Transport Phenomena in Biological Systems

Transport Phenomena in Biological Systems

SECOND EDITION

George A. Truskey

Fan Yuan

David F. Katz

Duke University, Durham, NC

PEARSON
Prentice Hall

ISBN-13: 978-0-13-156988-1

Upper Saddle River, New Jersey 07458

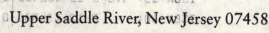

Library of Congress Cataloging-in-Publication Data
Truskey, George A.
Transport phenomena in biological systems / George A. Truskey, Fan Yuan, David F. Katz.—2nd ed.
p. cm.
Includes bibliographical references and index.
ISBN 0-13-156988-0 (alk. paper)
1. Biological transport. I. Yuan, Fan, Ph.D. II. Katz, David F. III. Title.
QH509.T78 2009
571.6'4—dc22

2008043589

Vice President and Editorial Director, ECS: *Marcia J. Horton*
Senior Editor: *Holly Stark*
Associate Editor: *Dee Bernhard*
Editorial Assistant: *William Opaluch*
Director of Team-Based Project Management: *Vince O'Brien*
Senior Managing Editor: *Scott Disanno*
Production Liaison: *Jane Bonnell*
Production Editor: *Bharath Parthasarathy, TexTech*
Manufacturing Manager: *Alan Fischer*
Manufacturing Buyer: *Lisa McDowell*
Marketing Manager: *Tim Galligan*
Marketing Assistant: *Mack Patterson*
Art Director: *Kenny Beck*
Cover Designer: *Kristine Carney*
Cover Images: left: *3D4Medical.com/Getty Images, Inc.*; top right: *Zephyr/Science Photo Library/Photo Research*; bottom right: *D. Phillips/ Science Photo Library/Photo Research*
Art Editor: *Greg Dulles*
Media Editor: *Daniel Sandin*
Media Project Manager: *John M. Cassar*
Composition/Full-Service Project Management: *TexTech International Pvt. Ltd.*

Pearson Education Ltd., *London*
Pearson Education Singapore, Pte. Ltd.
Pearson Education Canada, Inc.
Pearson Education–Japan
Pearson Education Australia PTY, Limited
Pearson Education North Asia, Ltd., *Hong Kong*
Pearson Educación de Mexico, S.A. de C.V.
Pearson Education Malaysia, Pte. Ltd.
Pearson Education, Upper Saddle River, New Jersey

PEARSON
Prentice Hall

ISBN-13: 978-0-13-156988-1
ISBN-10: 0-13-156988-0

19

To Anna, Peter, Nora, and the memory of Claire

GAT

To Qiyan Lou and Zuyue Yuan

FY

To Cindy, Helen, Frank, and Stanley

DFK

Contents

Part I Introduction to Physiological Fluid Mechanics 53

2 Conservation Relations and Momentum Balances 55

Part II Fundamentals and Applications of Mass Transport in Biological Systems 259

6 Mass Transport in Biological Systems 261

12 Cell Adhesion 582

Part IV Transport in Organs 623

13 Transport of Gases Between Blood and Tissues 625

Preface to the Second Edition

This textbook was developed to meet the needs of curricula in biomedical engineering and bioengineering. We have been pleased with the extent to which the book has been adopted in undergraduate and graduate courses. At the same time, we have received valuable feedback that has led to significant changes in this second edition. First, we have added a new chapter (Chapter 17) on Energy Transport in Biological Systems. This is an important topic present in many biotransport courses. We cover a range of relevant physiological and technological topics, focusing on heat transfer. Since our intent is to provide an introduction to this important area of biomedical engineering, we only briefly note more advanced topics that would be covered in advanced elective courses.

Second, we have added over 70 new problems at the ends of chapters. We find that the most important way for students to learn is by solving problems which apply the concepts presented in each chapter. We hope that these additional problems will stimulate thinking about transport phenomena and enhance understanding, so that students have an improved capability to use basic concepts of transport phenomena in complex bioengineering contexts.

Third, we have corrected a number of errors and clarified statements that students have found confusing, and we have eliminated some sections that were not found to be useful.

We thank the many colleagues and students who provided comments and suggestions about our book. These include: James Bryers, University of Washington; William Merryman, University of Alabama at Birmingham; Victor Baracos and Robert Tranquillo at the University of Minnesota; Edward Leonard, Columbia University; Brad Gelfand, University of Virginia; and John Tarbell at City University of New York. We give grateful acknowledgment, too, to the reviewers of the second edition: Ronald L. Miller, Colorado School of Mines; David Nelson, Michigan Technological University; and Aleksander S. Popel, Johns Hopkins University. Bonnie Lai and Steve Wallace were teaching assistants for the course at Duke and provided valuable insights. Finally, we thank all of our students in BME 207. Your feedback has helped shape this book as much as that by anyone else.

George A. Truskey
Fan Yuan
David F. Katz
Durham, NC

Preface to the First Edition

The transport of energy, mass, and momentum is essential to the function of living systems. Changes in these processes often underlie pathological conditions. Transport phenomena are also central to the operation of instrumentation used to analyze living systems, and to many of the technological interventions used to repair or improve tissues or organs. Transport processes are manifest from the smallest spatial scales of molecular dimensions to the large scales of whole organs and of organisms themselves. They also span the minute time scales of individual chemical events to the lifetimes of living systems. As scientific and technological advancements have shrunk the temporal and spatial scales of observation and understanding of living systems, attention to attendant transport processes at those scales has followed suit.

The objective study of biological transport phenomena began historically in the field of physiology and, indeed, helped define that field. Today, the engineering application of biological transport phenomena contributes to research advances in physiology, immunology, and cell and molecular biology. Thus, transport processes are important considerations in basic research related to molecule, organelle, cell and organ function; the design and operation of devices, such as filtration units for kidney dialysis, high density cell culture and biosensors; and applications including drug and gene delivery, biological signal transduction, and tissue engineering. Clearly, attention to the basic mechanisms of transport processes and, concomitantly, their biological and biomedical contexts, is important to curricula for educating biomedical engineers.

Teaching undergraduate and graduate students in bioengineering about transport phenomena in living systems is challenging. This teaching must integrate the development of fundamental principles of transport processes, the mathematical expression of these principles and the solution of transport equations, along with characterization of composition, structure and function of the living systems to which they are applied. The overwhelming majority of textbooks on transport processes are oriented primarily toward chemical and mechanical engineers, and lack applications and descriptions of biological and biomedical contexts. While many of these texts are excellent, there is a need for a book that integrates biological and engineering concepts in the development of the transport equations and, meanwhile, provides detailed and current applications. It is our goal that this text meets that need.

The materials in this textbook will help to develop both skills and contextual knowledge for engineers, enabling them to establish and critically analyze models of biological transport and reaction processes. We have sought to present engineering fundamentals and biological contexts in a unified way. The book covers topics in fluid mechanics, mass transport, and biochemical interactions and reactions. Inclusion of the latter has great biological and biomedical motivation, since so many relevant processes and technologies involve chemical reactions. Each engineering concept is motivated by specific biological problems. Immediately after the concept is developed, biological and/or biomedical applications are presented. In this way, the student can gain an understanding of the specific topic presented, as well as its application to important problems in biology and medicine. Each chapter contains a number of examples and homework problems, that either elaborate upon problems discussed in the text or address new biomedical questions. Problems and examples include analytical as well as numerical solutions. We emphasize analytical solutions because they often provide physical insights that are important for introductory material, even if such insights provide only first-order levels of understanding. More complex problems that require numerical solution are presented and the use of MATLAB in the solution of these problems is discussed. References to current literature are provided for those who are interested in more detailed analyses.

Our target audience consists of advanced undergraduates or graduate students. We assume some exposure to the study of biology, but Chapter 1 summarizes many basic concepts in cell biology and physiology. References to relevant texts in cell biology and physiology are provided. No previous exposure to mass and momentum transport, nor to chemical kinetics is assumed, but understanding of introductory material in chemistry is needed. Although students should be familiar with most of the mathematical concepts that are discussed in the text, we include an Appendix that provides a review of important concepts, and presents material about use of MATLAB in problem solving.

The text contains an introduction and four parts. The introduction describes the motivation for the text and its organization and provides a brief overview of transport processes at the cell and tissue level. The first two parts are based on the analogy in development of principles for momentum and mass transport. Balance relations are presented for momentum (Chapter 2) and mass (Chapter 6), and are applied to some simplified biomedical contexts that demonstrate the effects of geometry and boundary conditions. Next, we develop the conservation and constitutive relations in three dimensions for momentum (Chapter 3) and mass (Chapter 7), and apply them to more complex problems. Subsequent chapters in each part are devoted to specific applications. The third part examines biochemical interactions and effects of mass transport upon these interactions. The fourth part focuses on transport in organs and whole organisms. Applications in all parts range from molecular events within cells (Chapters 11 and 12) to biochemical reactions that affect transport between cells (Chapters 13 to 15). The final chapter of the book synthesizes these concepts by considering several examples of whole-body transport.

This book can be used for both introductory and advanced courses on transport phenomena in biomedical engineering. A one-semester course might include Chapters 1, 2, 3, 6, 7, and 10, and focus on the basic concepts of biological transport phenomena. We present physiological fluid mechanics before diffusion processes to provide a basis for understanding models describing the diffusion coefficient in Section 6.6. It is possible to present the analysis of diffusion (Chapter 6) prior to physiological fluid mechanics, by deferring discussion of Section 6.6 until low Reynolds number flow is presented. Time permitting, one or more of the application chapters could also be covered. An advanced course might use some or all of the chapters on advanced topics (Chapters 4, 5, 8, 9 and 11 through 16).

Contemporary biology, medical science, and biotechnology are replete with important transport problems yet to be solved. Such problems embody interrelationships amongst biological, chemical, and physical processes. By presenting these relationships in the context of biomedical applications, we hope, in this textbook, to provide students and researchers with knowledge and insights needed to address and solve these important problems.

The completion of this book would not be possible without the assistance of many colleagues and students who made many helpful suggestions. We are especially grateful to the following colleagues who reviewed large portions of the text or who used earlier drafts of the book in their courses: Robert Tranquillo and Victor Baracos at the University of Minnesota, John Tarbell at the City University of New York, Stanley Berger at the University of California, Berkeley, Norman Harris at the Pennsylvania State University. Roger Barr at Duke University, and Robert Roselli at Vanderbilt University provided helpful comments. We'd also like to thank the following graduate students who have reviewed the text or provided solutions to many problems: Bernard Chan, Joshua Henshaw, Joel Irick, Jeff Lamack, Priya Karani, Sarah Kieweg, Sarah McGuire, Valerie Mckinney, Mark Lavender, Sheng Tong, and Steve Wallace. Jennifer Peters provided drawings in Chapter 4. Diane Feldman provided expertise in editorial assistance and important suggestions that improved the clarity of the text. We greatly appreciate the financial support provided by the Textbook awards program at the Whitaker Foundation and the support and encouragement of Jack Linehan of the Foundation. Lastly, we thank the students of BME 207 who provided countless questions and who endured many drafts of the book as it evolved.

GEORGE A. TRUSKEY
FAN YUAN
DAVID F. KATZ
Durham, NC

Transport Phenomena in Biological Systems

Introduction

CHAPTER 1

1.1 The Role of Transport Processes in Biological Systems

The functioning of cells, organs, and tissues requires the efficient delivery of nutrients and regulators of growth and other processes of function. Organisms control the concentrations of molecules in their tissues and organs. Consequently, specialized mechanisms regulate the movement of molecules to, across, and within cells. These mechanisms are the subject of this book.

Many organs, such as the lungs, liver, and kidneys, are organized to enable the rapid exchange of molecules between the blood and tissues. In addition, various levels of biological organization—for example, the density of capillaries in various tissues and the size and structure of cells—can be explained, in part, by the rates of transport of molecules from their sources to their sites of delivery. Alterations in transport processes are important factors in a number of diseases, such as atherosclerosis, cancer, and kidney diseases.

Transport phenomena involve the integrated study of momentum, mass, and energy transfer, as well as the thermodynamics and kinetics of chemical reactions. For the bioengineer, a mechanistic understanding of transport processes is important for the characterization of physiological and cellular processes, the design and operation of a number of devices, and the development of new therapies. Examples of biomedical devices influenced by transport processes include kidney dialysis machines, heart–lung bypass machines, biosensors, and membrane oxygenators. Transport processes are critical in the removal of toxins from blood, the design of replacement tissues, and the delivery of drugs. Understanding and exploiting transport processes will be necessary in the application of molecular medicine. For example, the methods for delivering gene therapies must exploit biological transport pathways in order to successfully deliver the gene, in a functional form, to sites of action in the body.

Because biological systems have evolved unique adaptations to regulate molecular transport, the development and analysis of biomedical engineering transport problems requires an understanding of cellular biology and physiology. In this text, we focus upon descriptions of cellular and organ-level processes of transport, the formulation of the equations describing transport processes, and the solution of these equations for problems of interest to biomedical engineers.

The field of transport phenomena is sufficiently far advanced that the basic processes can be characterized mathematically. The predictive capabilities of models are quite good, even for complex biological systems. Analytical and numerical solutions are available for many problems pertaining to biological function and the design of many technologies. Nevertheless, there are many important biomedical transport problems that have not been solved. Such problems frequently demonstrate the interrelationships among biological, chemical, and physical processes.

1.2 Definition of Transport Processes

Two physical phenomena are involved in the transport of molecules: diffusion and convection. *Diffusion* is the random motion of molecules that arises from thermal energy transferred by molecular collisions. *Convection* is a mechanism of transport resulting from the bulk motion of fluids. The movement of energy and momentum in biological systems is influenced by these two mechanisms.

1.2.1 Diffusion

Collisions between molecules occur trillions of times per second. Each collision results in the random motion of solute and solvent molecules. This random motion gives rise to diffusion, which occurs in gases, in liquid solutions, in membranes, and in the interstitial spaces of tissues. The speed at which a molecule diffuses in a fluid or membrane depends upon its size and shape, the temperature, and the fluid viscosity, a property that reflects the resistance to flow.

In spite of the random nature of these collisions, net motion of molecules results. The term *random walk* is used to describe the net molecular motion arising from such collisions. A small sequence of a random walk is shown in Figure 1.1. Each change in direction is the result of collisions between the molecule of interest and the fluid molecules.

Figure 1.1 shows the random movement of a particle from the position marked by the filled circle to the position marked "X". The net movement between these two points can be characterized by the root-mean-square displacement $\langle r^2 \rangle^{1/2}$ which is calculated from a large number of random walks. The root-mean-square displacement is thus a measure of the distance traveled by a diffusing particle. The relation between random walks and diffusion is discussed in more detail in Section 6.5.

Random motion can be viewed macroscopically by observing the spreading of a dye droplet after it is added to water. Initially, dye molecules are close to the site of application. As time proceeds, the outer edges of the droplet become less distinct. Gradients in color appear, and the intensity decreases with distance away from the point of application of the dye. Eventually, the dye is uniformly distributed throughout the water.

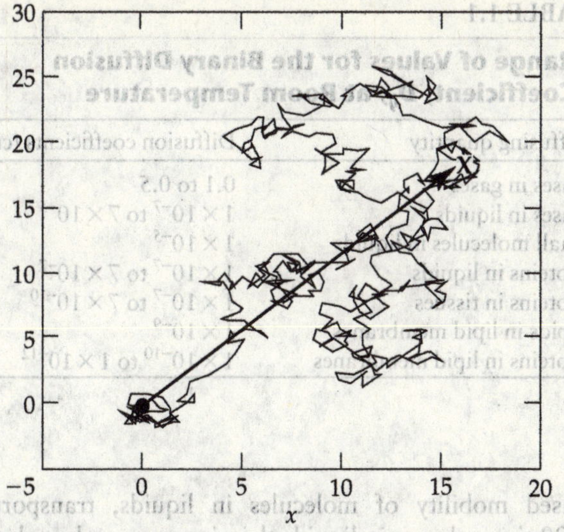

FIGURE 1.1 500-step random walk. The walk begins at coordinates $x = 0, y = 0$ denoted by the "•" and ends in the upper right hand quadrant denoted by "X". The arrow represents the displacement.

A macroscopic consequence of random molecular motion is that diffusing molecules move from regions of higher concentration to regions of lower concentration. These spatial differences are known as *concentration gradients*. The net movement of molecules through a unit area in a given direction per unit time is known as a *flux*. In general, a flux is defined, for any transported quantity, as the amount of the quantity passing through a unit area per unit time, and it may depend upon both position and time. Fluxes can be defined for mass, energy (including heat), and momentum. In addition, fluxes can be defined for groups of moving cells. Fluxes have a magnitude and direction and are thus vectors. The diffusion flux is proportional to the gradient of the concentration.

The relation between the diffusion flux and the concentration gradient (also known as a *constitutive equation*)[1] was first quantified in 1855 by Adolph Fick, a German physiologist, and is known as Fick's first law. Fick's first law is used widely in studies of diffusion in dilute solutions. Fick developed his relation from careful experimentation and by analogy with Fourier's law of heat conduction, which states that the flux of heat is proportional to the gradient of temperature.

The quantity that relates the diffusion flux to the concentration gradient is the *binary diffusion coefficient*, D_{ij}, where the subscript i refers to the solute and j to the solvent. The diffusion coefficient is a function of temperature and pressure. The magnitude of the diffusion coefficient depends upon the solute and the medium through which diffusion occurs—for example, gases, liquids, or tissues (Table 1.1). The diffusion coefficient is largest for gases, due to the relative unimportance of intermolecular forces and the relatively low density. Diffusing molecules travel distances much larger than the molecular size before colliding with other molecules. Because

[1]A constitutive equation is a specific relation of material properties. Unlike a conservation relation, which is universal, a constitutive equation applies to a specific class of materials or a specific range of conditions. For example, Fick's law is strictly applicable to the diffusion in dilute binary solutions. In practice, Fick's law has been used over a wider range of conditions.

TABLE 1.1

Range of Values for the Binary Diffusion Coefficient, D_{ij}, at Room Temperature

Diffusing quantity	Diffusion coefficients ($cm^2\ s^{-1}$)
Gases in gases	0.1 to 0.5
Gases in liquids	1×10^{-7} to 7×10^{-5}
Small molecules in liquids	1×10^{-5}
Proteins in liquids	1×10^{-7} to 7×10^{-7}
Proteins in tissues	1×10^{-7} to 7×10^{-10}
Lipids in lipid membranes	1×10^{-9}
Proteins in lipid membranes	1×10^{-10} to 1×10^{-12}

of the decreased mobility of molecules in liquids, transport by diffusion is 10,000–100,000 times slower in liquids than in gases, and a solute molecule diffusing in a liquid collides with liquid molecules when it has traveled a distance that is much smaller than its size. Diffusion in membranes is further reduced, relative to diffusion in liquids. In cell membranes, protein diffusion is slower than lipid diffusion, in part because proteins interact with the cytoskeleton as well as membrane lipids and other proteins.

Prior to developing the theory of relativity, Albert Einstein determined that the diffusion coefficient is related to the mean square distance a molecule moves during a random walk (Figure 1.1) [1]. For a random walk in two dimensions (x and y), he found that

$$D_{ij} = \frac{\langle x^2 \rangle + \langle y^2 \rangle}{4t}. \tag{1.2.1}$$

One important result from Equation (1.2.1) is that the time required to diffuse increases with the square of the distance over which diffusion occurs. Diffusion is a very rapid, and hence, efficient process when distances are small. For example, a typical cell is about 20 μm (= 0.002 cm) in diameter. The protein diffusion coefficient within the cell cytoplasm is about 1×10^{-7} $cm^2\ s^{-1}$. A protein can diffuse from the edge of the cell to the center in about 2.5 seconds. When a molecule must be transported large distances, however, diffusion becomes a very slow and inefficient process. The time for the same protein to diffuse through a tissue only 0.2 cm thick is 27.7 hours.

In complex structures, such as tissues, diffusion distances of molecules and other particles are greatly increased due to the presence of obstructions created by the extracellular matrix and cells. Further, these obstructions exert drag forces on diffusing molecules that slow their movement. However, Fick's law describes diffusion in tissues if the binary diffusion coefficient is replaced with an *effective diffusion coefficient* D_{eff}. The binary diffusion coefficient characterizes the diffusion of one molecule relative to another. The effective diffusion coefficient incorporates the effects of increased diffusion distances and the drag forces exerted by the extracellular matrix and cells.

1.2.2 Convection

One means of overcoming limits imposed by diffusion is to move molecules or cells of interest by fluid motion. Convection is a mechanism of transport resulting from the bulk motion of fluids. Gases and liquids flow following the application of forces such as gravity, pressure, or shearing forces. Effects of forces applied to a surface are characterized in terms of *stresses* (which have units of force per unit area). The stress depends upon the magnitude of the applied force, the direction of the force, and the surface to which the force is applied. *Shear stresses* result from forces applied tangentially to a surface, causing two contiguous parts of the material to slide relative to each other. Biological examples of shearing forces are those which occur in joints and in the eyelids. Stresses acting perpendicular (i.e., normal) to a surface can be compressive or tensile. Pressure, which helps propel the flow of blood through the circulatory system, is a compressive normal stress. The application of a shear stress and a pressure gradient can result in fluid motion.

The net motion of a fluid carries along solutes that are dissolved within it. Solute motion differs slightly from the local fluid motion because the solute molecules are also diffusing simultaneously. If fluid motion is slow relative to diffusion, then diffusion will dominate in causing solute transport. By contrast, if the fluid motion is fast relative to diffusion, the net fluid motion will be the dominant means of transport.

The fluid *viscosity* μ is a measure of the frictional resistance of a fluid to flow. The frictional force that must be applied in order to produce motion is proportional to the fluid viscosity. For a pure fluid, the viscosity is a thermodynamic function of temperature and pressure. Gases have relatively low viscosities (Table 1.2). Liquids are much more viscous and dense, due to the presence of intermolecular forces. The magnitudes of various liquid viscosities vary over four orders of magnitude, whereas the viscosities of gases vary by less than one order of magnitude.

Density is a material property that characterizes how closely molecules are arranged. Physically, the density is the ratio of the mass of the system to the volume of the system. The density of a pure fluid varies with temperature and pressure. Generally, solids are denser than liquids and liquids are denser than gases. One important exception is ice, which is less dense than water. For mixtures, the density is a function of temperature, pressure, and composition.

TABLE 1.2

Range of Values for Viscosity, Density, and Kinematic Viscosity at Room Temperature

	Viscosity, μ (g cm^{-1} s^{-1})	Density, ρ (g cm^{-3})	Kinematic viscosity, $\nu = \mu/\rho$ (cm^2 s^{-1})
Gases	10^{-4}	0.001	0.1
Liquids			
Water	0.01	1.0	0.01
Glycerol	10	1	10
Blood	0.03	1.2	0.025

The ratio of viscosity to density is known as the *kinematic viscosity* and is denoted by

$$\nu = \frac{\mu}{\rho}. \tag{1.2.2}$$

The kinematic viscosity is a measure of the efficiency of *momentum transport*. One conceptual view of the mechanics of fluid motion is that the applied forces transfer momentum to the fluid. The movement of momentum through the fluid can be regarded as analogous to diffusion. Although gases have low viscosities, they also have low densities. As a result, gases have a greater kinematic viscosity than some liquids (Table 1.2). The kinematic viscosity is analogous to the diffusion coefficient (and has the same units) and characterizes the "diffusion" of momentum [2]. Since the kinematic viscosity of air is greater than the kinematic viscosity of water, the same amount of momentum is transferred over longer distances in air than in water.

The units of shear stress and pressure are force per unit area (N m^{-2} [Pa] or dyne cm^{-2}). Since a force equals the rate of change of momentum with time, the shear stress can also be represented as the momentum per unit area per unit time.

Another way to conceive shear stress applied to fluids is to note that the shear stress transports momentum to the fluid adjacent to the site of application of the shear stress [3]. Because of molecular attractions, some fluid is drawn along while the other layers of the fluid resists this motion. The net result is that the next fluid layer moves more slowly than the layer adjacent to the site of application of the shear stress. Thus, the shear stress transports momentum to the fluid, resulting in a velocity gradient. In fact, the shear stress is commensurate with a momentum flux, a view developed by Bird, Stewart, and Lightfoot in their classic text *Transport Phenomena* [3].

There is an analogy among energy, mass, and momentum transport, which can be summarized as the following general relation:

$$\begin{pmatrix}\text{Flux of quantity}\\ \text{being transported}\end{pmatrix} \propto -\begin{pmatrix}\text{Gradient of quantity}\\ \text{being transported}\end{pmatrix}. \tag{1.2.3}$$

The negative sign in Equation (1.2.3) is used because transport occurs down a gradient—that is, from a greater to a lesser magnitude of the quantity being transported. In order to render the flux a positive quantity in the direction of transport, a negative sign is often applied.[2]

The relations for momentum, mass, and energy are summarized in Table 1.3. All of these transport processes are important in biological systems.

For simple fluids, the viscosity is the coefficient of proportionality between the shear stress and the velocity gradient. This relation, given by Equation (2.5.6), is known as *Newton's law of viscosity* and is another example of a constitutive relation. It applies to a number of common fluids with a single component, such as

[2]Note that, in mechanics, tensile stresses are positive (see Section 2.3.3) and compressive stresses are negative. As a result, a negative sign is not used in mechanics to relate the shear stress and velocity gradient. In spite of the analogy among mass, momentum, and energy transport, we adopt the convention used in mechanics.

TABLE 1.3

Relations between Fluxes and Gradients for Molecular Transport

Molecular transport mechanism	Flux	Gradient	Coefficient of proportionality
Momentum	Shear stress	Velocity	Viscosity
Mass	Mass or molar flux	Concentration[a]	Diffusion coefficient
Energy	Energy	Temperature	Thermal conductivity

[a]For charged molecules, transport is down an electrochemical gradient, defined as the sum of the concentration gradient plus the potential field gradient. This kind of transport is discussed in detail in Section 7.4.

water, alcohol, and air. For these fluids, the viscosity depends upon temperature and pressure only. For mixtures, such as polymer solutions and blood, the viscosity depends upon the velocity gradient, a quantity also known as the *shear rate*. Mixtures do not follow Newton's law of viscosity, and more complex constitutive relations must be applied.

When an applied force on a fluid in motion is removed, some time elapses before fluid motion ceases. The change in velocity with time arises from a balance between viscous and inertial forces. Viscous forces act to retard fluid motion, whereas inertial forces act to keep the fluid in motion. Because viscous forces are sensitive to viscosity, and inertia depends upon the mass or density, the relative contributions of these forces vary among different fluids. A dimensionless grouping of parameters known as the *Reynolds number* describes the ratio of inertial forces to viscous forces:

$$\mathrm{Re} = \frac{\text{inertial forces/volume}}{\text{viscous forces/volume}} = \frac{\rho v^2/L}{\mu v/L^2} = \frac{\rho L v}{\mu} \tag{1.2.4}$$

Here, L is a characteristic length and v is a characteristic velocity for the flow. For objects moving at the same speed, the relative significance of viscous and inertial forces depends upon size. For example, for a fish 50 cm long moving in water at the relatively slow speed of 1 cm s^{-1}, the Reynolds number is 5,000. Viscous forces are much less significant than inertial forces. If a white blood cell with a diameter of 10 μm moves at the speed of our hypothetical fish, then the Reynolds number is 0.1. Even though this is a relatively high speed for a cell, viscous forces dominate. Consistent with the analogy between energy, mass, and momentum transport, the Reynolds number also represents the ratio of momentum transport by convection to momentum transport by viscous diffusion [2].

One remarkable feature of fluid motion is that the character of the flow changes dramatically above a critical value of the Reynolds number. Flow can be characterized as *laminar* or *turbulent*. For steady laminar flow, the velocity at any given location does not change with time. When flow is turbulent, however, the velocity fluctuates randomly due to the formation and dissipation of eddies of fluid at high energy. The analysis of turbulent flow is more complex than that of laminar flow. For the most part, flow inside the body is laminar.

Biological tissues consist of water, cells, and an *extracellular matrix*—an interconnected network of proteins and *proteoglycans*, which are proteins containing significant amounts of polysaccharides. The matrix provides structural rigidity to the tissue. It is porous, and water can flow between the matrix molecules. On the scale of the pores (e.g., 0.01–10 μm), the local motion of fluid is described by relationships similar to those applied to fluid moving in a tube. Over a scale much larger than the dimensions of the matrix (e.g., 10–100 μm), however, the relationship is different. The average velocity of fluid moving through the tissue is proportional to the change in the pressure drop with distance, a constitutive relation known as *Darcy's law*. The constant of proportionality is the ratio of the *Darcy permeability* to the viscosity. The Darcy permeability represents the conductivity of the porous medium to flow and depends upon the detailed microstructure of the porous medium. Flow in tissues is examined in Chapter 8.

1.2.3 Transport by Binding Interactions

In addition to being affected by convection and diffusion, molecular transport is influenced by noncovalent interactions between two molecules. Many weak interactions between binding molecules can produce a net interaction that is quite stable. Such interactions are known as *binding* interactions and are specific. That is, for a given *receptor* molecule, only a small number of structurally similar molecules can bind to the receptor. These interactions can be described with methods of chemical kinetics and equilibrium.

Binding reactions provide a means of chemical recognition, selective transport, and cell signal amplification. For example, when a foreign molecule or cell enters the body, it binds to an antibody, which is a protein molecule of the immune system. Such foreign molecules that bind to antibodies are known as *antigens*. The antigen-bound antibody then binds to cells of the immune system via a portion of the antibody known as the Fc segment. The immune system cells can now digest the foreign molecule or cell. Thus, antibody binding enables the immune system to recognize and remove foreign molecules and cells.

Binding reactions are used to selectively transport molecules into cells. Specialized molecules in the cell membrane, known as *transporters*, bind to ions and small molecules, enabling them to pass across the cell membrane. Some binding interactions can enable cells to transport molecules against a concentration gradient. The process requires energy from the cell and is known as *active transport*. When hormones and proteins bind to receptors on the cell membrane, they are transported into the cell by a process involving the formation of vesicles derived from the plasma membrane. Such a process is referred to as *endocytosis*.

In some cases, molecules bound to cell membrane receptors exert their effects indirectly. Binding of the molecule, also known as a *ligand*, to the receptor produces a change in the three-dimensional spatial organization of the portion of the receptor exposed to the cytoplasm. Such changes in structure are known as *conformational changes*. The modified cytoplasmic side of the receptor is either directly or indirectly involved in chemical reactions producing molecules that can exert a biological response within the cell. Such a process of indirect action is known as *cell signaling*. One advantage of cell signaling is that a single molecule can initiate a reaction that produces many product molecules, which can then produce a biological response. This amplification of the signal can enable the ligand to be effective even at low concentrations.

1.3 Relative Importance of Convection and Diffusion

Length scales in biological systems range over eight orders of magnitude (Table 1.4). Clearly, no single transport process can function efficiently over these length scales. At short distances, diffusion can be quite rapid. As distance increases, the diffusion time increases as the square of this distance; thus, diffusion becomes increasingly less efficient. In biological systems, convection typically transports molecules over distances for which diffusion is too slow. For example, blood transports oxygen bound to hemoglobin in red blood cells (known as *erythrocytes*) over large distances in the larger blood vessels of the body by convection. In the smallest blood vessels, blood flow is very slow and oxygen is transported to local tissues by convection.

Because of the different length scales in biological systems, the relative importance of diffusion or convection varies with the specific situation. The significance of diffusion versus convection can be evaluated in two equivalent ways. One approach is to calculate the *Peclet* number, which represents the ratio of mass transport by convection to mass transport by diffusion. This number is given by

$$\mathrm{Pe} = \frac{\text{Mass transport by convection}}{\text{Mass transport by diffusion}} = \left(\frac{L^2}{D_{ij}}\right)\left(\frac{\mathrm{v}}{L}\right) = \frac{\mathrm{v}L}{D_{ij}}, \tag{1.3.1}$$

where v is a characteristic velocity and L is a characteristic length. When the Peclet number is much less than unity, diffusion is more rapid than convection. In many cases, the average velocity represents the characteristic velocity. Conversely, when the Peclet number is very large, convection is the dominant mode of transport. For a fixed-length scale and velocity, diffusion is most important for small molecules such as oxygen (Table 1.5). Convection is essential to transport larger molecules—such as proteins—and cells.

Over the dimensions of a cell (10 μm), diffusion is highly efficient for internal transport. Although diffusion is efficient at such length scales, concentration gradients may still arise if the molecule reacts very rapidly relative to the speed at which it is transported to the reacting site. Such reactions are said to be *diffusion limited*, because the rate of reaction depends upon the rate at which the reacting molecule is transported to the reaction site. Thus, some reactions of membrane-bound proteins

TABLE 1.4

Relevant Length Scales in Biological Systems

Quantity	Length scale (m)
Proteins and nucleic acids	10^{-8}
Organelles	10^{-7}
Cells	10^{-6} to 10^{-6}
Capillary spacing	10^{-4}
Organs	10^{-1}
Whole body	10^{0}

Source: From Ref. [4].

TABLE 1.5

Relative Importance of Diffusion and Convection

Molecule	MW (g mol^{-1})	D_{ij} ($cm^2\ s^{-1}$)	Diffusion time, L^2/D_{ij} (s)	Pe = Lv/D_{ij}
Oxygen	32	2×10^{-5}	5	0.05
Glucose	180	2×10^{-6}	50	0.50
Insulin	6,000	1×10^{-6}	100	1.0
Antibody	150,000	6×10^{-7}	167	1.67
Particle	**Diameter**	**D_{ij} ($cm^2\ s^{-1}$)**	**Diffusion time (s)**	**Pe**
Virus	0.1 μm	5×10^{-8}	2,000	20
Bacterium	1 μm	5×10^{-9}	20,000	200
Cell	10 μm	5×10^{-10}	200,000	2,000

Note: For L = 100 μm, and if v = 1 μm s^{-1}, the time for convection is always equal to L/v = 100 s for all molecules and particles.

(e.g., the reaction of oxygen with enzymes on the mitochondrial membrane) are often limited by the diffusion of reactants to the membrane.

An alternative way of comparing the relative roles of diffusion and convection is to compare the times required for a molecule to be transported by each process. The Peclet number is equivalent to computing the ratio of the diffusion time ($t_d = L^2/D_{ij}$) to the time for convection ($t_c = L/v$). Figure 1.2 is a log–log plot that shows how diffusion times and convection times change with distance transported. For short distances, convection is slower than diffusion, whereas for longer distances, diffusion is slower than convection.

The distance at which transport by the two processes becomes equal is inversely related to the diffusion coefficient. The diffusion coefficients shown in Figure 1.2 span the range of values for proteins and small solutes. For proteins, diffusion is an efficient process for dimensions on the order of the size of a cell or smaller (10 μm = 0.001 cm). For small molecules such as gases, glucose, or urea,

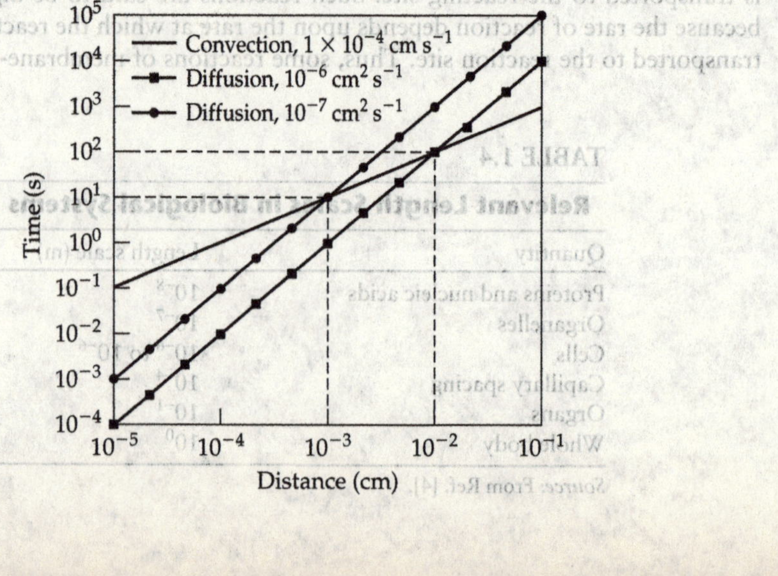

FIGURE 1.2 Effect of distance on diffusion and convection times.

diffusion is efficient for distances on the order of 100 μm. Interestingly, this distance is about the normal spacing between capillaries in many tissues. Chemical reactions lead to a decrease in the concentration of the diffusing molecule and reduce the distance over which diffusion is an efficient process.

The cardiovascular system uses convection to optimize oxygen delivery to the various organs. The dynamic response of other molecules (e.g., hormones) that are transported through the blood is limited by the oxygen-delivery requirements. The elapsed time for the delivery of these molecules is on the order of a few minutes, sufficient to meet normal demands. Body movement, however, requires a much faster response time than can be accomplished by convective transport through blood and diffusion in tissues. Therefore, the nervous system uses electrical conduction of signals through transmembrane ion movement and the release of neurotransmitters, at speeds as high as 500 m s^{-1}.

1.4 Transport Within Cells

In order to regulate the movement of molecules into and through cells, specialized mechanisms have developed in cells to transport molecules efficiently. The cells of higher animals are complex structures consisting of many organelles (Figure 1.3 and Table 1.6) that perform specialized functions such as protein synthesis (*endoplasmic reticulum*), protein degradation (*lysosomes*), export of proteins (*Golgi apparatus*), and cell division (*nucleus*).

Much of the organization of cells is a result of the need to regulate the transport of molecules between the cell and extracellular fluid and between the cytoplasm and various organelles. Ion gradients are maintained across cell membranes. Cells use the energy of the ions to drive chemical reactions and to transmit information. Endocytosis is the major mechanism to transport larger molecules that cannot permeate across the cell membrane. Such molecules may be involved in cellular nutrition or in the transfer of signals from the cell exterior to the cell interior. Proteins embedded in the cell membrane serve as the means to specifically transport these solutes and to transmit signals.

Transport within cells is highly regulated and complex. In addition to diffusion, transport is governed by interactions of solutes with specialized proteins. An extensive network of membrane-bound vesicles transports proteins to and from the extracellular fluid and between organelles. In addition, motor proteins can carry solutes over long distances. What follows is a summary of the mechanisms that exist to selectively transport and target molecules within the cell.

1.4.1 Transport Across the Cell Membrane

The cell membrane—or plasma membrane—consists of *lipids* and *proteins*. The lipids themselves are subdivided into three groups: *phospholipids*, *sphingolipids*, and *cholesterol*. By weight, phospholipids account for more than half of the lipid component of the plasma membrane. Sphingolipids account for 17% of the lipid weight, and cholesterol accounts for 28% of the weight. Phospholipids and sphingolipids are *amphipathic molecules*—containing both hydrophilic and hydrophobic

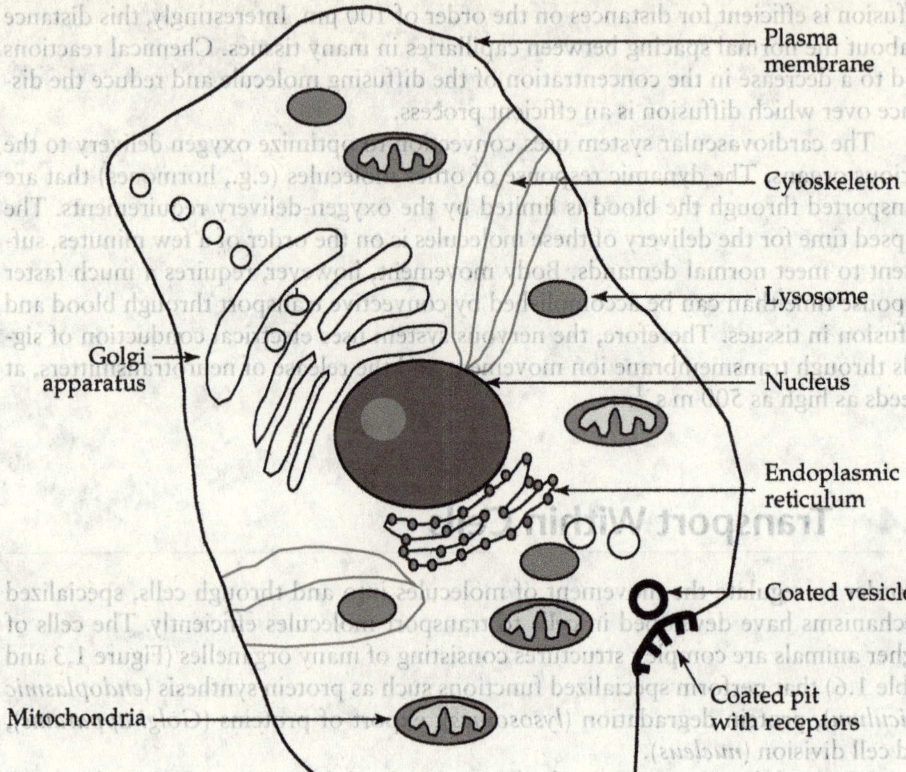

FIGURE 1.3 Schematic of a mammalian eukaryotic cell, showing the major organelles.

regions. Phospholipids and sphingolipids consist of long chains of nonpolar and hydrophobic hydrocarbons (the squiggly lines in Figure 1.4) and charged polar and hydrophilic head groups (the spheres attached to the squiggly lines in Figure 1.4). The proteins in cell membranes are classified as either *transmembrane proteins*, which span the membrane, or *peripheral proteins*. Peripheral proteins are present on only one side of the plasma membrane, either partially embedded in the plasma membrane or linked to the molecule's polar group or to another protein.

The polar portion of phospholipids and sphingolipids is on the outer portion of the membrane, in contact with extracellular or intracellular water. The two nonpolar hydrocarbon tails reside in the interior. The polar portion of cholesterol is also directed toward the exterior of the membrane. Cholesterol increases the rigidity of the membrane and decreases the permeability of the phospholipids. The lipid compositions on either side of the plasma membrane are different. There is a greater concentration of negatively charged lipids on the cytoplasmic side. The lipids can diffuse freely within the plane of the plasma membrane but cannot easily move from one side of the membrane to the other. Although some proteins can diffuse freely within the plasma membrane, many are segregated into distinct regions. This segregation

TABLE 1.6

Major Organelles of Eukaryotic Cells

Plasma Membrane consists of a lipid bilayer containing proteins that separates the cell from the external environment. The lipid bilayer is permeable to gases and small nonpolar molecules. The lipid membranes are impermeable to larger biological molecules. Protein channels and receptors permit selective transport.

Cytosol is the intracellular fluid in which all organelles reside. The cytosol consists of water, ions, dissolved gases, small polar molecules, proteins, and the cytoskeleton. The cytosol plus all organelles except the nucleus is called the cytoplasm.

Nucleus contains DNA in the form of chromosomes bound to proteins known as histones. Also present within the nucleus is the nucleolus, where ribosomes are assembled. Ribosomes are macromolecular complexes involved in the translation of RNA to protein. Communication between the nucleus and the cytoplasm occurs through pores, which consist of a protein complex that controls both the movement of proteins that regulate DNA transcription to messenger RNA (mRNA) and the transport of RNA from the nucleus to the cytoplasm.

Endoplasmic Reticulum is an extensive membranous network in contact with the plasma membrane. The endoplasmic reticulum is the site of synthesis of proteins and lipids found in the plasma and organelle membranes. Proteins for export are secreted into the endoplasmic reticulum. The *rough endoplasmic reticulum* contains *ribosomes* bound to the membrane and is the site of protein synthesis. RNA synthesized in the nucleus binds to ribosomes. Transfer RNA (tRNA) transports amino acids to the ribosomes. During translation of RNA into proteins, the growing peptide chain moves into the endoplasmic reticulum, where it undergoes conformational changes and chemical modifications. Lipid metabolism occurs in the *smooth endoplasmic reticulum.*

Golgi Apparatus is a membranous network involved in protein secretion and organelle formation. Vesicles bud from the Golgi and fuse with the plasma membrane or the membrane of other organelles, thereby releasing their protein contents.

Lysosome is an organelle that contains proteolytic enzymes at a low pH and is involved in protein hydrolysis. Vesicles containing proteins fuse with lysosomes, thereby preventing direct interaction between lysosomal contents and the cytoplasm.

Mitochondria are organelles that generate energy from glucose and oxygen. Mitochondria store energy in the form of adenosine triphosphate (ATP), which is used in various metabolic reactions.

Cytoskeleton represents networks of filamentous proteins (actin, tubulin, keratins, and vimentin) that provide structural support for the cell and regulate movement of proteins and organelles. Actin microfilaments are involved in cell motility and muscle contraction. Microtubules are polymers of tubulin that interact with the proteins kinesin and dynein to selectively transport molecules and organelles within cells. Microtubules are involved in nuclear division and cell mitosis. Intermediate filaments provide structural support for the cell.

Coated Pits and Vesicles are specialized regions of the cell membrane containing clathrin and other proteins. These molecules enable membranes to fold and form vesicles, which transport molecules into the cell from the extracellular fluid or from receptors present in the cell membrane.

Source: Adapted from Ref. [5].

Note: Prokaryotic cells have no nucleus or organelle. In contrast, eukaryotic cells have a nucleus and organelles.

results from protein–protein interactions in the membrane and protein binding to the cytoskeleton.

The lipid composition of plasma membranes varies among cell types and species. Similarly, the lipid composition of an organelle differs from the compositions of the membranes of other organelles. Proteins serve many important functions in the binding of molecules from the extracellular fluid, the regulated transport of ions and molecules across the cell, the attachment of cells to other cells or to the extracellular matrix, and the transmission of signals from the extracellular fluid to the cell interior. Both lipids and proteins contain charged sugar groups (referred to

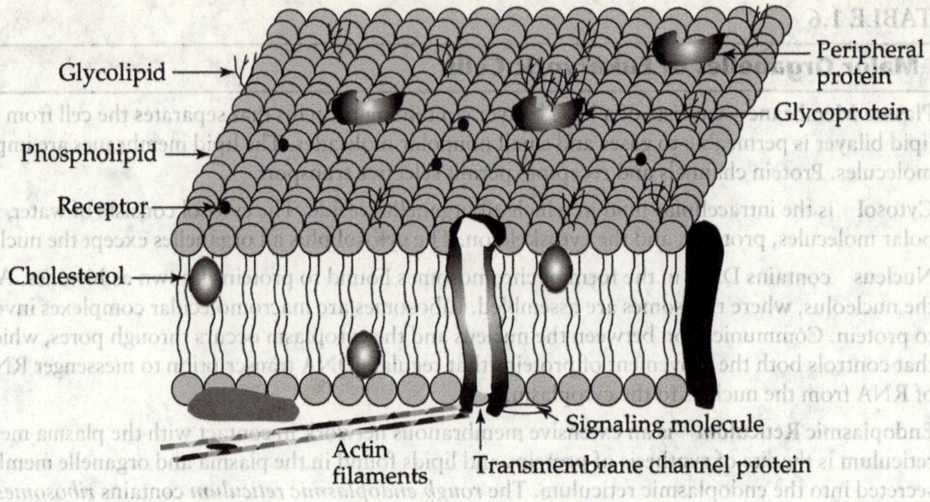

FIGURE 1.4 Schematic of a section of the cell plasma membrane.

as *glycolipids* and *glycoproteins*, respectively) that are involved in the binding of molecules and in intracellular signal generation.

Lipid bilayers are model membranes that exhibit limited permeability to molecules. Hydrophobic molecules, such as gases (O_2, N_2, and CO_2) and small organic molecules, have high permeability across lipid bilayers. Surprisingly, water exhibits a relatively high permeability across lipid bilayers. Small polar solutes (e.g., urea) have much lower permeability across lipid bilayers. The lipid membranes are virtually impermeable to ions. For small, uncharged polar molecules, such as ethanol, the permeability is proportional to the solubility of the molecule in organic solvents. Interestingly, the permeability of anesthetic gases correlates with the solubility of the gases in the interfacial polar head groups of the lipid bilayers. Due to the differences in the polarity and structure of the membrane, diffusion across the lipid bilayers is more complex than it is in simple fluids such as water [6].

Ions and small polar molecules (e.g., glucose, sodium, and amino acids) have low permeabilities across lipid bilayers and are transported across cell membranes by *transport proteins*, which are transmembrane proteins that are classified as either *carriers* or *channels*. Solutes bind to the carrier protein and are carried across the membrane due to a conformational change of the molecule. Transport across channels occurs when hydrophilic pores open within the proteins, allowing the solutes to diffuse through the proteins.

Carrier-Mediated Transport. Several features distinguish carrier-mediated transport from simple diffusion (see Chapter 14). Carrier-mediated transport produces a nonlinear relation between the rate of transport and concentration (Figure 1.5). At high concentrations, the rate of transport becomes independent of concentration in a process known as *saturation*. Saturation arises because the carrier sites become occupied as fast as they appear. Since transport occurs by the binding of solutes to carriers, transport can be blocked competitively, via a structurally similar solute, or noncompetitively, with a dissimilar solute.

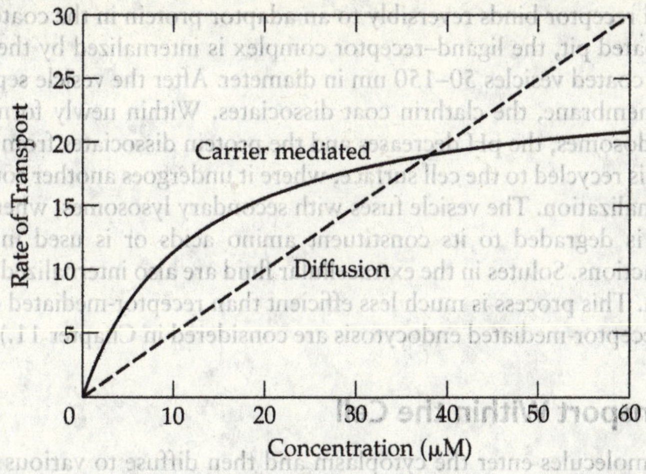

FIGURE 1.5 Comparison of carrier-mediated transport and diffusion across cell membranes.

Some carriers, such as the glucose carrier that is present on many cells, are known as *uniporters* and transport a single solute. Other carriers transport two solutes in the same direction (*symporters*) or in opposite directions (*antiporters*). Transport can be energy independent if the solutes are transported down an electrochemical gradient. Alternatively, active transport may be involved if the transport occurs against an electrochemical gradient.[3] The required energy often derives from the hydrolysis of the molecule adenosine triphosphate (ATP), coupled with the transport. The transport of a solute against an electrochemical gradient, coupled with ion transport down an electrochemical gradient, does not involve ATP hydrolysis and is referred to as *secondary active transport*.

Channel Transport. Channels enable transport by providing hydrophilic pores through which solutes move (see Chapter 14). Consequently, transport can occur only in the direction of the electrochemical gradient. The rate of transport through channels can be as much as a thousand times faster than transport rate by carriers. Channel proteins transport ions, and they are selective for specific ions. The channels are in either an open or a closed state. Channels can be opened (or *gated*) electrically, mechanically, or chemically. Ion channels play a major role in regulating transmembrane electrical potentials. These channels are critically important in electrically conductive cells such as nerve and cardiac cells.

Transport by Receptor-Mediated Endocytosis. The transport of large peptides and proteins into cells involves *receptor-mediated endocytosis*. The molecule binding to the receptor is often referred to as a *ligand*. The ligand binds reversibly to its receptor on the cell surface. Some receptors diffuse over the cell surface until they contact a region of the membrane that contains the protein *clathrin* on its cytoplasmic side. Such regions, referred to as *coated pits*, contain binding sites for receptors. The

[3] Transport of uncharged molecules is in the direction of the concentration gradient. For charged molecules, transport is affected by the electrical potential and concentration gradients. As discussed in Chapter 7, these two processes generate an electrochemical gradient.

ligand-bound receptor binds reversibly to an adaptor protein in the coated pit. Once within the coated pit, the ligand–receptor complex is internalized by the continuous formation of coated vesicles 50–150 nm in diameter. After the vesicle separates from the plasma membrane, the clathrin coat dissociates. Within newly formed vesicles, known as endosomes, the pH decreases and the protein dissociates from its receptor. The receptor is recycled to the cell surface, where it undergoes another round of binding and internalization. The vesicle fuses with secondary lysosomes, wherein the protein portion is degraded to its constituent amino acids or is used in subsequent metabolic reactions. Solutes in the extracellular fluid are also internalized during vesicle formation. This process is much less efficient than receptor-mediated endocytosis. (Models of receptor-mediated endocytosis are considered in Chapter 11.)

1.4.2 Transport Within the Cell

Many small molecules enter the cytoplasm and then diffuse to various parts of the cell. The organelles and cytoskeleton offer barriers to diffusion, increasing the time for molecules to reach them. Ions may associate nonspecifically with charged molecules or bind to specific sites on proteins, in order to perform specialized functions. For example, the calcium ion serves as a *second messenger*, relaying signals generated at the cell surface to various sites within the cell. A number of membrane proteins (e.g., calmodulin) bind calcium and serve as stores for that element. In addition, the calcium ion can be transported into the endoplasmic reticulum and stored there.

Polar hormones and drugs exert a number of actions by binding to receptors on the cell surface. This binding event initiates a cascade of biological reactions that affect cell function. For example, hormones can stimulate *signaling* by activating a protein found on the cytoplasmic face of the membrane known as a *G-protein* (Figure 1.6). The signaling process involves the activation of the enzyme *adenylate cyclase*, which catalyzes the formation of cyclic adenosine monophosphate (cAMP) from ATP. The cascade of events produces an amplification of the initial signal. For some signaling molecules, the rates of binding and reaction are limited by the rate of diffusion in the cell membrane. A large number of signaling pathways exist that can positively or negatively regulate ion transport and generation of other second messengers. In turn, these second messengers can stimulate cell division or regulate specific genes and subsequent protein synthesis. The cascade of reactions enables tight control of the various steps and amplification of signals.

Small and hydrophobic hormones, such as steroid hormones, retinoids, and vitamin D, bind to carrier proteins, which transport these hormone molecules across the cell membrane. Within the cell, they dissociate from the carrier, diffuse through the cytoplasm, and bind to receptors in the nucleus or the cytoplasm. The receptor–hormone complex enters the nucleus and binds to DNA. This hormone–protein–DNA complex can then regulate the expression of genes.

Newly synthesized proteins that are secreted from the cell are transported in vesicles by a pathway known as *exocytosis*. Proteins, destined for secretion or insertion into regions of the membrane, are synthesized at *ribosomes* attached to the endoplasmic reticulum. The proteins then move into the endoplasmic reticulum, which, together with the Golgi apparatus, forms a continuous network. Within the Golgi apparatus, the proteins undergo modifications and are sorted. The vesicles fuse with the plasma membrane and release their contents into the extracellular fluid.

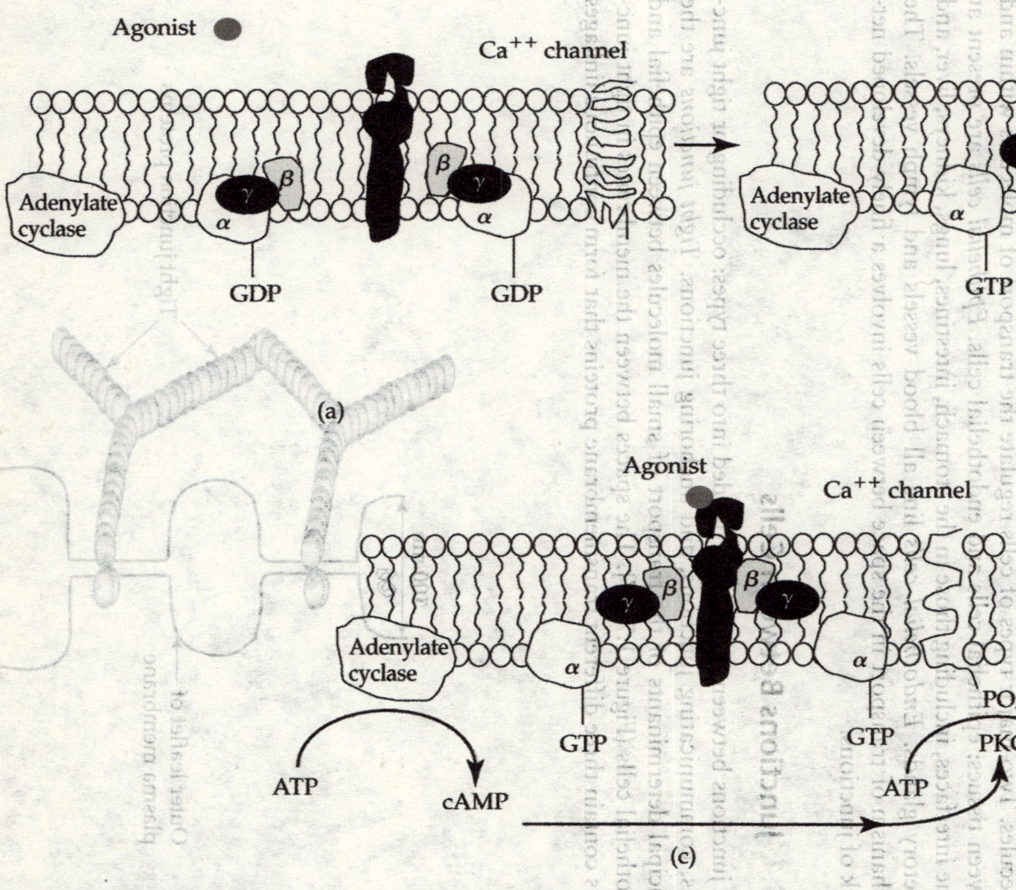

FIGURE 1.6 Schematic of G-protein signaling pathways in the cell membrane. (a) The G protein consists of α, β, and γ subunits. In the inactive state, the three subunits are bound together and diffuse in the membrane. (b) Binding of an agonist to a receptor produces a conformational change to the receptor, and the β subunit binds to this activated receptor. This activates the formation of guanosine triphosphate (GTP) from guanosine diphosphate (GDP), causing the complex to dissociate. As a result, the G protein dissociates from the receptor, and the agonist also dissociates. (c) The GTP-Gα complex diffuses in the membrane until it binds to adenylyl cyclase. This binding results in the production of cyclic adenosine monophosphate (cAMP) and hydrolysis of GTP to GDP. The complex diffuses until it binds to the $\beta\gamma$ complex. The G-protein complex is ready for another cycle. The cAMP activates the protein kinase C (PKC) pathway that leads to the addition of phosphate groups (i.e., phosphorylation) to proteins lining calcium channels. The channel opens, allowing Ca^{++} to enter the cell. Ca^{++} can then affect other functional changes. The channel is closed following removal of the phosphate group by the enzyme phosphodiesterase.

Nerve cells can have very long appendages, known as *axons*. These axons are needed to transmit electrical signals over long distances. Since diffusion is inefficient over such distances and convection does not occur, proteins and organelles are transported via microtubule *motor proteins*, which interact with microtubules and provide directed motion at speeds as high as 3–4.5 μm s^{-1}. The hydrolysis of ATP provides the energy to drive the motor proteins. In smaller cells, these motor proteins also serve to maintain the position of organelles.

1.5 Transcellular Transport

The movement of molecules between tissues must be finely regulated for the optimal concentration to be reached in the cells and to prevent the accumulation of toxic molecules. Two major types of cells regulate the transport of molecules within and between tissues: epithelial cells and endothelial cells. *Epithelial cells* are present at tissue interfaces, including those in the stomach, intestines, lungs, kidneys, liver, and secretory glands. *Endothelial cells* line all blood vessels and lymph vessels. The mechanism of transport in the space between cells involves a finely developed network of junctions.

1.5.1 Junctions Between Cells

The junctions between cells can be divided into three types: occluding or tight junctions, communicating junctions, and anchoring junctions. *Tight junctions* are the principal determinants of the transport of small molecules between epithelial and endothelial cells (Figure 1.7) [7]. The spaces between the membranes in tight junctions contain three different transmembrane proteins that form continuous linkages

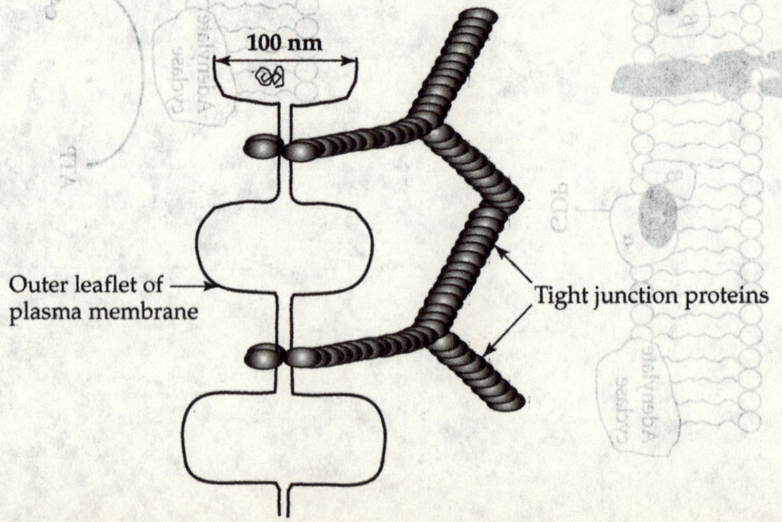

FIGURE 1.7 Schematic of the arrangement of the tight junction.

surrounding the cells. These proteins are also linked to the cytoskeleton. The number of such linkages influences the degree to which transport between cells is limited. In some epithelial cells, tight junctions enable the passage of specific ions. The permeability of tight junctions can be controlled by a number of stimuli that alter the interaction of the cytoskeleton with tight junction proteins.

There are two types of communicating junctions: gap junctions and synaptic junctions. *Gap junctions* consist of 2- to 4-nm-wide openings between two cells. The openings form a channel connecting two cells by means of a complex of protein molecules known as *connexons*. Connexons permit transcellular transport of electric currents and solutes with molecular weights less than 1,000 daltons. Variations in gap junction permeability among cells are due, in part, to differences in connexon structure. The proteins forming connexons can rotate to close the pores in the presence of a small number of calcium ions or a high pH. Although many cells, including epithelial cells, exhibit gap junctions, their functions are unclear.

Synaptic junctions are specialized forms of connections that nerve cells make with other nerve cells or muscle cells. Following stimulation by electrical signals in nerve axons, vesicles containing neurotransmitters fuse with the end of the synapse. Neurotransmitters rapidly diffuse across the small gaps between the nerve ending and target cells, where they interact with channels on the target cells. These channels transiently open, allowing ions to enter. The result is transmission of electrical impulses along the nerves or the activation of muscular contractions.

Anchoring junctions connect cells to the extracellular matrix or other cells. The connections involve transmembrane proteins linked to the cytoskeleton. For example, *adherens junctions* are linked to the cytoskeleton protein actin and can be found in the cell–cell adhesion belt and in cell–matrix focal contacts. A second set of junctions involves intermediate filaments.

1.5.2 Epithelial Cells

Epithelial cells line the cavities of organs such as the intestines, stomach, and lungs. These cells serve a number of functions, including secretion and absorption, and act as selective permeability barriers between fluid on different sides of cells. Epithelial cells are typically bound together in sheets, in different cell packing and orientation geometries (e.g., *columnar*, *cuboidal*, and *squamous*), alone, or adjacent to each other. The cells have a definite orientation, and basal, lateral, and apical surfaces can be described. The cell membrane in each of these regions has a specialized function. Cells with membranes that provide different functions, depending on the cell location, are considered *polarized*. The various types of epithelia with a transport function are listed in Table 1.7.

Epithelial cells that line the interior of the small intestine regulate nutrient transport into the body. The membrane properties differ on the surfaces that are exposed to the intestinal lumen and to the blood capillaries (known as *polarity*) in order to regulate nutrient transport to the blood (Figure 1.8). Like all epithelial cells, these cells consist of an apical side, which faces the intestinal lumen; a basal side, which is in contact with the extracellular space between the epithelium and blood vessels; and the lateral side, which is connected to other epithelial cells via tight junctions. The apical surface has a large number of membrane folds known as *microvilli*, which provide a large surface area for the exchange of solutes.

TABLE 1.7

Examples of Epithelial Cells Involved in Transport

1. Cells involved in absorption
   - brush border of intestine
   - striated duct cells of exocrine glands
   - gallbladder epithelial cells
   - brush border of proximal tubule of kidney
   - distal tubule cell of kidney
   - epididymal cells
2. Cells with an internal barrier function
   - type I pneumocyte—lines air spaces of lungs
   - duct cells of glands and exocrine organs
   - glomerular epithelium
   - collecting duct cells of kidney

Specific carriers transport glucose and amino acids from the intestinal lumen into the epithelium. Shown in Figure 1.8 is a glucose symporter that moves glucose and sodium ions (Na^+) from the lumen of the intestines into the cell. The energy for the transport comes from the movement of Na^+ down an electrochemical gradient. On the basal surface, glucose is transported to the extracellular fluid passively by means of a uniporter. In order to maintain low Na^+ concentrations within cells, an enzyme known as Na^+/K^+-ATPase hydrolyzes one ATP to adenosine diphosphate (ADP) and drives three Na^+ out of the cell and two potassium ions (K^+) into the cell. Solute transport between epithelial cells is blocked by the presence of tight junctions.

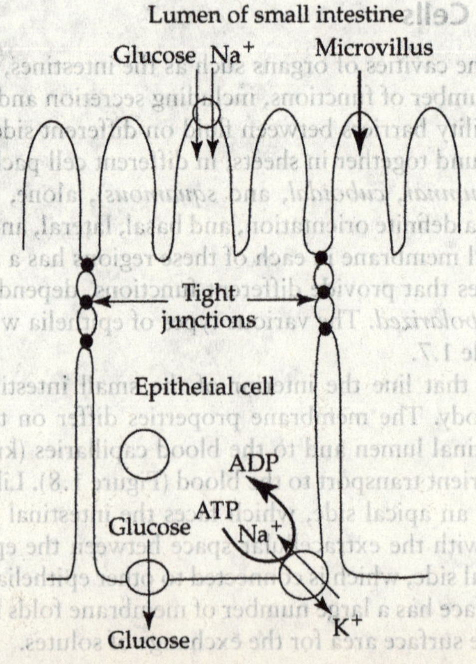

FIGURE 1.8 Schematic of a portion of an epithelial cell sheet showing directed transport of glucose from the intestinal lumen into the tissue.

1.5.3 Endothelial Cells

Endothelial cells line all blood vessels. The lining is called the *endothelium* and it regulates transport, much like epithelial cells. In addition, endothelial cells regulate coagulation and the adhesion of leukocytes.

There are three types of endothelial cells: continuous, fenestrated, and discontinuous (see Chapter 9). *Continuous endothelia* are found in all of the major blood vessels and capillary beds of the brain, muscles, heart, and lungs (see Figure 1.9). These endothelial cells are connected to each other by a system of gaps and tight junctions. The tight junctions of these cells are not as extensive as those found in epithelial cells. Except for the capillary endothelia in the brain, the junctions of continuous endothelial cells are permeable to solutes as large as 2 nm in radius. These cells often rest on a thin layer of extracellular matrix known as the *basal lamina*. Continuous endothelial cells make adherens junctions with the extracellular matrix. A prominent feature of continuous endothelia is an extensive number of vesicles.

Fenestrated endothelia are present in the capillary beds of endocrine glands and kidney glomerulus. Junctions similar to those found in continuous endothelium connect fenestrated endothelial cells. The *fenestrae* (from the Latin word for window) are openings in the plasma membrane between the capillary lumen and the underlying tissue or basement membrane (Figure 1.10). These are not transient structures, and the plasma membrane remains continuous. In some cases, the fenestrae contain a diaphragm that may serve as a molecular filter.

Discontinuous endothelia are not connected to each other. Their underlying extracellular matrix may be exposed to the capillary lumen, or another cell type may be positioned between the endothelial cells. Discontinuous endothelia are present in the liver, spleen, and bone marrow.

Molecules are transported across the endothelium in several ways. The junctions between cells permit the passage of molecules as large as 2 nm. Occasionally, very leaky junctions are observed, and agents that cause vessels to dilate or constrict

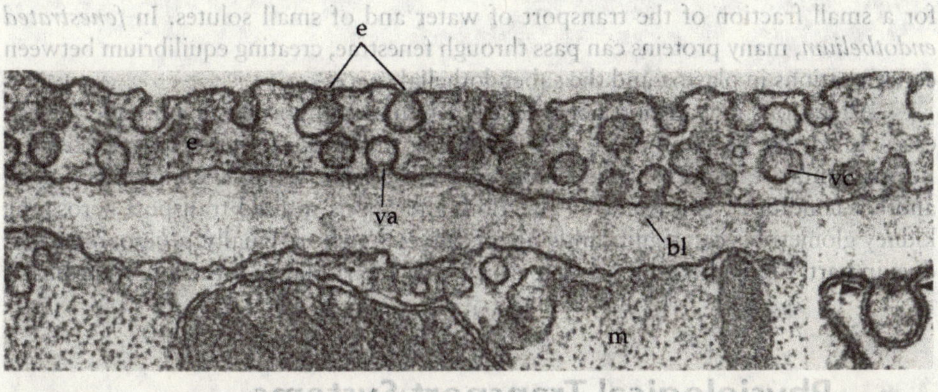

FIGURE 1.9 A continuous endothelial cell type from a vessel in the rat diaphragm. The endothelium (e) overlays a basal lamina (bl) and muscle cells (m) and exhibits a vesicle (va) in contact with the basal lamina. Vesicles are present throughout the endothelial cell (vc). (From Ref. [8], © 1984 by the American Physiological Society. Used by permission of Oxford University Press.)

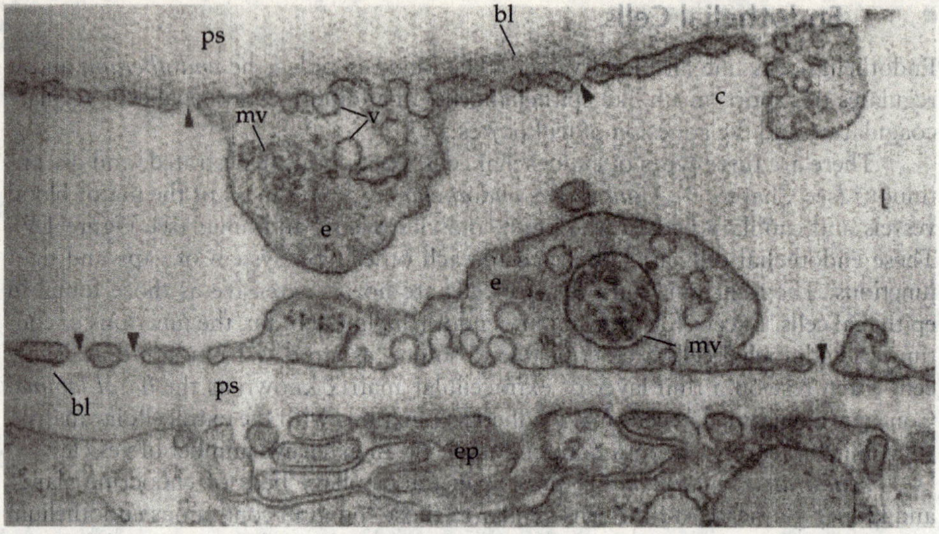

FIGURE 1.10 Example of fenestrated endothelium. The fenestrae (arrowheads) are near the peripheral portion of the endothelial cells (e). (ep) epithelial cell, (l) vessel lumen, (mv) multivesicular body, (ps) pericapillary space, (bl) basal lamina, (c) transendothelial channel. (From Ref. [8], © 1984 by the American Physiological Society. Used by permission of Oxford University Press.)

also increase the junction permeability. These changes in permeability involve coordinated interactions between the cytoskeleton and junction proteins [9].

Macromolecules are also transported across endothelium by vesicles. This process can occur in the fluid phase or by the binding of macromolecules to receptors on the endothelial cell surface. The distance traveled by vesicles is about 1 µm. In addition, vesicles have been observed to fuse, forming transient channels. Although vesicles can transport low-molecular-weight molecules, the vesicular pathway accounts for a small fraction of the transport of water and of small solutes. In *fenestrated endothelium*, many proteins can pass through fenestrae, creating equilibrium between concentrations in plasma and the subendothelial space.

Alterations in transport across endothelium are important in a number of diseases. In *edema*, for example, fluid accumulates in the extracellular space of tissues. Edema can arise from elevated blood pressure or damage to the endothelium. In another condition, as a result of infection, alterations in fluid transport across the kidney glomerulus can significantly affect kidney function. Finally, *atherosclerosis* is due, in part, to an alteration in the permeability of arterial endothelium to proteins.

1.6 Physiological Transport Systems

Most animals are composed of highly organized arrangements of cells known as tissues and organs. Often, many cells of a similar type are needed to perform a specific function and are grouped together in a tissue. Tissues themselves are not composed

TABLE 1.8

Examples of Organs and Organ Systems with Transport Functions

Organ or organ system	Transport functions
Respiratory system	Delivery of oxygen from the lungs to the blood and transport of carbon dioxide in the opposite direction
Cardiovascular system	Transport of oxygen within red blood cells
	Removal of carbon dioxide
	Delivery of antibodies and cells of immune system to sites of infection
	Thrombosis and hemostasis
Gastrointestinal tract	Digestion and absorption of nutrients
Liver	Carbohydrate storage and release
	Cholesterol metabolism and lipoprotein synthesis and metabolism
	Synthesis of plasma and transport proteins (e.g., albumin, transferring)
	Synthesis and export of molecules for tissue energy metabolism
	Urea synthesis
	Metabolism of toxins
Kidneys	Filtration of plasma
	Removal of urea and waste products
	Water reabsorption
	Maintenance of plasma volume and blood pH

solely of one type of cell, since they require a blood supply and innervation for communication, nourishment, and regulation. The major types of tissue are nerve, muscle, blood, lymphoid, epithelial, and connective tissue [5]. Organs are tissues grouped in an organized fashion to perform specialized functions. In turn, organs that have interrelated and coordinated functions are grouped together into organ systems. Examples include the digestive system, the respiratory system, and the reproductive system.

An important function of a number of organs and organ systems is the transport of molecules for growth, nutrition, repair, and communication (Table 1.8). The following discussion focuses upon five organs and organ systems that have significant transport functions: the cardiovascular system, the respiratory system, the gastrointestinal tract, the liver, and the kidneys, summarizing the transport functions and the relations between anatomy and physiology.

1.6.1 Cardiovascular System

The cardiovascular system consists of the heart, blood vessels, and blood. The primary function of the cardiovascular system is to transport oxygen from the lungs to the tissues. In addition, blood transports nutrients and hormones to various organs, brings waste products to the kidneys for filtration, transports molecules and cells of the immune system to sites of infection, and produces clots following injury to blood vessels.

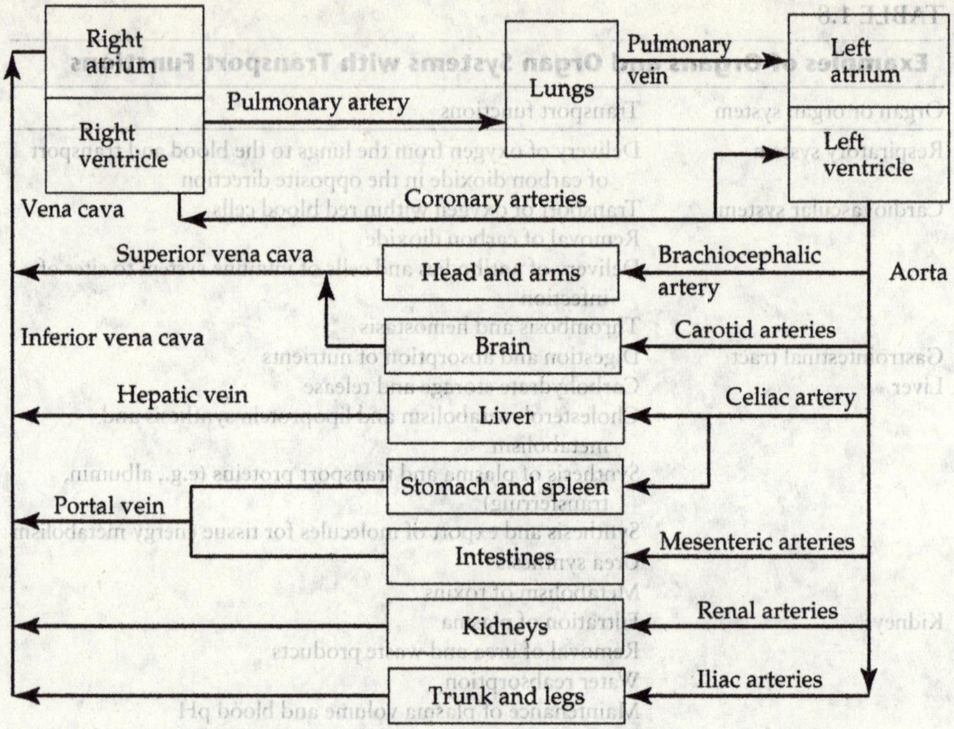

FIGURE 1.11 Simplified schematic of the blood flow distribution throughout the body.

The right side of the heart receives blood from the venous system and pumps it to the lungs (Figure 1.11). Blood then passes into the left side of the heart, where it is pumped to the body. Blood leaves the heart through arteries and returns via veins. Except for the pulmonary artery and vein, arterial blood is oxygenated, while venous blood has low oxygen concentrations.

The heart is a four-chambered pump consisting of muscle tissue that transports blood (Figure 1.12). It is divided into left and right parts, each with an *atrium* and a *ventricle*. Valves separate the atria and ventricles and maintain flow in one direction. The atria receive blood from the body. The ventricles pump blood to the organs. In order to reduce backflow, valves exist between the ventricles and blood vessels and between the atria and ventricles.

The beating of the heart is under neural and hormonal control. The atria and ventricles beat sequentially. Relaxation of the cardiac muscle causes the atria to fill. Once full, the atria contract, forcing blood into the ventricles. When the ventricular pressure exceeds atrial pressure, the *atrioventricular* valves close. The ventricles fill until the ventricular pressure exceeds the pressure on the arterial side. The aortic and pulmonary valves then open, and blood flows toward the organs. Because the left side of the heart pumps blood to all of the body except the lungs, it needs to generate a higher pressure. Consequently, the left ventricle is more muscular than the right ventricle.

The *cardiac output* is the amount of blood per minute that flows from the heart. The *stroke volume* is the amount of blood ejected during each heartbeat. Thus, the cardiac output (CO, measured in liters per minute) is the product of the

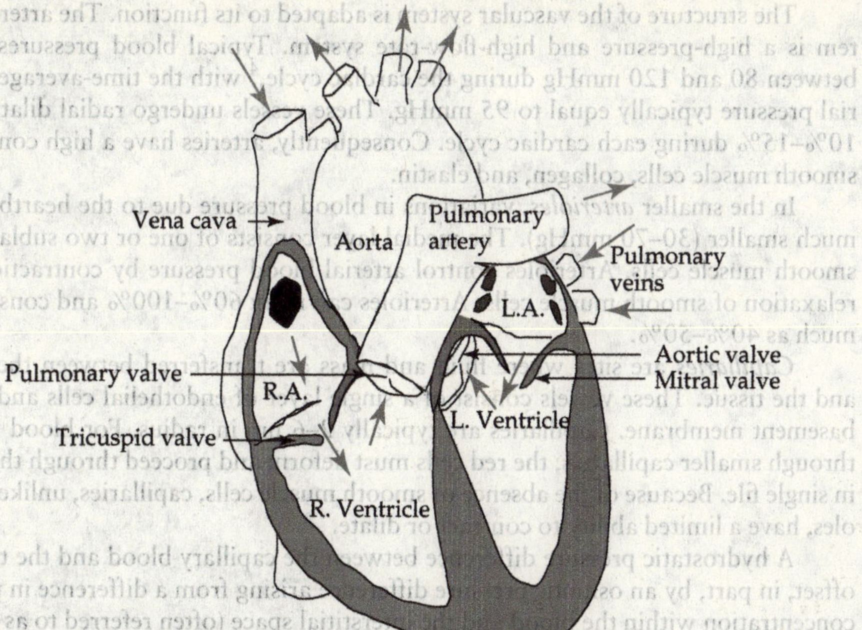

FIGURE 1.12 Diagram of the heart showing the four chambers, valves, and veins leading into and arteries leading away from the heart.

stroke volume (SV, measured in liters) and the heart rate (HR, measured in beats per minute); that is,

$$CO = SV \times HR. \tag{1.6.1}$$

The cardiac output in resting individuals is about 5 L min^{-1}, and the resting heart rate is between 60 and 72 beats per minute. During vigorous exercise, the heart rate can rise to 150 beats per minute, and the cardiac output can increase to as much as 25 L min^{-1}. The stroke volume also changes with exercise. During training, athletes are able to increase their stroke volume substantially, such that their heart rate increases only 50%–75%.

The system of blood vessels is divided into three main parts. The *arterial system* transports blood from the heart to the tissues. The *venous system* transports blood from the tissues to the heart. The *microcirculation* consists of small blood vessels in which solutes and solvents are exchanged with the tissue.

Blood vessels consist of endothelium, smooth muscle cells, and extracellular matrix. Arteries consist of three layers: the intima, the media, and the adventitia. The *intima* comprises the endothelium and a layer of extracellular matrix consisting of proteoglycan and collagen. The *media* consists of extracellular matrix and smooth muscle cells beneath the internal elastic lamina, providing structure and elasticity to the vessel. Collagen, elastin and glycosaminoglycans, present in the extracellular matrix, are responsible for the mechanical behavior of the blood vessel. Contraction and relaxation of the smooth muscle cells regulate the blood vessel diameter. The *adventitia* is a layer of loose connective tissue, smooth muscle cells, and fibroblasts. Capillaries and lymphatic vessels supplying the outer portion of the arterial wall are located within the adventitia.

The structure of the vascular system is adapted to its function. The arterial system is a high-pressure and high-flow-rate system. Typical blood pressures range between 80 and 120 mmHg during the cardiac cycle,[4] with the time-averaged arterial pressure typically equal to 95 mmHg. These vessels undergo radial dilations of 10%–15% during each cardiac cycle. Consequently, arteries have a high content of smooth muscle cells, collagen, and elastin.

In the smaller *arterioles*, variations in blood pressure due to the heartbeat are much smaller (30–70 mmHg). The medial layer consists of one or two sublayers of smooth muscle cells. Arterioles control arterial blood pressure by contraction and relaxation of smooth muscle cells. Arterioles can relax 60%–100% and constrict as much as 40%–50%.

Capillaries are sites where fluid and mass are transferred between the blood and the tissue. These vessels consist of a single layer of endothelial cells and a thin basement membrane. Capillaries are typically 2–6 μm in radius. For blood to pass through smaller capillaries, the red cells must deform and proceed through the tubes in single file. Because of the absence of smooth muscle cells, capillaries, unlike arterioles, have a limited ability to contract or dilate.

A hydrostatic pressure difference between the capillary blood and the tissue is offset, in part, by an osmotic pressure difference arising from a difference in protein concentration within the blood and the interstitial space (often referred to as *colloid osmotic pressure* or *oncotic pressure*). This net pressure difference causes fluid flow from the capillaries into the tissue and enhances convective transport of macromolecules in the interstitial space.

Due to a small pressure difference, fluid leaving the tissues is collected by a system of *lymphatic* vessels. Unlike the circulatory system, which operates in a loop, the lymphatic system is unidirectional. The lymph vessels connect to two main vessels that drain into the venous system. Although the flow of interstitial fluid in each tissue is quite low, the net drainage into all of the tissues amounts to several liters per day. In a number of diseases, there is a local accumulation of fluid in the interstitial space. This fluid accumulation, known as *edema*, can arise from damage to the lymph vessels, an elevation of the capillary pressure or capillary permeability, or a decrease in colloid osmotic pressure. Edema can occur in many kinds of tissue. Pulmonary edema, which can arise from mitral stenosis or heart failure, can severely impede breathing.

Blood leaves capillaries and flows into *venules*, which consist of an endothelial cell layer and a single layer of smooth muscle cells. Venules are exposed to pressures between 10 and 16 mmHg and serve as reservoirs for blood. About one-third of all the blood in a resting individual is stored in venules (Table 1.9). This volume can decrease in response to injury, such as hemorrhage. Veins are thin-walled vessels that receive blood from venules and transport blood to the right side of the heart. Veins are exposed to lower pressures than arteries and have thinner walls than arteries.

The mean arterial pressure $\overline{P}_a$ is the product of the cardiac output (CO) and the resistance (R) offered by the blood vessels of the circulatory system; that is,

$$\overline{P}_a = \mathrm{CO} \times R. \tag{1.6.2}$$

[4]Blood pressure is typically reported as mmHg. A pressure of 1 atm ($=1.0133 \times 10^5$ Pa $= 1.0133 \times 10^6$ dyn cm^{-2}) equals 760 mmHg.

TABLE 1.9

Distribution of Blood Volume

Region	Total (%)
Small veins and venules	45–53
Large veins	15
Lungs	10–12
Heart	8–11
Systemic arteries	10–12
Capillaries	4–5

TABLE 1.10

Vascular Resistance

Region	Total (%)
Small arteries	15
Arterioles	50
Capillaries	20
Venules	5
Veins	10

About 85% of the total *vascular resistance* occurs in the small arteries, arterioles, and capillaries (Table 1.10). Large arteries contribute negligibly to vascular resistance. Correspondingly, the largest pressure drop in the circulatory system occurs between the small arteries and the end of the capillary endothelium.

The blood-flow distribution to organs varies widely and is significantly altered during exercise (Table 1.11). This distribution is essential for the efficient transport of oxygen and metabolites. In resting individuals, organs involved in nutrient and fluid exchange, skeletal muscle, and the brain receive the greatest amount of blood flow. During exercise, the flow to digestive organs is reduced dramatically, while there is a significant increase in flow to skeletal muscle and the heart. Because of the importance of the brain, blood flow to the brain is maintained constant.

Blood pressure, heart rate, and hence cardiac output are under neural and hormonal control. Two types of nerves control heart function: *sympathetic nerves*, which stimulate the heart, and *parasympathetic nerves*, which slow down the heart. This innervation is responsible for maintaining blood pressure, blood volume, blood pH, oxygen levels, and carbon dioxide levels. Sympathetic nervous stimulation of the heart results in the local release of norepinephrine[5] from the synapses of nerve cells. The norepinephrine binds to receptors, increasing heart rate, conduction velocity, and contractility. (Receptors to which epinephrine and norepinephrine bind are known as *adrenergic receptors*.) Parasympathetic nervous stimulation of the heart results in the release of the neurotransmitter acetylcholine, which binds to receptors to reduce heart rate and conduction velocity.

[5]Norepinephrine is a neurotransmitter—a molecule released at the connection of a nerve cell to a muscle cell (*synapse*). Neurotransmitters transmit the signal generated by the nerve to the effector cell.

TABLE 1.11

Blood Flow Distribution during Rest and Heavy Exercise

	Rest		Heavy exercise	
Region	L min^{-1}	Percent of cardiac output	L min^{-1}	Percent of cardiac output
Digestive system	1.40	24	0.30	1
Renal	1.10	19	0.90	4
Brain	0.75	13	0.75	3
Heart	0.25	4	1.00	4
Skeletal muscle	1.20	21	22.00	85.5
Skin	0.50	9	0.60	2
Others	0.60	10	0.10	0.5
Cardiac output	5.80	100	25.65	100.0

Source: Adapted from Ref. [10].

Sympathetic nervous stimulation of most arteries, arterioles, venules, and veins releases norepinephrine, which binds β-adrenergic receptors and causes vessels to contract. Sympathetic nerves that release acetylcholine also innervate skeletal muscles. In skeletal muscle, acetylcholine binds to *cholinergic receptors* to produce vasodilation.

Blood flow in the microcirculation is under local regulation. Smooth muscle cells in blood vessels contract when stretched and relax after the stress is removed. Vessels respond rapidly to an increase or decrease in arterial or venous pressure and passively adjust local blood pressure toward the value present before the perturbation. The local release of molecules that cause blood vessels to dilate (*vasodilators*) or constrict (*vasoconstrictors*) influences the flow through the vessels. As a result, blood flow changes in response to various stimuli. Molecules that dilate or constrict blood vessels also alter the permeability of the endothelium to water and macromolecules [9].

Cells and molecules of the immune system are transported from sites of generation through the cardiovascular system to sites of infection. Red cells, white cells (*basophils, eosinophils, lymphocytes, mast cells, monocytes,* and *neutrophils*), and platelets form and, except for *T cells,* mature in the *bone marrow*. T cells mature and form subclasses in the thymus and migrate to the lymph nodes [11]. *B cells* reside in the spleen and lymph nodes and release antibodies that bind to molecules (antigens) that are present on pathogens (viruses, bacteria, yeast, and foreign molecules). T cells can bind to antibodies that are in turn bound to antigens. T cells and B cells respond to pathogens over periods of days to weeks. Neutrophils respond more rapidly, invading a site of infection and ingesting bacteria by a process known as *phagocytosis*. Over somewhat longer times, monocytes invade a site of infection. Once the monocytes are inside the tissue, chemical signals from the pathogen or other immune system cells cause the monocytes to transform into *macrophages*, which then bind to antibodies bound to antigens and ingest foreign molecules or cells by phagocytosis.

Cells of the immune system reach sites of infection via a sequential adhesion process [12]. At a site of infection, chemicals produced by the pathogens, or the pathogens themselves, diffuse to the endothelial cells and stimulate those cells to make proteins. These *adhesion receptors*, expressed on the endothelial cell surface, bind to counter-receptors on leukocytes. First, an initial adhesive contact occurs involving specialized sugar molecules on both endothelial cells and leukocytes known as *selectins*. Selectins bind rapidly to their counter-receptors. This initial contact causes the leukocytes to slow down and begin *rolling*, a process that involves the rapid formation and subsequent breakage of bonds as the bonds are stressed. The binding of selectins to their receptors activates a second class of molecules on leukocytes, known as *integrins*. These molecules bind to counter-receptors on endothelium and cause the leukocyte to stop rolling and become firmly arrested. Further interactions between integrins and their receptors cause the leukocytes to migrate between endothelial cells and enter the tissue, where they complete their immune function. (Models of adhesion and the role of transport processes are discussed in Chapter 12.)

Platelets produce clots following damage to blood vessels. A *clot* is an aggregate of red cells, platelets, and the protein fibrin. When portions of the extracellular matrix are exposed to blood after injury or damage to the endothelium, platelets bind rapidly to the extracellular matrix and release a number of molecules that promote the coagulation of blood cells and clotting proteins. The formation and growth of blood clots involves interplay among convection, diffusion, and chemical reaction.

1.6.2 Respiratory System

The respiratory system delivers oxygen to tissues, where it is metabolized, and removes carbon dioxide. The metabolism of oxygen produces ATP by a series of coupled reactions. ATP provides chemical energy for many biochemical reactions. Other functions of the respiratory system include the regulation of blood pH. The respiratory system consists of the nose, larynx, trachea, lungs, capillary bed, nervous system, and muscles (diaphragm), all working in a coordinated fashion to regulate breathing. Gas exchange occurs in the lungs (Figure 1.13a). The *trachea* terminates into two *bronchi*, each of which diverges into two *bronchioles*. This sequence of dividing airways continues for 23 generations (Figure 1.13b). The upper airways of the lung are involved in warming and humidifying the air. Beginning at the 15th generation of bronchioles, the gas exchange units, known as *alveoli*, appear. Capillaries surround each alveolus.

Respiration involves the intake of oxygen through the nose, followed by convective transport of the oxygen to the larynx and the lungs. Oxygen exchange between the air and blood occurs in the alveoli. Oxygen is transported in blood, dissolved in plasma,[6] and bound to *hemoglobin*, which is an oxygen-carrying protein within red blood cells. Because of its low solubility in plasma, most of the oxygen in blood is bound to hemoglobin. Convective transport through the blood carries the oxygen to tissues. At the lower blood-plasma concentrations of oxygen in tissues, oxygen dissociates from hemoglobin and diffuses into tissues and cells. Within cells, oxygen participates in the conversion of glucose to chemical energy in the form of ATP. The carbon dioxide produced by these reactions diffuses from the tissues into blood.

[6]Plasma is the liquid phase of blood and consists of water, ions, sugars, amino acids, and proteins.

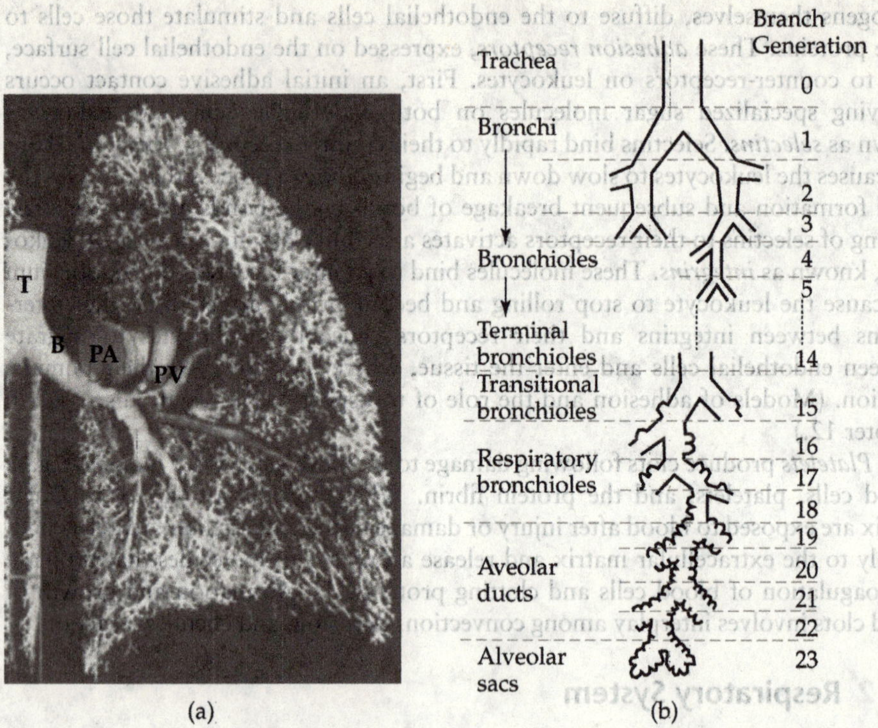

FIGURE 1.13 (a) Cast of a human lung, showing the trachea (T), one bronchus (B), the pulmonary artery (PA), and the pulmonary vein (PV). (b) Schematic of the organization of the airways in the human lung. (From Ref. [13], used with permission.)

The oxygen–hemoglobin dissociation curve is often reported as a function of the partial pressure of oxygen in the gas phase (Figure 1.14). Under normal conditions, the maximum partial pressure of oxygen in air (21 mol%) is 159.6 mmHg. The oxygen partial pressure, P_{O_2}, is related to the oxygen concentration in blood, C_{O_2}, through the formula

$$P_{O_2} = \frac{C_{O_2}}{H_{O_2}}, \tag{1.6.3}$$

where H_{O_2} represents the oxygen solubility in plasma. At 37°C, the solubility of oxygen in plasma is 1.4×10^{-6} mol L^{-1} mmHg^{-1}. The quantity P_{50} shown in Figure 1.14 is the oxygen partial pressure at which 50% of the sites on hemoglobin have bound oxygen. Under normal conditions, P_{50} for oxygen binding to hemoglobin is 26 mmHg. Arterial blood ranges from 95 to 100 mmHg pressure and is about 95%–97% saturated. Oxygen partial pressures in tissues are about 40 mmHg, and oxygen partial pressure in the pulmonary artery is typically 38 mmHg. Even at these partial pressures, hemoglobin saturation is about 70%. The total concentration of oxygen in blood is then

$$C_{O_2} = H_{O_2}P_{O_2}(1 - \text{Hct}) + \left(4C_{Hb}\bar{S} + H_{Hb}P_{O_2}\right)\text{Hct}, \tag{1.6.4}$$

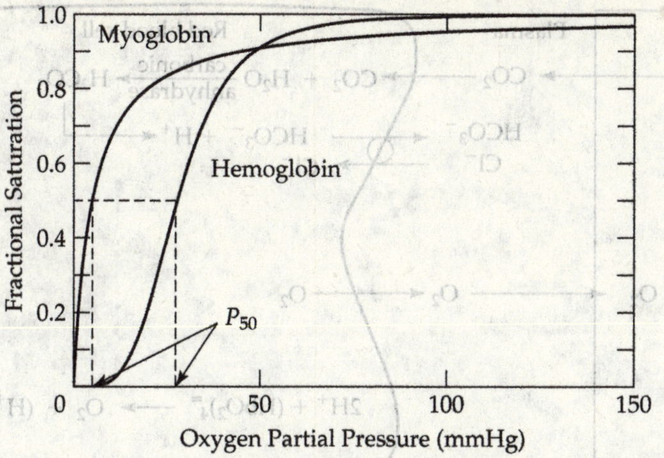

FIGURE 1.14
Oxygen–hemoglobin and oxygen–myoglobin dissociation curves. The fractional saturation is the relative amount of heme groups bound to molecular oxygen.

where C_{Hb} is the total concentration of heme groups in red blood cells, $\overline{S}$ is the average fractional saturation of hemoglobin, H_{Hb} is the oxygen solubility in hemoglobin solutions, and Hct is the hematocrit, the volume fraction of red blood cells. The fractional saturation is given by the following expression and is plotted in Figure 1.14.

$$\overline{S} = \frac{\left(P_{O_2}/P_{50}\right)^{2.6}}{1 + \left(P_{O_2}/P_{50}\right)^{2.6}}, \tag{1.6.5}$$

where $P_{50} = 26$ mmHg under normal conditions. The oxygen–hemoglobin dissociation curve is altered by a number of environmental changes. The curve can shift toward higher values for P_{50} by a decrease in pH or by an increase in temperature, carbon dioxide, or 2,3-diphosphogylcerate. In the presence of any of these factors, oxygen is more easily liberated from hemoglobin. Reduced pH is important in the normal release of oxygen and in the binding of carbon dioxide to hemoglobin in tissues, a phenomenon known as the *Bohr effect*.

Because oxygen partial pressures are low in the placental circulation (50 mmHg), fetal hemoglobin exhibits an oxygen–hemoglobin saturation curve in which P_{50} is shifted toward a lower value. Further, the hemoglobin content in fetal blood exceeds that in maternal blood. As a result, fetal blood has greater oxygen content than maternal blood.

Myoglobin is a protein that is present in muscle tissues and that contains a single heme group. Since myoglobin contains only one heme group, there is no cooperativity. Myoglobin binds oxygen very strongly ($P_{50} = 5.3$ mmHg) at the low oxygen partial pressures that appear in muscle during heavy exercise.

Carbon dioxide is present in three forms within blood (Figure 1.15) [14]. About 10% is dissolved at low levels in plasma. The solubility of carbon dioxide in plasma is more than 20 times the solubility of oxygen in plasma. Another 60% of carbon dioxide in blood is present in the plasma as bicarbonate ion. Although dissolved carbon dioxide reacts with water to form bicarbonate ion (HCO_3^-) and H^+, the reaction rate is greatly accelerated by the enzyme *carbonic anhydrase,* which is present in red cells. Although bicarbonate is produced in red cells, only about one-third of the bicarbonate remains within those cells. Although the red-cell membrane

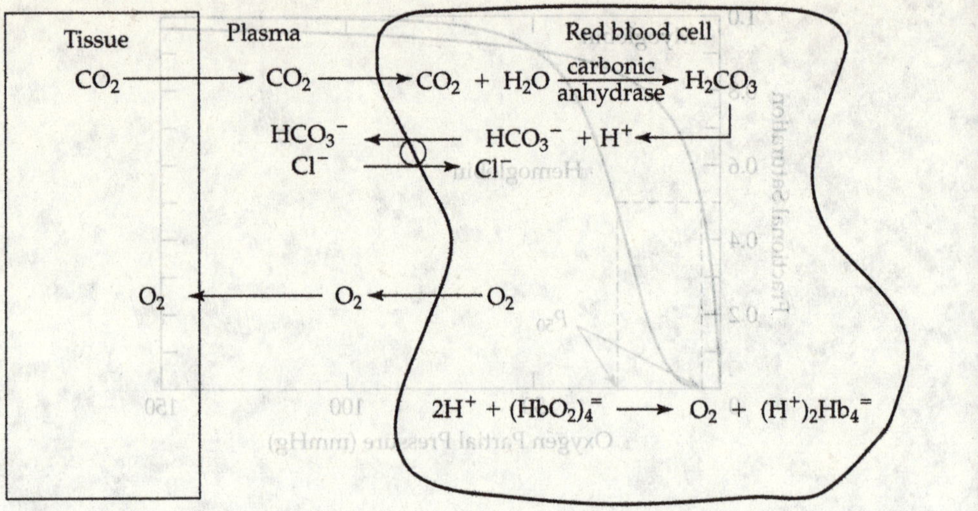

FIGURE 1.15 Schematic of transport and major reactions of oxygen and carbon dioxide with hemoglobin (Hb) in red blood cells. $(HbO_2)_4$ represents the fully oxygenated hemoglobin molecule.

is permeable to bicarbonate ion, transport is enhanced by a transmembrane transport protein, which exchanges one bicarbonate ion, between the interior of the red cell and the plasma, for one chloride ion, between the plasma and the interior of the red cell. Bicarbonate is one of several ions influencing the pH within the red cell, which, in turn, affects oxygen–hemoglobin binding.

The remaining 30% of the carbon dioxide in blood is bound to a modified form of hemoglobin in red cells. Carbon dioxide binds to uncharged amine groups ($-NH_2$) to form *carbamates*. Carbamate formation is enhanced by the conformational change that occurs upon the dissociation of oxygen from hemoglobin. Although carbamates can also form with plasma proteins, these carbamates account for a small fraction of those present in blood.

The lungs are an efficient system for gas exchange. The adult lung consists of 300 million alveoli that occupy a surface area of 130 m^2, an area equivalent to the size of two tennis courts! At 115 m^2, the capillary surface area is slightly smaller (Figure 1.16), and the capillary volume is 194 mL [15]. In order to reach the blood, oxygen diffuses through a surfactant film on the inner alveolar surface, across thin epithelial cells, across a narrow extracellular matrix, and, finally, across the capillary endothelium. The transport of oxygen and carbon dioxide across the layers of cells and matrix is rapid, because the layer is thin and these gases are soluble in the lipids of the cell membrane. Except during a disease, the diffusion of oxygen across the alveoli does not limit the transport of oxygen into blood. Under resting conditions, blood is completely oxygenated within one-third of its passage through the alveolar capillaries (Figure 1.16). During exercise, the blood-flow rate increases, due to the increase in cardiac output. As a result, blood travels farther along the length of alveolar capillaries before it is fully oxygenated. Nevertheless, blood is completely oxygenated before leaving the alveolar capillaries.

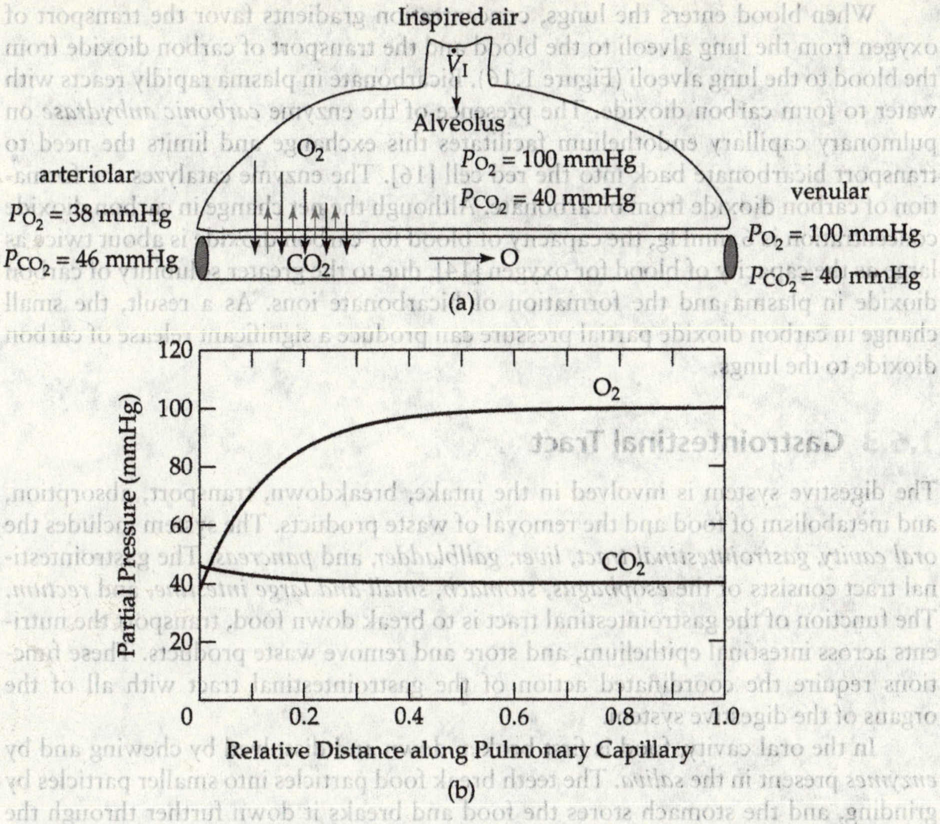

FIGURE 1.16 Oxygen and carbon dioxide exchange across the alveolar capillaries.

The net transport of gas between the lung and the blood vessels is obtained by applying a mass balance on the alveoli and capillaries (Figure 1.16). On equating the net gas flow into the alveolus, known as *ventilation*, with the net flow of gas between the arteriolar and venular ends of the capillary, known as *perfusion*, the resulting mass balance is

$$\dot{V}_{ALV}(C_I - C_{alv}) = Q(C_v - C_a), \tag{1.6.6}$$

where $\dot{V}_{ALV}$ is the alveolar ventilation rate, which is the volumetric flow rate of inspired air entering the alveoli, Q is the blood-flow rate through the lungs and equals the cardiac output, and the concentrations C_I, C_{alv}, C_v, and C_a refer to the gas concentration in the inspired air, alveoli, venular blood, and arteriolar blood, respectively. The gas phase concentrations can be obtained from the partial pressure using the ideal gas law. The alveolar volume equals the difference between the volume of inspired air minus the dead volume. $\dot{V}_{ALV}$ is the product of the alveolar volume multiplied by the breathing frequency (breaths per minute). Equation (1.6.6) applies to oxygen, carbon dioxide, or other gases used to measure perfusion and ventilation.

When blood enters the lungs, concentration gradients favor the transport of oxygen from the lung alveoli to the blood and the transport of carbon dioxide from the blood to the lung alveoli (Figure 1.16). Bicarbonate in plasma rapidly reacts with water to form carbon dioxide. The presence of the enzyme *carbonic anhydrase* on pulmonary capillary endothelium facilitates this exchange and limits the need to transport bicarbonate back into the red cell [16]. The enzyme catalyzes the formation of carbon dioxide from bicarbonate. Although the net change in carbon dioxide concentration is 6 mmHg, the capacity of blood for carbon dioxide is about twice as large as the capacity of blood for oxygen [14], due to the greater solubility of carbon dioxide in plasma and the formation of bicarbonate ions. As a result, the small change in carbon dioxide partial pressure can produce a significant release of carbon dioxide to the lungs.

1.6.3 Gastrointestinal Tract

The digestive system is involved in the intake, breakdown, transport, absorption, and metabolism of food and the removal of waste products. The system includes the *oral cavity, gastrointestinal tract, liver, gallbladder,* and *pancreas*. The gastrointestinal tract consists of the *esophagus, stomach, small and large intestine,* and *rectum*. The function of the gastrointestinal tract is to break down food, transport the nutrients across intestinal epithelium, and store and remove waste products. These functions require the coordinated action of the gastrointestinal tract with all of the organs of the digestive system.

In the oral cavity, food is first broken down and dissolved by chewing and by *enzymes* present in the *saliva*. The teeth break food particles into smaller particles by grinding, and the stomach stores the food and breaks it down further through the actions of digestive enzymes. The stomach can hold as much as 1 L of food. Food can reside in the stomach for periods ranging from several minutes to a few hours. The irregular surface of the stomach provides the mechanical means to disrupt and mix food. A number of specialized types of epithelial cells line different portions of the stomach and then release hydrochloric acid, enzymes, and proteins. A layer of mucus protects epithelial cells from the highly acidic pH of the stomach. Multiple layers of smooth muscle cells provide coordinated contractions to mix and move food through the stomach. Although most absorption occurs in the small intestine, water, alcohol, and lipid-soluble drugs are absorbed in the stomach.

Food moves sequentially through the small intestine in about three to four hours. The primary functions of the small intestine are absorption and secretion. The small intestine is about 7 m long and consists of the *duodenum* (0.25 m), *jejunum* (2.5–3 m), and *ileum* (4 m). After food leaves the stomach, it enters the duodenum, where the acids are neutralized. Pancreatic enzymes are involved in the breakdown of proteins, fats, carbohydrates, and polynucleotides into their constituents. These components are then absorbed in the jejunum by transport across the intestinal *mucosa*.

The inner surface of the intestine contains folds, termed *rugae*, that increase its surface area about fourfold (Figure 1.17). Each of these folds contains another set of folds, known as *villi*, which are typically several hundred microns long. Below the surface of the villi is an extensive network of capillaries. The combination of the large surface area of the villi and the abundant capillary network results in greater surface area for transport into the blood. The surface of the villi is covered with

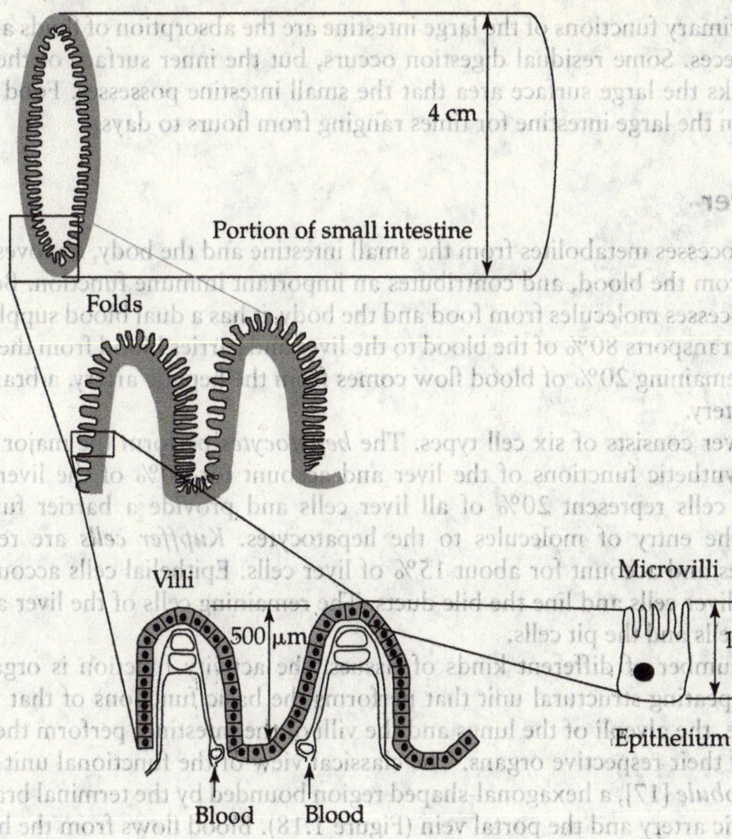

FIGURE 1.17 Schematic of the surface area of the inner intestinal mucosal surface.

epithelial cells, and the surface of those epithelial cells facing the intestinal lumen contains many finer folds, known as *microvilli*, in the cell membrane. Each cell contains several thousand microvilli, which are typically 0.5 μm long and 0.1 μm in diameter. Microvilli provide additional surface area to transfer mass across the cell membrane. The end result is a 300-fold increase in surface area.

In the lumen of the small intestine, hydrolytic enzymes break down carbohydrates, proteins, and fats to monosaccharides, amino acids, and triglycerides, which are subsequently transported across the epithelial membrane. The high concentrations of monosaccharides, amino acids, and triglycerides near the microvilli increase the rate of transport of these molecules across the epithelium.

Transport across epithelial cell membranes occurs by diffusion for lipid soluble molecules, such as alcohol, and for small polar molecules. Proteins, monosaccharides, and ions are transported across cell membranes in energy-dependent processes by carrier proteins. Ion transport is often linked to protein or sugar transport. In contrast, fatty acid triglycerides are emulsified by bile acids and diffuse across the epithelial cell membranes. Although a significant amount of water is removed from the digested food and secretions, osmotic gradients and passive diffusion drive water transport across epithelium. After entering the blood, sugar and amino acids are transported first to the liver. Fats are eventually transported to the liver in the form of large lipid particles called *chylomicrons*.

The primary functions of the large intestine are the absorption of fluids and the storage of feces. Some residual digestion occurs, but the inner surface of the large intestine lacks the large surface area that the small intestine possesses. Food waste may reside in the large intestine for times ranging from hours to days.

1.6.4 Liver

The liver processes metabolites from the small intestine and the body, removes toxic molecules from the blood, and contributes an important immune function. Because the liver processes molecules from food and the body, it has a dual blood supply. The portal vein transports 80% of the blood to the liver and carries blood from the intestines. The remaining 20% of blood flow comes from the hepatic artery, a branch of the celiac artery.

The liver consists of six cell types. The *hepatocytes* perform the major metabolic and synthetic functions of the liver and account for 60% of the liver cells. Endothelial cells represent 20% of all liver cells and provide a barrier function regulating the entry of molecules to the hepatocytes. *Kupffer cells* are resident macrophages and account for about 15% of liver cells. Epithelial cells account for 3%–5% of liver cells and line the bile ducts. The remaining cells of the liver are the fat-storing cells and the pit cells.

In a number of different kinds of tissues, the activity function is organized around a repeating structural unit that performs the basic functions of that tissue. For example, the alveoli of the lungs and the villi of the intestines perform the basic functions of their respective organs. The classical view of the functional unit of the liver is the *lobule* [17], a hexagonal-shaped region bounded by the terminal branches of the hepatic artery and the portal vein (Figure 1.18). Blood flows from the hepatic

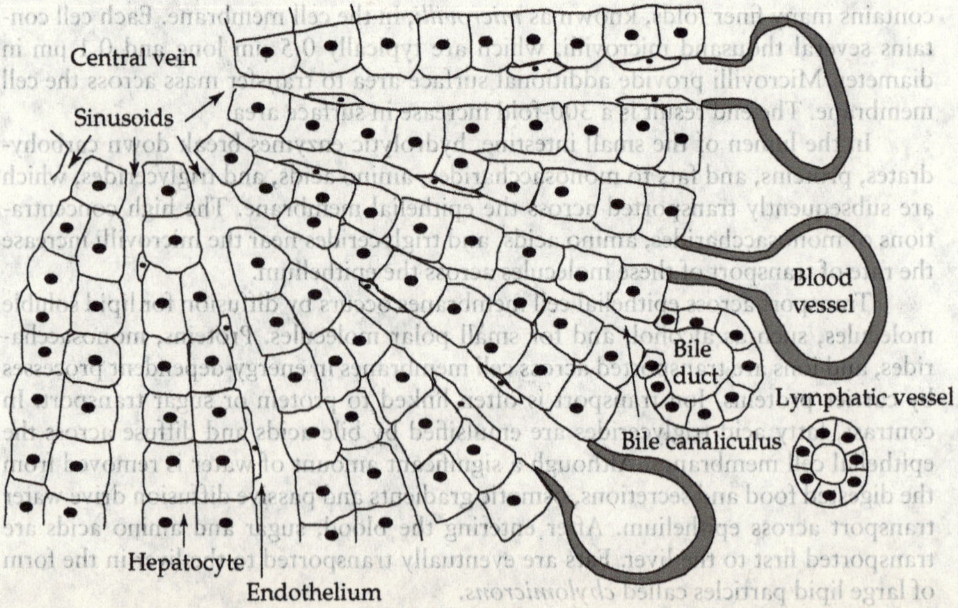

FIGURE 1.18 Schematic of a region of a hepatic lobule.

artery or portal vein through the sinusoids to the central vein. The sinusoids are analogous to the capillaries found in other tissues. The hepatocytes form a two-cell layer between sinusoids.

A thin layer of extracellular matrix, known as the *space of Disse*, lies between the endothelial cells and hepatocytes. Fat-containing cells are found in the space of Disse. Between the hepatocytes, bile collects in the bile *canaliculi*, which terminate in the bile duct. The liver lobules also contain lymphatic vessels and nerves.

Based upon recent studies of the structure and function of the liver, a different view of the functional unit of the liver has emerged. Sinusoids from different lobules have interconnections, and the state of oxygenation and metabolic function may vary within and between lobules. A conical segment of the classic lobule, such as that shown in Figure 1.18, represents the functional unit and is termed the *hepatic microvascular subunit*.

The liver endothelium is not continuous. *Kupffer cells* are present adjacent to and beneath endothelial cells. These Kupffer cells are in direct contact with endothelial cells. Sinusoidal endothelial cells contain numerous fenestrae that are 100 nm in diameter. Cytoskeletal proteins surrounding the fenestrae regulate the fenestrae diameters in response to various stimuli. Smaller molecules are transported across fenestrae without significant resistance. As a result, small molecules present in the space of Disse are probably in chemical equilibrium with the blood plasma. The fenestrae appear to prevent the passage of large particles.

The polarized surface of hepatocytes enables them to perform important metabolic and transport functions [18]. The basal surface of hepatocytes is in contact with sinusoidal blood through fenestrae and the extracellular space. The basal surfaces are exposed to high levels of circulating metabolites and proteins. Located on the basal surface are receptors for amino acids, glucose, and a number of metabolic proteins. The cytoplasmic side of this surface also contains molecules involved in formation and fusion of vesicles with the plasma membrane. This arrangement is important for the receptor-mediated uptake of proteins and release of synthesized molecules (e.g., albumin) by the hepatocytes. Bile acids, immunoglobulin A, and lipids are transported across the apical surface into the bile canaliculus. The lateral surfaces contain tight junctions that connect hepatocytes with each other and limit the transport of molecules between the hepatocytes. Bile is selectively secreted from the apical surface.

The major functions of hepatocytes are energy metabolism, bile synthesis and secretion, cholesterol and lipoprotein metabolism, detoxification, synthesis of vitamins and retinoids, and metabolism of copper, zinc, and iron ions. Hepatocytes process amino acids, fatty acids, and sugar molecules from the intestines and other organs and produce glucose, acetoacetate, triacylglycerols, and phospholipids [19]. Through the urea cycle, hepatocytes remove excess nitrogen and synthesize the essential amino acid arginine.

The liver synthesizes most of the cholesterol needed by the body and regulates plasma cholesterol and lipoprotein levels. Cholesterol is important for cell membranes and for myelin in nerves, and it is a precursor of steroid hormones and bile acids. Cholesterol is transported to the body tissues in the form of very-low-density lipoproteins (VLDLs) and low-density lipoproteins (LDLs). In turn, LDLs and high-density lipoproteins (HDLs) return cholesterol to the liver.

In the process of detoxification, hepatocytes modify molecules that are not normally synthesized or used by the body. Excessive levels of these molecules may be harmful. One important function of hepatocytes is to convert lipophilic molecules to a water-soluble form for removal by the intestines or kidneys. Otherwise, the molecule would mix with fat in fat tissue and remain for long periods of time. Enzymes involved in these reactions are concentrated in the liver.

1.6.5 Kidneys

The kidneys are responsible for the removal of waste products such as urea, for fluid and ion balances, and for the regulation of the blood pressure. *Filtration* of the blood plasma and *reabsorption* of water and ions remove waste products and balance fluids. The kidneys account for about 0.5% of the body weight, but receive 25% of the cardiac output (1.25 L min^{-1}). Large flow rates are needed in order to filter blood efficiently. Transport processes in the kidney are discussed in detail in Chapter 14. Kidney function is summarized here.

The kidneys receive blood from the renal arteries, which are branches of the abdominal aorta. The arteries branch repeatedly, ultimately forming interlobular arteries. The processes of filtration and reabsorption occur in the *nephron*, the functional unit of the kidney. The kidneys contain 1.2 million nephrons. Blood from the interlobular artery passes through *glomerular capillaries*. A net pressure difference between the capillary and *glomerulus* causes fluid and small solutes to flow across the glomerulus, a process known as *ultrafiltration*.

The glomerulus consists of a layer of fenestrated endothelium lining the capillary surface, a basement membrane, and a layer of epithelial cells (*podocytes*) (See Figure 14.6). The fenestrae permit unrestricted solute transport across the endothelium. The glomerular basement membrane is 300 nm thick and consists of collagen and negatively charged proteoglycans. Between the regions where the epithelial cells contact the membranes (termed *foot processes*) are narrow gaps through which solute and fluid are transported. The outer surfaces of the epithelial cells are coated with sialic acid, a negatively charged sugar group. These charged groups appear to play a role in the charge selectivity of the membrane.

The glomerulus is highly selective with respect to molecular size, as shown in Figure 1.19. Molecules with radii less than 1.5 nm are filtered across the membrane

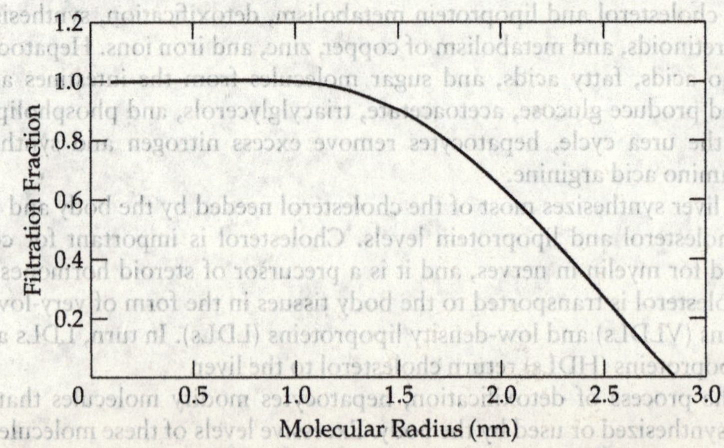

FIGURE 1.19 Fraction of solute filtered across the glomerulus as a function of molecular radius. The fraction filtered equals C_i^g/C_i^a, where C_i^g is the solute concentration in the glomerulus and C_i^a is the solute concentration in the afferent arterioles.

TABLE 1.12

Filtration and Excretion of Water, Electrolytes, and Solutes

Molecule	Filtration rate	Excretion rate
H_2O	180 L day^{-1}	1.44 L day^{-1}
Na^+	25,200 mmole day^{-1}	150 mmole day^{-1}
K^+	720 mmole day^{-1}	100 mmole day^{-1}
Ca^{++}	270 mmole day^{-1}	5 mmole day^{-1}
HCO_3^-	4,320 mmole day^{-1}	2 mmole day^{-1}
Cl^-	18,000 mmole day^{-1}	150 mmole day^{-1}
$C_6H_{12}O_6$ (glucose)	800 mmole day^{-1}	0.5 mmole day^{-1}
$CO(NH_2)_2$ (urea)	933 mmole day^{-1}	467 mmole day^{-1}

Source: Adapted from Ref. [20].

without any restriction. Solutes with molecular radii between 1.5 and 3.0 nm are transported across the membrane to a limited extent. The network structure of the membrane is not uniform. As a result, there are some openings between matrix molecules that allow larger molecules to penetrate. The frequency of these openings decreases with the size of the opening, limiting the passage of large solutes. The membrane also is charge selective: positively charged molecules pass through the membrane more easily than do uncharged molecules of the same molecular radius. In contrast, negatively charged molecules are filtered to a lesser extent than are uncharged molecules of the same size. A variety of diseases of the kidney result in protein in the urine, or *proteinurea,* a condition caused by a change in filtration due to an alteration in either the charge or the size selectivity of the glomerular membrane.

The filtrate flows from the glomerulus to the tubule network, where solutes are reabsorbed by active and passive transport across the epithelial cells to the blood. Water is passively transported to maintain an osmotic balance. Urea and other waste products are concentrated in the filtrate. To maximize *reabsorption* of fluid and solutes by the blood, there is a *countercurrent flow* of blood in the capillaries and the tubules. The extent to which ions and water are reabsorbed is summarized in Table 1.12. Blood leaves the nephrons through the interlobular veins, which merge into the renal vein. Urine from many nephrons enters the collecting ducts and flows into the bladder.

A mass balance summarizes the overall exchange of solutes and water in the kidney. Essentially, the mass entering the capillaries minus the mass leaving the capillaries equals the mass excreted in urine. Because solutes are flowing, the mass balance can be written as

$$C_i^a \text{RPF}^a - C_i^v \text{RPF}^v = C_i^u Q^u, \qquad (1.6.7)$$

where C_i^a and C_i^v are the concentrations of the solute i in the renal artery and vein, respectively, C_i^u is the concentration in the urine, and RPF^a and RPF^v are the rates of renal plasma flow in the artery and vein, respectively. $\text{RPF}^a = \text{RPF}^v + Q^u$.

In the *proximal tubule,* a significant fraction of the ions and water and most of the glucose and amino acids are transported from the tubule to the *peritubular* capillaries. A variety of antiporters, symporters, and uniporters transport ions, along with glucose, amino acids, lactate, or phosphate, from the tubule lumen into the epithelial cells, which contain a network of tight junctions that permit transport of water and Na^+ and Cl^- ions in the lower portion of the proximal tubule. In the *loop of Henle*, cations are transported between cells, although water is not. Sodium ion reabsorption and hydrogen ion excretion continue by a similar mechanism. In the distal tubule, Na^+ and Cl^- reabsorption continues with little transport of water. The driving mechanism in all of this transport is the Na^+/K^+-ATPase, which actively maintains low sodium concentrations within the cell. As a result, the symporters and antiporters function by transporting the ions down an electrochemical gradient.

1.6.6 Integrated Organ Function

The functions of tissues are regulated locally or systemically. Systemic regulation can occur through the release of molecules from one type of cell that are utilized by another type of cell. *Paracrine* regulation involves molecules with short biological half-lives, limiting the distances over which the molecules can be active. The release of vasodilators is a good example of paracrine regulation. These molecules are released by endothelium in response to local changes in blood flow or other agents and act on nearby smooth muscle cells. Alternatively, regulation can occur as the result of one type of cell binding to a second type. The binding of immune system cells to each other and to cells in various tissues is a good example of such interactions. In addition, cells can release signaling molecules that act on the cells secreting the molecules. Such autocrine regulation is important during development, for smooth muscle contraction, and for platelet aggregation.

Systemic regulation occurs via synaptic transmission and the endocrine system. Synaptic transmission involves the release of neurotransmitters from nerve cells upon *effector* cells and is relatively rapid. For example, acetylcholine release affects the speed of muscle cell contractions. Diffusion distances are very short, so that the release step does not produce a significant time delay. In order to limit the response, the *affinity* of the neurotransmitter for the receptor is low. For long-term regulation, hormones are secreted by various tissues of the endocrine system (e.g., thyroid and adrenal glands). These molecules are transported through blood to their effector sites. Hormone binding initiates a cascade of signaling events within cells that affect cell function.

1.7 Application of Transport Processes in Disease Pathology, Treatment, and Device Development

A quantitative analysis of transport processes is essential to the understanding of normal physiological and cell biological processes. Further, many disease processes result from alterations in transport pathways. Treatment often involves delivering a local agent to a site of disease or injury. This section provides examples of the variety of roles that transport processes play in the progress and treatment of disease. First, we describe how transport processes contribute to the development of atherosclerosis. Then, we examine the role of transport processes in a variety of therapies used in treating cancer. Finally, we see how transport processes apply in the development of devices or therapies that use biomolecules and cells.

1.7.1 Transport Processes and Atherosclerosis

Atherosclerosis is a disease of the large and midsized elastic and muscular arteries. Complications of the disease are the leading cause of death in the United States and many other developed countries. Atherosclerosis is characterized by the formation of lesions in the *arterial intima* consisting of lipids, plasma proteins, fibrous tissue, smooth muscle cells, and macrophages. When fully developed, lesions protrude into the lumen of the artery, reducing blood flow or causing platelets to adhere to the arterial wall, resulting in thrombosis or embolism. Complications arising from atherosclerosis (heart attack, heart failure, and stroke) are the leading cause of death in the United States, Canada, and Europe. Although a number of risk factors have been identified (e.g., elevated plasma cholesterol, hypertension, and diabetes), the mechanism of how the disease begins and progresses is poorly understood. Understanding these mechanisms could produce new therapies to treat the disease prior to the onset of clinical symptoms.

Characteristic features of early atherosclerotic lesions are lipid-filled macrophages and the presence of extracellular lipids. Lesions first appear around arterial branches or in highly curved arteries, possibly due to the alteration of endothelial cell function and monocyte transport by arterial fluid dynamics. The lipids present in atherosclerotic lesions are derived primarily from lipoproteins. Numerous epidemiological studies have demonstrated a correlation between plasma cholesterol levels and the risk of atherosclerosis. The mechanism by which plasma cholesterol influences the initiation and progression of the disease is poorly understood. Oxidation of lipoproteins appears to play a significant role in the progression of lesions. Oxidized lipoproteins can damage endothelium and lead to the unregulated accumulation of cholesterol within macrophages. Lipoproteins from patients with coronary artery disease show an increased susceptibility to oxidation. Thus, transport of lipoproteins into the arterial wall and their subsequent metabolism play an important role in the initiation and progression of atherosclerotic lesions.

LDLs are the major carriers of cholesterol in the blood. The LDL macromolecule is composed of a hydrophobic core of trigylceride and cholesterol, surrounded by phospholipid, cholesterol esters, and protein. The protein consists of two structurally similar units, known as *apoprotein B*, each of molecular weight 250,000. The apoprotein B molecule is entwined around the lipid and binds specifically to receptors on the cell surface. LDLs form in the blood as a result of the actions of the lipases that modify the lipids of VLDLs. Apoprotein B in the LDL molecules is essential for the receptor-mediated recognition of LDLs on the liver and peripheral tissues.

In normal vessels, the arterial endothelium presents a significant barrier to the entry of LDLs into the arterial wall. The major route of transport across normal endothelium is believed to be endocytosis. The LDL receptor does not appear to play a significant role in LDL transport, because the endothelial cells are exposed to high concentrations of LDLs in the plasma, reducing the number of LDL receptors on the cell surface. Regions of arteries where lesions develop exhibit increased endothelial permeability to macromolecules. The permeability of the arterial endothelium is 10–100 times greater than that of the surrounding regions. Endothelial cell junctions become transiently leaky and allow the entry of macromolecules. The cause of the transient permeability is not known, although indirect evidence suggests that the local fluid dynamics may play a role.

Many factors affect the rate of transport of macromolecules within the arterial wall, such as the macromolecule size and concentration, the macromolecular surface charge, the extracellular matrix composition and charge, transmural pressure differences, and the macromolecular binding to extracellular matrix constituents and cell surface sites. The fenestrated internal elastic lamina between the intima and the media may serve as a transport barrier. Within the media, smooth muscle cells metabolize LDLs. Both receptor-dependent and receptor-independent pathways are involved. As a result of convection, diffusion, binding to the matrix, and cellular metabolism in the media, there is a localized accumulation of lipoproteins in the intima.

Atherosclerotic plaques generally form in characteristic regions of the circulatory system. Often, highly lesion-prone areas are adjacent to lesion-resistant ones. However, the sites that are predisposed to the formation of lesions are fairly consistent in individuals. Because the plaques result from the metabolism and accumulation of cholesterol within the arterial wall, regional variations in permeability may explain the predisposition for plaque formation in specific locations.

The initiation and progression of the disease are influenced by monocyte transport to and accumulation within the vessel wall, as well as by the transport of LDLs and its subsequent chemical modification and metabolism. In studies with cholesterol-fed animals, the permeability of the endothelium has not altered as the disease begins. Increased LDL accumulation appears to be due to either binding to the extracellular matrix or accumulation within macrophages that have invaded the tissue. Binding of LDLs to the matrix may increase the likelihood that the LDLs are chemically modified by oxidants within the arterial wall. As the disease progresses, the permeability of the endothelium does increase, possibly due to the release of chemicals known as *cytokines* from macrophages or to the oxidation of LDLs. Cytokines and oxidized LDLs stimulate endothelium, altering their normal metabolic function and reducing the barrier function of junctions.

A possible sequence of events is shown in Figure 1.20. Following the elevation of plasma cholesterol levels (*hypercholesterolemia*), LDLs accumulate within the vessel wall at sites of elevated permeability. Elevated LDL concentrations stimulate the expression of adhesion molecules on endothelial cells, possibly as the result of oxidation. Adhesion molecule expression is further increased by local fluid dynamics. Receptors on blood monocytes bind to these adhesion molecules. Firm binding of the monocyte to the endothelium provides a resistive force that balances the shear forces produced by blood flow. (The transport and binding processes involved in leukocyte adhesion are discussed in Section 12.4.) Adherent monocytes migrate between the endothelial cells and enter the tissue. Macrophages do not have receptors for LDLs themselves; rather, they have receptors for modified forms of LDLs. The modified forms of LDLs bind to these *scavenger receptors*, and the molecules are internalized. Although the LDL receptors are regulated by cholesterol consumption, the scavenger receptors are not. The accumulation of cholesterol by these scavenger receptors is unregulated, and macrophages accumulate lipid deposits, which gives

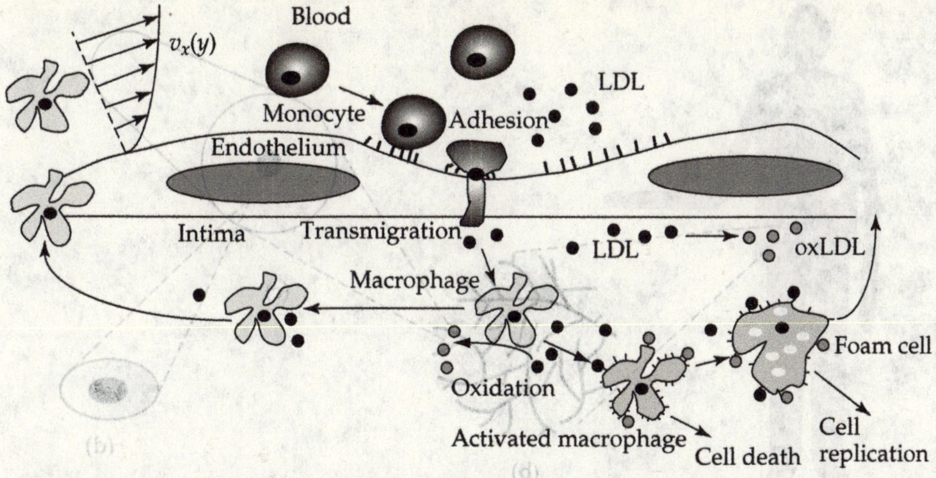

FIGURE 1.20 Schematic of events leading to macrophage accumulation and foam cell formation.

the cells a foamy appearance due to the lipid particles within the cells. Hence, the cells are given the name *foam cells*. Normally, macrophages perform their scavenger function, remove foreign material, and leave the tissue. However, if the plasma cholesterol remains high and LDLs continue to accumulate in the vessel wall, macrophages remain. If the underlying risk factors are not removed, the process acts as a positive feedback loop, resulting in excessive accumulation of foam cells. Over a period of years, these macrophages release enzymes that weaken the vessel wall, making rupture of the plaque likely. When the plaque interior is exposed to blood, a clot forms—occluding the vessel and leading to a heart attack or stroke.

1.7.2 Transport Processes and Cancer Treatment

Cancer is the second leading cause of death in the United States [21]. In 2007, more than 1,500 cancer patients died each day and about 1.45 million new cases were diagnosed. The major approaches to cancer treatment are surgery, radiation therapy, and chemotherapy. Among them, surgery and radiation therapy are used only for patients with a few large tumors in the body. When tumors have already *metastasized* (i.e., disseminated) to many locations in the body, patients have to be treated with chemical drugs. Currently, chemotherapy acts in the body by killing the dividing cells. However, some vital normal tissues (e.g., intestine, liver, and bone marrow) are also damaged during the treatment. Accordingly, there is great interest in developing more effective means to deliver drugs into tumors. Delivery is also an important issue in new approaches to cancer treatment (see Chapter 15).

In any approach based on therapeutic agents, it is the concentration at *all* target sites that determines therapeutic efficacy in cancer treatment. As a result, the delivery of therapeutic agents in solid tumors is critical. The efficiency of delivery depends on the physical and chemical properties of these agents, which can vary

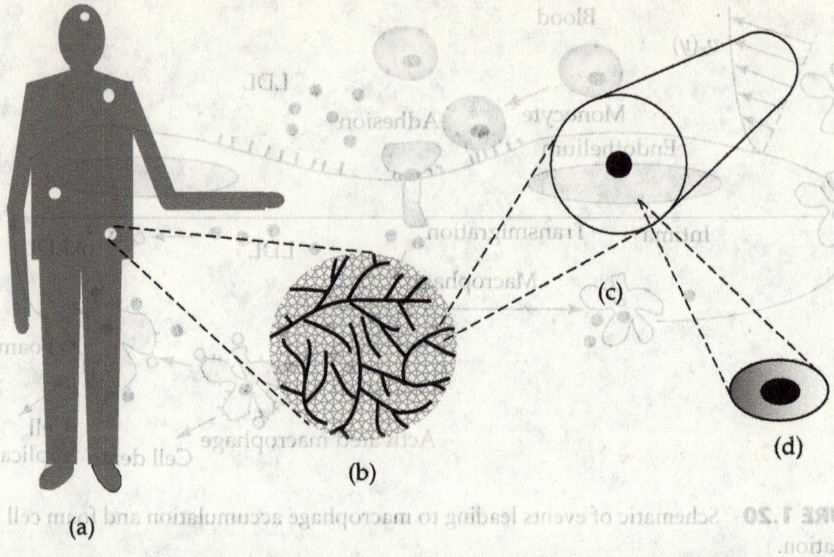

FIGURE 1.21 Schematic of drug delivery to tumor cells at four different levels: (a) the body level (~1 m), (b) the tissue level (~1 cm), (c) the microvessel level (~0.01 cm), and (d) the cell level (~10 μm). The open circles in (a) represent solid tumors. The solid curves in (b) represent tumor vessels. There are two concentric cylinders in (c). The inner cylinder represents a microvessel and the outer cylinder represents the region in tumor tissues that receives nutrients from this microvessel; (d) represents a tumor cell with a nucleus.

from small molecules to cells. For example, anticancer agents can be oxygen in radiation therapy, cytotoxic drugs in chemotherapy, photosensitizers in photodynamic therapy, viral or nonviral vectors in gene therapy, and effector cells or vaccines in immunotherapy. The delivery of therapeutic agents may involve tumor microcirculation, transvascular transport, interstitial transport, transport across the plasma membrane of cells, and intracellular transport. The details of some of these transport phenomena are discussed in Chapter 15; principles that govern intracellular and extracellular transport are also discussed in other chapters.

In general, drug and gene delivery to tumor cells can be studied at four different length scales (Figure 1.21): the body level (~1 m), the tissue level (~1 cm), the microvessel level (~0.01 cm), and the cell level (~10 μm). The whole-body distribution of drugs and genes must be analyzed in order to understand which organs are targeted by these agents. Determining local distributions of drugs and genes within a tumor or around a microvessel is important in understanding whether the drug concentrations around all tumor cells are higher than threshold levels that are required to kill tumor cells. If targets of drugs and genes are inside tumor cells or inside the nucleus, then transport in tumor cells becomes important. Barriers to intracellular transport include the plasma membrane, the membrane of vesicles in cells, the cytoskeleton, and the nuclear envelope. It is a challenge to facilitate drug and gene transport across these barriers.

1.7.3 Transport Processes, Artificial Organs, and Tissue Engineering

Since the 1960s, a number of *assist devices* have been developed, including circulatory support devices, heart–lung machines, oxygenators, and artificial kidneys. Some assist devices (e.g., heart–lung machines) are used in surgery or intensive-care units, whereas the kidney dialysis machine is used if the kidneys stop functioning. While the machine is quite efficient at removing low-molecular-weight waste products (see Section 7.8), patients on dialysis suffer long-term health problems because only part of the kidney function is replaced.

Tissue engineering involves the growth of cells to replace the functions of damaged organs and tissues [22]. One approach involves growing cells in a biodegradable scaffold and implanting the tissue construct. Important issues for *in vitro* culture include providing an adequate nutrient supply, obtaining enough cells, maintaining differentiated cell function, and using or not using stem cells. When multiple cell types are needed, the growth and spatial localization of each type must be regulated. Upon implantation, an adequate blood supply must be developed for the cells to survive. Tissues that are actively studied are skin, artery, cartilage, bladder, and liver. Tissue-engineered skin is available now for the treatment of burns and leg ulcers. The development of engineered tissues is affected by nutrient and growth factor transport [23].

Short of creating a complete tissue or organ that functions like the original organ, some researchers are developing *hybrid devices* that consist of cells and synthetic material. One approach is to *microencapsulate* cells in a polymeric material and inject them into various cavities (e.g., the peritoneum). The capsule prevents immune rejection and permits the use of nonhuman animal cells. Studies with animals show that microencapsulated pancreatic islet cells, which secrete insulin, can provide very good control of blood glucose levels. The capsule must be sturdy in order to prevent rupture and release of cells. The capsule size is limited by the metabolic demands of the cells. As a result, difficulties persist in providing sufficient numbers of cells to replace normal cell function.

Other assist devices being developed include the bioartificial kidney and artificial liver [24], which may serve as a bridge between organ failure and organ transplantation, although an artificial kidney [25] could in principle replace dialysis. These devices are extracorporeal. Within them (Figure 1.22), cells from the kidney or liver are coated onto one surface of a hollow fiber reactor, which keeps the cells isolated, limiting their immunological rejection. A pressure difference across the fiber forces the flow of plasma across the fiber and cells. The presence of the cells provides metabolic functions that are not achieved by simple filtration.

These three approaches hold much future promise, but many fundamental scientific and engineering issues remain to be addressed. Addressing these issues will require knowledge of cell biology, materials science, mechanical analysis, molecular biology, and transport processes.

The growing knowledge about intracellular organization and communication creates new opportunities to treat a variety of diseases in which signaling pathways are disrupted or altered. Such treatments involve the delivery of genes and promoters or inhibitors of signaling pathways. Because these molecules are not normally transported across cell membranes, various strategies are employed to enhance their entry

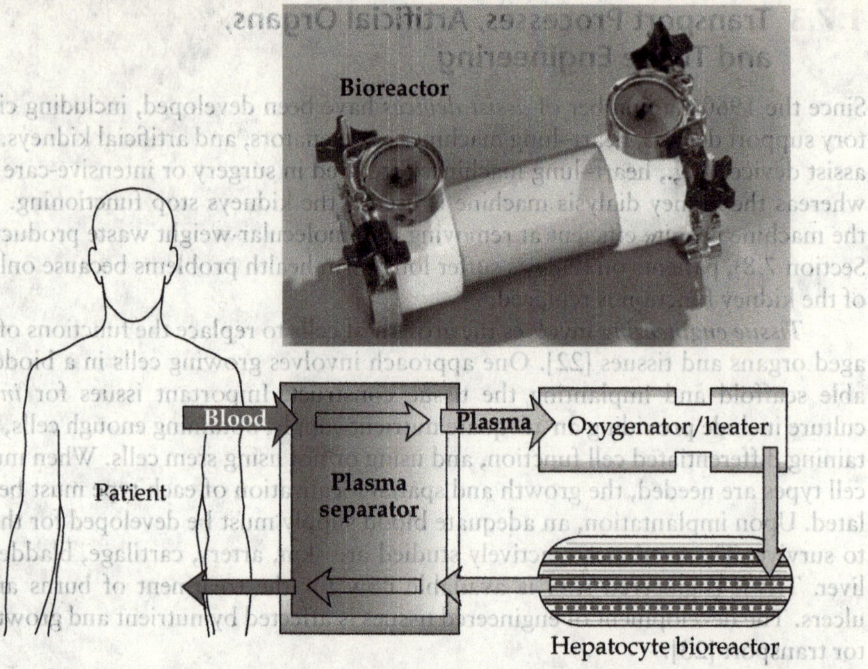

FIGURE 1.22 Schematic of flow circuit for extracorporeal artificial liver. (Reprinted with permission from Ref. [24]. © 2002, American Association for the Advancement of Science.)

into cells. Such strategies include using viruses, applying alternating electrical fields to transiently disrupt the cell membrane (a process known as *electroporation*), and linking drugs to molecules internalized by receptor-mediated endocytosis. A current challenge for biomedical engineers is to develop a mechanistic and quantitative understanding of transport routes into the cell, in order to develop efficient ways to deliver drugs and genes.

1.8 Relative Importance of Transport and Reaction Processes

As indicated in the preceding survey of cell biology and organ transport physiology, diffusion and convection often occur in conjunction with specialized cellular transport processes and chemical reactions. One of the tasks facing the bioengineer is to quantify the relative contribution of each of the different processes to the overall transport process. This is done using dimensional analysis.

Often, diffusing molecules pass through an endothelial or epithelial layer and then diffuse into tissue. In such cases, it is important to know the relative contribution of the endothelium and tissue to the diffusive resistance. The *Biot number* (Bi) is a dimensionless number that is a measure of the relative resistances of each process and is defined by

$$\mathrm{Bi} = \frac{\text{Mass transfer across a cell layer}}{\text{Mass transfer by diffusion through tissue}} = \frac{k_m L}{D_{\mathrm{eff}}}, \tag{1.8.1}$$

where k_m is the permeability of the cell layer, L is the distance over which diffusion in the tissues occurs, and D_{eff} is the effective diffusion coefficient. If the tissue is the major resistance, Bi is much greater than unity. If, however, the membrane is the limiting resistance, then Bi is much less than unity.

After a molecule reaches a site of action, it usually undergoes a chemical transformation, known as a *reaction*, or it interacts through nonspecific forces to generate a biological signal. For example, glucose is important in the oxidative phosphorylation reactions of glucose metabolism in the mitochondria of cells [26]. A series of chemical reactions transfers the energy in glucose to ATP. *Noncovalent* interactions, however, result in the formation of a complex that does not chemically change either of the molecules. The formation of complexes can produce changes in the conformation, or three-dimensional structure, of the molecules. These conformational changes can then initiate additional noncovalent interactions or covalent bond formation. This cascade of reactions can produce biological signals leading to alterations in cell function, replication, or protein synthesis.

Comparing timescales can provide useful information about the importance of various transport and reaction steps. One of the most commonly used approaches in modeling transport and reaction processes is the identification of the rate-limiting step. When the product is formed as the result of a sequence of reactions, the rate-limiting step is the slowest step in the sequence (see Sections 10.3 and 10.4). Consequently, the overall speed at which a process occurs equals the rate of this step. Identifying the rate-limiting step can simplify the reaction model significantly.

If the intrinsic reaction dynamics are very rapid, then the rate at which a reaction occurs will be affected by the rate at which the molecules are delivered to the reaction site. Reactions that are sensitive to the rate of transport are known as *transport-limited* reactions. The rate of reaction can be enhanced by decreasing the characteristic length if the process is limited by diffusion or by increasing the fluid velocity if the process is limited by flow.

1.9 QUESTIONS

1.1 What differences would you expect to observe between a random walk of a molecule in a gas and a random walk of a protein molecule in water?

1.2 If the Peclet number for the transport of a drug in tissue is 0.1, what is the dominant mechanism of transport?

1.3 Explain why an increase in hydrogen ion concentration would shift the oxygen–hemoglobin dissociation curve to the right.

1.4 Explain how each of the following passes from the extracellular fluid into cells: (1) oxygen, (2) calcium ion, (3) ethanol, (4) glucose, and (5) growth factors.

1.5 Why does the blood flow distribution to the various organs shift during heavy exercise?

1.6 Why is edema in the legs often seen in patients with chronic hypertension?

1.7 During hyperventilation, the arterial levels of carbon dioxide are less than 40 mmHg. What effect should hyperventilation have upon blood pH?

1.8 How does the Bohr effect influence the oxygen–hemoglobin dissociation curve?

1.9 How might living at high altitudes affect oxygen uptake and what adaptations might occur?

1.10 What is the function of folds, villi, and microvilli in the small intestine? Why are these structures absent in the large intestine?

1.11 Describe how the structure of the liver permits efficient exchange of molecules between the hepatocytes and blood.

1.12 Why does drinking a cup of coffee often result in increased production of urine?

1.13 In what ways can alterations in transport processes affect the growth of tumors?

1.14 How does the transport of nutrients affect the development of tissue-engineered blood vessels in the research laboratory?

1.10 PROBLEMS

1.1 The oxygen diffusion coefficient in tissue is about 1.1×10^{-5} cm^2 s^{-1}. The fluid filtration velocity is typically 1 μm s^{-1}.

(a) How large a distance between capillaries would be needed for convection to influence oxygen transport to tissues?

(b) Based upon reported values for the distance between capillaries, do you think that convection is an important mechanism for oxygen transport in tissues?

1.2 The solubility of oxygen in plasma at 37°C is 1.4×10^{-6} mol L^{-1} mmHg^{-1}. The heme concentration in red blood cells is 0.0203 mol L^{-1} = $4C_{Hb}$ [27]. The blood volume fraction (hematocrit) is typically 0.45 for men and 0.40 for women. Determine the fraction of oxygen in solution and that bound to hemoglobin in arterial and venous blood. For simplicity, assume that the solubility in red blood cells equals the solubility in plasma.

1.3 Compare the amount of oxygen taken up by blood and the amount of carbon dioxide released by the blood as blood passes through the lung capillaries. The difference in carbon dioxide (at standard temperature and pressure) between arteries and veins in the lung is 2.27 cm^3 per 100 cm^3 for plasma carbon dioxide and 1.98 cm^3 per 100 cm^3 for red-blood-cell carbon dioxide.

1.4 The alveolar epithelium, basement membrane, and lung capillary endothelium are typically 1 μm thick. Under resting conditions, hemoglobin binding with oxygen reaches a steady state (i.e., it does not change with time) in about 0.33 seconds. Is oxygen diffusion across the alveolus a significant factor in the time required for the hemoglobin to oxygenate as it traverses the capillary?

1.5 Data for the dimensions of the branches in the pulmonary arterial network are shown in Table 1.13. Calculate the volume of blood and the surface area of the vessel wall in each order of vessels and the cumulative volume and surface area from order 1 to order 11.

1.6 Morphometric data of a human lung are shown in Table 1.14 (see also Figure 1.13). The number of bronchi N equals 2^z, where z is the order. Calculate the volume of air and the surface area of the bronchial system in each order of airway, and calculate the cumulative volume and surface area from order 0 to order 23.

1.7 Using the data in Table 1.12 and a renal artery flow rate of 125 L min^{-1}, determine

(a) the fraction of water filtered across the glomerulus,

(b) the renal vein flow rate, and

(c) the concentration of sodium ion leaving the glomerulus and in the renal vein.

The blood volume fraction (hematocrit) is typically 0.45 for men and 0.40 for women. The sodium ion concentration in the blood entering renal arteries is 150 mM.

1.8 The permeability of normal rabbit arterial endothelium to low-density lipoproteins (LDL) is 5×10^{-9} cm s^{-1}. The diffusion coefficient of LDL in the rabbit arterial wall is 1×10^{-10} cm^2 s^{-1}. The rabbit aorta thickness is 150 μm.

(a) Determine the Biot number.

(b) What does this result indicate about the endothelium as a barrier to LDL transport?

TABLE 1.13

Morphometric Data of a Pulmonary Arterial Network

Order[a]	Mean vessel diameter (mm)	Mean vessel length (mm)	Number of vessels
1	0.024	0.116	300,358
2	0.044	0.262	97,519
3	0.073	0.433	31,662
4	0.122	0.810	9,736
5	0.192	1.510	2,925
6	0.352	2.720	774
7	0.533	4.600	202
8	0.875	8.190	49
9	1.519	14.260	12
10	2.486	11.870	4
11	5.080	25.000	1

Source: Adapted with kind permission of Springer Science + Business Media from Ref. [28].

[a]The smallest arterioles are vessels of order 1, and the pulmonary artery is of order 11.

TABLE 1.14

Morphometric Data for the Bronchi of a Human Lung

Order[a]	Mean diameter (mm)	Mean length (mm)
0	18.0	120.0
1	12.2	47.6
2	8.3	19.0
3	5.6	7.6
4	4.5	12.7
5	3.5	10.7
6	2.8	9.0
7	2.3	7.6
8	1.86	6.4
9	1.54	5.4
10	1.3	4.6
11	1.09	3.9
12	0.95	3.3
13	0.82	2.7
14	0.74	2.3
15	0.66	2.0
16	0.6	1.65
17	0.54	1.41
18	0.5	1.17
19	0.47	0.99
20	0.45	0.83
21	0.43	0.7
22	0.41	0.59
23	0.41	0.5

Source: Adapted from Ref. [14].

[a]The trachea is called the airway of order 1, and the alveoli are the airways of order 23.

1.9 For the following data, determine the oxygen consumption rate $\dot{V}_{O_2}$ under rest and exercise conditions.

	Rest	Exercise
Pulmonary blood flow (L min^{-1})	5.8	25
Arterial P_{O_2} (mmHg)	40	15
Venous P_{O_2} (mmHg)	100	100

Note that oxygen in blood is present in red blood cells bound to hemoglobin and freely dissolved in the red cell and the blood plasma. Equation (1.6.3) relates the partial pressure (mmHg) to the concentration in plasma and the total concentration of oxygen in blood is given by Equation (1.6.4), where $4C_{Hb}$ is 0.0203 M; the hematocrit or volume fraction of red cells, Hct, is typically 0.45 for males and 0.40 for females; H_{O_2} is 1.33×10^{-6} M $mmHg^{-1}$; and H_{Hb} is 1.50×10^{6} M $mmHg^{-1}$.

1.10 (a) Determine the rate of oxygen removal from the lung venules for the following conditions: inspired air containing oxygen at 21% partial pressure, alveolar oxygen partial pressure at 105 mmHg, blood flow rate of 5 L min^{-1}, respiration rate of 10 breaths per minute, and a respiratory volume of 0.56 L per breath. The alveolar volume equals the difference between the respiratory volume and the dead volume (0.15 L). Assume that $T = 37°C$ and $R = 0.08206$ L atm mol^{-1} K^{-1} and 1 atm = 760 mmHg.

(b) During exercise, the blood flow rate can rise to 25 L min^{-1}, the respiration rate rises to 30 breaths per minute and oxygen demand increases to 4 L min^{-1}. The partial pressure of oxygen in the pulmonary artery declines to 20 mmHg but remains at 100 mmHg in the pulmonary vein. The total concentration of oxygen in blood is given by Equation (1.6.4) where $4C_{Hb}$ is 0.0203 M; the hematocrit or volume fraction of red cells, Hct, is typically 0.45; H_{O_2} is 1.33×10^{-6} M $mmHg^{-1}$; and H_{Hb} is 1.50×10^{-6} M $mmHg^{-1}$. Determine the respiratory volume and oxygen consumption rate if the alveolar concentration of oxygen and dead volume are unchanged.

1.11 During exercise, the cardiac output can rise to 25 L min^{-1} from a resting level of 5 L min^{-1}. The heart rate of a well-trained athlete might rise from 60 beats per minute to 105 beats per minute and the mean arterial pressure may rise from 100 to 130 mmHg, whereas the heart rate of a sedentary person might rise from 72 beats per minute to 125 beats per minute and the mean arterial pressure may rise from 100 to 150 mmHg. Determine the volume of blood ejected during each heartbeat (stroke volume) and the peripheral resistance for an athlete and a sedentary person. Assess the power of the left side of the heart for the athlete and the sedentary person.

$$P = Wf = f \int \bar{p}_a dV,$$

where P is power, $\bar{p}_a$ is the mean arterial pressure, V is the ventricular volume, W is work, and f is heart rate in beats per second. Make sure that your units are consistent. Note: 1 mmHg = 133.3 Pa (1 Pa = 1 N m^{-2}).

1.12 At elevated altitudes, the body adapts to the reduced barometric pressure to extract sufficient oxygen to permit normal metabolic functions and do work. For example, at an altitude of 3,650 m (close to 12,000 feet above sea level) the barometric pressure drops to 485 mmHg. For an oxygen pressure drop in the lungs of 30 mmHg, determine the oxygen uptake rate for a respiration rate of 20 breaths per minute. Estimate the oxygen saturation in venous blood if the hematocrit rises to 0.60 and the partial pressure of oxygen blood is at a partial pressure equal to 98% of the alveolar level.

1.13 The basal metabolic rate for a resting individual is about 1,650 kcal per day. Using the results from Problem 1.11, determine the fraction of the metabolic energy used to pump blood through the body.

1.14 Use the data in Table 1.12 for Na^+, K^+, glucose, and urea to determine the ratio of the concentration in urine to the concentration in plasma. What would the ratio be if the solute were transported along with water?

1.15 The glomerular filtration rate is an important measure of kidney function. A common method to determine the glomerular filtration rate is to add a low molecular weight sugar, inulin, to the blood and measure its concentration in the blood and urine. Inulin is not reabsorbed. The plasma concentration of inulin was found to be 0.001 g/mL and the concentration in urine was 0.125 g mL^{-1}. Determine the glomerular filtration rate using a urine production rate of 1 mL min^{-1}.

1.11 REFERENCES

1. Einstein, A., *Investigations on the Theory of the Brownian Movement*, R. Furth, editor. 1956, New York: Dover Publications, Inc.
2. Probstein, R., *Physicochemical Hydrodynamics*. 1989, Boston: Butterworths.
3. Bird, R., Stewart, W., and Lightfoot, E., *Transport Phenomena*. 2d ed. 2002, New York: John Wiley and Sons.
4. Lightfoot, E., "The role of mass transfer in tissue function," in *The Biomedical Engineering Handbook*, J. Bronzino, editor. 1995, Boca Raton: CRC Press; IEEE Press. pp. 1656–1670.
5. Alberts, B., Bray, D., Lewis, J., Raff, M., Roberts, K., and Watson, J.D., *Molecular Biology of the Cell*. 3d ed. 1994, New York: Garland Publishing, Inc.
6. Paula, S., and Deamer, D., "Membrane permeability barriers to ionic and polar solutes." *Curr. Top. Membr.*, 1999. **48**: pp. 77–95.
7. Mitic, L.L., and Anderson, J.M., "Molecular architecture of tight junctions." *Ann. Rev. Physiol.*, 1998. **60**: pp. 121–142.
8. Simionescu, M., and Simionescu, N., "Ultrastructure of the microvascular wall: Functional correlations," in *Section 2: The Cardiovascular System. Volume IV: Microcirculation*, E. Renkin and C. Michel, editors. 1984, Bethesda: American Physiological Society. Oxford University Press, pp. 41–102.
9. Ogunrinade, O., Kameya, G.T., and Truskey, G.A., "Effect of fluid shear stress on the permeability of the arterial endothelium." *Ann. Biomed. Eng.*, 2002. **30**: pp. 1–17.
10. Rhoades, R., and Tanner, G., *Medical Physiology*. 1995, Boston: Little, Brown and Company.
11. Fuchs, E., "Cellular immunology," in *Section 14: Cell Physiology*, J. Hoffman and J. Jamieson, editors. 1997, New York: Oxford University Press. pp. 743–785.
12. Springer, T.A., "Adhesion receptors of the immune system." *Nature*, 1990. **346**: pp. 425–434.
13. Weibel, E., "Design of airways and blood vessels considered as branching trees," in *The Lung: Scientific Foundations*, R. Crystal, et al., editors. 1997, Philadelphia: Lippincott-Raven Publishers. pp. 1061–1071.
14. Rhoades, R., and Tanner, G., "Gas transfer and transport," in *Medical Physiology*. 1995, Boston: Little, Brown and Company. pp. 386–400.
15. Weibel, E., "Design and morphometry of the pulmonary gas exchanger," in *The Lung: Scientific Foundations*, R. Crystal, et al., editors. 1997, Philadelphia: Lippincott-Raven Publishers. pp. 1147–1157.
16. Klocke, R., "Carbon dioxide transport," in *The Lung: Scientific Foundations*, R. Crystal, et al., editors. 1997, Philadelphia: Lippincott-Raven Publishers. pp. 1633–1642.
17. McCuskey, R.S., "The hepatic microvascular system," in *The Liver: Biology and Pathobiology*, I. Arias, et al., editors. 1994, New York: Raven Press, Ltd. pp. 1089–1106.
18. Hubbard, A., Barr, V., and Scott, L., "Hepatocyte surface polarity," in *The Liver: Biology and Pathobiology*, I. Arias, et al., editors. 1994, New York: Raven Press, Ltd. pp. 189–214.
19. Seifter, S. and Englard, S., "Energy metabolism," in *The Liver: Biology and Pathobiology*, I. Arias, et al., editors. 1994, New York: Raven Press, Ltd. pp. 323–364.
20. Berne, R., and Levy, M., *Physiology*. 3d ed. 1993, St. Louis: Mosby Year Book.
21. American Cancer Society. *Cancer Facts & Figures 2007*. 2007, Atlanta: American Cancer Society.
22. Saltzman, W.M. *Tissue Engineering: Principles for the Design of Replacement Organs and Tissues*, 2004. New York: Oxford University Press.
23. Griffith, L.G., and Naughton, G., "Tissue engineering—current challenges and expanding opportunities." *Science*, 2002. **295**: pp. 1009–1014.
24. Strain, A.J., and Neuberger, J.N., "A bioartificial liver—state of the art." *Science*, 2002. **295**: pp. 1005–1009.
25. Humes, H., Buffington, D., Mackay, S., Funke, A., and Weitzel, W., "Replacement of renal function in uremic animals with a tissue-engineered kidney." *Nat. Biotechnol.*, 1999. **17**: pp. 451–455.
26. Lehninger, A., *Biochemistry*. 1975, New York: Worth Publishers, Inc.
27. Groebe, K., "Factors important in modeling oxygen supply to red muscle," in *Oxygen Transport in Biological Systems*, S. Eggington and H. Ross, editors. 1992, New York: Cambridge University Press. pp. 231–252.
28. Fung, Y., *Biomechanics: Motion, Flow, Stress, and Growth*. 3d ed. 1990, New York: Springer-Verlag.

Introduction to Physiological Fluid Mechanics

PART I

Liquids or gases are fluids; they may consist of a single component or be mixtures of different molecules. Whereas solids are rigid and can be formed into many shapes, fluids adopt the shapes of the containers in which they are placed. For the study of fluid mechanics, a more precise definition of a fluid is as follows [1]: A *fluid* is a material that deforms continuously when subjected to a force applied tangentially to a surface. Biological fluids are quite complex, exhibiting solid- and liquidlike behavior and deforming in a time-dependent fashion. Some fluids do not flow, but behave as solids when the shear stress applied to one surface is less than the threshold, or *yield*, value. Above this threshold value, the material begins to flow. Many biological fluids consist of a liquid phase and a cell phase. Examples of biological fluids include blood; synovial fluid in joints; lymph, which is produced by the filtration of blood plasma through tissues; and the vitreous fluid of the eye. The cell phase often exhibits complex time-dependent behavior. Tissues consist of cells embedded in an extracellular matrix. A fluid is present in the *interstitial space* between extracellular matrix molecules.

Fluid mechanics is the study of the motion of fluids in response to the application of stresses. Experiments, as well as analytical and numerical solutions of the conservation of linear momentum and mass, are used to determine the velocity distribution and the forces acting upon the fluid or solid surfaces for an imposed stress or flow. In turn, this information is used to calculate the pressure drop that occurs or the work that is required to move the fluid. In order to describe the movement of heat or mass in a fluid, the velocity field is needed to determine the local motion.

Fluid motion is determined from a balance of the forces acting on the fluid, together with the conservation of mass. The force balance describes the variation in velocity and pressure with time and space. In Chapter 2, we discuss how to formulate the conservation relations and solve simple problems. The importance of geometry on fluid flow and forces is examined. In Chapter 3, generalized forms of the conservation relations are developed and applied to two-dimensional flows and unsteady-flow problems. Chapters 4 and 5 examine specific biomedical applications of these results.

Biological fluid mechanics represents the full range of fluid dynamic phenomena exhibited by living organisms, including the flight of birds, the swimming of fish, and the movement of bacteria, as well as fluid flows in various circulatory systems such as blood and the phloem and xylem of plants. One goal of biological fluid dynamics is to explain the shape and function of organisms in terms of their adaptation to the flow of air and water surrounding them. The book by Vogel [2] is an excellent and readable introduction to this fascinating area.

In contrast, *physiological fluid mechanics* is concerned with fluid flow within organisms and the relation between fluid flow and physiological processes. Physiological fluid mechanics provides important insights into normal and pathological processes and into the design of many bioengineering applications, from artificial kidneys to vascular grafts to engineered tissues. The distinction between biological and physiological fluid mechanics deals essentially with the class of problems analyzed. This part of the book focuses on physiological fluid mechanics.

CHAPTER 2

Conservation Relations and Momentum Balances

2.1 Introduction

In this chapter, we analyze fluid motion and the forces exerted by and on moving fluids and apply the analyses to important, yet relatively simple, flow problems in different geometries. General relations for fluid velocity and acceleration are developed in Section 2.2. The conservation relations for mass and linear momentum are set forth in Section 2.3. The problems analyzed in Section 2.4 deal with fluid statics—the distribution of pressure and stress in a fluid at rest. Commonly used constitutive relations are presented in Section 2.5. The flow conditions typically encountered in the lungs and blood are examined in Section 2.6. Next, fluid motion in some simple flows is explored in Section 2.7. The focus is upon problems that can be solved analytically. (Some numerical solutions using MATLAB® are presented in later chapters.) Several different constitutive equations are used to show the effect of differences in fluid rheology on the flow profile and force. We conclude the chapter with a detailed examination of the properties of blood and their effects on steady blood flow.

2.2 Fluid Kinematics

Prior to discussing the forces acting on a system, we examine a number of results that characterize fluid motion in time and space, or *kinematics*. When describing motion, we must specify a frame of reference. Several choices are available. One choice is to use a fixed reference frame and to examine motion relative to this coordinate system. Alternatively, the reference frame may move, and all motion is assessed relative to the moving reference frame. For the most part, we will use a fixed reference frame, although occasionally the solution of problems becomes mathematically easier with a moving reference frame.

A qualitative description illustrates the difference between fixed and moving reference frames with regard to velocity. Imagine that you are recording the speed and direction of a small pleasure boat moving on a river. If you record the motion from the riverbank, you adopt a fixed reference frame. The velocity that you measure depends upon the speed and direction of the boat, which is influenced by the local speed and direction of the fluid. For instance, in a region of the river that narrows or is shallow, the speed of the water increases. If the boat moves in the direction of the river flow, the boat accelerates unless the driver adjusts the motor. If you move to a different location on the riverbank, your location relative to the boat changes. The speed of the boat is the same as in your previous location, but the direction of the motion is different. Observing motion from a fixed reference frame is referred to as an *Eulerian* description and is a common reference frame for fluid mechanics. Alternatively, you could measure the velocity of the river while you are on the boat. The speed you measure is the net motion of the boat relative to the motion of the river. The approach is known as a *Lagrangian* description of motion [3]. For the most part, the Eulerian approach is used to develop and apply conservation equations for mass and momentum. The Lagrangian approach is used to examine the motion of proteins and cells in a fluid.

2.2.1 Control Volumes

In fluid mechanics, a *control volume*—a defined region of space—is used to examine the flow of mass, momentum, and energy (Figure 2.1). Specification of a control volume makes the analysis and solution of a problem easier. Control volumes can be fixed in size or may change size with time. Further, control volumes may be fixed in space or move with some arbitrary velocity. The Eulerian view examines the control volume from a fixed frame of reference. In contrast, the Lagrangian view associates the control volume with a specific mass of fluid and follows the mass as it moves through the flow field.

In general, the control volume has a volume $V(t)$ and a surface $S(t)$. A unit normal vector **n** is defined at each point on the surface of the control volume. This vector is extremely useful in calculating velocities and stresses.

2.2.2 Velocity Field

Knowing the fluid velocity as a function of position and time enables one to calculate velocity gradients and many of the forces and stresses within the fluid. Again, the frame of reference used to define velocity must be carefully specified.

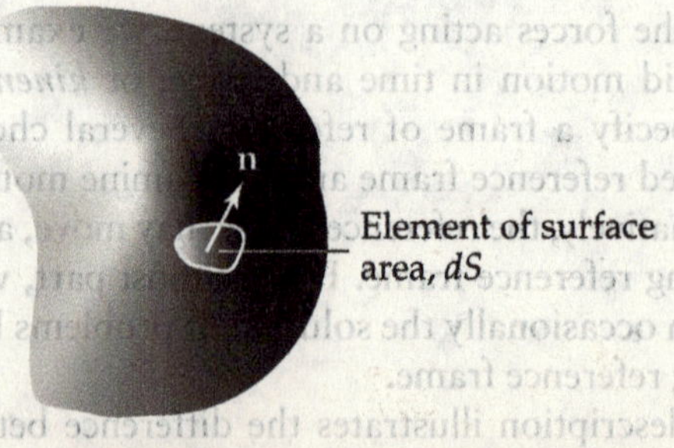

FIGURE 2.1 Control volume and unit normal vector.

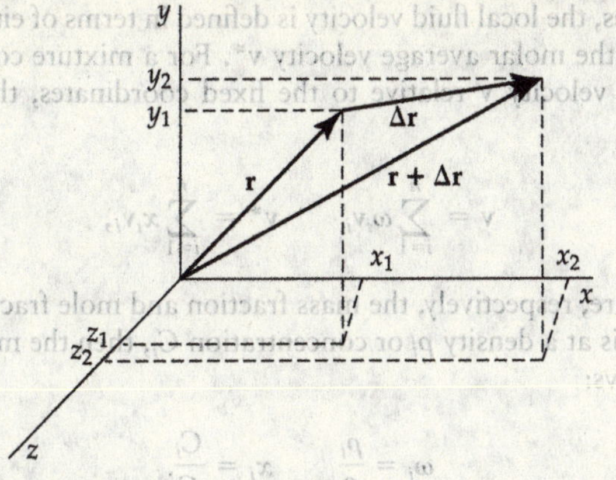

FIGURE 2.2 Motion of a differential fluid element.

A Lagrangian description examines motion of a differential fluid volume using a fixed reference frame (Figure 2.2). At time t, the particle is at position **r**, which has coordinates x_1, y_1, and z_1. At time $t + \Delta t$, the particle has moved to $\mathbf{r} + \Delta\mathbf{r}$, with coordinates x_2, y_2, and z_2. The velocity of a fluid element is the rate of change of the position of the fluid element with respect to time.

Velocity has a direction and a magnitude and thus is a vector quantity. For a rectangular coordinate system, the velocity is

$$\mathbf{v} = \frac{d\mathbf{r}}{dt} = \mathbf{e}_x \mathrm{v}_x + \mathbf{e}_y \mathrm{v}_y + \mathbf{e}_z \mathrm{v}_z = \mathbf{e}_x \frac{dx}{dt} + \mathbf{e}_y \frac{dy}{dt} + \mathbf{e}_z \frac{dz}{dt}, \tag{2.2.1}$$

where the vectors $\mathbf{e}_x$, $\mathbf{e}_y$, and $\mathbf{e}_z$ are unit vectors that point parallel to the x-, y-, and z-axes, respectively. Other notations commonly used in fluid mechanics for these unit vectors are $\mathbf{e}_1$, $\mathbf{e}_2$, and $\mathbf{e}_3$ and **i**, **j**, and **k** [4].

In the Eulerian description, a velocity is associated with each location in the fluid. That is, the observer examines fluid motion through a differential region, or window, fixed in space. Consider a differential cubic control volume fixed in space with origin at x, y, z (Figure 2.3). At a given instant in time, fluid elements enter the volume with one velocity and leave with another velocity. If the volume shrinks to a point, the velocity of a material point at that instant and at that location equals the local fluid velocity. Velocity is a function of location and time (i.e., $\mathbf{v} = \mathbf{v}(x, y, z, t)$).

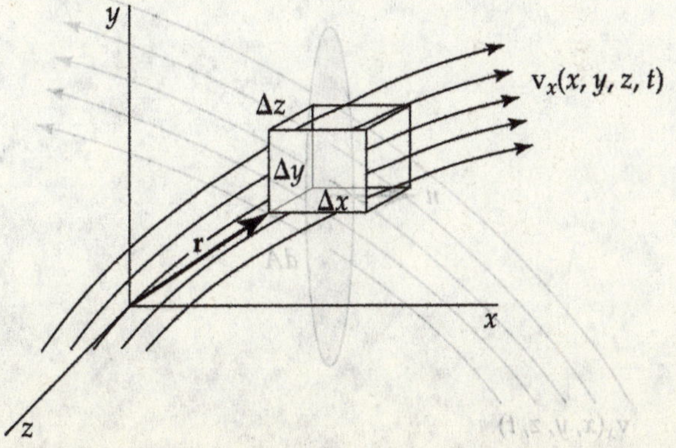

FIGURE 2.3 Motion of fluid through a volume $\Delta x \Delta y \Delta z$.

For mixtures, the local fluid velocity is defined in terms of either the mass average velocity v or the molar average velocity v*. For a mixture containing i components, each with velocity v relative to the fixed coordinates, these quantities are defined as

$$\mathbf{v} = \sum_{i=1}^{n} \omega_i \mathbf{v}_i \qquad \mathbf{v}^* = \sum_{i=1}^{n} x_i \mathbf{v}_i, \tag{2.2.2a,b}$$

where ω_i and x_i are, respectively, the mass fraction and mole fraction of the fluid. If each component is at a density ρ_i or concentration C_i, then the mass and mole fractions are as follows:

$$\omega_i = \frac{\rho_i}{\rho} \qquad x_i = \frac{C_i}{C}. \tag{2.2.3a,b}$$

The total density and concentration are summed over all components:

$$\rho = \sum_{i=1}^{n} \rho_i \qquad C = \sum_{i=1}^{n} C_i. \tag{2.2.4a,b}$$

For most applications with mixtures, the mass average velocity is used. The present section of the text focuses on the motion of pure fluids or the bulk behavior of mixtures. In Part II of the text, we examine the transport of individual components of dilute solutions.

2.2.3 Flow Rate

When the velocity is known, the average velocity and the flow rate can be determined. The average velocity $\langle \mathbf{v} \rangle$ is the integral of the velocity that passes through a reference area, divided by the cross-sectional area A (Figure 2.4); that is,

$$\langle \mathbf{v} \rangle = \frac{1}{A} \int_A \mathbf{v} \cdot \mathbf{n} dA, \tag{2.2.5}$$

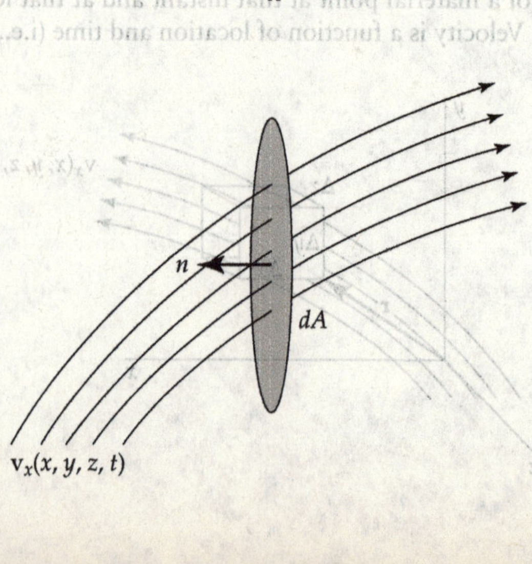

FIGURE 2.4 Differential area dA for calculating average velocity and flow rate.

where **n** is a unit vector normal to the surface area A. This average normal velocity is also referred to as the volumetric flux of fluid. The flow rate characterizes the amount of fluid that crosses a unit area of the flow field. The volumetric flow rate is

$$Q = \langle \mathbf{v} \rangle A = \int_A \mathbf{v} \cdot \mathbf{n} dA. \tag{2.2.6}$$

The mass flow rate is the integral of the product of the local density and the velocity:

$$M = \int_A \rho \mathbf{v} \cdot \mathbf{n} dA. \tag{2.2.7}$$

2.2.4 Acceleration

Acceleration is a vector defined as the rate of change of velocity. Fluid acceleration **a** results when the velocity changes with time or space (i.e., $\mathbf{a} = \mathbf{a}(x, y, z, t)$). The total derivative of the velocity is obtained by applying the chain rule to the changes in position in the x, y, and z directions (see Section A.1.E):

$$\mathbf{a} = \frac{\partial \mathbf{v}}{\partial t} + \frac{dx}{dt}\frac{\partial \mathbf{v}}{\partial x} + \frac{dy}{dt}\frac{\partial \mathbf{v}}{\partial y} + \frac{dz}{dt}\frac{\partial \mathbf{v}}{\partial z} = \frac{\partial \mathbf{v}}{\partial t} + \mathrm{v}_x\frac{\partial \mathbf{v}}{\partial x} + \mathrm{v}_y\frac{\partial \mathbf{v}}{\partial y} + \mathrm{v}_z\frac{\partial \mathbf{v}}{\partial z}. \tag{2.2.8a}$$

Using vector notation, we can write the second through fourth terms on the right-hand side of Equation (2.2.8a) in a more compact form that is independent of the particular coordinate system (see Section A.3.C):

$$\mathrm{v}_x\frac{\partial \mathbf{v}}{\partial x} + \mathrm{v}_y\frac{\partial \mathbf{v}}{\partial y} + \mathrm{v}_z\frac{\partial \mathbf{v}}{\partial z} = \mathbf{v} \cdot \nabla \mathbf{v}. \tag{2.2.8b}$$

Substituting Equation (2.2.8a) into Equation (2.2.8b) yields

$$\mathbf{a} = \frac{\partial \mathbf{v}}{\partial t} + \mathbf{v} \cdot \nabla \mathbf{v}. \tag{2.2.8c}$$

The first term on the right-hand side of Equation (2.2.8c) arises from the instantaneous change in velocity with time. This term is referred to as *local acceleration*. The second term is unique to fluid elements and represents fluid acceleration that arises from a spatial change in the velocity field. This term is called *convective acceleration*. For example, if fluid flowing in a tube passes through a narrow constriction, the fluid velocity must increase in order to maintain a constant flow rate. Physiologically, such a situation might arise during blood flow when there is a constriction, or *stenosis*, in a blood vessel due to a calcified valve or atherosclerosis. Conversely, a local expansion of the vessel, known as an *aneurysm*, can produce fluid deceleration. These conditions can cause significant complications.

Example 2.1 For the following velocity field, which describes high-speed flow of a fluid with negligible viscosity through a conical converging channel (Figure 2.5), compute the axial or x-component of the fluid acceleration.

$$v_x = \frac{V_0}{\left[1 + ((r_L - r_0)/r_0)(x/L)\right]^2} \qquad v_r = \frac{2V_0((r_L - r_0)/r_0)r}{L\left[1 + ((r_L - r_0)/r_0)(x/L)\right]^3}. \quad (2.2.9a,b)$$

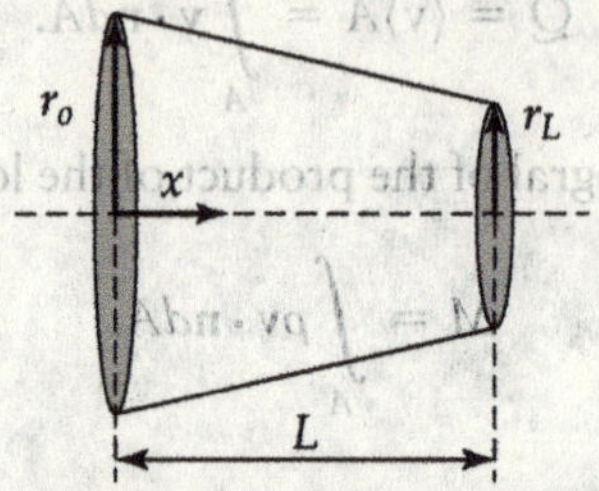

FIGURE 2.5 Flow through a conical channel.

Solution For steady flow, the x component of acceleration is, from Equation (2.2.8),

$$a_x = \mathbf{e}_x \cdot \mathbf{a} = \mathbf{e}_x \cdot \left(v_x \frac{\partial \mathbf{v}}{\partial x} + v_r \frac{\partial \mathbf{v}}{\partial r} \right) = v_x \frac{\partial v_x}{\partial x} + v_r \frac{\partial v_x}{\partial r}. \quad (2.2.10)$$

From Equations (2.2.9a) and (2.2.9b), the derivatives are

$$\frac{\partial v_x}{\partial x} = \frac{-V_0((r_L - r_0)/r_0)}{L\left[1 + ((r_L - r_0)/r_0)(x/L)\right]^3} \qquad \frac{\partial v_x}{\partial r} = 0, \quad (2.2.11a,b)$$

Combining terms,

$$a_x = -\frac{V_0^2\left[(r_L - r_0)/r_0\right]}{L\left[1 + ((r_L - r_0)/r_0)(x/L)\right]^5} \quad (2.2.12)$$

The result is general for a converging or diverging channel. For a converging channel, $r_L < r_0$ and the acceleration is positive.

2.2.5 Fluid Streamlines

Once the velocity field is known, there are several ways to display the information. One approach is to show the velocity vectors as arrows (Figure 2.6) in which the size of the vector is proportional to the magnitude of the velocity. This method of display works well for one-dimensional flows and some simple two-dimensional flows. Alternatively, for complex flows, one can display a curve that is everywhere tangent to the instantaneous velocity vectors (Figure 2.7). Such a curve is known as a *streamline.*

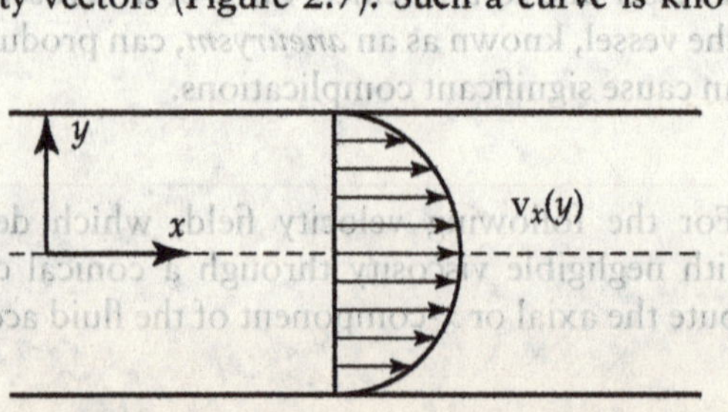

FIGURE 2.6 Vector representation of one-dimensional flow between parallel plates.

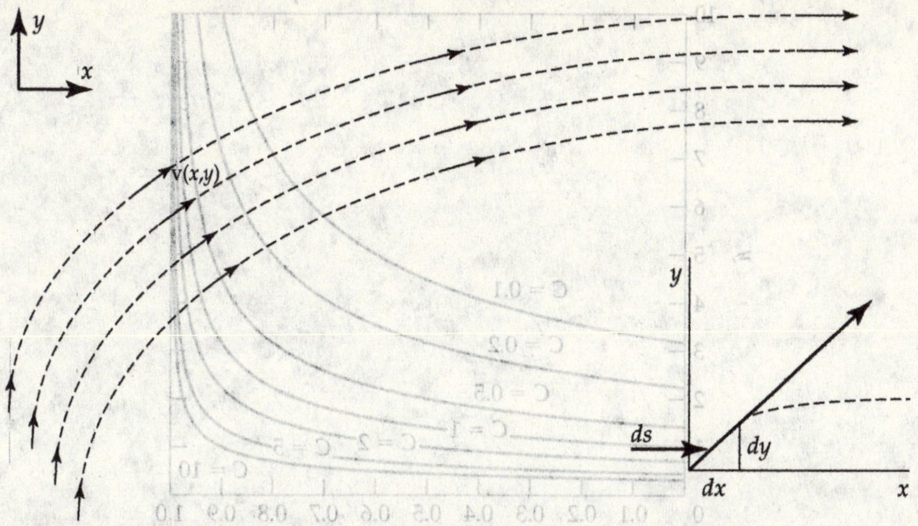

FIGURE 2.7 Streamlines (dotted lines) are everywhere tangential to the velocity vector. The axes on the right-hand side show the relation between a differential unit of length along the streamline, *ds*, and the differential components in the *x* and *y* directions and the velocity components.

For steady flow, a fluid element moves along streamlines. Since the streamlines are always tangent to the velocity vector, there is no flow normal to the streamline. This result is very useful in developing control volumes to analyze complex flows.

Because the velocity vector is everywhere tangential to the streamlines, the principal of similar triangles can be used to relate the velocity components to differential changes in position of the streamlines *dx*, *dy*, and *dz* (Figure 2.7):

$$\frac{d\mathrm{r}}{\mathrm{v}} = \frac{dx}{\mathrm{v}_x} = \frac{dy}{\mathrm{v}_y} = \frac{dz}{\mathrm{v}_z}. \tag{2.2.13}$$

The next example illustrates how to compute the streamlines with the use of Equation (2.2.13).

Example 2.2 Consider the two-dimensional steady velocity field

$$\mathrm{v}_x = y\left(1 - \frac{x}{L}\right) \qquad \mathrm{v}_y = \frac{y^2}{2L}, \tag{2.2.14a,b}$$

where $0 \le x \le L$. Determine the streamlines.

Solution Inserting Equations (2.2.14a) and (2.2.14b) into Equation (2.2.13) yields

$$\frac{dx}{y(1 - (x/L))} = \frac{2L\,dy}{y^2}. \tag{2.2.15a}$$

Rearranging gives

$$\frac{dx}{L(1 - (x/L))} = \frac{2dy}{y}. \tag{2.2.15b}$$

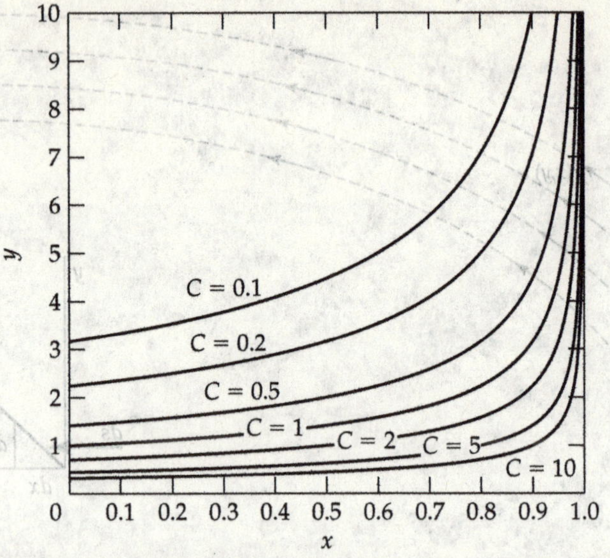

FIGURE 2.8 Streamlines represented by Equation (2.2.15d).

Integrating Equation (2.2.15b) produces

$$\ln\left(1 - \frac{x}{L}\right) = -2\ln y + \ln(C), \tag{2.2.15c}$$

or

$$y = \sqrt{\frac{C}{(1 - x/L)}}. \tag{2.2.15d}$$

A family of streamlines is plotted in Figure 2.8 for various values of C.

Numerous techniques are available to visualize fluid flows. Some visualization techniques trace *pathlines* if the motion of an individual particle is followed or *streaklines* if a continuous path is examined. Injecting individual bubbles or dye for a brief interval produces pathlines. Streaklines are produced by the continuous injection of dye at one location. Pathlines and streaklines represent the trajectories of particles (particle paths). Even if the agent that is used to visualize flow has the same density as the fluid, pathlines, streaklines, and streamlines are identical only for steady flow.

2.3 Conservation Relations and Boundary Conditions

A complete analysis of any transport process requires the specification of conservation relations, constitutive equations, and boundary conditions. These equations can be specified either in an integrated form that describes the average behavior of the fluid in a control volume or in a differential form that describes motion and forces at each point in the fluid. Both approaches are used in this text, although the differential form is emphasized. The resulting equations can be solved analytically or numerically, and the solutions can be used to address either a fundamental scientific issue, such as the mechanism of transport of molecules in a tissue, or a design issue, such as the particle size and amount of enzyme needed for a bioreactor. For flow problems, we need to specify the principles of conservation of mass and conservation of linear momentum. For fluid motion, conservation relations are first stated in words and individual terms are described. In subsequent sections, the conservation relations are applied to some simple, but important, situations. Such problems require a relation between the stress and the velocity gradient, or *shear rate*. Various constitutive equations are considered in Section 2.5. Then, in Chapter 3, the conservation of mass and the conservation of linear momentum are generalized through the derivation of differential equations that are valid in three dimensions for different coordinate systems. A similar approach is used in Part II of the text to describe mass transport.

2.3.1 Conservation of Mass

Consider a control volume of constant size that is fixed in space. A fluid mixture passes through the borders of the volume. Within the volume, chemical reactions can occur. Such a dynamic situation can lead to changes in the concentration of the various components of the mixture. If we examine an individual solute *i*, we can state the conservation relation in words:

$$\begin{bmatrix}\text{Rate of accumulation of}\\ \text{mass of } i \text{ in control volume}\end{bmatrix} = \begin{bmatrix}\text{Transport of } i \text{ into}\\ \text{control volume}\end{bmatrix} - \begin{bmatrix}\text{Transport of mass of } i\\ \text{out of control volume}\end{bmatrix} + \begin{bmatrix}\text{Gain or loss of mass } i\\ \text{due to chemical reaction}\end{bmatrix}. \tag{2.3.1}$$

In this section, only nonreacting systems are considered, and the third term on the right-hand side vanishes. We relax this assumption in the other three parts of the text. The system under study can be a single, pure component fluid or a mixture. At this point, we do not focus on the individual components of the system, but rather on the total mass entering or leaving the control volume. Since we are examining the total mass of the system, entry into and exit from the control volume can occur only by fluid flow. As a result, the mass balance can be restated as

$$\begin{bmatrix}\text{Rate of accumulation of}\\ \text{mass in control volume}\end{bmatrix} = \begin{bmatrix}\text{Flow of mass into}\\ \text{control volume}\end{bmatrix} - \begin{bmatrix}\text{Flow of mass out}\\ \text{of control volume}\end{bmatrix}. \tag{2.3.2}$$

2.3.2 Momentum Balances

Specification of the velocities and forces in a flowing system requires solving Newton's second law of motion. Before presenting a complete statement of Newton's second law for a fluid, we examine some specific, idealized applications using momentum balances. The momentum balances can be expressed in words as

$$\begin{bmatrix}\text{Rate of momentum}\\ \text{accumulation}\end{bmatrix} = \begin{bmatrix}\text{Rate of momentum flow}\\ \text{from control volume}\end{bmatrix} - \begin{bmatrix}\text{Rate of momentum flow}\\ \text{from control volume}\end{bmatrix} + \begin{bmatrix}\text{Sum of forces acting}\\ \text{on control volume}\end{bmatrix}. \tag{2.3.3}$$

In general, momentum is the product of mass and velocity ($m\mathbf{v}$). The rate of momentum accumulation is the rate of change of momentum, which equals the force. In fluid mechanics, the momentum per volume is an intensive variable (i.e., a variable that is independent of the mass or volume of the fluid) and equals the product of density and velocity ($\rho\mathbf{v}$). (The derivative of $\rho\mathbf{v}$ with respect to time equals the force per unit volume.) The flow rate of momentum is the product of momentum and the local volumetric flow rate. Rearranging Equation (2.3.3) so that all of the momentum terms are on the same side indicates that the net force acting on a system is simply the net change of momentum (i.e., the momentum accumulation minus the net rate of momentum flow across the control volume).

2.3.3 Forces

Forces acting on the control volume are divided into *body forces* and *surface forces*. Body forces, such as gravity and electromagnetic fields, act on the entire fluid mass throughout the control volume. For example, the net gravity force is

$$\mathbf{F}_g = \int_{V(t)} \rho \mathbf{g}\, dV, \tag{2.3.4}$$

where $\mathbf{g}$ is the acceleration due to gravity vector. These forces are generally independent of fluid motion.

Electromagnetic fields produce body forces that affect the motion of electrolytes in solution or the flow of ferromagnetic fluids. (Electrolyte transport is considered in Chapter 7.)

Forces per unit area, acting on control volume surfaces, are known as *stresses*, with units of force per unit area (e.g., Newton per square meter [or *pascals*, Pa] or dyne per square centimeter). Stresses are *tensors* (see Section A.3) and each element of the tensor has two directional components associated with it (Figure 2.9). Stresses are represented as $\boldsymbol{\sigma}$ or σ_{ij}, where the index i refers to the plane on which the stress acts and the index j refers to the direction in which the stress acts. The indices i and j can take the values for the three orthogonal directions (x, y, and z in Figure 2.9).

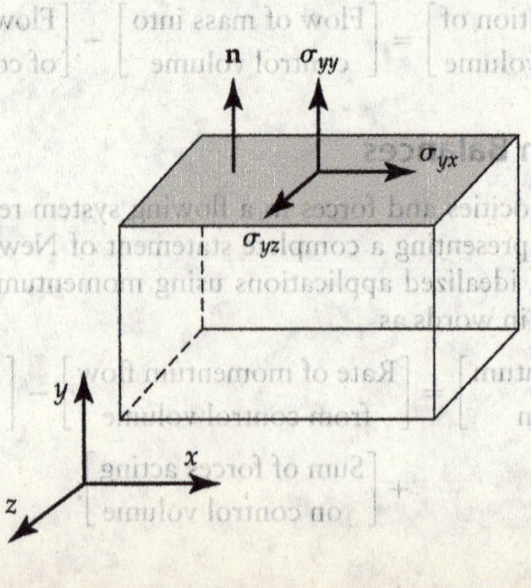

FIGURE 2.9 Cubic fluid element showing the stresses acting on a face of constant y. As shown, the normal stress σ_{yy} is tensile.

A cubic control volume is shown in Figure 2.9. The orientation of each surface is described in terms of a vector **n** of unit magnitude that is normal to, and directed away from, the surface (*the unit outward normal vector*). Stresses act normal or tangential to a control volume surface. Tangential stresses are also known as *shear stresses*. Thus, σ_{yy} is a normal stress acting on a plane of constant y in the y direction. Likewise σ_{yx} is a shear stress acting in the x direction on a plane of constant y.

The sign convention used in this book is that the stress is positive when exerted by the fluid in the direction of the unit outward normal vector. Since **n** points outward in the positive y direction from the control volume surface in Figure 2.9, σ_{yy} is positive and is tensile. A consequence of this sign convention is that compressive stresses such as pressure are negative, because these stresses point in the direction opposite that of the unit normal. Tensile stresses, by contrast, are positive, since they act in the direction of the unit normal. For shear stresses, the convention is adopted that the fluid on the face with the greater algebraic value exerts positive stresses on the face with the lesser value. Thus, σ_{yx}, σ_{yy}, and σ_{yz}, shown acting on the plane of constant y that represents the upper surface of the cube in Figure 2.9, are positive.

Since the stress is associated with two different directions (the local unit vector normal to the surface, and the direction of the local force at the surface), it is a tensor quantity with nine components. The following two important properties of stress tensors are valid for most fluids and are stated without proof (interested students should consult a reference on fluid mechanics for a proof of these statements [1,3,5]):

1. Stresses and torques on a material point are in equilibrium.
2. The stress tensor is symmetric (i.e., $\sigma_{ij} = \sigma_{ji}$).

The first property implies that the stresses acting on two different adjacent surfaces of control volumes are equal and opposite. This result is used in the boundary conditions discussed in Section 2.3.4. The second property implies that the number of stresses that must be specified is reduced from nine to six. Indeed, if the flow situation is two-dimensional, only four stresses need to be specified. (These statements apply to all fluids of biological interest.)

Another important fluid property is that *a fluid at rest cannot support a shear stress*. That is, if a shear stress is applied to a fluid at rest, it will begin to flow. Later, we will relax this restriction in considering fluids that also have solid-like properties resulting in the presence of a yield stress. Pressure is the only stress that acts on a fluid at rest. Pressure is compressive and acts normal to a surface. At a point, pressure is uniform in all directions (i.e., it is *isotropic*). Because of this distinction between stresses that can be supported at rest and stresses under motion, the stress tensor σ is divided into two components as follows:

$$\boldsymbol{\sigma} = -p\mathbf{I} + \boldsymbol{\tau}. \tag{2.3.5}$$

Here, p is the pressure, $\boldsymbol{\tau}$ is the so-called *deviatoric stress*, and $\mathbf{I}$ is the identity matrix. This matrix has diagonal elements of unit magnitude and off-diagonal elements equal to zero. The negative sign is introduced to be consistent with the sign convention, because pressure acts as a compressive normal stress. The deviatoric stress represents shear stresses arising from fluid stress.

The force of the fluid acting on a surface with a unit outward normal vector **n** is

$$\mathbf{F} = \int_S \mathbf{n} \cdot \boldsymbol{\sigma} dS. \qquad (2.3.6)$$

The term $\mathbf{n} \cdot \boldsymbol{\sigma}$ represents the vector acting on the surface S and is sometimes referred to as the *stress vector*.

2.3.4 Boundary Conditions

Solutions of the conservation of linear momentum and conservation of mass are used to determine the forces acting on the fluid if the flow is specified, or to determine the flow field for a given applied force. In order to solve such problems, boundary conditions are needed for either the stress or the velocity. The specific boundary conditions depend upon whether the boundary is an interface between two immiscible fluids or between a solid and a fluid. First consider a solid–fluid surface or interface. Either the stresses or the velocities at this interface need to be specified. If the stress or force applied to a solid surface is known, then the boundary condition is that the normal stresses are continuous across an interface. Hence,

$$\sigma_{ii}|_1 = \sigma_{ii}|_2. \qquad (2.3.7a)$$

In terms of the pressure and deviatoric stresses, this relation is

$$(\tau_{ii} - p)|_1 = (\tau_{ii} - p)|_2, \qquad (2.3.7b)$$

where the indices refer to one of the three orthogonal coordinates that are normal to the surface, and subscripts 1 and 2 refer to the surfaces that make up the interface.

Tangential to the surface, the stresses are also equal at an interface; that is,

$$\tau_{ij}|_1 = \tau_{ij}|_2, \qquad (2.3.7c)$$

where the direction i is normal to the interface and j is tangential to the interface. These stresses act in the same direction.

At the fluid–solid boundary, the "no-slip" condition for velocity is almost always used. This boundary condition expresses the fact that the fluid velocity tangent to the impermeable solid surface equals the velocity of the solid surface, or

$$\mathbf{V}_{\text{solid}} = \mathbf{v}. \qquad (2.3.7d)$$

This condition was first postulated in the 19th century without experimental proof. Subsequent experiments indicate that this relation is valid for almost all viscous fluids. Attractive forces between the fluid and the solid produce strong adhesion at the surface. The solid–fluid attractive forces are stronger than those between fluid molecules. Thus, when a stress is applied to the fluid, the fluid molecules above the layer contacting the surface move, whereas the fluid molecules contacting the surface are stationary.

If the solid is porous, fluid motion exists normal to the interface and relative to the solid. A porous medium consists of a fluid or void phase, through which flow occurs, and a solid phase. We will discuss flow in porous media in detail in Chapter 8.

The medium is characterized by a porosity or void volume fraction ε, which is the volume fraction of the medium through which the fluid flows. Fluid velocity in a porous medium is characterized by a superficial velocity, v_s, which is the ratio of the volumetric flow rate through the porous medium, divided by the cross-sectional area A of the medium normal to the direction of flow (this area includes both the solid and fluid phases):

$$v_s = \frac{Q}{A}. \tag{2.3.7e}$$

The interstitial velocity is the fluid velocity that occurs within the fluid phase of the porous medium. Since the volume available for fluid flow in that medium is reduced due to the presence of the solid phase, the interstitial velocity is greater than the superficial velocity. Because the flow rate into the porous medium equals the flow rate in the fluid phase in the porous medium, applying Equation (2.2.6) indicates that

$$v_s = \epsilon\langle v\rangle. \tag{2.3.7f}$$

Many biological tissues are porous. Flow through porous tissues is important in the delivery of high-molecular-weight solutes and the movement of lymph (see Chapter 8).

For a fluid–fluid interface, the stress boundary conditions of Equations (2.3.7a) to (2.3.7c) apply. Further, the velocity for each fluid is the same at the interface:

$$\mathbf{v}|_1 = \mathbf{v}|_2. \tag{2.3.7g}$$

2.4 Fluid Statics

2.4.1 Static Equilibrium

Although most of the transport problems encountered in biomedical engineering involve moving fluids, there are some important results under static conditions that are relevant to biomedical engineers. Under such conditions, the only forces acting on a fluid element are pressure ($\mathbf{F}_p$) and gravity ($\mathbf{F}_g$). A balance of forces yields

$$\mathbf{F}_g + \mathbf{F}_p = 0. \tag{2.4.1}$$

Applying this force balance to a cubic control volume (Figure 2.10) yields

$$x \text{ direction} \quad \left(p|_x - p|_{x+\Delta x}\right)\Delta y \Delta z = 0, \tag{2.4.2a}$$

$$y \text{ direction} \quad -\rho g \Delta x \Delta y \Delta z + \left(p|_y - p|_{y+\Delta y}\right)\Delta x \Delta z = 0, \tag{2.4.2b}$$

$$z \text{ direction} \quad \left(p|_z - p|_{z+\Delta z}\right)\Delta x \Delta y = 0. \tag{2.4.2c}$$

Because of the relation between stress and pressure given in Equation (2.3.5), the sign convention for pressure is opposite that used for stresses. Since pressure is compressive, it is positive when it acts on a surface in which the outward normal points in the negative direction and negative when it acts on a surface in which the outward

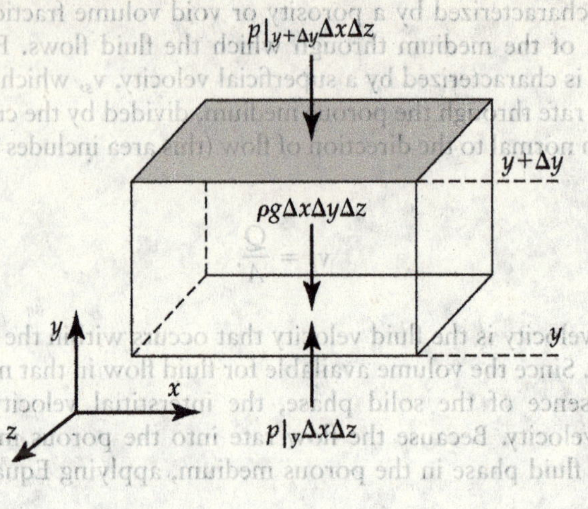

FIGURE 2.10 The y component of pressure.

normal points in the positive direction. Gravity acts in the negative y direction ($\mathbf{F}_g = -g\mathbf{e}_y$). Dividing each of the force balances by a volume element $\Delta x\ \Delta y\ \Delta z$ yields

$$\frac{(p|_x - p|_{x+\Delta x})}{\Delta x} = 0, \tag{2.4.3a}$$

$$-\rho g + \frac{(p|_y - p|_{y+\Delta y})}{\Delta y} = 0, \tag{2.4.3b}$$

and

$$\frac{(p|_z - p|_{z+\Delta z})}{\Delta z} = 0. \tag{2.4.3c}$$

Taking the limit as the volume shrinks to zero and using the definition of the derivative, we obtain the following differential balances:

$$-\frac{\partial p}{\partial x} = 0, \tag{2.4.4a}$$

$$-\rho g - \frac{\partial p}{\partial y} = 0, \tag{2.4.4b}$$

$$-\frac{\partial p}{\partial z} = 0. \tag{2.4.4c}$$

Equations (2.4.4a) to (2.4.4c) can be written in vector notation as follows:

$$\rho \mathbf{g} - \nabla p = 0. \tag{2.4.5}$$

The force balance represented by Equation (2.4.5) is used to find the pressure field and determine the forces on surfaces and submerged objects.

A useful application of this analysis is to calculate the pressure distribution in a column of fluid (Figure 2.11). Equations (2.4.4a) to (2.4.4c) indicate that the pressure is uniform in the x and z directions. Rearranging Equation (2.4.4b) yields

$$\frac{dp}{dy} = -\rho g. \tag{2.4.6}$$

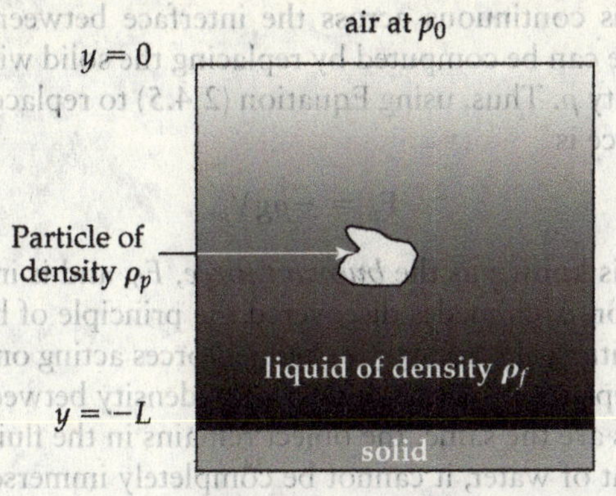

FIGURE 2.11 An arbitrarily shaped particle acted upon by a column of liquid.

Assuming constant density and assuming that $p = p_0$ at $y = 0$, we can integrate Equation (2.4.6) to yield

$$p = p_0 - \rho g y. \tag{2.4.7}$$

Since $y < 0$, the pressure increases with increasing depth in the fluid. The pressure difference with respect to atmospheric pressure, $p - p_0$, is known as the *gage pressure*. For water with a density of 1,000 kg m^{-3}, the pressure doubles every 10.34 m. The hydrostatic pressure difference from head to foot in a human 1.8 m tall is 18,660 N m^{-2} (or *pascals*, Pa), or 140 mmHg. Normal systolic blood pressure is only 120 mmHg, however, due to the location of the heart.

Knowing the distribution of forces in the fluid, one can calculate the force of the fluid acting on an arbitrarily shaped body or particle by a uniform column of liquid in equilibrium with air above it (Figure 2.11). The particle has a density ρ_p and a volume V_p. The fluid exerts a surface force due to the pressure on the submerged object. The surface force on the particle is the integral of the stress over the surface of the object:

$$\mathbf{F}_p = \int_S \mathbf{n} \cdot \boldsymbol{\sigma} dS. \tag{2.4.8}$$

Since the fluid is static ($\boldsymbol{\tau} = 0$), $\boldsymbol{\sigma} = -p\mathbf{I}$. The product $\mathbf{n} \cdot \mathbf{I}$ equals $\mathbf{n}$. Thus, the pressure force is

$$\mathbf{F}_p = -\int_S p\mathbf{n}\, dS. \tag{2.4.9}$$

The *divergence theorem* (Section A.3) relates surface and volume integrals:

$$\int_S p\mathbf{n}\, dS = \int_V \nabla p\, dV. \tag{2.4.10}$$

Thus, the pressure force can be written as

$$\mathbf{F}_p = -\int_V \nabla p\, dV. \tag{2.4.11}$$

Since the pressure is continuous across the interface between the liquid and the particle, the pressure can be computed by replacing the solid with an equivalent volume of fluid of density ρ. Thus, using Equation (2.4.5) to replace ∇p with $\rho \mathbf{g}$, we find that the pressure force is

$$\mathbf{F}_p = -\rho \mathbf{g} V_p. \tag{2.4.12}$$

This pressure force is known as the *buoyant force, F_b*, and is in the direction opposite the gravity vector. Archimedes discovered the principle of buoyancy over 2,300 years ago. The condition of stability (i.e., no net forces acting on an object that is initially submerged) depends upon the difference in density between the object and the fluid. If the densities are the same, the object remains in the fluid. If the object has a density less than that of water, it cannot be completely immersed in a liquid at rest. In that case, the surface and volume integrations are only over the submerged portion of the object ($V_p = V_{\text{submerged}}$). If the object's density exceeds the density of water, the object sinks to the bottom.

DNA, proteins, and other molecules dissolved in solution, as well as suspended cells, can be considered immersed objects with a specific density. (The density changes when the molecules interact with water molecules. The result is that some water becomes bound to the solute, and the solute is *hydrated*.) The resulting buoyant force is important in calculating the speed of settling of molecules and cells and the separation of molecules during *ultracentrifugation*, a process that produces centrifugal forces substantially greater than gravity and that enables molecules with small density differences to be separated. Instead of the particle density, its reciprocal, the *partial specific volume*, is often used. Typical values for the partial specific volume are between 0.70 and 0.75 $\text{cm}^3\ \text{g}^{-1}$ protein [6], which indicates that proteins are about 30% denser than water. In Chapter 6, we will revisit the computation of buoyant force on a molecule in an ultracentrifuge, as we analyze the role of fluid friction in the computation of diffusion coefficients.

2.4.2 Surface Tension

The behavior of molecules at the interface between two immiscible liquids or between a liquid and a gas is different from their behavior throughout the interior of the liquid. In the bulk-liquid phase, each liquid molecule is surrounded by other liquid molecules. Attractive forces on molecules are, on average, the same in all directions. At the interface, however, the environment surrounding individual molecules undergoes a sharp discontinuity. At a liquid–gas interface, the attractive forces among liquid molecules is much greater than any attractive force between the liquid molecules and the gas. As a result, the surface is under tension due to the discontinuity of attractive forces at the surface.

Surface tension affects soap-film bubbles, liposomes, and cell membranes. Surface tension, denoted by the symbol γ, has units of force per unit length or energy per unit area. Because surface tension also has units of energy per unit area, an alternative—and equally valid—view of surface tension is that it represents differential work needed to cause a differential increase in surface area of a fluid. Typical air–liquid surface tensions are presented in Table 2.1.

Due to surface tension, a pressure difference across an interface produces a curved surface. Pressure is higher on the concave side. Consider a spherical drop of

TABLE 2.1

Air–Liquid Surface Tension at 20°C

Liquid	Surface tension (N m^{-1})
Ethanol	0.0228
Mercury	0.4840
Methanol	0.0225
Water	0.0728

Source: Adapted from Ref. [7].

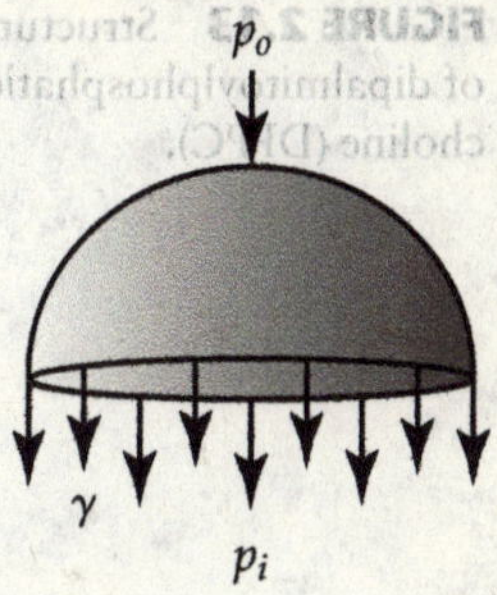

FIGURE 2.12 Forces acting on a spherical surface.

radius R, as shown in Figure 2.12. The external pressure is p_o, and the pressure inside the drop is p_i. A force balance yields

$$(p_i - p_o)\pi R^2 - \gamma 2\pi R = 0. \tag{2.4.13}$$

Defining $\Delta p = p_i - p_o$ and rearranging Equation (2.4.13) yields

$$\Delta p = \frac{2\gamma}{R}. \tag{2.4.14}$$

For an arbitrary curved surface, the following, more general, relation holds:

$$\Delta p = \gamma\left(\frac{1}{R_1} + \frac{1}{R_2}\right), \tag{2.4.15}$$

where R_1 and R_2 are the principal radii of curvature.

Equations (2.4.14) and (2.4.15) are known as the *law of Laplace*. Nonspherical shapes arise when a drop is on a surface. For a sphere, $R_1 = R_2$ and Equation (2.4.15) equals Equation (2.4.14). For a planar surface, R_1 and R_2 go to infinity and the pressure difference equals zero.

The air–liquid surface tension is reduced by the presence of *surfactants*—molecules that contain polar and nonpolar segments. Surfactants are *amphipathic* molecules, which means that the "molecule consists of parts, each of which has an affinity for a different phase" ([8], p. 254). The molecules are usually abbreviated RX, where R is a long-chain alkyl group, or H, and X is a polar or ionic group such as $-OH$, $-COOH$, SO_3^-, and NR_3^+. These molecules have low solubility in water due to the alkyl chain, and they concentrate at the air–water interface. The polar groups are in contact with the aqueous phase, and the alkyl chain is in contact with air. The exact orientation depends upon the surface density of the molecules.

Surface tension effects are significant in lung physiology and pathology. Alveoli are the sites of gas exchange from the lung to the blood. Alveoli are roughly spherical in shape and are connected to terminal bronchi in the lung. Alveolar size varies widely. Equation (2.4.14) indicates that the pressure to inflate smaller alveoli is greater than the pressure to inflate larger alveoli. Since there is a range of alveolar sizes, a surfactant is secreted to prevent the collapse of smaller alveoli and overdistension of larger alveoli, a condition known as *atelectasis* [9]. The surfactant is a lipoprotein-rich phospholipid whose primary lipid component is *dipalmitoylphosphatidylcholine* (DPPC; Figure 2.13).

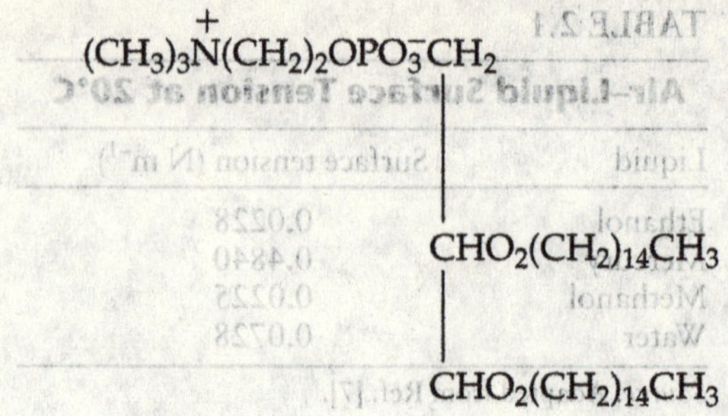

FIGURE 2.13 Structure of dipalmitoylphosphatidylcholine (DPPC).

The surface tension decreases as the area decreases. A decrease in area occurs during heavy exhalation. Since smaller alveoli go through a larger change in relative area during expiration, the surface tension of these alveoli decreases more than the surface tension of large alveoli, and collapse does not occur. When the alveoli expand during inhalation, surface tension increases. The mechanism underlying this effect is not well understood. Although it is widely held that, during compression, the surfactant becomes enriched in DPPC as other constituents are removed, recent data suggest that the surfactant undergoes a phase transition to a more ordered phase during compression [10].

In developing fetuses, surfactant is not produced until the last six weeks before normal delivery. Consequently, premature infants can develop infant respiratory distress syndrome due to lack of surfactant. Infants with this problem have labored breathing, edema, and incomplete expansion of the lungs. Death can result if lung damage is too severe. Efforts to reduce complications involve adding synthetic surfactants.

2.4.3 Membrane and Cortical Tension

White blood cells and other cells in suspension adopt a spherical shape, suggesting that the cell surface exerts a tension. This tension is much lower than the surface tension induced by a lipid surfactant layer because these cells have excess area present as membrane folds and ruffles. Beneath the cell membrane is a cytoskeletal network known as the *cortex*, which provides some mechanical rigidity to the membrane and is a site of transduction of signals [11]. The mechanical behavior of the cell surface is influenced by the properties of the membrane and the cortex.

Blood consists of red cells, white cells (or leukocytes), platelets, and plasma. Red cells are the most abundant and account for many of the unique properties of blood flow (Section 2.8). The concentration of cells in blood is shown in Table 2.2. Red cells are highly elastic and readily deform within capillaries. Leukocytes are substantially more rigid. Treating leukocytes with a variety of molecules that affect the cytoskeleton produces changes in the cortical tension. Increased leukocyte rigidity can produce an occlusion of small vessels under certain pathological conditions.

The mechanical properties of red cells and leukocytes can be analyzed with micropipet techniques (Figure 2.14). A suction pressure p_p is applied to draw a cell from a fluid solution into the micropipet. The external pressure in the bath is p_o and the pressure within the cells is p_i. Under a sufficiently high suction pressure, and for red cells in the presence of a low-osmolality solution, the cell outside the pipet

TABLE 2.2

Blood Cell Concentrations

Blood cell type	Range (cells mL^{-1})
Erthyrocytes	4.2 to 6.5 × 10^9
Leukocytes	
Neutrophils	4 to 10 × 10^6
Lymphocytes	2.5 to 5.0 × 10^6
Monocytes	1 to 10 × 10^5
Eosinophils	0 to 5 × 10^5
Basophils	0 to 1 × 10^5
Platelets	1.5 to 3.5 × 10^8

Source: Adapted from Ref. [12].

adopts a nearly spherical shape. Under these conditions, the membrane is under isotropic tension, and Equation (2.4.15) applies [13], where $\Delta p = p_o - p_p$. This pressure drop can be rewritten as $\Delta p = (p_i - p_p) - (p_i - p_o)$ to explicitly include the intracellular pressure. Applying the law of Laplace to the portion of the cell within the micropipet and the cell body yields

$$\Delta p = 2T_c\left(\frac{1}{R_p} - \frac{1}{R_c}\right), \tag{2.4.16}$$

where T_c is the cortical tension and represents contributions from the cell membrane and underlying cytoskeleton. The law of Laplace has been used to determine the bursting strength and area expansivity modulus of red cells and lipid vesicles and the cortical tension of leukocytes. Red-cell membranes and phospholipid vesicles are resistant to significant area dilation. Red cells and phospholipid vesicles rupture when the tension exceeds 10 N m^{-1}, suggesting that lipids in the membrane are the major structural component affecting red-cell rupture [14].

For a neutrophil, this cortical tension is approximately 0.024–0.035 mN m^{-1} [13] and causes the cells to maintain a spherical shape in solution. When a neutrophil is drawn into a micropipet and the tension exceeds the cortical tension, the cell flows freely in the micropipet.

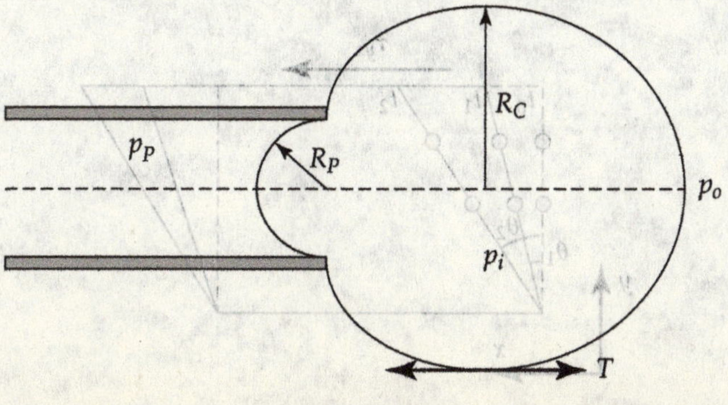

FIGURE 2.14 Schematic of pressures and tension acting on a cell drawn into a micropipet.

Example 2.3 For a white cell of radius 5 μm and a cortical tension of 0.03 mN m^{-1}, determine the pressure difference needed to draw the cell into a micropipet of radius 1 μm. What is the size of this pressure difference in terms of a column of mercury or water?

Solution From Equation (2.4.16), the pressure difference is 48 N m^{-2}. From Equation (2.4.7), the pressure force corresponds to the height h of a column of fluid and is given by $\Delta p = \rho_f gh$. The densities of mercury and water at 20°C are 13,800 and 1,000 kg m^{-3}, respectively. The corresponding heights are 3.55×10^{-4} m (or 0.0355 cm) and 4.90×10^{-3} m (or 0.49 cm). The smaller the micropipet radius, the greater is the pressure difference needed to draw the cell into the micropipet.

2.5 Constitutive Relations

2.5.1 Newton's Law of Viscosity

In order to use Newton's second law of motion to determine the velocity field and forces acting on a fluid, we must identify the variables that affect shear stress. In fluid mechanics, constitutive equations provide the needed relations between the shear stress and the fluid velocity. Unlike a conservation relationship, which is valid for all materials, a constitutive relationship is not universal and applies to a limited class of fluids. Experimental measurements are needed to derive constitutive relationships. What follows is a heuristic presentation of the relationship between shear stress and velocity gradient.

The application of a shear stress to a control volume produces a deformation of the control volume (Figure 2.15). During the time interval $\Delta t\ (= t_2 - t_1)$, the angle changes an amount $\Delta\theta\ (= \theta_2 - \theta_1)$. For a fixed time interval, increasing the shear stress results in an increase in the angle by which the control volume is deformed. Likewise, an increase in the shear stress produces the same change in angle in a shorter time interval. Thus, the shear stress is proportional to the rate of deformation, not the deformation itself. The rate of deformation is simply $d\theta/dt$, where θ is the angular change in the edge of the control volume.

Next, we show that the rate of deformation of a fluid element equals the velocity gradient. Consider the change in the location of material points in a small time

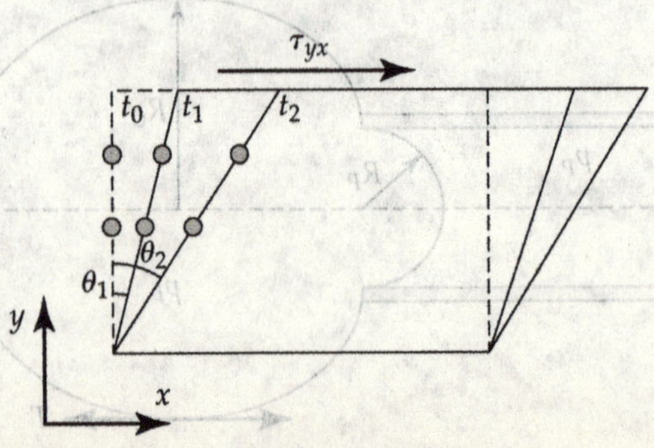

FIGURE 2.15 Continuous deformation of a fluid element exposed to a shear stress.

differential Δt after the application of a shear stress (Figure 2.16). From the geometry of the deformation, we have

$$\tan(\Delta\theta) = \frac{\Delta x}{\Delta y}. \tag{2.5.1}$$

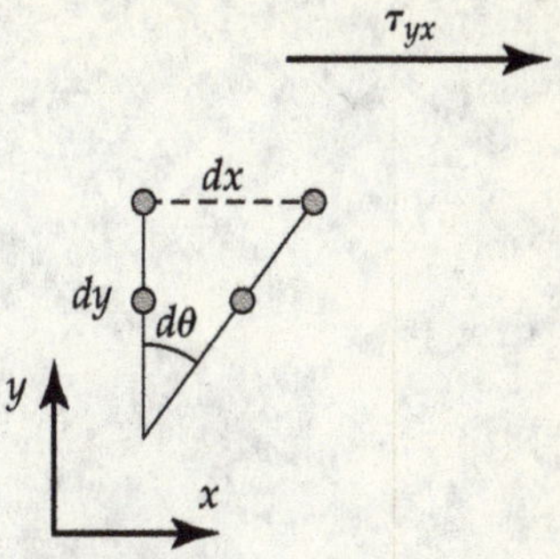

FIGURE 2.16 Location of a material point after a time interval dt in a fluid exposed to a shear stress τ_{yx}.

The quantity Δx represents the deformation of the fluid element in the x direction, and Δy is the corresponding deformation in the y direction. The finite *strain* in the x direction is the ratio of Δx to the undeformed length in the x direction. For differential changes in the angle (i.e., $\Delta\theta \ll 1$),

$$\tan(\Delta\theta) \approx \Delta\theta. \tag{2.5.2}$$

The x position of the fluid element at each x location is the product of the velocity v_x and the time interval Δt. Thus, from Equations (2.5.1) and (2.5.2), the change in angle can be represented as

$$\Delta\theta = \frac{v_x|_{y+\Delta y}\,\Delta t - v_x|_y\,\Delta t}{\Delta y} = \frac{\left(v_x|_{y+\Delta y} - v_x|_y\right)\Delta t}{\Delta y} = \frac{\Delta v_x \Delta t}{\Delta y}. \tag{2.5.3}$$

The rate of deformation can be written as

$$\frac{\Delta\theta}{\Delta t} = \frac{\Delta v_x\,\Delta t}{\Delta t\,\Delta y} = \frac{\Delta v_x}{\Delta y}. \tag{2.5.4a}$$

Taking the limit as Δt goes to zero results in the following relation between the rate of deformation and the velocity gradient:

$$\frac{d\theta}{dt} = \frac{dv_x}{dy}. \tag{2.5.4b}$$

The velocity gradient is known as the *shear rate*, $\dot{\gamma}_x$. Thus, constitutive relations take the following form:

$$\tau_{yx} = f(\dot{\gamma}_x). \tag{2.5.5}$$

Data are needed to specify the form of Equation (2.5.5) (i.e., the functional relationship) for a particular fluid. This simplest possible functional relationship, obeyed by many fluids, is a linear relation between shear stress and shear rate, given by

$$\tau_{yx} = \mu\dot{\gamma}_x = \mu\frac{dv_x}{dy}, \tag{2.5.6}$$

where the coefficient of proportionality is the *viscosity*, μ. Fluids that obey Equation (2.5.6) are known as *Newtonian fluids*, and the equation is often referred to as *Newton's law of viscosity*. That name is misleading, however, since the expression is not universal, unlike conservation equations and the laws of thermodynamics.

In some texts (e.g., Ref. [15]), a minus sign is introduced in the constitutive equation for a Newtonian fluid in order to preserve the analogy between momentum, mass, and energy transfer (see Equation (6.4.1)). Introducing the minus sign into Newton's law of viscosity implies that the stress is negative when it acts in the same direction as the unit normal. Further, compressive stresses are positive. Either sign convention is acceptable.

2.5.2 Non-Newtonian Rheology

Rheology is the branch of mechanics that studies the deformation of fluids. The distinction between Newtonian and non-Newtonian fluid behavior can be explained in terms of the shear stress and the shear rate. In general, the *apparent viscosity* is defined to represent the ratio of the shear stress to the shear rate; that is,

$$\eta_{\text{app}}(T, p, \dot{\gamma}_x) = \frac{\tau_{yx}}{\dot{\gamma}_x}. \tag{2.5.7}$$

The term *apparent viscosity* is used to indicate that, unlike a true fluid viscosity, as defined by Newton's law of viscosity, the ratio of the shear stress to the shear rate depends on the shear rate. The apparent viscosity η_{app} is determined at each shear rate from measured values of the shear rate and shear stress. For a Newtonian fluid, the apparent viscosity is a constant for all ratios of shear stress and shear rate. In general, however, the apparent viscosity changes with the shear rate. Methods for determining the relationship between shear stress and shear rate are discussed in Sections 2.7.4, 2.7.5, and 2.8.1.

Several general classes of fluids have been defined on the basis of the relationship between the shear stress and the shear rate (Figure 2.17) or the behavior of η_{app} as a function of shear rate (Figure 2.18).

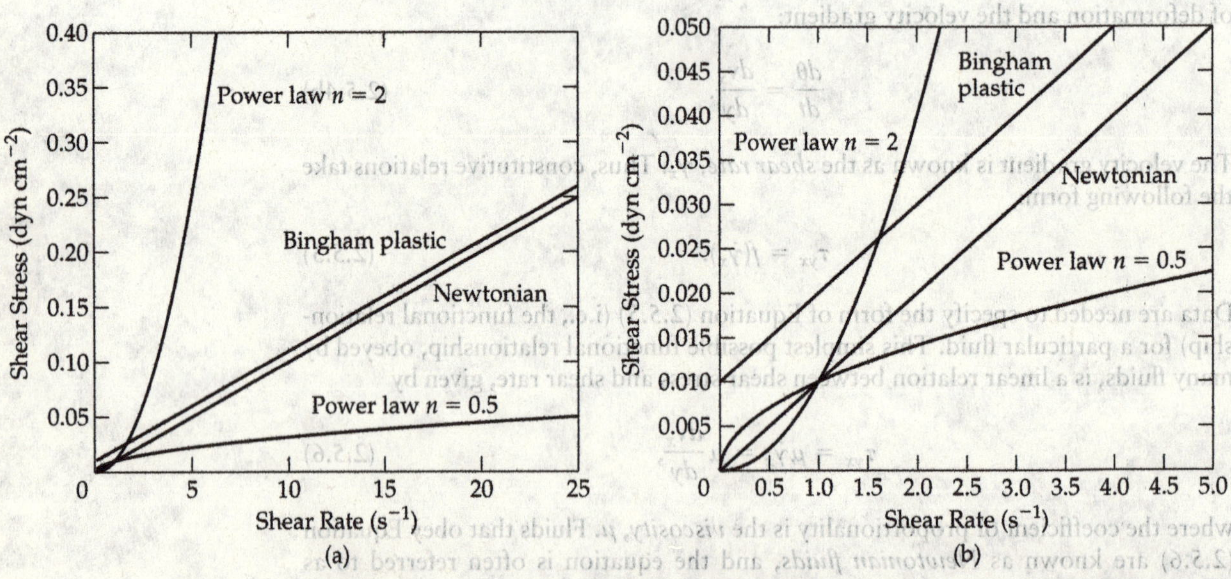

FIGURE 2.17 Relationship between the shear stress and shear rate for Newtonian and non-Newtonian fluids. Fluid properties are Newtonian fluid, $\mu = 0.01$ g cm^{-1} s^{-1}; Bingham plastic, $\tau_0 = 0.01$ dyn cm^{-2}, $\mu_0 = 0.01$ g cm^{-1} s^{-1}; shear-thickening fluid (dilatant), $m = 0.01$ g cm^{-1}, $n = 2.0$; and shear thinning fluid (pseudoplastic), $m = 0.01$ g cm^{-1} s$^{-1.5}$, $n = 0.5$.

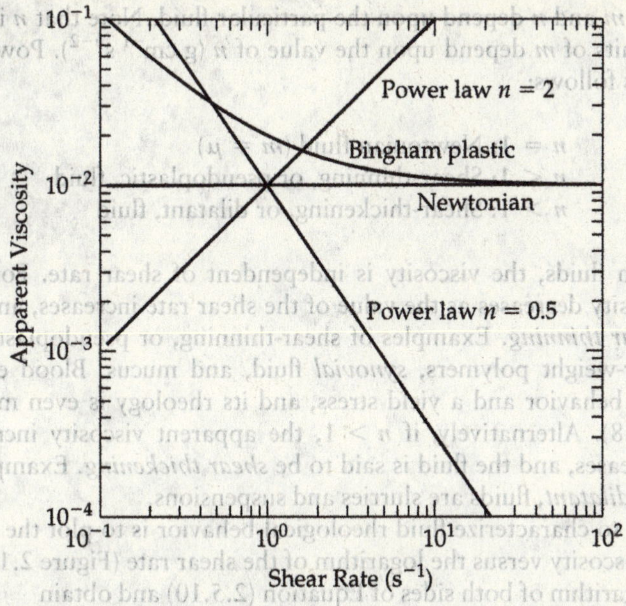

FIGURE 2.18 Apparent viscosity versus shear rate for fluids shown in Figure 2.17.

Newtonian Fluid. As previously noted, the shear stress is proportional to the shear rate and the apparent viscosity equals the true viscosity ($\eta_{app} = \mu$), given by the formula

$$\tau_{yx} = \mu \dot{\gamma}_x. \tag{2.5.8}$$

Bingham Plastic. A Bingham plastic is a material that has solid as well as fluid-like properties. It does not flow until the applied stress exceeds the *yield stress*, τ_0, for the material. Below the yield stress, the shear rate and velocity gradient vanish; that is,

$$|\tau_{yx}| < \tau_0 \qquad \dot{\gamma}_x = 0. \tag{2.5.9a}$$

In rectangular coordinates, a zero shear rate is equivalent to a constant velocity. Above the yield stress, the relationship between shear stress and shear rate is

$$|\tau_{yx}| > \tau_0 \qquad \tau_{yx} = \pm\tau_0 + \mu_0 \dot{\gamma}_x, \tag{2.5.9b}$$

where μ_0 depends upon the temperature and pressure and is independent of the shear rate. The sign in Equation (2.5.9b) is positive when the shear stress is positive and negative when the shear stress is negative. The equation is similar to Newton's law of viscosity, except that the fluid exhibits a yield stress.

Power Law Fluids. In general, a power law fluid is one type of fluid for which the *apparent viscosity* (i.e., the slope of the curve of shear stress versus shear rate at a particular value of $\dot{\gamma}_x$) is a function of the shear rate raised to a power. For power law fluids, the apparent viscosity can be written as

$$\eta_{app} = m|\dot{\gamma}_x|^{n-1}. \tag{2.5.10}$$

The quantities m and n depend upon the particular fluid. Note that n is dimensionless, but the units of m depend upon the value of n (g cm^{-1} s^{n-2}). Power law fluids are classified as follows:

$n = 1$: Newtonian fluid ($m = \mu$)
$n < 1$: Shear-thinning, or pseudoplastic, fluid
$n > 1$: Shear-thickening, or dilatant, fluid

For Newtonian fluids, the viscosity is independent of shear rate. For $n < 1$, the apparent viscosity decreases as the value of the shear rate increases, and the fluid is said to be *shear thinning*. Examples of shear-thinning, or pseudoplastic, fluids are high-molecular-weight polymers, *synovial* fluid, and mucus. Blood exhibits both shear-thinning behavior and a yield stress, and its rheology is even more complex (see Section 2.8). Alternatively, if $n > 1$, the apparent viscosity increases as the shear rate increases, and the fluid is said to be *shear thickening*. Examples of shear-thickening, or *dilatant*, fluids are slurries and suspensions.

One way to characterize fluid rheological behavior is to plot the logarithm of the apparent viscosity versus the logarithm of the shear rate (Figure 2.18). To do so, we take the logarithm of both sides of Equation (2.5.10) and obtain

$$\log(\eta_{app}) = \log(m) + (n - 1)\log(|\dot{\gamma}_x|). \tag{2.5.11a}$$

For a power law fluid, a plot of the logarithm of the apparent viscosity versus the logarithm of the shear rate is a straight line of slope $n - 1$, with an intercept equal to $\log(m)$. For a Newtonian fluid, the viscosity is independent of the shear rate ($n = 1$) and the line is horizontal (Figure 2.18). For $n > 1$, the slope is positive, and for $n < 1$, the slope is negative. Deviations from a power law fluid can be detected as nonlinearities in the plot (Figure 2.18). A log-log plot for Bingham plastics produces a nonlinear dependence of apparent viscosity that asymptotically approaches a constant value. This nonlinear behavior is due to the existence of a yield stress. Many biological fluids exhibit shear-thinning behavior. Sometimes the power law model, with or without a yield stress, accurately fits rheological measurements. This greatly simplifies subsequent fluid mechanical analysis. However, often the slope of the log-log plot of apparent viscosity versus shear rate is not a straight line. In those instances, more complex constitutive equations are often applied [16].

Example 2.4 For the following data, determine the type of fluid and the values of the properties characterizing the fluid:

Shear rate (s^{-1})	Shear stress (Nm^{-2})	Shear rate (s^{-1})	Shear stress (Nm^{-2})
0.1	0.0002	10	0.14
0.5	0.0019	50	1.77
1.0	0.0049	100	5.00
5.0	0.056	500	55.90

Solution The most straightforward approach is to plot the log of the shear stress versus the log of the shear rate. Newtonian and power law fluids exhibit linear behavior, whereas a Bingham plastic is nonlinear. A $\log_{10}$-$\log_{10}$ plot of the data is linear, with a slope of 1.48 and an intercept of -2.27. Substituting Equation (2.5.10) into Equation (2.5.7) and taking the base-10 logarithm of each side yields

$$\log(\tau_{yx}) = \log(m) + n\log(\dot{\gamma}_x). \tag{2.5.11b}$$

From the results of the fit, m equals 0.0053 kg m^{-1} s$^{-0.52}$ and n equals 1.48.

2.5.3 Time-Dependent Viscoelastic Behavior

Some fluids exhibit both viscous and elastic properties. In response to an applied force, such fluids exhibit both an instantaneous change in motion, due to elastic behavior, and a time-dependent deformation, due to viscous behavior. Fluids that exhibit such behavior are termed *viscoelastic*. As a result, the constitutive equation is a function of both the *strain* and the derivative of the strain with respect to time, or the shear rate.

For example, consider the flow of a fluid between two flat plates separated by a distance h. At time $t = 0$, the upper plate is suddenly set into motion. For a Newtonian fluid, the fluid velocity distribution changes with time and asymptotically approaches the steady-state value, with a characteristic time of $\rho h^2/\mu$. (Typically, the characteristic time represents the time for the solution to decay to a value equal to $1/e$ of the initial value.) A *viscoelastic* fluid may undergo a sudden, almost instantaneous, elastic displacement, followed by a gradual decay to the steady value, or the fluid might overshoot the steady velocity before declining to the steady state. In some cases, recoil may occur. Other transient phenomena associated with viscoelastic fluids include *stress relaxation* and *creep* [16].

For a viscous Newtonian fluid, the shear stress is proportional to the shear rate ($\dot{\gamma}_x$) as given by Equation (2.5.8). For a linear elastic (Hookean) material, the shear stress is proportional to the normalized change in position of material elements of the fluid, or the *strain*, (γ_x). The formula is

$$\tau_{yx} = G\gamma_x, \tag{2.5.12}$$

where G is the shear elastic modulus with units of Pa (Newton per square meter) or dyne per square centimeter. The shear rate (or strain rate) and the strain are related by

$$\gamma_x(t_0, t) = \int_{t_0}^{t} \dot{\gamma}_x(t')dt' \quad \text{or} \quad \frac{d\gamma_x(t)}{dt} = \dot{\gamma}_x(t). \tag{2.5.13}$$

One commonly used constitutive equation to describe viscoelastic fluids is the Maxwell model, given by

$$\tau_{yx} + \frac{\mu}{G}\frac{\partial \tau_{yx}}{\partial t} = \mu\dot{\gamma}_x, \tag{2.5.14a}$$

where G is the elastic modulus. The ratio μ/G has units of time and is often referred to as the *relaxation time* λ. Assuming that the stress in the fluid is finite at $t = -\infty$, then the solution of Equation (2.5.14a) can be represented as [16]

$$\tau_{yx} = \int_{-\infty}^{t} \left[\frac{\mu}{\lambda} \exp(-(t - t')/\lambda) \right] \dot{\gamma}_x(t')dt'. \quad (2.5.14b)$$

Example 2.5 Determine the shear stress of a Maxwell fluid for application of an instantaneous change in shear rate at time 0 from 0 to $\dot{\gamma}_{x0}$.

Solution Since the shear rate is 0 for time less than 0, the integral in Equation (2.5.14b) is nonzero for times greater than 0. For a constant shear rate, that equation becomes

$$\tau_{yx} = \frac{\mu \dot{\gamma}_{x0}}{\lambda} \exp(-t/\lambda) \int_{0}^{t} \exp(t'/\lambda)dt' = \mu \dot{\gamma}_{x0} \exp(-t/\lambda) \exp(t'/\lambda)|_0^t. \quad (2.5.14c)$$

Evaluating the function between the limits yields

$$\tau_{yx} = \mu \dot{\gamma}_{x0} \exp(-t/\lambda)(\exp(t/\lambda) - 1) = \mu \dot{\gamma}_{x0}(1 - \exp(-t/\lambda)). \quad (2.5.14d)$$

For a Maxwell fluid, the shear stress does not rise instantly to the steady-state value, but approaches steady state with a time constant λ. A Newtonian fluid can be viewed as a special case of a Maxwell fluid for which $G = 0$. For this case, note that the time constant λ is infinite and $\tau_{yx} = \mu \dot{\gamma}_{x0}$. Thus, stress in a Maxwell fluid exhibits a time-dependent response to an applied shear rate. By contrast, stress in a Newtonian fluid does not vary with time.

Only a limited number of real fluids behave similarly to an ideal Maxwell fluid. Although a number of other constitutive relations have been proposed to describe viscoelastic fluids [16], perhaps the following generalized expression for a linear viscoelastic fluid is most useful:

$$\tau_{yx} = \int_{-\infty}^{t} G(t - t')\dot{\gamma}_x(t')dt'. \quad (2.5.15a)$$

In this equation, the *relaxation modulus* $G(t - t')$ is a property of the fluid, whereas the shear rate is a property of the flow field. For a Maxwell fluid, a comparison of Equations (2.5.14b) and (2.5.15a) indicates that

$$G(t - t') = \frac{\mu}{\lambda} \exp(-(t - t')/\lambda). \quad (2.5.15b)$$

A variety of different transients are applied experimentally to study viscoelastic behavior. In addition to a step input, the response to an imposed oscillatory shear rate (or the velocity of flow of a viscous fluid through narrow channels) is the most common input for the quantitative characterization of biological and nonbiological fluids. The most general approach is to represent an imposed oscillation with the complex representation

$$\exp(i\omega t) = \cos(\omega t) + i\sin(\omega t). \tag{2.5.16}$$

Thus,

$$\dot{\gamma}_x(t) = \dot{\gamma}_x^0 \exp(i\omega t), \tag{2.5.17a}$$

and

$$\gamma_x(t) = \frac{\dot{\gamma}_x^0 \exp(i\omega t)}{i\omega}. \tag{2.5.17b}$$

Using complex notation, we find that the shear stress is

$$\tau_{yx} = (\eta' + i\eta'')\dot{\gamma}_x^0, \tag{2.5.18a}$$

$$\tau_{yx} = (G' + iG'')\gamma_x^0, \tag{2.5.18b}$$

where η' and η'' are, respectively, the real and imaginary components of the complex viscosity, and G' and G'' are, respectively, the real and imaginary components of the complex modulus of elasticity. Since γ_x and γ_0 are related by Equations (2.5.17a) and (2.5.17b), Equations (2.5.18a) and (2.5.18b) can also be written as

$$\tau_{yx} = i\omega(\eta' + i\eta'')\gamma_x \tag{2.5.19}$$

and

$$G' = -\omega\eta'' \qquad G'' = \omega\eta'. \tag{2.5.20a,b}$$

G' and G'' are also known as the *storage* and *loss moduli*, respectively. The quantity G' represents the storage of elastic energy, and G'' represents the dissipation of energy by viscous forces. For a perfectly elastic material, G'equals the elastic modulus and G'' equals zero. Likewise, for a viscous fluid, η' equals the viscosity and η'' equals zero. As a result, G'' is nonzero. Since viscous forces dissipate energy, G'' is known as the loss modulus. The ratio of G' and G'' (or η'' and η') indicates the relative importance of viscous and elastic contributions to material behavior and is referred to as the *loss tangent*.

Synovial fluid, a biological viscoelastic fluid that lubricates joints and absorbs compressive loads, provides a good illustration of these concepts. The fluid consists of electrolytes and proteins. The major protein consists of a protein core linked to hyaluronic acid.

Rheological data for synovial fluid (Figure 2.19) show that synovial fluid exhibits viscoelastic behavior. At low frequencies, the loss modulus is much larger than the storage modulus ($G'' \gg G'$), and the fluid behavior is largely viscous. As the frequency increases, both moduli increase in magnitude. For frequencies greater than $2\,\text{rad}\,\text{s}^{-1}$, the storage modulus exceeds the loss modulus, and elastic behavior dominates. Consistent with this behavior, steady-flow measurements in a viscometer indicate that synovial fluid exhibits shear-thinning behavior. Frequencies associated with walking ($3\,\text{rad}\,\text{s}^{-1}$) and running ($15.7\,\text{rad}\,\text{s}^{-1}$) lie in a range in which elastic forces

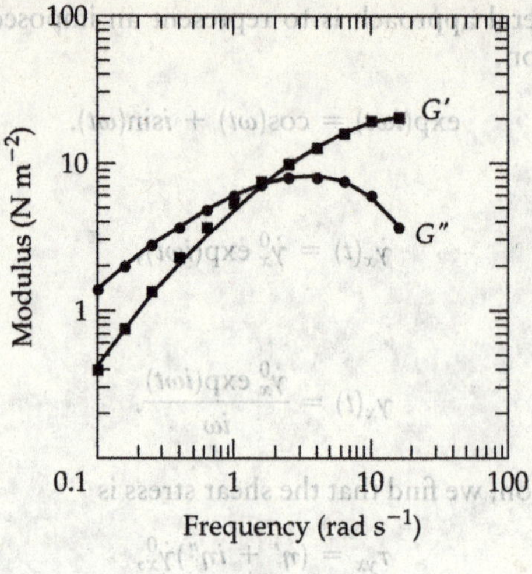

FIGURE 2.19 Storage and loss moduli for human synovial fluid. (Adapted with permission from Ref. [17].)

dominate. Elastic behavior may be beneficial at these frequencies, whereas viscous behavior may be helpful in bearing loads without motion. With age, both the storage and loss moduli decrease, and G'' contributes more significantly at all frequencies. Other viscoelastic biological fluids include the mucus in the intestine, the respiratory system, and the uterine cervix, and the vitreous of the eye.

2.6 Laminar and Turbulent Flow

A remarkable feature of fluid flow is that fluid motion undergoes a transition in its behavior at higher flow rates. Such a change in fluid motion is evident to anyone who has watched smoke rise from a fire or water stream from a hose or faucet. In the late 19th century, Osborne Reynolds documented this transition for flow in cylindrical tubes. Consider the steady flow of a Newtonian fluid in a cylindrical tube, as shown in Figure 2.20a. A dye is injected continuously into the fluid. For low flow rates, the dye traces a straight-line path. (A more quantitative statement about what constitutes a low flow rate is considered shortly.) The width of the dye band broadens due to molecular diffusion as the dye moves farther downstream. Since this is a steady flow, the dye pathline traces a fluid streamline. Flow in which the fluid travels in a steady, time-independent manner at each location is known as *laminar* flow. Fluid motion is said to occur in laminae—in this case, concentric cylinders around the centerline.

Now consider the case in which the flow rate increases. At first, the dye pattern is the same. But at some flow rate, depending on the radius of the cylinder and the

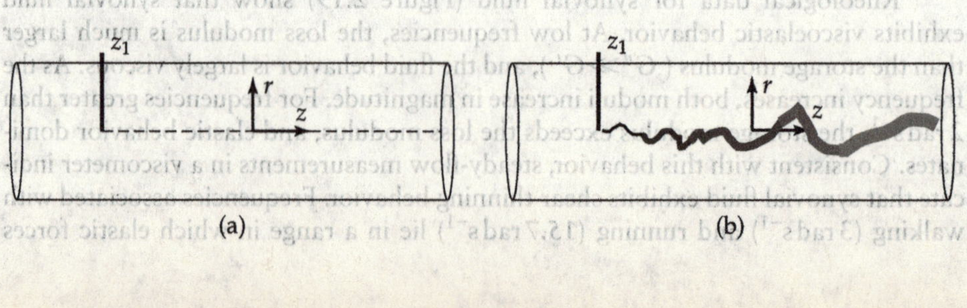

FIGURE 2.20 Release of dye into a stream of flowing Newtonian liquid and demonstration of the transition from laminar (a) to turbulent (b) flow.

properties of the fluid, the flow pattern undergoes a transition. The motion of the dye in the radial direction becomes more and more random. This movement is much faster than diffusion. At first, the behavior occurs intermittently, and, for a period, the flow resumes its laminar behavior. As the flow rate increases further, however, the intermittent random and chaotic behaviors of the dyes become more prevalent, until the behavior is completely chaotic (Figure 2.20b). Dye is dispersed throughout the entire radius of the tube. Such fluid flow is called *turbulent*.

In steady laminar flow, the velocity at a given location does not change with time. A significant feature of turbulent flow is that the velocity at a given location changes with time even when the overall flow is steady, as shown in Figure 2.21. However, the time-averaged velocity at a given location is a well-defined quantity, as shown by the dashed line in Figure 2.21. Often, the velocity in turbulent flow is characterized as the sum of a time-averaged component and a fluctuating component with a zero time average.

Turbulence affects many transport phenomena. For example, at low flow rates, the pressure drop is generally proportional to the average velocity. For turbulent flow in a smooth cylindrical tube, the pressure drop is proportional to the velocity raised to the power of 1.75. For biological organisms, such a change in the functional relationship between pressure drop and the velocity affects the energy needed to create the pressure drop. Turbulence affects the transport of molecules, often producing more rapid mixing than what is observed under laminar-flow conditions. Turbulent stresses can, however, damage biological molecules and cells. (The effect of turbulence in heart valve design is considered in Section 5.8.)

Whether a flow is laminar or turbulent can be determined simply by computing the ratio of inertial to viscous forces, a quantity known as the *Reynolds number* and defined as

$$\mathrm{Re} = \frac{\rho v L}{\mu}, \tag{2.6.1}$$

where ρ is the fluid density, μ is the fluid viscosity, v is a characteristic velocity, and L is a characteristic length. For fluid flow in a tube, the characteristic velocity is the average velocity and the characteristic length is the tube diameter. The diameter is

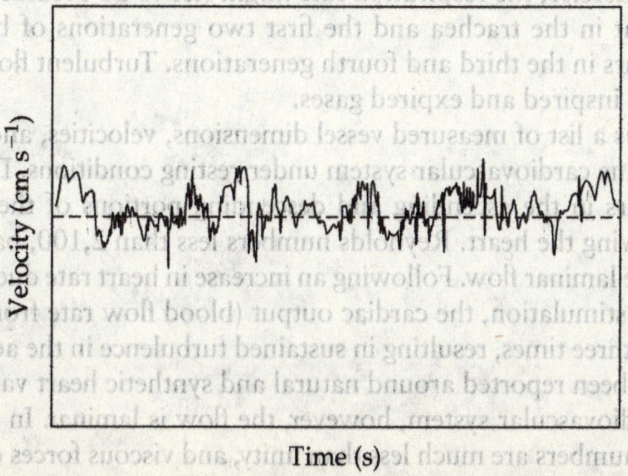

FIGURE 2.21 Velocity fluctuations in turbulent flow at one location in Figure 2.20b.

chosen rather than the length, because that is the distance over which velocity gradients occur. For noncircular tubes, the characteristic length is the hydraulic diameter D_h.

$$D_h = \frac{4(\text{Cross-sectional area})}{\text{Perimeter}}. \tag{2.6.2}$$

For a rectangular channel of width w and height h, $D_h = 2wh/(w+h)$.

For flow in cylindrical tubes, flow is always laminar for Re less than 2,100. Under conditions in which vibrations are eliminated and the inner walls of the cylinder are extremely smooth, laminar flow can be produced for Re as high as 10^5. Typically, though, turbulent flow begins around a Re of 4,000 and is sensitive to local roughness on the inner walls of the cylinder. The intermediate region between laminar and turbulent flow is known as *transitional flow*. In this region, flow is unstable and exhibits intermittent turbulence.

Besides roughness, the onset of turbulence is influenced by local changes in geometry, such as constrictions and bends.

The dynamics of fluid flow are strongly influenced by the relative importance of inertial and viscous forces, which explains the importance of the Reynolds number. Inertial forces arise from the convective acceleration of fluid and are proportional to ρv^2, with units of force per unit area. Viscous forces per unit area can be represented by the viscous shear stress, which, for a Newtonian fluid, is proportional to $\mu v/L$ (Equation (2.5.6)). The ratio of inertial to viscous forces is

$$\frac{\text{Inertial forces/volume}}{\text{Viscous forces/volume}} = \frac{\rho v^2}{\mu v/L} = \frac{\rho v L}{\mu} = \text{Re}. \tag{2.6.3}$$

The characterization of laminar and turbulent flows in tubes in terms of a single dimensionless variable, the Reynolds number, is a rather remarkable result. As we show in Chapter 3, other dimensionless groups become important if flow involves a gas–liquid interface or surface tension. For many biomedical applications, the Reynolds number is the dominant dimensionless group that characterizes flow.

Reynolds numbers for airflow in the lungs are listed in Table 2.3. The human lung consists of 23 generations of branching tubes (see Figure 1.13b). During quiet breathing, which corresponds to roughly 12 breaths per minute, transitional flow may arise in the trachea. Otherwise, flow is laminar. During vigorous breathing, such as that during exercise, the respiration rate might rise to 30 breaths per minute. Airflow is turbulent in the trachea and the first two generations of branches. Transitional flow occurs in the third and fourth generations. Turbulent flow leads to better mixing between inspired and expired gases.

Table 2.4 is a list of measured vessel dimensions, velocities, and Reynolds numbers for the canine cardiovascular system under resting conditions. During peak flow, turbulence occurs in the ascending and descending portions of the aorta, the main blood vessel leaving the heart. Reynolds numbers less than 2,100, based on the mean velocity, indicate laminar flow. Following an increase in heart rate due to physical exertion or nervous stimulation, the cardiac output (blood flow rate from the heart) may increase two to three times, resulting in sustained turbulence in the aorta. Further, turbulent flow has been reported around natural and synthetic heart valves. Throughout most of the cardiovascular system, however, the flow is laminar. In the microcirculation, Reynolds numbers are much less than unity, and viscous forces dominate.

TABLE 2.3

Human Airway Dimensions, Velocities, and Reynolds Numbers

		Quiet breathing ($Q = 0.5 \times 10^{-3}$ m^3 s^{-1})			Vigorous breathing ($Q = 2.0 \times 10^{-3}$ m^3 s^{-1})	
Generation	Internal diameter (cm)	Length (cm)	$\bar{v}$ (cm s^{-1})	Re	$\bar{v}$ (cm s^{-1})	Re
Trachea	1.80	12.0	197	2,325	790	9,324
1	1.22	4.76	215	1,719	859	6,876
2	0.83	1.90	235	1,281	941	5,124
3	0.56	0.76	250	921	1,002	3,684
4	0.45	1.27	202	594	809	2,376
5	0.35	1.07	161	369	643	1,476
10	0.13	0.46	38	32	151	127
15	0.066	0.20	4.4	1.9	17.8	7.6
20	0.045	0.083	0.3	0.09	1.2	0.37

Source: Reproduced from Ref. [18], with permission of www.annualreviews.org © 1977.

TABLE 2.4

Dimensions, Velocities, and Reynolds Numbers in the Canine Cardiovascular System

Vessel	Internal diameter (cm)	Length (cm)	Peak blood velocity (cm s^{-1})	Re_{peak}	Mean blood velocity (cm s^{-1})	Re_{mean}
Ascending Aorta	1.5	5	120	4,500	20	750
Descending Aorta	1.3	20	105	3,400	20	648
Abdominal Aorta	0.9	15	55	1,250	15	341
Femoral Artery	0.4	10	100	1,000	10	100
Arteriole	0.005	0.15	75	0.09	0.5–1	0.0006–0.0012
Capillary	0.0006	0.06	7	0.001	0.02–0.17	$2.86–24.3 \times 10^{-6}$
Venule	0.004	0.15	35	0.035	0.2–0.5	0.0002–0.0005
Inferior Vena Cava	1.0	30	25	700		
Main Pulmonary Artery	1.7	3.5	70	3,000	0.15	6.43

Source: Adapted from Ref. [19], with permission.

2.7 Application of Momentum Balances

The number of flow problems that can be solved analytically is limited. Often, it is necessary to make simplifying assumptions about the character of the flow or the boundary conditions. Nevertheless, the simplified problems are important because the assumptions can be approximately satisfied in many different types of flow. Further, information about these simple flows can provide some insight into more complex flow.

In this section, a number of steady, one-dimensional laminar flow problems of biological interest are examined. These problems treat the effect of geometry, the manner in which flow is generated, the constitutive equations upon the flow field, and the resulting forces acting on the fluid. The assumptions used in each problem are stated, and conditions are identified for which these assumptions are valid.

2.7.1 Flow Induced by a Sliding Plate

As our first example of momentum balances, we consider flow between two parallel plates, as shown in Figure 2.22a. The lower plate is stationary and the upper plate moves at a fixed speed V. Under idealized conditions, this flow represents the type of flow between moving surfaces. Biological examples include a red cell moving near a capillary wall and the relative motion of two joints separated by a thin layer of synovial fluid. Both of these situations are far more complex than this simple model suggests. Nevertheless, an analysis of such a simple flow field provides insight into the relationship between the motion of the plate and the forces acting on the plate.

For simplicity, consider the steady motion of the upper plate. The only forces arising are those due to shear stress exerted by the fluid. Assume that far upstream and downstream, the fluid is exposed to the atmosphere, so no pressure gradients arise. If the gap is thin and the plates are large, and edge effects are negligible, the fluid velocity is constant in the x direction and change only in the y direction.

Before analyzing the problem mathematically, we establish a coordinate system and sketch the velocity profile. The origin of the rectangular coordinate system is set on the surface of the lower plate. Other appropriate choices are the upper plate and the midplane between the plates. Because of the no-slip condition, the velocity at the lower plate ($y = 0$) is zero and the velocity of the fluid in contact with the upper plate ($y = h$) is V. The velocity profile is drawn as linear, although at this point in the analysis, it is not clear that the profile is linear.

A rectangular control volume is shown in Figure 2.22b. Shear stresses act on the surfaces at y and $y + \Delta y$. Momentum enters the control volume at x and leaves at $x + \Delta x$. In other words, the momentum balance is

$$\begin{bmatrix}\text{Sum of forces acting} \\ \text{in the } x \text{ direction}\end{bmatrix} = \begin{bmatrix}\text{Rate of momentum} \\ \text{flow out at } x + \Delta x\end{bmatrix} - \begin{bmatrix}\text{Rate of momentum} \\ \text{flow in at } x\end{bmatrix} \tag{2.7.1}$$

Applying a force or momentum balance to a small volume of fluid $\Delta x \Delta y \Delta z$ yields

$$(\tau_{yx}|_{y+\Delta y} - \tau_{yx}|_{y})\,\Delta x \Delta z = (\rho v_x v_x|_{x+\Delta x} - \rho v_x v_x|_{x})\,\Delta y \Delta z. \tag{2.7.2}$$

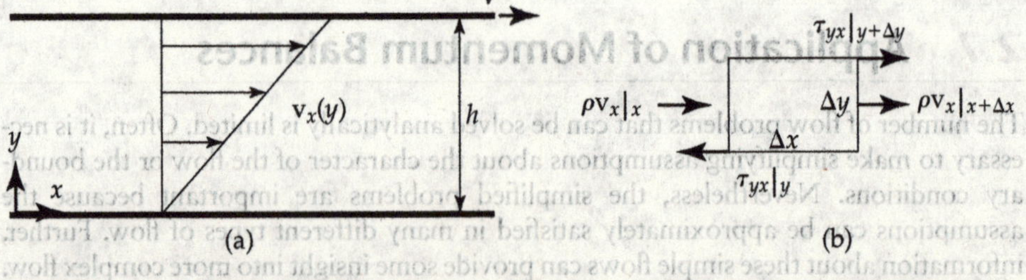

FIGURE 2.22 (a) Schematic of flow between two plates separated by a gap of thickness h containing a Newtonian fluid of viscosity μ. (b) Momentum balance applied to a cubic control volume of fluid.

Note the sign convention used for the shear stress. The unit outward normal on the surface at y points in the minus y direction, so the shear stress is negative. Conversely, the unit outward normal on the surface at $y + \Delta y$ points in the positive y direction, and the shear stress is positive.[1] The rate of momentum flow per unit volume is the product of the local flow rate $v_x \Delta y \Delta z$ and the local momentum ρv_x. The fluid velocity does not change in the x direction. As a result, $\rho v_x|_x = \rho v_x|_{x+\Delta x}$, and the right-hand side of Equation (2.7.2) equals zero.

Dividing Equation (2.7.2) by $\Delta x \Delta y \Delta z$ yields

$$\frac{\tau_{yx}|_{y+\Delta y} - \tau_{yx}|_y}{\Delta y} = 0. \tag{2.7.3}$$

Next, take the limit as Δy goes to zero and apply the definition of the derivative. Recall from introductory calculus that the definition of the derivative is

$$\lim_{\Delta t \to 0} \frac{f(t + \Delta t) - f(t)}{\Delta t} = \frac{df}{dt}. \tag{2.7.4}$$

Therefore, Equation (2.7.3) becomes

$$\frac{d\tau_{yx}}{dy} = 0. \tag{2.7.5}$$

Due to the absence of an applied pressure, the gradient in shear stress is zero. This result also means that the shear stress does not vary in the y direction. In terms of momentum, Equation (2.7.5) indicates that there is no momentum transport in the y direction.

The stresses on either surface are not known and need to be determined. The fluid velocity on the surfaces is as follows:

$$y = 0 \qquad v_x = 0, \tag{2.7.6a}$$

$$y = h \qquad v_x = V. \tag{2.7.6b}$$

Integrating Equation (2.7.5) yields $\tau_{yx} = \text{constant} = C_1$. Thus, the shear stress is independent of location and is uniform everywhere. To determine the constant, substitute into Newton's law of viscosity:

$$\tau_{yx} = \mu \frac{dv_x}{dy} = C_1. \tag{2.7.7}$$

Integrating Equation (2.7.7) yields

$$v_x = \frac{C_1}{\mu} y + C_2. \tag{2.7.8}$$

Equation (2.7.8) contains two constants of integration that are determined by evaluating the boundary conditions. Applying the boundary condition at $y = 0$ requires that C_2 equal zero. Next, applying the boundary condition at $y = h$ requires that

$$C_1 = \mu \frac{V}{h}. \tag{2.7.9}$$

[1]If a negative sign is used in Equation (2.5.6) to maintain the analogy with mass and energy transfer, then the shear stress is positive at y points and negative on the surface at $y + \Delta y$.

The fluid velocity field is therefore

$$v_x = V\frac{y}{h}. \tag{2.7.10}$$

The shear stress acting on the plate is

$$\tau_{yx} = \mu \frac{dv_x}{dy}\bigg|_{y=h} = \mu\frac{V}{h}. \tag{2.7.11}$$

The shear stress acts in the positive x direction. The plate creates a shear stress that causes the fluid to move. Thus, the direction of the shear stress and the fluid motion are the same in this case.

Example 2.6 Two parallel plates are separated by a thin gap of thickness h containing a Bingham plastic, as shown in Figure 2.22a. A force per unit area, τ_{yx}, is exerted on the upper plate. Under what conditions does the upper plate move?

Solution Equation (2.7.5) is valid for a fluid moving between parallel plates. Equation (2.7.7) indicates that the shear stress is uniform throughout the fluid. The upper plate moves when $\tau_{yx} > \tau_0$. For $\tau_{yx} < \tau_0$, v_x is constant. The lower plate is not moving, so $v_x\ (y = 0) = 0$. Since v_x is a constant everywhere between the two plates and v_x must equal zero at $y = 0$, v_x must equal zero at all points in the range $0 < y < h$. Therefore, the upper plate moves only when the yield stress is exceeded. The velocity field is

$$\tau_{yx} < \tau_0 \qquad v_x = 0, \tag{2.7.12a}$$

$$\tau_{yx} > \tau_0 \qquad v_x = \frac{(\tau_{yx} - \tau_0)y}{\mu_0}. \tag{2.7.12b}$$

Note that, for a Bingham plastic, the velocity of the upper plate is less than the velocity that would exist for a Newtonian fluid (i.e., $\tau_0 = 0$).

2.7.2 Pressure-Driven Flow through a Narrow Rectangular Channel

The next two cases characterize flow through channels. Flow is induced by a pressure gradient, which could be produced by a pump or gravity. These examples also examine the effect of geometry upon the flow and shear stress. In the first example, we examine flow through a narrow rectangular channel of height h and width w (Figure 2.23). Flow through this kind of channel is important in many biomedical devices, such as hemodialyzers and ultrafiltration units. Parallel-plate channels are used widely to study the effect of flow on cell adhesion and cell function.

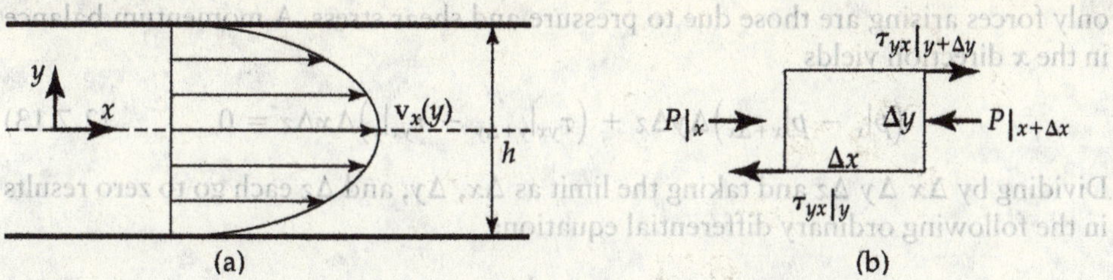

FIGURE 2.23 (a) Pressure-driven flow through a rectangular channel. Note that the velocity profile is symmetric around the centerline. The channel thickness is h and the channel width is w. (b) Momentum balance applied to a cubic control volume.

The velocity profile is sketched in Figure 2.23a. Due to the no-slip boundary condition, the velocity is zero at either surface ($y = h/2$). In order to analyze this flow, we must make the following assumptions:

1. The pressure varies only in the direction of flow.
2. The fluid density is constant, which indicates that the fluid is incompressible.
3. The flow is steady; that is, pressure, shear stress, and velocity do not change with time.
4. The fluid is Newtonian.
5. Flow is fully developed. This means that the channel length is much longer than the entrance length L_e where the velocity depends upon axial distance along the channel, $L_e \ll L$.
6. Edge effects are neglected. To meet this assumption, we require a long, wide channel; that is, $h/w \ll 1$ and $h/L \ll 1$, where w is the width and L is the length of the channel.
7. The flow is laminar.

The first and fifth assumptions imply that there is only one shear stress, namely, τ_{yx}. The pressure is balanced by shear stress acting in the x direction. With only one shear stress acting on the control volume, there is only one velocity component, v_x, and the flow is considered *fully developed*. This means that the shear stresses and velocity field do not change along the x direction, so v_x is a function of y only. Thus, the net flow of momentum in the x direction is zero.

Because the plates are not moving, the velocity of the fluid in contact with the upper and lower plates is zero. Further, the velocity and shear stress must be symmetric about the midplane. For this reason, the coordinate system in the y direction has its origin along the midplane between the plates. Other possible choices for the origin in the y direction include the upper and lower surfaces of the channel. The proposed velocity profile is consistent with the expected symmetry about the midplane and no-slip boundary conditions.

A control volume for analysis is shown in Figure 2.23b. Since there is no net momentum flow and the flow is steady, the sum of all forces must equal zero. The only forces arising are those due to pressure and shear stress. A momentum balance in the x direction yields

$$\left(p|_x - p|_{x+\Delta x}\right)\Delta y \Delta z + \left(\tau_{yx}|_{y+\Delta y} - \tau_{yx}|_y\right)\Delta x \Delta z = 0. \tag{2.7.13}$$

Dividing by $\Delta x\ \Delta y\ \Delta z$ and taking the limit as Δx, Δy, and Δz each go to zero results in the following ordinary differential equation:

$$\frac{dp}{dx} = \frac{d\tau_{yx}}{dy}. \tag{2.7.14}$$

The pressure changes only in the x direction (i.e., $p = f(x)$), and the shear stress changes in the y direction ($\tau_{yx} = g(y)$). Thus,

$$\frac{df(x)}{dx} = \frac{dg(y)}{dy}. \tag{2.7.15}$$

The left- and right-hand sides of Equation (2.7.15) can be equal only if *the derivatives* are each equal to a constant C_1. Hence, the left-hand side of Equation (2.7.14) can be integrated to yield

$$p = C_1 x + C_2. \tag{2.7.16}$$

The pressure can be specified at two locations, away from both the entrance and the exit. Thus, at $x = x_0$, $p = p_0$, and at $x = x_L$, $p = p_L$. Defining $\Delta p = p_0 - p_L$ and $x_L - x_0 = L$, we find that the pressure is

$$p = p_0 + \frac{\Delta p}{L}(x_0 - x), \tag{2.7.17}$$

and Equation (2.7.14) becomes

$$-\frac{\Delta p}{L} = \frac{d\tau_{yx}}{dy}. \tag{2.7.18}$$

Integrating Equation (2.7.18) results in

$$\tau_{yx} = -\frac{\Delta p}{L}y + C_3. \tag{2.7.19}$$

The stresses are not specified on any boundary, but the velocity is known. To find the constant of integration, determine the velocity and apply the boundary condition that $v_x = 0$ at $y = \pm h/2$. To find the velocity, insert Newton's law of viscosity, Equation (2.5.6), into Equation (2.7.19) to yield

$$\mu\frac{dv_x}{dy} = -\frac{\Delta p}{L}y + C_3. \tag{2.7.20}$$

After integrating Equation (2.7.20), we have

$$v_x = -\frac{\Delta p}{2\mu L}y^2 + \frac{C_3}{\mu}y + C_4. \tag{2.7.21}$$

Applying the boundary conditions results in $C_3 = 0$ and $C_4 = \Delta p h^2/8\mu L$. The velocity profile is

$$v_x = \frac{\Delta p h^2}{8\mu L}\left(1 - \frac{4y^2}{h^2}\right). \tag{2.7.22}$$

Note that v_x is symmetric around $y = 0$, which indicates that the velocity gradient and shear stress are zero at $y = 0$. The *symmetry condition* that the velocity gradient is zero along the centerline is used in subsequent examples and problems.

Before examining the stress, we need to compute several results with respect to the velocity and flow rate. Equation (2.7.22) describes a parabola and the velocity is a maximum at $y = 0$ with a value of

$$v_{max} = \frac{\Delta p h^2}{8\mu L}, \tag{2.7.23}$$

and Equation (2.7.22) can be written as

$$v_x = v_{max}\left(1 - \frac{4y^2}{h^2}\right). \tag{2.7.24}$$

The volumetric flow rate is the integral of the velocity over the cross-sectional area through which fluid flows; that is,

$$Q = \int_{-h/2}^{h/2}\int_0^w v_x dz dy = v_{max} w \int_{-h/2}^{h/2}\left(1 - \frac{4y^2}{h^2}\right)dy \tag{2.7.25}$$

or

$$Q = v_{max} w\left(y - \frac{4y^3}{3h^2}\right)\Bigg|_{-h/2}^{h/2} = \frac{2v_{max} wh}{3}. \tag{2.7.26}$$

The average velocity through the cross-sectional area A is

$$\langle v\rangle = \frac{1}{A}\int_A v_x dA = \frac{1}{wh}\int_{-h/2}^{h/2}\int_0^w v_x dz dy, \tag{2.7.27}$$

The volumetric flow rate can be written as

$$Q = \langle v\rangle wh. \tag{2.7.28}$$

To relate the average and maximum velocities for flow through a channel, equate Equations (2.7.26) and (2.7.28)

$$\langle v\rangle = \tfrac{2}{3}v_{max}. \tag{2.7.29}$$

Since $C_3 = 0$, the shear stress, Equation (2.7.19), is

$$\tau_{yx} = -\frac{\Delta p}{L}y. \tag{2.7.30}$$

The shear stress distribution τ_{yx} is sketched in Figure 2.24. The shear stress is positive for $y < 0$ and negative for $y > 0$. This change in sign is due to the convention for the sign of the shear stress and to the location of the origin of the coordinate

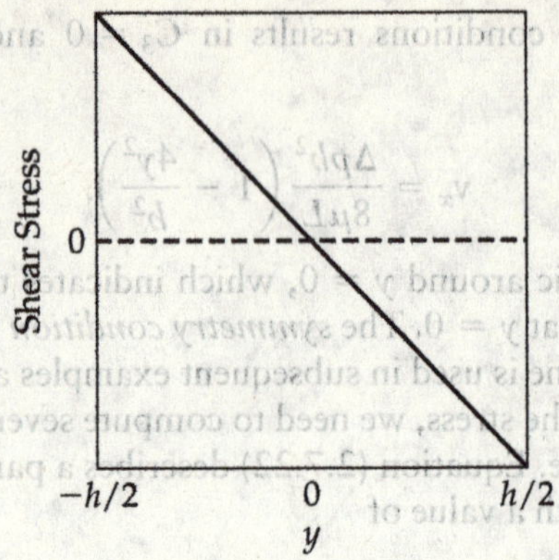

FIGURE 2.24 Shear stress distribution for flow between parallel plates.

system. For $y < 0$, a positive shear stress points in the negative x direction. For $y > 0$, the normal is in the positive y direction, and the negative sign indicates that the shear stress acts in the negative x direction.

2.7.3 Pressure-Driven Flow through a Cylindrical Tube

In this example, we consider the pressure-driven laminar flow of a Newtonian fluid through a cylindrical tube. Such flows arise in many biomedical applications, such as flow in ultrafiltration and dialysis units, bioreactors, needles, infusion systems, and capillary tube viscometers. In vivo, this type of flow occurs in the cardiovascular system, although the flow is often unsteady, and blood behaves as a non-Newtonian fluid (see Chapter 5). In fact, the original flow studies by Poiseuille were motivated by flow in the microcirculation. Given problems with blood clotting, Poiseuille conducted studies with water flowing through glass tubes [20].

Because of the geometry, cylindrical coordinates are a simpler system to use than rectangular coordinates. The fluid is Newtonian; flow is steady and laminar. For this problem, we make many of the assumptions we made for the last problem. We expect the velocity to be independent of the azimuthal angular direction. The flow is fully developed, so v_z is a function of r only. Channel flow is expected to exhibit radial symmetry about the centerline. A velocity profile consistent with these ideas is shown schematically in Figure 2.25a. From the expected velocity profile, the origin for the radial direction is chosen as the centerline and the z-axis is along the centerline.

Control volume and momentum balances are shown in Figure 2.25b. Since there is no net momentum flow due to the assumption of fully developed flow, and since the flow is steady, the sum of all forces must equal zero. The only forces arising are those due to pressure and shear stress, so we have

$$\left(p|_z - p|_{z+\Delta z}\right) r\Delta\theta\Delta r + \left((r+\Delta r)\tau_{rz}|_{r+\Delta r} - r\tau_{rz}|_r\right)\Delta\theta\Delta z = 0. \quad (2.7.31)$$

Dividing by $r\Delta\theta\Delta r\Delta z$ yields

$$-\frac{p|_{z+\Delta z} - p|_z}{\Delta z} + \frac{(r+\Delta r)\tau_{rz}|_{r+\Delta r} - r\tau_{rz}|_r}{r\,\Delta r} = 0. \quad (2.7.32)$$

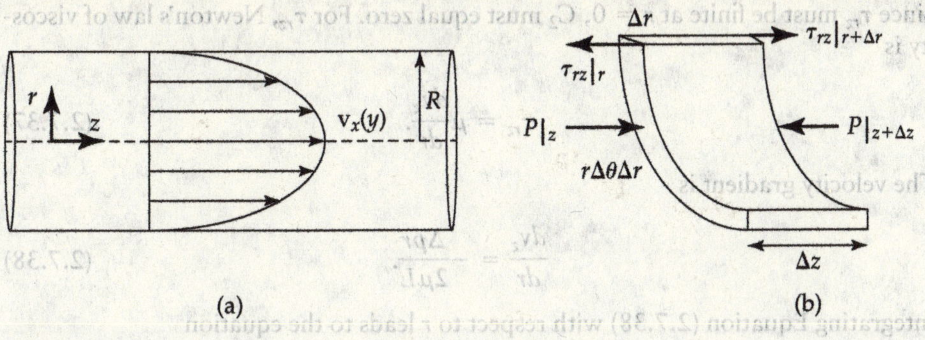

FIGURE 2.25 (a) Laminar flow of a Newtonian fluid through a cylinder of radius R and (b) momentum balance on a differential volume $r\Delta\theta\Delta r\Delta z$.

Taking the limit as Δz and Δr approach zero, and using the definition of the derivative (Equation (2.7.4)), we find that Equation (2.7.31) reduces to the following differential balance:

$$\frac{dp}{dz} = \frac{1}{r}\frac{d(r\tau_{rz})}{dr}. \tag{2.7.33}$$

Equation (2.7.33) is the analog of Equation (2.7.14) for cylindrical coordinates. The left-hand side of Equation (2.7.33) is a function of z, and the right-hand side is a function of r only. Employing the same argument as that used to solve Equation (2.7.14), each side of Equation (2.7.33) equals a constant. Thus, for the pressure gradient,

$$\frac{dp}{dz} = C_1. \tag{2.7.34a}$$

The constant is simply equal to the pressure drop over the length L of the tube, or

$$-\frac{dp}{dz} = \frac{\Delta p}{L}, \tag{2.7.34b}$$

where $\Delta p = p_0 - p_L$ and L is the distance between the locations on the z-axis at which p_0 and p_L are measured. Substituting Equation (2.7.34b) into Equation (2.7.33) results in the following expression:

$$\frac{1}{r}\frac{d(r\tau_{rz})}{dr} = -\frac{\Delta p}{L}. \tag{2.7.35}$$

Multiplying both sides of Equation (2.7.35) by r, integrating with respect to r, and dividing the result by r yields

$$\tau_{rz} = -\frac{\Delta p r}{2L} + \frac{C_2}{r}. \tag{2.7.36}$$

Since τ_{rz} must be finite at $r = 0$, C_2 must equal zero. For τ_{rz}, Newton's law of viscosity is

$$\tau_{rz} = \mu \frac{dv_z}{dr}. \tag{2.7.37}$$

The velocity gradient is

$$\frac{dv_z}{dr} = -\frac{\Delta p r}{2\mu L}. \tag{2.7.38}$$

Integrating Equation (2.7.38) with respect to r leads to the equation

$$v_z = -\frac{\Delta p r^2}{4\mu L} + C_3. \tag{2.7.39}$$

Applying the no-slip boundary condition at $r = R$ and solving for C_3 yields

$$C_3 = \frac{\Delta p R^2}{4\mu L}. \tag{2.7.40}$$

Inserting Equation (2.7.40) for C_3 into Equation (2.7.39) and rearranging terms yields the velocity profile:

$$v_z = \frac{\Delta p R^2}{4\mu L}\left(1 - \frac{r^2}{R^2}\right). \tag{2.7.41}$$

The velocity is a parabolic function of radial position with a maximum value at $r = 0$ equal to

$$v_{max} = \frac{\Delta p R^2}{4\mu L}. \tag{2.7.42}$$

The definition of the average velocity in cylindrical coordinates is the integral of the velocity over the cross-sectional area through which flow occurs, divided by the cross-sectional area, or

$$\langle v \rangle = \frac{1}{A}\int_A v_z dA = \frac{1}{\pi R^2}\int_0^R \int_0^{2\pi} v_z r d\theta dr. \tag{2.7.43}$$

Using Equations (2.7.41) to (2.7.43), the average and maximum velocities are related as follows:

$$\langle v \rangle = \tfrac{1}{2} v_{max}. \tag{2.7.44}$$

Note that this result differs from the result obtained for flow through a rectangular slit (Equation (2.7.29)), for which the average velocity is two-thirds of the maximum velocity. This difference is due to the different geometries.

The volumetric flow rate Q equals $\langle v \rangle \pi R^2$. From Equations (2.7.42), (2.7.43), and (2.7.44), the flow rate is

$$Q = \frac{\Delta p \pi R^4}{8\mu L}. \tag{2.7.45}$$

Equation (2.7.45) is known as *Poiseuille's law*. Originally, the functional relationship was determined from experimental measurements independently by Hagen and Poiseuille [20]. The most significant aspect of the equation is the fourth-power dependence upon radius. For example, a 10% decrease in tube radius at a constant pressure drop leads to a 46% drop in flow rate.

Quantitative statements can be made about entrance effects (assumption 5 in Section 2.7.2) and conditions for laminar flow (assumption 6). As the fluid enters a cylindrical tube, the velocity changes in the r and z directions. Variations in the z direction become insignificant for lengths greater than the entrance length L_e. This length has been measured experimentally for Newtonian fluids and found to be

$$L_e = 0.058D\mathrm{Re}, \tag{2.7.46}$$

where D is the inner diameter of the cylinder. For rectangular slits, the entry length must be greater than $0.04h\mathrm{Re}$, where $\mathrm{Re} = \rho\langle v\rangle D_h/\mu$ and D_h is defined in Equation (2.6.2).

For laminar flow in cylindrical tubes, the Reynolds number must be less than 2,100.

2.7.4 Pressure-Driven Flow of a Power Law Fluid in a Cylindrical Tube

The constitutive equation affects the relationship between pressure and flow. As an example, consider the flow of a power law fluid through a cylindrical tube. The momentum balance (Equation (2.7.35)) and boundary conditions are still valid. Since the shear stress must be finite at $r = 0$, the integrated momentum balance is

$$\tau_{rz} = -\frac{\Delta p r}{2L}. \tag{2.7.47}$$

The constitutive equation for power law fluids is

$$\tau_{rz} = m\left|\frac{dv_z}{dr}\right|^{n-1}\left(\frac{dv_z}{dr}\right). \tag{2.7.48}$$

The velocity is a maximum at the centerline and declines toward zero with increasing radial distance for distances less than the inner radius of the cylinder. Thus, the velocity gradient is negative, and

$$\left|\frac{dv_z}{dr}\right| = -\frac{dv_z}{dr}. \tag{2.7.49}$$

Equation (2.7.48) becomes

$$\tau_{rz} = -m\left|\frac{dv_z}{dr}\right|^{n} = -\frac{\Delta p r}{2L}. \tag{2.7.50}$$

Rearranging Equation (2.7.50) in terms of the velocity gradient yields

$$\left|\frac{dv_z}{dr}\right| = \left(\frac{\Delta p}{2mL}\right)^{1/n} r^{1/n}. \tag{2.7.51}$$

Using Equation (2.7.49) to eliminate the absolute-value sign in Equation (2.7.51), we get

$$\frac{dv_z}{dr} = -\left(\frac{\Delta p}{2mL}\right)^{1/n} r^{1/n}. \tag{2.7.52}$$

Integrating Equation (2.7.52) with respect to r yields

$$v_z = -\left(\frac{\Delta p}{2mL}\right)^{1/n}\left(\frac{r^{1+1/n}}{1+1/n}\right) + C_5. \tag{2.7.53}$$

The constant of integration can be found by applying the no-slip condition at $r = R$. The resulting velocity profile is

$$v_z = \left(\frac{\Delta p}{2mL}\right)^{1/n}\left(\frac{R^{1+1/n}}{1+1/n}\right)\left[1 - \left(\frac{r}{R}\right)^{1+1/n}\right]. \tag{2.7.54}$$

For a Newtonian fluid, $m = \mu$, $n = 1$, and Equation (2.7.54) becomes equal to Equation (2.7.41).

Normalized velocity profiles for power law fluids are shown in Figure 2.26. A Newtonian fluid ($n = 1$) serves as a reference. For a shear-thinning fluid ($n < 1$), the apparent viscosity declines as the shear stress decreases. Thus, the viscosity is lower near the wall than near the centerline. As a result of the reduced viscosity near the wall, the fluid velocity is greater than it would be for a Newtonian fluid. Conversely, for $n > 1$, the viscosity increases with increasing shear stress, and the velocity is less than the corresponding velocity for a Newtonian fluid.

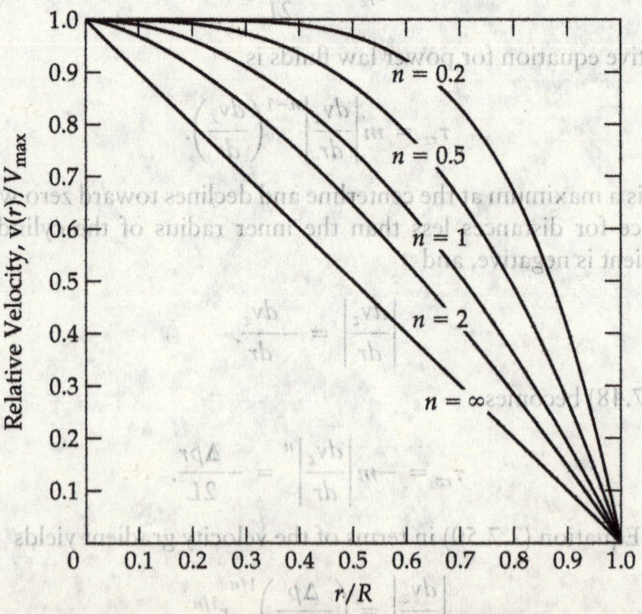

FIGURE 2.26 Velocity profiles for power law fluids flowing through a cylindrical tube.

The flow rate is obtained by multiplying the average velocity by the cross-sectional area πR^2 through which flow occurs:

$$Q = \frac{n\pi}{3n+1}\left(\frac{\Delta P}{2Lm}\right)^{1/n} R^{(3n+1)/n}. \tag{2.7.55}$$

For shear-thinning fluids, the dependence of the flow rate on the radius is greater than for a Newtonian fluid, although a smaller pressure drop is needed.

2.7.5 Flow between Rotating Cylinders

As another example of flow in cylindrical coordinates, consider the angular flow of a Newtonian fluid of viscosity μ between two cylinders of radii R and εR (Figure 2.27). The gap between the cylinders is narrow ($1 - \varepsilon \ll 1$), and the length $L \gg R$. The outer cylinder is rotating with a rotational speed of Ω. Note that this flow is similar to the flow induced by a sliding plate. The primary difference is that the surfaces are curved. Such a flow is used to measure fluid viscosity by calculating the torque required to maintain the outer cylinder at a constant rotational speed.

The solution that is developed is valid when inertial forces are small and flow is essentially one-dimensional. These assumptions are valid for thin gaps, long cylinders, and low rotational speeds. The only nonzero velocity component is v_θ, the angular velocity, which ranges from zero at the inner radius εR to ΩR at the outer radius R (Figure 2.28). The velocity v_θ varies only with r and not with θ. If the velocity v_θ varied in the θ direction, then the flow rate at one angle would exceed the flow rate at another angle. While this can occur at higher rotational speeds due to secondary flows in the z direction, at low rotational speeds v_θ is a function of r only.

The only nonzero viscous stress component is $\tau_{r\theta}$ which acts in the θ direction on a face of constant r. A schematic of the control volume is shown in Figure 2.29. There are four components to the stress in the r direction, $\tau_{r\theta}$ acting on the surfaces

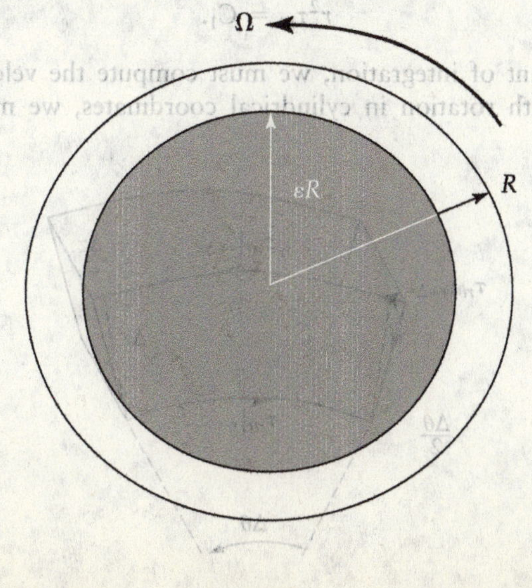

FIGURE 2.27 Flow between rotating cylinders.

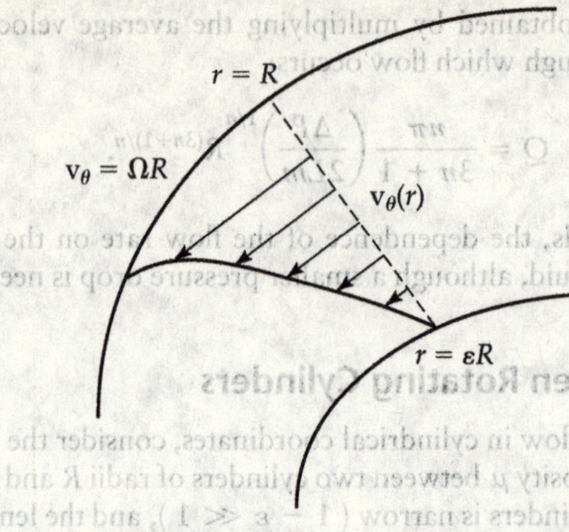

FIGURE 2.28 Sketch of velocity profile for flow between rotating cylinders.

at r and $r + \Delta r$ and the components of $\tau_{\theta r}$ on a face of constant θ acting in the r direction [21]. Thus,

$$\tau_{r\theta}|_{r+\Delta r}(r + \Delta r)\Delta\theta\Delta z - \tau_{r\theta}|_r r\Delta\theta\Delta z + 2\tau_{r\theta}|_{r+\Delta r}\Delta r\Delta z \sin(\Delta\theta/2) = 0, \tag{2.7.56a}$$

where we have used the symmetry property of the stress tensor. Dividing by the volume element $r\Delta\theta\ \Delta z\ \Delta r$ and collecting terms,

$$\frac{\tau_{r\theta}|_{r+\Delta r} r - \tau_{r\theta}|_r r + \tau_{r\theta}|_{r+\Delta r}\Delta r(1 + 2\sin(\Delta\theta/2))}{r\Delta r} = 0. \tag{2.7.56b}$$

Since the angle is small, $\sin(\Delta\theta/2) \approx \Delta\theta/2$. Taking the limit as Δr goes to zero yields

$$\frac{d\tau_{r\theta}}{dr} + 2\frac{\tau_{r\theta}}{r} = \frac{1}{r^2}\frac{d(r^2\tau_{r\theta})}{dr} = 0. \tag{2.7.57}$$

Integrating Equation (2.7.57) gives

$$r^2\tau_{r\theta} = C_1. \tag{2.7.58}$$

To find the constant of integration, we must compute the velocity profile. Since we are dealing with rotation in cylindrical coordinates, we must determine the

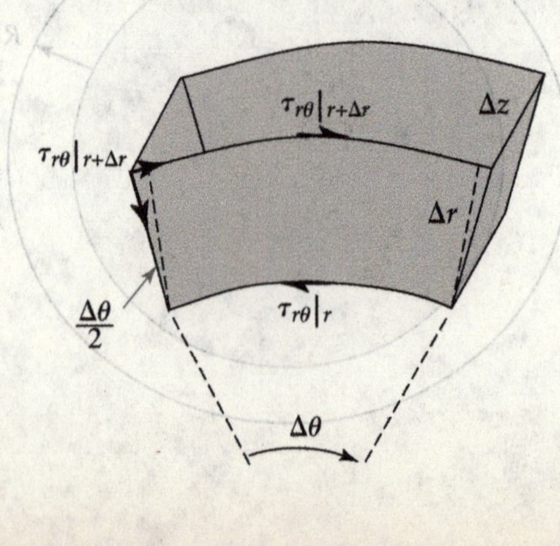

FIGURE 2.29 Stresses acting on the fluid element for flow between rotating cylinders.

relationship between the shear stress component $\tau_{r\theta}$ and the shear rate. As before (see Equations (2.5.1) to (2.5.6)), the shear stress is proportional to the angular rate of deformation, or

$$\tau_{r\theta} = \mu \frac{d\theta}{dt} \tag{2.7.59}$$

and

$$d\theta \approx \frac{ds}{dr}, \tag{2.7.60}$$

where

$$ds = r d\Omega dt = r d\left(\frac{v_\theta}{r}\right) dt. \tag{2.7.61}$$

Substituting for $d\theta$ and ds yields the following expression relating $\tau_{r\theta}$ and the velocity gradient:

$$\tau_{r\theta} = \mu \frac{d}{dt}\left(\frac{ds}{dr}\right) = \mu r \frac{d(v_\theta/r)\, dt}{dr dt}, \tag{2.7.62a}$$

or

$$\tau_{r\theta} = \mu r \frac{d}{dr}\left(\frac{v_\theta}{r}\right). \tag{2.7.62b}$$

Substituting Equation (2.7.62b) into Equation (2.7.58) gives

$$\mu r^3 \frac{d}{dr}\left(\frac{v_\theta}{r}\right) = C_1. \tag{2.7.63}$$

Integrating Equation (2.7.63) yields

$$v_\theta = -\frac{C_1}{2\mu r} + C_2 r. \tag{2.7.64}$$

The boundary conditions are as follows:

$$r = \varepsilon R \qquad v_\theta = 0, \tag{2.7.65}$$

$$r = R \qquad v_\theta = R\Omega. \tag{2.7.66}$$

Applying the boundary conditions yields the following expression for the velocity profile:

$$v_\theta = \frac{\Omega R \varepsilon}{(1 - \varepsilon^2)}\left(\frac{r}{\varepsilon R} - \frac{\varepsilon R}{r}\right). \tag{2.7.67}$$

The torque T required to maintain the outer cylinder at a constant rotational speed equals the product of the force of the fluid on the cylinder and the outer radius. The force on the outer cylinder equals the product of the shear stress at $r = R$ and the surface area of the cylinder, $2\pi RL$, where L is the cylinder length. In vector notation,

$$\mathbf{F}|_{r=R} = \int_A \mathbf{n}\cdot\boldsymbol{\tau}dA = \mathbf{n}\cdot\mathbf{e}_r\mathbf{e}_\theta \int_0^{2\pi}\int_0^L \tau_{r\theta}|_{r=R}Rdzd\theta = \mathbf{e}_\theta 2\pi RL\tau_{r\theta}|_{r=R} \quad (2.7.68)$$

$$\mathbf{T} = \mathbf{F}|_{r=R} \times \mathbf{e}_r R = \tau_{r\theta}|_{r=R}(2\pi RL)R(\mathbf{e}_\theta \times \mathbf{e}_r) = \tau_{r\theta}|_{r=R}2\pi R^2 L\mathbf{e}_z. \quad (2.7.69)$$

The shear stress at $r = R$ is found by substituting Equation (2.7.67) into Equation (2.7.62b). The result is

$$\tau_{r\theta}|_{r=R} = \frac{2\mu\Omega\varepsilon^2}{(1-\varepsilon^2)}, \quad (2.7.70)$$

and the magnitude of the torque is

$$T = 4\pi\mu L\Omega R^2\frac{\varepsilon^2}{(1-\varepsilon^2)}. \quad (2.7.71)$$

For a Newtonian fluid, a plot of the torque versus the rotational speed is proportional to the viscosity, as given by Equation (2.7.71). Thus, this flow geometry is the basis for a class of devices used to measure viscosity. Viscometers based upon rotating coaxial cylinders are known as *Couette viscometers.*

Two related problems can also be solved: the case when the inner cylinder rotates at a constant speed and the outer cylinder is stationary and the case when both cylinders rotate at different speeds in the same or opposite directions. Rotation of both cylinders in the same direction at the same rotational speed produces solid-body rotation, for which $\tau_{r\theta}$ and the torque are both zero. In most Couette viscometers, only the outer cylinder rotates, because this rotation produces the most stable flow with minimal secondary flows.

The preceding analysis is strictly valid for laminar flow without vortices or other secondary flows. The Reynolds number for the system is

$$\mathrm{Re} = \frac{\rho R^2\Omega}{\mu}. \quad (2.7.72)$$

When the outer cylinder rotates, the transition Reynolds number is a function of $1-\varepsilon$, reaching a minimum of 50,000 when $1-\varepsilon = 0.05$. (Interestingly, if the Reynolds number is defined in terms of the distance $\mathrm{Re}_\varepsilon = \rho\varepsilon R^2\Omega/\mu$ over which viscous forces change, then this transition value is 2,500 for $\varepsilon = 0.05$, close to the value obtained for flow through a cylindrical tube.) When the inner cylinder rotates, the transition occurs when the Reynolds number defined by Equation (2.7.72) equals $41.3\varepsilon(1-\varepsilon)^{-1.5}$[15].

Even before the onset of turbulence, the velocity becomes much more complicated than the simple flows we have described. The foregoing analysis was based on the assumptions that v_θ is the only velocity component and that v_θ is a function of the radius only. Flow conditions for which these assumptions are valid depend on the Reynolds numbers at the inner and outer cylinders and upon the gap thickness ε. These parameters are combined in the *Taylor number,* defined as [22]

$$\mathrm{Ta} = \frac{\rho\varepsilon R^2\Omega}{\mu}\sqrt{\varepsilon} = \mathrm{Re}_\varepsilon\sqrt{\varepsilon}. \quad (2.7.73)$$

When Ta < 41.3, the flow is laminar and is well described by the preceding analysis. When both cylinders are rotating in the same direction, and for Taylor numbers between 41.3 and 400, flow is laminar, but *Taylor vortices* arise. These are counter-rotating flows with axes along the circumference. A number of flow patterns have been characterized as a function of the Reynolds numbers at the inner and outer cylinders [23].

2.8 Rheology and Flow of Blood

2.8.1 Measurement of Blood Viscosity

Fluid viscosity is determined by measuring the applied force (or force per unit area) and flow rate in one of several devices known as ***viscometers***. Several different types of viscometers that are in common use include the cone-and-plate viscometer, the Couette viscometer, and the capillary-tube viscometer. Another class of devices known as *rheometers* can measure dynamic properties of fluids and solids. Several common designs require fluid volumes ranging from 0.5 to 3 mL including the cone-and-plate viscometer, the Couette viscometer (Equation (2.7.71)), the rotating-disk viscometer, and the capillary-tube viscometer. For smaller sample volumes and for rheological measurements with cells, magnetic beads that oscillate or rotate within the fluid can be used to measure viscosity.

A general feature of viscometers and rheometers is the independent determination of the shear stress and shear rate. For Couette, cone-and-plate, and rotating-disk viscometers, one surface rotates at a constant speed, and the torque needed to maintain constant speed is measured. The shear rate is proportional to the angular velocity, and the shear stress is proportional to the torque. For capillary-tube viscometers, the shear rate is proportional to the flow rate and the shear stress is proportional to the pressure drop per unit length.

Capillary-tube viscometers are especially useful for two-phase fluids such as blood, since settling effects are minimized if flow is horizontal. These viscometers consist of one or more capillary tubes, often with radii greater than 300 μm, connecting two reservoirs under different hydrostatic pressures. The shear stress can be determined from Equations (2.7.37) to (2.7.41) and from measurements of the pressure drop between two points away from the ends of the tube. An order-of-magnitude estimate of the shear rate at the tube wall is

$$\dot{\gamma}_r = \frac{dv_z}{dr} \approx \frac{\langle v \rangle}{2R} = \frac{Q}{2\pi R^3}. \tag{2.8.1}$$

Equation (2.8.1) states that the ratio of the average velocity to the tube radius should be proportional to the shear rate. The ratio $Q/2\pi R^3$ is often known as the *reduced velocity*, $\overline{U}$. Thus, $\dot{\gamma}_r = k\overline{U}$, where k is a constant. For a Newtonian fluid, $k = 8$.

The optimal way to present data is to create a log-log plot of the shear stress versus the reduced velocity. If the data fall on a straight line, then the fluid is a power law fluid and the slope is equal to n. If the slope is equal to unity, then the fluid is Newtonian. To see this, substitute Equation (2.8.1) into Equation (2.5.10) for a cylindrical geometry. Since the velocity decreases with increasing radius, the shear rate is always negative. Thus,

$$\tau_w = m(\dot{\gamma}_r)^n = mk^n\overline{U}^n. \tag{2.8.2}$$

Taking the logarithm of Equation (2.8.2) yields

$$\log(\tau_w) = n\log(k) + n\log(\overline{U}) + \log(m). \tag{2.8.3}$$

In order to determine the constant k and m, we must assume a specific fluid behavior. If the fluid exhibits a yield stress, a plot of $\log(\tau_w)$ against $\log(\overline{U})$ will not fall on a straight line.

2.8.2 Rheology of Blood Flow in Large Vessels

Whole blood consists of an aqueous phase (plasma), containing salts, sugars, and proteins (Table 2.5), and a cellular phase, which includes red blood cells (RBCs), white cells, and platelets. RBCs constitute the dominant cells in blood and occupy 40%–50% of the blood volume.

Plasma is a Newtonian fluid with a viscosity between 1.16 and 1.35 mPa s (also known as centipoises, or cp) at 37°C. By comparison, water has a viscosity of 0.69 mPa s at 37°C. Approximately 98% of the increase in the plasma viscosity above that of water is due to the presence of proteins [24]. This effect of proteins upon viscosity can be interpreted with reference to the viscosity of a dilute solution of particles [8]. Representing the volume fraction as ϕ, we obtain the suspension viscosity

$$\eta = \mu_{sol}(1 + 2.5\phi). \tag{2.8.4}$$

This relation was first developed by Albert Einstein in 1906 to describe the viscosity of dilute suspensions of rigid spherical particles. Equation (2.8.4) is valid for ϕ less than 0.1.

In contrast to plasma, whole blood exhibits significant non-Newtonian behavior (Figure 2.30a), due to the presence of RBCs. The fluid exhibits a yield stress that depends upon the *hematocrit*—the volume fraction of blood. The yield stress is given by [25]

$$\tau_0 \approx (\text{Hct} - \text{Hct}_c)^3, \tag{2.8.5}$$

where Hct is the hematocrit and Hct_c (= 0.04) is the critical or limiting hematocrit below which a yield stress is not observed. The yield stress is highly sensitive to the square of the fibrinogen concentration.

Above the yield stress, but below 100 s^{-1}, blood behaves as a power law fluid. Above 100 s^{-1}, blood behaves as a Newtonian fluid. A number of constitutive equations have been proposed, the simplest of which is the Casson equation,

$$\tau_w{}^{1/2} = \tau_0{}^{1/2} + \eta_N{}^{1/2}(-\dot{\gamma}_r)^{1/2}, \tag{2.8.6}$$

where η_N is the viscosity at high shear rates. Data in Figure 2.30a are replotted in Figure 2.30b as the square root of the shear stress versus the square root of the shear

TABLE 2.5

Properties of Whole Blood

Parameters	Values	Solute	meq L^{-1}
pH	7.35 – 7.40	Na^+	144
$\mu/\mu H_2O$ (37°C)	~3.0	K^+	5
$\rho/\rho H_2O$ (25°C/4°C)	1.0546	Ca^{++}	2.5
Surface tension	~75 dyn cm^{-1}	Mg^{++}	1.5
Venous hematocrit	Male 0.47	Cl^-	107
	Female 0.42	$HCO3^-$	27
Whole blood volume	~78 mL kg^{-1} body wt.	$HPO_4^=$	2
Plasma or serum colloid osmotic pressure	~330 mmH_2O	$H_2PO_4^-$	2
		$SO_4^=$	0.5
Water content	~93% by volume	Amino acids	2
Cellular components	cells mL^{-1} whole blood	Creatinine	0.2
Erythrocyes	Male 5.4×10^9	Lactate	1.2
	Female 4.8×10^9	Glucose	5.6
Leukocytes	$\sim 7.4 \times 10^6$	Protein	1.2
Platelets	$\sim 2.8 \times 10^8$	Urea	1

rate. The yield stress is typically quite low (≈ 0.02 dyn cm^{-2}). The physiological significance of the yield stress has not been identified. The viscosity parameter

$$\eta_N = \eta_p(1 + 2.5\text{Hct} + 7.35\text{Hct}^2) \tag{2.8.7}$$

is a function of the hematocrit [25].

The Casson equation was derived for a two-phase medium and adequately represents the rheology of blood (Figure 2.30). At low flow rates, the predicted velocity profile is blunted (Figure 2.31), typical of a fluid exhibiting a yield stress. At higher flow rates, the profile approaches the parabolic shape of Newtonian flow. On the basis of the range of shear rates in the circulation (Table 2.4), blood is often assumed

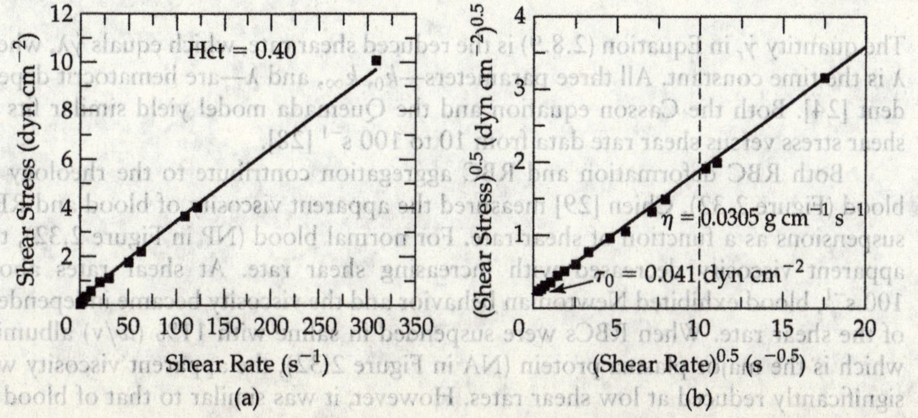

FIGURE 2.30 (a) Shear stress versus shear rate for blood at a hematocrit of 0.40. (b) Data in (a) replotted according to Equation (2.8.6). (Adapted from Ref. [25].)

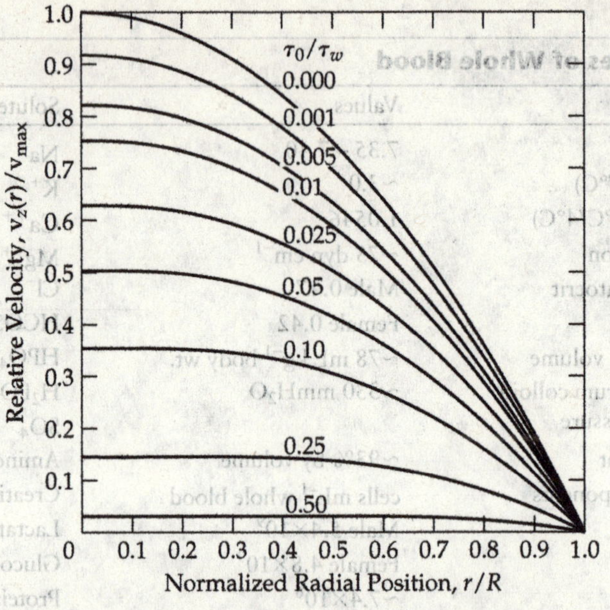

FIGURE 2.31 Velocity profiles for a fluid that obeys Equation (2.8.6). (From Ref. [26].)

to behave as a Newtonian fluid in the major arteries and veins. Nevertheless, the shear-thinning behavior does have a quantitative effect on both the velocity profile and the shear stresses exerted by the fluid on the vessel wall.

A number of other constitutive equations have been proposed to describe the rheological properties of blood [27]. Some of these models do not include a yield stress. Another commonly used model is the Quemada model,

$$\eta_{app} = \eta_p(1 - 0.5k\text{Hct})^{-2}, \tag{2.8.8}$$

where

$$k = \frac{k_0 + k_\infty \dot{\gamma}_r^{1/2}}{1 + \dot{\gamma}_r^{1/2}}. \tag{2.8.9}$$

The quantity $\dot{\gamma}_r$ in Equation (2.8.9) is the reduced shear rate, which equals $\dot{\gamma}\lambda$, where λ is the time constant. All three parameters—k_0, k_∞, and λ—are hematocrit dependent [24]. Both the Casson equation and the Quemada model yield similar fits of shear stress versus shear rate data from 10 to 100 s^{-1} [28].

Both RBC deformation and RBC aggregation contribute to the rheology of blood (Figure 2.32). Chien [29] measured the apparent viscosity of blood and RBC suspensions as a function of shear rate. For normal blood (NP in Figure 2.32), the apparent viscosity decreased with increasing shear rate. At shear rates above 100 s^{-1}, blood exhibited Newtonian behavior and the viscosity became independent of the shear rate. When RBCs were suspended in saline with 11% (w/v) albumin, which is the major plasma protein (NA in Figure 2.32), the apparent viscosity was significantly reduced at low shear rates. However, it was similar to that of blood at

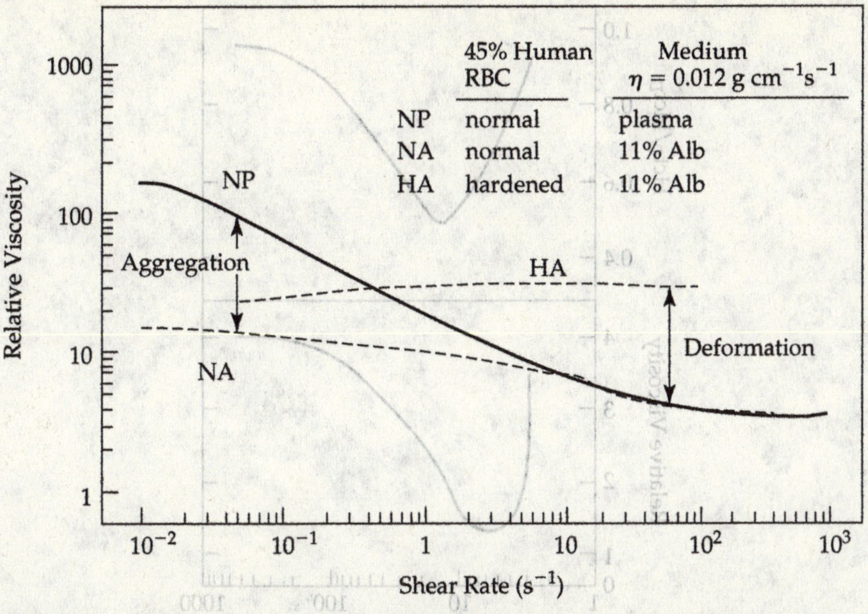

FIGURE 2.32 The apparent viscosity of RBC suspensions (divided by the plasma viscosity) as a function of shear rate. NP = normal RBC in plasma; NA = normal RBC in isotonic saline containing 11% albumin in order to make the liquid viscosity equal the plasma viscosity; HA = glutaraldehyde-fixed RBC in the same saline solution. (Reprinted with permission from Ref. [29], © 1970 American Association for the Advancement of Science.)

shear rates above 10 s^{-1}. This result suggests that some protein in blood interacts with RBCs to increase the blood viscosity above that of RBCs in saline. Subsequent studies showed that fibrinogen binds to RBCs, causing them to aggregate. At higher shear rates, the aggregates break due to forces acting on them, and the apparent viscosity decreases. When the red cells were made rigid by fixation in glutaraldehyde and were suspended in saline (HA in Figure 2.32), the fluid was approximately Newtonian, but with a much higher viscosity than that of blood at high shear rates. Thus, the lower blood viscosity at high shear rates is due to deformation of the RBCs, which reduces the overall work that the heart would perform relative to a fluid with a suspension of rigid cells.

2.8.3 Blood Flow in Small Tubes

Experiments performed in glass capillary tubes with diameters less than 300 μm show that blood rheology is influenced by the two-phase nature of blood. In these experiments, measurements of hematocrit levels are taken in the tube at its point of entry (Hct_T) or at its point of discharge (Hct_D).

Two related effects occur due to the relative motion between blood cells and the fluid (Figure 2.33a). One of these is the *Fahreus effect,* in which the volume fraction of blood in the tube (i.e., the hematocrit) is lower than the volume fraction at the point of discharge ($Hct_T < Hct_D$). Note further that the discharge hematocrit is

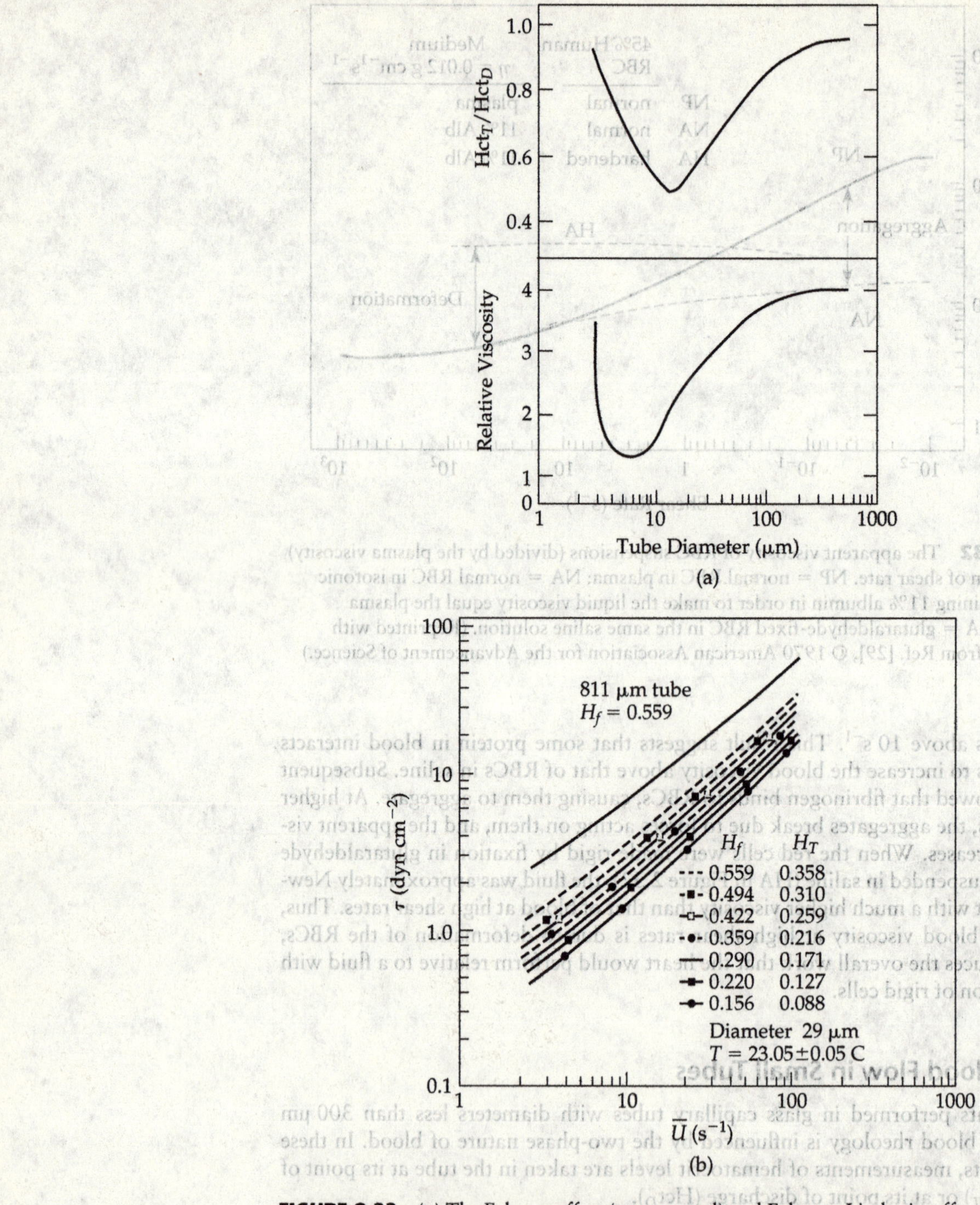

FIGURE 2.33 (a) The Fahreus effect (upper panel) and Fahreus–Lindquist effect (lower panel) (Reprinted from Ref. [30], with permission from IOS Press.) (b) The effect of tube size and hematocrit on the relation between the wall shear stress and the reduced velocity. (Redrawn from Ref. [31], with permission from Elsevier.)

equal to the volume fraction at the point of entry. At first glance, this puzzling result appears to contradict the conservation of mass; that is, more blood appears to be leaving the tube than is entering the tube!

The Fahreus effect arises because the tube hematocrit (Hct_T) and discharge hematocrit (Hct_D) are really two very different quantities. They are respectively defined as

$$Hct_T = \frac{2}{R_T^2}\int_0^{R_T} Hct_T(r)rdr \tag{2.8.10a}$$

and

$$Hct_D = \frac{\int_0^{R_T} Hct_T(r)v_z(r)rdr}{\int_0^{R_T} v_z(r)rdr}, \tag{2.8.10b}$$

where $Hct_T(r)$ is the local hematocrit. In general, the local hematocrit is greater near the center of the tube and declines to zero at $r = R_T$.

The tube hematocrit Hct_T is the spatial average over the cross section of the tube. The higher local hematocrit ($Hct_T(r)$) near the center of the tube contributes less to the overall value of Hct_T. In contrast, the hematocrit at the point of discharge (Hct_D) is weighted by the local velocity. When fluid is collected in the discharge, the faster-moving fluid elements are weighted more than the slower-moving ones near the tube wall. As a result, the sampling at the point of discharge is weighted toward the red-cell-rich regions, in contrast to the average given by Equation (2.8.10a). Problem 2.24 is a specific test of this model.

Obviously, if the local hematocrit is uniform across the vessel wall, then the discharge and tube hematocrit are equal. A nonuniform hematocrit arises, in part because, in flowing blood, there is a narrow region near the tube wall that is free of red cells. The red-cell-free region results from exclusion of cells near the surface. In large tubes, the cell-free region is a small fraction of the cross-sectional area, and the hematocrit is approximately uniform. But as the tube size decreases, the cell-free region contributes more to the volume because the cell-free region is a larger fraction of the cross-sectional area. Consequently, the tube hematocrit decreases with decreasing tube radius.

A related phenomena is the *Fahreus–Lindquist effect,* in which the apparent viscosity of blood decreases in tubes with diameters between 30 and 300 μm. Barbee and Cokelet [31] showed that the shear-stress–shear-rate curves for flow in smaller tubes could be predicted from the results in larger tubes if the volume fraction of hematocrit in the larger tube equaled the volume fraction of hematocrit in the smaller tube (Figure 2.33). Thus, the reduction in fluid viscosity in the tubes arose simply from the decreased volume of hematocrit in the tube.

For tube diameters less than 10 μm, the tube hematocrit and the blood viscosity in the tube increase with decreasing tube diameter. This is because the tube radius approaches the size of the RBCs. The cells pass through the tube in single file, and, in smaller tubes, the individual cells are deformed. Red cells cannot enter tubes smaller than 2.7 μm in diameter, so the hematocrit is not defined for vessels 2.7 μm in diameter or smaller.

2.8.4 Blood Flow in Capillaries

The resistance to flow in arterioles and capillaries is determined from the following modified form of Equation (1.6.2):

$$\Delta P = RQ. \tag{2.8.11}$$

Here, ΔP is the pressure drop between the inlet and outlet to the vessel, R is the resistance, and Q is the volumetric flow rate. If the pressure drop is held constant, then the resistance can be determined from the velocity of red cells [32].

The Fahreus–Lindquist effect indicates that the flow resistance in arterioles and capillaries should be less than that which is predicted on the basis of the bulk viscosity of blood. Measurements of the resistance to flow in capillaries indicate, however, that the resistance offered by the vessels is about two times greater than expected and that the hematocrit is lower than expected [32]. This change in resistance is not explained simply by the difference in hematocrit. Instead, what accounts for the difference [32] is the irregular shape of the inner surface of the arterioles and capillaries due to protrusion of the endothelial cell nuclei, vessel branches, the presence of white blood cells, and the presence of an "endothelial surface layer" (Figure 2.34). Tapering, due

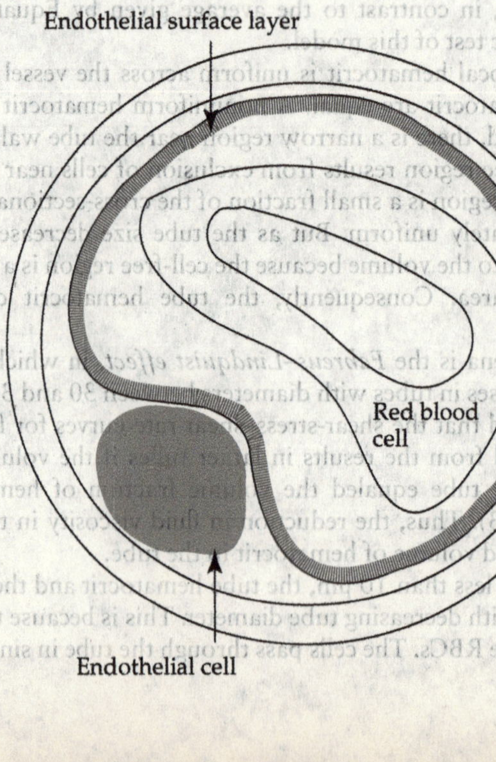

FIGURE 2.34 Cross-sectional view of a capillary showing the effect upon red blood cells of vessel tapering due to the presence of the capillary endothelial cell and the endothelial surface layer.

to the presence of endothelial cell nuclei, is most significant in capillaries and can increase resistance twofold.

The endothelial surface layer consists of the glycocalyx—a layer of glycoproteins and glycolipids on the surface of the endothelium—and a thicker layer of bound protein. Flow resistance measurements (see Problem 2.27) and morphological measurements indicate that the endothelial surface layer is 0.5–1.0 μm thick [33]. This layer is deformable but restricts red cells from approaching the endothelium when the red cells are moving at 20 μm s^{-1}. The more rigid white cells deform the endothelial surface layer and exhibit only a minor change in resistance, due to the presence of the endothelial surface layer. The resistance offered by the endothelial surface layer is reduced if the vessels are treated with an enzyme that breaks down the glycocalyx (e.g., neuraminidase or heparinase). Even partial disruption of the glycocalyx by enzymes causes a significant change in the overall thickness of the endothelial surface layer.

2.8.5 Regulation of Blood Flow

For a Newtonian fluid, the flow rate for a given pressure drop increases with the radius to the fourth power (Equation (2.7.45)). For blood, the use of Equation (2.8.6) as the constitutive equation yields

$$Q = \frac{\pi R^4 \Delta P}{8\eta_N L}\left[1 + \frac{11}{21}\left(\frac{\tau_0}{\tau_w}\right)^4 - \frac{16}{7}\left(\frac{\tau_0}{\tau_w}\right)^{1/2} + \frac{8}{3}\left(\frac{\tau_0}{\tau_w}\right)\right]. \quad (2.8.12)$$

Note that this result treats blood in the small vessels as a single-phase fluid and ignores the separation of blood in red-cell-rich and cell-free layers. For the same pressure drop per unit length, blunting of the velocity profile results in a reduction of the flow rate relative to that of a Newtonian fluid. For the biologically relevant values of $\tau_0 = 0.041$ dyn cm^{-2} and an average wall shear stress of 15 dyn cm^{-2}, this results in an 11.2% reduction in blood flow relative to that of a Newtonian fluid with a viscosity of η_N. Although the flow rate is affected by the non-Newtonian rheology, the leading effect of vessel radius on flow is given by the R^4 dependence for a Newtonian fluid. To calculate resistance (the limiting value of Equation (2.8.12)), Equation (2.7.45) is used for a Newtonian fluid.

Equation (2.8.12) explains in part why small arteries, arterioles, and capillaries contribute 85% of the vascular resistance (Table 1.9). Venules make a small contribution because they are more extensible than arterioles. Blood vessels are dynamic structures and respond to pressure and flow. As a result, short- and long-term *changes* to blood vessels alter either the vessel radius or the vessel thickness, altering the vessel resistance and the wall shear stress.

Blood vessels respond actively to a sudden change in blood flow rate or pressure. An increase in the pressure difference across the artery wall leads to constriction of the vessel. In the absence of any active behavior of arterial smooth muscle cells, an increase in the pressure in the fluid phase leads to an increase in the vessel diameter analogous to the filling of a balloon. Active stimulation of arterial smooth muscle cells by the circumferential stress, generated by an increase in blood pressure, causes the smooth muscle to contract in what is known as a *myogenic response*. This phenomenon is explained [34] by assuming that the blood vessels exhibit viscoelastic properties, that vessel diameters change with applied pressure, and that the diameters change in the same manner in passive tissue as in actively contracting vessels. Further, the function of smooth muscle cells that are present in arterioles, arteries, and veins is controlled by nerves that connect to the smooth muscle cells, as well as by compounds that are secreted by endothelial cells. Some of these compounds relax smooth muscle cells, causing an increase in diameter (*vasodilators*). Others cause smooth muscle cells to contract, resulting in a decrease in diameter (*vasoconstrictors*). Such neural regulation can produce local changes in blood flow in response to physiological changes (e.g., exercise, fear, and eating).

An increase in blood flow in an artery or arteriole results in the release of the vasodilators *nitric oxide* [35] and *prostaglandin* [36] by endothelial cells. Nitric oxide (NO) is a potent vasodilator that is released from endothelium in response to an increase in flow.[2] Nitric oxide diffuses through the blood vessel and enters the smooth muscle cells, where it then binds with guanylyl cyclase to stimulate the production of cyclic guanine monophosphate, which relaxes smooth muscle cells. In addition to being stimulated by shear stress, nitric oxide release is stimulated by acetylcholine, histamine, adenosine triphosphate, adenosine diphosphate, and *hypoxia* (low blood concentrations of oxygen). The result is a greater increase in blood vessel diameter than would be expected due to the passive change that results from the increased pressure across the artery wall. Vessel dilation leads to a local reduction in resistance to flow. Removal of the endothelium abolishes flow-dependent vasodilation. Conversely, a decrease in blood flow causes the release of endothelin, which produces smooth muscle cell contraction and also a decrease in vessel diameter. Flow leads to an increase in mRNA and protein levels for nitric oxide synthase and a decrease in mRNA and protein levels of the precursor to endothelin [37]. Although fluid shear stresses activate a number of signaling pathways and genes [38,39], the manner in which fluid stresses are transduced into biological responses is poorly understood.

Changes in pressure or blood flow lead to long-term changes in vessel properties, a process known as *remodeling* [40]. Increased pressure produces an increase in the thickness of the media, due to an increase in the size of smooth muscle cells (*hypertrophy*). In *hypertension*, chronic elevation of blood pressure produces an increased stiffness of the vessel wall. A chronic increase in blood flow to a vessel leads to an increase in its diameter, and a chronic decrease in blood flow leads to long-term vessel constriction. These phenomena are important in a number of conditions. A pathological constriction of a blood vessel (stenosis) can sometimes develop. Downstream from the stenosis, blood vessel dilation is observed. This dilation is induced by a chronically elevated release of nitric oxide due to altered flow conditions [41]. Differences in the ability of arterial and venous endothelium to release nitric oxide may explain the improved survival of artery bypass grafts in treatment of blocked coronary arteries. Thus, far from passively containing blood, arteries respond dynamically to changes in blood flow conditions.

[2] The 1998 Nobel Prize in medicine was awarded to Robert Furchgott, Louis Ignarro, and Ferid Murad for their discovery of the biological function of nitric oxide.

2.9 QUESTIONS

2.1 Imagine that people are measuring the speed of a boat on a river. One person stands on the riverbank and records speeds as the boat moves past her field of view. A second person is in a boat that moves at the speed of the river and records the speed of the first boat at the same time as the observer on the riverbank does. Describe conditions under which the observer in the boat will measure speeds that are faster and slower than the speeds measured by the observer on the riverbank.

2.2 An embolism is a blood clot that detaches and lodges farther downstream. Discuss whether a clot that forms immediately downstream from a stenosis (a narrowing of a blood vessel) will be more or less likely to form an embolism.

2.3 Explain the difference between body and surface forces.

2.4 Why does a cell that is detached from a surface adopt a spherical shape?

2.5 If liquid mercury is poured onto a slanted surface, the liquid will break up into beads. Water has a greatly reduced tendency to do this. Explain.

2.6 Explain how the equations of fluid statics can be used to measure differences in pressure.

2.7 Red blood cells do not exhibit a critical membrane tension and will flow into a micropipet for all pressures. Explain the functional significance of this observation.

2.8 Suppose you wished to perform a momentum balance on flow through a tapered vessel. At any axial position z, the cross-sectional area is $A(z)$. State the momentum balance.

2.9 Discuss the effect of a narrow constriction in an artery upon the acceleration of blood.

2.10 Suppose that the apparent viscosity of blood is measured in a Couette viscometer. Discuss how the apparent viscosity at one shear rate might change with time.

2.11 While measuring the viscosity of blood in a capillary tube, you add a small amount of fibrinogen. How does the apparent viscosity change with time?

2.10 PROBLEMS

2.1 For the following velocity vector in centimeters per second and unit normal, determine the flow rate for $0 < x < 2$ and $0 < y < 3$.

$$\mathbf{v} = 3x\mathbf{e}_x + 6y\mathbf{e}_y + 5z^2\mathbf{e}_z, \qquad (2.10.1)$$

$$\mathbf{n} = \frac{\mathbf{e}_x}{\sqrt{2}} + \frac{\mathbf{e}_y}{\sqrt{2}}. \qquad (2.10.2)$$

2.2 The unit normal vector is directed outward from a surface and has unity magnitude. The unit normal can be resolved into three vectors parallel to the axes of the coordinate system (Figure 2.35). For a rectangular coordinate system given by $\mathbf{n} = a\mathbf{e}_x + a\mathbf{e}_y + a\mathbf{e}_z$, determine the magnitude of the coefficient.

2.3 Prove the following vector relation:

$$\nabla \cdot (\rho \mathbf{v}\mathbf{v}) = \rho \mathbf{v} \cdot \nabla \mathbf{v} + \mathbf{v}\nabla \cdot (\rho \mathbf{v}). \qquad (2.10.3)$$

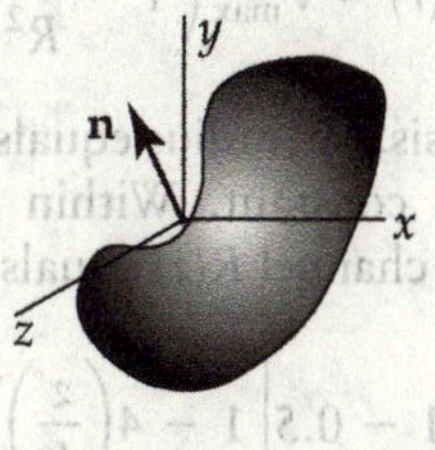

FIGURE 2.35

2.4 Consider the following velocity field:

$$\mathbf{v} = U_0(x^2 - y^2 + x)\mathbf{e}_x - U_0(2xy + y)\mathbf{e}_y.$$

(a) Compute the acceleration at $y = 1$ and $x = 0$ and $x = 2$. $U_0 = 2\ \text{m s}^{-1}$.

(b) Compute the volumetric flow rate through the plane at $x = 5$ and extending from $y = 0$ to $y = 5$

and width 3 in the z direction. The unit normal vector is in the positive x direction.

2.5 For steady flow through a conical nozzle of length L, the velocity in the direction of flow, x direction, is approximately

$$v_x = U_0\left(1 - \frac{x}{L}\right)^{-2} \qquad (2.10.4)$$

where U_0 is the velocity at the entrance to the nozzle.

(a) Determine the acceleration in the x direction at $x = 1$ m for $U_0 = 5$ m s^{-1} and $L = 2$ m.

(b) The velocity given by Equation (2.10.4) is approximate. Identify two locations along the nozzle where the velocity will not be correct.

2.6 A stenosis is a narrowing of a blood vessel or valve. Stenosis of a blood vessel arises during atherosclerosis and could occlude an artery, depriving the tissue downstream of oxygen. Further, the fluid shear stresses acting on the endothelial cells lining blood vessels may affect the expression of genes that regulate endothelial cell function. Such gene expression can influence whether the stenosis grows or not.

Consider a symmetric stenosis as shown in Figure 2.36. Assume that the velocity profile within the stenosis of radius $R_i(z)$ has the same shape as the profile outside the stenosis and is represented as

$$v_z(r) = v_{max}\left(1 - \frac{r^2}{R^2}\right) \qquad (2.10.5)$$

Outside the stenosis, the radius equals R_0 and the maximum velocity is constant. Within the stenosis, the radius of the fluid channel $R(z)$ equals

$$R(z) = R_0\left\{1 - 0.5\left[1 - 4\left(\frac{z}{L}\right)^2\right]^{1/2}\right\} \qquad (2.10.6)$$

The origin of the z axis is the midpoint of the stenosis.

(a) Develop an expression for v_{max} in a stenosis in terms of the volumetric flow rate Q, cylindrical tube of radius R_0, and distance along the stenosis z/L.

(b) Compute the shear stress acting on the surface of the stenosis ($r = R_i$) at $z = 0$ relative to the value outside the stenosis.

2.7 A parallel-plate flow chamber is being designed to study the adhesion of leukocytes to endothelium. The channel width is 2 cm. Determine channel heights that can be used to generate wall shear stresses (i.e., $\tau_{yx}(y = -h/2) = \tau_w$) as high as 20 dyn cm^{-2} with flow rates less than 2 cm^3 s^{-1}. The viscosity of the tissue culture liquid medium is 0.0087 g cm^{-1} s^{-1}.

2.8 In micropipet systems that measure membrane and cortical tension, the pressure difference $\Delta p = p_p - p_0$ is produced by a column of fluid. Thus, $\Delta p = \rho g h$. The fluid is water ($\rho = 1$ g cm^{-3}), and the minimum column height that can be maintained is 2.5 μm.

(a) Determine the minimum pressure that can be generated.

(b) For a cell of radius 6 μm and a micropipet radius of 1.5 μm, determine the smallest cortical tension that can be measured.

2.9 The law of Laplace (Equation (2.4.16)) describes the balance of forces when a cell is drawn into a pipet. For this equilibrium situation, a hemispherical cellular cap of the dimensions of the pipet radius is formed. When the pressure drop just exceeds the value calculated from Equation 2.4.16, the cell is drawn into the pipet and behaves like a liquid droplet. This equation can be used to determine the smallest size capillary into which a white blood cell can enter.

(a) Consider a monocyte of radius 6.5 μm and a cortical tension of 0.06 mN m^{-1}. Determine the smallest capillary that this cell can enter for a capillary pressure drop of 0.2 mmHg. Note that the density of mercury (Hg) is 13.6 g cm^{-3}. Ignore the possibility that the capillary can deform.

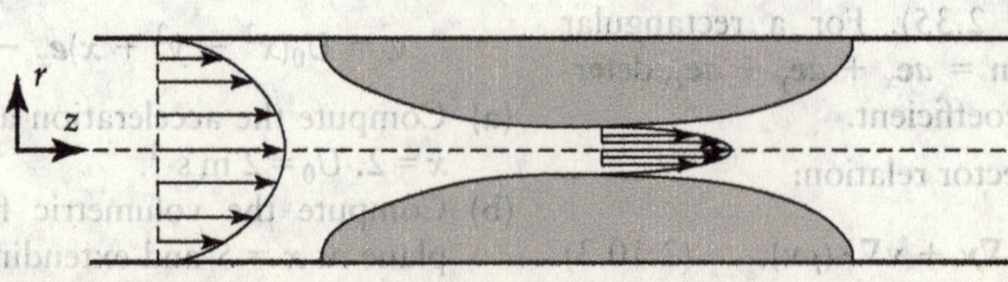

FIGURE 2.36 Flow through a stenosis.

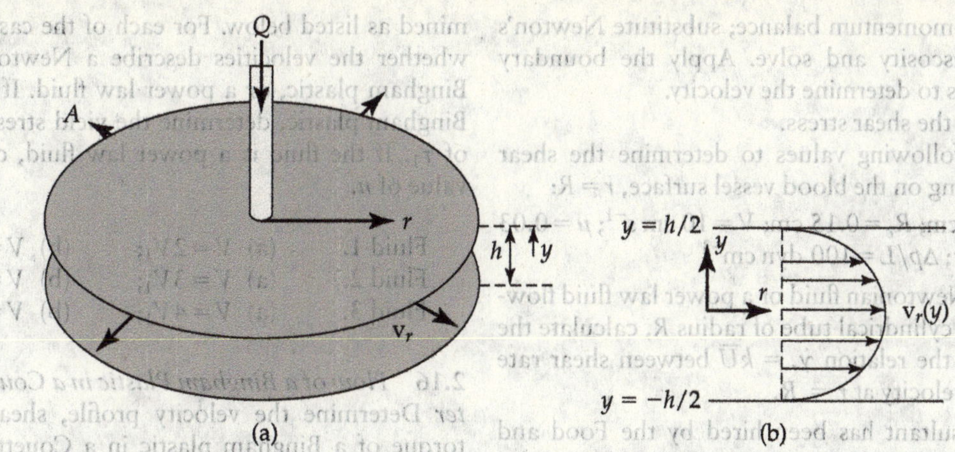

FIGURE 2.37 Schematic of a radial flow channel of height h. (a) Top view; (b) side view.

(b) Could a cell of radius 3.0 μm and the same cortical tension enter the capillary determined in part (a)? Explain.

2.10 A number of different types of flow chambers have been designed to study adhesion between cells in vitro. One important design feature is the ability to have a range of shear stresses. This ability allows a wide range of data to be collected in a single experiment, reducing experimental variability. One such device is a radial-flow channel, shown in Figure 2.37. In this device, fluid flows from a source in a radial direction and the volumetric flow rate at any radial position r is Q. The flow itself is driven by an unspecified pressure gradient. Solve for the velocity profile. Use your solution and the constant-flow-rate condition to show that the wall shear stress at $y = -h/2$ is

$$\tau_w = \frac{3\mu Q}{\pi r h^2}. \tag{2.10.7}$$

2.11 A hollow-fiber dialysis unit is one design that provides a large surface area between blood and dialysate to permit efficient mass exchange. Total blood flow rates in the device are typically 200 mL min^{-1}, similar to the flow rate in kidneys. A typical hollow-fiber unit is 30 cm long with 250 fibers, each of diameter 200 μm. The fluid density is 1.05 g cm^{-3} and the viscosity is 0.03 g cm^{-1} s^{-1}. Assuming that blood behaves as a Newtonian fluid, determine the entrance length for each fiber and assess whether it is small relative to the length of the hollow-fiber unit.

2.12 Consider a catheter of radius R_c placed in a small artery of radius R as shown in Figure 2.38. The catheter moves at a constant speed V. In addition, blood flows through the annular region between R_c and R under a pressure gradient $\Delta p/L$ that only varies in the z direction. We want to determine the effect of the catheter upon the shear stress at $r = R$. Assume steady, fully developed flow of a Newtonian fluid.

(a) State the momentum balance and boundary conditions.

(b) Sketch the velocity profile and provide a justification for its shape.

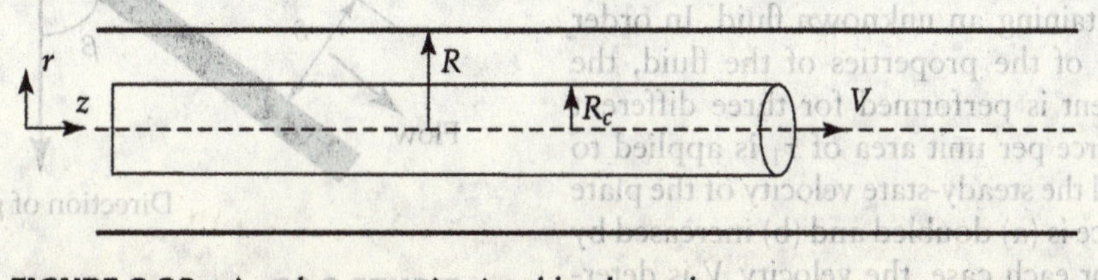

FIGURE 2.38 A catheter moving in a blood vessel.

(c) Solve the momentum balance; substitute Newton's law of viscosity and solve. Apply the boundary conditions to determine the velocity.
(d) Calculate the shear stress.
(e) Use the following values to determine the shear stress acting on the blood vessel surface, $r = R$:
$R = 0.17$ cm; $R_c = 0.15$ cm; $V = 10$ cm s^{-1}; $\mu = 0.03$ g cm^{-1} s^{-1}; $\Delta p/L = 100$ dyn cm^{-3}

2.13 For a Newtonian fluid or a power law fluid flowing through a cylindrical tube of radius R, calculate the constant k in the relation $\dot{\gamma}_r = k\overline{U}$ between shear rate and reduced velocity at $r = R$.

2.14 A consultant has been hired by the Food and Drug Administration (FDA) Catsup Division to evaluate a manufacturer's advertisement claims. The OrgoFoods company has developed a new pesticide-free catsup that they claim is "20% slower" than the leading competitor's brand. When pressed for details, the OrgoFoods scientists say that this claim means that the yield stress of their product is 20% higher than that of the competitor's product.

In order to evaluate OrgoFoods's claim, FDA scientists measured the yield stress of the two brands of catsup using a "Bostwich viscometer," which consists of flow down an inclined plane, as shown in Figure 2.39. Treat catsup as an incompressible Bingham plastic. Assume that the flow is fully developed and the plate width is very wide, such that the flow is one-dimensional. The thickness of the catsup is 0.5 cm and the density is 2 g cm^{-3}. The FDA scientists measured the angle at which the catsup first started to flow. Using the following data, determine the yield stress and evaluate the validity of the manufacturer's claim.

Catsup	Angle at which catsup begins to flow
OrgoFoods	5°
Competitor	20°

2.15 Two parallel plates are separated by a thin gap of thickness H containing an unknown fluid. In order to determine some of the properties of the fluid, the following experiment is performed for three different fluids. A known force per unit area of τ_1 is applied to the upper plate, and the steady-state velocity of the plate is V_1. Next, the force is (a) doubled and (b) increased by a factor of four. For each case, the velocity V is determined as listed below. For each of the cases, determine whether the velocities describe a Newtonian fluid, a Bingham plastic, or a power law fluid. If the fluid is a Bingham plastic, determine the yield stress τ_0 in terms of τ_1. If the fluid is a power law fluid, determine the value of n.

Fluid 1.	(a) $V = 2V_1$;	(b) $V = 4V_1$
Fluid 2.	(a) $V = 3V_1$;	(b) $V = 7V_1$
Fluid 3.	(a) $V = 4V_1$;	(b) $V = 16V_1$

2.16 *Flow of a Bingham Plastic in a Couette Viscometer* Determine the velocity profile, shear stress, and torque of a Bingham plastic in a Couette viscometer. Explain how μ_0 and τ_0 can be determined.

2.17 *Flow of a Power Law Fluid in a Couette Viscometer* Determine the velocity profile, shear stress, and torque of a power law fluid in a Couette viscometer. Explain how m and n can be determined.

2.18 *Stress Relaxation Function* An alternative and perhaps more intuitively appealing derivation of Equation (2.5.19) is as follows: The step function can be considered to occur over the time interval from $t_0 - \varepsilon$ to t_0, where $\varepsilon \ll t_0$ and $t_0 = 0$ (Figure 2.40). During this time, the strain changes from 0 to γ_0. The shear rate or strain rate is zero for $t < t_0 - \varepsilon$ and $t > t_0$. During the interval from $t_0 - \varepsilon$ to t_0, the shear rate is γ_0/ε.

(a) Use Equation (2.5.15) to determine an expression for τ_{yx} as a function of γ_0, ε, and the integral of the relaxation function.

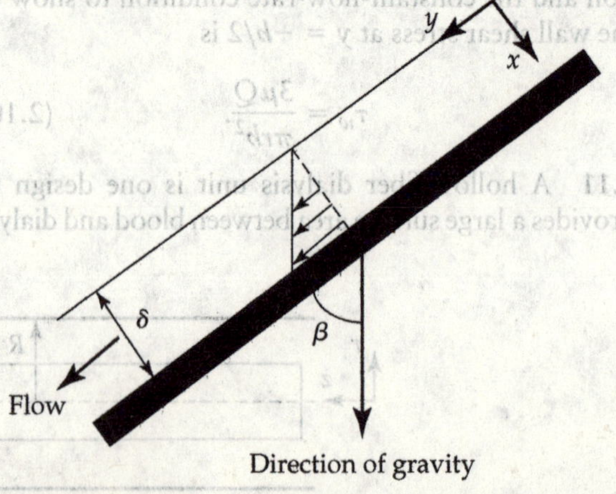

FIGURE 2.39

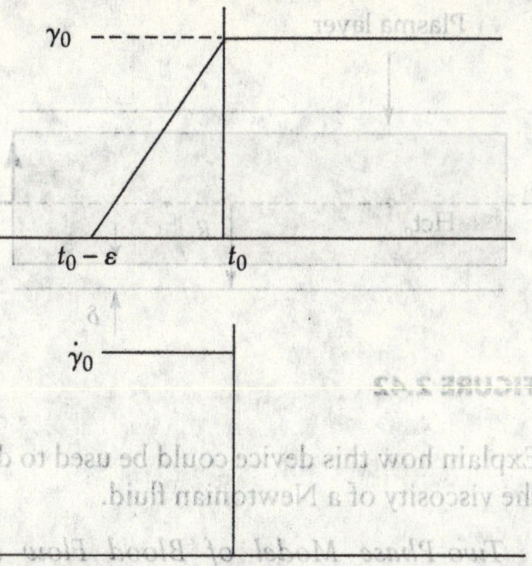

FIGURE 2.40 Approximation of the step function as a ramp over interval ε.

(b) Now take the limit as ε goes to zero. Use L'Hôpital's rule to find the limit (see Section A.1.B). You need to use Leibniz's rule when you differentiate the integral (see Section A.1.H).

2.19 *Rheology of Human Neutrophils* From studies of the flow of neutrophils in micropipets, it is possible to measure the apparent viscosity of the cytoplasm. Using the data obtained by Tsai et al. [42], determine the rheological relationship that best describes the behavior of the cytoplasmic viscosity.

Shear rate (s^{-1})	Apparent viscosity (Pa s)
0.111	450 ± 93
1	137 ± 13
2.70	97 ± 7
5.75	58 ± 4
7	58 ± 2

2.20 *Rheology of Hyaluronic Acid* Hyaluronic acid is a high-molecular polymer of the repeating disaccharide units glucuronic acid and N-acetylglucosamine. Present in all tissues and usually linked with protein in the form of proteoglycans, hyaluronic acid is a component of the synovial fluid that is present between joints and is thought to play a role in joint lubrication and resisting compressive forces.

(a) The data in the table at the bottom of this page were obtained at 25°C in a cone-and-plate viscometer for solutions of 5 mg mL^{-1} hyaluronic acid in phosphate-buffered saline with a pH of 7.2. Determine whether the fluid behaves as a power law fluid, and if so, determine the parameters m and n.

(b) The hyaluronic acid solution was treated with the enzyme hyaluronidase for 5 minutes at 25°C, and the data on the right-hand side of the table were recorded. Determine the effect of the enzyme upon the rheology of hyaluronic acid.

(c) Explain why the enzyme might change the rheology of hyaluronic acid.

2.21 Consider a Bingham plastic flowing in a cylindrical tube of radius R. Flow is induced by the pressure gradient $\Delta P/L$. The tube is horizontal and flow is steady and fully developed.

(a) Determine a criterion for the onset of flow.

(b) Determine an expression for the flow rate once flow begins. Compare with a Newtonian fluid of the same viscosity.

2.22 *Neglect of Curvature in a Couette Viscometer* An analytical solution of equations for data obtained from a Couette viscometer cannot be found when the rheological equations are complex. Show that if ε is

Hyaluronic acid		Hyaluronic acid plus hyaluronidase	
Shear rate (s^{-1})	Shear stress (dyn cm^{-2})	Shear rate (s^{-1})	Shear stress (dyn cm^{-2})
3.75	2.39	375	3.31
7.50	4.28	450	3.93
11.3	6.23	525	4.48
15.0	8.05	600	5.10
22.5	11.2	750	6.37
30.0	14.1	900	7.79
37.5	17	1,125	9.61

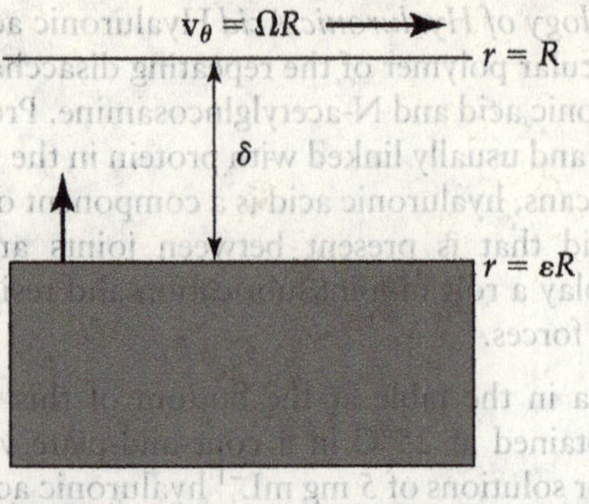

FIGURE 2.41 Schematic of a Couette viscometer when $\delta \ll R$ or ε is approximately equal to 1.

close to unity, curvature can be neglected and the problem can be treated as flow between parallel plates separated by a gap of thickness $\delta = R(1 - \varepsilon) \ll R$ (see Figure 2.41).

(a) Determine the relationship between torque and viscosity for a Newtonian fluid.
(b) Calculate the error in the shear stress when curvature is neglected for $\delta/R = 0.005$, 0.01, and 0.03 (or $\varepsilon = 0.995$, 0.99, and 0.97).
(c) Develop a relationship for determining m and n for a power law fluid under the conditions when curvature can be neglected.

2.23 A new type of viscometer has been developed that involves rotating a cylinder of radius R and length L at a constant rotational speed of ω in a large volume of liquid. The cylinder is very long ($L \gg R$), and edge effects can be neglected. The walls of the fluid container are far enough away from the rotating cylinder such that they do not alter the velocity field produced by the cylinder. Far from the cylinder surface, the fluid is not moving and the pressure is uniform. Thus, you can assume that the fluid is semi-infinite such that the velocity equals zero as the radial distance approaches ∞.

(a) Using cylindrical coordinates, provide an expression for the θ component of linear momentum and a brief justification. Indicate any assumptions made.
(b) Sketch the velocity profile for a Newtonian fluid.
(c) For a Newtonian fluid, determine the velocity profile for $v_\theta(r)$. State the boundary conditions.
(d) Compute the torque acting on the cylinder by the fluid. Note: $e_\theta \times e_r = -e_z$.

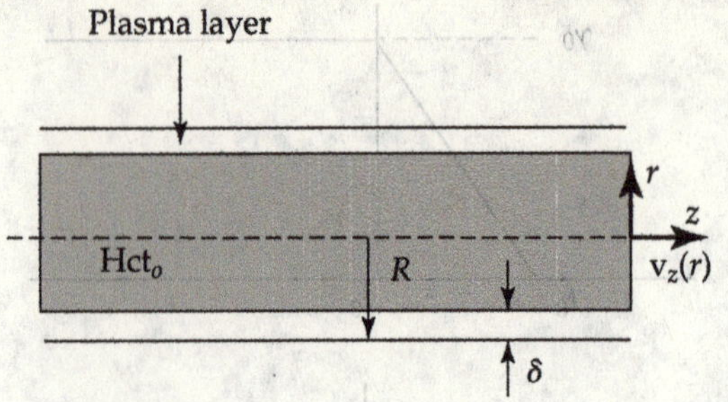

FIGURE 2.42

(e) Explain how this device could be used to determine the viscosity of a Newtonian fluid.

2.24 *Two-Phase Model of Blood Flow in Small Tubes* One explanation for the reduced hematocrit in a tube postulates that, in small capillary tubes (15 μm < R_T < 150 μm), blood separates into a cell-free plasma layer of thickness δ (typically, 5 μm) near the tube wall and a central core enriched in red blood cells (Figure 2.42). The hematocrit of blood before entering the tube is Hct_F. The tube hematocrit Hct_T is a function of location, with a value of Hct_0 for $0 < r < R_T - \delta$ and 0 for $r > R_T - \delta$. (Hct_0 is the core hematocrit.)

(a) Given the definitions of the tube and discharge hematocrits (Equations (2.8.10a) and (2.8.10b), respectively) and the observation that $Hct_F = Hct_D$, derive a relationship between Hct_0 and Hct_F. Assume that $v_z(r)$ is a simple parabolic velocity profile for a single-phase fluid.
(b) Show that, for tubes less than 500 μ in diameter and δ equal to 5 μ, $Hct_T < Hct_F$.

2.25 When the blood microvessel dimensions approach the dimensions of the red cell, the red cell squeezes through capillaries and the viscosity increases. To model this and predict an increase in viscosity, we will assume that the red cells pass through the capillaries as a continuous train equal to the capillary length L. The cells have a radius R_c. The plasma flows in a thin gap between the cell and the vessel wall (Figure 2.43). The gap thickness is $\delta = R - R_c$ and the viscosity is μ. Because the gap is so thin relative to the vessel wall thickness, assume that the velocity profile is linear and equals $V_c y/\delta$ where $y = R - r$ and $\delta = R - R_c$.

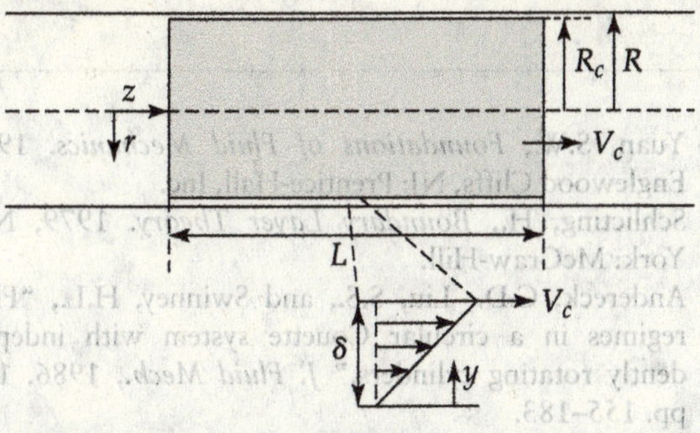

FIGURE 2.43 Schematic of flow between a red blood cell and the capillary wall.

(a) Determine an expression for the shear stress acting on the cells if the velocity in the gap can be approximated as $v_z = V_c y/\delta$.
(b) Perform a force balance on the cell to relate the pressure and shear stresses. Use this result to relate the cell velocity to $\Delta p/L$.
(c) Relate the average velocity to the cell velocity.
(d) Use the results from parts (b) and (c) to derive an expression for the effective viscosity of the red cell suspension μ_{eff} by matching the red cell velocity to the mean velocity for Poiseuille flow in a tube.

$$\langle v \rangle = \frac{R^2}{8\mu_{\text{eff}}}\frac{\Delta p}{L} \tag{2.10.8}$$

(e) Show that the ratio μ_{eff}/μ increases as the ratio δ/R increases for $0.1 < \delta/R < 0.4$, or R_c/R varies from 0.6 to 0.9.

2.26 *Velocity Profiles for Blood* The constitutive relation for the flow of blood in a cylindrical tube of radius R is

$$|\tau_{rz}| > \tau_0 \qquad (\tau_{tz})^{1/2} = (\tau_0)^{1/2} + (\eta_N)^{1/2}\left(\frac{dv_z}{dr}\right)^{1/2} \tag{2.10.9a}$$

and

$$|\tau_{rz}| \le \tau_0 \qquad \frac{dv_z}{dr} = 0, \tag{2.10.9b}$$

where τ_0 is the yield stress and η_N is the Newtonian viscosity at high shear rates.

(a) Show that the velocity profile is

$$r_c < r < R \quad v_z = \frac{\Delta PR^2}{4\eta_N L}\left[\left(1-\frac{r^2}{R^2}\right) - \frac{8}{3}\frac{r_c^{1/2}}{R^{1/2}}\left(1-\frac{r^{3/2}}{R^{3/2}}\right) + \frac{2r_c}{R}\left(1-\frac{r}{R}\right)\right], \tag{2.10.10a}$$

$$0 < r < r_c \quad v_z = \frac{\Delta PR^2}{4\eta_N L}\left[1 - \frac{8}{3}\frac{r_c^{1/2}}{R^{1/2}} + \frac{2r_c}{R} - \frac{r_c^2}{3R^2}\right], \tag{2.10.10b}$$

where

$$r_c = \frac{\tau_0 2L}{\Delta p} \tag{2.10.11}$$

is the critical radius at which the yield stress equals the magnitude of the shear stress.

(b) Show that the ratio of the wall shear stress ($\tau_w = |\tau_{rz}(r = R)|$) to the yield stress can be written as

$$\frac{\tau_w}{\tau_0} = \frac{R}{r_c}. \tag{2.10.12}$$

(c) Plot the ratio of the velocity to the maximum value for a Newtonian fluid ($\Delta pR^2/4\eta_N L$) as a function of r/R for ratios of τ_w/τ_0 equal to 0.001, 0.01, 0.1, 0.2, and 0.4.
(d) On the basis of the preceding results, comment on the deviation in the wall shear stress and velocity profile in a large artery for which τ_w equals 10–15 dyn cm^{-2} ($\tau_0 - 0.02$ dyn cm^{-2}) and a venule for which τ_w equals 0.2–2 dyn cm^{-2}.

2.27 The glycocalyx layer cannot be observed with a light microscope, but its presence can be inferred and its thickness estimated from the resistance to flow of a red cell [32]. It was found that treatment of the vessel with the enzyme heparinase reduced the resistance by 14%. The vessel diameter, which represents the distance from the endothelial cell surface, equaled 29 μm. Using Equation (2.7.45) to relate the resistance to the vessel diameter, estimate the thickness of the glycocalyx. Assume that the vessel diameter is uniform.

2.11 REFERENCES

1. Deen, W.M., *Analysis of Transport Phenomena*. 1998, New York: Oxford University Press.
2. Vogel, S., *Life in Moving Fluids*. 1981, Princeton, NJ: Princeton University Press.
3. Batchelor, G., *An Introduction to Fluid Dynamics*. 1970, Cambridge, UK: Cambridge University Press.
4. Whitaker, S., *Introduction to Fluid Mechanics*. 1981, Malabar, FL: Krieger Publishing Company.
5. Aris, R., *Vectors, Tensors, and the Basic Equations of Fluid Mechanics*. 1962, Englewood Cliffs, NJ: Prentice-Hall, Inc.
6. Creighton, T.E., *Proteins, Structures and Molecular Principles*. 1984, New York: W.H. Freeman.
7. White, F., *Fluid Mechanics*. 1986, New York: McGraw-Hill Book Company.
8. Hiemenz, P., *Principles of Colloid and Surface Chemistry*. 1977, New York: Marcel Dekker, Inc.
9. Rhoades, R., and Tanner, G., "Gas transfer and transport," in *Medical Physiology*. 1995, Boston: Little, Brown and Company. pp. 386–400.
10. Piknova, B., Schram, V., and Hall, S.B., "Pulmonary surfactant: Phase behavior and function." *Curr. Opin. Struct. Biol.*, 2002. **12**: pp. 487–494.
11. Alberts, B., Bray, D., Lewis, J., Raff, M., Roberts, K., and Watson, J.D., *Molecular Biology of the Cell*. 3d ed. 1994, New York: Garland Publishing, Inc.
12. Rhoades, R., and Tanner, G., *Medical Physiology*. 1995, Boston: Little, Brown and Company.
13. Waugh, R., and Hochmuth, R., "Mechanics and deformability of hematocytes," in *The Biomedical Engineering Handbook*, J. Bronzino, editor. 1995, Boca Raton, FL: CRC Press. pp. 474–486.
14. Fung, Y., *Biomechanics, Mechanical Properties of Living Tissues*. 3d ed. 1993, New York: Springer-Verlag.
15. Bird, R., Stewart, W., and Lightfoot, E., *Transport Phenomena*. 2d ed. 2002, New York: John Wiley and Sons.
16. Bird, R., Armstrong, R., and Hassager, O., *Dynamics of Polymeric Liquids*. 2d ed. *Vol. 1. Fluid Mechanics*. 1987, New York: John Wiley and Sons.
17. Balazs, E., and Gibbs, D., "The rheological properties and biological function of hyaluronic acid," in *Chemistry and Biology of the Intercellular Matrix*, E. Balazs, editor. 1970, New York: Academic Press. pp. 1241–1253.
18. Pedley, T., "Pulmonary fluid dynamics." *Ann. Rev. Fluid Mech.*, 1977. **9**: pp. 229–274.
19. Pedley, T.J., *The Fluid Mechanics of Large Blood Vessels*. 1980, Cambridge, UK: Cambridge University Press.
20. Sutera, S., and Skalak, R., "The history of Poiseuille's law." *Ann. Rev. Fluid Mech.*, 1993. **25**: pp. 1–19.
21. Yuan, S.W., *Foundations of Fluid Mechanics*. 1967, Englewood Cliffs, NJ: Prentice-Hall, Inc.
22. Schlicting, H., *Boundary Layer Theory*. 1979, New York: McGraw-Hill.
23. Andereck, C.D., Liu, S.S., and Swinney, H.L., "Flow regimes in a circular Couette system with independently rotating cylinders." *J. Fluid Mech.*, 1986. **164**: pp. 155–183.
24. Cokelet, G., "The rheology and tube flow of blood," in *Handbook of Bioengineering*, R. Skalak and S. Chien, editors. 1987, New York: McGraw-Hill. pp. 14.1–14.17.
25. Merrill, E., "Rheology of blood." *Physiol. Rev.*, 1969. **49**: pp. 863–888.
26. Merrill, E.W., Benis, A.M., Gilliland, E.R., Sherwood, R.K., and Salzmann, E.W., "Pressure–flow relations of human blood in hollow fibers at low flow rates." *J. Appl. Physiol.*, 1965. **20**: pp. 954–967.
27. Easthope, P.L. and Brooks, D.E., "A comparison of rheological constitutive functions for whole human blood." *Biorheology*, 1980. **17**: pp. 235–347.
28. Kleinstreuer, C., Buchanan, J.R., Lei, M., and Truskey, G.A., "Computational analysis of particle-hemodynamics and prediction of the onset of arterial diseases," in *Cardiovascular Techniques. Biomechanical Systems Techniques and Applications*, C. Leondes, editor. 2001, Boca Raton, FL: CRC Press. pp. 1–1 to 1–68.
29. Chien, S., "Shear dependence of effective cell volume as a determinant of blood viscosity." *Science*, 1970. **268**: pp. 977–989.
30. Gaehtgens, P., "Flow of blood through narrow capillaries: Rheological mechanisms determining capillary hematocrit and apparent viscosity." *Biorheology*, 1980. **17**: pp. 183–189.
31. Barbee, J.H., and Cokelet, G.R., "Prediction of blood flow in tubes with diameters as small as 29 μm." *Microvasc. Res.*, 1971. **3**: pp. 277–297.
32. Pries, A.R., Secomb, T.W., Jacobs, H., Sperandio, M., Osterloh, K., and Gaehtgens, P., "Microvascular blood flow resistance: role of endothelial surface layer." *Am. J. Physiol.*, 1997. **273**: pp. H2272–H2279.
33. Pries, A.R., Secomb, T.W., and Gaehtgens, P., "The endothelial surface layer." *Pflugers Arch.*, 2000. **440**: pp. 653–666.
34. Lee, S., and Schmid-Schonbein, G.W., "Biomechanical model for the myogenic response in the microcirculation: Part I—formulation and initial testing." *J. Biomech. Eng.*, 1996. **118**: pp. 145–151.
35. Buga, G.M., Gold, N.E., Fukuto, J.M., and Ignarro, L.J., "Shear stress-induced release of nitric oxide from

endothelial cells grown on glass beads." *Hypertension,* 1991. **17**: pp. 187–193.

36. Frangos, J.A., Eskin, S.G., McIntire, L.V., and Ives, C.L., "Flow effects on prostacyclin production by cultured human endothelial cells." *Science,* 1985. **227**: pp. 1477–1479.

37. Kuchan, M.J., and Frangos, J.A., "Shear stress regulates endothelin-1 release via protein kinase C and cGMP in cultured endothelial cells." *Am. J. Physiol.,* 1993. **264**: pp. H150–H158.

38. Malek, A. and Izumo, S., "Physiological fluid shear stress causes downregulation of endothelin-1 mRNA in bovine aortic endothelium." *Am. J. Physiol.,* 1992. **263**: pp. C389–C396.

39. Davies, P.F., "Flow-mediated endothelial mechanotransduction." *Physiol. Rev.,* 1995. **75**: pp. 519–560.

40. Langille, B.L., "Blood flow-induced remodeling of the artery wall," in *Flow-Dependent Regulation of Vascular Function,* J.A. Bevan, G. Kaley, and G.M. Rubanyi, editors. 1995, New York: Oxford University Press. pp. 277–299.

41. Calvo, W.J., Hajduczok, G., Russell, J.A., and Diamond, S.L., "Inhibition of nitric oxide but not prostacyclin prevents poststenotic dilatation in rabbit femoral artery." *Circulation,* 1999. **99**: pp. 1069–1076.

42. Tsai, M.A., Frank, R.S., and Waugh, R.E., "Passive mechanical behavior of human neutrophils: effect of cytochalasin B." *Biophys J.,* 1993. **66**: pp. 2166–2172.

CHAPTER 3

Conservation Relations for Fluid Transport, Dimensional Analysis, and Scaling

3.1 Introduction

The problems analyzed in Chapter 2 involved applying a momentum balance from first principles for each flow situation. In this chapter, we develop a more efficient process for analyzing problems that involve steady and unsteady flow in one, two, or three dimensions. Rather than developing momentum balances for each problem, we derive more general, three-dimensional forms of the equations for the conservation of mass and linear momentum. Both equations are differential equations and are derived for a differential unit cube (i.e., in rectangular coordinates) that is fixed in space. The dimensions of this cube are shrunk to zero in the process of derivation, leading to the differential nature of the final equations, which apply generally to any point in the fluid. Indeed, the resulting equations are sometimes referred to as *point equations*, to contrast them with equations that represent mass or momentum balances over finite volumes of fluid. Next, we use vector calculus to generalize the results to any coordinate system. In Section 3.2, the equation of conservation of mass is derived, resulting in a single expression. Because the variable is mass, which is a scalar, the result is referred to as a *scalar equation*. Then, in Section 3.3, the general conservation of linear momentum is developed. Since momentum is a vector, this result is written either in vector notation as a single equation (a *vector equation*) or in the form of three equations, one for each of the vector components. The form of the equation of conservation of linear momentum for Newtonian fluids—the *Navier–Stokes* equation—is presented, along with applications. The conservation equations are summarized in tabular form for easy reference, and biomedical examples are provided for the use of these equations in this and subsequent chapters. Several applications of the Navier–Stokes equations are considered in Section 3.4. In order to present results in a useful way and to simplify the complex features of many flow problems, it is often useful to relate the dependent and independent variables by suitable reference quantities. This process of *scaling* often is used to recast the conservation equation in terms of nondimensional variables. Accordingly, in Section 3.5, the concept of *dimensional analysis* is introduced and the Navier–Stokes equations are cast in a dimensionless form. Dimensional analysis is used to correlate data and to simplify the Navier–Stokes equation. Section 3.6 involves the examination of one such simplification, the case of low-Reynolds-number flow.

3.2 Differential Form of the Equation of Conservation of Mass in Three Dimensions

3.2.1 General Form of the Equation of Conservation of Mass

To establish the conservation of mass, consider a fixed cubic volume through which a pure fluid passes (see Figure 3.1). Transport is by the flow of fluid only. In words, the conservation of mass for a pure fluid is

$$\begin{bmatrix}\text{Rate of accumulation of}\\ \text{mass in control volume}\end{bmatrix} = \begin{bmatrix}\text{Flow rate of mass}\\ \text{into control volume}\end{bmatrix} - \begin{bmatrix}\text{Flow rate of mass}\\ \text{from control volume}\end{bmatrix}. \tag{3.2.1}$$

Equation (3.2.1) is also valid for the total mass of a mixture. The validity of the preceding statement is demonstrated in Chapter 7.

Equation (3.2.1) is developed mathematically as follows: For a cubic control volume (Figure 3.1), the mass is the product of the density and the volume element $\Delta x \Delta y \Delta z$. The mass flow rate (mass/time) is the product of the density ρ, the local velocity, and the cross-sectional area through which flow occurs. Assume that, for each direction, mass enters the control volume at the smaller value of the coordinate (e.g., x for a surface of constant x and area $\Delta y \Delta z$) and exits at the larger value of the coordinate (e.g., $x + \Delta x$). The mass flow rate in a given direction is the product of the density and the velocity component in that direction. Accounting for mass flow into and from the control volume on each surface, we can write Equation (3.2.1) as

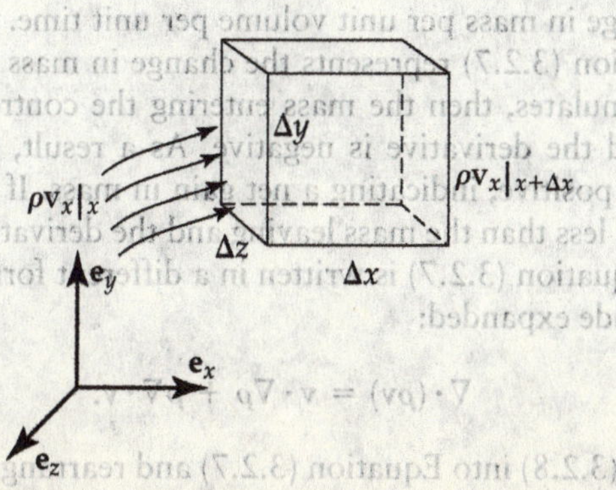

FIGURE 3.1 Flow across a surface of constant x for a rectangular control volume.

$$\Delta x \Delta y \Delta z \frac{\partial \rho}{\partial t} = [\rho v_x|_x - \rho v_x|_{x+\Delta x}]\Delta y \Delta z + [\rho v_y|_y - \rho v_y|_{y+\Delta y}]\Delta x \Delta z + [\rho v_z|_z - \rho v_z|_{z+\Delta z}]\Delta x \Delta y. \quad (3.2.2)$$

Dividing each term by the volume element $\Delta x \Delta y \Delta z$ yields

$$\frac{\partial \rho}{\partial t} = \frac{(\rho v_x|_x - \rho v_x|_{x+\Delta x})}{\Delta x} + \frac{(\rho v_y|_y - \rho v_y|_{y+\Delta y})}{\Delta y} + \frac{(\rho v_z|_z - \rho v_z|_{z+\Delta z})}{\Delta z}. \quad (3.2.3)$$

Taking the limit of Equation (3.2.3) as Δx, Δy, and Δz approach zero and using the definition of the derivative gives

$$\frac{\partial \rho}{\partial t} = -\left(\frac{\partial \rho v_x}{\partial x} + \frac{\partial \rho v_y}{\partial y} + \frac{\partial \rho v_z}{\partial z}\right). \quad (3.2.4)$$

Equation (3.2.4) is one form of the conservation of mass (in Cartesian coordinates). The equation can be expressed in a more compact form by using vector notation. As described in Section A.3.C, the gradient operator in Cartesian coordinates is

$$\nabla = \mathbf{e}_x \frac{\partial}{\partial x} + \mathbf{e}_y \frac{\partial}{\partial y} + \mathbf{e}_z \frac{\partial}{\partial z}, \quad (3.2.5)$$

where $\mathbf{e}_x$, $\mathbf{e}_y$, and $\mathbf{e}_z$ are unit vectors in the x, y, and z directions, respectively. The divergence of the mass flow rate per unit area, $\rho \mathbf{v}$, is (see Equation (A.3.10))

$$\nabla \cdot (\rho \mathbf{v}) = \frac{\partial \rho v_x}{\partial x} + \frac{\partial \rho v_y}{\partial y} + \frac{\partial \rho v_z}{\partial z}. \quad (3.2.6)$$

Equation (3.2.4) can thus be written more compactly as

$$\frac{\partial \rho}{\partial t} = -\nabla \cdot (\rho \mathbf{v}). \quad (3.2.7)$$

Equation (3.2.7) is a general form of the conservation of mass, often referred to as the *continuity equation*, and is valid for any coordinate system. Table 3.1 lists the continuity equation in rectangular, cylindrical, and spherical coordinates.

The physical interpretation of Equation (3.2.7) is as follows: The term $\partial \rho / \partial t$ represents the change in mass per unit volume per unit time. The term on the right-hand side of Equation (3.2.7) represents the change in mass flow rate per unit volume. If mass accumulates, then the mass entering the control volume exceeds the mass leaving it and the derivative is negative. As a result, the right-hand side of Equation (3.2.7) is positive, indicating a net gain in mass. If mass is removed, then the mass entering is less than the mass leaving and the derivative is positive.

Sometimes Equation (3.2.7) is written in a different form, with the derivatives on the right-hand side expanded:

$$\nabla \cdot (\rho \mathbf{v}) = \mathbf{v} \cdot \nabla \rho + \rho \nabla \cdot \mathbf{v}. \quad (3.2.8)$$

Inserting Equation (3.2.8) into Equation (3.2.7) and rearranging terms yields

TABLE 3.1

The Conservation of Mass (Continuity Equation)	
Rectangular coordinates (x, y, z)	$\frac{\partial \rho}{\partial t} = -\left(\frac{\partial \rho v_x}{\partial x} + \frac{\partial \rho v_y}{\partial y} + \frac{\partial \rho v_z}{\partial z}\right)$
Cylindrical coordinates (r, θ, z)	$\frac{\partial \rho}{\partial t} = -\left(\frac{1}{r}\frac{\partial(\rho r v_r)}{\partial r} + \frac{1}{r}\frac{\partial(\rho v_\theta)}{\partial \theta} + \frac{\partial(\rho v_z)}{\partial z}\right)$
Spherical coordinates (r, θ, ϕ)	$\frac{\partial \rho}{\partial t} = -\left(\frac{1}{r^2}\frac{\partial(\rho r^2 v_r)}{\partial r} + \frac{1}{r\sin\theta}\frac{\partial(\rho v_\theta \sin\theta)}{\partial \theta} + \frac{1}{r\sin\theta}\frac{\partial(\rho v_\phi)}{\partial \phi}\right)$

$$\frac{\partial \rho}{\partial t} + \mathbf{v} \cdot \nabla \rho = -\rho \nabla \cdot \mathbf{v}. \tag{3.2.9}$$

Both terms on the left-hand side of Equation (3.2.9) represent the accumulation of mass per unit volume. The first term represents the change in density with time. Such a change might occur if the fluid is being heated uniformly. The second term represents the change in density due to the flow of mass across the surface of the control volume. This will occur if the density varies spatially. The two terms sometimes appear as

$$\frac{D}{Dt} = \frac{\partial}{\partial t} + \mathbf{v} \cdot \nabla, \tag{3.2.10}$$

where D/Dt is known as the *substantial derivative*. This quantity represents a reference frame that moves with the local fluid velocity. Thus, the conservation of mass can be written as

$$\frac{D\rho}{Dt} = -\rho \nabla \cdot \mathbf{v}. \tag{3.2.11}$$

3.2.2 Conservation of Mass for Incompressible Fluids

A special case of the continuity equation applies to an *incompressible fluid*, whose density does not vary with pressure and temperature (ρ = constant). As a result, Equations (3.2.7) and (3.2.11) reduce to

$$\nabla \cdot \mathbf{v} = 0. \tag{3.2.12}$$

At constant temperature, all single-component liquids behave as incompressible fluids. Liquid solutions are incompressible if the components are well mixed. Gases can be treated as incompressible when their fluid velocity is less than the speed of sound. (This relationship can be developed with reference to the *Mach number*, which is the ratio of the fluid velocity to the speed of sound in the fluid. Gases behave as incompressible fluids when the Mach number is less than unity.) If significant temperature gradients are present, the assumption of incompressibility is not valid. For example, gas motions around a fire or a heated pipe arise from density variations that are

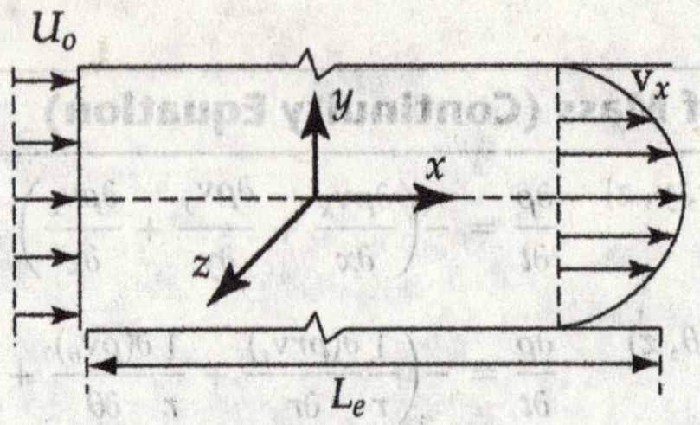

FIGURE 3.2 Flow into a rectangular channel.

induced by heat. Similarly, in the ocean, significant gradients in salinity and density arise from temperature variations, and these variations can generate fluid motion. For most biomedical problems, incompressibility can be assumed without any apprecia-ble error. Therefore, Equation (3.2.12) is typically used for the conservation of mass.

An important result that arises from an analysis of the conservation of mass is that if a velocity component changes in the direction of flow, then there is a corre-sponding change in a velocity component in another direction. This other velocity component is often referred to as a *secondary flow*, because it occurs in a direction other than the direction of the main flow and is frequently smaller than the main flow. For example, consider the flow of an incompressible fluid in the entry region of a nar-row rectangular channel, as shown in Figure 3.2. At the entrance, flow is uniform. At the end of the entry length, the velocity profile is a parabola, as we determined in Sec-tion 2.7.2. Since the channel width is much greater than the channel height, velocity variations in the z direction are negligible. As the uniform flow enters the channel, the fluid near the surfaces decelerates and fluid near the centerline accelerates to maintain a constant flow rate. As a result, v_x changes in the x and y directions. For this case, the continuity equation in rectangular coordinates for an incompressible fluid reduces to

$$\frac{\partial v_x}{\partial x} + \frac{\partial v_y}{\partial y} = 0. \tag{3.2.13}$$

Because v_x is a function of x and y in the entry region, the derivative of v_x with respect to x is nonzero and there is a velocity component v_y that varies with y. As fluid moves along the x direction, the variation of v_x in the x direction decreases, with a corresponding decline in v_y. When the flow is fully developed, v_x is a function of y only, and v_y vanishes. For fully developed flow, each term in Equation (3.2.13) is identically zero.

Example 3.1 Consider a two-dimensional inviscid flow (i.e., $\mu = 0$) that impinges on a surface as shown in Figure 3.3. The velocity in the y direction is $v_y = Uy$. Apply the conservation of mass for an incompressible fluid to determine v_x.

Solution Applying the two-dimensional form of the conservation of mass for an incompressible fluid (Equation (3.2.13)) to this particular flow yields

$$\frac{\partial v_x}{\partial x} = -U. \tag{3.2.14}$$

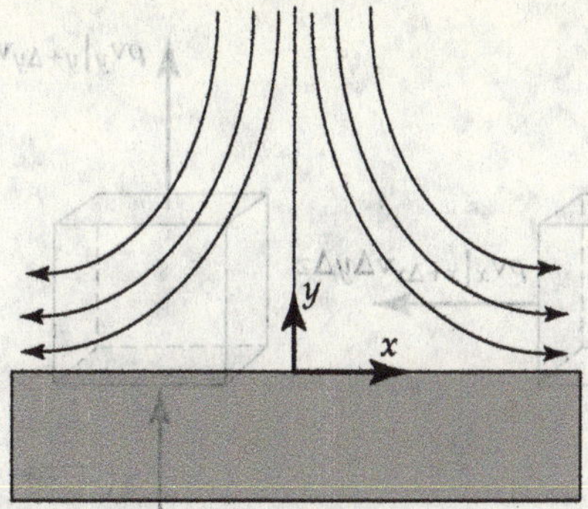

FIGURE 3.3 Two-dimensional flow toward a surface.

Integrating Equation (3.2.14) with respect to x gives the following expression for the velocity, with one constant of integration:

$$v_x = -Ux + C. \tag{3.2.15}$$

Since v_x is zero at $x = 0$, C is zero. Thus,

$$v_x = -Ux. \tag{3.2.16}$$

Note that this result is not valid near $y = 0$. Even for fluids with low viscosity and high velocities, we must account for viscous forces near the surface.

3.3 Differential Form of the Conservation of Linear Momentum and the Navier–Stokes Equations in Three Dimensions

3.3.1 General Form of the Equation of Conservation of Linear Momentum

The derivation of the differential form of the equation of conservation of linear momentum is similar to the derivation of the conservation of mass, except that momentum is a vector quantity, whereas mass is a scalar. The conservation of linear momentum can be stated as[1]

$$\begin{bmatrix}\text{Rate of momentum}\\ \text{accumulation}\end{bmatrix} = \begin{bmatrix}\text{Rate of momentum}\\ \text{flow in}\end{bmatrix} - \begin{bmatrix}\text{Rate of momentum}\\ \text{flow out}\end{bmatrix} + \sum \text{forces}. \tag{3.3.1}$$

Again, we can use a rectangular coordinate system to recast Equation (3.3.1) into a mathematical form. A cubic control volume for the application of the

[1]Alternatively, Equation (3.3.1) can be written as $ma = \Sigma$ forces, where m is the mass of fluid in the control volume and a is the fluid acceleration. For a fluid, the acceleration includes both the rate of change of the fluid velocity with respect to time and the convective acceleration (Equation (2.2.8c)).

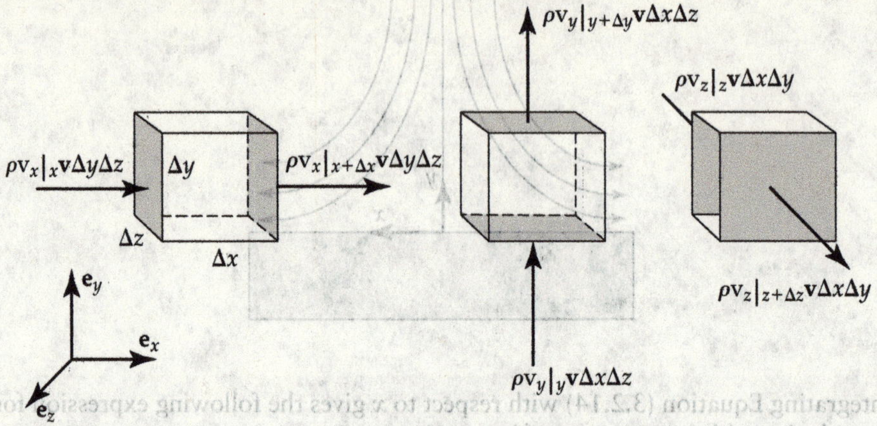

FIGURE 3.4 Momentum transport across surfaces of cubic control volume.

conservation of linear momentum is shown in Figure 3.4. The rate of momentum accumulation is

$$\frac{\partial \rho \mathbf{v}}{\partial t}\Delta x \Delta y \Delta z. \tag{3.3.2a}$$

The flow of momentum arises from the motion of the fluid. The momentum flow rate per unit volume is the product of the flow rate (velocity times area) and the momentum per unit volume ($\rho\mathbf{v}$). By analogy with mass, momentum flows into and leaves the control volume through surfaces of constant x, y, and z. The total flow rate of momentum in the x, y, and z directions is

$$(\rho\mathbf{v}\mathrm{v}_x|_x - \rho\mathbf{v}\mathrm{v}_x|_{x+\Delta x})\Delta y\Delta z + (\rho\mathbf{v}\mathrm{v}_y|_y - \rho\mathbf{v}\mathrm{v}_y|_{y+\Delta y})\Delta x\Delta z$$
$$+ (\rho\mathbf{v}\mathrm{v}_z - \rho\mathbf{v}\mathrm{v}_z|_{z+\Delta z})\Delta x\Delta y. \tag{3.3.2b}$$

Substituting Equations (3.3.2a) and (3.3.2b) into the momentum balance yields

$$\Delta x\Delta y\Delta z\frac{\partial \rho\mathbf{v}}{\partial t} = \Delta x\Delta y\Delta z\left[\frac{(\rho\mathbf{v}\mathrm{v}_x|_x - \rho\mathbf{v}\mathrm{v}_x|_{x+\Delta x})}{\Delta x} + \frac{(\rho\mathbf{v}\mathrm{v}_y|_y - \rho\mathbf{v}\mathrm{v}_y|_{y+\Delta y})}{\Delta y}\right.$$
$$\left. + \frac{(\rho\mathbf{v}\mathrm{v}_z|_z - \rho\mathbf{v}\mathrm{v}_z|_{z+\Delta z})}{\partial z}\right] + \sum \mathbf{F}. \tag{3.3.2c}$$

Taking the limit as Δx, Δy, and Δz approach zero produces

$$dV\frac{\partial \rho\mathbf{v}}{\partial t} = -dV\left[\frac{\partial \rho \mathrm{v}_x\mathbf{v}}{\partial x} + \frac{\partial \rho \mathrm{v}_y\mathbf{v}}{\partial y} + \frac{\partial \rho \mathrm{v}_z\mathbf{v}}{\partial z}\right] + \sum \mathbf{F}, \tag{3.3.3}$$

where dV is the differential volume element, $dxdydz$. Equation (3.3.3) is a vector relation. By analogy with Equations (3.2.6) and (3.2.7), we obtain

$$\nabla\cdot(\rho\mathbf{v}\mathbf{v}) = \left[\frac{\partial(\rho \mathrm{v}_x\mathbf{v})}{\partial x} + \frac{\partial(\rho \mathrm{v}_y\mathbf{v})}{\partial y} + \frac{\partial(\rho \mathrm{v}_z\mathbf{v})}{\partial z}\right]. \tag{3.3.4}$$

Equation (3.3.3) now becomes

$$dV\left[\frac{\partial \rho \mathbf{v}}{\partial t} + \nabla \cdot (\rho \mathbf{v}\mathbf{v})\right] = \sum \mathbf{F}. \tag{3.3.5}$$

The vector expression $\nabla \cdot (\rho \mathbf{v}\mathbf{v})$ can be rearranged as

$$\nabla \cdot (\rho \mathbf{v}\mathbf{v}) = \rho \mathbf{v} \cdot \nabla \mathbf{v} + \mathbf{v}\nabla \cdot (\rho \mathbf{v}). \tag{3.3.6}$$

The left-hand side of Equation (3.3.5) can be rearranged by substituting Equation (3.3.6) and expanding the derivative of the momentum with respect to time:

$$dV\left[\rho\frac{\partial \mathbf{v}}{\partial t} + \rho \mathbf{v} \cdot \nabla \mathbf{v} + \mathbf{v}\frac{\partial \rho}{\partial t} + \mathbf{v}\nabla \cdot (\rho v)\right] = \sum \mathbf{F}. \tag{3.3.7}$$

The third and fourth terms in brackets on the left-hand side of Equation (3.3.7) are equal to zero, since they represent the product of the velocity and the conservation of mass (Equation (3.2.7)). Thus, Equation (3.3.7) simplifies to

$$dV\left[\rho\frac{\partial \mathbf{v}}{\partial t} + \rho \mathbf{v} \cdot \nabla \mathbf{v}\right] = \sum \mathbf{F}. \tag{3.3.8}$$

Note that the sum in brackets in Equation (3.3.8) can be recast as the product of the density and the substantial derivative of the velocity. For velocity, the substantial derivative represents the convective fluid acceleration, while the derivative with respect to time represents acceleration due to changes in velocity with time. The term $\mathbf{v} \cdot \nabla \mathbf{v}$ represents the *convective acceleration*—the fluid acceleration that arises from changes in velocity with position (Equation (2.2.8c)).

To complete the derivation of the conservation of linear momentum, the forces acting on the control volume must be specified. These forces are either *body forces* ($\mathbf{F}_b$) or *surface forces* ($\mathbf{F}_s$). Thus, $\mathbf{F} = \mathbf{F}_b + \mathbf{F}_s$. Body forces act on the entire body. Examples are gravity and electromagnetic forces. Here, we consider only gravity, for which the body force is expressed as

$$\mathbf{F}_b = \mathbf{F}_g = \rho \mathbf{g}\Delta x \Delta y \Delta z. \tag{3.3.9}$$

Surface forces act on the surfaces of the control volume and include pressure and viscous stresses. Pressure acts normal to the surface and represents the stress acting on a fluid element at rest. Viscous stresses arise from the motion of the fluid and act both normally and tangentially to the surfaces (see Figure 3.5). The total stress tensor is the sum of the pressure and the viscous stresses, or

$$\boldsymbol{\sigma} = \boldsymbol{\tau} - \mathbf{I}p. \tag{3.3.10}$$

The reason for writing the stress in this manner is that the pressure at a given point is the same in all directions (i.e., it is *isotropic*), whereas the viscous stresses (referred to in Section 2.3 as the *deviatoric* stress) are not generally the same in all directions. The negative sign in Equation (3.3.10) arises because a compressive stress is considered to be negative, whereas pressure is a positive quantity.

Another commonly used notation for tensors is the *Einstein convention*, in which the tensor is specified in terms of subscripts (e.g., $\boldsymbol{\sigma} = \sigma_{ij}$). The subscripts indicate that

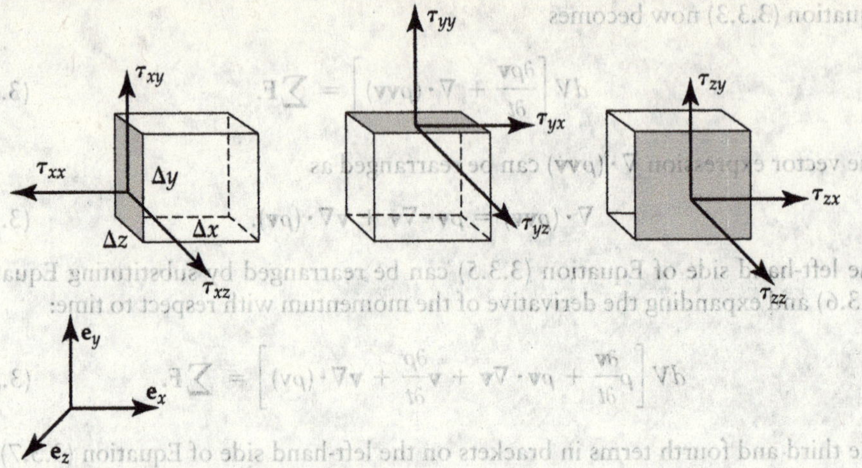

FIGURE 3.5 Schematic of components of the viscous stress tensor.

the stress acts in the j direction on a face normal to the i-axis. As noted in Chapter 1, an important property of the stress tensor is that it is symmetric (i.e., $\sigma_{ij} = \sigma_{ji}$).

The pressure is normal to every surface and directed inward. The viscous stress term has nine components. For each surface of constant x, y, or z, three viscous stresses act, as shown in Figure 3.5. Thus, for a surface of constant y, the fluid stresses acting in the x, y, and z directions are τ_{yx}, τ_{yy}, and τ_{yz}, respectively. In rectangular coordinates, the components of the stress tensor are

$$\sigma_{ij} = \begin{vmatrix} \tau_{xx} - p & \tau_{yx} & \tau_{zx} \\ \tau_{xy} & \tau_{yy} - p & \tau_{zy} \\ \tau_{xz} & \tau_{yz} & \tau_{zz} - p \end{vmatrix} \tag{3.3.11}$$

A net force on the fluid element arises from a gradient in the stress tensor. This gradient is due to a difference in the stresses across two or more planes normal to one of the axes. Thus, the surface force acting in the x direction derives from a difference in the pressure that is acting on the surfaces $\Delta y \Delta z$ at x and $x + \Delta x$ and from the viscous stresses τ_{xx}, τ_{yx}, and τ_{zx}. In this text, we adopt the convention that the fluid on the face with the greater algebraic value of the space variable defining it exerts positive stresses on the face that has the lesser value. This convention is consistent with the sign convention for pressure. The surface forces in the x direction are

$$\begin{aligned} F_{S_x} = &-[p|_{x+\Delta x} - p|_x]\Delta y \Delta z + [\tau_{xx}|_{x+\Delta x} - \tau_{xx}|_x]\Delta y \Delta z \\ &+[\tau_{yx}|_{y+\Delta y} - \tau_{yx}|_y]\Delta x \Delta z + [\tau_{zx}|_{z+\Delta z} - \tau_{zx}|_z]\Delta x \Delta y. \end{aligned} \tag{3.3.12a}$$

Likewise, the surface forces in the y and z directions are

$$\begin{aligned} F_{S_y} = &-[p|_{y+\Delta y} - p|_y]\Delta x \Delta z + [\tau_{xy}|_{x+\Delta x} - \tau_{xy}|_x]\Delta y \Delta z \\ &+[\tau_{yy}|_{y+\Delta y} - \tau_{yy}|_y]\Delta x \Delta z + [\tau_{zy}|_{z+\Delta z} - \tau_{zy}|_z]\Delta x \Delta y \end{aligned} \tag{3.3.12b}$$

and

$$\begin{aligned} F_{S_z} = &-[p|_{z+\Delta z} - p|_z]\Delta x \Delta y + [\tau_{xz}|_{x+\Delta x} - \tau_{xz}|_x]\Delta y \Delta z \\ &+[\tau_{yz}|_{y+\Delta y} - \tau_{yz}|_y]\Delta x \Delta z + [\tau_{zz}|_{z+\Delta z} - \tau_{zz}|_z]\Delta x \Delta y. \end{aligned} \tag{3.3.12c}$$

Each of the force components is now rewritten as the product of the force per unit volume and the volume element. As a result, Equations (3.3.12a) through (3.3.12c) become, respectively,

$$F_{S_x} = \Delta x \Delta y \Delta z \left[-\frac{p|_{x+\Delta x} - p|_x}{\Delta x} + \frac{\tau_{xx}|_{x+\Delta x} - \tau_{xx}|_x}{\Delta x} + \frac{\tau_{yx}|_{y+\Delta y} - \tau_{yx}|_y}{\Delta y} + \frac{\tau_{zx}|_{z+\Delta z} - \tau_{zx}|_z}{\Delta z} \right], \quad (3.3.13a)$$

$$F_{S_y} = \Delta x \Delta y \Delta z \left[-\frac{p|_{y+\Delta y} - p|_y}{\Delta y} + \frac{\tau_{xy}|_{x+\Delta x} - \tau_{xy}|_x}{\Delta x} + \frac{\tau_{yy}|_{y+\Delta y} - \tau_{yy}|_y}{\Delta y} + \frac{\tau_{zy}|_{z+\Delta z} - \tau_{zy}|_z}{\Delta z} \right], \quad (3.3.13b)$$

and

$$F_{S_z} = \Delta x \Delta y \Delta z \left[-\frac{p|_{z+\Delta z} - p|_z}{\Delta z} + \frac{\tau_{xz}|_{x+\Delta x} - \tau_{xz}|_x}{\Delta x} + \frac{\tau_{yz}|_{y+\Delta y} - \tau_{yz}|_y}{\Delta y} + \frac{\tau_{zz}|_{z+\Delta z} - \tau_{zz}|_z}{\Delta z} \right]. \quad (3.3.13c)$$

Next, taking the limit as Δx, Δy, and Δz approach zero, and writing the expression in vector form, we obtain

$$\begin{aligned} \mathbf{F} &= \mathbf{e}_x F_x + \mathbf{e}_y F_y + \mathbf{e}_z F_z \\ &= -dV\left(\mathbf{e}_x \frac{\partial p}{\partial x} + \mathbf{e}_y \frac{\partial p}{\partial y} + \mathbf{e}_z \frac{\partial p}{\partial z} \right) + dV\mathbf{e}_x \left(\frac{\partial \tau_{xx}}{\partial x} + \frac{\partial \tau_{yx}}{\partial y} + \frac{\partial \tau_{zx}}{\partial z} \right) \\ &\quad + dV\mathbf{e}_y \left(\frac{\partial \tau_{xy}}{\partial x} + \frac{\partial \tau_{yy}}{\partial y} + \frac{\partial \tau_{zy}}{\partial z} \right) + dV\mathbf{e}_z \left(\frac{\partial \tau_{xz}}{\partial x} + \frac{\partial \tau_{yz}}{\partial y} + \frac{\partial \tau_{zz}}{\partial z} \right). \end{aligned} \quad (3.3.14)$$

Equation (3.3.14) can be written in a more compact notation. The terms involving pressure are simply the gradient of the pressure, ∇p. The viscous stress terms are the divergence of the viscous stress tensor, $\nabla \cdot \boldsymbol{\tau}$. Inserting the forces per unit volume into Equation (3.3.8) yields

$$dV\left[\rho \frac{\partial \mathbf{v}}{\partial t} + \rho \mathbf{v} \cdot \nabla \mathbf{v} \right] = dV[-\nabla p + \nabla \cdot \boldsymbol{\tau} + \rho \mathbf{g}]. \quad (3.3.15)$$

Dividing both sides of Equation (3.3.15) by the differential volume element yields the conservation of linear momentum:

$$\rho \frac{\partial \mathbf{v}}{\partial t} + \rho \mathbf{v} \nabla \cdot \mathbf{v} = -\nabla p + \nabla \cdot \boldsymbol{\tau} + \rho \mathbf{g}. \quad (3.3.16)$$

Equation (3.3.16) is another form of Newton's second law of motion, expressed per unit of volume. The equations of conservation of linear momentum in rectangular, cylindrical, and spherical coordinates are summarized in Table 3.2.

In the absence of fluid motion, $\mathbf{v} = 0$ and $\boldsymbol{\tau} = 0$. Therefore, the equation of fluid statics (Equation (2.4.5)) is recovered:

$$\nabla p = \rho \mathbf{g}. \tag{3.3.17}$$

Equation (3.3.17) states that gravity gives rise to a pressure gradient. This result can be used to calculate the force that is acting upon submerged objects, which we studied in Chapter 2.

An examination of the equations for conservation of mass (Equation (3.2.9)) and linear momentum (Equation (3.3.16)) shows that there are four equations and

TABLE 3.2

Conservation of Linear Momentum

Rectangular coordinates

x component

$$\rho\left[\frac{\partial v_x}{\partial t} + v_x\frac{\partial v_x}{\partial x} + v_y\frac{\partial v_x}{\partial y} + v_z\frac{\partial v_x}{\partial z}\right] = \rho g_x - \frac{\partial p}{\partial x} + \left[\frac{\partial \tau_{xx}}{\partial x} + \frac{\partial \tau_{yx}}{\partial y} + \frac{\partial \tau_{zx}}{\partial z}\right] \tag{3.3.18a}$$

y component

$$\rho\left[\frac{\partial v_y}{\partial t} + v_x\frac{\partial v_y}{\partial x} + v_y\frac{\partial v_y}{\partial y} + v_z\frac{\partial v_y}{\partial z}\right] = \rho g_y - \frac{\partial p}{\partial y} + \left[\frac{\partial \tau_{xy}}{\partial x} + \frac{\partial \tau_{yy}}{\partial y} + \frac{\partial \tau_{zy}}{\partial z}\right] \tag{3.3.18b}$$

z component

$$\rho\left[\frac{\partial v_z}{\partial t} + v_x\frac{\partial v_z}{\partial x} + v_y\frac{\partial v_z}{\partial y} + v_z\frac{\partial v_z}{\partial z}\right] = \rho g_z - \frac{\partial p}{\partial z} + \left[\frac{\partial \tau_{xz}}{\partial x} + \frac{\partial \tau_{yz}}{\partial y} + \frac{\partial \tau_{zz}}{\partial z}\right] \tag{3.3.18c}$$

Cylindrical coordinates

r component

$$\rho\left[\frac{\partial v_r}{\partial t} + v_r\frac{\partial v_r}{\partial r} + \frac{v_\theta}{r}\frac{\partial v_r}{\partial \theta} - \frac{v_\theta^2}{r} + v_z\frac{\partial v_r}{\partial z}\right] = \rho g_r - \frac{\partial p}{\partial r} + \left[\frac{1}{r}\frac{\partial (r\tau_{rr})}{\partial r} + \frac{1}{r}\frac{\partial \tau_{\theta r}}{\partial \theta} - \frac{\tau_{\theta\theta}}{r} + \frac{\partial \tau_{zr}}{\partial z}\right] \tag{3.3.19a}$$

θ component

$$\rho\left[\frac{\partial v_\theta}{\partial t} + v_r\frac{\partial v_\theta}{\partial r} + \frac{v_\theta}{r}\frac{\partial v_\theta}{\partial \theta} + \frac{v_r v_\theta}{r} + v_z\frac{\partial v_\theta}{\partial z}\right] = \rho g_\theta - \frac{\partial p}{\partial \theta} + \left[\frac{1}{r^2}\frac{\partial (r^2\tau_{r\theta})}{\partial r} + \frac{1}{r}\frac{\partial \tau_{\theta\theta}}{\partial \theta} + \frac{\partial \tau_{z\theta}}{\partial z}\right] \tag{3.3.19b}$$

z component

$$\rho\left[\frac{\partial v_z}{\partial t} + v_r\frac{\partial v_z}{\partial r} + \frac{v_\theta}{r}\frac{\partial v_z}{\partial \theta} + v_z\frac{\partial v_z}{\partial z}\right] = \rho g_z - \frac{\partial p}{\partial z} + \left[\frac{1}{r}\frac{\partial (r\tau_{rz})}{\partial r} + \frac{1}{r}\frac{\partial \tau_{\theta z}}{\partial \theta} + \frac{\partial \tau_{zz}}{\partial z}\right] \tag{3.3.19c}$$

Spherical coordinates

r component

$$\rho\left[\frac{\partial v_r}{\partial t} + v_r\frac{\partial v_r}{\partial r} + \frac{v_\theta}{r}\frac{\partial v_r}{\partial \theta} + \frac{v_\phi}{r\sin\theta}\frac{\partial v_r}{\partial \phi} - \frac{v_\theta^2 + v_\phi^2}{r}\right] = \rho g_r - \frac{\partial p}{\partial r} + \left[\frac{1}{r^2}\frac{\partial (r^2\tau_{rr})}{\partial r} + \frac{1}{r\sin\theta}\frac{\partial (\tau_{\theta r}\sin\theta)}{\partial \theta} + \frac{1}{r\sin\theta}\frac{\partial \tau_{r\phi}}{\partial \phi} - \frac{\tau_{\theta\theta} + \tau_{\phi\phi}}{r}\right] \tag{3.3.20a}$$

TABLE 3.2

(Continued)

θ **component**

$$\rho\left[\frac{\partial v_\theta}{\partial t} + v_r\frac{\partial v_\theta}{\partial r} + \frac{v_\theta}{r}\frac{\partial v_\theta}{\partial \theta} + \frac{v_\phi}{r\sin\theta}\frac{\partial v_\theta}{\partial \phi} + \frac{v_r v_\theta}{r} - \frac{v_\phi^2\cot\theta}{r}\right] = \rho g_\theta - \frac{\partial p}{\partial \theta}$$
$$+\left[\frac{1}{r^2}\frac{\partial(r^2\tau_{r\theta})}{\partial r} + \frac{1}{r\sin\theta}\frac{\partial(\tau_{\theta\theta}\sin\theta)}{\partial\theta} + \frac{1}{r\sin\theta}\frac{\partial\tau_{\theta\phi}}{\partial\phi} + \frac{\tau_{r\theta}}{r} - \frac{\cot\theta}{r}\tau_{\phi\phi}\right] \tag{3.3.20b}$$

ϕ **component**

$$\rho\left[\frac{\partial v_\phi}{\partial t} + v_r\frac{\partial v_\phi}{\partial r} + \frac{v_\theta}{r}\frac{\partial v_\phi}{\partial \theta} + \frac{v_\phi}{r\sin\theta}\frac{\partial v_\phi}{\partial \phi} + \frac{v_r v_\phi}{r} + \frac{v_\phi v_\theta\cot\theta}{r}\right] = \rho g_\phi - \frac{1}{r\sin\theta}\frac{\partial p}{\partial \phi}$$
$$+\left[\frac{1}{r^2}\frac{\partial(r^2\tau_{r\phi})}{\partial r} + \frac{1}{r}\frac{\partial\tau_{\theta\phi}}{\partial\theta} + \frac{1}{r\sin\theta}\frac{\partial\tau_{\phi\phi}}{\partial\phi} + \frac{\tau_{r\phi}}{r} + \frac{2\cot\theta}{r}\tau_{\theta\phi}\right] \tag{3.3.20c}$$

13 unknowns (pressure, three velocity components, and nine components of the viscous stress tensor). We can reduce the number of unknowns by three by noting that the viscous stress tensor is symmetric ($\tau_{ij} = \tau_{ji}$) for most fluids. Nevertheless, the only way in which we can further reduce the number of unknowns is to specify a relationship between viscous stresses and the fluid velocity. For all fluids, viscous stresses arise from velocity gradients; as noted earlier, the relationship between these two quantities is the *constitutive equation* (e.g., Newton's law of viscosity).

3.3.2 The Navier–Stokes Equation for an Incompressible Newtonian Fluid

The following generalized constitutive relationship has been developed for an incompressible Newtonian fluid:

$$\tau_{ij} = \mu\left(\frac{\partial v_i}{\partial x_j} + \frac{\partial v_j}{\partial x_i}\right). \tag{3.3.21a}$$

(A more general relation is available for compressible fluids [1,2]. Since most problems in physiological fluid mechanics deal with incompressible fluids, this loss of generality will not present any problems here.) Equation (3.3.21a) is consistent with the statement of Newton's law of viscosity that we used previously for simple one-dimensional flow in rectangular coordinates (Equation (2.5.6). Further, Equation (3.3.21a) is consistent with the property of symmetry that is expected for the stress tensor. Expressed in vector form, Equation (3.3.21a) becomes

$$\boldsymbol{\tau} = \mu(\nabla\mathbf{v} + (\nabla\mathbf{v})^T), \tag{3.3.21b}$$

where $(\nabla\mathbf{v})^T$ is the *transpose* of the tensor $\nabla\mathbf{v}$. Recall that the transpose represents exchanging elements ij for elements ji in Equation (3.3.21a). For an incompressible Newtonian fluid, expressions for $\boldsymbol{\tau}$ in rectangular, cylindrical, and spherical coordinates are listed in Table 3.3. More general forms are available as well [3].

TABLE 3.3

Shear-Stress Tensor for an Incompressible Newtonian Fluid

Rectangular coordinates

$$\tau_{xx} = 2\mu \frac{\partial v_x}{\partial x} \quad (3.3.22a)$$

$$\tau_{yx} = \tau_{yx} = \mu\left(\frac{\partial v_x}{\partial y} + \frac{\partial v_y}{\partial x}\right) \quad (3.3.22b)$$

$$\tau_{zx} = \tau_{xz} = \mu\left(\frac{\partial v_x}{\partial z} + \frac{\partial v_z}{\partial x}\right) \quad (3.3.22c)$$

$$\tau_{yy} = 2\mu \frac{\partial v_y}{\partial y} \quad (3.3.22d)$$

$$\tau_{zy} = \tau_{yz} = \mu\left(\frac{\partial v_y}{\partial z} + \frac{\partial v_z}{\partial y}\right) \quad (3.3.22e)$$

$$\tau_{zz} = 2\mu \frac{\partial v_z}{\partial z} \quad (3.3.22f)$$

Cylindrical coordinates

$$\tau_{rr} = 2\mu \frac{\partial v_r}{\partial r} \quad (3.3.23a)$$

$$\tau_{r\theta} = \tau_{\theta r} = \mu\left(r\frac{\partial}{\partial r}\left(\frac{v_\theta}{r}\right) + \frac{1}{r}\frac{\partial v_r}{\partial \theta}\right) \quad (3.3.23b)$$

$$\tau_{zr} = \tau_{rz} = \mu\left(\frac{\partial v_r}{\partial z} + \frac{\partial v_z}{\partial r}\right) \quad (3.3.23c)$$

$$\tau_{\theta\theta} = 2\mu\left(\frac{1}{r}\frac{\partial v_r}{\partial \theta} + \frac{v_r}{r}\right) \quad (3.3.23d)$$

$$\tau_{z\theta} = \tau_{\theta z} = \mu\left(\frac{\partial v_\theta}{\partial z} + \frac{1}{r}\frac{\partial v_z}{\partial \theta}\right) \quad (3.3.23e)$$

$$\tau_{zz} = 2\mu \frac{\partial v_z}{\partial z} \quad (3.3.23f)$$

Spherical coordinates

$$\tau_{rr} = 2\mu \frac{\partial v_r}{\partial r} \quad (3.3.24a)$$

$$\tau_{r\theta} = \tau_{\theta r} = \mu\left(r\frac{\partial}{\partial r}\left(\frac{v_\theta}{r}\right) + \frac{1}{r}\frac{\partial v_r}{\partial \theta}\right) \quad (3.3.24b)$$

$$\tau_{r\phi} = \tau_{\phi r} = \mu\left(r\frac{\partial}{\partial r}\left(\frac{v_\phi}{r}\right) + \frac{1}{r\sin\theta}\frac{\partial v_r}{\partial \phi}\right) \quad (3.3.24c)$$

$$\tau_{\theta\theta} = 2\mu\left(\frac{1}{r}\frac{\partial v_r}{\partial \theta} + \frac{v_r}{r}\right) \quad (3.3.24d)$$

$$\tau_{\theta\phi} = \tau_{\phi\theta} = \mu\left(\frac{\sin\theta}{r}\frac{\partial}{\partial \theta}\left(\frac{v_\phi}{\sin\theta}\right) + \frac{1}{r\sin\theta}\frac{\partial v_\theta}{\partial \phi}\right) \quad (3.3.24e)$$

$$\tau_{\phi\phi} = 2\mu\left(\frac{1}{r\sin\theta}\frac{\partial v_\phi}{\partial \phi} + \frac{v_r}{r} + \frac{v_\theta \cot\theta}{r}\right) \quad (3.3.24f)$$

The vector form of the Navier–Stokes equation is obtained by substituting Equation (3.3.21b) into Equation (3.3.16):

$$\rho \frac{\partial \mathbf{v}}{\partial t} + \rho \mathbf{v} \cdot \nabla \mathbf{v} = -\nabla p + \mu \nabla^2 \mathbf{v} + \rho \mathbf{g}. \tag{3.3.25}$$

Table 3.4 presents the scalar components of the Navier–Stokes equation in rectangular, cylindrical, and spherical coordinates. The table is useful in formulating complex flow problems.

TABLE 3.4

Navier–Stokes Equation for an Incompressible Fluid

Rectangular coordinates

x direction

$$\rho\left(\frac{\partial v_x}{\partial t} + v_x \frac{\partial v_x}{\partial x} + v_y \frac{\partial v_x}{\partial y} + v_z \frac{\partial v_x}{\partial z}\right) = -\frac{\partial p}{\partial x} + \mu\left[\frac{\partial^2 v_x}{\partial x^2} + \frac{\partial^2 v_x}{\partial y^2} + \frac{\partial^2 v_x}{\partial z^2}\right] + \rho g_x \tag{3.3.26a}$$

y direction

$$\rho\left(\frac{\partial v_y}{\partial t} + v_x \frac{\partial v_y}{\partial x} + v_y \frac{\partial v_y}{\partial y} + v_z \frac{\partial v_y}{\partial z}\right) = -\frac{\partial p}{\partial y} + \mu\left[\frac{\partial^2 v_y}{\partial x^2} + \frac{\partial^2 v_y}{\partial y^2} + \frac{\partial^2 v_y}{\partial z^2}\right] + \rho g_y \tag{3.3.26b}$$

z direction

$$\rho\left(\frac{\partial v_z}{\partial t} + v_x \frac{\partial v_z}{\partial x} + v_y \frac{\partial v_z}{\partial y} + v_z \frac{\partial v_z}{\partial z}\right) = -\frac{\partial p}{\partial z} + \mu\left[\frac{\partial^2 v_z}{\partial x^2} + \frac{\partial^2 v_z}{\partial y^2} + \frac{\partial^2 v_z}{\partial z^2}\right] + \rho g_z \tag{3.3.26c}$$

Cylindrical coordinates

r direction

$$\rho\left(\frac{\partial v_r}{\partial t} + v_r \frac{\partial v_r}{\partial r} + \frac{v_\theta}{r}\frac{\partial v_r}{\partial \theta} - \frac{v_\theta^2}{r} + v_z \frac{\partial v_r}{\partial z}\right) = -\frac{\partial p}{\partial r} + \mu\left[\frac{\partial}{\partial r}\left(\frac{1}{r}\frac{\partial (r v_r)}{\partial r}\right) + \frac{1}{r^2}\frac{\partial^2 v_r}{\partial \theta^2} - \frac{2}{r^2}\frac{\partial v_\theta}{\partial \theta} + \frac{\partial^2 v_r}{\partial z^2}\right] + \rho g_r \tag{3.3.27a}$$

θ direction

$$\rho\left(\frac{\partial v_\theta}{\partial t} + v_r \frac{\partial v_\theta}{\partial r} + \frac{v_\theta}{r}\frac{\partial v_\theta}{\partial \theta} + \frac{v_r v_\theta}{r} + v_z \frac{\partial v_\theta}{\partial z}\right) = -\frac{1}{r}\frac{\partial p}{\partial \theta} + \mu\left[\frac{\partial}{\partial r}\left(\frac{1}{r}\frac{\partial (r v_\theta)}{\partial r}\right) + \frac{1}{r^2}\frac{\partial^2 v_\theta}{\partial \theta^2} + \frac{2}{r^2}\frac{\partial v_r}{\partial \theta} + \frac{\partial^2 v_\theta}{\partial z^2}\right] + \rho g_\theta \tag{3.3.27b}$$

z direction

$$\rho\left(\frac{\partial v_z}{\partial t} + v_r \frac{\partial v_z}{\partial r} + \frac{v_\theta}{r}\frac{\partial v_z}{\partial \theta} + v_z \frac{\partial v_z}{\partial z}\right) = -\frac{\partial p}{\partial z} + \mu\left[\frac{1}{r}\frac{\partial}{\partial r}\left(r\frac{\partial v_z}{\partial r}\right) + \frac{1}{r^2}\frac{\partial^2 v_z}{\partial \theta^2} + \frac{\partial^2 v_z}{\partial z^2}\right] + \rho g_z \tag{3.3.27c}$$

(continued)

TABLE 3.4

(Continued)

Spherical coordinates

r direction

$$\rho\left[\frac{\partial v_r}{\partial t} + v_r\frac{\partial v_r}{\partial t} + \frac{v_\theta}{r}\frac{\partial v_r}{\partial \theta} + \frac{v_\phi}{r\sin\theta}\frac{\partial v_r}{\partial \phi} - \frac{v_\theta^2 + v_\phi^2}{r}\right] = -\frac{\partial p}{\partial r} + \rho g_r$$
$$+ \mu\left[\left(\frac{1}{r^2}\frac{\partial}{\partial r}\left(r^2\frac{\partial v_r}{\partial r}\right)\right) + \frac{1}{r^2\sin\theta}\frac{\partial}{\partial \theta}\left(\sin\theta\frac{\partial v_r}{\partial \theta}\right) + \frac{1}{r^2\sin^2\theta}\frac{\partial^2 v_r}{\partial \phi^2} - 2\frac{v_r}{r^2} - \frac{2}{r^2}\frac{\partial v_\theta}{\partial \theta} - \frac{2}{r^2}v_\theta\cot\theta - \frac{2}{r^2\sin^2\theta}\frac{\partial v_\phi}{\partial \phi}\right] \tag{3.3.28a}$$

θ direction

$$\rho\left(\frac{\partial v_\theta}{\partial t} + v_r\frac{\partial v_\theta}{\partial r} + \frac{v_\theta}{r}\frac{\partial v_\theta}{\partial \theta} + \frac{v_\phi}{r\sin\theta}\frac{\partial v_\theta}{\partial \phi} + \frac{v_\theta v_r}{r} - \frac{v_\phi^2\cot\theta}{r}\right) = -\frac{1}{r}\frac{\partial p}{\partial \theta} + \mu\left[\frac{1}{r^2}\frac{\partial}{\partial r}\left(r^2\frac{\partial v_\theta}{\partial r}\right)\right.$$
$$\left. + \frac{1}{r^2\sin\theta}\frac{\partial}{\partial \theta}\left(\sin\theta\frac{\partial v_\theta}{\partial \theta}\right) + \frac{1}{r^2\sin^2\theta}\frac{\partial^2 v_\theta}{\partial \phi^2} + \frac{2}{r^2}\frac{\partial v_r}{\partial \theta} - \frac{v_\theta}{r^2\sin^2\theta} - \frac{2\cos\theta}{r^2\sin^2\theta}\frac{\partial v_\phi}{\partial \phi}\right] + \rho g_\theta \tag{3.3.28b}$$

ϕ direction

$$\rho\left(\frac{\partial v_\phi}{\partial t} + v_r\frac{\partial v_\phi}{\partial r} + \frac{v_\theta}{r}\frac{\partial v_\phi}{\partial \theta} + \frac{v_\phi}{r\sin\theta}\frac{\partial v_\phi}{\partial \phi} + \frac{v_\phi v_r}{r} + \frac{v_\theta v_\phi}{r}\cot\theta\right) = -\frac{1}{r\sin\theta}\frac{\partial p}{\partial \phi} + \mu\left[\frac{1}{r^2}\frac{\partial}{\partial r}\left(r^2\frac{\partial v_\phi}{\partial r}\right)\right.$$
$$\left. + \frac{1}{r^2\sin\theta}\frac{\partial}{\partial \theta}\left(\sin\theta\frac{\partial v_\phi}{\partial \theta}\right) + \frac{1}{r^2\sin^2\theta}\frac{\partial^2 v_\phi}{\partial \phi^2} + \frac{2}{r^2\sin\theta}\frac{\partial v_r}{\partial \phi} - \frac{v_\phi}{r^2\sin^2\theta} + \frac{2\cos\theta}{r^2\sin^2\theta}\frac{\partial v_\theta}{\partial \phi}\right] + \rho g_\phi \tag{3.3.28c}$$

Example 3.2 Here we reexamine the problem analyzed in Section 2.7.3 and shown schematically in Figure 3.6. Consider the steady, fully developed, one-dimensional, laminar flow of an incompressible Newtonian fluid in a circular tube of radius R. In this derivation, the Navier–Stokes equations in cylindrical coordinates, Equations (3.3.27a) through (3.3.27c), are used.

Solution Since the flow is steady, the derivatives with respect to time vanish (i.e., $\partial/\partial t = 0$). The fluid is incompressible and ρ is constant. Assuming that the

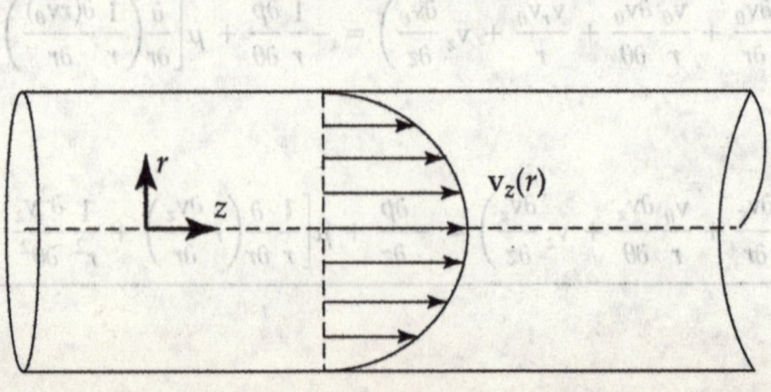

FIGURE 3.6 Laminar flow of an incompressible fluid through a cylindrical tube of radius R for a pressure gradient of $\Delta P/L$.

tube is much longer than the entrance length, flow is one-dimensional for lengths longer than the entrance length. As a result, $v_r = v_\theta = 0$. The conservation of mass for an incompressible fluid (Table 3.1) reduces to

$$\frac{\partial v_z}{\partial z} = 0. \tag{3.3.29}$$

Integrating Equation (3.3.29) indicates that v_z is not a function of z and varies in the radial direction only. Thus, the assumption of a fully developed flow (i.e., that v_z depends only on r) is a consequence of the assumption that the flow is one-dimensional.

With these results, the component of the Navier–Stokes equation for an incompressible Newtonian fluid in cylindrical coordinates simplifies to

$$\rho\left(\overbrace{\frac{\partial v_r}{\partial t}}^{0\text{ steady state}} + \underbrace{v_r\frac{\partial v_r}{\partial r}}_{0,\text{ since } v_r=0} + \frac{v_\theta}{r}\frac{\partial v_r}{\partial \theta} - \underbrace{\frac{v_\theta^2}{r}}_{0,\text{ since } v_\theta=0} + \overbrace{v_z\frac{\partial v_r}{\partial z}}^{0,\text{ since } v_r=0}\right) = -\frac{\partial p}{\partial r} + \mu\left(\underbrace{\frac{\partial}{\partial r}\left(\frac{1}{r}\frac{\partial (r v_r)}{\partial r}\right)}_{0,\text{ since } v_r=0}\right.$$
$$\left. + \underbrace{\frac{1}{r^2}\frac{\partial^2 v_r}{\partial \theta^2}}_{0,\text{ since } v_r=0} - \underbrace{\frac{2}{r^2}\frac{\partial v_\theta}{\partial \theta}}_{0,\text{ since } v_\theta=0} + \underbrace{\frac{\partial^2 v_r}{\partial z^2}}_{0,\text{ since } v_r=0}\right) + \overbrace{\rho g_r}^{0,\text{ since } g=0}. \tag{3.3.30}$$

Because the flow is steady, one-dimensional, and fully developed, Equation (3.3.30) simplifies to

$$\frac{\partial p}{\partial r} = 0 \quad \text{or} \quad p \neq f(r). \tag{3.3.31a,b}$$

In the θ direction, the conservation of linear momentum yields

$$\rho\left(\overbrace{\frac{\partial v_\theta}{\partial t}}^{0\text{ steady state}} + \underbrace{v_r\frac{\partial v_\theta}{\partial r}}_{\substack{0,\text{ since } v_r=0\\ v_\theta=0}} + \underbrace{\frac{v_\theta}{r}\frac{\partial v_\theta}{\partial \theta}}_{0,\text{ since } v_\theta=0} + \underbrace{\frac{v_r v_\theta}{r}}_{\substack{0,\text{ since } v_r=0\\ v_\theta=0}} + \underbrace{v_z\frac{\partial v_\theta}{\partial z}}_{0,\text{ since } v_\theta=0}\right) = -\frac{1}{r}\frac{\partial p}{\partial \theta} + \mu\left(\underbrace{\frac{\partial}{\partial r}\left(\frac{1}{r}\frac{\partial (r v_\theta)}{\partial r}\right)}_{0,\text{ since } v_\theta=0}\right.$$
$$\left. + \underbrace{\frac{1}{r^2}\frac{\partial v_\theta}{\partial \theta}}_{0,\text{ since } v_\theta=0} + \underbrace{\frac{2}{r^2}\frac{\partial v_r}{\partial \theta}}_{0,\text{ since } v_r=0} + \underbrace{\frac{\partial^2 v_\theta}{\partial z^2}}_{0,\text{ since } v_\theta=0}\right) + \overbrace{\rho g_\theta}^{0,\text{ since } g_\theta=0} \tag{3.3.32}$$

which reduces to

$$\frac{\partial p}{\partial \theta} = 0 \quad \text{or} \quad p \neq f(\theta). \tag{3.3.33a,b}$$

In the z direction, the conservation of linear momentum yields

$$\rho\left(\overbrace{\frac{\partial v_z}{\partial t}}^{\text{0, steady state}} + \underbrace{v_r\frac{\partial v_z}{\partial r}}_{\text{0, since } v_r = 0} + \underbrace{\frac{v_\theta}{r}\frac{\partial v_z}{\partial \theta}}_{\text{0, since } v_z(r) \text{ only}, \; v_\theta = 0} + \underbrace{v_z\frac{\partial v_z}{\partial z}}_{\text{0, since } v_z(r) \text{ only}}\right) = -\frac{\partial p}{\partial z} + \mu\left(\frac{1}{r}\frac{\partial}{\partial r}\left(r\frac{\partial v_z}{\partial r}\right)\right.$$

$$\left. + \underbrace{\frac{1}{r^2}\frac{\partial^2 v_z}{\partial \theta^2}}_{\text{0, since } v_z(r) \text{ only}} + \underbrace{\frac{\partial^2 v_z}{\partial z^2}}_{\text{0, since } v_z(r) \text{ only}}\right) + \underbrace{\rho g_z}_{\text{0, since } g_z \text{ negligible}}, \tag{3.3.34}$$

which reduces to

$$\frac{\mu}{r}\frac{d}{dr}\left(r\frac{dv_z}{dr}\right) = \frac{dP}{dz}. \tag{3.3.35}$$

From Equations (3.3.31b) and (3.3.33b), p is a function of z only. This result is identical to Equation (2.7.38), except that the derivative is replaced with $-\Delta p/L$. The solution of Equation (3.3.35) proceeds as discussed earlier.

3.4 Fluid Motion with More Than One Dependent Variable

The flows that we have studied so far have involved a single independent variable. This simplification enabled us to solve several classical problems, such as laminar flow in a cylindrical tube. These flows have important biological applications. However, most flows depend upon more than one independent variable. Some of these problems can be solved analytically, but many require numerical solutions. Here, we consider two examples that can be solved analytically: two-dimensional flow in a rectangular channel and the time that is required to reach a steady state of a flow in a rectangular channel.

3.4.1 Two-Dimensional Flow in a Channel

Flow in rectangular channels is important for many biomedical engineering devices. Applications include flat-plate kidney dialysis units and gas exchangers and parallel-plate flow channels to apply a well-defined fluid force on cultured cells. A complete understanding of fluid dynamics is needed to design a channel that meets the requirements for specific applications. The analysis of the laminar flow of a Newtonian fluid in a narrow channel in Section 2.7.2 assumed that the channel width was much greater than the channel height ($w \gg h$). In reality, the channel width is finite, and the velocity in the direction of flow (the x direction) varies in both the y and z directions due to the presence of solid boundaries. If the width is greater than the height, the change in the velocity profile in the z direction is confined to a region near the sidewalls.

In this analysis, conditions are developed for which we can assume that the velocity profile and the shear stress do not significantly vary with position in the z direction. First we solve the problem for flow in a two-dimensional channel of height h and finite width w (see Figure 3.7). Then we use this solution to determine the effect of the channel width on the velocity profile and the shear stress on the surfaces $y = \pm h/2$.

For the problem at hand, the flow is steady and fully developed (i.e., v_x does not change in the direction of flow). The only nonzero component of the velocity is v_x. Since $v_y = v_z = 0$ and the pressure varies only in the x direction, which is the direction of flow, Equations (3.3.26a) through (3.3.26c) reduce to

$$0 = -\frac{\partial p}{\partial x} + \mu\left(\frac{\partial^2 v_x}{\partial y^2} + \frac{\partial^2 v_x}{\partial z^2}\right). \tag{3.4.1}$$

The boundary conditions are that, on each surface, the velocity is 0; thus,

$$y = \pm h/2 \qquad v_x(\pm h/2, z) = 0, \tag{3.4.2a,b}$$

$$z = \pm w/2 \qquad v_x(y, 0) = v_x(y, w) = 0. \tag{3.4.3a,b}$$

Further, the velocity profiles are symmetric around the centerline plane; hence,

$$y = 0 \qquad \frac{\partial v_x}{\partial y} = 0, \tag{3.4.4}$$

$$z = 0 \qquad \frac{\partial v_x}{\partial z} = 0. \tag{3.4.5}$$

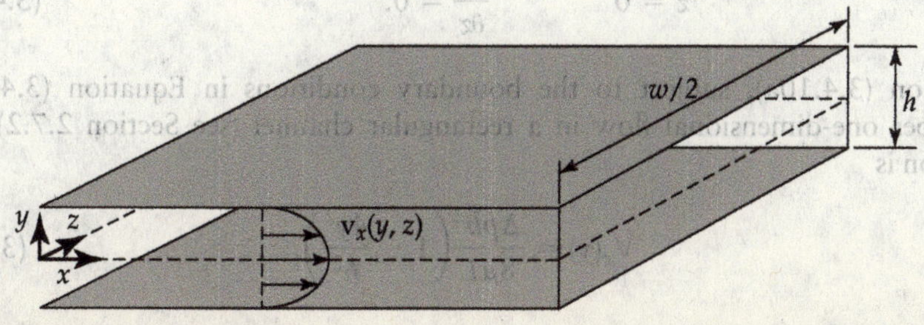

FIGURE 3.7 Flow in a rectangular channel of height h and width w. The section is through the plane $z = 0$.

Because the pressure is a function of distance in the direction of flow (the x-axis), the pressure gradient is constant. The pressure gradient can be defined in terms of the measured pressure drop over a distance L, as follows:

$$-\frac{\partial p}{\partial x} = \frac{\Delta p}{L}. \tag{3.4.6}$$

As a result, Equation (3.4.1) is written as

$$0 = \frac{\Delta p}{L} + \mu\left(\frac{\partial^2 v_x}{\partial y^2} + \frac{\partial^2 v_x}{\partial z^2}\right). \tag{3.4.7}$$

Equation (3.4.7) is nonhomogeneous and cannot be solved with the separation-of-variables method (see Section A.2) without a transformation. Since the velocity depends upon the y-coordinate only when $w \gg h$, the velocity v_x can be written as the sum of two terms, one that depends upon the y-coordinate only and a second that depends upon the y- and z-coordinates:

$$v_x(y, z) = V_x(y) + \phi(y, z). \tag{3.4.8}$$

With this transformation, Equation (3.4.8) can be rewritten as

$$0 = \frac{\Delta p}{L} + \mu\frac{d^2 V_x}{dy^2} + \mu\left(\frac{\partial^2 \phi}{\partial y^2} + \frac{\partial^2 \phi}{\partial z^2}\right). \tag{3.4.9}$$

Equation (3.4.9) can be divided into two equations:

$$0 = \frac{\Delta p}{L} + \mu\frac{d^2 V_x}{dy^2} \tag{3.4.10a}$$

$$0 = \frac{\partial^2 \phi}{\partial y^2} + \frac{\partial^2 \phi}{\partial z^2}. \tag{3.4.10b}$$

Equations (3.4.10a) and (3.4.10b) must satisfy the boundary conditions given by Equations (3.4.2a) and (3.4.2b). Equation (3.4.10b) must also satisfy the boundary conditions given by Equations (3.4.3a) and (3.4.3b). Hence, we have

$$y = \pm h/2 \qquad V_x = 0 \qquad \phi = 0, \tag{3.4.11a}$$

$$z = \pm w/2 \qquad \phi = -V_x, \tag{3.4.11b}$$

$$z = 0 \qquad \frac{\partial \phi}{\partial z} = 0. \tag{3.4.11c}$$

Equation (3.4.10a), subject to the boundary conditions in Equation (3.4.11a), describes one-dimensional flow in a rectangular channel (see Section 2.7.2). The solution is

$$V_x(y) = \frac{\Delta p h^2}{8\mu L}\left(1 - \frac{4y^2}{h^2}\right). \tag{3.4.12}$$

Equation (3.4.10b) is homogeneous (see Section A.2) and can be solved using the separation-of-variables method. The solution can be posited in the form

$$\phi(y, z) = Y(y)Z(z). \tag{3.4.13}$$

Substituting Equation (3.4.13) into Equation (3.4.10b) and rearranging terms yields

$$\frac{1}{Y}\frac{d^2Y}{dy^2} + \frac{1}{Z}\frac{d^2Z}{dz^2} = 0. \tag{3.4.14}$$

Boundary conditions in terms of Y and Z can be stated as follows:

$$y = 0 \qquad \frac{dY}{dy} = 0, \tag{3.4.15a}$$

$$y = \pm h/2 \qquad Y = 0, \tag{3.4.15b}$$

$$z = 0 \qquad \frac{dZ}{dz} = 0. \tag{3.4.15c}$$

The boundary condition at $z = \pm w/2$ cannot be stated in terms of Y or Z alone and is applied in terms of the variable ϕ.

Each of the terms in Equation (3.4.14) must equal a constant if the assumption of the separation of variables is to be applicable. This constant is defined as

$$\lambda^2 = -\frac{1}{Y}\frac{d^2Y}{dy^2} = \frac{1}{Z}\frac{d^2Z}{dz^2}. \tag{3.4.16}$$

The negative sign was chosen for the y-direction term because the boundary conditions are homogeneous in that direction and we expect a periodic solution. For a finite domain, the solutions for Y and Z are

$$Y = A_1\sin(\lambda y) + A_2\cos(\lambda y), \tag{3.4.17a}$$

$$Z = B_1\sinh(\lambda z) + B_2\cosh(\lambda z), \tag{3.4.17b}$$

From the symmetry condition at $y = 0$, we have

$$\frac{dY}{dy} = A_1\lambda\cos(\lambda 0) - A_2\lambda\sin(\lambda 0) = 0. \tag{3.4.18}$$

The constant A_1 must equal zero, since $\cos(0) = 1$. Satisfying the boundary conditions at $y = \pm h/2$ requires that

$$\cos(\lambda h/2) = 0. \tag{3.4.19}$$

(Note that $\cos(-x) = \cos(x)$.) Equation (3.4.19) can be satisfied if $\lambda h/2$ equals $\pi/2, 3\pi/2, \ldots$. Thus, many values of λ can satisfy the solution:

$$\lambda_{n'} = n'\pi/h \quad n' = 1, 3, 5, \ldots. \tag{3.4.20a}$$

An alternative way of writing the values of λ is

$$\lambda_n = (2n + 1)\pi/h \quad n = 0, 1, 2, 3, \ldots \tag{3.4.20b}$$

In the z direction, the boundary condition is

$$\frac{dZ}{dz} = B_1\lambda_n \cosh(\lambda_n 0) + B_2\lambda_n \sinh(\lambda_n 0) = 0. \tag{3.4.21}$$

Since $\cosh(0) = 1$, the constant B_1 equals zero. From these three conditions, the solution $\phi(y, z)$ is

$$\phi = \sum_{n=0}^{\infty} A_n \cosh(\lambda_n y)\cos(\lambda_n y). \tag{3.4.22}$$

At $z = \pm w/2$, we have

$$\phi(y, \pm w/2) = \sum_{n=0}^{\infty} A_n \cosh\left(\frac{\lambda_n w}{2}\right)\cos(\lambda_n y) = -\frac{\Delta p h^2}{8\mu L}\left(1 - \frac{4y^2}{h^2}\right). \tag{3.4.23}$$

The coefficients A_n are obtained by multiplying each side of Equation (3.4.23) by $\cos((2m + 1)\pi y/h)$ and integrating from $y = -h/2$ to $y = h/2$ (see Equations (A.2.11) and (A.2.12)),

$$-\int_{-h/2}^{h/2} \frac{\Delta p h^2}{8\mu L}\left(1 - \frac{4y^2}{h^2}\right)\cos\left(\frac{(2m + 1)\pi y}{h}\right)dy =$$

$$\sum_{n=0}^{\infty} A_n \int_{-h/2}^{h/2} \cosh(\lambda_n w)\cos\left(\frac{(2n + 1)\pi y}{h}\right)\cos\left(\frac{(2m + 1)\pi y}{h}\right)dy. \tag{3.4.24}$$

Only one integral with $n = m$ on the right-hand side of Equation (3.4.24) is nonzero. Thus, the coefficients A_n are determined as

$$A_n = -\frac{\dfrac{\Delta p h^2}{8\mu L}\displaystyle\int_{-h/2}^{h/2} (1 - 4y^2/h^2)\cos\big((2n + 1)\pi y/h\big)dy}{\cosh\big((2n + 1)\pi w/2h\big)\displaystyle\int_{-h/2}^{h/2} \cos^2\big((2n + 1)\pi y/h\big)dy} \quad n = 0, 1, 2, \ldots \tag{3.4.25}$$

or

$$A_n = -\frac{(-1)^n(\Delta p h^2/8\mu L)(32/(2n + 1)^3\pi^3)}{\cosh\big((2n + 1)\pi w/2h\big)} \quad n = 0, 1, 2, \ldots \tag{3.4.26}$$

and the solution for the velocity field is

$$v_x(y, z) = \frac{\Delta p h^2}{8\mu L}\left(1 - \frac{4y^2}{h^2}\right) - \frac{\Delta p h^2}{8\mu L}\sum_{n=0}^{\infty}\frac{32(-1)^n \cosh\big((2n + 1)\pi z/h\big)\cos\big((2n + 1)\pi y/h\big)}{(2n + 1)^3\pi^3\cosh\big((2n + 1)\pi w/2h\big)}. \tag{3.4.27}$$

This result can be used to determine conditions for which the channel will be sufficiently wide for flow to be one-dimensional. As the ratio of channel width to height increases, the velocity profile approaches the parabolic profile for an infinitely wide channel (see Figure 3.8). At the centerline, the velocity for $w/h = 5$ is indistinguishable from the parabolic profile (see Figure 3.8a). However, closer to $z = \pm w/2$, even a ratio of $w/h = 10$ gives a velocity that lies below the value for an infinitely wide channel. Thus, for z near $\pm w/2$, a very large ratio of w/h is needed to approach the velocity for an infinitely wide channel.

There are two ways to calculate the shear stress acting upon the surfaces $y = \pm h/2$. In one approach, the average shear stress on the surface at $y = -h/2$ can be found by differentiating the velocity field in the y direction and integrating in the z direction (note that $\sinh(-x) = -\sinh(x)$)

$$\tau_{yx} = \frac{\mu}{w}\int_{w/2}^{w/2}\left(\frac{\partial v_x}{\partial y}\bigg|_{y=-h/2}\right)dz$$

$$= \frac{\Delta p h}{2L}\left[1 - 16\left(\frac{h}{w}\right)\sum_{n=0}^{\infty}\frac{(-1)^n \tanh\big((2n+1)\pi w/2h\big)}{(2n+1)^3\pi^3}\right]. \quad (3.4.28)$$

Using this result, we find that the wall shear stress is equal to 95% of the value for an infinitely wide plate when $w/h > 10$ (see Figure 3.9) and 99% of the value for an infinitely wide plate when $w/h = 50$.

Alternatively, the shear stress is computed from the flow rate (see Problem 3.4). Since $Q = \langle v_x \rangle wh = (2/3)v_{max}wh$ for an infinitely wide plate (Equation (2.7.26)), the shear stress acting upon the surface $y = -h/2$ is

$$\tau_w = \frac{6\mu Q}{wh^2}. \quad (3.4.29)$$

The agreement between the two calculations for the w/h ratio required for the shear stress to be within 1% of the value for an infinitely wide plate is very good.

3.4.2 Time Required to Establish a Steady Flow in a Rectangular Channel

In Section 2.7.2, we considered the steady, pressure-driven flow between two parallel plates that are separated by a narrow gap h that is much smaller than the channel width ($h \ll w$). Flow is often examined at steady state, for which the velocity and shear stress do not change with time. However, it may be important to determine the time required for the flow to change from rest to a new steady-state velocity when a pressure gradient is applied. Assuming that the flow is laminar, v_x is a function of y and t only and the channel is horizontal, Equations (3.3.26a) through (3.3.26c) reduce to

$$\rho\frac{\partial v_x}{\partial t} = -\frac{\partial p}{\partial x} + \mu\frac{\partial^2 v_x}{\partial y^2}. \quad (3.4.30)$$

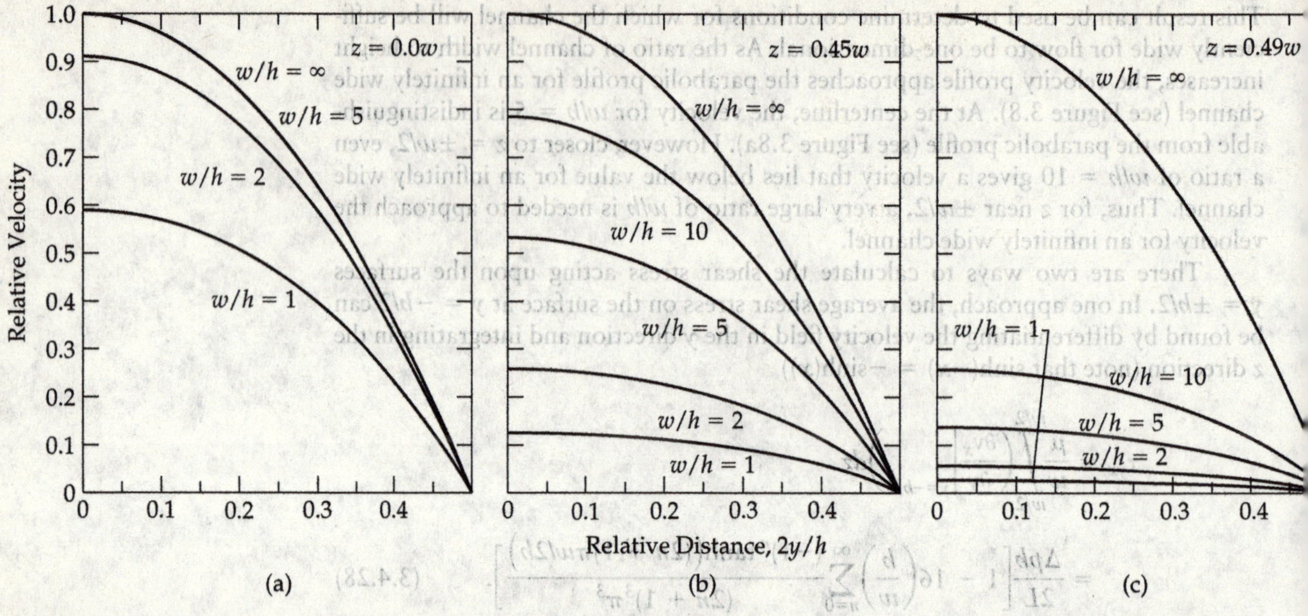

FIGURE 3.8 Velocity profiles in the y direction as a function of the ratio w/h and distance in the z direction normal to the directi of flow.

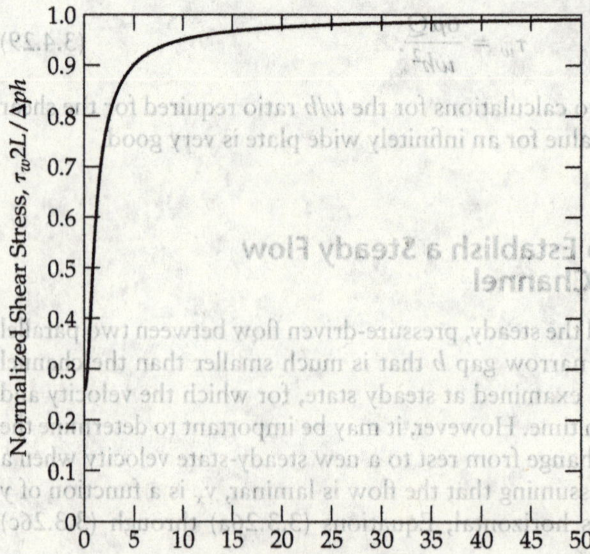

FIGURE 3.9 Effect of the channel width on the ratio of the shear stress at $y = -h/2$ to the value calculated for an infinitely wide plate.

The coordinate axes are centered about the midplane between the two plates. The boundary conditions are

$$y = \pm h/2 \qquad v_x = 0, \tag{3.4.31a}$$

$$y = 0 \qquad \frac{\partial v_x}{\partial y} = 0. \tag{3.4.31b}$$

The initial condition is

$$t = 0 \qquad v_x = 0. \tag{3.4.32}$$

The pressure varies only in the x direction, and the pressure gradient equals $-\Delta p/L$. As a result, Equation (3.4.30) is not homogeneous. To solve it, note that a steady-state solution exists, Equation (2.7.22). So the velocity can be divided into two functions—one that represents the steady-state velocity V_{ss} and the other, $\psi(y, t)$, that depends upon time and position:

$$v_x = V_{ss} + \psi(y, t). \tag{3.4.33}$$

Substituting for the pressure gradient and v_x yields

$$\rho\frac{\partial\psi}{\partial t} = \frac{\Delta p}{L} + \mu\frac{d^2V_{ss}}{dx^2} + \mu\frac{\partial^2\psi}{\partial y^2}. \tag{3.4.34}$$

As a result, Equation (3.4.34) is divided into two separate equations. For V_{ss}, we have

$$0 = \frac{\Delta p}{L} + \mu\frac{d^2V_{ss}}{dx^2}. \tag{3.4.35}$$

The boundary conditions are obtained by setting terms with V_{ss} and $\psi(y, t)$ to 0:

$$y = \pm h/2 \qquad V_{ss} = 0 \tag{3.4.36a}$$

$$y = 0 \qquad \frac{\partial V_{ss}}{\partial y} = 0. \tag{3.4.36b}$$

The solution of Equation (3.4.35), subject to the boundary conditions in Equations (3.4.36a) and (3.4.36b), was obtained in Section 2.7.2 (see also Equation (3.4.12)).

For the time-dependent component of the velocity, we have the following homogeneous partial differential equation:

$$\rho\frac{\partial\psi}{\partial t} = \mu\frac{\partial^2\psi}{\partial y^2}. \tag{3.4.37}$$

The boundary conditions and initial condition are

$$y = 0 \qquad \frac{\partial\psi}{\partial y} = 0, \tag{3.4.38a}$$

$$y = \pm h/2 \qquad \psi = 0, \tag{3.4.38b}$$

$$t = 0 \qquad \psi = -V_{ss}, \tag{3.4.38c}$$

As $t \to \infty$, $\psi = 0$. To solve Equation (3.4.37), express ψ as a product of functions that depend on y alone and t alone:

$$\psi = Y(y)T(t). \tag{3.4.39}$$

With these variables and the definition of the kinematic viscosity, $\nu = \mu/\rho$, Equation (3.4.37) becomes

$$\frac{1}{\nu T}\frac{dT}{dt} = \frac{1}{Y}\frac{d^2Y}{dy^2}. \tag{3.4.40a}$$

Because T depends upon time only and Y depends upon the y-coordinate only, the two terms in Equation (3.4.40a) must equal a constant. We choose a negative sign in anticipation that the y direction has homogeneous boundary conditions:

$$\frac{1}{\nu T}\frac{dT}{dt} = \frac{1}{Y}\frac{d^2Y}{dy^2} = -\lambda^2. \tag{3.4.40b}$$

The solutions for T and Y are

$$T = A\exp(-\lambda^2 \nu t), \tag{3.4.41a}$$

$$Y = B_1\sin(\lambda y) + B_2\cos(\lambda y). \tag{3.4.41b}$$

Boundary conditions for Y are

$$y = 0 \qquad \frac{dY}{dy} = 0, \tag{3.4.42a}$$

$$y = \pm h/2 \quad Y = 0. \tag{3.4.42b}$$

As in the last problem, the symmetry condition implies that $B_1 = 0$. Likewise, satisfying the boundary conditions at $y = \pm h/2$ requires that $\lambda_n = (2n + 1)\pi/h$. Thus, ψ can be written as

$$\psi = \sum_{n=0}^{\infty} C_n \exp(-\lambda_n^2 \nu t)\cos(\lambda_n y). \tag{3.4.43}$$

Orthogonality is used in applying the initial condition. The coefficients are

$$C_n = -\frac{\dfrac{\Delta p h^2}{8\mu L}\displaystyle\int_{-h/2}^{h/2}(1 - 4y^2/h^2)\cos\big((2n+1)\pi y/h\big)dy}{\displaystyle\int_{-h/2}^{h/2}\cos^2\left(\frac{(2n+1)\pi y}{h}\right)dy}$$

$$= -(-1)^n\frac{\Delta p h^2}{8\mu L}\left(\frac{32}{(2n+1)^3\pi^3}\right) \qquad n = 0,1,2,\ldots. \tag{3.4.44}$$

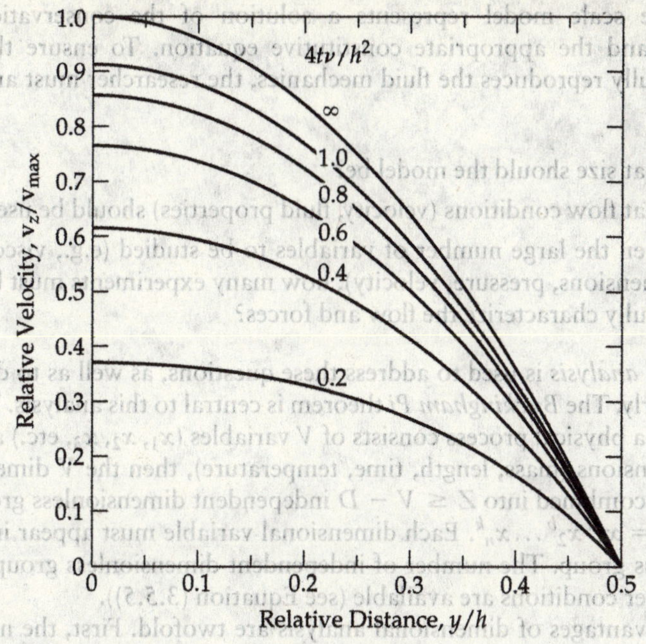

FIGURE 3.10 Relative velocity profile as a function of time after the initiation of flow in a rectangular channel of width w and height h.

The velocity profile is

$$v_x(y,t) = \frac{\Delta p h^2}{8\mu L}\left(1 - \frac{4y^2}{h^2}\right) - \frac{\Delta p h^2}{8\mu L}\sum_{n=0}^{\infty}\frac{32(-1)^n \exp\left(-(2n+1)^2\pi^2 \nu t/h^2\right)\cos\left((2n+1)\pi y/h\right)}{(2n+1)^3\pi^3}. \quad (3.4.45)$$

Notice that, for this problem, the time is normalized by $h^2/4\nu$. For short times after the start of flow, the velocity profile around the centerline is relatively flat (see Figure 3.10). As this normalized time increases, the velocity approaches its steady-state value. The velocity profile approaches within 1% of the steady-state value when $\nu t/h^2 = 5$.

3.5 Dimensional Analysis and Dimensionless Groups

3.5.1 Dimensional Analysis

Often, a problem in fluid dynamics or mass transfer is too complicated to be readily solved by the conservation and constitutive equations. An alternative approach is to perform experiments on scale models. An example of this approach is the testing of scale models of flow in blood vessels or the lungs. When the experiment is performed properly, the scale model represents a solution of the conservation of linear momentum and the appropriate constitutive equation. To ensure that the scale model faithfully reproduces the fluid mechanics, the researcher must answer several questions:

1. What size should the model be?
2. What flow conditions (velocity, fluid properties) should be used?
3. Given the large number of variables to be studied (e.g., viscosity, density, dimensions, pressure, velocity), how many experiments must be performed to fully characterize the flow and forces?

Dimensional analysis is used to address these questions, as well as to design experiments properly. The *Buckingham Pi* theorem is central to this analysis. This theorem states that if a physical process consists of V variables (x_1, x_2, x_3, etc.) and D fundamental dimensions (mass, length, time, temperature), then the V dimensional variables can be combined into $Z \leq V - D$ independent dimensionless groups, each of the form $Z_i = x_1{}^a x_2{}^b \ldots x_n{}^k$. Each dimensional variable must appear in at least one dimensionless group. The number of independent dimensionless groups can be less than Z if other conditions are available (see Equation (3.5.5)).

The advantages of dimensional analysis are twofold. First, the net number of variables to be studied is reduced. Second, if the dimensionless variables are the same for two different systems of the same dimensions, then the flow is the same for the two different systems. One limitation of dimensional analysis is that the dimensionless groups are not unique. This limitation is alleviated, to some extent, by several conventions that have been adopted.

As an application of dimensional analysis, consider the steady flow of a Newtonian fluid in a cylinder of inner radius R and length L. We wish to find the relationship between the pressure drop Δp and the other remaining variables: viscosity (μ), density (ρ), length (L), diameter (D), and average velocity ($\langle \mathrm{v} \rangle$). This relationship can be stated functionally as

$$\Delta p = f(\rho, \mu, L, D, \langle \mathrm{v} \rangle). \tag{3.5.1}$$

Previously, we solved this problem for laminar flow. By substituting Equation (2.7.44) into Equation (2.7.42) and using the relation between the radius and diameter, we have

$$\Delta p = \frac{32\mu \langle \mathrm{v} \rangle L}{D^2}. \tag{3.5.2}$$

Equation (3.5.2) is valid only for the laminar flow of a Newtonian fluid and is therefore not a universal relation. No analytical solution has been obtained that is valid for all flow conditions.

For the general case of flow of a Newtonian fluid in a cylindrical tube, there are six dimensional variables and three fundamental dimensions (mass, length, and time). According to the Buckingham Pi theorem, there are then three dimensionless groups. One is L/D, the ratio of the length scales. The other two groups are less obvious. One hint is to note that Δp has units of force per unit area (or mass length^{-1}

time^{-2}). Two other groups that have similar units are $\rho\langle \mathrm{v}\rangle^2$ and $\mu\langle \mathrm{v}\rangle/D$. By convention, the dimensionless grouping $\rho\langle \mathrm{v}\rangle D/\mu$ is known as the *Reynolds number* (Re) and represents the ratio of inertial ($\rho\langle \mathrm{v}\rangle^2$) to viscous ($\mu\langle \mathrm{v}\rangle/D$) forces per unit area (see Chapter 2). This is the second dimensionless group. The third grouping involves the pressure term and depends upon the type of flow. For most applications, the Reynolds number is greater than unity and inertial forces are highly significant. As a result, Δp is most appropriately scaled by the inertial force per unit area ($\rho\langle \mathrm{v}\rangle^2$). The dimensionless form of Equation (3.5.1) is

$$\frac{\Delta p}{\rho\langle \mathrm{v}\rangle^2} = g\left(\frac{\rho\langle \mathrm{v}\rangle D}{\mu}, \frac{L}{D}\right). \tag{3.5.3}$$

In order to apply the Buckingham Pi theorem to this problem, we first choose a *basis group* consisting of D dimensional variables. Each variable is raised to a power. Then we multiply each remaining dimensional variable by the basis group. For each group, the exponents are adjusted to create a dimensionless group. Each dimensional variable must be used at least once. There is no unique basis group; however, the dependent variable in the functional relation is usually not included in the basis group, because we want a specific relation for the dependent variable.

To determine the dimensionless groups that arise from Equation (3.5.1), choose the basis group $\langle \mathrm{v}\rangle^a\rho^bD^c$. To obtain a dimensionless pressure, multiply pressure by the basis group:

$$\Delta p\langle \mathrm{v}\rangle^a\rho^bD^c. \tag{3.5.4a}$$

In terms of the units (mass m, length l, and time t), we have

$$(\mathrm{ml^{-1}t^{-2}})(\mathrm{lt^{-1}})^a(\mathrm{ml^{-3}})^b(\mathrm{l})^c. \tag{3.5.4b}$$

For each of the dimensions, the exponents must sum to zero to yield a dimensionless group. For mass,

$$0 = 1 + b. \tag{3.5.4c}$$

Correspondingly, the expressions for length and time are

$$0 = -1 + a - 3b + c, \tag{3.5.4d}$$

$$0 = -2 - a. \tag{3.5.4e}$$

Solving Equations (3.5.4c) through (3.5.4e), we find that $a = -2$, $b = -1$, and $c - 0$. The resulting dimensionless group is thus $\Delta p/\rho\langle \mathrm{v}\rangle^2$. We can repeat this process to identify the other dimensionless groups.

We can simplify Equation (3.5.3) further and deduce the functional dependence upon L. Holding all other variables fixed, if the length of the tube is doubled, then we expect that the pressure drop will also double. Alternatively, for two tubes of different lengths, but with all other parameters held fixed, we expect the pressure drop per unit length, $\Delta p/L$, to be fixed. Thus, Δp is proportional to L. Because L appears in only one dimensional grouping in Equation (3.5.3), Δp is proportional to L/D, and it follows that

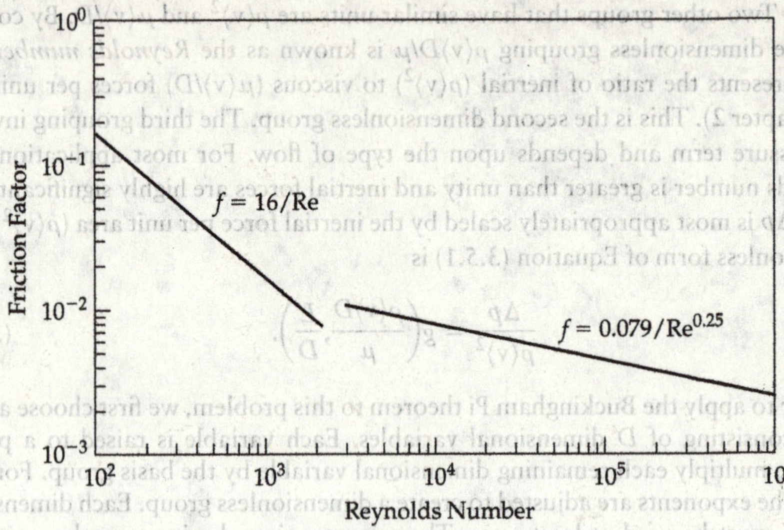

FIGURE 3.11 The Fanning friction factor versus the Reynolds number.

$$\frac{\Delta p}{\rho\langle v\rangle^2} = \frac{L}{D} g\left(\frac{\rho\langle v\rangle D}{\mu}\right). \tag{3.5.5}$$

A dimensionless group that is commonly used in the engineering literature to characterize the pressure drop is the *Fanning friction factor*, defined as [3,4]

$$f = \frac{\Delta p}{2\rho\langle v\rangle^2}\frac{D}{L}. \tag{3.5.6}$$

From Equations (3.5.5) and (3.5.6), we see that the Fanning friction factor depends only upon the Reynolds number (see Figure 3.11). For laminar flow, $f = 16/\text{Re}$ (Re < 2,100 for pipe flow). For turbulent flow (Re > 4,300), the friction factor depends also upon the roughness of the pipes.

The advantages of dimensional analysis are apparent by comparing Equations (3.5.1) and (3.5.5). The pressure drop given in Equation (3.5.1) depends upon five variables. If 10 different values of each variable were obtained, a total of 10^5 measurements would be needed. Further, many different graphs would be needed to present the data. The use of Equation (3.5.6) reduces the number of dependent variables to one, and the results can be shown on a single graph. Although application of the Buckingham Pi theorem may seem a bit arcane, it has proven valuable in studying many problems in transport theory.

3.5.2 Dimensionless Form of the Navier–Stokes Equation

Mathematically, the independent and dependent variables in the equations of motion can be transformed into dimensionless counterparts via division by parameters of the same dimension. Those parameters, referred to as *characteristic values*, typically are chosen such that the magnitudes of the new dimensionless variables are never greater than about unity. The resulting dimensionless equations can often be solved more easily; doing so involves inspecting the magnitudes of the various terms and neglecting those terms which are multiplied by coefficients (themselves dimensionless), the magnitudes of which are small. In Chapter 4, several examples of this process are presented. Here, the mathematical exercise of creating dimensionless variables and equations is introduced. A direct consequence of this exercise is the observation that flow problems with similar geometries, but different dimensions, may, nonetheless, exhibit *dynamic similarity*, in which a single mathematical solution applies to the problems. This similarity relates to the concepts of dimensional analysis, presented in the previous section.

We now generalize the foregoing results by casting the Navier–Stokes equation (3.3.25) into dimensionless form. We will show this explicitly for the x component of the two-dimensional form of the equation. The results can be generalized to three dimensions for arbitrary coordinate systems. In dimensional form, Equation (3.3.26a) is

$$\rho\left(\frac{\partial v_x}{\partial t} + v_x\frac{\partial v_x}{\partial x} + v_y\frac{\partial v_x}{\partial y}\right) = -\frac{\partial p}{\partial x} + \mu\left(\frac{\partial^2 v_x}{\partial x^2} + \frac{\partial^2 v_x}{\partial y^2}\right) + \rho g_x. \quad (3.5.7)$$

In this example, we assume that the characteristic lengths and velocities in the x and y directions are the same. This restriction is not absolutely necessary but simplifies the presentation. In Chapter 4, we present examples that use different length and velocity scales in the x and y directions. Thus, the characteristic length is L. The characteristic velocity is the average velocity $\langle v_x\rangle$, and the characteristic pressure is $\rho\langle v_x\rangle^2$. Time is made dimensionless by a characteristic time T. We define the following dimensionless variables:

$$\begin{aligned} x^* &= x/L & y^* &= y/L,\\ v_x^* &= v_x/\langle v_x\rangle & v_y^* &= v_y/\langle v_y\rangle,\\ p^* &= p/\rho\langle v_x\rangle^2 & g_x^* &= g_x/g,\\ t^* &= t/T \end{aligned}$$

The derivatives are made dimensionless as follows:

$$\frac{\partial v_x}{\partial x} = \frac{\partial v_x^*\langle v_x\rangle}{\partial(x^*L)} = \frac{\langle v_x\rangle}{L}\frac{\partial v_x^*}{\partial x^*}. \quad (3.5.8a)$$

A similar relation is written for the derivatives with respect to y. For the second derivatives, we have

$$\frac{\partial^2 v_x}{\partial x^2} = \frac{\partial^2 v_x^*\langle v_x\rangle}{\partial(x^*L)^2} = \frac{\langle v_x\rangle}{L^2}\frac{\partial^2 v_x^*}{\partial x^{*2}}. \quad (3.5.8b)$$

With these definitions of the dimensionless variables, Equation (3.5.7) becomes

$$\frac{\rho\langle v_x\rangle^2}{L}\left(\frac{L}{\langle v_x\rangle T}\frac{\partial v_x^*}{\partial t^*} + v_{x^*}\frac{\partial v_x^*}{\partial x^*} + v_{y^*}\frac{\partial v_x^*}{\partial y^*}\right)$$
$$= -\frac{\rho\langle v_x\rangle^2}{L}\frac{\partial p^*}{\partial x^*} + \frac{\mu\langle v\rangle}{L^2}\left(\frac{\partial^2 v_x^*}{\partial x^{*2}} + \frac{\partial^2 v_x^*}{\partial y^{*2}}\right) + \rho g_{x^*}. \quad (3.5.9)$$

Multiplying both sides of Equation (3.5.9) by $L/\rho\langle v_x\rangle^2$ yields

$$\left(\mathrm{St}\frac{\partial v_x^*}{\partial t^*} + v_x^*\frac{\partial v_x^*}{\partial x^*} + v_y^*\frac{\partial v_x^*}{\partial y^*}\right) = -\frac{\partial p^*}{\partial x^*} + \frac{1}{\mathrm{Re}}\left(\frac{\partial^2 v_x^*}{\partial x^{*2}} + \frac{\partial^2 v_x^*}{\partial y^{*2}}\right) + \frac{1}{\mathrm{Fr}}g_x^*, \quad (3.5.10)$$

where $\mathrm{St} = L/T\langle v_x\rangle$ is the *Strouhal number*, $\mathrm{Re} = \rho\langle v_x\rangle L/\mu$ is the Reynolds number, and $\mathrm{Fr} = \langle v_x\rangle^2/gL$ is the *Froude number*. The Strouhal number is the ratio of the characteristic time that is required to move the fluid through a distance L to the characteristic time of flow. The Froude number is the ratio of inertial forces ($\rho\langle v_x\rangle^2$) to gravitational forces (ρgL) and is important for *free-surface* flows (e.g., with an air–liquid interface). Because the Froude number is unimportant when free surfaces are not present, it is not often used in problems of human physiology. It can, however, be important in problems concerning medical devices with air–liquid interfaces and high Reynolds numbers. Note that Equation (3.5.10) can be written in three dimensions as

$$\mathrm{St}\frac{\partial \mathbf{v}^*}{\partial t^*} + \mathbf{v}^* \cdot \nabla^*\mathbf{v}^* = -\nabla^* p^* + \frac{1}{\mathrm{Re}}\nabla^{*2}\mathbf{v}^* + \frac{1}{\mathrm{Fr}}\mathbf{g}^*. \quad (3.5.11)$$

In cardiovascular blood flow, the characteristic time is the reciprocal of the frequency ω of the heartbeat. This time scale is used because the velocity profile repeats itself after every heartbeat. Instead of the Strouhal number, a related parameter, known as the *Womersley number* $\alpha = \sqrt{\omega R^2/v}$, is typically used. The quantity R^2/v represents the characteristic time for viscous forces to change in the radial direction in a circular tube and ω is the frequency in radians ($\omega = 2\pi f$ where f is beats per second). Typical values of α in the arterial system range from 2 to 10. The significance of the Womersley number upon arterial flows is examined in Chapter 5.

A two-dimensional flow can be completely characterized in terms of the three dimensionless groups St, Re, and Fr. Thus, if two different physical situations have the same values for these three groups and all boundary and initial conditions are the same, then they have the same solution of Equation (3.5.10).

3.5.3 Dimensional Analysis and Dynamic Similarity

Two criteria must be met to perform experiments properly with scale models. The first criterion, known as *dimensional similarity*, is that all dimensions of the scale and original models must be in the same ratio. For example, if the body of a scale model of an aircraft is one-hundredth the original length, then all other dimensions must be reduced by a factor of 100.

The second criterion, known as *dynamic similarity*, is that the value of each relevant dimensionless group must be the same for the model and the original. Thus, if the Reynolds number and Froude number are important, then we must have $\mathrm{Re}_o = \mathrm{Re}_s$ and $\mathrm{Fr}_o = \mathrm{Fr}_s$, where the subscript "$o$" represents the original model and "s" represents the scale model.

Example 3.3 A laboratory model of steady flow in the coronary arteries is being developed. All dimensions are three times larger than the original dimensions. The working fluid is a glycerol–water mixture that has the same physical properties as blood, a density of 1.1 g cm^{-3}, and a viscosity of 0.03 g cm^{-1} s^{-1}. Determine the velocity of the scale model relative to that of the original model.

Solution Because the flow is steady, the Strouhal number (or the Womersley parameter) is not relevant. To maintain dynamic similarity, we must have $Re_o = Re_s$, or

$$\frac{\rho_o \langle v_o \rangle D_o}{\mu_o} = \frac{\rho_s \langle v_s \rangle D_s}{\mu_s}. \tag{3.5.12}$$

Since $\rho_o = \rho_s$, $\mu_o = \mu_s$, and $D_s = 3D_o$, we have

$$\langle v_s \rangle = \frac{\langle v_o \rangle}{3}. \tag{3.5.13}$$

Example 3.4 Now consider unsteady, oscillatory flow in a coronary artery. If the frequencies of the scale model and the original are the same, can the fluid from Example 3.3 be used?

Solution In addition to requiring that the Reynolds numbers be the same, we also require that the Wormersley parameters be the same, or

$$3R_o\sqrt{\frac{\omega_s}{\nu_s}} = R_o\sqrt{\frac{\omega_o}{\nu_o}}. \tag{3.5.14}$$

Rearranging Equation (3.5.14) indicates that the ratio of kinematic viscosities is

$$\frac{\nu_s}{\nu_o} = 9. \tag{3.5.15}$$

Thus the fluid does not have the same properties as blood.

Substituting Equation (3.5.15) and $D_s = 3D_o$ into Equation (3.5.12) yields the following relation for the velocities:

$$\langle v_s \rangle = 3\langle v_o \rangle. \tag{3.5.16}$$

Example 3.5 Determine the fluid requirements for a $\frac{1}{5}$ scale model used to examine the fluid dynamics of flow down an inclined plane.

Solution During flow down an inclined plane, the liquid is in contact with air. For this case, we must meet two requirements: $Re_o = Re_s$ and $Fr_o = Fr_s$. In terms of dimensional variables, we have

$$\frac{\rho_o \langle v_o \rangle L_o}{\mu_o} = \frac{\rho_s \langle v_s \rangle L_o}{5\mu_s}, \tag{3.5.17}$$

$$\frac{\langle v_o \rangle^2}{L_o} = \frac{5\langle v_s \rangle^2}{L_o}. \tag{3.5.18}$$

Solving for μ_s yields

$$\mu_s = 0.089 \frac{\mu_o}{\rho_o} \rho_s. \tag{3.5.19}$$

Thus, the fluid used for the scale model must have a lower viscosity than the fluid used in the actual physical situation.

Dimensional analysis is also used to correlate results from experiments and simulations in a compact way. One physiological application of dimensional analysis is the calculation of the additional pressure drop produced by leukocytes that adhere to the surface of endothelial cells [5]. Leukocytes are a class of blood cells that are involved in the body's response to injury and infection. In the presence of a stimulus, such as an infection or a wound, endothelial cells express specific receptors for leukocytes. The leukocytes have counter-receptors that bind to these receptors, causing the leukocytes to roll and arrest on the endothelial surface (see Figure 3.12).

The motion of the blood exerts a drag force on the adherent leukocyte. Because the leukocyte diameter d is comparable to the venule diameter D, there is an increase in the pressure drop compared with that for flow in a cylindrical tube without the adherent leukocyte. The leukocyte obstructs a region of the tube, as shown in Figure 3.12. A larger pressure drop is then required to maintain the same flow rate. The following correlation was derived from experiments and computational fluid dynamics for adherent leukocytes in a cylindrical tube [5]:

$$f^* = \frac{1}{\mathrm{Re}} \exp\left[2.877 + 4.63\left(\frac{d}{D}\right)^4\right]. \tag{3.5.20}$$

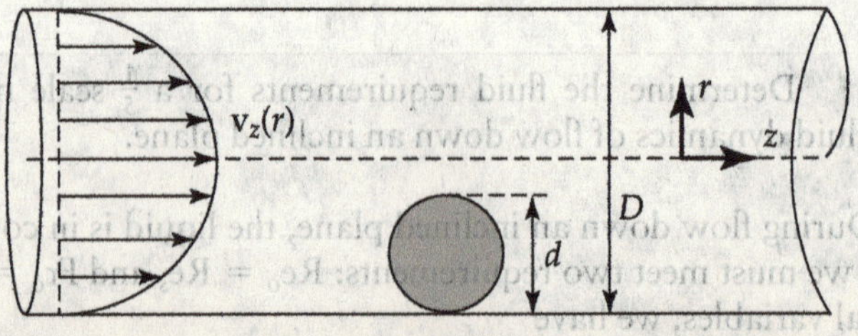

FIGURE 3.12 Schematic of a leukocyte adhering to endothelium in a venule.

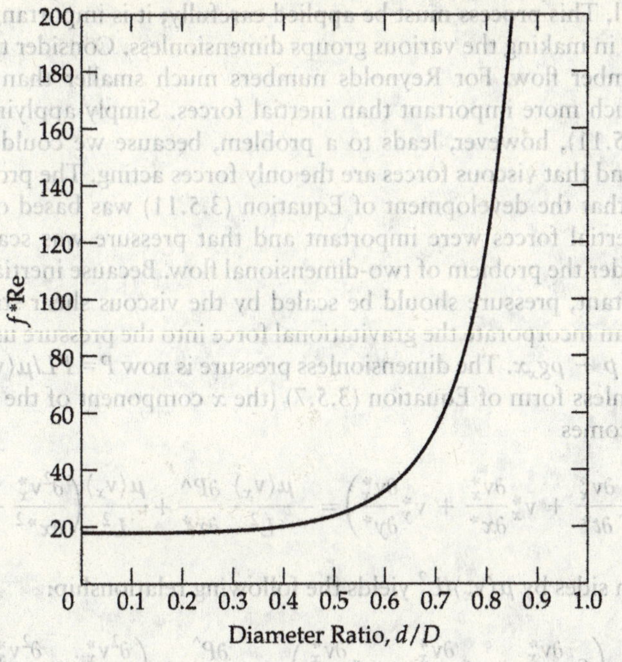

FIGURE 3.13 Friction factor (Equation (3.5.20)) for flow through a cylindrical tube of diameter D with an adherent leukocyte of diameter d.

In this equation, d is the leukocyte diameter, D is the venule diameter, and $\mathrm{Re} = \rho\langle \mathrm{v}\rangle D/\mu$. The parameter

$$f^* = \frac{\Delta p'}{2\rho\langle \mathrm{v}\rangle^2}\frac{D}{L_e} \tag{3.5.21}$$

is a modified form of the friction factor, where p'/L_e is the pressure drop per unit length due to the adherent leukocyte.

When the leukocyte diameter is much less than the blood vessel diameter, the modified friction factor given by Equation (3.5.20) approaches 16/Re, the limit for laminar flow. The small deviation may reflect experimental error. As the ratio of the leukocyte diameter to the blood vessel diameter increases, the friction factor increases, indicating a greater resistance to flow, even at the same Reynolds number. By plotting f^*Re against the ratio of the leukocyte diameter to the blood vessel diameter, we can clearly discern the effect of the diameter ratio (see Figure 3.13). The increased flow resistance, which is due to an adherent leukocyte, can be quite significant when the leukocyte diameter is more than one-half the diameter of the blood vessel.

3.6 Low-Reynolds-Number Flow

3.6.1 Conservation Relations for Low-Reynolds-Number Flow

One application of the dimensionless form of the conservation relations is the simplification of these equations that can result if one or more of the dimensionless groups are very small. This process must be applied carefully; it is important to use appropriate scaling in making the various groups dimensionless. Consider the case of low-Reynolds-number flow. For Reynolds numbers much smaller than unity, viscous forces are much more important than inertial forces. Simply applying this logic to Equation (3.5.11), however, leads to a problem, because we could conclude that $\nabla^{*2}\mathbf{v}^* = 0$ and that viscous forces are the only forces acting. The problem with this reasoning is that the development of Equation (3.5.11) was based on the assumptions that inertial forces were important and that pressure was scaled by inertial forces. Consider the problem of two-dimensional flow. Because inertial forces are no longer important, pressure should be scaled by the viscous shear stresses $\mu\langle v_x\rangle/L$. Further, we can incorporate the gravitational force into the pressure using a modified pressure $P = p - \rho g_x x$. The dimensionless pressure is now $P = PL/\mu\langle v_x\rangle$. As a result, the dimensionless form of Equation (3.5.7) (the x component of the Navier–Stokes equation) becomes

$$\frac{\rho\langle v_x\rangle^2}{L}\left(\frac{L}{\langle v_x\rangle T}\frac{\partial v_x^*}{\partial t^*} + v_x^*\frac{\partial v_x^*}{\partial x^*} + v_y^*\frac{\partial v_x^*}{\partial y^*}\right) = -\frac{\mu\langle v_x\rangle}{L^2}\frac{\partial P^\wedge}{\partial x^*} + \frac{\mu\langle v_x\rangle}{L^2}\left(\frac{\partial^2 v_x^*}{\partial x^{*2}} + \frac{\partial^2 v_x^*}{\partial y^{*2}}\right) \quad (3.6.1)$$

Dividing both sides by $\mu\langle v_x\rangle/L^2$ yields the following relationship:

$$\mathrm{Re}\left(\mathrm{St}\frac{\partial v_x^*}{\partial t^*} + v_x^*\frac{\partial v_x^*}{\partial x^*} + v_y^*\frac{\partial v_x^*}{\partial y^*}\right) = -\frac{\partial P^\wedge}{\partial x^*} + \left(\frac{\partial^2 v_x^*}{\partial x^{*2}} + \frac{\partial^2 v_x^*}{\partial y^{*2}}\right). \quad (3.6.2)$$

If inertial forces are negligible, then Re and Re·St are much less than unity and the left-hand side of Equation (3.6.2) vanishes. Equation (3.6.2) then reduces to

$$0 = -\frac{\partial P^\wedge}{\partial x^*} + \left(\frac{\partial^2 v_x^*}{\partial x^{*2}} + \frac{\partial^2 v_x^*}{\partial y^{*2}}\right). \quad (3.6.3a)$$

We now rearrange and use vector notation to generalize to three-dimensional flow in any coordinate system. The result is the relation

$$\nabla P = \mu\nabla^2\mathbf{v}, \quad (3.6.3b)$$

where ∇^2 is the Laplacian operator (see Equation (A.3.11)). Equation (3.6.3b) is often referred to as the *Stokes equation*.

Equation (3.6.3b) indicates that in the limit of no inertial forces, the viscous stresses balance the pressure term. This means that at very low Reynolds numbers, the flow responds instantaneously to a change in conditions (e.g., a pressure drop in the flow rate). Such a flow is known as a *quasi-steady flow*.[2] Detailed calculations indicate that the assumption of low-Reynolds-number flow is valid for Re less than 0.1 [6].

Flows with low Reynolds numbers are also known as *creeping flows* or *Stokes flows*. Low Reynolds numbers result from low fluid velocities, small values of the characteristic dimension, and/or a very large viscosity. Stokes flows are important factors in a number of biomedical transport problems that involve cells, proteins,

[2] If the boundary conditions are time dependent, then time appears as a parameter in the solution.

and micro-fluidic devices with effective diameters D ranging from 0.5×10^{-7} cm to 10^{-3} cm [7]. That is, in these problems the characteristic dimensions are small. The viscosity is close to the value of that for water (1 mPa s at 20°C), and velocities can range from 0.1×10^{-6} cm s^{-1} to 20 cm s^{-1}. Reynolds numbers range from 5×10^{-13} to 2.

3.6.2 Low-Reynolds-Number Flow Around a Sphere

To study the effect of low-Reynolds-number flow over proteins or cells, we determine the force acting on a rigid sphere that is exposed to a uniform flow field under conditions when the Reynolds number is extremely small. A slightly different problem—a sphere moving at a constant velocity in a quiescent (i.e., nonmoving) fluid—yields a different velocity field, but the same force [6]. The results from this problem are used to analyze the viscous and pressure forces that act upon biological molecules and cells.

For low-Reynolds-number flow around a sphere of radius R, the velocity varies in the radial and angular directions above and below the sphere's equator (see Figure 3.14). Spherical coordinates are most appropriate for this problem, because the solid boundary is spherical. These coordinates simplify the application of the boundary conditions. The velocity components far from the sphere are

$$r \rightarrow \infty \qquad \mathrm{v}_r = U_0 \cos\theta, \tag{3.6.4a}$$

$$\mathrm{v}_\theta = -U_0 \sin\theta. \tag{3.6.4b}$$

Far from the surface of the sphere, the pressure is uniform at p_0. On the surface, the no-slip condition applies and the velocities vanish; that is,

$$r = R \qquad \mathrm{v}_r = \mathrm{v}_\theta = 0. \tag{3.6.5a,b}$$

Our goals are to determine the velocity profile, to calculate the shear stress, and then to determine the force on the sphere. There is symmetry about the equator and the midplane through the sphere that is normal to the equatorial plane. Thus, there is no variation in the ϕ direction. For steady flow, the continuity equation for an incompressible fluid (see Table 3.1) reduces to

$$\frac{1}{r^2}\frac{\partial(r^2 \mathrm{v}_r)}{\partial r} + \frac{1}{r\sin\theta}\frac{\partial(\mathrm{v}_\theta \sin\theta)}{\partial\theta} = 0. \tag{3.6.6}$$

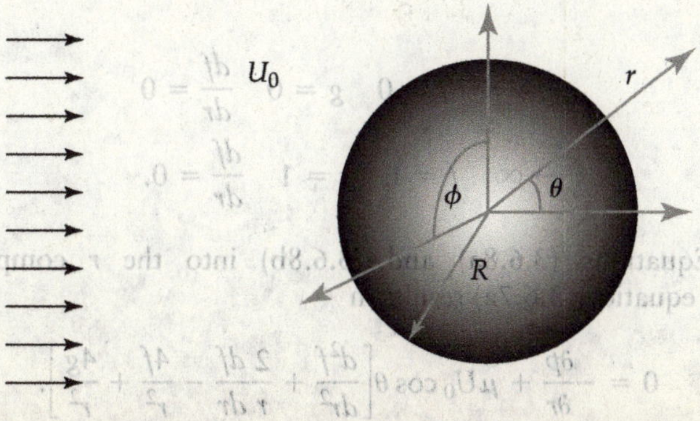

FIGURE 3.14 Uniform flow past a sphere of radius R at low Reynolds numbers.

For steady flow, neglecting variations in p around the sphere due to gravity and neglecting inertial terms in the Navier–Stokes equation (i.e., Equation (3.6.3b) applies), the r and θ components of the Navier–Stokes equation (Equations (3.3.28a) and (3.3.28b)) are

$$0 = -\frac{\partial p}{\partial r} + \mu\left(\frac{1}{r^2}\frac{\partial}{\partial r}\left(\left(r^2\frac{\partial v_r}{\partial r}\right)\right) + \frac{1}{r^2 \sin\theta}\frac{\partial}{\partial\theta}\left(\sin\theta\frac{\partial v_r}{\partial\theta}\right) - 2\frac{v_r}{r^2} - \frac{2}{r^2}\frac{\partial v_\theta}{\partial\theta} - \frac{2}{r^2}v_\theta\cot\theta\right) \tag{3.6.7a}$$

and

$$0 = -\frac{1}{r}\frac{\partial p}{\partial\theta} + \mu\left(\frac{1}{r^2}\frac{\partial}{\partial r}\left(r^2\frac{\partial v_\theta}{\partial r}\right) + \frac{1}{r^2\sin\theta}\frac{\partial}{\partial\theta}\left(\sin\theta\frac{\partial v_\theta}{\partial\theta}\right) + \frac{2}{r^2}\frac{\partial v_r}{\partial\theta} - \frac{v_\theta}{r^2\sin^2\theta}\right). \tag{3.6.7b}$$

Based on the behavior of the velocity far from the sphere, the following forms are assumed for the velocity components in spherical coordinates:

$$v_r = U_0 f(r)\cos\theta, \tag{3.6.8a}$$

$$v_\theta = -U_0 g(r)\sin\theta, \tag{3.6.8b}$$

$$v_\phi = 0. \tag{3.6.8c}$$

Here, $f(r)$ and $g(r)$ are functions that must be determined. There are four boundary conditions for the function $f(r)$ and two for $g(r)$. At $r = R$, v_r and v_θ must equal zero. This requires that $f(r)$ and $g(r)$ each equal zero. Far from the surface of the sphere, as r goes to infinity, $f(r)$ and $g(r)$ must each equal unity.

To solve for the velocity and pressure fields, we apply the continuity equation to relate $f(r)$ to $g(r)$:

$$\frac{d}{dr}(r^2 f) = 2rg \quad \text{or} \quad g = f + \frac{r}{2}\frac{df}{dr}. \tag{3.6.9a,b}$$

The continuity equation provides the two remaining constraints on the function f. At $r = R$, $g = f = 0$, and the derivative of f with respect to r equals zero. Likewise, as r goes to infinity, the same derivative goes to zero. Summarizing these conditions, we have

$$r = R \quad f = 0 \quad g = 0 \quad \frac{df}{dr} = 0 \tag{3.6.10a}$$

$$r \to \infty \quad f = 1 \quad g = 1 \quad \frac{df}{dr} = 0. \tag{3.6.10b}$$

Substituting Equations (3.6.8a) and (3.6.8b) into the r component of the Navier–Stokes equation (3.6.7a) results in

$$0 = -\frac{\partial p}{\partial r} + \mu U_0\cos\theta\left[\frac{d^2 f}{dr^2} + \frac{2}{r}\frac{df}{dr} - \frac{4f}{r^2} + \frac{4g}{r^2}\right]. \tag{3.6.11}$$

To ensure that the function f depends upon radial position only, the pressure is divided as follows:

$$p = p_\infty + P(r)\cos\theta. \tag{3.6.12}$$

Here, p_∞ is the pressure far from the surface. With this transformation for the pressure, Equation (3.6.11) becomes

$$0 = -\frac{dP}{dr} + \mu U_0\left[\frac{d^2f}{dr^2} + \frac{2}{r}\frac{df}{dr} - \frac{4f}{r^2} + \frac{4g}{r^2}\right]. \tag{3.6.13}$$

Likewise, after substituting for p, v_r, and v_θ, Equation (3.6.7b) is

$$0 = \frac{P}{r} + \mu U_0\left[-\frac{d^2g}{dr^2} - \frac{2}{r}\frac{dg}{dr} + \frac{2g}{r^2} - \frac{2f}{r^2}\right]. \tag{3.6.14}$$

Equations (3.6.9b), (3.6.13), and (3.6.14) are three equations in the three variables P, f, and g. After differentiation of Equation (3.6.14) and some algebraic manipulation, these three equations can be combined into the following equation:

$$0 = r\frac{df}{dr} - r^2\frac{d^2f}{dr^2} - r^3\frac{d^3f}{dr^3} - \frac{r^4}{8}\frac{d^4f}{dr^4}. \tag{3.6.15}$$

Equation (3.6.15) is known as *Euler's equation* [8] and has the following general solution:

$$f = C_1 + \frac{C_2}{r} + \frac{C_3}{r^3} + C_4r^2. \tag{3.6.16}$$

The boundary condition as r approaches infinity requires that C_4 equals zero and C_1 equals unity. When we apply the two boundary conditions at $r = R$, we find that $C_2 = -3R/2$ and $C_3 = R^3/2$. Thus, $f(r)$, $g(r)$, and $P(r)$ can be found. The resulting expressions for the velocity and pressure are

$$v_r = U_0\left[1 - \frac{3}{2}\frac{R}{r} + \frac{1}{2}\frac{R^3}{r^3}\right]\cos\theta, \tag{3.6.17}$$

$$v_\theta = -U_0\left[1 - \frac{3}{4}\frac{R}{r} - \frac{1}{4}\frac{R^3}{r^3}\right]\sin\theta, \tag{3.6.18}$$

and

$$p = p_\infty - \rho g x - \frac{3\mu U_0}{2R}\left(\frac{R}{r}\right)^2\cos\theta, \tag{3.6.19}$$

where the x-direction is in the vertical direction (Figure 3.15). The velocity and pressure fields are used to compute the force exerted by the fluid on the sphere.

Superimposing a rectangular coordinate system on the sphere, with the z-direction parallel to the direction of flow, we find that the force exerted by the fluid on the sphere has only a z-component (see Figure 3.15). That force is given by

$$F_z = \mathbf{F}\cdot\mathbf{e}_z = -\mathbf{e}_z\cdot\int_S \mathbf{n}\cdot\boldsymbol{\sigma}\,dS. \tag{3.6.20}$$

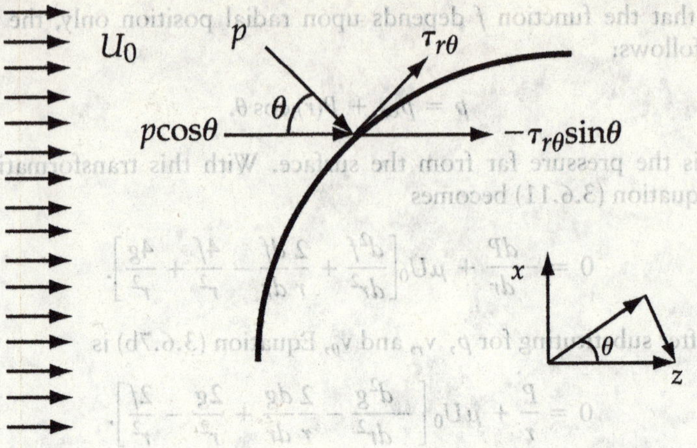

FIGURE 3.15 Components of the pressure and shear stress acting in the direction of flow, the z-direction.

In spherical coordinates, the differential surface area on the sphere is $R^2 \sin\theta d\theta d\phi$.

The pressure and shear stress that act upon the surface contribute to the stress. The nonzero shear stress component on the surface of the sphere is $\tau_{r\theta}$. The resolution of unit vectors in the z direction is shown in Figure 3.15. The force in the direction of flow can be written as

$$F_z = \int_0^{2\pi}\int_0^{2\pi} p|_{r=R} \cos\theta R^2 \sin\theta d\theta d\phi + \int_0^{2\pi}\int_0^{\pi} -\tau_{r\theta}|_{r=R} \sin\theta R^2 \sin\theta d\theta d\phi. \quad (3.6.21)$$

If we consider a particle settling under gravity with the gravity vector parallel to the z-axis, the pressure term at $r = R$ in Equation (3.6.21) can be rewritten as

$$p|_{r=R} = p_\infty - \rho g R\cos\theta - \frac{3\mu U_0}{2R}\cos\theta. \quad (3.6.22)$$

Evaluating the shear stresses at the sphere surface yields

$$\tau_{r\theta}|_{r=R} = -\frac{3\mu U_0}{2R}\sin\theta. \quad (3.6.23)$$

We insert Equations (3.6.22) and (3.6.23) into Equation (3.6.21) and integrate to obtain the following expression for the surface forces that are exerted by the moving fluid on the stationary sphere:

$$F_z = \frac{4}{3}\pi R^3 \rho g + 2\pi\mu U_0 R + 4\pi\mu U_0 R. \quad (3.6.24)$$

The first term is the buoyant force. The second term is known as the *form drag*; it arises from the nonuniform pressure field around the sphere. The third term, known as the *friction drag*, is caused by the viscous stresses that act upon the sphere. The sum of the form drag and the friction drag is known as the *drag force* and is given by

$$F_D = 6\pi\mu U_0 R. \quad (3.6.25)$$

Equation (3.6.25) is known as *Stokes's law* and is valid when $\text{Re} = \rho U_0 D/\mu$ is less than approximately 0.1.

The definition of the friction factor (Equation (3.5.6)) can be generalized in terms of the force F that is exerted by the fluid on an object, whether it is a submerged object or the wall of a tube. The generalized equation is

$$f = \frac{F}{\frac{1}{2}\rho U_0^{\,2} A}, \qquad (3.6.26)$$

where A is the projected area normal to the flow and U_0 is a characteristic velocity. For a sphere, the projected area is πR^2. Substituting Equation (3.6.25) for the force into Equation (3.6.26) yields

$$f = \frac{24}{\text{Re}} \quad \text{Re} < 0.1. \qquad (3.6.27)$$

For Reynolds numbers between 2 and 500, the following correlation fits the data well:

$$f = \frac{18.5}{\text{Re}^{0.6}} \quad 2 < \text{Re} < 500. \qquad (3.6.28)$$

Equations (3.6.27) and (3.6.28) were plotted in Figure 3.16, and the results were used to determine the force on a nonmoving sphere that is exposed to a fluid moving at a uniform velocity U_0. The same forces would be determined for a sphere moving at velocity U_0 in a nonmoving fluid [6]. As a result, this analysis has general utility; it can be used to calculate the drag on a particle moving at velocity U_0 at low Reynolds numbers.

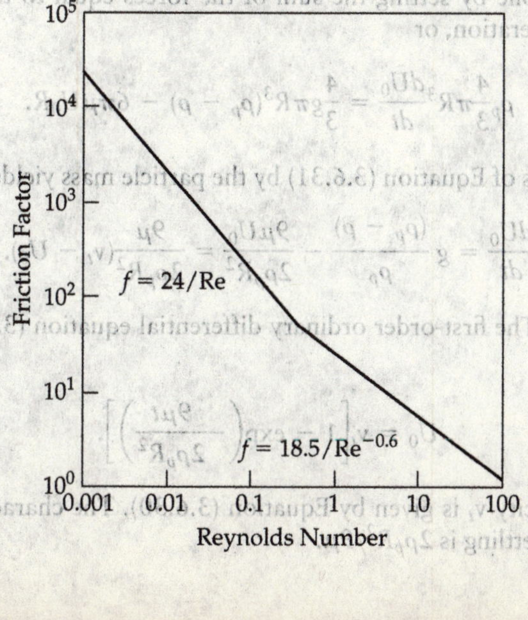

FIGURE 3.16 Friction factor as a function of the Reynolds number for a laminar flow.

Example 3.6 Determine the Reynolds number of a bacterium 10 μm in diameter that is traveling through water ($\nu = 0.01\ \text{cm}^2\ \text{s}^{-1}$) at 10 μm s^{-1}.

Solution From the data given,

$$\text{Re} = \frac{(10^{-3}\text{cm s}^{-1})(10^{-3}\text{cm})}{0.01\ \text{cm}^2\ \text{s}^{-1}} = 10^{-4}.$$

Clearly, inertial forces are negligible for the motion of cells and biological molecules.

The foregoing calculations can be used to determine the velocity on a small particle of radius R as it moves under gravity. In the steady state, the vertical component of the force balance yields (see Figure 3.17)

$$0 = F_g - F_b - F_d, \tag{3.6.29}$$

where $F_g = (4/3)\pi R^3 \rho_p g$ is the gravitational force, in which ρ_p is the particle density.

Substituting for each of the forces and rearranging terms result in the following expression for the velocity:

$$U_0 = \frac{2gR^2}{9\mu}(\rho_p - \rho). \tag{3.6.30}$$

The steady-state value U_0 of the settling velocity is often referred to as the *terminal velocity* v_t.

A short time is needed for the velocity to attain its steady-state value. We can estimate the time that is required to reach steady state by solving the unsteady-state velocity. This is done by setting the sum of the forces equal to the product of the mass and the acceleration, or

$$\rho_p \frac{4}{3}\pi R^3 \frac{dU_0}{dt} = \frac{4}{3} g\pi R^3(\rho_p - \rho) - 6\pi\mu U_0 R. \tag{3.6.31}$$

Dividing both sides of Equation (3.6.31) by the particle mass yields

$$\frac{dU_0}{dt} = g\frac{(\rho_p - \rho)}{\rho_p} - \frac{9\mu U_0}{2\rho_p R^2} = \frac{9\mu}{2\rho_p R^2}(\text{v}_t - U_0). \tag{3.6.32}$$

Initially, $U_0 = 0$. The first-order ordinary differential equation (3.6.32) has the following solution:

$$U_0 = \text{v}_t\left[1 - \exp\left(-\frac{9\mu t}{2\rho_p R^2}\right)\right]. \tag{3.6.33}$$

The terminal velocity v_t is given by Equation (3.6.30). The characteristic time constant for particle settling is $2\rho_p R^2/9\mu$.

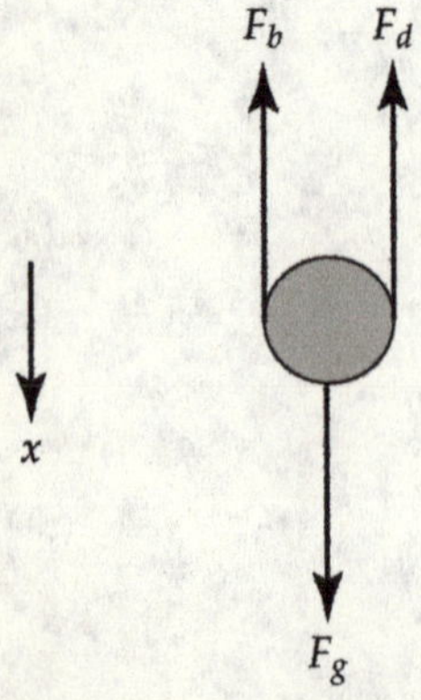

FIGURE 3.17 Schematic of the forces acting on a sphere of radius R settling under the influence of gravity.

Example 3.7 Determine the terminal velocity and the time that is required to reach steady state for a sphere of density 1.05 g cm^{-3} and radius 10.35 mm that is settling in water (ρ = 1 g cm^{-3} and μ = 0.01 g cm^{-1} s^{-1}).

Solution From Equation (3.6.30), the settling velocity is

$$v_t = \frac{2(980\,\text{cm s}^{-2})(0.001035\,\text{cm})^2(1.05 - 1.00\,\text{g cm}^{-3})}{9(0.01\,\text{g cm}^{-1}\text{s}^{-1})} = 0.00117\,\text{cm s}^{-1}.$$

The time constant for settling is

$$t_c = \frac{2(1.05\text{ g cm}^{-3})(0.001035\text{ cm})^2}{9(0.01\text{ g cm}^{-1}\text{ s}^{-1})} = 2.5 \times 10^{-5}\text{ s}.$$

After five time constants, the velocity is within 1% of the terminal velocity (see Figure 3.18).

The Reynolds number for the particle at its terminal velocity is 2.54×10^{-4}, so this flow is clearly one in which inertial forces are negligible. The time constant for settling is certainly a small number, but does its magnitude indicate that we can neglect acceleration?

One way to answer this question is to calculate the distance moved during the time that it takes for the particle to reach within 1% of its terminal velocity and then to compare this distance with the size of the particle. The distance that

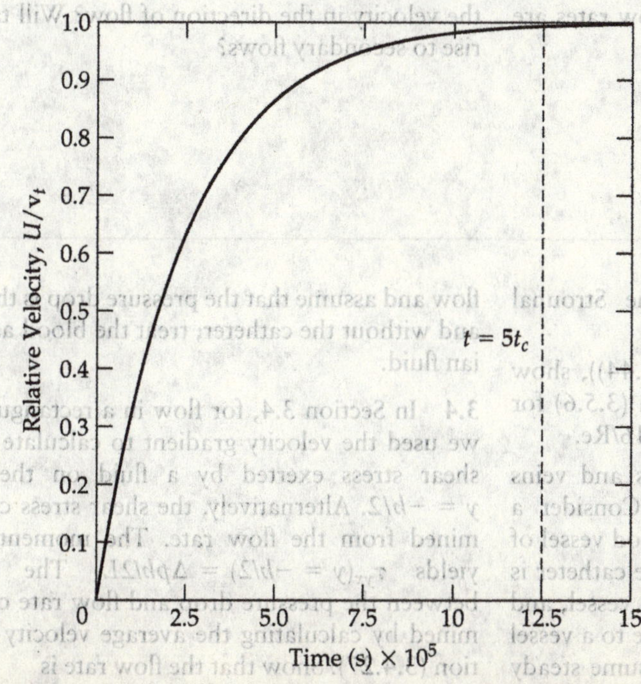

FIGURE 3.18 Approach to the terminal velocity for a sphere of radius 10.35 mm settling under the influence of gravity.

the particle moves can be determined by integrating the velocity with respect to time. The result is

$$x = v_t\left[t - \frac{2\rho_p R^2}{9\mu}\left(1 - \exp\left(-\frac{9\mu t}{2\rho_p R^2}\right)\right)\right]. \tag{3.6.34}$$

For the given values for the fluid and particle, the particle moves a distance of 1.17×10^{-7} cm during the time it takes to reach within 1% of the terminal velocity. This distance is about 0.01% of the radius of the particle. Hence, because the distance is small relative to the particle dimensions, we can safely neglect fluid acceleration and assume that the particle instantaneously adapts to a change in velocity.

3.7 QUESTIONS

3.1 Discuss whether the assumption of incompressibility is suitable for problems concerning gas transport in the lung.

3.2 What restrictions are placed on the analysis of the time that is required for steady flow to develop in a rectangular channel (Section 3.4.2)?

3.3 Discuss the advantages of using dimensional analysis to characterize the flow in a branching blood vessel in which the diameters and flow rates are varied.

3.4 Provide a qualitative explanation as to why the flow at extremely low Reynolds numbers is not explicitly dependent on time.

3.5 Use the analysis of low-Reynolds-number flow to determine how to separate molecules of different sizes and densities in a high-speed centrifuge.

3.6 The major arteries taper slightly with distance away from the heart. What effect does tapering have on the velocity in the direction of flow? Will tapering give rise to secondary flows?

3.8 PROBLEMS

3.1 Derive the relationship between the Strouhal number St and the Womersley number α.

3.2 Using Poiseuille's law (Equation (2.7.44)), show that the friction factor defined in Equation (3.5.6) for laminar flow in a cylindrical tube is simply 16/Re.

3.3 Catheters that are placed in arteries and veins will alter the flow through the vessel. Consider a catheter of radius εR that is placed in a blood vessel of radius R (see Figure 3.19). Assume that the catheter is concentric with the centerline of the blood vessel, and determine the reduction in flow rate relative to a vessel of the same radius without a catheter. Assume steady flow and assume that the pressure drop is the same with and without the catheter; treat the blood as a Newtonian fluid.

3.4 In Section 3.4, for flow in a rectangular channel, we used the velocity gradient to calculate the average shear stress exerted by a fluid on the surface at $y = -h/2$. Alternatively, the shear stress can be determined from the flow rate. The momentum balance yields $\tau_{yx}(y = -h/2) = \Delta p h/2L$. The relationship between the pressure drop and flow rate can be determined by calculating the average velocity from Equation (3.4.27). Show that the flow rate is

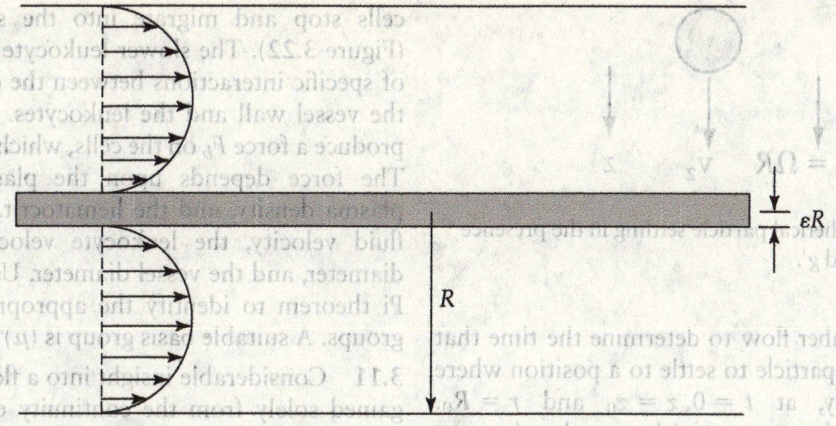

FIGURE 3.19 Sketch of the velocity profile for steady flow around a catheter placed in the center of a blood vessel.

$$Q = \frac{\Delta p w h^3}{12\mu L}$$

$$\times\left[1 - 192\left(\frac{h}{w}\right)\sum_{n=0}^{\infty}\frac{(-1)^n\tanh((2n+1)\pi w/2h)}{(2n+1)^5\pi^5}\right]. \quad (3.8.1)$$

3.5 For a rectangular flow channel with $w = 2.0$ cm and $h = 0.05$ cm, determine the time required for the flow to reach within 1% of steady state for a kinematic viscosity equal to 0.007 cm^2 s^{-1}.

3.6 Parallel-plate flow channels similar to the system shown schematically in Figure 3.6 are used to study the response of endothelial cells to fluid shear stresses. It has been proposed [9] that temporal, shear-stress gradients may affect endothelial cell function.

(a) Using the results from Section 3.4.2, calculate the derivative of the shear stress with respect to time.

(b) Plot the shear-stress gradient normalized by the maximum value.

(c) Determine flow rates and channel dimensions for shear stresses of 1 and 10 dyn cm^{-2} and gradients differing by an order of magnitude for each case. Use v = 0.007 or 0.07 cm^2 s^{-1} and w = 2 cm.

3.7 The velocity profile for the transient change in velocity in a circular tube of radius R following the onset of flow is given by

$$v_z = \frac{\Delta P R^2}{4\mu L}\left(1 - \frac{r^2}{R^2}\right)$$

$$- \frac{2\Delta P R^2}{\mu L}\sum_{n=1}^{\infty}\frac{J_0(\lambda_n r/R)\exp(-\lambda_n^2 vt/R^2)}{\lambda_n^3 J_1(\lambda_n)}, \quad (3.8.2)$$

where $J_v(x)$ is the Bessel function of order v (see Figure A.3 and Equations (A.2.31) and (A.2.34)) and λ_n are the roots of $J_0(\lambda_n) = 0$. Bessel functions and their roots can be simply calculated in MATLAB. For example, $J_n(x)$ is simply besselj(n,x).

Use the above equation and write a MATLAB program to determine the time for the velocity profile to reach within 1% of the steady-state value. Explain why this result differs from that obtained for a rectangular channel.

3.8 One way to isolate protein molecules in a mixture is to allow them to settle in a solution with a density gradient. Settling under the influence of gravity is rather slow. A high-speed centrifuge increases the gravitational acceleration, which reduces the time that is required to reach steady state. Consider a protein molecule of radius a and density ρ_π that is undergoing ultracentrifugation in a liquid solution. The solution has a density variation given by $\rho(z) = \rho_0 + \alpha z$, a viscosity μ, and a gravitational acceleration $g' = \Omega^2 R$, where R is the radial distance (see Figure 3.20). Use the results for the settling velocity of a particle under conditions of

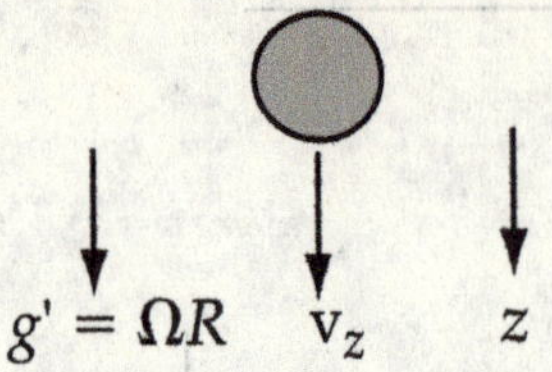

FIGURE 3.20 A spherical particle settling in the presence of a gravitational field g'.

low-Reynolds-number flow to determine the time that is required for the particle to settle to a position where $\rho(z) = \rho_\pi$. Initially, at $t = 0, z = z_0$ and $r = R_0$. Neglect effects due to particle acceleration or deceleration.

3.9 One method of measuring the cytoplasmic viscosity of cells is to aspirate a portion of the cell into a long micropipet of radius R_p, as shown in Figure 3.21. For such flows, Re $\ll$ 1. The characteristic velocity is equal to the time rate of change of the cell length, $L' = dL/dt$, in the micropipet. This length of change depends upon the pressure drop ΔP across the cell, the micropipet radius R_p, the viscosity μ_c of the cell cytoplasm, and the radius R_c of the cell. Use the Buckingham Pi theorem to determine the dimensionless groups that are involved.

3.10 In postcapillary venules, leukocytes are close to the vessel wall, rolling or moving more slowly than the average fluid velocity V_f. At sites of injury or infection, the leukocyte velocity V_c slows further, and the cells stop and migrate into the surrounding tissue (Figure 3.22). The slower leukocyte velocity is a result of specific interactions between the endothelial cells of the vessel wall and the leukocytes. These interactions produce a force F_b on the cells, which slows them down. The force depends upon the plasma viscosity, the plasma density, and the hematocrit, as well as on the fluid velocity, the leukocyte velocity, the leukocyte diameter, and the vessel diameter. Use the Buckingham Pi theorem to identify the appropriate dimensionless groups. A suitable basis group is $(\mu)^a(V_f)^b(d_t)^c$.

3.11 Considerable insight into a flow problem can be gained solely from the continuity equation. Consider laminar flow over a flat plate, as shown in Figure 3.23. Fluid approaches the plate with a uniform velocity U. When the fluid is in contact with the surface, flow in the x-direction decelerates within a thin region of thickness δ. This *boundary layer* grows with distance along the plate (i.e., $\delta = f(x)$). Using the no-slip condition along the plate ($y = 0$, $v_x = 0$, and $v_y = 0$) and $v_x = U$ (a constant outside the boundary layer), prove the following statements, using the conservation of mass for an incompressible fluid:

(a) $v_y(x, y)$ is everywhere positive for $0 \leq y \leq \delta$.

(b) v_y has a parabolic shape with respect to y very close to the plate.

(c) v_y reaches a positive maximum at $y = \delta$.

For part (b), assume that v_x can be represented in terms of a series (e.g., $v_x = a_1(x)y + a_2(x)y^2 + \cdots$).

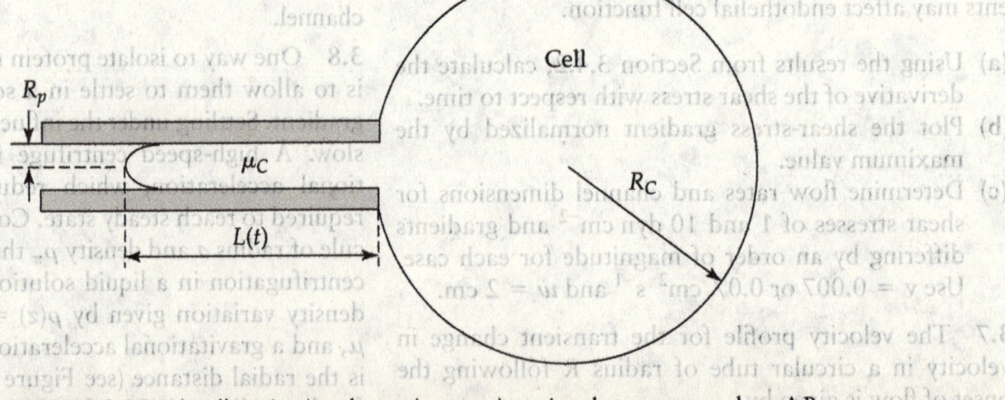

FIGURE 3.21 A cell projection drawn into a micropipet by a pressure drop ΔP.

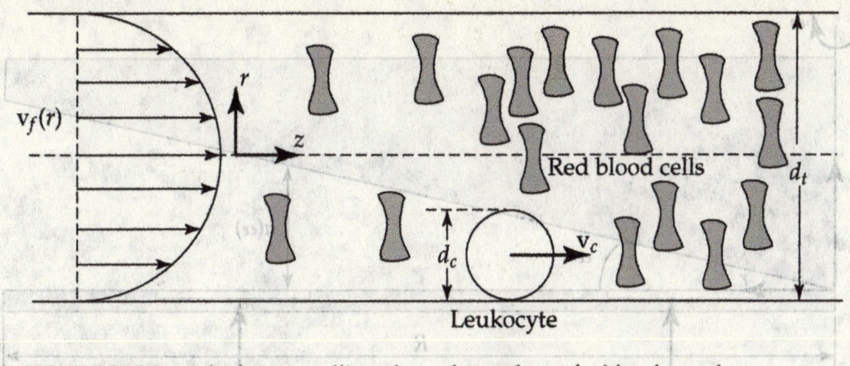

FIGURE 3.22 A leukocyte rolling along the surface of a blood vessel.

3.12 Endothelial cells line blood vessels and play an important role in protein transport, the maintenance of vascular tone, hemostasis, and the immunological response of tissues. Endothelial cells are exposed to unsteady shear stresses between 10 and 200 dyn cm^{-2}. These shear stresses are believed to affect the function of the cells. One approach that has been used to study the response of endothelial cells in vitro has been to grow the cells on surfaces and then expose the cells to laminar shear stresses in a cone-and-plate viscometer, as shown in Figure 3.24. The plate is stationary, and the cone rotates with an angular speed ω (rad s^{-1}). The objective of this problem is to analyze the conditions under which the velocity can be represented by the simple relationship

$$v_\theta = \frac{\omega r z}{h(\alpha)}, \tag{3.8.3}$$

where α is the angle between the cone and the plate (typically, less than 3°, or 0.0524 rad), $h = r \tan\alpha$ is the local gap height, and the coordinates r and z are as shown shortly. For such small angles, $\tan\alpha \approx \alpha$. Boundary conditions are $z = 0$, $v_r = v_\theta = v_z = 0$ and $z = 4$, $v_r = v_z = 0$, $v_\theta = \omega r$.

In general, flow is symmetric in the θ direction. Because the angle $\alpha \ll 1$, $v_z \ll v_r$ and $v_z \ll v_\theta$. Consequently, $\partial/\partial r \ll \partial/\partial z$. The equation of continuity and the Navier–Stokes equation in cylindrical coordinates, respectively, reduce to

$$\frac{1}{r}\frac{\partial r v_r}{\partial r} + \frac{v_z}{z} = 0 \tag{3.8.4}$$

and

$$-\frac{\rho v_\theta^{\,2}}{r} = -\frac{\partial p}{\partial r} + \mu\frac{\partial^2 v_r}{\partial z^2}, \tag{3.8.5a}$$

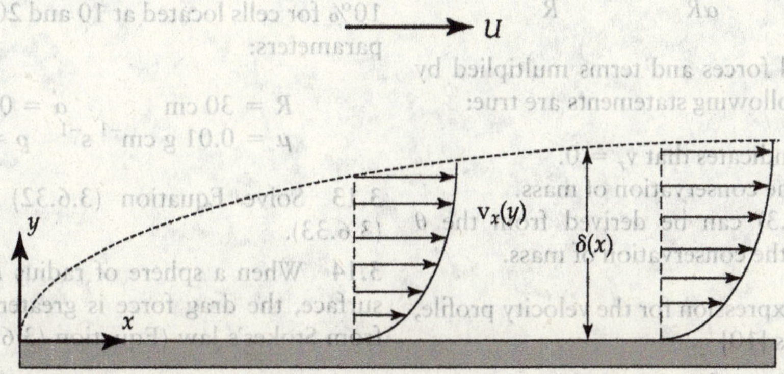

FIGURE 3.23 Schematic of flow over a flat plate.

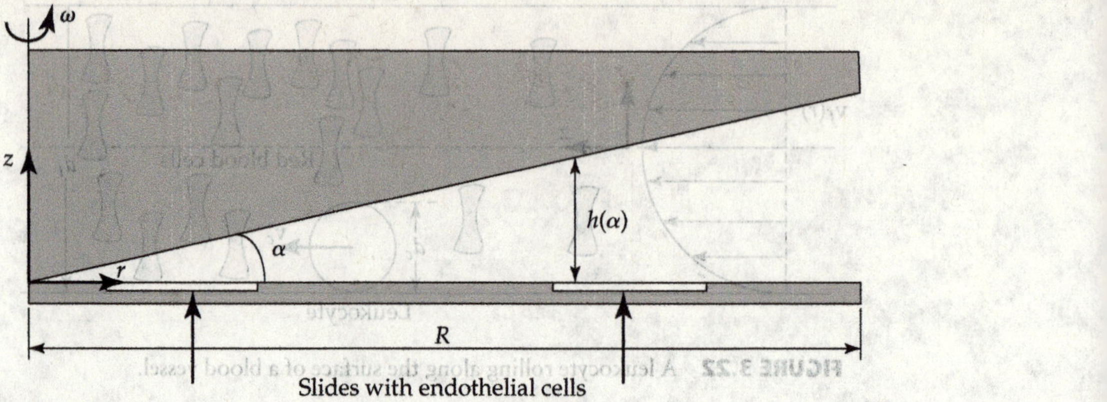

FIGURE 3.24 Schematic of cone-and-plate viscometer.

$$\theta, \quad \rho\left[\mathrm{v}_r\frac{\partial \mathrm{v}_\theta}{\partial r} + \mathrm{v}_z\frac{\partial \mathrm{v}_\theta}{\partial z}\right] = \mu\frac{\partial^2 \mathrm{v}_\theta}{\partial z^2}, \quad (3.8.5b)$$

$$z, \quad 0 = -\frac{\partial p}{\partial z} + \mu\frac{\partial^2 \mathrm{v}_z}{\partial z^2}. \quad (3.8.5c)$$

(a) Use the following dimensionless variables to cast Equations (3.8.5a) through (3.8.5c) into dimensionless form and to develop a criterion for neglecting inertial forces:

$$\mathrm{v}_z^* = \frac{\mathrm{v}_z}{\omega\alpha R} \qquad \mathrm{v}_r^* = \frac{\mathrm{v}_r}{\omega R} \qquad \mathrm{v}_\theta^* = \frac{\mathrm{v}_\theta}{\omega R}$$

$$P^* = \frac{P\alpha}{\mu\omega} \qquad z^* = \frac{z}{\alpha R} \qquad r^* = \frac{r}{R}.$$

(b) Neglecting inertial forces and terms multiplied by α, show that the following statements are true:

1. r-momentum indicates that $\mathrm{v}_r = 0$.
2. $\mathrm{v}_z = 0$, from the conservation of mass.
3. Equation (3.8.3) can be derived from the θ component of the conservation of mass.

(c) A more accurate expression for the velocity profile, valid for small α, is [10]

$$\mathrm{v}_\theta = \omega r\left\{\left(\frac{z}{r\alpha}\right) + \varepsilon^2\left[-3.294 + 10^{-3}\left(\frac{z}{r\alpha}\right) + 2.778 \times 10^{-3}\left(\frac{z}{r\alpha}\right)^4 + 2.50 \times 10^{-3}\left(\frac{z}{r\alpha}\right)^5 - 1.984 \times 10^{-3}\left(\frac{z}{r\alpha}\right)^7\right]\right\}, \quad (3.8.6)$$

where

$$\varepsilon = \frac{\rho r^2\omega\alpha^2}{\mu}.$$

Use this result to determine when the shear stress on the plate surface $[\tau_{z\theta}(z = 0)]$ would differ by more than 10% for cells located at 10 and 20 cm for the following parameters:

$R = 30$ cm $\qquad \alpha = 0.05236$ radian(3°)
$\mu = 0.01$ g cm^{-1} s^{-1} $\qquad \rho = 1$ g cm^{-3}

3.13 Solve Equation (3.6.32) to obtain Equation (3.6.33).

3.14 When a sphere of radius R moves near a solid surface, the drag force is greater than that computed from Stokes's law (Equation (3.6.25)). This additional drag force is due to the presence of the surface. In such a case, the drag force is often written as

$$F_D = 6\pi\mu R v F^*(H/R), \qquad (3.8.7)$$

where H is the height of the sphere above the surface, v is the fluid velocity at height H ($v = \dot{\gamma}H$, where $\dot{\gamma}$ is the shear rate), and $F^*(H/R)$ is a function that accounts for the increased force on the sphere. This force is tabulated [11] and ranges from 1.69 when H/R is unity to unity when H/R is infinite. Compute the drag force acting on a leukocyte of radius 6 μm that adheres to the endothelium in a flow channel. The average velocity in the channel is 2000 $\mu m\ s^{-1}$, the channel height is 250 μm, and the channel width is 2 cm. The fluid viscosity is 0.009 $g\ cm^{-1}\ s^{-1}$.

3.15 The velocity field surrounding a sphere of radius R moving at a velocity U_0 through a Newtonian fluid of density r and viscosity μ is given by [6]

$$v_r = \frac{U_0}{2}\cos\theta\left(\frac{R}{r}\right)^2\left(\frac{3r}{R} - \frac{R}{r}\right) \qquad (3.8.8a)$$

and

$$v_\theta = \frac{U_0}{4}\sin\theta\left(\frac{R}{r}\right)\left(\left(\frac{R}{r}\right)^2 + 3\right). \qquad (3.8.8b)$$

(a) Using Equations (3.8.8a), (3.8.8b), (3.6.7a), and (3.6.7b), determine $p(r, \theta)$.

(b) Show that the drag force acting on the sphere is $6\pi\mu R U_0$.

3.16 In flow-chamber assays of leukocyte adhesion to the endothelium, the leukocytes are suspended in a tissue culture medium, and they flow through a parallel-plate flow channel of length L, height h, and width w (Figure 3.25). The width is much greater than the height ($w \gg h$). Endothelial cells are grown on glass slides that make up the lower surface of the flow chamber. Leukocytes are slightly more dense than the tissue culture medium and must settle under gravity to adhere. Find the distance a leukocyte travels through the flow chamber before it reaches the lower surface of the chamber.

(a) For the following cell and channel dimensions and fluid and cell properties, and for a wall shear stress of 1 $dyn\ cm^{-2}$ acting on the endothelial cells, show that the Reynolds number for the leukocytes is much less than unity:

$w = 2$ cm $\quad h = 0.025$ cm $\quad R_{leukocyte} = 6\ \mu m$

$L = 7.5$ cm $\quad \rho_{leukocyte} = 1.07\ g\ cm^{-3}$

$\rho = 1.01\ g\ cm^{-3} \quad \mu = 0.0085\ g\ cm^{-1}\ s^{-1}$

(b) State force balances acting on the cell in the x and y directions. Neglect any acceleration of the cell.

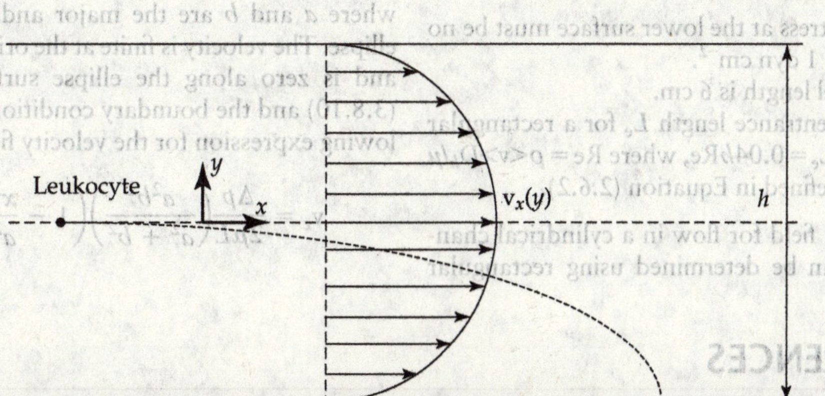

FIGURE 3.25 Sketch of the trajectory of a leukocyte moving in a velocity field $v_x(y)$ and settling under gravity.

(c) Solve the force balances to determine the distance traveled by a cell that is initially at the centerline.

(d) Discuss the effect of cell settling on the measurement of the adhesion of leukocytes to endothelium under flow conditions.

3.17 You have been hired by Biomicrofluidics, Inc., to design a microfluidic device to capture tumor cells present in blood. These captured cancer cells will be analyzed in a microarray system to develop gene profiles of cancer for treatment and assess the success of therapies. Basically, red cells are separated from the other cells and the solution of white cells and cancer cells flows through a series of parallel-plate channels in which the lower surface of each channel is coated with antibody to cancer cells. There are 20 identical parallel channels of equal height, width, and length. The cancer cells adhere if the shear stress is below a critical value.

(a) Use the design conditions below to determine the channel height to permit efficient capture of the cells on the surface of the slide.

(b) Determine whether the total pressure drop ($\Delta P = n\Delta P_{\text{channel}}$) is less than 5,000 Pa and the flow is laminar (1 Pa = 10 dyn cm^{-2}).

Data and design conditions:

1. Channel width w is 20 times the height h to ensure that the flow is one-dimensional.
2. The total flow rate must be 10 cm^3 h^{-1}.
3. The viscosity is 0.008 g cm^{-1} s^{-1} and the density is 1 g cm^{-3}.
4. The shear stress at the lower surface must be no higher than 1 dyn cm^{-2}.
5. The channel length is 6 cm.
6. Where the entrance length L_e for a rectangular channel is $L_e = 0.04h\text{Re}$, where $\text{Re} = \rho\langle v\rangle D_h/\mu$ and D_h is defined in Equation (2.6.2).

3.18 The velocity field for flow in a cylindrical channel of radius R can be determined using rectangular coordinates using the following approach. Consider steady, fully developed laminar flow of a Newtonian fluid in a circular channel of radius R and length L. Show that for rectangular coordinates

$$-\frac{\Delta p}{L} = \mu\left(\frac{\partial^2 v_z}{\partial x^2} + \frac{\partial^2 v_z}{\partial y^2}\right) \quad (3.8.9)$$

The radius is related to the x and y coordinates as $R^2 = x^2 + y^2$.

(a) State the boundary conditions.

(b) The following equation is proposed for the solution:

$$v_z = A + B(x^2 + y^2)$$

Show that this equation satisfies the boundary condition at $x = 0$, $y = 0$ and Equation (3.8.9). Determine the value of B.

(c) Find the value of A from the boundary condition on the surface of the cylindrical tube $R = \sqrt{x^2 + y^2}$. Show that the result is identical to Equation (2.7.22).

3.19 Consider steady, fully developed laminar flow of a Newtonian fluid in a channel of elliptical cross-section and length L. The ellipse is centered at the origin and is described by the following relation:

$$\frac{x^2}{a^2} + \frac{y^2}{b^2} = 1, \quad (3.8.10)$$

where a and b are the major and minor axes of the ellipse. The velocity is finite at the origin ($x = 0$ and $y = 0$) and is zero along the ellipse surface. Use Equation (3.8.10) and the boundary conditions to obtain the following expression for the velocity field;

$$v_z = \frac{\Delta p}{2\mu L}\left(\frac{a^2b^2}{a^2 + b^2}\right)\left(1 - \frac{x^2}{a^2} - \frac{y^2}{b^2}\right) \quad (3.8.11)$$

3.9 REFERENCES

1. Whitaker, S., *Introduction to Fluid Mechanics.* 1981, Malabar, FL: Krieger Publishing Company.
2. Bird, R., Armstrong, R., and Hassager, O., *Dynamics of Polymeric Liquids.* 2d ed. Vol. 1. *Fluid Mechanics.* 1987, New York: John Wiley and Sons.
3. Bird, R., Stewart, W., and Lightfoot, E., *Transport Phenomena.* 2d ed. 2002, New York: John Wiley and Sons.
4. White, F., *Fluid Mechanics.* 1986, New York: McGraw-Hill Book Company.

5. Chapman, G.B., and Cokelet, G.R., "Model studies of leukocyte–endothelium–blood interactions." *Biorheology*, 1997. **34**: pp. 37–56.
6. Happel, J., and Brenner, H., *Low Reynolds Number Hydrodynamics*. 1983, Boston: Martinus Nijhoff Publishers.
7. Brody, J.P., Yager, P., Goldstein, R.E., and Austin, R.H., "Biotechnology at low Reynolds number." *Biophys J.*, 1996. **71**: pp. 3430–3441.
8. Denn, M.M., *Process Fluid Mechanics*. 1980, Englewood Cliffs, NJ: Prentice-Hall, Inc.
9. Bao, X., Clark, C.B., and Frangos, J.A., "Temporal gradient in shear-induced signaling pathway: involvement of MAP kinase, c-fos, and connexin43." *Am. J. Physiol.*, 2000. **278**: pp. H1598–H1605.
10. Sdougos, H.P., Bussolari, S.R., and Dewey, C.F., "Secondary flow and turbulence in a cone-and-plate device." *J. Fluid Mech.*, 1984. **138**: pp. 379–404.
11. Goldman, A.J., Cox, R.G., and Brenner, H., "Slow viscous motion of a sphere parallel to a plane wall: I. Motion through a quiescent fluid." *Chem. Eng. Sci.*, 1967. **22**: pp. 637–651.

CHAPTER 4

Approximate Methods for the Analysis of Complex Physiological Flow

4.1 Introduction

Many problems in physiological fluid mechanics are so complex that detailed analytical or numerical solutions of the differential forms of the conservation of mass and conservation of linear momentum are not feasible. Examples include blood flow through the heart and red-cell motions in capillaries. For such problems, two different types of simplification can be employed. One, instead of using the differential form of the conservation relations, integral average equations can be derived. These equations do not describe the properties at each point in the system but, rather provide an average description of the fluid motion and forces acting on the system. The resulting equations are often easier to solve than the complete set of differential relations. The trade-off is that the integral analysis often requires some simplifications that lead to approximate results. Nevertheless, an integral analysis can provide important information about the magnitudes of forces and flows.

Two approaches are used to develop the integral forms of the conservation relations. One is to derive these integral forms de novo from an analysis of general flow situations. The other is to use the differential relations and integrate over a control volume. The latter approach is used in Sections 4.2 and 4.3 to derive integral, or macroscopic, forms of the conservation of mass and linear momentum. In Section 4.4, a simplified form of the integral analysis is derived. The result, known as *Bernoulli's equation*, provides important physical insights into many types of flows.

The second approach to simplification of the analysis of complex physiological flow problems is the use of order-of-magnitude estimates of the continuity equation and the Navier–Stokes equation. An appropriate scaling of these equations can result in the elimination of terms that do not contribute significantly to fluid motion or forces. Such an approach was used in Section 3.6 to derive the equations for low-Reynolds-number flow. Similar scaling arguments based on physical dimensions and the relative magnitudes of viscous and inertial forces can lead to an approximate analysis of fluid motion for high-speed flows or cases of complex geometries. Boundary layer analysis affords key insights into scaling based on the relative importance of forces and is used in the simplification of many flow and mass transfer problems. Section 4.5 examines the development of the boundary layer equations to determine the viscous forces near a surface exposed to a high-speed flow. This kind of analysis is used to examine some complex flows. In Section 4.6, boundary layer results are used to develop a criterion for flow separation, a phenomenon in which a region of the flow field moves in a direction opposite that of the main flow. In Section 4.7, a scaling analysis is applied to flows that pass through narrow spaces, an approach known as *lubrication theory.*

4.2 Integral Form of the Equation of Conservation of Mass

As a first step in deriving the integral form of the equation of conservation of mass, consider the differential form of Equation (3.2.7):

$$\frac{\partial \rho}{\partial t} = -\nabla \cdot (\rho \mathbf{v}). \tag{4.2.1}$$

To obtain the average behavior of the fluid, we integrate over the control volume V:

$$\int_V \frac{\partial \rho}{\partial t} dV = -\int_V \nabla \cdot (\rho \mathbf{v}) dV. \tag{4.2.2}$$

For a constant control volume, the order of differentiation and integration are interchangeable. The integral of the density throughout the volume is simply the mass m of the control volume. Thus, the term on the left-hand side of Equation (4.2.2) becomes

$$\int_V \frac{\partial \rho}{\partial t} dV = \frac{\partial}{\partial t}\left(\int_V \rho dV\right) = \frac{dm}{dt}. \tag{4.2.3}$$

Next, we apply the divergence theorem to the right-hand side of Equation (4.2.2) (Section A.3.C). This theorem allows us to convert the volume integral into an integral over the control surface:

$$\int_V \nabla \cdot (\rho \mathbf{v}) dV = \int_S \rho \mathbf{v} \cdot \mathbf{n} dS. \tag{4.2.4}$$

As before (Section 2.2.1), the vector $\mathbf{n}$ is the unit normal directed from the control surface into the fluid. The surface integral on the right-hand side of Equation (4.2.4) represents the rate of mass flow across the surface S.

Using Equations (4.2.3) and (4.2.4) to replace the two integrals in Equation (4.2.2), we obtain the following version of the integral form of the conservation of mass:

$$\frac{dm}{dt} = -\int_S \rho \mathbf{v} \cdot \mathbf{n} dS. \tag{4.2.5}$$

In words, Equation (4.2.5) states that the change in mass of the control volume equals the integral over the surface of the mass flow rates. When the velocity is in the same direction as the unit normal, there is a net flow of mass from the control volume. The integral in Equation (4.2.5) is positive, and the right-hand side of that equation is negative, representing a decrease in mass with time. Likewise, when the velocity is in a direction opposite that of the unit outward normal, there is a net flow of mass *into* the control volume. Consequently, the integral is negative and the right-hand side of Equation (4.2.5) is positive, resulting in an accumulation of mass within the control volume.

By examining specific flow situations, the integral in Equation (4.2.5) can be expressed in a way that explicitly shows the individual mass flow rates. Consider, for example, flow through a branching tube, as shown in Figure 4.1. Such a situation corresponds to flow through the bronchi of the lung or a branching blood vessel. The region depicted in the figure is the control volume. For convenience, the control surface is divided into five regions. Surface 1 is the cross-sectional area of the inlet channel normal to the direction of flow. Surface 2 is the surface bounding the walls of the main channel. Surface 3 bounds the walls of the two daughter channels. Surfaces 4 and 5 are the cross-sectional areas normal to flow in each of the daughter channels. On surface 1, the product $\mathbf{v} \cdot \mathbf{n}$ is negative, since the normal vector and the velocity vector point in opposite directions. On surfaces 2 and 3, $\mathbf{v} \cdot \mathbf{n} = 0$, because there is no flow normal to the surface. The product $\mathbf{v} \cdot \mathbf{n}$ over surfaces 4 and 5 is positive, since the velocity vector and the surface normal vector point in the same direction. Thus, we need focus only upon the flow over surfaces 1, 4, and 5.

The integral of the product of the fluid density and the differential flow rate is the mass flow rate and is denoted $\dot{m}$. Thus, for the flow shown in Figure 4.1, Equation (4.2.5) can be written as

$$\frac{dm}{dt} = \dot{m}_1 - \dot{m}_4 - \dot{m}_5. \tag{4.2.6}$$

Thus, the mass accumulates within the control volume if the mass flow into the control volume exceeds the mass flow out of the control volume.

For a system with p inlet channels and q outlet channels, Equation (4.2.5) becomes

$$\frac{dm}{dt} = \sum_{i=1}^{p} \dot{m}_i - \sum_{j=1}^{q} \dot{m}_j. \tag{4.2.7}$$

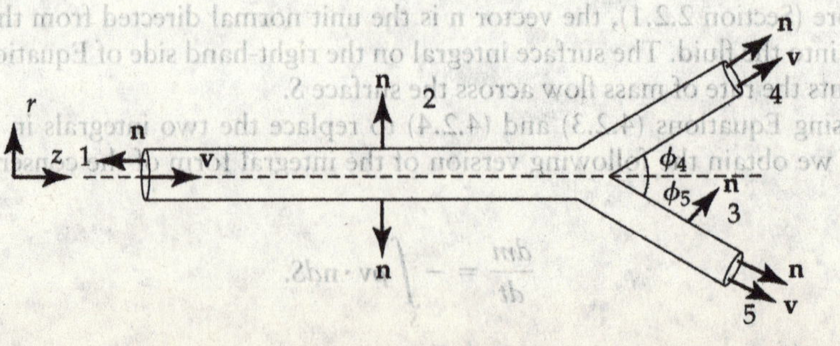

FIGURE 4.1 Flow through a branching tube, showing the various control surfaces.

If the density is constant, then the accumulation is zero, and the rate of mass flow in equals the rate of mass flow out. Alternatively, this result can be written in terms of volumetric flow rates $Q = \dot{m}/\rho$, or

$$\left(\sum_{i=1}^{p} Q_i\right)_{\text{in}} = \left(\sum_{j=1}^{q} Q_j\right)_{\text{out}}. \tag{4.2.8}$$

Example 4.1 In the kidney glomerulus, water and small-molecular-weight solutes permeate across the glomerular capillaries into the Bowman's space (Figure 4.2). The flow rate entering the afferent arteriole is 3.0 nL min^{-1}, and the flow rate filtering across the endothelium is 0.45 nL min^{-1}. Determine the volumetric flow rate leaving the capillary and entering the efferent arteriole. Assume steady flow.

Solution Divide the surface of the control volume into three surfaces: surface 1, the inlet in the afferent arteriole normal to the direction of flow; surface 2, a surface along the walls of the capillary; and surface 3, an outlet on the efferent arteriolar side normal to the direction of flow. The capillary endothelium is permeable to water and low-molecular-weight solutes, so flow through surface 2 is nonzero. Thus,

$$Q_1 = Q_2 + Q_3. \tag{4.2.9}$$

From the data provided, Q_3 is 2.55 nL min^{-1}. About 15% of the fluid entering the capillary is filtered across the glomerular membrane. Most of this fluid entering the glomerulus is reabsorbed by the blood in the proximal tubule. Nevertheless, the loss of fluid from the capillaries does increase the concentration of solutes leaving the glomerular capillaries. This solute concentration affects solute and water transport across the capillary. The fluid velocity is not constant and varies with distance along the capillary. A simple model of glomerular filtration is presented in Example 4.7 and is discussed in more detail in Chapter 14.

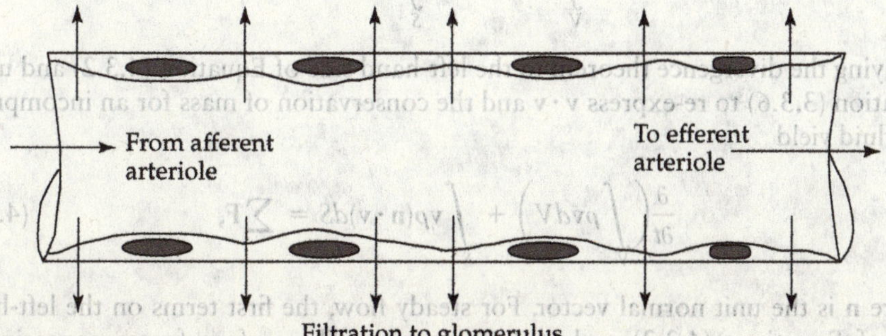

FIGURE 4.2 Schematic of flow in the capillaries of the kidney glomerulus.

4.3 Integral Form of the Equation of Conservation of Linear Momentum

To develop integrated forms of the equation of conservation of linear momentum, we start with the differential form of the conservation of linear momentum, Equation (3.3.16):

$$\rho\left(\frac{\partial \mathbf{v}}{\partial t} + \mathbf{v}\cdot\nabla\mathbf{v}\right) = -\nabla p + \nabla\cdot\boldsymbol{\tau} + \rho\mathbf{g}. \tag{4.3.1}$$

Integrating each term of this equation over the control volume V yields

$$\int_{v} \rho\left(\frac{\partial \mathbf{v}}{\partial t} + \mathbf{v}\cdot\nabla\mathbf{v}\right)dV = \int_{V} (-\nabla p + \nabla\cdot\boldsymbol{\tau} + \rho\mathbf{g})dV. \tag{4.3.2}$$

The terms on the left-hand side of Equation (4.3.2) are the rate of change with time of the total momentum $d\mathbf{P}/dt$ of the system, due to the accumulation and flow of momentum. The right-hand side of Equation (4.3.2) represents the different forces acting on the system. Thus, Equation (4.2.2) can be explicitly written as Newton's second law of motion for a fluid:

$$\frac{d\mathbf{P}}{dt} = \sum \mathbf{F}. \tag{4.3.3}$$

Equation (4.3.3) represents one form of the integrated, or macroscopic, form of the conservation of linear momentum. The summation is performed over the entire surface for surface forces (pressure and viscous stresses) and over the entire control volume for body forces (e.g., gravity). In many problems, the inlet and outlet surfaces for the flows are one-dimensional, and the fluid properties are assumed to be constant over each inlet and outlet.

The forces are described by converting the volume integral for each force into a surface integral. The divergence theorem (Section A.3.C) transforms a volume integral into a surface integral:

$$\int_{V} \nabla\cdot\mathbf{v}dV = \int_{S} \mathbf{n}\cdot\mathbf{v}dS. \tag{4.3.4}$$

Applying the divergence theorem to the left-hand side of Equation (4.3.2) and using Equation (3.3.6) to re-express $\mathbf{v}\cdot\mathbf{v}$ and the conservation of mass for an incompressible fluid yield

$$\frac{\partial}{\partial t}\left(\int_{V} \rho\mathbf{v}dV\right) + \int_{S} \mathbf{v}\rho(\mathbf{n}\cdot\mathbf{v})dS = \sum \mathbf{F}, \tag{4.3.5}$$

where $\mathbf{n}$ is the unit normal vector. For steady flow, the first terms on the left-hand side of Equations (4.3.2) and (4.3.5) are zero. The surface forces appearing in

Equation (4.3.2) are usually rewritten by applying the divergence theorem. For the second term on the right-hand side of that equation, the divergence theorem gives

$$\int_V \nabla \cdot \boldsymbol{\tau} dV = \int_S \mathbf{n} \cdot \boldsymbol{\tau} dS. \tag{4.3.6}$$

Thus, the right-hand side of Equation (4.3.5) becomes

$$\sum \mathbf{F} = -\int_S p\mathbf{n} dS + \int_S \mathbf{n} \cdot \boldsymbol{\tau} dS + m\mathbf{g}, \tag{4.3.7}$$

where m is the total mass of the system. Combining Equations (4.3.5) and (4.3.7) yields the following relationship:

$$\frac{\partial \int_V \rho \mathbf{v} dV}{\partial t} + \int_S \mathbf{v}\rho(\mathbf{n} \cdot \mathbf{v}) dS = -\int_S p\mathbf{n} dS + \int_S \mathbf{n} \cdot \boldsymbol{\tau} dS + m\mathbf{g}. \tag{4.3.8}$$

Equation (4.3.8) is the integral form of the equation of conservation of linear momentum. It is a vector equation and can be resolved into components in each of the three orthogonal axes of an appropriate coordinate system. Thus, there is one equation for each of the directions in which flow is present.

Example 4.2 For steady, laminar flow through the branching vessels shown in Figure 4.1, determine the net force exerted by the fluid on the solid surface. The two branches are positioned in the horizontal midplane of the entry tube, and outflow surfaces 4 and 5 are inclined at an angle of 30 degrees relative to the midplane of the tube. Assume that flow in each of the branches is far enough downstream that a fully developed flow is reestablished. Use the following conditions: $Q_5 = 0.8\, Q_4$; $R_4 = R_5 = 0.9R_1$.

Solution From the conservation of mass for steady flow, Equation (4.2.8) becomes

$$Q_1 = Q_4 + Q_5 = 1.8\, Q_4. \tag{4.3.9}$$

As a result, $Q_4 = 0.556\, Q_1$, and $Q_5 = 0.444\, Q_1$.

Since flow is steady, the time derivative in Equation (4.3.8) is zero. Further, the gravitational force is negligible because the branches are in a horizontal plane and flow is not driven by gravity. Thus, Equation (4.3.8) reduces to

$$\int_S \mathbf{v}\rho(\mathbf{n} \cdot \mathbf{v}) dS = -\int_S p\mathbf{n} dS + \int_S \boldsymbol{\tau} \cdot \mathbf{n} dS. \tag{4.3.10}$$

The product $\mathbf{n} \cdot \mathbf{v}$ is nonzero only over surfaces 1, 4, and 5. Since the flow is fully developed and laminar, the velocities are given by Equation (2.7.41). The maximum velocity can be written as $2Q/\pi R^2$. From the information provided, the term on the left-hand side of Equation (4.3.10) is

$$\int_S \mathbf{v}\rho(\mathbf{n}\cdot\mathbf{v})dS = -\int_{S_1} \rho v_1^2 \mathbf{e}_z dS + \int_{S_4} \rho v_4^2(\mathbf{e}_z \cos\phi_4 + \mathbf{e}_r \sin\phi_4)dS$$

$$+ \int_{S_2} \rho v_5^2(\mathbf{e}_z \cos\phi_5 + \mathbf{e}_r \sin\phi_5)dS. \qquad (4.3.11a)$$

For the z component of Equation (4.3.11a),

$$\mathbf{e}_z\cdot\int_S \mathbf{v}\rho(\mathbf{n}\cdot\mathbf{v})dS = -\rho\left(\frac{2Q_1}{\pi R_1^2}\right)^2 \int_{r_1=0}^{R_1}\int_{\theta=0}^{2\pi} r_1\left(1-\frac{r_1^2}{R_1^2}\right)^2 dr_1 d\theta$$

$$+ \rho\cos\phi_4\left(\frac{2Q_4}{\pi R_4^2}\right)^2 \int_{r_4=0}^{R_4}\int_{\theta=0}^{2\pi} r_4\left(1-\frac{r_4^2}{R_4^2}\right)^2 dr_4 d\theta$$

$$+ \rho\cos\phi_5\left(\frac{2Q_5}{\pi R_5^2}\right)^2 \int_{r_5=0}^{R_5}\int_{\theta=0}^{2\pi} r_5\left(1-\frac{r_5^2}{R_5^2}\right)^2 dr_5 d\theta. \qquad (4.3.11b)$$

Each integral on the right-hand side of Equation (4.3.11b) equals $\pi R^2/3$. Thus, Equation (4.3.11b) reduces to

$$\mathbf{e}_z\cdot\int_S \mathbf{v}\rho(\mathbf{n}\cdot\mathbf{v})dS = \frac{4\rho}{3\pi}\left[-\left(\frac{Q_1}{R_1}\right)^2 + \cos\phi_4\left(\frac{Q_4}{R_4}\right)^2 + \cos\phi_5\left(\frac{Q_5}{R_5}\right)^2\right]. \qquad (4.3.11c)$$

Substituting for Q_4 and Q_5 in terms of Q_1 and for R_4 and R_5 in terms of R_1 yields

$$\mathbf{e}_z\cdot\int_S \mathbf{v}\rho(\mathbf{n}\cdot\mathbf{v})dS = \frac{4\rho}{3\pi}\left(\frac{Q_1}{R_1}\right)^2\left\{-1 + \cos\phi_4\left(\frac{0.556}{0.9}\right)^2 + \cos\phi_5\left(\frac{0.444}{0.9}\right)^2\right\}$$

$$= -0.1947\rho\left(\frac{Q_1}{R_1}\right)^2. \qquad (4.3.11d)$$

The z component of the convective acceleration is negative, due to deceleration of the fluid resulting from the increase in total cross-sectional area through the expansion. For the r component, the integral through surface 1 is zero. As a result,

$$\mathbf{e}_r\cdot\int_S \mathbf{v}\rho(\mathbf{n}\cdot\mathbf{v})dS = \frac{4\rho}{3\pi}\left(\frac{Q_1}{R_1}\right)^2\left\{0 + \sin\phi_4\left(\frac{0.556}{0.9}\right)^2 + \sin\phi_5\left(\frac{0.444}{0.9}\right)^2\right\}$$

$$= 0.1326\rho\left(\frac{Q_1}{R_1}\right)^2. \qquad (4.3.11e)$$

The right-hand side of Equation (4.3.11e) represents the net force. The pressure integral over surfaces 2 and 3 is zero, leaving only the terms at surfaces 1, 4, and 5.

Assuming that the pressure is uniform over each region, the first term on the right-hand side of Equation (4.3.8) is

$$-\int_S p\mathbf{n}dS = \mathbf{e}_z p_1 \pi R_1^2 - (\mathbf{e}_z \cos\phi_4 + \mathbf{e}_r \sin\phi_4)p_4\pi R_4^2$$
$$- (\mathbf{e}_z \cos\phi_5 + \mathbf{e}_r \sin\phi_5)p_5\pi R_5^2. \quad (4.3.12a)$$

This equation can be rewritten as

$$-\int_S p\mathbf{n}dS = \mathbf{e}_z \pi R_1^2[p_1 - 0.81p_4\cos\phi_4 - 0.81p_5\cos\phi_5]$$
$$-\mathbf{e}_r 0.81\pi R_1^2[p_4 \sin\phi_4 + p_5\sin\phi_5]. \quad (4.3.12b)$$

For flow through surfaces 1, 4, and 5, the force due to stresses of the fluid passing through cross-sectional areas parallel to the flow is generally small and is neglected [1]. The fluid stresses acting on the solid surfaces are nonzero and cannot be neglected. Thus, the momentum equation reduces to the pair of equations

$$-0.1947\rho\left(\frac{Q_1}{R_1}\right)^2 = \pi R_1^2[p_1 - 0.81p_4\cos\theta_4 - 0.81p_5\cos\theta_5] + \mathrm{F}_{fs_z}$$
$$= F_z(z \text{ component}) \quad (4.3.13a)$$

and

$$0.1326\rho\left(\frac{Q_1}{R_1}\right)^2 = -0.81\pi R_1^2[p_4\sin\theta_4 + p_5\sin\theta_5] + \mathrm{F}_{fs_r}$$
$$= F_r(r \text{ component}), \quad (4.3.13b)$$

where F_{fs} is the fluid shear force acting on the solid surfaces. The integration of shear stresses on surfaces 1, 4, and 5 is zero. The net force on the solid surface is simply

$$F = \sqrt{F_r^2 + F_z^2} = 0.2356\rho\left(\frac{Q_1}{R_1}\right)^2. \quad (4.3.13c)$$

Specifying the pressure enables one to determine the fluid stresses acting on the solid surface. Note that for a straight tube, both the acceleration and the net force F are zero. The pressure and shear forces thus balance.

4.4 Bernoulli's Equation

A special case of the conservation of linear momentum, known as Bernoulli's equation, can be derived for steady frictionless flow along a streamline. Recall that a streamline is a line whose tangent is everywhere parallel to the velocity vector v

(Section 2.2.5). For frictionless flow at steady state, the equation of conservation of linear momentum reduces to

$$\rho\mathbf{v}\cdot\nabla\mathbf{v} = -\nabla p + \rho\mathbf{g}. \tag{4.4.1}$$

Equation (4.4.1) indicates that the force arising from convective acceleration is balanced by pressure and gravity forces. Gravity acts in the negative z-direction. It is more helpful to express gravity in terms of the potential $\phi = zg$, or $\rho\mathbf{g} = -\rho\nabla\phi$. Thus,

$$0 = \rho\mathbf{v}\cdot\nabla\mathbf{v} + \nabla p + \rho\nabla\phi. \tag{4.4.2}$$

Applying this result along a streamline (Figure 4.3), we obtain the component of Equation (4.4.2) tangent to a streamline. The unit vector $\mathbf{e}_t$ tangent to all the streamlines is represented as

$$\mathbf{e}_t = \frac{\partial\mathbf{r}}{\partial s}, \tag{4.4.3}$$

where $\mathbf{r} = x\mathbf{e}_x + y\mathbf{e}_y + z\mathbf{e}_z$ is the direction vector and s is the local distance along the streamline.

The component of Equation (4.4.2) tangent to a streamline is the dot product of that equation and $\mathbf{e}_t$, so

$$0 = \mathbf{e}_t\cdot(\rho\mathbf{v}\cdot\nabla\mathbf{v} + \nabla p + \rho\nabla\phi). \tag{4.4.4}$$

The dot product of $\mathbf{e}_t$ and the first term on the right-hand side of Equation (4.4.4) is

$$\mathbf{e}_t\cdot\rho\mathbf{v}\cdot\nabla\mathbf{v} = \rho\frac{\partial\mathbf{r}}{\partial s}\cdot\left(\mathrm{v}_x\frac{\partial\mathbf{v}}{\partial x} + \mathrm{v}_y\frac{\partial\mathbf{v}}{\partial y} + \mathrm{v}_z\frac{\partial\mathbf{v}}{\partial z}\right). \tag{4.4.5}$$

To express the dot product on the right-hand side of Equation (4.4.5), note that $s = f(x, y, z)$. Consequently, the differential operator is

$$\frac{\partial}{\partial s} = \frac{\partial x}{\partial s}\frac{\partial}{\partial x} + \frac{\partial y}{\partial s}\frac{\partial}{\partial y} + \frac{\partial z}{\partial s}\frac{\partial}{\partial z}. \tag{4.4.6}$$

As a result, it can be shown [2] that Equation (4.4.5) becomes

$$\mathbf{e}_t\cdot\rho\mathbf{v}\cdot\nabla\mathbf{v} = \rho\mathrm{v}\frac{\partial\mathrm{v}}{\partial s}, \tag{4.4.7}$$

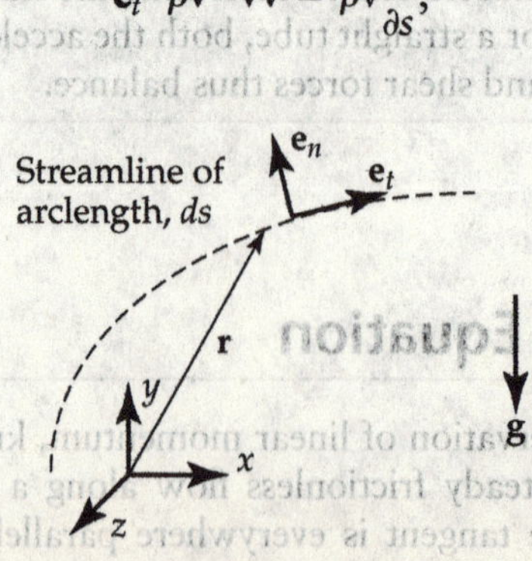

FIGURE 4.3 Flow streamline showing normal and tangent vectors.

where

$$v = \sqrt{v_x^2 + v_y^2 + v_z^2} \tag{4.4.8}$$

is the magnitude of the velocity.

Lastly, the right-hand side of Equation (4.4.7) can be expressed as follows:

$$\rho v \frac{\partial v}{\partial s} = \frac{\rho}{2} \frac{\partial v^2}{\partial s}. \tag{4.4.9}$$

The dot products of the second and third terms on the right-hand side of Equation (4.4.4) can be found from Equation (4.4.6) thus:

$$\frac{\partial \mathbf{r}}{\partial s} \cdot \nabla = \frac{\partial x}{\partial s} \frac{\partial}{\partial x} + \frac{\partial y}{\partial s} \frac{\partial}{\partial y} + \frac{\partial z}{\partial s} \frac{\partial}{\partial z} = \frac{\partial}{\partial s}. \tag{4.4.10}$$

Hence, along a streamline, Equation (4.4.4) can be rewritten as

$$\mathbf{e}_t \cdot (\rho \mathbf{v} \cdot \nabla \mathbf{v} + \nabla p + \rho \nabla \phi) = \frac{\rho}{2} \frac{\partial v^2}{\partial s} + \frac{\partial p}{\partial s} + \rho \frac{\partial \phi}{\partial s} = 0. \tag{4.4.11a}$$

Substituting gz for ϕ and multiplying both sides of Equation (4.4.11a) by ds gives

$$\frac{\rho}{2} dv^2 + dp + \rho g dz = 0. \tag{4.4.11b}$$

Integrating Equation (4.4.11b) yields

$$\frac{1}{2} \rho v^2 + p + \rho g z = \text{constant.} \tag{4.4.11c}$$

The constant of integration is different for each streamline. Equations (4.4.11b) and (4.4.11c) are, respectively, the differential and integrated forms of Bernoulli's equation and are extremely helpful in characterizing velocity in many flow situations. These forms of Bernoulli's equation are valid for isothermal, steady flow with negligible viscous losses or chemical reactions. (A more general form of Bernoulli's equation is presented in Section 4.4.2.) Viscous losses are significant near solid surfaces where velocity gradients are large. In addition, viscous losses are significant when the flow conditions change dramatically, as in expansions, contractions, or branches. If the changes are gradual, then viscous losses are minimal and Bernoulli's equation applies. The next example applies Bernoulli's equation. The choice of the streamline is important in solving the problem and avoiding situations in which viscous losses are significant.

Example 4.3 Consider steady laminar flow through a cylindrical channel expansion in a horizontal channel, shown in Figure 4.4. The channel diameter increases gradually and frictional forces can be neglected. Determine a relationship between the pressure drop and velocity at locations 1 and 2 (i.e., the velocity is a function of y alone).

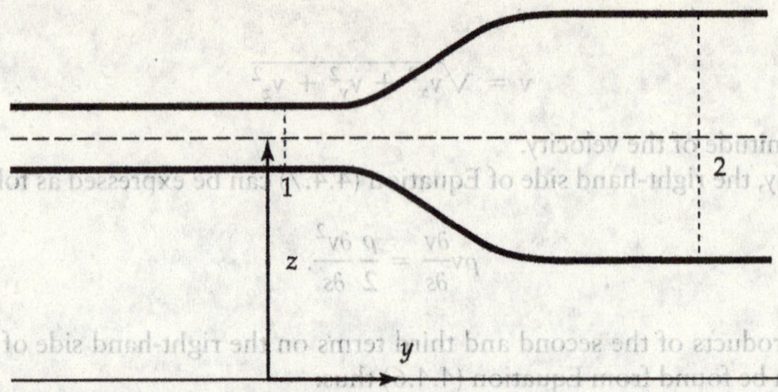

FIGURE 4.4 Flow through an expanding channel.

Solution The most appropriate streamline to choose is the one along the centerline of the inlet tube, shown by the dashed line in the figure. This is one streamline that does not change vertical position. As a result, the gravity term vanishes. Equation (4.4.10b) along the centerline streamline reduces to

$$\frac{\rho}{2}dv^2 = -dp. \tag{4.4.12}$$

The opening is gradual so that flow recirculation does not occur.

To solve Equation (4.4.12), choose two points along the streamline, one upstream of the expansion and the other downstream from the expansion. From the conservation-of-mass equation, the flow rate entering at point 1 equals the flow rate at any point along the flow channel. Applying Equation (4.2.5) for steady flow, we have

$$\langle v_1 \rangle A_1 = \langle v_2 \rangle A_2. \tag{4.4.13}$$

Since the cross-sectional area at location 2 is greater than the cross-sectional area at location 1, the velocity decreases upon entering the expansion. Consequently, the change in velocity with distance along the streamline is negative (i.e., $dv < 0$).

Integrating Equation (4.4.12) between positions 1 and 2, we obtain the following expression:

$$\frac{\rho}{2}(v_1^2 - v_2^2) = p_2 - p_1. \tag{4.4.14}$$

The velocity along the centerline streamline at location 1 or 2 equals the maximum velocity or twice the average velocity. Using Equation (4.4.13) to relate the velocities at lines 1 and 2, and solving for p_2, we have

$$p_2 = p_1 + \frac{\rho}{2}v_1^2\left(1 - \left(\frac{A_1}{A_2}\right)^2\right). \tag{4.4.15}$$

Equation (4.4.15) indicates that the rise in pressure in the expansion results from fluid deceleration. The pressure change along the flow path is sketched in Figure 4.5. After the fluid enters the expansion, the pressure increases, reaches a maximum, and then decreases. Once fully developed flow is reestablished downstream from the expansion, the pressure decreases.

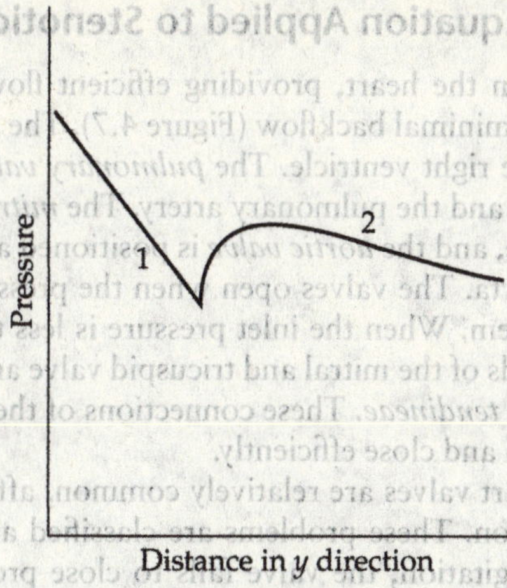

FIGURE 4.5 Pressure changes for flow through an expanding channel.

Example 4.4 Consider high-speed turbulent flow through a narrow constriction, as shown in Figure 4.6. Assuming that the velocity profiles are uniform in turbulent flow (i.e., the time-averaged velocity v is a constant and does not vary with distance from the wall), determine v_2 for a known pressure difference.

Solution For steady flow, Bernoulli's equation is applicable. Again, the flow is horizontal and $z_1 = z_2$. For uniform flow, the conservation of mass yields

$$v_1 A_1 = v_2 A_2. \tag{4.4.16}$$

Solving Equation (4.4.16) for v_1, substituting into Equation (4.4.11c), and rearranging yields the following expression for v_2:

$$v_2 = \sqrt{\frac{2(p_1 - p_2)}{\rho\left(1 - \left(A_2^2/A_1^2\right)\right)}}. \tag{4.4.17}$$

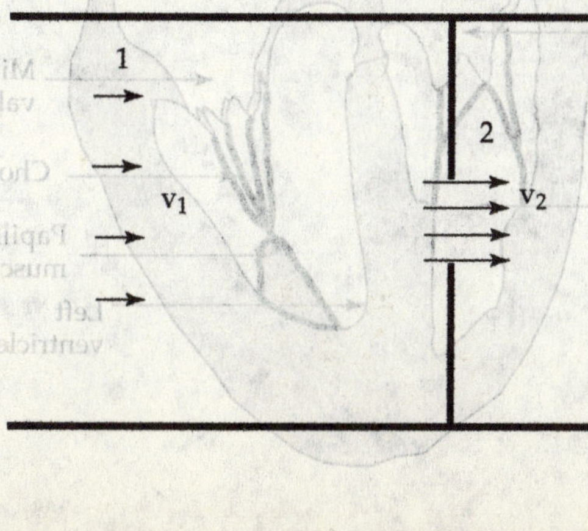

FIGURE 4.6 High-speed flow through a narrow constriction.

4.4.1 Bernoulli's Equation Applied to Stenotic Heart Valves

There are four valves in the heart, providing efficient flow of blood through the organ's chambers, with minimal backflow (Figure 4.7). The *tricuspid valve* separates the right atrium and the right ventricle. The *pulmonary valve* lies between the outflow of the right atrium and the pulmonary artery. The *mitral valve* lies between the left atrium and ventricle, and the *aortic valve* is positioned at the exit of the left ventricle as it meets the aorta. The valves open when the pressure above them exceeds the pressure beneath them. When the inlet pressure is less than the outlet pressure, the valves close. The ends of the mitral and tricuspid valve are connected to papillary muscles via the *chordae tendineae*. These connections of the valves to the heart wall allow the valves to open and close efficiently.

Pathologies of heart valves are relatively common, affecting a fairly large percentage of the population. These problems are classified as either *regurgitation* or *stenosis*. In valve regurgitation, the valve fails to close properly, causing blood to flow back into the atria or ventricles. Valve regurgitation is most common in the aortic valve, followed by the mitral valve. This pathology is a result of degeneration of the aortic valve, rheumatic heart disease, bacterial endocarditis, trauma, or damage to the papillary muscle connected to the mitral valve [3]. Mild forms of regurgitation may cause few symptoms, but more severe regurgitation causes shortness of breath and abnormal heartbeats.

A stenosis is a narrowing within a flow channel. With respect to valves, the term *stenosis* refers to a decreased area of opening of the valve. This pathology can arise from degeneration and calcification of the valves for unknown reasons, congenital abnormalities, bacterial endocarditis, or rheumatic heart fever. In some cases, the valve structure is normal, but the aorta downstream from the valve is constricted due to a congenital malformation.

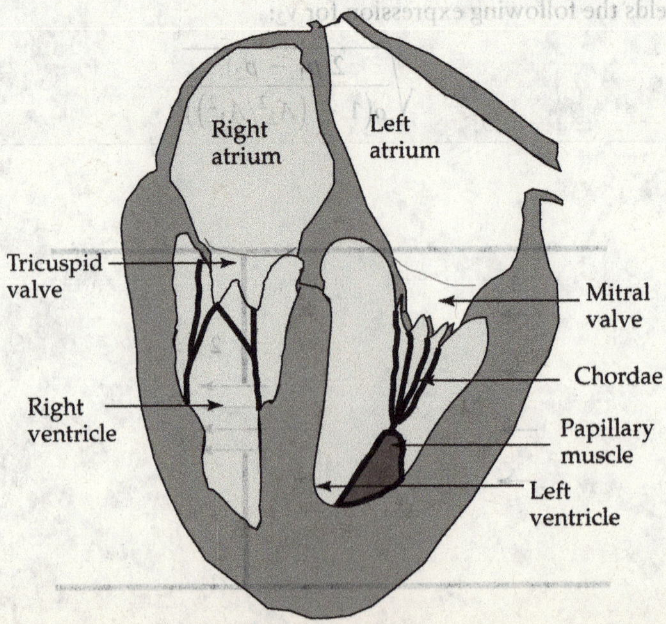

FIGURE 4.7 A cross-section through the heart, showing the mitral and tricuspid valves. These valves are connected to the ventricles via chordae attached to papillary muscles.

Valvular stenosis causes the heart to expend more work in pumping blood through the body. The heart compensates for the narrowing by increasing its muscle mass (hypertrophy); however, this often leads to secondary complications that degrade the overall function of the organ. This decrease in heart function is known as heart failure. Ultimately, valvular stenosis can produce serious health problems and premature death.

Mild cases of stenosis or regurgitation can be treated with drugs to regulate heart contraction and limit blood clot formation. About 90,000 operations are performed each year to treat valvular disorders. Most involve replacement of the aortic or mitral valves. About 19,600 deaths per year in the United States are attributed to complications arising from valvular heart disease, often congestive heart failure [4]. The hemodynamics of replacement valves is discussed in Chapter 5.

A modification of Bernoulli's equation is used to determine the pressure drops across normal and stenotic heart valves. The severity of the stenosis is determined by the pressure gradient across the valve. In turn, this pressure gradient is a measure of the additional work that the heart must do to pump blood through the valve. Stenotic valves can be diagnosed noninvasively[1] or invasively. The most common noninvasive method now in use is ultrasound, referred to as *echocardiography* when it is used to diagnose anatomic and functional changes to the heart. *Doppler ultrasound* is utilized to determine the fluid velocity, which is obtained from the frequency shift between the transmitted and received signals. Bernoulli's equation is used to relate the velocity through the valve to the pressure drop across the valve.

Equation (4.4.11c) is the most common form of Bernoulli's equation used to relate velocity and pressure. Small changes in height are neglected so that the gravity term is ignored. For flow through valves, site 1 is upstream in the atrium or ventricle and site 2 is at the valve opening (Figure 4.8). Using a value of 1,070 kg m^{-3} in SI units for the density of blood and a velocity in meters per second, we find that the pressure drop in mmHg (1 mmHg = 133.32 Pa) is related to the velocity as [5,6]

$$p_1 - p_2 = 4(v_2^2 - v_1^2). \tag{4.4.18}$$

Since physicians commonly report pressure in units of mmHg, the constant 4 has units of square seconds per square meter mmHg. When the velocity through the valve is much greater than the velocity in the heart chamber, this equation can be simplified to

$$p_1 - p_2 = 4v_{max}^2, \tag{4.4.19}$$

where v_{max} is the maximum jet velocity through the valve.

Alternatively, pressure can be measured directly with the use of cardiac catheterization, an invasive procedure in which a catheter is inserted into a vein or artery in the leg and guided to the heart. Cardiac catheterization is a widely used procedure and is often performed on an outpatient basis. While it is generally safe, there is a small risk of complications. Such a procedure is warranted to assess the complications of stenotic

[1]Noninvasive medical procedures are those that do not require surgery or entry into the body by incision. Examples of noninvasive procedures include most imaging procedures, such as ultrasound and magnetic resonance imaging scanning. Minimally invasive procedures involve incisions and the sampling of fluids or insertion of catheters or cameras.

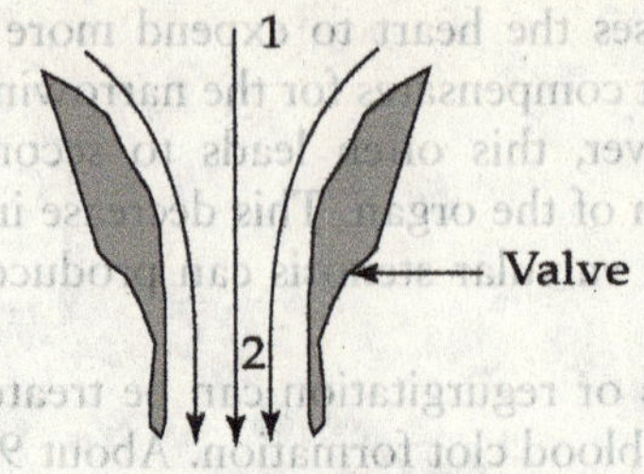

FIGURE 4.8 Schematic of open valve showing location of points 1 and 2 along streamline for application of Bernoulli's equation.

valves, but noninvasive echocardiography can provide initial confirmation of the presence and severity of stenosis.

4.4.2 The Engineering Bernoulli Equation: The Effects of Viscous Losses and Time-Dependent Energy Changes

Bernoulli's equation applies when viscous losses and work on the fluid are negligible. Often, this is not the case. An examination of Equations (4.4.11b) and (4.4.11c) indicates that each term has units of energy per unit volume. Thus, Bernoulli's equation is an energy balance, and a more general form of the equation has been developed from the conservation of mechanical energy [1,7]. In this section, an approximate approach is used to generalize Bernoulli's equation and to estimate the effect of viscous losses and changes in energy with time.

In an *open system,*[2] in which mass and energy are exchanged across system boundaries, energy accumulates within the system due to a net energy generation (or loss) and energy transfer across the system's boundaries. The system energy consists of the kinetic, potential, electrical, chemical, and all other forms of energy. The first law of thermodynamics states that the energy change of a system equals the work done on the system, plus the heat transferred to the system. The energy added or removed is transported by mass, which enters or exits the system. In differential form, the rate of change of the system energy is

$$\frac{dE}{dt} + \int_S \rho \hat{E} \mathbf{v} \cdot \mathbf{n} dS = \frac{dQ}{dt} + \frac{dW}{dt}, \tag{4.4.20}$$

where E is the total system energy and $\hat{E}$ is the energy per unit mass. The total system energy represents the sum of the internal (E_i), mechanical (E_M), electrical (E_e), and chemical (E_c) energy of the system. Work done by or on the system includes mechanical work[3] (W_M, sometimes referred to as shaft work), pressure–volume work (W_{PV}), and other forms of work. The fluid does work to overcome viscous-energy losses (W_V). This work represents an irreversible loss of energy that is converted to heat. Work is positive when done by the surroundings on the system and negative when

[2]A system is identical to a control volume.

[3]Mechanical work is work done by moving parts, such as turbines and propellers. Within biological systems, work is generated by pressure–volume work during the contraction of heart and other tissues and the conversion of chemical energy into mechanical energy, such as the movement of muscle by the interaction of actin and myosin.

done by the system on the surroundings. Likewise, heat is positive when released by the surroundings and added to the system and negative when released by the system and added to the surroundings.

For an isothermal, incompressible fluid with no chemical reactions and negligible viscous heating, only kinetic and potential energy are important. These two kinds of energy are given by

$$E_K = \int_V \frac{1}{2}\rho v^2 dV \quad \text{and} \quad E_P = \int_V \rho gz dV, \qquad (4.4.21a,b)$$

where v is the magnitude of the velocity vector. The work done by or on the system is mechanical work, pressure–volume work, or work performed to overcome viscous stresses. With these assumptions, the first law of thermodynamics for an incompressible fluid becomes

$$\frac{d(E_K + E_P)}{dt} = -\int_S \rho\left(\frac{1}{2}v^2 + gz\right)\mathbf{v}\cdot\mathbf{n}dS + \frac{dW_M}{dt} + \frac{dW_{PV}}{dt} + E_V. \qquad (4.4.22)$$

The pressure–volume work due to the deformation of the system's boundaries is

$$\frac{dW_{PV}}{dt} = -\int_S p\mathbf{v}\cdot\mathbf{n}dS. \qquad (4.4.23)$$

The mechanical-work term is specific to the type of process that is occurring. The following expression can be developed for the rate of energy loss to viscous forces [1,7]:

$$E_V = \int_V (\boldsymbol{\tau}\cdot\nabla)\cdot\mathbf{v}dV. \qquad (4.4.24)$$

Expressions exist for energy losses due to viscous dissipation in laminar and turbulent flow [1], but these terms can often be difficult to determine in a given situation. Empirical correlations of viscous energy losses are available for entrance regions, bends, and other geometries in pipes and ducts [1,7].

Inserting Equation (4.4.23) into Equation (4.4.22) and collecting terms integrated over the surface yields

$$\frac{d(E_K + E_P)}{dt} = -\int_S \left(\frac{1}{2}\rho v^2 + \rho gz + p\right)\mathbf{v}\cdot\mathbf{n}dS + \frac{dW_M}{dt} + E_V. \qquad (4.4.25)$$

Equation (4.4.25) represents what is known as the engineering Bernoulli equation for an isothermal system without chemical reactions. Bernoulli's equation itself is obtained from Equation (4.4.25) by assuming a steady flow without mechanical work or viscous dissipation and integrating along a tube surface formed by streamlines. Then, let tube diameter go to zero.

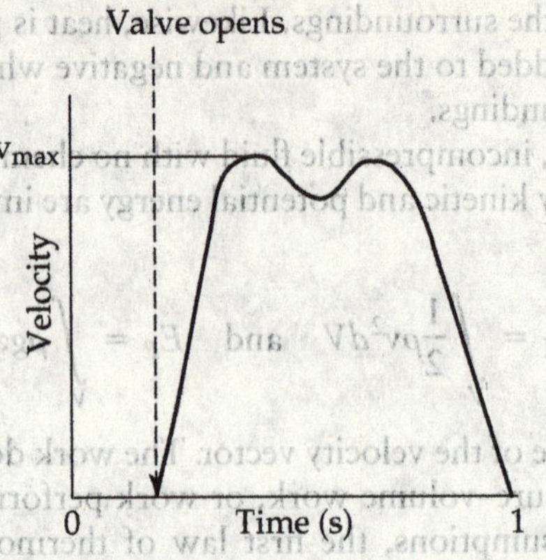

FIGURE 4.9 Schematic of velocity variation across a stenotic valve. (Adopted from Ref. [6].)

Equation (4.4.25) has been used to assess possible errors arising from viscous losses and unsteady flow during the rise in pressure and flow as a valve opens (systole) and during the period when the velocity and pressure decrease (diastole). A typical velocity profile during a pressure pulse is shown in Figure 4.9. For flow between the valves, no mechanical work is done (although valve movement contributes to viscous-energy losses), and height changes lead to minimal changes in the gravity and potential-energy terms. Thus, only the kinetic energy, pressure, and viscous-loss terms are important. Then, Equation (4.4.25) becomes

$$\frac{d}{dt}\left(\int_V \frac{1}{2}\rho v^2 dV\right) = -\int_S \left(\frac{1}{2}\rho v^2 + p\right)\mathbf{v}\cdot\mathbf{n}dS + E_V. \qquad (4.4.26)$$

Next, Equation (4.4.26) is applied along a streamline that passes through the valve opening. In this case, the volume and surface area are constant, and the time derivative can be rewritten as

$$\frac{d}{dt}\left(\int_V \frac{1}{2}\rho v^2 dV\right) = \int_V \frac{1}{2}\rho\frac{dv^2}{dt}dV = \rho\int_V v\frac{dv}{dt}dV. \qquad (4.4.27)$$

Inserting Equation (4.4.27) into Equation (4.4.26) and integrating along the streamline results in the following expression:

$$p_1 - p_2 = \frac{\rho}{2}\left(v_2{}^2 - v_1{}^2\right) + \rho\int_{l=1}^{2}\frac{dv}{dt}dl + \frac{E_V}{\int_S \mathbf{v}\cdot\mathbf{n}dS}. \qquad (4.4.28)$$

The second term on the right-hand side of Equation (4.4.28) represents acceleration due to a time-varying change in velocity between locations 1 and 2 in Figure 4.9. The viscous losses E_V can be estimated from the boundary layer theory described in Section 4.5. For mitral stenosis, the boundary layer is quite thin and viscous forces are negligible (see Problem 4.6).

The differential distance along the streamline is represented by dl. The magnitudes of the terms in Equation (4.4.28) have been estimated for the stenotic mitral valve [5]. Maximum velocities in the valve range from 1 to 3 m s^{-1}. Typical peak velocities in the left atrium are 0.2 m s^{-1}. The pressure drop due to convective acceleration (the first term on the right-hand side of the equation) ranges from 3.8 to 36 mmHg. As the valve opens and closes, the acceleration due to time-varying velocities increases with the square of the distance [8,9]. Thus, the integral in Equation (4.2.28) can be estimated as

$$\rho \int_{l=1}^{2} \frac{dv}{dt} dl \approx \frac{\rho l}{3} \frac{dv_2}{dt}. \tag{4.4.29}$$

The time-dependent component of the fluid acceleration is typically 30 − 40 m s^{-2}, and the length from the center of the atrium to the valve opening is 0.05 m. The resulting pressure generated by fluid acceleration during the opening and closing of the valve is 4–5 mmHg. This pressure is quite significant relative to the pressure that is due to convective acceleration. During the period of peak velocity, the time-varying acceleration term in the equation drops to about 0.2 mmHg, and this term does not contribute significantly to the total pressure drop.

Overall, Equations (4.4.18) and (4.4.19) can be used to characterize pressures in normal and stenotic aortic and mitral valves in adults. Errors can arise when this approach is used in children, due to the smaller size of a child's chamber. The equations do not work well when there is significant backflow because of valve damage (e.g., regurgitation). Echocardiography is often used in conjunction with catheterization of the heart to help identify the locations of highest and lowest pressures during catheterization, even when the strict applicability of Equations (4.4.18) and (4.4.19) cannot be justified.

4.5 Boundary Layer Theory

4.5.1 Background to Boundary Layer Theory

High-Reynolds-number flows occur in the lungs and large arteries (Tables 2.3 and 2.4). Due to the importance of such flows, it is critical to describe them and to determine the forces acting on surfaces. The analysis of high-Reynolds-number flows is complicated because viscous forces affect the flow and forces near the surface of an object. Failure to account for viscous forces leads to erroneous estimates of the forces exerted by the fluid on solid surfaces.

Initial attempts to analyze high-Reynolds-number flows began in the late 19th century. The initial approach was to assume that, at high Reynolds numbers, viscous forces could be neglected throughout the entire flow domain. The resulting equation, known as the *inviscid flow equation,* or Euler equation, is

$$\rho \frac{\partial \mathbf{v}}{\partial t} + \rho \mathbf{v} \cdot \nabla \mathbf{v} = -\nabla p + \rho \mathbf{g}. \tag{4.5.1}$$

For high-Reynolds-number flows, the inviscid flow equations provide an accurate description of the flow and pressure field away from surfaces. However, the drag forces on submerged objects computed with inviscid flow theory did not agree with measured values. The central insight that resolved the disagreement between theory and experiment, developed by the German physicist Ludwig Prandtl[4] in 1904, was that viscous forces were not negligible everywhere. Viscous forces could not be neglected in a very thin region near the solid surface, known as a boundary layer. From a mathematical point of view, this makes sense, since ignoring viscous forces (i.e., the term $\mu \nabla^2 \mathbf{v}$ in the Navier–Stokes equation (3.3.25)) means removing the second derivatives in velocity. This assumption reduces the conservation of linear momentum from a system of second-order partial differential equations to a system of first-order partial differential equations. As a result, it is not possible to satisfy all of the boundary conditions. Boundary layer theory resolves this problem by dividing the flow region into two domains. In a thin region, or boundary layer, near the surface, viscous and inertial forces are of similar magnitude. Hence, two different length scales exist in the boundary layer: Normal to the surface, the dominant length scale is the boundary layer thickness δ. Parallel to the surface, the length scale is the characteristic size L of the object.

The objectives of boundary layer theory are (1) to describe the flow field near the surface, (2) to estimate the boundary layer thickness, and (3) to determine the drag force exerted on the surface. The upcoming sections provide an introduction to the basic developments of the boundary layer equations and their applications to some simple flows [10,11]. Boundary layer flows arise in arteries and the lungs. The concepts employed help to explain a number of phenomena observed in curved and branched arteries. Understanding the arguments used to perform the scaling analysis is valuable in analyzing other flows in which the equations of continuity and conservation of linear momentum are simplified.

4.5.2 Derivation of the Boundary Layer Equations

Consider flow at a uniform velocity U_0 approaching the stationary submerged object shown in Figure 4.10. A local coordinate system is chosen in which the x-coordinate is everywhere parallel to the surface and the y-coordinate is normal to the surface. At the surface of the object, the velocity in the direction tangent to the surface is zero. Far from the surface (in a region known as the free stream), viscous forces are very small and inertial forces dominate. Viscous and inertial forces are of comparable magnitude over a boundary layer of thickness $\delta \ll L$. Within the boundary layer, fluid decelerates in the x direction and causes a flow in the y direction away from the surface. The fluid velocity normal to the surface increases over the thickness of the boundary layer and ultimately equals the free-stream velocity.

Boundary layer theory permits the equations of conservation of mass and linear momentum to be simplified and the boundary layer thickness $\delta(x)$ to be estimated within an order of magnitude. The conservation of mass and linear momentum for steady two-dimensional flows is as follows:

$$\frac{\partial v_x}{\partial x} + \frac{\partial v_y}{\partial y} = 0 \quad \text{(continuity)}, \tag{4.5.2}$$

[4]Prandtl's research on boundary layer theory and the design of wings provided the basis for aerodynamics theory.

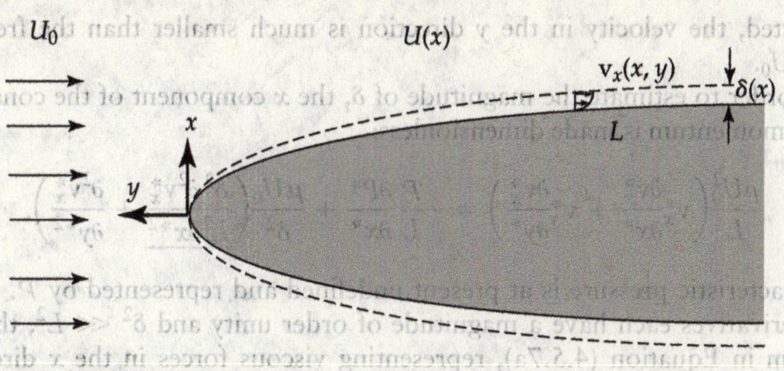

FIGURE 4.10 Generalized boundary layer flow.

$$\rho\left(v_x\frac{\partial v_x}{\partial x}+v_y\frac{\partial v_x}{\partial y}\right)=-\frac{\partial P}{\partial x}+\mu\left(\frac{\partial^2 v_x}{\partial x^2}+\frac{\partial^2 v_x}{\partial y^2}\right)\quad (x\text{ momentum}),\qquad (4.5.3a)$$

$$\rho\left(v_x\frac{\partial v_y}{\partial x}+v_y\frac{\partial v_y}{\partial y}\right)=-\frac{\partial P}{\partial y}+\mu\left(\frac{\partial^2 v_y}{\partial x^2}+\frac{\partial^2 v_y}{\partial y^2}\right)\quad (y\text{ momentum}).\qquad (4.5.3b)$$

The pressure is the sum of the thermodynamic pressure and the gravitational term ($P = p - \rho gz$). Outside the boundary layer, the velocity field is obtained by solving Equation (4.5.1).

To simplify Equations (4.5.2), (4.5.3a), and (4.5.3b), we need to cast them in dimensionless form and identify those terms which can be neglected. Proper scaling of the dimensional variables is necessary because the characteristic fluid velocity and length are different in the x and y directions. Appropriate length scales in those directions are L and $\delta(x)$, respectively. The characteristic velocity in the x direction is U_0. The characteristic velocity in the y direction is not apparent at this point in the analysis and is represented by the unknown variable V. Thus, the following dimensionless groups are defined:

$$x^* = \frac{x}{L} \qquad y^* = \frac{y}{\delta}, \qquad (4.5.4a,b)$$

$$v_x^* = \frac{v_x}{U_0} \qquad v_y^* = \frac{v_y}{V}. \qquad (4.5.4c,d)$$

Recasting the equation of continuity in dimensionless form yields

$$\frac{U_0}{L}\frac{\partial v_x^*}{\partial x^*}+\frac{V}{\delta}\frac{\partial v_y^*}{\partial y^*}=0. \qquad (4.5.5)$$

The dimensionless derivatives have been appropriately scaled, and each term has a magnitude of about unity. Since the flow is truly two-dimensional, V/δ is of the same order of magnitude as U_0/L. (Such a relation is denoted with the symbol $\sim$.) Although the characteristic velocity in the y direction is much smaller than U_0, the velocity gradients in the x and y directions are comparable. Rearranging terms, we obtain the following expression for V:

$$V \sim U_0\frac{\delta}{L} \ll U_0. \qquad (4.5.6)$$

As expected, the velocity in the y direction is much smaller than the free-stream velocity U_0.

In order to estimate the magnitude of δ, the x component of the conservation of linear momentum is made dimensionless:

$$\frac{\rho U_0^2}{L}\left(v_x^*\frac{\partial v_x^*}{\partial x^*} + v_y^*\frac{\partial v_x^*}{\partial y^*}\right) = -\frac{\mathcal{P}}{L}\frac{\partial P^*}{\partial x^*} + \frac{\mu U_0}{\delta^2}\left(\underline{\frac{\delta^2}{L^2}\frac{\partial^2 v_x^*}{\partial x^{*2}}} + \frac{\partial^2 v_x^*}{\partial y^{*2}}\right). \quad (4.5.7a)$$

The characteristic pressure is at present undefined and represented by $\mathcal{P}$. Since the scaled derivatives each have a magnitude of order unity and $\delta^2 \ll L^2$, the underlined term in Equation (4.5.7a), representing viscous forces in the x direction, is much smaller than viscous forces in the y direction and can be neglected.

After the term multiplied by δ^2/L^2 is neglected, further rearrangement of Equation (4.5.7a) yields

$$\frac{\rho U_0 \delta^2}{\mu L}\left(v_x^*\frac{\partial v_x^*}{\partial x^*} + v_y^*\frac{\partial v_x^*}{\partial y^*}\right) = -\frac{\mathcal{P}\delta^2}{\mu U_0 L}\frac{\partial P^*}{\partial x^*} + \frac{\partial^2 v_x^*}{\partial y^{*2}}. \quad (4.5.7b)$$

In the boundary layer, viscous forces are comparable to inertial forces. Consequently, the group of dimensional quantities multiplying the inertial terms on the left-hand side of Equation (4.5.7b) must be first-order quantities. Rearranging and solving for δ yields

$$\delta \sim \left(\frac{\mu L}{\rho U_0}\right)^{1/2} \sim L\mathrm{Re}_L^{-1/2}, \quad (4.5.7c)$$

where $\mathrm{Re}_L = \rho U_0 L/\mu$ is the Reynolds number based upon the characteristic dimensions of the object. Thus, the boundary layer thickness scales with the square root of the reciprocal of the Reynolds number. As inertial forces become greater, the boundary layer thickness becomes smaller. The numerical value of the constant of proportionality in Equation (4.5.7c) is obtained by solution of the boundary layer equations.

Now that the magnitude of δ has been estimated, the characteristic pressure can be estimated. The pressure gradient in the x direction drives the flow past the object and cannot be neglected. Hence, $\mathcal{P}\delta^2/\mu U_0 L$, the term multiplying the dimensionless pressure in Equation (4.5.7b), must be first order. Replacing δ with Equation (4.5.7c) and solving for $\mathcal{P}$ yields

$$\mathcal{P} \sim \rho U_0^2. \quad (4.5.7d)$$

Therefore, the characteristic pressure is proportional to the inertial forces.

Applying the same logic of non-dimensionalization to the y component of the conservation of linear momentum indicates that the pressure gradient does not change in the y direction. In dimensionless form, the y component of the conservation-of-linear-momentum equation is

$$\frac{\rho U_0^2 \delta}{L^2}\left(v_x^*\frac{\partial v_y^*}{\partial x^*} + v_y^*\frac{\partial v_y^*}{\partial y^*}\right) = -\frac{\rho U_0^2}{\delta}\frac{\partial P^*}{\partial y^*} + \frac{\mu U_0}{\delta L}\left(\frac{\delta^2}{L^2}\frac{\partial^2 v_y^*}{\partial x^2} + \frac{\partial^2 v_y^*}{\partial y^{*2}}\right). \quad (4.5.8a)$$

Since the second derivative of v_y with respect to the x direction is multiplied by $(\delta/L)^2$, which is small, viscous forces in the x direction are smaller than those in the y direction and can be neglected. Using Equation (4.5.7c) to replace the boundary layer thickness yields

$$\frac{\partial P^*}{\partial y^*} = \left(\frac{\delta}{L}\right)^2 \left(\frac{\partial^2 v_y^*}{\partial y^{*2}} - \frac{L}{\delta}\left(v_x^* \frac{\partial v_y^*}{\partial x^*} + v_y^* \frac{\partial v_y^*}{\partial y^*}\right)\right). \tag{4.5.8b}$$

The right-hand side of Equation (4.5.8b) is multiplied by $(\delta/L)^2$, which is very small. Consequently, the magnitude of the right-hand side is much less than unity, and the equation reduces to

$$\frac{\partial P^*}{\partial y^*} \approx 0. \tag{4.5.8c}$$

This result indicates that the pressure varies only in the direction of flow ($P = P(x)$ only). In the y direction, the inertial and viscous forces balance.

Since the pressure does not depend upon the y direction, the pressure gradient in the boundary layer is equal to the pressure gradient in the free stream. Viscous forces are negligible in the free stream, and the pressure and velocity can be related in terms of Bernoulli's equation. Choosing a streamline that does not vary in the vertical direction, we integrate Bernoulli's equation and obtain

$$P(x) + \frac{1}{2}\rho U(x)^2 = \text{constant}. \tag{4.5.9a}$$

Note that $U(x)$ is assumed to be known. Taking the derivative of P with respect to x and rearranging yields the following expression for the pressure gradient:

$$\frac{dP}{dx} = -\rho U(x)\frac{dU(x)}{dx}. \tag{4.5.9b}$$

Since $U = U(x)$ only, a total derivative was taken. As a result, the pressure is expressed in terms of the known velocity $U(x)$ in the free stream.

To summarize, an order-of-magnitude analysis of Equations (4.5.2), (4.5.3a), and (4.5.3b) results in the following two equations, known as the boundary layer equations:

$$\frac{\partial v_x}{\partial x} + \frac{\partial v_y}{\partial y} = 0, \tag{4.5.10a}$$

$$\rho\left(v_x \frac{\partial v_x}{\partial x} + v_y \frac{\partial v_x}{\partial y}\right) = \rho U(x)\frac{dU(x)}{dx} + \mu\frac{\partial^2 v_x}{\partial y^2}. \tag{4.5.10b}$$

The foregoing analysis has reduced the number of variables from three (P, v_x, and v_y) to two (v_x and v_y). Correspondingly, the number of equations was also reduced from three to two.

4.5.3 Integral Momentum Equations for Boundary Layer Flows

Although Equations (4.5.10a) and (4.5.10b) can be solved numerically, highly accurate approximate solutions can be found with the use of an integral analysis. Integrating the two equations from $y = 0$ to $y = \infty$ and noting that $v_y(0) = 0$ yields

$$v_y(\infty) = -\int_0^\infty \frac{\partial v_x}{\partial x}dy, \tag{4.5.11a}$$

$$\int_0^\infty \rho\left(v_x\frac{\partial v_x}{\partial x} + v_y\frac{\partial v_x}{\partial y}\right)dy = \int_0^\infty \left(\rho U(x)\frac{dU(x)}{dx} + \mu\frac{\partial^2 v_x}{\partial y^2}\right)dy. \tag{4.5.11b}$$

Equations (4.5.11a) and (4.5.11b) are simplified as follows: The second term on the right-hand side of Equation (4.5.11b) is integrated to yield the shear stress at $y = 0$ (i.e., the wall shear stress, τ_w):

$$\int_0^\infty \mu\frac{\partial^2 v_x}{\partial y^2}dy = \int_0^\infty \mu\frac{\partial}{\partial y}\left(\frac{\partial v_x}{\partial y}\right) = \mu\frac{\partial v_x}{\partial y}\bigg|_{y=0}^{y=\infty} = -\tau_w. \tag{4.5.12}$$

As y approaches ∞, the velocity approaches U_0 and the velocity gradient approaches zero.

Next, the second term on the left-hand side of Equation (4.5.11b) is integrated by parts (Section A.1.A) to yield

$$\int_0^\infty v_y\frac{\partial v_x}{\partial y}dy = v_y v_x\bigg|_0^\infty - \int_0^\infty v_x\frac{\partial v_y}{\partial y}dy. \tag{4.5.13a}$$

At $y = 0$, $v_x = 0$ and $v_y = 0$; at $y = \infty$, $v_x = U(x)$ and $v_y(\infty)$ is obtained from Equation (4.5.11a). Furthermore, from the conservation-of-mass equation (4.5.10a), $\partial v_y/\partial y$ is equal to $-\partial v_x/\partial x$. As a result, Equation (4.5.13a) becomes

$$\int_0^\infty v_y\frac{\partial v_x}{\partial y}dy = -U\int_0^\infty \frac{\partial v_x}{\partial x}dy + \int_0^\infty v_x\frac{\partial v_x}{\partial x}dy. \tag{4.5.13b}$$

Substituting Equations (4.5.12) and (4.5.13b) into Equation (4.5.11b) and rearranging yields

$$-\tau_w = \rho\int_0^\infty \left(2v_x\frac{\partial v_x}{\partial x} - U\frac{\partial v_x}{\partial x} - U\frac{dU}{dx}\right)dy. \tag{4.5.14}$$

Note that Equation (4.5.14) depends upon one unknown: v_x. $U(x)$ is known once the geometry is specified. The equation is often rewritten in a slightly different form by noting the following:

$$2v_x\frac{\partial v_x}{\partial x} = \frac{\partial v_x^2}{\partial x}, \tag{4.5.15a}$$

$$\frac{\partial(Uv_x)}{\partial x} = U\frac{\partial v_x}{\partial x} + v_x\frac{dU}{dx}. \tag{4.5.15b}$$

Substituting for $2v_x\,\partial v_x/\partial x$ and $U\partial v_x/\partial x$ and rearranging yields

$$-\tau_w = \rho\int_0^{\infty}\left(\frac{\partial v_x(v_x - U)}{\partial x} + (v_x - U)\frac{dU}{dx}\right)dy. \tag{4.5.16}$$

Switching the order of integration and derivative gives the following equation, which is known as the *von Karman integral momentum equation*:

$$\tau_w = \rho\frac{\partial}{\partial x}\left[\int_0^{\infty} v_x(U - v_x)dy\right] + \rho\frac{dU}{dx}\int_0^{\infty}(U - v_x)dy. \tag{4.5.17}$$

In order to solve Equation (4.5.17) and calculate the shear stress, it is necessary to specify v_x. This is often done by assuming a velocity profile that satisfies the boundary conditions. The following example is illustrative:

Example 4.5 Flow Over a Flat Plate

A uniform flow field with velocity U_0 approaches a flat plate of length L that is oriented in the direction of flow, as shown in Figure 4.11. Using boundary layer theory, determine the thickness of the boundary layer and the wall shear stress. For this flow, $U(x) = U_0$.

Solution Since $U(x) = U_0$, Equation (4.5.17) reduces to

$$\tau_w = \rho\frac{d}{dx}\left[\int_0^{\infty} v_x(U - v_x)dy\right]. \tag{4.5.18}$$

In general, the velocity v_x approaches U_0 asymptotically as y approaches ∞. The physical picture is simplified by assuming that at $y = \delta$, the boundary layer thickness, $v_x = U_0$. Thus, any proposed velocity field must satisfy the following two boundary conditions:

$$y = 0 \qquad v_x = 0, \tag{4.5.19a}$$

$$y = \delta \qquad v_x = U_0. \tag{4.5.19b}$$

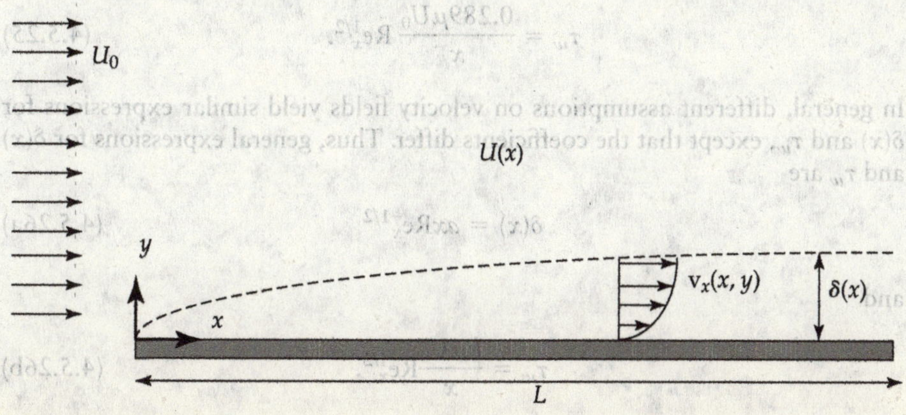

FIGURE 4.11 Flow over a flat plate.

An additional condition used for more complex functional representations of v_x is

$$y = \delta \qquad \frac{dv_x}{dy} = 0. \tag{4.5.19c}$$

As a result of this assumption, the upper limit on the integral in Equation (4.5.18) becomes δ, and we have

$$\tau_w = \rho \frac{d}{dx}\left[\int_0^\delta v_x(U_0 - v_x)dy\right]. \tag{4.5.20}$$

The simplest velocity field that satisfies Equations (4.5.19a) and (4.5.19b) (but not Equation (4.5.19c)) is

$$v_x = \frac{U_0 y}{\delta}. \tag{4.5.21}$$

Inserting Equation (4.5.21) into Equation (4.5.20) and integrating yields the following expression for τ_w in terms of $\delta(x)$:

$$\tau_w = \frac{\rho U_0^2}{6}\frac{d\delta}{dx}. \tag{4.5.22}$$

τ_w can also be obtained by calculating the velocity gradient at $y = 0$. The result is a first-order ordinary differential equation for $\delta(x)$:

$$\tau_w = \mu \frac{\partial v_x}{\partial y}\bigg|_{y=0} = \frac{\mu U_0}{\delta} = \frac{\rho U_0^2}{6}\frac{d\delta}{dx}. \tag{4.5.23}$$

In order to solve Equation (4.5.23), we need to specify an initial condition, namely, that at $x = 0$, $\delta(x) = 0$. Solving Equation (4.5.23) for $\delta(x)$ yields

$$\delta(x) = \sqrt{\frac{12\mu x}{\rho U_0}} = 3.464 x Re_x^{-1/2}, \tag{4.5.24}$$

where $\text{Re}_x = \rho U_0 x/\mu$. Substituting for $\delta(x)$ in Equation (4.5.23) and solving for the wall shear stress, we get

$$\tau_w = \frac{0.289\mu U_0}{x}\text{Re}_x^{1/2}. \tag{4.5.25}$$

In general, different assumptions on velocity fields yield similar expressions for $\delta(x)$ and τ_w, except that the coefficients differ. Thus, general expressions for $\delta(x)$ and τ_w are

$$\delta(x) = ax\text{Re}_x^{-1/2} \tag{4.5.26a}$$

and

$$\tau_w = \frac{b\mu U_0}{x}\text{Re}_x^{1/2}. \tag{4.5.26b}$$

TABLE 4.1

Coefficients in Equations (4.5.26a) and (4.5.26b) for Various Approximations of v_x

$v_x/U_0 = f(y/\delta)$	a	b
exact	5.00	0.332
y/δ	3.5	0.289
$(3y/2\delta) - (y^3/2\delta^3)$	4.6	0.323
$\sin(\pi y/2\delta)$	4.8	0.327

Table 4.1 lists various values of a and b for different velocity fields that satisfy Equations (4.5.19a), (4.5.19b), and, in some cases, (4.5.19c). The exact solution was obtained numerically [11]. Clearly, a simple approximation for v_x in the boundary layer can lead to highly accurate estimates of the wall shear stress.

4.6 Flow Separation

Boundary layer theory applies to the flow over submerged objects, as well as in channels and blood vessels, as long as the boundary layer thickness is much less than the thickness of the channel. The theory affords an insight into a number of important flows that arise in channels in which the cross-sectional area changes, such as flow through an expanding channel. Under the appropriate conditions, regions of *flow reversal* and *recirculation* arise in which some of the fluid moves in a reverse direction. The location at which flow first reverses direction is known as the *separation point*, and the downstream location where all of the fluid again moves in the same direction is known as the *reattachment point* (Figure 4.12). Flow recirculation produces additional stresses on the solid walls and requires a greater pressure drop for the same flow rate. Further, recirculation affects the transfer of mass between the fluid and the walls of the chamber. Systems in which flow separation is minimized or

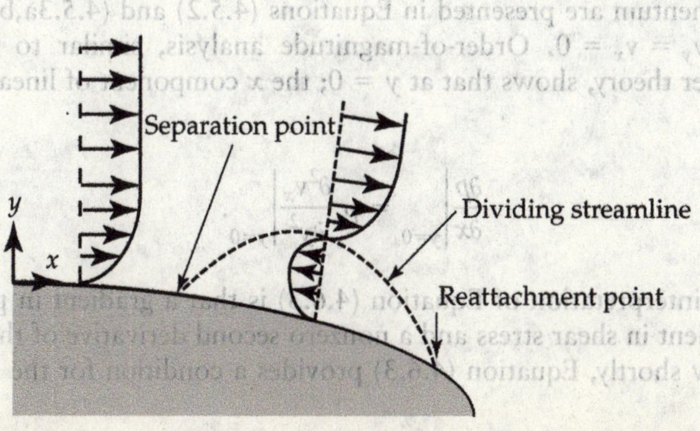

FIGURE 4.12 Schematic of region of flow separation for flow along a curved surface.

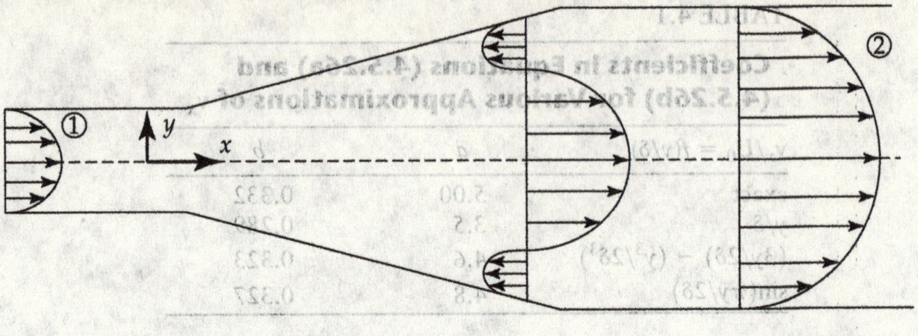

FIGURE 4.13 Flow through an expansion.

eliminated are highly desirable. Understanding flow separation is important in the design of aircraft and ships. Flow separation and reversal arise in the lungs and in blood flow. Flow reversal and separation in arteries are implicated in atherosclerosis, the underlying pathology behind heart disease and stroke.

In order to identify the conditions that produce flow separation, consider the differential form of Bernoulli's equation for flow in a region with a varying cross-sectional area (Figure 4.13):

$$\rho v_x \frac{dv_x}{dx} + \frac{dp}{dx} + \rho g \frac{dy}{dx} = 0. \tag{4.6.1}$$

If the channel is horizontal, $dy/dx = 0$. Equation (4.6.1) then simplifies to

$$\rho v_x \frac{dv_x}{dx} + \frac{dp}{dx} = 0. \tag{4.6.2}$$

Conservation of mass dictates that as the cross-sectional area increases, the velocity decreases. Thus, an increase in cross-sectional area causes fluid deceleration ($dv_x/dx < 0$). According to Equation (4.6.2), the pressure must rise ($dp/dx > 0$). This positive or adverse pressure gradient is a necessary, but not sufficient, condition for flow separation.

The onset of flow separation can be explained by considering the case of the steady two-dimensional flow of an incompressible fluid passing through a rectangular expansion (Figure 4.13). In this case, the conservation of mass and the conservation of linear momentum are presented in Equations (4.5.2) and (4.5.3a,b). At the surface, $y = 0$, $v_y = v_x = 0$. Order-of-magnitude analysis, similar to that used in boundary layer theory, shows that at $y = 0$; the x component of linear momentum reduces to

$$\left.\frac{\partial p}{\partial x}\right|_{y=0} = \mu \left.\frac{\partial^2 v_x}{\partial y^2}\right|_{y=0}. \tag{4.6.3}$$

The physical interpretation of Equation (4.6.3) is that a gradient in pressure gives rise to a gradient in shear stress and a nonzero second derivative of the velocity. As we shall show shortly, Equation (4.6.3) provides a condition for the onset of flow separation.

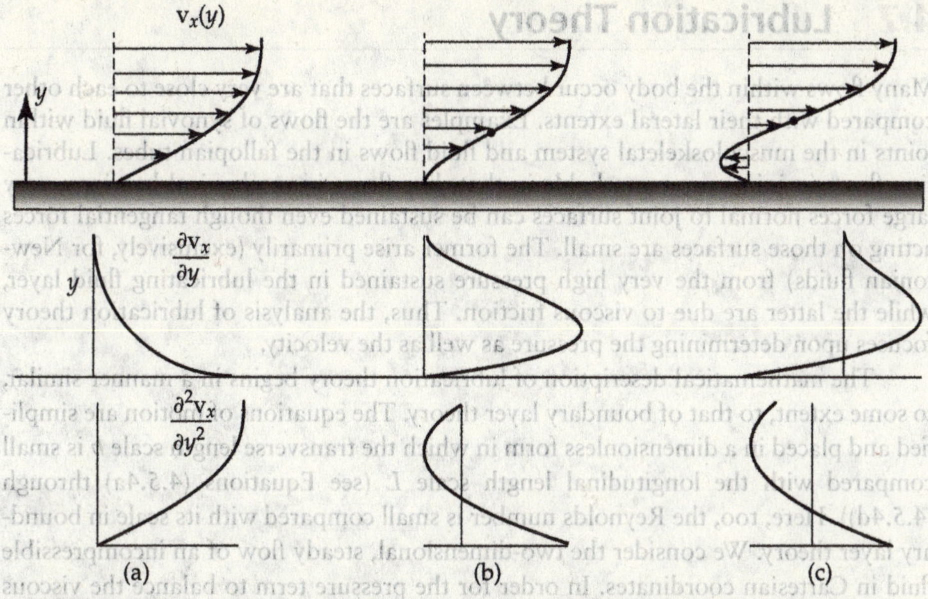

FIGURE 4.14 Schematic of velocity field (v_x), velocity gradient ($\partial v_x/\partial y$), and second derivative of velocity $\partial^2 v_x/\partial y^2$. Dot in (b) and (c) represents the location of an inflection point.

For flow in constant cross-sectional areas, such as a narrow channel of constant thickness, the pressure gradient $\partial p/\partial x$ is negative, the velocity gradient is negative, and $\partial^2 v_x/\partial y^2|_{y=0}$ is negative (Figure 4.14a). Consequently, flow separation does not occur. Due to an increase in cross-sectional area, the pressure gradient is positive at $y = 0$, and $\partial^2 v_x/\partial y^2|_{y=0}$ is also positive. Away from the wall, the velocity gradient decreases in magnitude with increasing y. As a result, $\partial^2 v_x/\partial y^2$ must pass through an inflection point ($\partial^2 v_x/\partial y^2 = 0$) where the velocity gradient is a maximum. Thus, $\partial^2 v_x/\partial y^2$ becomes negative at large distances from the wall. If the adverse pressure gradient is small, then $\partial^2 v_x/\partial y^2$ is small enough such that the velocity gradient is always positive (Figure 4.14b). As a result, this adverse pressure gradient is not large enough to induce flow separation. At a somewhat larger adverse pressure gradient, $\partial v_x/\partial y|_{y=0} = 0$, and the wall shear stress is zero. This is the condition for flow separation. At higher adverse pressure gradients, $\partial v_x/\partial y|_{y=0}$ is negative, and flow reversal occurs. Thus, flow reversal occurs when the adverse pressure gradient is large enough to overcome the viscous forces at the wall.

The region of flow separation is bounded by locations at which the velocity gradient is zero at the surface. Correspondingly, the shear stress is also zero at these points. The size of the region of flow reversal depends on the geometry of the expansion and the upstream Reynolds number. The size of the recirculation zone and the magnitude of the Reynolds number for onset of separation grow as the steepness of the expansion increases. For a given geometry, the size of the recirculation region grows with the Reynolds number. This analysis is limited to two-dimensional flows.

4.7 Lubrication Theory

Many flows within the body occur between surfaces that are very close to each other compared with their lateral extents. Examples are the flows of synovial fluid within joints in the musculoskeletal system and fluid flows in the fallopian tubes. Lubrication flows in joints are remarkable in that, like flows in mechanical bearings, very large forces normal to joint surfaces can be sustained even though tangential forces acting on those surfaces are small. The former arise primarily (exclusively, for Newtonian fluids) from the very high pressure sustained in the lubricating fluid layer, while the latter are due to viscous friction. Thus, the analysis of lubrication theory focuses upon determining the pressure as well as the velocity.

The mathematical description of lubrication theory begins in a manner similar, to some extent, to that of boundary layer theory. The equations of motion are simplified and placed in a dimensionless form in which the transverse length scale h is small compared with the longitudinal length scale L (see Equations (4.5.4a) through (4.5.4d)). Here, too, the Reynolds number is small compared with its scale in boundary layer theory. We consider the two-dimensional, steady flow of an incompressible fluid in Cartesian coordinates. In order for the pressure term to balance the viscous stress term in the x equation of motion, the reference pressure must be $P = \mu UL/h^2$. The resulting equations of motion, neglecting terms multiplied by h/L and $(h/L)^2$, are

$$\frac{\partial p}{\partial x} = \frac{\partial}{\partial y}\left(\mu \frac{\partial \mathrm{v}_x}{\partial y}\right) \qquad (4.7.1a)$$

and

$$\frac{\partial p}{\partial y} = 0. \qquad (4.7.1b)$$

Note that the y velocity $\mathrm{v}_y \ll \mathrm{v}_x$ and can therefore be neglected; note also that, as in boundary layer theory, the pressure varies only in the x direction. Let us consider flow between two surfaces: a lower surface that is a stationary horizontal plane and an upper surface given by $y = h(x)$ and all of whose points move with velocity U in the x direction (Figure 4.15). For simplicity, we shall restrict this analysis to the case in which there is no movement of the upper surface in the y direction, although that case can also be developed analytically.

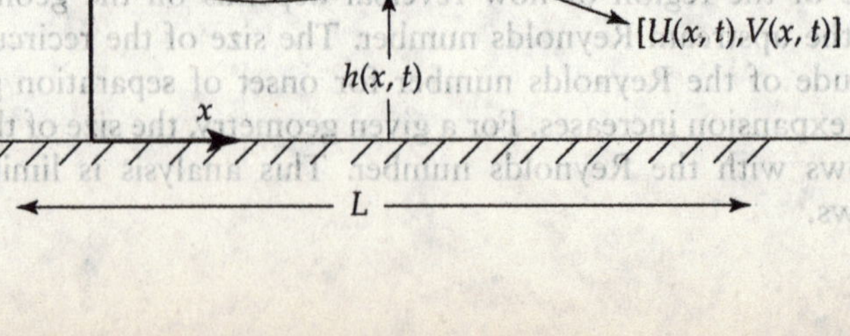

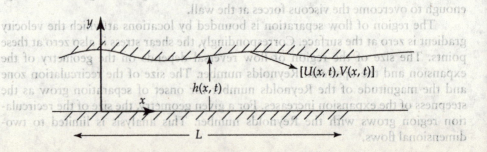

FIGURE 4.15 Schematic of lubrication flow through a narrow channel.

The boundary conditions are

$$v_x = 0 \qquad y = 0, \tag{4.7.2a}$$

$$v_x = U \qquad y = h. \tag{4.7.2b}$$

For a Newtonian fluid, the solution is

$$v_x = \frac{1}{2\mu}\frac{dp}{dx}[y(y-h)] + U\frac{y}{h}. \tag{4.7.3}$$

The volumetric fluid flux is given by

$$Q_x = \int_0^{h(x)} v_x dy = -\frac{h^3}{12\mu}\frac{dp}{dx} + \frac{Uh}{2}. \tag{4.7.4}$$

We now integrate the continuity equation over the thickness of the lubrication layer, noting that $v_y \approx 0$. Then, using Leibniz's rule for differentiating an integral, we have

$$0 = \int_0^{h(x)} \frac{dv_x}{dx} dx = \frac{d}{dx}\int_0^{h(x)} v_x dx - v_x(x,h)\frac{dh}{dx} = \frac{dQ_x}{dx} - U\frac{dh}{dx}. \tag{4.7.5}$$

Now, combining Equations (4.7.4) and (4.7.5), we obtain

$$\frac{1}{\mu}\frac{d}{dx}\left(h^3\frac{dp}{dx}\right) = -6U\frac{dh}{dx}. \tag{4.7.6}$$

This is the famous Reynolds equation of lubrication theory, as applied to our problem. A more general form of the Reynolds equation that allows for the velocity of the upper surface to have a y component V and that also allows U and V to vary with x is

$$\frac{1}{\mu}\frac{d}{dx}\left(h^3\frac{dp}{dx}\right) = 6\left[h\frac{dU}{dx} - U\frac{dh}{dx} + 2V\right]. \tag{4.7.7}$$

Given information about $h(x)$, $U(x)$, and $V(x)$ and boundary conditions on pressure at upstream and downstream ends of the flow region, Equation (4.7.7) can be solved for $p(x)$. The net normal (lubricating) forces on the upper and lower surfaces can then be readily obtained.

Example 4.6 Lubrication Forces Due to a Thin Film between Two Surfaces

We shall consider the case of a lubrication layer between two flat surfaces, one of which is inclined and moves (i.e., slides) relative to the other (Figure 4.16). In this case,

$$h(x) = h_1 - \frac{h_1 - h_2}{L}x. \tag{4.7.8}$$

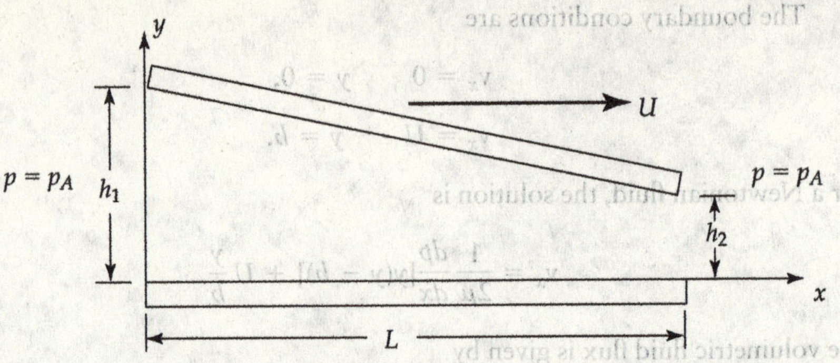

FIGURE 4.16 Lubrication flow for an angled sliding plate.

This example demonstrates the remarkably high normal forces that can be generated in lubrication layers and the very low relative friction that occurs in the longitudinal direction. The flow geometry is that of a slider bearing in machine design. An obvious physiological reference is the lubrication of diarthrodial joints within the body (e.g., those in the knee, hip, shoulder, ankle, or elbow). Such biological lubrication is a biomechanically sophisticated and complex phenomenon [12,13]. The lubricating fluid is synovial fluid, a dialysate of blood plasma, which contains a small cellular component and hyaluronic acid, a long-chain polymer that is found at many places within the body (e.g., the vitreous fluid of the eye and the cumulus oophorus that encapsulates the egg at the time of ovulation). Structurally, the presence of the hyaluronic acid is like a network of macromolecules, the interstices of which are filled by a low-viscosity fluid. The presence of hyaluronic acid in synovial fluid gives rise to complex rheological behavior, including shear thinning. Overall, the lubrication mechanism in diarthrodial joints depends not simply upon the properties of synovial fluid, but also upon the roles of the adjacent cartilage and other mechanisms [12]. Our model lubrication problem here focuses exclusively upon viscous lubrication effects by a thin film of fluid. Moreover, we consider a Newtonian fluid. Exact analytic solutions for non-Newtonian fluids in lubrication problems are rare. Nonetheless, this example demonstrates how movements of apposing surfaces in our diarthrodial joints can support forces much higher than, for example, our body weights.

In this example, the pressure is obtained by combining Equation (4.7.8) with the Reynolds Equation (4.7.6) and integrating. Applying the boundary conditions that $p(x) = P_A$ at $h = h_1$ and $h = h_2$, we obtain

$$p(x) - p_A = \frac{6\mu U}{\alpha(h_1 + h_2)}\left[\frac{(h_2 - h)(h - h_1)}{h^2}\right], \tag{4.7.9a}$$

where

$$\alpha = \frac{h_1 - h_2}{L}. \tag{4.7.9b}$$

The vertical force or *load* F_N acting on the upper surface is obtained by resolving the pressure stress into the vertical direction and integrating along the length L.

The depth of the surface in the z direction is W. The integration is facilitated by changing the variable of integration from x to h:

$$F_N = \frac{-W}{(1+\alpha^2)^{\frac{1}{2}}}\int_0^L (p - p_A)dx \quad (4.7.10a)$$

$$= \frac{W}{\alpha(1+\alpha^2)^{\frac{1}{2}}}\int_{h_1}^{h_2} (p - p_A)dh \quad (4.7.10b)$$

$$= \frac{1}{(1+\alpha^2)^{\frac{1}{2}}}\left(\frac{6\mu UW}{\alpha^2}\right)\left(\ln\beta - 2\frac{\beta - 1}{\beta + 1}\right). \quad (4.7.10c)$$

Here, $\beta = h_2/h_1$. Quite often, $\alpha \ll 1$, so, for algebraic simplicity, we shall neglect the factor $(1+\alpha)^{-1/2}$ in the rest of the analysis.

The tangential force F_F acting on the upper, sliding surface has two components. One, F_V, is obtained by integrating the viscous stress $\tau_{yx}(x, h)$ along the length of that surface:

$$\tau_{yx}(x, h) = \mu\frac{dv_x(x, h)}{dy} \quad (4.7.11a)$$

$$= \frac{\mu U}{h} + \frac{h}{2}\frac{dp}{dx} \quad (4.7.11b)$$

$$= \frac{\mu U}{h} + \frac{4h}{2}\frac{dp}{dh} \quad (4.7.11c)$$

$$= 2\mu U\left(\frac{3h_1h_2}{h_1 + h_2}\frac{1}{h_2} - \frac{1}{h}\right) \quad (4.7.11d)$$

$$F_V = W\int_0^L \tau_{yx}\,dx \quad (4.7.12a)$$

$$= -W\frac{1}{\alpha}\int_{h_1}^{h_2} \tau_{yx}dh \quad (4.7.12b)$$

$$= \frac{2\mu U}{\alpha}W\left(3\frac{\beta - 1}{\beta + 1} - \ln\beta\right). \quad (4.7.12c)$$

Strictly speaking, there is a component of the pressure in the x direction that contributes a force F_P to F_F. This force is given by

$$F_P = \frac{W\alpha}{(1+\alpha^2)^{\frac{1}{2}}}\int_0^L (p - p_A)dx \quad (4.7.13a)$$

$$F_p = \frac{W}{(1+\alpha^2)^{\frac{1}{2}}}\int_{h_1}^{h_2}(p - p_A)dh. \tag{4.7.13b}$$

Since $\alpha^2 \ll 1$, Equation (4.7.13b) reduces to

$$F_p = W\int_{h_1}^{h_2}(p - p_A)dh \tag{4.7.13c}$$

$$= \frac{6\mu UW}{\alpha}\left(\ln\beta - 2\frac{\beta-1}{\beta+1}\right). \tag{4.7.13d}$$

Overall, then, we find that

$$F_F = F_v + F_p, \tag{4.7.14a}$$

and

$$F_F = \frac{2\mu UW}{\alpha}\left(2\ln\beta - 3\frac{\beta-1}{\beta+1}\right). \tag{4.7.14b}$$

The ratio of the tangential to the normal force is thus

$$R = \frac{\alpha}{3}\left[\frac{2(\beta+1)\ln\beta - 3(\beta-1)}{(\beta+1)\ln\beta - 2(\beta-1)}\right]. \tag{4.7.15}$$

As a numerical example, consider a layer of dimensions $h_1 = 1.2$ mm, $h_2 = 1$ mm, and $L = 5$ cm. The ratio $R = 0.07$, a very low frictional effect.

Example 4.7 Flow Through a Thin Channel with Porous Walls: A Model of Filtration

We next consider another variation of the two-dimensional channel problem solved initially in Chapter 2. In this example, we allow for a nonzero component v_w of the y velocity at the wall. The problem is a model of physiological filtration processes. For example, there is filtration of blood in the kidneys. As the blood flows through the glomerular capillaries, about one-fifth of its water content passes across the surfaces of the capillaries to enter the proximal portions of the renal tubules. The filtration pressure that drives this process is the fluid pressure within the capillaries, minus the pressure in the capsule space, minus the osmotic pressure of the plasma proteins (see Chapter 14). We can analyze the flow as longitudinal fluid flow along the axis of a capillary, together with transverse flow of fluid across the capillary wall, which is modeled as a porous medium (Figure 4.17). A more complete analysis that includes the effect of osmotic pressure is given in [14]. Here, we consider the two-dimensional, steady, laminar flow of an incompressible Newtonian fluid in a channel of length L and height $2h$. We assume that

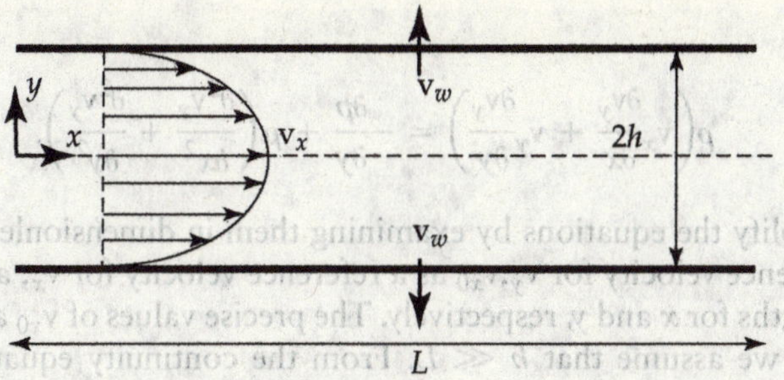

FIGURE 4.17 Flow through a rectangular channel with porous walls.

there is continuity of the normal component of velocity and of pressure between the fluid at the wall and the fluid within the porous wall at its surface with the channel. We retain the no-slip boundary condition for the longitudinal component of velocity at the channel wall. Within the porous wall, velocity is related to pressure by Darcy's law, which is discussed in Chapter 8. For now, we denote $\widetilde{v}$ and $\widetilde{p}$ as the velocity and pressure, respectively, in fluid within the wall. Then,

$$v_w|_{\text{fluid at wall}} = \widetilde{v}|_{\text{within wall at outer surface}} \tag{4.7.16a}$$

$$= K\nabla \widetilde{p}|_{\text{at outer surface of wall}} \tag{4.7.16b}$$

Here, K is the *hydraulic conductivity* for the wall (see Equation (8.3.2)). If we assume that the wall has a thickness l, then we have

$$v_w = \frac{K}{l}(\widetilde{p}_s - \widetilde{p}_i) \tag{4.7.17a}$$

$$= \frac{K}{l}(p|_{\text{fluid at wall}} - \widetilde{p}_i), \tag{4.7.17b}$$

where $\widetilde{p}_s$ and $\widetilde{p}_i$ are the pressures within the wall at its outer and inner surfaces, respectively. With reference to the kidney, therefore, $\widetilde{p}_i$ includes the pressure in the capsule space plus the osmotic pressure. In this example, we make the assumption that $\widetilde{p}_i$ is constant. Note that Equation (4.7.17b) is the boundary condition for the normal component of channel velocity at the upper wall; since p is a function of the axial position x along the channel, v_w is now a function of x. At the lower channel wall, $y = -h$, we have an analogous boundary condition, only now the normal wall velocity is $-v_w$.

The continuity, x-momentum, and y-momentum equations are, respectively,

$$\frac{\partial v_x}{\partial x} + \frac{\partial v_y}{\partial y} = 0, \tag{4.7.18a}$$

$$\rho\left(v_x\frac{\partial v_x}{\partial x} + v_y\frac{\partial v_x}{\partial v_y}\right) = -\frac{\partial p}{\partial x} + \mu\left(\frac{\partial^2 v_x}{\partial x^2} + \frac{\partial^2 v_x}{\partial y^2}\right), \tag{4.7.18b}$$

and

$$\rho\left(v_x\frac{\partial v_y}{\partial x} + v_y\frac{\partial v_y}{\partial y}\right) = -\frac{\partial p}{\partial y} + \mu\left(\frac{\partial^2 v_y}{\partial x^2} + \frac{\partial^2 v_y}{\partial y^2}\right). \tag{4.7.18c}$$

We now simplify the equations by examining them in dimensionless form, using v_{w0} as a reference velocity for v_y,v_{x0} as a reference velocity for v_x, and L and h as reference lengths for x and y, respectively. The precise values of v_{x0} and v_{w0} do not matter, since we assume that $h \ll L$. From the continuity equation, we must have $v_{w0} \ll v_{x0}$, and this is sufficient to help develop the simplification of the equations of motion. If we examine the x-momentum equation in dimensionless form, we find that the inertial terms are multiplied by $\rho v_{x0}h^2/\mu L$. We can regard this quantity as the effective Reynolds number for our problem. That is, a Reynolds number based on the channel height h alone is reduced by the factor h/L. We shall assume that the Reynolds number is small and that inertial forces are negligible. It follows that the reference pressure is $\mu v_{x0}L/h^2$. The solution for v_x is thus the same as that obtained earlier in Section 2.7.2, or

$$v_x = -\frac{h^2}{2\mu}\frac{\partial p}{\partial x}\left[1 - \frac{y^2}{h^2}\right]. \tag{4.7.19}$$

(Note that, in developing Equation (2.7.22), the channel height was h, not $2h$ as here.)

We next integrate the continuity equation to obtain v_y:

$$v_y = -\int_0^y \frac{\partial v_x}{\partial x}\,dy + C_1, \tag{4.7.20}$$

$$\frac{\partial v_x}{\partial x} = -\frac{h^2}{2\mu}\frac{\partial^2 p}{\partial x^2}\left(1 - \frac{y^2}{h^2}\right), \tag{4.7.21}$$

$$v_y = \frac{h^2}{2\mu}\frac{\partial^2 p}{\partial x^2}\int_0^y\left(1 - \frac{y^2}{h^2}\right)dy + C_1, \tag{4.7.22}$$

$$v_y = \frac{h^2}{2\mu}\frac{\partial^2 p}{\partial x^2}\left(y - \frac{y^3}{3h^2}\right) + C_1. \tag{4.7.23}$$

We obtain the value of the constant of integration, $C_1 = 0$, by applying a symmetry boundary condition on the centerline of the channel, namely, $v_y = 0$ at $y = 0$. Then

$$v_y = \frac{h^2}{2\mu}\frac{\partial^2 p}{\partial x^2}\left(y - \frac{y^3}{3h^2}\right). \tag{4.7.24}$$

Applying the boundary condition $v_y = v_w$ at $y = h$ at the wall, we obtain

$$v_w = \frac{h^2}{2\mu}\frac{\partial^2 p}{\partial x^2}\left(h - \frac{h}{3}\right) = \frac{h^3}{3\mu}\frac{\partial^2 p}{\partial x^2} \tag{4.7.25}$$

and

$$v_y = \frac{h^3}{3\mu}\frac{\partial^2 p}{\partial x^2}\left(\frac{3y}{2h} - \frac{y^3}{2h^3}\right) = \frac{v_w}{2}\left(\frac{3y}{h} - \frac{y^3}{h^3}\right). \quad (4.7.26)$$

Substituting Equation (4.7.17b) into Equation (4.7.25), we obtain an equation for $p(x)$:

$$\frac{d^2p}{dx^2} - \frac{3\mu K}{lh^3}p = -\frac{3\mu K}{lh^3}\widetilde{p}_i. \quad (4.7.27)$$

Equation (4.7.27) is analogous to the Reynolds equation (4.7.6) or (4.7.7) developed in the previous section.

Suppose now that we take $p = p_0$ at $x = 0$ and $p = p_L$ at $x = L$. Then we can integrate Equation (4.7.27) to obtain the pressure as a function of x. Before doing so, however, it is convenient to define a transformed pressure as

$$p^* = \frac{p_0 - p}{p_0 - p_L}. \quad (4.7.28)$$

The resulting equation for pressure and the associated boundary conditions are as follows:

$$\frac{d^2p^*}{dx^2} - \frac{3\mu K}{lh^3}p^* = \frac{3\mu K}{lh^3}\left(\frac{\widetilde{p}_i - p_0}{p_0 - p_L}\right), \quad (4.7.29)$$

$$p^* = 0 \quad \text{at } x = 0, \quad (4.7.30a)$$

$$p^* = 1 \quad \text{at } x = L. \quad (4.7.30b)$$

The general solution of Equation (4.7.29) is

$$p^*(x) = C_1 \sinh\sqrt{\frac{3\mu K}{lh^3}}\,x + C_2 \cosh\sqrt{\frac{3\mu K}{lh^3}}\,x - \frac{\widetilde{p}_i - p_0}{p_0 - p_L}, \quad (4.7.31)$$

where C_1 and C_2 are constants of integration, and the third term is the particular solution of the ordinary differential equation resulting from its nonzero right-hand side. Applying the boundary conditions, we obtain, after some algebraic manipulations,

$$p(x) = p_0 - \left[(p_0 - p_L) + (\widetilde{p}_i - p_0)\left(1 - \cosh\sqrt{(3\mu K/lh^3)}L\right)\right]\frac{\sinh\sqrt{(3\mu K/lh^3)}x}{\sinh\sqrt{(3\mu K/lh^3)}L}$$

$$+ (\widetilde{p}_i - p_0)\left(1 - \cosh\sqrt{(3\mu K/lh^3)}x\right). \quad (4.7.32)$$

4.8 Peristaltic Pumping

There are a number of physiological phenomena in which tubular structures propel their contents by organized contractions of longitudinal and circular muscle fibers. These contractions produce localized reductions in the diameters of the tubes, which are organized into waves that propagate along the tube axes. Such "peristalsis" can consist of individual or entire wave trains of contractions. Together with pressure differences applied to the ends of the tubes, peristaltic waves can result in the net transport of fluid within the tubes, cells, and other material suspended within the fluid. Examples of peristaltic pumping include transport in the esophagus, intestines, ureter, and fallopian tubes. Abnormalities of peristalsis can lead to discomfort (e.g., esophageal acid reflux), infection (e.g., retrograde flow of bacteria in the ureter), and possibly infertility (e.g., abnormal sperm or egg transport in the fallopian tubes). A full biophysical understanding of peristaltic pumping must take into account the mechanics of the contractions themselves. Here, we focus only upon the resulting flow of fluid within a tube undergoing peristalsis of its wall [15]. The fluid mechanics of peristaltic pumping can be described to varying degrees of sophistication and mathematical complexity. Different representations of the anatomy and physiology of peristalsis can lead to different conclusions about its fluid mechanical details. In this section, we present a relatively simple mathematical analysis, based directly on the work of Shapiro et al. [16] and Zien and Ostrach [17]. Further mathematical treatments and physiological discussions of peristaltic pumping are found in [18].

Our analysis of peristaltic pumping characterizes the net flow rate through the tube (i.e., the pumping capacity) and also the potential for reflux in some regions of flow (which can lead to pathology, as in the retrograde transport of bacteria in the ureter). In addition to the contractions of the tube wall, there may exist an externally applied pressure gradient acting along the tube. The existence, direction, and magnitude of such pressure gradients are found to vary in different physiological contexts. Pressure gradients can be created when there is contractile activity of organs or spaces at one or both ends of the tubes undergoing peristalsis. For example, at the uterine end of the fallopian tube there is a sphincterlike *utero-tubal junction* that can open and close. The ovarian end of this tube increases in diameter into a *fimbria* that opens into the *peritoneal cavity,* within which pressure can vary due to contractions of a number of pelvic structures. Overall, the fluid mechanics of peristaltic pumping depend upon interactions between detailed properties of the contractile waves and the externally applied pressure gradients.

We consider a Newtonian fluid contained within a circular tube whose length is very long compared with its diameter, $2a$ (Figure 4.18). We assume that the Reynolds number for the flow is low enough that fluid inertia can be neglected. A sinusoidal traveling-wave train with a length comparable to the tube length (i.e., with an effectively infinite number of individual waves) propagates along the tube wall with constant wave speed c, wave amplitude b, and wavelength λ. Following Shapiro et al. [16], we note that the investigation of this problem is greatly facilitated by undertaking analyses in both a fixed "laboratory" frame of reference (Figure 4.18a) and also a moving "wave" frame that translates to the right with speed c (Figure 4.18b). In the wave frame, the crests of the waves appear stationary. As in

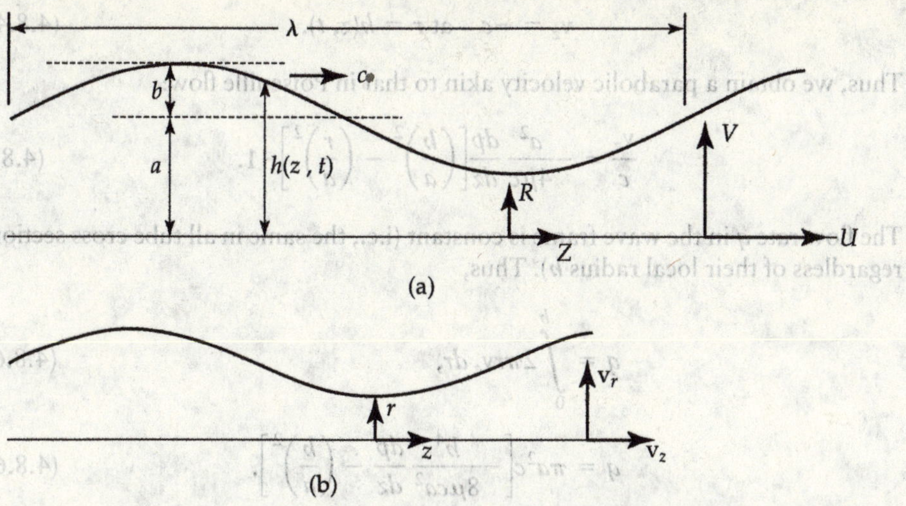

FIGURE 4.18 (a) Schematic of peristaltic wave in the moving coordinate system. The wave crests appear stationary. (b) Schematic of peristaltic wave in the fixed coordinate system.

the case of Poiseuille flow in a tube, we shall work toward developing a relationship between the flow rate and the pressure within the tube.

Here, we have introduced cylindrical coordinate systems for the fixed frame (R,Z) and the moving frame (r,z). The radial and transverse velocities in the two frames are denoted U,V and v_z,v_r, respectively. Note that

$$z = Z - ct, \tag{4.8.1a}$$

$$r = R, \tag{4.8.1b}$$

$$v_z(z, r) = U(Z - ct, R) - c, \tag{4.8.2a}$$

$$v_r(z, r) = V(Z - ct, R). \tag{4.8.2b}$$

There are three characteristic lengths in this problem: a, b, and λ. We assume that $\lambda \gg a$. We do not restrict the magnitude of the wave amplitude b, except that it cannot exceed the tube radius; that is, $b \leq a$. Under these conditions, the local slope of the tube wall is small. Our problem then has certain similarities in scaling to the lubrication theory analysis of the previous section. As a result, the transverse velocities v_r and V are small compared with the longitudinal velocities v_z and U, respectively, and the gradients of pressure in the transverse direction are small compared with those in the longitudinal direction. That is, the pressure is instantaneously uniform over each cross section of the tube.

Initially, we set up the problem in the wave frame. The equation of motion for the axial velocity v_z and the associated boundary conditions are as follows:

$$\frac{\mu}{r}\frac{\partial}{\partial r}\left(r\frac{\partial v_z}{\partial r}\right) = \frac{dp}{dz}, \tag{4.8.3}$$

$$\frac{\partial v_z}{\partial r} = 0 \quad \text{at } r = 0, \tag{4.8.4a}$$

$$v_z = -c \quad \text{at } r = h(z, t). \tag{4.8.4b}$$

Thus, we obtain a parabolic velocity akin to that in Poiseuille flow:

$$\frac{v_z}{c} = -\frac{a^2}{4\mu c}\frac{dp}{dz}\left[\left(\frac{h}{a}\right)^2 - \left(\frac{r}{a}\right)^2\right] - 1. \tag{4.8.5}$$

The flow rate q in the wave frame is constant (i.e., the same in all tube cross sections, regardless of their local radius h). Thus,

$$q = \int_0^h 2\pi r v_z\, dr, \tag{4.8.6a}$$

$$q = \pi a^2 c\left[-\frac{h^4}{8\mu c a^2}\frac{dp}{dz} - \left(\frac{h}{a}\right)^2\right]. \tag{4.8.6b}$$

In the laboratory frame, the flow rate is

$$Q = \int_0^h 2\pi R U dR = q + \pi h^2 c. \tag{4.8.7}$$

We now introduce the sinusoidal wave train

$$h(z, t) = a + b\sin\frac{2\pi}{\lambda}(z - ct). \tag{4.8.8}$$

In order to evaluate the pumping capacity of peristalsis, we now obtain $\overline{Q}$, the time-averaged value of Q over one period T of the sine wave contractions:

$$\overline{Q} = \frac{1}{T}\int_0^T Q dt \tag{4.8.9a}$$

$$\overline{Q} = q + \frac{\pi a^2 c}{T}\int_0^T \left(\frac{h}{a}\right)^2 dt \tag{4.8.9b}$$

$$\overline{Q} = q + \pi a^2 c\left(1 + \frac{1}{2}\phi^2\right). \tag{4.8.9c}$$

Here, $\phi = b/a$. We can use Equation (4.8.6b) to study pressure as a function of position:

$$\frac{dp}{dz} = -8\mu\left(\frac{q}{\pi h^4} + \frac{c}{h^2}\right). \tag{4.8.10}$$

Note that Equation (4.8.9) applies to any type of wall deformation, provided that the slope is small.

Substituting Equation (4.8.8) into Equation (4.8.10) enables us to analyze p as a function of z. We shall focus upon the change in pressure over one wavelength, Δp_λ. We find that

$$\Delta p_\lambda = \int_0^\lambda \frac{dp}{dz}\, dz \tag{4.8.11a}$$

$$\Delta p_\lambda = -8\mu c \int_0^\lambda \frac{dz}{h^2} - \frac{8}{\pi}\mu q \int_0^\lambda \frac{dz}{h^4} \tag{4.8.11b}$$

$$\Delta p_\lambda = \frac{4\mu c \lambda}{a^2}\left\{\frac{8\phi^2\left(1 - \frac{1}{16}\phi^2\right) - 4\phi\left(1 - \frac{1}{4}\phi\right)\left(1 - \frac{3}{2}\phi^2\right)\Theta}{(1-\phi^2)^{\frac{7}{2}}}\right\}, \tag{4.8.11c}$$

where

$$\Theta = \frac{\overline{Q}}{\pi a^2 c\left(2\phi - \frac{\phi^2}{2}\right)}. \tag{4.8.12}$$

Equations (4.8.11c) and (4.8.12) relate the pressure drop per wavelength along the tube to the time-averaged flow rate $\overline{Q}$. The latter in turn is given by

$$\overline{Q} = -\frac{\pi a^4}{8\mu}\frac{\Delta p_\lambda}{\lambda}\frac{(1-\phi^2)^{\frac{7}{2}}}{(1+\frac{3}{2}\phi^2)} + \pi a^2 c \frac{4\phi^2(1-\frac{1}{16}\phi^2)}{(1+\frac{3}{2}\phi^2)}. \tag{4.8.13}$$

Note that, in the absence of peristalsis (i.e., $\phi = 0$), Equation (4.8.13) reduces to Equation (2.7.45) for pressure-driven flow through a cylindrical tube. Equations (4.8.11) and (4.8.13) indicate that, in peristaltic pumping, a finite mean flow rate is associated both with a favorable (negative) mean pressure gradient and with the pressure gradient generated by the contractions of the tube walls. Indeed, the pressure gradient generated by peristalsis can overcome adverse (positive) pressure gradients below a limiting value. This property is demonstrated as follows: Let $\Delta p_{\lambda 0}$ be the pressure change per wavelength at which there is zero mean flow (i.e., $\overline{Q} = 0$). Then

$$\Delta p_{\lambda 0} = \frac{32\mu c\lambda}{a^2}\frac{\phi^2(1-\frac{1}{16}\phi^2)}{(1-\phi^2)^{\frac{7}{2}}}. \tag{4.8.14}$$

For $|\Delta p_1| < |\Delta p_{\lambda 0}|$, there is a positive net mean flow.

4.9 QUESTIONS

4.1 What effect does the assumption of an incompressible fluid have upon the conservation of mass?

4.2 For each of the following situations, explain the relative advantages and disadvantages of a differential or integral approach:

(a) Determination of the shear stress acting on an adherent cell
(b) Estimation of the work done by the heart
(c) Determination of the filtration flow rate and pressure drop of fluid through tissues
(d) Estimation of the pressure drop across heart valves

4.3 State the two assumptions used to derive Bernoulli's equation.

4.4 Would Bernoulli's equation be applicable at the location where the superior mesenteric artery branches from the abdominal aorta?

4.5 In boundary layer theory, the boundary layer thickness δ is much less than the characteristic length L in the x direction ($\delta \ll L$). What does this assumption imply about the pressure and the characteristic velocity in the y direction?

4.6 For each of the following locations, state whether boundary layers are present:

(a) At the inflow to the coronary arteries
(b) In a capillary
(c) In a lymphatic vessel
(d) In the entrance tubing to a blood oxygenator (inner diameter 2.5 cm) operating at 5 L min^{-1}

4.7 Specify what conditions are necessary for flow separation to arise.

4.8 What effect does flow separation have upon the overall pressure drop for flow through a channel?

4.10 PROBLEMS

4.1 Use Leibniz's rule for differentiating an integral (Section A.1.H) to determine an integral form of the conservation of mass when the volume is a function of time.

4.2 Use an integral analysis to examine the effect of tapering on the pressure drop in a blood vessel of length L, Assume that the outlet radius R_2 is related to the inlet radius by $R_2 = ER_1$, where $E < 1$.

4.3 Consider the entrance region of a cylindrical tube of radius R, as shown in Figure 4.19. The inlet velocity profile is uniform, with velocity U_0. The fluid is Newtonian and the flow is steady. Within the entrance region, the velocity field v_z is two dimensional. Downstream, the velocity field is fully developed and depends on r only. Use the integral forms of the conservation of linear momentum and mass for steady flow to relate the drag force exerted by the fluid on the walls of the tube to the pressure drop in the entrance region.

4.4 For the force balance developed in Example 4.2, plot the net force as a function of the angle of each branch from 0° to 180°.

4.5 *Application of boundary layer theory to estimate the entrance length for channel flow* Developing flow within a channel can be examined with boundary layer

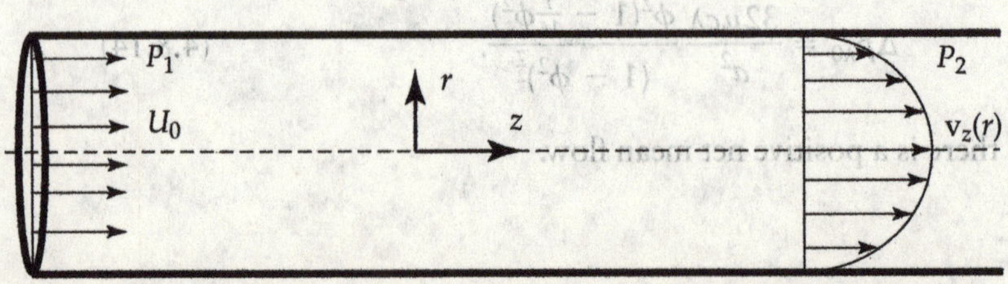

FIGURE 4.19 Flow in the entrance region of a cylindrical tube of radius R.

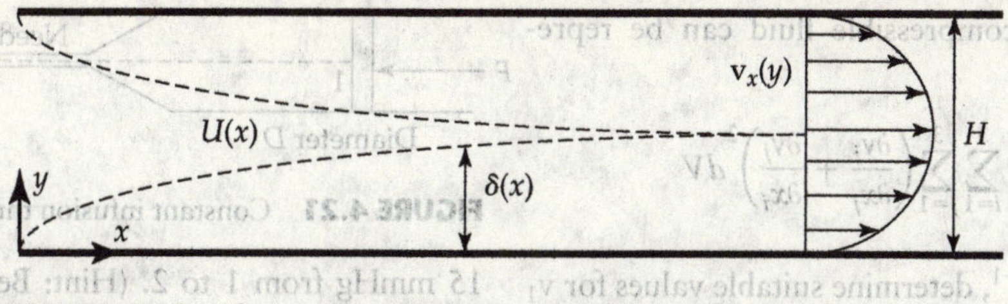

FIGURE 4.20 Flow in the entrance region of a narrow rectangular channel.

theory. Consider a rectangular channel of height H and width w such that $w \gg H$ (Figure 4.20). The velocity field in the entrance region depends upon the x and y directions. As shown in Figure 4.20, a boundary layer develops as flow enters the channel:

Once the boundary layer has grown to equal $H/2$, the flow is fully developed.

(a) As a first approximation, assume that the boundary layer is described with the use of results for flow over a flat plate, that is, $\delta(x) = 5.00x\ \mathrm{Re}_x^{-1/2}$ where $\mathrm{Re}_x = \rho U x/\mu$. Develop an expression for the entrance length in terms of the channel Reynolds number $\mathrm{Re}_x = 2\rho\langle v\rangle H/\mu = 2\rho Q/w\mu$, where $\langle v\rangle$ is the average velocity in the channel.

(b) The analysis in part (a) assumes that $U(x) = U_0$. In fact, the free-stream velocity changes as the boundary layer grows in the channel.

1. Use the result that the flow rate Q is constant to develop a relation for $U(x)$.
2. Use a linear velocity profile $v_x = U(x)\, y/\delta$ in the boundary layer to derive an expression for the growth of the boundary layer. For this problem, the von Karman momentum integral equation is

$$\tau_w = \rho \frac{\partial}{\partial x}\left[\int_0^\infty v_x(U - v_x)dy\right] + \rho \frac{dU}{dx}\int_0^\infty (U - v_x)dy.$$

3. Numerically or analytically integrate your result from part (b.2) to obtain an expression for the entrance length.
4. Explain why the result you obtained in part (b.3) differs from the one obtained in part (a).

MATLAB can be used to do numerical integration. Create an M-file as follows:

```
function dx=entrance(t,x)
global h Q w;
%These variables pass between program
and subroutine.
dx=[%Specify the function to be
integrated, where delta, boundary layer
thickness equals x(1)];
```

To integrate, the following commands are executed in the Command Window:

```
global h Q w;
h = provide value;
Q = provide value;
w = provide value;
[X,Y] = ode15s('entrance',[0 10],[0]);
%The first term specifies the limits of
integration for x and the second provides
an initial value for delta. The value of
10 is a guess. You will want to continue
integration until delta equals h.
```

The global command permits the specification of the values external to the M-file. Ode15s is a numerical integration routine for "stiff" equations—those that change their magnitude rapidly over a short time scale.

4.6 Using a boundary layer approximation, determine the contribution of viscous stresses to the pressure drop in the modified Bernoulli equation (4.4.28) for flow through a stenotic mitral valve. The valve thickness is 0.1 cm and the area ranges from 0.25 to 2 cm^2. Assume a maximum jet velocity of 200 cm s^{-1} and a constant viscosity of 0.03 g cm^{-1} s^{-1}. Assume also that v_x/U_0 can be represented by $\sin(\pi y/2\delta)$. The viscous loss term for an incompressible fluid can be represented as [7]:

$$E_V = \frac{\mu}{2}\int_V \sum_{i=1}^{3}\sum_{j=1}^{3}\left(\frac{\partial v_i}{\partial x_j} + \frac{\partial v_j}{\partial x_i}\right)^2 dV$$

4.7 For $v_2 = 2$ m s^{-1}, determine suitable values for v_1 such that Equation (4.4.19) can be used with no error greater than 10%.

4.8 Pulmonary valve stenosis is suspected in an infant with poor blood oxygenation. The right ventricle is underdeveloped, with a maximum cross-sectional area of 2 cm^2. From echocardiography, the velocities in the right ventricle and across the pulmonary valve are 0.5 and 1.3 m s^{-1}, respectively. Estimate the pressure drop across the valve and the cross-sectional area of the valve.

4.9 A patient is being examined by echocardiography for possible mitral valve stenosis. The left atrial velocity averaged 0.5 m s^{-1} while the valve was open. The maximum velocity through the valve was 1.8 m s^{-1}. The cross-sectional diameter in the left atrium where the velocity was recorded was 1.5 cm.

(a) Determine the pressure drop between the left atrium and the valve.

(b) Calculate the cross-sectional open area of the mitral valve.

4.10 For flow through a valve, estimate the viscous losses for an upstream centerline velocity equal to $v_1 = 80$ cm s^{-1} and $R_1 = 1.25$ cm and a downstream velocity $v_2 = 500$ cm s^{-1}. Use the following relation to approximate the viscous losses

$$E_V = \tau_{w2} - \tau_{w1}$$

If the flow is laminar, use a parabolic velocity profile. If turbulent, use

$$v_z = v_{max}\left(1 - \frac{r^a}{R^a}\right) \quad a = 5$$

4.11 Pulmonary banding is performed in infants with congenital heart defects that make them susceptible to pulmonary hypertension. A band is placed around the pulmonary artery to induce a stenosis and thereby reduce flow and pressure in the pulmonary artery. For the following conditions, determine the area reduction and radius of the stenosis needed to produce a pressure drop of 15 mmHg from 1 to 2. (Hint: Be sure to calculate the Reynolds number in each region.)

Flow rate	2000 cm^3 min^{-1}
Upstream pulmonary artery diameter	1.3 cm
Density of blood	1,070 kg m^{-3}
Conversion factor	133.32 Pa mmHg^{-1}
Viscosity of blood	0.035 g cm^{-1} s^{-1}

4.12 In designing a hollow-fiber hemodialyzer, assess whether the constitutive equation affects the net force acting on the walls of a cylindrical channel in the entrance region. Consider steady laminar flow of the following power law fluid into a cylindrical tube of radius R and length L ($R \ll L$). The velocity profile is

$$v_z = v_{max}\left[1 - \left(\frac{r}{R}\right)^3\right]$$

Flow enters the tube at a constant velocity U_0. Use an integral balance to determine the net force acting on the tube walls and show that it is less than the value obtained for a Newtonian fluid, $\pi R^2 \rho \langle v\rangle^2/3$.

4.13 Use an integral analysis to determine the net force acting on the walls of a tapered blood vessel of length L. Assume that the outlet radius R_2 is related to the inlet radius R_1 as $R_2 = ER_1$ where $E < 1$. Assume that the velocity profiles for the inlet and outlet are parabolic.

4.14 A drug is being injected at a constant infusion rate of Q from a syringe of diameter D into a needle of diameter d (Figure 4.21). The fluid viscosity is μ and density ρ. Determine the force on the syringe. Assume that the flow through the catheter is laminar and that the blood pressure is p. State all other assumptions made.

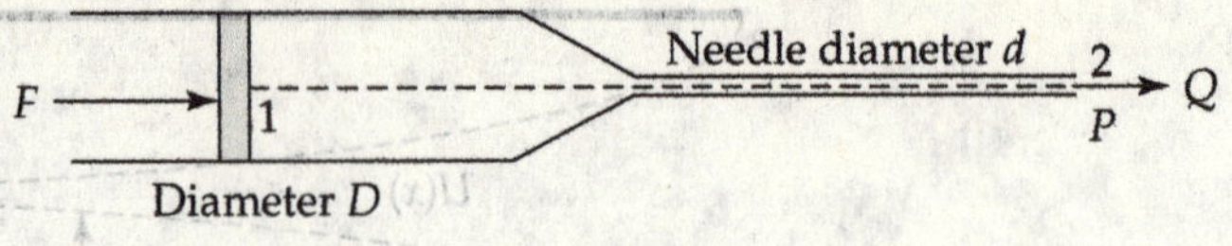

FIGURE 4.21 Constant infusion through a syringe.

4.15 A patient had a mitral valve that failed to close completely. The velocity through the centerline of the valve opening was well-represented by the following expression:

$$v_2 = 500 \sin(2\pi t) \text{ cm s}^{-1}$$

The velocity at a distance of 5 cm upstream was $v_1 = 50\sin(2\pi t)$ in units of centimeter per second. The fluid density is 1,050 kg m^{-3}. Estimate whether fluid acceleration can be neglected.

4.16 Fluid forces acting on endothelium can detach the cells and expose the underlying tissue to blood. Measuring this adhesion force under a number of biological conditions can provide important information about the extent to which endothelial cells might detach. Because the force is sensitive to the surface to which cells attach and the biological state of the cells, the measurement should be performed on endothelium adhering to blood vessels in their native state.

One approach to measuring the adhesion force is to remove a blood vessel and open it to expose the endothelial layer. A jet of fluid is applied to create a force on the tissue (Figure 4.22). This fluid force is resisted by the cells and represents the adhesive force as shown in Figure 4.22.

(a) Use an integral balance to determine an expression for the force in terms of the flow rate from the jet, Q. Assume that the velocity in the jet is uniform.

(b) For the following data, determine the pipette diameter needed to produce a force as high as 0.05 dyne (5×10^{-7} N) with a pressure below 30 mmHg ($\rho_{Hg} = 13.6$ g cm^{-3} and g = 980 cm s^{-2}); $\rho = 1.05$ g cm^{-3}, $\mu = 0.009$ g cm^{-1} s^{-1}, $L_p = 10$ cm (pipette length). Assume fully developed laminar flow in the pipette.

4.17 Due to the ready availability of mouse strains lacking specific genes, the mouse has become an important model for studying atherosclerosis. While mice do not normally develop atherosclerosis, deletion of specific genes involved in cholesterol metabolism leads to atherosclerosis that resembles the disease in humans.

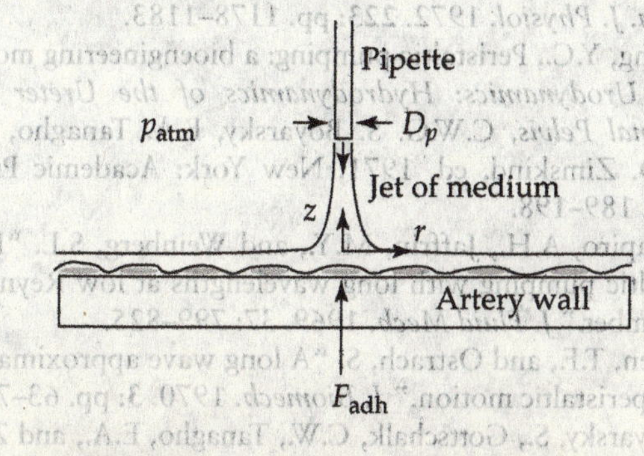

FIGURE 4.22 A fluid jet impinging on endothelium.

Since arterial hemodynamics, and the arterial wall shear stress in particular, are implicated in the disease, there is interest in comparing the mean shear stresses in the mouse and human.

(a) A reference shear stress is based on the flow from the aortic root using Poiseuille's law. Using this result (Equation (2.7.45)), show that the wall shear stress can be written as

$$\tau_w = \frac{8\mu^2 Re}{\rho D^2}$$

(b) Using the following data, determine the reference values for τ_w and α, the Womersley number ($\alpha = R\sqrt{\omega/\nu}$ where the frequency ω is in radians per second). Assume that the viscosity (0.035 g cm^{-1} s^{-1}) and density (1.05 g cm^{-3}) are the same for both species. Based on this result, do you think that the hemodynamic conditions are similar or different between the human and the mouse?

	Human	Mouse
Cardiac output (L min^{-1})	5.0	0.012
Heart rate (beats per minute)	60	600
Aortic root diameter (cm)	3.0	0.1

4.18 (a) For the approximate velocity profile $v_x/U_0 = a_0 + a_1(y/\delta) + a_3(y/\delta)^3$ and the following boundary conditions, determine the second expression for velocity in Table 4.1.

$$y = 0 \quad v_x = 0$$

$$y = \delta \quad v_x = U_0$$

$$\frac{\partial v_x}{\partial y} = 0$$

(b) Use this result in the von Karman momentum integral to determine expressions for the coefficients a and b in Equations (4.5.26a) and (4.5.26b), respectively.

4.19 Introducing the appropriate dimensionless variables, derive Equations (4.7.1a) and (4.7.1b).

4.20 Explore the magnitudes of forces in thin layers, as given in Equations (4.7.10c) and (4.7.14b). Start with the dimensions of the lubricating layer set forth in Example 4.6. The force magnitudes also depend upon the sliding velocity U, the fluid viscosity μ, and the depth W of the layer. Indeed, the forces are proportional to the product μUW. Compare a water-like fluid with viscosity $\mu = 1$ centipoise $= 10^{-3}$ Pa s against a fluid with viscosity $\mu = 10$ poise $= 1$ Pa s. If $W = L$, how large must sliding velocities be to support a weight of, say, 10 lbf = 44.48 N?

4.21 Analyze the flow in a two-dimensional channel with porous walls similar to those in the problem solved in Example 4.7, except now assume that the normal velocity v_w at the wall is constant. Show that the effect of the flow into the wall is to reduce the pressure gradient. Sketch the velocity profiles for this problem in comparison with that for a solid wall.

4.22 Rework the analysis of peristaltic pumping (Equations (4.8.3) through 4.8.13)) for two-dimensional flow in a channel of mean height $2a$. Now investigate the conditions under which reflux flow can occur. There will be reflux when some portions of the velocity profile for the axial velocity v_x versus y are positive and some portions are negative. Obtain an expression $v_x(y, t)$, and then take its time average $\bar{v}$, producing

$$\frac{\bar{v}}{c} = \frac{3}{2}\left\{1 - (1-\phi^2)^{\frac{-1}{2}} + \frac{3}{2}\phi^2(1-\phi^2)^{\frac{-5}{2}}\left(\frac{y}{a}\right)^2 + \bar{Q}\left[(1-\phi^2)^{\frac{-1}{2}} - \left(1+\frac{\phi^2}{2}\right)(1-\phi^2)^{\frac{-5}{2}}\left(\frac{y}{a}\right)^2\right]\right\}.$$

Rearrange terms in the preceding equation to obtain

$$\frac{\bar{v}}{c} = \frac{3}{2}(1-\phi^2)^{\frac{-1}{2}}\left\{\bar{Q} - \left[1 - (1-\phi^2)^{\frac{1}{2}}\right]\right\} + \frac{3}{2}\left(1+\frac{\phi^2}{2}\right)(1-\phi^2)^{\frac{-5}{2}}\left(\frac{3\phi^2}{2+\phi^2} - \bar{Q}\right)\left(\frac{y}{a}\right)^2.$$

By inspection of this equation, determine two conditions involving Q versus ϕ in which there are positive and negative values of v_x for different values of y. Sketch the velocity profiles for these two cases.

4.11 REFERENCES

1. Bird, R., Stewart, W., and Lightfoot, E., *Transport Phenomena*. 2d ed. 2002, New York: John Wiley and Sons.
2. Whitaker, S., *Introduction to Fluid Mechanics*. 1968, Malabar, FL: Krieger Publishing Company.
3. Berkow, R., Fletcher, A.J., and Bondy, P.K., *The Merck Manual of Diagnosis and Therapy*. 16th ed. 1992, Rahway, NJ: Merck Research Laboratories.
4. American Heart Association, *Heart and Stroke Statistical Update*. 2003, Dallas: American Heart Association.
5. Hatle, L., Brubakk, A., Tromsdal, A., and Angelsen, B. "Noninvasive assessment of pressure drop in mitral stenosis by Doppler ultrasound." *British Heart Journal*, 1978. 40: pp. 131–140.
6. Yoganathan, A.P., Cape, E.G., Sung, H.-W., Williams, F.P., and Jimoh, A. "Review of hydrodynamic principles for the cardiologist: applications to the study of blood flow and jets by imaging techniques." *J. Amer. Coll. Cardiol.* 1988. 12: pp. 1344–1353.
7. White, F., *Fluid Mechanics*. 1986, New York: McGraw-Hill Book Company.
8. Labovitz, A.J., and Williams, G.A., *Doppler Echocardiography: The Quantitative Approach*. 1992, Philadelphia: Lea and Febiger.
9. Snider, A.R., Serwer, G.A., and Ritter, S.B., *Echocardiography in Pediatric Heart Disease*. 1997, St. Louis, MO: Mosby.
10. Denn, M.M., *Process Fluid Mechanics*. 1980, Englewood Cliffs, NJ: Prentice-Hall, Inc.
11. Schlicting, H., *Boundary Layer Theory*. 1979, New York: McGraw-Hill.
12. Mow, V.C., and Ateshian, G.A., "Lubrication and wear of diarthrodial joints," in *Basic Orthopaedic Biomechanics*, V.C. Mow and W.C. Hayes, ed. 1997, Philadelphia: Lippincott-Raven. pp. 275–315.
13. Moore, D.F., *Principles and Applications of Tribology*. 1975, New York: Pergamon.
14. Deen, W.M., Robertson, C.R., and Brenner, B.M. "A model of glomerular ultrafiltration in the rat." *Am. J. Physiol.* 1972. 223: pp. 1178–1183.
15. Fung, Y.C., Peristaltic pumping: a bioengineering model, in *Urodynamics: Hydrodynamics of the Ureter and Renal Pelvis*, C.W.G. S. Boyarsky, E.A. Tanagho, and P.D. Zimskind, ed. 1971, New York: Academic Press. pp. 189–198.
16. Shapiro, A.H., Jaffrin, M.Y., and Weinberg, S.L. "Peristaltic pumping with long wavelengths at low Reynolds number." *J. Fluid Mech.* 1969. 37: 799–825.
17. Zien, T.F., and Ostrach, S. "A long wave approximation to peristaltic motion." *J. Biomech.* 1970. 3: pp. 63–75.
18. Boyarsky, S., Gottschalk, C.W., Tanagho, E.A., and Zimskind, P.D., eds. *Urodynamics: Hydrodynamics of the Ureter and Renal Pelvis*. 1971, New York: Academic Press.

Fluid Flow in the Circulation and Tissues

CHAPTER 5

5.1 Introduction

Blood flow throughout the cardiovascular system plays a critical role in the delivery of oxygen to tissues and in the removal of carbon dioxide. Recent research demonstrates that red blood cells and certain plasma proteins are important in the transport of nitric oxide (NO), which is a key factor in regulating the diameter of blood vessels and blood flow to tissues. In addition, blood is involved in the transport of other nutrients, metabolites (e.g., glucose, iron, and cholesterol), and hormones, as well as the transport of white cells and immunoglobulins involved in the immune response. Cancer cells are also transported through blood as they spread from one tissue to another in a process known as *metastasis*.

As noted in Section 1.6.1, a number of mechanisms regulate the distribution of blood to different organs and tissues. Such regulation occurs rapidly during exercise or after a meal. Blood flow regulation involves the dilation or contraction of arterioles by hormones secreted by the kidney (angiotensin), nerve cells (epinephrine), and endothelial cells (NO, a vasodilator, and endothelin, a vasoconstrictor). On a longer timescale, blood pressure is regulated by changes in the thickness and inner diameter of vessels through the synthesis of extracellular matrix.

Blood flow is altered in a number of pathological conditions. During a heart attack or a stroke, there is partial or complete occlusion of blood vessels due to the formation of a blood clot that reduces or halts blood flow. Unless blood flow is reestablished, the tissue perfused by the occluded blood vessel can die from oxygen starvation (a condition known as an *infarct*). Brain and heart tissue cannot regenerate. If an infarct occurs, the complications can be severe, ranging from the loss of a specific brain function (as in paralysis, memory loss, or aphasia) to reduced cardiac output and even death. A number of treatments are used to reestablish blood flow. In *angioplasty*, the occlusion is disrupted mechanically, usually with a balloon catheter. Alternatively, enzymes such as tissue plasminogen activator can dissolve the clot and have been proven to have a major effect upon improving the outcome for patients suffering from heart attacks and strokes. Less-severe occlusions due to progressive atherosclerosis produce a transient disruption of blood flow and reduce oxygen levels in tissue (a condition known as *ischemia*). Ischemic episodes in the heart, resulting from increased exertion, lead to *angina*. Atherosclerosis can also occur in arteries of the leg, leading to ischemic episodes in the lower limbs during routine physical activity (a condition known as *claudication*). Other diseases in which blood vessel occlusion is significant include sickle-cell anemia, diabetes, deep-vein thrombosis, and various disorders of the clotting cascade.

In addition to pathologies that can compromise blood flow by occluding vessels, elevated blood pressure (*hypertension*) alters the rigidity of vessel walls and the responses of the walls to agents that dilate blood vessels (*vasodilators*). Although a number of biological molecules that affect blood pressure are known, the manner in which these agents affect hypertension is poorly understood. Hypertension is a risk factor for cardiovascular disease. Hypertension also leads to an increased pressure difference across the capillary endothelium, resulting in the accumulation of fluid in tissue, a condition known as *edema*. In the lungs, edema resulting from damage to the heart by a heart attack can compromise breathing and lead to death.

A quantitative characterization of blood flow is important in understanding physiological and pathological processes, the interface of devices that perfuse blood (e.g., blood oxygenators and hemodialyzers), and artificial heart valves. Flow in the arterial system differs from the steady laminar flow of a Newtonian fluid in the following ways:

(a) The rheology of blood is non-Newtonian. The corresponding effects on blood flow were described in Chapter 2.
(b) Blood flow is pulsatile.
(c) Vessel entrance lengths are not sufficiently long for flow to become fully developed.
(d) Curved and branched vessels affect the flow.
(e) Arterial walls are viscoelastic.

The effects of time-varying pressure pulses on the velocity profile and shear stress at the wall are described in Section 5.2. *Entrance lengths* are the lengths of the entrance regions of vessels over which the velocity depends upon the axial coordinate. Entrance lengths are sensitive to curvature and unsteady flow and are presented in Section 5.3. The influences of vessel curvature and branching are examined in Sections 5.4 and 5.5. Flow in specific blood vessels is considered in Section 5.6. The concluding sections of the chapter focus on the role of blood flow in specific biological conditions. Section 5.7 examines the effect of arterial fluid dynamics in atherosclerosis. In Section 5.8, hemodynamic factors affecting heart valves are discussed. In Section 5.9, we briefly examine flow in the venous system and the impact of low pressure on the venous side on the design of shunts to treat congenital defects of the heart. We do not consider pulse wave propagation, a topic studied in more advanced texts [1,2].

5.2 Oscillating Flow in a Cylindrical Tube

Blood fills the ventricles and atria of the heart as the heart muscle relaxes (Figure 5.1). Contraction of the heart muscle results in the flow of blood from the right ventricle into the pulmonary artery and from the left ventricle into the aorta. Heart valves

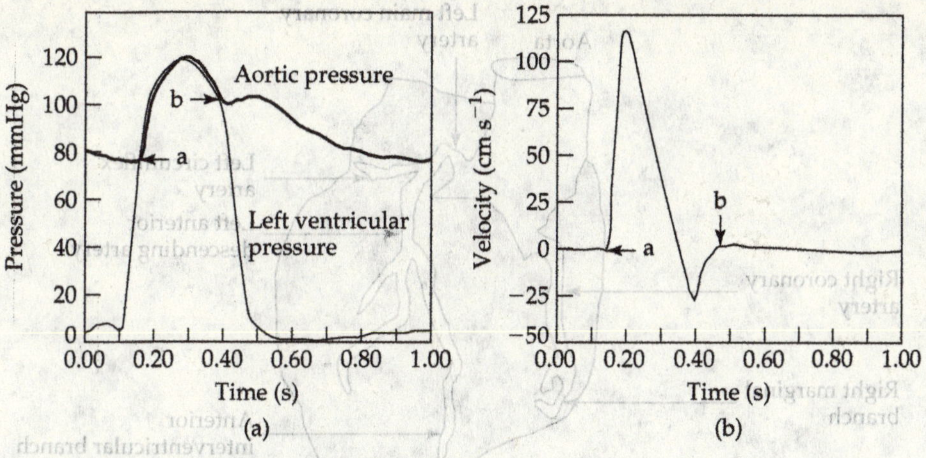

FIGURE 5.1 (a) Pressure in the aorta and left ventricle during a single heartbeat. Point a represents the opening of the aortic valve and point b represents the closing of the valve. (b) Average velocity in the aorta. Flow begins when the valve opens. The velocity increases rapidly until the pressure starts to decline. (Adapted from Ref. [3], with permission of Cambridge University Press.)

prevent the backflow of blood. The result is that the pressure and flow vary with time over the period of heart relaxation and contraction.

Blood flow is divided into two phases: *systole*, during which blood is pumped from the heart, and *diastole*, during which no blood is pumped from the heart and the ventricles fill with blood. As the heart contracts, left ventricular pressure rises rapidly from about 5 to 120 mmHg in about 0.2 seconds (Figure 5.1a). Aortic pressure typically ranges from 80 to 120 mmHg. When the ventricular pressure exceeds the aortic pressure, the aortic valve opens and blood flows into the aorta (point a in Figures 5.1a and 5.1b). Aortic flow reaches a maximum when the local pressure derivative begins to decrease prior to the point of maximum pressure. As the pressure decreases, the valve starts to close. The small amount of backflow is due to bulging of the valves and flow into the coronary arteries [3]. The aortic velocity is approximately zero until the left ventricle begins to contract again. Thus, the aorta maintains a relatively high pressure, even when there is no flow. The pressure and velocity pulse change with distance from the heart [1], the maximum pressure declining and the pulse width broadening with increasing distance. In the arterioles, the pressure change is greatly attenuated. Pressure pulses are almost nonexistent in the venous system, and flow is steady. The velocity in arteries downstream from the aorta is affected by the resistance downstream from the branch. In most vessels, the velocity does not go to zero during diastole, although in some vessels (e.g., the external carotid artery and the femoral artery), there is a period of negative velocity.

The coronary arteries branch from the aorta and supply blood to the heart muscle (Figure 5.2). There are two main branches: the left coronary artery, which supplies the left side of the heart, and the right coronary artery, which supplies the right side. The major arteries lie near the surface, with part of the coronary arteries exposed (although sometimes lying beneath a layer of fatty tissue) and part of each vessel buried in the heart muscle tissue. Flow in the coronary arteries is influenced by the motion of the heart, and the resulting velocity differs substantially from that found in the aorta. During systole, the heart muscle contracts to expel blood from the ventricles. Contraction compresses the coronary arteries, reducing flow through those vessels (Figure 5.3). During diastole, the heart muscle relaxes and the coronary arteries fill with blood.

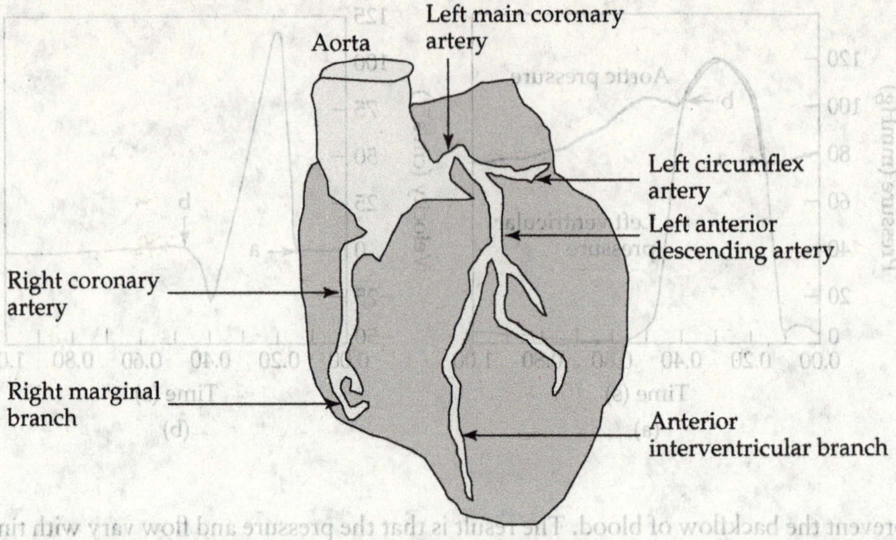

FIGURE 5.2 Diagram of the coronary arteries.

Mathematically, the pressure and velocity waveforms can be decomposed into Fourier series. Thus, if the relationship between pressure and velocity is known for a pressure waveform that varies as a sine or cosine function, the entire velocity field can be determined by multiplying the various solutions by the Fourier coefficients. A minimum of 6 harmonics are needed to describe the pressure waveform accurately, and 10 harmonics are needed to describe the flow waveform [3]. The velocity field in response to an oscillating pressure field was determined in a manner that permitted practical application by Womersley [5]. In this section, the velocity distribution as a function of radial position and time is determined for fully developed flow in a rigid circular tube, and major features of the velocity field are discussed. While these conditions represent a vast simplification of the actual flow in arteries, the results do provide some insight into the manner in which the velocity is affected by the frequency and Reynolds number.

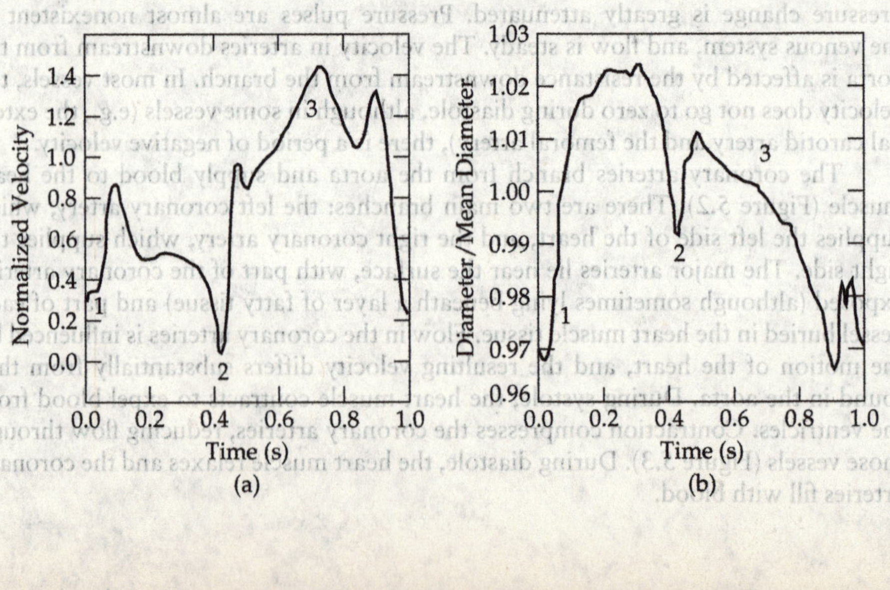

FIGURE 5.3 (a) Waveform of the velocity in the canine left main coronary artery. (b) Associated change in vessel diameter. (Adapted from Ref. [4], with permission.) Times marked as 1, 2 and 3 are discussed in Section 5.6.4.

For unsteady flow, a dimensional analysis indicates that the velocity depends upon two dimensionless groups (Section 3.5): the Reynolds number and the Strouhal number ($\mathrm{St} = L_x/T\langle v_x\rangle$) For oscillatory flow, the characteristic time is proportional to the reciprocal of the frequency ω (rad s^{-1}). In place of the Strouhal number, the Womersley number α is used. This dimensionless group scales the characteristic time $1/\omega$ of the oscillating flow with the characteristic time of momentum transport R^2/ν, where R is the vessel radius and ν is the kinematic viscosity. The operative equation is

$$\alpha = R\sqrt{\omega/\nu}. \tag{5.2.1}$$

The frequency used is the dominant frequency in blood flow arising from the heartbeat. Representative values of the Womersley number presented in Table 5.1 indicate that the Womersley number declines in smaller arteries. These differences in the magnitude of the Womersley number have a significant effect on the shape of the velocity distribution.

To determine the velocity distribution, consider the laminar flow of a Newtonian fluid in a rigid, long, circular blood vessel exposed to an oscillating pressure field. The problem is similar to that discussed in Section 2.7.3, except that the flow is time dependent. Flow is only in the axial direction ($v_r = v_\theta = 0$) and is fully developed. (v_z is a function of radial position only.) As a result, only the z component of the Navier–Stokes equation in cylindrical coordinates (Equation (3.3.27c)) is nonzero. The resulting terms are given in the formula

$$\rho\frac{\partial v_z}{\partial t} = -\frac{\partial p}{\partial z} + \frac{\mu}{r}\frac{\partial}{\partial r}\left(r\frac{\partial v_z}{\partial r}\right). \tag{5.2.2}$$

The pressure gradient oscillates in time with frequency ω (rad s^{-1}) and can be represented as

$$-\frac{\partial p}{\partial z} = \frac{\Delta p}{L}\cos(\omega t) = \mathrm{Re}[A^* e^{i\omega t}], \tag{5.2.3}$$

where $i = \sqrt{-1}$ and $\mathrm{Re}[e^{i\omega t}]$ denotes the real part of $e^{i\omega t} = \cos(\omega t) + i\sin(\omega t)$. In general, A^* may be complex, although in the analysis that follows, it is assumed to be a real constant. The complex notation for the pressure field simplifies the solution of

TABLE 5.1

Vessel Radius, Reynolds Number, and Womersley Number for Various Human Arteries Under Resting Conditions (60 Beats per Minute)

Vessel	Radius (cm)	$\mathrm{Re}_{\mathrm{mean}}$	α
Proximal aorta [1,3]	1.5	1,500	21.7
Femoral artery [1]	0.27	180	3.9
Left main coronary artery [1,6]	0.425	270	6.15
Left anterior descending coronary artery [6]	0.17	80	2.4
Right coronary artery [7]	0.097	233	1.82
Terminal arteries [3]	0.05	17	0.72

Equation (5.2.2). Using the general complex representation for the pressure gradient and substituting yields the following relation:

$$-\rho\frac{\partial v_z}{\partial t} + \frac{\mu}{r}\frac{\partial}{\partial r}\left(r\frac{\partial v_z}{\partial r}\right) = -A^* e^{i\omega t}. \tag{5.2.4}$$

Boundary conditions for this problem are as follows:

$$r = 0 \qquad \frac{\partial v_z}{\partial r} = 0 \qquad v_z \text{ is finite,} \tag{5.2.5a}$$

$$r = R \qquad v_z = 0. \tag{5.2.5b}$$

Equation (5.2.4) can be solved as a function of time by assuming that the velocity can be represented as the product of two functions, one of which, u, is a function of radial position and the other of which is equal to $e^{i\omega t}$:

$$v_z = ue^{i\omega t}. \tag{5.2.6}$$

Substituting Equation (5.2.6) into Equation (5.2.4) yields

$$\frac{i\omega ue^{i\omega t}}{\nu} + \frac{e^{i\omega t}}{r}\frac{d}{dr}\left(r\frac{du}{dr}\right) = \frac{A^*}{\mu}e^{i\omega t} \tag{5.2.7}$$

Since each term in the foregoing equation is multiplied by $e^{i\omega t}$, the time dependence cancels. Rearranging terms yields

$$\frac{d^2u}{dr^2} + \frac{1}{r}\frac{du}{dr} - \frac{i\omega u}{\nu} + \frac{A^*}{\mu} = 0. \tag{5.2.8}$$

Substituting Equation (5.2.6) into Equations (5.2.5a) and (5.2.5b), we obtain the boundary conditions for u:

$$r = 0 \qquad \frac{du}{dr} = 0 \ (u \text{ is finite}), \tag{5.2.9a}$$

$$r = R \qquad u = 0. \tag{5.2.9b}$$

The solution of Equation (5.2.8) for $A^* = 0$ is Bessel's equation (A.2.31), with χ equal to zero and $m = i^{3/2}\alpha R$. The general solution of Equation (5.2.8) is

$$u = \frac{A^*}{i\omega\rho} + C_1 J_0(i^{3/2}\sqrt{\omega/\nu}r) + C_2 Y_0(i^{3/2}\sqrt{\omega/\nu}r), \tag{5.2.10}$$

where $J_0(i^{3/2}\sqrt{\omega/\nu}r)$ and $Y_0(i^{3/2}\sqrt{\omega/\nu}r)$ are, respectively, Bessel functions of the first and second kind of order zero. These functions are graphed in Figure A.3. At $r = 0$, the velocity is finite. Since $Y_0(r)$ goes to negative infinity as r goes to zero, C_2 must equal zero. After evaluating the boundary condition at $r = R$ and rearranging terms, we find that the velocity is

$$v_z = \frac{A^*}{i\omega\rho}\left(1 - \frac{J_0(i^{3/2}\alpha r/R)}{J_0(i^{3/2}\alpha)}\right)e^{i\omega t}. \tag{5.2.11}$$

Because the pressure gradient is the real part of $A^*\exp(i\omega t)$, the velocity is the real part of Equation (5.2.11). Also, because the Bessel functions depend upon i, the real part of the velocity depends upon the magnitude of the argument $\alpha r/R$.

The solution is graphed by creating the following MATLAB M-file to compute v_z:

```
function z=oscill(t);
global A rho al r f;
pi=3.14159;
n=-1:0.01:1;
r=n;
z=A/(i*2*pi*f*rho)*(1-...
  besselj(0,i*sqrt(i)*al*n)/besselj(0,i*sqrt(i)*al))*...
  exp(i*2*pi*f*t);
```

Figure 5.4 is a graph of v_z as a function of radial position and phase of the cycle (ωt is divided into 360 degrees) for a value of $\alpha = 3.34$. For this particular value of α, the velocity is not in phase with the pressure waveform. The pressure is a maximum at 0°, whereas the velocity at the centerline is largest at around 75°. At zero time, when the pressure gradient is a maximum, the profile deviates significantly from a parabolic velocity profile observed with steady flow. This deviation is the result of inertial forces. As time progresses, the magnitude of the velocity increases while the pressure gradient decreases. Fluid acceleration becomes less significant, and between 60° and 100°, the velocity profile tends toward a parabolic shape. As the velocity decreases, the velocity profile deviates from the parabolic shape and secondary minima in the velocity appear as inertial forces, again affecting the slowest velocities ($\alpha r/R$ is largest). As the fluid begins to decelerate, negative velocities appear near the tube walls. These regions of flow reversal are due to the positive pressure gradient. The slower velocity components near the wall are affected first, but eventually all of the velocities are negative. At 180°, the velocity profile has the same shape as the profile at 0°, except that the direction of the velocity has reversed.

The phase shift and distortion of the velocity profile are dependent upon the value of α. At lower values of α, the velocity profile is in phase with the pressure waveform, and the shape of the velocity is parabolic at all times (Figure 5.5). Conceptually, this behavior is consistent with the zero-frequency limit in which the velocity profile is parabolic.

The convergence of the velocity profile to a parabolic profile for small values of α can be shown as follows. For small values of the argument, the Bessel function $J_0(x)$ can be expanded into the following power series:

$$J_0(x) = 1 - \frac{x^2}{2^2} + \frac{x^4}{2^4 2!} - \frac{x^6}{2^6(3!)^2} + \cdots. \qquad (5.2.12)$$

Truncating Equation (5.2.12) at the first term and substituting into Equation (5.2.11) yields the following expression:

$$v_z = \frac{A^*}{i\omega\rho}\left(1 - \left[\frac{1 - \frac{1}{4}(i^3\alpha^2(r/R)^2)}{1 - \frac{1}{4}(i^3\alpha^2)}\right]\right)[\cos(\omega t) + i\sin(\omega t)]. \qquad (5.2.13)$$

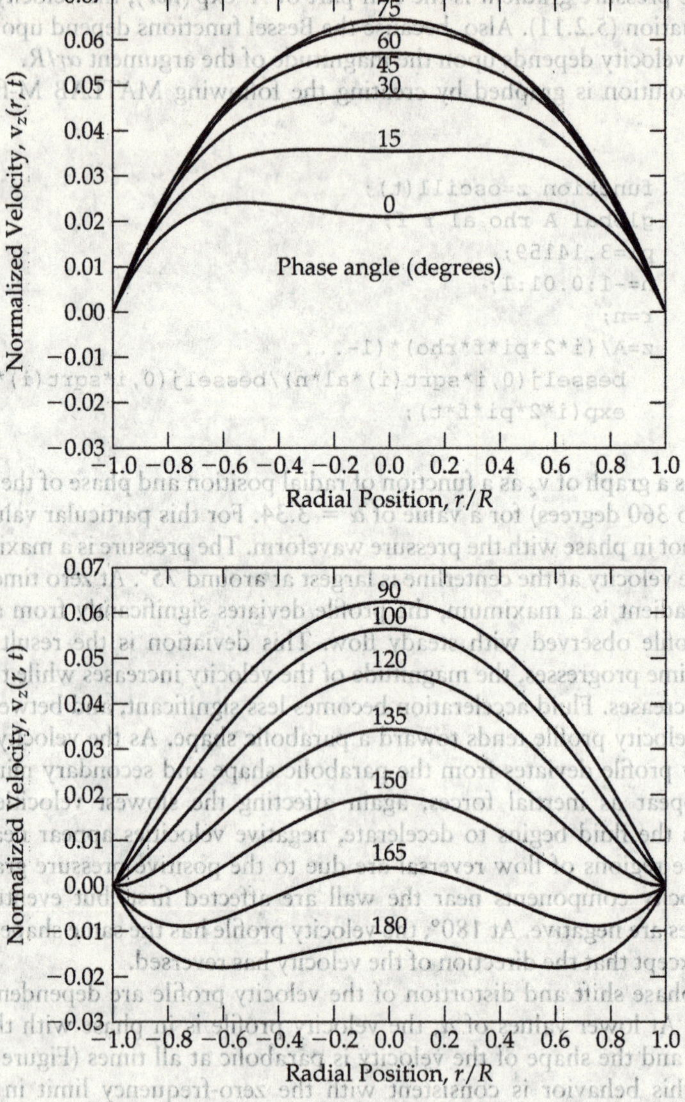

FIGURE 5.4 Velocity profiles for the oscillatory pressure gradient given by Equation (5.2.3) with $A^* = 1$ dyn cm^{-1}, $\rho = 1$ g cm^{-3}, $\omega = 3$ cycles per second, $s = 6\pi$ rad s^{-1}, and $\alpha = 3.34$. These values correspond to values obtained for a dog femoral artery [1,5]. Velocity is normalized by A^*/ρ and a frequency of 1 Hz.

Note that i^3 equals $-i$ and $1/i = i/(ii) = -i$. Multiplying the numerator and denominator of the term in brackets by the complex conjugate of the denominator, $1 + i\alpha^2/4$, and collecting terms yields the following expression:

$$v_z = \frac{A^*R^2}{\alpha^2\mu}\left(\frac{4\alpha^2(1 - (r/R)^2) - i\alpha^4(1 + (r/R)^2)}{16 + \alpha^4}\right)[\cos(\omega t) + i\sin(\omega t)]. \qquad (5.2.14)$$

Taking the real part of Equation (5.2.14) and assessing the limit as α goes to zero yields a parabolic velocity profile (e.g., Equation (2.7.41)) that is in phase with the pressure pulse; that is,

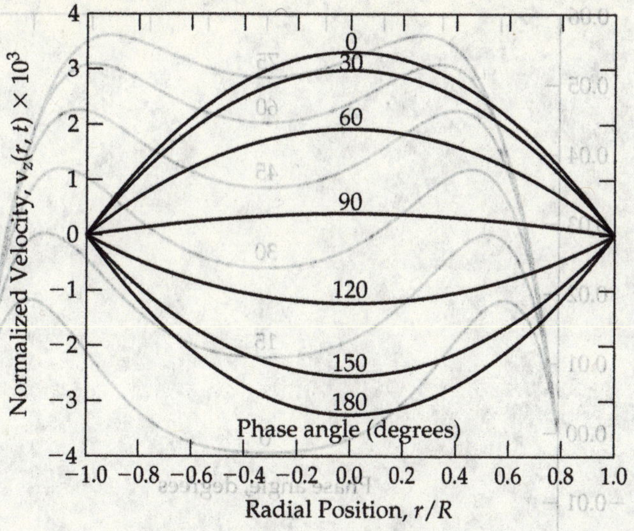

FIGURE 5.5 Velocity profile for $\alpha = 0.5$. Other parameters are as specified in the legend to Figure 5.4.

$$v_z = A^* \frac{R^2}{4\mu}\left(1 - \left(\frac{r}{R}\right)^2\right)\cos(\omega t), \tag{5.2.15}$$

where

$$A^* = \frac{\Delta P}{L}. \tag{5.2.16}$$

When the velocity profile shape does not change with time and the magnitude is in phase with the pressure wave, the velocity is said to be *quasi steady*. The velocity adapts rapidly to the change in the pressure gradient, and the acceleration term in Equation (5.2.4) is small. This analysis is valid when α is 0.2 or less.

As the Womersley number increases in magnitude, inertial forces dominate, the profile becomes blunter (Figures 5.6 and 5.7), and the phase shift increases. At higher values of α, the profile is blunter near the centerline, and the shape appears similar at all times, with local maxima in the velocity near the walls (Figure 5.7).

The foregoing discussion indicates that the Womersley number does not simply represent the characteristic time for oscillatory flow, but also includes a scaling of viscous and inertial forces. To see this, multiply and divide Equation (5.2.1) by the square root of the average velocity:

$$\alpha = \sqrt{R^2\omega\langle v\rangle/\langle v\rangle\nu} = \sqrt{\mathrm{Re}R\omega/2\langle v\rangle} = \sqrt{\mathrm{Re\,St}/2}. \tag{5.2.17}$$

Thus, the Womersley number equals the square root of the product of the Reynolds number and the Strouhal number. Low values of α represent low frequencies, or low values of the Reynolds number.

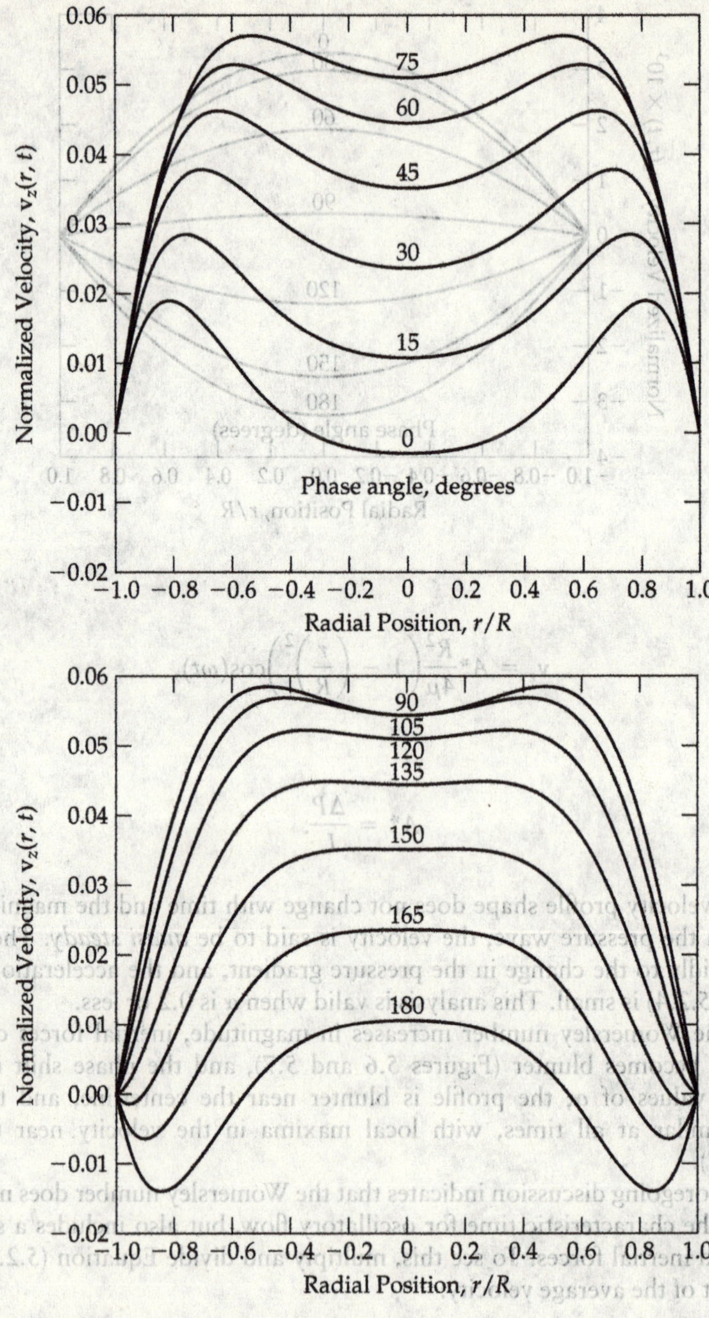

FIGURE 5.6 Velocity profile for $\alpha = 6.5$. Other parameters are as specified in the legend to Figure 5.4.

Implicit in this analysis is the assumption that the velocity depends only upon one dimensionless group, α. To ascertain that this is indeed the case, cast Equation (5.2.8) into dimensionless form. The characteristic velocity is the average velocity obtained for steady flow in a circular tube, $A^* R^2/8\mu$, and the dimensionless velocityis $u^* = u8\mu/A^* R^2$. The dimensionless radial position is $r^* = r/R$. As a

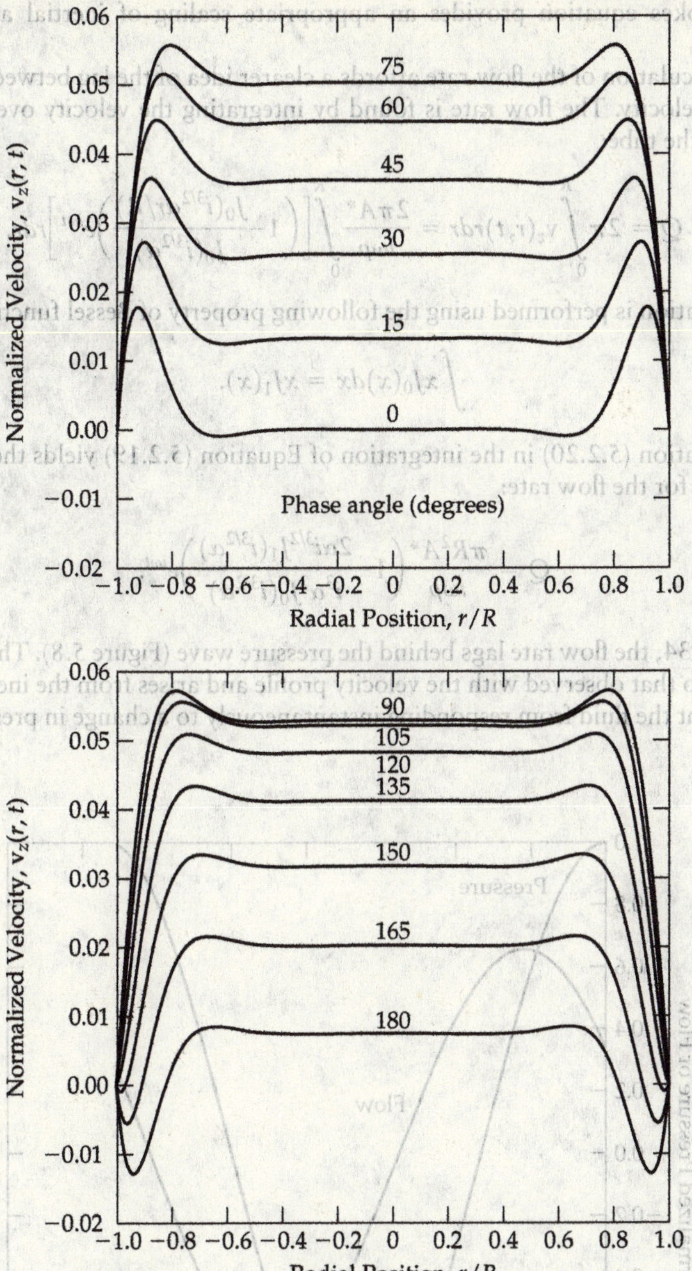

FIGURE 5.7 Velocity profile for $\alpha = 15$. Other parameters are as specified in the legend to Figure 5.4.

result, Equation (5.2.8) becomes

$$\frac{d^2u^*}{dr^{*2}} + \frac{1}{r^*}\frac{du^*}{dr^*} - i\alpha^2 u^* + 8 = 0. \tag{5.2.18}$$

Thus, u^* is a function of only α and the dimensionless position. The velocity v_z also depends upon ωt. The dimensionless form of the axial component of the

Navier–Stokes equation provides an appropriate scaling of inertial and viscous forces.

A calculation of the flow rate affords a clearer idea of the lag between the pressure and velocity. The flow rate is found by integrating the velocity over the cross section of the tube:

$$Q = 2\pi \int_0^R v_z(r,t)rdr = \frac{2\pi A^*}{i\omega\rho} \int_0^R \left[\left(1 - \frac{J_0(i^{3/2}\alpha r/R)}{J_0(i^{3/2}\alpha)}\right)e^{i\omega t}\right] rdr. \quad (5.2.19)$$

The integration is performed using the following property of Bessel functions:

$$\int xJ_0(x)dx = xJ_1(x). \quad (5.2.20)$$

Using Equation (5.2.20) in the integration of Equation (5.2.19) yields the following expression for the flow rate:

$$Q = \frac{\pi R^2 A^*}{i\omega\rho}\left(1 - \frac{2\alpha i^{3/2} J_1(i^{3/2}\alpha)}{i^3\alpha^2 J_0(i^{3/2}\alpha)}\right)e^{i\omega t}. \quad (5.2.21)$$

For $\alpha = 3.34$, the flow rate lags behind the pressure wave (Figure 5.8). The phase lag is similar to that observed with the velocity profile and arises from the inertial forces that prevent the fluid from responding instantaneously to a change in pressure.

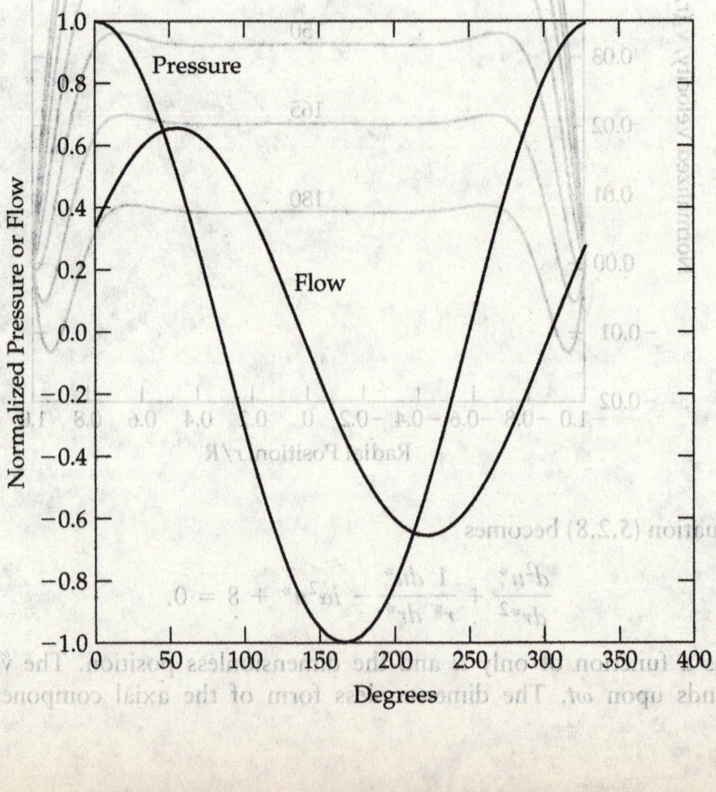

FIGURE 5.8 Relationship between pressure and flow rate for $\alpha = 3.34$.

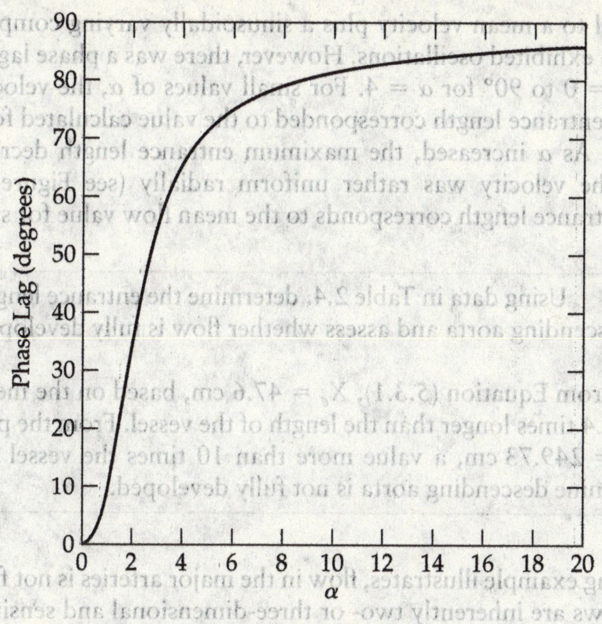

FIGURE 5.9 Phase lag between pressure and flow as a function of α.

The phase lag as a function of α is [8],

$$\text{Phase lag} = \tan^{-1}[\text{Imaginary}(Q(t=0))/\text{Real}(Q(t=0))]. \qquad (5.2.22)$$

For $\alpha = 0$, the phase lag vanishes, as the preceding analysis suggests (Equation (5.2.15)). As α increases, the velocity lags behind the pressure (Figure 5.9). The phase lag increases with increasing values of α, reaching a maximum of 90° for larger values of α.

5.3 Entrance Lengths

When flow conditions change due to a change in tube dimensions, there is a distance known as the *entrance length* over which the velocity adjusts to the change (see Section 2.7.3). Within this entrance region, flow depends upon both axial and radial position. Beyond the entrance region, the flow is fully developed and does not change with axial distance. Since the velocity profile asymptotically approaches the fully developed velocity profile, the entrance length is typically taken as the distance over which the velocity is within 1% of the fully developed velocity [9]. For steady flow, the entrance length for a tube of radius R depends upon the Reynolds number according to the formula,

$$X_e = 0.113R\text{Re}. \qquad (5.3.1)$$

In order to determine the entrance length for unsteady flow in a cylindrical tube, the equation for continuity and the Navier–Stokes equation were solved numerically for a velocity equal to a mean velocity plus a sinusoidally varying component [9]. The entrance length exhibited oscillations. However, there was a phase lag that increased from 0° for $\alpha = 0$ to 90° for $\alpha = 4$. For small values of α, the velocity was quasi-steady and the entrance length corresponded to the value calculated for the instantaneous velocity. As α increased, the maximum entrance length decreased. At high values of α, the velocity was rather uniform radially (see Figure 5.7), and for $\alpha > 12$, the entrance length corresponds to the mean flow value for steady flow.

Example 5.1 Using data in Table 2.4, determine the entrance length for flow in the canine descending aorta and assess whether flow is fully developed.

Solution From Equation (5.3.1), $X_e = 47.6$ cm, based on the mean velocity, a value about 2.4 times longer than the length of the vessel. From the peak Reynolds number, $X_e = 249.73$ cm, a value more than 10 times the vessel length. Thus, flow in the canine descending aorta is not fully developed.

As the preceding example illustrates, flow in the major arteries is not fully developed. Thus, these flows are inherently two- or three-dimensional and sensitive to the inlet flow conditions.

5.4 Flow in Curved Vessels

Many blood vessels are curved, and blood vessel curvature has a profound effect on flow. In the aortic arch and the Circle of Willis, vessels branch to distribute the blood to various tissues (Figure 5.10). The thoracic and abdominal aorta are tethered to the spine and assume its curvature. Other examples of vessels with significant curvature include the coronary arteries and the inferior vena cava (IVC). In general, the vessel curvature is in three dimensions, with complex effects on flow. The analysis that follows is restricted to the simpler case in which the curvature of a cylindrical tube occurs in a plane (a case known as *planar curvature*). Three-dimensional effects are then briefly discussed. More realistic situations are considered in Section 5.6.

For constant planar curvature, flow depends on the radius a and the radius of curvature, R (Figure 5.11). The angle θ originates at the outer wall in the midplane that bisects the tube. Inertial forces push fluid toward the outer wall (Figure 5.11), generating fluid motion in the radial and angular directions (i.e., v_r and v_θ are nonzero).

Flow in a curved cylindrical tube is governed by the ratio of the tube radius a to the radius of curvature, R, and by the Dean number De:

$$\delta = \frac{a}{R} \qquad \mathrm{De} = \sqrt{\delta}\,\mathrm{Re}. \tag{5.4.1a,b}$$

The Dean number represents a scaling of the Reynolds number by the ratio of the tube radius to the radius of curvature. (Several definitions of the Dean number are used, so it is important to check the definition carefully [10].) For a straight tube, the radius of curvature is infinite, and δ and De are zero. The curvature is the reciprocal

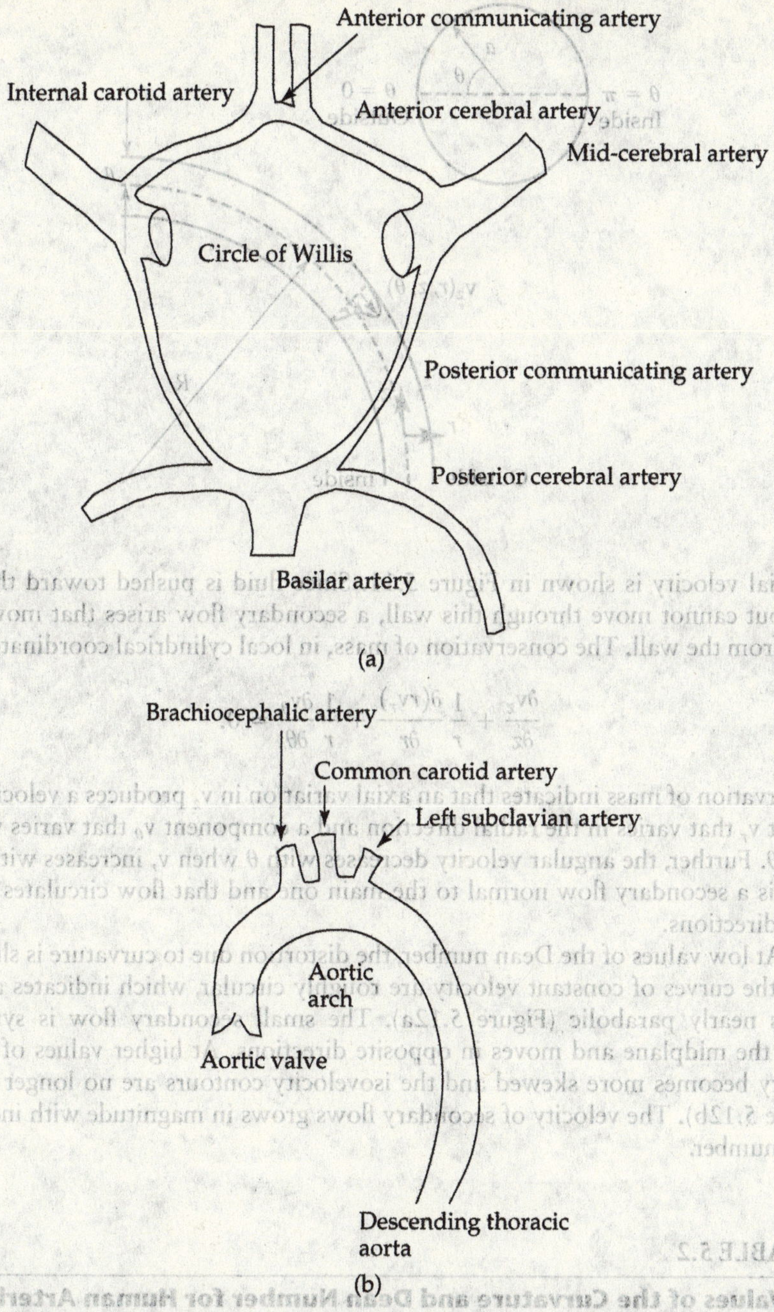

FIGURE 5.10 (a) The cerebral arteries that form the Circle of Willis. (b) Curvature of branches of the aortic arch.

of the radius of curvature. As the curvature increases, both δ and De increase in magnitude. Thus, for a fixed value of Re, inertial forces arising from curvature (*centrifugal forces*) become more important. Dean numbers for curved arteries range from 23 to 700 (Table 5.2).

The inertial forces cause the maximum in the axial velocity to shift toward the outer wall. The inner wall has lower velocities and shear stresses. A schematic of

FIGURE 5.11 Geometry of a curved vessel.

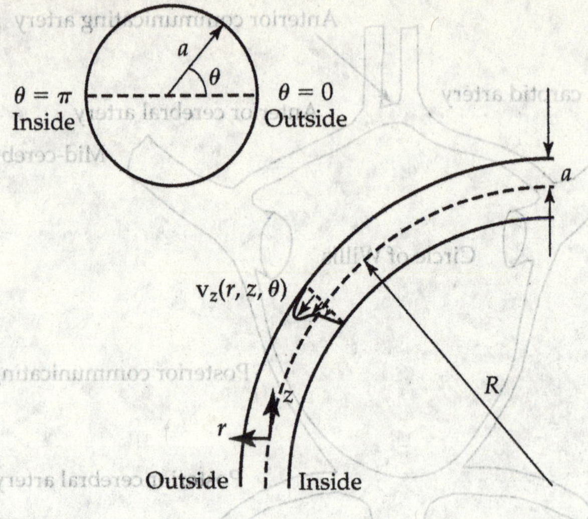

the axial velocity is shown in Figure 5.11. Since fluid is pushed toward the outer wall, but cannot move through this wall, a secondary flow arises that moves fluid away from the wall. The conservation of mass, in local cylindrical coordinates, is

$$\frac{\partial v_z}{\partial z} + \frac{1}{r}\frac{\partial (r v_r)}{\partial r} + \frac{1}{r}\frac{\partial v_\theta}{\partial \theta} = 0. \tag{5.4.2}$$

Conservation of mass indicates that an axial variation in v_z produces a velocity component v_r that varies in the radial direction and a component v_θ that varies with the angle θ. Further, the angular velocity decreases with θ when v_r increases with r. The result is a secondary flow normal to the main one and that flow circulates in the r and θ directions.

At low values of the Dean number, the distortion due to curvature is slight and all of the curves of constant velocity are roughly circular, which indicates a profile that is nearly parabolic (Figure 5.12a). The small secondary flow is symmetric about the midplane and moves in opposite directions. At higher values of De, the velocity becomes more skewed and the isovelocity contours are no longer circular (Figure 5.12b). The velocity of secondary flows grows in magnitude with increasing Dean number.

TABLE 5.2

Values of the Curvature and Dean Number for Human Arteries

Vessel	Radius (cm)	$\delta = a/R$	Re_{mean}	De
Aortic arch [1]	1.5	0.22	1,500	707
Left main coronary artery [6]	0.425	0.10	150	47.4
Left anterior descending coronary artery [6]	0.17	0.082	80	22.9
Right coronary artery [7]	0.097	0.024	233	36

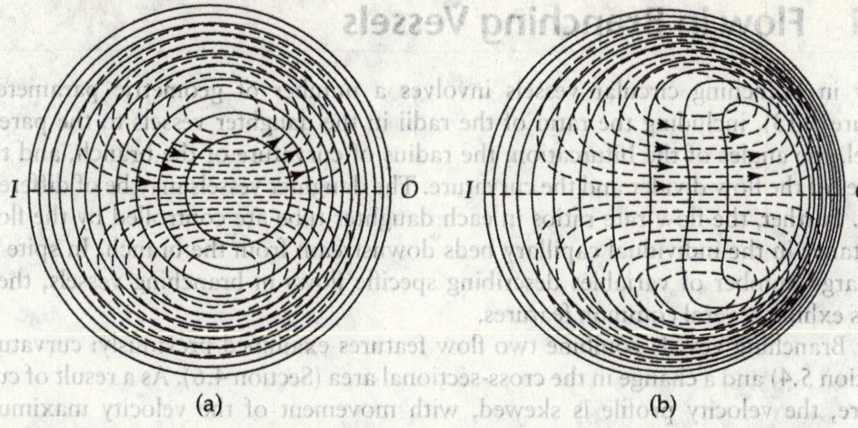

FIGURE 5.12 Lines of constant velocity (solid lines) and streamlines (dashed lines) for Dean numbers of (a) 34 and (b) 214. The view is a cross section normal to the direction of axial flow. *I*, inner wall; *O*, outer wall. (Adapted from Ref. [10], with permission of www.annualreviews.org © 1977.)

In general, the detailed velocity profiles for steady flow in curved tubes can be determined only by a numerical solution of the Navier–Stokes equation or by experiment [3]. However, for values of De less than 34, and $\delta \ll 1$, an approximate analytical solution is available [3]. For this solution, the wall shear-stress values are [3]

$$\tau_{rz}\big|_{r=a} = \frac{\mu\langle \mathrm{v}\rangle}{4a}\left[1 + \frac{\mathrm{De}^2 \cos\theta}{73{,}728}\right] \tag{5.4.3}$$

and

$$\tau_{r\theta}\big|_{r=a} = \frac{\mu\langle \mathrm{v}\rangle}{4a}\left(\frac{\mathrm{De}\sin\theta}{72}\right). \tag{5.4.4}$$

The secondary shear stress $\tau_{r\theta}$ arises from the secondary flow and is smaller than the value for the main flow component. The wall shear stress τ_{rz} due to the primary flow is greatest at the outer wall and lowest at the inner wall.

The effect of a fully developed pulsatile flow has been examined numerically [11,12] for sinusoidally varying pressure, using a curvature representative of the aortic arch. At low values of the Reynolds number, the flow behaves similarly to that described for steady flow. Secondary flows near the centerline move toward the outer wall, and fluid near the wall moves from the outer wall to the inner wall. For Re above 300, the flow changes. Secondary vortices arise and disappear throughout the pulse cycle. These vortices move in a direction opposite to that of the main secondary flow. The axial flow is reversed along the inner wall. The largest shear-stress magnitudes are located along the inner wall, not the outer wall. However, the time-averaged shear stresses are close to the steady-flow shear stresses. Due to flow reversal along the inner wall, the time-averaged values along the outer wall are greater than the values along the inner wall.

5.5 Flow in Branching Vessels

Flow in branching circular vessels involves a number of geometric parameters (Figure 5.13), including the ratio of the radii in the daughter vessels to the parent vessel, the angles of the bifurcation, the radius of curvature of the branch, and the shapes of the flow divider and the curvature. The daughter vessels may be of different sizes. Further, the flow rate ratios in each daughter tube are controlled by the flow resistance in the individual capillary beds downstream from the branch. In spite of the large number of variables describing specific flows in branching vessels, these flows exhibit several common features.

Branching vessels combine two flow features examined previously: curvature (Section 5.4) and a change in the cross-sectional area (Section 4.6). As a result of curvature, the velocity profile is skewed, with movement of the velocity maximum toward the flow divider. Secondary flows, similar to those observed in curved vessels, arise in the angular direction. The increase in cross-sectional area due to branching produces adverse pressure gradients that may give rise to flow reversal. In a Y-type branch (Figure 5.13), such separation arises along the outer wall of the daughter vessels. Examples of Y-type branches include the iliac arteries, which branch from the end of the abdominal aorta, and the carotid artery bifurcation.

In a T-type branch (Figure 5.14), such as in the renal arteries, there is a smaller side branch at a right angle to the main branch. Downstream from the location at which the vessel branches, the vessel diameter may be the same size or smaller than the diameter before the branch. Flow recirculation in the parent vessel does not occur below a critical Reynolds number based on the upstream velocity. The recirculation zone grows with increasing Reynolds number. Steady-flow studies in the rabbit [13] and dog [14] aortoceliac branch exhibit a critical Reynolds number that is dependent upon the flow into the side branch (Figure 5.15). Below the critical Reynolds number, inertial forces are not sufficiently large to produce an adverse pressure gradient with a magnitude large enough to cause the shear gradient to change directions (see Section 4.6). Decreasing the flow into the side branch suppresses the formation of the recirculation zone, presumably by decreasing the magnitude of the adverse pressure gradient. Under unsteady flow conditions, the Reynolds number based on the instantaneous velocity, rather than the average velocity, often

FIGURE 5.13 (a) Major geometric variables involved in a vessel branch. (b) Flow features present at a vessel branch. Note that for branching vessels, the notation of inner and outer walls differs from the convention used with curved vessels (see Figure 5.11).

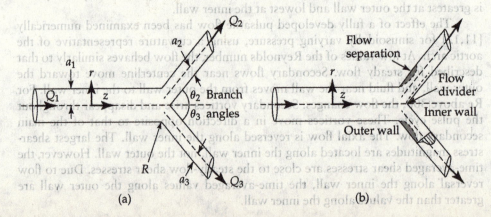

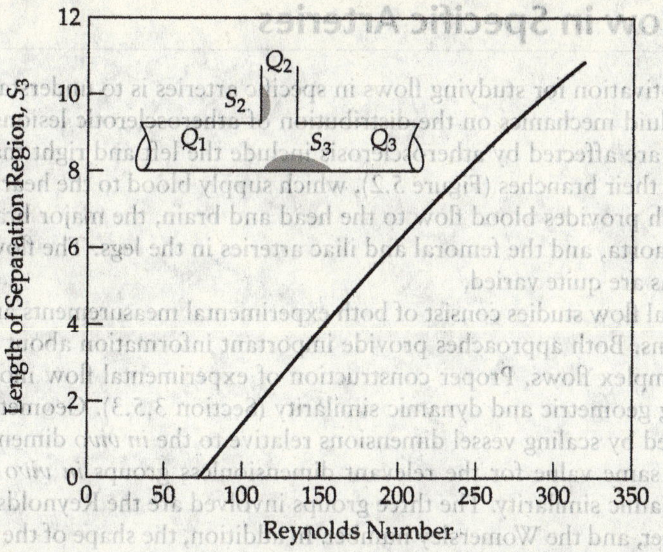

FIGURE 5.14 Growth of recirculation zone in a *T*-branch. (Based on results in Ref. [13,14].)

drops below the critical Reynolds number. As a result, flow recirculation is confined to a portion of the cardiac cycle.

The cross-sectional area of many arteries is not constant. Before the branch arises, the cross-sectional area increases slightly. Beyond the branch, the cross-sectional area of the daughter vessels decreases with distance from the branch [15]. This tapering may aid in limiting the onset, and the size of the region, of flow reversal.

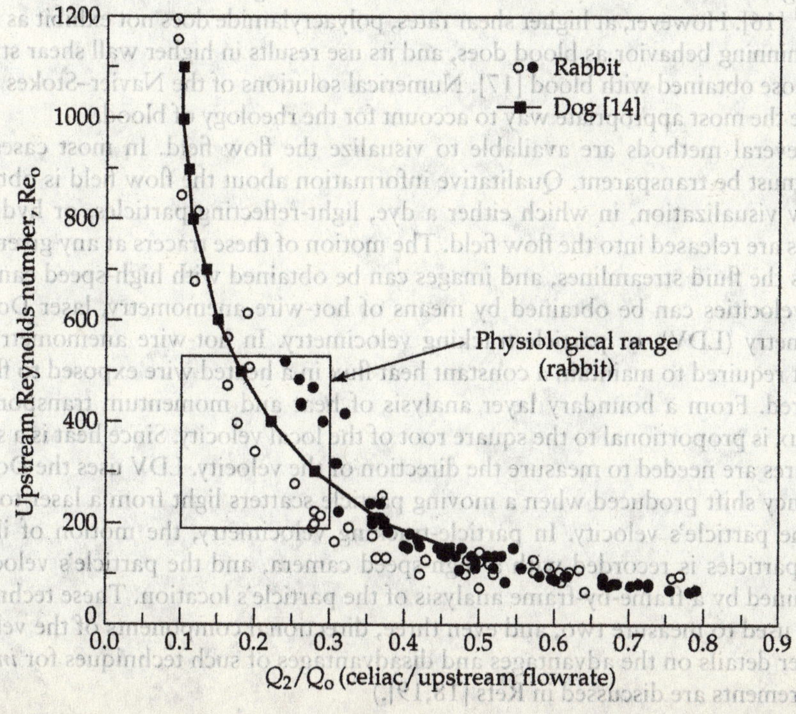

FIGURE 5.15 Critical Reynolds number for the onset of flow recirculation on the lateral sides of the aortoceliac branch for the dog [14] and rabbit (Adapted from Ref. [13], with permission from Elsevier.)

5.6 Flow in Specific Arteries

A prime motivation for studying flows in specific arteries is to understand the effect of arterial fluid mechanics on the distribution of atherosclerotic lesions. The major vessels that are affected by atherosclerosis include the left and right main coronary arteries and their branches (Figure 5.2), which supply blood to the heart, the carotid artery, which provides blood flow to the head and brain, the major branches of the abdominal aorta, and the femoral and iliac arteries in the legs. The flows in each of these regions are quite varied.

Arterial flow studies consist of both experimental measurements and numerical computations. Both approaches provide important information about the behavior of these complex flows. Proper construction of experimental flow models requires maintaining geometric and dynamic similarity (Section 3.5.3). Geometric similarity is maintained by scaling vessel dimensions relative to the *in vivo* dimensions. Maintaining the same value for the relevant dimensionless groups *in vivo* and *in vitro* ensures dynamic similarity. The three groups involved are the Reynolds number, the Dean number, and the Womersley number. In addition, the shape of the velocity profile as a function of time must be simulated.

Many *in vitro* arterial flow studies have used Newtonian fluids, because blood is Newtonian for shear rates above 100 s^{-1}. There is no fluid that mimics the full rheological behavior of blood. It is not practical to use blood in such studies because the cells can interfere with the methods that are employed to measure the velocity and streamlines. While non-Newtonian properties have a quantitative effect on fluid velocities and shear stresses, flow studies with non-Newtonian fluids are qualitatively similar to those obtained with the use of Newtonian fluids. Polyacrylamide solutions show good agreement with the behavior of blood in the region of shear rate from 1 to 100 s^{-1} [16]. However, at higher shear rates, polyacrylamide does not exhibit as much shear-thinning behavior as blood does, and its use results in higher wall shear stresses than those obtained with blood [17]. Numerical solutions of the Navier–Stokes equation are the most appropriate way to account for the rheology of blood.

Several methods are available to visualize the flow field. In most cases, the vessel must be transparent. Qualitative information about the flow field is obtained by flow visualization, in which either a dye, light-reflecting particles, or hydrogen bubbles are released into the flow field. The motion of these tracers at any given time follows the fluid streamlines, and images can be obtained with high-speed cameras. Fluid velocities can be obtained by means of hot-wire anemometry, laser Doppler velocimetry (LDV), or particle-tracking velocimetry. In hot-wire anemometry, the current required to maintain a constant heat flux in a heated wire exposed to flow is measured. From a boundary layer analysis of heat and momentum transport, the heat flux is proportional to the square root of the local velocity. Since heat is a scalar, two wires are needed to measure the direction of the velocity. LDV uses the Doppler frequency shift produced when a moving particle scatters light from a laser to measure the particle's velocity. In particle-tracking velocimetry, the motion of illuminated particles is recorded with a high-speed camera, and the particle's velocity is determined by a frame-by-frame analysis of the particle's location. These techniques can be used to measure two, and even three, directional components of the velocity. (Further details on the advantages and disadvantages of such techniques for *in vitro* measurements are discussed in Refs [18,19].)

Although hot-wire anemometry has been attempted *in vivo*, none of the three methods is well suited for such measurements. Color Doppler ultrasound is widely used to measure flow clinically, but is difficult to apply for detailed velocity field information. Magnetic resonance imaging (MRI) shows considerable promise for *in vivo* determination of velocity, and in vitro measurements for sinusoidal flow in a tube agree reasonably well with theoretical values [20]. Nevertheless, the large sample volumes needed for MRI limit its applicability, especially in regions of stenosis [21].

Initially, researchers used vessels that represented a composite of vessels from many different individuals or an idealization of typical geometries. This approach was very helpful in elucidating the basic features of the flow field around specific branches. Correlations between hemodynamics and lesions were then developed. Examination of the vessel geometries and the lesion distribution patterns from many individuals indicated that composite models were insufficient to explain individual lesion distribution data [22]. In recent years, noninvasive imaging methods such as MRI have been applied to obtain data from living subjects, including the actual vessel geometry and the variation of flow with respect to geometry. Coupled with new advances to identify different types of lesions *in vivo*, these methods may soon make it possible to obtain a vastly improved understanding of the role of arterial hemodynamics in atherosclerosis.

Several recent reviews have surveyed the experimental and computational studies of blood flow in arteries [21,23–25]. These papers should be consulted for more information and additional references.

5.6.1 Carotid Artery

The carotid artery is the major artery that provides oxygenated blood to the brain and head. The vessel arises as a branch of the aortic arch (Figure 5.10b). In the neck, the common carotid branches into the internal and external carotid arteries in the shape of a Y (Figure 5.16a). The internal carotid has a unique structure in which there is a bulge just beyond the bifurcation. A pressure sensor is located here. The clinical interest in the carotid arteries is that they are sites of atherosclerosis. Blood clots that form at the internal carotid artery break free (*embolize*) and lodge in smaller vessels of the brain, causing a stroke.

In humans, a typical mean Re for flow in the common carotid is 400, and 70% of the flow from the common carotid goes into the internal carotid artery. In the region of the internal carotid artery where the bulge is largest, the cross-sectional area equals 120% of the area of the common carotid artery; it then declines to 50% of the area of the common carotid artery. The external carotid artery has a cross-sectional area equal to 32% of the area of the common carotid artery.

The large increase in cross-sectional area in the internal carotid artery leads to flow separation. *In vitro* steady-flow experiments with composite models show flow recirculation at the outer wall of the internal carotid artery [27] (the dashed lines in Figure 5.16a). The flow divider is a region of high shear stress. The flow cycle in the carotids consists of a brief period of high flow during contraction of the ventricles, followed by a longer period in which the flow is much lower and varies little with time (Figure 5.16b). Flow reversal occurs during part of the cycle [26]. During fluid acceleration (Figure 5.16b), flow separation in the internal carotid artery does not occur. The increase in velocity during the acceleration phase compensates for the

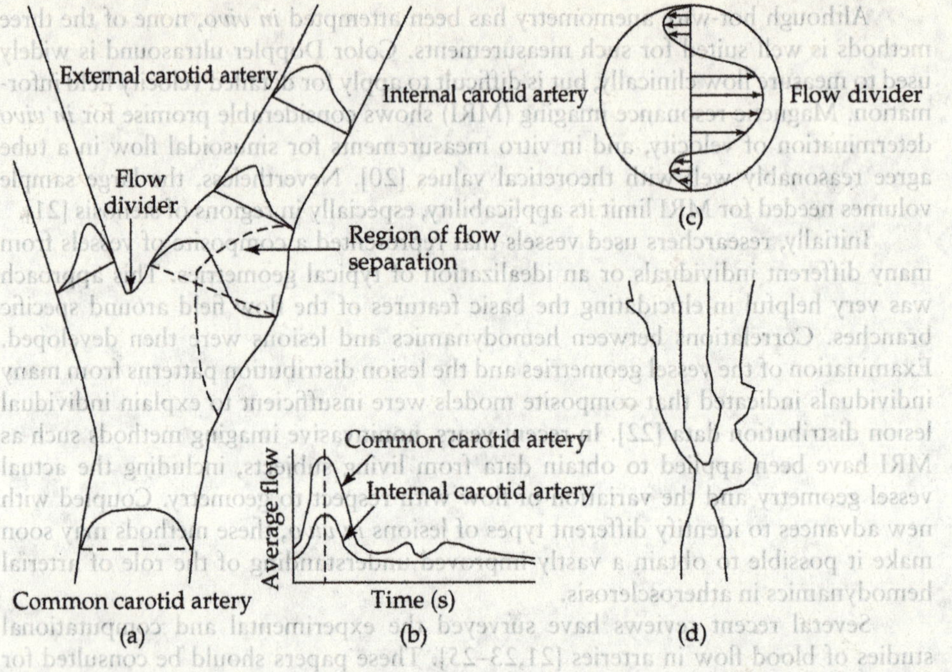

FIGURE 5.16 (a) Composite carotid artery geometry showing the region of flow separation [26]. (b) The dashed vertical line shows the velocity profile at the location of maximum flow [24,26]. (c) Radial flow profile at a cross section in the separation region of the internal carotid artery at the point of maximum velocity [24]. (d) Schematic of the actual internal geometry in a diseased carotid vessel [24].

adverse pressure gradient arising from the increase in cross-sectional area. Flow reversal begins after the fluid starts to decelerate. Although flow reversal does occur at the same general location as that observed under steady flow, the location of the region of reversal moves with time. The maximum size of the region of flow reversal is larger under pulsatile flow than steady flow, possibly due to fluid deceleration that arises during unsteady flow. As a result, the magnitudes of the shear stresses are greater for pulsatile than for steady flow. The size and direction of secondary flows vary with position (Figure 5.16c). Away from the bulge, the secondary flows are similar to those observed in curved tubes. At the bulge in the internal carotid artery, the secondary flow exhibits a wide region of flow toward the flow divider, with a narrow region near the wall in which fluid recirculates toward the inner wall (Figure 5.16c).

5.6.2 Aorta

The aortic arch, thoracic aorta, and abdominal aorta have different geometries, and the resulting flows are all different. Flow entering the ascending aorta is significantly affected by flow through the aortic valve. The velocity profile leaving the valves is skewed toward the outer wall. The arch exhibits considerable three-dimensional curvature as it bends around the esophagus and trachea [28]. The flow exhibits the general features expected for unsteady flow in curved tubes. The velocity maximum is shifted toward the inner wall [29]. The three-dimensional nature of the curvature results in helical flow patterns that extend into the descending thoracic aorta, which is attached to the spine and exhibits a small amount of curvature. Helical flows that arise in the arch persist in the descending aorta.

Blood that enters the abdominal aorta supplies the abdomen and the legs. About 54% of the cardiac output reaches the abdominal aorta (Figure 5.17). Of this

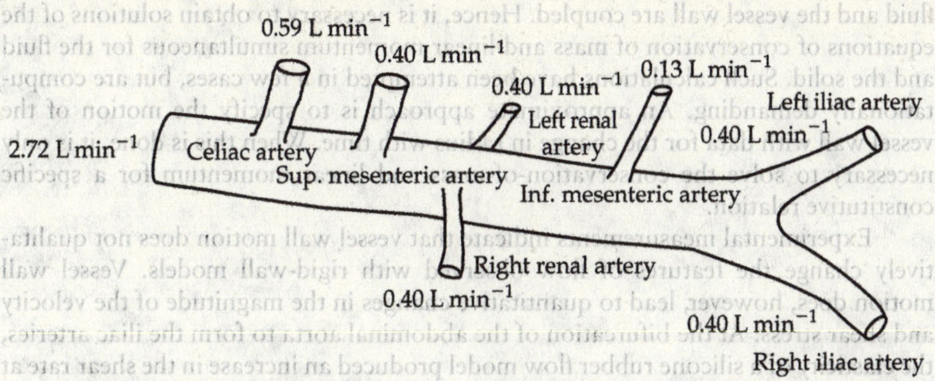

FIGURE 5.17 Schematic of the major blood vessels in the abdominal aorta and the reported flow rates in the vessels in the human under resting conditions. (Adapted from Ref. [20], with permission.)

percentage, more than two-thirds supplies the intestines, stomach, pancreas, spleen, and kidneys. The kidneys receive about 20% of the cardiac output. The close spacing among the various branches in the abdominal aorta (Figure 5.17) influences flow in at least two ways: The distance between branches in the abdominal aorta is shorter than the entrance lengths required for flows to redevelop, and the significant flow to the branches causes fluid deceleration in the mid-abdominal aorta.

Flow visualization studies have examined the general features of fluid flow in rigid models of the abdominal aortas of dogs, rabbits, and humans. Under steady-flow conditions, two thin, paired recirculation zones along the lateral celiac and superior mesenteric flow dividers occur in the dog [14] and rabbit [13] models. Barakat et al. [30] reported helical flow lateral to the celiac and superior mesenteric ostia in a rabbit model, with flow separation on the dorsal wall opposite the superior mesenteric artery. In pulsatile flow studies in dog and human aortas, transient secondary flows that appear during systolic deceleration were described as horseshoe vortices near the branch orifices [31]. Under pulsatile flow, regions of flow recirculation may oscillate or may exist only during part of the cycle [32].

The flow field in the human abdominal aorta has been examined *in vitro* by flow visualization [33] and MRI [34] and *in vivo* by MRI [20]. While transient flow reversal does occur on the lateral sides of the celiac and superior mesenteric arteries, flow above the renal arteries is relatively orderly. At the level of the renal arteries and below, however, a series of vortices arises at various points in the cardiac cycle. In the study, these vortices were confined to the posterior wall of the aorta. Flow reversal occurs at the iliac artery branch as fluid decelerates during late systole. Under exercise conditions, flow through the abdominal aorta more than doubles and blood flow to the branches drops [33]. As a result, secondary flows are much smaller in magnitude.

5.6.3 Effect of Vessel Wall Elasticity

The influence of vessel wall elasticity upon flow has been difficult to study experimentally, because materials are not available that accurately match the elastic behavior of arteries. The stress–strain curve is nonlinear, and vessel wall properties vary spatially. Thus, any material that is used differs from the properties of the vessel wall, and results must be interpreted cautiously. Further, errors may arise in measuring the velocity near the vessel wall. Simulation is difficult because the motion of the fluid and the vessel wall are coupled. Hence, it is necessary to obtain solutions of the equations of conservation of mass and linear momentum simultaneous for the fluid and the solid. Such calculations have been attempted in a few cases, but are computationally demanding. An approximate approach is to specify the motion of the vessel wall with data for the change in radius with time. When this is done, it is only necessary to solve the conservation-of-mass and linear momentum for a specific constitutive relation.

Experimental measurements indicate that vessel wall motion does not qualitatively change the features of flow observed with rigid-wall models. Vessel wall motion does, however, lead to quantitative changes in the magnitude of the velocity and shear stress. At the bifurcation of the abdominal aorta to form the iliac arteries, the elasticity of a silicone rubber flow model produced an increase in the shear rate at the flow divider and a decrease along the outer wall [35]. The influence of vessel wall motion on the shear rates at distinct locations is dependent upon the phase shift between the pressure and velocity waveforms. The larger the phase shift, the less is the effect of convective acceleration upon overcoming adverse pressure gradients. As a result, flow reversal is more pronounced with higher velocities near the flow divider [36]. Quantitatively, vessel strains of about 4% produce a 10% change in the wall shear rate.

5.6.4 Coronary Arteries

The coronary arteries supply blood to the heart. The left and right coronary arteries originate at either side of the root of the aorta and are partially embedded in the outer surface of the heart. The vessels curve along the myocardial surface and undergo branching (Figure 5.2). In addition, the coronary arteries are subjected to compression by the motion of the heart. As a result, the diameter changes by about 5% during the cardiac cycle (Figure 5.3b).

Computational and experimental studies of flow in coronary arteries show that secondary flows arise due to vessel curvature. In a numerical study, Qui and Tarbell [4] used a canine coronary artery geometry and published flow waveforms for the canine coronary arteries (Figure 5.3a). Vessel wall motion was included in the model by using the measured change in diameter of the canine vessel (Figure 5.3b). Newtonian behavior was assumed for the fluid. The inlet velocity profile was assumed to be uniform, and the mean value was obtained from the flow data. Velocity profiles are shown in Figure 5.18 for three time points noted in Figures 5.3(a) and 5.3(b) (0.05, 0.4, and 0.65 seconds). The axial velocity was greater along the outer wall, except in the region of the vessel where there is an increase in cross-sectional area. At $t = 0.05$ seconds, the vessel wall is expanding. During this period, flow resembles the developing flow observed for curved tubes discussed in Section 5.4. Higher shear stresses are present on the outer wall of the curved vessels, due to centrifugal forces. Shear stresses along the inner wall are about 68% lower than along the outer wall and only a few percent smaller than values obtained for steady laminar flow through a straight tube of similar radius [4]. Variations in shear stress occur along each wall due to the out-of-plane curvature. At 0.4 seconds, the flow is at a global minimum and the diameter is at a local minimum. The magnitudes of all velocities are much lower, but the behavior is similar to that observed at 0.05 seconds. At 0.65 seconds, the flow is nearing a maximum, but the diameter is decreasing.

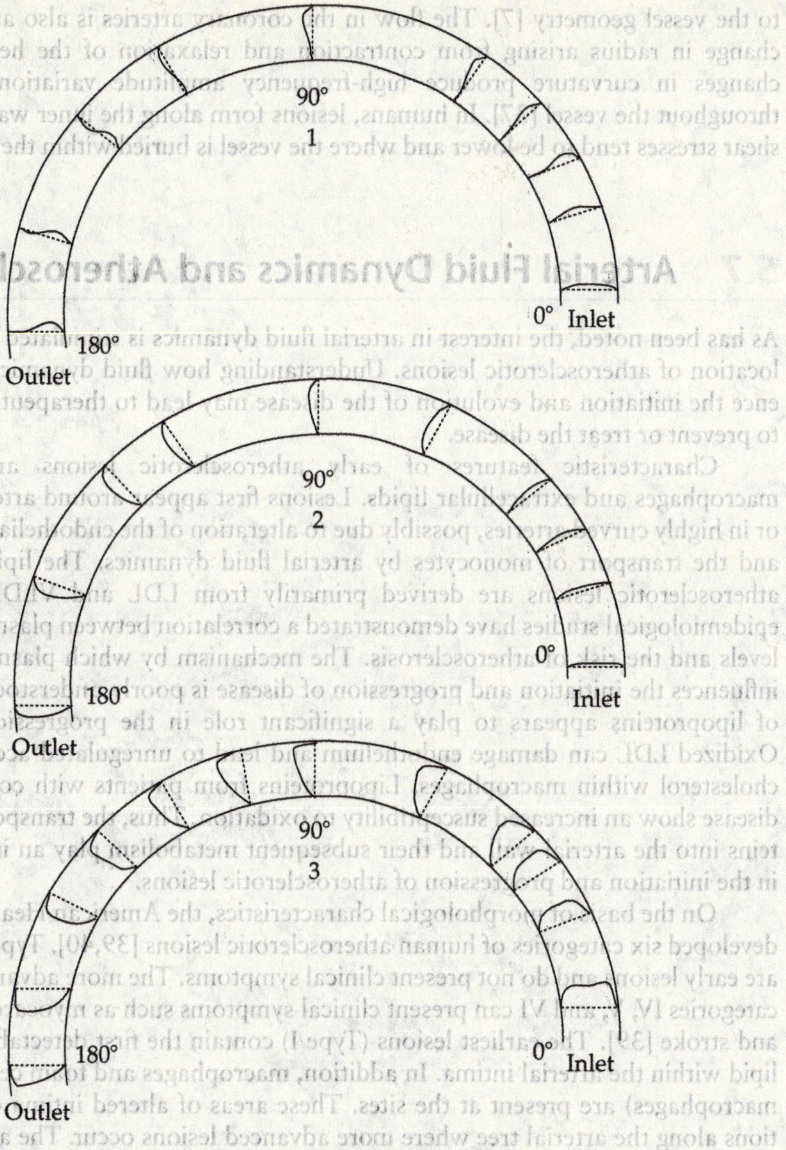

FIGURE 5.18 Velocity profiles in the left main coronary artery. The displayed time points correspond to those in Figure 5.3. Point 1 is 0.05s, point 2 is 0.4 s, and point 3 is 0.65 s. Profiles were obtained by numerical solution of a three-dimensional form of the Navier–Stokes equation. (Adapted from Ref. [4], with permission.)

During the first 90° of the curved vessel, the velocity shifts toward the outer wall. Flow reversal occurs along the inner wall of the distal portion of the vessels due to the small component of backflow arising from the arterial input waveform and expansion of the vessels during diastole [4]. Secondary flows exhibited symmetric vortices similar to predicted flows in curved tubes. As expected for flow in curved tubes, the mean wall shear stress was greater along the outer wall than along the inner wall. Interestingly, the wall shear stress amplitudes were similar for the outer and inner walls.

Along the right coronary artery, the local curvature is quite significant, resulting in oscillations in the time-averaged wall shear stress along the length of the vessel [7]. The wall shear-stress distribution is not very sensitive to the inlet waveform, due to the vessel geometry [7]. The flow in the coronary arteries is also affected by the change in radius arising from contraction and relaxation of the heart. Dynamic changes in curvature produce high-frequency amplitude variations that differ throughout the vessel [37]. In humans, lesions form along the inner wall [38], where shear stresses tend to be lower and where the vessel is buried within the myocardium.

5.7 Arterial Fluid Dynamics and Atherosclerosis

As has been noted, the interest in arterial fluid dynamics is stimulated by the spatial location of atherosclerotic lesions. Understanding how fluid dynamics might influence the initiation and evolution of the disease may lead to therapeutic approaches to prevent or treat the disease.

Characteristic features of early atherosclerotic lesions are lipid-filled macrophages and extracellular lipids. Lesions first appear around arterial branches or in highly curved arteries, possibly due to alteration of the endothelial cell function and the transport of monocytes by arterial fluid dynamics. The lipids present in atherosclerotic lesions are derived primarily from LDL and VLDL. Numerous epidemiological studies have demonstrated a correlation between plasma cholesterol levels and the risk of atherosclerosis. The mechanism by which plasma cholesterol influences the initiation and progression of disease is poorly understood. Oxidation of lipoproteins appears to play a significant role in the progression of lesions. Oxidized LDL can damage endothelium and lead to unregulated accumulation of cholesterol within macrophages. Lipoproteins from patients with coronary artery disease show an increased susceptibility to oxidation. Thus, the transport of lipoproteins into the arterial wall and their subsequent metabolism play an important role in the initiation and progression of atherosclerotic lesions.

On the basis of morphological characteristics, the American Heart Association developed six categories of human atherosclerotic lesions [39,40]. Types I, II, and III are early lesions and do not present clinical symptoms. The more advanced lesions in categories IV, V, and VI can present clinical symptoms such as myocardial infarction and stroke [39]. The earliest lesions (Type I) contain the first detectable deposits of lipid within the arterial intima. In addition, macrophages and foam cells (lipid-filled macrophages) are present at the sites. These areas of altered intima occur at locations along the arterial tree where more advanced lesions occur. The altered intimas are considered precursors of these lesions.

Type II lesions are visible to the unaided eye as yellow-colored streaks [40]. They are more commonly called fatty streaks and consist primarily of intimal macrophages and smooth muscle cell containing lipid droplets. Type II lesions develop in young children. Autopsies performed on children who died between the ages 2 and 15 years revealed that 99% had Type II lesions in the aorta [41]. Sixty-five percent of young people who died between the ages of 12 and 14 years and who were examined at autopsy had Type I or II lesions along the left proximal coronary arteries, and 8% had more advanced lesions there.

Intermediate atherosclerotic lesions (Type III) in humans are composed of extracellular lipid pools, fibrous tissue, smooth muscle cells, macrophages, and lipid-filled foam cells. These lesions represent an advanced form of intimal thickening, can lead to the clinically relevant advanced lesions (Type IV), and are considered intermediate between Type II and the more advanced lesions (Types IV–VI). The advanced lesions contain all the elements of Type III lesions, but also include a central core of lipid (cholesterol esters and cholesterol crystals) and a fibrous cap just under the endothelium. Type IV lesions first appear in men during the fourth decade of life [39], are prone to rupture, and may be responsible for sudden cardiac death.

Type V lesions are advanced atherosclerotic lesions consisting of all the elements of Type IV lesions, but also including layers of fibrous tissue [39]. Type V lesions can be calcified lesions or can contain fibrous connective tissue with little or no lipid present. Type VI lesions represent advanced lesions with a fissure and blood clot [39]. These lesions may have ruptured, but have not caused clinical symptoms.

5.7.1 Hemodynamic Variables Associated with Atherosclerosis

In an attempt to identify the hemodynamic factors influencing atherosclerosis, researchers examined the relationship between intimal thickening (Type III lesions) and various measures of the fluid dynamics. The normal intima is a relatively thin acellular layer of the vessel wall. Intimal thickening is easily measured from tissue specimens. While intimal thickening does represent the presence of a developing lesion, its clinical significance is unclear.

Three hemodynamic variables that are of particular interest are the time-averaged wall shear stress, the peak shear stress, and the difference between the maximum and minimum shear stresses. The influence of spatial gradients in shear stress can be assessed with the use of a time-averaged wall shear-stress gradient. In order to characterize the relative contribution of flow reversal, the oscillatory shear index (OSI) [42], representing the fraction of the cardiac cycle in which flow is reversed, is used. The relevant equation is

$$\text{OSI} = \frac{\int_0^T |\boldsymbol{\tau}_w^* \cdot \mathbf{n}|\,dt}{\int_0^T |\boldsymbol{\tau}_w \cdot \mathbf{n}|\,dt}, \tag{5.7.1}$$

where $\boldsymbol{\tau}_w$ is the shear stress evaluated at the aortic wall and $\boldsymbol{\tau}_w^*$ is the stress component acting in the direction opposite that of the time-averaged shear stress. The OSI varies with location along the vessel wall and is zero if there is no reversed flow. Larger values of the OSI represent longer periods of flow reversal or larger magnitudes of shear stress during reversed flow.

Intimal thickening in human carotid arteries [42], in the infrarenal abdominal aorta [33], and at the aortic bifurcation [43] is strongly correlated with the reciprocal of the mean and maximum wall shear stresses and with the OSI [42]. These results suggest that lesion-prone regions are associated with regions where flow undergoes reversal during part of the cardiac cycle. In the coronary arteries, however, the OSI is greater along the inner wall, but is much less in coronary arteries than in the carotid bifurcation or the abdominal aorta bifurcation [4].

Studies with an induced stenosis have confirmed a role for arterial fluid mechanics in atherosclerosis. Creating a stenosis in the aorta of animals produces high shear-stress regions within the stenosis and a region of reversed flow downstream.

Endothelial cell replication is increased upstream and downstream of a coarctation, but not in the stenosis [44]. Lipids are deposited downstream in regions where flow reversal occurs [45]. Changes in the shapes of endothelial cells were not correlated with fluid shear stress [45], suggesting that mechanical deformations of the vessel wall, produced when the stenosis was created, also affect the endothelial cell response to the coarctation. By inserting a plug into the vessel wall of rats that partially occludes flow, but does not alter the vessel dimensions, a stenosis was produced without deforming the vessel [46]. In this model, lesions containing macrophages and lipids that were induced in hypercholesterolemic rats were associated with regions of low shear stress. These results suggest that arterial fluid mechanics influences several key events in the initiation of lesions, including endothelial permeability and monocyte adhesion (see Figure 1.20).

The lack of consistency among the hemodynamic variables in different blood vessels may indicate a complex influence of hemodynamics upon atherosclerosis. Different physiological and pathological processes may be affected by fluid stresses in different ways. When these processes all contribute to the formation of lesions, the effect of arterial fluid mechanics upon the initiation and progression of disease is difficult to isolate by correlating the formation of lesions with hemodynamic variables. An alternative approach is to examine the relationship between hemodynamics and individual events that occur in the vessel wall during atherosclerosis. Since such studies cannot be done in humans, animal models and cell culture studies are crucial to an understanding of how fluid mechanics influences the individual steps in the disease process.

In order to examine the effect of hemodynamics upon individual steps in the formation of lesions, the spatial distribution of two of the earliest events in atherosclerosis—monocyte adhesion and the accumulation and elevated permeability to lipoproteins—were examined in the rabbit aorta. In the normocholesterolemic rabbit aorta, the density of intimal macrophages, derived from blood monocytes, was dramatically higher in the arch and about lesion-prone regions of major orifices in the abdominal aorta than in nonbranch regions of the aorta [47] (Figure 5.19). In rabbits fed a hypercholesterolemic diet, macrophage densities were greater near intercostal orifices than in nonbranch regions [48]. A 0.25%-cholesterol diet in rabbits for two weeks led to a rapid increase in intimal macrophages around intercostal arteries and the major abdominal branches [49].

Flow visualization in models of the rabbit aortoceliac branch indicated that flow reversal began earliest lateral to the branch, and these regions exhibited low and oscillating shear stresses [13]. Proximal to the branch and at the flow divider, shear stresses were higher. Computational three-dimensional fluid dynamics was used to examine the relationship between the wall shear stress, the OSI and wall shear stress gradient, and regions of elevated permeability and monocyte accumulation in the normocholesterolemic rabbit aorta [50]. Consistent with flow visualization [13], regions of transient flow reversal and elevated OSI were observed on the lateral sides of the branches. However, the regions of largest OSI were opposite the branch and were found to have little monocyte adhesion. The sites of highest LDL permeability were associated with regions of elevated shear-stress gradient, but not all sites of elevated permeability were associated with high shear-stress gradients. Elevated permeability may be influenced by the release of the vasodilator nitric oxide from endothelial cells [51]. Cell culture studies indicate that nitric oxide synthesis and enzyme activity are enhanced by fluid shear stress.

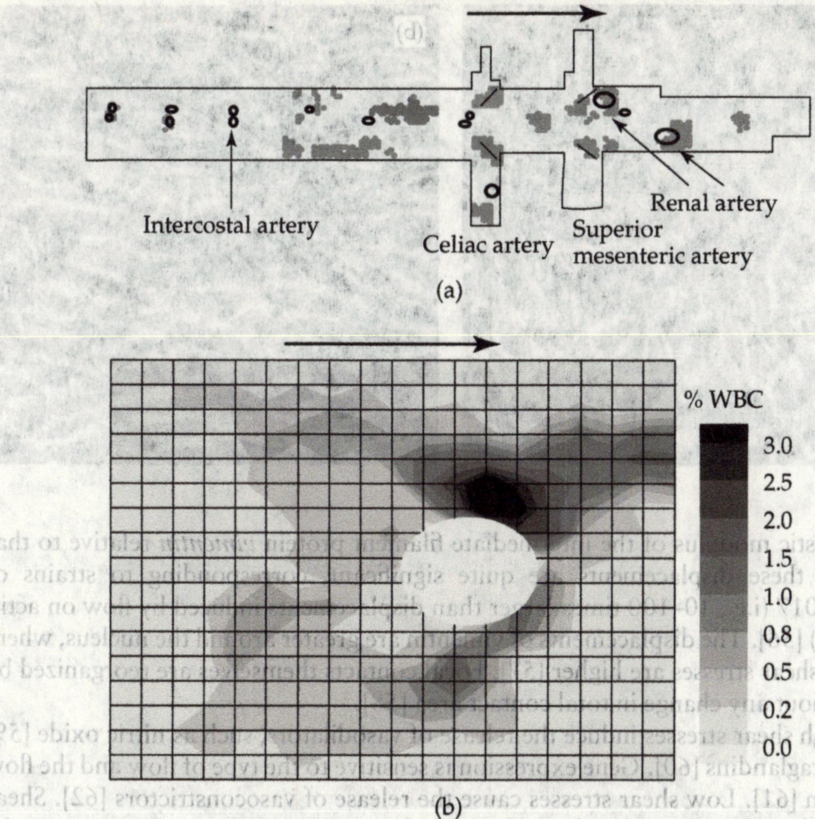

FIGURE 5.19 Distribution of intimal macrophages (i.e., the shaded regions) in the normal rabbit aorta (a) and around the aorto-celiac branch (b) [47]. Horizontal arrows in each panel indicate direction of mean flow.

Of the three hemodynamic parameters, only the wall shear-stress gradient exhibited a significant correlation with the density of intimal monocytes. A comparison of hemodynamic variables with lesion growth data [52] indicated that some regions where lesions formed were associated regions of low and oscillating shear stresses. However, low shear stresses and OSI could not explain lesions distal to the aortoceliac orifice. The results suggest that different hemodynamic factors may affect monocyte transport, endothelial cell function, and macromolecular transport.

5.7.2 Effect of Hemodynamics upon Endothelial Cell Function

In vitro and *in vivo* studies indicate that fluid shear stress affects endothelial cell function. Away from vessel branches, endothelial cells are elongated. Near branches, where flow undergoes reversal, the cell shape is rounder and there is more variation in shape among the cells. Exposure of cultured endothelial cells to shear stresses from 5 to 25 dyn cm^{-2} for 24 hours leads to cell elongation in the direction of flow (Figure 5.20). When cells are exposed to recirculating flow *in vitro* or to a stenosis *in vivo*, cell orientation is associated with the local shear stress acting on the cells. The application of flow leads to a significant decrease in F-actin within 5 minutes of the onset of flow [54]. In the longer term, the actin cytoskeleton remodels itself, orienting toward the direction of flow and forming stress fibers [55]. Due to the

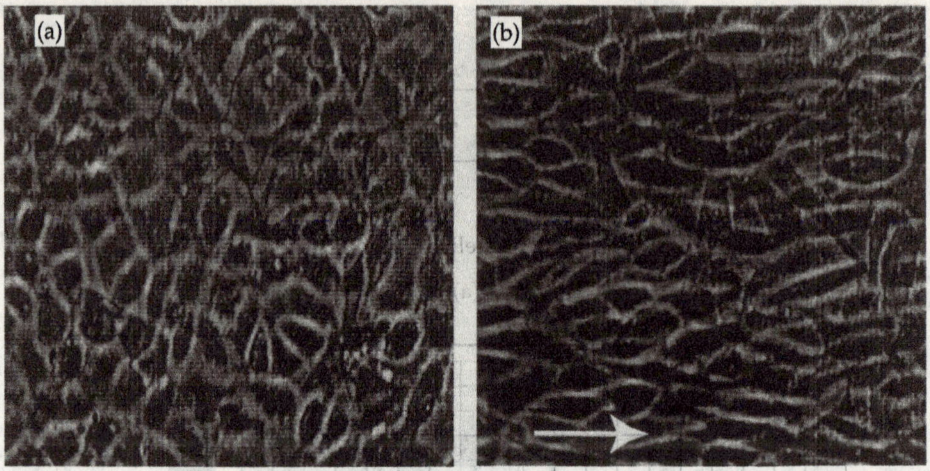

FIGURE 5.20 (a) Bovine aortic endothelial cells cultured under static conditions. (b) Bovine aortic endothelial cells after exposure to 16.7 dyn cm^{-2} for 24 hours. Arrow represents the direction of flow. (From Ref. [53], reprinted with permission.)

lower elastic modulus of the intermediate filament protein *vimentin* relative to that of actin, these displacements are quite significant, corresponding to strains of 0.004–0.017 (i.e., 10–100 times larger than displacements induced by flow on actin filaments) [56]. The displacements of vimentin are greater around the nucleus, where the fluid shear stresses are higher [57]. Focal contacts themselves are reorganized by flow without any change in total contact area [58].

High shear stresses induce the release of vasodilators, such as nitric oxide [59] and prostaglandins [60]. Gene expression is sensitive to the type of flow and the flow waveform [61]. Low shear stresses cause the release of vasoconstrictors [62]. Shear stress is likely to be translated into biological responses by interactions among the cytoskeleton, ion channels, and membrane receptors [63].

Endothelial cells are not flat. The nucleus protrudes from the cell body about 3.5 μm under static conditions. Exposure of the endothelium to flow causes a decrease in cell height to 1.8 μm [64]. Although these cell heights are small relative to the diameter of an artery, they have a significant effect upon flow. Barbee et al. [64] solved the Navier–Stokes equation to obtain the velocity field and stress distribution surrounding a confluent layer of endothelial cells. The shape of the cells was obtained by atomic force microscopy and served as one surface in a finite-element model. A linear velocity profile was imposed, and the equations were solved for steady-state conditions. Barbee et al. [64] found that, for confluent endothelium, the local shear stress is proportional to the local cell height and can be represented as

$$\tau/\tau_w = 1 + \kappa(h - h_{avg}), \tag{5.7.2}$$

where τ_w is the wall shear stress in the absence of the cell and $h - h_{avg}$ is the variation in surface height about the mean over a single cell. A representative fit for bovine aortic endothelial cells exposed to 16.7 dyn cm^{-2} for 24 hours is shown in Figure 5.21. While there is considerable scatter in the data, the data fit well with Equation (5.7.2). The slope κ is sensitive to flow conditions and is 0.120 μm^{-1} for cells not exposed to flow and 0.08 μm^{-1} after a 24-hour exposure of cells to flow [64]. Shear stresses can vary as much as threefold over the surface of an endothelial cell. The nonuniform distributions of shear stresses may provide cues for the cells to

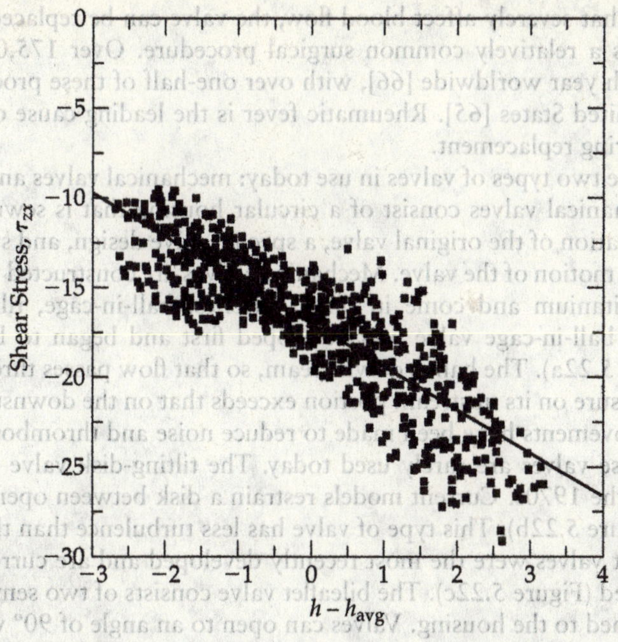

FIGURE 5.21 Relationship between shear stress at distinct locations on the cell and the deviation of the cell height from the average cell height for cells exposed to 16.7 dyn cm^{-2} for 24 hours. The unit of $h - h_{avg}$ is μm [53].

respond to a flow. Consistent with this view, displacements induced in the cytoskeleton are greater in the nuclear region [57].

Elucidating the mechanism by which the cells sense shear stress is an active area of research. While endothelial cells have viscoelastic properties, the fluid physiological shear stresses are not large enough to cause significant deformation of the cells. Rather, the flow may deform molecules on the surface of the cell or cytoskeletal proteins, thereby activating a chain of biochemical and genetic processes that cause the cell to adapt to the flow.

5.8 Heart-Valve Hemodynamics

Flow around natural and artificial heart valves is complex, due to the valves' opening and closing. Disturbances in flow through normal heart valves caused by an underlying valve pathology can be detected by a stethoscope. Flow through artificial valves generates complex flows that can affect the behavior of blood.

5.8.1 Artificial Heart Valves

Improperly functioning heart valves occur as a result of congenital defects, rheumatic fever, infection, and calcification or other aging-related disorders. The effect on valve function is incomplete opening of the valve (*stenosis*) or improper closing, leading to fluid leakage (*regurgitation*). The aortic and mitral valves are the most commonly affected valves [65]. Often, the first sign of a valve disease is a murmur or fatigue. In some cases, the valve disease can be treated with drugs to regulate the heart rate. For valve defects that severely affect blood flow, the valve can be replaced. Heart valve replacement is a relatively common surgical procedure. Over 175,000 valves are implanted each year worldwide [66], with over one-half of these procedures occurring in the United States [65]. Rheumatic fever is the leading cause of severe valve damage requiring replacement.

There are two types of valves in use today: mechanical valves and tissue valves [66,67]. Mechanical valves consist of a circular housing that is sewn to the vessel wall at the location of the original valve, a specific valve design, and struts or hinges to restrain the motion of the valve. Mechanical valves are constructed from pyrolytic carbon and titanium and come in three varieties: ball-in-cage, tilting-disk, and bileaflet. The ball-in-cage valve was developed first and began to be used in the 1960s (Figure 5.22a). The ball is downstream, so that flow passes through the valve when the pressure on its upstream portion exceeds that on the downstream portion. Various improvements have been made to reduce noise and thrombosis. Because of their size, these valves are rarely used today. The tilting-disk valve was originally developed in the 1970s. Current models restrain a disk between opened and closed positions (Figure 5.22b). This type of valve has less turbulence than the ball-in-cage valve. Bileaflet valves were the most recently developed and are currently the most commonly used (Figure 5.22c). The bileaflet valve consists of two semicircular disks that are attached to the housing. Valves can open to an angle of 90° with respect to the valve plane, thereby minimizing obstruction caused by the valve surface.

The major problems with mechanical heart valves are thrombus formation, red-cell damage (hemolysis), and infection. These problems arise because of the need to use nonbiological materials and the significant difference between normal and mechanical valve function. Anticoagulant therapy is required to minimize the formation of clots. In spite of these potential problems, tilting-disk and bileaflet valves function very well. Mechanical failure is rare. When it does occur, failure can be lethal. The use of anticoagulant therapy minimizes clot formation, but does pose the risk of bleeding.

Tissue valves were developed to overcome the shortcomings of mechanical valves. Tissue valves are preserved with glutaraldehyde at high pressures and mounted on stents [67]. Low levels of glutaraldehyde improve valve performance and minimize calcification. The most common source of these valves is the pig, although human grafts are available from cadavers. Preserving tissue architecture is

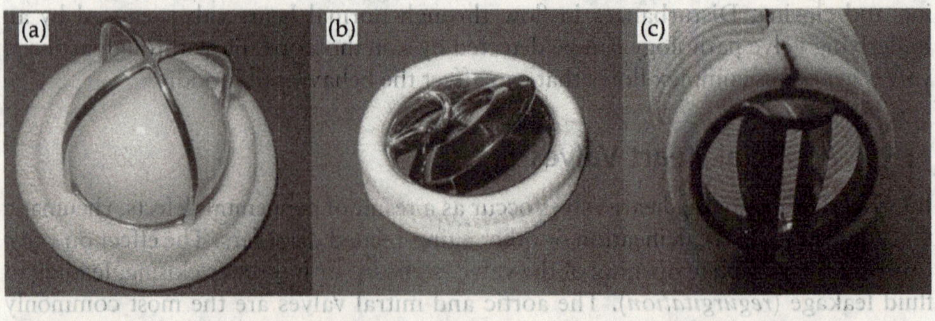

FIGURE 5.22 Mechanical heart valves. (a) The ball-in-cage or Starr Edwards valve. (b) The tilting-disk valve. (c) The bileaflet valve.

very important. Tissue valves exhibit reduced thrombosis relative to mechanical valves. Long-term anticoagulation is not needed. Problems, however, include leaflet wear and calcification and the formation of turbulent jets. The fixation process is necessary to maintain the valves and minimize rejection, but fixation alters valve properties. Valve failure in tissue valves arises from mechanical failure of the tissue graft. Valve failure becomes more common after 10 years of implantation. Because of the shorter lifetime of tissue valves, they are given primarily to patients 70 years of age or older.

5.8.2 Turbulent Flow Around Heart Valves

Turbulent flow generated during the opening and closing of artificial valves has significant effects on biological function. In this section, the essential features of turbulent flow are summarized. These concepts are then used in the characterization of valves. In laminar flow, the velocity at a particular location is predictable. If the velocity at a point is measured under steady-flow conditions, the velocity does not change with time. If the flow oscillates, the same velocity occurs at the same part of each cycle. During unsteady flow, the velocity changes continuously with time, and the change is a smoothly varying function. Turbulent flow behaves differently. As a result, the velocity at a given instant of time varies with time, even for steady flow. For oscillatory turbulent flow, the velocity will not be exactly the same for the same part of each cycle.

Turbulent flow arises when inertial forces are much greater than viscous forces. In laminar flow, small fluctuations in fluid motion arising from vibrations in the fluid or small nonuniformities in the boundaries are rapidly dissipated due to viscous forces. At large enough Reynolds numbers, however, inertial forces are more significant, resulting in an amplification of these fluctuations. In turbulence, the result is that fluctuations in energy lead to local variations in velocities and eddies that interact with one another. Whereas there is a single characteristic dimension in laminar flow, often equal to the width or diameter of the flow channel, many length scales exist in turbulent flow. At the larger length scales of turbulence, inertial forces dominate and give rise to the eddies, which are unstable and break into smaller eddies. The smallest eddies are dominated by viscous forces. As a result, the energy generated in turbulence is ultimately dissipated as heat through the action of those forces. In general, turbulence can exist throughout the fluid, or it can be localized to a region. Examples of localized turbulence include turbulent flow separation over a ball moving though air and turbulence arising from a mechanical heart valve. Turbulence appears to be unique to mechanical valves and does not generally occur with native tissue valves.

Although the velocity does vary at a given point, it is not completely random. The velocity component v_i can be described in terms of a time-averaged velocity $\overline{V}_i$ and a random fluctuating velocity v_i':

$$v_i = \overline{V}_i + v_i'. \tag{5.8.1}$$

The fluctuating component has a time average of zero. The pressure can be cast in a similar form. These expressions for the velocity and pressure can be inserted into the conservation-of-mass equation and the Navier–Stokes equation for an incompressible

Newtonian fluid, and the time-average value of these equations can then be obtained [68]. In performing the time average, the order of differentiation and integration can be interchanged. Thus,

$$\frac{1}{T}\int_T \nabla\cdot\mathbf{v}dt = \nabla\cdot\left(\frac{1}{T}\int_T \mathbf{v}dt\right) = 0. \tag{5.8.2}$$

Thus,

$$\nabla\cdot\mathbf{v} = \nabla\cdot\overline{\mathbf{V}} = 0. \tag{5.8.3}$$

Assuming that gravity does not contribute significantly to the flow, the Navier–Stokes equation becomes

$$\rho\left(\frac{\partial\overline{\mathbf{V}}}{\partial t} + \overline{\mathbf{V}}\cdot\nabla\overline{\mathbf{V}} + \overline{\mathbf{v}'\cdot\nabla\mathbf{v}'}\right) = -\nabla\overline{p} + \mu\nabla^2\overline{\mathbf{V}}, \tag{5.8.4}$$

where the overbar represents a time average. The third term on the left-hand side can be rewritten as

$$\overline{\mathbf{v}'\cdot\nabla\mathbf{v}'} = \overline{\nabla\cdot\mathbf{v}'\mathbf{v}'}, \tag{5.8.5}$$

since $\nabla\cdot\mathbf{v}' = 0$.

Making this substitution and rearranging the term with the fluctuating velocity yields

$$\rho\left(\frac{\partial\overline{\mathbf{V}}}{\partial t} + \overline{\mathbf{V}}\cdot\nabla\overline{\mathbf{V}}\right) = -\nabla\overline{p} + \mu\nabla^2\overline{\mathbf{V}} - \rho\overline{\nabla\cdot\mathbf{v}'\mathbf{v}'}. \tag{5.8.6}$$

The term $\rho\overline{\mathbf{v}'\mathbf{v}'}$ is known as the *Reynolds stress tensor* and represents the average momentum flux due to the fluctuating component of velocity [68]. For fully developed turbulence away from solid surfaces, Reynolds forces can be much larger than viscous stresses. Near a solid surface, viscous stresses are still dominant.

The design of artificial valves requires suitable materials that are durable for long periods of repetitive motion and impact forces and that interact minimally with blood cells. Hemodynamic requirements are that there be a minimal pressure drop, little or no regurgitation, no stagnation or flow separation, little turbulence, and moderate shear stresses. The first three requirements are intended to minimize the work on the heart. Maximizing the effective orifice area (EOA) minimizes the pressure drop across the valve [66]. The EOA is determined by the relation

$$\text{EOA} = \frac{Q_{\text{RMS}}}{51.6\sqrt{\Delta\overline{p}}}, \tag{5.8.7}$$

where Q_{RMS} is the root-mean-square flow rate through the valve, in units of $\text{cm}^3\,\text{s}^{-1}$, cubic centimeters per second, and $\Delta\overline{p}$ is the mean pressure drop across the valve in mmHg. The EOA, in units of cm^2, square centimeters, represents a functional measure of valve performance, not a true area. Thus, the larger the EOA, the lower the transvalvular pressure gradient and the better the performance. Note that the EOA for the ball-in-cage valves is much lower than that for other valves (Table 5.3).

TABLE 5.3

Turbulent Stresses Arising Downstream from Artificial Aortic Valves [66,69]

Valve type	Effective orifice area, EOA (cm^2)	Percent regurgitation	Peak velocity ($cm\ s^{-1}$)	Peak turbulent shear stress, ($dyn\ cm^{-2}$)
Ball in cage	1.6	3–7	180	3,500
Tilting disk	3.1	10–13	210	1,500
Bileaflet	3.2	10	210	1,700
Tissue	1.9–3.2	1.5	175	2,950

Note: Typical values for a 2.5-cm diameter valve.

The percent regurgitation represents the fraction of the volume flowing through the valve that returns to the heart before the valve closes. Normal native valves have a percent regurgitation between 0.5% and 1.0%. Replacement valves have much higher levels of regurgitation than native valves, although tissue valves come the closest to native valves (Table 5.3).

Turbulent stresses are generated throughout the cardiac cycle while the valve opens and closes. These stresses are present as a jet that occupies a small region of fluid volume downstream from the valve itself. The peak stresses can be quite significant (Table 5.3).

In vitro studies have provided important information on the flow fields around artificial valves [66]. In all cases, the flow is time dependent, with rapid changes in the velocity and stress. In the ball-and-cage design, two symmetric jets form on either side of the ball and move along the vessel wall. Turbulent stresses are maximal about 3 cm downstream. A region of flow reversal forms behind the ball. The jets produced by the tilting-disk valve are asymmetric, due to the open area produced as the valve tilts. During the period of maximum flow from the heart, a region of flow reversal forms behind the valve near the smaller jet. When the bileaflet valve opens, flow occurs between the two valves or on each side of them. Although the velocity tends to be larger on the lateral sides, the flow is more uniform than with the other designs. Turbulent stresses are greatest about 1 cm downstream from the valve. Tissue valves also exhibit rather uniform velocities, although flow reversal can occur near the vessel wall during the deceleration phase of the cardiac cycle.

Turbulent shear stresses above 500 dyn cm^{-2} can damage red blood cells. Platelets are activated by shear stresses in the range of 100 dyn cm^{-2}. Thus, the stresses generated by artificial valves can damage blood. Damage is a function of the shear stress and the length of time that the cells are exposed to turbulent shear stresses. The shorter the duration of exposure to a shear stress, the higher the shear stress must be to cause damage. Since the period of turbulence in the heart valves is short, it is likely that platelet activation or red blood cell damage results from multiple exposures of the cells to the turbulent environment. Further, the threshold for damage is lowered when the cells contact a surface.

Turbulence may damage cells through the smallest eddies—those in which the turbulent stresses are dissipated by viscous forces. The length scale of these eddies is known as the *Kolmogorov length scale* and is given by [68]

$$\eta = \left(\frac{\nu^3}{\bar{\varepsilon}}\right)^{1/4} \sim L\mathrm{Re}_L^{-3/4}, \tag{5.8.8}$$

where ν is the kinematic viscosity, L is the largest length scale for turbulence, and $\bar{\varepsilon}$ is the rate of energy dissipation per unit mass [68]. The quantity $\bar{\varepsilon}$ has units of length squared per time cubed. The largest length scale is the half-width of the turbulent jet. Thus, as the Reynolds number gets larger, the Kolmogorov length scale gets smaller. For a valve diameter of 2.5 cm, the jet width, which is used as an estimate of L, ranges from 0.01 to 0.05 cm. For a peak velocity of 210 cm s^{-1}, the Kolmogorov length scale ranges from 4 to 6 μm. This length is the same order of magnitude as that of a red cell. Thus, if a red cell is trapped transiently in the smallest eddy when it dissipates its energy, the energy could stretch the red cell, causing damage.

5.9 Fluid Dynamic Effects of Reconstructive Surgery for Congenital Heart Defects

Congenital heart defects are the most common congenital abnormality, affecting about 3% of newborns. These defects include openings between the atria or ventricles (*atrial–septal* or *atrial–ventricular* defects), transposition of major arteries, a malfunctioning connection between the aorta and the pulmonary artery (patent *ductus arteriosus*), fused valves (*atresia*), and only a single functioning ventricle. Often, these defects occur in isolation, but in some pathologies (e.g., Down's syndrome and Tetralogy of Fallot), there are multiple defects. These defects are believed to arise during embryonic development, due to missing or altered cues between cardiac cells or their precursors, resulting in improper cell organization and malformation of the resulting tissue structure. Both environmental and genetic factors play a role.

Atrial–septal and atrial–ventricular defects are the most common congenital cardiac malformations and can range from a mild atrial–septal defect that may not be noticed until middle age to a life-threatening atrial–ventricular defect that requires surgery shortly after birth. Fortunately, septal defects can be treated and the prognosis is excellent. More difficult to treat are the cases of multiple defects, tricuspid valve atresia, or a single functioning ventricle. In some cases, the only treatment is a heart transplant. This is a difficult choice for parents and surgeons because of the limited availability of donor hearts and the requirement of lifelong immunosuppressant therapy. Thus, there is considerable interest in alternative approaches, including those involving tissue engineering.

One approach that has emerged in the treatment of either tricuspid atresia or a single functioning ventricle is the *Fontan procedure*, or *cavopulmonary connection*. Normally, blood enters the right atrium through the vena cava, which is formed by joining the inferior vena cava (IVC) and the superior vena cava (SVC). The IVC supplies venous blood from the lower extremities and abdomen, whereas the SVC supplies blood from the head, neck, and arms. When the right ventricle is not functional, the Fontan procedure involves connecting the IVC and the SVC directly to the pulmonary arteries, which bring unoxygenated blood to the lung. This connection eliminates any mixing of venous and arterial blood that would occur with a single ventricle. Flow arises from the venous side into the pulmonary arteries, due to the small pressures present in the veins feeding into the right side of the heart. Although patients do suffer from a number of complications, the procedure has enabled many recipients to survive to adulthood and bear children when they otherwise would have died before reaching school age [70].

Because the venous pressure is low, it is important to minimize any pressure losses that might occur from poor mixing of the blood as it passes from the vena cava into the pulmonary arteries. In order to characterize the effect of different pressure losses, the overall rate of energy loss is determined from Equation (4.4.25) for the case of no mechanical work, negligible potential-energy or height variations, and negligible viscous losses:

$$\frac{dE_K}{dt} = -\int_S \left(\frac{1}{2}\rho v^2 + p\right)\mathbf{v}\cdot\mathbf{n}dS. \tag{5.9.1}$$

The pressures and velocities can be estimated *in vivo* from MRI flow measurements. *In vitro* flow modeling or computational simulations can examine the effect of a wide range of flows and geometries and identify those which would lead to minimal energy losses. While the anatomy and flow requirements limit conditions that can be varied during a surgical procedure, the analysis has identified key factors that can affect energy losses.

Flow visualization and numerical solution of the Navier–Stokes equation are employed to examine energy losses. Unlike arterial flow, venous flow can be treated as steady. Most of the variation of the flow with time disappears as the flow enters the venous system.

A realistic, albeit simplified, connection geometry is shown in Figure 5.23 [71]. The SVC is smaller than the IVC and accounts for 40% of the flow into the pulmonary artery connection. Because of its smaller size, the velocities in the SVC are higher than those in the IVC. The flow enters from the SVC and IVC and mixes before exiting by either the right pulmonary artery (RPA) or the left pulmonary artery (LPA). Because of the higher velocities in the SVC, a jet forms as fluid enters the pulmonary artery connection. The jet creates vortices on either side. Flow from the SVC curves into the RPA.

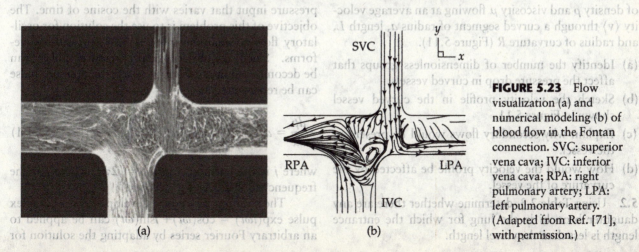

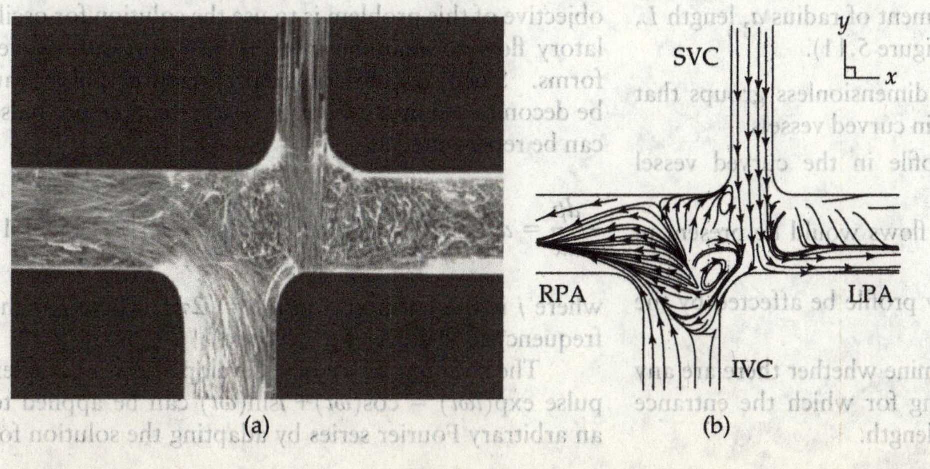

FIGURE 5.23 Flow visualization (a) and numerical modeling (b) of blood flow in the Fontan connection. SVC: superior vena cava; IVC: inferior vena cava; RPA: right pulmonary artery; LPA: left pulmonary artery. (Adapted from Ref. [71], with permission.)

From flow studies, a number of important conclusions have emerged that have helped guide surgery. Energy losses are greatest when there is no offset between the IVC and SVC. Flaring of the connections reduces energy losses. Physiologically accurate models produce about 150% more energy loss than idealized models. When the SVC and IVC are offset, the maximum energy loss occurs in the pulmonary artery downstream from where the fluids mix. Actual connections examined by MRI show poor mixing due to deviation from the idealized geometry [72]. These studies can help to guide both surgery and postsurgical care.

5.10 QUESTIONS

5.1 For flow through branching vessels, flow separation occurs along the outer wall.

(a) Give a one- or two-sentence explanation of why flow separation occurs.

(b) Explain why the separation occurs along the outer wall.

5.2 Describe the effect of vessel curvature on flow.

5.3 What is the physical significance of the Dean number?

5.4 Define quasi–steady flow, and discuss the conditions for which the flow field can be approximated as quasi steady.

5.5 Explain why flow reversal occurs along the inner wall, not the outer wall, of a curved tube subjected to an oscillating pressure gradient.

5.6 For a Y-branch, discuss the effect of the cross-sectional area of the daughter vessels on flow separation.

5.7 Discuss several ways in which arterial fluid mechanics may affect the process of atherosclerosis.

5.8 Which features in an artificial valve affect the generation of turbulence?

5.9 How do turbulent eddies affect red blood cells and vascular endothelium?

5.11 PROBLEMS

5.1 The pressure drop in a curved vessel is affected by the curvature of the vessel. Consider a Newtonian fluid of density ρ and viscosity μ flowing at an average velocity $\langle v \rangle$ through a curved segment of radius a, length L, and radius of curvature R (Figure 5.11).

(a) Identify the number of dimensionless groups that affect the pressure drop in curved vessels.

(b) Sketch the velocity profile in the curved vessel shown in Figure 5.11.

(c) Explain why secondary flows would be present in the vessel.

(d) How would the velocity profile be affected by the curvature of the vessel.

5.2 Using Table 2.3, determine whether there are any daughter bronchi in the lung for which the entrance length is less than the vessel length.

5.3 *Pulsatile Flow of Newtonian Fluid in Arteries* [5] We have discussed the solution of unsteady flow for a pressure input that varies with the cosine of time. The objective of this problem is to use the solution for oscillatory flow to examine more realistic pressure waveforms. Since a physiological pressure pulse can be decomposed into a Fourier series, the pressure pulse can be represented as

$$-\frac{dp}{dx} = a_0 + \sum_{j=1}^{n}\left[a_j \cos(\omega t) + b_j \sin(\omega t)\right], \qquad (5.11.1)$$

where j is the harmonic and $\omega = j2\pi f$, where f is the frequency in Hz (cycles per second).

The solution procedure developed for a complex pulse $\exp(i\omega t) = \cos(\omega t) + i\sin(\omega t)$ can be applied to an arbitrary Fourier series by adapting the solution for the different terms in the series and the change in the Womersley parameter as

$$\alpha_j = R\sqrt{\frac{j2\pi f\rho}{\mu}}. \tag{5.11.2}$$

(a) Determine the velocity profile for a pressure gradient represented by Equation (5.11.1) with the Womersley number for the harmonics represented by Equation (5.11.2).

(b) The paper by Womersley [5] has data for the first six Fourier components of flow in the femoral artery of a dog ($R = 0.15$ cm, $\mu = 0.04\ \text{g cm}^{-1}\,\text{s}^{-1}$, $\rho = 1.05\ \text{g cm}^{-3}$, and $f = 3$ cycles per second). Data for the coronary arteries are available from Holenstein and Nerem [73]. The Fourier coefficients need to be determined, and one approach is presented shortly. Using these results, determine how well the Womersley results predict the flow waveform for the given pressure waveform.

(c) Using the exact pressure pulses, compare the velocity profiles of the vessels examined in part (b), and contrast them with results obtained from using the leading coefficient in the Fourier series. How do the higher harmonics affect the velocity profiles for these conditions?

To obtain a representation of the pressure gradient pulse, take discrete points from the pressure-versus-time graph at two locations. Note the period p_i at each value of x. Enter your data as a MATLAB data file x.data. Compute the fast Fourier transform in MATLAB with the command y = fft(x). In general, y is a complex number. Compute the inverse fast Fourier transform with the command z = ifft(y). You can verify the accuracy of the result by plotting z as a function of p_i. The coefficients in the Fourier series can be determined as follows:

$$a_0 = \frac{1}{N}y(1), \tag{5.11.3}$$

$$a_i = \frac{2}{N}\text{real}[y(i+1)], \tag{5.11.4}$$

and

$$b_i = \frac{2}{N}\text{imag}[y(i+1)]. \tag{5.11.5}$$

5.4 *Oscillatory Flow of a Casson Fluid* The rheology of blood is non-Newtonian. Blood exhibits both a small, but finite, yield stress and power law behavior. At high shear rates, blood behaves as a Newtonian fluid. The objective of this project is to analyze the oscillatory flow of blood in a rigid cylindrical tube. The conservation of linear momentum for this problem is

$$\rho\frac{\partial v_z}{\partial t} = -\frac{\partial p}{\partial z} + \frac{1}{r}\frac{\partial}{\partial r}(r\tau_{rz}). \tag{5.11.6}$$

The constitutive equation is

$$(-\tau_{rz})^{1/2} = (-\tau_0)^{1/2} + \left(-m\frac{\partial v_z}{\partial r}\right)^{1/2}$$

$$|\tau_{rz}| \geq |\tau_0|. \tag{5.11.7a}$$

The negative signs are introduced because the shear rate and shear stress are negative for $0 < r < R$. As before, the velocity is constant for shear stresses less than the yield shear stress. Hence,

$$\frac{\partial v_z}{\partial r} = 0 \qquad |\tau_{rz}| < |\tau_0|. \tag{5.11.7b}$$

The pressure wave varies as

$$-\frac{\partial p}{\partial z} = A\cos(\omega t) = A\text{Re}\left[exp(i\omega t)\right]. \tag{5.11.8}$$

Substituting Equations (5.11.7a) and (5.11.8) into Equation (5.11.6) gives the following expression for the velocity [74]:

$$\rho\frac{\partial v_z}{\partial t} = -\frac{\partial P}{\partial z} + \frac{1}{r}\frac{\partial}{\partial r}\left[r\left(\tau_0 + 2{\tau_0}^{1/2}\left(m\frac{\partial v_z}{\partial r}\right)^{1/2} + \frac{\partial v_z}{\partial r}\right)\right]. \tag{5.11.9}$$

Since we are interested in assessing the deviations from Newtonian behavior, we will examine conditions under which non-Newtonian behavior is present but not significant. These conditions hold when the Womersley parameter is less than unity. The following approximate solution has been derived [74]:

$$v_z = \frac{R^2A}{4m}\left[\exp(i\omega t)\left(1 - \frac{r^2}{R^2}\right) - \frac{8}{3}\left(\frac{2\tau_0}{AR}\exp(i\omega t)\right)^{1/2}\left(\frac{r^{3/2}}{R^{3/2}} - 1\right) + \frac{\tau_0}{AR}\left(\frac{r}{R} - 1\right)\right]$$
$$+ \frac{AR^2}{m}\alpha^2\sin(\omega t)F(r,t). \tag{5.11.10}$$

In this equation,

$$F(r,t) = \left(\frac{r}{4R}\right)^2 \left(\left(\frac{r}{2R}\right)^2 - 1\right) - \left(\frac{r_c}{R}\right)^{1/2} \left[\frac{r^2}{3R^2}\left(\frac{4}{49}\left(\frac{r}{R}\right)^{3/2} - \frac{1}{4}\right) + \frac{1}{8}\left(\frac{r}{R}\right)^{3/2}\left(\frac{1}{7}\left(\frac{r}{R}\right)^2 - \frac{2}{3}\right)\right] + \frac{r_c}{9R}\left(\frac{r}{R}\right)^{3/2}\left(\frac{2}{7}\left(\frac{r}{R}\right)^{3/2} - 1\right) + \frac{3}{64} + \frac{5r_c}{63R} - \frac{143}{1176}\left(\frac{r_c}{R}\right)^{1/2}. \quad (5.11.11)$$

The velocity is the real part of the preceding expression.

(a) For $\alpha < 1$, compare the velocity profiles for the Newtonian and non-Newtonian fluids. Use the following values:

$$\tau_0 = 0.02 \text{ dyn cm}^{-2}$$
$$R = 0.10 \text{ cm}$$
$$\mu = 0.03 \text{ g cm}^{-1}\text{ s}^{-1}$$
$$f = 1 \text{ cycle per second}$$
$$A = 10 \text{ mmHg cm}^{-1} \text{ (convert to dyn cm}^{-3})$$

(b) Assess the phase shift between the pressure and flow waveforms for the Newtonian and non-Newtonian cases. Explain any deviation between the two flow rates and velocities.

(c) Evaluate the shear stress at $r = R$, and determine the difference between the results for the Newtonian and non-Newtonian cases.

5.5 Assuming steady flow, use Equations (5.4.3) and (5.4.4) to estimate the wall shear stress acting on the artery walls in the left anterior descending coronary artery.

5.6 Determine the maximum shear stress acting on a cultured endothelial cell exposed to a shear stress of 15 dyn cm^{-2} (1) when the cell has not been exposed to flow in culture and (2) after exposure to flow for 24 hours.

5.7 Suppose that the surface area of an endothelial cell is represented as half of an ellipsoid with a long-axis half-length of 20 μm, a half-width of 5 μm, and a height of 2.5 μm. Determine the force acting on the cell, assuming that Equation (5.7.2) applies and that κ for a cell exposed to flow can be used. Compare this result with the force acting on an elliptical disk of the same length and width.

5.8 Determine the value of α for which the amplitude and phase shift for the flow rate are within 5% of the quasi-steady state value. The amplitude can be determined from the following relation

$$|Q| = \sqrt{Q\bar{Q}},$$

where $\bar{Q}$ is the complex conjugate.

5.9 Based on the results of the previous problem and the following data, determine whether flow in the mouse coronary artery is quasi-steady. The heart rate is 600–700 bpm and the coronary artery diameter is 0.04 cm.

5.10 Murray's law provides a relationship between flow rate and radius that minimizes the overall power for steady flow of a Newtonian fluid [75]. Murray posited that a cost function for the overall power of the circulatory system represented a balance between the power to pump blood and the metabolic consumption rate. The power of pumping blood equals the rate of work done to overcome viscous resistance. This power is equal to the product of the average velocity times the viscous force acting on the vessel wall ($r = R$).

(a) Using this relation, show that for a Newtonian fluid, the pumping power equals

$$\Delta p Q = (8\mu L Q^2)/(\pi R^4).$$

(b) The metabolic power is assumed to be equal to the product of the metabolic energy per unit volume of blood times the blood volume. Simply treating the blood as a tube of radius R and length L, then the cost function F is

$$F = \Delta p Q + E_m \pi R^2 L.$$

From the first derivative of F with respect to R, determine the relationship between Q and the vessel radius. Using the second derivative, show that this is a maximum.

(c) Relate the shear stress at the vessel wall to the flow rate and show that the result from part (b), Murray's law, requires that the wall shear stress be constant.

5.11 Most branching blood vessels consist of a parent vessel and two daughter vessels [76]. Let the radius of the parent vessel be R_0 and the radius of each daughter vessel is R_1 and R_2. The area ratio is defined as

$$\beta = \frac{R_1^2 + R_2^2}{R_0^2}$$

and the bifurcation index is $\alpha = R_2/R_1$.

(a) Using the results from part (b) of problem 5.10, show that

$$\beta = \frac{1 + \alpha^2}{(1 + \alpha^3)^{2/3}}.$$

(b) Determine the value of α that leads to a minimum in the value of β. What does this result imply about the area ratio?

5.12 REFERENCES

1. Nichols, W., and O'Rourke, M., *McDonald's Blood Flow in Arteries*. 1998, New York: Arnold and Oxford University Press.
2. Fung, Y.C., *Biomechanics: Circulation*. 1997, New York: Springer.
3. Pedley, T.J., *The Fluid Mechanics of Large Blood Vessels*. 1980, Cambridge, UK: Cambridge University Press.
4. Qui, J., and Tarbell, J.M., "Flow in a compliant curved tube model of a coronary artery." *J. Biomech. Eng.*, 2000. 122: pp. 77–85.
5. Womersley, J.R., "Method for the calculation of velocity, rate of flow and viscous drag in arteries when the pressure gradient is known." *J. Physiol.*, 1955. 127: pp. 553–563.
6. He, X., and Ku, D.N., "Pulse flow in the human left coronary artery bifurcation: average conditions." *J. Biomech. Eng.*, 1996. 118: pp. 74–82.
7. Myers, J.G., Moore, J.A., Ojha, M., Johnston, K.W., and Ethier, C.R., "Factors influencing blood flow patterns in the human right coronary artery." *Ann. Biomed. Eng.*, 2001. 29: pp. 109–120.
8. Stephanopoulos, G., *Chemical Process Control*. 1984, Englewood Cliffs, NJ: Prentice-Hall.
9. He, X., and Ku, D.N., "Unsteady entrance flow development in a straight tube." *J. Biomech. Eng.*, 1994. 116: pp. 74–82.
10. Berger, S.A., Talbot, L., and Yao, L.S., "Flow in curved pipes." *Ann. Rev. Fluid Mech.*, 1983. 15: pp. 461–512.
11. Chang, L.J., and Tarbell, J.M., "Numerical simulation of fully developed sinusoidal and pulsatile (physiological) flow in curved tubes." *J. Fluid Mech.*, 1985. 161: pp. 175–198.
12. Hamakiotes, C.C., and Berger, S.A., "Fully developed pulsatile flow in a curved pipe." *J. Fluid Mech.*, 1988. 195: pp. 23–55.
13. Malinauskas, R.A., Sarraf, P., Barber, K.M., and Truskey, G.A., "Association between secondary flow in models of the aorto-celiac junction and subendothelial macrophages in the normal rabbit." *Atherosclerosis*, 1998. 140: pp. 121–134.
14. Karino, T., Motomiya, M., and Goldsmith, H.L., "Flow patterns at the major T-junctions of the dog descending aorta." *J. Biomech.*, 1990. 23: pp. 537–548.
15. MacLean, N.F., Kratky, R.G., Macfarlane, T.W.R., and Roach, M.R., "Taper: an important feature of Y-bifurcations in porcine renal arteries and human cerebral arteries." *J. Biomech.*, 1992. 25: pp. 1047–1052.
16. Liepsch, D.W., and Moravec, S., "Pulsatile flow of a non-Newtonian fluid in distensible models of human arteries." *Biorheology*, 1984. 23: pp. 571–586.
17. Mann, D.E., and Tarbell, J.M., "Flow of non-Newtonian blood analog fluids in rigid curved and straight artery models." *Biorheology*, 1990. 27: pp. 711–733.
18. Chew, Y.T., Chew, T.C., Low, H.T., and Lim, W.L., "Techniques in the determination of the flow effectiveness of prosthetic heart valves," in *Biomechanical Systems: Techniques and Applications*, C.T. Leonedes, editor. 2001, Boca Raton, FL: CRC Press LLC. pp. 2.1–2.48.
19. Kleppe, J.A., Olin, J.G., and Menon, R.K., "Point velocity measurements," in *The Measurement, Instrumentation and Sensors Handbook*, J.G. Webster, editor. 1999, Boca Raton, FL: CRC Press LLC. pp. 29.1–29.55.
20. Moore, J.E., and Ku, D., "Pulsatile velocity measurements in a model of the human abdominal aorta under resting conditions." *J. Biomech. Eng.*, 1994. 116: pp. 337–346.
21. Ku, D.N., "Blood flow in arteries." *Ann. Rev. Fluid Mech.*, 1997. 29: pp. 399–434.
22. Friedman, M.H., Deters, O.J., Mark, F.F., Bargeron, C.B., and Hutchins, G.M., "Arterial geometry affects hemodynamics: a potential risk factor for atherosclerosis." *Atherosclerosis*, 1983. 46: pp. 225–231.
23. Lou, Z., and Yang, W.-J., "Biofluid dynamics at arterial bifurcations." *Crit. Rev. Biomed. Eng.*, 1992. 19: pp. 455–493.
24. Berger, S.A., and Jou, L.-D., "Flow in stenotic vessels." *Ann. Rev. Fluid Mech.*, 2000. 32: pp. 347–382.
25. Kleinstreuer, C., Hyun, S., Buchanan, J.R., Longest, P.W., Archie, J.P., and Truskey, G.A., "Hemodynamic

parameters and early intimal thickening in branching blood vessels." *Crit. Rev. Biomed. Eng.*, 2001. **29**: pp. 1–64.

26. Ku, D.N. and Giddens, D.P., "Laser Doppler anemometer measurements of pulsatile flow in a model carotid bifurcation." *J. Biomech.*, 1987. **20**: pp. 349–362.
27. Zarins, C.K., Giddens, D.P., Bharadvaj, B.K., Sottiurai, V.S., Mabon, R.F., and Glagov, S., "Carotid bifurcation atherosclerosis: quantitative correlation of plaque localization with flow velocity profiles and wall shear stress." *Circ. Res.*, 1983. **53**: pp. 501–514.
28. Chandran, K.B., "Flow dynamics in the human aorta." *J. Biomech. Eng.*, 1993. **115**: pp. 611–616.
29. Chandran, K.B., "Flow dynamics in the human aorta: techniques and applications," in *Biomechanical Systems: Techniques and Applications*, C.T. Leonedes, ed. 2001, Boca Raton, FL: CRC Press. pp. 5.1–5.26.
30. Barakat, A.I., Uhthoff, P.A.F., and Colton, C.K., "Topographical mapping of sites of enhanced HRP permeability in the normal rabbit aorta." *ASME J. Biomech. Eng.*, 1992. **114**: pp. 283–292.
31. Fukushima, T., Homma, T., and Harakawa, K., "Flow separation and horseshoe vortex in a tube with side branches during pulsatile flow," in *Role of Blood Flow in Atherogenesis*, R.M. Nerem, Yasuo Yoshida, T. Yamaguchi, C.G. Caro, S. Glagov, eds. 1988, Springer-Verlag New York, LLC. pp. 81–90.
32. Lutz, R.J., Hsu, L., Menawat, A., Zrubek, J., and Edwards, K., "Comparison of steady and pulsatile flow in a double branching arterial model." *J. Biomech.*, 1983. **16**(9): pp. 753–766.
33. Moore, J.E., Ku, D.N., Zarins, C.K., and Glagov, S., "Pulsatile flow visualization in the abdominal aorta under differing physiologic conditions: implications for increased susceptibility to atherosclerosis." *J. Biomech. Eng.*, 1992. **114**: pp. 391–397.
34. Moore, J. and Ku, D., "Wall shear stress measurements in a model of the human abdominal aorta under exercise conditions." *Adv. Bioeng.*, 1992. **22**: pp. 289–291.
35. Duncan, D.D., Bargeron, C.B., Borchardt, S.E., Deters, O.J., Gearhart, S.A., Mark, F.F., and Friedman, M.H., "The effect of compliance on wall shear in casts of a human aortic bifurcation." *J. Biomech. Eng.*, 1990. **112**: pp. 183–188.
36. Dutta, A., Wang, D.M., and Tarbell, J.M., "Numerical analysis of flow in an elastic artery model." *J. Biomech. Eng.*, 1992. **114**: pp. 26–33.
37. Moore, J.E., Weydahl, E.S., and Santamarina, A., "Frequency dependence of dynamic curvature effects on flow through coronary arteries." *J. Biomech. Eng.*, 2001. **123**: pp. 129–133.
38. Fox, B., and Seed, W.A., "Location of early atheroma in the human coronary arteries." *J. Biomech. Eng.*, 1981. **103**: pp. 208–212.
39. Stary, H.C., Chandler, A.B., Dinsmore, R.E., Fuster, V., Glagov, S., Insull, W.J., Rosenfeld, M.E., Schwartz, C.J., Wagner, W.D., and Wissler, R.W., "A definition of advanced types of atherosclerotic lesions and a histological classification of atherosclerosis." *Arterioscler. Thromb. Vasc. Biol.*, 1995. **15**: pp. 1512–1531.
40. Stary, H.C., Chandler, A.B., Glagov, S., Guyton, J.R., Insull, W.J., Rosenfeld, M.E., Schaffer, S.A., Schwartz, C.J., Wagner, W.D., and Wissler, R.W., "A definition of initial, fatty streak, and intermediate lesions of atherosclerosis. A report from the Committee on Vascular Lesions of the Council on Arteriosclerosis, American Heart Association." *Arterioscler. Thromb.*, 1994. **14**: pp. 840–856.
41. Stary, H.C., "Evolution and progression of atherosclerotic lesions in coronary artereis." *Arteriosclerosis*, 1989. 9(Suppl. I): pp. I-19–I-23.
42. Ku, D.N., Giddens, D.P., Zarins, C.K., and Glasgov, S., "Pulsatile flow and atherosclerosis in the human carotid bifurcation: positive correlation between plaque location and low and oscillating shear stress." *Arteriosclerosis*, 1985. **5**: pp. 293–302.
43. Friedman, M.H., Deters, O.J., Bargeron, C.B., Hutchins, G.M., and Mark, F.F., "Shear-dependent thickening of the human arterial intima." *Atherosclerosis*, 1986. **60**: pp. 161–171.
44. Langille, B.L., Reidy, M.A., and Kline, R.L., "Injury and repair of endothelium at sites of flow disturbances near abdominal aortic coarctations in rabbits." *Arteriosclerosis*, 1986. **6**: pp. 146–154.
45. Zand, T., Nunnari, J.J., Hoffman, A.H., Savilonis, B.J., Macwilliams, B., Majno, G., and Joris, I., "Endothelial adaptations in aortic stenosis." *Am. J. Pathol.*, 1988. **133**: pp. 407–418.
46. Zand, T., Hoffman, A.H., Savilonis, B.J., Underwood, J.M., Nunnari, J.J., Majno, G., and Joris, I., "Lipid deposition in rat aortas with intraluminal hemispherical plug stenosis." *Am. J. Pathol.*, 1999. **155**: pp. 85–92.
47. Malinauskas, R.A., Herrmann, R.A., and Truskey, G.A., "The distribution of intimal white blood cells in the normal rabbit aorta." *Atherosclerosis*, 1995. **115**: pp. 147–163.
48. Back, M.R., Carew, T.E., and Schmid-Schoenbein, G.W., "Deposition pattern of monocytes and fatty streak development in hypercholesterolemic rabbits." *Atherosclerosis*, 1995. **116**: pp. 103–115.
49. Truskey, G., Herrmann, R., Kait, J., and Barber, K., "Focal increases in VCAM-1 and intimal macrophages at atherosclerosis-susceptible sites in the rabbit aorta after short-term cholesterol feeding." *Arterioscler. Thromb. Vasc. Biol.*, 1999. **19**: pp. 393–401.
50. Buchanan, J., Kleinstreuer, C., Truskey, G., and Lei, M., "Relation between non-uniform hemodynamics and sites

of altered permeability and lesion growth." *Atherosclerosis*, 1999. **143**: pp. 27–40.

51. Forster, B.A., and Weinberg, P.D., "Changes with age in the influence of endogenous nitric oxide on transport properties of the rabbit aortic wall near branches." *Arterioscler. Thromb. Vasc. Biol.*, 1997. **17**: pp. 1361–1368.
52. Zeindler, C.M., Kratky, R.G., and Roach, M.R., "Quantitative measurements of early atherosclerotic lesions on rabbit aortae from vascular casts." *Atherosclerosis*, 1989. **76**: pp. 245–255.
53. Xiao, Y., "The response of endothelial cells in recirculating flow." Ph.D. dissertation. 1997, Duke University, Durham, NC.
54. Morita, T., Kurihara, H., Macmura, K., Yoshizumi, M., and Yazaki, Y., "Disruption of cytoskeletal structures mediates shear stress-induced endothelin-1 gene expression in cultured porcine aortic endothelial cells." *J. Clin. Invest.*, 1993. **92**: pp. 1706–1712.
55. Dewey, C.F., Bussolari, S.R., Gimbrone, M.A., and Davies, P.F., "The dynamic response of vascular endothelial cells to fluid shear stress." *J. Biomech. Eng.*, 1981. **103**: pp. 177–185.
56. Satcher, R.L., and Dewey, C.F., "Theoretical estimates of mechanical properties of the cell cytoskeleton." *Biophys. J.*, 1996. **71**: pp. 109–118.
57. Helmke, B.P., Thakker, D.B., Goldman, R.D., and Davies, P.F., "Spatiotemporal analysis of flow-induced intermediate filament displacement in living endothelial cells." *Biophys. J.*, 2001. **80**: pp. 184–194.
58. Davies, P.F., Robotewskyj, A., and Griem, M.L., "Quantitative studies of endothelial cell adhesion: directional remodeling of focal adhesion sites in response to flow forces." *J. Clin. Invest.*, 1994. **93**: pp. 2031–2038.
59. Buga, G.M., Gold, N.E., Fukuto, J.M., and Ignarro, L.J., "Shear stress–induced release of nitric oxide from endothelial cells grown on glass beads." *Hypertension*, 1991. **17**: pp. 187–193.
60. Frangos, J.A., Eskin, S.G., McIntire, L.V., and Ives, C.L., "Flow effects on prostacyclin production by cultured human endothelial cells." *Science*, 1985. **227**: pp. 1477–1479.
61. Garcia-Cardena, G., Comander, J., Anderson, K.R., Blackman, B.R., and Gimbrone, M.A., "Biomechanical activation of vascular endothelium as a determinant of its functional phenotype." *Proc. Natl. Acad. Sci.*, 2001. **98**: pp. 4478–4485.
62. Kuchan, M.J., and Frangos, J.A., "Shear stress regulates endothelin-1 release via protein kinase C and cGMP in cultured endothelial cells." *Am. J. Physiol.*, 1993. **264**: pp. H150–H158.
63. Davies, P.F., "Flow-mediated endothelial mechanotransduction." *Physiol. Rev.*, 1995. **75**: pp. 519–560.
64. Barbee, K.A., Davies, P.F., and Lal, R., "Shear stress–induced reorganization of the surface topography of living endothelial cells imaged by atomic force microscopy." *Circ. Res.*, 1994. **74**: pp. 163–171.
65. American Heart Association, *Heart and Stroke Statistical Update*. 2003, Dallas: American Heart Association.
66. Yoganathan, A.P., "Cardiac valve prostheses," in *The Biomedical Engineering Handbook*, J.D. Bronzino, editor. 2000, Boca Raton: CRC Press LLC. pp. 127.1–127.23.
67. Sapirstein, J.S., and Smith, P.K., "The "ideal" replacement heart valve." *Am. Heart J.*, 2001. **141**: pp. 856–860.
68. Mathieu, J., and Scott, J., *An Introduction to Turbulent Flow*. 2000, New York: Cambridge University Press.
69. Chandran, K.B., "Dynamic behavior analysis of mechanical heart valve prostheses," in *Biomechanical Systems: Techniques and Applications*, C.T. Leonedes, editor. 2001, Boca Raton, FL: CRC Press LLC. pp. 3.1–3.31.
70. Mair, D.D., Puga, F.P., and Danielson, G.K., "The Fontan procedure for tricuspid atresia: early and late results of a 25-year experience with 216 patients." *J. Amer. Coll. Cardiol.*, 2001. **37**: pp. 933–939.
71. Ryu, K., Healy, T.M., Ensley, A.E., Sharma, S., Lucas, C., and Yoganathan, A.P., "Importance of accurate geometry in the study of the total cavopulmonary connection: computational simulations and in vitro experiments." *Ann. Biomed. Eng.*, 2001. **29**: pp. 844–853.
72. Sharma, S., Ensley, A.E., Hopkins, K., Chatzimavroudis, G.P., Healy, T.M., Tam, V.K.H., Kanter, K.R., and Yoganathan, A.P., "In vivo flow dynamics of the total cavopulmonary connection from three-dimensional multislice magnetic resonance imaging." *Ann. Thorac. Surg.*, 2001. **71**: pp. 889–898.
73. Holenstein, R., and Nerem, R.M., "Parametric analysis of flow in the intramyocardial circulation." *Ann. Biomed. Eng.*, 1990. **18**: pp. 347–365.
74. Rohlf, K., and Tenti, G., "The role of the Womersley number in pulsatile blood flow: A theoretical study of the Casson model." *J. Biomech.*, 2001. **34**: pp. 141–148.
75. Murray, C.D. 1926. The physiological principle of minimum work. I. The vascular system and the cost of blood volume. *Proc. Natl. Acad. Sci.* 12: 207–214.
76. Zamir, M. 2000. *The Physics of Pulsatile Flow. American Institute of Physics Press*, Springer-Verlag, New York. 220 pp.

Fundamentals and Applications of Mass Transport in Biological Systems

PART II

A fundamental requirement of all organisms is the efficient transport of solutes and solvent. Such transport supplies chemicals for cellular metabolism and energy generation, as well as signals to change cellular function. At the cellular level, the changes include cell division and the activation of various functional properties. These cellular changes produce dramatic changes in the function of multicellular organisms and even in some single-celled organisms such as the amoeba. The changes can be long term (involving fetal development, puberty, long-term muscle conditioning, or cancer) or short term (changes in the heart rate during excitement).

At the cellular level, molecules move rapidly by diffusion. This movement assists in the delivery of nutrients and metabolites from blood capillaries to cells. Because many biological functions require nonuniform distributions of molecules, specialized mechanisms have evolved to localize the concentrations of molecules. Cells and organelles have phospholipid membranes, which are soluble to small nonpolar molecules, but insoluble to small polar molecules and ions. Transmembrane protein channels permit the selective movement of ions and small polar molecules (e.g., glucose) in response to specific stimuli. Larger molecules enter cells by binding to specific membrane-bound receptors and by endocytosis. Within cells, motor proteins transport molecules to specific locations, sometimes moving these molecules against concentration gradients.

Depending upon the size of the molecule, diffusion becomes inefficient at distances greater than 100–200 μm As a result, convection and electrical conduction are needed for rapid transport. Convection occurs in the flow of molecules and cells within the blood, in the lymphatic system, on the surfaces of many epithelia, and within the kidneys and the intestines. Electrical conduction transports signals efficiently over larger distances (1–200 cm) on a more rapid time scale (10–100 m).

Part II of the text is concerned with transport by diffusion and convection. In Chapter 6, the classical continuum approach is used to develop constitutive and conservation relations and to solve a number of problems concerning steady and unsteady diffusion that are relevant to biomedical engineering. Next, in Chapter 7, results from Part I and Chapter 6 are used in problems in which convection and diffusion occur. We examine transport properties from a molecular and microscopic viewpoint, and we describe current theories of transport in porous media such as tissues. The material in this section provides the groundwork for examining important biomedical problems in which transport occurs in conjunction with biochemical reactions. In Part III, we consider the effect of transport processes on reaction rates, and we discuss biological mechanisms for localizing molecules. In Part IV, we examine the application of these concepts at the organ and the whole-body level.

CHAPTER 6

Mass Transport in Biological Systems

6.1 Introduction

In this chapter, we introduce basic concepts related to steady and unsteady diffusion in one dimension. Biological examples are given throughout. Sections 6.2, 6.3, and 6.4 present relations for the solute flux in a mixture, the conservation relations for a component of a mixture, and Fick's law of diffusion, respectively. Section 6.5 introduces the diffusion process as a random walk. The molecular basis of diffusion and the relation between fluid mechanics and diffusion are considered in Section 6.6. Sections 6.7 and 6.8, respectively, examine steady and unsteady diffusion in one dimension. Section 6.9 considers diffusion-limited reactions, which are so fast that the rate depends only upon the diffusion process.

6.2 Solute Fluxes in Mixtures

To begin a consideration of mass transport, the velocity and flux of the various components of a mixture must be defined precisely. The velocity of each component ($\mathbf{v}_i$) of a mixture comprises two factors: the *diffusion velocity* ($\mathbf{v}_d$) and the *bulk velocity* of the mixture ($\mathbf{v}$). Mathematically, then,

$$\mathbf{v}_i = \mathbf{v}_d + \mathbf{v}. \tag{6.2.1}$$

The diffusion velocity varies randomly with time due to collisions between the solute and solvent molecules. The diffusion velocity is considered in more detail in Section 6.5.

The bulk velocity represents a *mixture velocity*, which is a weighting of the velocities of the various components of the mixture. In most applications, the weightings are based on mass or moles. The bulk velocity in Equation (6.2.1) is the *mass average velocity*. This velocity is defined as

$$\mathbf{v} = \frac{1}{\rho}\sum_{i=1}^{n}\rho_i\mathbf{v}_i = \sum_{i=1}^{n}\omega_i\mathbf{v}_i, \tag{6.2.2a}$$

where ρ_i is the density of component i, $\rho = \sum_{i=1}^{n}\rho_i$ is the total density of the mixture, and $\omega_i = \rho_i/\rho$ is the weight fraction of the solute. The *molar average velocity* is defined as

$$\mathbf{v}^* = \frac{1}{C}\sum_{i=1}^{n}C_i\mathbf{v}_i = \sum_{i=1}^{n}x_i\mathbf{v}_i, \tag{6.2.2b}$$

where C_i is the molar concentration (in units of moles per volume) of the solute, $C = \sum_{i=1}^{n}C_i$ is the total concentration of the solution, and x_i is the mole fraction C_i/C. A less commonly used mixture velocity is the volume-averaged velocity.

A *flux* is the amount of material crossing a unit area normal to the direction of transport in a given unit of time (see Figure 6.1). As such, fluxes are vectors. The vector direction represents the direction of net transport. Fluxes occur in the transport of heat, energy, or, indeed, any quantity. As noted in Section 1.2, fluid shear stress can be considered a momentum flux [1].

Fluxes are defined for fixed ($\mathbf{n}_i$, $\mathbf{N}_i$) or moving ($\mathbf{j}_i$, $\mathbf{J}_i$) coordinates and in mass ($\mathbf{n}_i$, $\mathbf{j}_i$) or molar ($\mathbf{N}_i$, $\mathbf{J}_i$) units. For fixed coordinates, the *mass* and the *molar* fluxes of component i are, respectively,

$$\mathbf{n}_i = \rho_i\mathbf{v}_i \tag{6.2.3a}$$

and

$$\mathbf{N}_i = C_i\mathbf{v}_i. \tag{6.2.3b}$$

Inserting the definition of the mass flux into Equation (6.2.2a) shows that the sum of the mass fluxes equals the product of the total mixture density times the mass average velocity:

$$\rho\mathbf{v} = \sum_{i=1}^{n}\mathbf{n}_i. \tag{6.2.4a}$$

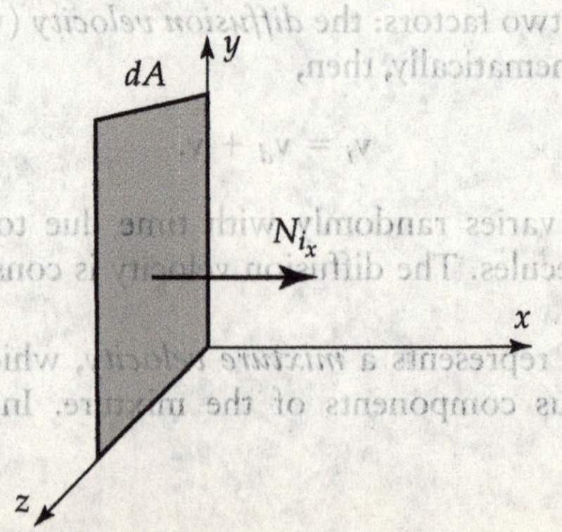

FIGURE 6.1 Flux in the x direction.

Likewise, the molar average velocity is related to the sum of the molar fluxes as follows:

$$C\mathbf{v}^* = \sum_{i=1}^{n} \mathbf{N}_i. \tag{6.2.4b}$$

The mass average velocity is the most commonly used reference for *diffusive mass flux* ($\mathbf{j}_i$) or *diffusive molar flux* ($\mathbf{J}_i$), based upon moving coordinates. The respective formulas are

$$\mathbf{j}_i = \rho_i(\mathbf{v}_i - \mathbf{v}) = \rho_i \mathbf{v}_d \tag{6.2.5a}$$

and

$$\mathbf{J}_i = C_i(\mathbf{v}_i - \mathbf{v}) = C_i \mathbf{v}_d. \tag{6.2.5b}$$

Other driving forces that produce component velocities that differ from the bulk velocity are discussed in Chapter 7.

The *mass flux* and *molar flux* based upon the molar average velocity $\mathbf{v}^*$ are, respectively,

$$\mathbf{j}_i^* = \rho_i(\mathbf{v}_i - \mathbf{v}^*) \tag{6.2.5c}$$

and

$$\mathbf{J}_i^* = C_i(\mathbf{v}_i - \mathbf{v}^*). \tag{6.2.5d}$$

These four fluxes represent the movement of a component of a mixture by diffusion when the mixture is moving. Fluxes based upon fixed ($\mathbf{n}_i$, $\mathbf{N}_i$) and moving ($\mathbf{j}_i$, $\mathbf{J}_i$, $\mathbf{j}_i^*$, $\mathbf{J}_i^*$) coordinates are related as follows:

$$\mathbf{j}_i = \mathbf{n}_i - \rho_i \mathbf{v}, \tag{6.2.6a}$$

$$\mathbf{J}_i = \mathbf{N}_i - C_i \mathbf{v}, \tag{6.2.6b}$$

$$\mathbf{j}_i^* = \mathbf{n}_i - \rho_i \mathbf{v}^*, \tag{6.2.6c}$$

$$\mathbf{J}_i^* = \mathbf{N}_i - C_i \mathbf{v}^*. \tag{6.2.6d}$$

Example 6.1 Fluxes for Binary Systems

For a binary mixture (i = 1, 2), begin with Equation (6.2.6d) and derive an explicit expression for the molar flux of component 1 relative to fixed coordinates, or $\mathbf{N}_1$.

Solution Rearranging Equation (6.2.6d) in terms of $\mathbf{N}_1$ yields

$$\mathbf{N}_1 = \mathbf{J}_1^* + C_1 \mathbf{v}^*. \tag{6.2.7}$$

Equation (6.2.4b) can be rearranged to express the molar average velocity in terms of the fluxes:

$$v^* = \frac{1}{C}(N_1 + N_2). \quad (6.2.8)$$

Using Equation (6.2.8) to replace v* in Equation (6.2.7) produces

$$N_1 = J_1^* + x_1(N_1 + N_2). \quad (6.2.9)$$

Solving explicitly for N_1 gives the final result:

$$N_1 = \frac{1}{1 - x_1}(J_1^* + x_1 N_2). \quad (6.2.10)$$

Any applied force that increases the velocity of one or more solutes relative to the motion of the bulk fluid produces a flux. A flux can arise from a concentration gradient, from an applied electrical field, from pressure (pressure diffusion), and from temperature gradients (temperature diffusion). Here, the total flux is specified as the sum of the fluxes arising from ordinary diffusion J^{*D}_i, applied electrical fields J^{*e}_i, pressure diffusion J^{*p}_i, and temperature diffusion, J^{*T}_i; that is,

$$J^*_i = J^{*D}_i + J^{*e}_i + J^{*p}_i + J^{*T}_i. \quad (6.2.11)$$

To study the function of a protein, of DNA, or of another biological macromolecule, it is often necessary to separate the molecule from a mixture that contains a large number of components. The methods cannot be harsh, or the three-dimensional structure (the *conformation*) may be altered, or *denatured*. The fluxes in Equation (6.2.11) have been exploited in a variety of separation techniques. Electrical potential gradients are used to separate molecules on the basis of their electrical charge (electrophoresis) or their pH (isoelectric focusing). Ultracentrifugation separates macromolecules by size through the differential effects of pressure gradients produced by rotation within the ultracentrifuge. Due to the sensitivity of biological molecules to temperature, thermal gradients are not routinely used to purify such molecules. In addition to the diffusive flux, fluxes arising from applied electric fields are important in charged solute transport across membranes and in electrophoresis (see Section 7.4).

6.2.1 The Dilute-Solution Assumption

The preceding expressions for fluxes are general. The components of most biological solutions are dilute, and several simplifications can therefore be made. A dilute solution is defined as a solution in which the solvent is the dominant component, whether measured on a mass or a molar basis. We label C_S and x_S as the solvent concentration and mole fraction, respectively, the *dilute-solution assumption* being that

$$C_s \gg C_i \text{ or } x_s \approx 1 \gg x_i \quad \text{for } i = 1, 2, \ldots, n, \quad i \neq s. \quad (6.2.12)$$

Similar statements can be made for the mass fraction and mass concentration. Further, the solution density is approximately constant because the solutes are extremely dilute. The major simplification that arises from the dilute-solution assumption is that the mass average velocity equals the molar average velocity, because both are approximately equal to the solvent velocity. That is, because

$$\mathbf{v}^* \approx \mathbf{v}_s \quad \text{and} \quad \mathbf{v} \approx \mathbf{v}_s, \tag{6.2.13a}$$

it follows that

$$\mathbf{v} \approx \mathbf{v}^*. \tag{6.2.13b}$$

Consequently, $\mathbf{J}_i^* = \mathbf{J}_i$, and the molar flux relative to fixed coordinates simplifies to

$$\mathbf{N}_i = \mathbf{J}_i + C_i\mathbf{v}_s. \tag{6.2.14}$$

Note that Equation (6.2.10) reduces to the dilute-solution limit if the solvent equals component 2, since $x_i \ll 1$ and $\mathbf{N}_2 \approx C\mathbf{v}_s$.

There are two important conclusions to be drawn from Equation (6.2.14). First, as already noted, the average fluid velocity, whether on a mass or molar basis, is approximately equal to the solvent velocity. Second, for multicomponent transport in which the solvent is the dominant component, the only interactions that are important are those between the solute and the solvent. Thus for a dilute multicomponent solution, the transport of each solute through the solvent can be studied as though the mixture were binary.

The dilute-solution approximation is used in many transport problems in biological systems and can be justified as follows: The most common solvent is water, which has a density of 1 g cm^{-3}. One mole of water weighs 18 g, to yield a concentration of 55.56 M. Dissolved-salt concentrations range from 1 μM to 300 mM. The mole fractions of these salts range from 1.80×10^{-8} to 0.0054. Protein concentrations range from 1 nM to 1 mM, with corresponding mole fractions between 1.80×10^{-11} and 1.80×10^{-5}. Under most physiological conditions then, the dilute-solution approximation is reasonable. There are some cases, however, in which the approximation may not be appropriate: gas diffusion in the lung, the purification of biological molecules in high salt concentrations (1–5 M), and fermentation processes. While the dilute-solution approximation does simplify the analysis by reducing the problem to a pseudobinary system, the presence of other molecules in solution can produce deviations in the diffusion coefficient from the value in the pure solvent.

6.3 Conservation Relations

6.3.1 Equation of Conservation of Mass for a Mixture

In this section, we generalize the equation of conservation of mass that was developed in Section 2.3 for mixtures and chemical reactions. Biological mass transport studies frequently involve the concentration of the component under study and the reaction rate or the rate of transport at a given location. These quantities are obtained by solving the conservation of mass for the particular biological situation. Although transport problems can rely upon coordinate systems that are either fixed or moving, a fixed reference is frequently used. In words, the conservation relation, or *material balance*, is

$$\begin{bmatrix} \text{Rate of} \\ \text{accumulation} \\ \text{of } i \text{ within volume} \end{bmatrix} = \begin{bmatrix} \text{Moles of } i \\ \text{entering across} \\ \text{surface of area } S \end{bmatrix} - \begin{bmatrix} \text{Moles of } i \\ \text{leaving across} \\ \text{surface of area } S \end{bmatrix} + \begin{bmatrix} \text{Rate of production of } i \\ \text{by chemical reaction} \\ \text{within volume} \end{bmatrix}. \tag{6.3.1}$$

The rate of transport across the surfaces is described in terms of fluxes.

For most applications, the flux relative to fixed coordinates ($\mathbf{N}_i$ or $\mathbf{n}_i$) is used to describe the rate of transport across the surface of a system. Solutes can enter and leave the control volume by diffusion or convection. Chemical reactions that occur at surface boundaries are considered as boundary conditions. Equation (6.3.1) can be stated in a differential or integral form.

In this chapter, steady-state and unsteady one-dimensional diffusion and convection problems are examined. In the problems considered in this chapter, no chemical reactions are occurring. In Chapter 7, steady-state two- and three-dimensional problems are considered, as well as steady-state diffusion and convection problems. Reaction and diffusion problems are considered in Chapter 10.

6.3.2 Boundary Conditions

Several common boundary conditions are used in mass transport analyses. The concentration or flux may be known at a boundary. Unlike temperature or velocity, concentrations are not necessarily continuous across a fluid–fluid interface. This property is most readily apparent at a gas–liquid interface. Gases have a finite, and usually low, solubility in liquids. For example, at 20°C and a pressure of 1.033×10^5 Pa, the oxygen concentration in water is 1.38 mM when liquid water is in equilibrium with oxygen. This relation is often expressed as a *Henry's law constant* H_i in the formula

$$C_i|_1 = H_i C_i|_2. \tag{6.3.2a}$$

Solute equilibrium at the interface between two immiscible liquids is described in terms of a *partition coefficient*, Φ_i, or solubility:

$$C_i|_1 = \Phi_i C_i|_2. \tag{6.3.2b}$$

Both H_i and Φ_i can be derived from thermodynamic equilibrium relations. The Henry's law constant is a dilute-solution limit that applies when a gas is sparingly soluble in a liquid. The *partition coefficient* is the ratio of the solute activity coefficients in the two phases (i.e., $\Phi_i = \gamma_i|_2/\gamma_i|_1$) and is usually a function of the concentrations in each phase. The partition coefficient is also used to describe solute equilibrium between a porous medium and a fluid (see Section 8.2).

If the boundary is impermeable, there is no flux across it. This condition often occurs at a solid–fluid boundary and is expressed in the equation

$$N_{i_x}|_2 = 0. \tag{6.3.3a}$$

If the solid surface is permeable, Equations (6.3.2a) and (6.3.2b) do not apply and the following relation is used:

$$N_{i_x}|_1 = k(\Phi_i C_i|_2 - C_i|_1). \tag{6.3.3b}$$

In this formula, k is the permeability. Equation (6.3.3b) is used at the interface between a fluid and a permeable solid, such as a membrane, or a single-cell layer, such as epithelium or endothelium.

At a gas–liquid boundary or a boundary between two immiscible liquids, the fluxes in each fluid must be equal. For example, if transport is in the x direction between fluids 1 and 2, then at the fluid boundary,

$$N_{i_x}|_1 = N_{i_x}|_2. \tag{6.3.4a}$$

If a chemical reaction occurs at a surface, then the flux to the surface equals the rate of reaction:

$$N_{i_x}|_1 = R_{i_x}|_2. \tag{6.3.4b}$$

The sign of the reaction rate depends upon whether the reaction produces ($R_{i_x} > 0$) or consumes ($R_{i_x} < 0$) component i. For example, if the reacting surface is at $x = 0$ and transport occurs in the positive x direction, a positive reaction rate indicates that the component is being produced at the surface. The concentration decreases with increasing distance in the x direction, yielding a negative concentration gradient and a positive flux.

6.4 Constitutive Relations

The conservation relation depends upon the fluxes of the solute. To solve the conservation relation for a particular problem, constitutive relations are needed to relate fluxes with concentrations. The most commonly used relation, Fick's law of diffusion, was developed for dilute solutions with nonreacting solutes. Fick developed this relationship from experiments and by analogy with Fourier's law of heat conduction. Fick's law works well for most biomedical problems and is used in most of the subsequent analysis. Departures from Fick's law, such as concentration-dependent diffusion coefficients and multicomponent diffusion, are briefly considered.

Adolph Fick was one of the first biomedical engineers. In addition to developing the most commonly used constitutive relation for diffusion, Fick is known for his numerous contributions to physiology [2]. Fick's doctoral work in 1851 was concerned with astigmatism. He made additional contributions in the area of vision. In hemodynamics, he developed a relation known as Fick's law, which is used to calculate cardiac output from arteriovenous oxygen differences. Fick made significant contributions in the area of muscle dynamics and heat generation as well. His analyses were quantitative and thorough. Fick also made several instruments for physiological measurements, including the first device to measure intraocular pressure and a manometer to measure arterial pressure gradients.

6.4.1 Fick's Law of Diffusion for Dilute Solutions

In his study of diffusion, Fick developed two relations that are sometimes referred to as Fick's first and second laws of diffusion. *Fick's first law* is a constitutive equation that is valid for dilute solutions. *Fick's second law* describes non-steady-state diffusion in a dilute binary solution without any chemical reactions.

Fick's first law states that the diffusive flux J_{i_x} (or N_{i_x} for $v = 0$) of a dissolved solute equals the negative product of the diffusion coefficient and the concentration gradient; in one dimension,

$$J_{i_x} = -D_{ij}\frac{dC_i}{dx}, \tag{6.4.1}$$

where D_{ij} is the binary diffusion coefficient of the solute i in the solvent j. For dilute binary mixtures, $D_{ij} = D_{ji}$. The negative sign in Equation (6.4.1) arises because net solute diffusion proceeds from a region of higher concentration to a region of lower concentration. In other words, in the net direction of diffusion, the derivative is negative and the flux is positive.

Although Fick's first law of diffusion was developed from experiments, it is possible to derive the relation heuristically. Consider the diffusive motion of the molecules in a dilute solution. The presence of convection is removed by focusing on the diffusive flux J_{i_x}. A macroscopic concentration difference is applied so that a steady-state flux arises; the flux at time t is the same at time $t + \Delta t$. The amount of material transported through a small one-dimensional differential region in the x direction during this interval is $AJ_{i_x}\Delta t$. Over the interval Δt, half of the solute molecules move a distance Δx. Thus, the flux in the positive x direction is

$$AJ_{i_x}\Delta t = -\frac{1}{2}(n_i(x + \Delta x) - n_i(x)). \tag{6.4.2}$$

The minus sign indicates that material is transported from x to $x + \Delta x$. Rearranging terms, we obtain

$$AJ_{i_x} = -\frac{n_i(x + \Delta x) - n_i(x)}{2\Delta t}. \tag{6.4.3}$$

As noted in Section 1.2 and discussed further in Section 6.5, diffusion arises from the random motion of molecules. For one-dimensional transport $D_{ij} = (\Delta x)^2/2\Delta t$ or $2\Delta t = (\Delta x)^2/D_{ij}$. Using this result to replace Δt, Equation (6.4.3) can be rewritten explicitly for the flux as

$$J_{i_x} = -D_{ij}\left(\frac{n_i(x + \Delta x) - n_i(x)}{(A\Delta x)\Delta x}\right), \tag{6.4.4}$$

where $A\Delta x$ is the differential volume ΔV. When we shrink the volume element to zero and use the definition of the local concentration, $C_i(x) = n_i/\Delta V$, the result is Fick's first law of diffusion:

$$J_{i_x} = \lim_{\Delta x \to 0} - D_{ij}\left(\frac{C_i(x) - C_i(x + \Delta x)}{\Delta x}\right) = -D_{ij}\frac{dC_i}{dx}. \tag{6.4.5}$$

Equation (6.4.5) can be generalized to three dimensions to yield

$$J_i = -D_{ij}\nabla C_i. \tag{6.4.6a}$$

The molar flux relative to fixed coordinates can now be written for dilute solutions. A similar, but more general, result can be obtained for ideal gases, for which case, however, the density of the mixture may change as the composition changes. A more general form of Fick's law is

$$J_i = -CD_{ij}\nabla x_i. \tag{6.4.6b}$$

When convection is present, Equation (6.2.14) is used for the molar flux relative to fixed coordinates. Substituting Equation (6.4.6a) into Equation (6.2.14) yields

$$N_i = -D_{ij}\nabla C_i + C_i\mathbf{v}_s. \tag{6.4.7}$$

This result is used in problems with convection and diffusion.

6.4.2 Diffusion in Concentrated Solutions

Fick's law applies strictly to a dilute binary solution. For dilute solutions, the diffusion coefficient is a function of temperature and pressure. Diffusion arises from solute–solvent interactions, and solute–solute interactions are negligible. As a result, the diffusion coefficient is independent of the solute concentration. For concentrated solutions, however, Fick's law for binary solutions must be amended to account for nonideal behavior arising from solute–solute interactions. This is done by noting that the driving force for solute fluxes is not the concentration gradient, but the gradient in chemical potential, μ_i, [3]. That is,

$$J_{i_x} = -\frac{D_{ij}C_i}{RT}\nabla\mu_i, \tag{6.4.8}$$

where the chemical potential is [3]

$$\mu_i = \mu_i^0 + RT\ln(\gamma_i x_i), \tag{6.4.9}$$

in which μ_i^0 is the reference chemical potential and γ_i is the activity coefficient that accounts for solute–solute interactions. The reference chemical potential is independent of the composition, and the activity coefficient is a function of the mole fraction only. As x_i approaches zero, γ_i approaches unity. Taking the gradient of Equation (6.4.9) and substituting into Equation (6.4.8) yields

$$J_{i_x} = -D_{ij}C_i\nabla\ln(\gamma_i x_i). \tag{6.4.10}$$

Using the chain rule (Section A.1.E), we can rewrite Equation (6.4.10) as

$$J_{i_x} = -\frac{D_{ij}C_i}{\gamma_i x_i}\nabla\gamma_i x_i = -\frac{D_{ij}C_i}{\gamma_i x_i}[x_i\nabla\gamma_i + \gamma_i\nabla x_i]. \tag{6.4.11}$$

Since γ_i is a function of x_i, applying the chain rule to $\nabla\gamma_i$ results in

$$\nabla\gamma_i = \frac{d\gamma_i}{dx_i}\nabla x_i. \tag{6.4.12}$$

Substituting Equation (6.4.12) into Equation (6.4.11) and noting that $C_i = x_iC$ yields

$$J_{i_x} = -\frac{D_{ij}C}{\gamma_i}\left[x_i\frac{d\gamma_i}{dx_i}\nabla x_i + \gamma_i\nabla x_i\right] = -D_{ij}C\left(1 + \frac{d\ln\gamma_i}{d\ln x_i}\right)\nabla x_i. \tag{6.4.13}$$

For solutions of constant density, $C\nabla x_i = \nabla C_i$. In studies with concentrated solutions, an *apparent diffusivity* is often defined as

$$D_{\text{app}} = D_{ij}\left(1 + \frac{d\ln\gamma_i}{d\ln x_i}\right). \tag{6.4.14}$$

In the limit as the solute mole fraction goes to zero, the derivative in Equation (6.4.14) also goes to zero and D_{app} approaches D_{ij}. (This limiting value of D_{app} is sometimes referred to as the *diffusion coefficient at infinite dilution* and is represented as D_{ij}^0.) A limitation to the use of Equation (6.4.14) is that activity coefficients are generally not known for many solutes as a function of composition, and even when they are known, values of D_{app} computed from Equation (6.4.14) do not always agree with experimental values [4].

6.5 Diffusion as a Random Walk

Before solving problems by using the equation of conservation of mass for dilute solutions, let us examine the molecular basis of diffusion. Such an approach provides an alternative explanation for the diffusion equations and relates the diffusion coefficient to fundamental properties that describe molecular motion. Diffusion on a molecular scale involves a small number of molecules that undergo random interactions. As a result, probabilities are used to describe the average position or state of the molecules.

In liquids and gases, diffusion arises from molecular collisions. Thermal energy is transferred to kinetic energy. During a collision, the probability that a molecule has an energy E_i is $p(E_i)$. This probability is given by summing over the number n of molecules [5]:

$$p(E_i) = \frac{\exp(-E_i/k_BT)}{\sum_{j=1}^{n}\exp(-E_j/k_BT)}. \tag{6.5.1}$$

In Equation (6.5.1), k_B is *Boltzmann's constant*, 1.38×10^{-23} J molecule^{-1} K^{-1}. Because the distribution is exponential, most of the energy transfer involves lower energy states. A molecule travels a distance δ over a time interval τ before encountering another collision. Because the angle of collision is random, the direction of displacement is also random. The energy transferred follows Equation (6.5.1), so the distance traveled between collisions and the time between collisions are not constant.

The macroscopic demonstration of the random motion of molecules was documented in 1828 by the botanist Robert Brown, who observed the movement of pollen after it was placed in a liquid. Hence, the term *Brownian motion* is used. Subsequent measurements by Brown and others showed that the speed of the fluctuations increases with decreasing fluid viscosity and that the speed is inversely related to the particle size [6].

The following analysis of diffusion as a random walk, first presented by Albert Einstein [6], relates the diffusion coefficient to molecular events. The book by Howard Berg [7] provides an interesting presentation of random-walk analysis with applications to biology.

Consider a molecule or microscopic particle in a quiescent fluid that undergoes a three-dimensional displacement δ during a time interval τ (see Figure 6.2; this analysis can be extended to a moving fluid as long as the reference frame coincides with the bulk motion of the fluid). The origin of the coordinate system is placed at the particle location at time zero, the beginning of the period of observation. Each position $\mathbf{r}_i$ is related to the previous position $\mathbf{r}_{i-1}$ by the formula

$$\mathbf{r}_i = \mathbf{r}_{i-1} + \delta, \tag{6.5.2}$$

where $\mathbf{r}_{i-1}$ and $\mathbf{r}_i$ are vectors that extend from the origin to the particle location at steps $i-1$ and i, respectively. Because the motion is random, the displacement δ is not the same during each step.

Random walks are examined first in one dimension, and then the vector components are summed in each direction to obtain the result for three dimensions. For simplicity, assume that N particles move a fixed distance of $\pm\delta$ during each step. Then

$$x_i(n) = x_i(n-1) \pm \delta. \tag{6.5.3}$$

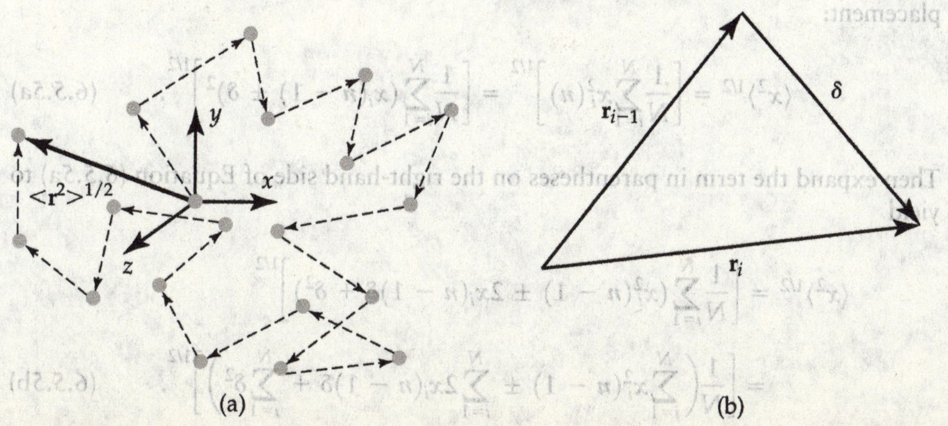

FIGURE 6.2 (a) Random motion of a particle. The particle moves a distance δ in time interval τ. (b) Locations of vectors $\mathbf{r}_i$, $\mathbf{r}_{i-1}$, and δ.

The ± sign indicates that a particle could move either forward or backward. The average location of N particles after n steps is [7]

$$\langle x(n)\rangle = \frac{1}{N}\sum_{i=1}^{N}[x_i(n-1) \pm \delta]. \tag{6.5.4a}$$

Since a time τ elapses during each step, the average over many steps is equivalent to averaging over time. Summing each term on the right-hand side of Equation (6.5.4a) separately yields

$$\langle x(n)\rangle = \frac{1}{N}\sum_{i=1}^{N}[x_i(n-1) \pm \delta] = \frac{1}{N}\sum_{i=1}^{N}x_i(n-1) + \frac{1}{N}\sum_{i=1}^{N} \pm \delta. \tag{6.5.4b}$$

Then we repeatedly substitute for each step until the right-hand side is in terms of δ only:

$$\langle x(n)\rangle = \frac{1}{N}\sum_{i=1}^{N}x_i(n-1) + \frac{1}{N}\sum_{i=1}^{N} \pm \delta$$
$$= \frac{1}{N}\sum_{i=1}^{N}x_i(n-2) + \frac{2}{N}\sum_{i=1}^{N} \pm \delta = \cdots = \frac{n}{N}\sum_{i=1}^{N} \pm \delta. \tag{6.5.4c}$$

For large numbers of particles, the summation of $\pm\delta$ on the right-hand side of Equation (6.5.4b) equals zero, because a particle is equally likely to move forward and backward. Thus, the mean displacement of N particles is zero ($\langle x(n)\rangle = \langle x\rangle = 0$). In the same way that many tosses of a coin are likely to result in heads 50% of the time, the average displacement approaches zero for large numbers of steps. For small values of N, the mean distance that is traveled may be nonzero, because the number of steps in the positive and negative directions may not balance.

Although the mean displacement is zero, the region of space that is sampled by the diffusing molecules grows with time. This is an important distinction that distance can be quantified in terms of the root-mean-square displacement $\langle r^2\rangle^{1/2}$. For a one-dimensional random walk, this quantity is $\langle x^2\rangle^{1/2}$. To see that this term is nonzero, substitute Equation (6.5.3) into the definition of the root-mean-square displacement:

$$\langle x^2\rangle^{1/2} = \left[\frac{1}{N}\sum_{i=1}^{N}x_i^2(n)\right]^{1/2} = \left[\frac{1}{N}\sum_{i=1}^{N}(x_i(n-1) \pm \delta)^2\right]^{1/2}. \tag{6.5.5a}$$

Then expand the term in parentheses on the right-hand side of Equation (6.5.5a) to yield

$$\langle x^2\rangle^{1/2} = \left[\frac{1}{N}\sum_{i=1}^{N}(x_i^2(n-1) \pm 2x_i(n-1)\delta + \delta^2)\right]^{1/2}$$
$$= \left[\frac{1}{N}\left(\sum_{i=1}^{N}x_i^2(n-1) \pm \sum_{i=1}^{N}2x_i(n-1)\delta + \sum_{i=1}^{N}\delta^2\right)\right]^{1/2}. \tag{6.5.5b}$$

Using the argument developed to show that the mean displacement is zero, we can also show that the second term on the right-hand side of Equation (6.5.5b) is zero as well. As a result, the right-hand side of Equation (6.5.5b) reduces to

$$\langle x^2\rangle^{1/2} = \left[\frac{1}{N}\left(\sum_{i=1}^{N} x_i^2(n-1) + \sum_{i=1}^{N}\delta^2\right)\right]^{1/2} = [\langle x_i^2(n-1)\rangle + \delta^2]^{1/2}. \quad (6.5.5c)$$

After substitution for $x(n-1)$, the same procedure results in

$$\langle x^2\rangle^{1/2} = [\langle x(n-1)^2\rangle + \delta^2]^{1/2} = [\langle x(n-2)^2\rangle + 2\delta^2]^{1/2}. \quad (6.5.5d)$$

Repeating this process of substitution gives the following relation between the root-mean-square displacement and the number of steps:

$$\langle x^2\rangle^{1/2} = [n\delta^2]^{1/2} = n^{1/2}\delta \quad (6.5.6)$$

A single time step equals τ. After n steps, the elapsed time t equals $n\tau$. Thus, $n = t/\tau$, and from Equation (6.5.6), the mean-square displacement in one direction can be written as

$$\langle x^2\rangle = \delta^2 t/\tau. \quad (6.5.7)$$

Defining the binary diffusion coefficient for a one-dimensional random walk[1] as $D_{ij} = \delta^2/2\tau$, we have

$$\langle x^2\rangle = 2D_{ij}t. \quad (6.5.8)$$

Equation (6.5.8) indicates that the mean-squared distance that is sampled by a molecule or particle increases linearly with the square root of time. (The factor of two accounts for particle movement in either the positive x direction or the negative x direction.)

For a three-dimensional random walk,

$$\langle r^2\rangle = \langle x^2\rangle + \langle y^2\rangle + \langle z^2\rangle. \quad (6.5.9)$$

Because the motion is random, the random walk in the other two directions yields expressions similar to Equation (6.5.8) (i.e., $\langle x^2\rangle = \langle y^2\rangle = \langle z^2\rangle$). Thus, the mean-square displacement is

$$\langle r^2\rangle = 6D_{ij}t. \quad (6.5.10)$$

Likewise, for a two-dimensional random walk, $\langle r^2\rangle = 4D_{ij}t$. Similar results can be obtained if the displacement δ varies from step to step (see Problem 6.3 at the end of the chapter).

[1]The displacement δ and the binary diffusion coefficient depend upon both the diffusing molecule and the solvent. The same kinetic energy transferred from the solvent to the solute produces a smaller displacement for a larger molecule than for a smaller molecule.

This molecular view of diffusion can be used to describe non-steady-state diffusion and to determine the probability distribution for a particle as a function of position and time. In so doing, we derive what is commonly termed the *diffusion equation*. The probability p that a particle is at location x, y, z equals the average of the probabilities that particles at locations $x \pm \delta, y \pm \delta, z \pm \delta$ are displaced to location x, y, z; that is,

$$p(x, y, z; t + \tau) = \frac{1}{6}\begin{bmatrix} p(x + \delta, y, z; t) + p(x - \delta, y, z; t) + p(x, y + \delta, z; t) \\ + p(x, y - \delta, z; t) + p(x, y, z + \delta; t) + p(x, y, z - \delta; t) \end{bmatrix}. \tag{6.5.11}$$

The sum of probabilities over all locations is unity, or

$$1 = \int_{-\infty}^{\infty} p(x, y, z; t)dV. \tag{6.5.12}$$

Each term in Equation (6.5.11) is approximated by a series expansion in time. For example, $p(x, y, z; t + \tau)$, $p(x + \delta, y, z; t)$, and $p(x - \delta, y, z; t)$ are

$$p(x, y, z; t + \tau) = p(x, y, z; t) + \frac{\partial p}{\partial t}\tau + \cdots, \tag{6.5.13a}$$

$$p(x + \delta, y, z; t) = p(x, y, z; t) + \frac{\partial p}{\partial x}\delta + \frac{\partial^2 p}{\partial x^2}\frac{\delta^2}{2} + \cdots, \tag{6.5.13b}$$

$$p(x - \delta, y, z; t) = p(x, y, z; t) - \frac{\partial p}{\partial x}\delta + \frac{\partial^2 p}{\partial x^2}\frac{\delta^2}{2} + \cdots. \tag{6.5.13c}$$

Because δ is small, terms higher than δ^2 can be neglected. Substituting the results of the Taylor series expansion into Equation (6.5.11) yields

$$p(x, y, z; t + \tau) = p(x, y, z; t) + \frac{\partial p}{\partial t}\tau = p(x, y, z; t) + \frac{\delta^2}{6}\left[\frac{\partial^2 p}{\partial x^2} + \frac{\partial^2 p}{\partial y^2} + \frac{\partial^2 p}{\partial z^2}\right]. \tag{6.5.13d}$$

Rearranging terms in Equation (6.5.13d), we obtain

$$\frac{\partial p}{\partial t}\tau = \frac{\delta^2}{6}\left[\frac{\partial^2 p}{\partial x^2} + \frac{\partial^2 p}{\partial y^2} + \frac{\partial^2 p}{\partial z^2}\right]. \tag{6.5.14}$$

The quantity $\delta^2/6\tau$ equals the diffusion coefficient D_{ij} for a three-dimensional random walk. The probabilities can be rewritten in generalized vector notation as

$$\frac{\partial p}{\partial t} = D_{ij}\nabla^2 p. \tag{6.5.15}$$

An equation similar to Equation (6.5.15) can be developed for the solute concentration using the concentration distribution and Fick's law (Equation (7.3.13)). The

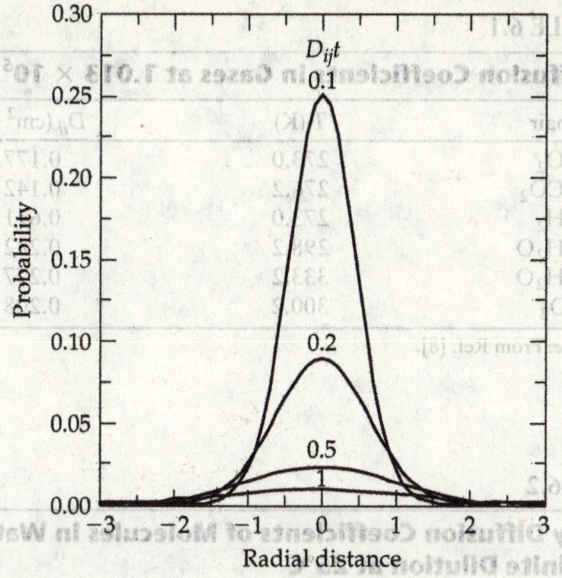

FIGURE 6.3 The probability distribution for a particle diffusing as a function of time.

solution (see Section 6.8) for the case in which a diffusing particle begins at at time $t = 0$ is

$$p(r, t) = \frac{1}{8(\pi D_{ij}t)^{3/2}} \exp(-r^2/4D_{ij}t), \tag{6.5.16}$$

where $r = (x^2 + y^2 + z^2)^{1/2}$ Equation (6.5.16) is simply the three-dimensional form of the normal distribution with a mean value of zero. The variance, which equals the root-mean-square displacement, is $\sqrt{6D_{ij}t}$. As time increases, the distribution becomes broader and particles are found at more locations, each with lower probability (Figure 6.3).

6.6 Estimation of Diffusion Coefficients in Solution

In this section, the random-walk results and the drag on a sphere at a low Reynolds number (Section 3.6) are used to estimate the diffusion coefficient of solutes, particularly macromolecules. The random-walk calculations that we discussed in Section 6.5 relate the diffusion coefficient to displacements and time. However, the diffusion coefficient is affected by differences in the solute size and noncovalent interactions (e.g., van der Waals forces) with the solvent. In general, the diffusion coefficient is a function of temperature, fluid viscosity, and solute size. For example, diffusion coefficients of gases in gases are very rapid. At 25°C, the diffusion coefficient of carbon dioxide in nitrogen is 0.165 cm^2 s^{-1}. Typical values of diffusion coefficients in air are given in Table 6.1. In contrast, diffusion coefficients of small solutes in water

TABLE 6.1

Diffusion Coefficients in Gases at 1.013 × 10⁵ Pa

Gas pair	T (K)	D_{ij} (cm^2 s^{-1})
Air–O_2	273.0	0.1775
Air–CO_2	276.2	0.142
Air–H_2	273.0	0.611
Air–H_2O	298.2	0.282
Air–H_2O	333.2	0.277
N_2–O_2	300.2	0.208

Source: From Ref. [8].

TABLE 6.2

Binary Diffusion Coefficients of Molecules in Water at Infinite Dilution at 25°C

Solute	$D^0 \times 10^{-5}$ (cm^2 s^{-1})
Oxygen	2.10
Sucrose	0.52
Urea	1.38
Glycine	1.06
Ethanol	0.840
Albumin	0.061
Hemoglobin	0.069
Fibrinogen	0.020

Source: From Ref. [4].

Note: Infinite dilution refers to the case in which each solute molecule is completely surrounded by solvent molecules and does not interact with other solute molecules.

are about 10,000 times smaller than diffusion coefficients in gases (see Table 6.2). Diffusion coefficients of proteins in liquids are about two orders of magnitude smaller than diffusion coefficients of small solutes in liquids.

6.6.1 Transport Properties of Proteins

In dilute solutions, such as those most commonly found in biological systems, the motion of solutes is caused by thermal energy that is transferred from the solvent molecules during random collisions between them and solute molecules. This transfer of energy leads to fluid motion that is retarded by the drag of the solvent on the solute. To determine the diffusion coefficient, it is necessary to understand the effects of the drag force. For solute molecules with a characteristic dimension about 10 times greater than that of the solvent molecules, the continuum approximation applies and this drag force can be approximated using low-Reynolds-number hydrodynamics [9,10] (Section 3.6). On the basis of this size constraint, such an approach can be applied to molecules larger than sucrose. In this section, several general results for transport properties are developed and applied to proteins.

6.6.2 The Stokes–Einstein Equation

In the previous section, we related the diffusion coefficient to the root-mean-square displacement of a particle or molecule. The displacement is a complex function of the shape of the particle and the energy transferred during each collision. This energy can be expressed in terms of the thermal energy of the system. From a force balance on the particles, we then relate the root-mean-square displacement to the drag acting on the particles and the thermal energy of the system. In so doing, we derive an expression for the diffusion coefficient.

For low-Reynolds-number flow, the drag force on a large molecule or particle is proportional to the particle velocity. For an arbitrarily shaped particle, the force and velocity are related as

$$\mathbf{F}_D = \mathbf{K}\cdot\mathbf{v}, \tag{6.6.1}$$

where **K** is the translation tensor [9]. Equation (6.6.1) is a generalization of Stokes Law. In the analysis of diffusion, the force is specified and the velocity is found. As a result, the velocity is computed as

$$\mathbf{v} = \mathbf{K}^{-1}\cdot\mathbf{F}_D, \tag{6.6.2}$$

where $\mathbf{K}^{-1}$ is the inverse of the translation tensor.

The components of **K** are referred to as *friction coefficients* f_{ij} [10]. **K** is a symmetric tensor (i.e., $f_{ij} = f_{ji}$) that can be expressed in terms of the principal friction coefficients f_1, f_2, and f_3 [9]. The principal friction coefficients are the *eigenvalues* obtained by solving $\det[\mathbf{K} - f\mathbf{I}] = 0$. For an isotropic body, $f_1 = f_2 = f_3 = f$, and

$$\mathbf{F}_D = f\mathbf{v}. \tag{6.6.3}$$

For a sphere of radius R, $f = 6\pi\mu R$ (see, e.g., Equation (3.6.25)).

In general, the frictional drag coefficients are sensitive to the shape and orientation of the particle. In the diffusive transport of nonspherical shapes, the particle orientation changes rapidly due to each collision. Accordingly, to apply Equation (6.6.2) to the case of diffusion, it is necessary to average over all orientations. When this is done, the inverse of the translation tensor is replaced by its average, $\bar{f}$, and the equation then becomes

$$\mathbf{v} = \frac{\mathbf{F}_D}{\bar{f}}. \tag{6.6.4}$$

The average value of $\bar{f}$ is defined as the harmonic mean [9]

$$\frac{1}{\bar{f}} = \frac{1}{3}\left(\frac{1}{f_1} + \frac{1}{f_2} + \frac{1}{f_3}\right). \tag{6.6.5}$$

TABLE 6.3

Values of the Mean Frictional Drag Coefficient for Different Shapes [9,10]

Shape	Frictional drag coefficient
Sphere of radius R	$f = 6\pi\mu R$
Prolate ellipsoid, $p = a/b > 1$, where a is a major axis, b is a minor axis	$\bar{f} = \dfrac{6\pi\mu b(p^2 - 1)^{1/2}}{p^{1/3}\ln[p + (p^2 - 1)^{1/2}]}$
Oblate ellipsoid, $p = a/b < 1$	$\bar{f} = \dfrac{6\pi\mu b(1 - p^2)^{1/2}}{p^{1/3}\tan^{-1}[1 - p^2)^{1/2}p^{-1}]}$
Thin circular disk of radius a	$\bar{f} = 16\mu a$
Cylinder of radius a and length L	$\bar{f} \approx \dfrac{4\pi\mu L}{\ln(L/a) + 0.193}$

Source: From Refs [9,10].

Table 6.3 lists the frictional drag coefficients for several representative shapes, computed with low-Reynolds-number hydrodynamics.

Two approaches have been used to determine the dependence of the diffusion coefficient upon the energy of the system and the resistance to flow. Einstein employed a thermodynamic approach [6] that is summarized at the end of the chapter (Section 6.10). The alternative approach uses a force balance that is averaged over periods that are long compared with the collision time, and is presented here [10,11]. Particles accelerate due to the difference between the drag force (F_D) and the force arising from collisions (F_T) (Figure 6.4). In one dimension, the force balance is

$$m_p\frac{d^2x}{dt^2} = -\bar{f}\frac{dx}{dt} + F_T, \tag{6.6.6a}$$

where m_p is the mass of the particle. Equation (6.6.6a) is referred to as the *Langevin equation* [10,11]. In order to express it in terms of the mean-square displacement, each term is multiplied by x and averaged over a period that is long compared with the collision time:

$$\left\langle m_p x\frac{d^2x}{dt^2}\right\rangle = -\left\langle \bar{f}x\frac{dx}{dt}\right\rangle + \langle xF_T\rangle. \tag{6.6.6b}$$

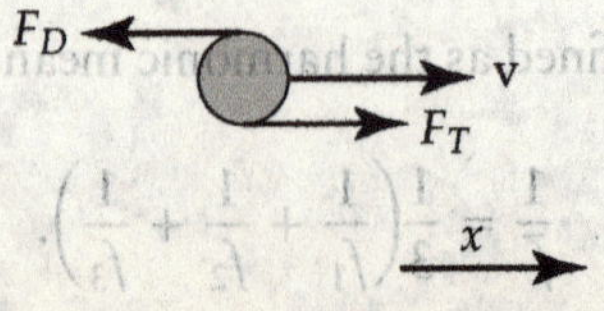

FIGURE 6.4 Forces acting on a solute molecule.

Since F_T is a random variable with a zero mean, the average $\langle xF_T\rangle$ is also zero. The derivative on the left-hand side of Equation (6.6.6b) can be rewritten as

$$x\frac{d^2x}{dt^2} = \frac{d}{dt}\left(x\frac{dx}{dt}\right) - \left(\frac{dx}{dt}\right)^2. \qquad (6.6.7)$$

Substituting Equation (6.6.7) into Equation (6.6.6b) and noting that $xdx = dx^2/2$, we obtain

$$\left\langle \frac{1}{2}m_p\frac{d}{dt}\left(\frac{dx^2}{dt}\right)\right\rangle - \left\langle m_p\left(\frac{dx}{dt}\right)^2\right\rangle = -\left\langle \frac{\bar{f}}{2}\frac{dx^2}{dt}\right\rangle. \qquad (6.6.8)$$

Replacing dx/dt with the particle velocity and rearranging terms in Equation (6.6.8) yields

$$m_p\langle \mathrm{v}_x{}^2\rangle = \frac{\bar{f}}{2}\frac{d\langle x^2\rangle}{dt} + \frac{1}{2}m_p\frac{d^2\langle x^2\rangle}{dt^2}. \qquad (6.6.9)$$

The kinetic energy of a particle, $m_p\langle \mathrm{v}_x{}^2\rangle$, moving in one dimension equals the thermal energy, k_BT. Making this substitution and solving Equation (6.6.9) subject to $\langle x^2\rangle$ and $d\langle x^2\rangle/dt$ each being equal to zero at time zero yields

$$\langle x^2\rangle = \frac{2k_BT}{\bar{f}}t - \frac{2m_pk_BT}{\bar{f}^2}(1 - \exp(-\bar{f}t/m_p)). \qquad (6.6.10)$$

Equation (6.6.10) indicates that, following the application of random forces, the particle velocity changes with a time constant equal to $m_p/\bar{f}$. At long times, the mean-square displacement increases linearly with time. Replacing the long-time behavior of $\langle x^2\rangle$ with Equation (6.5.8) and rearranging terms yields the following expression for the diffusion coefficient:

$$D_{ij} = \frac{k_BT}{\bar{f}}. \qquad (6.6.11)$$

This expression is widely known as the *Stokes–Einstein equation*. Note that the same result is obtained if we consider motion in three dimensions.

Often, the frictional drag coefficient is reported as the ratio $\bar{f}/f_0$ relative to the frictional coefficient of an equivalent sphere, where $f_0 = 6\pi\mu R_{eq}$, in which the equivalent radius is

$$R_{eq} = \left(\frac{3\bar{V}}{4\pi N_A}\right)^{1/3}, \qquad (6.6.12)$$

where N_A is Avogadro's number and $\bar{V}$ is the partial molar volume of the protein under study (cm^3 (g mol)$^{-1}$).

Alternatively, the equivalent radius can be measured from X-ray crystallography data [9].

There are several ways to measure diffusion coefficients from Equation (6.6.11) and/or random-walk calculations. The frictional coefficient can be determined from particle sedimentation rates in an ultracentrifuge and then the diffusion coefficient is calculated from Equation (6.6.11). The advantage of this approach is that detailed information on the particle shape is not needed. Drawbacks of this approach are that the experiments are lengthy due to the time needed to produce measurable separation and measurements are limited to the solution phase.

The development of a method that measures light and particle displacements with high accuracy permits the measurement of the movement of groups of particles or individual particles. In dynamic light scattering, a laser passes through a solution of macromolecules and scattering is measured as a function of time. The scattering electromagnetic field is an exponential function of the scattering vector which has magnitude $q \equiv (4\pi/\lambda)\ \sin\ (\theta/2)$ where θ is the angle of scattering relative to the detector. The particle displacement over time τ is proportional to the reciprocal of the scattering vector, $\Delta x \sim q^{-1}$ [12]. Since the correlation among the various scattering particles declines exponentially over the timescale t, the diffusion coefficient is determined from the correlation as a function of time and the Stokes–Einstein equation [12]. Information about the molecular shape is needed in order to identify the appropriate relation for the friction coefficient.

Single-particle tracking methods are used to track the movement of fluorescently labeled proteins or lipids or organelles within cells or on cell membranes. One interesting result of these studies is that the diffusion of a number of molecules or organelles does not follow the linear relation between mean-square displacement and time [13]. In some cases, a convective flow arises and the root-mean-square displacement is proportional to time. Such flows may arise from active displacement of membranes or organelles. More interesting are cases for which the mean-square displacement varies with t^{α} where $\alpha < 1$. This case is known as anomalous diffusion and arises due to the presence of obstructions. A key feature of this type of diffusion is that the measured results depend on the length scale over which diffusion is measured. Over short timescales, the diffusing molecule or organelle does not contact the obstructions and diffusion appears to conform to the random-walk model (Equation (6.5.8)). Over longer timescales, the diffusing particles encounter the obstructions which increase the path length and the diffusion time. As a result, the mean-square displacement grows more slowly with time [13]. Conversely, localized domains in cell membranes have a density of barriers (*corrals*), slowing diffusion on short timescales. Once the diffusing molecule escapes from the corral, its mean-square displacement increases linearly with time.

6.6.3 Estimation of Frictional Drag Coefficients

The results presented in Section 6.6.1 indicate that the frictional drag coefficients for molecules can be determined from diffusion measurements or from sedimentation in an ultracentrifuge. Both approaches have been employed to characterize the size and molecular weight M of globular proteins. We can use the resulting measurements to assess the accuracy of the hydrodynamic approach to estimating frictional drag.

Data from sedimentation and diffusion can be utilized to estimate molecular weight as follows: assuming solid-body rotation in an ultracentrifuge, we obtain the sedimentation coefficient [14][2]

$$s = \frac{1}{\omega^2 r}\frac{dr}{dt} = \frac{M(1-\bar{v}\rho)}{N_A \bar{f}}, \tag{6.6.13}$$

where ρ is the density of the solution and $\bar{v}$ is the *partial specific volume*[3] of the protein ($cm^3\ g^{-1}$). Using Equation (6.6.11) to solve for $\bar{f}$ and substituting into Equation (6.6.13) yields the following expression for M:

$$M = \frac{sN_A k_B T}{D_i(1-\bar{v}\rho)}. \tag{6.6.14}$$

The molecular weight of a protein can be found from Equation (6.6.14) by using independent measurements of s, D_i, and $\bar{v}$. The accuracy in the value of M depends upon the accuracy of the value of the partial specific volume of the protein [14].

For a number of proteins, the diffusion and sedimentation coefficients at 20°C, the molecular weights, X-ray crystallographic data for shapes, and measured values of $\bar{f}/f_0$ are known (see Table 6.4). One way to assess the accuracy of estimates of the hydrodynamic properties is to compare the molecular weights obtained by the hydrodynamic analysis with those values computed from a knowledge of the structure of the molecules. The molecular weights measured from Equation (6.6.14) agree very well with the values that have been determined from the composition of the molecule (the amino acid sequence plus glycosylation). The consistency between the two measurements of molecular weight indicates that the values of $\bar{f}/f_0$ obtained from sedimentation and diffusion are the same. This result validates the use of Equation (6.6.14) to determine molecular weights and indicates that measurements of sedimentation and diffusion coefficients provide important information about the molecular shape. The value of $\bar{f}/f_0$ and the dimensions found from analysis of crystal structure indicate that the proteins cannot be adequately treated as spheres.

Although molecular weights can be measured from equilibrium concentration gradients in an ultracentrifuge (an approach that does not require a value for $\bar{f}$), hydrodynamic measurements are not used extensively for this purpose. Gel filtration

[2]Equation (6.6.13) is derived from a force balance that is applied to the particle. Solid-body rotation occurs in the angular direction θ. At steady state, the particle is exposed to a centrifugal force, a buoyant force, and a drag force according to the formula

$$0 = fv_r - m_p\omega^2 r + \rho V_p \omega^2 r,$$

where m_p is the particle mass, V_p is the particle volume, and ρV_p is the fluid mass that is displaced by the particle. Noting that the partial specific volume of the protein $\bar{v}$ equals V_p/m_p and that v_r equals dr/dt, we can rearrange the force balance to yield

$$f\frac{dr}{dt} = m_p\omega^2 r(1-\bar{v}\rho).$$

The particle mass is simply M/N_A. Making this substitution and further rearranging the force balance results in Equation (6.6.13).

[3]The partial specific volume represents the volume change in solution that arises from interactions between the protein and the solvent water [5] $\bar{v} = \bar{V}/M$.

TABLE 6.4

Hydrodynamic, Transport, and Structural Properties of Selected Proteins

Protein	s^0_{20w} $(S)^a$	D^0_{20w} $(10^7 cm^2 s^{-1})$	$\bar{v}$ $(cm^3 g^{-1})$	Molecular weight Structure	Molecular weight Measured[b]	$\bar{f}/f_0$ [b]	Dimensions from crystal structure (nm)
Pancreatic trypsin inhibitor (bovine)	1.0	12.9	0.718	6,520	6,670	1.321	2.9 × 1.9 × 1.9
Cytochrome C (equine)	1.83	13.0	0.715	12,310	11,990	1.116	2.5 × 2.5 × 3.7
Ribonuclease A (bovine)	1.78	10.7	0.704	13,690	13,600	1.290	3.8 × 3.8 × 2.2
Lysozyme (hen)	1.91	11.3	0.703	14,320	13,800	1.240	4.5 × 3.0 × 3.0
Trypsin (bovine)	2.5	9.3	0.727	23,200	23,890	1.187	5.0 × 4.0 × 4.0
Carbonic anhydrase (human)	3.23	10.7	0.729	28,800	27,020	1.053	4.7 × 4.1 × 4.1
Superoxide dismutase (bovine)	3.35	8.92	0.729	33,900	33,600	1.132	7.2 × 4.0 × 3.8
Phosphoglycerate kinase (yeast)	3.09	6.38	0.749	45,800	46,800	1.377	7.0 × 4.5 × 3.5
Concanavalin A	3.8	6.34	0.732	51,260	54,240	1.299	8.0 × 4.5 × 3.0
Hemoglobin (human)	4.48	6.9	0.746	64,500	62,300	1.18	9.0 × 8.8 × 5.5
Oxyhemoglobin (equine)	4.22	6.02	0.750	64,610	67,980	1.263	7.0 × 5.5 × 5.5
Malate dehydrogenase (porcine)	4.53	5.76	0.742	74,900	73,900	1.344	6.4 × 6.4 × 4.5
Plasma vitronectin (human)[c]	4.2–4.6	5.7	0.711	75,000–78,000	70,000	1.37	not determined
Alcohol dehydrogenase (equine)	5.08	6.23	0.750	79,870	79,070	1.208	4.5 × 5.5 × 11.0
Lactate dehydrogenase (dogfish)	7.54	4.99	0.74	146,200	141,000	1.273	7.4 × 7.4 × 8.4
Fibrinogen (human)[d]	7.9	2.0	0.715	340,000	336,000	2.34	47.5 × 9.0 (cylinder)[e]

Source: From Ref. [14] with permission from Elsevier, except [c]from Ref. [15] and [d]from Ref. [16].

[a]S = Svedberg = 10^{-13} second.

[b]Determined from independent measurements of sedimentation and diffusion using Equation (6.6.14).

[e]Combination of electron microscopy and X-ray crystallography of fibrinogen after limited proteolysis. Data represent the length and diameter of the molecule, respectively.

is now the method most commonly used to determine molecular weight. This technique involves passing the protein solution through a column of packed porous beads. Molecules smaller than the pore size of the beads pass into the beads and take longer to travel through the column than do other molecules. Thus, the smaller the molecule, the more time it takes for it to pass through the column. By calibrating the time required to pass through a column with the molecular weight of a set of standards of known molecular weight, the molecular weight of the protein can be determined. This method is easier to apply than ultracentrifugation, and, further, the samples need not be pure [14].

Predicting $\bar{f}$ from X-ray crystallographic methods and estimates of the shape of the molecule leads to values that deviate from the measured values of $\bar{f}$ by 15%–30% (see Table 6.4). Thus, the simple theory used to estimate D_{ij} (Equation (6.5.8)) fails to account for all factors affecting $\bar{f}$ and D_{ij}. Two other factors must be considered: the actual shape of the surface and the hydration of the molecule.

6.6.4 The Effects of Actual Surface Shape and Hydration

Table 6.5 compares experimental and computed values of the diffusion coefficient for a lysozyme and for the tobacco mosaic virus. The shape of a molecule may differ substantially from one of the idealized shapes listed in the table. A space-filling model[4] of lysozyme (see Figure 6.5) shows that modeling the molecule as an ellipse of rotation is quite approximate. Accurate inclusion of the three-dimensional shape requires a sophisticated analysis of low-Reynolds-number hydrodynamics [17].

In solution, a protein is hydrated due to electrostatic and electrodynamic interactions. Water that is firmly associated with the protein increases the size of the molecule, altering the frictional coefficient. If hydration were uniform over the protein and occurred in discrete layers, then hydration would result in larger molecular dimensions and no special consideration would be needed. However, hydration is not uniform over the molecule; it is greater near charged groups [18]. Hydration increases the molecular volume V in accordance with the relationship

$$V = \frac{M}{N_A}(\bar{v} + \delta_h/\rho), \qquad (6.6.15)$$

where δ_h is the extent of hydration (g water/g protein) and ρ is the density of water. Values for δ_h range from 0.2 to 0.6 g water/g protein. The quantity δ_h is not always known independently and must be inferred.

TABLE 6.5

Comparison of Experimental and Computed Values of the Diffusion Coefficient

	Diffusion coefficient, $D^0_{20w} \times 10^{-6}$ cm^2 s^{-1}				
Object	Measured	Sphere	Hydrated sphere	Ellipsoid	Computed [17]
Lysozyme	1.13 ± 0.05	1.35	1.245	1.232	1.17 ± 0.04
Tobacco mosaic virus	0.044 ± 0.001			0.050	

[4]Several representations are used to show the shape of proteins. Space-filling models show the entire volume occupied by each component of the molecule. These models provide information about the overall three-dimensional shape of the molecule. Conversely, ball-and-stick models provide information about the arrangement of the individual amino acids within proteins (the secondary structure).

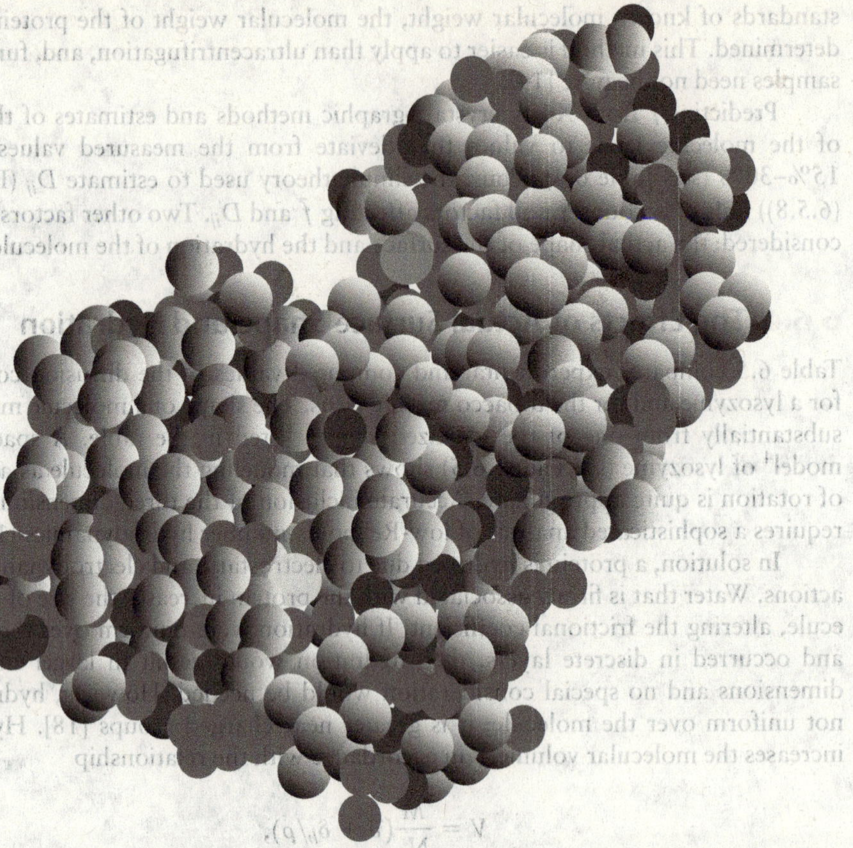

FIGURE 6.5 A space-filling representation of lysozyme. (Image from the Brookhaven National Laboratory protein data bank.)

An interesting analysis serves to illustrate the accuracy of existing correlations and more sophisticated computational techniques. The translational diffusion coefficient of lysozyme was computed from Equation (6.5.8), with a variety of shapes used to estimate the friction coefficient [17]. Four cases were considered:

1. An equivalent sphere based upon the X-ray crystallography data ($R_{eq} = 1.59 \times 10^{-7}$ cm)
2. A hydrated sphere (with δ_h assumed equal to 0.2 g water/g protein and $R_{eq} = 1.72 \times 10^{-7}$ cm)
3. A prolate ellipsoid of revolution
4. A numerical computation based on the three-dimensional shape shown in Figure 6.5

An examination of Table 6.5 indicates that accounting for the shape more accurately improves the estimated diffusion coefficient. For the protein lysozyme, the error for an ellipsoid is 8.85%, and the value from the numerical computations of Brune and Kim [17] is within the experimental error. A second example is the tobacco mosaic virus, which is roughly 300 nm long, with a diameter of 18 nm. Both a prolate ellipsoid and a cylinder yielded similar values that were within 16% of the measured value. The numerical computation of Brune and Kim treated the virus as a 16-sided prism and obtained a value for the diffusion coefficient that was in exact agreement with data. The major conceptual result from this analysis is that accurate estimates of the shape are needed to obtain reliable estimates of the diffusion coefficient. A drawback is that the method used to obtain the computed results relies on detailed structural information of the molecule and requires a significant computational effort. Thus, as a practical matter, approximate methods must be employed to estimate the diffusion coefficients of proteins. The best approach is to use the most reliable shape data, realizing that the error in the estimated diffusion coefficient is about 10%.

The diffusion coefficient developed for a random walk is stated in terms of the displacement δ and the time τ between collisions (see Equation (6.5.7)). In order to determine these two quantities, one additional relation is needed. This relation is obtained from the definition of the diffusion velocity v_d as the average speed of the particle as it moves a distance δ in a time τ, or

$$v_d = \delta/\tau. \tag{6.6.16}$$

Generalizing Equation (6.5.7) to three dimensions, we have $\langle r^2\rangle/t = \delta^2/\tau$. Solving for τ and inserting the definition of $D_{ij}(=\delta^2/6\tau)$ for a three-dimensional random walk yields

$$\delta = 6D_{ij}/v_d \tag{6.6.17}$$

The velocity can be determined by equating the kinetic energy $(m_p v_d^2/2)$ of the molecule or particle to its thermal energy $(3k_BT/2)$. In three dimensions, the result is

$$v_d = \sqrt{3k_BT/m_p}, \tag{6.6.18}$$

where m_p is the mass of the molecule or particle and k_B equals 1.38×10^{-16} erg molecule^{-1} K^{-1}. The molecular mass equals the weight of molecules per mole (M) divided by Avogadro's number (N_A). Substituting Equations (6.6.11) and (6.6.18) into Equation (6.6.17) and solving for δ yields

$$\delta = \frac{1}{\bar{f}}\sqrt{12 m_p k_B T} = \frac{1}{\bar{f}}\sqrt{\frac{12 M k_B T}{N_A}}. \tag{6.6.19}$$

Example 6.2 Using the data in Table 6.4, determine the time to move a root-mean-square displacement of 1 μm and the time between collisions for the protein lysozyme at 20°C.

Solution For lysozyme, Equation (6.6.11) and the measured value for the diffusion coefficient yield a value of $\bar{f} = 3.58 \times 10^{-8}$ g s^{-1}. Using this result in Equation (6.6.19) yields $\delta = 2.95 \times 10^{-9}$ cm. From the relation $D_{ij} = \delta^2/6\tau$ for a three-dimensional random walk, the time τ between collisions is found to be 1.28×10^{-12} s. From Equation (6.5.10), lysozyme moves a root-mean-square distance of 1 μm in 0.0015 s, which requires 1.15×10^9 time intervals of duration τ. Such short distances and interaction times during each step are due to the high mass and large frictional coefficient of the protein.

Polypeptides and polynucleic acids (RNA and DNA) differ from globular proteins structurally and are consequently often treated as random coils [14]. *Random coils* can adopt a number of configurations, and the root-mean-square end-to-end distance, $\langle d_0^2\rangle^{1/2}$, can be described in terms of a modified random walk. For a random-coil molecule that can rotate freely, the root-mean-square end-to-end distance is

$$\langle d_0^2\rangle^{1/2} = n^{1/2}l, \qquad (6.6.20)$$

where n is the number of bonds and l is the length of each segment. The contour length L equals the product nl and represents the end-to-end length of the molecule if it were stretched. The contour length is the maximum possible length the molecule can assume.

This model describing the root-mean-square end-to-end distance is known as the *random-flight chain* or the *free-draining chain*, to imply that solvent within the domain of the molecule moves freely with respect to the molecule. In reality, the bonds cannot rotate freely, and the finite molecular shapes restrict the movement of the components of the molecule. (These restrictions are known as *steric* restrictions.) As a result, natural and synthetic random-coil macromolecules satisfy Equation (6.6.20) only in solvents with which the molecule does not interact. Further, due to constraints on the orientation of intersegment bonds, l differs from the length of the segment and must be determined experimentally [5].

For polypeptides and polynucleotides in water, there is significant interaction between the solvent and the molecule, and considerable solvent is associated with the macromolecule. In this case, Equation (6.6.20) becomes

$$\langle d_0^2\rangle^{1/2} = (C_n n)^{1/2}\ l, \qquad (6.6.21)$$

where C_n is known as the *characteristic ratio*. This ratio approaches a limiting value C_∞ for large chain lengths. In general, $C_n = l_K/l$, where l_K is known as the *Kuhn statistical segment length* [19]. Polypeptides without the amino acids glycine or proline have a value of C_∞ equal to 130, and their mean-square end-to-end distance is given in nanometers. The large value of C_∞ indicates significant interaction of the amino acid side chains with water and stiffening of the segments. For polyglycine, a polypeptide consisting of repeating units of glycine, however, C_∞ equals 0.9, indicating that it behaves close to a random-flight chain [14].

For DNA, l_K equals 150 nm and l equals 0.34 nm, which indicates that DNA is a stiff chain. For a bacterial DNA with 1.5×10^6 base pairs, the root-mean-square end-to-end distance is 8.75 μm, which is larger than the size of the bacterium! Clearly, the organization within the bacterium is such that DNA is not behaving according to Equation (6.6.20).

For polyamino acids or polynucleotides, the friction coefficient can be calculated under the assumption that the random coil behaves as a sphere with a radius equal to the radius of gyration, R_G. The diffusion coefficient depends upon the radius of gyration as follows [5]:

$$D_{ij} = \frac{0.196 k_B T}{\sqrt{6}\mu R_G}. \qquad (6.6.22)$$

For open-ended coils,

$$R_G = \frac{\langle d_0{}^2\rangle^{1/2}}{\sqrt{6}}. \tag{6.6.23a}$$

DNA plasmid is a circular piece of DNA found in bacteria and used in molecular biology and gene delivery. For a circular random coil, which represents a model for DNA plasmid,

$$R_G = \frac{\langle d_0{}^2\rangle^{1/2}}{\sqrt{12}}. \tag{6.6.23b}$$

For dilute solutions of DNA, this analysis indicates that the diffusion coefficient is inversely proportional to the square root of the number of base pairs, a result that has been confirmed experimentally [20,21].

Example 6.3 Determine the diffusion coefficient at 37°C for a 10,000-base-pair plasmid DNA used in gene delivery studies.

Solution Using $l_K = 150$ nm and $l = 0.34$ nm, we find that R_G is 0.206 μm. The diffusion coefficient is then

$$D_{ij} = \frac{0.196(1.38 \times 10^{-16}\ \text{erg K}^{-1})(310\ \text{K})}{\sqrt{6}(0.01\ \text{g cm}^{-1}\ \text{s}^{-1})(0.206 \times 10^{-4}\ \text{cm})} = 1.66 \times 10^{-8}\ \text{cm}^2\ \text{s}^{-1}. \tag{6.6.24}$$

6.6.5 Correlations

A number of correlations aid in estimating diffusion coefficients in liquid solutions [4,8]. Two of the most relevant for solute diffusion in aqueous solution are the Wilkie–Chang correlation and the Stokes–Einstein equation. *The Wilkie–Chang correlation* is the semiempirical relation

$$D_{ij} = 7.4 \times 10^{-10}\left(\frac{T(\phi M)^{0.5}}{\mu \overline{V}_0{}^{0.6}}\right), \tag{6.6.25}$$

where $\overline{V}_0$ is the molar volume of the solute (cm^3 mol^{-1}) at its normal boiling point, μ is the viscosity of the solution (g cm^{-1} s^{-1}), T is the absolute temperature in K, M is the molecular weight of the solvent, and ϕ is an association parameter for the solvent. The recommended value of ϕ for water is 2.26. Values for $\overline{V}_0$ are listed in Reid et al. [8]. The Wilkie–Chang correlation applies both to solutes with sizes similar to those of the solvent and to liquid–liquid diffusion. The equation is accurate to within 10%.

6.7 Steady-State Diffusion in One Dimension

In this and subsequent sections, we apply a continuum approach to the analysis of transport problems. The continuum approach does not work with the fine details of molecular interactions, but it is used to develop conservation relations that describe molecular motion. This surprisingly robust approach is useful at length scales of 1 nm or greater. In the remainder of this section, we examine steady-state diffusion in one dimension without chemical reaction. The analysis demonstrates the effects of geometry and boundary conditions on the concentration distribution and flux. In Section 6.8, we consider non-steady-state diffusion. Later chapters consider diffusion with convection and the influence of convection (Chapter 7) and diffusion (Chapter 11) on chemical reactions.

6.7.1 Diffusion in Rectangular Coordinates

As a first example, consider one-dimensional diffusion in rectangular coordinates (see Figure 6.6). These coordinates apply to diffusion through a variety of tissues, such as skin and liver, when the concentrations are uniform and constant over a surface, when curvature is not significant, and when the width and length are much greater than the thickness. Diffusion occurs normal to a surface with an area A and through a volume element $A\Delta x$. A material balance on component i entering at x and leaving at $x + \Delta x$ yields

$$\frac{\partial C_i}{\partial t} A\Delta x = (N_{ix}|_x - N_{ix}|_{x+\Delta x})A + R_i A\Delta x. \tag{6.7.1}$$

Equation (6.7.1) is a general statement regarding one-dimensional transport in rectangular coordinates. Dividing each term of the equation by the volume element $A\,\Delta x$ leads to the formula

$$\frac{\partial C_i}{\partial t} = \frac{(N_{ix}|_x - N_{ix}|_{x+\Delta x})}{\Delta x} + R_i. \tag{6.7.2}$$

When we take the limit as $\Delta x \to 0$ and use the definition of the derivative, the result is the following differential expression for the conservation of mass for one-dimensional transport in rectangular coordinates:

$$\frac{\partial C_i}{\partial t} = -\frac{\partial N_{ix}}{\partial x} + R_i. \tag{6.7.3}$$

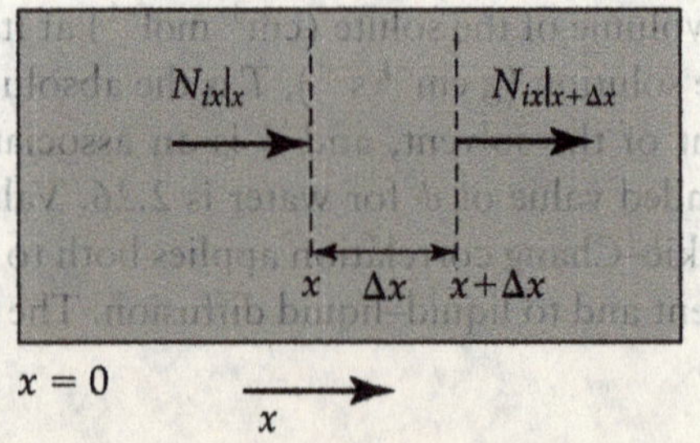

FIGURE 6.6 Diffusion through a small rectangular volume of area A and thickness Δx.

For the case of diffusion only, $v_x = 0$ and $N_{i_x} = J_{i_x}$. Substituting Fick's law, $N_{ix} = -D_{ij}(\partial C_i/\partial x)$ into Equation (6.7.3) yields

$$\frac{\partial C_i}{\partial t} = D_{ij}\frac{\partial^2 C_i}{\partial x^2} + R_i. \tag{6.7.4a}$$

When no chemical reactions occur, $R_i = 0$, and the result is known as *Fick's second law*:

$$\frac{\partial C_i}{\partial t} = D_{ij}\frac{\partial^2 C_i}{\partial x^2}. \tag{6.7.4b}$$

For steady-state transport, the derivative with respect to time is zero.

Steady-State Diffusion Across Membranes. Membranes permit the selective transport between two fluid phases. Biological membranes are made from phospholipid bilayers (see Figure 1.5), which surround cells and many organelles. Synthetic membranes consist of a polymer matrix that provides mechanical rigidity and irregular openings through which solute enters. As a first approximation, thin, uniform layers of cells, such as endothelial or epithelial cells, can be treated as membranes. Membranes often consist of two phases; usually the solute is able to diffuse through only one phase. Because the structure of membranes is complex, the binary diffusion coefficient is not an appropriate description of diffusion. Rather, an *effective diffusion coefficient* $D_{i,\text{eff}}$ is used in Fick's equation in place of the binary diffusion coefficient. The effective diffusion coefficient depends upon the diffusion-path length, upon the volume fraction available to the solvent, and upon the hydrodynamic interactions between the solute and the solid matrix. Factors affecting $D_{i,\text{eff}}$ are considered in Chapter 8.

Example 6.4 Consider steady-state transport across a thin membrane of thickness L (see Figure 6.7). At $x = 0$, $C_m = \Phi C_0$, and at $x = L$, $C_m = \Phi C_L$, where Φ is the partition coefficient. Develop an expression for the flux.

Solution If convection and chemical reactions are unimportant, then Equation (6.7.4a) reduces to

$$0 = D_{i,\text{eff}}\frac{d^2 C_m}{dx^2}, \tag{6.7.5}$$

where C_m is the solute concentration in the membrane. Boundary conditions are as follows:

$$x = 0 \quad C_m = \Phi C_0 \tag{6.7.6a}$$

$$x = L \quad C_m = \Phi C_L. \tag{6.7.6b}$$

The solution of Equation (6.7.5) is

$$C_m = Ax + B \tag{6.7.7}$$

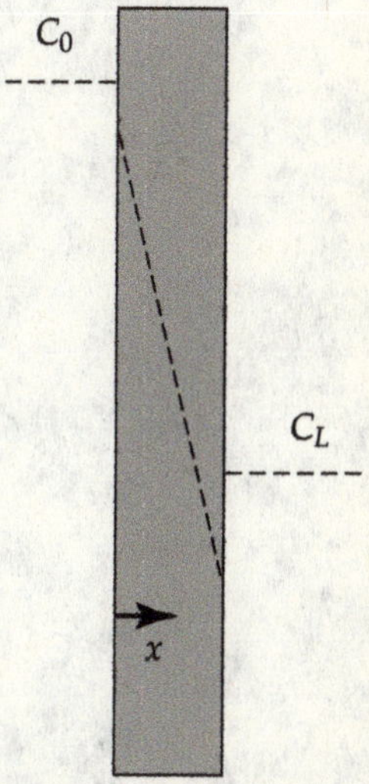

Membrane of thickness L

FIGURE 6.7 Schematic of steady diffusion across a membrane of thickness L that separates two well-mixed solutions. For this situation $\Phi < 1$.

Application of the boundary conditions yields

$$x = 0 \quad \Phi C_0 = B, \tag{6.7.8a}$$

$$x = L \quad \Phi C_L = AL + \Phi C_0, \tag{6.7.8b}$$

or

$$A = -\frac{\Phi(C_0 - C_L)}{L}. \tag{6.7.8c}$$

The resulting concentration is

$$C_m = \Phi C_0 - \Phi(C_0 - C_L)\frac{x}{L}. \tag{6.7.9}$$

From the definition of the diffusive flux, the flux of solute across the membrane is

$$N_{ix|x=0} = J_{ix} = -D_{i,\text{eff}}\frac{dC_m}{dx} = \frac{D_{i,\text{eff}}\Phi}{L}(C_0 - C_L). \tag{6.7.10}$$

For one-dimensional diffusion in a rectangular coordinate system, the flux is a constant and is independent of position. If $C_0 > C_L$, the solute flux is in the positive x direction; if $C_0 < C_L$, the solute flux is in the negative x direction.

One important result arising from Equation (6.7.10) is that membrane transport depends upon the product of the diffusion coefficient and the partition coefficient. When $\Phi < 1$, the surface concentrations in the membrane are less than the concentrations in the fluid, and the flux is reduced (see Figure 6.7). The product $D_{i,\text{eff}}\Phi/L$ is often referred to as the *permeability*. The permeability depends upon the membrane thickness and is not an intrinsic property of the membrane. However, it has practical value since it can be measured. We discuss this further in Section 6.8.4.

Homogeneous phospholipid membranes serve as models for biological membranes, and solute transport has been studied extensively in these membranes. Biological membranes consist of a number of phospholipids, as well as cholesterol and proteins. Gases and small organic solvents are soluble in purified phospholipid membranes and diffuse readily across them. For a large range of solutes, the permeability is proportional to the partition coefficient (Figure 6.8). Deviations from the linear relation are due to differences in the diffusion coefficients.

Example 6.5 Diffusion Through a Two-Phase medium

Many biological tissues have layers with different extracellular matrix components and different orientations of these components. As a result, diffusion coefficients vary from region to region. (Even within a region, the diffusion coefficient depends upon the direction of transport, a situation known as *anisotropy*.) Consider the steady-state, one-dimensional diffusion of a protein across a tissue that consists of

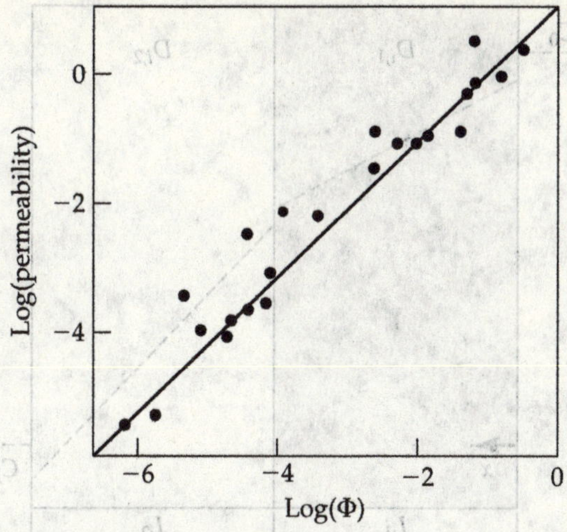

FIGURE 6.8 Relationship between the permeability coefficient and the partition coefficient for solute transport across phospholipid membranes. (Adapted from Ref. [22], with permission from Springer Science + Business Media.)

an acellular phase and a cellular phase (see Figure 6.9). There are no chemical reactions. An example of such a tissue is an elastic artery that contains an elastic lamina of thickness L_1 and a layer of smooth muscle cells of thickness L_2. (Actually, an artery consists of repeating layers of elastin and smooth muscle cells, but for the present discussion, consider a single layer of each.) The protein diffusion coefficients of the two layers are $D_{i,1}$ and $D_{i,2}$, respectively. The concentration of the protein at the surface (at $x = 0$) is C_0, and the concentration at $x = L_1 + L_2 = L$ is C_L. Determine the concentration as a function of position x and the flux across the tissue.

Solution For one-dimensional steady-state diffusion in rectangular coordinates with no chemical reaction, Equation (6.7.5) reduces to

$$\frac{d^2C_i}{dx^2} = 0. \tag{6.7.11}$$

This equation is valid for both the cellular and acellular phases. Thus, we really have two problems with two solutions:

$$C_1 = A_1 + B_1x, \tag{6.7.12a}$$

and

$$C_2 = A_2 + B_2x. \tag{6.7.12b}$$

There are four constants of integration—A_1, A_2, B_1, and B_2— and four boundary conditions. Two of the boundary conditions are given in the statement of the problem:

$$x = 0 \quad C_1 = \Phi_1C_0, \tag{6.7.13a}$$

$$x = L \quad C_2 = \Phi_2C_L. \tag{6.7.13b}$$

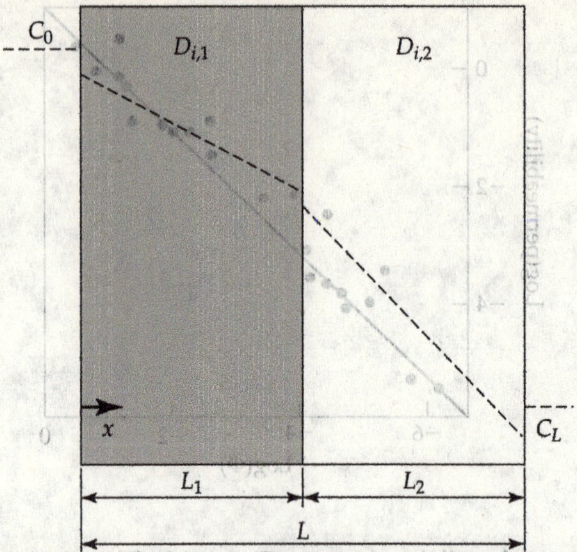

FIGURE 6.9 Diffusion through a two-layer laminate. Each layer is described by a separate diffusion coefficient.

The other two conditions are at $x = L_1$ and are

$$C_1/\Phi_1 = C_1/\Phi_2 \quad \text{and} \quad N_{1x} = N_{2x}. \tag{6.7.13c,d}$$

Applying the condition at $x = 0$ yields $A_1 = \Phi_1 C_0$. From the second boundary condition, we get $A_2 = \Phi_2 C_L - B_2 L$. Next, we apply Equations (6.7.13c) and (6.7.13d) to yield the following two equations with two unknowns:

$$C_0 + B_1 L_1/\Phi_1 = C_L - B_2 L_2/\Phi_2, \tag{6.7.14a}$$

$$D_{i,1} B_1 = D_{i,2} B_2. \tag{6.7.14b}$$

Solving for the unknowns yields expressions for the concentration profiles and the flux:

$$0 < x < L_1 \quad C_1 = \Phi_1 C_0 - \frac{D_{i,2}\Phi_1\Phi_2(C_0 - C_L)x}{D_{i,2}\Phi_2 L_1 + D_{i,1}\Phi_1 L_2}, \tag{6.7.15a}$$

$$L_1 < x < L_2 \quad C_2 = \Phi_2 C_L - \frac{D_{i,1}\Phi_1\Phi_2(C_0 - C_L)(x - L)}{D_{i,2}\Phi_2 L_1 + D_{i,1}\Phi_1 L_2}, \tag{6.7.15b}$$

$$N_{1x} = N_{2x} = \frac{D_{i,1} D_{i,2}\Phi_1\Phi_2(C_0 - C_L)}{D_{i,2}\Phi_2 L_1 + D_{i,1}\Phi_1 L_2}. \tag{6.7.15c}$$

As was the case in the previous example, the flux is independent of location in the region between $x = 0$ and $x = L$. For a single-phase medium, $D_{i,1} = D_{i,2}$. Equation (6.7.15c) reduces to Equation (6.7.10). The flux can be rewritten as

$$N_{1x} = N_{2x} = \frac{\Phi_{\text{eff}} D_{\text{eff}}}{L}(C_0 - C_L), \tag{6.7.16a}$$

from which it follows that

$$\frac{L}{\Phi_{\text{eff}}D_{\text{eff}}} = \frac{L_1}{\Phi_1 D_{i,1}} + \frac{L_2}{\Phi_2 D_{i,2}}. \tag{6.7.16b}$$

The terms $L_j/\Phi_j D_{i,j}$ represent diffusive resistances. For diffusion through two media in series, diffusion resistances are additive, like electrical resistances acting in series. Generalizing to n layers in series yields

$$\frac{L}{\Phi_{\text{eff}}D_{\text{eff}}} = \frac{L_1}{\Phi_1 D_{i,1}} + \frac{L_2}{\Phi_2 D_{i,2}} + \cdots + \frac{L_n}{\Phi_n D_{i,n}} = \sum_{j=1}^{N} \frac{L_j}{\Phi_j D_{i,j}}. \tag{6.7.17}$$

Diffusion in Gases. Diffusion in gases is more complex than diffusion in liquids because the dilute-solution assumption cannot be invoked. In general, a flux of one component of a mixture induces a flux of another component of the mixture, even in the absence of bulk motion. As a result, the flux must be specified in terms of Equations (6.2.9) and (6.4.7) as follows:

$$\mathbf{N}_1 = -CD_{ij}\nabla x_1 + x_1(\mathbf{N}_1 + \mathbf{N}_2). \tag{6.7.18}$$

The relationship between the fluxes can be deduced from the physics of the problem. To see this, consider a situation such as that depicted in Figure 6.10. In the figure, a narrow capillary tube of length L connects two large gas reservoirs. The concentrations in each reservoir are kept constant so that, at steady state, $C_A = C_{A_0}$ and $C_B = 0$ at $x = 0$ and $C_A = 0$, and $C_B = C_{B_0}$ at $x = L$. A constant pressure is maintained throughout, and there are no chemical reactions. The steady-state material balance for each component is

$$\frac{dN_{A_x}}{dx} = 0 \qquad \frac{dN_{B_x}}{dx} = 0. \tag{6.7.19a,b}$$

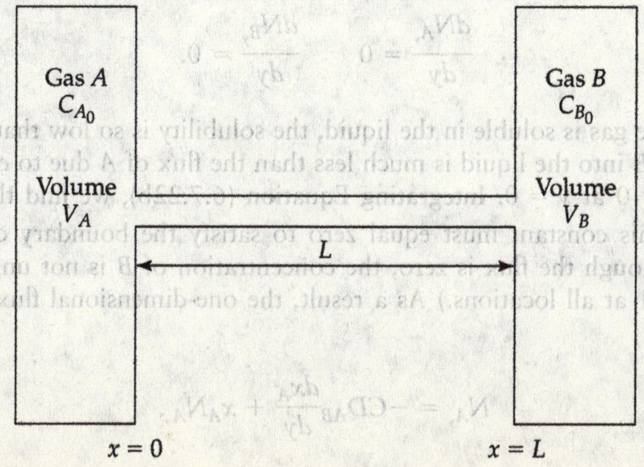

FIGURE 6.10 Diffusion of two gases. The concentrations in each reservoir are kept constant such that at steady state, $C_A = C_{A_0}$ and $C_B = 0$ at $x = 0$ and $C_A = 0$ and $C_B = C_{B_0}$ at $x = L$.

For one-dimensional diffusion, Equation (6.7.18) reduces to the following relationship for components A and B:

$$N_{A_x} = -CD_{AB}\frac{dx_A}{dx} + x_A(N_{A_x} + N_{zB_x}), \tag{6.7.20a}$$

$$N_{B_x} = -CD_{AB}\frac{dx_B}{dx} + x_B(N_{A_x} + N_{B_x}). \tag{6.7.20b}$$

(In Equations (6.7.20a) and (6.7.20b), we have assumed that $D_{AB} = D_{BA}$.)

Because pressure is uniform throughout, the movement of 1 mole of component A in one direction is balanced by that of 1 mole of B in the opposite direction ($N_{B_x} = -N_{A_x}$), and Equations (6.7.20a) and (6.7.20b) reduce to the single equation

$$N_{A_x} = -CD_{AB}\frac{dx_A}{dx} = -N_{B_x}. \tag{6.7.21}$$

Assuming ideal-gas behavior, we have $C_A = P_A/RT$, where P_A is the partial pressure of A. Because the total pressure is constant everywhere, C is constant. The resulting problem is analogous to diffusion through a membrane, except that $\Phi_i = 1$. This special case of gas diffusion is known as *equimolar counterdiffusion.*

A more complex situation arises when we consider the steady-state evaporation of a liquid of component A into a stagnant layer of gas B. The assumption of a stagnant layer means that there is no bulk flow of the gas.

Example 6.6 Consider the steady-state evaporation from a pure liquid A into a stagnant column of length L (see Figure 6.11). At $y = L$, the column is connected to a larger reservoir of gas B at a concentration C_{B_0}. Gas B is insoluble in liquid A, and the concentration of A at $y = L$ is zero. The pressure is uniform. Determine the steady-state distribution of A in the column.

Solution Material balances yield

$$\frac{dN_{A_y}}{dy} = 0 \qquad \frac{dN_{B_y}}{dy} = 0. \tag{6.7.22a,b}$$

Although the gas is soluble in the liquid, the solubility is so low that the flux of component B into the liquid is much less than the flux of A due to evaporation. Thus, $N_{B_y} \approx 0$ at $y = 0$. Integrating Equation (6.7.22b), we find that N_{B_y} is a constant. This constant must equal zero to satisfy the boundary condition at $y = 0$. (Although the flux is zero, the concentration of B is not uniform, since $x_A + x_B = 1$ at all locations.) As a result, the one-dimensional flux expression becomes

$$N_{A_y} = -CD_{AB}\frac{dx_A}{dy} + x_A N_{A_y}, \tag{6.7.23}$$

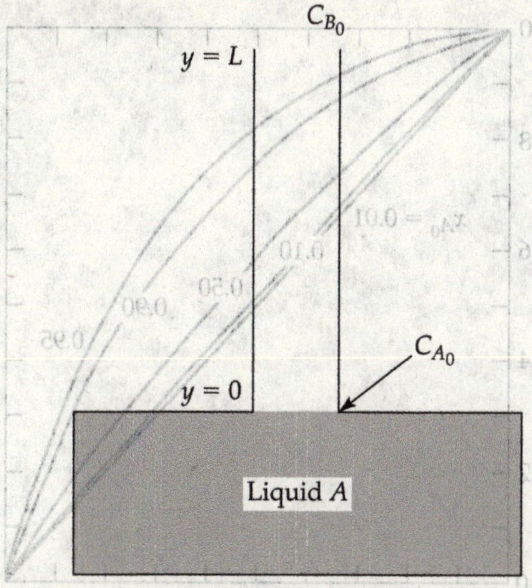

FIGURE 6.11 Evaporation of liquid A and diffusion through a stagnant layer of gas B.

or, after solving for N_{A_y},

$$N_{A_y} = -\frac{CD_{AB}}{1 - x_A}\frac{dx_A}{dy}. \tag{6.7.24}$$

Substituting Equation (6.7.24) into Equation (6.7.22a) yields

$$\frac{d}{dy}\left(\frac{1}{1 - x_A}\frac{dx_A}{dy}\right) = 0. \tag{6.7.25}$$

The boundary conditions are $x_A = x_{A_0}$ at $y = 0$ and $x_A = 0$ at $y = L$. Integrating Equation (6.7.25) twice yields

$$\ln(1 - x_A) = -C_1 y + C_2. \tag{6.7.26}$$

From the boundary condition at $y = 0$, $C_2 = \ln(1 - x_A)$. Applying the boundary condition at $y = L$, we find that $C_1 = \ln(1 - x_A)/L$. The result is

$$\ln(1 - x_A) = \ln(1 - x_{A_0}) - (y/L)\ln(1 - x_{A_0}), \tag{6.7.27a}$$

or

$$x_A = 1 - (1 - x_{A_0})^{1-y/L}. \tag{6.7.27b}$$

For large values of x_{A_0}, significant deviations from linearity are observed (Figure 6.12). For liquids with low vapor pressures, x_{A_0} is small, and the concentration profile is linear with distance. This linear behavior can be recovered from Equation (6.7.27b) by expanding the second term on the right-hand side when $x_{A_0} \ll 1$. We obtain

$$(1 - x_{A_0})^{1-y/L} \approx 1 - (1 - y/L)x_{A_0}, \tag{6.7.28a}$$

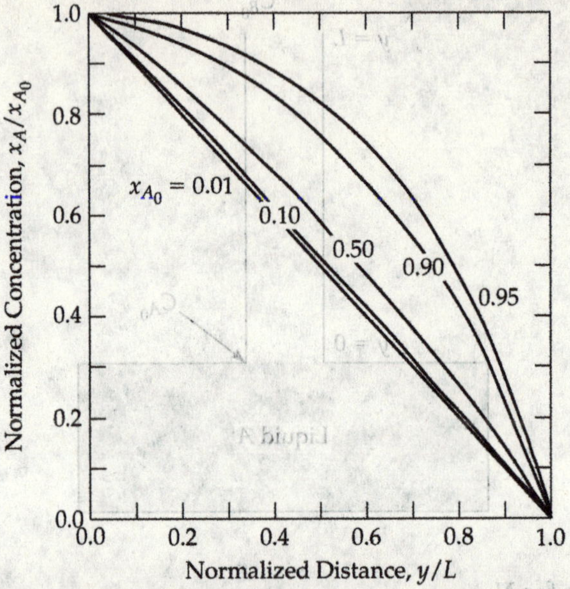

FIGURE 6.12 Concentration profiles for liquid evaporation and diffusion through a stagnant layer of gas as described by Equation (6.7.27b).

and

$$x_A \approx x_{A_0}(1 - y/L), \tag{6.7.28b}$$

which is the linear profile predicted for the dilute-solution limit.

The oxygen mole fraction in the lung is 0.21, which is close to the value in air. Thus, an analysis of oxygen diffusion in the lung must use Equation (6.7.23) for the flux. Anesthetics and other gases (e.g., NO, CO, CO_2, and Ar) are sometimes present in the lung, usually in very low concentrations—often much less than 1%. The diffusion of these gases can be analyzed by using the assumption of dilute solutions.

6.7.2 Radial Diffusion in Cylindrical Coordinates

Many tissues and biomedical devices have curved surfaces. Examples include blood vessels, the tubule system of the kidney, and hollow-fiber membranes. Diffusion through a curved region differs from diffusion through a rectangular region. For example, if transport is in the radial direction only, then the surface area through which transport occurs changes with the radial distance. This changing surface area affects the flux and concentration distribution. Instead of rectangular coordinates, cylindrical coordinates should be used. Consider the general case of one-dimensional radial diffusion, as shown in Figure 6.13. Choosing a thin annular shell from r to $r + \Delta r$ and of length L and examining the flux of material in the radial and axial directions yields the following expression:

$$\frac{\partial C_i}{\partial t} 2\pi r L \Delta r = (N_{ir}|_r(r) - N_{ir}|_{(r+\Delta r)}(r + \Delta r))2\pi L + R_i 2\pi r L \Delta r. \tag{6.7.29}$$

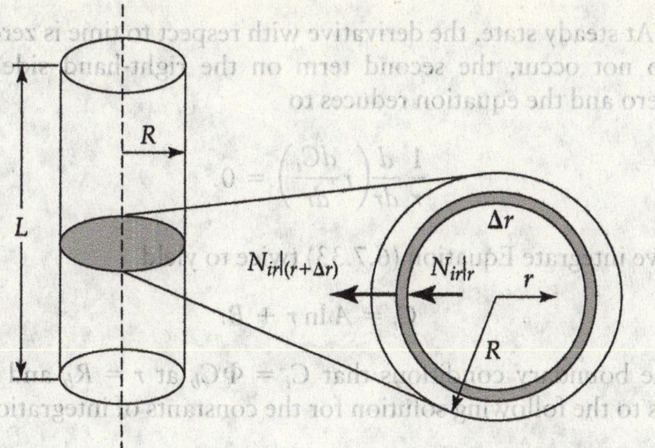

FIGURE 6.13 Radial diffusion through a cylindrical shell.

Dividing both sides of Equation (6.7.29) by the volume element $2\pi Lr\Delta r$ results in

$$\frac{\partial C_i}{\partial t} = \frac{(N_{ir}|_r r - N_{ir}|_{(r+\Delta r)}(r + \Delta r))}{r\Delta r} + R_i. \tag{6.7.30}$$

Taking the limit as $\Delta r \rightarrow 0$ and using the definition of the derivative results in the following differential equation:

$$\frac{\partial C_i}{\partial t} = -\frac{1}{r}\frac{\partial (rN_{ir})}{\partial r} + R_i. \tag{6.7.31}$$

Substituting Fick's law for diffusion without convection, $N_{ir} = -D_{ij}\partial C_i/\partial r$, into Equation (6.7.31) gives

$$\frac{\partial C_i}{\partial t} = \frac{D_{ij}}{r}\frac{\partial}{\partial r}\left(r\frac{\partial C_i}{\partial r}\right) + R_i. \tag{6.7.32}$$

Example 6.7 Steady-State Radial Diffusion Through a Blood Vessel

Balloon angioplasty is a procedure that is used to open blocked coronary arteries. A balloon catheter is inserted into the femoral artery and guided to the constricted artery, where it is inflated. In some cases, the vessel closes again due to the formation of a clot or the excessive growth of smooth muscle cells, a condition called *restenosis*. A drug that treats restenosis after balloon angioplasty is infused into the circulation long enough for the concentration of the drug in the blood to be constant at C_b. Measurements in coronary arteries have found that the drug concentration in the vessel is C_o in the outer part of the artery (known as the *adventitia*). If the inner radius of the artery is R_b and the outer radius of the vessel is R_o, determine the concentration profile and flux of the drug into the artery at R_b. Neglect chemical reactions.

Solution At steady state, the derivative with respect to time is zero. If chemical reactions do not occur, the second term on the right-hand side of Equation (6.7.32) is zero and the equation reduces to

$$\frac{1}{r}\frac{d}{dr}\left(r\frac{dC_i}{dr}\right) = 0. \tag{6.7.33}$$

To find C_i, we integrate Equation (6.7.33) twice to yield

$$C_i = A \ln r + B. \tag{6.7.34}$$

Applying the boundary conditions that $C_i = \Phi C_b$ at $r = R_b$ and $C_i = \Phi C_o$ at $r = R_o$ leads to the following solution for the constants of integration:

$$A = \frac{\Phi(C_b - C_o)}{\ln(R_b/R_o)} \qquad B = \Phi C_b - \frac{\ln R_b \Phi(C_b - C_o)}{\ln(R_b/R_o)}. \tag{6.7.35a,b}$$

The concentration is then

$$C = \Phi C_b - \frac{\ln(R_b/r)\Phi(C_b - C_o)}{\ln(R_b/R_o)}, \tag{6.7.36}$$

and the flux into the artery is

$$N_i = -D_{ij}\frac{dC_i}{dr} = -\frac{D_{ij}\Phi(C_b - C_o)}{r\ln(R_b/R_o)}. \tag{6.7.37}$$

Note that the absolute value of the magnitude of the flux decreases as r increases. The reason for this change with respect to position is that the cross-sectional area through which transport occurs increases with increasing radial distance. Because the same amount of solute must be transported through each cross section, increasing the cross-sectional area leads to a reduced flux.

For thin vessels ($R_b/R_o \approx 1$), the curvature can be neglected, and the flux approaches that for transport across a membrane. To see this, note the following expansion of the logarithm for a small parameter ε:

$$\ln(1 + \varepsilon) \approx \varepsilon - 0.5\varepsilon^2 + \cdots. \tag{6.7.38}$$

Rewriting R_b/R_o as $1 + R_b/R_o - 1$, where $R_b/R_o - 1 = \varepsilon$, yields

$$\ln\left(\frac{R_b}{R_o}\right) \approx \frac{R_b}{R_o} - 1 - 0.5\left(\frac{R_b}{R_o} - 1\right)^2. \tag{6.7.39}$$

Substituting Equation (6.7.39) (without the quadratic term) into Equation (6.7.37) and approximating r as R_o gives

$$N_i \approx D_{ij}\frac{\Phi(C_b - C_o)}{R_o - R_b}, \tag{6.7.40}$$

where $R_o - R_b$ is the thickness of the vessel. The relative error incurred in using the linearized form of Equation (6.7.37) is listed in Table 6.6. For a 10% increase in radius between the inner and outer walls, the error due to linearization is certainly acceptable.

TABLE 6.6

Relative Percent Error Due to the Linearization of Flux

R_{Ob}/R_b	Percent error
1.01	0.50
1.05	2.48
1.10	4.92
1.20	9.70
1.50	23.3

6.7.3 Radial Diffusion in Spherical Coordinates

In some cases, diffusion occurs through spherical objects; white blood cells, beads containing immobilized proteins, and spherical aggregates of cultured cells are examples. A schematic of diffusion in a spherical volume is shown in Figure 6.14. To perform a material balance, we assume that concentration varies only in the radial direction and that concentrations are independent of the angle of orientation (i.e., $C = C(r)$ only). For an analysis of radial diffusion, consider a thin spherical shell with an inner surface area of $4\pi r^2$ and an outer surface area of $4\pi(r + \Delta r^2)$. The volume of this shell is $4\pi r^2 \Delta r$. A material balance on the solute entering, leaving, or accumulating within this volume is

$$\frac{\partial C_i}{\partial t} 4\pi r^2 \Delta r = (N_{ir}|_r r^2 - N_{ir}|_{(r+\Delta r)}(r + \Delta r)^2)4\pi + R_i 4\pi r^2 \Delta r. \tag{6.7.41}$$

Dividing the terms in the material balance by $4\pi r^2 \Delta r$ yields

$$\frac{\partial C_i}{\partial t} = \frac{(N_{ir}|_r r^2 - N_{ir}|_{(r+\Delta r)}(r + \Delta r)^2)}{r^2 \Delta r} + R_i. \tag{6.7.42}$$

Taking the limit as $\Delta r \rightarrow 0$ and dividing by r^2 results in

$$\frac{\partial C_i}{\partial t} = \frac{1}{r^2}\frac{\partial}{\partial r}(N_{ir} r^2) + R_i. \tag{6.7.43}$$

Substituting Fick's law for diffusion without convection, $N_{ir} = -D_{ij}(\partial C_i/\partial r)$, into Equation (6.7.43) gives

$$\frac{\partial C_i}{\partial t} = \frac{D_{ij}}{r^2}\frac{d}{dr}\left(r^2 \frac{\partial C_i}{\partial r}\right) + R_i. \tag{6.7.44}$$

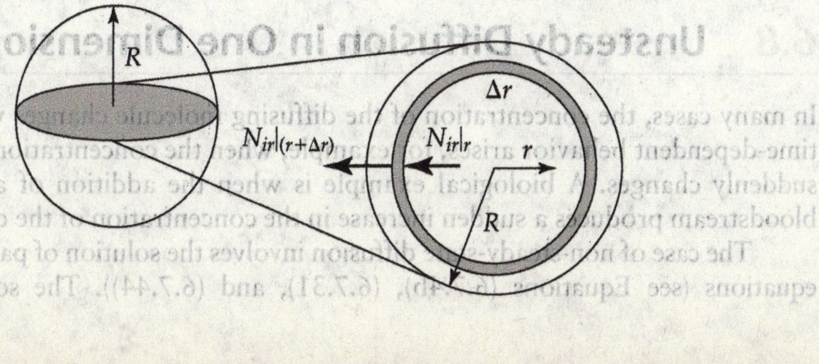

FIGURE 6.14 Diffusion through a spherical shell.

Example 6.8 Dissolution and Release of a Drug

A porous spherical polymer of radius R contains a drug to treat brain tumors. Experiments were performed *in vitro* to characterize the release of the drug. At the surface of the polymer ($r = R$), the drug concentration is ΦC_0. Far from the surface ($r \rightarrow \infty$), the drug concentration drops to zero. Determine the steady-state release rate from the polymer.

Solution At steady state with no chemical reactions, Equation (6.7.43) reduces to

$$0 = \frac{1}{r^2}\frac{d}{dr}\left(r^2\frac{dC_i}{dr}\right). \tag{6.7.45}$$

Integrating twice results in the following expression:

$$C_i = B - \frac{A}{r}. \tag{6.7.46}$$

From the boundary condition at $r \rightarrow \infty$, $B = 0$. From the boundary condition at $r = R$, $A = -RC_0$. The resulting concentration profile and flux are, respectively,

$$C_i = \frac{\Phi C_0 R}{r} \tag{6.7.47}$$

and

$$N_{ir} = -D_{ij}\frac{dC_i}{dr} = D_{ij}\frac{\Phi C_0 R}{r^2}. \tag{6.7.48}$$

As the distance from the surface increases, the flux decreases, due to the increase in the cross-sectional area through which transport occurs. The steady-state release rate (in moles per second) at the polymer surface is simply the product of the flux and the surface area:

$$\text{Release Rate} = N_{ir}|_{r=R}4\pi R^2 = 4\pi D_{ij}\Phi C_0 R. \tag{6.7.49}$$

Note that this analysis is approximate because the drug concentration at the surface does not remain constant but decreases with time as the drug leaves the polymer.

6.8 Unsteady Diffusion in One Dimension

In many cases, the concentration of the diffusing molecule changes with time. Such time-dependent behavior arises, for example, when the concentration at a boundary suddenly changes. A biological example is when the addition of a drug into the bloodstream produces a sudden increase in the concentration of the drug in tissues.

The case of non-steady-state diffusion involves the solution of partial differential equations (see Equations (6.7.4b), (6.7.31), and (6.7.44)). The solution of such equations is more complex than that of the simple ordinary differential equations encountered in steady-state diffusion. In this section, three approaches are presented for solving non-steady-state diffusion problems. Under some conditions, the diffusion problem can be simplified by assuming that diffusion occurs in a semi-infinite medium. A change of variables transforms the partial differential equation into an ordinary differential equation. This simplification is discussed in Section 6.8.1, and a criterion for its applicability is developed. Next, the solution of non-steady-state diffusion problems by separation of variables is discussed in Section 6.8.2. Finally, in some problems, the surface concentration changes continuously with time. For such problems, Laplace transform methods are used. A sample problem is considered in Section 6.8.3.

6.8.1 One-Dimensional Diffusion in a Semi-Infinite Medium

Consider a semi-infinite material containing a solute at a concentration C_0 (see Figure 6.15a). Solute freely diffuses in this material and no chemical reactions are occurring. At time zero, the surface at $x = 0$ is placed in contact with a well-mixed liquid reservoir containing solute maintained at a concentration C_1. We want to determine the flux of the solute at the surface, $x = 0$.

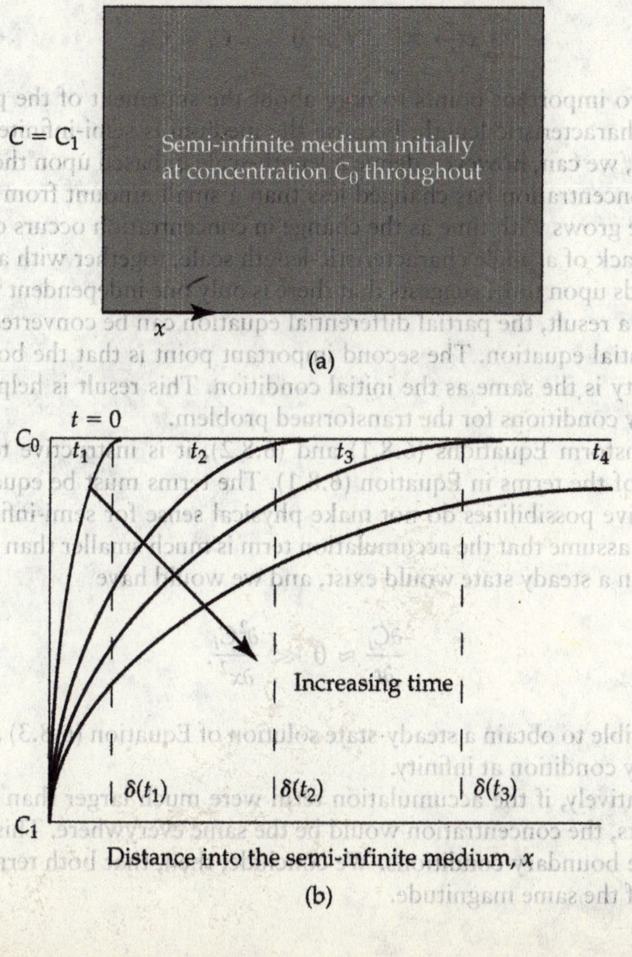

FIGURE 6.15 (a) A schematic of the physical domains for diffusion in a semi-infinite medium. (b) A sketch of the change in concentration with time.

At first glance, this problem may seem to be an impractical one, with little relevance to biological transport processes. Certainly, dimensions of cells, tissues, and organs are finite. If, however, the distance over which diffusion occurs is small relative to the characteristic dimension of the domain, then the system does behave as if it were semi-infinite. Only when changes in concentration occur throughout the entire domain is the finite nature of the domain apparent. Thus, the analysis is limited to times that are small relative to the time required for a solute to diffuse throughout the entire domain. This analysis provides an analytical expression for the flux. At the end of the analysis, more accurate statements about the applicability of this model are made. Because we are dealing with non-steady-state diffusion without convection or chemical reactions, Equation (6.7.4b) applies, with R_i equal to zero. Hence, we have

$$\frac{\partial C_i}{\partial t} = D_{ij}\frac{\partial^2 C_i}{\partial x^2}. \tag{6.8.1}$$

The boundary and initial conditions are as follows:

$$x \geq 0 \qquad t \leq 0 \qquad C_i = C_0, \tag{6.8.2a}$$

$$x = 0 \qquad t \geq 0 \qquad C_i = C_1, \tag{6.8.2b}$$

$$x \rightarrow \infty \qquad t \geq 0 \qquad C_i = C_0. \tag{6.8.2c}$$

There are two important points to note about the statement of the problem. First, there is no characteristic length, because the medium is semi-infinite. As shown in Figure 6.15b, we can, however, define a length-scale δ, based upon the distance over which the concentration has changed less than a small amount from C_0 (e.g., 1%). This distance grows with time as the change in concentration occurs over larger distances. The lack of a single characteristic-length scale, together with a distance scale δ that depends upon time, suggests that there is only one independent variable in this problem. As a result, the partial differential equation can be converted into an ordinary differential equation. The second important point is that the boundary condition at infinity is the same as the initial condition. This result is helpful in defining the boundary conditions for the transformed problem.

To transform Equations (6.8.1) and (6.8.2), it is instructive to estimate the magnitudes of the terms in Equation (6.8.1). The terms must be equal, because the two alternative possibilities do not make physical sense for semi-infinite media. To see this, first assume that the accumulation term is much smaller than the gradient in the flux. Then a steady state would exist, and we would have

$$\frac{\partial C_i}{\partial t} \approx 0 \ll \frac{\partial^2 C_i}{\partial x^2}. \tag{6.8.3}$$

It is not possible to obtain a steady-state solution of Equation (6.8.3) and still satisfy the boundary condition at infinity.

Alternatively, if the accumulation term were much larger than the concentration gradients, the concentration would be the same everywhere. This case also fails to satisfy the boundary conditions. We conclude, then, that both terms in Equation (6.8.1) are of the same magnitude.

To perform an order-of-magnitude analysis, approximate the left-hand and right-hand sides of Equation (6.8.1) with the use of characteristic scaling. For the concentration, the characteristic scale is the concentration difference $C_1 - C_0$. Length is scaled by δ and time is scaled by t. Next, we determine the relation between these two variables by examining the magnitude of each term in Equation (6.8.1). We have

$$\frac{\partial C_i}{\partial t} \sim \frac{C_1 - C_0}{t} \tag{6.8.4a}$$

and

$$D_{ij}\frac{\partial^2 C_i}{\partial x^2} \sim D_{ij}\frac{C_1 - C_0}{\delta^2}. \tag{6.8.4b}$$

Recall that the symbol ~ means "of the same order of magnitude" and that $C_1 - C_0$ is the characteristic concentration. Accordingly, the two expressions are not equalities, and we can approximate the magnitude of the derivatives of the terms on the right-hand side. Each term is replaced by its characteristic quantity. Because the characteristic length is unknown, we replace it with δ, which is the unspecified length scale in the x direction. Equating Equations (6.8.4a) and (6.8.4b) yields the following expression for the characteristic-length scale δ:

$$\delta \sim \sqrt{D_{ij}t}. \tag{6.8.5}$$

Equation (6.8.5) reveals two interesting features of diffusion in semi-infinite media. First, the characteristic-length scale grows with the square root of time, and second, there is only one independent variable in the problem. Consequently, we can transform Equation (6.8.1) from a partial differential equation into an ordinary differential equation.

To use these insights in the solution to our problem, we must scale our dimensional quantities (concentration, time, and space) by their characteristic values as follows:

$$\theta = \frac{C_i - C_0}{C_1 - C_0}, \tag{6.8.6a}$$

$$\eta = \frac{x}{2\delta} = \frac{x}{\sqrt{4D_{ij}t}}. \tag{6.8.6b}$$

Because the concentrations are scaled by $C_1 - C_0$, they vary from zero to unity. The x-coordinate is scaled by δ, the characteristic length. The 2 is introduced into the denominator of Equation (6.8.6b) to simplify the resulting ordinary differential equation. Since the characteristic length contains time, and we anticipate only one independent variable, it is not necessary to scale time separately. Note that the variables listed in Equations (6.8.6a) and (6.8.6b) are dimensionless.

Next, we write the derivatives in Equation (6.8.1) in terms of the scaled dimensionless variables. To do this, we apply the chain rule (Section A.1.E), first to the time derivative:

$$\frac{\partial C_i}{\partial t} = (C_1 - C_0)\frac{d\theta}{d\eta}\frac{\partial \eta}{\partial t}. \tag{6.8.7a}$$

Taking the derivative of η with respect to t yields

$$\frac{\partial \eta}{\partial t} = \frac{\partial}{\partial t}\left[\frac{x}{\sqrt{4D_{ij}t}}\right] = -\frac{x}{2t\sqrt{4D_{ij}t}} = -\frac{\eta}{2t}. \tag{6.8.7b}$$

Substituting Equation (6.8.7b) into Equation (6.8.7a) yields

$$\frac{\partial C_i}{\partial t} = -\frac{\eta(C_1 - C_0)}{2t}\frac{d\theta}{d\eta}. \tag{6.8.7c}$$

The spatial derivatives are transformed as follows:

$$\frac{\partial C_i}{\partial x} = (C_1 - C_0)\frac{d\theta}{d\eta}\frac{\partial \eta}{\partial x}, \tag{6.8.8a}$$

$$\frac{\partial^2 C_1}{\partial x^2} = (C_1 - C_0)\frac{\partial}{\partial x}\left[\frac{d\theta}{d\eta}\frac{\partial \eta}{\partial x}\right]. \tag{6.8.8b}$$

Equation (6.8.8b) is simplified by noting that $\partial\eta/\partial x = 1/\sqrt{4D_{ij}t}$ is independent of x. Thus,

$$\frac{\partial^2 C_i}{\partial x^2} = \frac{(C_1 - C_0)}{\sqrt{4D_{ij}t}}\frac{\partial}{\partial x}\left[\frac{d\theta}{d\eta}\right] = \frac{(C_1 - C_0)}{\sqrt{4D_{ij}t}}\frac{d}{d\eta}\left[\frac{\partial \theta}{\partial x}\right]. \tag{6.8.8c}$$

Applying the chain rule to the partial derivative of θ with respect to x yields

$$\frac{\partial \theta}{\partial x} = \frac{d\theta}{d\eta}\frac{\partial \eta}{\partial x} = \frac{1}{\sqrt{4D_{ij}t}}\frac{d\theta}{d\eta}. \tag{6.8.8d}$$

Substituting Equation (6.8.8d) into Equation (6.8.8c) and rearranging terms yields

$$\frac{\partial^2 C_i}{\partial x^2} = \frac{(C_1 - C_0)}{4D_{ij}t}\frac{d^2\theta}{d\eta^2}. \tag{6.8.8e}$$

Now substituting Equations (6.8.7c) and (6.8.8e) into Equation (6.8.1), we obtain the following scaled ordinary differential equation describing diffusion in a semi-infinite medium:

$$-\frac{\eta(C_1 - C_0)}{2t}\frac{d\theta}{d\eta} = \frac{(C_1 - C_0)}{4D_{ij}t}\frac{d^2\theta}{d\eta^2}. \tag{6.8.9a}$$

Canceling identical terms on the left- and right-hand sides of Equation (6.8.9a) and rearranging terms produces

$$-2\eta\frac{d\theta}{d\eta} = \frac{d^2\theta}{d\eta^2}. \tag{6.8.9b}$$

The boundary conditions are

$$\eta = 0 \quad \theta = 1 \ (x = 0), \tag{6.8.10a}$$

$$\eta \to \infty \quad \theta = 0 \ (x \to \infty \text{ and } t = 0). \tag{6.8.10b}$$

Thus, for diffusion in a semi-infinite medium, we have succeeded in transforming the unsteady diffusion partial differential equation (6.8.1) into a second-order ordinary differential equation. This was possible because the problem depended only upon a single independent variable η.

Equation (6.8.9b) is solved by first rewriting the derivatives as follows:

$$u = \frac{d\theta}{d\eta} \tag{6.8.11a}$$

and

$$\frac{du}{d\eta} = \frac{d^2\theta}{d\eta^2}. \tag{6.8.11b}$$

Equation (6.8.9b) then becomes

$$-2\eta u = \frac{du}{d\eta}. \tag{6.8.12a}$$

Integrating Equation (6.8.12a) yields

$$u = \frac{d\theta}{d\eta} = Ae^{-\eta^2}. \tag{6.8.12b}$$

Integrating Equation (6.8.12b) gives the following expression for θ in terms of two constants A and B:

$$\theta = A\int_0^{\eta} (-z^2)dz + B. \tag{6.8.13}$$

Here, z is a dummy variable of integration.

Applying the boundary condition at $\eta = 0$, Equation (6.8.10a) yields $B = 1$. Applying the boundary condition at $\eta = \infty$, Equation (6.8.10b) gives a value for the constant A:

$$A = \frac{-1}{\int_0^{\infty} \exp(-z^2)dz}. \tag{6.8.14}$$

The dimensionless concentration

$$\theta = 1 - \frac{\int_0^{\eta} \exp(-z^2)dz}{\int_0^{\infty} \exp(-z^2)dz}. \tag{6.8.15}$$

The value of the integral in the denominator of Equation (6.8.15) is $\sqrt{\pi}/2$. Equation (6.8.15) is rewritten as

$$\theta = 1 - \frac{2}{\sqrt{\pi}}\int_0^{\eta} \exp(-z^2)dz. \tag{6.8.16}$$

The integral in Equation (6.8.16) appears frequently in mathematical physics, and is known as the *error function*, erf(η),

$$\text{erf}(\eta) = \frac{2}{\sqrt{\pi}} \int_0^{\eta} (-z^2)dz. \tag{6.8.17}$$

The solution can then be written as

$$\theta = 1 - \text{erf}(\eta). \tag{6.8.18}$$

Equation (6.8.18) is shown in Figure 6.16. Some useful values of the error function are

$$\text{erf}(\eta = 0) = 0.0, \tag{6.8.19a}$$

$$\text{erf}(\eta = 3) \approx 1.00, \tag{6.8.19b}$$

and

$$\text{erfc}(\eta) = 1 - \text{erf}(\eta). \tag{6.8.19c}$$

where erfc(η) is known as the *complementary error function*.

In many calculations, the flux into a semi-infinite medium (i.e., $N_{i_x}(x = 0)$) is needed. This flux can be obtained from the dimensionless concentration as follows:

$$N_{ix}(x = 0) = -D_{ij}\frac{\partial C_i}{\partial x}\bigg|_{x=0} = -D_{ij}(C_1 - C_0)\frac{\partial \theta}{\partial x}\bigg|_{x=0} \tag{6.8.20a}$$

and

$$N_{ix}(x = 0) = -D_{ij}(C_1 - C_0)\frac{d\theta}{d\eta}\frac{\partial \eta}{\partial x}\bigg|_{x=0} = \frac{D_{ij}(C_1 - C_0)}{\sqrt{4D_{ij}t}}\frac{d\theta}{d\eta}\bigg|_{\eta=0}. \tag{6.8.20b}$$

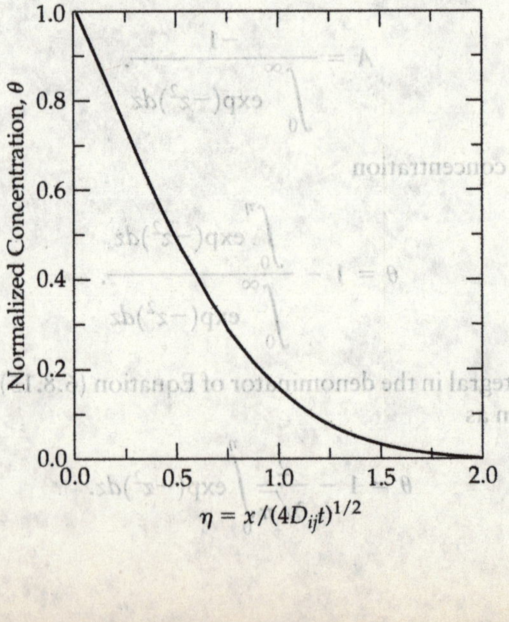

FIGURE 6.16 Dimensionless concentration (Equation (6.8.18)) as a function of η for unsteady diffusion in a semi-infinite medium.

The derivative in Equation (6.8.20b) can be calculated using Leibnitz's rule for differentiating an integral (Section A.1.H) to yield

$$\left.\frac{d\theta}{d\eta}\right|_{\eta=0} = \left[-\frac{2\exp(-\eta^2)}{\sqrt{\pi}}\right]_{\eta=0} = -\frac{2}{\sqrt{\pi}}. \tag{6.8.21}$$

The flux at $x = 0$ is

$$N_{ix}(x = 0) = \sqrt{\frac{D_{ij}}{\pi t}}(C_1 - C_0). \tag{6.8.22}$$

The flux is proportional to the reciprocal of the square root of time. This dependence upon time arises because the solute concentration at the surface decreases and solute within the material must diffuse a distance δ to reach the surface.

The major advantage of the semi-infinite solution is the simplicity of the flux expression, Equation (6.8.22). As is shown in the next section, the corresponding solution for a finite medium is a trigonometric series, the analysis of which requires programming on a computer. We can simplify the solution if we apply the semi-infinite medium approximation to a finite medium.

As an example, consider a long, thin rectangular tissue section of thickness L. Initially, the concentration is zero throughout. At $t = 0$, the surface at $x = 0$ is in equilibrium, with a concentration C_0, but the surface at $x = L$ is maintained at a concentration of zero. The assumption of a semi-infinite medium is not valid when the concentration at $x = L$ becomes greater than zero. An estimate of this time can be made by setting $\eta = 3$, for which the error function is approximately unity and the concentration is close to zero. The distance x equals L. Thus,

$$3 = \frac{L}{\sqrt{4D_{ij}t}}. \tag{6.8.23}$$

Rearranging terms yields

$$t = \frac{L^2}{36D_{ij}}. \tag{6.8.24}$$

The solution for diffusion in a semi-infinite medium is valid for times that are less than $L^2/36D_{ij}$.

Protein Adsorption to Biomaterials. The function of biomaterials is strongly dependent upon the adsorption of proteins upon their surfaces. The kinetics and amount adsorbed often depend upon the surface properties. In many experimental situations, the initial kinetics of adsorption are independent of surface properties and are *diffusion limited*, even for a solution such as blood plasma, which contains many different proteins. In diffusion-limited kinetics, the time required for a molecule to diffuse to the surface is much longer than the time required for the molecule to react at the surface. In effect, the molecules react as soon as they reach the surface. Thus, the concentration of the adsorbing molecule in solution is zero at the surface. Further, diffusion-limited kinetics are worth studying because, in the absence of mixing of the

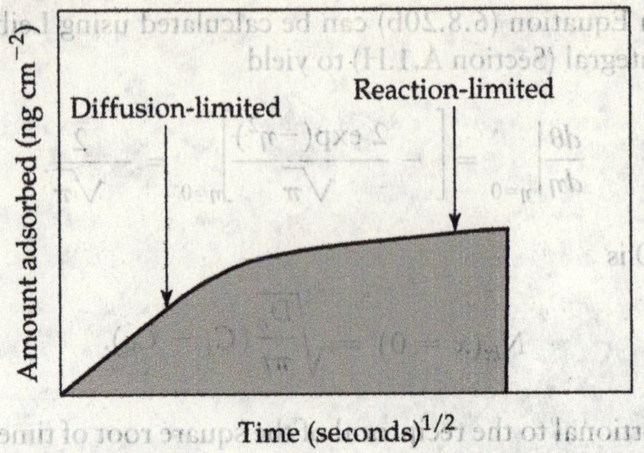

FIGURE 6.17 A schematic of the kinetics of protein adsorption in a static experiment. (Adapted from Ref. [23], with permission.)

fluid, the diffusion-limited case represents the upper limit for the molecule's reaction rate (see Figure 6.17). That is, the rate at which the reaction proceeds on the surface can be no greater than the rate of delivery of molecules to the surface.

When a material is exposed to a solution containing a number of proteins, diffusion-limited kinetics prevail during the first few minutes, as proteins adsorb to the surface. This length of time is often sufficient for proteins to completely cover the surface. Following the diffusion-limited period, weakly bound molecules desorb and are replaced by molecules that adhere more firmly in a process known as *competitive binding*. As a result, there is a rearrangement of the composition on the surface.

The rate of adsorption and the amount adsorbed can be determined for the case of diffusion-limited kinetics. The assumption that is most commonly used is that the amount adsorbed is such a small fraction of the total amount of protein in solution that the concentration of the solution far from the surface does not change. As a result, we can treat the fluid as a semi-infinite medium in which the concentration is depleted near the surface (see Figure 6.18).

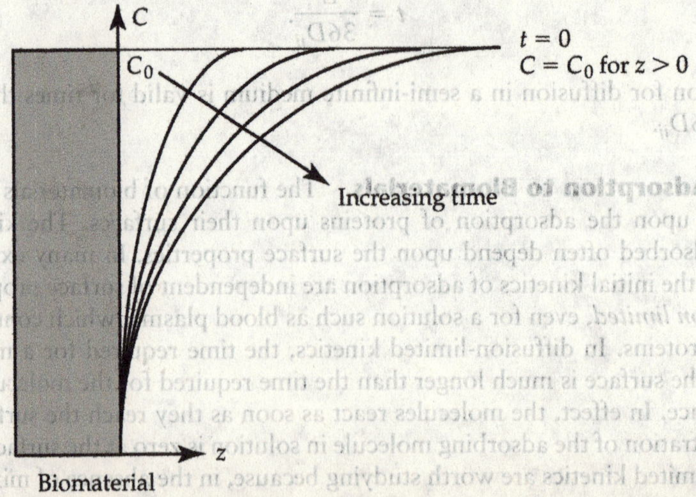

FIGURE 6.18 A schematic of the changing solution concentration near the surface of a biomaterial.

Example 6.9 Determine the diffusion-limited rate of adsorption and the surface concentration for protein adsorption to a surface.

Solution The problem is similar to the case of diffusion in a semi-infinite medium, except that the boundary condition at $z = 0$ is different. We have

$$z \geq 0 \quad t \leq 0 \quad C_i = C_0, \tag{6.8.25a}$$

$$z = 0 \quad t \geq 0 \quad C_i = 0, \tag{6.8.25b}$$

$$z \to \infty \quad t \geq 0 \quad C_i = C_0. \tag{6.8.25c}$$

Since $C_1 = 0$, the dimensionless concentration is defined as $\theta' = C_i/C_0 = 1 - \theta$. On the basis of the preceding analysis, the solution of Equation (6.8.1), subject to the boundary conditions (Equations (6.8.25a) through (6.8.25c)), is

$$\theta' = \frac{2}{\sqrt{\pi}} \int_0^{\eta} \exp(-z^2) dz. \tag{6.8.26}$$

The flux of protein at the surface $z = 0$ is

$$N_{iz}(z = 0) = \sqrt{\frac{D_{ij}}{\pi t}} C_0. \tag{6.8.27}$$

If the process is diffusion limited, then the rate of adsorption equals the flux at the surface:

$$\frac{dC_{ads}}{dt} = N_{iz}(z = 0). \tag{6.8.28}$$

The adsorbed concentration C_{ads} is in units of mass per unit area. Substituting Equation (6.8.27) into Equation (6.8.28) and integrating gives the surface concentration as a function of time for diffusion-limited adsorption:

$$C_{ads} = 2C_0 \sqrt{\frac{D_{ij} t}{\pi}}. \tag{6.8.29}$$

The amount adsorbed thus increases with the square root of the product of the diffusion coefficient and time.

Equation (6.8.29) has been used to identify the major proteins in blood that adsorb to surfaces. According to that equation, the larger $C_0\sqrt{D_{ij}}$ is, the more protein adsorbs to a surface. Values of $C_0\sqrt{D_{ij}}$ are tabulated in Table 6.7. The data indicate that diffusion-limited kinetics favor albumin adsorption.

The kinetics of protein adsorption are diffusion limited only during the first few minutes of adsorption. At later times, adsorption is dominated by the affinity of the particular molecule for the surface. For all concentrations, the first minute or two are dominated by diffusion-limited kinetics. At low plasma dilutions (0.25%), the amount of fibrinogen adsorbed increases monotonically, suggesting that the surface is

TABLE 6.7

Major Plasma Proteins Involved in Diffusion-Limited Adsorption

Protein	Concentration C_0, mg ml^{-1}	Molecular weight	$D_{ij} \times 10^7$ cm^2 s^{-1}	$C_0\sqrt{D_{ij}} \times 10^7$, M cm s$^{-1/2}$
Albumin	40	66,000	6.1	4.73
IgG	8–17	150,000	4.0	0.38–0.72
LDL	4	2,000,000	2.0	0.0089
HDL	3	170,000	4.6	0.12
α-macroglobulin	2.7	725,000	2.4	0.018
Fibrinogen	2–3	340,000	2.0	0.026–0.039
Transferrin	2.3	77,000	5.0	0.21
α-antitrypsin	2	54,000	5.2	0.27
Haptoglobins	2	100,000	4.7	0.14
C_3	1.6	180,000	4.5	0.060
IgA	1–4	150,000	4.0	0.042–0.17

Source: Adapted from Ref. [23], with permission.

incompletely covered and that a number of different molecules can adsorb. As the plasma concentration increases, however, proteins cover the surface completely. As a result, a weakly adsorbing protein such as fibrinogen is displaced by a protein with a higher affinity. This effect is more pronounced with less-dilute plasma concentrations.

6.8.2 One-Dimensional Unsteady Diffusion in a Finite Medium

Diffusion in a semi-infinite medium leads to an analytical solution that is valid for values of time that are less than or equal to those listed in Equation (6.8.24). However, we seek a result that describes diffusion in a finite medium and is valid until a steady state is reached. Three cases are considered in this section: unsteady diffusion in rectangular coordinates, in spherical coordinates, and from a point source.

Unsteady Diffusion in Rectangular Coordinates. In this case, a rectangular slab of thickness $2L$ contains a diffusing material at a concentration C_0. At time zero, the surfaces at $y = L$ and $y = -L$ are raised to a concentration C_1. We want to find the concentration within the slab as a function of time and position. Reaction and convection do not occur.

The conservation of mass for one-dimensional unsteady diffusion is

$$\frac{\partial C_i}{\partial t} = D_{ij}\frac{\partial^2 C_i}{\partial y^2}. \tag{6.8.30}$$

The boundary conditions are as follows:

$$-L \leq y \leq L \quad t \leq 0 \quad C_i = C_0, \tag{6.8.31a}$$

$$y = 0 \quad t \geq 0 \quad \frac{\partial C_i}{\partial y} = 0, \tag{6.8.31b}$$

$$y = \pm L \quad t \geq 0 \quad C_i = C_1. \tag{6.8.31c}$$

Note that the boundary condition at the center of the slab, $y = 0$, is a symmetry condition because both sides of the domain are identical. Thus, it is necessary to solve the problem over only one-half of the domain, $0 \leq y \leq L$. The solution to this problem can be generalized by casting the problem in a dimensionless form. The following dimensionless variables are used:

$$\eta = \frac{y}{L} \qquad \theta = \frac{C_i - C_0}{C_1 - C_0} \qquad \tau = \frac{tD_{ij}}{L^2}.$$

The dimensionless concentration θ is defined such that it varies from zero in the center to unity at the surface $y = L$. Recalling that we need to solve the problem over only half of the region, we can restate Equation (6.8.30) and the associated boundary conditions as

$$\frac{\partial \theta}{\partial \tau} = \frac{\partial^2 \theta}{\partial \eta^2}, \tag{6.8.32}$$

$$0 \leq \eta \leq 1 \quad \tau \leq 0 \quad \theta = 0, \tag{6.8.33a}$$

$$\eta = 0 \quad \tau \geq 0 \quad \frac{\partial \theta}{\partial \eta} = 0, \tag{6.8.33b}$$

and

$$\eta = 1 \quad \tau \geq 0 \quad \theta = 1. \tag{6.8.33c}$$

To apply separation of variables, both boundary conditions need to be homogeneous. They can be made so with the following transformation:

$$\theta'(\eta,\tau) = 1 - \theta(\eta,\tau). \tag{6.8.34}$$

Equations (6.8.32) and (6.8.33) now become

$$\frac{\partial \theta'}{\partial \tau} = \frac{\partial^2 \theta'}{\partial \eta^2}. \tag{6.8.35}$$

The boundary conditions are

$$0 \leq \eta \leq 1 \quad \tau \leq 0 \quad \theta' = 1, \tag{6.8.36a}$$

$$\eta = 0 \quad \tau \geq 0 \quad \frac{\partial \theta'}{\partial \eta} = 0, \tag{6.8.36b}$$

and

$$\eta = 1 \quad \tau \geq 0 \quad \theta' = 0. \tag{6.8.36c}$$

Equation (6.8.35) and the associated initial and boundary conditions can now be solved by the method of separation of variables (see Section A.2). Assume a solution of the form $\theta'(\eta, \tau) = X(\eta)T(\tau)$. Substituting this solution into Equation (6.8.35) and rearranging terms yields

$$\frac{1}{T}\frac{dT}{d\tau} = \frac{1}{X}\frac{d^2X}{d\eta^2}. \tag{6.8.37}$$

The left-hand side of Equation (6.8.37) is a function of τ only, and the right-hand side is a function of η only. This can be true if and only if each side equals a constant $\pm\lambda^2$. To ensure that a characteristic-value problem results, the sign of λ^2 is chosen such that

$$\frac{d^2X}{d\eta^2} = -\lambda^2 X \qquad \frac{dT}{d\tau} = -\lambda^2 T. \tag{6.8.38a,b}$$

The negative sign ensures that T decreases with time. If the sign were positive, the concentration would increase without limit. The solutions of these two equations are

$$\theta' = XT = (A\sin(\lambda\eta) + B\cos(\lambda\eta))\exp(-\lambda^2\tau). \tag{6.8.39}$$

Applying the boundary condition at zero yields

$$\left.\frac{\partial\theta'}{\partial\eta}\right|_{\eta=0} = 0 = \lambda(A\cos(\lambda\eta) - B\sin(\lambda\eta))\exp(-\lambda^2\tau)\Big|_{\eta=0}. \tag{6.8.40}$$

For $\eta = 0$, the sine term is zero, but the cosine term is unity. Therefore, A must equal zero to satisfy this boundary condition. Equation (6.8.40) reduces to

$$\theta' = B\cos(\lambda\eta)\exp(-\lambda^2\tau). \tag{6.8.41}$$

At $\eta = 1$, $\theta' = 0$. If B were to equal zero, then the result would be trivial: the concentration would not change. We know that this is not the case. Alternatively, the cosine term is zero when λ is $\pi/2$, $3\pi/2$, $5\pi/2$, and so on. This condition can be written as

$$\lambda = (n + 1/2)\pi \qquad n = 0,1,2,3,\ldots. \tag{6.8.42}$$

Substituting Equation (6.8.42) into the expression for θ (Equation (6.8.41)) and summing over all values of n yields

$$\theta' = \sum_{n=0}^{\infty} B_n \cos\left[(n + 1/2)\pi\eta\right]\exp\left[-(n + 1/2)^2\pi^2\tau\right]. \tag{6.8.43}$$

The terms B_n are evaluated from the initial condition and the orthogonality relations for the cosine function over the domain [0, 1]. At time $\tau = 0$,

$$1 = \sum_{n=0}^{\infty} B_n \cos[(n + 1/2)\pi\eta]. \tag{6.8.44}$$

To obtain the orthogonality condition, multiply both sides by $\cos[(m + 1/2)\pi\eta]$ and integrate from $\eta = 0$ to $\eta = 1$.

The result is

$$\int_0^1 \cos[(m + 1/2)\pi\eta]d\eta = \sum_{n=0}^{\infty} B_n \int_0^1 \cos[(m + 1/2)\pi\eta]\cos[(m + 1/2)\pi\eta]d\eta. \tag{6.8.45}$$

The right-hand side of Equation (6.8.45) is nonzero only when $m = n$. Performing the integrations leads to the following values for B_n:

$$B_n = \frac{2(-1)^n}{(n + 1/2)\pi}. \tag{6.8.46}$$

The final result for θ is

$$\theta = 1 - \theta' = 1 - 2\sum_{n=0}^{\infty} \frac{(-1)^n}{(n + 1/2)\pi} \cos\Big[(n + 1/2)\pi\eta\Big] \exp\Big[-(n + 1/2)^2\pi^2\tau\Big]. \tag{6.8.47}$$

The most convenient way to present this result is in graphical form. Shown in Figure 6.19 is the solution for rectangular coordinates. Note that the time for the concentration at $y = 0$ to reach 99% of the final concentration is approximately $2L^2/D_{ij}$, which justifies the use of this quantity as the characteristic diffusion time.

Unsteady Diffusion in Spherical Coordinates. Problems involving unsteady diffusion in spherical coordinates can be easily solved in a manner analogous to that used to solve problems involving unsteady diffusion in rectangular coordinates. Consider unsteady diffusion in a sphere of radius R. Initially, the concentration within the sphere is uniform at C_0. At time zero, the concentration at the surface of the sphere ($r = R$) is raised to C_1. In dimensional form, the non-steady-state diffusion equation with no chemical reaction is

$$\frac{\partial C_i}{\partial t} = \frac{D_{ij}}{r^2}\frac{\partial}{\partial r}\left(r^2 \frac{\partial C_i}{\partial r}\right). \tag{6.8.48}$$

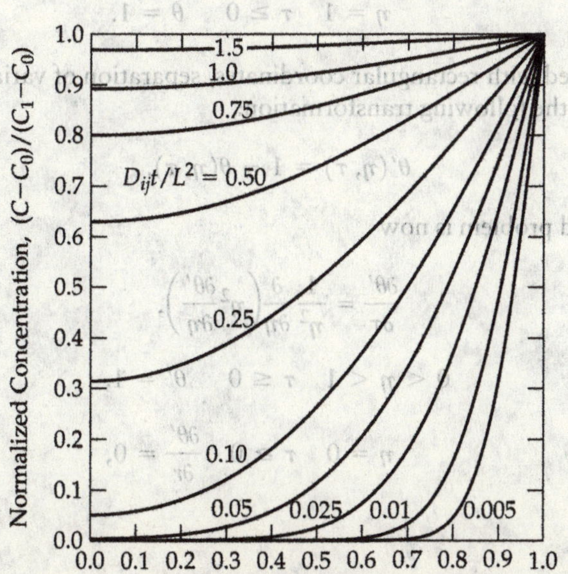

FIGURE 6.19 Unsteady diffusion in a rectangular slab of thickness $2L$ for various values of $D_{ij}t/L^2$.

The initial condition is

$$0 < r < R \quad t \le 0 \quad C = C_0. \tag{6.8.49a}$$

The boundary conditions are

$$r = 0 \quad t \ge 0 \quad \frac{\partial C_i}{\partial r} = 0 \tag{6.8.49b}$$

and

$$r = R \quad t \ge 0 \quad C = C_1. \tag{6.8.49c}$$

The following dimensionless variables are used:

$$\eta = \frac{r}{R} \qquad \theta = \frac{C_i - C_0}{C_1 - C_0} \qquad \tau = \frac{tD_{ij}}{R^2}.$$

Equations (6.8.48) and (6.8.49) can now be cast into dimensionless form:

$$\frac{\partial \theta}{\partial \tau} = \frac{1}{\eta^2}\frac{\partial}{\partial \eta}\left(\eta^2 \frac{\partial \theta}{\partial \eta}\right), \tag{6.8.50}$$

$$0 < \eta < 1 \quad \tau \le 0 \quad \theta = 0, \tag{6.8.51a}$$

$$\eta = 0 \quad \tau \ge 0 \quad \frac{\partial \theta}{\partial r} = 0, \tag{6.8.51b}$$

and

$$\eta = 1 \quad \tau \ge 0 \quad \theta = 1. \tag{6.8.51c}$$

As was performed with rectangular coordinates, separation of variables can be used if we first apply the following transformation:

$$\theta'(\eta, \tau) = 1 - \theta(\eta, \tau). \tag{6.8.52}$$

The transformed problem is now

$$\frac{\partial \theta'}{\partial \tau} = \frac{1}{\eta^2}\frac{\partial}{\partial \eta}\left(\eta^2 \frac{\partial \theta'}{\partial \eta}\right), \tag{6.8.53}$$

$$0 < \eta < 1 \quad \tau \le 0 \quad \theta' = 1, \tag{6.8.54a}$$

$$\eta = 0 \quad \tau \ge 0 \quad \frac{\partial \theta'}{\partial r} = 0, \tag{6.8.54b}$$

and

$$\eta = 1 \quad \tau \ge 0 \quad \theta' = 0. \tag{6.8.54c}$$

The solution of Equation (6.8.53) can proceed with separation of variables if we define the new variable f as

$$f = \theta'\eta. \tag{6.8.55}$$

Making this substitution, Equation (6.8.53) then becomes

$$\frac{\partial f}{\partial \tau} = \frac{\partial^2 f}{\partial \eta^2}. \tag{6.8.56}$$

The solution of Equation (6.8.56) is identical to Equation (6.8.35). Thus,

$$\theta' = \left(\frac{A}{\eta}\sin(\lambda\eta) + \frac{B}{\eta}\cos(\lambda\eta)\right)\exp(-\lambda^2\tau). \tag{6.8.57}$$

The boundary condition at $\eta = 0$ can now be restated as

$$\left.\frac{\partial \theta'}{\partial \eta}\right|_{\eta=0} = -\left(\frac{A(\sin(\lambda\eta) - \eta\lambda\cos(\lambda\eta))}{\eta^2} + \frac{B(\cos(\lambda\eta) - \eta\lambda\sin(\lambda\eta))}{\eta^2}\right)$$
$$\times \exp(-\lambda^2\tau) = 0. \tag{6.8.58}$$

Applying L'Hôpital's rule (Section A.1.B), the constant B must equal zero in order for the derivative to equal zero. (Note that this boundary condition is the same as $f(\eta = 0) = 0$.)

Application of the boundary condition at $\eta = 1$ yields

$$\theta' = A\sin(\lambda) = 0, \tag{6.8.59}$$

which is satisfied for

$$\lambda = n\pi \quad n = 1,2,3,\ldots \tag{6.8.60}$$

Hence, the solution for θ' becomes a series with coefficients A_n; that is,

$$\theta' = \sum_{n=1}^{\infty}\frac{A_n}{\eta}\sin(n\pi\eta)\exp(-n^2\pi^2\tau). \tag{6.8.61}$$

The constants A_n can be evaluated from the initial condition

$$\eta = \sum_{n=1}^{\infty}A_n\sin(n\pi\eta), \tag{6.8.62a}$$

and from the orthogonality relation for the sine:

$$\int_0^1 \eta\sin(m\pi\eta)d\eta = \sum_{n=1}^{\infty}A_n\int_0^1 \sin(n\pi\eta)\sin(m\pi\eta)d\eta. \tag{6.8.62b}$$

The right-hand side of Equation (6.8.62b) is nonzero when $m = n$. Then

$$A_n = \frac{-2(-1)^n}{n\pi}. \tag{6.8.62c}$$

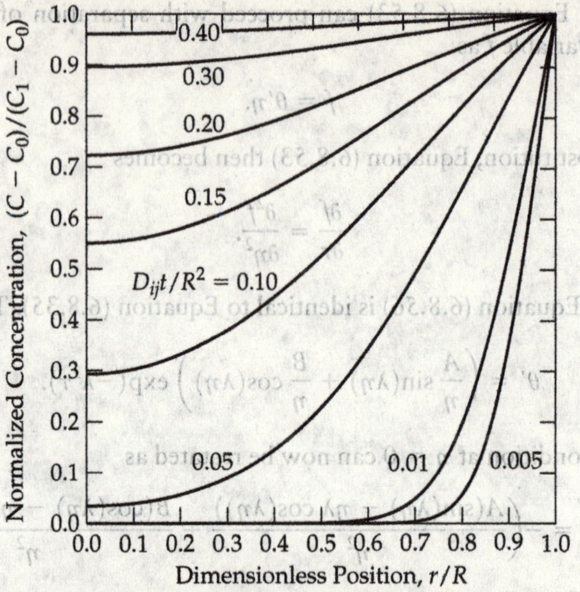

FIGURE 6.20 Unsteady diffusion in a sphere of radius *R*.

The resulting solution for the dimensionless concentration is

$$\theta = 1 + \sum_{n=1}^{\infty} \frac{2(-1)^n}{\eta(n\pi)} \sin(n\pi\eta) \exp\left(-n^2\pi^2\tau\right). \qquad (6.8.63)$$

Figure 6.20 contains plots of Equation (6.8.63) for various values of τ. For the same dimensionless time, the concentration at a given relative distance from the center is greater for the sphere than for the rectangular slab. (A similar result is also valid for the cylinder [24].) The reason for this difference is that the cross-sectional area changes with radial distance for a sphere (and a cylinder). For a sphere, the dimensionless time required for the concentration at $r = 0$ to reach 99% of the final concentration is 0.55, which is 3.6 times shorter than the corresponding dimensionless time for a rectangular slab.

Diffusion from a Point Source. An important class of diffusion problems is those in which a finite concentration is added at a given point or in which a constant flux of solute is released at a given point. Such problems provide insight into the behavior of diffusing molecules and can serve in an analysis of more complex problems. Although, in reality, the concentration or source must be applied over a finite region, conditions can be designed to approximate a point source for the sake of analysis.

First, consider diffusion away from a point source in one dimension. Equation (6.8.30) applies for rectangular coordinates. A finite mass of solute is added to the origin at time zero. Far from the source, the concentration is zero. Diffusion is symmetric around the origin, so the flux there is always zero. The boundary and initial conditions are

$$t = 0 \quad x > 0 \quad C_i = 0, \qquad (6.8.64a)$$

$$t > 0 \quad x \to \infty \quad C_i = 0. \qquad (6.8.64b)$$

At any time, the total amount of solute per unit area that is initially added at $x = 0$, M_0, is found by integrating over the entire domain:

$$t > 0 \quad M_0 = \int_{-\infty}^{\infty} C_i(x, t)dx = 2\int_0^{\infty} C_i(x, t)dx. \tag{6.8.64c}$$

The term on the far right-hand side of Equation (6.8.64c) arises because diffusion is symmetric around the origin.

Because the problem has no characteristic length, it cannot be solved by separation of variables. Although this problem appears similar to the case of diffusion in a semi-infinite medium, the initial condition prevents any application of the separation of variables. Instead, the problem is most easily solved by the method of *Laplace transforms* [25]. The Laplace transform of Equation (6.7.4b), subject to Equations (6.8.64a) through (6.8.64c), is,

$$s\overline{C} = D_{ij}\frac{d^2\overline{C}}{dx^2}. \tag{6.8.65}$$

This equation has the following solution:

$$\overline{C} = A\exp\left(-x\sqrt{s/D_{ij}}\right) + B\exp\left(x\sqrt{s/D_{ij}}\right). \tag{6.8.66}$$

From the boundary condition at infinity, B equals zero. The constant A is found by taking the Laplace transform of Equation (6.8.64c):

$$\frac{M_0}{s} = 2\int_0^{\infty} \overline{C}(x, t)dx. \tag{6.8.67}$$

Substituting Equation (6.8.66) into Equation (6.8.67) and solving for A results in the following expression for $\overline{C}(x, t)$:

$$\overline{C}(x, t) = \frac{M_0}{2\sqrt{D_{ij}s}}\exp(-x\sqrt{s/D_{ij}}). \tag{6.8.68}$$

Equation (6.8.68) can be inverted with the use of a table of Laplace transforms [25]. The result is

$$\overline{C}(x, t) = \frac{M_0}{2\sqrt{\pi D_{ij}t}}\exp\left(-\frac{x^2}{4D_{ij}t}\right). \tag{6.8.69}$$

This result is simply the normal distribution.

The one-dimensional solution can be generalized to radial diffusion in three dimensions. For this geometry, the problem can then be restated as

$$\frac{\partial C_i}{\partial t} = \frac{D_{ij}}{r^2}\frac{\partial}{\partial r}\left(r^2\frac{\partial C_i}{\partial r}\right), \tag{6.8.70}$$

$$t = 0 \quad r > 0 \quad C_i = 0, \tag{6.8.71a}$$

$$t > 0 \quad r \to \infty \quad C_i = 0, \tag{6.8.71b}$$

and

$$t > 0 \quad M = \int_V C_i(x,t)dV = 4\pi \int_0^\infty C_i(x,t)r^2 dr, \tag{6.8.71c}$$

where M is the total amount (in moles, molecules, or mass) of solute released at $r = 0$ at time zero. Using the transformation $f = Cr$, we can recast Equation (6.8.70) as

$$\frac{\partial f}{\partial t} = D_{ij}\frac{\partial^2 f}{\partial r^2}. \tag{6.8.72}$$

Taking the Laplace transform of Equation (6.8.72) and solving for $\bar{f}$ yields the following expression for the Laplace transform of the concentration $\overline{C}$:

$$\overline{C} = \frac{A}{r}\left(-r\sqrt{s/D_{ij}}\right) + \frac{B}{r}\exp\left(r\sqrt{s/D_{ij}}\right). \tag{6.8.73}$$

As r goes to infinity, the term $\exp\left(r\sqrt{s/D_{ij}}\right)$ grows faster than $1/r$. Therefore, B must equal zero. The constant A is determined by substituting the result into the Laplace transform of Equation (6.8.71c). We have

$$\frac{M}{s} = 4\pi \int_0^\infty \frac{A\left(-r\sqrt{s/D_{ij}}\right)}{r} r^2 dr. \tag{6.8.74}$$

Solving for A yields the following solution for $\overline{C}$:

$$\overline{C} = \frac{M}{4\pi D_{ij} r}\exp\left(-r\sqrt{s/D_{ij}}\right). \tag{6.8.75}$$

Equation (6.8.75) can be inverted with the use of a table of Laplace transforms [25]. The result is,

$$C_i = \frac{M}{8(\pi D_{ij} t)^{3/2}}\left(-r^2/4D_{ij}t\right). \tag{6.8.76}$$

This is the solution presented earlier for the probability density function for a random walk (see Equation (6.5.16) and Figure 6.3).

6.8.3 Model of Diffusion of a Solute into a Sphere from a Well-Stirred Bath

There is considerable interest among bioengineers in growing cells within supports of polymeric matrices or gels. These supports, or scaffolds, may ultimately be used to replace damaged or diseased tissues. Encapsulation is employed to immobilize cells for *in vivo* injection or to provide easy separation of cells and cell products. *In vivo* encapsulated cells can evade detection by the immune system; nonhuman animal cells can be used as a potential artificial liver or pancreas [26]. The capsule must permit the transport of low-molecular-weight nutrients, but must inhibit the diffusion of immunoglobulins. A number of important bioengineering issues arise in the design of such replacements, including material strength and biocompatibility, the ability of the material to regulate cell function, the transport of nutrients and macromolecules, genetic manipulation to produce cells that synthesize large quantities of desired proteins, and the development of a blood supply. In this section, we examine one way to characterize the diffusion of macromolecules through polymeric gels.

One approach to measuring diffusion in gels is to study diffusion from a well-mixed solution into gel, as shown in Figure 6.21 [27]. Spherical beads of gel are prepared. At time zero, the beads are added to the solution, which contains a peptide or a protein. At various times, samples are drawn from the solution and the concentration of the peptide or protein is measured. (This can be done by measuring the absorbance at a particular wavelength or by tagging the molecule with a fluorescent or radioactive molecule.) Knowing the size and number of beads added and the solution volume and concentration as a function of time, we want to determine the diffusion coefficient of the peptide or protein in the gel.

The diffusion coefficient of a molecule diffusing through a gel can be found from measurements of the concentration of the solution as a function of time. The measurements are fit to a model for diffusion from a well-mixed bath into a sphere of radius R. Because many particles are used in the experiments, we assume that the particle density is sufficiently low and that the solution is well mixed such that diffusion into one bead does not affect what is happening to another bead.

This problem involves solving for diffusion into the beads with a time-varying surface concentration. The surface concentration changes because the volume of the solution is finite. The measured variable is the solution concentration that decreases as the solute molecules diffuse into the beads. The solute concentration in the beads,

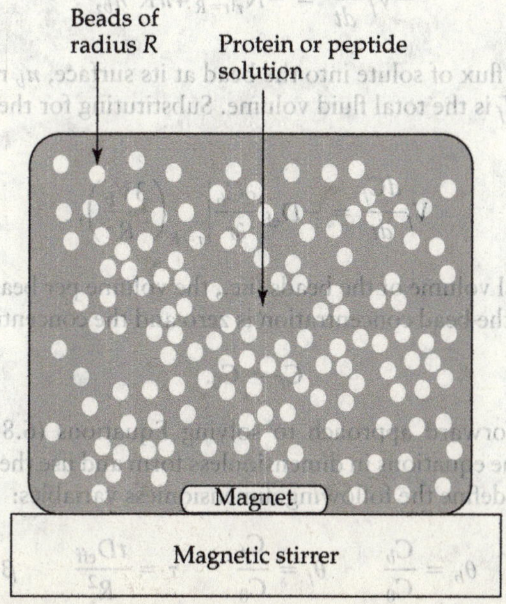

FIGURE 6.21 Diffusion of peptide or protein from a well-mixed solution into beads of gel.

C_b, and the concentration in the fluid, C_f, as a function of time are determined as follows:

For the beads, the conservation of mass for unsteady one-dimensional radial diffusion in spherical coordinates is

$$\frac{\partial C_b}{\partial t} = \frac{D_{\text{eff}}}{r^2}\frac{\partial}{\partial r}\left(r^2\frac{\partial C_b}{\partial r}\right). \tag{6.8.77}$$

The initial condition is

$$0 < r < R \quad t \leq 0 \quad C_b = 0, \tag{6.8.78a}$$

$$r = 0 \quad t \geq 0 \quad \frac{\partial C_b}{\partial r} = 0, \tag{6.8.78b}$$

and

$$r = R \quad t \geq 0 \quad C_b = \Phi C_f(t). \tag{6.8.78c}$$

where Φ is the partition coefficient. In the formulation of the mass balances, we have implicitly neglected the volume fraction that is occupied by the gel itself. Because the gel occupies 2% or 4% of the volume of the beads, this error is not significant.

An additional relation is needed to determine C_f as a function of time. This relation is obtained by performing a mass balance on the solute. The result can be stated as

$$\begin{bmatrix}\text{Change in solute}\\ \text{concentration in bath}\end{bmatrix} = \begin{bmatrix}\text{Rate of transport}\\ \text{into beads}\end{bmatrix}. \tag{6.8.79}$$

Mathematically, this mass balance is

$$-V_f\frac{dC_f}{dt} = -N_r|_{r=R}\,4\pi R^2 n_b, \tag{6.8.80}$$

where $N_r|_{r=R}$ is the flux of solute into the bead at its surface, n_b represents the number of beads, and V_f is the total fluid volume. Substituting for the flux and rearranging terms yields

$$V_f\frac{dC_f}{dt} = -D_{\text{eff}}\left.\frac{\partial C_b}{\partial r}\right|_{r=R}\left(\frac{3V_b}{R}\right), \tag{6.8.81}$$

where V_b is the total volume of the beads (i.e., the volume per bead times the number of beads). Initially, the bead concentration is zero and the concentration in the fluid is

$$C_f = C_0. \tag{6.8.82}$$

The most straightforward approach to solving Equations (6.8.77), (6.8.78), and (6.8.81) is to cast the equations in dimensionless form and use the method of Laplace transforms [1]. We define the following dimensionless variables:

$$\rho = \frac{r}{R} \qquad \theta_b = \frac{C_b}{C_0} \qquad \theta_f = \frac{C_f}{C_0} \qquad \tau = \frac{tD_{\text{eff}}}{R^2} \qquad \beta = \frac{V_f}{\Phi V_b}.$$

The solution proceeds as follows: Equation (6.8.77) becomes

$$\frac{\partial \theta_b}{\partial \tau} = \frac{1}{\rho^2}\frac{\partial}{\partial \rho}\left(\rho^2 \frac{\partial \theta_b}{\partial \rho}\right). \tag{6.8.83}$$

The Laplace transform of Equation (6.8.83) is

$$s\overline{\theta}_b = \frac{1}{\rho^2}\frac{d}{d\rho}\left(\rho^2 \frac{d\overline{\theta}_b}{d\rho}\right). \tag{6.8.84}$$

Equation (6.8.84) is solved by defining the new variable $\overline{\theta}_b = f/\rho$. The resulting equation is

$$sf = \frac{d^2 f}{d\rho^2}. \tag{6.8.85}$$

After solving Equation (6.8.85) and substituting for $\overline{\theta}_b$, we have

$$\overline{\theta}_b = \frac{A}{\rho}\sinh(\sqrt{s}\rho) + \frac{B}{\rho}\cosh(\sqrt{s}\rho). \tag{6.8.86}$$

From the boundary condition at $\rho = 0$, B equals zero. The constant A can be obtained from the boundary condition at the surface and the mass balance relation. At the surface, $\rho = 1$, and we have

$$\overline{\theta}_b = A\sinh(\sqrt{s}) = \Phi\overline{\theta}_f \tag{6.8.87}$$

and

$$\beta\Phi\frac{d\theta_f}{d\tau} = -3\left.\frac{\partial \theta_b}{\partial \rho}\right|_{\rho=1}. \tag{6.8.88}$$

The Laplace transform of Equation (6.8.88) is

$$\beta\Phi(s\overline{\theta}_f - 1) = -3\left.\frac{\partial \overline{\theta}_b}{\partial \rho}\right|_{\rho=1}. \tag{6.8.89}$$

Substituting Equation (6.8.87) into Equation (6.8.89) for $\overline{\theta}_f$ and solving for $\overline{\theta}_b$ yields

$$\overline{\theta}_b(\rho = 1) = \frac{\Phi}{s} - \frac{3}{\beta s}\left.\frac{\partial \overline{\theta}_b}{\partial \rho}\right|_{\rho=1}. \tag{6.8.90}$$

Solving Equation (6.8.90) for $\overline{\theta}_b$ and using Equation (6.8.81) to find $\overline{\theta}_f$ yields

$$\overline{\theta}_f = \frac{\beta\tanh(\sqrt{s})}{(s\beta - 3)\tanh(\sqrt{s}) + 3\sqrt{s}}. \tag{6.8.91}$$

The final operation is to take the inverse Laplace transform of Equation (6.8.91). Inversion requires evaluating the following integral:

$$\theta_f = \frac{1}{2\pi i}\lim_{L\to\infty}\int_{\gamma - iL}^{\gamma + iL} e^{ts}\overline{\theta}_f ds. \tag{6.8.92}$$

Equation (6.8.92) is evaluated by the method of residues [25]. The integrand has a double pole at $s = 0$ and poles at $s = -\lambda_i^2$, where the λ_i are roots of the equation

$$\tan \lambda_i = \frac{3\lambda_i}{3 + \beta\lambda_i^2}. \tag{6.8.93}$$

Shown in Figure 6.22 is a graph of the first three roots of Equation (6.8.93) for $\beta = 0.1$, 0.5, 1.0, and 2.0. The residues are in the following formula:

$$\theta_f = \text{Res}(s = 0) + \sum_{i=1}^{\infty} \text{Res}(s = -\lambda_i^2). \tag{6.8.94}$$

To find the residues at $s = 0$, we expand $\tanh(\sqrt{s})$ in terms of a Taylor series (Section A.1.C):

$$\tanh(\sqrt{s}) \approx \sqrt{s} - s^{3/2}/3. \tag{6.8.95}$$

Then we substitute Equation (6.8.95) into Equation (6.8.91) and take the limit as s goes to zero. The result is $\text{Res}(s = 0) = \beta/(1 + \beta)$, which represents the dilution of the solution by the beads. The residues at $s = -\lambda_i^2$ can be determined by noting that $e^{ts}\overline{\theta}_f$ can be written as $P(s)/Q(s)$. The residues are determined from the following formula:

$$\text{Res}(s = -\lambda_i^2) = \frac{P(s = -\lambda_i^2)}{dQ/ds|_{s=-\lambda_i^2}}. \tag{6.8.96}$$

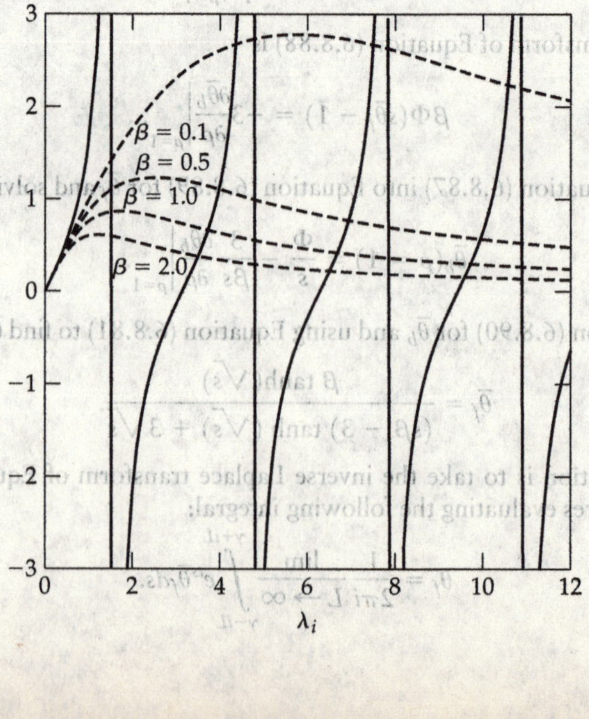

FIGURE 6.22 Roots of Equation (6.8.93). The right-hand side of Equation (6.8.93) is shown as the dashed lines and the left-hand side is shown as the solid lines for $\beta = 0.1$, 0.5, 1.0, and 2.0.

After some algebraic manipulation, the concentration in the fluid is found to be

$$\theta_f = \frac{\beta}{\beta + 1} + \sum_{i=1}^{\infty} \frac{6\beta \exp(-\lambda_i^2 \tau)}{9(1+\beta) + \beta^2 \lambda_i^2} \tag{6.8.97}$$

The dimensionless concentration in the fluid phase is calculated by summing Equation (6.8.97) over a large number of values of λ_i.

The dimensionless fluid-phase concentration depends upon two variables: the dimensionless time τ and the ratio of fluid to bead volume β. Decreasing β, due to an increase in bead volume or a decrease in fluid volume, leads to large changes in concentration that occur rapidly (see Figure 6.23). Because D_{eff} affects the temporal change in the fluid concentration, the optimal choice of β depends upon the rate at which the concentration can be measured.

6.8.4 Quasi-Steady Transport Across Membranes

Because membranes function to control transport, one of the most important properties of biological and synthetic membranes is the permeability. As previously noted (Equation (6.7.10)), the permeability is not an intrinsic property, but it is the product of the solute partition coefficient and the diffusion coefficient divided by the thickness of the membrane. A complete characterization of the transport properties of a membrane requires the measurement of the membrane diffusion coefficient of the solute. In general, the surface concentration changes with time, since the reservoir volumes are finite. Although solutions can be obtained, the expressions are often complex (see Problem 6.18 at the end of the chapter). If the timescale by which the surface concentration changes is much slower than the timescale for diffusion across the membrane, a *quasi-steady-state* approach is used. In such an approach, the unsteady form of the conservation relation is not solved directly. Rather, a simple

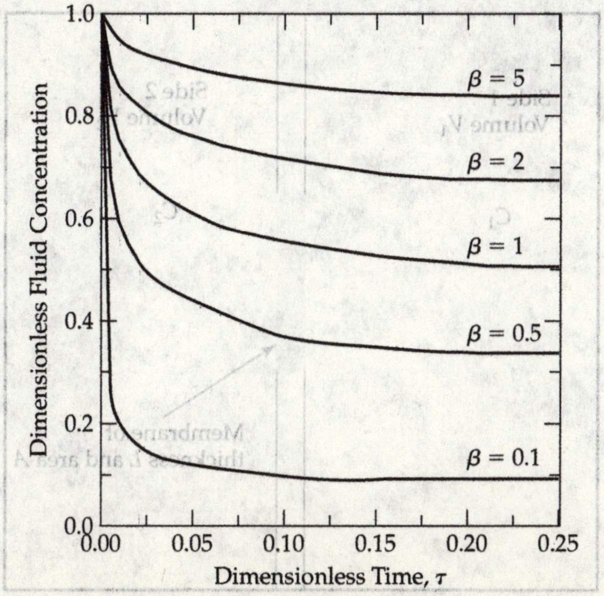

FIGURE 6.23 Dimensionless fluid concentration as a function of dimensionless time and the dimensionless ratio of fluid to bead volume.

approximation is derived for which it can be assumed that the diffusion in the membrane is much faster than the diffusion that leads to the change in the concentrations at the boundaries. This analysis produces a simple analytical expression that is accurate when these assumptions are valid.

As an application, consider a thin membrane of thickness L and area A that separates two solutions of volumes V_1 and V_2 (see Figure 6.24). For times less than zero, there are no solutes in the reservoirs or the membrane. At time zero, the concentrations of solute are $C_1 = C_0$ and $C_2 = 0$. At various times, the concentration C_2 is measured.

To apply the quasi–steady-state analysis, assume that the time for the solute to diffuse across the membrane ($t_{D,\text{memb}}$) is much faster than the time for the solute concentration to change on either side of the membrane (t_c). Quantitatively,

$$t_{D,\text{memb}} \ll t_c. \tag{6.8.98}$$

Based upon the results of Section 6.8.2, the diffusion time is proportional to L^2/D_{ij}. Although we do not yet have an expression for the time required for the concentration in either reservoir to change appreciably (t_c), we may suspect that Equation (6.8.98) is satisfied if the volume of the membrane is much less than the volume of the reservoir. After the analysis, we reexamine this assumption. If Equation (6.8.98) is satisfied, we can assume steady-state transport across the membrane. The diffusion equation is

$$\frac{\partial C_m}{\partial t} \approx 0 \approx D_m \frac{d^2 C_m}{dx^2}. \tag{6.8.99}$$

We use Equation (6.7.10) as the solution to Equation (6.8.99). The solution is

$$N_{i_x} = -D_m \frac{dC_m}{dx} = \frac{D_m \Phi}{L}(C_1 - C_2). \tag{6.8.100}$$

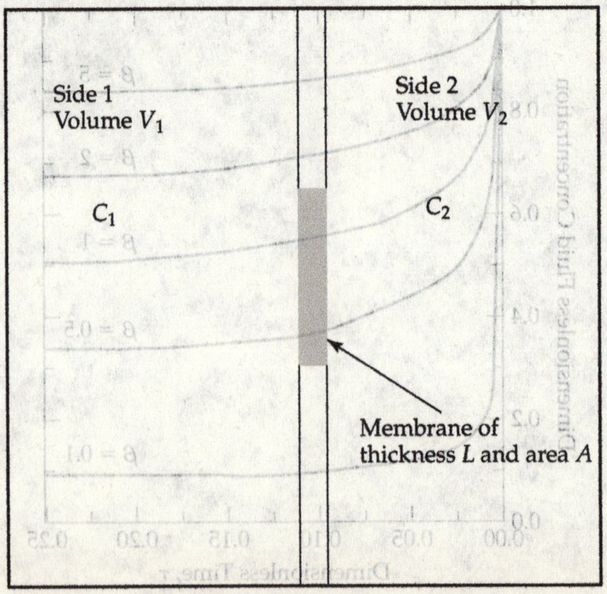

FIGURE 6.24 Schematic of a system for measurement of diffusion coefficients across membranes.

Next, an unsteady mass balance is applied to the solute on each side of the membrane. In words, the mass balance is

$$\begin{bmatrix}\text{Moles of solute leaving}\\ \text{side 1 per unit time}\end{bmatrix} = \begin{bmatrix}\text{Moles of solute transported}\\ \text{across membrane}\end{bmatrix}. \quad (6.8.101)$$

A similar expression can be written for the solute entering side 1. The number of moles transported across the membrane per unit time is simply the product of the flux and the area. Expressed mathematically, the mass balance is

$$-V_1\frac{dC_1}{dt} = A_m D_m \Phi \frac{(C_1 - C_2)}{L}. \quad (6.8.102)$$

The concentrations C_1 and C_2 can be related by noting that after the solute leaves side 1, either it is in the membrane or it is on side 2. The loss of solute from side 1 is balanced by the gain of solute in the membrane or on side 2; that is,

$$V_1\frac{dC_1}{dt} = -\left(\frac{V_m}{\Phi}\frac{dC_m}{dt} + V_2\frac{dC_2}{dt}\right). \quad (6.8.103)$$

If $V_1 = V_2 = V$ and $V_m \ll V$, then the first term on the right-hand side is much less than the other two terms. This means that the amount of solute in the membrane is small relative to the amount of solute in either reservoir. As a result, Equation (6.8.103) can be simplified to

$$\frac{dC_1}{dt} = -\frac{dC_2}{dt}. \quad (6.8.104)$$

Using the initial conditions $C_1 = C_0$ and $C_2 = 0$, we can integrate Equation (6.8.104) to yield

$$C_1 - C_0 = -C_2 \quad (6.8.105a)$$

or

$$C_2 = C_0 - C_1. \quad (6.8.105b)$$

Substituting Equation (6.8.105b) into Equation (6.8.102) leads to a differential equation in one variable:

$$-V\frac{dC_1}{dt} = A_m D_m \Phi \frac{(2C_1 - C_0)}{L}. \quad (6.8.106)$$

Integrating Equation (6.8.106) and applying the initial condition that $C_1 = C_0$, leads to the following result:

$$\ln\left(\frac{2C_1 - C_0}{C_0}\right) = -\frac{2A_m D_m \Phi t}{VL}. \quad (6.8.107)$$

The concentration thus decreases exponentially with time. The steady-state concentrations equal $C_0/2$ in each reservoir. A graph of $\ln[(2C_1 - C_0)/C_0]$ versus t has a slope equal to $-2A_mD_m\Phi/VL$. Equation (6.8.107), then, provides a straightforward method for determining the diffusion coefficient.

Although the membrane thickness L can be determined quite easily for synthetic membranes, the thickness of biological tissues is often more difficult to determine, and it may be variable for a given specimen. Thus, it is often convenient to work with the permeability $P = D_m\Phi/L$ instead. The slope of a graph of $\ln[(2C_1 - C_0)/C_0]$ versus t equals $-2A_mP/V$.

From Equation (6.8.107), we see that the characteristic time for the concentration on side 1 or side 2 to change is given by $VL/A_mD_m\Phi$. Thus, Equation (6.8.98) becomes

$$\frac{L^2}{D_m} \ll \frac{VL}{2D_mA_m\Phi}. \tag{6.8.108}$$

Rearranging terms, we see that Equation (6.8.108) reduces to

$$2\Phi LA_m \ll V. \tag{6.8.109}$$

This is essentially the criterion we anticipated to ensure that the membrane is in a quasi–steady state.

The quasi-steady-state approach can be used to analyze a wide range of problems in which the timescales are very different. The simplification the approach provides often leads to an analytical solution that gives physical insight into the roles of the variables controlling the transport or flow process.

Example 6.10 Endothelial cells have been grown on porous membranes to measure the permeability across the cells and to identify the cellular basis for transport. (For an example, see Ref. [28].) In these experiments, the endothelial cells are grown on one side of a membrane of thickness L and area A. When the cells completely cover the surface (a condition known as *confluence*), the cells form a monolayer. Identical solutions containing salts and a macromolecule of interest are placed on both sides of the membrane, except that the solution on one side (side 1 as depicted in Figure 6.24) contains the labeled macromolecule. The change in the concentration of the labeled molecule on side 2 is monitored as a function of time. The initial concentration of the labeled molecule on side 1 was 10 μg mL^{-1}, and the initial concentration on side 2 was zero. The cells are grown on circular membranes of radius 0.65 cm, and the volumes on each side of the membrane are each 2 cm^3. The following data for albumin were obtained in separate experiments using membranes with and without a monolayer of cells:

Concentration of Labeled Albumin on Side 2 (μg mL^{-1})		
Time (s)	Without a monolayer of endothelial cells	With a monolayer of endothelial cells
0	0	0
1,200	0.39	0.010
2,400	0.71	0.020
4,800	1.30	0.035

Determine the permeability of the endothelial monolayer to albumin.

Solution For each time point of each data set, the quantity $\ln(2C_1/C_0 - 1)$ is determined. A linear regression of this quantity on time yields a straight line for each condition. The permeability is found by using Equation (6.8.10) and the definition of permeability ($P_m = D_m\Phi/L_m$). Without the cells, the permeability is 4.83×10^{-5} cm s^{-1}. With the endothelium present, the permeability is 1.09×10^{-6} cm s^{-1}. To determine the permeability of the endothelium, note that the endothelium and the membrane act in series. For this case, Equation (6.7.16b) can be written as

$$\frac{1}{P_{\text{Tot}}} = \frac{1}{P_m} + \frac{1}{P_e} \tag{6.8.11b}$$

For the values given, P_e is 1.12×10^{-6}. The endothelium represents almost 98% of the total resistance across the system.

6.9 Diffusion-Limited Reactions

Chemical reactions involve the formation of covalent or ionic bonds. Such reactions are essential to many biological processes, such as the metabolism of sugars. Equally important, many biochemical and cellular events are affected by noncovalent interactions between molecules. These interactions involve electrostatic and electrodynamic interactions and produce changes in the chemical nature of the molecules. Examples of such binding events include antigen–antibody and hormone–receptor binding.

If the reaction or binding rates are very fast, chemical reactions and binding events are limited by the rates of transport of the interacting molecules. In the extreme case of diffusion-limited reactions, the intrinsic reaction rate is so fast that reactants are consumed as soon as they reach the reaction site. In Example 6.9, we considered the diffusion-limited adsorption of proteins to a surface. In that example, the reaction rate equaled the diffusive flux to the surface. Thus, a hallmark of diffusion-limited reactions is that they depend on the diffusion coefficient and are independent of the intrinsic chemical process. In the examples that follow, we examine three important cases of diffusion-limited reactions: reactions in solutions, reactions between a molecule in solution and a molecule bound to a cell, and reactions on surfaces. We estimate diffusion-limited rate constants. In Chapter 10, the combined effect of chemical reactions and transport processes is considered.

6.9.1 Diffusion-Limited Binding and Dissociation in Solution

In this simplified analysis, we consider the diffusion of molecules toward a reacting, central molecule. The diffusing molecules, or *ligands*, are assumed to be much smaller than the central molecule, or *receptor*. We assume that ligands diffuse toward a large nondiffusing receptor of radius $R_R(D_R \ll D_L)$ and interact at the receptor surface. The steady-state diffusion equation is

$$\frac{D_L}{r^2}\frac{d}{dr}\left(r^2\frac{dC_L}{dr}\right) = 0. \qquad (6.9.1)$$

Note that since the reaction is heterogeneous, it does not appear in the conservation relation. The following boundary conditions are used. At the receptor surface, all of the ligand is consumed and the concentration of the ligand is zero:

$$r = R_R \quad C_L = 0. \qquad (6.9.2a)$$

This boundary condition corresponds to the diffusion-limited case because the ligands react as soon as they reach $r = R_R$.

Far from the surface of the receptor, the ligand concentration is uniform; that is,

$$r \to \infty \quad C_L = C_{L_0}. \qquad (6.9.2b)$$

For this condition to be applicable, the receptor concentration must be dilute so that the distance between receptors is large.

The solution of Equation (6.9.1), subject to the boundary conditions given by Equation (6.9.2a) and (6.9.2b), is

$$C_L = C_{L_0}\left(1 - \frac{R_R}{r}\right). \qquad (6.9.3)$$

The diffusion-limited forward rate constant is defined in terms of a reaction rate based upon the concentration far from the receptor surface:

$$\text{Rate} = 4\pi R_R^2 N_r|_{r=s} = k_+ C_{L_0}. \qquad (6.9.4)$$

The reaction rate per receptor (mole per second or molecule per second)[5] equals the product of the flux at the surface and the surface area of the molecule. For simplicity, the receptor is treated as a sphere. The units of k_+ are either cubic centimeter per second or cubic centimeter per mole per second (or per molar per second) depending upon whether the rate constant is reported on a per molecule basis or on a molar basis. The interconversion of these two notations is made by using Avogadro's number (6.02×10^{23} molecules/mole) and noting that the receptor radius R_R is in units of centimeter per molecule.

Using Equation (6.9.4) to evaluate the flux and inserting into Equation (6.9.4) yield the following expression for k_+

$$k_+ = 4\pi R_R D_L. \qquad (6.9.5a)$$

[5]The overall reaction is bimolecular. Thus, the overall rate (mole per cubic centimeter per second or molars per second) can be expressed as $R = k_+ C_L C_R$, where C_R is the concentration of the centrally reacting molecule. The rate given by Equation (6.9.4) is for a single centrally reacting receptor molecule.

If both the receptor and ligand can diffuse, then Equation (6.9.5a) is

$$k_+ = 4\pi R_R(D_L + D_R). \tag{6.9.5b}$$

Diffusion-limited association represents the upper limit for the association rate constant in the absence of mixing. Likewise, diffusion-limited dissociation represents the fastest rate of dissociation, because the unbinding step is rapid, however slow diffusion is. For this problem, Equation (6.9.1) applies, but the boundary conditions are now

$$r = R_R \quad C_L = C_{L_0}, \tag{6.9.6a}$$

$$r \to \infty \quad C_L = 0, \tag{6.9.6b}$$

where C_{L_0} is the moles or number of ligand molecules in solution immediately adjacent to the receptor per volume of receptor. This is the same problem we solved in Example 6.8. The solution is

$$C_L = \frac{C_{L_0}R_R}{r}. \tag{6.9.7}$$

To compute the diffusion-limited dissociation rate constant k_-, we note that the rate equals $kC_{L_0}V$, where V is the volume of a spherical receptor. This rate equals the flux times the surface area of the receptor:

$$N_r|_{r=R_R}4\pi R_R^2 = k_-C_{L_0}\frac{4}{3}\pi R_R^3. \tag{6.9.8a}$$

The rate coefficient k_- is a first-order rate coefficient (Section 10.2), with units of per second. Using Equation (6.9.7) to evaluate the flux, we obtain the following result:

$$k_- = 3\frac{D_L}{R_R^2}. \tag{6.9.8b}$$

For macromolecules, D_L ranges between 10^{-6} and 10^{-7} cm^2 s^{-1} and R_R is about 2 nm [29], corresponding to values of k_+ between 2.5×10^{-13} and 2.5×10^{-12} cm^3 molecule^{-1} s^{-1},or 1.51×10^8 and 1.51×10^9 M^{-1}s^{-1}. Corresponding values of k_- range from 7.5×10^6 s^{-1} to 7.5×10^7 s^{-1}. Thus, these estimates represent the upper limit for the association and dissociation rate constants in the absence of mixing.

6.9.2 Diffusion-Limited Binding Between a Cell Surface Protein and a Solute

Many important cellular functions are controlled by the binding of molecules to receptor molecules on cell membrane surfaces (see Figure 6.25). Many receptors are transmembrane proteins with a portion of the protein protruding into the cell cytoplasm. Receptors function in a variety of ways. In some cases, the binding event alters the function of the cytoplasmic portion of the molecule (the *cytoplasmic tail*, in the jargon of cell biologists), initiating a series of biochemical interactions

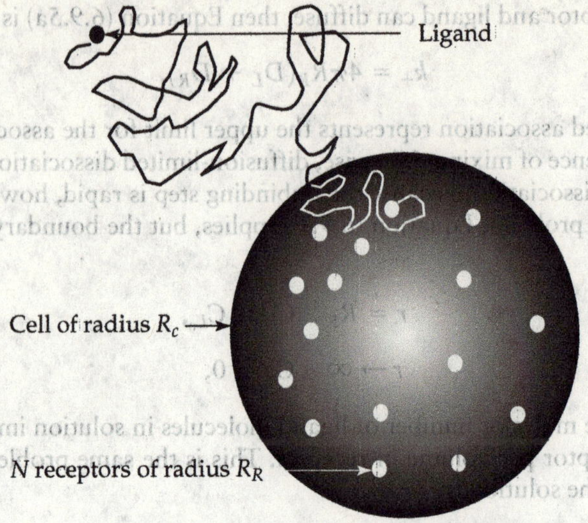

FIGURE 6.25 Diffusion of ligand to a cell containing N receptors.

that lead to altered cell function. Examples include the binding of cell adhesion molecules to their receptors. (Cell adhesion is discussed extensively in Chapter 12.) In other cases, binding opens a transmembrane channel that permits the flow of ions across the cell (as described in Section 1.4.1). In still others, the bound molecule is internalized within the cell. After it is in the cell, the molecule exerts its function. Internalization of the receptor may occur continuously, regardless of the presence of the ligand (a process known as *constitutive endocytosis*), as is the case with the receptor involved in the internalization of low-density lipoprotein that supplies cholesterol to cells. Alternatively, internalization of the receptor is triggered by binding of the receptor to its ligand. An example of this type of internalization is the binding of epidermal growth factor to its receptor. Often, the binding of receptor to ligand may initiate multiple events that lead to alterations in cell function. One example is receptor-mediated endocytosis, which is considered in more detail in Chapter 11.

In many tissues, the rate of ligand binding to receptors on the cell surface is limited by the rate of diffusion of the ligands to the surface. The result obtained in Section 6.9.1 cannot be used when individual receptors are immobilized on cells in suspension, except for the limiting case when the entire cell is covered with a single type of receptor. For the more general case of ligand diffusion and reaction with N individual receptors on the surface of suspended cells, an approximate analysis was developed by Berg and Purcell [7,30]. This analysis is based on the assumption that receptors are far enough apart that the binding of one ligand to a receptor does not deplete the concentration of ligands near other receptors. As a result, binding events on one receptor are independent of those occurring on another receptor. Typically, cells contain hundreds of different receptors. For each receptor, the number of copies on the cell surface ranges from 5,000 to 1,000,000, with an average of around 100,000 copies. A typical spherical cell radius is 10 μm. If receptors are treated as disks with a typical radius of 2 nm, the fractional surface area that is covered by receptors ranges from 5×10^{-5} to 1×10^{-2}. Thus, based on receptor surface density, the assumption that receptors do not interact appears reasonable.

The low coverage of individual receptors on the cell surface permits the cell to contain many different types of receptors. This raises a question about how efficient binding can be. Is the probability of binding low because of the reduced coverage of receptors on the surface? Certainly, this would be the case if the receptors were to diffuse freely in solution. However, because they are limited to the surface of the cell, the binding process is surprisingly efficient.

Berg and Purcell addressed this question by noting that the diffusion-limited rate constant for binding to N receptors on a cell is less than the value for binding to a cell that is completely covered with receptors ($k_+ = 4\pi R D_L$). The diffusion-limited rate constant is inversely proportional to the time for a molecule to diffuse to the reacting site. For N receptors on the surface, this time is the sum of the time required to reach the surface of a cell plus the time in which a molecule near the surface diffuses to a receptor [31]. The time to reach the surface is analogous to that which was calculated for the case of diffusion to a uniformly reactive sphere. Since the cell is not uniformly reactive, however, additional time is required while the molecule is near the cell surface and searches to find a receptor. Mathematically, we have

$$\frac{1}{k_+} = \frac{1}{4\pi R D_L} + \frac{1}{k_+}\Big|_{\text{Binding to } N \text{ receptors}}. \tag{6.9.9}$$

The problem of steady-state binding to N individual receptors can be solved by considering an isolated receptor of radius R_R as shown in Figure 6.26. A cylindrical coordinate system is used. Far from the surface, the concentration is uniform at C_0. On the disk surface, the ligand concentration is zero due to rapid reaction. On the remaining portion of the cell surface, the membrane is impermeable to the ligand and the flux is zero. Mathematically, the statement of the problem is as follows:

$$\frac{1}{r}\frac{\partial}{\partial r}\left(r\frac{\partial C_L}{\partial r}\right) + \frac{\partial^2 C_L}{\partial z^2} = 0, \tag{6.9.10}$$

$$z \to \infty \quad C_L = C_0, \tag{6.9.11a}$$

$$z \to 0 \quad 0 \le r \le R_R \quad C_L = 0, \tag{6.9.11b}$$

$$z \to 0 \quad r > R_R \quad \frac{\partial C_L}{\partial r} = 0. \tag{6.9.11c}$$

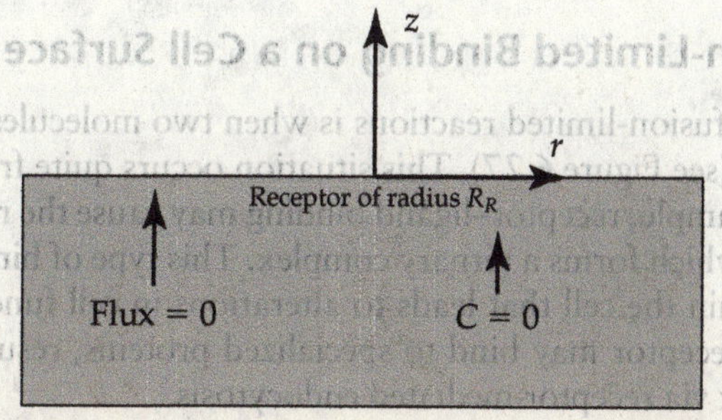

FIGURE 6.26 A schematic of diffusion-limited binding to a disk-like receptor.

This problem is known as *Weber's disk*; the solution can be found in Crank [24]. Of significance to this problem is the flux at the surface of the disk, given by

$$-D_L \frac{\partial C_L}{\partial z}\Big|_{z=0} = \frac{2D_L C_0}{\pi\sqrt{(R_R^2 - r^2)}} \qquad 0 \le r \le R_R. \tag{6.9.12}$$

The diffusion-limited rate constant k_+ is obtained by equating the reaction rate (which is based upon the concentration far from the surface) with the rate of diffusion to the disk:

$$-\int_0^{R_R} D_L \frac{\partial C_L}{\partial z}\Big|_{z=0} 2\pi r dr = k_+ C_0. \tag{6.9.13}$$

Substituting Equation (6.9.12) into Equation (6.9.13) and integrating yields the following relation for the diffusion-limited rate constant for binding to N receptors:

$$k_+|_{\text{Binding to } N \text{ receptors}} = Nk_+ = 4NR_R D_L. \tag{6.9.14}$$

Inserting Equation (6.9.14) into Equation (6.9.9) and rearranging yields the overall rate constant for binding to N receptors in the cell

$$k_+ = 4\pi R_c D_L \left(\frac{NR_R}{NR_R + \pi R_c}\right). \tag{6.9.15}$$

The units of k_+ are square centimeters per second per molecule. The term in parentheses represents the extent to which distributing the receptors over the cell surface reduces the rate constant relative to a uniformly reactive cell. Equation (6.9.15) can be used to assess the receptor density needed such that k_+ is 50% of the value for a cell that is completely covered by receptors. Setting the term in parentheses equal to 0.5, and using $R_R = 2$ nm and $R = 10$ μm, we find that $N = 15{,}708$. This value is quite surprising, given that the receptor density we have assumed covers only 0.063% of the cell surface. The reason that the receptor number is so low is that the time for a ligand to reach the surface of the cell is much longer than the time to explore a region near the cell surface. Thus, when a ligand reaches the surface, it is more likely to explore regions near the surface than simply to diffuse away. As a result, the probability of forming a bond increases significantly. By a similar argument for dissociation from a receptor on the cell surface, the diffusion-limited rate constant for dissociation is

$$k_- = \frac{3D_L}{R_R^2}\left(\frac{NR_R}{NR_R + \pi R_c}\right). \tag{6.9.16}$$

6.9.3 Diffusion-Limited Binding on a Cell Surface

Another case of diffusion-limited reactions is when two molecules on the cell surface bind to each other (see Figure 6.27). This situation occurs quite frequently in biological systems. For example, receptor–ligand binding may cause the receptor to bind to a signaling protein, which forms a ternary complex. This type of binding initiates a cascade of events within the cell that leads to alterations in cell function. Alternatively, the ligand-bound receptor may bind to specialized proteins, resulting in internalization of the complex via receptor-mediated endocytosis.

FIGURE 6.27 Diffusion of two molecules on the cell surface.

Because diffusion is occurring on the cell surface, the previous analysis must be modified. Instead of diffusing in three dimensions toward a central receptor, the receptors diffuse in two dimensions. Since the density of the receptors is generally sparse, the distance between receptors is typically small enough that the curvature of the cell can be neglected. For this problem, a cylindrical coordinate system with radial diffusion is the most appropriate coordinate system to use. Receptors are modeled as disks of radius R_R. Diffusion coefficients in membranes are much smaller than diffusion coefficients in solution. Values are typically 10^{-9} cm^2 s^{-1} for lipids diffusing in lipid membranes. Diffusion coefficients for proteins range from 10^{-10} to 10^{-12} cm^2 s^{-1}. At the receptor surface, the rate of binding is so fast that the ligand concentration is zero. Far from the surface of the receptor, the ligand concentration is uniform at C_0. Because the surface of the cell is finite, we must specify a distance at which the concentration equals C_0. The most reasonable choice is the half-distance between two receptors, represented as b.

Mathematically, the problem can be stated as

$$\frac{D_L}{r}\frac{d}{dr}\left(r\frac{dC_L}{dr}\right) = 0, \tag{6.9.17}$$

$$r = R_R \quad C = 0, \tag{6.9.18a}$$

$$r = b \quad C = C_0. \tag{6.9.18b}$$

The solution of Equation (6.9.17), subject to the boundary conditions given by Equations (6.9.18a) and (6.9.18b), is

$$C = C_0\frac{\ln(r/R_R)}{\ln(b/R_R)}. \tag{6.9.19}$$

The diffusion-limited association constant is found by equating the rate of diffusion along the perimeter of the receptor to the diffusion-limited reaction rate:

$$2\pi R_R N_r|_{r=R_R} = k_+ C_0. \tag{6.9.20}$$

Evaluating the flux with Equation (6.9.19) and solving Equation (6.9.20) for k_+ yields

$$k_+ = \frac{2\pi D_L}{\ln(b/R_R)}. \tag{6.9.21}$$

The diffusion-limited dissociation constant is found by solving Equation (6.9.17), subject to the following boundary conditions:

$$r = R_R \quad C = C_0, \tag{6.9.22a}$$

$$r = b \qquad C = C_b. \tag{6.9.22b}$$

The result is

$$k_- = \frac{2D_L}{R_R^2 \ln(b/R_R)}. \tag{6.9.23}$$

For a receptor radius of 2 nm, a density of 1×10^5 molecules cell^{-1}, a diffusion coefficient of 1×10^{-10} cm^2 s^{-1}and a cell radius of 10 μm, k_+ is 1.82×10^{-10}cm^2 s^{-1} and k_- is 1,449 s^{-1}. Direct comparisons between values for k_+ in solution and on the surface are difficult, but the value for the diffusion-limited dissociation constant is less for molecules on cell membranes than for those in solution.

6.10 A Thermodynamic Derivation of the Stokes–Einstein Equation

This section gives a summary of the derivation of the Stokes–Einstein equation that was developed by Einstein [6]. Diffusion represents the random motion of molecules, driven by collisions between molecules. In a dilute system, collisions between solute molecules are very infrequent. Thus, the driving force for diffusion is due mainly to the thermal molecular motion of solvent molecules. To simplify the discussion, the derivation that follows assumes that molecular motion is one dimensional (i.e., only in the x direction). However, the Stokes–Einstein equation is also valid for three-dimensional diffusion in a dilute system.

Consider N_o particles suspended in a solvent and bounded between $x = 0$ and $x = L$. The particles are much larger than solvent molecules, but smaller than the length scale of macroscopic diffusion. The concentration distribution of the particles is $C(x)$.

In the state of thermodynamic equilibrium, the concentration profile is uniform, unless an external force $-F_x(x)$ acts on each particle such that the diffusion flux is balanced by a negative convective flux $-\mathrm{v}_x C$ across any cross-section. We can state this condition mathematically as

$$\mathrm{v}_x C + D_{ij}\frac{dC}{dx} = 0, \tag{6.10.1}$$

where v_x is numerically equal to the average velocity of the particles, induced by the thermal motion of solvent molecules. For low-Reynolds-number flow, v_x is proportional to F_x. Thus,

$$F_x = f\mathrm{v}_x, \tag{6.10.2}$$

where f is the frictional drag coefficient. Substituting Equation (6.10.2) into Equation (6.10.1), we have

$$\frac{F_x}{f}C + D_{ij}\frac{dC}{dx} = 0. \tag{6.10.3}$$

In order to find F_x, we need to apply the concept of "virtual displacement" that is widely used in stress–strain analysis in solid materials. A virtual displacement $\delta u(x)$ of a particle is defined as an arbitrary displacement that (a) does not affect the equilibrium state of the system and (b) vanishes at the boundary of the field. Thus, virtual displacement is an equilibrium process. As a result, the Gibbs free energy G should not change, and we have

$$\delta G = \delta H - T\delta S = 0, \tag{6.10.4}$$

where H is the enthalpy, T is the temperature, S is the entropy, and δ indicates the virtual change in the corresponding physical quantity. Furthermore, δH is equal to the change in the internal energy of the particles, since all particles are bounded between $x = 0$ and $x = L$ and the volume of the system does not expand. In this case, δH is equal to the virtual work done by F_x, or

$$\delta H = \int_0^L (F_x \delta u) CA dx, \tag{6.10.5}$$

where A is the cross-sectional area in the x direction. During a virtual displacement, δS is calculated on the basis of the Boltzmann equation.

Now consider a particle system with a volume of V_1. When the volume is increased to V_2, the entropy change obeys the Boltzmann equation

$$\Delta S = k_B n_0 \ln (V_2/V_1), \tag{6.10.6}$$

where n_0 is the number of particles in the system and k_B is the Boltzmann constant. Consider a small volume $A\Delta x$ between $x = 0$ and $x = L$. Thus, $V_1 = A\Delta x$. Within V_1, $n_0 = CA\Delta x$. At the end of the virtual displacement, the volume changes to

$$V_2 = A(\delta u|_{x+\Delta x} - \delta u|_x + \Delta x) \approx A\left(1 + \frac{\partial \delta u}{\partial x}\right)\Delta x = V_1\left(1 + \frac{\partial \delta u}{\delta x}\right). \tag{6.10.7}$$

Substituting Equation (6.10.7) into Equation (6.10.6) and assuming that $\partial\delta u/\partial x \ll 1$, we have

$$\Delta S = k_B n_0 \ln\left(1 + \frac{\partial \delta u}{\partial x}\right) \approx k_B n_0 \frac{\partial \delta u}{\partial x}. \tag{6.10.8}$$

The virtual change in the system entropy can be calculated by integrating Equation (6.10.8):

$$\delta S = \int_0^L k_B CA \frac{\partial \delta u}{\partial x} dx = k_B A[C\delta u|_0^L - \int_0^L \delta u \frac{\partial C}{\partial x} dx] = -k_B A \int_0^L \delta u \frac{\partial C}{\partial u} dx. \tag{6.10.9}$$

In Equation (6.10.9), we use the boundary conditions of the virtual displacement (i.e., it vanishes at $x = 0$ and $x = L$). Combining Equations (6.10.4), (6.10.5), and

(6.10.9), we have

$$\int_0^L \left(F_x C + k_B T \frac{\partial C}{\partial x} \right) \delta u dx = 0. \qquad (6.10.10)$$

The virtual displacement δu is arbitrary. Thus, the integral in Equation (6.10.10) is zero if the integrand is zero. Therefore, we have

$$F_z C + k_B T \frac{\partial C}{\partial x} = 0. \qquad (6.10.11)$$

Substituting F_x from Equation (6.10.11) into Equation (6.10.3), we obtain

$$D_{ij} = \frac{k_B T}{f}. \qquad (6.10.12)$$

In the derivation of the Stokes–Einstein equation, we have assumed that the suspended-particle system is dilute and the particle size is much larger than the solvent size. Quantitatively, it has been shown that the former should be at least five times larger than the latter [4]. For highly viscous solutions, then, D_{ij} is no longer inversely related to μ [4].

6.11 QUESTIONS

6.1 Using the data in Table 6.7, show that the major protein components of blood represent a dilute solution.

6.2 Tissues consist of an extracellular matrix and a fluid phase that contains water, ions, sugars, and proteins. What effect does the extracellular matrix have on the random motion and diffusion coefficient of molecules in the fluid phase?

6.3 Explain how the random motion of molecules produces a concentration gradient and a net movement of solute.

6.4 What factors affect the frictional drag coefficient?

6.5 Choose one of the proteins listed in Table 6.4, and examine the effect of three-dimensional shape on the value of $\bar{f}/f_0$. Does one of the idealized shapes listed in Table 6.3 accurately predict this ratio?

6.6 What is the accuracy of protein molecular weights that are obtained by using measurements of transport properties?

6.7 How is the time to reach steady state affected by a doubling of the thickness of a slab of thickness $2L$?

6.8 What effect does the partition coefficient have on permeability?

6.9 Discuss the strengths and limitations of using cell cultures to study the permeability across an endothelial cell monolayer.

6.10 Why does it take less time to reach steady state in diffusion through a sphere of radius R than it does in diffusion through a rectangular region of thickness R?

6.11 Describe experiments that could be used to determine the diffusion coefficient and partition coefficient of an antibody in a tumor.

6.12 Discuss how the analysis presented in Section 6.8.3 can be used to obtain the diffusion coefficient of a peptide or an oligonucleotide diffusing into a polymer bead.

6.13 Would k_+ be greater or less than the result for N individual receptors (Equation (6.9.15)) if all of the receptors aggregated into a single patch?

6.12 PROBLEMS

6.1 Using the Stokes–Einstein equation and the Wilkie–Chang correlation, estimate the diffusion coefficient of oxygen in water at 298 K. Take the molecular diameter to have a value of 0.3467 nm. For oxygen, the partial molar volume is 25.6 $cm^3\ mol^{-1}$. Compare your answer for the diffusion coefficient with the measured value of $2.10 \times 10^{-5}\ cm^2\ s^{-1}$.

6.2 A new protein has been isolated by molecular biologists. Before they can measure its diffusion coefficient at 25°C, their instrument breaks and repairs will take six months! The researchers are in a panic because they need this information for a paper they are writing. The biologists do know that the molecule is spherical and has a radius of 12 nm. Before the instrument broke, they measured the diffusion coefficient of albumin (R = 3 nm and MW = 68,000) as $6.8 \times 10^{-7}\ cm^2\ s^{-1}$. Calculate the diffusion coefficient of the new protein.

6.3 Using recursion relations that were presented in Section 6.5, develop a program to describe a two-dimensional random walk.

(a) Develop an expression analogous to Equation (6.5.6) for a one-dimensional random walk when δ is not constant, but is a random number between −1 and 1.

(b) For a one-dimensional random walk, use a spreadsheet (e.g., Excel) or MATLAB to show that the root-mean-square displacement after 500 steps is approximated by the theoretical relation developed in the first part and is proportional to the absolute value of the x position after 500 steps.

6.4 Estimate the diffusion coefficient of fibrinogen, assuming that it is

(a) a prolate ellipsoid or

(b) a cylindrical rod.

Which shape is a more accurate model of the shape of fibrinogen?

6.5 Consider steady one-dimensional diffusion through a funnel of varying cross section (Figure 6.28). Unlike a tube of a constant cross section, the concentration varies nonlinearly with distance.

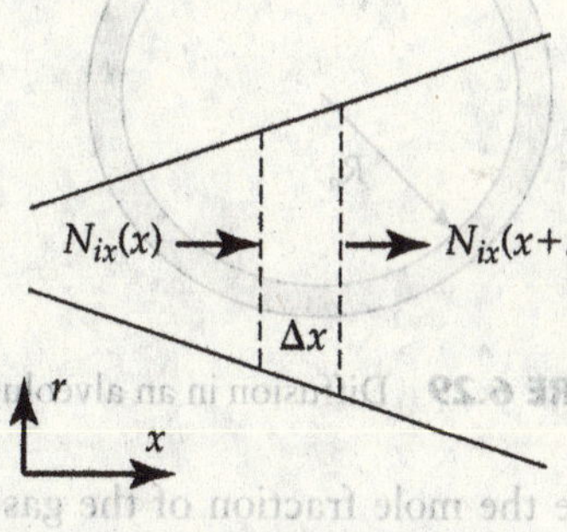

FIGURE 6.28 Diffusion through a funnel.

(a) Neglect chemical reactions and convection and derive the following material balance for steady one-dimensional diffusion through a region of varying cross section (i.e., $A = A(x)$):

$$\frac{d(N_{ix}A(x))}{dx} = 0. \tag{6.12.1}$$

Integrate this equation and explain briefly what the result indicates.

(b) The radius varies linearly with distance along the funnel according to the formula

$$r(x) = r_0\left(1 + \frac{x}{L}\right). \tag{6.12.2}$$

Use Fick's law, the result from part (a) of this problem, and Equation (6.12.2) to show that the steady-state flux is

$$N_{ix} = \frac{2D_{ij}\,(C_0 - C_L)}{L\ (1 + x/L)^2}. \tag{6.12.3}$$

The boundary conditions are

$$x = 0 \quad C_i = C_0$$
$$x = L \quad C_i = C_L.$$

6.6 Nitric oxide (NO) is an extremely potent vasodilator that is used to treat newborns who have pulmonary hypertension and adults who have undergone certain operative procedures. We want to examine the transport of the gas through an alveolus and into the capillaries. The gas is added at a concentration of less than 100 parts per million (ppm). The alveolus is modeled as a sphere of radius R_a (see Figure 6.29).

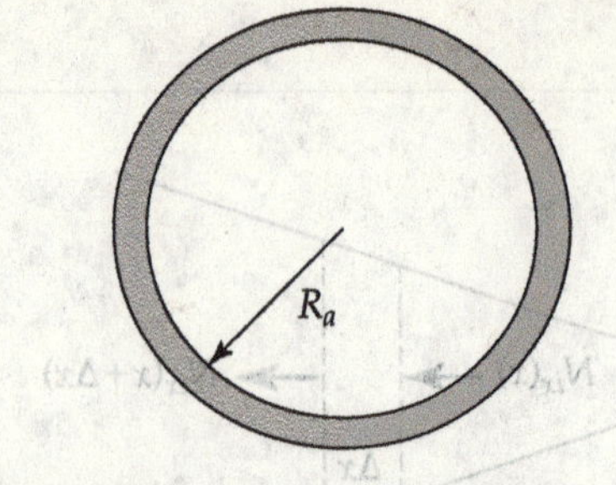

FIGURE 6.29 Diffusion in an alveolus.

(a) Determine the mole fraction of the gas and assess the maximum error in the flux, assuming that the gas is a dilute solution of NO.

(b) In each inspiration, the concentration of NO in each alveolus is 30 ppm and the gas is initially uniformly mixed. Is the gas completely removed between breaths? Use the following data:

alveolus radius = 50 μm $\quad D_{NO} = 0.2\ \text{cm}^2\ \text{s}^{-1}$

gas concentration in blood = 0

6.7 An uncharged membrane separates two aqueous salt solutions that contain a protein at concentrations of C_L and C_R, with $C_L > C_R$ (see Figure 6.30). Stirring the solutions reduces, but does not eliminate, mass transfer effects near the membrane surface. The salt concentrations are the same for both solutions, so the potential differences are negligible. Figure 6.30 shows concentration distributions for the protein solutions.

For each sketch, briefly discuss whether the concentration profiles are physically possible, and if they are, determine whether the partition coefficient $\Phi(= C_m/C_L$ or $C_m/C_R)$ is greater than, less than, or equal to unity.

6.8 In Section 6.7, diffusion through multiple layers of tissue arranged in series was examined. Now consider steady-state diffusion through two media arranged parallel to each other (see Figure 6.31). Assume that diffusion is one dimensional. At $x = 0$, $C_1 = \Phi_1 C_0$, and $C_2 = \Phi_2 C_0$. At $x = L$, $C_1 = \Phi_1 C_L$ and $C_2 = \Phi_2 C_L$. Develop an expression for the steady-state flux across the two media. Show that the diffusive resistances act in parallel.

6.9 Consider a rectangular laminate consisting of two layers, as shown in Figure 6.9. Assume that $\Phi_1 = \Phi_2 = 1$.

(a) For the following values, determine the effective diffusion coefficient:

$$D_{i,1} = 5 \times 10^{-6}\ \text{cm}^2\text{s}^{-1} \qquad L_1 = 20\ \mu\text{m}$$

$$D_{i,2} = 7 \times 10^{-7}\ \text{cm}^2\text{s}^{-1} \qquad L_2 = 80\ \mu\text{m}$$

(b) Determine conditions for which the two-layer model behaves as an effective one-layer model.

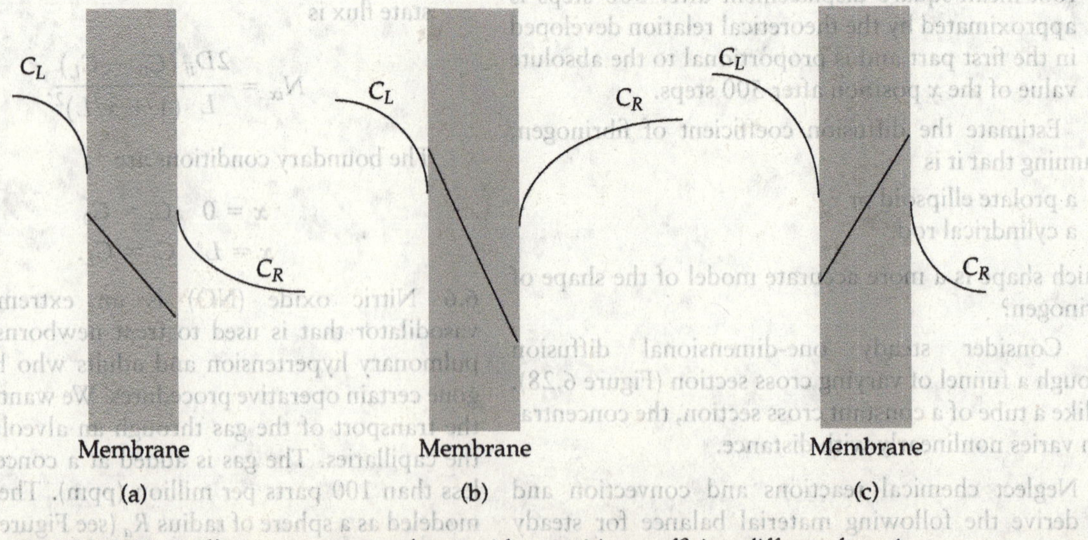

FIGURE 6.30 Diffusion across a membrane with a partition coefficient different from 1.

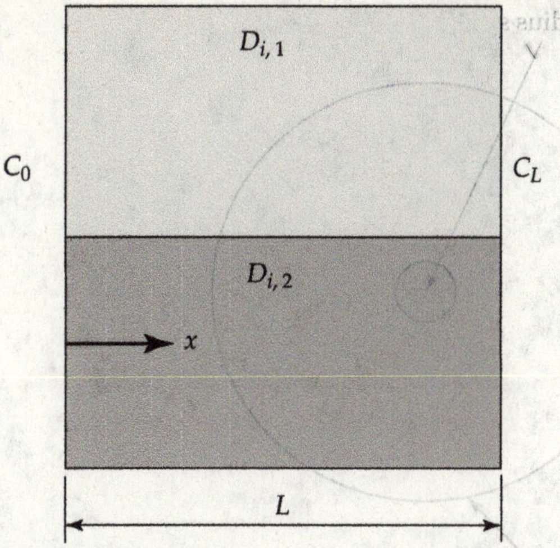

FIGURE 6.31 Diffusion through two media arranged in parallel.

6.10 Consider a two-layer model of an artery, as shown in Figure 6.32. The layers are of thickness $R_0 - R_1$ and $R_1 - R_i$. The inner layer has the diffusion coefficient D_i and the outer layer has the diffusion coefficient D_0. The solute concentration in the lumen (i.e., $0 < r < R_i$) is C_i, and the concentration at R_0 is C_0. Calculate the effective diffusion coefficient.

6.11 Beginning with Equation (6.8.101), derive a generalized quasi–steady-state relation for transport across a thin membrane when the volumes on the two sides of the membrane differ. Show that the result reduces to Equation (6.8.107) when $V_1 = V_2$.

6.12 Low-density lipoprotein (LDL) is the major cholesterol-carrying lipoprotein in the body. Its entry into cells occurs when LDL binds to receptors that are localized on specialized regions of the cell surface known as *coated pits.* (The name arises from the electron-dense appearance of the membrane in electron micrographs.) A coated pit contains proteins that regulate the binding of receptors and the formation of vesicles. When a coated pit forms a vesicle, LDL molecules are transported to lysosomes. In the lysosome, the cholesterol is esterified and enters the cell cytoplasm; the protein portion is degraded to amino acid.

Determine the rate constant for the diffusion-limited dissociation of LDL receptors from binding sites in coated pits. Binding and dissociation of LDL receptors to coated-pit proteins occurs independently of LDL binding to its receptor. Assume that coated pits have a radius s and are separated by a distance $2b$ (see Figure 6.33), and use the following data to determine k_- for the dissociation of a receptor from a ternary complex in coated pits on the cell membrane surface:

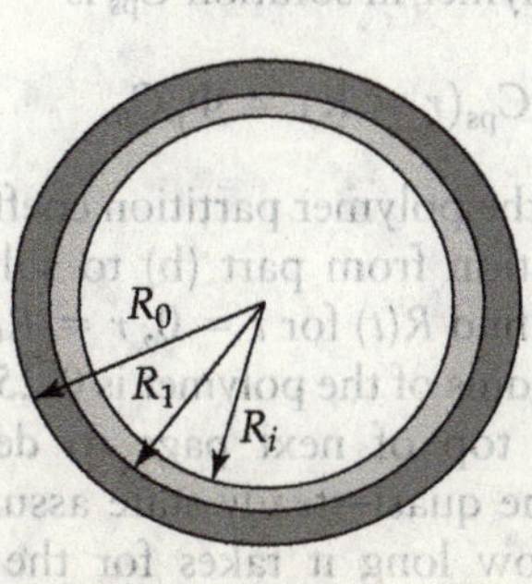

FIGURE 6.32 Diffusion through a cylindrical laminate.

P	0.30 coated pit μm^{-2}	Number density of coated pits
N	100,000 receptors $cell^{-1}$	Number of receptors per cell
A	5,000 μm^2	Surface area of cell
R_R	1 nm	Receptor radius
s	0.10 μm	Radius of a coated pit
D_R	4.5×10^{-11} $cm^2\ s^{-1}$	Diffusion coefficient of receptor
λ	0.20 min^{-1}	Rate constant for vesicle formation
b	1.0 μm	Half of the separation distance between two coated pits

6.13 There is considerable interest in culturing endothelial cells onto polymeric surfaces to provide a small-diameter vascular graft that resists clotting. The adhesion of endothelial cells is greatly enhanced if the cells are grown on surfaces that contain fibronectin, a preadsorbed protein. One approach to adsorbing fibronectin onto the surface is to incubate the graft for a short period of time with the patient's plasma, which contains fibronectin. The other choice is to incubate the graft with a purified solution of fibronectin. In all cases,

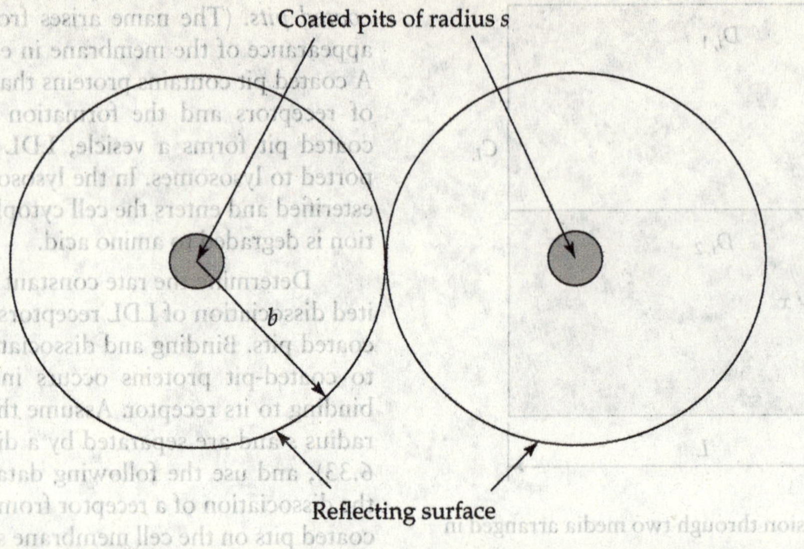

FIGURE 6.33 Schematic of two coated pits and the geometry for receptor diffusion.

the volume of fluid is sufficiently large that the change in concentration of any single protein is very small. Fibronectin adsorption is rapid and can be assumed to be diffusion limited.

(a) When plasma is incubated with the graft surface, how long does it take to cover the surface with (1) albumin and (2) fibronectin? Treat plasma as consisting of albumin (ALB) at 50 mg mL^{-1} and fibronectin (FN) at 0.200 mg mL^{-1}. Take

$$D_{FN} = 1.5 \times 10^{-7}\ \mathrm{cm^2\ s^{-1}}$$
$$D_{ALB} = 6.0 \times 10^{-7}\ \mathrm{cm^2\ s^{-1}}$$

Albumin concentration required to completely cover the surface = 0.388 μg cm^{-2}.

Fibronectin concentration required to completely cover the surface = 0.364 μg cm^{-2}.

(b) If plasma is incubated with the graft for 20 seconds, which proteins adsorb to the graft? Assume that adsorption is irreversible.

(c) What is the potential advantage of incubating the graft surface with a purified solution of fibronectin at 200 μg mL^{-1}, instead of with plasma?

6.14 Example 6.8 examined the slow, steady release of a drug from a polymer. The erosion of the polymer was neglected, on the grounds that polymer erosion is slow relative to the rate of transport of the drug. We can test this assumption quantitatively by using the following model for erosion: Erosion results from the polymer dissolving in the solvent and drug leaves the polymer as a result of erosion. Assume that the rate of polymer erosion at the surface equals the rate of diffusion of the polymer from the surface.

(a) Show that a material balance leads to the following relation for the polymer:

$$-C_p \frac{dR^3}{dt} = 3R^2 N_{pr}|_{r=R}. \quad (6.12.4)$$

C_p is the polymer concentration in the undissolved state.

(b) Determine the flux, assuming steady-state diffusion of the polymer away from the surface. The concentration of polymer in solution C_{ps} is

$$C_{ps}(r = R) = \Phi_p C_p, \quad (6.12.5)$$

where Φ_p is the polymer partition coefficient.

(c) Use the solution from part (b) to solve Equation (6.12.4) and find $R(t)$ for $t = 0$, $r = R_0$.

(d) The initial radius of the polymer is 0.25 cm. Use the table at the top of next page to determine the validity of the quasi–steady-state assumption, and determine how long it takes for the polymer to dissolve.

Property	Polymer	Drug
Diffusion coefficient ($cm^2\ s^{-1}$)	1×10^{-10}	2×10^{-6}
Solution solubility ($mol\ cm^{-3}$)	$C_{ps}(r = R) = 1 \times 10^{-10}$	$C_0 = 1 \times 10^{-6}$
Concentration ($mol\ cm^{-3}$)	$C_p = 1 \times 10^{-8}$	$C_d = 1 \times 10^{-7}$
Partition coefficient	$\Phi_p = 0.01$	$\Phi_d = 1.00$

6.15 Determine the time required for a lipid-soluble molecule (e.g., oxygen) to become uniformly distributed within a cell. The cell is modeled as a sphere of radius R. The lipid membrane is modeled as a thin layer with permeability P. For times less than zero, the concentration inside the cell (C_i) is C_0. For times greater than zero, the concentration outside the cell is raised to C_1.

(a) Show that the solution to this problem is

$$\theta' = \sum_{n=1}^{\infty} \frac{2(\sin\lambda_n - \lambda_n\cos\lambda_n)}{\lambda_n\eta(\lambda_n - \cos\lambda_n\sin\lambda_n)}\sin(\lambda_n\eta)\exp(-\lambda_n^2\tau), \quad (6.12.6)$$

where

$$\eta = \frac{r}{R} \qquad \theta' = \frac{C_1 - C_i}{C_1 - C_0} \qquad \tau = \frac{tD}{R^2},$$

and the terms λ_n are roots of the transcendental equation

$$\frac{\lambda_n}{1 - B_i} = \tan\lambda_n, \quad (6.12.7)$$

where $B_i = PR/D_{ij}$.

(b) Determine the time to reach steady state for the following set of parameters:

$$P = 1{,}10^{-2}{,}10^{-4}\ cm\ s^{-1},\ D_{ij} = 1 \times 10^{-6}\ cm^2 s^{-1},\ R = 10\ \mu m.$$

(*Note:* Newton's method can be used to solve the transcendental equation that you obtain after rearranging it to the form $f(\lambda_n) = 0$.)

6.16 A patch has been designed to release a drug to treat airsickness. The patch is applied an hour before the flight and is worn until the flight is over. The patch is a cylindrical disk 3-mm thick with a radius of 1 cm. The drug is initially at a concentration of 50 mM. Assume that the concentration in the adjacent skin is much less than the concentration in the patch. The diffusion coefficient of the drug in the patch is $1 \times 10^{-8}\ cm^2\ s^{-1}$. The manufacturer must have the release rate above $0.5 \times 10^{-10}\ mol\ s^{-1}$ to ensure that there is enough drug in the bloodstream for the drug to be effective.

Using the data given, assess whether the desired release rate can be achieved for a 14-hour flight from San Francisco to Beijing. (*Hint*: Determine whether the semi-infinite medium approximation is valid.) If the criterion is not met, suggest possible ways to produce a higher flux.

6.17 Show that Equation (6.8.56) results after making the substitution of Equation (6.8.55) for θ' into Equation (6.8.53).

6.18 Consider unsteady diffusion across a membrane of thickness L. Initially, the concentration of solute in the membrane is zero. At time zero, the concentration at $x = 0$ is raised to C_0 and the concentration at $x = L$ is maintained at zero.

(a) Use the following dimensionless variables:

$$\eta = \frac{x}{L} \qquad \theta = \frac{C_m}{C_0} \qquad \tau = \frac{tD_m}{L^2}.$$

Show that the solution for θ can be rewritten as

$$\theta(\eta,\tau) = \psi(\eta) + \phi(\eta,\tau), \quad (6.12.8)$$

where $\psi(\eta)$ is the steady-state solution. State boundary conditions for the problem. Show that

$$\psi(\eta) = 1 - \eta. \quad (6.12.9)$$

(b) Show that $\phi(\eta, \tau)$ can be solved by separation of variables, and obtain the following solution for $\theta(\eta, \tau)$:

$$\theta = 1 - \eta - \frac{2}{\pi}\sum_{n=1}^{\infty}\frac{\sin(n\pi\eta)}{n}\exp(-n^2\pi^2\tau). \quad (6.12.10)$$

(c) Use Equation (6.12.10) to assess the time required for the membrane to reach steady state. Compare this result with the result obtained from the quasi–steady-state analysis.

(d) The following relation is derived from a mass balance:

$$V\frac{dC_1}{dt} = AD_m \frac{\partial C_m}{\partial x}\bigg|_{x=0}, \qquad (6.12.11)$$

where C_1 is the concentration on side 1 of the membrane, as shown in Figure 6.24. At $t = 0$, $C_1 = C_0$. Show that the solution for C_1 is

$$\frac{C_1}{C_0} = 1 - \frac{AD_m}{VL}\left[t + \frac{2L^2}{D_m\pi^2}\sum_{n=1}^{\infty}\frac{1}{n^2} \times (1 - \exp(-n^2\pi^2\tau))\right]. \qquad (6.12.12)$$

(e) Use the result from part (d) to assess the accuracy of the quasi–steady-state analysis.

6.19 Suppose a cell of radius 15 μm contains 100,000 receptors, each 1.5 nm in radius. How close is the diffusion-limited rate constant k_+ to the maximum value for the case when the surface is completely covered with receptors?

6.20 Derive the expression for the diffusion-limited dissociation constant for receptor–ligand dissociation on cell membranes (see Equation (6.9.23)).

6.21 Prove the following relation:

$$erfc(\eta) = \frac{2}{\sqrt{\pi}}\int_{\eta}^{\infty} e^{-x^2}dx \qquad (6.12.13)$$

6.22 Single particle tracking methods enable the determination of molecular diffusion coefficients from the analysis of the mean-square displacement or a random walk. This approach is very useful in measuring diffusion within cells when it is not possible to establish a concentration gradient. The molecule can be labeled with a fluorescent dye and its motion tracked over time. For three-dimensional molecular motion, determine the diffusion coefficient inside the cell for the following data.

Time (s)	Mean-square displacement (μm)
720	4.32
1440	8.64

6.23 Drug-eluting stents release molecules into blood vessels to prevent smooth muscle cells from dividing. A stent manufacturer wants to know the diffusion coefficient of the drug in the tissue in order to evaluate whether the drug can rapidly reach the smooth muscle cells in the vessel wall. To determine the diffusion coefficient of the drugs in the tissue, blood vessels are removed from animals and the lumen is filled with drug. The drug is allowed to diffuse into the tissue and the total amount of drug entering the tissue is measured. The drug concentration in the lumen is assumed constant. For this analysis, the curvature of the blood vessel is ignored (i.e., it is treated as a rectangular region because the vessel thickness is less than the curvature) and diffusion is assumed to be semi-infinite.

(a) Since the flux represents the rate of release per unit area, determine the uptake M over an area A by integrating the product of the flux times cross-sectional area from $t = 0$ to t'. Use this result to determine an expression for the diffusion coefficient as a function of time, the uptake, and area.

(b) Use the following data to determine the diffusion coefficient.

(c) Assess whether the assumption of a semi-infinite tissue is appropriate for this experiment.

M	5×10^{-13} moles
C_0	1×10^{-6} M (mol L^{-1})
t'	25 minutes
A	0.75 cm^2
L (tissue thickness)	0.004 cm

6.24 A polymer sphere of radius 0.2 cm containing a therapeutic protein at 0.1 mg mL^{-1} is placed in a large volume of buffer and the protein diffuses from the polymer to the buffer. The buffer volume is large enough that the protein concentration can be considered zero. The protein diffusion coefficient is 5×10^{-10} cm^2 s^{-1}. Determine the time required for the protein concentration at the center of the sphere to equal 5% of the initial concentration in the polymer.

6.25 A new antibody therapy is being developed to neutralize the human immunodeficiency virus (HIV).

The antibody is added to a polymeric gel and binds to HIV. In order to interpret data in which antibody binding is measured, experiments are performed to measure the diffusion coefficient of the antibody. In these experiments, the gel is uniformly distributed on a microscope slide. To minimize out of focus fluorescence, the gel thickness is 5 μm. A total of 1 μg of the fluorescent antibody is added in a very narrow region that can be considered a line source so that diffusion is one-dimensional. The following data were obtained at a position of 50 μm from the line source.

Distance from source (μm)	C_i/M_o at 3,000 s	C_i/M_o at 6,000 s
500	1.2×10^{-4}	1.6
350	5.2	285
250	770	3700

Use the result for diffusion from a one-dimensional point source in rectangular coordinates to determine the diffusion coefficient.

6.26 One way to produce a constant rate of drug delivery is to have the drug in a reservoir at a high concentration and diffusing across a membrane of thickness l. Such a system is shown schematically in Figure 6.34. Assuming that the drug diffuses rapidly through the skin and enters the blood, then the drug concentration in the skin can be assumed to be zero. For this case, the concentration distribution in the membrane is

$$\frac{C}{\Phi C_0} = 1 - \eta' - 2\sum\frac{\sin(n\pi\eta')}{n\pi}\exp(-n^2\pi^2\tau)$$

where $\eta' = y/l$ and $\tau = tD/l^2$. Here we assume that the drug concentration is constant and steady state is reached for $t = 0.5$.

(a) Show that at steady state, the release rate of the drug at $y = L$ is independent of time. Note that the release rate is the flux times the membrane area, A.

(b) For $D_{ij} = 5\times10^{-9}$ cm^2 s^{-1} $\Phi = 0.1$, and $A = 2.0$ cm^2, determine the thickness and concentration C_0 to ensure that steady state is reached in 10 minutes and the release rate is 0.15 mg h^{-1}.

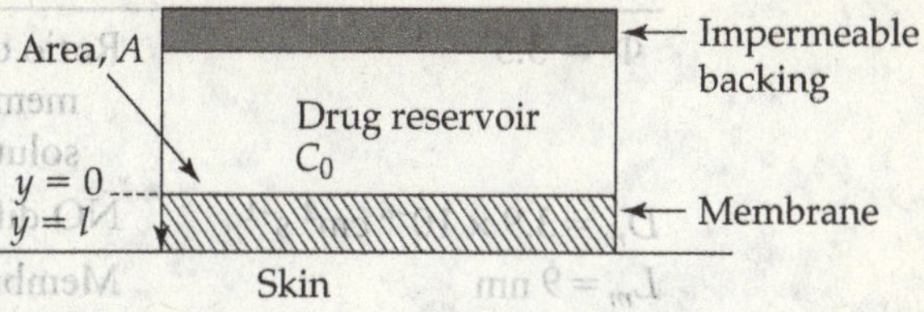

FIGURE 6.34 Diffusion of a drug from a reservoir across a thin membrane.

(c) The concentration of drug can be treated as constant at $y = 0$ if concentration of drug in the reservoir changes by less than 3%. Is this criterion met after the drug delivery system is worn for 10 hours?

6.27 Nitric oxide (NO) plays key roles in regulating the blood vessel diameter and as a neurotransmitter. You are asked to evaluate whether NO transport across the cell membrane is a significant resistance to the overall transport of NO into cells. For this analysis, consider the cell to be a rectangular composite consisting of the cell membrane and cell cytoplasm. Using the data that follows determine the contribution of the cell membrane to the overall permeability.

6.28 Consider the steady-state diffusion of a solute through a membranous sphere as shown in Figure 6.35. Such a situation occurs when drugs are released from liposomes. The membrane is defined by the region between R_0 and R_i. The solute concentration is highest at the core region. The spheres are placed in a large reservoir and the solute concentration in the bath is zero. Assume that the partition coefficient is one.

(a) State the steady-state form of the differential mass balance for diffusion in a sphere with no convection or reaction.

(b) State boundary conditions.

(c) Solve to obtain the concentration distribution in the membrane.

(d) Determine the flux at $r = R_0$.

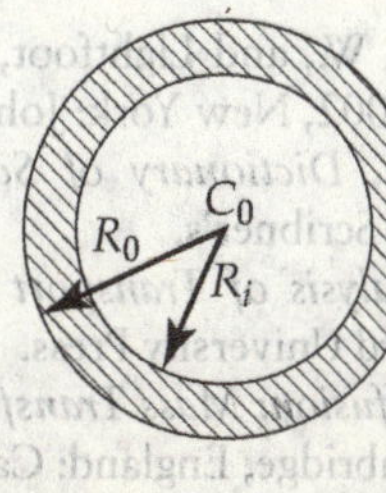

FIGURE 6.35

$\Phi = 3.5$	Ratio of equilibrium concentration of NO in lipid membrane to NO concentration in aqueous solution
$D_m = 3.9 \times 10^{-6}$ cm^2 s^{-1}	NO diffusion coefficient in cell membrane
$L_m = 9$ nm	Membrane thickness
$D_c = 3.3 \times 10^{-5}$ cm^2 s^{-1}	NO diffusion coefficient in cell
$L_c = 5$ μm	Half-thickness of cell
$\Phi_c = 1.0$	NO partition coefficient in cell

6.29 The diffusion of a protein into tissue is studied at short times by measuring the uptake of labeled protein into the tissue of thickness L. Initially, there is no labeled protein in the tissue. At time zero, the surface at $x = 0$ is placed at concentration C_1.

(a) Assuming that the tissue can be treated as a semi-infinite media, what is the flux into the tissue at $x = 0$.

(b) Determine the uptake M by the tissue with surface area A, which is defined as

$$M(t) = A\int_0^t -D\frac{\partial c}{\partial x}\bigg|_{x=0} dt.$$

(c) Determine the diffusion coefficient for the following data:

$$M(t = 600\text{ s}) = 8.74 \times 10^{-13}\text{ mole}$$
$$C_1 = 1 \times 10^{-9}\text{ mol cm}^{-3}$$
$$L = 0.015\text{ cm}$$
$$A = 1\text{ cm}^2$$

(d) Assess whether the semi-infinite medium assumption is valid.

6.30 A gene therapy procedure is being developed to treat muscular dystrophy. As part of this study, the investigators wish to know the diffusion coefficient of DNA in muscle tissue. A small piece of tissue is placed in a diaphragm separating two reservoirs of equal volume.

(a) Initially $C_1 = 0.0015$ M and $C_2 = 0$ and $V_1 = V_2 = V = 100$ mL. At steady state, the concentrations on either side are equal. The concentration in the tissue equals 2×10^{-5} M. Determine the partition coefficient between the membrane and tissue.

(b) After 48 hours from the start of the experiment, the concentration on side 1 has dropped to 10.5×10^{-4} M. Using the data below, determine the diffusion coefficient.

$$A_m = 50\text{ cm}^2$$
$$L = 40\text{ μm}$$

6.13 REFERENCES

1. Bird, R., Stewart, W., and Lightfoot, E., *Transport Phenomena*. 2d ed. 2002, New York: John Wiley and Sons.
2. Gillispie, C., ed., *Dictionary of Scientific Biography*. 1990, New York: Scribner's.
3. Deen, W.M., *Analysis of Transport Phenomena*. 1998, New York: Oxford University Press.
4. Cussler, E.L., *Diffusion: Mass Transfer in Fluid Systems*. 2d ed. 1997, Cambridge, England: Cambridge University Press.
5. Tinico, I., Sauer, K., Wang, J.C., and Puglisi, J.D., *Physical Chemistry: Principles and Applications in Biological Sciences*. 4th ed. 2002, Upper Saddle River, NJ: Prentice Hall.
6. Einstein, A., *Investigations on the Theory of the Brownian Movement*, R. Furth, editor. 1956, New York: Dover Publications, Inc.
7. Berg, H., *Random Walks in Biology*. 1993, Princeton, NJ: Princeton University Press.

8. Reid, R., Prausnitz, J., and Sherwood, T., *The Properties of Gases and Liquids*. 3d ed. 1980, New York: McGraw-Hill.
9. Happel, J., and Brenner, H., *Low Reynolds Number Hydrodynamics*. 1983, Boston: Martinus Nijhoff Publishers.
10. Probstein, R., *Physicochemical Hydrodynamics*. 1989, Boston: Butterworths.
11. Feynman, R.P., Leighton, R.B., and Sands, M., *The Feynman Lectures on Physics. Vol. 1*. 1963, Reading, MA: Addison-Wesley.
12. Lomakin, A., Teplow, D.B., and Benedek, G.B., "Quasi-elastic Light Scattering for Protein Assembly Studies" in *Methods in Molecular Biology, vol. 299: Amyloid Proteins: Methods and Protocols*, E. M. Sigurdsson, editor. 2005, Totowa, NJ: Humana Press Inc.
13. Saxton, M.J., and Jacobson, K. "Single-particle tracking: applications to membrane dynamics." *Annu. Rev. Biophys. Biomol. Struct*, 1997. **26**: pp. 373–99.
14. Squire, P.G., and Himmel, M.E., "Hydrodynamics and protein hydration," *Arch. Biochem. Biophys*., 1979. **196**: 165–177.
15. Preissner, K.T., "Structure and biological role of vitronectin." *Annu. Rev. Cell Biol*., 1991. **7**: pp. 275–310.
16. Doolittle, R., "Fibrinogen and fibrin," in *Hemostasis and Thrombosis*, A. Bloom, et al., editors. 1994, Edinburgh: Churchill Livingstone. pp. 491–513.
17. Brune, D., and Kim, S., "Predicting protein diffusion coefficients." *Proc. Natl. Acad. Sci. USA*, 1993. **90**: pp. 3835–3839.
18. Teller, D., Swanson, E., and De Haen, C., "The translational friction coefficient of proteins." *Methods Enzymol*., 1979. **61**: pp. 103–124.
19. Eisenberg, D., and Crothers, D., *Physical Chemistry with Applications to the Life Sciences*. 1979, Menlo Park, CA: Benjamin/Cummings Publishing Co.
20. Smith, D.E., Perkins, T.T., and Chu, S., "Dynamical scaling of DNA diffusion coefficients." *Macromolecules*, 1996. **29**: pp. 1372–1373.
21. Pluen, A., Netti, P.A., Jain, R.K., and Berk, D.A., "Diffusion of macromolecules in agarose gels: comparison of linear and globular configurations." *Biophys. J*., 1999. **77**: pp. 542–552.
22. Walter, A., and Gutknecht, J., "Permeability of small nonelectrolytes through lipid bilayer membranes," *J. Membr. Biol*., 1986. **90**: 207–217.
23. Andrade, J.D., and Hlady, V., "Plasma protein adsorption: the big twelve." *Ann. NY Acad. Sci*., 1987. **516**: pp. 158–172.
24. Crank, J., *The Mathematics of Diffusion*. 2d ed. 1975, Oxford: Clarendon Press.
25. Hildebrand, F.B., *Advanced Calculus for Applications*. 2d ed. 1976, Englewood Cliffs, NJ: Prentice-Hall. pp. 186–268.
26. Lysaght, M., and Aebischer, P., "Encapsulated cells as therapy." *Sci. Am*., 1999. **280**: pp. 76–82.
27. Øyaas, J., Storrø, I., Svendsen, H., and Levine, D., "The effective diffusion coefficient and the distribution constant for small molecules in calcium-alginate gel beads." *Biotechnol. Bioeng*., 1995. **47**: pp. 492–500.
28. Jo, H., Dull, R.O., Hollis, T.M., and Tarbell, J.M., "Endothelial albumin permeability is shear dependent, time-dependent, and reversible." *Am. J. Physiol*., 1991. **260**: pp. H1992–H1996.
29. Lauffenburger, D.A., and Linderman, J.J., *Receptors: Models for Binding, Trafficking, and Signaling*. 1993: Oxford University Press.
30. Berg, H., and Purcell, E., "Physics of chemoreception." *Biophys. J*., 1977. **20**: pp. 193–219.
31. Hammer, D., "Transport analysis of receptor–ligand interactions," in *Chemical Engineering Problems in Biotechnology*, M. Shuler, editor. 1990, New York: American Institute of Chemical Engineers. pp. 81–105.

CHAPTER 7

Diffusion with Convection or Electrical Potentials

7.1 Introduction

In the previous chapter, steady- and unsteady-state diffusion in one dimension was examined. While there are important physiological examples to which such an analysis applies, in many other cases, diffusion occurs in two and three dimensions or solute transport results from driving forces other than concentration gradient. In this chapter, more general relations are developed and applied to transport in dilute solutions. In Sections 7.2 and 7.3, conservation relations are developed for solute transport in three dimensions. As with the analysis of momentum and forces in Chapter 3, this generalized approach provides a framework for analyzing a wide range of problems, after casting the conservation relations into dimensionless form (Section 7.4).

In Section 7.5, we develop fluxes for the case of electrolyte transport in dilute solutions and apply those fluxes to transport across charged and uncharged membranes. A solute flux can result from a number of driving forces, including gradients in concentration, pressure, temperature, electrical potential, and momentum. Because organisms function at a constant temperature and pressure, evolution has exploited the other three gradients for two purposes. First, electrical potential or momentum gradients are used to transport molecules more rapidly over distances greater than a few hundred microns. Such rapid transport is essential for the rapid transmission of signals in prokaryotes. Second, maintaining electrical or chemical gradients across membranes in cells is an important mechanism for storing energy. Electrical potential and momentum gradients are also extremely important in the isolation and characterization of biological molecules.

The remainder of the chapter focuses upon transport by diffusion and convection. Classical results for transport to membranes or reacting surfaces are presented in Section 7.6. Macroscopic forms of the conservation of mass for solutions are presented in Section 7.7. Mass transfer coefficients, presented in Section 7.8, represent an alternative approach to analyzing convection diffusion problems. Finally, in Section 7.9, these concepts are applied to the analysis of mass exchange devices, such as those used in hemodialysis and blood gas oxygenators.

7.2 Fick's Law of Diffusion and Solute Flux

In Chapter 6, we developed the following general form of Fick's law for a binary system based upon the molar flux relative to the molar average velocity (Equation (6.4.6b)):

$$\mathbf{J}_i = \mathbf{J}_i^* = -CD_{ij}\nabla x_i. \tag{7.2.1}$$

Here, $x_i = C_i/C$ is the mole fraction of component i, C is the total concentration of the solution, and D_{ij} is the binary diffusion coefficient. (Note that $C = \sum_{i=1}^{n} C_i$, where n equals 2 for a binary system.) Fick's law states that a solute moves in the direction of decreasing concentration. In general, the diffusion coefficient is a function of temperature, fluid viscosity, and the chemical activity of the solute. An analogous relation can be stated for $\mathbf{j}_i$ in terms of the mass fraction $\omega_i = \rho_i/\rho$.

When convection and other driving forces for transport are present, Fick's law is often stated with a flux $\mathbf{N}_i$, based upon fixed coordinates, replacing $\mathbf{J}_i^*$. Substituting Equation (7.2.1) into Equation (6.2.6d) and rearranging terms produces

$$\mathbf{N}_i = -CD_{ij}\nabla x_i + C_i\mathbf{v}^*. \tag{7.2.2}$$

Inserting the relation between $\mathbf{v}^*$ and the molar fluxes (Equation (6.2.4b)) into Equation (7.2.2) yields

$$\mathbf{N}_i = -CD_{ij}\nabla x_i + x_i\sum_{i=1}^{n}\mathbf{N}_i. \tag{7.2.3}$$

For dilute solutions, the total mixture concentration is approximately equal to the solvent concentration ($C \approx C_{\text{solv}}$), the density is constant, and the diffusion coefficient is independent of the concentration of the solute. Furthermore, the mass average and molar average velocity are each approximately equal to the solvent velocity ($\mathbf{v} \approx \mathbf{v}^* \approx \mathbf{v}^*_{\text{solv}}$). As a result, Equations (7.2.1) and (7.2.2) become

$$\mathbf{J}_i^* = -D_{ij}^0\nabla C_i, \tag{7.2.4a}$$

$$\mathbf{N}_i = -D_{ij}^0\nabla C_i + C_i\mathbf{v}, \tag{7.2.4b}$$

where D_{ij}^0 is the binary diffusion coefficient at infinite dilution of component i relative to component j. Although Equations (7.2.4a) and (7.2.4b) were derived for a binary solution, they are valid for dilute multicomponent solutions, as is shown in Section 7.3.1. Note that for subsequent analysis in which a dilute solution is assumed, the superscript "0" is omitted from the diffusion coefficient.

7.3 Conservation of Mass for Dilute Solutions

Having specified the solute fluxes when diffusion and convection occur, we now develop conservation relations for transport in three dimensions. These relations are stated for the mass and molar fluxes and are related to the conservation of mass derived in Chapter 3.

In words, the conservation of mass for component i in a multicomponent system is

$$\begin{bmatrix}\text{Rate of accumulation}\\ \text{of } i\end{bmatrix} = \begin{bmatrix}\text{Net rate of flow of}\\ i \text{ across surface } S\end{bmatrix} + \begin{bmatrix}\text{Rate of production of } i\\ \text{by chemical reaction}\end{bmatrix}. \tag{7.3.1}$$

If the component i is a reactant and is consumed by the reaction, then the rate of production is negative. The flow rates are shown in Figure 7.1 for molar fluxes. Mass fluxes would be similar.

In order to relate the conservation of mass for a mixture to the results obtained in Chapter 3 for a pure component (Equation (3.2.7)), the conservation relations are first developed using mass fluxes based on fixed coordinates. In terms of mass fluxes, Equation (7.3.1) is stated as

$$\Delta x \Delta y \Delta z \frac{\partial \rho_i}{\partial t} = \left[n_{ix}|_x - n_{ix}|_{x+\Delta x}\right] \Delta y \Delta z + \left[n_{iy}|_y - n_{iy}|_{y+\Delta y}\right] \Delta x \Delta z + \left[n_{iz}|_z - n_{iz}|_{z+\Delta z}\right] \Delta x \Delta y + r_i\, \Delta x \Delta y \Delta z, \tag{7.3.2}$$

where r_i is the reaction rate based upon mass (mass/volume/time). Dividing each term by the volume element $\Delta x \Delta y \Delta z$ yields

$$\frac{\partial \rho_i}{\partial t} = \frac{n_{ix}|_x - n_{ix}|_{x+\Delta x}}{\Delta x} + \frac{n_{iy}|_y - n_{iy}|_{y+\Delta y}}{\Delta y} + \frac{n_{iz}|_z - n_{iz}|_{z+\Delta z}}{\Delta z} + r_i. \tag{7.3.3}$$

Taking the limit as Δx, Δy, and Δz approach zero, we have

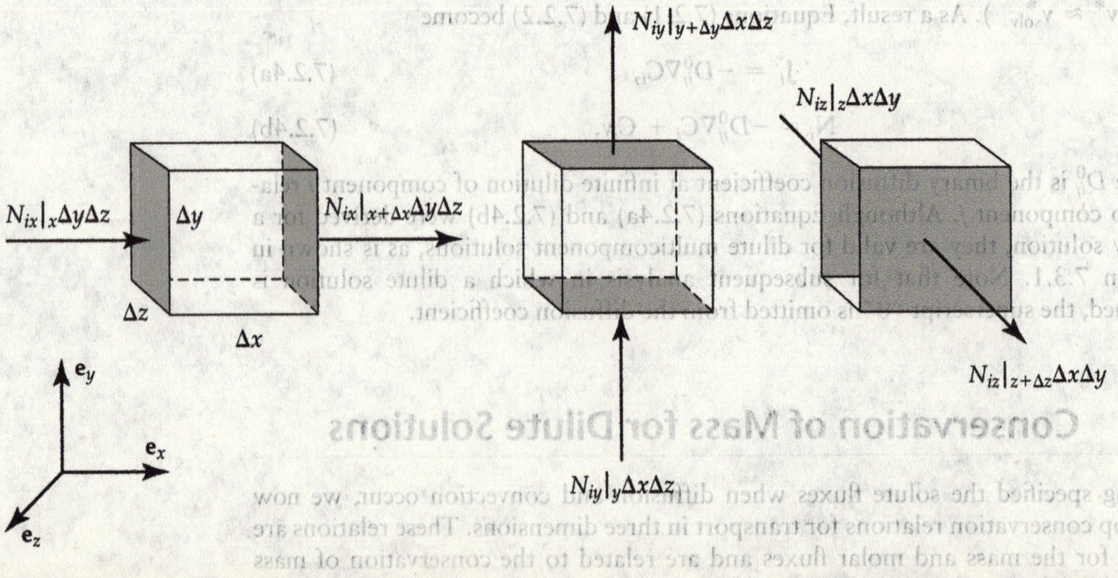

FIGURE 7.1 Schematic of components of solute flux.

$$\frac{\partial \rho_i}{\partial t} = -\left(\frac{\partial n_{ix}}{\partial x} + \frac{\partial n_{iy}}{\partial y} + \frac{\partial n_{iz}}{\partial z}\right) + r_i. \tag{7.3.4}$$

Equation (7.3.4) is one form of the conservation of mass for a multicomponent system. Generalizing this result by using vector notation for the terms in parentheses in Equation (7.3.4) gives

$$\frac{\partial \rho_i}{\partial t} = -\nabla \cdot \mathbf{n}_i + r_i. \tag{7.3.5}$$

Equation (7.3.5) is related to Equation (3.2.7) for a pure system as follows; summing Equation (7.3.5) for all components of the mixture yields

$$\frac{\partial \rho}{\partial t} = -\nabla \cdot \sum_{i=1}^{n} \mathbf{n}_i + \sum_{i=1}^{n} r_i. \tag{7.3.6}$$

The summation of the mass fluxes in Equation (7.3.6) is $\rho\mathbf{v}$ (cf. Equation (6.2.4a)). The sum of the mass reaction rates for all components is zero, since mass is neither created nor destroyed. Equation (7.3.6) can be rewritten in a more compact form as

$$\frac{\partial \rho}{\partial t} = -\nabla \cdot (\rho \mathbf{v}). \tag{7.3.7}$$

Thus, the conservation of mass is valid for both a pure component and all components in a mixture.

In a manner similar to the derivation of Equation (7.3.5), the conservation of mass for a mixture based upon the molar flux relative to fixed coordinates is

$$\frac{\partial C_i}{\partial t} = -\nabla \cdot \mathbf{N}_i + R_i. \tag{7.3.8}$$

We use Equation (7.3.8) as the basis for all subsequent analyses. Table 7.1 lists the equations for rectangular, cylindrical, and spherical coordinates. Note that these are general results.

TABLE 7.1

Conservation of Mass Using Molar Fluxes Based on Fixed Coordinates

Rectangular	$\frac{\partial C_i}{\partial t} = -\left(\frac{\partial N_{i_x}}{\partial x} + \frac{\partial N_{i_y}}{\partial y} + \frac{\partial N_{i_z}}{\partial z}\right) + R_i$	(7.3.9a)
Cylindrical	$\frac{\partial C_i}{\partial t} = -\left(\frac{1}{r}\frac{\partial (rN_{i_r})}{\partial r} + \frac{1}{r}\frac{\partial N_{i_\theta}}{\partial \theta} + \frac{\partial N_{i_z}}{\partial z}\right) + R_i$	(7.3.9b)
Spherical	$\frac{\partial C_i}{\partial t} = -\left(\frac{1}{r^2}\frac{\partial (r^2 N_{i_r})}{\partial r} + \frac{1}{r\sin\theta}\frac{\partial (N_{i\theta}\sin\theta)}{\partial \theta} + \frac{1}{r\sin\theta}\frac{\partial N_{i_\phi}}{\partial \phi}\right) + R_i$	(7.3.9c)

In most problems considered in this text, the solutions are dilute and the fluid is incompressible. Using Equation (7.2.4b) for the molar flux based on fixed coordinates, we may write Equation (7.3.8) as

$$\frac{\partial C_i}{\partial t} = \nabla\cdot(D_{ij}\nabla C_i - C_i\mathbf{v}) + R_i. \tag{7.3.10a}$$

Rearranging Equation (7.3.10a) results in the following expression:

$$\frac{\partial C_i}{\partial t} + \nabla\cdot(C_i\mathbf{v}) = D_{ij}\nabla^2 C_i + R_i. \tag{7.3.10b}$$

The second term on the left-hand side of Equation (7.3.10b) can be expressed as

$$\nabla\cdot(C_i\mathbf{v}) = C_i\nabla\cdot\mathbf{v} + \mathbf{v}\cdot\nabla C_i = \mathbf{v}\cdot\nabla C_i, \tag{7.3.11}$$

where we have used the relation $\nabla\cdot\mathbf{v} = 0$ for an incompressible fluid. Thus, the general form of the conservation of mass for dilute solutions that are incompressible and that follow Fick's law is

$$\frac{\partial C_i}{\partial t} + \mathbf{v}\cdot\nabla C_i = D_{ij}\nabla^2 C_i + R_i. \tag{7.3.12}$$

Note that the left-hand side of Equation (7.3.12) is the substantial derivative of the concentration (cf. Equation (3.2.10)). Table 7.2 expresses Equation (7.3.12) in rectangular, cylindrical, and spherical coordinates.

The next example considers the effect of convection upon the concentration profile and solute flux. Biological membranes and synthetic membranes are permeable to water as well as to the solute. Pressure differences across the membranes produce fluid flow, which gives rise to convective transport of solutes across the membrane. Due to the pressure difference between capillaries and lymphatic vessels in tissues, fluid and proteins are convected into tissues (see Chapters 8 and 9). Alterations in fluid flow across endothelium resulting from the action of a number of compounds, such as histamine, that affect endothelial structures can produce a condition known as *edema*, in which fluid accumulates in the tissues.

TABLE 7.2

Conservation Relations for Dilute Solutions

Rectangular	$\frac{\partial C_i}{\partial t} + v_x\frac{\partial C_i}{\partial x} + v_y\frac{\partial C_i}{\partial y} + v_z\frac{\partial C_i}{\partial z} = D_{ij}\left(\frac{\partial^2 C_i}{\partial x^2} + \frac{\partial^2 C_i}{\partial y^2} + \frac{\partial^2 C_i}{\partial z^2}\right) + R_i$	(7.3.13a)
Cylindrical	$\frac{\partial C_i}{\partial t} + v_r\frac{\partial C_i}{\partial r} + \frac{v_\theta}{r}\frac{\partial C_i}{\partial \theta} + v_z\frac{\partial C_i}{\partial z} = D_{ij}\left(\frac{1}{r}\frac{\partial}{\partial r}\left(r\frac{\partial C_i}{\partial r}\right) + \frac{1}{r^2}\frac{\partial^2 C_i}{\partial \theta^2} + \frac{\partial^2 C_i}{\partial z^2}\right) + R_i$	(7.3.13b)
Spherical	$\frac{\partial C_i}{\partial t} + v_r\frac{\partial C_i}{\partial r} + \frac{v_\theta}{r}\frac{\partial C_i}{\partial \theta} + \frac{v_\phi}{r\sin\theta}\frac{\partial C_i}{\partial \phi}$ $= D_{ij}\left(\frac{1}{r^2}\frac{\partial}{\partial r}\left(r^2\frac{\partial C_i}{\partial r}\right) + \frac{1}{r^2\sin\theta}\frac{\partial}{\partial \theta}\left(\sin\theta\frac{\partial C_i}{\partial \theta}\right) + \frac{1}{r^2\sin^2\theta}\frac{\partial^2 C_i}{\partial \phi^2}\right) + R_i$	(7.3.13c)

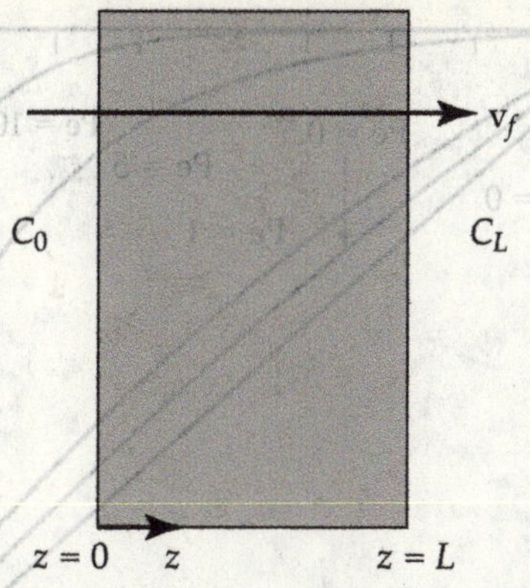

FIGURE 7.2 Schematic of transport across a membrane with convection.

Example 7.1 Consider steady-state one-dimensional diffusion and convection across a membrane of thickness L (Figure 7.2). Such a membrane may be a synthetic membrane or a cellular layer. No chemical reactions are occurring. At $z = 0$, $C = \Phi C_0$ and at $z = L$, $C = \Phi C_L$, where Φ is the partition coefficient of the solute in the membrane. Determine the concentration profile and flux across the membrane, and compare it with results for diffusion only.

Solution From the statement of the problem, R_i equals zero, and the only nonzero velocity is in the z direction. The velocity v_z is constant and equal to v_f. With these simplifications, Equation (7.3.13a) reduces to

$$v_f \frac{dC_i}{dz} = D_{\text{eff}} \frac{d^2C_i}{dz^2}, \tag{7.3.14}$$

where v_f is called the filtration velocity across the membrane.[1] Integrating Equation (7.3.14) yields the following solution:

$$C_i = A \exp\left(\frac{v_f}{D_{\text{eff}}} z\right) + B. \tag{7.3.15}$$

Applying the boundary conditions at $z = 0$ and $z = L$, we find the constants A and B. The resulting concentration, as a function of position, is

$$C_i = \Phi C_0 - \Phi(C_0 - C_L)\left[\frac{1 - \exp(\text{Pe}_L(z/L))}{1 - \exp(\text{Pe}_L)}\right], \tag{7.3.16}$$

where the Peclet number $\text{Pe}_L = v_f L/D_{\text{eff}}$, represents the relative importance of diffusion and convection (Section 1.3). Equation (7.3.16) is plotted in Figure 7.3 in terms of $(C_i - \Phi C_L)/(\Phi C_0 - \Phi C_L)$. For small values of the Peclet number, diffusion is much more important than convection, and the profile is linear. This linear profile at low values of the Peclet number can be obtained by expanding the

[1]For larger solutes, the solute velocity lags behind the fluid velocity due to hydrodynamic drag forces resulting from the membrane matrix. This effect is incorporated by multiplying v_f by W, the hydrodynamic resistance factor for convection. See Section 14.4 and Equation (14.4.6).

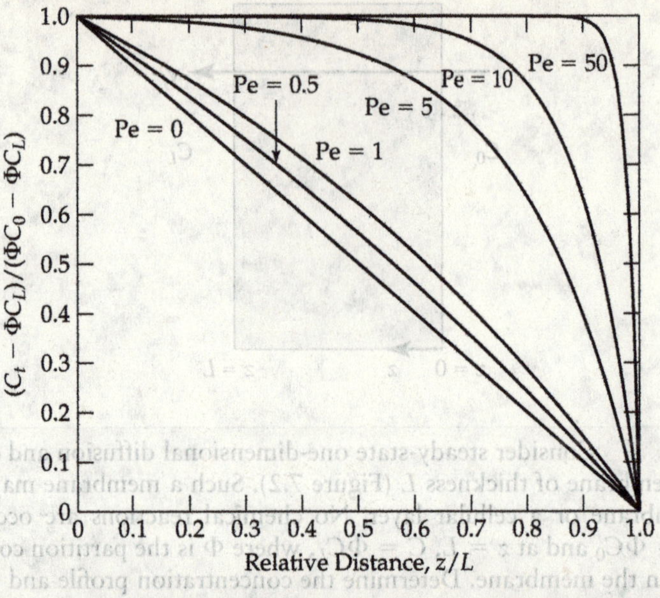

FIGURE 7.3 Effect of convection upon concentration distribution across membrane.

exponential as $\exp(x) \approx 1 + x \ldots$, which is valid for small values of the argument ($x < 0.1$). For larger values of the Peclet number, convection becomes increasingly important and the concentration decreases slowly near $z = 0$. Thus, the curve exhibits a distinct concave shape.

At larger values of the Peclet number, the profile is flat throughout most of the membranes until close to $z = L$, at which point the concentration drops rapidly to ΦC_L.

The flux at any location is

$$N_{i_z} = -D_{ij}\frac{dC_i}{dz} + C_i v_f = \frac{\Phi D_{ij}}{L}\mathrm{Pe}_L\left[C_0 - \frac{(C_0 - C_L)}{1 - \exp(\mathrm{Pe}_L)}\right]. \quad (7.3.17)$$

In the limit as the Peclet number goes to zero, the flux approaches the diffusion limit, Equation (6.7.10). Conversely, for very large values of the Peclet number, the flux approaches the limiting value, $\Phi C_0 v_f$, for a process dominated by convection. In the high Peclet number limit, the flux is controlled solely by the filtration velocity and the concentration at $z = 0$.

7.3.1 Transport in Multicomponent Mixtures

The previous results apply strictly to dilute solutions, which is a reasonable assumption for many biological applications. In general, the flux of a component of a mixture depends upon the concentration gradient of all components, as well as on gradients in temperature, electrical potential, and gravity. While most biological and biotechnological systems operate isothermally, electrical potential gradients exist across cells, and electrical potential gradients and gravity are used to separate biological macromolecules. Thus, to gain a wider appreciation of the manner by which a solute can undergo transport, we briefly examine the multicomponent transport equations.

Although there are many solutes in a typical biological solution, the solutes are dilute. The question then arises as to whether Fick's law can be used for these mixtures. To address this issue, we first consider the general relation for multicomponent transport and then determine what simplifications can be made for a dilute solution.

In the general case of a multicomponent mixture containing n components, the movement in one component[2] of the mixture is significant enough to produce movement in other components of the mixture. As a result, the component fluxes are coupled. Under isothermal conditions, the diffusion flux can be generalized to [1,2][3]

$$\mathbf{J}_i = -C_i \sum_{j=1}^{n} \mathcal{D}_{ij}\mathbf{d}_j, \tag{7.3.18}$$

where $\mathcal{D}_{ij}$ are the multicomponent diffusion coefficients and $\mathbf{d}_j$ represents the diffusion driving force of component j. For gases, the multicomponent diffusion coefficients equal the binary diffusion coefficients ($\mathcal{D}_{ij} = D_{ij}$). For binary liquid solutions and dilute liquid solutions, the multicomponent and binary diffusion coefficients are equivalent; for concentrated multicomponent mixtures, the equality is an approximation [1]. The multicomponent (and binary) diffusion coefficients are symmetric ($\mathcal{D}_{ij} = \mathcal{D}_{ji}$). The diffusion driving force is [1,2]

$$CRT\mathbf{d}_i = C_i\nabla\mu_i + (C_i\overline{V}_i - \omega_i)\nabla P - \rho_i\left(\mathbf{g}_i - \sum_{k=1}^{n}\omega_k\mathbf{g}_k\right), \tag{7.3.19}$$

where R is the universal gas law constant (8.314 J mol^{-1} K^{-1}), T is the absolute temperature (K), μ_i is the chemical potential of component i, $\overline{V}_i$ is the partial molar volume of component i, ω_i is the mass fraction, and $\mathbf{g}_i$ represents the body forces (including gravity and electrical potentials) acting on component i.

In Equation (7.3.19), the diffusion driving force $\mathbf{d}_i$ represents the sum of the forces acting on each component of the mixture, and $CRT\mathbf{d}_i$ has units of force per unit volume. In Section 6.6.2, diffusion is discussed as a balance between thermal forces and drag forces on the solute. This analysis can be generalized if other forces are present. Drag forces retard the motion generated by other forces. Thus, Equation (7.3.19) represents a macroscopic generalization of the various forces that cause the selective molecular motion. From thermodynamic arguments [1,2], the sum of all the $\mathbf{d}_i$ is zero.

The application of irreversible thermodynamics to multicomponent mixtures results in the following relation between the diffusion flux and the diffusion velocities:

$$\mathbf{d}_i = \sum_{\substack{j=1 \\ j\neq k}}^{n} \frac{x_i x_j}{\mathrm{D}_{ij}}(\mathbf{v}_j - \mathbf{v}_k), \tag{7.3.20}$$

where k is an arbitrary number between 1 and n.

[2]Generally, the solvent is assumed to be the most abundant component. With very concentrated solutions, it may not be feasible to make such a clear distinction between solute and solvent.

[3]Irreversible thermodynamics is used in the derivation of Equation (7.3.18). Conservation relations are developed for entropy and energy, assuming local equilibrium. In addition, the fluxes are assumed to be linearly related to gradients in concentration, electrical potential, pressure, temperature, stresses, and body forces. Further details and references are provided in [1].

The relations represented by Equation (7.3.20) for $n-1$ components of the mixture are known as the generalized *Stefan–Maxwell equations*. Since the $\mathbf{d}_i$ sum to zero, only $n-1$ equations are needed.

Example 7.2 Show that the generalized Stefan–Maxwell equations for a binary mixture reduce to the expressions that were developed for a dilute binary mixture—that is, Equation (6.4.7)—when only diffusion occurs.

Solution For a binary mixture, only one relation is needed and $k = 1$. Equation (7.3.20) becomes

$$\mathbf{d}_1 = \frac{x_1 x_2}{D_{12}}(\mathbf{v}_2 - \mathbf{v}_1), \tag{7.3.21}$$

where 1 is the solute and 2 is the solvent. For a dilute solution, $x_2 \approx 1$, $C_2 \approx C$, and $x_1 C_2 \approx C_1$. Using the definition of the flux $\mathbf{N}_1$ based on fixed coordinates (Equation (6.2.3b)), Equation (7.3.21) becomes

$$C D_{12}\mathbf{d}_1 = C_1\mathbf{v}_2 - \mathbf{N}_1. \tag{7.3.22}$$

In the absence of body and surface forces, Equation (7.3.19) simplifies to

$$CRT\mathbf{d}_i = C_i\nabla\mu_i = Cx_i\nabla\mu_i. \tag{7.3.23}$$

The chemical potential is defined by Equation (6.4.9). For dilute solutions, the activity coefficient γ_i is unity, and Equation (7.3.23) simplifies to

$$CRT\mathbf{d}_i = Cx_i\nabla\mu_i = CRTx_i\nabla \ln x_i = CRT\nabla x_i. \tag{7.3.24}$$

Thus, $CD_{12}\,\mathbf{d}_1 = CD_{12}\nabla x_1 = -\mathbf{J}_1$. Substituting this result into Equation (7.3.23) and rearranging terms yields Equation (6.4.7), assuming that, for dilute binary solutions, the multicomponent diffusion coefficients equal the binary diffusion coefficients.

For dilute solutions, the Stefan–Maxwell relations indicate that the diffusion of each solute in a multicomponent mixture can be examined separately. To derive this result, we set the subscript n in Equation (7.3.20) to be the solvent and the subscripts i and k to unity. The diffusion driving force can then be written as

$$\mathbf{d}_1 = -\frac{x_1 x_s}{D_{1s}}(\mathbf{v}_1 - \mathbf{v}_s) + \sum_{j=2}^{n-1}\frac{x_1 x_j}{\mathrm{D}_{1j}}(\mathbf{v}_j - \mathbf{v}_1). \tag{7.3.25}$$

Because the solvent is the dominant component ($x_s \approx 1$, $x_j \ll 1$), $x_1 x_j \ll x_1$. As a result, the second term on the right-hand side of Equation (7.3.25) is much smaller than the first term and can be neglected. Thus, from Example 7.2, each of the fluxes can be written in terms of Equation (7.2.4b) as

$$\mathbf{N}_i = -D_{is}\nabla C_i + C_i\mathbf{v}_s. \tag{7.3.26}$$

This result is analogous to Equation (6.4.7) and indicates that dilute multicomponent solutions behave as a series of binary solutions in which the flux of each solute depends only on transport relative to the solvent. (Such a solution is known as a *pseudobinary mixture.*) The result provides the justification for treating dilute biological solutions as binary mixtures of the individual solutes diffusing in the solvent liquid. To use a biomedical example, the flux of an individual plasma protein in blood depends only upon the diffusion and convection of each protein relative to the solvent.

7.4 Dimensional Analysis

Applying an analysis similar to that used in Chapter 3, we cast Equation (7.3.12) into dimensionless form. The characteristic concentration is the initial concentration C_0. The characteristic length is L. The characteristic velocity is the average velocity $\langle \mathrm{v} \rangle$. The diffusion time is L^2/D_{ij}. The corresponding dimensionless quantities are

$$C^* = \frac{C_i}{C_0} \qquad \mathbf{x}^* = \frac{\mathbf{x}}{L} \qquad \mathbf{v}^* = \frac{\mathbf{v}}{\langle \mathrm{v} \rangle} \qquad t^* = \frac{tD_{ij}}{L^2},$$

where $\mathbf{x}^*$ is the dimensionless position vector and t^* is known as the *Fourier number*, which scales time with the characteristic diffusion time. Applying the same methods used with the conservation of mass and the Navier–Stokes equation, we obtain the dimensionless form of Equation (7.3.12):

$$\frac{\partial C_i^*}{\partial t^*} + \frac{\langle \mathrm{v} \rangle L}{D_{ij}} \mathbf{v}^* \cdot \nabla^* C_i^* = \nabla^{*2} C_i^* + \frac{R_i L^2}{C_0 D_{ij}}. \tag{7.4.1}$$

Casting Equation (7.3.12) into the dimensionless form (7.4.1) produces two new dimensionless groups. The first of these, known as the Peclet number, is given by

$$\mathrm{Pe} = \frac{\langle \mathrm{v} \rangle L}{D_{ij}} \tag{7.4.2}$$

As discussed above and in Chapter 1, the Peclet number represents the ratio of the diffusion time L^2/D_{ij} to the convection time $L/\langle \mathrm{v} \rangle$. If the diffusion time is much less than the convection time, diffusion is the faster method of transport, and the Peclet number is much less than unity. In this case, Equation (7.4.1) reduces to

$$\frac{\partial C_i^*}{\partial t^*} = \nabla^{*2} C_i^* + \frac{R_i L^2}{C_0 D_{ij}}. \tag{7.4.3}$$

If the diffusion time is approximately equal to the convection time, the Peclet number is about unity. If the diffusion time is long relative to the time required to transport the molecule by convection, then convection is the faster method of transport, and the Peclet number is much greater than unity.

The Peclet number can be rewritten as the product of the Reynolds number and the Schmidt number (ν/D_{ij}):

$$\mathrm{Pe} = \mathrm{ReSc}. \tag{7.4.4}$$

The Schmidt number represents the ratio of viscous transport to diffusive transport and is a property of the fluid and the solute. For gases, the Schmidt number is about unity. For the diffusion of low-molecular-weight solutes in water, D_{ij} is about 1×10^{-5} cm^2 s^{-1}, and the Schmidt number is about 1,000. For protein diffusion in aqueous solutions, the Schmidt number ranges from 10^4 to 10^5. Thus, even for relatively low-Reynolds-number flow in aqueous solutions, the Peclet number may be significant.

For high-Peclet-number problems, diffusion may still be important, although not in the direction of convective transport. Separate length scales exist, resulting in different contributions of diffusion and convection in each direction. To understand the balance between convection and diffusion under these circumstances, consider flow of a solution in a cylindrical tube in which a solute reacts on the surface of the tube at $r = R$. The velocity is fully developed and is a function of r only. Velocities and concentrations in the angular direction are uniform. There are no chemical reactions in the tube. For solute transport, Equation (7.3.13b) simplifies to

$$\frac{\partial C_i}{\partial t} + v_z \frac{\partial C_i}{\partial z} = D_{ij}\left(\frac{1}{r}\frac{\partial}{\partial r}\left(r\frac{\partial C_i}{\partial r}\right) + \frac{\partial^2 C_i}{\partial z^2}\right). \tag{7.4.5}$$

There are two length scales: $2R$ in the radial direction and L in the axial direction, where R is much smaller than L. The z and r are made dimensionless by using L and R, respectively. That is, $z^* = z/L$ and $r^* = r/R$, because we expect that diffusion is important in the r direction. The dimensionless time t^* and the Peclet number are defined using R instead of L. The dimensionless time and radial Peclet number are $t^* = tD_{ij}/4R^2$ and $\mathrm{Pe}_R = \langle v \rangle 2R/D_{ij}$. In dimensionless form, Equation (7.4.5) becomes

$$\frac{\partial C^*}{\partial t^*} + \mathrm{Pe}_R \frac{2R}{L} v_z^* \frac{\partial C^*}{\partial z^*} = 4\left[\frac{1}{r^*}\frac{\partial}{\partial r^*}\left(r^*\frac{\partial C^*}{\partial r^*}\right) + \left(\frac{R}{L}\right)^2 \frac{\partial^2 C^*}{\partial z^{*2}}\right], \tag{7.4.6}$$

where v_z^* is the z component of v^* as previously defined.

Since $R \ll L$, the second term on the right-hand side of Equation (7.4.6) is smaller than the first term and can be neglected. Thus, at steady state, axial diffusion is much less than axial convection, and axial convection is balanced by radial diffusion.

Example 7.3 Estimate the Peclet number for the axial transport of a protein in a blood capillary.

Solution From the data in Table 2.4, the average velocity in a capillary is 0.02–0.17 cm s^{-1}, and the length is 0.06 cm. The kinematic viscosity is 0.03 cm^2 s^{-1} and the protein diffusion coefficient is 6.0×10^{-7} cm^2 s^{-1}.

For these conditions, the Reynolds number is between 0.04 and 0.34. The Schmidt number is 5×10^4. The axial Peclet number ranges from 2.0×10^3 to 1.7×10^4, suggesting that axial diffusion is unimportant. Although convection is a significant means of transport, radial diffusion cannot be neglected, as discussed in Section 7.6.

The second dimensionless group is the dimensionless reaction rate, given by

$$R_i^* = \frac{R_i L^2}{C_0 D_{ij}}. \tag{7.4.7}$$

This dimensionless group is the ratio of the diffusion time L^2/D_{ij} to the reaction time C_0/R_i. If the diffusion time is rapid relative to the reaction time, the chemical reaction does not affect the concentration profile and $R_i^* \ll 1$, which means that this term can be neglected. Alternatively, if the diffusion time is slow relative to the reaction time, then the reactant is rapidly consumed, and significant concentration gradients may arise.

Several limiting cases of Equation (7.3.12) arise. If convection (Pe = 0) and chemical reactions ($R_i = 0$) are unimportant, then Equation (7.3.12) reduces to

$$\frac{\partial C_i}{\partial t} = D_{ij} \nabla^2 C_i. \tag{7.4.8}$$

This is the unsteady diffusion equation, also known as *Fick's second law*. Solutions of this equation were discussed in Chapter 6. When the concentration no longer changes with time (i.e., a steady state is reached), Equation (7.4.8) reduces to

$$\nabla^2 C_i = 0. \tag{7.4.9}$$

Equation (7.4.9) is known as *Laplace's equation*. Some simple one-dimensional examples of Equation (7.4.8) were solved in Chapter 6. A two-dimensional case of Laplace's equation (3.4.10b) was considered when we examined fluid flow in a channel of finite width. Equations (7.4.8) and (7.4.9) are homogeneous and can often be solved by the method of separation of variables. A large set of solutions of Laplace's equation has been compiled for diffusion [3].

In addition to the dimensionless groups for mass transport introduced previously, a number of other dimensionless groups that arise in problems involving mass transfer and chemical reactions are listed in Table 7.3. In general, these groups can be viewed as ratios of transport times. Many of the groups are discussed in subsequent chapters.

7.5 Electrolyte Transport

Many biological molecules are charged, and their transport is influenced by electrical potential gradients. Such gradients arise naturally due to differences in ion concentration across cell membranes, as well as to localization of charges on cell membrane surfaces and the extracellular matrix. In addition, a number of techniques used to characterize biological molecules in a solution apply an electric field to separate DNA or proteins on the basis of their charge and size. The most commonly used method involves gel electrophoresis.

TABLE 7.3

Dimensionless Groups Arising in Mass Transfer and Chemical Reactions

Group	Definition	Physical interpretation	Applications
Schmidt number	$Sc = \dfrac{\nu}{D_{ij}}$	$\dfrac{\text{Momentum transport}}{\text{Diffusive transport}}$	Convective-diffusion problems
Fourier number	$t^* = \dfrac{tD_{ij}}{L^2}$	$\dfrac{\text{Time}}{\text{Diffusion time}}$	Unsteady diffusion
Dimensionless residence time	$\tau = \dfrac{t\langle v\rangle}{L}$	$\dfrac{\text{Time}}{\text{Residence time}}$	Flow problems
Peclet number	$Pe = \dfrac{\langle v\rangle L}{D_{ij}} = ReSc$	$\dfrac{\text{Diffusion time}}{\text{Convection time}}$	Convective-diffusion problems
Sherwood number	$Sh = \dfrac{k_f L}{D_{ij}}$	$\dfrac{\text{Mass transfer}}{\text{Diffusion}}$	Convective-diffusion problems
Biot number	$Bi = \dfrac{k_f L}{D_{\text{eff}}}$	$\dfrac{\text{Mass transfer}}{\text{Internal diffusion}}$	Interphase mass transfer
Damkohler number	$Da = \dfrac{kL}{k_f}$	$\dfrac{\text{Chemical reaction}}{\text{Mass transfer}}$	Mass transfer and surface reaction
Thiele modulus	$\phi = \sqrt{\dfrac{k_n L^2 C^{n-1}}{D_{ij}}}$	$\dfrac{\text{Diffusion time}}{\text{Reaction time}}$	Chemical reactions and diffusion
Reaction rate modulus	$R_i^* = \dfrac{R_i L^2}{C_0 D_{ij}}$	$\dfrac{\text{Diffusion time}}{\text{Reaction time}}$	Chemical reactions and diffusion

k_f is the mass transfer coefficient (length time^{-1}).
k is the rate coefficient for the first-order reaction (time^{-1}).
D_{eff} is the effective diffusion coefficient in the solid phase (length^2 time^{-1}).
k_n is the reaction rate coefficient for a reaction of order n ((volume/moles)$^{n-1}$ time^{-1}).

7.5.1 Nernst–Planck Equation

When an ion in a dilute solution is subjected to an electric field, the flux produced is the sum of the mobilities induced by the electrical field, diffusion, and bulk convection. Transport resulting from each of these driving forces may not be in the same direction. The flux due to the electrical potential gradient is simply the product of the concentration and the velocity of ion migration $\mathbf{v}_{mi}$:

$$\mathbf{J}_i^{*e} = C_i \mathbf{v}_{mi}. \tag{7.5.1}$$

Since the movement of a molecule occurs at low Reynolds numbers (when the Reynolds number is based on the size of the molecule), the migration velocity is simply the force per mole of molecules divided by the frictional coefficient multiplied by Avogadro's number $\mathbf{F}/(\bar{f} N_A)$ The electrical force acting on a mole of ion is the product of the electrical potential gradient, ($\nabla\psi$), and the charge and Faraday's constant F (96,487 C mol^{-1}) [4].[4] The expression for the flux due to an applied electrical potential is

$$\mathbf{J}_i^{*e} = -\frac{C_i z_i F}{\bar{f} N_A} \nabla\psi, \tag{7.5.2}$$

where z_i is the net charge on the molecule and ψ is the electrical potential (in volts). The flux and the potential gradient are in the same direction for anions and in opposite directions for cations.

Equation (7.5.2) is simplified by using the relation between the frictional coefficient and the diffusion coefficient (Equation (6.6.11), $\bar{f} = k_B T/D_{ij}$); the universal gas constant R is the product of Avogadro's number and the Boltzmann constant k_B. Substituting these expressions into Equation (7.5.2) yields

$$\mathbf{J}_i^{*e} = -\frac{D_{ij} C_i z_i F}{RT} \nabla\psi. \tag{7.5.3}$$

This result is analogous to diffusion in that the flux is proportional to a gradient. On the basis of fixed coordinates, the flux becomes

$$\mathbf{N}_i = -D_{ij}\nabla C_i - \frac{D_{ij} C_i z_i F}{RT}\nabla\psi + C_i\mathbf{v}. \tag{7.5.4}$$

Equation (7.5.4) is referred to as the *Nernst–Planck equation*.[5] This equation is valid for dilute solutions. For more concentrated solutions, the fluxes and ionic currents interact [1,5].

[4]Faraday's constant is the product of the elementary charge (1.602×10^{-19} coulombs) and Avogadro's number.

[5]An alternative derivation uses the diffusion flux for dilute solutions [1]. In Equation (7.3.19) the term ρg_i represents the body force per unit volume. When an electrical potential is applied, the body force acting on solute i per unit volume is $z_i C_i F \nabla\psi$. Since the mass fraction is related to the mole fraction as

$$x_i = \frac{\omega_i/M_i}{\sum_{i=1}^{n} \omega_i/M_i},$$

where M_i is the molecular weight of component i, Equation (7.3.19) simplifies to

$$\mathbf{J}_i = -D_{ij} C \mathbf{d}_i = -D_{ij} C \nabla x_i + D_{ij}\frac{z_i C_i F}{RT}\nabla\psi\left(1 - \frac{M_i}{z_i C \sum_{i=1}^{n}(\omega_i/M_i)}\sum_{k=1}^{n} z_i C_i\right).$$

By electroneutrality (Equation (7.5.5)), the summation of the products of the charges and the concentrations is zero. Since $\mathbf{N}_i = \mathbf{J}_i + C\mathbf{v}_i$, this result is equivalent to Equation (7.5.4).

Two other relations are needed to characterize electrolyte transport fully. The first of these is called *electroneutrality* and requires that the sum of the charges in solution equals zero. For n ions, electroneutrality is written as

$$\sum_{i=1}^{n} C_i z_i = 0. \tag{7.5.5}$$

Near a surface, electroneutrality may not be satisfied. In this case, the sum in Equation (7.5.5) is defined as the charge density.

The second relation is that the current is the sum of the ion fluxes, or

$$\mathrm{i} = F\sum_{i=1}^{n} \mathrm{N}_i z_i. \tag{7.5.6}$$

Inherent in this relation (and subsequent ones) is the assumption that the ion fluxes are independent of one another.

As an application of electrolyte transport, consider a 1:1 electrolyte (M^+X^-) solution between two electrodes, as shown in Figure 7.4. When an electrical potential $\Delta\psi$ is applied across the electrodes, the positively charged ion (the cation) M^+ migrates to the negative electrode (the cathode) at $z = 0$, where it reacts with an electron e^- to form the metal M as follows:

$$z = 0 \quad M^+ + e^- \rightarrow M. \tag{7.5.7a}$$

Free electrons are generated at the anode ($z = L$), where the metal dissolves according to the reaction

$$z = L \quad M \rightarrow M^+ + e^-. \tag{7.5.7b}$$

Due to electroneutrality, the negatively charged anion concentration is equal to the cation concentration at all locations within the solution; that is,

$$C_{M^+} = C_{X^-} = C. \tag{7.5.8}$$

Although the concentrations are the same, the fluxes are not, as is shown next.

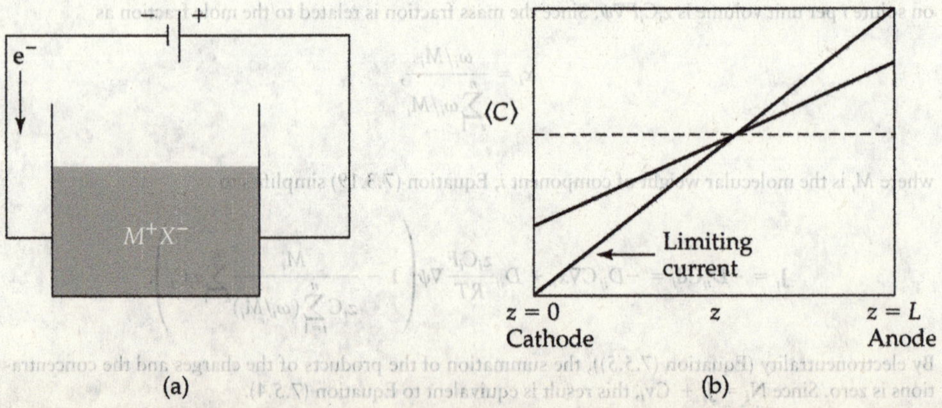

FIGURE 7.4 (a) Schematic of electrical potential difference $\Delta\psi$ applied to a solution of the 1:1 electrolyte M^+X^-. (b) Graph of steady-state profile of cation concentration $\langle C \rangle$.

Consider the case of ion movement in an electrolyte to which a current has been applied under steady-state conditions, with no bulk flow ($v = 0$) and no homogeneous reactions ($R_i = 0$). The conservation relations for the cation and anion are

$$\frac{\partial N_{M^+z}}{\partial z} = \frac{\partial N_{X^-z}}{\partial z} = 0, \tag{7.5.9}$$

where the fluxes represent the sum of the fluxes due to diffusion and electrical potentials. Integrating Equation (7.5.9) indicates that the fluxes of the anion and cation are each equal to a constant:

$$N_{M^+z} = C_1, \tag{7.5.10a}$$

$$N_{X^-z} = C_2. \tag{7.5.10b}$$

Since the anion does not react at $z = 0$ and $z = L$, $C_2 = 0$. The constant C_1 is related to the current density by substituting Equations (7.5.10a) and (7.5.10b) into Equation (7.5.6), yielding

$$i = FN_{M^+z} = FC_1, \tag{7.5.11}$$

where i is the magnitude of current in the z-direction.

Since the applied voltage and the resistance of the solution are known in principle, the current density i can be considered to be known. Using the Nernst–Planck equation for the fluxes, we have

$$N_{M^+z} = \frac{i}{F} = -D_+\frac{dC}{dz} - \frac{D_+CF}{RT}\frac{d\psi}{dz}, \tag{7.5.12a}$$

$$N_{X^-z} = 0 = -D_-\frac{dC}{dz} + \frac{D_-CF}{RT}\frac{d\psi}{dz}. \tag{7.5.12b}$$

The potential gradient is determined from Equation (7.5.12b) and is

$$\frac{d\psi}{dz} = \frac{RT}{CF}\frac{dC}{dz}. \tag{7.5.13}$$

Next, we substitute Equation (7.5.13) into Equation (7.5.12a) to yield

$$\frac{i}{F} = -2D_+\frac{dC}{dz}. \tag{7.5.14}$$

Integrating Equation (7.5.14) and applying the boundary conditions $C = C_0$ at $z = 0$ and $C = C_L$ at $z = L$ yields

$$C = C_0 - \frac{z}{L}(C_0 - C_L). \tag{7.5.15}$$

The current density is

$$i = 2FD_+\frac{(C_0 - C_L)}{L}. \tag{7.5.16}$$

The maximum or limiting current occurs when C_0 vanishes:

$$i_{max} = \frac{2FD_+C_L}{L}. \tag{7.5.17}$$

Lastly, the potential difference across the solution can be found by integrating Equation (7.5.13). The result is

$$\Delta\psi = \frac{RT}{F}\ln\left(\frac{C_L}{C_0}\right). \tag{7.5.18}$$

Note that the potential difference is infinite at the limiting current. Equation (7.5.18) is known as the *Nernst equation* and also represents the potential difference across a membrane.

The previous results can be generalized for an arbitrary *binary electrolyte*. Such an electrolyte can be represented as $M_{v+}^{z+}\ X_{v-}^{z-}$, where the superscripts $z+$ and $z-$ refer to the positive and negative charges, respectively, and the subscripts $v+$ and $v-$ refer to the number of cations and anions per molecule, respectively. For example, the salt calcium chloride is $CaCl_2$, where $z+$ is 2, $z-$ is -1, $v+$ is 1, and $v-$ is 2. If the salt concentration is C, it is related to the ion concentrations as

$$C = \frac{C_+}{v_+} = \frac{C_-}{v_-}. \tag{7.5.19}$$

Substituting Equation (7.5.4) into Equation (7.3.8) and assuming no chemical reactions, we obtain separate conservation relations for the cation and anion. Using Equation (7.5.19) we obtain

$$\frac{\partial C}{\partial t} + \mathbf{v}\cdot\nabla C = D_+\nabla^2 C + \frac{D_+z_+F}{RT}\nabla\cdot(C\nabla\psi) \tag{7.5.20a}$$

and

$$\frac{\partial C}{\partial t} + \mathbf{v}\cdot\nabla C = D_-\nabla^2 C + \frac{D_-z_-F}{RT}\nabla\cdot(C\nabla\psi), \tag{7.5.20b}$$

where D_+ and D_- are the binary diffusion coefficients of the cation and anion, respectively, in water. Subtracting Equation (7.5.20b) from Equation (7.5.20a) results in the following relation between the potential and concentration gradients:

$$\nabla\cdot(C\nabla\psi) = \frac{RT}{F}\left(\frac{D_- - D_+}{z_+D_+ - z_-D_-}\right)\nabla^2 C. \tag{7.5.21}$$

Substituting Equation (7.5.21) into either Equation (7.5.20a) or (7.5.20b) gives the following conservation relation for binary electrolytes:[6]

$$\frac{\partial C}{\partial t} + \mathbf{v}\cdot\nabla C = D_{\text{eff}}\nabla^2 C. \tag{7.5.22}$$

[6] *RT/F* has units of volts. *R* is 8.3144 J mol^{-1} K^{-1}, *T* is in Kelvin, and *F* is 96,487 coulombs mol^{-1}. One coulomb equals 1 amp s, and 1 volt equals 1 J s^{-1}A^{-1}. At 25°C, *RT/F* equals 25.69 mV.

In this equation,

$$D_{\text{eff}} = \frac{(z_+ - z_-)D_+D_-}{z_+D_+ - z_-D_-} \tag{7.5.23}$$

is the effective diffusion coefficient. Equation (7.5.22) for binary electrolytes is identical to the result for uncharged molecules (Equation (7.3.12)) except that the binary diffusion coefficient is replaced with the effective diffusion coefficient.

7.5.2 Electrolyte Transport Across Membranes

Because ion transport across membranes is important in biological and biotechnological applications, it is worth examining in some detail. Ion transport is affected by the charge of the membrane and the manner in which ions are transported across the membrane. Transport across uncharged membranes is a straightforward extension of previous concepts, and an example is provided next. For charged membranes, both the concentration gradient and potential gradient must be determined.

Example 7.4 Electrolyte Transport Across an Uncharged Membrane

Consider the steady-state one-dimensional transport of a binary electrolyte across an uncharged membrane of thickness L. There are no chemical reactions, convection is unimportant, and there is no applied electrical potential. However, a potential arises across the membrane due to differences in the diffusion coefficient of ions. Derive an expression for this potential.

Solution For steady-state one-dimensional transport in rectangular coordinates, Equation (7.3.9a) reduces to

$$\frac{dN_{iz}}{dz} = 0. \tag{7.5.24}$$

Equation (7.5.24) indicates that the anion and cation fluxes are constant. Since there is no current across the membrane, Equation (7.5.6) implies that $z_+N_+ = -z_-N_-$, and it follows that

$$z_+D_+\left[\frac{dC_+}{dz} + \frac{z_+C_+F}{RT}\frac{d\psi}{dz}\right] = -z_-D_-\left[\frac{dC_-}{dz} + \frac{z_-C_-F}{RT}\frac{d\psi}{dz}\right]. \tag{7.5.25}$$

Rearranging terms in Equation (7.5.25) and using the electroneutrality condition yields the following relation between the potential gradient and the concentration gradient:

$$\frac{d\psi}{dz} = -\left(\frac{D_+ - D_-}{z_+D_+ - z_-D_-}\right)\frac{RT}{F}\frac{1}{C_+}\frac{dC_+}{dz}. \tag{7.5.26}$$

Integrating Equation (7.5.26) from $z = 0$ to $z = L$ produces the following expression for the potential difference across the membrane:

$$\psi(L) - \psi(0) = \left(\frac{D_+ - D_-}{z_+ D_+ - z_- D_-}\right)\frac{RT}{F}\ln\left(\frac{C_0}{C_L}\right), \qquad (7.5.27)$$

where C_0/C_L is the ratio of concentrations between $z = 0$ and $z = L$ for either anion or cation. The electrical potential represented by Equation (7.5.27) is known as the *diffusion potential.* This potential arises because the diffusion coefficients of the anion and cation are not identical.

The term in parenthesis on the right-hand side of Equation (7.5.27) represents a charge-weighted ratio of diffusion coefficients. The greater the difference in the diffusion coefficients, the greater is the potential that arises. Table 7.4 lists the diffusion coefficients for a number of anions and cations. Except for the hydrogen and hydroxyl ions, most ion diffusion coefficients are between 1×10^{-5} cm^2 s^{-1} and 2×10^{-5} cm^2 s^{-1}. The hydrogen (H^+) and hydroxyl (OH^-) ions have higher diffusion coefficients because protons (i.e., hydrogen ions) are transferred from solvent to ions via hydrogen bond connections [6]. Hydroxyl ion mobility then arises as a result of this rapid transfer of hydrogen ions. Such rapid transport is important in many biochemical reactions. Because of the rapidity of proton transfer, these reactions are often diffusion limited.

For many salts (e.g., NaCl), the anion and cation diffusivities are not the same. These small differences in diffusion coefficients can give rise to significant diffusion potentials, according to Equation (7.5.26). For example, consider a membrane-separating solution of either NaCl or KCl. If the concentrations differ by a factor of 10, then the diffusion potentials are 12.27 mV for NaCl and 1.11 mV for KCl. (Note that the sign of the potential depends upon whether C_0 is greater or less than C_L.)

The analysis of ion transport across charged membranes is complicated because the charge distribution within the membrane affects the concentration and

TABLE 7.4

Diffusion Coefficients of Anions and Cations at 25°C

Cation	Charge, z_+	$D_+ \times 10$ cm^2 s^{-1}	Anion	Charge, z_-	$D_- \times 10$ cm^2 s^{-1}
H^+	+1	9.312	OH^-	−1	5.260
Na^+	+1	1.334	Cl^-	−1	2.032
K^+	+1	1.957	NO_3^-	−1	1.902
NH_4^+	+1	1.954	HCO_3^-	−1	1.105
Mg^{++}	+2	0.7063	HCO_2^-	−1	1.454
Ca^{++}	+2	0.7920	$SO_4^{=}$	−2	1.065
Cu^{++}	+2	0.72	HSO_4^-	−1	1.33

Source: Adapted from Ref. [5]. Used with permission.

potential distributions. This is true for biological membranes, even though ion transport occurs through distinct channels that occupy a fraction of the membrane area. Although the potential differences across cell membranes are small (e.g., 100 mV), the electric field gradients are enormous (e.g., 200,000 V cm^{-1}), because the plasma membrane is very thin (5 nm) [7]. As a result, it is often assumed that the potential varies linearly with position within the membrane. For lipids, the charge density cannot be so large that higher order terms are negligible. (This is not the case for channels with charges distributed across them [7].)

In order to develop an analytical solution for the concentration distribution of an ion in a charged membrane, three assumptions are made: (1) The electric potential varies linearly across the membrane; (2) each ion behaves independently; and (3) the properties are uniform across the membrane. For a membrane of thickness L, the potential distribution is approximated as [8] (Figure 7.5)

$$\frac{d\psi}{dz} \approx -\frac{\psi_0 - \psi_L}{L} = \frac{V_m}{L}. \tag{7.5.28}$$

This simplification is commonly referred to as the *constant-field assumption.* The transmembrane potential V_m represents the difference of intracellular potential and the extracellular potential. Assuming no convection, a steady state, and one-dimensional transport, insertion of the Nernst–Planck equation into the conservation relation for each ion yields

$$0 = -D_{ij}\frac{d^2C_i}{dz^2} - \frac{D_{ij}z_iF}{RT}\frac{V_m}{L}\frac{dC_i}{dz}. \tag{7.5.29}$$

Since the system is at steady state, the flux is constant. Integrating across the membrane from $z = 0$, where $C_i = \Phi_i C_0$, to $z = L$, where $C_i = \Phi_i C_L$, we have the following expressions for the concentration distribution and flux:

$$C_i = -\left(\frac{\Phi_i(C_L - C_0)[\exp(-V_m z_i F z/RTL) - 1]}{1 - \exp(-V_m z_i F/RT)}\right) + \Phi_i C_0, \tag{7.5.30a}$$

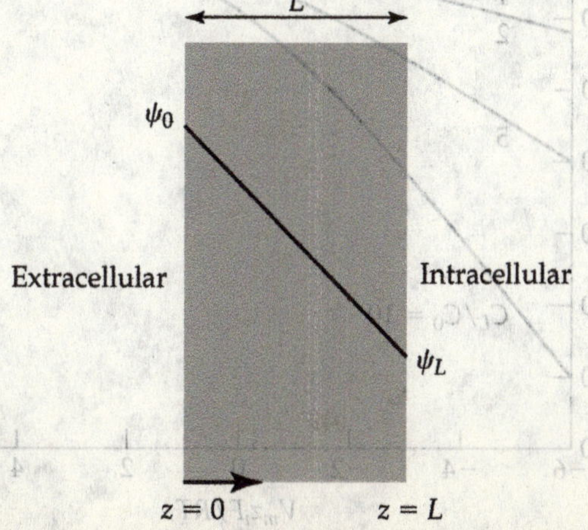

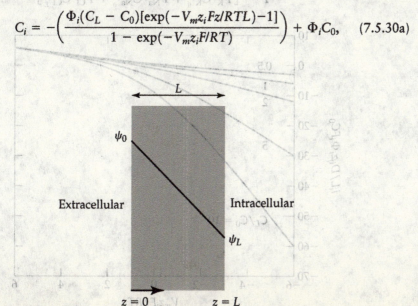

FIGURE 7.5 Potential difference across a charged cellular membrane. The transmembrane potential V_m equals the potential inside the cell minus the potential outside the cell, $\psi_i - \psi_0$.

$$N_{i_z} = -\frac{\Phi_i D_{ij} z_i F V_m}{RTL}\left(\frac{C_0\exp(-V_m z_i F/RT) - C_L}{\exp(-V_m z_i F/RT) - 1}\right), \tag{7.5.30b}$$

where Φ_i is the partition coefficient of ion i.

Although the potential difference across the membrane is constant, the concentration is a nonlinear function of z.

The normalized current $i_i L/D_{ij} z_i F \Phi_i C_0$ depends on the normalized voltage $V'_m = V_m z_i F/RT$ and the concentration ratio C_L/C_0 (Figure 7.6). When C_L/C_0 equals unity, the flux and ionic current are linearly related to the transmembrane potential. For large negative transmembrane potentials $\psi_L < \psi_0$, the normalized current asymptotically approaches $V'_m C_L/C_0$ and is directed into the cell. Conversely, at high positive voltages, the normalized current asymptotically approaches V'_m, and all the curves approach each other.

The transmembrane potential is the second major result that can be obtained from an analysis of transport across charged membranes. The three principal ions transported across channels are K^+, Na^+, and Cl^-. Although Ca^{++} plays a crucial role in transferring intracellular signals, its concentration is much less than the concentration of the other three ions, and it does not contribute substantially to the transmembrane potential.

In order to determine the transmembrane potential, note that the net current across the membrane is zero. As a result,

$$0 = N_{K^+} + N_{Na^+} - N_{CL^-}. \tag{7.5.31}$$

Substituting Equation (7.5.30b) into Equation (7.5.31) for each ion and solving for the transmembrane potential yields

$$V_m = -\frac{RT}{F}\ln\left[\frac{P_{K^+}C_{K^+_L} + P_{Na^+}C_{Na^+_L} + P_{Cl^-}C_{Cl^-_0}}{P_{K^+}C_{K^+_0} + P_{Na^+}C_{Na^+_0} + P_{Cl^-}C_{Cl^-_L}}\right], \tag{7.5.32}$$

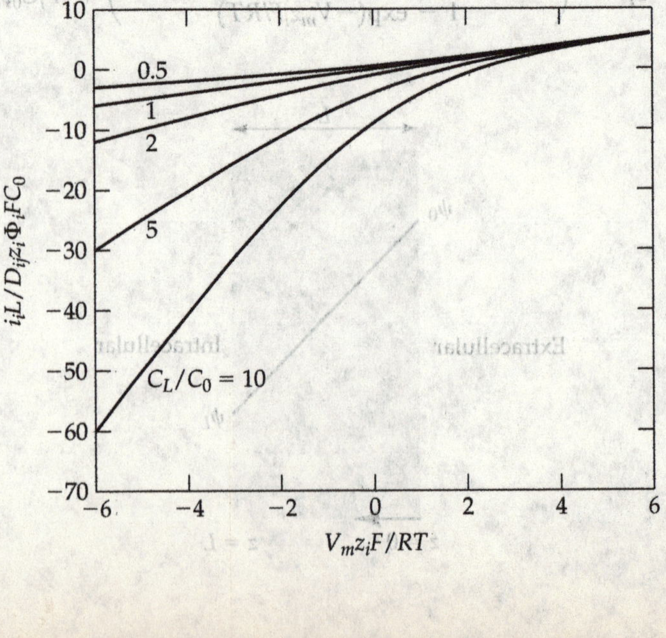

FIGURE 7.6 Relationship between normalized outward ionic current and normalized transmembrane voltage for different values of C_L/C_0.

where the permeabilities P_i equal $\Phi_i D_{ij}/L$. For a single ion, Equation (7.5.32) reduces to the Nernst equation (7.5.18).

The preceding analysis assumes that ions permeate across the entire membrane. Actually, ion transport is strictly controlled by ion channels—transmembrane proteins with an aqueous pore that selectively permits the transport of ions. Thus, the permeabilities depend upon the fractional area of the membrane that is occupied by the respective channels.

Channels are specific to a given ion. Channels are opened or closed, depending upon whether a stimulus is present. A channel opening, or gating, is controlled by the voltage, ligand, or stretch [9]. From a sequence and structural perspective, channels can be classified into families that have similarities in structure. Such a classification is extremely helpful in understanding the function and evolution of various channels [9].

Channels exhibit a number of properties that raise questions about the applicability of the constant-field assumption and a continuum approach (reviewed by Hille [9]). The gating mechanism does not follow the simple relation shown for a homogeneous membrane. Channels can be selectively blocked by pharmacological agents. The charge distribution across the membrane is not uniform, suggesting that the electrical potential gradient is not constant. Further, the flux is *saturable*. That is, at high concentrations, the flux reaches a maximum and does not increase further with increasing concentration. These behaviors are consistent with transport by either a carrier or a channel. The difference is that a carrier binds the ion and then moves the ion across the membrane, whereas a channel consists of an aqueous pore through which ions move by diffusion and electrostatic interactions.

Before the crystal structure of a potassium channel was determined in 1998, showing the exact pore structure [10], available kinetic evidence indicated that ions were transported by pores within the channels. Ion transport rates typically range between 10^7 and 10^8 ions per second. In contrast, carriers, which move molecules by binding, can transport molecules only at speeds ranging from 10^2 to 5×10^4 molecules per second [9]. The function of specific potassium channel blockers requires that the channel be open, and the effect of the blockers can be reduced by increasing the extracellular concentration of potassium. Further, most pores show decreasing rates of transport as the size of the ion increases, an observation consistent with a pore. Much like binding reactions exhibit a maximum level, often referred to as *saturation*, the rate of transport through pores reaches a plateau at high ion concentrations, due to the narrow channel dimensions that permit only single-file motion of ions. As a result, raising the extracellular ion concentration does not increase the flux once the ions are as closely spaced as possible within the channel pore.

The crystal structure of the bacterial potassium channel KcsA was obtained at a 0.25-nm resolution and reveals some important features about the mechanism of ion transport (Figure 7.7). Since the amino acid sequence is related to voltage-gated potassium channels found in mammals, structural information from this molecule may be applicable to other ion channels. The pore is 4.5 nm long, and the diameter is variable, but does not appear to be responsible for the selective behavior of the channel. Rather, the intracellular and extracellular entrance regions are negatively charged, thereby excluding anions from the channel. Near the extracellular surface, the uncharged polar and nonpolar amino acids—glycine, tyrosine, valine, and threonine—act as filters that select potassium over sodium. The carbonyl groups, rather than the

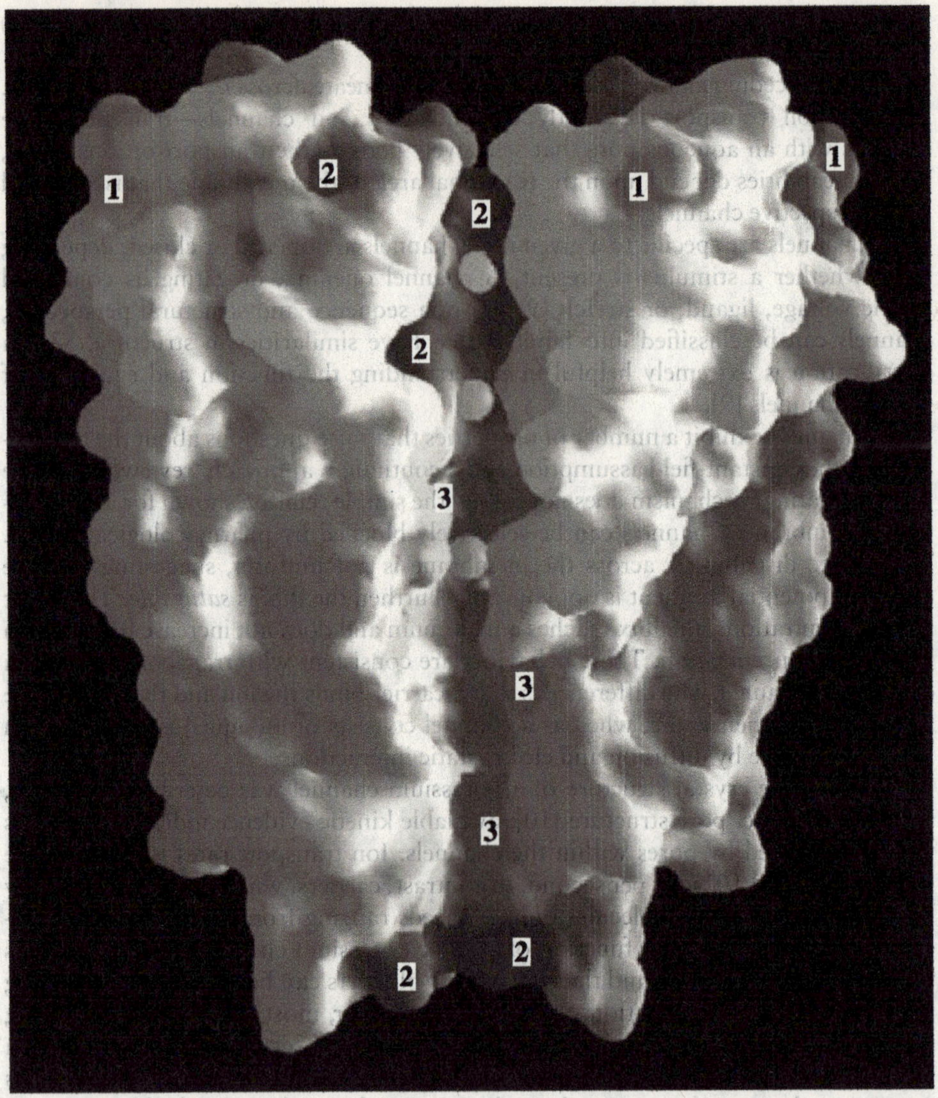

FIGURE 7.7 Surface contour of three of the four tetrameric units of the bacterial potassium channel KcsA from Doyle et al. [10]. Positively charged regions are denoted by the number 1, negatively charged regions by the number 2, and hydrophobic surfaces by the number 3. Reprinted with permission from AAAS.

side groups, point into the pore. This structure stabilizes the potassium ion and its analogs, rubidium and cesium, better than it does the smaller sodium ion.

Electrophysiologists are interested in understanding how the structure of the pore relates to its properties. To address this issue, they have used models to compute the flux through the pores. One approach involves applying the Nernst–Planck equation to nonuniform pore geometry and charge distributions [11]. In order to obtain the potential field in these situations, the following Poisson's equation is solved for the charge distribution [5],

$$\nabla^2\psi = -\frac{F}{\varepsilon}\sum_i z_i C_i, \qquad (7.5.33)$$

where ε is the permittivity of the solvent. The solution of Equation (7.5.33) provides the charge distribution around a central ion. An important parameter arising from the solution of Poisson's equation is the Debye length, which is

$$\lambda = \left(\frac{\varepsilon RT}{F^2 \sum_i z_i^2 C_i} \right)^{1/2}. \qquad (7.5.34)$$

The Debye length represents the characteristic distance over which counterion concentration is elevated around the central ion. The Debye length ranges from 1.0 nm for 100-mM KCl to 0.56 nm for 300-mM KCl, a range that is comparable to the channel dimensions. With small numbers of ions passing through the channel, the concentration distribution cannot be treated as continuous [12]. The Nernst–Planck equations do not account for the discrete distribution of ions and overestimate the screening produced by counterions [13].

In order to determine the conditions under which continuum models accurately model ion transport in channels, transport through the pores has been modeled by both a continuum model and molecular dynamics simulations [12,13]. Molecular dynamics involves the solution of Newton's laws of motion for a charged particle moving through a solvent, which is treated as a continuum. Molecular dynamics can handle nonuniformity in charge distribution around charged or polar regions of the pore wall.[7] In the absence of fixed or induced charges on the pore walls, the two approaches agree, even for channel radii as small as 0.3 nm. When induced or fixed charges are present, the solution of the Nernst–Planck equation overestimates charge shielding. Shielding becomes more important as the channel radius increases, and the agreement between molecular dynamics and the solution of the Nernst–Planck equation improves. These results indicate that the most accurate analysis of ion transport in pores occurs when we consider them as individual ions [12]. The practical impact is that the solution of the Nernst–Planck equation yields unreasonably low values for ion permeability across channels [14].

The gating of channels by appropriate stimuli is a critical feature of ionic conductance across membranes and explains the electrical behavior of excitable tissues such as the heart or brain. Since the extracellular and intracellular ion concentrations are relatively stable, the potential and the current change with time, due to opening and closing of the channels. Under these conditions, the current flow due to ion i is [8,15]

$$i_i = g(t)(\psi - \psi_i), \qquad (7.5.35)$$

where $g(t)$ is the time-dependent conductance, ψ is the membrane potential, and ψ_i is the Nernst or equilibrium potential of ion i (see Equation (7.5.18)). A number of kinetic models have been developed to describe the kinetic behavior of the conductance [8,15].

[7] In molecular dynamics, the equations of motion are solved for the ion and solvent, but current computational facilities can reasonably solve only the simplest problems.

7.6 Diffusion and Convection

In the microcirculation, mass transport occurs by both diffusion and convection. Endothelial cells release a number of bioactive molecules that affect several local functions. For example, fluid flow stimulates the release of nitric oxide, which causes vessels to dilate. Nitric oxide diffuses into the tissue and into the vessel lumen. As a result, the local release of nitric oxide may affect vasodilation downstream. In addition, the local concentrations of molecules that affect blood clotting are influenced by the flow rate. Thus, high flow rates may reduce the local concentration of some clotting molecules, reducing blood clot formation (thrombosis).

High flow rates minimize the importance of diffusion in the direction of flow. Normal to the direction of flow, however, diffusion is still important. To see this, consider steady, one-dimensional flow and mass transport in a rectangular, two-dimensional channel of length L and height H, with $H < L$:

$$\mathrm{v}_x \frac{\partial C_i}{\partial x} = D_{ij}\left(\frac{\partial^2 C_i}{\partial x^2} + \frac{\partial^2 C_i}{\partial y^2}\right). \tag{7.6.1}$$

The x-coordinate can be made dimensionless by normalizing relative to $L(x^* = x/L)$, whereas the y coordinate is scaled by $H(y^* = y/H)$. The concentration can be normalized by a reference concentration C_0. ($C^* = C_i/C_0$, where C_0 could be the concentration at the entrance to the channel.) The velocity in the x direction is scaled by the average channel velocity $\langle \mathrm{v} \rangle (\mathrm{v}_x^* = \mathrm{v}_x/\langle \mathrm{v} \rangle)$. We obtain

$$\mathrm{Pe}_H \frac{H}{L} \mathrm{v}_x^* \frac{\partial C_i^*}{\partial x^*} = \frac{H^2 \partial^2 C_i^*}{L^2 \partial x^{*2}} + \frac{\partial^2 C_i^*}{\partial y^{*2}}, \tag{7.6.2}$$

where $\mathrm{Pe}_H = \langle \mathrm{v} \rangle H/D_{ij}$. Since $H/L < 1$ and the dimensionless variables are all of order one, terms multiplied by $(H/L)^2$ are much less than unity. Thus, the diffusion term in the x direction is negligible. The convective term remains, since $\mathrm{Pe}_H H/L$ can be larger than unity. As a result, convective transport in the direction of flow is balanced by diffusion. To expand on these arguments, two cases are examined: the release of material from the wall of a vessel and mass transfer for flow over a flat plate.

7.6.1 Release from the Walls of a Channel: A Short-Contact-Time Solution

As noted previously, endothelial cells release biologically active molecules into the blood in response to many stimuli. In addition, viruses or bacteria that enter tissue may pass across the endothelial cell barrier to enter the blood. In this section, we analyze how the concentration of a molecule that is released into a Newtonian fluid is affected by convection and diffusion. The additional effect of mixing by red blood cells is not examined.

Consider a cylindrical tube of radius R through which the flow of a Newtonian fluid with an average velocity $\langle \mathrm{v} \rangle$ is fully developed. A biologically active molecule is released into the fluid at $r = R$ (Figure 7.8). The concentration of the molecule at R is C_0. The solute concentration entering the channel is zero. The concentration C_0 is

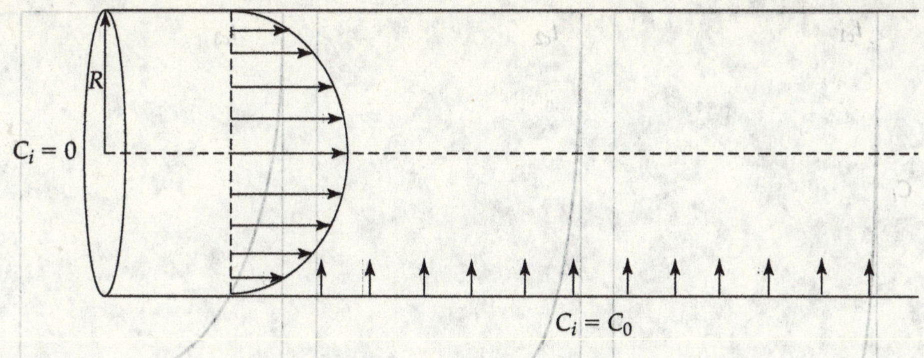

FIGURE 7.8 Mass transfer of a solute from the walls of a cylindrical tube of radius R into a Newtonian fluid undergoing laminar flow.

sufficiently small that dilute solutions can be assumed and Fick's law is applicable. For this problem, the conservation relation yields

$$v_z \frac{\partial C_i}{\partial z} = D_{ij}\left(\frac{\partial^2 C_i}{\partial z^2} + \frac{1}{r}\frac{\partial}{\partial r}\left(r\frac{\partial C_i}{\partial r}\right)\right). \tag{7.6.3}$$

The boundary conditions for the problem are as follows:

$$z = 0 \quad C_i = 0, \tag{7.6.4a}$$

$$r = R \quad C_i = C_0, \tag{7.6.4b}$$

$$r = 0 \quad \frac{\partial C_i}{\partial r} = 0. \tag{7.6.4c}$$

Although general solutions of Equation (7.6.3), subject to the boundary conditions in Equations (7.6.4), are available using separation of variables method [1,5,16], the analysis presented here is restricted to the important limiting case of short contact times between the fluid and the walls. The contact time t_c equals $z/\langle v_z \rangle$, where $\langle v_z \rangle$ is the average velocity of the fluid. In this case, the change in solute concentration in the fluid is confined to a narrow region near the tube walls. The contact time increases with axial distance. The concentration profiles at various values of the contact time is sketched in Figure 7.9. The concentration profiles are similar to those observed for diffusion in a semi-infinite medium (Section 6.8.1); thus, a similar analysis is used. Consequently, the extent of the change in concentration depends, not simply on the axial position z, but on the contact time as well. To see this, note that if the flow rate is very slow, then, at a given position z, there is sufficient time for the solute to diffuse radially. However, if the flow rate is much greater, the solute concentration profile changes much less. Thus, the concentration profile is very different when examined at the same axial distance, but at two different velocities.

The foregoing restriction, sometimes referred to as the *Lévêque approximation*[8] in honor of the individual who first developed the analysis, allows us to make three important simplifications to the problem. First, by focusing on short contact time, the concentration change occurs near $r = R$ and the fluid appears to be semi-infinite.

[8]Lévêque was a French scientist who published this analysis in the 1920s.

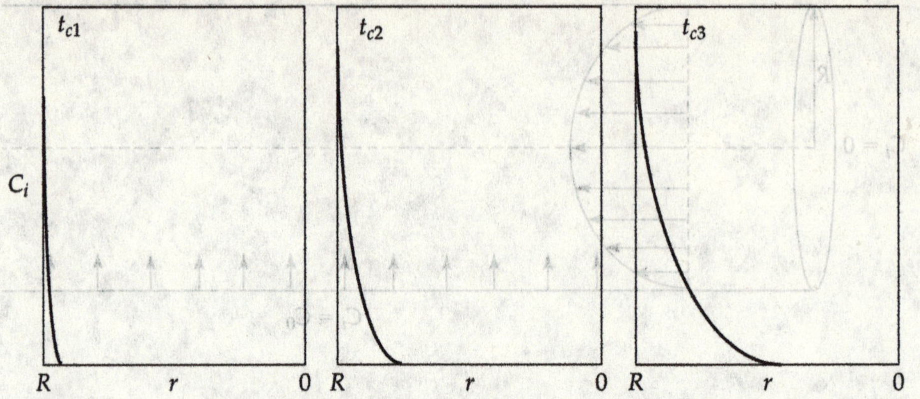

FIGURE 7.9 Sketch of the concentration as a function of position for several values of the contact time.

Second, the velocity profile can be simplified by defining a new coordinate $y = R - r$. We begin by expanding the velocity profile as

$$v_z = v_{max}\left(1 - \frac{r^2}{R^2}\right) = v_{max}\left(1 - \frac{r}{R}\right)\left(1 + \frac{r}{R}\right). \tag{7.6.5}$$

Next, we replace the radial coordinate r with $R - y$. Since $y/R \ll 1$, the velocity profile can be simplified to

$$v_z = v_{max}\frac{y}{R}\left(2 - \frac{y}{R}\right) \approx 2v_{max}\frac{y}{R}. \tag{7.6.6}$$

Substituting Equation (7.6.6) for the velocity profile into Equation (7.6.3) and neglecting axial diffusion results in

$$2v_{max}\frac{y}{R}\frac{\partial C_i}{\partial z} = \frac{D_{ij}}{r}\frac{\partial}{\partial r}\left(r\frac{\partial C_i}{\partial r}\right). \tag{7.6.7}$$

The third simplification is that curvature can be neglected by replacing r with $R - y$. The second derivative in the radial direction becomes

$$\frac{1}{r}\frac{\partial}{\partial r}\left(r\frac{\partial C_i}{\partial r}\right) = \frac{1}{(R-y)}\frac{\partial}{\partial y}\left((R-y)\frac{\partial C_i}{\partial y}\right) = \frac{\partial^2 C_i}{\partial y^2} - \frac{1}{(R-y)}\frac{\partial C_i}{\partial y}. \tag{7.6.8}$$

Because R is much larger than y, the second term on the right side of Equation (7.6.8) is smaller than the first term on the right side and can be neglected. As a result, Equation (7.6.8) reduces to

$$2v_{max}\frac{y}{R}\frac{\partial C_i}{\partial z} = D_{ij}\frac{\partial^2 C_i}{\partial y^2}. \tag{7.6.9}$$

The boundary conditions are as follows:

$$z = 0 \quad C_i = 0, \tag{7.6.10a}$$

$$y = 0 \quad C_i = C_0, \tag{7.6.10b}$$

$$y \to \infty \quad C_i = 0. \tag{7.6.10c}$$

Since there is no natural length scale in the y direction, we expect, on the basis of our analysis of unsteady diffusion in semi-infinite media, that the problem can be recast in terms of a single variable (e.g., a similarity solution). To assess the form of this variable, we perform an order-of-magnitude analysis (see Section 6.8.1). The derivative of C_i with respect to z can be scaled as

$$\frac{\partial C_i}{\partial z} \sim \frac{C_0}{z}. \tag{7.6.11a}$$

The characteristic-length scale in the y direction is δ, the distance over which the concentration changes from zero to C_0. The diffusion term in Equation (7.6.9) becomes

$$D_{ij}\frac{\partial^2 C_i}{\partial y^2} \sim D_{ij}\frac{C_0}{\delta^2}. \tag{7.6.11b}$$

Substituting Equations (7.6.11a) and (7.6.11b) into Equation (7.6.9) yields

$$2\mathrm{v}_{\max}\frac{\delta}{R}\frac{C_0}{z} \sim D_{ij}\frac{C_0}{\delta^2}. \tag{7.6.12a}$$

Rearranging terms and solving for δ gives

$$\delta \sim \left(\frac{RzD_{ij}}{2\mathrm{v}_{\max}}\right)^{1/3}. \tag{7.6.12b}$$

Defining a single variable $\eta \sim y/\delta$ for transport in the y and z directions, we have

$$\eta = y\left(\frac{2\mathrm{v}_{\max}}{9RD_{ij}z}\right)^{1/3}. \tag{7.6.13}$$

The derivatives in Equation (7.6.9) are transformed into functions of η as follows: The derivative in the z direction is

$$\frac{\partial C_i}{\partial z} = \frac{dC_i}{d\eta}\frac{\partial \eta}{\partial z} = -\frac{1}{3}\frac{y}{z}\left(\frac{2\mathrm{v}_{\max}}{9RD_{ij}z}\right)^{1/3}\frac{dC_i}{d\eta} = -\frac{1}{3}\frac{\eta}{z}\frac{dC_i}{d\eta}. \tag{7.6.14}$$

As we did in the development of Equation (6.8.8c), we can show that

$$\frac{\partial^2 C_i}{\partial y^2} = \left(\frac{2\mathrm{v}_{\max}}{9HD_{ij}z}\right)^{2/3}\frac{d^2 C_i}{d\eta^2}. \tag{7.6.15}$$

We substitute Equations (7.6.14) and (7.6.15) into Equation (7.6.9) and rearrange terms to obtain

$$-3y\left(\frac{2\mathrm{v}_{\max}}{9RD_{ij}z}\right)\left(\frac{2\mathrm{v}_{\max}}{9RD_{ij}z}\right)^{-2/3}\eta\frac{dC_i}{d\eta} = \frac{d^2 C_i}{d\eta^2}. \tag{7.6.16}$$

After we rearrange and collect terms once more, Equation (7.6.9) becomes a function of one variable η:

$$-3\eta^2\frac{dC_i}{d\eta} = \frac{d^2 C_i}{d\eta^2}. \tag{7.6.17}$$

The boundary conditions are

$$\eta = 0(y = 0) \qquad C_i = C_0, \tag{7.6.18a}$$

$$\eta \to \infty(y \to \infty \text{ and } z = 0) \qquad C_i = 0. \tag{7.6.18b}$$

To solve Equation (7.6.17), we define the variable

$$u = \frac{dC_i}{d\eta}. \tag{7.6.19}$$

As a result, Equation (7.6.17) is rewritten as

$$-3\eta^2 u = \frac{du}{d\eta}. \tag{7.6.20}$$

The solution of Equation (7.6.20) is

$$u = \frac{dC_i}{d\eta} = A\exp(-\eta^3). \tag{7.6.21}$$

Integrating again yields

$$C_i = A\int_0^{\eta} \exp(-z^3)dz + B. \tag{7.6.22}$$

After evaluating the boundary conditions at $\eta = 0$ and $\eta \to \infty$, we have

$$C = C_0\left[1 - \frac{\int_0^{\eta} \exp(-z^3)dz}{\int_0^{\infty} \exp(-z^3)dz}\right] = C_0\frac{\int_{\eta}^{\infty} \exp(-z^3)dz}{\int_0^{\infty} \exp(-z^3)dz}. \tag{7.6.23}$$

This result can be simplified, since the integral in the denominator has a known value. First, we express Equation (7.6.23) in terms of gamma functions that are defined as

$$\Gamma(n) = \int_0^{\infty} w^{n-1}e^{-w}dw. \tag{7.6.24}$$

(Values for the gamma function can be obtained directly from MATLAB.) Next, setting $w = z^3$, we find that the integral in the denominator is

$$\int_0^{\infty} \exp(-z^3)dz = \left(\frac{1}{3}\right)\Gamma\left(\frac{1}{3}\right). \tag{7.6.25}$$

Noting that $n\Gamma(n) = \Gamma(n + 1)$, we see that the integral evaluates to $\Gamma(4/3) = 0.8930$. Equation (7.6.23) is then

$$C = 1.1198C_0\int_{\eta}^{\infty} \exp(-z^3)dz. \tag{7.6.26}$$

Figure 7.10 is a graph of the concentration as a function of η. The concentration declines rapidly with η, reaching a value of 0.0049 at $\eta = 1.5$.

The flux of solute leaving the surface at $y = 0$ is determined from Leibniz's rule (Appendix, A.1.H). We have

$$N_{i_y}(y=0) = -D_{ij}\frac{\partial C_i}{\partial y}\bigg|_{y=0} = -D_{ij}\frac{\partial \eta}{\partial y}\frac{dC_i}{d\eta}\bigg|_{\eta=0}$$

$$= -1.1198D_{ij}\left(\frac{2\text{v}_{\max}}{9RD_{ij}z}\right)^{1/3} C_0\frac{d}{d\eta}\left[\int_{\eta}^{\infty}\exp(-z^3)dz\right]\Bigg|_{\eta=0} \tag{7.6.27}$$

$$N_{i_y}(y=0) = 0.6783D_{ij}C_0\left(\frac{2\langle\text{v}\rangle}{RD_{ij}z}\right)^{1/3} = \frac{0.6783D_{ij}C_0}{R}\left(\text{Pe}\frac{R}{z}\right)^{1/3}, \tag{7.6.28}$$

where $\langle \text{v}\rangle = \text{v}_{\max}/2$ and $\text{Pe} = \langle \text{v}\rangle 2R/D_{ij}$.

Restricting the deviation between the linear velocity profile (Equation (7.6.6)) and the true parabolic profile to no more than 10%, the concentration profile and the flux are valid for values of $zD_{ij}/\text{v}_{\max}R^2 < 0.01$. If we consider the dimensions of arterioles, venules, and capillaries (Table 2.4), we can estimate whether these results are applicable along the lengths of the vessels. For arterioles, the values of z must be less than 0.0625–1.25 cm for $D_{ij} = 1 \times 10^{-6}\ \text{cm}^2\ \text{s}^{-1}$ to $1 \times 10^{-7}\ \text{cm}^2\ \text{s}^{-1}$. Likewise, for venules, the corresponding lengths range from 0.016 to 0.4 cm for the two diffusion coefficients. These values are comparable to the length of the arterioles or venules, suggesting that the approximation is not valid for mass transport through the entire lengths of these microvessels. In capillaries, the approximation is valid for diffusion coefficients smaller than $6 \times 10^{-10}\ \text{cm}^2\ \text{s}^{-1}$. For larger diffusion coefficients, the length over which the assumption is valid is less than the length of the capillary.

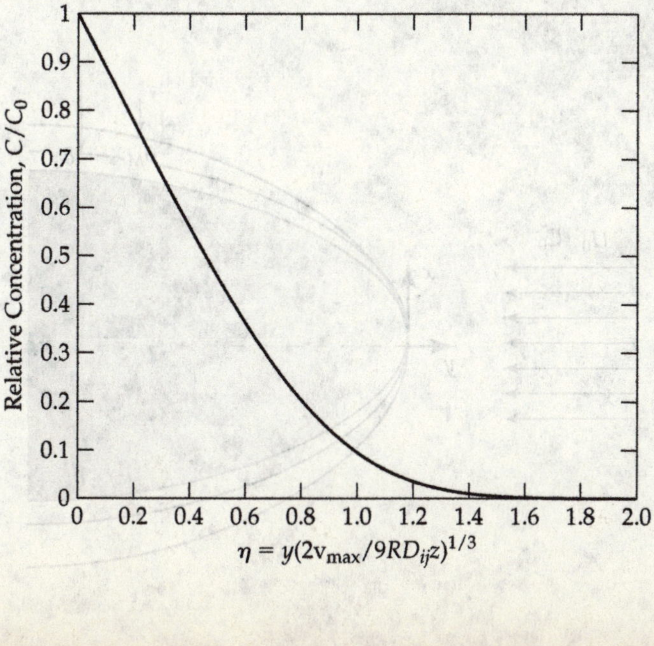

FIGURE 7.10 Graph of Equation (7.6.26), relative concentration versus η for short contact times.

For the flux (Equation (7.6.28)), corrections were developed that extend the range of accuracy of the short-contact-time solution [5]. For values of $zD_{ij}/v_{max}R^2 < 0.1$, the following expression can be used:

$$N_{iy}(y = 0) = \frac{D_{ij}C_0}{R}\left[0.6783\left(\text{Pe}\frac{R}{z}\right)^{1/3} - 0.6 - 0.1485\left(\text{Pe}\frac{R}{z}\right)^{-1/3}\right]. \quad (7.6.29)$$

The short-contact-time solution (Equation (7.6.28)) is applicable to protein transport in capillaries, but the corrected equation is applicable to smaller solutes such as oxygen (see Problem 7.7).

7.6.2 Momentum and Concentration Boundary Layers

If a fluid flowing at high velocities encounters a surface that is reactive or soluble, then boundary layers for momentum and solute transport arise. Such a situation occurs in larger arteries and blood oxygenators during the adhesion of platelets or the formation of a thrombus. These boundary layers can be important in mass transport across membranes for various dialysis procedures or in the oxygenation of blood or bioreactors.

The analysis of concentration boundary layers incorporates that of momentum boundary layers. Consider a fluid with a solute at concentration C_0 and a uniform velocity U_0 coming into contact with an arbitrarily shaped object, as shown in Figure 7.11. Assuming that the surface is reactive and the solute concentration is zero at the surface, a concentration boundary is produced due to the balance between convection and diffusion. To distinguish between these two boundary layers, we denote the momentum boundary layer thickness as δ_M and the concentration boundary layer thickness as δ_C. In general, the thicknesses of these two boundary layers will not be the same, and their relative sizes will depend upon the magnitude of the Schmidt number. To determine the order of magnitude of the concentration

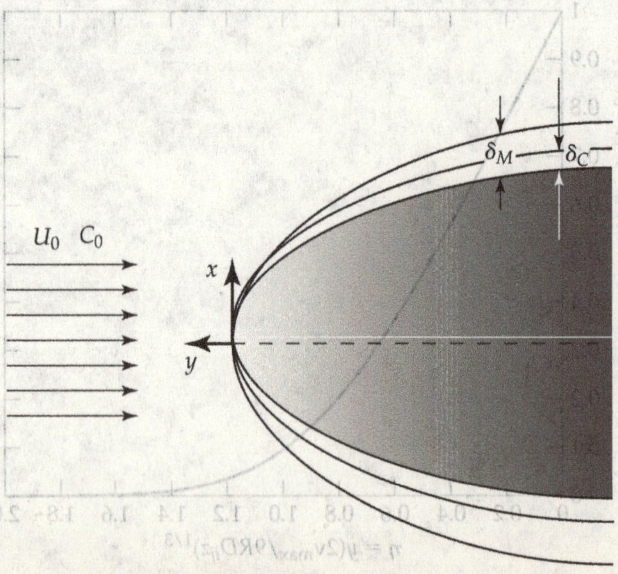

FIGURE 7.11 Schematic of momentum and concentration boundary layers.

boundary layer and the flux at the surface, we begin with the steady-state two-dimensional form of the conservation relation for dilute solutions:

$$v_x \frac{\partial C_i}{\partial x} + v_y \frac{\partial C_i}{\partial y} = D_{ij}\left(\frac{\partial^2 C_i}{\partial x^2} + \frac{\partial^2 C_i}{\partial y^2}\right). \tag{7.6.30}$$

In order to scale the velocity appropriately, it is necessary to estimate the magnitude of the concentration boundary layer relative to the momentum boundary layer. The governing dimensionless group is the Schmidt number. For most solutes in water, the Schmidt number is 1,000 or larger. Since the Schmidt number ν/D_{ij} measures the importance of momentum transport relative to that of diffusive transport, a larger value indicates that momentum transport is greater than diffusive transport. Consequently, the concentration gradient near the surface is confined to a narrower region than the momentum gradient is. That is, $\delta_C \ll \delta_M$ for $\mathrm{Sc} \gg 1$. As a result, the velocity in the concentration boundary layer is linear with y, and y is on the order of δ_C:

$$v_x = U_0 \frac{y}{\delta_M} \sim U_0 \frac{\delta_C}{\delta_M}. \tag{7.6.31}$$

Applying the conservation of mass and using Equation (7.6.31) yields an estimate for the magnitude of v_y:

$$v_y = \int \frac{\partial v_x}{\partial x} dy \sim U_0 \frac{y^2}{\delta_M x} \sim U_0 \frac{\delta_C^2}{\delta_M L}. \tag{7.6.32}$$

The concentration scales as C_0, the y coordinate scales as δ_C, and the x coordinate scales as L. Using the scaling for v_x and v_y, we obtain the dimensionless form of Equation (7.6.30):

$$U_0 \frac{\delta_C}{\delta_M} \frac{C_0}{L} v_x^* \frac{\partial C^*}{\partial x^*} + U_0 \frac{{\delta_C}^2}{\delta_M L} \frac{C_0}{\delta_C} v_y^* \frac{\partial C^*}{\partial y^*} = D_{ij}\left(\frac{C_0}{L^2} \frac{\partial^2 C^*}{\partial x^{*2}} + \frac{C_0}{{\delta_C}^2} \frac{\partial^2 C^*}{\partial y^{*2}}\right). \tag{7.6.33}$$

Collecting terms, we find that (7.6.33) becomes

$$\frac{U_0}{L} \frac{\delta_C}{\delta_M}\left(v_x^* \frac{\partial C^*}{\partial x^*} + v_y^* \frac{\partial C^*}{\partial y^*}\right) = \frac{D_{ij}}{{\delta_C}^2}\left(\frac{{\delta_C}^2}{L^2} \frac{\partial^2 C^*}{\partial x^{*2}} + \frac{\partial^2 C^*}{\partial y^{*2}}\right). \tag{7.6.34}$$

Since $\delta_C \ll L$, diffusion in the x direction can be neglected. After rearrangement of dimensional terms, Equation (7.6.34) reduces to

$$\frac{U_0}{D_{ij} L} \frac{{\delta_C}^3}{\delta_M}\left(v_x^* \frac{\partial C^*}{\partial x^*} + v_y^* \frac{\partial C^*}{\partial y^*}\right) = \frac{\partial^2 C^*}{\partial y^{*2}}. \tag{7.6.35}$$

In order for diffusion and convection to balance, the dimensionless group in Equation (7.6.35) must be of order one, or

$$\frac{U_0}{D_{ij} L} \frac{{\delta_C}^3}{\delta_M} \sim 1. \tag{7.6.36}$$

Since δ_M scales as $L\mathrm{Re}^{-1/2}$ and since $U_0L/D_{ij} = \mathrm{Re}\,\mathrm{Sc}$, the concentration boundary layer scales as

$$\delta_C \sim L\mathrm{Sc}^{-1/3}\mathrm{Re}^{-1/2} \sim \delta_M\mathrm{Sc}^{-1/3}. \tag{7.6.37}$$

Since the Schmidt number is large, our original assumption that $\delta_C \ll \delta_M$ is valid.

Using the expression for δ_C, we can now estimate the flux at the surface $y = 0$ as

$$N_{i_y}(y = 0) = -D_{ij}\frac{\partial C_i}{\partial y} \sim D_{ij}\frac{C_0}{\delta_C} \sim \frac{D_{ij}C_0}{L}\mathrm{Sc}^{1/3}\mathrm{Re}^{1/2}. \tag{7.6.38}$$

Applying boundary layer theory then permits us to determine the constant relating the flux to the terms on the right-hand side of Equation (7.6.38).

7.7 Macroscopic Form of Conservation Relations for Dilute Solutions

Just as we developed macroscopic balances for the conservation of mass for a single-component system and for the conservation of linear momentum (Section 4.3), macroscopic forms of the conservation of mass for dilute multicomponent solutions are obtained by integrating Equation (7.3.8) over the control volume:

$$\int_V \frac{\partial C_i}{\partial t}dV = -\int_V \nabla\cdot\mathbf{N}_i dV + \int_V R_i dV. \tag{7.7.1}$$

Note that such a transformation is possible because the control volumes are fixed in space and not changing.

To simplify Equation (7.7.1), we define the following volume average concentrations and reaction rates:

$$\langle C_i\rangle = \frac{1}{V}\int_V C_i dV \qquad \langle R_i\rangle = \frac{1}{V}\int_V R_i dV. \tag{7.7.2a,b}$$

With these definitions, Equation (7.7.1) becomes

$$V\frac{\partial\langle C_i\rangle}{\partial t} = -\int_V \nabla\cdot\mathbf{N}_i dV + V\langle R_i\rangle. \tag{7.7.3}$$

The integral of the flux can be computed by applying the divergence theorem (Section A.3.C). We obtain

$$\int_V \nabla\cdot\mathbf{N}_i dV = \int_S \mathbf{N}_i\cdot\mathbf{n}\, dS. \tag{7.7.4}$$

As a result of this transformation, the macroscopic form of the conservation relation for dilute multicomponent solutions is

$$\frac{\partial\langle C_i\rangle}{\partial t} = -\frac{1}{V}\int_S \mathbf{N}_i\cdot\mathbf{n}dS + \langle R_i\rangle, \tag{7.7.5}$$

where **n** is the unit normal vector that points out from the surface of the control volume. An alternative way of expressing this relationship is to divide the surfaces into n parts and average the flux over each surface. In this case, the average is based on the surface area. Then, Equation (7.7.5) can be written as

$$\frac{d\langle C_i\rangle}{dt} = -\frac{1}{V}\sum_{j=1}^{n}(\langle \mathbf{N}_i \cdot \mathbf{n}\rangle S)_j + \langle R_i\rangle. \tag{7.7.6}$$

A common application of Equation (7.7.6) is mass transfer in cylindrical tubes and ducts. This situation is applicable to many devices involved in biomedical mass transfer applications (e.g., blood oxygenators and hemodialyzers), as well as to filtration across capillaries. The channel walls may be permeable to solute and solvent. There is flow through the channel and transport across its permeable walls (Figure 7.12). The system can be divided into three parts: the inlet (1), the outlet (2), and the permeable walls (3). For flow through the inlet and outlet, diffusion is negligible, so that, for each region, $N_{iz} = C_i(r, z)v_z(r)$. The velocity-weighted average concentration can be defined as

$$\overline{C}_i = \frac{\int_0^{2\pi}\int_0^R C_i v_z r\,dr\,d\theta}{\int_0^{2\pi}\int_0^R v_z r\,dr\,d\theta}. \tag{7.7.7}$$

Neglecting diffusion in the axial direction, we obtain the average flux at the entrances and exits:

$$\langle N_{iz}\rangle = \frac{\int_0^{2\pi}\int_0^R N_{iz} r\,dr\,d\theta}{\pi R^2} = \frac{\int_0^{2\pi}\int_0^R C_i v_z r\,dr\,d\theta}{\pi R^2} = \overline{C}_i\langle v_z\rangle. \tag{7.7.8}$$

The average flux across the walls at $r = R$ can be due to diffusion or convection, depending on the mode of transport across the permeable surface. This flux is simply $\langle N_{iz}\rangle|_3$. Equation (7.7.6) can then be written as

$$\frac{d\langle C_i\rangle}{dt} = \frac{S}{V}\overline{C}_i\langle v_z\rangle\Big|_1 - \frac{S}{V}\overline{C}_i\langle v_z\rangle\Big|_2 - \frac{S}{V}\langle \mathbf{N}_i \cdot \mathbf{n}\rangle\Big|_3 + \langle R_i\rangle. \tag{7.7.9a}$$

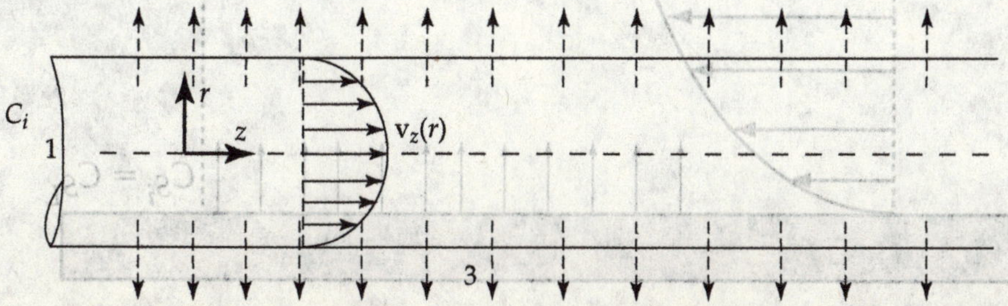

FIGURE 7.12 Application of macroscopic balances to flow through a tube with permeable walls.

One difficulty with applying Equation (7.7.9a) to unsteady conditions and in the presence of chemical reactions is that the two averages for the solute concentration (Equations (7.7.2a) and (7.7.7)) are not identical. Consequently, Equation (7.7.9a) is generally used only for steady-state conditions without homogeneous reactions ($\langle R_i \rangle = 0$). Under these conditions, Equation (7.7.9a) reduces to

$$\overline{C}_i\langle v_z\rangle|_1 - \overline{C}_i\langle v_z\rangle|_2 = \frac{2L}{R}\langle \mathbf{N}_i \cdot \mathbf{n}\rangle|_3, \tag{7.7.9b}$$

where L is the tube length. The flux can be specified with the use of mass transfer coefficients, as described in the next section.

7.8 Mass Transfer Coefficients

In many mass transfer problems, we are interested in the flux of a particular component of a mixture or in the rate of chemical reactions. Many mass transfer problems, however, are too complicated to solve analytically. An alternative approach is to develop correlations of experimental data based upon mass transfer coefficients. The mass transfer coefficient can then be used to calculate the flux of the solute.

The mass transfer coefficient can be defined by considering the example presented in Figure 7.13. In the figure, a soluble material enters the fluid by transport from the surface S into a moving fluid. The surface may be a solid, in which the case the solute dissolves into the fluid. Further, the surface may be the surface of a submerged object that is dissolving or the walls of an enclosed channel [17]. The concentration of the solute in solution adjacent to the solid is C_{S_i}. For the case of a dissolving material, C_{S_i} is a constant. The average flux of solute into the fluid is

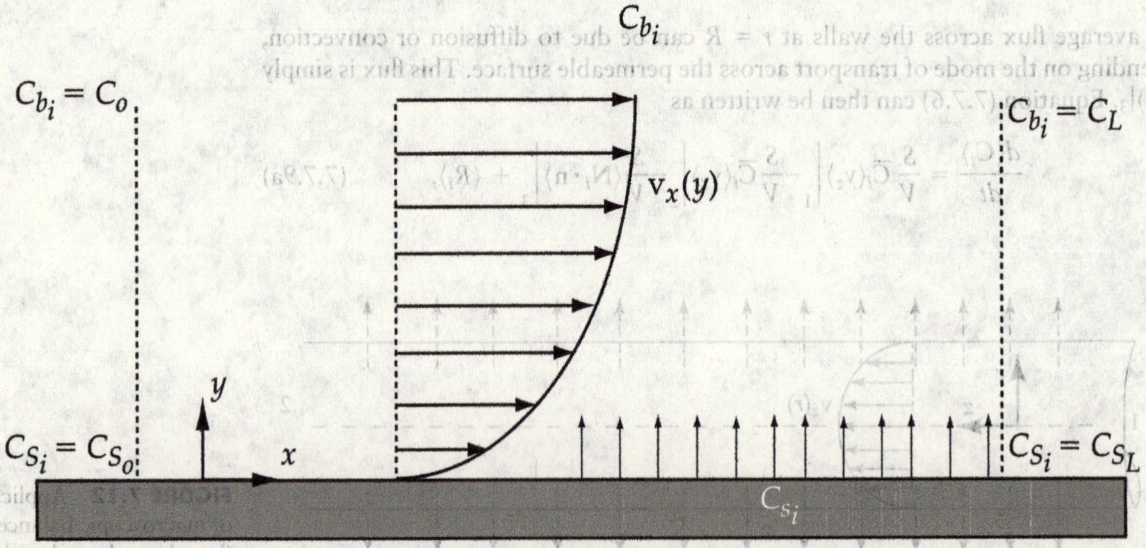

FIGURE 7.13 Mass transfer from a solid surface into a moving fluid.

$$\langle N_{iy}(y=0)\rangle = -\frac{1}{S}\int_A D_{ij}\left(\frac{\partial C_i}{\partial y}\right)\bigg|_{y=0} dS, \tag{7.8.1}$$

where S is the total surface area of the solid in contact with the fluid. The mass transfer coefficient k_f is defined as

$$\langle N_{iy}(y=0)\rangle = k_f \Delta C_i, \tag{7.8.2}$$

where ΔC_i is a suitable reference concentration difference [17].

For flow over submerged objects, the concentration far from the surface is constant and represented by C_∞ [2,17]. Thus, the concentration difference represents the difference between the mean surface concentration C_{S_i} and C_∞. For this case, the mass transfer coefficient is referred to as the mean mass transfer coefficient k_m and

$$k_m = -\frac{1}{S(C_{S_i}-C_\infty)}\int_S D_{ij}\left(\frac{\partial C_i}{\partial y}\right)\bigg|_{y=0} dS. \tag{7.8.3}$$

The mass transfer coefficient defined in Equation (7.8.2) is an *average* mass transfer coefficient. If the fluid and surface concentrations change with location, one could perform a differential mass balance and define a *local* mass transfer coefficient as

$$N_{iy}(y=0) = k_{\text{loc}}(C_{S_i} - C_{b_i}), \tag{7.8.4}$$

where C_{b_i} is the bulk concentration. For confined flow in channels, it is defined as the concentration weighted by the velocity passing through a cross-sectional area A normal to flow:

$$C_{b_i} = \frac{\int_A C_i \mathbf{v}\cdot\mathbf{n}dA}{\int_A \mathbf{v}\cdot\mathbf{n}dA} = \frac{\int_A C_i v_z dA}{\int_A v_z dA}. \tag{7.8.5}$$

In general, C_{b_i} is a function of distance along the flow direction.

Equation (7.8.4) is often difficult to use because average values at the inlet and outlet of the channel are often determined and the spatial variation of k_{loc}, C_{S_i}, and C_{b_i} may not be known. Consequently, an important average value used in many correlations is the log-mean concentration difference [2]:

$$\langle N_{iy}(y=0)\rangle = k_{\ln}\left(\frac{(C_{S_0}-C_0)-(C_{S_L}-C_L)}{\ln\left((C_{S_0}-C_0)/(C_{S_L}-C_L)\right)}\right). \tag{7.8.6}$$

The mass transfer coefficient is a function of the concentration gradient, which, in turn, depends upon the flow field and transport properties. Dimensional analysis is useful in displaying which dimensionless groups influence the mass transfer coefficient. Equating Equations (7.8.1) and (7.8.2) gives the relationship between the mass transfer coefficient and the concentration gradient:

$$k_f = -\frac{1}{S\Delta C_i}\int_S D_{ij}\left(\frac{\partial C_i}{\partial y}\right)\bigg|_{y=0} dS. \tag{7.8.7}$$

The mass transfer coefficient is a function of the concentration gradient, which, in turn, depends upon the flow field and transport properties. Dimensional analysis is useful in displaying which dimensionless groups influence the mass transfer coefficient. By defining

$$C^* = \frac{C_i}{\Delta C_i} \quad \text{and} \quad y^* = \frac{y}{L},$$

we can rewrite Equation (7.8.3) in dimensionless form as

$$\frac{k_f L}{D_{ij}} = \frac{1}{S}\int_S \left(\frac{\partial C^*}{\partial y^*}\right)_{y^*=0} dS. \tag{7.8.8}$$

The dimensionless group on the left-hand side of Equation (7.8.8) is known as the *Sherwood number* ($\text{Sh} = k_f L/D_{ij}$), a dimensionless mass transfer coefficient. This group is also referred to as the mass transfer *Nusselt number*. On the basis of the dimensional analysis performed in Section 7.4, the dimensionless concentration gradient—and hence the Sherwood number—depends upon the Reynolds number and the Schmidt number according to the relation

$$\text{Sh} = f(\text{Re}, \text{Sc}). \tag{7.8.9}$$

The Sherwood number has been determined for a number of experimental systems. Correlations are presented in Table 7.5. The general form of these correlations is

$$\text{Sh} = a\text{Re}^b\text{Sc}^c. \tag{7.8.10}$$

The values of the exponents in Equation (7.8.10) can be deduced from analytical solutions for idealized cases, as discussed in Section 7.6. For instance, an order-of-magnitude solution for boundary flows predicts that the Sherwood number is proportional to the square root of the Reynolds number and the one-third root of the Schmidt number (Equation (7.6.38)). This is, in fact, what is found experimentally.

Example 7.5 For steady laminar flow through a cylindrical blood vessel with radius 1.0 cm, determine the extent to which fluid convection enhances solute transport to the vessel wall for $\langle v \rangle = 10$ cm s^{-1} and $D_{ij} = 1 \times 10^{-6}$ cm^2 s^{-1}. Use $\nu = 0.03$ cm^2 s^{-1}.

Solution

The Sherwood number represents the ratio of mass transfer to diffusive transport. From the information given, $\text{Sc} = 3 \times 10^4$ and $\text{Re} = 333.33$. From the correlation given in Table 7.5 for steady laminar flow through a cylindrical tube, $\text{Sh} = 1055$. Clearly, these flow conditions greatly enhance the transport of the solute.

TABLE 7.5

A Selection of Mass Transfer Correlations

Physical system	Correlation	Comments
Laminar flow along a flat plate	$\frac{k_{loc}z}{D_{ij}} = 0.323\ Re_{loc}^{1/2}\ Sc^{1/3}$	$Re_{loc} = \langle v \rangle z/\nu$ z = distance along plate $\langle v \rangle$ = average fluid velocity
Laminar flow in a circular pipe	$\frac{k_{ln}D}{D_{ij}} = 1.86\left(ReSc\frac{D}{L}\right)^{1/3}$	$Re = \langle v \rangle D/\nu$ D = diameter $2(z/R)/(Re\ Sc) < 0.02$
	$\frac{k_{loc}D}{D_{ij}} = 3.657$	$2(z/R)/(Re\ Sc) > 0.05$
Laminar flow in a channel	$\frac{k_{loc}4H}{D_{ij}} = 3.1058\left(\frac{\langle v \rangle H^2}{4D_{ij}z}\right)^{1/3}$	$2H$ = channel height $\frac{4D_{ij}z}{\langle v \rangle H^2} < 0.01$
	$\frac{k_{loc}2H}{D_{ij}} = 3.770$	$\frac{4D_{ij}z}{\langle v \rangle H^2} > 0.02$
Forced convection around solid sphere	$\frac{k_m D}{D_{ij}} = 2.0 + 0.6Re^{1/2}Sc^{1/3}$	D = sphere diameter $Re = \langle v \rangle D/\nu$
Spinning disk	$\frac{k_m D}{D_{ij}} = 0.62Re^{1/2}Sc^{1/3}$	$Re = D^2\omega/\nu$ ω = angular speed (rad s^{-1})
Packed bed	$\frac{k_m}{v^o} = 0.62Re^{-0.42}Sc^{-0.67}$	$Re = D_p v^o/\nu$ v^o = superficial velocity D_p = particle diameter

$Sc = \frac{\nu}{D_{ij}}$.

7.9 Mass Transfer Across Membranes: Application to Hemodialysis

An important application of biomedical engineering to the treatment of chronic diseases was the development of various treatment modalities to replace some of the functions of kidneys. End-stage renal disease is diagnosed when the kidneys have lost over 90% of their function of filtering blood and producing urine. About three-quarters of the cases of chronic kidney failure are due to complications arising from diabetes, hypertension, or glomerulonephritis. In 2005, approximately 485,000 people in the United States had end-stage renal disease. The incidence of new cases was 107,000. Deaths resulting from complications due to the treatment of end-stage renal disease were almost 86,000 in 2005.[9] The incidence of the disease is increasing with the aging of the population and the increased prevalence of type II diabetes. In 1997, the total cost for all forms of dialysis treatment in the United States was $15.6 billion.

At present, there are two treatments for end-stage renal disease: transplantation and dialysis. Dialysis can be performed by connecting a dialysis machine, or dialyzer, to the bloodstream (hemodialysis and hemodiafiltration) or by draining fluid from the peritoneal cavity (peritoneal dialysis). Although kidney transplants are extremely successful (the five-year survival rate following living-donor transplant is 90.1%), only 17,429 transplants were performed in 2005, an increase of 40% from 1997, but far below the need. The majority of the patients receive some form of hemodialysis. The problem is that although dialysis can lead to an improved quality of life, the five-year survival rate is only 35.8%. Further, the current mode of dialysis can affect the quality of life. In addition to treating end-stage renal disease, hemodialysis is used on individuals suffering from acute renal failure. Such treatment can save the individual's life while the cause of organ failure is reversed.

The major goal of hemodialysis is to provide limited kidney function by removing toxic solutes accumulating in the blood. Although the specific toxins responsible for death from kidney failure are incompletely characterized, all of the solutes have molecular weights less than 500 and can be removed by dialysis. These solutes include sodium, potassium, hydrogen ion, urea, creatinine, and phosphate. The dialysate, the fluid that is on the other side of dialysis, must be maintained precisely. Some solutes present in dialysate are listed in Table 7.6.

There are several different methods in use today, all falling under the general heading of "hemodialysis." In the oldest approach, dialysis involves mass transfer across a membrane due to a difference in concentration. This method was first demonstrated in animals in 1913. The first successful demonstration of dialysis in humans was performed by Wilhelm Kolff in the Netherlands in 1944. As factors limiting successful outcomes with dialysis were eliminated in the 1960s, the treatment became more widely available.

Hemodialysis units are either flat-plate or hollow-fiber units. The flat-plate design consists of a stack of membrane sheets separated by thin gaps to allow the passage of fluid. The blood and dialysate flow in alternating layers, providing both a larger surface area for exchange and short diffusion distances. The flat-plate design affords a greater surface area, but the volume of blood required is also larger. Hollow-fiber units contain as many as 11,000 fibers with an inner diameter of 200 μm. Blood flows within the hollow fibers, and the dialysate flows around the fibers (often referred to as the shell side of the dialyzer). Hollow-fiber dialyzers are often used in hemofiltration.

[9]Detailed statistics are available at the United States Renal Data System website, www.usrds.org. Data from U.S. Renal Data System, USRDS 2007 Annual Data Report: Atlas of Chronic Kidney Disease and End-Stage Renal Disease in the United States, National Institutes of Health, National Institute of Diabetes and Digestive and Kidney Diseases, Bethesda, MD, 2007.

TABLE 7.6

Solutes Present in Dialysate

Solute	Concentration (meq L^{-1})
Sodium	135–145
Potassium	0–4
Chloride	102–106
Bicarbonate	30–39
Acetate	2–4
Calcium	0–3.5
Magnesium	0.5–1.0
Dextrose	11
H^+	$0.5–0.79 \times 10^{-7}$

Source: From Ref. [22].
meq, milliequivalent.

The second major method of removing solutes involves augmenting the concentration difference across membranes with a pressure difference. This procedure, known as hemodiafiltration [18], produces a convective flow of water from the blood to the dialysate. Because the membrane permits the transport of low-molecular-weight solutes, these solutes are also transported by convection, enhancing their overall flux. High-molecular-weight solutes are excluded and accumulate near the inner membrane surface. (Hemodiafiltration is a form of *ultrafiltration*, in which a pressure gradient enhances solvent flow and excludes macromolecular transport across a membrane.)

The major mass transfer design goal is to maximize the removal of solute from the blood during each pass through the dialyzer. The variables that can be controlled are the flow rates of the blood and the dialysate, the total membrane exchange area (typically $0.8-2.1\ m^2$), the length of the exchange unit, the difference in concentration across the membrane, the relative directions of blood and dialysate flow, and, for hemofiltration, the transmembrane pressure difference. The discussion that follows addresses how mass balances are applied to predict the performance of the various devices that are used in dialysis. The kinetics of solute removal also depend on the distribution of the solute (e.g., urea) throughout the body. A complete description of the dynamics of solute removal requires a pharmacokinetic model (Chapter 16).

Two types of membranes are in widespread use: cellulose and hollow fibers. Cellulose derivatives are used in hemodialyzers. The cellulose membranes all contain the sugar group cellobiose in a polymeric form. These membranes are thin (approximately 8 μm when dry, $20-30$ μm when hydrated) and are highly permeable to solutes with a molecular weight less than 500. A major advance achieved over the past 30 years was to reduce the thickness of the membranes, thereby improving flux across them. More recently, the pore size and permeability to water (often referred to as the *hydraulic permeability*) of cellulose membranes have been increased due to the replacement of hydroxyl groups. Hemofiltration is generally performed with hollow-fiber membranes. These membranes are asymmetric and have a thin skin (typically 1 μm) on the inner surface and a relatively thick, but porous, foam structure that provides mechanical support. Hemofiltration membranes are composed of cellulose derivatives, polysulfone, or polyamides.

7.9.1 Cocurrent Exchange

In hemodialysis, there are four modes in which the flow of blood and dialysate can operate: (a) The two flow streams run in the same direction (*cocurrent* flow—see Figure 7.14a); (b) the flow streams run in opposite directions (*countercurrent* flow—see Figure 7.14b); (c) a well-mixed dialysate flows at a uniform concentration; and (d) for hollow-fiber systems, the flows run normal to each other. In this section, we analyze cocurrent exchange. Performing a differential material balance on fluid and solutes entering and leaving the device results in the following expressions for the molar flow rate $\dot{M}_i$ (moles per second) of component i:

$$\text{Blood} \quad d\dot{M}_i = -Q_B dC_{i_B}, \tag{7.9.1}$$

$$\text{Dialysate} \quad d\dot{M}_i = Q_D dC_{i_D}. \tag{7.9.2}$$

Here, Q_B and Q_D are the flow rates of the blood and dialysate, respectively. Equations (7.9.1) and (7.9.2) indicate that the loss of solute from the blood is balanced by the gain of that solute in the dialysate. In addition, the exchange across the membrane can be described with a mass transfer coefficient. We have

$$d\dot{M}_i = k_0(C_{i_B} - C_{i_D})dA_m, \tag{7.9.3}$$

where k_0 is the overall mass transfer coefficient, which includes fluid phase conductivities on the blood (k_B) and dialysate sides (k_D) and permeability across the membrane (P_m). For a flat-plate dialyzer,

$$\frac{1}{k_0} = \frac{1}{k_D} + \frac{1}{P_m} + \frac{1}{k_B}. \tag{7.9.4a}$$

The terms in Equation (7.9.4a) are sometimes referred to as mass transfer resistances, since, like electrical resistances, these resistances are additive in series:

$$R_0 = R_D + R_m + R_B. \tag{7.9.4b}$$

Equations (7.9.1) through (7.9.3) are used to develop a global expression for the molar flow rate in terms of the concentrations entering and leaving on the blood and

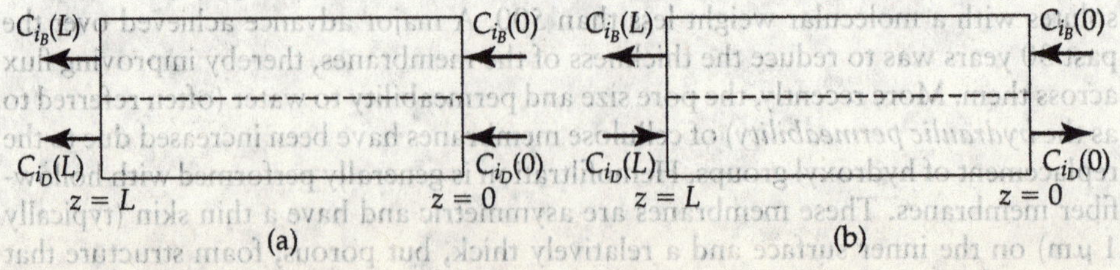

FIGURE 7.14 Schematic of inflows and outflows for (a) cocurrent and (b) countercurrent hemodialysis units.

dialysate sides. This is done by first rearranging terms in Equations (7.9.1) and (7.9.2) such that dC_{i_D} is subtracted from dC_{i_B}:

$$dC_{i_B} - dC_{i_D} = -d\dot{M}_i\left(\frac{1}{Q_D} + \frac{1}{Q_B}\right). \tag{7.9.5}$$

Equation (7.9.3) is then used to replace $d\dot{M}_i$ in Equation (7.9.5), yielding

$$dC_{i_B} - dC_{i_D} = -k_0 dA_m\left(\frac{1}{Q_D} + \frac{1}{Q_B}\right)(C_{i_B} - C_{i_D}). \tag{7.9.6}$$

Integrating from the inlet to the outlet results in the following expression:

$$\ln\left(\frac{C_{i_B}(0) - C_{i_D}(0)}{C_{i_B}(L) - C_{i_D}(L)}\right) = k_0 A_m\left(\frac{1}{Q_D} + \frac{1}{Q_B}\right). \tag{7.9.7}$$

The flow rates can be replaced by integrating Equations (7.9.1) and (7.9.2) from the inlet to the outlet. We obtain

$$\dot{M}_i = Q_B(C_{i_B}(0) - C_{i_B}(L)) = -Q_D(C_{i_D}(0) - C_{i_D}(L)) \tag{7.9.8}$$

or

$$Q_B = \frac{\dot{M}_i}{(C_{i_B}(0) - C_{i_B}(L))} \qquad Q_D = \frac{\dot{M}_i}{(C_{i_D}(L) - C_{i_D}(0))}. \tag{7.9.9a,b}$$

Substituting Equations (7.9.9a) and (7.9.9b) into Equation (7.9.7) and solving for the molar flow rate yields

$$\dot{M}_i = k_0 A_m \frac{[C_{i_B}(0) - C_{i_D}(0)] - [C_{i_B}(L) - C_{i_D}(L)]}{\ln\left(\dfrac{C_{i_B}(0) - C_{i_D}(0)}{C_{i_B}(L) - C_{i_D}(L)}\right)}. \tag{7.9.10}$$

Equation (7.9.10) indicates that the molar flow rate depends on a log-mean concentration difference. The equation also provides an adequate definition for the average mass transfer coefficient and is applicable to cocurrent flow.

7.9.2 Countercurrent Exchange

For countercurrent flow (Figure 7.14b), the flow rate of the dialysate is in the negative z direction. As a result, Equation (7.9.2) becomes

$$d\dot{M}_{i_D} = -Q_D dC_{i_D}. \tag{7.9.11}$$

Consequently, Equation (7.9.5) becomes

$$dC_{i_B} - dC_{i_D} = -d\dot{M}_i\left(\frac{1}{Q_B} - \frac{1}{Q_D}\right). \tag{7.9.12}$$

Equation (7.9.3) is then used to replace $d\dot{M}_i$ in Equation (7.9.5), resulting in

$$dC_{i_B} - dC_{i_D} = -k_0 dA_m\left(\frac{1}{Q_B} - \frac{1}{Q_D}\right)(C_{i_B} - C_{i_D}). \quad (7.9.13)$$

Integrating from the inlet to the outlet results in the following expression:

$$\ln\left(\frac{C_{i_B}(0) - C_{i_D}(0)}{C_{i_B}(L) - C_{i_D}(L)}\right) = k_0 A_m\left(\frac{1}{Q_B} - \frac{1}{Q_D}\right). \quad (7.9.14)$$

Since the design goal is to remove the largest amount of solute in each pass, countercurrent exchange is superior to cocurrent exchange. The reason for this is that, with cocurrent exchange, the exit concentration on the dialysate side ($C_{i_D}(L)$) is slightly less than the outlet concentration on the blood side ($C_{i_B}(L)$) and much less than the inlet concentration on the blood side ($C_{i_B}(0)$). In contrast, with countercurrent exchange, dialysate enters at $z = L$ and exits at $z = 0$. The outlet concentration on the dialysate side ($C_{i_D}(0)$) can then exceed the outlet concentration on the blood side ($C_{i_B}(L)$) and approach the inlet concentration on the blood side ($C_{i_B}(0)$) (Figure 7.15).

The preceding results can be used to characterize the efficiency of a dialysis unit. This is done by defining the clearance K as the mass transfer rate divided by the concentration difference at the inlet to the unit [19]:

$$K = \frac{\dot{M}_i}{C_{i_B}(\text{inlet}) - C_{i_D}(\text{inlet})}. \quad (7.9.15)$$

The location of the inlet for the dialysate depends upon whether flow is cocurrent ($C_{i_D}(\text{inlet}) = C_{i_D}(0)$) or countercurrent ($C_{i_D}(\text{inlet}) = C_{i_D}(L)$). Using Equations (7.9.9a) and (7.9.9b) to substitute for the mass flow rate, we obtain the clearance for cocurrent exchange:

$$K = Q_B\frac{C_{i_B}(0) - C_{i_B}(L)}{C_{i_B}(0) - C_{i_D}(0)} = Q_D\frac{C_{i_D}(L) - C_{i_D}(0)}{C_{i_B}(0) - C_{i_D}(0)}. \quad (7.9.16)$$

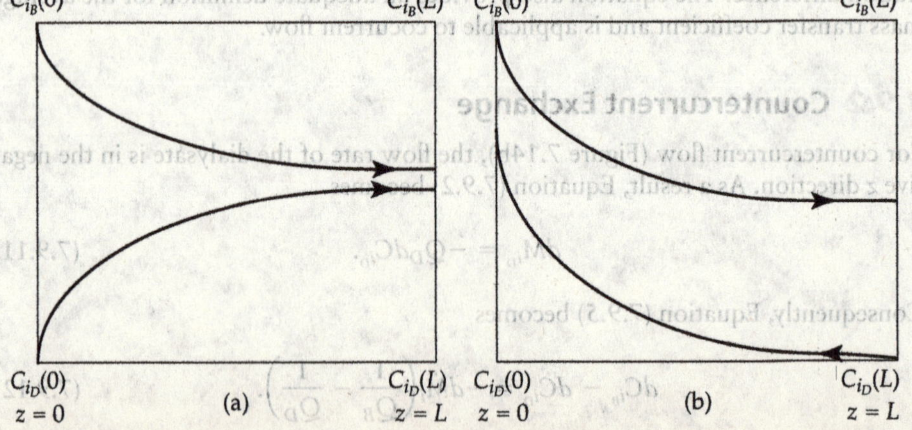

FIGURE 7.15 Schematics of (a) cocurrent and (b) countercurrent exchange of solute from blood to dialysate.

The extraction fraction $E = K/Q_B$ represents the fraction of solute that is removed from blood. From Equation (7.9.16), the fraction for cocurrent exchange is given by the relation

$$E = \frac{K}{Q_B} = \frac{C_{i_B}(0) - C_{i_B}(L)}{C_{i_B}(0) - C_{i_D}(0)} = \frac{Q_D}{Q_B}\frac{(C_{i_D}(L) - C_{i_D}(0))}{C_{i_B}(0) - C_{i_D}(0)}. \quad (7.9.17)$$

With the aid of Equations (7.9.9a), (7.9.9b), (7.9.10a), and (7.9.10b), the extraction ratio for cocurrent and countercurrent exchange is written as a function of the two variables FR and Z [19], which represent the ratio of mass transfer rate to blood flow rate and the ratio of blood to dialysate flow rates, respectively,

$$E = \frac{1}{1+Z}\{1 - \exp[-FR(1+Z)]\} \text{ (cocurrent exchange)}, \quad (7.9.18a)$$

$$E = \frac{1 - \exp[FR(1-Z)]}{Z - \exp[FR(1-Z)]} \text{ (countercurrent exchange)}. \quad (7.9.18b)$$

In the preceding two equations,

$$Z = \frac{Q_B}{Q_D} \quad \text{and} \quad FR = \frac{k_0 A_M}{Q_B}. \quad (7.9.19a,b)$$

Equations (7.9.18a) and (7.9.18b) are the primary results used in the design optimization of dialysis exchange. In order to apply those equations, it is necessary to obtain values for the mass transfer coefficients. The overall mass transfer resistance across the membrane is the sum of the resistances on the blood side, across the membrane, and on the dialysate side. On the blood side, the Lévêque approximation for short contact time is generally not applicable. In general, the blood-side Sherwood number Sh_B for mass transfer across a membrane in a rectangular channel of thickness H and length L is a function of two variables: $x = zD_{eff}/\langle v\rangle H^2$, where z is the axial dimension and D_{eff} is the effective diffusion coefficient in blood; and the transmembrane Sherwood number $Sh_m = k_m H/D_{eff} = P_m H/D_{eff}$ [20] (note that $x = z/\text{Pe}\,H$).

For flat-plate hemodialyzers, the value of x is sufficiently large that the blood-side Sherwood number Sh_B approaches its asymptotic value of approximately 4.30 for permeable membranes [18]. The same result is also used for hollow fibers; the diameter of the fiber replaces the channel thickness.

The diffusion coefficient of solutes in blood is affected by the volume of blood cells and the increased mixing arising from collisions among red cells. The effective diffusion coefficient consists of two terms. The first term represents the reduction in diffusion coefficient due to the presence of red cells. The second term represents mixing and is proportional to the shear rate. The effective diffusion coefficient can be written as

$$D_{eff} = 0.53D_{ij} + 5.292 \times 10^{-9}\dot{\gamma}_w. \quad (7.9.20)$$

This equation would need to be modified if the solute binds to serum proteins.

The membrane mass transfer coefficient is simply equal to the membrane permeability. The value of the solute diffusion coefficient in the membrane is much less than its value in aqueous solution. The extent of this reduction is a strong function of the solute radius [18]. The mass transfer coefficient on the dialysate side is a strong function of the specific geometry used, and few general relations are available.

Example 7.6 **Mass Transfer Resistances for Hemodialysis**

For urea (molecular weight MW = 60), vitamin B_{12} (MW = 1,355), and albumin (MW = 66,000), determine the resistances offered by the blood, membrane, and dialysate for the following conditions for hollow-fiber membranes:

Urea diffusion coefficient at 37°C	1.81×10^{-5} cm^2 s^{-1}
D_m/D_{ij} for urea	0.16
Vitamin B_{12} diffusion coefficient at 37°C	4.5×10^{-6} cm^2 s^{-1}
D_m/D_{ij} for vitamin B_{12}	0.05
Albumin diffusion coefficient at 37°C	6.34×10^{-7} cm^2 s^{-1}
D_m/D_{ij} for albumin	2×10^{-5}
Dialysate mass transfer coefficient (cm min^{-1})	$k_D = 0.874 Q_D^{0.8}(D_{ij})^{2/3}$
Dialysate flow rate, Q_D	500 cm^3 min^{-1}
Surface area of exchanger	1.5 m^2
Shear rate	250 s^{-1}
Membrane thickness	20 μm
Inner diameter of hollow fiber d	180 μm

Solution For each of the terms in Equation (7.9.4b), the resistance equals the reciprocal of the mass transfer coefficient. For the blood phase, this resistance is

$$R_B = \frac{1}{k_B} = \frac{d}{4.30(0.53D_{ij} + 5.292 \times 10^{-9}\dot{\gamma}_w)}. \tag{7.9.21a}$$

The membrane resistance is simply

$$R_m = \frac{L_m}{D_m}. \tag{7.9.21b}$$

Finally, the dialysate resistance is [18]

$$R_D = \frac{1}{126.103(D_{ij})^{2/3}}. \tag{7.9.21c}$$

Results are summarized in Table 7.7.

TABLE 7.7

Mass Transfer Resistances for Urea, Vitamin B$_{12}$, and Albumin in a Hemodialyzer

Molecule	R_B (min cm^{-1})	R_m (min cm^{-1})	R_D (min cm^{-1})	R_0 (min cm^{-1})	R_m/R_0
Urea	6.39	11.52	11.5	29.41	0.39
Vitamin B_{12}	18.8	148	29.1	195.9	0.76
Albumin	42	2.63×10^6	107.5	2,630,149.5	0.99994

As expected, the absolute resistance increases with increasing size of the molecule. The membrane is most sensitive to the molecular size and becomes a greater fraction of the total mass transport resistance as the size increases. For albumin, the membrane accounts for almost all of the mass transport resistance. The resistance values reported in Table 7.7 for urea and vitamin B_{12} are similar to values reported for a number of commercial devices [21].

Example 7.7 Using the results in Example 7.6 for urea and vitamin B_{12}, determine the extraction ratio for cocurrent and countercurrent exchange. Use the following flow rates for Q_B and Q_D: 50, 100, 150, 200, 250, 300, and 500 mL min^{-1}.

Solution The range of flow rates given and the values of k_0 and A_m from Example 7.6 correspond to a flow rate ratio Z ranging from 0.1 to 10 and FR ranging from 0.153 to 10.2. Extraction is greatest for smaller values of the flow ratio and larger values of FR (Figure 7.16). Thus, the optimum membrane has a rapid rate of transport across its surface relative to the rate in blood and a slower flow rate in the blood phase than in the dialysate phase. Although countercurrent exchange leads to higher values of the extraction ratio for a given value of flow ratio and of permeability flow ratio, the effect is most significant for larger values of FR and for values of Z between 0.5 and 3. Typical operating conditions (Table 7.8) indicate that Z is about 0.5. For the data given and for Q_B of 250 mL min^{-1}, FR is 2.04 for urea and 0.306 for vitamin B_{12}. The clearance K (see Equation (7.9.15)) represents the rate of removal of a particular solute. Clearance values for urea and vitamin B_{12} are reported in Table 7.9 for cocurrent and countercurrent exchange.

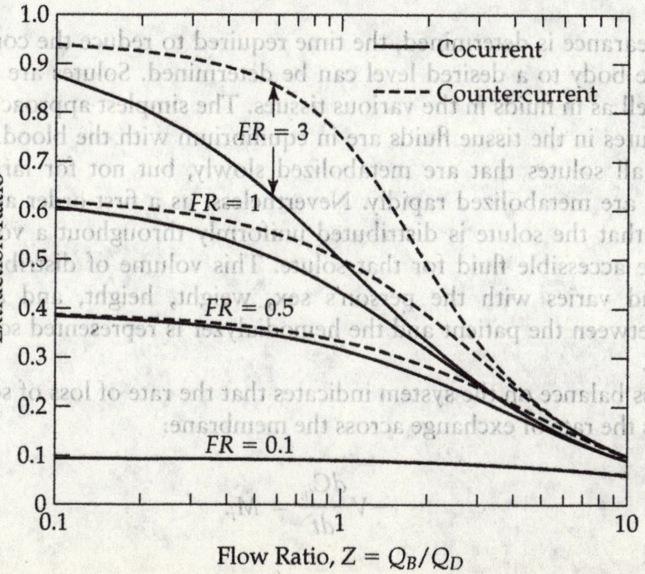

FIGURE 7.16 The extraction ratio as a function of the flow ratio (Z) and permeability flow ratio (FR) for cocurrent and countercurrent exchange.

TABLE 7.8

Typical Operating Conditions for Hemodialyzers with Cellulose Membranes

Blood flow rate (mL min^{-1})	250
Dialysate flow rate (mL min^{-1})	500
$k_0 A$, for urea (mL min^{-1})	300–500
Urea clearance (mL min^{-1})	<200
Urea clearance/body weight (mL min^{-1} kg^{-1})	<3
Vitamin B_{12} clearance (mL min^{-1})	30–60

Source: From Ref. [19].

TABLE 7.9

Clearance *K* (mL min^{-1}) for Cocurrent and Countercurrent Exchange

Solute	Cocurrent exchange	Countercurrent exchange
Urea	158.9	195.0
Vitamin B_{12}	61.3	62.1

Due to urea's larger rate of transmembrane exchange, the clearance of urea is greater than the clearance of vitamin B_{12}. Note that these calculations are consistent with reported values for commercial devices (Table 7.8).

Once the clearance is determined, the time required to reduce the concentration of solute in the body to a desired level can be determined. Solutes are present in the blood, as well as in fluids in the various tissues. The simplest approach is to assume that the solutes in the tissue fluids are in equilibrium with the blood. This is likely true for small solutes that are metabolized slowly, but not for larger solutes or solutes that are metabolized rapidly. Nevertheless, as a first-order approximation, we assume that the solute is distributed uniformly throughout a volume V equal to all of the accessible fluid for that solute. This volume of distribution for urea is 40 L and varies with the person's sex, weight, height, and age [22]. The exchange between the patient and the hemodialyzer is represented schematically in Figure 7.17.

A mass balance on the system indicates that the rate of loss of solute from the body equals the rate of exchange across the membrane:

$$-V\frac{dC_{i_B}}{dt} = \dot{M}_i. \tag{7.9.22}$$

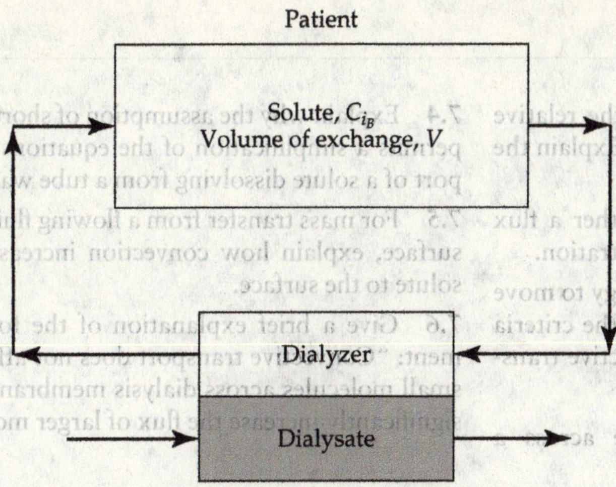

FIGURE 7.17 Schematic of solute exchange between the patient and the hemodialyzer.

Several expressions can be used for the mass flow rate. In the analysis presented here, the definition of the clearance given in Equation (7.9.15) is most useful. Consequently, Equation (7.9.22) becomes

$$-V\frac{dC_{i_B}}{dt} = K(C_{i_B}(0) - C_{i_D}(\text{inlet})). \tag{7.9.23}$$

For the solutes of interest (urea and vitamin B_{12}), the initial dialysate concentration is zero. The inlet concentration to the dialyzer equals the solute concentration of the whole body. As a result, integration of Equation (7.9.23) yields

$$C_{i_B}(t) = C_{i_{B_0}}\exp\left(-\frac{Kt}{V}\right), \tag{7.9.24}$$

where $C_{i_{B_0}}$ is the initial concentration of the solute in the blood.

For a typical dialysis time of 4 hours, the fractional reduction $C_{i_B}(t)/C_{i_{B_0}}$ for urea and for vitamin B_{12} is listed in Table 7.10. Because of the large distribution volume of solutes, considerable time is required to cause an appreciable reduction in solute concentration in blood. For solutes larger than urea and vitamin B_{12}, the change in concentration during this interval is quite small.

TABLE 7.10

Fractional Reduction in Solute After Four Hours of Cocurrent or Countercurrent Hemodialysis

Solute	Cocurrent exchange	Countercurrent exchange
Urea	0.386	0.310
Vitamin B_{12}	0.692	0.689

7.10 QUESTIONS

7.1 What dimensionless group describes the relative importance of convection versus diffusion? Explain the physical basis of this group.

7.2 For a charged molecule, explain whether a flux could occur from a lower to a higher concentration.

7.3 Active transport processes require energy to move a molecule across a membrane. Determine the criteria that can be applied to determine whether active transport is occurring in the following cases:

(a) Transport of an uncharged molecule across a membrane
(b) Transport of ions across a membrane

7.4 Explain why the assumption of short contact times permits a simplification of the equations for the transport of a solute dissolving from a tube wall into fluid.

7.5 For mass transfer from a flowing fluid to a reactive surface, explain how convection increases the flux of solute to the surface.

7.6 Give a brief explanation of the following statement: "Convective transport does not affect the flux of small molecules across dialysis membranes, but it does significantly increase the flux of larger molecules."

7.11 PROBLEMS

7.1 For the following electrolytes, compute the effective diffusion coefficients and diffusion potentials when a 1-mM solution and a 0.1-mM solution are separated by an uncharged membrane at 25°C:

(a) $CuSO_4$
(b) $MgCl_2$

7.2 The permeability across a thin rectangular region of tissue is defined as

$$P_i = \frac{N_{i_z}}{C_0 - C_L}, \tag{7.11.1}$$

where C_0 is the solute concentration in the blood, C_L is the concentration in the lymph, and L is the thickness of the membrane. Assume that $C_L = 0$.

(a) Use this definition and the flux across a membrane (when fluid flows across the membrane) ($\Phi = 1$) to find an expression for the permeability in the presence of convective transport across the membrane.
(b) Show that for diffusion-limited transport ($\mathrm{Pe} \to 0$), $P_i = D_{ij}/L$. Note that $\exp(x) \approx 1 + x$ for $x < 0.01$.
(c) For the following conditions, find the error that would result in the calculated permeability if convection were neglected as a transport mechanism:

$$D_{ij} = 1 \times 10^{-10}\,\mathrm{cm^2\,s^{-1}}$$
$$\mathrm{v} = 1 \times 10^{-6}\,\mathrm{cm\,s^{-1}}$$
$$L = 0.001\,\mathrm{cm}$$
$$C_L = 0$$
$$C_0 = 1 \times 10^{-7}\,\mathrm{M}$$

7.3 Derive Equations (7.5.30a) and (7.5.30b) for ion transport across charged membranes.

7.4 Use Equation (7.5.30b) to find the outward ionic current when the transmembrane potential is zero. Explain how this result can be used to determine the ion permeability.

7.5 For the following conditions, find the resting potential across a typical nerve:

$$C_{K^+_0} = 4\,\mathrm{mM} \qquad C_{K^+_L} = 140\,\mathrm{mM}$$
$$C_{Na^+_0} = 150\,\mathrm{mM} \qquad C_{Na^+_L} = 12\,\mathrm{mM}$$
$$C_{Cl^-_0} = 120\,\mathrm{mM} \qquad C_{Cl^-_L} = 4\,\mathrm{mM}$$
$$P_{Cl^-}/P_{K^+} = 0.1 \qquad P_{Na^+}/P_{K^+} = 0.1$$

7.6 For the conditions listed in Problem 7.5, what is the concentration of negatively charged protein needed to maintain electroneutrality inside the cell and in the extracellular fluid?

7.7 Determine whether the short-contact-time analysis for mass transport is valid for oxygen (1.10×10^{-5} cm^2 s^{-1}) in capillaries and venules.

7.8 Assuming that an acceptable error in the linear approximation for the velocity profile is 10%, find the range of validity of Equation (7.6.6). Use this result together with the criterion that $zD_{ij}/v_{max}R^2 < 0.01$ to determine the smallest relative concentration that can be at the exit of a tube of length L in order for the assumption to be valid.

7.9 Consider the short-contact-time release of a soluble molecule from the walls of a rectangular channel of length L and height H. Making the same assumptions that were made with regard to short-contact-time analysis in a cylindrical tube, show that the steady-state form of the conservation relation can be approximated as

$$6\langle v\rangle \frac{y'}{H}\frac{\partial C_i}{\partial z} = D_{ij}\frac{\partial^2 C_i}{\partial y'^2}, \qquad (7.11.2)$$

where $y' = H/2 - y$.

7.10 For the flow of a fluid over a flat membrane of length 10 cm, determine the length-average mass transfer coefficient. The relevant properties of the system are $v = 0.01$ cm^2 s^{-1}, $D_{ij} = 5 \times 10^{-6}$ cm^2 s^{-1}, and $\langle v\rangle = 5$ cm s^{-1}.

7.11 Calculate the effect of red-cell mixing on the diffusion coefficients of urea and albumin at shear rates of 1, 10, and 1,000 s^{-1}.

7.12 For the flow of a fluid over a flat membrane with permeability P_i, determine an expression for the flux in terms of the bulk fluid concentrations on either side of the membrane.

7.13 Derive Equations (7.9.10) and (7.9.18b) for countercurrent flow in a hemodialysis unit. Note that, for countercurrent flow, the following relationships apply:

$$C_{i_D}(L) < C_{i_D}(0)$$

$$\dot{M}_i = Q_B(C_{i_B}(0) - C_{i_B}(L)) = Q_D(C_{i_D}(0) - C_{i_D}(L))$$

7.14 Some intermediate molecular weight (IMW) molecules may be responsible for toxic effects associated with kidney dialysis. However, these molecules bind to plasma proteins, principally albumin, reducing their transport across the dialysis membrane. A novel membrane has been developed that produces high clearance at high flow rates of dialysate.

(a) Use the data given below to compute the IMW extraction fraction for countercurrent exchange with and without protein in solution and compare with the result for urea.

Q_D	741 mL min^{-1}	Dialysate flow rate
Q_B	208 mL min^{-1}	Blood flow rate
K_{urea}	200 mL min^{-1}	Urea clearance
K_{IMW}	136 mL min^{-1}	Intermediate molecular weight clearance without protein
K_{IMW}	24 mL min^{-1}	Intermediate molecular weight clearance with protein

(b) As noted, many of these intermediate-molecular-weight molecules bind to albumin that is too big to move across the membrane. Thus, the total concentration of these molecules (C_{iB_T}) equals the sum of the free molecule (C_{iB_f}) and the albumin-bound molecule ($C_{iB_{bound}}$). The loss of IMW molecules from the blood is based on the total amount of IMW molecules, $dM_i = -Q_B dC_{iB_T}$. Assuming that the number of binding sites for the IMW molecules is large, the bound concentration is proportional to the free concentration, i.e., $C_{iB_{bound}} = k_A C_{iB_f}$. Since only the free molecule can move across the membrane, membrane transport depends only on the free molecule (i.e., $dM_i = k_0(C_{iB_f} - C_{iD})dA_m$). Use these results to show that an equation similar to Equation (7.9.13) can be derived for $dC_{iB} - dC_{iD}$, except that Q_B is replaced with $Q_B\,(1+K_A)$.

(c) Assuming that Q_B is replaced with $Q_B\,(1+K_A)$, determine the value of K_A for the intermediate molecular weight protein.

7.15 A flat-plate dialyzer is being used to extract a toxin from blood. To enhance flow, the unit is connected to the femoral artery in the leg and a new high-permeability membrane is being used. There are 90 channels for blood and 90 channels for dialysate, and the unit is operated in countercurrent exchange. Each channel has a height $2H$, length L, and width W, with $W \gg 2H$. The dialysate concentration in the inlet is zero. The toxin level is C_0.

(a) Using data below and appropriate correlations in Table 7.5, determine the time required to reduce the toxin below 10^{-12} mol cm^{-3}. Note V is the volume of distribution.

(b) You have been asked to determine whether you could use a maximum of 50 L of dialysate to obtain the same final toxin level.

1. Does recycling the dialysate help? (Hint: Determine the equilibrium concentration in blood and dialysate.)
2. Is there another way this can be done to reduce the toxin levels in less than 6 hours?

$Q_B = 500$ mL min^{-1}
$Q_D = 750$ mL min^{-1}
$D_{ij} = 5 \times 10^{-6}$ cm^2 s^{-1}
$v_{blood} = 0.03$ cm^2 s^{-1}
$v_{dialysate} = 0.009$ cm^2 s^{-1}
$C_0 = 3 \times 10^{-9}$ mol cm^{-3}
$H = 0.005$ cm
$W = 10$ cm
$L = 25$ cm
$P_m = 0.005$ cm s^{-1}
$V = 10$ L
$Re = \langle v \rangle D_h / v$
$D_h = 4WH/(W + 2H)$

7.16 Concentration polarization arises during fluid ultrafiltration when high molecular proteins accumulate near the membrane surface because the membrane restricts transport. While an exact solution can be obtained by numerical integration of the solute conservation relations, we will perform an approximate analysis as follows. Consider a protein solution flowing over a membrane permeable to water and low-molecular-weight solutes (panel A in Figure 7.18). A pressure difference exists across the membrane driving fluid flow. The filtration velocity is v_f. We will simplify the analysis by assuming that the concentration change is confined to a thin boundary layer near the membrane surface of thickness δ (panel B in Figure 7.18).

(a) For steady-state transport across the membrane, show that the flux is

$$N_s = -D_{ij}\frac{dC}{dy} - v_f C = -v_f C_f, \qquad (7.11.3)$$

where C_f is the filtrate concentration.

(b) Integrate Equation (7.11.3) from $y = 0$ to $y = d$ to obtain the polarization ratio $(C_w - C_f)/(C_b - C_f)$.

(c) Let the mass transfer coefficient be defined as $k_f = D_{ij}/\delta$. For the long contact time solution in a cylindrical tube of diameter D, $Sh = k_f D/D_{ij} = 3.657$. Determine the polarization ratio for $Pe = v_f D/D_{ij} = 10$.

(d) This analysis assumes that the filtration velocity is constant. In general, the filtration velocity depends on the membrane properties and the difference in hydrostatic and osmotic pressure differences. Assume that the filtration velocity can be represented as

$$v_f = L_p(\Delta p_c - \sigma(\pi - \pi_f)), \qquad (7.11.4)$$

where π_f is the osmotic pressure in the filtrate, and the osmotic pressure can be described by Equation (9.3.18). Discuss briefly the effect of concentration polarization on the filtrate velocity.

7.17 Peritoneal dialysis is an alternative method of dialysis in which solute exchange occurs across the peritoneal membrane that lines the abdominal cavity. The approach avoids connecting a patient to an external hemodialyzer several times per week for 3–4 hours per dialysis. In peritoneal dialysis, a volume V_D of a glucose solution that is at the same salt concentration as peritoneal fluid is injected into the peritoneal cavity through an indwelling catheter. After 5 hours, the fluid volume V_D is withdrawn from the peritoneum and replaced with an equal volume of fresh fluid. The process is shown schematically in Figure 7.19.

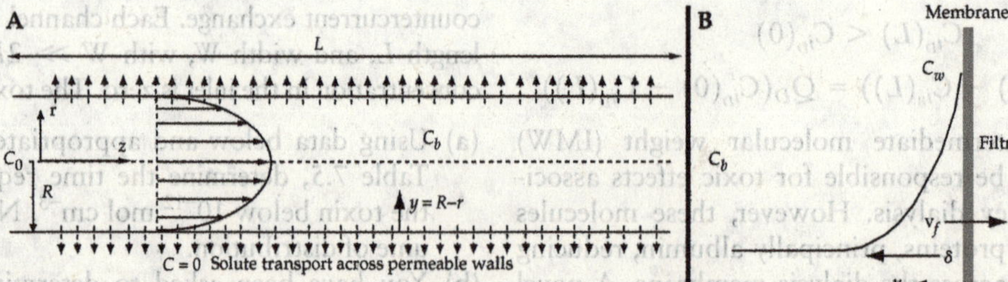

FIGURE 7.18 (A) Solute transport across an ultrafiltration membrane. (B) Schematic of concentration profile during concentration polarization at the membrane surface.

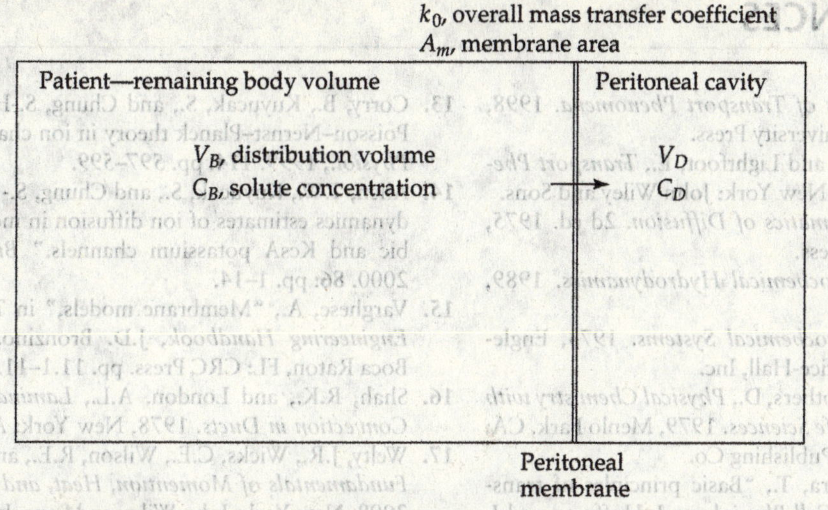

FIGURE 7.19 Schematic of peritoneal dialysis.

(a) For a constant dialysate volume V_D, perform a mass balance on the solute in the dialysate C_D to describe the solute accumulation between periods of transfer of peritoneal fluid. The fluid in both compartments is well mixed. Mass transfer across the peritoneum is described in terms of an overall mass transfer coefficient k_o and the membrane exchange area is A_m.

(b) Solve the resulting differential equation, assuming that the concentration in the body C_B is constant and C_D is initially zero.

(c) For the data below, determine the concentration of urea and vitamin B_{12} at the end of the exchange period as a fraction of the concentration in the body.

$t = 5$ hours		Exchange period
$V_D = 2$ L	$k_oA_m = 21$ cm^3 min^{-1}	Urea
$V_B = 40$ L	$k_oA_m = 5$ cm^3 min^{-1}	Vitamin B_{12}

7.18 To predict the change in urea concentration during the period between dialyses, the model describing the change in plasma levels of molecules removed by dialysis needs to be modified to include a term for production as well as residual removal by the kidney. The modified mass balance is

$$V\frac{dC_i}{dt} = G - (K + K_R)C_i, \qquad (7.11.5)$$

where G is the constant production rate and K_R is the residual clearance of the kidneys.

(a) For $G = 0.00022$ mol min^{-1}, $V = 40$ L, $K_R = 6$ mL min^{-1}, $K = 167$ mL min^{-1}, and a initial urea concentration of 37 mM, solve Equation (7.11.5) and determine the urea concentration after 4 hours of dialysis. Determine the effect of residual production and synthesis on the level of urea in the plasma during this 4-hour period of dialysis.

(b) Now consider the rise in the urea concentration during the period between dialysis treatments. There is no clearance by dialysis ($K = 0$), but there is residual removal from the kidney and production. Modify the solution from part (a) to determine the steady-state concentration and the time to reach steady state.

(c) Calculate the dialysate concentration in blood at the beginning of the next dialysis, which is three days later. Determine the number of cycles for the concentrations until the concentration rise between cycles is the same. Give the final levels before and after dialysis.

7.12 REFERENCES

1. Deen, W.M., *Analysis of Transport Phenomena*. 1998, New York: Oxford University Press.
2. Bird, R., Stewart, W., and Lightfoot, E., *Transport Phenomena*. 2d ed. 2002, New York: John Wiley and Sons.
3. Crank, J., *The Mathematics of Diffusion*. 2d ed. 1975, Oxford: Clarendon Press.
4. Probstein, R., *Physicochemical Hydrodynamics*. 1989, Boston: Butterworths.
5. Newman, J.S., *Electrochemical Systems*. 1973, Englewood Cliffs, NJ: Prentice-Hall, Inc.
6. Eisenberg, D., and Crothers, D., *Physical Chemistry with Applications to the Life Sciences*. 1979, Menlo Park, CA: Benjamin/Cummings Publishing Co.
7. Macey, R., and Moura, T., "Basic principles of transport," in *Section 14, Cell Physiology*, J. Hoffman and J. Jamieson, editors. 1997, New York: American Physiological Society. pp. 181–260.
8. Plonsey, R., and Barr, R.C., *Bioelectricity: A Quantitative Approach*. 2d ed. 2000, New York: Plenum Press.
9. Hille, B., *Ion Channels of Excitable Membranes*. 3d ed. 2001, Sunderland, MA: Sinauer Associates, Inc.
10. Doyle, D.D., Cabral, J.M., Pfuetzner, R.A., Kuo, A., Gulbis, J.M., Cohen, S.L., Chait, B.T., and MacKinnon, R., "The structure of the potassium channel molecular basis of K^+ conduction and selectivity." *Science*, 1998. 280: pp. 69–77.
11. Levitt, D.G., "Modeling of ion channels." *J. Gen. Physiol.*, 1999. 113: pp. 789–794.
12. Corry, B., Kuyucak, S., and Chung, S.-H., "Tests of continuum theories as models of ion channels. II. Poisson–Nernst–Planck theory versus Brownian dynamics." *Biophys. J.*, 2000. 78: pp. 2364–2381.
13. Corry, B., Kuyucak, S., and Chung, S.-H., "Test of the Poisson–Nernst–Planck theory in ion channels." *J. Gen. Physiol.*, 1999. **114**: pp. 597–599.
14. Allen, T.W., Kuyucak, S., and Chung, S.-H., "Molecular dynamics estimates of ion diffusion in model hydrophobic and KcsA potassium channels." *Biophys. Chem.*, 2000. 86: pp. 1–14.
15. Varghese, A., "Membrane models," in *The Biomedical Engineering Handbook*, J.D. Bronzino, editor. 2000, Boca Raton, FL: CRC Press. pp. 11.1–11.23.
16. Shah, R.K., and London, A.L., *Laminar Flow Forced Convection in Ducts*. 1978, New York: Academic Press.
17. Welty, J.R., Wicks, C.E., Wilson, R.E., and Rorrer, G.L., *Fundamentals of Momentum, Heat, and Mass Transfer*. 2008, New York: John Wiley and Sons, Inc.
18. Yeun, J.Y., and Depner, T.A., "Principles of hemodialysis," in *Dialysis and Transplantation*, W.F. Owen, B.J.G. Pereira, and M.H. Sayegh, editors. 2000, Philadelphia: W.B. Saunders. pp. 1–32.
19. Galletti, P.M., Colton, C.K., and Lysaght, M.J., "Artificial kidney," in *The Biomedical Engineering Handbook*, J.D. Bronzino, editor. 2000, Boca Raton, FL: CRC Press. pp. 130–1 to 130–25.
20. Colton, C.K., Smith, K.A., Stroeve, P., and Merrill, E.W., "Laminar flow mass transfer in a flat duct with permeable walls." *AIChE J.*, 1971. 17: pp. 773–780.
21. Sargent, J.A., and Gotch, F.A., "Principles and biophysics of dialysis," in *Replacement of Renal Function by Dialysis*, C. Jacobs, et al., editors. 1996, Boston: Kluwer Academic Publishers. pp. 35–102.
22. Colton, C.K., "Analysis of membrane processes for blood purification." *Blood Purification*, 1987. 5: pp. 202–251.

Transport in Porous Media

CHAPTER

8

8.1 Introduction

Porous media are solid materials with internal pore structures. The pores can be either empty or filled with fluids. Porous structures vary significantly among different media (see Figure 8.1). A *structure with a regular array of cylindrical pores* can be found in micro- or nanofabricated materials. A *foam structure* is composed of a continuous solid phase with interconnected channels or isolated pores and is often observed as a sponge. A *granular structure*, exhibited by a pile of sand, consists of solid particles and the void space between them. A *fiber matrix* is the primary structure in polymeric gels. Biological tissues can contain several of these structures simultaneously.

There are three compartments in biological tissues: blood and lymph vessels, cells, and interstitium (see Figure 8.2). The interstitial space can be further divided into the extracellular matrix and the interstitial fluid. Although the volume fraction of each compartment is tissue-dependent, it is generally less than 10% for the vascular space, which is smaller than those for the other two compartments. The extravascular region (cells and the interstitium) can be considered a porous medium, with pores saturated with interstitial fluid.

There are two different populations of pores in biological tissues. One exists between cells as part of a granular structure; the other exists between extracellular fibrous molecules as part of a fiber matrix. The fiber-matrix structure is embedded in the granular structure to form a composite material (see Figure 8.2). The interstitial pores are either isolated or connected to form hydrophilic channels that are critical to the transport of nutrients, metabolites, growth factors, inhibitors, modulators, and other signaling molecules in tissues.

The extracellular matrix is a network of protein and polysaccharide macromolecules that are either synthesized by the surrounding cells or originated from circulating blood. Examples of matrix molecules include proteoglycans, collagens, elastin, fibronectin, and laminin. The extracellular matrix functions as a mechanical

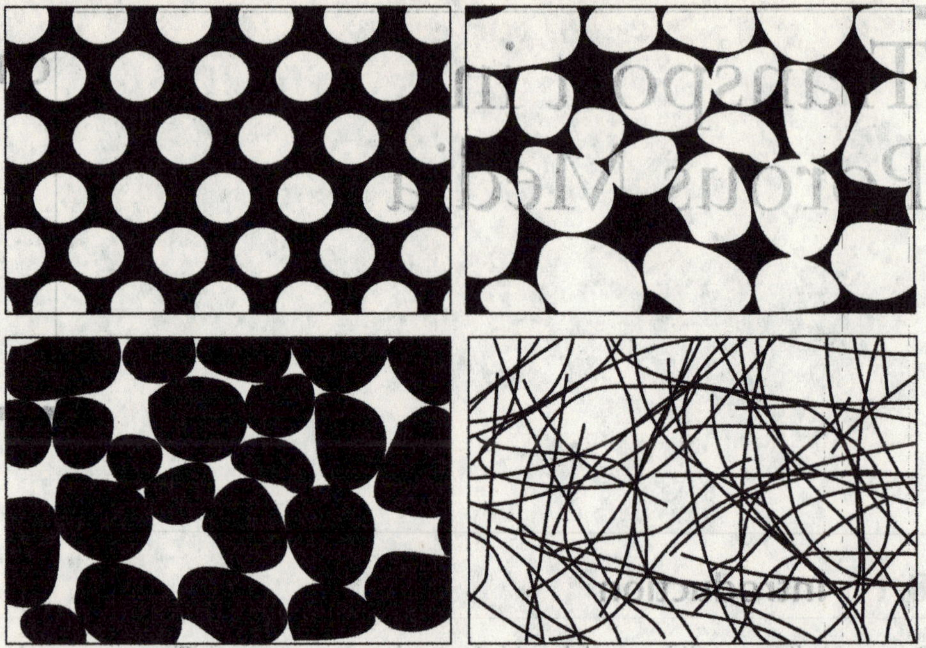

FIGURE 8.1 Examples of porous structures. Upper left, a regular array of cylindrical pores; upper right, a foam structure; lower left, a granular structure; and lower right, a fiber matrix. In all examples, the white regions represent void spaces or the fluid phases of the media, and the black regions represent the solid phases.

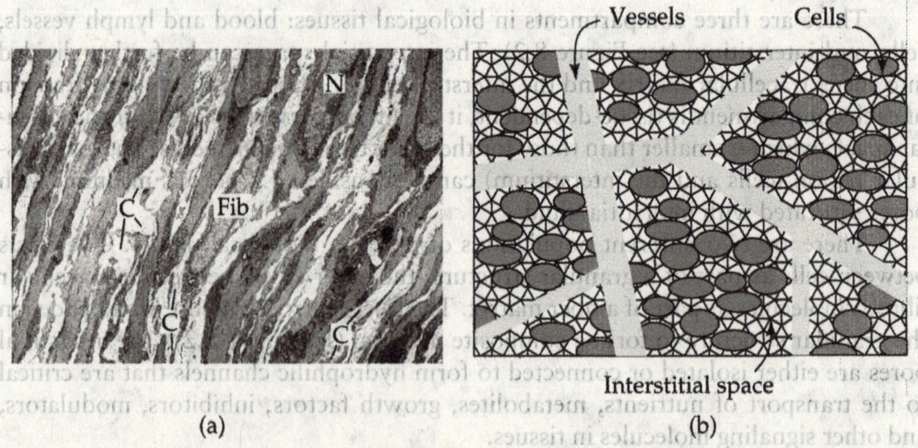

FIGURE 8.2 Compartments in biological tissues. (a) An electron micrograph of smooth muscle tissues, where Fib indicates fibroblast, N indicates the nucleus of smooth muscle cells, and C indicates collagen fibrils. Blood vessels are not shown in this figure (modified from Plate 35 in Ref. [1], with permission). (b) A schematic of biological tissues. The vessels can be either blood or lymph vessels. The cells include all populations in the tissue.

scaffold of tissues (e.g., collagens), a substrate for cell adhesion and migration (e.g., fibronectin), a depot for growth factors (e.g., fibronectin), or a depot for water (e.g., glycosaminoglycans). Biologically, cell–matrix interactions regulate cell physiology, proliferation, differentiation, and apoptosis. In this chapter, we will focus on transport-related properties of the extracellular matrix.

8.2 Porosity, Tortuosity, and Available Volume Fraction

Porous media can be characterized by their *specific surface* and *porosity*, respectively defined as

$$s = \frac{\text{Total interface area}}{\text{Total volume}} \tag{8.2.1}$$

and

$$\varepsilon = \frac{\text{Void volume}}{\text{Total volume}}. \tag{8.2.2}$$

Note that the unit of s is one over length and that ε is dimensionless. The void volume is the total volume of the void space in a porous medium; the interface is the border between solid and void spaces. Both s and ε depend on the structure of pores.

Some porous media, such as biological tissues, are deformable under mechanical loads. Material deformation can change the spatial distribution of the porosity. Thus, the local porosity may vary both spatially and temporally. If a porous material is homogenous, its porosity can be easily calculated. Biological tissues are heterogeneous, in which the porosity is equal to the volume fraction of local interstitial fluid. If the total volume in Equation (8.2.2) is based on the interstitial space, ε is, in general, larger than 0.9. If cells and vessels need to be considered (i.e., if the total volume is based on the entire tissue), ε can vary from 0.06 in the brain [2] to 0.30 in the skin [3]. In some tumor tissues, ε can be as high as 0.60 [2] (see Chapter 15).

Example 8.1 Determine the specific surface and the porosity of a porous medium with uniformly distributed cylindrical pores. Assume that the pores are parallel to each other (see Figure 8.3), the diameter of pores is d, and the number of cylinders per unit cross-sectional area is n_A.

Solution The void volume is the space within the cylinders. To calculate the specific surface and the porosity, we must first determine the total interface area and the void volume. We have

$$\text{Total interface area} = \pi d L N, \tag{8.2.3}$$

$$\text{Void volume} = \pi \frac{d^2}{4} L N, \tag{8.2.4}$$

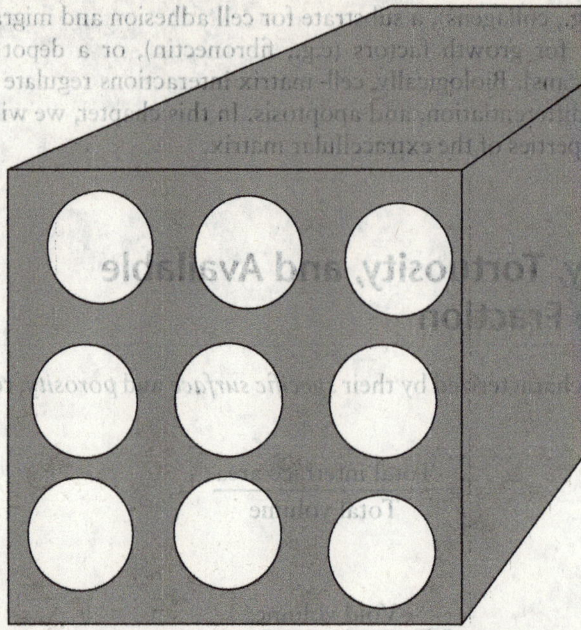

FIGURE 8.3 A porous medium with cylindrical pores.

where L is the length of the pores and N is the total number of cylinders in the material. Numerically, N is equal to the product of the cross-sectional area A and n_A. The specific surface and the porosity can be calculated by substituting Equations (8.2.3) and (8.2.4) into Equations (8.2.1) and (8.2.2), respectively. The results are

$$s = \frac{\text{Total interface area}}{\text{Total volume}} = \frac{\pi dLN}{AL} = \pi d n_A \tag{8.2.5}$$

and

$$\varepsilon = \frac{\text{Void volume}}{\text{Total volume}} = \frac{\pi(d^2/4)LN}{AL} = \pi\frac{d^2}{4}n_A. \tag{8.2.6}$$

Example 8.2 Determine the porosity of a porous material consisting of uniformly distributed solid spheres (see Figure 8.4). The space between spheres is filled with a solution. Assume that the diameter of the spheres is d and that the number of spheres per unit volume is n_V.

Solution In contrast to the situation in Example 8.1, we focus on the solid phase in the calculation of the total interface area and the void volume. They are

$$\text{Total interface area} = \pi d^2 N \tag{8.2.7}$$

and

$$\text{Void volume} = V - \pi\frac{d^3}{6}N, \tag{8.2.8}$$

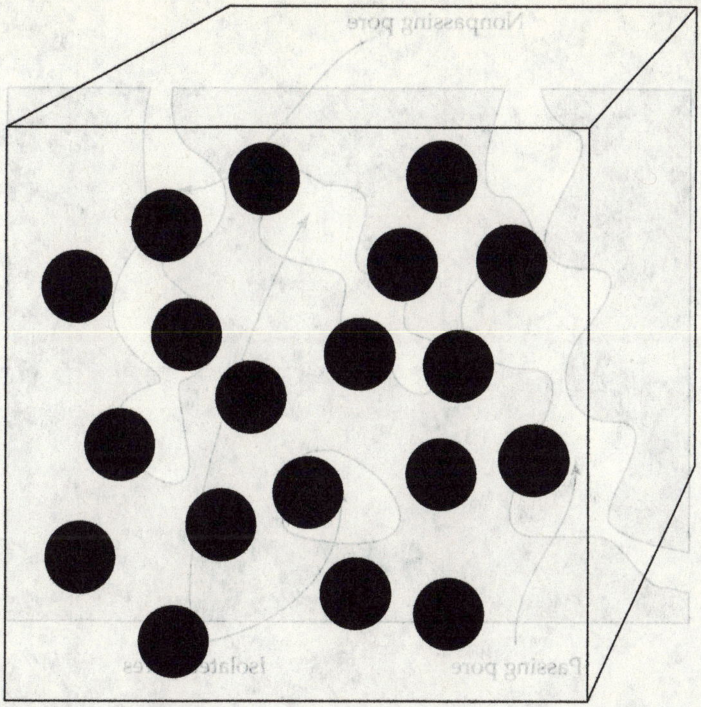

FIGURE 8.4 A porous medium with distributed solid spheres.

where N is the total number of spheres in the material and V is the total volume of the porous material. Substituting Equations (8.2.7) and (8.2.8) into Equations (8.2.1) and (8.2.2), respectively, we get

$$s = \frac{\text{Total interface area}}{\text{Total volume}} = \pi d^2 n_V \tag{8.2.9}$$

and

$$\varepsilon = \frac{\text{Void volume}}{\text{Total volume}} = 1 - \pi \frac{d^3}{6} n_V, \tag{8.2.10}$$

The porosity is a measure of the average void volume fraction in a specific region of porous medium. It does not provide any information on how different pores are connected or on how many pores are available for water and solute transport. Therefore, we divide the pores into the following three categories (see also Figure 8.5):

- pores
  - penetrable pores
    - passing pores (1)
    - nonpassing pores (2)
  - isolated pores (3)

Based on this classification of pores, the porosity can be expressed as the sum,

$$\varepsilon = \varepsilon_i + \varepsilon_p + \varepsilon_n, \tag{8.2.11}$$

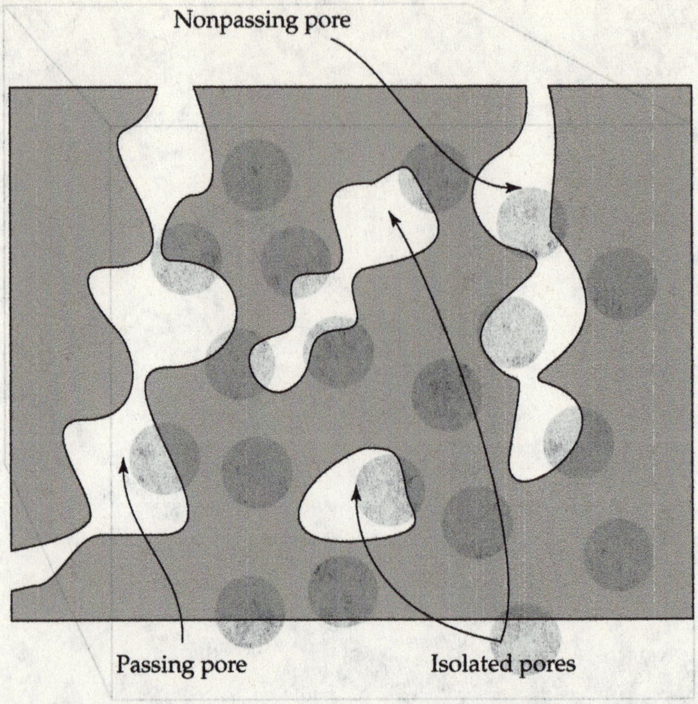

FIGURE 8.5 Classification of pores based on their connectedness. A pore is defined as a *passing* component when it connects to at least two subdomains of the outer surface of finite porous media. For the porous structure shown in this figure, the passing pores connect to two boundaries of the rectangular material. A *nonpassing* pore connects to only one subdomain of the outer surface. Passing and nonpassing pores together are called *penetrable* pores. *Isolated* pores have no connections to the outer surface of the porous media.

where the subscripts i, p, and n indicate isolated, passing, and nonpassing pores, respectively. Isolated pores are not accessible to external solvents and solutes; therefore, they can sometimes be considered as part of the solid phase in transport analysis. In this case, the void volume is defined as the total volume of penetrable pores.

The path length between points A and B in a porous medium is measured by the distance between these points through connected pores. The shortest path length, L_{min}, can be characterized by the *geometric tortuosity*

$$T = \left(\frac{L_{min}}{L}\right)^2, \qquad (8.2.12)$$

where L is the straight-line distance between A and B. The tortuosity depends on the locations of A and B and on the structures of porous media. By definition, T is always greater than or equal to unity.

Not all penetrable pores are accessible to solutes. Such accessibility will depend upon the molecular properties of the solutes. For example, a pore will be inaccessible to a solute if the solute molecule is larger than the pore or if the pore is surrounded by other pores that are smaller than the solute molecule [4]. The portion of accessible volume that can be occupied by the solute is called the available volume. For a

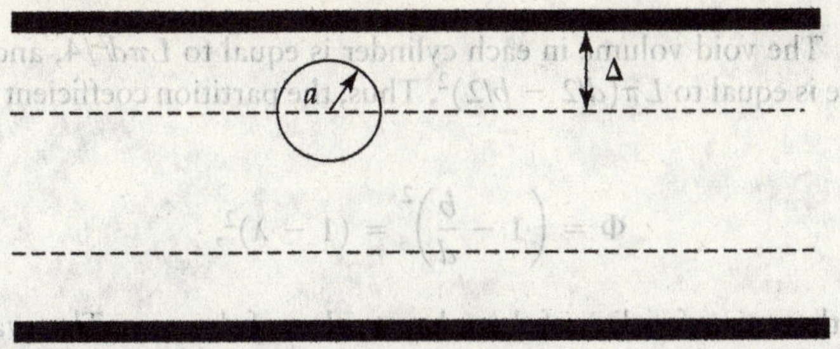

FIGURE 8.6 Steric exclusion of solutes near the surface of solid phase in porous media. The circle represents the solute molecule. Δ is the distance between the solute and the surface.

solute, the ratio of the available volume to the total volume is defined as the *available volume fraction*:

$$K_{AV} = \frac{\text{Available volume}}{\text{Total volume}}. \tag{8.2.13}$$

By definition, K_{AV} is molecule dependent and always smaller than the porosity ε. For example, K_{AV} in a rat tumor tissue is 0.1 for a dextran molecule with a molecular weight of 70,000, while ε is 0.5 [5]. There are three possible scenarios that can cause K_{AV} to be less than ε. First, the centers of the solute molecules cannot reach the solid surface in the void space. In this case, the difference between the total void volume and the available volume can be estimated as the product of the area of the solid surface and the distance Δ between the solute and the surface (see Figure 8.6). The distance Δ is equal to the radius of the solute if the solute can come into contact with the surface. However, many biological surfaces are electrically charged. Solutes that have the same charge as the solid surface does may not be able to reach the surface. In such cases, Δ is larger than the radius of the solute.

The second scenario that can cause K_{AV} to be less than ε is the situation in which some of the void space is smaller than the solute molecules. The third scenario is the inaccessibility of large penetrable pores surrounded by pores that are smaller than the solutes. In general, K_{AV} decreases with the size of the solutes. This molecular size dependence can be influenced by the charge on the solutes and solid surfaces and by the structures of porous media.

The ratio of the available volume to the void volume is defined as the *partition coefficient* Φ of the solutes. The partition coefficient is a measure of solute partitioning at equilibrium between external solutions and the void space in porous media. This parameter is different from that with the same name used in chemistry, in which the partition coefficient characterizes the partitioning of solutes between two liquid phases, such as oil and water (see Section 6.3.2). By definition, the partition coefficient in porous media is related to K_{AV} and ε via the formula

$$\Phi = \frac{K_{AV}}{\varepsilon}. \tag{8.2.14}$$

Example 8.3 Determine the partition coefficient and the available volume fraction of a spherical solute in the same porous medium as that in Example 8.1. Assume that the diameter of the solute is b and that electric charge–charge interactions are negligible.

Solution The void volume in each cylinder is equal to $L\pi d^2/4$, and the available volume is equal to $L\pi(d/2 - b/2)^2$. Thus, the partition coefficient can be calculated as

$$\Phi = \left(1 - \frac{b}{d}\right)^2 = (1 - \lambda)^2, \tag{8.2.15}$$

where λ is the ratio of radius of the solute to that of the pore. The available volume fraction is equal to the product of the partition coefficient and the porosity, which was determined in Example 8.1. Thus,

$$K_{AV} = \pi\frac{d^2}{4}n_A(1 - \lambda)^2, \tag{8.2.16}$$

if $\lambda \ll 1$, $\Phi \approx 1$, and $K_{AV} \approx \varepsilon$.

Example 8.3 assumes that the molecules are spheres. However, polymer molecules are flexible chains with different conformations and configurations. For example, the conformation of a linear polymer chain is closer to a random coil rather than a sphere (see Figure 8.7).

The size of flexible molecules is characterized by the *radius of gyration*, R_g, which is given by

$$R_g^2 = \frac{1}{N}\sum_i r_{ic}^2 = \frac{1}{2N^2}\sum_i\sum_j r_{ij}^2, \tag{8.2.17}$$

where r_{ic} is the average distance between segment i and the center of mass of the polymer chain, r_{ij} is the average distance between segments i and j in the polymer chain, and N is the total number of segments in the polymer chain. All of the equations that we have derived for spherical molecules can be used approximately for chain molecules if we replace the radius with the radius of gyration. For a more precise calculation of the partition coefficient of flexible polymer chains, investigators have used statistical models for polymer partitioning between an external solution and the void space in porous media with different pore structures [6,7]. When the flexible polymer can be modeled

FIGURE 8.7 The random coil model of a linear polymer chain.

as a linear chain of rigid rods (or segments) and the length of a segment (ℓ) is much less than the pore radius (i.e., $\ell/a \ll 1$), analytical expressions are available for pores with simple geometries [6,7]. If the pore is a sphere, then we have

$$\Phi_{\text{sphere}} = \frac{6}{\pi^2}\sum_{n=1}^{\infty}\frac{1}{n^2}\exp[-n^2\pi^2\lambda_s^2], \tag{8.2.18}$$

where λ_s is the ratio of R_g to the radius of the spherical pore. If the pore is between two parallel plates, then the relationship is

$$\Phi_{\text{plate}} = \frac{8}{\pi^2}\sum_{n=0}^{\infty}\frac{1}{(2n+1)^2}\exp\left[-\frac{(2n+1)^2\pi^2}{4}\lambda_p^2\right], \tag{8.2.19}$$

where λ_p is the ratio of R_g to the half-distance between plates. If the pore is a circular cylinder, then we have

$$\Phi_{\text{cylinder}} = 4\sum_{n=1}^{\infty}\frac{1}{\alpha_n^2}\exp[-\alpha_n^2\lambda_c^2], \tag{8.2.20}$$

where λ_c is the ratio of R_g to the radius of the cylindrical pore and α_n are the roots of $J_0(\alpha_n) = 0$, in which J_0 is the Bessel function of the first kind, of order zero (see Section A.2). The roots can be solved easily with commercial software. The partition coefficient of flexible polymer chains in a cylindrical pore can also be estimated with Equation (8.2.15), by replacing λ with λ_c. The results are plotted in Figure 8.8. The estimation is close to that predicted by Equation (8.2.20). Note that we previously mentioned that Equation (8.2.20) is derived by assuming that $\ell/a \ll 1$. If this condition is not satisfied, the equation underestimates Φ_{cylinder} [7].

Some porous media, such as polymeric gels or the interstitial space in tissues, are fiber matrices. Therefore, the space inside and near the surface of fibers will not be available to solutes. (In other words, the solutes are excluded from this space; see Figure 8.9.) The size of the space is known as the *exclusion volume*. Around a single fiber,

$$\text{Exclusion volume} = \pi(r_f + r_s)^2 L, \tag{8.2.21}$$

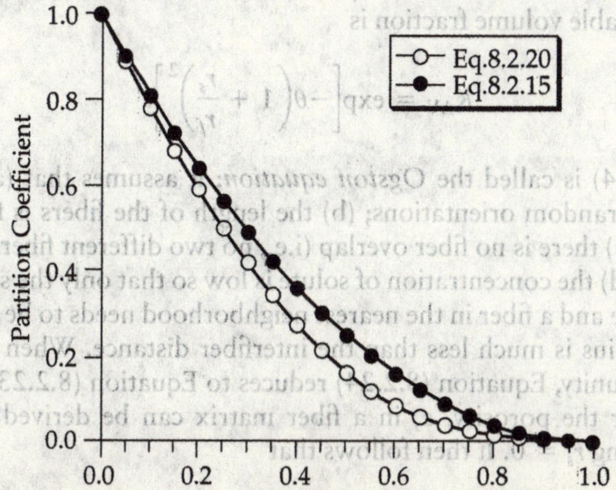

FIGURE 8.8 Comparison of the partition coefficients of flexible polymer chains calculated by Equations (8.2.15) and (8.2.20), respectively, for the same ratio of radius of gyration of polymers versus radius of the pore.

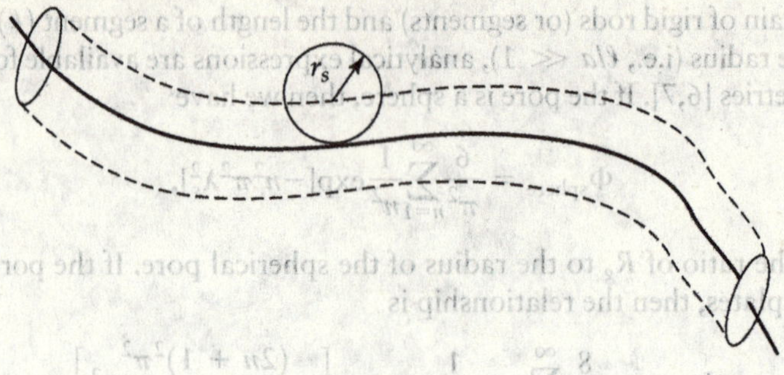

FIGURE 8.9 The exclusion volume around a fiber. The exclusion volume is indicated by the cylinder around the fiber. The radius of the cylinder is equal to the radius of solutes, r_s, plus the radius of fibres, r_f.

where r_f and r_s are the radii of the fiber and solute, respectively, and L is the length of fiber.

For a matrix of fibers, the expression of the exclusion volume can easily be derived if the minimum distance between fibers is larger than $2(r_f + r_s)$. This requirement is satisfied when the fiber density is very low or when the fibers are arranged in a parallel manner. In such cases, the exclusion volume fraction is stated simply as

$$\text{Exclusion volume fraction} = \frac{\pi(r_f + r_s)^2 LN}{V} = \theta\left(\frac{r_s}{r_f} + 1\right)^2, \quad (8.2.22)$$

where θ is the volume fraction of fibers (i.e., the weight concentration times the effective specific volume of fibers). By definition,

$$K_{AV} = 1 - \text{exclusion volume fraction} = 1 - \theta\left(\frac{r_s}{r_f} + 1\right)^2. \quad (8.2.23)$$

The calculation of K_{AV} is complicated if the minimum distance between fibers in the matrix is smaller than $2(r_f + r_s)$. Ogston has developed a statistical model to investigate molecular exclusion in a randomly oriented fiber matrix [8]. On the basis of this model, the available volume fraction is

$$K_{AV} = \exp\left[-\theta\left(1 + \frac{r_s}{r_f}\right)^2\right]. \quad (8.2.24)$$

Equation (8.2.24) is called the *Ogston equation*; it assumes that (a) the fibers are rigid rods with random orientations; (b) the length of the fibers is finite, but much larger than r_s; (c) there is no fiber overlap (i.e., no two different fibers share any portion of space); (d) the concentration of solute is low so that only the steric interaction between a solute and a fiber in the nearest neighborhood needs to be considered; and (e) the fiber radius is much less than the interfiber distance. When $\theta(1 + r_s/r_f)^2$ is much less than unity, Equation (8.2.24) reduces to Equation (8.2.23). The void volume fraction or the porosity, ε, in a fiber matrix can be derived from Equation (8.2.24) by letting $r_s = 0$. It then follows that

$$\varepsilon = \exp[-\theta]. \quad (8.2.25)$$

When θ is much less than unity,

$$\varepsilon = 1 - \theta, \tag{8.2.26}$$

which can also be derived from Equation (8.2.23) by letting $r_s = 0$.

A close examination of Equations (8.2.24) and (8.2.25) reveals two paradoxes. One is that Equation (8.2.24) should reduce to Equation (8.2.23) instead of Equation (8.2.25) when r_s/r_f is much less than unity, so that the minimum distance between fibers is larger than $2(r_f + r_s)$. Another paradox is that the porosity in Equation (8.2.25) approaches $\exp(-1) = 0.37$ instead of zero when θ approaches unity. These paradoxes follow from the assumption that the fiber radius is much less than the interfiber distance, an assumption that is valid only in matrices with a low volume fraction of fibers. When θ is high, the assumption is unwarranted. As a result, Equations (8.2.24) and (8.2.25) become invalid. In most polymeric gels, θ is less than 10%. Under such conditions, Equation (8.2.25) can be approximated by Equation (8.2.26), and Equation (8.2.24) will reduce to Equation (8.2.23) if r_s/r_f is much less than unity. The paradoxes then disappear.

By definition, the partition coefficient of solutes in the liquid phase of the fiber-matrix material is

$$\Phi = \frac{\exp\left[-\theta\left(1 + \left(\frac{r_s}{r_f}\right)\right)^2\right]}{1 - \theta}. \tag{8.2.27}$$

Equations (8.2.24), (8.2.26), and (8.2.27) indicate that the partition coefficient, the porosity, and the available volume fraction in fiber matrices can be predicted theoretically if one knows the radii of the fibers and the solutes and the concentration of fibers. Note that the value of $\theta(1 + r_s/r_f)^2$ is not necessarily smaller than unity, even though θ is much less than unity. This is because r_s/r_f can be much larger than unity.

The foregoing discussion is limited to neutral solutes under dilute conditions in fiber matrices. In biological tissues, the presence of cells and charge–charge interactions between different solutes or between solutes and fibers render the theoretical analysis of solute exclusion much more complicated. Various models have been proposed in the literature for analyzing the effects of charge and concentration of solutes on K_{AV} in polymeric gels [9,10]. However, the effects of cell arrangement and cell density on K_{AV} have received little attention [11].

The partition coefficients in different organs of the body can be estimated by measuring the equilibrium concentration ratios of solutes between the interstitial fluid in each organ and the blood. For albumin, the partition coefficient is observed to be 0.50 in the liver, 0.61 in the skin, and 0.90 in the gut [3]. The porosity of tissues can be estimated by measuring the volume fraction of the interstitial fluid. The porosity is 0.163, 0.302, and 0.094 in the liver, skin, and gut, respectively [3]. Once the partition coefficient and the porosity are known, the available volume fraction of albumin can be calculated, according to Equation (8.2.14). The figures are 0.082, 0.184, and 0.085 in the liver, skin, and gut, respectively. In this calculation, one can see that the available volume fraction of albumin in the gut is similar to that in the liver, although the partition coefficient of albumin in the gut is higher than that in the liver. The available volume fraction of other macromolecules in different tissues has been reviewed in [12].

8.3 Fluid Flow in Porous Media

8.3.1 Darcy's Law

Fluid flow in porous media has been studied for more than a century [13]. The interaction between solid and liquid phases in porous media was first quantified by Darcy in 1856, nearly at the same time that Fick developed his theory of molecular diffusion. In a study of water percolating through sand, Darcy discovered that the flow rate was proportional to the pressure gradient. This empirical relationship, called *Darcy's law*, is found to be valid in many porous media and was theoretically derived by other investigators later, on the basis of mechanical analyses of fluid flow in porous media [13,14]. The derivation shows that Darcy's law is invalid for non-Newtonian fluids, for Newtonian liquids at high velocity, and for gases at very low and very high velocities. The derivation also reveals that Darcy's law neglects the friction within the fluid and the exchange of momentum between the fluid and solid phases. Except for the friction within the fluid, which is discussed in Section 8.3.2, these exceptional cases of Darcy's law have rarely been observed in the interstitium of biological tissues. Therefore, Darcy's law has been widely used in the analysis of interstitial fluid flow.

The movement of fluid molecules in porous media follows tortuous pathways in the void space (see Figure 8.10). To describe the fluid flow in porous media, two approaches can be used. One is to numerically solve the governing equations for fluid flow in individual pores if the structures of pore networks are known. The other approach is to assume that a porous medium is a uniform material. In this so-called *continuum approach* proposed by Darcy in 1856, there are three length scales. The

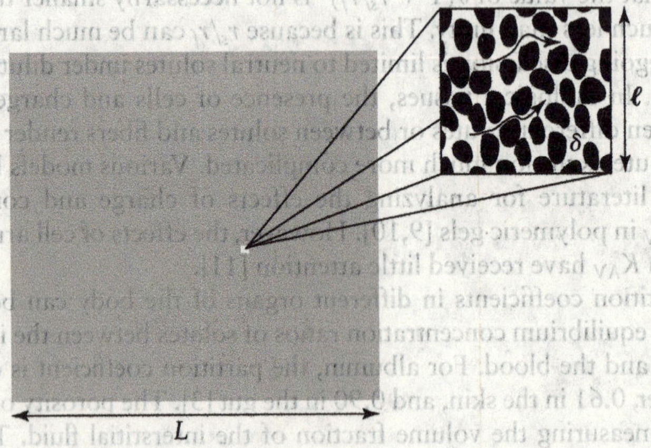

FIGURE 8.10 A sketch of fluid and solute transport in a porous medium. The medium is shown as a filled gray area. The medium can have any structure, but it is represented by a granular structure in this figure. The solid phase is shown as the black areas in the insert. Fluid and solutes can move between solid particles. The characteristic size of pores, δ, is equal to the average distance between adjacent particles. A small volume with a dimension, ℓ, is shown as the insert. The characteristic distance of transport, L, is the size of the gray area. The curved arrows in the insert indicate examples of transport pathways.

first one is the average size δ of the pores. The second is the distance L over which macroscopic changes of physical quantities (e.g., fluid velocity and pressure) must be considered. In most cases, L is chosen to be the characteristic linear dimension of the porous medium (e.g., the size of tissues or the distance between adjacent blood vessels). The continuum approach requires that L be at least two orders of magnitude larger than δ so that there may exist a third length scale, ℓ, between δ and L. To define ℓ, we consider a volume V_i of dimension ℓ in the porous medium. The volume fraction of the void space is then the volumetric porosity ε, and the volume fraction of the solid phase is equal to $1 - \varepsilon$. When ℓ is close to δ, the porosity ε is highly sensitive to the value of ℓ. When ℓ is increased gradually, the fluctuation in ε will decrease. There exists a value ℓ_0 of ℓ beyond which ε is a smooth function of ℓ, although it still fluctuates with a very small amplitude due to the random distribution of pore size in the volume V_i. The continuum approach requires that $\delta \ll \ell_0 \ll L$. In biological tissues, $\delta < 0.1$ μm, $\ell_0 \sim 1$ μm, and $L \sim 100$ μm to 10 cm. Thus, transport in biological tissues can be studied with the continuum approach.

The volume with dimension ℓ_0 is called the *representative elementary volume* (REV) [14]. In porous media, the REV can be taken to be a point, called a material (or physical) point, since $\ell_0 \ll L$ (see Figure 8.10). In that case, the details of pore structures are neglected, and each spatial point simultaneously contains two phases: a void phase with a volume fraction of ε and a solid phase with a volume fraction of $1 - \varepsilon$. In this book, we consider only porous media with pores filled with fluid. Thus, we do not specifically distinguish the void versus fluid phases. At each material point (i.e., in each REV), any physical quantity can be defined as the volume average of the same quantity defined in a *pure* medium. There are two different ways of averaging over a volume. One is based on the volume of each phase (i.e., the fluid and the solid phase); another is based on the total volume of REV. For example, the fluid velocity $\mathbf{v}_f$ at a material point can be defined as the velocity of each fluid particle in the fluid phase, averaged over the volume of the fluid phase. Alternatively, the fluid velocity $\mathbf{v}$ at a material point can be defined as the velocity of each fluid particle in the fluid phase, averaged over the REV. On the basis of these definitions, $\mathbf{v} = \varepsilon \mathbf{v}_f$.

Fluid transport in porous media must satisfy the law of mass conservation (see Chapters 2 and 3). For a pure, incompressible fluid, the mass balance equation in the liquid phase states simply that the divergence of the fluid velocity is equal to zero (see Equation (3.2.12)). The same equation is also valid in a porous medium, if there is no fluid production (know as a *source*) or fluid consumption (known as a *sink*) in the medium. However, sources and sinks are often present in biological tissues. For example, fluid is exchanged between interstitial space and the blood or lymph vessels. Thus, the mass balance equation (3.2.12) needs to be modified by adding a source term and a sink term:

$$\nabla \cdot \mathbf{v} = \phi_B - \phi_L. \tag{8.3.1a}$$

Here, $\mathbf{v}$ is the fluid velocity averaged in the REV. The mass balance equation can also be written as

$$\nabla \cdot (\varepsilon \mathbf{v}_f) = \phi_B - \phi_L, \tag{8.3.1b}$$

where $\mathbf{v}_f$ is the fluid velocity averaged in the volume of fluid phase. In Equations (8.3.1a) and (8.3.1b), ϕ_B and ϕ_L are rates of volumetric flow in sources and sinks, respectively, per unit volume of a porous medium. In biological tissues, they represent the rate of fluid flow per unit volume from blood vessels into the interstitial space and from the interstitial space into lymph vessels, respectively. The values of ϕ_B and ϕ_L are determined by Starling's law, which is discussed in Chapter 9. In solid tumors, where there are no functional lymph vessels, $\phi_L = 0$. In dead tissues, where there is no flow in blood vessels or lymph vessels, $\phi_B = \phi_L = 0$. In this case, Equations (8.3.1a) and (8.3.1b) reduce to Equation (3.2.12).

The equation for momentum balance in porous media is Darcy's law, introduced earlier. In a homogeneous and isotropic medium, Darcy's law can be written as

$$\mathbf{v} = -K\nabla p, \tag{8.3.2}$$

where ∇p is the gradient of the hydrostatic pressure and K is a constant defined as the *hydraulic conductivity*. Note that p is defined as the average quantity within the fluid phase in the REV. For nonisotropic and heterogeneous media, K is a tensor and it depends upon the location in the medium.

Substituting Equation (8.3.2) into Equation (8.3.1a), we get

$$-\nabla\cdot(K\nabla p) = \phi_B - \phi_L. \tag{8.3.3}$$

Equations (8.3.2) and (8.3.3) are the governing equations for fluid flow in rigid porous media. Later, we will show that these equations are also valid for steady-state flow in deformable tissues. A special case of Equation (8.3.3) is

$$\nabla^2 p = 0, \tag{8.3.4}$$

when K is a constant and $\phi_B = \phi_L = 0$. In this case, the interstitial fluid pressure is governed by a Laplace equation.

Example 8.4 Consider one-dimensional flow through a porous medium with hydraulic conductivity K (see Figure 8.11). The thickness of the medium is h. The pressures at $x = 0$ and $x = h$ are p_1 and p_2, respectively. Determine the pressure and velocity distributions in the medium.

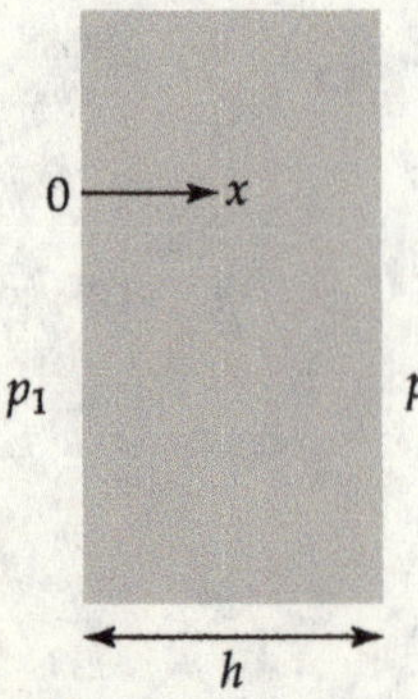

FIGURE 8.11 One-dimensional flow through a porous medium.

Solution To solve any problem of fluid flow, one can start from the mass balance and momentum equations. In this problem, $\phi_B = \phi_L = 0$ and the flow is one dimensional. Equations (8.3.2) and (8.3.3) then become, respectively,

$$\mathrm{v}_x = -K\frac{dp}{dx} \tag{8.3.5}$$

and

$$\frac{d^2p}{dx^2} = 0. \tag{8.3.6}$$

Integrating Equation (8.3.6) twice, we get

$$p = a_1x + a_2, \tag{8.3.7}$$

where a_1 and a_2 are constants that can be determined from the boundary conditions of the pressure:

$$a_1 = \frac{p_2 - p_1}{h} \quad \text{and} \quad a_2 = p_1. \tag{8.3.8}$$

Thus,

$$p = \frac{p_2 - p_1}{h}x + p_1. \tag{8.3.9}$$

The velocity profile can be obtained by substituting Equation (8.3.9) into Equation (8.3.5). The result is

$$v_x = K\frac{p_1 - p_2}{h}. \tag{8.3.10}$$

For one-dimensional flow, one can see that the pressure decreases linearly as a function of distance and the velocity is a constant in the medium.

Example 8.5 Therapeutic agents can be infused directly into solid tumors. Intratumoral infusion enhances the convective transport of drugs, which is critical to the delivery of macromolecules and nanoparticles. Macromolecules and nanoparticles diffuse very slowly in biological tissues. It may take several months for these agents to diffuse a millimeter's distance, whereas the time constant for convective transport over the same distance can be as short as a few minutes. To simplify the analysis of convective transport in a tumor, we consider only the fluid flow and assume that the tumor is removed from the body. Thus, the tumor has no blood and lymph circulation, and the pressure at its surface is equal to zero. The needle tip is inserted into the center of the tumor, after which a small fluid cavity forms around the tip. The radius of the cavity is δ, which can be approximated by the radius of the needle. We assume that solid tumors are spherical, that the radius of the tumor is a, and that the hydraulic conductivity is K. If the infusion rate is a constant Q, determine the pressure profile in the tumor at steady state.

Solution The fluid flow is spherically symmetric about the tip of the infusion needle (see Figure 8.12). Therefore, the velocity is unidirectional along the radial direction. It is convenient to use the spherical coordinate system with its origin at the needle tip, in which the pressure and the velocity depend only on r and are independent of the other two coordinates.

As in the last example, we start the solution procedure with the mass balance and momentum equations. In spherical coordinates, Equations (8.3.2) and (8.3.3) can be written, respectively, as

$$v_r = -K\frac{dp}{dr}, \tag{8.3.11}$$

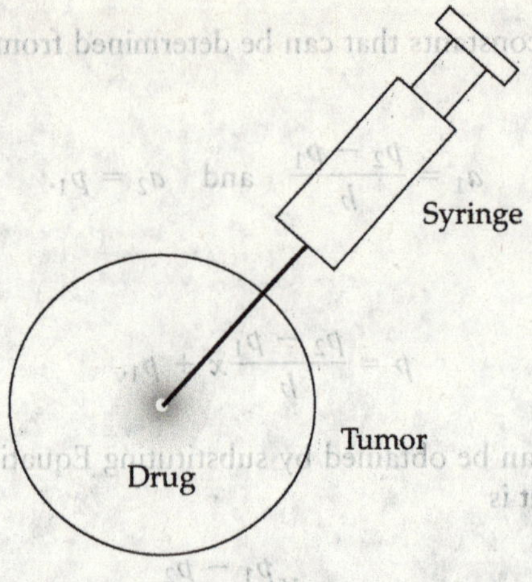

FIGURE 8.12 Intratumoral infusion of a fluid via a needle connected to the syringe. The infusion creates a fluid cavity (open circle) at the center of the tumor. The gray area indicates the distribution volume of the infused fluid.

$$\frac{1}{r^2}\frac{d}{dr}\left(r^2\frac{dp}{dr}\right) = 0, \qquad (8.3.12)$$

where $\phi_B = \phi_L = 0$, since there is no source or sink in the tissue after the tumor is removed from the body. Integrating Equation (8.3.12) twice, we get

$$p = -\frac{c_1}{r} + c_2. \qquad (8.3.13)$$

The boundary condition is $p = 0$ at $r = a$. Thus,

$$c_1 = ac_2. \qquad (8.3.14)$$

The boundary condition at $r = \delta$ is unknown, but the infusion rate Q is known. Because of the requirement of mass balance, the infusion rate is equal to the rate of flow across any spherical surface. Hence,

$$Q = \int_0^{2\pi}\int_0^{\pi} v_r r^2 \sin\theta d\theta d\phi. \qquad (8.3.15)$$

Substituting Equation (8.3.13) into Equation (8.3.11) to obtain v_r and then substituting the resulting expression for v_r into Equation (8.3.15), we get

$$Q = -4\pi K c_1. \qquad (8.3.16)$$

Solving Equations (8.3.14) and (8.3.16) simultaneously yields

$$c_1 = -\frac{Q}{4\pi K} \quad \text{and} \quad c_2 = -\frac{Q}{4\pi K a}. \qquad (8.3.17)$$

Thus,

$$p = \frac{Q}{4\pi K}\left(\frac{1}{r} - \frac{1}{a}\right) \qquad (8.3.18)$$

and

$$v_r = \frac{Q}{4\pi r^2}. \quad (8.3.19)$$

The velocity profile can also be derived directly from the mass balance equation because the flow rate across any spherical surface is a constant and equal to Q. Consequently,

$$Q = v_r(4\pi r^2). \quad (8.3.20)$$

Rearranging terms in Equation (8.3.20) gives the same result as Equation (8.3.19). Note that the fluid velocity and pressure are independent of δ at any location between δ and a.

If the gravitational force is not negligible, then Darcy's law must be modified to read

$$\mathbf{v} = -K(\nabla p - \rho \mathbf{g}), \quad (8.3.21)$$

where ρ is the density of the fluid and $\mathbf{g}$ is the acceleration due to gravity. Equation (8.3.21) indicates that the gravitational force can be neglected only when $\Delta p/L \gg \rho g$, where L is the distance over which the change in pressure is Δp.

The hydraulic conductivity is inversely proportional to the viscosity of the fluid (μ). The product of K and μ is defined as the *specific hydraulic permeability* k and depends only upon microscopic structures of the porous medium. For porous media with simple structures, K or k can be predicted theoretically.

Example 8.6 Determine the hydraulic conductivity and specific hydraulic permeability in the same porous medium as that discussed in Example 8.1. Assume that the liquid is a Newtonian fluid.

Solution Fluid flow in a circular cylinder is governed by Poiseuille's law (see Chapter 2), which predicts the dependence of the flow rate q on the pressure gradient:

$$q = -\frac{\pi d^4}{128\mu}\frac{dp}{dx}. \quad (8.3.22)$$

The total flow rate Q across the porous medium is equal to the sum of the flow rates across all of the cylinders; that is, $Q = An_A q$, where A is the cross-sectional area of the porous medium and n_A is the number of pores per unit area. By definition, the velocity v in the porous medium in the axial direction of cylinders is equal to the total flow rate per unit area. Thus,

$$v = -\frac{n_A \pi d^4}{128\mu}\frac{dp}{dx}. \quad (8.3.23)$$

Comparing this equation with Darcy's law (Equation (8.3.2)), we derive

$$K = \frac{n_A \pi d^4}{128\mu} \quad \text{and} \quad k = \frac{n_A \pi d^4}{128}. \tag{8.3.24}$$

In Example 8.6, we assumed the cross section of pores to be circular. Analytical solutions of fluid flow in noncircular, cylindrical pores have been derived by Kozeny (1927), on the basis of the Navier–Stokes equation [14]. Kozeny found that

$$K = \frac{c\varepsilon^3}{\mu s^2}, \tag{8.3.25}$$

where ε is the porosity, s is the *specific surface*, defined by Equation (8.2.1), and c is a shape factor, also called the *Kozeny constant*. Some examples of c are given in Table 8.1.

For noncylindrical pores (see Figure 8.1), a more general equation for calculating K is the *Kozeny–Carman equation*,

$$K = \frac{\varepsilon^3}{G\mu s_0^2(1-\varepsilon)^2}, \tag{8.3.26}$$

where G is also called the *Kozeny constant* in the literature. The variable s_0 is the *Carman-specific surface*, defined as the area of the surface that is exposed to the fluid per unit volume of the solid phase. For porous media with parallel, cylindrical pores, $s = s_0(1 - \varepsilon)$ and G is equal to $1/c$. In this case, Equation (8.3.26) reduces to Equation (8.3.25). In other porous media, G is equal to $1/(Tc)$ (which is less than $1/c$), where T is the tortuosity defined in Equation (8.2.12) [14]. For fiber-matrix materials, $s_0 = 2\pi r_f L_f/\pi r_f^2 L_f = 2/r_f$, here r_f is the fiber radius and L_f is the total fiber length. In this case, Equation (8.3.26) becomes

$$K = \frac{r_f^2\varepsilon^3}{4G\mu(1-\varepsilon)^2}. \tag{8.3.27}$$

In fiber-matrix materials, the value of G depends on the porosity. When the fibers are randomly oriented [15],

$$G(\varepsilon) = \frac{2\varepsilon^3}{3(1-\varepsilon)}\left\{\frac{1}{-2\ln(1-\varepsilon) - 3 + 4(1-\varepsilon) - (1-\varepsilon)^2} + \frac{2}{-\ln(1-\varepsilon) - [1-(1-\varepsilon)^2]/[1+(1-\varepsilon)^2]}\right\}. \tag{8.3.28}$$

Equation (8.3.28) predicts that G is approximately equal to 5 if $\varepsilon < 0.7$ and that G approaches $-5/3(1-\varepsilon)\ln(1-\varepsilon)$ as $\varepsilon \to 1$. Empirically, it has been observed that K in polymeric gels is a power function of the volume fraction of polymer fibers, θ:

$$K = \frac{m}{\mu}\theta^n. \tag{8.3.29}$$

TABLE 8.1

Values of c for Different Shapes of Cylindrical Pores

Shape	c
Circle	0.5
Square	0.5619
Equilateral triangle	0.5974
Strip	0.6667

Here, m and n are constants that depend on the type, size, and cross-link density of the fibers, as well as on the range of θ [16–18]. For agarose gels, $K\mu = k = 0.0244\theta^{-2.45}$ nm^2 [16].

In biological tissues, the dependence of the hydraulic conductivity upon the composition and structures of the tissues is complicated. The complication is mainly caused by cells. The distance between adjacent cells may vary from a few nanometers to several micrometers (i.e., from the diameter of a fiber molecule to the size of a cell). In Section 8.3.2, it will be shown that the plasma membranes of cells may provide a major resistance to fluid flow in the interstitial space—a resistance that is absent in polymeric gels.

Experimental studies have demonstrated that, in biological tissues, K is dependent upon the temperature, whereas k is independent of the temperature [19,20]. This suggests that the temperature dependence of K is mediated through the viscosity of fluid. Thus,

$$\frac{K_{T_1}}{K_{T_2}} = \frac{\mu_{T_2}}{\mu_{T_1}}, \tag{8.3.30}$$

where the two subscripts, T_1 and T_2, indicate two different temperatures.

8.3.2 Brinkman Equation

The interstitial space can be considered a network of channels filled with porous media. Fluid flow in such channels may not be modeled correctly by Darcy's law, because the fluid velocity in Darcy's model does not satisfy the no-slip boundary condition on the channel wall (see Section 2.3.4). From the physical point of view, Darcy's law assumes that the viscous resistance at the fluid–solid interface is much larger than that within the fluid. This assumption is valid, however, only when the specific permeability k of porous media is low. That condition in turn implies a high fiber concentration in a fiber matrix. The viscous stress within the fluid may not be negligible when k is large. In this case, the momentum equation must be derived again, using the Stokes equation for low-Reynolds-number flow (Equation (3.6.3b)). The result, called the *Brinkman equation*, is

$$\mu\nabla^2\mathbf{v} - \frac{1}{K}\mathbf{v} - \nabla p = 0. \tag{8.3.31}$$

Darcy's law, Equation (8.3.2), can be considered a special case of the Brinkman equation (8.3.31) wherein the first term can be neglected. In this case, the specific permeability k of the porous medium is much less than the square of the characteristic length L over which macroscopic changes in fluid velocity must be considered.

Example 8.7 Assume that the interstitial space between two cells can be considered to be a porous channel bounded by two parallel plates (see Figure 8.13). The effective hydraulic conductivity in the channel (K_{channel}), defined as the ratio of the fluid flux to the pressure gradient, depends on the specific hydraulic permeability

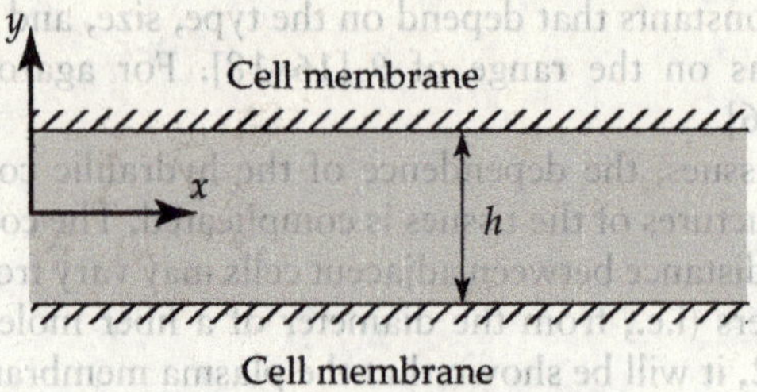

FIGURE 8.13 Fluid flow in a channel that separates two cells. The channel represents the interstitial space filled with extracellular matrix and interstitial fluid. It can be modeled as a uniform porous medium. The height of the channel is denoted by h.

in the porous medium (k) and the interaction of fluid with the channel wall. The channel height h is much smaller than the size of cells. Therefore, the flow can be assumed to be unidirectional. Derive the velocity profile and the expression of K_{channel} as a function of k, h, and the viscosity of the fluid (μ).

Solution The flow is unidirectional in the channel and is governed by the mass balance equation and the Brinkman equations. With $\phi_B = \phi_L = 0$, Equations (8.3.1a) and (8.3.31) become, respectively,

$$\frac{\partial \mathrm{v}_x}{\partial x} = 0, \tag{8.3.32}$$

$$-\frac{\partial p}{\partial y} = 0, \tag{8.3.33}$$

and

$$\mu \frac{\partial^2 \mathrm{v}_x}{\partial y^2} - \frac{\mu}{k}\mathrm{v}_x - \frac{\partial p}{\partial x} = 0. \tag{8.3.34}$$

Equations (8.3.32) and (8.3.33) indicate that v_x is independent of x and p is independent of y, respectively. Thus, Equation (8.3.34) becomes

$$\mu \frac{d^2 \mathrm{v}_x}{dy^2} - \frac{\mu}{k}\mathrm{v}_x = \frac{dp}{dx}. \tag{8.3.35}$$

The right-hand side of the equation is a function of x, whereas the left-hand side of the equation is a function of y. Therefore, the only possibility is that both sides are independent of x and y (i.e., they are equal to a constant). If we call this constant B and rearrange terms in Equation (8.3.35), we get

$$\frac{d^2 \mathrm{v}_x}{dy^2} - \frac{1}{k}\mathrm{v}_x = \frac{1}{\mu}B. \tag{8.3.36}$$

The general solution of Equation (8.3.36) is

$$\mathrm{v}_x = c_1 \sinh\left(\frac{y}{\sqrt{k}}\right) + c_2 \cosh\left(\frac{y}{\sqrt{k}}\right) - \frac{k}{\mu}B, \tag{8.3.37}$$

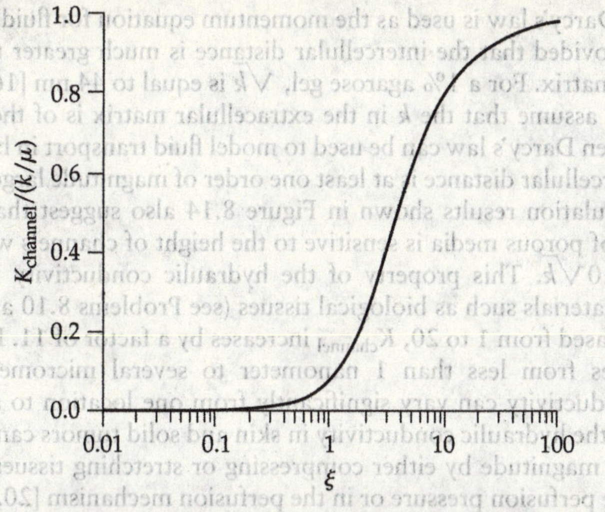

FIGURE 8.14 The dependence of K_{channel} on $\xi = h/\sqrt{k}$.

where c_1 and c_2 are constants that can be determined from the boundary conditions of v_x, namely,

$$\frac{dv_x}{dy} = 0 \quad \text{at} \quad y = 0 \tag{8.3.38a}$$

and

$$v_x = 0 \quad \text{at} \quad y = h/2. \tag{8.3.38b}$$

Substituting the boundary conditions into Equation (8.3.37), we get

$$v_x = -\frac{k}{\mu}B\left[1 - \frac{\cosh(y/\sqrt{k})}{\cosh(h/(2\sqrt{k}))}\right]. \tag{8.3.39}$$

The fluid flux q is equal to the flow rate per unit cross-sectional area:

$$q = \frac{1}{h}\int_{-h/2}^{h/2} v_x\,dy = -\frac{k}{\mu}B\left[1 - \frac{2\sqrt{k}}{h}\tanh\left(\frac{h}{2\sqrt{k}}\right)\right]. \tag{8.3.40}$$

The effective hydraulic conductivity of the channel is defined as the ratio of the fluid flux to the pressure gradient. Thus,

$$K_{\text{channel}} = \frac{k}{\mu}\left[1 - \frac{2\sqrt{k}}{h}\tanh\left(\frac{h}{2\sqrt{k}}\right)\right]. \tag{8.3.41}$$

K_{channel} can be normalized by k/μ, which is the hydraulic conductivity of the porous medium within the channel. Equation (8.3.41) indicates that the ratio of K_{channel} to k/μ depends only on $h/\sqrt{k}$. This dependence is plotted in Figure 8.14. The plot demonstrates that the effect of the channel wall on the fluid flow is less than 10% if $h/\sqrt{k} > 20$. In that case, $K_{\text{channel}} \approx k/\mu$, and the momentum equation can be approximated by Darcy's law. From this example, one can observe that $\sqrt{k}$ is a characteristic-length scale that determines the validity of Darcy's law.

In summary, Darcy's law is used as the momentum equation for fluid flow in biological tissues, provided that the intercellular distance is much greater than $\sqrt{k}$ in the extracellular matrix. For a 1% agarose gel, $\sqrt{k}$ is equal to 44 nm [16] (see Equation 8.3.29). If we assume that the k in the extracellular matrix is of the same order of magnitude, then Darcy's law can be used to model fluid transport in biological tissues when the intercellular distance is at least one order of magnitude larger than 44 nm.

The simulation results shown in Figure 8.14 also suggest that the hydraulic conductivity of porous media is sensitive to the height of channels when that height is less than $10\sqrt{k}$. This property of the hydraulic conductivity is important in deformable materials such as biological tissues (see Problems 8.10 and 8.11). When $h/\sqrt{k}$ is increased from 1 to 20, K_{channel} increases by a factor of 11. In biological tissues, h ranges from less than 1 nanometer to several micrometers. Thus, the hydraulic conductivity can vary significantly from one location to another. In perfused tissues, the hydraulic conductivity in skin and solid tumors can be altered over five orders of magnitude by either compressing or stretching tissues through slight changes in the perfusion pressure or in the perfusion mechanism [20,21].

8.3.3 Squeeze Flow

Squeeze flow refers to the fluid flow caused by the relative movement of solid boundaries toward each other. The boundaries can be either external ones for any fluid flow or internal ones for fluid flow in porous media. The squeeze flow problem has been analyzed for both Newtonian and non-Newtonian fluids. The applications of squeeze flow in biological systems are numerous. Here is a list of some examples: (a) designing a rheometer to measure the viscosity or the apparent viscosity of fluids [22]; (b) determining mechanical and fluid transport properties of biological tissues [23–25]; (c) optimizing the design of mechanical heart valves [26,27]; (d) analyzing the spreading of drug delivery formulations within body cavities [28]; and (e) understanding deformation-induced fluid flow in biological tissues [29–34].

Tissue deformation causes changes in the volume fraction of interstitial space, since cells are nearly incompressible. The changes in the interstitial space in turn lead to interstitial fluid flow. To illustrate this phenomenon, we will consider fluid flow between two adjacent cells during the compression of a tissue. The interstitial space between these cells is filled with interstitial fluid and extracellular matrix molecules. The distance between adjacent cells is, in general, much smaller than the diameter of the cells. Thus, the curvature of the cell membrane can be neglected here, and the deformation-induced convection can be modeled as a squeeze flow in a porous medium between two parallel plates (Figure 8.15).

In a cylindrical coordinate system, the continuity equation for axisymmetric flow is derived from Equation (8.3.1a) with $\phi_B = \phi_L = 0$. In that case, the equation is

$$\frac{1}{r}\frac{\partial(r\mathrm{v}_r)}{\partial r} + \frac{\partial \mathrm{v}_z}{\partial z} = 0, \qquad (8.3.42a)$$

where v_r and v_z are velocity components in the r and z directions, respectively. The integral form of the continuity equation can be written as

$$\langle \mathrm{v}_r \rangle_{\text{at } r=R} = \frac{RV_h}{2h}, \qquad (8.3.42b)$$

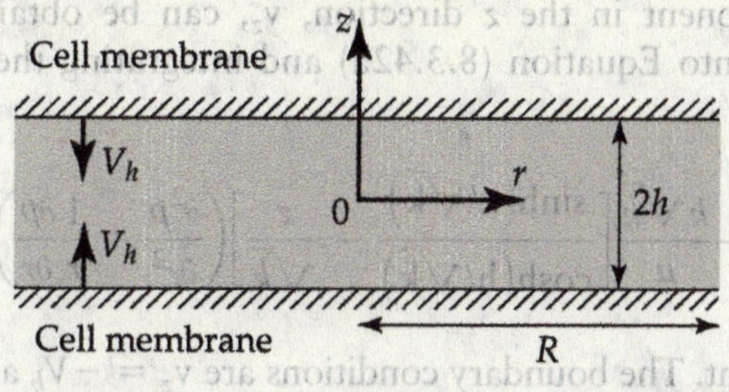

FIGURE 8.15 Squeeze flow in a porous medium between two parallel plates. The channel represents the interstitial space filled with extracellular matrix and interstitial fluid. V_h is the velocity of cell membrane movement and h is the half-distance between cells; R is the radius of the plate, which is on the same order of magnitude as cell radius; and z and r are cylindrical coordinates.

where the angle brackets, $\langle\rangle$, indicate the spatial average in the z direction. We assume that $h \ll R$. Thus, Equation (8.3.42b) requires that $\langle v_r \rangle_{r=R} \gg V_h$. The orders of magnitude of v_r and v_z can be estimated as $\langle v_r \rangle_{r=R}$ and V_h, respectively, indicating that $|v_r| \gg |v_z|$. This relationship is accurate in most regions, except near $r = 0$ *and* $z = \pm h$.

In a porous medium with solid boundaries, the transport of fluid momentum is governed by the Brinkman equation (8.3.31). In cylindrical coordinates, this equation becomes

$$\frac{\partial p}{\partial z} = 0, \tag{8.3.43}$$

$$\mu \frac{\partial^2 v_r}{\partial z^2} - \frac{\mu}{k} v_r - \frac{\partial p}{\partial r} = 0. \tag{8.3.44}$$

The derivation of Equation (8.3.43) is based on the fact that $\partial p/\partial z$ and $\partial p/\partial r$ are of the same orders of magnitude as v_z and v_r, respectively. Thus, $\partial p/\partial z$ is negligible compared with $\partial p/\partial r$, since $|v_z| \ll |v_r|$. The derivation of Equation (8.3.44) is also based on the condition that $|v_z| \ll |v_r|$. (The higher order terms have been neglected.)

Integrating Equation (8.3.44), we obtain

$$v_r = c_1 \sinh\left(\frac{z}{\sqrt{k}}\right) + c_2 \cosh\left(\frac{z}{\sqrt{k}}\right) - \frac{k}{\mu}\frac{\partial p}{\partial r}, \tag{8.3.45}$$

where c_1 and c_2 are constants that are determined by the boundary conditions: $v_r = 0$ at $z = h$ and $\partial v_r/\partial r = 0$ at $z = 0$. Substituting these boundary conditions into Equation (8.3.45) yields

$$v_r = \frac{k}{\mu}\frac{\partial p}{\partial r}\left[\cosh\left(\frac{z}{\sqrt{k}}\right)/\cosh\left(\frac{h}{\sqrt{k}}\right) - 1\right]. \tag{8.3.46}$$

The velocity component in the z direction, v_z, can be obtained by substituting Equation (8.3.46) into Equation (8.3.42a) and integrating the resulting equation. The final solution is

$$v_z = -\frac{k\sqrt{k}}{\mu}\left[\frac{\sinh(z/\sqrt{k})}{\cosh(h/\sqrt{k})} - \frac{z}{\sqrt{k}}\right]\left(\frac{\partial^2 p}{\partial r^2} + \frac{1}{r}\frac{\partial p}{\partial r}\right) + c_3, \tag{8.3.47}$$

where c_3 is a constant. The boundary conditions are $v_z = -V_h$ at $z = h$ and $v_z = 0$ at $z = 0$. Substituting the boundary conditions into Equation (8.3.47), we have $c_3 = 0$ and

$$\frac{V_h}{(k\sqrt{k}/\mu)\left[\tanh(h/\sqrt{k}) - h/\sqrt{k}\right]} = \left(\frac{\partial^2 p}{\partial r^2} + \frac{1}{r}\frac{\partial p}{\partial r}\right). \tag{8.3.48}$$

Note that Equation (8.3.48) corresponds to the Reynolds equation in lubrication theory, described in Chapter 4 (Equation (4.7.6)). Substituting Equation (8.3.48) into Equation (8.3.47), we have

$$v_z = -V_h\frac{\sinh(z/\sqrt{k})/\cosh(h/\sqrt{k}) - (z/\sqrt{k})}{\left[\tanh(h/\sqrt{k}) - (h/\sqrt{k})\right]}. \tag{8.3.49}$$

Integrating Equation (8.3.48) twice gives the pressure distribution in the porous medium:

$$p = \frac{r^2}{4}A + c_4 \ln r + c_5. \tag{8.3.50}$$

Here, A is the term on the left-hand side of Equation (8.3.48), and c_4 and c_5 are constants. The boundary conditions are $\partial p/\partial r = 0$ at $r = 0$ and $p = p_0$ at $r = R$. Substituting the boundary conditions into Equation (8.3.50) gives

$$p = p_0 + \frac{V_h}{(4k\sqrt{k}/\mu)\left[\tanh(h/\sqrt{k}) - h/\sqrt{k}\right]}(r^2 - R^2). \tag{8.3.51}$$

Substituting the pressure profile into Equation (8.3.46), we have

$$v_r = \frac{V_h r}{2\sqrt{k}}\frac{\left[\cosh(z/\sqrt{k})/\cosh(h/\sqrt{k}) - 1\right]}{\left[\tanh(h/\sqrt{k}) - (h/\sqrt{k})\right]}. \tag{8.3.52}$$

Given the velocity $V_h(t)$ as a function of time, one can determine $h(t)$ by directly integrating $V_h(t)$. We get

$$h = h_0 - \int_0^t V_h(\xi)\,d\xi, \tag{8.3.53}$$

where h_0 is the initial distance between the adjacent cells and ξ is a dummy variable of integration. When both $h(t)$ and $V_h(t)$ are known, the fluid velocity and pressure profiles can be determined by Equations (8.3.52), (8.3.49), and (8.3.51).

The preceding analysis is similar to that for a squeeze flow of pure Newtonian fluid between two parallel plates [22]. From the analysis, one can observe that velocity and pressure profiles are implicitly dependent on time. The time dependence is introduced through V_h and h in Equation (8.3.53).

In the analysis, we have assumed that $|v_r| \gg |v_z|$. This assumption is not valid near the boundaries $r = 0$ and $z = \pm h$, on which v_r is equal to zero while v_z may differ from zero. However, the velocity boundary conditions are satisfied exactly at $r = 0$ and $z = \pm h$, and $|v_r| \gg |v_z|$ is true away from these boundaries. Thus, the velocity and pressure profiles determined by Equations (8.3.52), (8.3.49), and (8.3.51) should not differ significantly from the exact solutions of the *Brinkman equation*. Another assumption made in the derivation of these equations is that the specific hydraulic permeability k is a constant. This assumption is valid only for small tissue deformation. When the deformation is large, one must consider the deformation-induced changes in k [20,35,36]. As an exercise, Problem 8.10 shows how to include the changes in k in the analysis of squeeze flow.

The rate of deformation, $V_h(t)$, can be related to forces on the cell membranes. To derive such a relationship, one needs to know the mechanical properties of porous media. If the medium is a poroelastic material (i.e., the solid phase of the medium is an elastic material) and tissue deformation is infinitesimal, then the forces on the cell membranes can be calculated on the basis of the poroelastic theory discussed in Section 8.5. One special case is that the normal forces on the cell membrane due to the elastic compression of the extracellular matrix are negligible compared with the pressure within the matrix if the interstitial space is highly compressible. In this case, the total force on the cell membrane is equal to the integration of the pressure, given in Equation (8.3.51), over the entire surface of the cell (see Problem 8.14). A general treatment of tissue-deformation-induced fluid flow is beyond the scope of this text [23,29–34].

8.4 Solute Transport in Porous Media

8.4.1 General Considerations

The transport of solutes in porous media is governed by the same principles as those governing transport in solutions. Within the liquid phase of a medium, Fick's law still applies for diffusion (see Chapter 6), and the Nernst–Planck equation governs the migration of electrolytes in an electric field (see Section 7.4). However, analyses of solute transport can be complicated if the continuum approach (see Section 8.3.1) is used. Among the challenges facing the continuum approach are the following:

(i) The diffusion of solutes is characterized by the effective diffusion coefficient D_{eff} inside the pores. This coefficient is related to the diffusion coefficient of the same solutes in solutions, the available volume fraction of the solutes, hydrodynamic effects, the binding of the solutes to pore surfaces, and the tortuosity of pathways for diffusion (see Section 8.4.2). The tortuosity defined by Equation (8.2.12) can be nonisotropic and is, in general, a second-order tensor, as is the effective diffusion coefficient [14,37]. The absolute value of D_{eff} is smaller than that in solutions, because of all the preceding factors pertaining to D_{eff}. Another important factor that can influence diffusion in porous media is the connectedness of pores at different locations. In general, the probability that two pores are connected through void space decreases when the distance between the pores increases. We can imagine a porous material that consists of many layers of identical pores. In each layer, the number of pores is 100. If only 99% of the pores in each layer are connected to those in the previous layer, then the numbers of pores that are connected to the first layer are 99, 98,..., 37 in the 2nd, 3rd,..., 100th layers, respectively. As a result, the available space for diffusion and D_{eff} in this material decrease exponentially with the distance of diffusion [4].

(ii) The convective velocities of solutes are not necessarily equal to the solvent velocities, because solutes are more hindered by porous structures than solvents are. The ratio of the solute velocity v_s to the solvent velocity v_f is defined as the *retardation coefficient*:

$$f = \frac{v_s}{v_f}. \tag{8.4.1}$$

The value of f is between zero and unity. In the literature, $1 - f$ is defined as the *reflection coefficient* σ, a phenomenological parameter for characterizing the hindrance of convective transport across a membrane (see Chapter 9). The retardation coefficient in biological tissues has received little study. It depends on the fluid velocity, the solute size, and pore structures. Nevertheless, the flux N_s of convective transport across a tissue is proportional to v_s and the local concentration of solutes, C, according to the formula

$$N_s = v_s C = f v_f C. \tag{8.4.2}$$

The fluid velocity is independent of the concentration of solute in dilute systems and can be solved independently, using the equations described in Section 8.3. Note that C is the average concentration defined over the entire REV shown in Figure 8.10, whereas v_f is the average velocity defined only in the fluid phase of the REV (see Equation (8.3.1b)).

(iii) According to the continuum approach, the fluid flow causes not only solute movement in the direction of the fluid velocity but also dispersion of solutes in other directions in porous media. Solute dispersion is characterized by the dispersion coefficient, which depends on the diffusion coefficient of the solute, the average velocity defined by the Darcy's law (see Equation 8.3.2), and the average size of the pores. Dispersion analysis is beyond the scope of this textbook; readers can find relevant information in various books on transport phenomena in porous media [13, 14,37]. In this text, we always neglect the dispersion phenomenon.

(iv) The concentrations of solutes in the continuum approach can be discontinuous at the interfaces between solutions and porous media or between different porous media. Consequently, the boundary conditions at the interfaces of any two regions (say, regions 1 and 2, denoted by subscripts) should be modified to

$$N_1 = N_2 \tag{8.4.3}$$

and

$$\frac{C_1}{K_{AA_1}} = \frac{C_2}{K_{AA_2}}, \tag{8.4.4}$$

where N is the flux of solute, C is the solute concentration, and K_{AA} is the area fraction available at the interface for solute transport. The K_{AA} can be approximated by K_{AV} in macroscopically isotropic porous materials.

When all these issues are considered, the equation governing the transport of a neutral molecule is

$$\frac{\partial C}{\partial t} + \nabla \cdot (f \mathrm{v}_f C) = D_{eff} \nabla^2 C + \Phi_B - \Phi_L + Q, \tag{8.4.5}$$

(see also Section 15.5.1), where Q is the rate of solute production per unit volume due to chemical reactions, and Φ_B and Φ_L represent the rates of solute transport per unit volume from blood vessels into the interstitial space and from the interstitial space into lymph vessels, respectively. Blood and lymph vessels are sources and sinks not only for fluid transport, as previously discussed, but also for solute transport in biological tissues. The rates of solute transport Φ_B and Φ_L are governed by the Kedem–Kachalsky equation, which will be described in Chapter 9. Equation (8.4.5) assumes that the effective diffusion coefficient D_{eff} is isotropic and uniform and that the dispersion coefficient is much smaller than D_{eff}. If the dispersion effect is to be taken into account, then D_{eff} in Equation (8.4.5) should be replaced by the sum of D_{eff} and the dispersion coefficient [13,14,37]. In Equation (8.4.5), C, Φ_B, Φ_L, and Q are defined as average quantities per unit tissue volume, whereas f, v_f, and D_{eff} are defined as the average quantities per unit volume of the fluid phase in tissues (i.e., the interstitial fluid space). For molecules that can diffuse freely through cells, D_{eff} is averaged over the volume of the extravascular space in tissues.

8.4.2 Effective Diffusion Coefficient in Hydrogels

As mentioned in the previous section, the effective diffusion coefficient D_{eff} depends on many factors. Here, we consider three of them relating to the diffusion of uncharged solutes in hydrogels (i.e., water containing polymeric gels): the diffusion coefficient of solutes in water (D_0), hydrodynamic interactions between the solute and the surrounding solvent molecules, and the tortuosity of diffusion pathways due to the steric exclusion of solutes in the fiber matrix. Under this consideration, we can write

$$D_{eff} = D_0 F S, \tag{8.4.6}$$

where F and S account for effects of hydrodynamic interactions and pathway tortuosity on the rate of diffusion, respectively. To illustrate how to determine F and S, we will discuss the hindered diffusion of spherical molecules (e.g., globular proteins) in hydrogels. In this case, F is defined as the ratio of friction coefficients of the solute in a porous medium and water, respectively. The value of the friction coefficient in water is simply $6\pi\mu r_s$ (see Chapter 3), where μ is the viscosity of water and r_s is the radius of the molecule. On the basis of this definition, F is a measure of the enhancement of drag on the solute molecule due to the presence of polymeric fibers in water. To determine F, two different approaches have been used. The first one assumes that the hydrogel is a uniform medium in which the movement of a spherical solute molecule with a constant velocity is governed by the Brinkman equation (8.3.31). This analysis is also called the *effective-medium* or *Brinkman-medium* approach in the literature [18,38]. By simultaneously solving the Brinkman equation (8.3.31) and the continuity equation (8.3.1a), with $\phi_B = \phi_L = 0$, we obtain

$$F = \left[1 + \frac{r_s}{\sqrt{k}} + \frac{1}{9}\left(\frac{r_s}{\sqrt{k}}\right)^2\right]^{-1}. \tag{8.4.7}$$

The derivation of Equation (8.4.7) also requires a continuous boundary condition for the velocity at the interface between the medium and the solute molecule.

The second approach assumes that hydrogels can be modeled as three-dimensional spaces filled with water and randomly placed cylindrical fibers [18,39]. The friction coefficient on a moving sphere between fibers with a given configuration is determined by solving the Stokes equation (3.6.3b) numerically. The numerical solutions are then ensemble averaged over many fiber configurations to obtain the average friction coefficient, which is equal to F after normalization by $6\pi\mu r_s$. In this case, F depends on the ratio α of fiber to solute radii and the volume fraction of fiber, θ, in hydrogels. The dependence can be fitted to the stretched exponential function

$$F(\alpha, \theta) = \exp(-a_1\theta^{a_2}), \tag{8.4.8}$$

where

$$a_1 = 3.727 - 2.460\alpha + 0.822\alpha^2 \tag{8.4.9}$$

and

$$a_2 = 0.358 + 0.366\alpha - 0.0939\alpha^2. \tag{8.4.10}$$

The tortuosity factor S depends on an adjusted volume fraction parameter

$$f_a = \left(1 + \frac{1}{\alpha}\right)^2 \theta. \tag{8.4.11}$$

The parameter f_a is equal to the excluded volume fraction of solute in hydrogels if the fiber density is very low or if the fibers are arranged in a parallel manner (see Equation (8.2.22)). Numerical simulations have demonstrated that the dependence of S on f_a can also be fitted by a stretched exponential function if $f_a < 0.7$ [18,38]. In that case,

$$S(\alpha, \theta) = \exp[-0.84 f_a^{1.09}]. \tag{8.4.12}$$

8.4.3 Effective Diffusion Coefficient in a Liquid-Filled Pore

Biological tissues in transport studies can be modeled as networks of liquid-filled pores if the cellular resistance to diffusion is much larger than that caused by the extracellular matrix. As in diffusion in hydrogels, the effective diffusion coefficient of uncharged

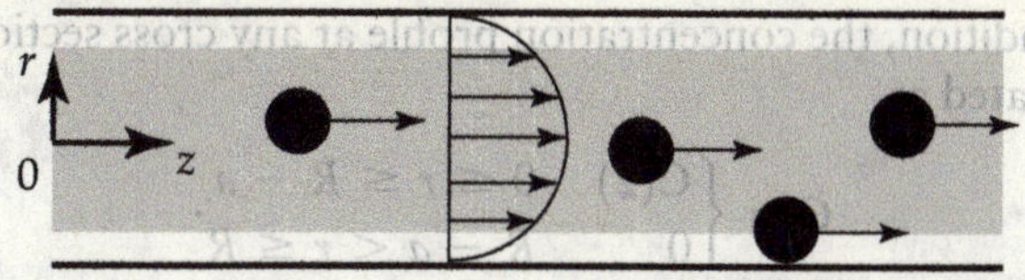

FIGURE 8.16 Geometry of mathematical models for solute transport in a cylindrical tube. Solid spheres represent solutes. The gray area in the middle of the pore represents the distribution volume of solutes in the pore. The fluid flow in the pore is fully developed and thus has a parabolic profile.

solutes in liquid-filled pores also depends on the diffusion coefficient D_0 of solutes in water, hydrodynamic interactions between the solute and the surrounding solvent molecules within the pores, and the steric exclusion of solutes near the walls of pores.

A general analysis of the effective diffusion coefficient of uncharged solutes in a network of pores is complicated. To simplify the analysis, we will consider only the diffusion of spherical solutes in a cylindrical pore, as shown in Figure 8.16. In addition, we include the convection of solutes in the cylinder, since it can easily be treated in the model. As a result, both the effective diffusion coefficient and the retardation coefficient of solutes will be determined simultaneously [9,40]. (The retardation coefficient is discussed further in Section 14.4.2.)

In our analysis, we assume that the entrance effect is negligible (see Section 2.7.3). Thus, the velocity of fluid has a parabolic profile when the fluid enters the pore. In cylindrical coordinates,

$$v_z = 2v_m\left[1 - \left(\frac{r}{R}\right)^2\right], \tag{8.4.13}$$

where v_z is the axial fluid velocity, v_m is the mean velocity in the pore, and r is the radial coordinate. We also assume that the distribution of solute concentration is uniform at the entrance to the pore. Furthermore, we assume that the volume fraction of spherical solutes in the cylinder is less than $2\lambda/3$, where $\lambda = a/R$, in which a and R are the radii of the solute and the cylinder, respectively. Under this assumption, solute–solute interactions are negligible [40]. Thus, we can separately analyze the movement of individual solutes in the pore.

When $\lambda \to 0$, solute–pore interactions can be neglected. In this case, solute transport is one dimensional, and the flux is

$$N_s = -D_0\frac{\partial C}{\partial z} + Cv_z, \tag{8.4.14}$$

where D_0 is the diffusion coefficient in water, C is the concentration of solutes, and z is the axial coordinate. In general, C depends on r and z, although the concentration profile has been assumed to be uniform at the entrance to the pore. The dependence of C on r is caused by the nonuniformity of the fluid velocity. The parabolic profile makes the solute concentration higher near the center than at the edge of the pore. This phenomenon, called *Taylor dispersion*, has been well studied [22]. Taylor dispersion is negligible if longitudinal dispersion is much slower than diffusion [40].

Under such a condition, the concentration profile at any cross section in the cylinder can be approximated as

$$C = \begin{cases} C(z) & 0 \le r \le R - a \\ 0 & R - a < r \le R \end{cases}. \tag{8.4.15}$$

Therefore, Equation (8.4.14) becomes

$$N_s = -D_0 \frac{dC}{dz} + Cv_z, \tag{8.4.16}$$

for $0 \le r \le R - a$ and $N_s = 0$ for $R - a < r \le R$. When λ does not approach zero, solute–pore interactions must be considered. Such interactions will alter the fluid resistance to solute movement and the velocity profile of the fluid. These alterations are taken into account through the introduction of two coefficients: the enhanced friction coefficient, K, and the lag coefficient, G, respectively [9,40]. In general, K and G are functions of λ and r/R. The fluid resistance is characterized by the frictional coefficient, which is inversely proportional to the diffusion coefficient of solute in the Stokes–Einstein equation (see Equation (6.6.11)). Changing the fluid resistance by a factor of K is equivalent to changing the diffusion coefficient by a factor of $1/K$. Thus, Equation (8.4.16) can be modified by including K and G in the diffusion and convection terms, respectively, to model the solute flux in the pore, that is

$$N_s = -\frac{D_0}{K}\frac{dC}{dz} + GCv_z. \tag{8.4.17}$$

The flux averaged over the entire cross-sectional area of the cylinder is defined as

$$\overline{N_s} = \frac{2}{R^2}\int_0^R N_s r\,dr. \tag{8.4.18}$$

Integrating Equation (8.4.17) gives

$$\overline{N_s} = -HD_0\frac{dC}{dz} + WCv_m, \tag{8.4.19}$$

where

$$H = \frac{2}{R^2}\int_0^{R-a} \frac{1}{K} r\,dr \tag{8.4.20}$$

and

$$W = \frac{4}{R^2}\int_0^{R-a} G\left[1 - \left(\frac{r}{R}\right)^2\right] r\,dr. \tag{8.4.21}$$

H and W are called *hydrodynamic resistance coefficients*. W is also called the *retardation coefficient*. On the basis of Equation (8.4.19), the effective diffusion coefficient of solute within the pore is defined as

$$D_{\text{eff}} = HD_0. \tag{8.4.22}$$

The values of H and W depend on K and G, which, in general, are determined by numerically solving the Stokes equation (3.6.3b) for fluid flow in a cylindrical pore with only one sphere located at a radial position r. The numerical simulation is repeated for different r and different solute radii, giving K and G as functions of λ and r/R. It has been observed that the dependence of K and G on r/R is less significant than their dependence on λ in calculating H and W. Thus, $K(\lambda, r/R)$ and $G(\lambda, r/R)$ can be approximated by $K(\lambda, 0)$ and $G(\lambda, 0)$, respectively. This approximation, known as the *centerline approximation* in the literature [9,40], assumes that all spheres are distributed on the centerline position in the pore. For $\lambda < 0.4$, K and G can be approximated [40], respectively, by

$$K^{-1}(\lambda, 0) = 1 - 2.1044\lambda + 2.089\lambda^3 - 0.948\lambda^5 \tag{8.4.23}$$

and

$$G(\lambda, 0) = 1 - \frac{2}{3}\lambda^2 - 0.163\lambda^3. \tag{8.4.24}$$

Substituting Equations (8.4.23) and (8.4.24) into Equations (8.4.20) and (8.4.21), respectively, we have

$$H(\lambda) = \phi(1 - 2.1044\lambda + 2.089\lambda^3 - 0.948\lambda^5) \tag{8.4.25}$$

and

$$W(\lambda) = \phi(2 - \phi)\left(1 - \frac{2}{3}\lambda^2 - 0.163\lambda^3\right), \tag{8.4.26}$$

where $\phi = (1 - \lambda)^2$ is the partition coefficient of solute in the pore (see Equation (8.2.15)). As $\lambda \rightarrow 0$, interactions between the pore and solutes become negligible. Thus, both H and W approach unity. The numerical results of H and W based on Equations (8.4.25) and (8.4.26), respectively, are plotted in Figure 8.17.

For $\lambda > 0.4$, Equations (8.4.23) through (8.4.26) are invalid. In this case, K, G, H, and W need to be expressed as more complicated functions of λ or plotted as functions of λ on the basis of their numerical values [9] (see Section 14.4.2).

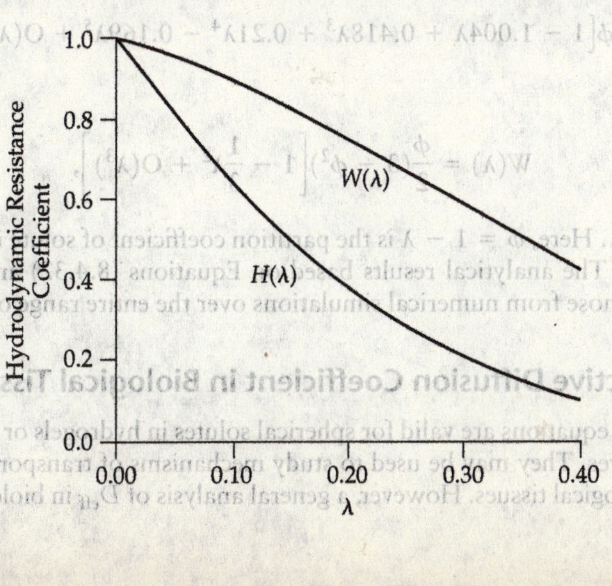

FIGURE 8.17 The dependence of H and W on λ.

In addition to treating circular cylinder geometry, the literature presents a complete analysis of the convection and diffusion of spherical molecules between two parallel plates (i.e., through a slit pore) [9,41]. This analysis is similar to that of cylindrical pores previously discussed. We will not repeat it, but we do list the results here. The expression for the average flux over the cross-sectional area of the slit is the same as Equation (8.4.19) in a circular cylinder, or

$$\overline{N}_s = -HD_0\frac{dC}{dz} + WCv_m. \tag{8.4.27}$$

However,

$$\overline{N}_s = \frac{1}{h}\int_0^h N_s dy, \tag{8.4.28}$$

$$H = \frac{1}{h}\int_0^{h-a} \frac{1}{K}dy, \tag{8.4.29}$$

and

$$W = \frac{3}{2h}\int_0^{h-a} G\left[1 - \left(\frac{y}{h}\right)^2\right]dy. \tag{8.4.30}$$

(Note that h is the half-width of the slit.) As in the case of a circular cylinder, the effective diffusion coefficient of solute within the pore is also (see Equation (8.4.22))

$$D_{\text{eff}} = HD_0. \tag{8.4.31}$$

The values of K and G depend on λ (i.e., a/h) and y/h, and they can be determined by numerically solving the Stokes equation (3.6.3b) for fluid flow in a slit pore [41]. Alternatively, analytical expressions for K and G can be obtained by using the centerline approximation mentioned earlier. On the basis of $K(\lambda, 0)$ and $G(\lambda, 0)$, integrating Equations (8.4.29) and (8.4.30) gives

$$H(\lambda) = \phi\left[1 - 1.004\lambda + 0.418\lambda^3 + 0.21\lambda^4 - 0.169\lambda^5 + O(\lambda^6)\right] \tag{8.4.32}$$

and

$$W(\lambda) = \frac{\phi}{2}(3 - \phi^2)\left[1 - \frac{1}{3}\lambda^2 + O(\lambda^3)\right], \tag{8.4.33}$$

for $0 < \lambda < 1$. Here, $\phi = 1 - \lambda$ is the partition coefficient of solute in the pore (see Problem 8.2). The analytical results based on Equations (8.4.32) and (8.4.33) are very close to those from numerical simulations over the entire range of λ [9].

8.4.4 Effective Diffusion Coefficient in Biological Tissues

The preceding equations are valid for spherical solutes in hydrogels or in liquid-filled, cylindrical pores. They may be used to study mechanisms of transport of uncharged solutes in biological tissues. However, a general analysis of D_{eff} in biological tissues is complicated and tissue dependent. Empirically, D_{eff} has been related approximately to the molecular weight M_r of solutes via the power function

$$D_{\text{eff}} = b_1(M_r)^{-b_2}, \tag{8.4.34}$$

where b_1 and b_2 are functions of the charge and shape of solutes, as well as of the structures of tissues [2]. The effects of tissue structures on D_{eff} increase with the size of solutes. For example, the ratio of diffusion coefficients in tissue versus water is close to unity for oxygen [42], but reduced to 0.1–0.3 for albumin and IgG in tumor tissues [43,44]. The ratio will approach zero as the size of solutes approaches the cutoff sizes of pores in tissues.

8.5 Fluid Transport in Poroelastic Materials

Biological tissues are deformable, and the deformation can be nonlinear. For deformable materials, both the mass and momentum balance equations must be modified [45,46]. As an introduction to deformable porous media, we will limit the scope of our analysis to infinitesimal deformation of poroelastic materials and neglect all nonlinear effects. The solid phase of the medium is modeled as a Hookean material (i.e., it is elastic and the strains are infinitesimal). In this case, stresses are linear functions of strains, and we have

$$\boldsymbol{\sigma} = \boldsymbol{\tau} - p\mathbf{I} = 2\mu_G\mathbf{E} + \mu_\lambda e\mathbf{I} - p\mathbf{I}, \tag{8.5.1}$$

where $\boldsymbol{\sigma}$ and $\boldsymbol{\tau}$ are the effective and partial solid stress tensors, respectively, $\mathbf{E}$ is the strain tensor of the tissue, $\mathbf{I}$ is the identity tensor, p is the fluid pressure, μ_G and μ_λ are called the *Lamé constants* and characterize the elastic property of the solid phase in the tissue, and

$$e = \text{Tr}(\mathbf{E}) = \nabla\cdot\mathbf{u} \tag{8.5.2}$$

is the *volume dilatation*, which is equal to the divergence of the displacement of the solid phase. In Equation (8.5.2), Tr indicates the trace operation and $\mathbf{u}$ is the displacement. For infinitesimal deformation, the strain tensor is a linear function of the displacement gradient:

$$\mathbf{E} = \frac{1}{2}\left[\nabla\mathbf{u} + (\nabla\mathbf{u})^{\text{T}}\right]. \tag{8.5.3}$$

Here, T indicates the transpose operation.

To derive mass and momentum balance equations, we assume that the intercellular space is much larger than the square root of the specific hydraulic permeability of the interstitium, that the porous media are saturated with liquid, and that the materials are isotropic. Under these assumptions, the mass conservation in the fluid phase requires that

$$\frac{\partial(\rho_f\varepsilon)}{\partial t} + \nabla\cdot(\rho_f\varepsilon\mathbf{v}_f) = \rho_f(\phi_B - \phi_L), \tag{8.5.4}$$

where ε is the porosity or the volume fraction of fluid, ρ_f is the fluid mass density, and ϕ_B and ϕ_L are the same source and sink terms as those in Equations (8.3.1a) and (8.3.1b) for fluid exchange between the interstitial space and the blood and lymph vessels, respectively. The terms ϕ_B and ϕ_L are defined as rates of volumetric fluid exchange per unit volume of porous materials. In Equations (8.5.1) through (8.5.4), τ and $\mathbf{u}$ are defined as the average quantities within the solid phase in the REV, whereas p and $\mathbf{v}_f$ are defined as the average quantities within the fluid phase in the REV (see Section 8.3.1). The mass conservation in the solid phase requires that

$$\frac{\partial[\rho_s(1-\varepsilon)]}{\partial t} + \nabla\cdot\left[\rho_s(1-\varepsilon)\frac{\partial \mathbf{u}}{\partial t}\right] = 0, \tag{8.5.5}$$

where ρ_s is the solid mass density. If both the fluid and the solid mass densities are constants [45], then summing Equations (8.5.4) and (8.5.5) gives

$$\nabla\cdot\left((1-\varepsilon)\frac{\partial \mathbf{u}}{\partial t} + \varepsilon\mathbf{v}_f\right) = \phi_B - \phi_L. \tag{8.5.6}$$

Momentum balances in both the fluid and the solid phases are governed by the generalized Biot law [30]. Transport in biological tissues is, in general, slow and within a small region. Thus, body forces (e.g., gravity and inertial forces) on a unit volume in both the fluid and the solid phases are negligible compared with the pressure gradient. In this case, the generalized Biot law can be simplified to

$$\varepsilon\left(\mathbf{v}_f - \frac{\partial \mathbf{u}}{\partial t}\right) = -K\nabla p, \tag{8.5.7}$$

$$\nabla\cdot\boldsymbol{\sigma} = 0, \tag{8.5.8}$$

where K is the hydraulic conductivity. Note that Equation (8.5.7) is similar to Darcy's law, except that the pressure gradient in deformable materials is proportional to the fluid velocity relative to the solid phase. In the steady state of tissue deformation, $\partial\mathbf{u}/\partial t = 0$. Therefore, Equations (8.5.6) and (8.5.7) reduce to Equations (8.3.1b) and (8.3.2), respectively.

Substituting Equations (8.5.1) through (8.5.3) into Equation (8.5.8), we get

$$\mu_G\nabla^2\mathbf{u} + (\mu_G + \mu_\lambda)\nabla e - \nabla p = 0. \tag{8.5.9}$$

The divergence of Equation (8.5.9) (see Section A.3.C) gives

$$(2\mu_G + \mu_\lambda)\nabla^2 e = \nabla^2 p. \tag{8.5.10}$$

The pressure term in Equation (8.5.10) can be replaced by the volume dilatation e if ε and K are homogeneous in the tissue. The procedure is as follows: taking the divergence of Equation (8.5.7), we get

$$\varepsilon\left(\nabla\cdot\mathbf{v}_f - \frac{\partial e}{\partial t}\right) = -K\nabla^2 p. \tag{8.5.11}$$

Combining Equation (8.5.11) with Equation (8.5.6), we get

$$\frac{\partial e}{\partial t} = K\nabla^2 p + \phi_B - \phi_L. \tag{8.5.12}$$

Substituting Equation (8.5.10) into Equation (8.5.12) yields

$$\frac{\partial e}{\partial t} = K(2\mu_G + \mu_\lambda)\nabla^2 e + \phi_B - \phi_L. \tag{8.5.13}$$

We call the factor $K(2\mu_G + \mu_\lambda)$ the *coefficient of consolidation*, c. Mathematically, it is equivalent to the diffusion coefficient if e is the concentration [45].

Example 8.8 In Example 8.5, we analyzed intratumoral infusion without taking tissue deformation into account. Here, we assume that the tumor tissue is a poroelastic material and that tissue deformation is infinitesimal. Determine the displacement in the tumor and the radius of the cavity at the steady state.

Solution The flow is spherically symmetric. In the steady state, Equation (8.5.12) predicts that

$$\nabla^2 p = 0, \tag{8.5.14}$$

since there is no source or sink in the tumor when it is removed from the body (see Example 8.5). Thus, the pressure profile in Equation (8.3.18) is still valid in poroelastic materials in the steady state:

$$p = \frac{Q}{4\pi K}\left(\frac{1}{r} - \frac{1}{a}\right).$$

For spherically symmetric flow, Equation (8.5.9) becomes

$$\mu_G\left[\frac{1}{r^2}\frac{\partial}{\partial r}\left(r^2\frac{\partial u_r}{\partial r}\right) - \frac{2}{r^2}u_r\right] + (\mu_G + \mu_\lambda)\frac{\partial e}{\partial r} - \frac{\partial p}{\partial r} = 0. \tag{8.5.15}$$

Substituting Equations (8.3.18) and (8.5.2) into Equation (8.5.15), we obtain

$$\frac{\partial}{\partial r}\left(\frac{1}{r^2}\frac{\partial(r^2 u_r)}{\partial r}\right) = -\frac{Q}{4c\pi}\frac{1}{r^2}. \tag{8.5.16}$$

Integrating Equation (8.5.16) twice, we get

$$u_r = \frac{Q}{8c\pi} + b_1 r + \frac{b_2}{r^2}, \tag{8.5.17}$$

where b_1 and b_2 are constants that can be determined by the boundary conditions. Assuming that there are no partial solid stresses (τ_{rr}) on the surfaces of the fluid cavity ($r = \delta$) or the tumor ($r = a$), it follows that

$$\tau_{rr} = (2\mu_G + \mu_\lambda)\frac{\partial u_r}{\partial r} + 2\mu_\lambda\frac{u_r}{r} = 0 \quad \text{at } r = \delta \text{ and } a. \tag{8.5.18}$$

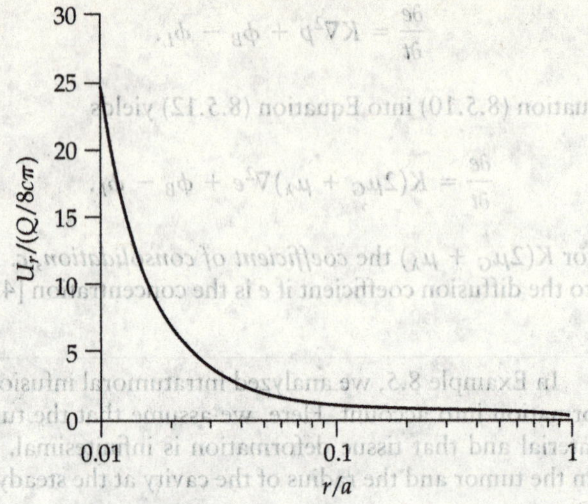

FIGURE 8.18 The displacement of tumor tissues in the radial direction. The displacement is non-dimensionalized by $Q/(8c\pi)$ and the radial distance is normalized by the radius of the tumor, a. At the surface of the fluid cavity, $r/a = 0.01$.

Substituting Equation (8.5.17) into Equation (8.5.18) yields

$$-\delta^3(2\mu_G + 3\mu_\lambda)b_1 + 4\mu_G b_2 = \frac{\mu_\lambda Q}{4c\pi}\delta^2 \tag{8.5.19}$$

and

$$-a^3(2\mu_G + 3\mu_\lambda)b_1 + 4\mu_G b_2 = \frac{\mu_\lambda Q}{4c\pi}a^2. \tag{8.5.20}$$

Analytic expressions for b_1 and b_2 can be derived from these equations. To simplify the expressions for b_1 and b_2, we assume that $\delta/a \ll 1$. Then,

$$b_1 \approx -\frac{\mu_\lambda Q}{4c\pi(2\mu_G + 3\mu_\lambda)a}\left(1 - \frac{\delta^2}{a^2}\right) \tag{8.5.21}$$

and

$$b_2 \approx \frac{\mu_\lambda Q a^2}{16\mu_G c\pi}\frac{\delta^2}{a^2}. \tag{8.5.22}$$

Substituting, Equations (8.5.21) and (8.5.22) into Equation (8.5.17) yields

$$u_r = \frac{Q}{8c\pi}\left[1 - \frac{2\mu_\lambda}{2\mu_G + 3\mu_\lambda}\left(1 - \frac{\delta^2}{a^2}\right)\frac{r}{a} + \frac{\mu_\lambda}{2\mu_G}\frac{\delta^2}{r^2}\right]. \tag{8.5.23}$$

The radius of the cavity at the steady state, δ_s, is equal to $\delta + (u_r)_{\text{at } r=\delta}$. Thus,

$$\delta_s = \delta + \frac{Q}{8c\pi}\left[\frac{2\mu_G + \mu_\lambda}{2\mu_G} - \frac{2\mu_\lambda}{2\mu_G + 3\mu_\lambda}\left(1 - \frac{\delta^2}{a^2}\right)\frac{\delta}{a}\right]. \tag{8.5.24}$$

For some solid tumors, $\mu_G = 15$ mmHg, $\mu_\lambda = 700$ mmHg, and $\delta/a = 0.01$ [46]. The ratio of u_r to $Q/(8c\pi)$ as a function of r/a is plotted in Figure 8.18.

8.6 PROBLEMS

8.1 There are about 6 L of blood in a 70-kg person. Assuming that the mass density of the human body is 1 kg L^{-1}, estimate the volume fraction of the vascular compartment of the body.

8.2 Consider the partitioning of spherical solutes between a laminated porous medium and external solutions. Assume that the pores are formed between adjacent parallel plates (i.e., slit pores). The half-width of the pores is h, and the number of pores per unit thickness of the material in the normal direction of the plates is n_L. The diameter of the solute is b and there are no charge–charge interactions. Determine the partition coefficient and the available volume fraction of the solute.

8.3 Determine the partition coefficient and the available volume fraction of a spherical solute in the same porous medium as that in Example 8.2. Assume that the diameter of the solute is b and that there are no charge–charge interactions.

8.4 Biological tissues are deformable. Therefore, the porosity ε depends on the volume dilatation e. Assuming that the total volume of cells and extracellular matrix molecules do not change during tissue deformation and that the initial porosity is ε_0, derive the relationship between ε and e.

8.5 A piece of avascular tissue is composed of cells, the extracellular matrix, and the interstitial fluid, as shown in Figure 8.19. Assume that the diameters of the cells and the fibers are d_c and d_f, respectively, the fiber length per unit volume is ℓ, the number of cells per unit volume is n_v, and the Kozeny constant is G.

(a) Find the porosity of the tissue.
(b) Find the hydraulic conductivity of the tissue using the Kozeny–Carman equation.

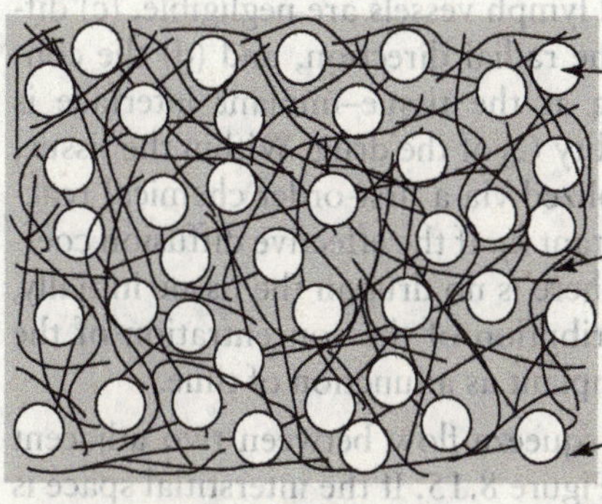

FIGURE 8.19 The composition of an avascular tissue. The cells are white spheres and the extracellular matrix is represented by black fibers.

8.6 Plot the velocity profile given in Example 8.7 for different values of $h/\sqrt{k}$. Under what conditions does the velocity profile approach a parabolic curve or a uniform profile? Plot the ratio of the flow rate q to $-h^2B/(12\mu)$ as a function of $h/\sqrt{k}$, and discuss the results.

8.7 A Newtonian fluid flows through a cylindrical pipe filled with a porous medium. The length of the pipe is L. The flow is one dimensional and at steady state. The pressures at two ends of the pipe are p_1 and p_2. The specific hydraulic permeability of the porous medium is k and the viscosity of the fluid is μ.

(a) Determine the velocity profile.
(b) Plot the nondimensionalized velocity $v/(kB/\mu)$ as a function of r/R, where $(P_1 - P_2)/L = B$. Assume that $k = 400\ nm^2$ and $R = 20$, 60, 100, and 400 nm, in turn.
(c) Under what conditions can viscous effects be neglected? That is, under what conditions is the error in the flow rate that is predicted by Darcy's law less than 10% of the flow rate that is determined by Brinkman's equation?

8.8 Consider fluid flow in a pipe with a porous wall (e.g., a blood vessel) that is permeable to the fluid. The radius and the length of the pipe are R and L, respectively. In general, the flow is axisymmetric, the fluid velocity has both radial and axial components, and these components have to be solved numerically. However, there exist two approximate approaches to this problem. One is to use lubrication theory to determine the relationship between the flow rate and the pressure gradient in the pipe if $R \ll L$ (see Chapter 4). Another is to estimate the axial velocity component independently by solving a problem of unidirectional flow in a pipe with an impermeable wall. In the second approach, the no-slip boundary condition at the wall is replaced by the equation

$$\sqrt{k}\frac{\partial v}{\partial r} = -\alpha v,$$

where k is the specific hydraulic permeability of the wall, v is the axial velocity of the fluid, r is the radial coordinate, and α is a dimensionless quantity that depends on the microstructure of the porous wall. The value of α can vary from 0.1 to 10 and may correlate with the average pore size in the pipe wall [47].

(a) Determine the velocity profile.

(b) Estimate the slip effect α on the pressure drop if the flow rate through the pipe is fixed.

8.9 Drugs are infused into an isolated solid tumor via a constant infusion pressure p_0. There is no blood and lymph circulation in the tumor, and the pressure at the surface of the tumor is equal to zero. Assume that the shape of solid tumors is spherical, the radii of the tumor and the fluid cavity are a and δ, respectively, and the hydraulic conductivity in tumors is K. Determine

(a) the pressure and velocity profiles in the tumor at the steady state and

(b) the infusion rate.

8.10 In the derivation of Equations (8.3.49), (8.3.51), and (8.3.52), we assume that the specific hydraulic permeability k is a constant. In general, k depends on local structures of tissues. Empirically,

$$k = \beta\left(\frac{\varepsilon}{1-\varepsilon}\right)^n,$$

where ε is the porosity and β and n are empirical constants [36]. When tissue is deformed, ε changes. If both the solid and fluid phases are assumed to be incompressible, ε can be related to the volume dilatation e (see Problem 8.4). For one-dimensional deformation, shown in Figure 8.15,

$$e = \frac{h - h_0}{h_0}.$$

We also assume that $V_h = V_{h_0}\exp(-\alpha t)$, where $\alpha = V_{h_0}/(\varepsilon_0 h_0)$ and t is the time. The baseline values of the constants of the model are $\beta = 0.003$ nm^2, $n = 3.2$, $\varepsilon_0 = 0.98$, $h_0 = 0.1$ μm, and $R = 5$ μm.

(a) Determine the velocity v_r as a function of z at $r = R$ for $\alpha t = 0$, 0.1, 1, and 3.

(b) Determine the pressure p as a function of αt at $r = 0$.

(c) Compare the pressure and velocity profiles you found in parts (a) and (b) with those determined by Equations (8.3.51) and (8.3.51), respectively, with $k = k_0 = \beta[\varepsilon_0/(1-\varepsilon_0)]^n$.

8.11 Assume that the interstitial space between two cells can be modeled as a porous medium bounded by two parallel plates (see Figure 8.13). Determine how the effective hydraulic conductivity in the channel (K_{channel}) changes with h if the interstitial space is deformed by following the procedure described in Example 8.7 and using the results from Problem 8.10.

8.12 Polymer implants have been used for the controlled release of drugs and genes in biological tissues. Consider a polymer membrane implanted in a tissue. To simplify the analysis of controlled drug release, we assume that the tissue is much larger than the thickness of the implant. In addition, we assume that the clearance of a drug in the tissue is very slow, so that the concentration of the drug at the tissue–implant interface is equal to the solubility C_0 of the drug. Also, the transport of drug in the tissue is one dimensional and in the normal direction of the membrane. Finally, we assume that convection and drug clearance through blood and lymph vessels are negligible. Thus, drug transport in the tissues involves only diffusion. If the effective diffusion coefficient is D_{eff} and there is no drug in the tissue initially, determine the distribution of the concentration of the drug outside the implant as a function of time.

8.13 Consider another situation of drug release from polymer implants. In this case, the implant is a sphere with radius a. Assume that (a) the tissue is much larger than the implant, (b) convection and drug clearance through blood and lymph vessels are negligible, (c) diffusion is only in the radial direction, and (d) the concentration of drug at the tissue–implant interface is equal to the solubility C_0 of the drug. Within the tissue, the drug is metabolized via a first-order chemical reaction with rate constant k_f. If the effective diffusion coefficient is D_{eff} and there is no drug in the tissue initially, determine the distribution of the concentration of the drug outside the implant as a function of time.

8.14 Consider a squeeze flow between two adjacent cells, as shown in Figure 8.15. If the interstitial space is highly compressible, determine the total forces on the cell membranes.

8.7 REFERENCES

1. Ross, M.H., Reith, E.J., and Romrell, L.J., *Histology: A Text and Atlas*. 2d ed. 1989, Baltimore: Williams & Wilkins.
2. Jain, R.K., "Transport of molecules in the tumor interstitium: a review." *Cancer Res.*, 1987. 47: pp. 3039–3051.
3. Khor, S.P., and Mayersohn, M., "Potential error in the measurement of tissue to blood distribution coefficients in physiological pharmacokinetic modeling: residual tissue blood. I. Theoretical considerations." *Drug Metab. Disp.*, 1991. 19: pp. 478–485.
4. Yuan, F., Krol, A., and Tong, S., "Available space and extracellular transport of macromolecules: effects of pore size and connectedness." *Ann. Biomed. Eng*, 2001. 29: pp. 1150–1158.
5. Krol, A., Maresca, J., Dewhirst, M.W., and Yuan, F., "Available volume fraction of macromolecules in a fibrosarcoma: implications for drug delivery." *Cancer Res.*, 1999. 59: pp. 4136–4141.
6. Casassa, E.F., "Equilibrium distribution of flexible polymer chains between a macroscopic solution phase and small voids." *Polym. Lett.*, 1967. 5: pp. 773–778.
7. Davidson, M.G., Suter, U.W., and Deen, W.M., "Equilibrium partitioning of flexible macromolecules between bulk solution and cylindrical pores." *Macromolecules*, 1987. 20: pp. 1141–1146.
8. Ogston, A.G., "The spaces in a uniform random suspension of fibers." *Trans. Farad Soc.*, 1958. 54: pp. 1754–1757.
9. Deen, W.M., "Hindered transport of large molecules in liquid-filled pores." *AIChE J.*, 1987. 33: pp. 1409–1425.
10. Buck, K.K., Gerhardt, N.I., Dungan, S.R., and Phillips, R.J., "The effect of solute concentration on equilibrium partitioning in polymeric gels." *J. Colloid Interface Sci.*, 2001. 234: pp. 400–409.
11. El-Kareh, A.W., Braunstein, S.L., and Secomb, T.W., "Effect of cell arrangement and interstitial volume fraction on the diffusivity of monoclonal antibodies in tissue." *Biophys. J.*, 1993. 64: pp. 1638–1646.
12. McGuire, S., Zaharoff, D., and Yuan, F. "Interstitial transport of macromolecules: implications for nucleic acid delivery in solid tumors." in *Pharmaceutical Perspectives of Nucleic Acid-Based Therapeutics*, R.I. Mahato and S.W. Kim, editors. 2002. London: Taylor & Francis Books, Ltd., pp. 434–454.
13. de Boer, R., *Theory of Porous Media: Highlights in the Historical Development and Current State*. 2000, Berlin: Springer.
14. Bear, J., *Dynamics of Fluids in Porous Media*. 1988, New York: Dover Publications.
15. Happel, J., and Brenner, H., *Low Reynolds Number Hydrodynamics with Special Application to Particulate Media*. 1983, The Hague: Martinus Nijhoff Publishers.
16. Johnson, E.M., and Deen, W.M., "Hydraulic permeability of agarose gels." *AIChE J.*, 1996. 42: pp. 1220–1224.
17. Tong, J., and Anderson, J.L., "Partitioning and diffusion of proteins and linear polymers in polyacrylamide gels." *Biophys. J.*, 1996. 70: pp. 1505–1513.
18. Phillips, R.J., "A hydrodynamic model for hindered diffusion of proteins and micelles in hydrogels." *Biophys. J.*, 2000. 79: pp. 3350–3354.
19. Swabb, E.A., Wei, J., and Gullino, P.M., "Diffusion and convection in normal and neoplastic tissues." *Cancer Res.*, 1974. 34: pp. 2814–2822.
20. Zhang, X.-Y., Luck, J., Dewhirst, M.W., and Yuan, F., "Interstitial hydraulic conductivity in a fibrosarcoma." *Am. J. Physiol.*, 2000. 279: pp. H2726–H2734.
21. Guyton, A.C., Scheel, K., and Murphree, D., "Interstitial fluid pressure: III. Its effect on resistance to tissue fluid mobility." *Circ. Res.*, 1966. 19: pp. 412–419.
22. Bird, R.B., Stewart, W.E., and Lightfoot, E.N., *Transport Phenomena*. 2d ed. 2002, New York: John Wiley & Sons.
23. Netti, P.A., Berk, D.A., Swartz, M.A., Grodzinsky, A.J., and Jain, R.K., "Role of extracellular matrix assembly in interstitial transport in solid tumors." *Cancer Res.*, 2000. 60: pp. 2497–2503.
24. Mow, V.C., Kuei, S.C., Lai, W.M., and Armstrong, C.G., "Biphasic creep and stress relaxation of articular cartilage in compression: theory and experiments." *J. Biomech. Eng.*, 1980. 102: pp. 73–84.
25. Frank, E.H., and Grodzinsky, A.J., "Cartilage electromechanics. I. Electrokinetic transduction and the effect of electrolyte pH and ionic strength." *J. Biomech.*, 1987. 20: pp. 615–627.
26. Maymir, J.C., Deutsch, S., Meyer, R.S., Geselowitz, D.B., and Tarbell, J.M., "Mean velocity and Reynolds stress measurements in the regurgitant jets of tilting disk heart valves in an artificial heart environment." *Ann. Biomed. Eng.*, 1998. 26: pp. 146–156.
27. Makhijani, V.B., Siegel, J.M.J., and Hwang, N.H., "Numerical study of squeeze-flow in tilting disc mechanical heart valves." *J. Heart Valve Dis.*, 1996. 5: pp. 97–103.
28. Katz, D.F., Henderson, M.H., Owen, D.H., Plenys, A.M., and Walmer, D.K. "What is needed to advance vaginal formulation technology?" In *Vaginal Microbicide Formulation Workshop*, W.F. Rencher, editor. 1998. Philadelphia: Lippencott-Raven, pp. 90–99.

29. Lai, W.M., Mow, V.C., Sun, D.D., and Ateshian, G.A., "On the electric potentials inside a charged soft hydrated biological tissue: streaming potential versus diffusion potential." *J. Biomech. Eng.*, 2000. 122: pp. 336–346.
30. Zienkiewicz, O.C., and Shiomi, T., "Dynamic behavior of saturated porous media: the generalized Biot formulation and its numerical solution." *Int. J. Num. Anal. Meth. Geomech.*, 1984. 8: pp. 71–96.
31. Steckwzy, R., Niedererz, P., and Knothe Tate, M.L., "A finite element analysis for the prediction of load-induced fluid flow and mechanochemical transduction in bone." *J. Theor. Biol.*, 2003. 220: pp. 249–259.
32. Cowin, S.C., "Bone poroelasticity." *J. Biomech.*, 1999. 32: pp. 217–238.
33. Mak, A.F., "The apparent viscoelastic behavior of articular cartilage: The contributions from the intrinsic matrix viscoelasticity and interstitial flows." *J. Biomech. Eng.*, 1986. 108: pp. 123–130.
34. Buschmann, M.D., Kim, Y.J., Wong, M., Frank, E., Hunziker, E.B., and Grodzinsky, A.J., "Stimulation of aggrecan synthesis in cartilage explants by cyclic loading is localized to regions of high interstitial fluid flow." *Arch. Biochem. Biophys.*, 1999. 366: pp. 1–7.
35. White, J.A., and Deen, W.M., "Agarose–dextran gels as synthetic analogs of glomerular basement membrane: water permeability." *Biophys. J.*, 2002. 82: pp. 2081–2089.
36. Gu, W.Y., Yao, H., Huang, C.Y., and Cheung, H.S., "New insight into deformation-dependent hydraulic permeability of gels and cartilage, and dynamic behavior of agarose gels in confined compression." *J. Biomech.*, 2003. 36: pp. 593–598.
37. Adler, P.M., *Porous Media: Geometry and Transport*. 1992, Boston: Butterworth-Heinemann.
38. Johnson, E.M., Berk, D.A., Jain, R.K., and Deen, W.M., "Hindered diffusion in agarose gels: test of effective medium model." *Biophys. J.*, 1996. 70: pp. 1017–1723.
39. Clague, D.S., and Phillips, R.J., "Hindered diffusion of spherical macromolecules through dilute fibrous media." *Phys. Fluids*, 1996. 8: pp. 1720–1731.
40. Anderson, J.L., and Quinn, J.A., "Restricted transport in small pores: a model for steric exclusion and hindered particle motion." *Biophys. J.*, 1974. 14: pp. 130–150.
41. Weinbaum, S., "Strong interaction theory for particle motion through pores and near boundaries in biological flows at low Reynolds number." *Lect. Math. Life Sci.*, 1981. 14: pp. 119–192.
42. Bentley, T.B., and Pittman, R.N., "Influence of temperature on oxygen diffusion in hamster retractor muscle." *Am. J. Physiol.*, 1997. 272: pp. H1106–H1112.
43. Pluen, A., Boucher, Y., Ramanujan, S., McKee, T.D., Gohongi, T., di Tomaso, E., Brown, E.B., Izumi, Y., Campbell, R.B., Berk, D.A., and Jain, R.K., "Role of tumor-host interactions in interstitial diffusion of macromolecules: cranial vs. subcutaneous tumors." *Proc. Natl. Acad. Sci. USA*, 2001. 98: pp. 4628–4633.
44. Berk, D.A., Yuan, F., Leunig, M., and Jain, R.K., "Direct in vivo measurement of targeted binding in a human tumor xenograft." *Proc. Natl. Acad. Sci. USA*, 1997. 94: pp. 1785–1790.
45. Basser, P.J., "Interstitial pressure, volume, and flow during infusion into brain tissue." *Microvasc. Res.*, 1992. 44: pp. 143–165.
46. Netti, P.A., Baxter, L.T., Boucher, Y., Skalak, R., and Jain, R.K., "Macro- and microscopic fluid transport in living tissues: application to solid tumors." *AIChE J.*, 1997. 43: pp. 818–834.
47. Beavers, G.S., and Joseph, D.D., "Boundary conditions at a naturally permeable wall." *J. Fluid Mech.*, 1967. 30: pp. 197–207.

Transvascular Transport

CHAPTER

9

9.1 Introduction

Nutrients and medicines absorbed in the intestine are delivered to individual organs in the body through the systemic blood circulation. In the microcirculation of each organ, the molecules that are involved escape from the lumen of microvessels to enter the interstitial space and then reach individual cells as a result of diffusion and convection. Metabolites and wastes that are released from cells enter the microvessels through either the same pathways (but in the opposite direction) or the lymphatic system. The exchange of molecules between the blood and the interstitium occurs mainly in two specific microvessels—capillaries and postcapillary venules—because the permeability of these vessels is much higher than that of other vessels.

The rates of transport across a microvessel wall are characterized by the hydraulic conductivity and the microvascular permeability coefficient of solutes. These quantities are phenomenological constants that depend upon the ultrastructure of the vessel wall, which in turn is modulated by chemical and physical microenvironments in tissues. The microvascular permeability coefficient also depends upon the physical and chemical properties of the solutes.

The capillary wall consists of three layers—the *glycocalyx, endothelium,* and *basement membrane*—as well as an incomplete layer of *pericytes* (see Figure 9.1). The glycocalyx is a layer of extracellular matrix composed of glycoproteins with a negative charge. The reported range of thickness is between 100 nm and 400 nm. (The thickness of this layer is still controversial, because it has been measured in different ways.) The functions of the glycocalyx are not fully understood. The structure is known to be a molecular sieve that resists the transport of macromolecules across the microvessel wall. The glycocalyx may also influence leukocyte–endothelial adhesion, since adhesion molecules on endothelial cells are embedded within it.

Endothelial cells are bound together by junctional proteins to form a single-cell tube around the glycocalyx. At the cell junctions, a variety of structures can exist, depending upon the type of junctional proteins that are present (see Figure 9.2). Linker proteins form *adhesion junctions* that connect the cytoskeleton of two adjacent cells

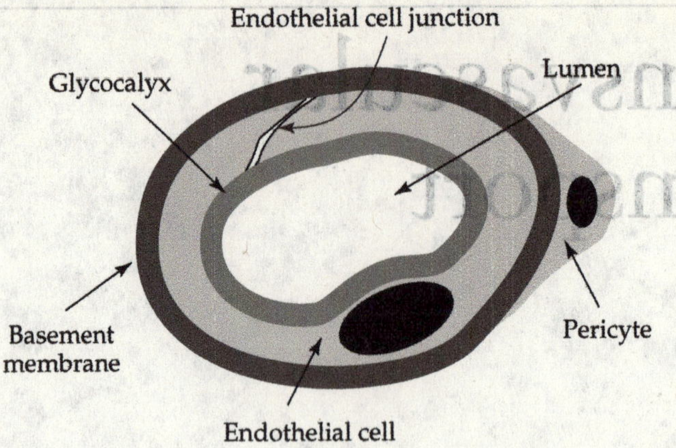

FIGURE 9.1 Structures of the microvessel wall.

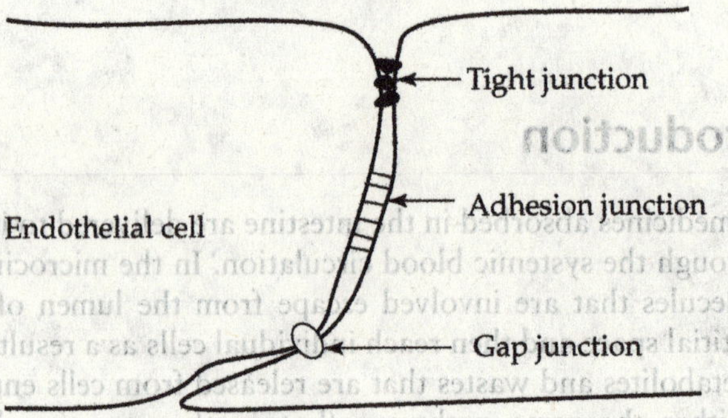

FIGURE 9.2 Three types of endothelial cell junctions in the microvessel wall.

(e.g., cadherins). These proteins provide mechanical strength to the cell–cell junctions. *Gap junctions* are formed by protein assemblies (connexons) that provide channels for molecular transport between the cytoplasm of adjacent cells. *Tight junctions,* found in brain capillaries (i.e., the blood–brain barrier), are formed by tight-junctional proteins (see Figure 1.8). The tight junctions are impermeable to hydrophilic molecules, so bloodborne solutes can cross the brain capillary wall only through facilitated or active transport (see Chapters 1 and 14). In all microvessels, the endothelial cell layer presents a major resistance to the transport of hydrophilic solutes.

The basement membrane is an electron-dense fiber-matrix layer (i.e., it appears as a dark layer when observed under an electron microscope) that contains type-IV collagen, proteoglycan (e.g., perlecan), laminin, fibronectin, and glycoproteins. The basement membrane provides little resistance to the transport of small molecules, but it must be considered in the study of the transport of macromolecules or nanoparticles across the microvessel wall.

9.2 Pathways for Transendothelial Transport

Pathways for the transport of solvents and solutes across the endothelium are not well understood. With an electron microscope, we can observe the structures of potential pathways. On the basis of these structures, capillaries can be divided into three categories: continuous, discontinuous, and fenestrated.

9.2.1 Continuous Capillaries

Continuous capillaries are found in most normal organs, such as the muscle, and the skin. Endothelial cells in these vessels are connected via junctional proteins, and the hydrophilic pores in the junctional clefts are formed due to missing proteins in the junctional protein strands. There are four different mechanisms of passive transport across the wall of continuous capillaries: direct diffusion through endothelial cells, diffusion and convection through endothelial junctions, diffusion on the cell membrane, and vesicle-mediated transport. The relative contributions of each mechanism to solute transport under specific conditions are still controversial.

Direct diffusion through cells (see Figure 9.3a) is a transport mechanism for small hydrophobic molecules or very small gas molecules (e.g., oxygen). The transport of most hydrophilic molecules across the microvessel wall relies on diffusion and convection through endothelial junctions (see Figure 9.3a). Diffusion on the cell membrane (see Figure 9.3b) is not well understood, but it may play a role in the transport of some large hydrophilic molecules across the endothelium.

Many vesicles exist in endothelial cells, and they may mediate the transendothelial cell transport of solutes. The size of these vesicles is 50 ~ 70 nm (see Figure 1.10). Vesicle-mediated transport is more complicated than the other three mechanisms discussed. Vesicles may facilitate transport in three ways (see Figure 9.4): Some vesicles may serve as shuttles for solute transport, some may transport solutes by means of fusion and fission, and others may form transendothelial cell channels [1]. Previous studies have shown that most vesicles cannot move freely between luminal

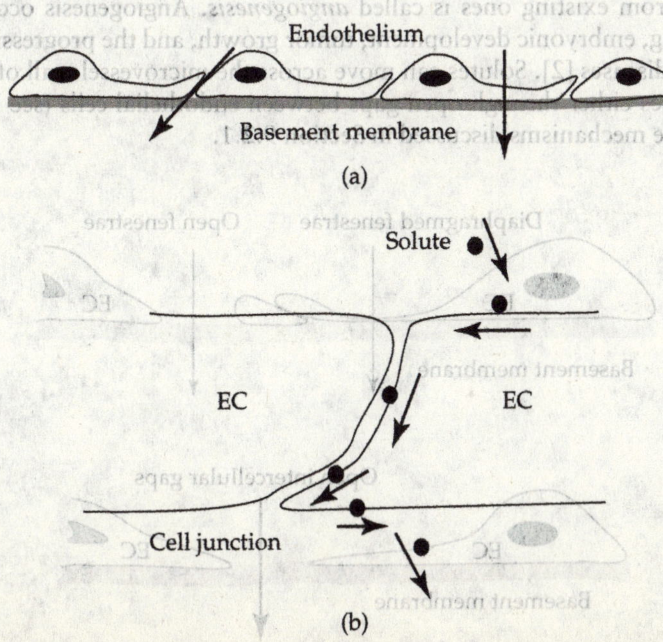

FIGURE 9.3 (a) Solute transport through endothelial junctions or through endothelial cells. (b) Diffusion of solutes along the cell membrane. The solute may first bind to receptors on the luminal side of the cell membrane, diffuse along with the receptor to the abluminal side of the cell, and then be released from the receptor. (EC = endothelial cell.)

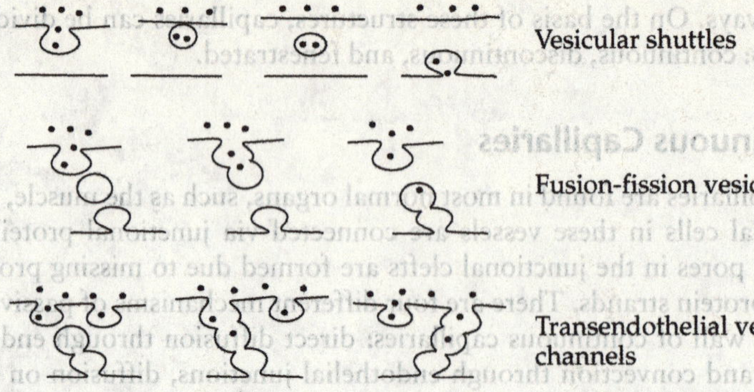

FIGURE 9.4 Vesicle-mediated transport of solutes across the endothelium in the microvessel wall. The solutes are represented by solid circles. (Adapted from Ref. [1], with permission.)

and abluminal surfaces of endothelial cells. Thus, the shuttle mechanism rarely occurs in transendothelial cell transport.

9.2.2 Fenestrated Capillaries

Fenestrated capillaries, found in the liver, the kidney, and tumors, are characterized by structures that contain many openings, or *fenestrae,* inside endothelial cells (see Figures 1.11 and 9.5). Some of the fenestrae are covered with membrane diaphragms, while others are completely open, allowing the transport of macromolecules, or even bloodborne cells, through the endothelial layer. In addition to the fenestrae, the mechanisms of transport across the wall of continuous capillaries exist in fenestrated capillaries.

9.2.3 Discontinuous Capillaries

Discontinuous capillaries are found in newly formed vessels. The process of forming new vessels from existing ones is called *angiogenesis*. Angiogenesis occurs during wound healing, embryonic development, tumor growth, and the progression of various vascular diseases [2]. Solutes can move across the microvessel wall of discontinuous capillaries either through open gaps between endothelial cells (see Figure 9.6) or through the mechanisms discussed in Section 9.2.1.

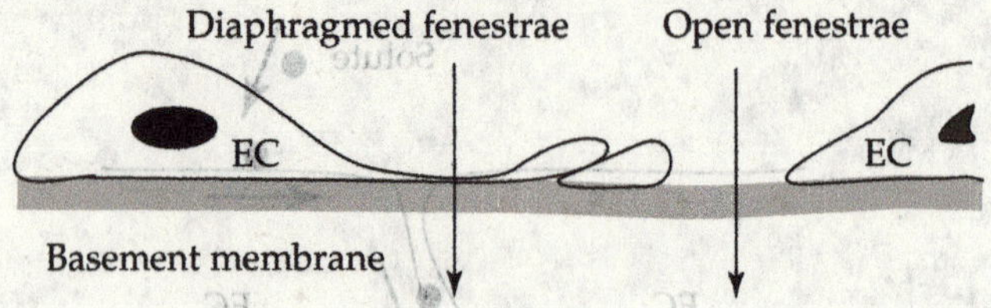

FIGURE 9.5 Diaphragmed and open fenestrae in the wall of fenestrated capillaries. The arrows indicate the pathways of transport.

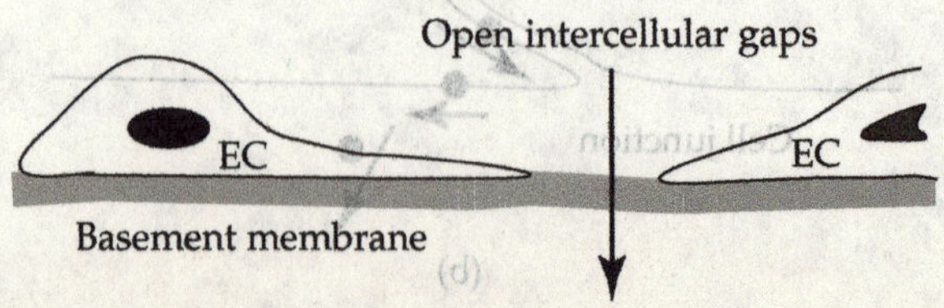

FIGURE 9.6 Open intercellular gaps in the wall of discontinuous capillaries. The arrow indicates the pathway of transport.

9.3 Rates of Transvascular Transport

The transport of molecules across the microvessel wall has been studied extensively in the field of microcirculatory research [3]. The microvessel wall is a laminated material, consisting of the glycocalyx, the endothelium, the basement membrane, and the pericytes (see Figure 9.1). However, most theories of transport treat the microvessel wall as a uniform membrane. The single-membrane model of the microvessel wall may be inappropriate for studying fluid absorption into microvessels at the steady state, but it is a reasonable approximation in the analysis of filtration and transient absorption (see Section 9.5). Therefore, we will assume in this book that the microvessel wall is a porous membrane with a uniform structure. Before discussing the rates of transvascular transport, we will first introduce the concept of *osmotic pressure,* since its gradient is an important driving force for fluid transport.

9.3.1 Osmotic Pressure

The concept of osmotic pressure is best demonstrated by the experiment illustrated in Figure 9.7. Before the system reaches equilibrium, the solvent has a tendency to move from the low-concentration to the high-concentration sides of the membrane if both chambers are open to the atmosphere. This phenomenon, called *osmosis,* was first observed by a French physicist, Jean-Antoine Nollet, in 1748. Molecular mechanisms of osmosis are not yet completely understood [4], although they have been investigated extensively in previous studies based on microhydrodynamic theories [5].

The mechanism of osmosis at the macroscopic level is related to the chemical potential of the solvent, which decreases when the concentration of solutes increases, due to an increase in the entropy. Therefore, the chemical potential of pure solvent is

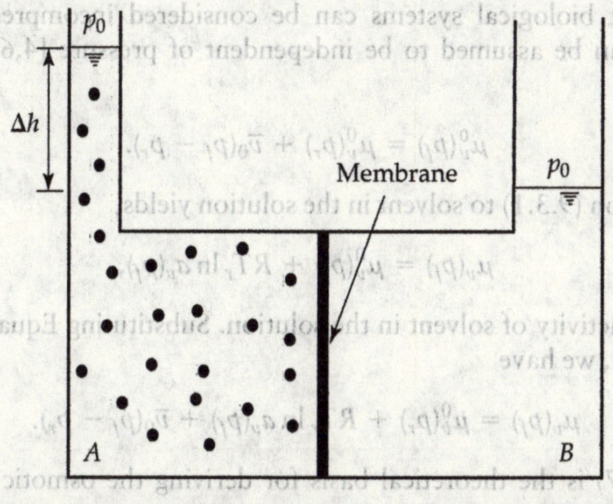

FIGURE 9.7 An apparatus for demonstrating the osmotic pressure in solutions. Two chambers, *A* and *B*, are separated by a membrane that is permeable only to water. Chamber *A* contains a solution; chamber *B* contains pure water. The water has a tendency to move from chamber *B* to chamber *A*. At equilibrium, the difference in the fluid levels between chambers *A* and *B* is Δh. Both chambers are open to the atmosphere. Thus, the pressures on both fluid surfaces are equal to p_0, the atmospheric pressure.

greater than the chemical potential of the solvent in solutions at the same pressure and temperature. Quantitatively, the chemical potential μ_i, of a substance i in a solution at the pressure p_r and the temperature T_r is related to the activity a_i of the substance (see Chapter 6) [4,6] via the formula

$$\mu_i = \mu_i^0 + RT_r \ln a_i, \tag{9.3.1}$$

where R is the gas constant, μ_i^0 is the chemical potential of the pure substance i at a hypothetical reference state in which a_i is equal to unity, and the pressure and temperature are at p_r and T_r, respectively. To simplify the discussion, we will first focus on a binary system (i.e., a system in which the solution contains only one solute in the solvent).

The chemical potential μ_v^0 of a pure solvent is a function of the pressure p, temperature T, and number of moles of solvent n_v. The dependence of μ_v^0 on p can be expressed, in general, as [6]

$$\left(\frac{\partial \mu_v^0}{\partial p}\right)_T = \overline{v}_0, \tag{9.3.2}$$

where

$$\overline{v}_0 = \frac{V_v}{n_v} \tag{9.3.3}$$

is the molar volume of the pure solvent and V_v is the solvent volume. Integrating Equation (9.3.2) from p_r to a final pressure p_f gives

$$\mu_v^0(p_f) = \mu_v^0(p_r) + \int_{p_r}^{p_f} \overline{v}_0 dp. \tag{9.3.4}$$

Liquids in most biological systems can be considered incompressible materials. Therefore, $\overline{v}_0$ can be assumed to be independent of pressure [4,6], and Equation (9.3.4) becomes

$$\mu_v^0(p_f) = \mu_v^0(p_r) + \overline{v}_0(p_f - p_r). \tag{9.3.5}$$

Applyng Equation (9.3.1) to solvent in the solution yields,

$$\mu_v(p_f) = \mu_v^0(p_f) + RT_r \ln a_v(p_f), \tag{9.3.6}$$

where a_v is the activity of solvent in the solution. Substituting Equation (9.3.5) into Equation (9.3.6), we have

$$\mu_v(p_f) = \mu_v^0(p_r) + RT_r \ln a_v(p_f) + \overline{v}_0(p_f - p_r). \tag{9.3.7}$$

Equations (9.3.7) is the theoretical basis for deriving the osmotic pressure in any solution.

For the specific situation shown in Figure 9.7, the chemical potential of solvent at any location in chamber A is governed by Equation (9.3.7), which we recast into the form

$$\mu_A(p_A) = \mu_v^0(p_r) + RT_r \ln a_A(p_A) + \overline{v}_0(p_A - p_r), \tag{9.3.8}$$

where the subscript A indicates chamber A and p_A is location dependent. The chemical potential of solvent at any location in chamber B is governed by Equation (9.3.5), or, in recast form,

$$\mu_B^0(p_B) = \mu_v^0(p_r) + \bar{v}_0(p_B - p_r), \tag{9.3.9}$$

where the subscript B indicates chamber B. At equilibrium, the chemical potential difference across the membrane must be equal to zero at any location along the membrane i.e.,

$$\mu_A(p_A) = \mu_B^0(p_B). \tag{9.3.10}$$

Thus,

$$-RT_r \ln a_A(p_A) = \bar{v}_0(p_A - p_B). \tag{9.3.11}$$

In general, the *osmotic pressure* is defined as the pressure that must be applied to a solution to keep its solvent in equilibrium with pure solvent at the same temperature [6]. On the basis of this definition, the osmotic pressure π in the solution contained in chamber A, shown in Figure 9.7, is equal to the equilibrium pressure difference across the membrane. Therefore,

$$\pi = p_A - p_B = \rho_f g \Delta h, \tag{9.3.12}$$

where ρ_f is the fluid mass density, g is the acceleration due to gravity, and Δh is the difference in fluid levels between chambers A and B (see Figure 9.7). Equation (9.3.12) is often used to determine the osmotic pressure in the solution if ρ_f is known and Δh is measured.

Substituting Equation (9.3.12) into Equation (9.3.11), we have

$$\pi = -\frac{RT}{\bar{v}_0} \ln a. \tag{9.3.13}$$

Equation (9.3.13) is broadly valid, so the subscripts r and A have been removed. The equation indicates that the osmotic pressure in a solution depends upon the temperature and pressure of the solution, as well as on the molar volume and the activity of the solvent at the same temperature and pressure.

In dilute binary systems, the mole fraction of solute, x_s, in the solution is much less than unity. In this case, the activity of solvent a is approximately equal to x_v, which is the mole fraction of solvent and is equal to $1 - x_s$. As a result, Equation (9.3.13) becomes

$$\pi = \frac{RT}{\bar{v}_0}[x_s + O(x_s^2)]. \tag{9.3.14}$$

In dilute systems, $\bar{v}_0$ is approximately equal to V/n_v (see Equation (9.3.3)), where V is the solution volume and $x_s = n_s/(n_s + n_v) \approx n_s/n_v$. Substituting these relationships into Equation (9.3.14) and neglecting high-order terms, we have

$$\pi = CRT, \tag{9.3.15}$$

where $C = n_s/V$ is the molar concentration of the solute. The relationship in Equation (9.3.15) was first observed empirically by van't Hoff. Thus, it is called the *van't Hoff* equation. The equation indicates that the osmotic pressure is proportional to the concentration of the solute, but is independent of the type of solute in dilute binary systems. In solutions with multiple solutes, Equation (9.3.15) is still valid, as long as the sum of mole fractions of all solutes is much less than unity. In this case, C in that equation is the total concentration of solutes; that is,

$$C = \sum_{j=1}^{N} C_j, \tag{9.3.16}$$

where N is the total number of different solutes in the solution and C_j is the molar concentration of the jth solute. Similar to the hydrostatic pressure, the osmotic pressure is a relative quantity. The reference value is selected at the point where the concentration of solute is equal to zero. One important application of Equation (9.3.15) is the measurement of a solute's molecular weight, given that the molar concentration is related to the mass concentration C_ρ, (gram per liter) through the molecular weight of solutes, M_r. Using C_ρ, we obtain, from Equation (9.3.15),

$$\pi M_r = C_\rho RT. \tag{9.3.17}$$

If π and T of a solution can be measured, and if C_ρ is known during the experimental preparation of the solution, M_r can be calculated from Equation (9.3.17).

In nondilute solutions of a neutral solute, the osmotic pressure can be expressed as a series expansion of C_ρ, or

$$\pi = RT\left(\frac{C_\rho}{M_r} + B_2 C_\rho^2 + B_3 C_\rho^3 + \cdots\right), \tag{9.3.18}$$

where B_2 and B_3 are called the second and the third ***virial coefficients***, respectively [7]. The first virial coefficient, B_1, is equal to $1/M_r$. B_2 and B_3 represent two- and three-body interactions of solutes, respectively. In general, the virial coefficients depend on temperature and on the chemical potential of the solvent, which in turn depends on solute–solute and solute–solvent interactions [7]. When solute–solute interactions in dilute systems are negligible, Equation (9.3.18) reduces to Equation (9.3.17). In addition to the dependence on concentration, the osmotic pressure depends on the charge of the solutes. Discussions of charge effects on the osmotic pressure can be found in the literature [7,8].

Example 9.1 Consider a glucose solution with concentration of 0.1 mg mL^{-1}. The temperature of the solution is maintained at 20°C. Determine the osmotic pressure of the solution.

Solution The molecular weight of glucose, $C_6H_{12}O_6$, is 180. Thus, the molar concentration of the glucose solution is 0.1 g L^{-1}/180 = 5.56×10^{-4} M. Substituting the molar concentration into Equation (9.3.15), we get

$$\pi = 5.56 \times 10^{-4} \times 8.3144 \times 10^{7} \times 293 = 1.35 \times 10^{4}\ \mathrm{g\ s^{-2}\ cm^{-1}}. \tag{9.3.19}$$

The same result can be obtained by substituting the mass concentration into Equation (9.3.17). In biomedical studies, people tend to use mmHg or cmH_2O as the pressure unit. The osmotic pressure in Equation (9.3.19) is equal to 10.1 mmHg, or 13.8 cmH_2O.

If the glucose in Example 9.1 is replaced by albumin with the same mass concentration, then the osmotic pressure is reduced to 0.0376 cmH_2O, since the molecular weight of albumin is 66,000. At high concentrations, Equation (9.3.15) is invalid. In this case, the osmotic pressure can be determined by either Equation (9.3.18), if all virial coefficients are known, or by an empirical relationship obtained by fitting a curve to experimental data. For albumin, the empirical relationship is

$$\pi = 0.345C_p + 2.657 \times 10^{-3}C_p^2 + 2.26 \times 10^{-5}C_p^3, \tag{9.3.20}$$

where the units of C_p and π are milligrams per milliliter and cmH_2O, respectively [9]. The constants in Equation (9.3.20) may vary slightly in different studies [10]. When $C_p = 1$ mg mL^{-1}, the predictions of π from Equation (9.3.15) and Equation (9.3.20) are 0.376 and 0.348 cmH_2O, respectively. The difference is only 8%. When $C_p = 50$ mg mL^{-1}, the difference increases to 30%.

9.3.2 Rate of Fluid Flow and Starling's Law of Filtration

Fluid flux across the microvessel wall has been studied for more than 100 years [11]. In previous chapters, we showed that fluid flow in pure solutions or porous media can be driven by shear stresses, pressure gradients, or both. The effect of shear stress on fluid flow across a membrane is absent, because the thickness of a membrane is negligible. However, the hydrostatic and the osmotic pressure differences, as driving forces for fluid flow across a semipermeable membrane, must be considered.

The effect of an osmotic pressure difference on fluid flow across a microvessel wall was first observed by Starling in 1896 (see Ref. [11]). He found that the rate of fluid flow across the microvessel wall was proportional to the hydrostatic pressure difference Δp minus the osmotic pressure difference $\Delta\pi$, or

$$J_v = L_p S(\Delta p - \Delta\pi), \tag{9.3.21}$$

where J_v is the rate of fluid flow, S is the surface area of the endothelium, and L_p is the *hydraulic conductivity* of the vessel wall. Equation (9.3.21) is called *Starling's law of filtration*. Later studies by Kedem and Katchalsky [12], based on irreversible thermodynamics, demonstrated that Starling's law must be modified if the microvessel wall is permeable to those solutes (e.g., albumin) which contribute significantly to the maintenance of the osmotic pressure difference across the microvessel wall. In this case, the modified form of Starling's law of filtration is

$$J_v = L_p S(\Delta p - \sigma_s \Delta\pi), \tag{9.3.22}$$

where σ_s, first introduced by Staverman [13], is called the *osmotic reflection coefficient*. The value of σ_s varies between zero and unity, depending on the solutes and the structures of the microvessel wall. When σ_s is equal to unity, the vessel wall is impermeable to the solutes. Consequently, Equation (9.3.22) reduces to Equation (9.3.21). If the permeability of the microvessel wall is extremely high, σ_s approaches zero. In the body, the osmotic pressure difference across a microvessel wall is maintained by the plasma proteins. For continuous capillaries, σ_s is between 0.7 and 0.9.

Starling's law can be compared with Darcy's law (see Equation (8.3.2)) in the analysis of fluid transport across a porous membrane, where L_p is equivalent to K per unit thickness of the membrane and J_v/S is the average velocity of fluid across the membrane. The only difference between the two equations is the presence of osmotic pressure, which is neglected in Darcy's law. In general, the osmotic pressure gradient needs to be considered in the analysis of fluid flow in porous media if it is comparable to or larger than the hydrostatic pressure gradient.

9.3.3 Rate of Solute Transport and the Kedem–Katchalsky Equation

In their study of fluid transport across the membrane (see Equation (9.3.22)), Kedem and Katchalsky also modeled solute transport across the membrane. They found that

$$J_s = J_v(1 - \sigma_f)\overline{C}_s + PS\Delta C, \tag{9.3.23}$$

where J_s is the rate of solute transport and ΔC is the difference in concentration across the membrane [12].[1] σ_f, P, and $\overline{C}_s$ are phenomenological constants: σ_f is the filtration reflection coefficient, which is generally not equal to σ_s; P is the microvascular permeability coefficient; and $\overline{C}_s$ is an average molar concentration of solutes in the membrane. For dilute solutions,

$$\overline{C}_s = \frac{\Delta C}{\ln(C_L/C_i)}, \tag{9.3.24}$$

where C_L is the concentration of solutes in the lumen of vessels and C_i is the concentration immediately outside the microvessel wall. The Taylor expansion (see Section A.1.C) of $\ln(x)$ is

$$\ln(x) = 2\left[\frac{x-1}{x+1} + \frac{1}{3}\left(\frac{x-1}{x+1}\right)^3 + \cdots + \frac{1}{2n+1}\left(\frac{x-1}{x+1}\right)^{2n+1} + \cdots\right]. \tag{9.3.25}$$

If $1 < C_L/C_i < 3$, $\ln(C_L/C_i)$ can be approximated (with a 10% error) by the first term in the Taylor expansion; that is,

$$\ln(C_L/C_i) \approx 2\frac{C_L/C_i - 1}{C_L/C_i + 1} = \frac{\Delta C}{(C_L + C_i)/2}. \tag{9.3.26}$$

[1]Note that we use J in most chapters of this book to represent the flux of solute diffusion (see Section 6.2). However, we also use J to represent the total rate of transport across a *membrane*, in order to be consistent with the notation in the literature.

The average concentration in the membrane then becomes

$$\overline{C}_s = \frac{C_L + C_i}{2}. \tag{9.3.27}$$

A different approach has been used by Patlak et al. [14] to derive the flux of solute transport across the microvessel wall. In this approach, the microvessel wall is assumed to be a uniform porous membrane. Transport across the membrane is considered to be one dimensional and involves both convection and diffusion. At the steady state,

$$J_s = J_v(1 - \sigma_f)C_m - D_{\text{eff}}S\frac{\partial C_m}{\partial x} = a_1, \tag{9.3.28}$$

where C_m is the concentration of solutes within the membrane, D_{eff} is the effective diffusion coefficient, and a_1 is a constant. Note that the reflection coefficient cannot be defined in porous media with finite thickness. Thus, $1 - \sigma_f$ in Equation (9.3.28) should be interpreted as the retardation coefficient f (see Equation (8.4.1)). Integrating Equation (9.3.28), we have

$$C_m = \frac{a_1}{J_v(1 - \sigma_f)} + a_2\exp\left(\text{Pe}\frac{x}{h}\right), \tag{9.3.29}$$

where h is the thickness of the membrane, a_2 is a constant, and Pe is the Peclet number, defined as

$$\text{Pe} = \frac{J_v(1 - \sigma_f)}{PS}, \tag{9.3.30}$$

which describes the ratio of convection to diffusion across the microvessel wall (see Section 7.3). Here, $P = D_{\text{eff}}/h$ is the microvascular permeability coefficient of solutes. The constants a_1 and a_2 can be determined by the boundary conditions:

$$C_m = C_L \quad \text{at } x = 0, \tag{9.3.31}$$

$$C_m = C_i \quad \text{at } x = h. \tag{9.3.32}$$

Substituting Equations (9.3.31) and (9.3.32) into Equation (9.3.29), we get

$$a_1 = \frac{C_i - C_L\exp(\text{Pe})}{1 - \exp(\text{Pe})}J_v(1 - \sigma_f), \tag{9.3.33}$$

$$a_2 = \frac{C_L - C_i}{1 - \exp(\text{Pe})}, \tag{9.3.34}$$

and

$$C_m = \frac{C_L[\exp(\text{Pe}\,x/h) - \exp(\text{Pe})] + C_i[1 - \exp(\text{Pe}x/h)]}{1 - \exp(\text{Pe})}. \tag{9.3.35}$$

Substituting Equation (9.3.33) into Equation (9.3.28) yields

$$J_s = J_v(1 - \sigma_f)\frac{C_i - C_L\exp(\text{Pe})}{1 - \exp(\text{Pe})}. \tag{9.3.36}$$

Equation (9.3.36) is called the *Patlak equation*. Three other forms have also been used in the literature:

$$J_s = J_v(1 - \sigma_f)\left(C_L - \frac{\Delta C}{1 - \exp(\text{Pe})}\right), \tag{9.3.37}$$

$$J_s = J_v(1 - \sigma_f)C_L + PS\Delta C\frac{\text{Pe}}{\exp(\text{Pe}) - 1}, \tag{9.3.38}$$

$$J_s = J_v(1 - \sigma_f)\frac{C_L + C_i}{2} + \frac{1}{2}\left\{\frac{J_v(1 - \sigma_f)[\exp(\text{Pe}) + 1]\,\Delta C}{\exp(\text{Pe}) - 1}\right\}. \tag{9.3.39}$$

In these equations, $\Delta C = C_L - C_i$. The three equations can be derived by rearranging terms in Equation (9.3.36). A special case of Equations (9.3.36) through (9.3.39) with $\sigma_f = 0$ has been derived in Example 7.1. When convection is slower than diffusion (i.e., $\text{Pe} < 1$), the Patlak equation reduces to the Kedem–Katchalsky equation

$$J_s = J_v(1 - \sigma_f)\overline{C}_s + PS\Delta C, \tag{9.3.40}$$

with $\overline{C}_s = (C_L + C_i)/2$. The first term in Equation (9.3.40) is identical to that in Equation (9.3.39); the difference in the second terms of Equations (9.3.39) and (9.3.40) is less than 10% if $\text{Pe} < 1.2$.

The derivation of the Patlak equation is based upon various implicit and explicit assumptions. For example, transport is assumed to be at the steady state. The one-dimensional convective diffusion Equation (9.3.28) assumes implicitly that the solution is dilute. A more general discussion of the mathematical modeling of solute transport across a uniform membrane can be found in [3].

9.4 Phenomenological Constants in the Analysis of Transvascular Transport

We have introduced four phenomenological constants in the analysis of transvascular transport: the hydraulic conductivity L_P; the permeability coefficient of solutes, P; the osmotic reflection coefficient σ_s; and the filtration reflection coefficient σ_f. Macroscopically, these constants can be explained as follows:

$$L_P = \left(\frac{J_v/S}{\Delta p}\right)_{\Delta\pi=0} = \left(\frac{\text{Fluid flux}}{\text{Hydrostatic pressure difference}}\right)_{\Delta\pi=0}, \tag{9.4.1}$$

$$P = \left(\frac{J_s/S}{\Delta C}\right)_{J_v=0} = \left(\frac{\text{Solute flux}}{\text{Concentration difference}}\right)_{J_v=0}, \tag{9.4.2}$$

$$\sigma_s = \left(\frac{\Delta p}{\Delta\pi}\right)_{J_v=0} = \left(\frac{\text{Hydrostatic pressure difference}}{\text{Osmotic pressure difference}}\right)_{J_v=0}, \tag{9.4.3}$$

$$\sigma_f = 1 - \left(\frac{J_s/S}{(J_v C_0)/S}\right)_{\Delta C=0} = 1 - \left(\frac{\text{Solute flux across vessel wall}}{\text{Solute flux in the solution}}\right)_{\Delta C=0}. \quad (9.4.4)$$

These equations form the theoretical basis for the experimental quantification of the phenomenological constants. Microscopically, these constants are related to the molecular properties of solutes and to the structures of pores in the membrane [5,15]. Compared with the phenomenological constants introduced in Chapter 8 in regard to three-dimensional transport in porous media, L_P, P, and σ_f are equivalent to the hydraulic conductivity (K) per unit thickness, the effective diffusion coefficient (D_{eff}) per unit thickness, and 1 minus the retardation coefficient (f), respectively, provided that the membrane can be treated as a uniform, porous material with a finite thickness and that transport in the material is one dimensional. The difference between the two reflection coefficients is that σ_s is used to characterize the membrane resistance to transport of solutes that maintain the osmotic pressure difference across the membrane, whereas σ_f is used to characterize the membrane resistance to convective transport of the solute of interest in the study. When both solutes are the same, σ_s is approximately equal to σ_f.

In theory, we can measure the permeability coefficient if we know the flux of solute transport and also the concentration difference across the microvessel wall under conditions in which the solvent flux is zero. The zero-flux condition is very difficult to achieve experimentally, however, because it requires a delicate balance of hydrostatic and osmotic pressures across the microvessel wall. Therefore, only the *apparent permeability coefficient*, defined as

$$P_{\text{app}} = \left(\frac{J_s/S}{\Delta C}\right)_{J_v \neq 0} \quad (9.4.5)$$

has been directly quantified in experiments (see Problem 9.7). The apparent permeability coefficient depends upon both diffusion and convection. Thus, P_{app} is a function of the pressure difference across the microvessel wall. When the pressure difference approaches zero, P_{app} is equal to P.

Values of P_{app} have been measured in capillaries in the frog mesentery [16]. In these experiments, the pressure outside the lumen is approximately equal to zero. Thus, the pressure difference across the microvessel wall is equal to the pressure in the lumen (i.e., the perfusion pressure). P_{app} is found to be an approximately linear function of the perfusion pressure (see Figure 9.8). Extrapolation of the linear curve to zero pressure provides an estimation of the permeability coefficient in these vessels.

The data shown in Figure 9.8 demonstrate that the microvascular permeability can be influenced by the ingredients in the perfusate. Other factors that can affect the microvascular permeability are (a) the size, charge, shape, and polarity of the solutes; (b) the size and density of the pores in the microvessel wall; and (c) the fiber size and charge densities of the extracellular matrix (e.g., the glycocalyx and the basement membrane). In general, the microvascular permeability decreases when the molecular weight of the solutes increases (see Figure 9.9). The dependence on molecular weight is nonlinear and is not fully understood [16].

Transendothelial transport depends on the charge of the solutes, because the vessel wall is negatively charged [17,18]. Therefore, the permeability of cationic

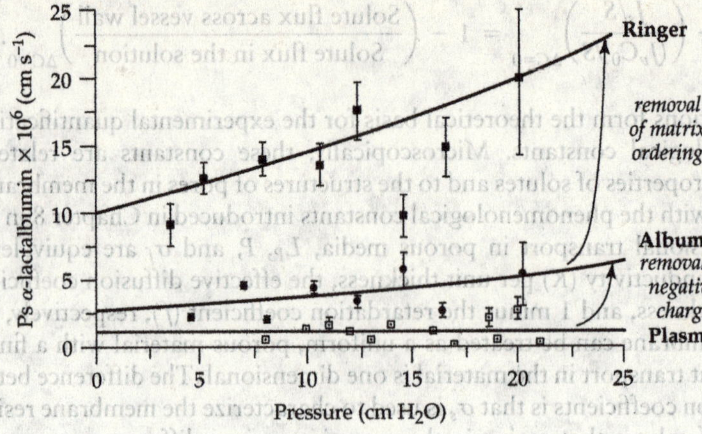

FIGURE 9.8 Dependence of the apparent microvascular permeability of α-lactalbumin on the perfusion pressure. The microvessels in the frog mesentery are perfused with either ringer solution, albumin solution, or plasma. The microvascular permeability of α-lactalbumin depends on the ingredients in the perfusate. (Reprinted from Ref. [16], with permission.)

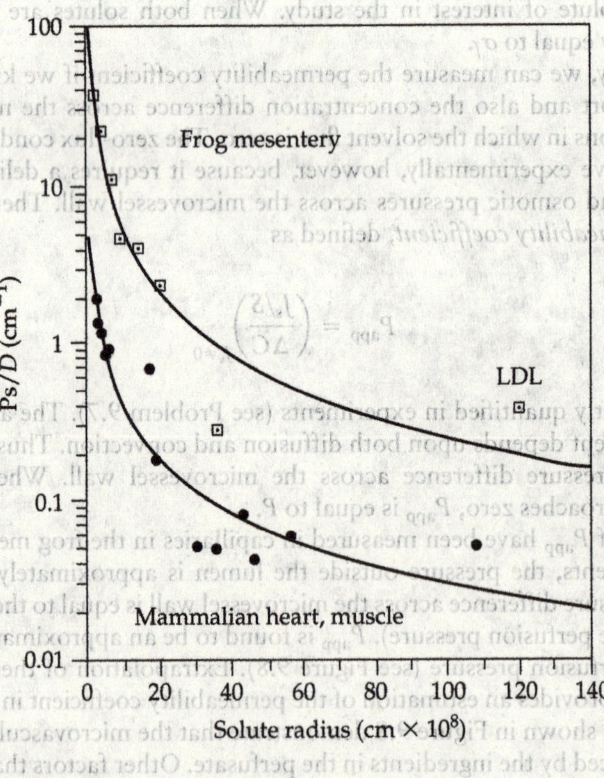

FIGURE 9.9 Dependence of the apparent microvascular permeability on the size of solutes in the frog mesentery and the muscle tissues. The permeability has been normalized by the diffusion coefficient of the same solutes in water. (Reprinted from Ref. [16], with permission.)

molecules is relatively higher than that of anionic ones if other properties of the solutes (e.g., the size) are approximately the same. For example, the molecular weight of *α-lactalbumin* is 14,176, which is close to that of ribonuclease (13,683) under physiological conditions. The net charges on *α-lactalbumin* and ribonuclease are −10 and +4, respectively. The permeability coefficient of *α-lactalbumin* in the frog mesentery is 2.1×10^{-6} cm s^{-1}, which is only 49% of the permeability coefficient of ribonuclease [17]. The dependence of the permeability on charge can be analyzed with the modified Nernst–Planck equation introduced in Chapter 7. The details of the analysis are described in Chapter 14, in discussing the glomerular filtration of charged solutes.

9.5 A Limitation of Starling's Law

Starling's law governs fluid transport across membranes. For membranes with macroscopically uniform structures, Starling's law accurately predicts the flux of fluid, based on the pressure differences across the membrane and the hydraulic conductivity. The capillary wall can be treated as a membrane because its thickness is one order of magnitude smaller than the circumference of the vessel. However, the structure of the microvessel wall is nonuniform, containing three barriers to fluid transport: the glycocalyx, the endothelium, and the basement membrane. Even within the endothelium, the structure is nonuniform, with endothelial cells, interendothelial clefts, and junctional protein strands. These nonuniform structures across the microvessel wall can render Starling's law inconsistent with some experimental data.

9.5.1 Fluid Filtration in the Steady State

The following example is often described in physiology textbooks: Consider blood in a capillary with a hydrostatic pressure p_B of 43 mmHg and an osmotic pressure π_B of 28 mmHg in the arterial end. In the interstitial space, the fluid pressure p_i is −2 mmHg, and the osmotic pressure π_i is 1 mmHg. The osmotic reflection coefficient σ_s is 0.9. The net pressure difference across the capillary wall is

$$p_{\text{net}} = (p_B - p_i) - \sigma_s(\pi_B - \pi_i) = 20.7 \text{ mmHg}, \tag{9.5.1}$$

which is the driving force for fluid transport across the wall. According to Starling's law, water in the lumen should filter through the vessel wall to enter the interstitial space. In the venous end of the capillary, the blood pressure p_B drops to 15 mmHg, whereas other pressures remain the same. Thus, the net pressure difference across the capillary wall changes to

$$p_{\text{net}} = (p_B - p_i) - \sigma_s(\pi_B - \pi_i) = -7.3 \text{ mmHg}. \tag{9.5.2}$$

In this case, water in the interstitial space should be reabsorbed into the lumen (see Figure 9.10).

The filtration and reabsorption hypothesis has been challenged by a classical experiment of single-vessel perfusion, performed by Michel and Phillips [19]. Microvessels in the frog mesentery were cannulated with glass micropipettes and perfused with albumin solution (50 mg mL^{-1}). After perfusion, the vessel was occluded with a glass restraining rod, while the pressure was maintained. The flux of water across the vessel wall, J_v/A, was then quantified by measuring the movement of red blood cells in the vessel. In this experiment, the osmotic pressure difference across the capillary wall was maintained at 22 cmH_2O, whereas the hydrostatic pressure difference Δp varied from 10 to 50 cmH_2O. A graph of J_v/A versus Δp is shown in Figure 9.11.

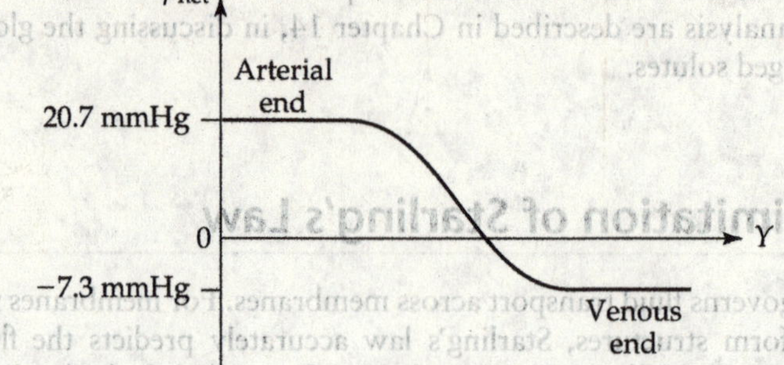

FIGURE 9.10 A classical view of Starling's law. Fluid is filtered in the arterial end and reabsorbed in the venous end of capillaries, due to the difference in the net pressure difference, p_{net}, across the capillary wall. Y indicates the distance along the axis of the capillary. The net pressure difference drops from 20.7 to −7.3 mmHg in the middle of the capillary.

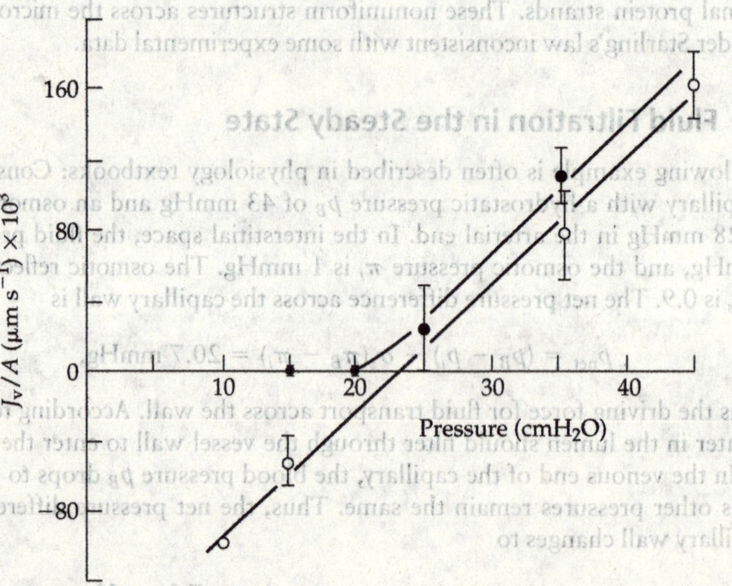

FIGURE 9.11 Single-vessel perfusion experiment. A capillary is perfused with albumin solution (50 mg mL^{-1}). The osmotic pressure of the perfusate is 22 cmH_2O. The perfusion pressure varies from 10 to 50 cmH_2O. The hydrostatic and osmotic pressures in the interstitial space are close to zero. In the transient experiment (open circles), the vessel is perfused for 3–5 seconds. In the steady-state experiment (closed circles), it is perfused for 2–5 minutes. J_v/A is the fluid flux, which is the same as J_v/S used in this chapter. (Reprinted from Ref. [19], with permission from Wiley publishers.)

These results demonstrate that Starling's law is valid only in the transient experiment, in which J_v/A is a linear function of Δp. The relationship between J_v/A and Δp in the steady-state experiment is nonlinear. There is no absorption of water in the steady state (i.e., J_v/A is always greater than zero), even when Δp is less than $\Delta\pi(=22$ cm$H_2O)$. Furthermore, a recent experiment demonstrates that J_v/A is independent of $\Delta\pi$ [10]. These data contradict the predictions of Starling's law.

9.5.2 A New View of Starling's Law

To explain the contradiction between Starling's law and experiment, Hu and Weinbaum proposed a nonlinear, three-dimensional model of fluid and solute transport through the microvessel wall [10,20]. This model considers four different regions in the wall and surrounding tissues: the glycocalyx, a junctional cleft between endothelial cells with a junctional protein strand in the middle, a semicircular region for albumin mixing at the exit of the cleft, and the extravascular space between the semicircular region and the midplane of two microvessels.

The glycocalyx is treated as a uniform membrane across which Starling's law is applied for determining the rate of fluid transport. The application of Starling's law yields

$$J_v = L_p S(\Delta p_{BG} - \sigma_s \Delta\pi_{BG}), \tag{9.5.3}$$

where $\Delta p_{BG} = p_B - p_G$ and $\Delta\pi_{BG} = \pi_B - \pi_G$, in which p_G and π_G are the hydrostatic and osmotic pressures, respectively, at the interface between the glycocalyx and endothelial cells. $\Delta\pi_{BG}$ can be either lower or higher than $\Delta\pi = \pi_B - \pi_i$, depending on the experimental conditions. In the junctional cleft region, the new model assumes that the pathways for water and albumin transport are the breaks in the junctional strand. The subendothelial regions are treated as uniform porous media to which the mass and momentum balance equations discussed in Chapter 8 are applied to determine the distributions of fluid velocity, pressure, and albumin concentration (or the osmotic pressure). The details of the model are omitted here; we recommend that readers consult the original papers [10,20].

The model by Hu and Weinbaum predicts that the major barriers to fluid and protein transport across the microvessel wall in both directions are the glycocalyx layer and the junctional protein strand. The latter is impermeable to fluid and solutes, except at the breaks caused by missing proteins. To summarize the main predictions of the model, we consider three typical cases:

Case 1. p_B is 35 cmH_2O, π_B is 27.2 cmH_2O, and π_i and p_i are equal to zero. In this case, there is fluid filtration across the capillary wall, since $\Delta p > \Delta\pi$. In the break regions in the junctional strand, convection is much faster than diffusion. As a result, albumin molecules cannot accumulate behind the junctional protein strand after they pass through the breaks. In addition, $\pi_G \approx 0$ at the steady state, indicating that the major osmotic pressure drop across the microvessel wall occurs in the glycocalyx layer.

Case 2. p_B, p_i, and π_B are the same as in Case 1, but π_i is increased to 27.2 cmH_2O (i.e., $\Delta\pi = 0$). In this case, π_G is increased slightly, but is still much less than π_B at the steady state. As a result, $\Delta\pi_{BG}$ is still close to 27 cmH_2O, although $\Delta\pi$ is equal to zero. Equation (9.5.3) predicts that there is a minimal increase in J_v compared with that in Case 1. This prediction is consistent with the experimental data obtained at the steady state and shown in [10]. However, Equation (9.3.21) predicts that J_v should be significantly higher than that in Case 1, since $\Delta\pi$ is reduced to zero.

Case 3. p_i, π_B, and π_i are the same as those in Case 1 or Case 2 and the microvessel has been perfused to establish the steady-state hydrostatic and osmotic pressure distributions in the microvessel wall. Then, p_B is suddenly decreased from 35 cmH_2O to 10 cmH_2O. During the initial, short period, transient fluid absorption occurs in the microvessel wall, since the osmotic pressure distribution in the wall, determined by the local concentration of albumin, cannot be changed quickly. However, fluid absorption cannot be sustained with a gradual accumulation of albumin in the region between the glycocalyx and the junctional protein strand. The accumulation is caused by the reduction in Δp, which in turn results in a significant decrease in the amount of albumin molecules being washed away by the convective flow through the break regions. Back diffusion of albumin from the subendothelial region also contributes to the accumulation behind the glycocalyx layer if π_i is maintained at 27.2 cmH_2O. As a result, $\Delta\pi_{BG}$ will decrease gradually with time, although $\Delta\pi$ is maintained. When the distributions of both osmotic and hydrostatic pressures reach new steady states, the driving force for fluid transport, $\Delta p_{BG} - \sigma_s \Delta\pi_{BG}$, is lower than in either Case 1 or Case 2, but it is still positive. There is no fluid absorption.

Further discussions of the predictions of the model can be found in [10,20]. Although this model has not been experimentally validated through the direct measurement of the distribution of albumin concentration in the microvessel wall, the model's predictions are consistent with the data in the literature [10,19]. Note that the analysis of Hu and Weinbaum does not indicate that Starling's law cannot be applied to the transmembrane transport of fluid; rather, it demonstrates that Starling's law may not apply directly to membranes with nonuniform structures, such as those found in the capillary wall. In nonuniform membranes, the osmotic pressure distribution is nonlinear and depends on the transport of solutes across the membranes.

9.6 PROBLEMS

9.1 Derive Equations (9.3.37) through (9.3.39) on the basis of the Patlak equation (9.3.36).

9.2 The permeability coefficient in a membrane, P, is defined as the flux of the solute, J_s, divided by the concentration difference ($\Delta C = C_1 - C_2$) across the vessel wall: $P = J_s/\Delta C$ (see Figure 9.12). If the effective diffusion coefficient of a nonelectrolyte molecule A in the membrane is D_A, the thickness of the membrane is L, and the available volume fraction of A in the membrane is K_A, determine the permeability coefficient of this molecule as a function of D_A, K_A, and L.

9.3 Consider the transport of an electrolyte molecule B across the membrane described in Problem 9.2. Assume that the net charge on B is z_B, $C_2 = 0$, the temperature of the membrane is T, the available volume fraction of B in the membrane is K_B, and the electric potential in the membrane (Ψ) is a linear function of x (see Figure 9.13). Determine

(a) the concentration distribution in the membrane;

(b) the permeability coefficient of molecule B as a function of $\Delta\Psi = \Psi_1 - \Psi_2$, D_B, K_B, L, T, and z_B.

9.4 The capillary wall can be modeled as a membrane with pores (see Figure 9.14). The diameter and the length of the pores are d and h, respectively. The density of pores is n_A, defined as the number of pores per unit surface area. The diffusion flux of molecules across the vessel wall can be described quantitatively by two methods: (a) one-dimensional diffusion in pores or (b) the Kedem–Katchalsky equation, with zero convective velocity ($J_s/S = P\Delta C$, where P is the vascular permeability coefficient and ΔC is the concentration difference across the vessel wall). If diffusion through different pores is identical and the solute size is much less than d, find P, using the intrinsic transport parameters d, h, n_A, and D (the diffusion coefficient in the pore).

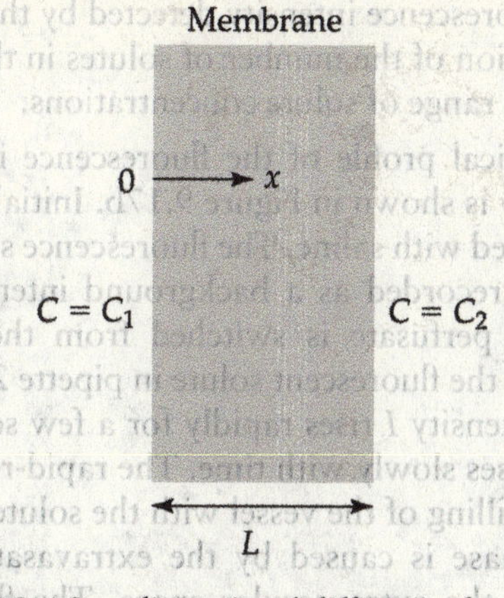

FIGURE 9.12 One-dimensional diffusion of a nonelectrolyte molecule across a membrane.

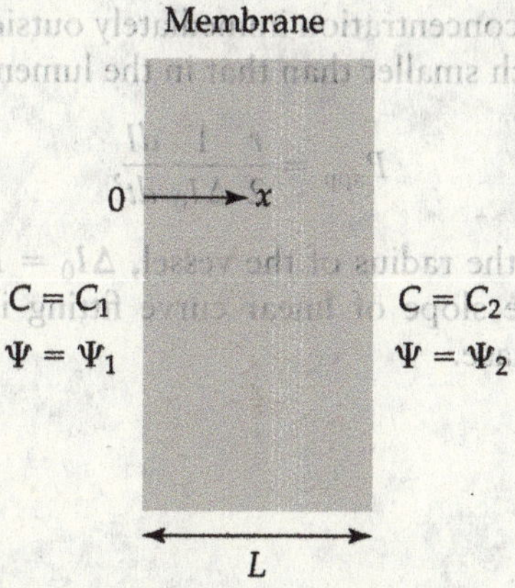

FIGURE 9.13 One-dimensional diffusion of an electrolyte molecule across a membrane.

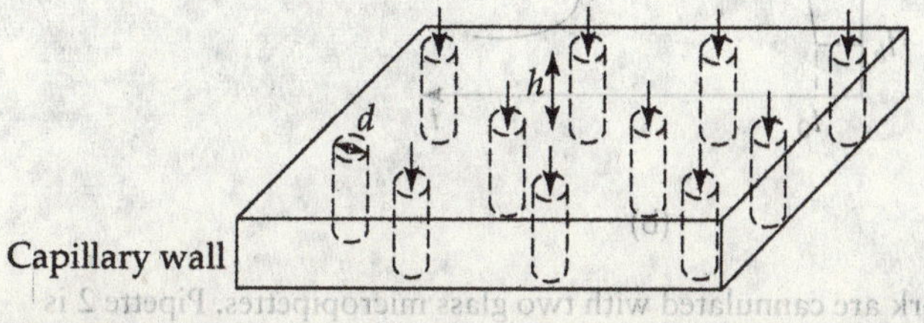

FIGURE 9.14 Transport of molecules across the capillary wall through the pores. The arrows indicate the direction of transport.

9.5 Fluid can cross the capillary wall through porous clefts between endothelial cells. The clefts can be considered as channels that contain random cylindrical fibers (see Figure 9.15). Assume that the area fraction of the porous clefts on the endothelial surface is A_p, the channel length is h, the average radius of the fibers is r_f, and the fiber length per unit volume of the porous region is ℓ.

(a) Find the porosity of the clefts.

(b) Find the hydraulic conductivity of the vessel wall, using the Kozeny–Carman theory.

(c) If the transmural pressure difference is Δp and the wall thickness is Δx, what is the fluid flux across the capillary wall?

9.6 Atherosclerosis is a cardiovascular disease that involves the accumulation of cholesterol in the arterial wall in regions with damaged endothelial cells. Assume that the damaged region is a circular hole with radius b, as shown in Figure 9.16; the thickness of the endothelial layer can be neglected. The endothelial layer is impermeable to cholesterol and plasma, except in the damaged region. The subendothelial space can be treated as a semi-infinite porous material with hydraulic

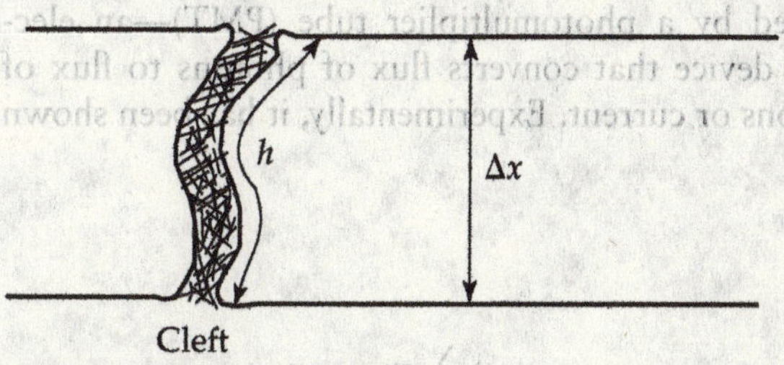

FIGURE 9.15 Solute transport across the capillary wall through the endothelial cleft.

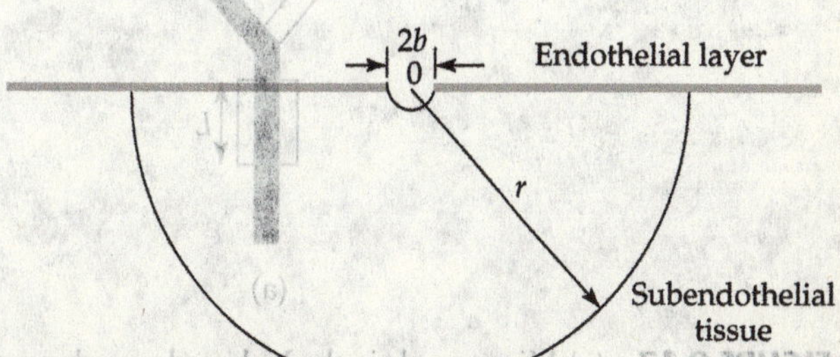

FIGURE 9.16 Transport of cholesterol in the subendothelial space.

conductivity K. The transport of fluid and cholesterol is in the radial direction and is symmetric about the center of the hole. The diffusion coefficient of cholesterol in the subendothelial tissue is D. The concentrations of cholesterol and the hydrostatic pressure at $r = b$ are approximately equal to those in the plasma, which are C_p and p_0, respectively. The concentration and the pressure at $r \rightarrow \infty$ are assumed to be zero. Determine the distributions of pressure, fluid velocity, and cholesterol concentration in the subendothelial space in the steady state.

9.7 *Experimental quantification of the apparent microvascular permeability* [16]. The single-vessel cannulation technique has been used to determine the apparent microvascular permeability P_{app}. The experimental details are as follows:

A Y-shaped vascular network in biological tissue (e.g., mesentery) is cannulated with two glass micropipettes (see Figure 9.17a). One pipette is used to perfuse vessels with fluorescent solute (e.g., rhodamine-labeled albumin), the other is used to wash out the dye in the vessel with physiological saline. The vascular network is prepared under a fluorescence microscope, and the intensity of the fluorescence in a rectangular window is monitored as a function of time. The intensity is detected by a photomultiplier tube (PMT)—an electronic device that converts flux of photons to flux of electrons or current. Experimentally, it has been shown that the fluorescence intensity detected by the PMT is a linear function of the number of solutes in the window over a large range of solute concentrations.

A typical profile of the fluorescence intensity in the window is shown in Figure 9.17b. Initially, the vessel is perfused with saline. The fluorescence signal in the window is recorded as a background intensity I_b. At $t = t_0$, the perfusate is switched from the saline in pipette 1 to the fluorescent solute in pipette 2. The fluorescence intensity I rises rapidly for a few seconds and then increases slowly with time. The rapid-rise phase is due to the filling of the vessel with the solute; the slow-increase phase is caused by the extravasation of the solute into the extravascular space. The fluorescence intensity at the time when I changes from the rapid- to the slow-increase phase is recorded as I_0. When the perfusate is switched back to saline, the fluorescence intensity drops to the background level.

Assume that, in short-term perfusion experiments, the concentration immediately outside the vessel wall is much smaller than that in the lumen. Show that

$$P_{app} = \frac{r}{2}\frac{1}{\Delta I_0}\frac{dI}{dt},$$

where r is the radius of the vessel, $\Delta I_0 = I_0 - I_b$, and dI/dt is the slope of linear curve fitting in the slow-increase phase.

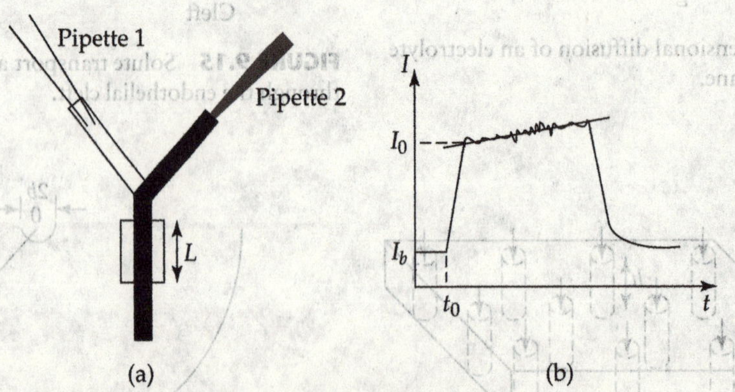

FIGURE 9.17 (a) Microvessels in the Y-shaped vascular network are cannulated with two glass micropipettes. Pipette 2 is used to perfuse vessels with fluorescent solutes and Pipette 1 is used to wash out the dye in the vessel with physiological saline. The fluorescence intensity in the rectangular window with length L is monitored as a function of time. (b) A typical profile of the fluorescence intensity in the window is shown as a function of time.

9.7 REFERENCES

1. Michel, C.C., "The transport of albumin: a critique of the vesicular system in transendothelial transport." *Am. Rev. Respir. Dis.*, 1992. **146**: pp. S32–S36.
2. Folkman, J., "Angiogenesis in cancer, vascular, rheumatoid and other disease." *Nat. Med.*, 1995. **1**: pp. 27–31.
3. Curry, F.E., ed., "Mechanisms and thermodynamics of transcapillary exchange." *Handbook of Physiology*, Sect. 2, "The Cardiovascular System," vol. IV, *The Microcirculation*, E.M. Renkin and C.C. Michel, editors. 1984, Bethesda, MD: American Physiological Society, pp. 309–374.
4. Chang, R., *Physical Chemistry for the Chemical and Biological Sciences*. 2000, Sausalito, CA: University Science Books.
5. Yan, Z.-Y., Weinbaum, S., and Pfeffer, R., "On the fine structure of osmosis including three-dimensional pore entrance and exit behavior." *J. Fluid Mech.*, 1986. **162**: pp. 415–438.
6. Tinoco, I.J., Sauer, K., and Wang, J.C., *Physical Chemistry: Principles and Applications in Biological Sciences*. 3d ed. 2002, Upper Saddle River, NJ: Prentice Hall.
7. Yousel, M.A., Datta, R., and Rodgers, V.G.J., "Understanding nonidealities of the osmotic pressure of concentrated bovine serum albumin." *J. Colloid Interface Sci.*, 1998. **207**: pp. 273–282.
8. Freeman, B. "Osmosis," *in Encyclopedia of Applied Physics*, G.L. Trigg, editor. 1995. New York: VCH Publishers, pp. 59–71.
9. McDonald, J.N. and Levick, J.R., "Effect of extravascular plasma protein on pressure–flow relations across synovium in anaesthetized rabbits." *J. Physiol.* (Lond.), 1993. **465**: pp. 539–559.
10. Hu, X., Adamson, R.H., Liu, B., Curry, F.E., and Weinbaum, S., "Starling forces that oppose filtration after tissue oncotic pressure is increased." *Am. J. Physiol.*, 2000. **279**: pp. H1724–H1736.
11. Michel, C.C., "Starling: The formulation on his hypothesis of microvascular fluid exchange and its significance after 100 years." *Exp. Physiol.*, 1997. **82**: pp. 1–30.
12. Kedem, O., and Katchalsky, A., "Thermodynamic analysis of the permeability of biological membranes to nonelectrolytes." *Biochim. Biophs. Acta.*, 1958. **27**: pp. 229–246.
13. Staverman, A.J., "The theory of measurement of osmotic pressure." *Rec. Trav. Chim.*, 1951. **70**: pp. 344–352.
14. Patlak, C.S., Goldstein, D.A., and Hoffman, J.F., "The flow of solute and solvent across a two-membrane system." *J. Theor. Biol.*, 1963. **5**: pp. 426–442.
15. Ganatos, P., Weinbaum, S., Fischbarg, J., and Liebovitch, L., "A hydrodynamic theory for determining the membrane coefficients for the passage of spherical molecules through an intercellular cleft." *Advances in Bioengineering* (ASME), 1980. pp. 193–196.
16. Curry, F.E., "Regulation of water and solute exchange in microvessel endothelium: studies in single perfused capillaries." *Microcirculation*, 1994. **1**: pp. 11–26.
17. Adamson, R.H., Huxley, V.H., and Curry, F.E., "Single capillary permeability to proteins having similar size but different charge." *Am. J. Physiol.*, 1988. **254**: pp. H304–H312.
18. Dellian, M., Yuan, F., Trubetskoy, V., Torchilin, V.P., and Jain, R.K., "Effect of molecular charge on microvascular permeability in a human tumor xenograft." *Brit. J. Cancer*, 2000. **82**: pp. 1513–1518.
19. Michel, C.C., and Phillips, M.E., "Steady-state fluid filtration at different capillary pressures in perfused frog mesenteric capillaries." *J. Physiol.*, 1987. **388**: pp. 421–435.
20. Hu, X., and Weinbaum, S., "A new view of Starling's hypothesis at the microstructural level." *Microvasc Res*, 1999. **58**: pp. 281–304.

The Effect of Mass Transport upon Biochemical Interactions

PART III

Biochemical interactions between molecules can be classified as either *chemical reactions* or *chemical interactions*. Chemical reactions involve the breaking and forming of covalent bonds. The product of the reaction has properties different from those of either reactant. Such reactions are important in all metabolic and synthetic reactions within the body. These reactions are often catalyzed by enzymes, a type of protein designed to break or form specific bonds between molecules. The second class of interactions involves noncovalent chemical interactions, such as electrostatic, electrodynamic, or hydrophobic interactions. In these interactions, electrons are not shared and covalent bonds are neither broken nor formed. Antigen–antibody binding and receptor–ligand interactions, as well as DNA and RNA base-pair binding, are noncovalent interactions. Noncovalent interactions are weaker than covalent bonds, but the presence of multiple noncovalent interactions can result in large net bond strengths. The region where the two molecules bind is known as the *binding region*. Because binding molecules must have complementary structures, only very specific combinations of molecules can bind.

Many of these diverse chemical processes share similar kinetic features, and a general approach can be taken to understand the dynamics of the chemical interactions and mechanisms involved. This chapter presents a survey of important biochemical interactions; it is not exhaustive, but does provide a sufficient background for more advanced study. Chapter 10 gives an introduction to the kinetics of biochemical interactions and the effect of mass transport on those interactions. In Chapter 11, we examine important, special cases of these interactions on the cell membrane and within cells. We develop expressions commonly used to analyze data, and we discuss how reaction sequences can generate properties needed for efficient cellular responses to an altered environment. In Chapter 12, we examine the effect of force on biochemical kinetics. These concepts are applied to cell adhesion.

Mass Transport and Biochemical Interactions

CHAPTER

10

10.1 Introduction

In this chapter, we examine the various kinetic expressions describing biochemical reactions, and we determine the effect of mass transport on the rates of reactions. Basic concepts in chemical kinetics and some simple reaction schemes are presented in Section 10.2. Section 10.3 follows with a discussion of a simple sequence of reactions and the use of an important simplification to examine complex reactions: the *quasi–steady-state assumption.* In Section 10.4, the concepts of Section 10.3 are extended to the case of enzyme kinetics. Enzyme reactions are regulated *in vivo* in order to control the output of the product(s) more precisely. Section 10.5 considers chemical means of regulation in enzyme reactions, and Section 10.6 examines the effect of diffusion and convection on the reaction rate of biomolecules.

10.2 Chemical Kinetics and Reaction Mechanisms

10.2.1 Reaction Rates

Reaction kinetics are described in terms of the *reaction rate*, which represents the amount of reactant consumed or product produced per unit time. Rates are often expressed per unit volume or per unit surface area. For N_i moles of reactant or product, the reaction rates are written as

$$R_i = \frac{1}{V}\frac{dN_i}{dt} = \frac{dC_i}{dt} \qquad (10.2.1a)$$

and

$$R_i^s = \frac{1}{S}\frac{dN_i}{dt}, \tag{10.2.1b}$$

where R_i is the reaction rate per unit volume (moles per unit volume per unit time), R_i^s is the reaction rate per unit surface area (moles per unit area per unit time), V is the volume in which the reaction occurs, S is the surface area when reactions occur at the interface between two phases, t is time, and C_i is the concentration (moles per unit volume) of the reactant or product. Reaction rates are expressed per unit volume when the reaction occurs through a particular region, such as a tissue. The reaction rate is positive for products and negative for reactants and may vary with location. In contrast, the reaction rate is specified per unit surface area when the reaction occurs at the interface between two different regions. For example, reactions occurring on the surface of endothelial cells (cf. such reactions as adenosine triphosphate [ATP] hydrolysis and the reactions of the coagulation cascade) are specified in terms of the interfacial surface area.

Homogeneous reactions are chemical reactions that occur within a single phase. Examples include polymerization reactions and acid–base reactions in aqueous solutions. Homogeneous reaction rates are often expressed per unit volume. Chemical reactions that occur at the interface between two phases are known as *heterogeneous reactions*. Examples include the hydrolysis of water and enzyme reactions on the surface of blood vessels. Heterogeneous reaction rates are expressed either per unit surface area or per unit weight of catalyst or tissue. The distinction between homogeneous and heterogeneous reactions is important when mass transfer limits the rate of reaction. Since homogeneous reactions occur throughout a single phase, the reaction rates appear in the conservation relations. Heterogeneous reactions occur at an interface and appear in the boundary conditions.

The *reaction stoichiometry* (i.e., the number of moles of individual molecules) relates the reaction rates of reactants and products. For example, consider the following expression for a reaction between two A molecules and three B molecules:

$$2A + 3B \rightleftharpoons C. \tag{10.2.2}$$

Equation (10.2.2) does not describe the mechanism of the reaction, for it is extremely unlikely that two molecules of A and three of B collide in just the right fashion to produce one molecule of C. Rather, Equation (10.2.2) expresses a balance of moles among reactants and products. Thus, for every 2 moles of A that react with 3 moles of B, 1 mole of C is produced. That is, the rate of appearance of C is half the rate of disappearance of A and one-third the rate of disappearance of B. Alternatively,

$$R_C = -\frac{1}{2}R_A = -\frac{1}{3}R_B. \tag{10.2.3}$$

The minus sign is needed because the concentrations of A and B decrease, and the concentration of C increases, as the reaction proceeds.

This result can be generalized by using υ_i to represent the stoichiometric coefficient of component i in the reaction. This coefficient is negative for reactants and positive for products. In general,

$$\frac{1}{\upsilon_i}R_i = \frac{1}{\upsilon_i}R_j, \tag{10.2.4}$$

where i and j represent specific reactants or products.

Example 10.1 Urea (NH_2CONH_2) decomposes in acidic solutions to form ammonium ion (NH_4^+) and carbonate (CO_3^{2-}) according to the following stoichiometry:

$$NH_2CONH_2 + 2H_2O \rightarrow 2NH_4^+ + CO_3^{2-}$$

Find the relation between the urea reaction rate and the rates of formation of ammonium and carbonate.

Solution Based upon Equation (10.2.4), the rates are related as follows:

$$\frac{1}{\nu_{NH_2CONH_2}} R_{NH_2CONH_2} = \frac{1}{\nu_{NH_4^+}} R_{NH_4^+} = \frac{1}{\nu_{CO_3^{2-}}} R_{CO_3^{2-}}. \qquad (10.2.5)$$

The stoichiometric coefficients are $\nu_{NH_2CONH_2} = -1$, $\nu_{NH_4^+} = 2$, and $\nu_{CO_3^{2-}} = 1$.

The reaction rate can be estimated by approximating the derivative as a finite change in the concentration over a period Δt

$$R_i = \frac{dC_i}{dt} \approx \frac{C_i(t + \Delta t) - C_i(t)}{\Delta t} = f(C_i(t)). \qquad (10.2.6)$$

The reaction rate is a function of the concentrations of reactants, products, or catalysts. Such *rate expressions* must be determined experimentally. One goal of biochemical kinetics is to determine the rate expression from data and then develop a mechanism consisting of a series of *elementary reactions* that are consistent with the observed rate expression. Elementary reactions are simple (often bimolecular) reactions between molecules.

Several approaches are used to develop rate expressions: direct differentiation of concentration-versus-time data to obtain rates (Figure 10.1), direct determination of the rate in flowing systems, and the use of the integrated form of postulated rate

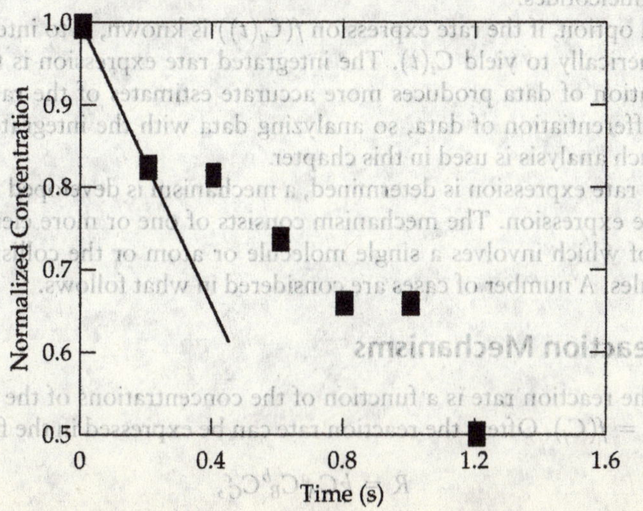

FIGURE 10.1 The initial reaction rate equals the slope of the solid line.

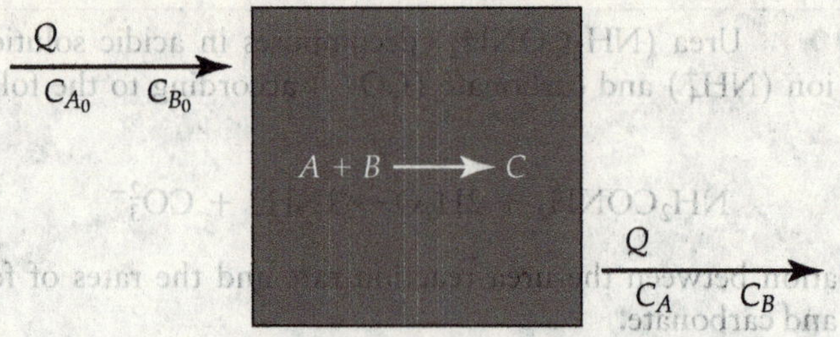

FIGURE 10.2 Schematic of a flowing and reacting system.

expressions. Direct determination of the rate by differentiation of the data is difficult, especially if the data have appreciable errors. Further, the concentration and reaction rate change with time. One way to get around this problem is to measure the initial rate of the reaction. When the reaction first begins, very little reactant is consumed and little product formed. Thus, the concentrations can be assumed to equal their initial values.

Alternatively, one can measure the reaction rate directly with a continuously flowing and reacting system. Reactants flow into the reactor, and the reaction proceeds in the reactor, which may have a catalyst or an elevated temperature to permit the reaction to occur. The outlet stream represents the concentrations of reactants or products in the reactor, since no further reaction occurs in the outlet (Figure 10.2). When the reaction is at steady state, a mass balance on the system yields

$$-R_i = \frac{Q}{V}(C_{i_0} - C_i), \tag{10.2.7}$$

where Q is the volumetric flow rate, V is the reactor volume, C_{i_0} is the inlet concentration of component i, and C_i is the concentration of i in the reaction and outlet. The ratio V/Q is known as the *residence time*—the characteristic time that a reactant resides in the reactor. The reaction rate multiplied by the reactor volume is equal to the flow rate times the difference in concentration. The advantage of this approach is that the reaction rate is determined directly without taking derivatives. While desirable, the approach requires sufficient quantities of reactant to permit flow. Because of the limited supply of biological molecules, this approach is not often used with proteins, peptides, or nucleotides.

A third option, if the rate expression $f(C_i(t))$ is known, is to integrate analytically or numerically to yield $C_i(t)$. The integrated rate expression is then fit to the data. Integration of data produces more accurate estimates of the rate coefficients than does differentiation of data, so analyzing data with the integrated rate law is preferred. Such analysis is used in this chapter.

After a rate expression is determined, a mechanism is developed to explain the observed rate expression. The mechanism consists of one or more elementary reactions, each of which involves a single molecule or atom or the collision of two or three molecules. A number of cases are considered in what follows.

10.2.2 Reaction Mechanisms

In general, the reaction rate is a function of the concentrations of the reactants and products, $R_i = f(C_j)$. Often, the reaction rate can be expressed in the form

$$R = kC_A^{\,a}C_B^{\,b}C_C^{\,c}, \tag{10.2.8}$$

where C_A, C_B, and C_C represent concentrations of reactants, products, or catalysts, and the indices a, b, and c represent the order of the reaction for molecules A, B, and C, respectively. Thus, the reaction is an ath-order reaction in A, a bth-order reaction in B, and a cth-order reaction in C. The overall order of the reaction is $a + b + c$, and k is known as the *rate coefficient*. This coefficient is independent of the concentration but is a function of temperature. The units of k depend upon the order of the reaction and the volume or area basis for the reaction.

We now examine several elementary reactions that occur often in the analysis of biochemical reaction kinetics. To distinguish these elementary reactions from reaction stoichiometries, the rate coefficient is used only in the elementary reactions. Note that the individual elementary reactions must sum to the overall reaction stoichiometry and must yield the observed rate expression. Another useful quantity that describes the kinetics of a reaction is the half-time $t_{1/2}$, the time over which the reactants decline to 50% of their initial value or the products reach 50% of the final value.

Three factors influence whether the collision of two reacting molecules produces a new molecule. First, the molecules must diffuse toward each other before the reaction can occur. If the molecules react quickly once they are in contact, then the reaction could be limited by how fast the molecules diffuse toward each other. Such a reaction is said to be *diffusion limited* (see Section 6.9). Second, the molecules must be oriented appropriately to react. Third, a reaction intermediate may form in order for the reaction to proceed to completion. The rate of formation of such an intermediate may be limited by an energy barrier, often known as an *activation energy*. This barrier slows the overall rate of reaction.

Each elementary step in a reaction sequence describes a rate expression. The *rate law* represents the sum of the rates of the individual elementary steps. A complicated rate expression indicates that the reaction consists of a number of elementary steps. Often, it is necessary to make assumptions about the intermediate steps in a reaction mechanism in order to obtain the resulting rate expression. It is easy to disprove a mechanism by showing that it fails to conform to the rate expression. However, the converse is not necessarily true: That a mechanism conforms to the rate law merely indicates that the mechanism is plausible; other information besides kinetics is needed to validate the mechanism.

While there are a number of possible elementary reactions, a few cases appear quite frequently and form the basis of most complicated reaction mechanisms. These include first-order and second-order irreversible and reversible reactions. We consider each in turn.

10.2.3 First-Order Reactions

In a first-order reaction, a reacting molecule decays to one or more products:

$$A \xrightarrow{k_1} B + C. \tag{10.2.9}$$

Examples of first-order reactions include radioisotope decay and fluorescence quenching. The rate of disappearance of A is

$$-\frac{dC_A}{dt} = k_1 C_A. \tag{10.2.10}$$

The subscript 1 for the rate coefficient denotes a first-order reaction, and the units of k_1 are inverse time (e.g., s^{-1}). If the initial concentration of C_A is C_{A_0}, then integrating Equation (10.2.10) yields

$$C_A = C_{A_0}\exp(-k_1 t). \tag{10.2.11}$$

The half-time for the reaction is determined by setting C_A/C_{A_0} to 0.5 and solving for time:

$$t_{1/2} = \frac{-\ln 0.5}{k_1}. \tag{10.2.12}$$

Example 10.2 Table 10.1 displays data for the disappearance of A as a function of time. Show that the reaction is first order, and determine the magnitude of k_1 and the half-time.

Solution Taking the logarithm of Equation (10.2.11) yields

$$\ln(C_A) = \ln(C_{A_0}) - k_1 t. \tag{10.2.13}$$

Equation (10.2.13) expresses a linear relationship between $\ln(C_A/C_{A_0})$ and time. The data in Table 10.1 follow this relationship (Figure 10.3). Linear regression of the data yields $k_1 = 0.05\ s^{-1}$. The half-time is 13.9 s.

TABLE 10.1

Data for Example 10.2

Time (s)	Concentration (μM)
0	15
5	10.68
10	9.10
20	5.52
40	2.03
60	0.05

10.2.4 Second-Order Irreversible Reactions

Second-order reactions are common and involve the collision of two molecules to produce a third. The rate expressions of such reactions depend on whether the two colliding molecules are the same type or different types.

When the molecules are the same type, the reaction is written as

$$A + A \xrightarrow{k_2} A_2 \tag{10.2.14}$$

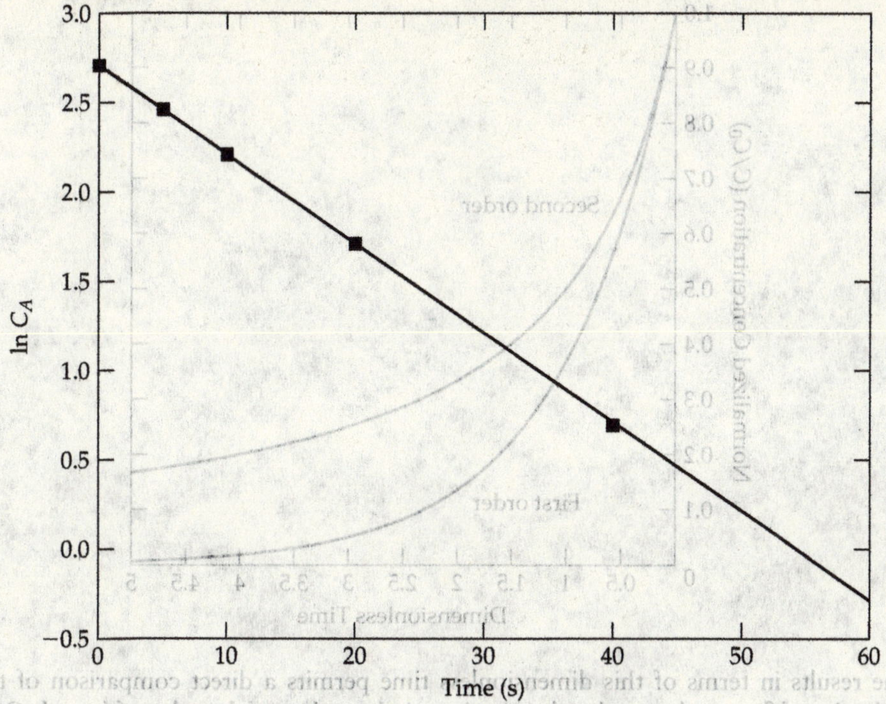

FIGURE 10.3 Logarithm of C_A versus time.

The rate expression is[1]

$$-\frac{dC_A}{dt} = k_2 C_A^2. \tag{10.2.15}$$

The rate coefficient k_2 has units of $M^{-1}\ s^{-1}$. Integrating Equation (10.2.15) subject to the initial condition that $C_A = C_{A_0}$ at $t = 0$ yields

$$\frac{1}{C_A} - \frac{1}{C_{A_0}} = k_2 t. \tag{10.2.16a}$$

To compare the kinetics of first- and second-order reactions, Equation (10.2.16a) can be rewritten as

$$\frac{C_A}{C_{A_0}} = \frac{1}{1 + k_2 C_{A_0} t}. \tag{10.2.16b}$$

Letting t' equal $k_1 t$ for a first-order reaction and $k_2 C_{A_0} t$ for a second-order reaction, we can plot the two reactions on the same graph, as shown in Figure 10.4. (Presenting

[1]Based on the stoichiometry, $-1/2 dC_A/dt = dC_{A_2}/dt$. The rate expression, Equation (10.2.15), could be written as $-1/2 dC_A/dt = k_2' C_A^2$. The two rate coefficients are related as $k_2 = 2k_2'$.

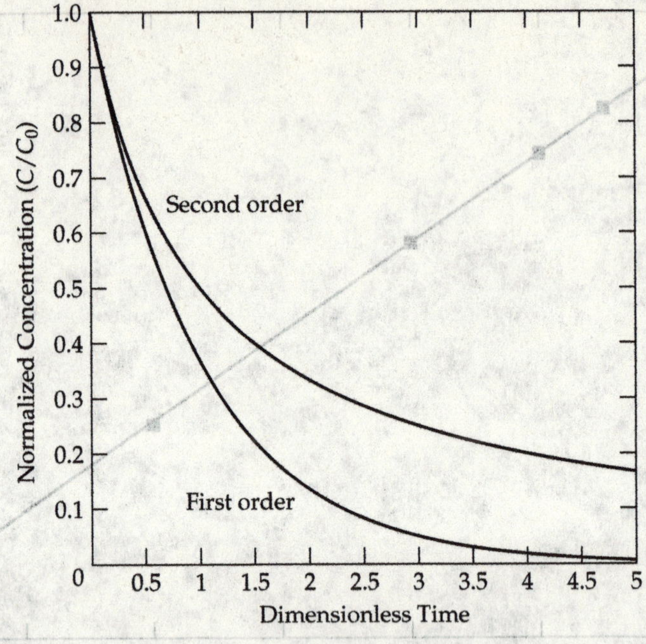

FIGURE 10.4 Comparison of first- and second-order reactions versus dimensionless time, $k_1 t$, for a first-order reaction and $k_2 C_{A_0} t$ for a second-order reaction.

the results in terms of this dimensionless time permits a direct comparison of the behavior of first- and second-order reactions, independent of the value of k_1 or $k_2 C_{A_0}$.) Clearly, in terms of dimensionless time, a second-order reaction decays more slowly than a first-order reaction. A semilogarithmic plot shows that only a first-order reaction is linear (Figure 10.5).

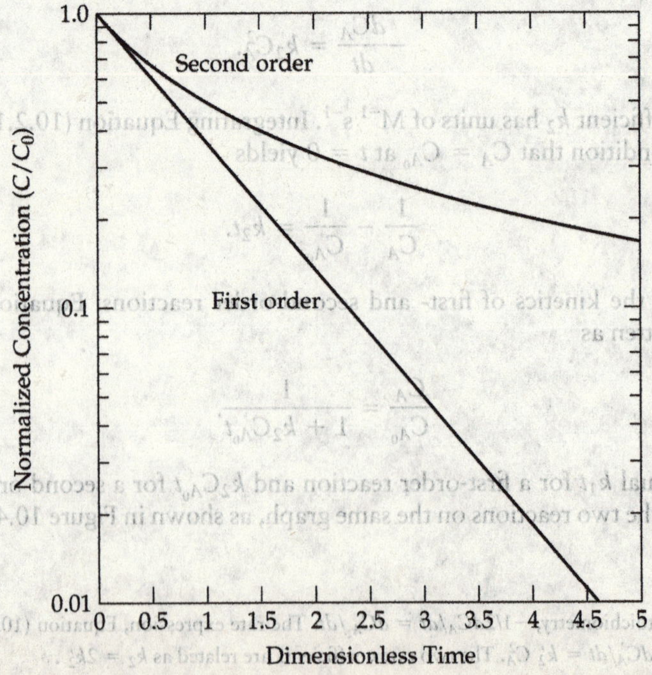

FIGURE 10.5 Semilogarithmic plot of first- and second-order reactions versus dimensionless time.

Example 10.3 **Reassociation Kinetics of DNA**

Consider a DNA solution consisting of double-stranded nucleotide sequences, each of the same length (typically a few hundred base pairs). Such small nucleotide chains are produced from long DNA chains by gentle disruption via sound waves generated by a vibrating rod, a process known as sonication. Next, the DNA sample is heated to separate the two strands. Upon cooling, the two strands reassociate. If the original DNA chain consists of a repeating sequence, then the kinetics of reassociation for the two complementary strands, S_1 and S_1', are modeled as a second-order irreversible reaction:

$$S_1 + S_1' \xrightarrow{k} S_1 - S_1' \tag{10.2.17}$$

Develop expressions for the fraction of DNA that is single stranded and the reaction half-time.

Solution This is a second-order reaction with rate coefficient k. In general, the concentrations of S_1 and S_1' are the same ($C_{S_1} = C_{S_1'} = C_S$), with an initial value of C_{S_0}. The concentration of S is usually given as the number of moles of nucleotide per liter of solution. The solution of the second-order reaction in which the reactants have the same concentration (Equation (10.2.16b)) can be written in terms of the fraction of DNA that is single stranded, $f = C_S/C_{S_0}$:

$$f = \frac{C_S}{C_{S_0}} = \frac{1}{1 + kC_{S_0}t}. \tag{10.2.18}$$

The half-time is

$$t_{1/2} = \frac{1}{C_{S_0}k}. \tag{10.2.19}$$

Unlike first-order reactions, the half-time is not fixed and depends upon the initial concentration of the reactant.

DNA consists of repeating and nonrepeating sequences. The kinetic reassociation of nonrepeating DNA sequences follows the expected trends for second-order kinetics given by Equation (10.2.18) (Figure 10.6). The curves in Figure 10.6 were obtained by isolating nonrepeating sequences and then breaking the DNA into 400 base pair chains. The chains were heated and allowed to reassociate. The differences in the curves reflect differences in the nonrepeating sequence length, or *sequence complexity*, of the DNA as well as differences in the value of the second-order rate coefficient. The sequence complexity, X, is the total number of base pairs of the longest nonrepeating sequences present in the DNA genome [1]. The complexity equals the total number of nucleotides in a given organism's genome when there are no repeating sequences. The presence of repeating sequences reduces the sequence complexity below the genome size.

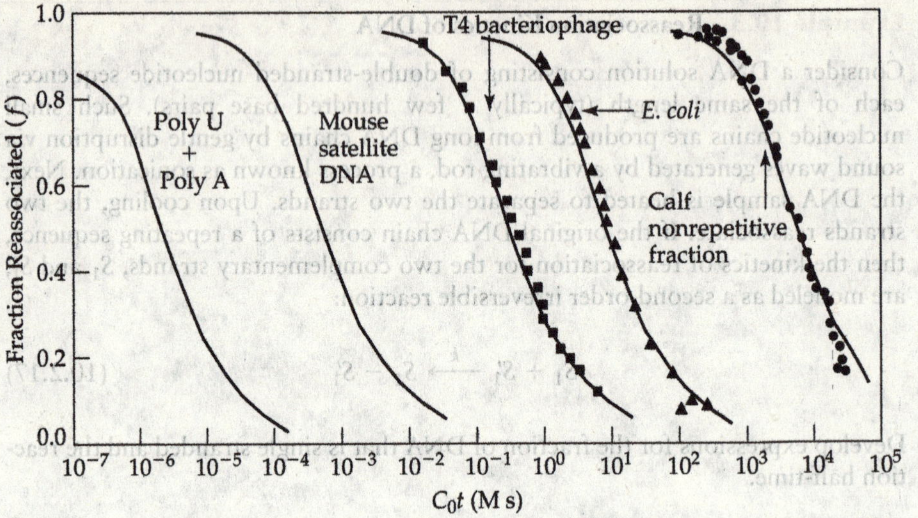

FIGURE 10.6 Reassociation kinetics for various types of nonrepeating DNA sequences. (Adapted from Ref. [2], with permission.)

The concentration, C_{S_0}, represents the molar concentration of DNA based on the average weight of a nucleotide. This is a straightforward quantity to determine but does not represent the actual molar concentration of DNA strands. For the case of two different nonrepeating sequences of different complexity at the same total DNA concentration, C_{S_0}, the solution of lower complexity has more DNA molecules per unit volume because its genome is shorter. Thus, the concentration of the fragments is greater for the sequence of lower complexity and the strands reassociate faster. For nonrepeating sequences, the concentration of the nucleotide chains C_0 is related to the total DNA concentration, C_{S_0}, as follows:

$$C_0 = \frac{C_{S_0}}{X} \tag{10.2.20}$$

For example, the differences in the reassociation curves of T4 bacteriophage and the calf nonrepetitive fraction shown in Figure 10.6 indicate that the complexities of the two DNA sequences differ by a factor of 10,000.

Reassociation kinetics of eukaryotic DNA are more complicated due to the different frequency of sequence repetition. Each population of sequences can be described by Equation (10.2.18), and the overall dissociation curve represents the sum of the reassociation curve for each subpopulation multiplied by the fraction of the sequences represented by that population (Figure 10.7). The measured complexity from such experiments, often referred to as the *kinetic complexity*, represents a weighted average of the complexity of the individual population complexities.

Alternatively, the two colliding molecules in a second-order reaction may be two *different* molecular species. In that case, the collision can be expressed as

$$A + B \xrightarrow{k_2} AB. \tag{10.2.21}$$

Examples of second-order reactions include antigen–antibody reactions. The reaction rate is

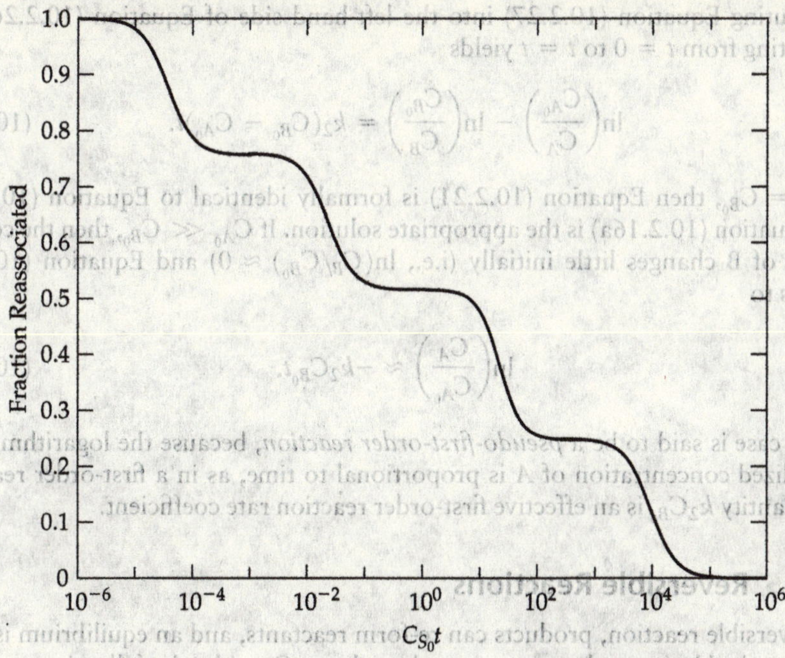

FIGURE 10.7 Reassociation kinetics for double-stranded DNA with various concentrations of repeating sequences.

$$-\frac{dC_A}{dt} = k_2 C_A C_B. \tag{10.2.22}$$

The rate of disappearance of B is the same as the rate of disappearance of A. Thus,

$$\frac{dC_A}{dt} = \frac{dC_B}{dt}. \tag{10.2.23}$$

For the initial condition that $C_A = C_{A_0}$, $C_B = C_{B_0}$ and $C_{AB} = 0$ at $t = 0$, integrating Equation (10.2.23) yields

$$C_{A_0} - C_A = C_{B_0} - C_B. \tag{10.2.24}$$

Substituting Equation (10.2.24) into Equation (10.2.22) gives

$$-\frac{dC_A}{dt} = k_2 C_A (C_{B_0} - C_{A_0} + C_A). \tag{10.2.25}$$

This first-order ordinary differential equation (Appendix, A.1.D) can be rewritten as

$$\frac{dC_A}{C_A(C_{B_0} - C_{A_0} + C_A)} = -k_2 dt. \tag{10.2.26}$$

The left-hand side of Equation (10.2.26) is expanded as follows:

$$\frac{dC_A}{C_A(C_{B_0} - C_{A_0} + C_A)} = \frac{1}{C_{B_0} - C_{A_0}}\left(\frac{dC_A}{C_A} - \frac{dC_A}{C_{B_0} - C_{A_0} + C_A}\right). \tag{10.2.27}$$

Substituting Equation (10.2.27) into the left-hand side of Equation (10.2.26) and integrating from $t = 0$ to $t = t$ yields

$$\ln\left(\frac{C_{A_0}}{C_A}\right) - \ln\left(\frac{C_{B_0}}{C_B}\right) = k_2(C_{B_0} - C_{A_0})t. \tag{10.2.28}$$

If $C_{A_0} = C_{B_0}$, then Equation (10.2.21) is formally identical to Equation (10.2.14), and Equation (10.2.16a) is the appropriate solution. If $C_{A_0} \ll C_{B_0}$, then the concentration of B changes little initially (i.e., $\ln(C_B/C_{B_0}) \approx 0$) and Equation (10.2.28) reduces to

$$\ln\left(\frac{C_A}{C_{A_0}}\right) \approx -k_2 C_{B_0} t. \tag{10.2.29}$$

Such a case is said to be a *pseudo-first-order reaction*, because the logarithm of the normalized concentration of A is proportional to time, as in a first-order reaction. The quantity $k_2 C_{B_0}$ is an effective first-order reaction rate coefficient.

10.2.5 Reversible Reactions

In a reversible reaction, products can re-form reactants, and an equilibrium is eventually reached between the reactants and products. Consider the following reversible reaction in which a complex is formed in a bimolecular reaction and decays via a first-order process:

$$A + B \underset{k_{-2}}{\overset{k_2}{\rightleftharpoons}} C. \tag{10.2.30}$$

The reaction rate is

$$\frac{dC_C}{dt} = k_2 C_A C_B - k_{-2} C_C. \tag{10.2.31}$$

The forward reaction is second order and the reverse reaction is first order. The *association rate coefficient* is k_2 and the *dissociation rate coefficient* k_{-2}. A common set of initial conditions are $C_A = C_{A_0}$, $C_B = C_{B_0}$, and $C_C = 0$ at $t = 0$. Defining the equilibrium constant as $K = k_2/k_{-2}$, we can rewrite Equation (10.2.31) as

$$\frac{dC_C}{dt} = k_2\left[(C_{A_0} - C_C)(C_{B_0} - C_C) - \frac{1}{K}C_C\right]. \tag{10.2.32}$$

Equation (10.2.32) was obtained using the conservation relations $C_{A_0} - C_A = C_C$ and $C_{B_0} - C_B = C_C$. The solution of Equation (10.2.32) is

$$\ln\left[\frac{2C_C/(b + \sqrt{b^2 - 4a}) - 1}{2C_C/(b - \sqrt{b^2 - 4a}) - 1}\right] = k_2\left(\sqrt{b^2 - 4a}\right)t, \tag{10.2.33}$$

where

$$a = C_{A_0}C_{B_0} \quad \text{and} \quad b = C_{A_0} + C_{B_0} + 1/K. \tag{10.2.34a,b}$$

Although there are a large number of other reactions that can be analyzed (for examples, see Refs [3] and [4]), most reactions can be considered a collection of first- and second-order reversible or irreversible reactions. The elucidation of reaction mechanisms requires stating the various elementary steps and then making assumptions about the kinetics involved.

10.3 Sequential Reactions and the Quasi–Steady-State Assumption

In this section, we explore the effect of the magnitude of the rate coefficients on the overall kinetics of a simple sequence of chemical reactions. Depending upon the magnitude of the rate coefficients, it is possible to simplify a reaction sequence. One of the most significant assumptions is the *quasi–steady-state assumption*, often applied to reaction intermediates when the time for the reactant to undergo reaction is much faster than the time to form the reactant.

A *sequential reaction* is a set of elementary reactions that proceed consecutively. Many biochemical reactions, such as those occurring during the oxidative metabolism of glucose, are sequential reactions. Some sequential reactions contain branch points at which two different reactions occur. Such reactions are catalyzed by enzymes and have complex kinetics. Before we examine these complex reactions, it is instructive to examine some simpler ones first. The simplest reaction sequence is the following, in which a reactant undergoes an irreversible first-order reaction to form a product that, in turn, irreversibly reacts in a first-order reaction to form a second product:

$$A \xrightarrow{k_1} B \xrightarrow{k_2} C. \tag{10.3.1}$$

The rates of change of A, B, and C are, respectively,

$$\frac{dC_A}{dt} = -k_1 C_A, \tag{10.3.2}$$

$$\frac{dC_B}{dt} = k_1 C_A - k_2 C_B, \tag{10.3.3}$$

and

$$\frac{dC_C}{dt} = k_2 C_B. \tag{10.3.4}$$

At $t = 0$, $C_A = C_{A_0}$ and $C_B = C_C = 0$. The solutions of these equations are, respectively,

$$C_A = C_{A_0}\exp(-k_1 t), \tag{10.3.5}$$

$$C_B = \frac{k_1 C_{A_0}}{k_2 - k_1}[\exp(-k_1 t) - \exp(-k_2 t)], \tag{10.3.6}$$

and

$$C_C = \frac{k_1 k_2 C_{A_0}}{k_2 - k_1}\left[\frac{1-\exp(-k_1 t)}{k_1} - \frac{1-\exp(-k_2 t)}{k_2}\right]. \tag{10.3.7}$$

The solution for $k_1 = 1\ \text{s}^{-1}$, $k_2 = 2\ \text{s}^{-1}$, and $C_{A_0} = 1\ \mu\text{M}$ is shown in Figure 10.8. The concentration of A drops exponentially with time, as expected. A good estimate of the time required for A to disappear is five time constants (i.e., $k_1 t = 5$, so that $\exp(-5) = 0.0067$). The concentration of B rises and reaches a maximum when the derivative of C_B with respect to time equals zero. This corresponds to a time of

$$t_{C_{B_{max}}} = \frac{-\ln(k_1/k_2)}{k_2 - k_1}. \tag{10.3.8}$$

Let us now consider some limiting behavior for the cases in which the second reaction is very fast or very slow compared with the first reaction. Suppose that the second reaction occurs much faster than the first reaction ($k_2 \gg k_1$). Then, because the first step is slower than the second step, the first reaction is said to be *rate limiting*. For very large differences between the values of k_1 and k_2, the concentrations of B and C, respectively, become

$$C_B \approx \frac{k_1 C_{A0}}{k_2}\exp(-k_1 t) = \frac{k_1 C_A}{k_2} \ll C_A \tag{10.3.9}$$

and

$$C_C \approx C_{A_0}\left[1 - \exp(-k_1 t)\right]. \tag{10.3.10}$$

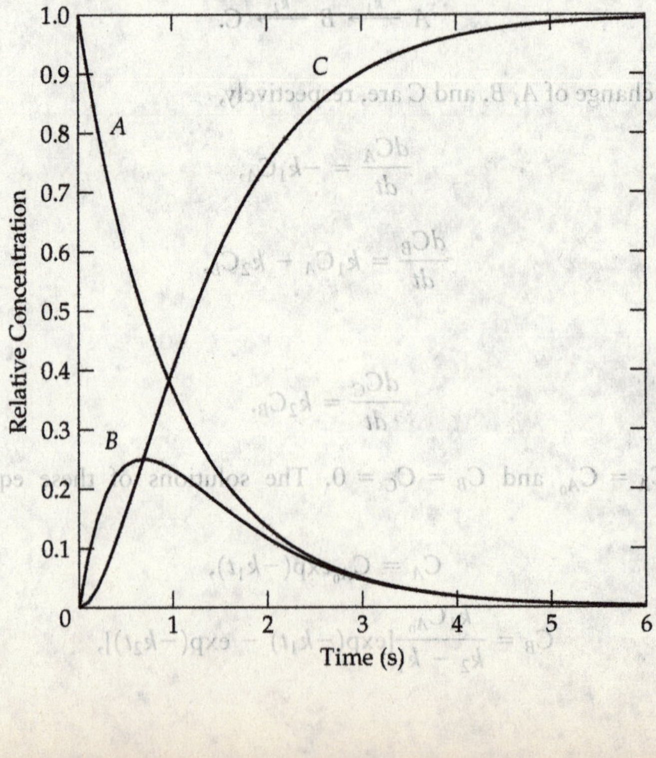

FIGURE 10.8 The change in the concentrations of A, B, and C for the sequential reaction shown in Equation (10.3.1) for $k_1 = 1\ \text{s}^{-1}$, $k_2 = 2\ \text{s}^{-1}$, and $C_{A_0} = 1\ \mu\text{M}$.

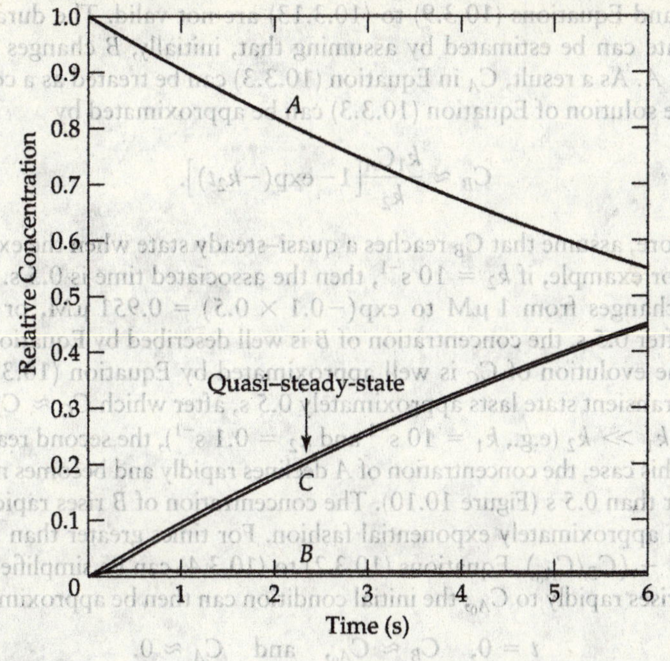

FIGURE 10.9 Sequential reactions for $k_1 = 0.1\ s^{-1}$, $k_2 = 10\ s^{-1}$, and $C_{A_0} = 1\ \mu M$. Curve A is Equation (10.3.5), Curve B is Equation (10.3.6), Curve C is Equation (10.3.7), and Quasi–steady-state refers to Equation (10.3.10).

Solutions of Equations (10.3.9) and (10.3.10) for $k_1 = 0.1\ s^{-1}$ and $k_2 = 10\ s^{-1}$ are shown in Figure 10.9. Also shown is the complete solution for C_C, Equation (10.3.7). For this particular set of values for k_1 and k_2, Equation (10.3.10) is a very good approximation of the exact solution. Because C_B is very small due to the rapid conversion to C, the reaction behaves as if A were converted directly to C.

Equations (10.3.9) and (10.3.10) can also be derived from Equations (10.3.3) and (10.3.4) if we assume that the concentration of B is in a quasi–steady state, that is, if B reacts to form C as soon as B is formed. Thus,

$$\frac{dC_B}{dt} \approx 0 \tag{10.3.11}$$

and

$$C_B \approx \frac{k_1}{k_2} C_A. \tag{10.3.12}$$

Substituting Equation (10.3.12) into Equation (10.3.4) yields

$$\frac{dC_C}{dt} \approx k_1 C_A. \tag{10.3.13}$$

The solution of Equation (10.3.13) is given by Equation (10.3.10). The assumption that the concentration of B is in a *quasi–steady-state* (Equation (10.3.11)) implies that the disappearance of A is approximately equal to the appearance of $C(C_C \approx C_{A_0} - C_A)$. There is a brief transient period during which C_B rises from zero to C_A and Equations (10.3.9) to (10.3.13) are not valid. The duration of this transient state can be estimated by assuming that, initially, B changes much more rapidly than A. As a result, C_A in Equation (10.3.3) can be treated as a constant. For this case, the solution of Equation (10.3.3) can be approximated by

$$C_B \approx \frac{k_1 C_A}{k_2}\left[1-\exp(-k_2 t)\right]. \tag{10.3.14}$$

Now, as before, assume that C_B reaches a quasi–steady state when the exponential is $\exp(-5)$. For example, if $k_2 = 10\ \text{s}^{-1}$, then the associated time is 0.5 s. During this period C_A changes from 1 μM to $\exp(-0.1 \times 0.5) = 0.951$ μM, or a little less than 5%. After 0.5 s, the concentration of B is well described by Equation (10.3.14). Likewise, the evolution of C_C is well approximated by Equation (10.3.10) (Figure 10.9). The transient state lasts approximately 0.5 s, after which $C_C \approx C_{A_0} - C_A$.

When $k_1 \gg k_2$ (e.g., $k_1 = 10\ \text{s}^{-1}$ and $k_2 = 0.1\ \text{s}^{-1}$), the second reaction is rate limiting. In this case, the concentration of A declines rapidly and becomes negligible for times greater than 0.5 s (Figure 10.10). The concentration of B rises rapidly and then decays in an approximately exponential fashion. For times greater than about 0.2 s, $C_C/C_{A_0} \approx 1 - (C_B/C_{A_0})$. Equations (10.3.2) to (10.3.4) can be simplified by assuming that C_B rises rapidly to C_{A_0}; the initial condition can then be approximated as

$$t = 0, \quad C_B \approx C_{A_0}, \quad \text{and} \quad C_A \approx 0. \tag{10.3.15}$$

For the simulated case, C_B/C_A reaches a maximum of ≈ 0.96 at $t = 0.465$ s, so the error is about 5%. In general, when $k_1 \gg k_2$, the problem simplifies to

$$\frac{dC_B}{dt} = -k_2 C_B, \tag{10.3.16}$$

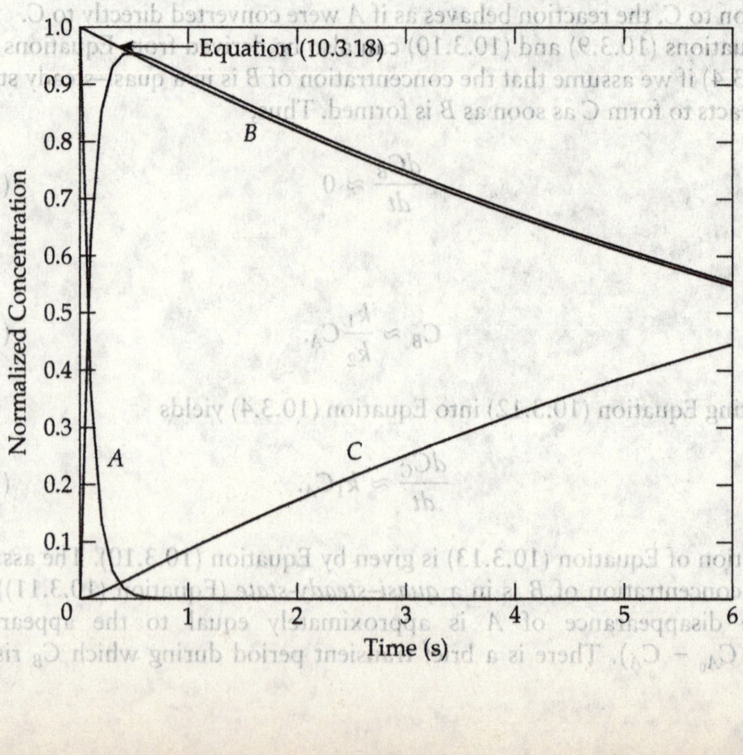

FIGURE 10.10 Sequential reactions for $k_1 = 10\ \text{s}^{-1}$, $k_2 = 0.1\ \text{s}^{-1}$, and $C_{A_0} = 1$.

$$\frac{dC_C}{dt} = k_2 C_B, \tag{10.3.17}$$

and $C_B \approx C_{A_0}$ at $t = 0$. The solution to this problem is

$$C_B \approx C_{A_0}\exp(-k_2 t), \tag{10.3.18}$$

$$C_C \approx C_{A_0}[1 - \exp(-k_2 t)]. \tag{10.3.19}$$

The approximate result for C_B is also shown in Figure 10.10. Certainly, at early times, this expression does not work well. But after 0.4 s, the two results (Equations (10.3.6) and (10.3.18)) are similar. This period corresponds to the time during which the concentration of A changes. Note that the solution of this case when the second reaction is rate limiting does not involve applying the quasi–steady-state assumption. The reason is that reactant A is not produced by any reaction.

The foregoing two cases show that the identification of an appropriate *rate-limiting step* can simplify the analysis of reaction sequences. For the sequential reactions studied here, the reaction sequence can be reduced from two steps to one. The actual kinetics of the formation of the product, however, depend upon which step is rate limiting. The slower step can be identified by examining the magnitude of the rate coefficient: The larger the rate coefficient, the faster is the step and the shorter is the time constant. In the case we examined in this section, the time constant $t_c = 1/k_i$, where i is 1 or 2. When the rate-limiting step is identified, the kinetic expression can be simplified by assuming a rapid rate of reaction. Since the reaction rate changes with time, this assumption is valid as long as the duration of the faster step is much less than the duration of the slower step and the concentrations of the other reactants do not change appreciably.

10.4 Enzyme Kinetics

Enzymes are proteins, RNA, or DNA that catalyze biological reactions. Enzymes have three important properties. First, they increase the rate of biochemical reactions by 10^4 to 10^{12} times the rate in the absence of the catalyst ([5], p. 420). Many important biochemical reactions do not occur at an appreciable rate without enzymes. Second, enzymes are specific to the reactant molecules, also known as the *substrates*, which interact at a specific site on the enzyme, often with high affinity. The high-affinity binding enables enzymes to function effectively in a solution containing a large number of biological molecules at low concentrations. Third, the activity of enzymes can be regulated in a number of ways, offering control over the rate and amount of product formed. Examples of regulation include cofactor, which binds to the enzyme. Alternatively, the activity of the enzyme can be controlled by a reaction product that inhibits the reaction.

Enzymes catalyze biochemical reactions following the binding of one or more substrates to the *active site* of the enzyme, a region that interacts specifically with the substrate (Figure 10.11). The active site provides appropriate orientations of the reacting molecules and alters the local electrodynamic environment to make the occurrence of the reaction more favorable. Specific amino acid side chains serve as

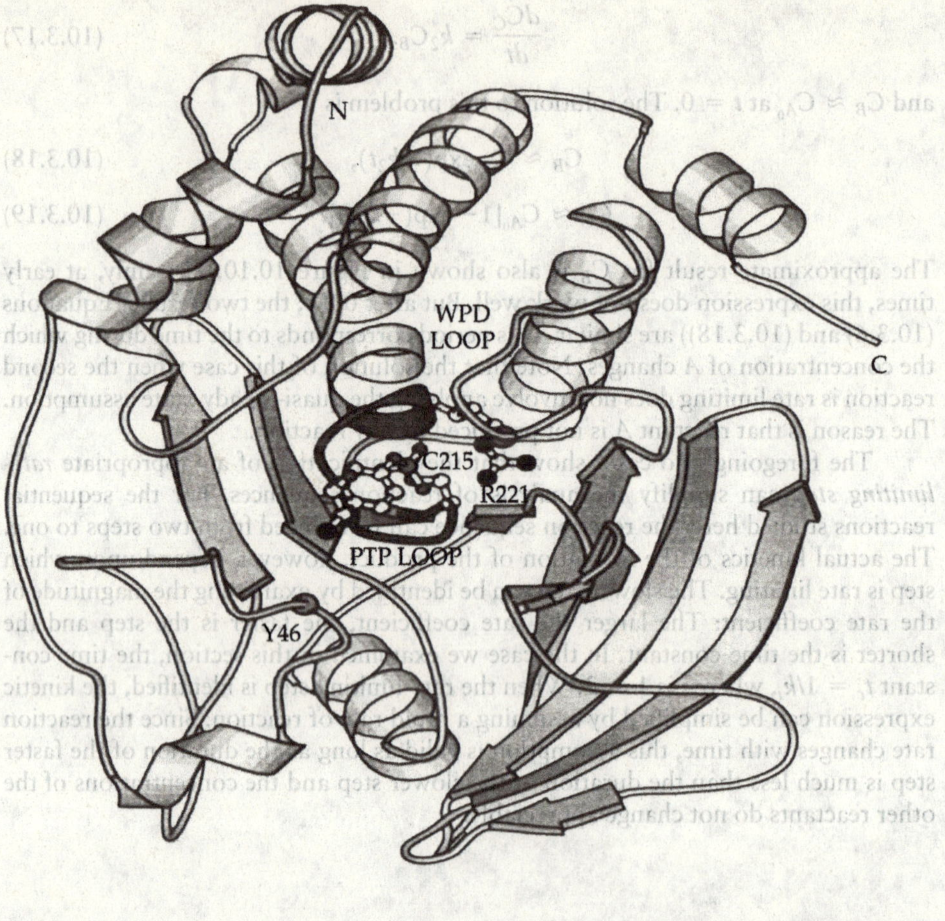

FIGURE 10.11 Structure of protein tyrosine phosphatase. C refers to the carboxyl terminus, N refers to the amino terminus. The amino acids tyrosine 46 and cysteine 215 stabilize the substrate in the active site, and arginine 221 is the catalytic group in the active site. (Adapted with permission from Ref. [7].)

catalytic agents facilitating bond breakage or the formation of the substrate. Techniques of molecular biology (e.g., *site-directed mutagenesis*) are used to determine the effect of single amino acid substitution on enzyme activity and to identify specific structure–function relations [6].

Although thermodynamically favorable ($\Delta G < 0$), many reactions are limited by the energy barrier needed to form an activated state ([7], pp. 357–360). As a result, these reactions do not occur in the absence of added heat or a catalyst. The enzyme provides an alternative reaction pathway that produces an activated state of the reactants with a lower energy barrier (Figure 10.12). As a result, the rate of reaction can increase significantly, but the overall change in energy between the reactants and products is not altered.

The overall reaction of an enzyme with a single reactant, or substrate, is written as

$$E + S \longleftrightarrow P + E, \tag{10.4.1}$$

where E is the enzyme, S is the substrate, and P is the product. The enzyme accelerates the rate of reaction but is not consumed in the reaction. The enzyme affects the *rate of reaction* but does not alter the *equilibrium*.

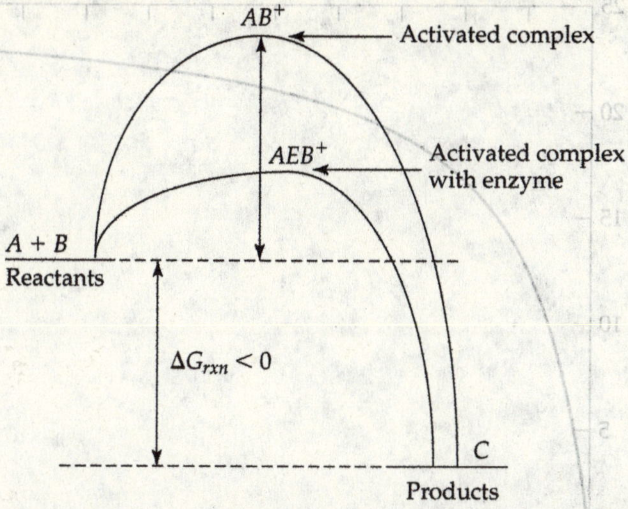

FIGURE 10.12 Energy pathway of the effect of an enzyme on a reaction [5,8]. In the absence of a catalyst, there is a large energy barrier to reaction in the activated state AB^+. The enzyme creates a different activated state AEB^+ with a much lower energy barrier, facilitating the conversion of A and B to form C.

For many enzymes involving a single substrate, experiments have shown that the rate of consumption of the substrate, R_S, can be written as

$$R_S = \frac{R_{\max} C_S}{K_M + C_S}, \tag{10.4.2}$$

where R_S is the magnitude of the rate of disappearance of the substrate (or reactant), $R_{\max}$ is the maximum reaction rate, C_S is the concentration of the substrate, and K_M is the *Michaelis constant*—the concentration of the substrate at which the reaction is half of its maximum rate. The rate expression has limiting behavior between a first-order reaction, when C_S is much less than K_M, and a reaction that does not depend upon the reactant concentration (zero-order kinetics) when C_S is much greater than K_M (Figure 10.13). Equation (10.4.2) is known as the *Michaelis–Menten* equation.

10.4.1 Derivation of Michaelis–Menten Kinetics

The following multistep mechanism explains the complex rate expression for enzymatic reactions (Equation (10.4.2)). The first step in the enzyme–substrate reaction is the formation of a complex between the enzyme and substrate in a reversible reaction:

$$E + S \underset{k_{-1}}{\overset{k_1}{\rightleftharpoons}} \text{ES}. \tag{10.4.3}$$

The complex is an unstable, short-lived intermediate in which the substrate is non-covalently attached to the enzyme prior to the chemical reorientation of the substrate to form a product. The complex then dissociates, yielding the enzyme and the reaction product P:

$$\text{ES} \xrightarrow{k_2} E + P. \tag{10.4.4}$$

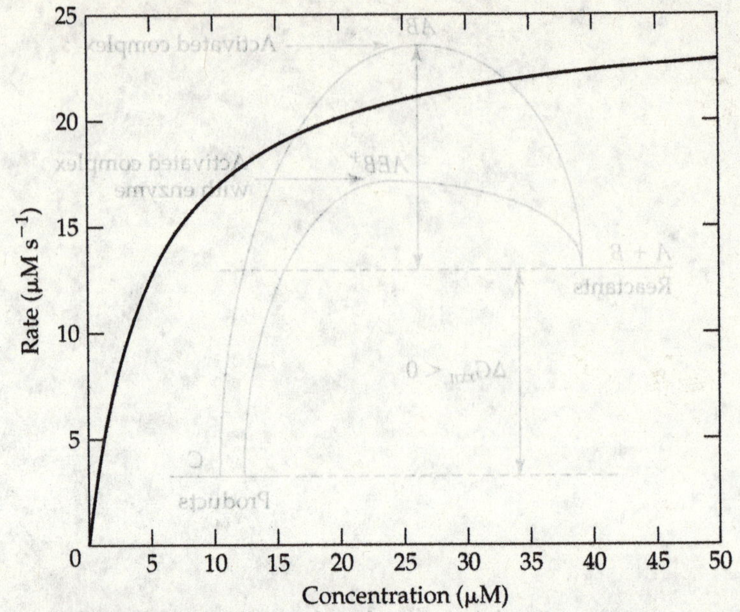

FIGURE 10.13 Rate of substrate consumption for an enzyme reaction as a function of substrate concentration for $K_M = 5$ μM and $R_{max} = 25$ μM s^{-1}.

The rate of disappearance of the substrate is

$$-R_S = \frac{dC_S}{dt} = -k_1 C_S C_E + k_{-1} C_{ES}. \tag{10.4.5}$$

The first term on the right-hand side of Equation (10.4.5) represents the rate of formation of the enzyme–substrate complex; the second term represents the rate of decay of the complex to free enzyme and unreacted substrate.

The rate of formation of the enzyme–substrate complex is

$$\frac{dC_{ES}}{dt} = k_1 C_E C_S - (k_{-1} + k_2) C_{ES}. \tag{10.4.6}$$

Because the enzyme is not consumed in the reaction, the total enzyme concentration is constant and equals C_{E_0}. The total enzyme concentration is equal to the sum of the concentrations of the free enzyme and the enzyme–substrate complex:

$$C_{E_0} = C_E + C_{ES}. \tag{10.4.7}$$

Finally, the rate of formation of the product is

$$\frac{dC_P}{dt} = k_2 C_{ES}. \tag{10.4.8}$$

The key assumption for simplifying Equations (10.4.5) to (10.4.8) is that the rate of formation of the product is much faster than the rate of formation of the complex. Consequently, the complex breaks down rapidly, and the rate of its accumulation is approximately zero:

$$\frac{dC_{ES}}{dt} \approx 0. \tag{10.4.9}$$

Equation (10.4.9) is the expression of the quasi–steady-state assumption that is used to simplify the analysis of the sequential reactions in Section 10.3. With this assumption, the concentration of the complex is

$$C_{ES} = \frac{k_1 C_E C_S}{k_{-1} + k_2}. \tag{10.4.10}$$

Substituting Equation (10.4.10) into Equation (10.4.7) and rearranging the resulting equation allows us to express the free enzyme and enzyme–substrate complex concentrations in terms of the total enzyme concentration:

$$C_E = \frac{(k_{-1} + k_2)C_{E_0}}{(k_{-1} + k_2) + k_1 C_S}, \tag{10.4.11}$$

$$C_{ES} = \frac{k_1 C_{E_0} C_S}{(k_{-1} + k_2) + k_1 C_S}. \tag{10.4.12}$$

Upon substituting Equations (10.4.11) and (10.4.12) into Equation (10.4.5) and rearranging terms, we obtain the following expression for the rate of disappearance of substrate:

$$R_S = -\frac{dC_S}{dt} = \frac{k_1 k_2 C_{E_0} C_S}{(k_{-1} + k_2) + k_1 C_S}. \tag{10.4.13}$$

Dividing the numerator and denominator of Equation (10.4.13) by k_1 yields

$$R_S = -\frac{dC_S}{dt} = \frac{k_2 C_{E_0} C_S}{\left[(k_{-1} + k_2)/k_1\right] + C_S}. \tag{10.4.14}$$

Comparing Equations (10.4.2) and (10.4.14) indicates that

$$R_{\max} = k_2 C_{E_0}, \tag{10.4.15a}$$

and

$$K_M = \frac{k_{-1} + k_2}{k_1}. \tag{10.4.15b}$$

Figure 10.13 is a plot of Equation (10.4.14) for $K_M = 5\ \mu\text{M}$ and $R_{\max} = 25\ \mu\text{M s}^{-1}$. As previously noted, two limiting extremes of kinetic behavior can be distinguished. At low substrate concentrations (i.e., $C_S \ll K_M$), the rate of consumption of reactant is proportional to the reactant concentration and the reaction is first order; that is,

$$R_S = \frac{R_{\max} C_S}{K_M}. \tag{10.4.16a}$$

By contrast, at high substrate concentrations (i.e., $C_S \gg K_M$), the rate of consumption of reactant is independent of the reactant concentration and the reaction is zero order; that is,

$$R_S = R_{\max}. \tag{10.4.16b}$$

Although the kinetic model represented by Equations (10.4.3) and (10.4.4) and the application of the quasi–steady-state assumption predict the observed rate law for enzymes, the mechanism is not unique. In 1903, Henri, and in 1913, Michaelis and Menten assumed that the binding and dissociation of enzyme to substrate was in equilibrium with a dissociation constant [10][2]

$$K_{D_1} = \frac{C_E C_S}{C_{ES}} = \frac{k_{-1}}{k_1}. \tag{10.4.17}$$

Consequently,

$$\frac{dC_S}{dt} \approx 0 \tag{10.4.18a}$$

and

$$\frac{dC_{ES}}{dt} = \frac{dC_P}{dt} = k_2 C_{ES}. \tag{10.4.18b}$$

Using Equations (10.4.7) and (10.4.18b) to solve for the concentration of the enzyme–substrate complex, we have

$$C_E = \frac{C_{E_0} K_{D_1}}{K_{D_1} + C_S}. \tag{10.4.19a}$$

The concentration of the enzyme–substrate complex is

$$C_{ES} = \frac{C_{E_0} C_S}{K_{D_1} + C_S}. \tag{10.4.19b}$$

The overall rate of reaction is

$$\frac{dC_P}{dt} = \frac{k_2 C_{E_0} C_S}{K_{D_1} + C_S}. \tag{10.4.20}$$

This result has the same form as Equation (10.4.2). The assumption represented by Equation (10.4.18a) that the substrate concentration is constant must be incorrect, since the substrate concentration must decline with time. Further, the rate of product formation closely parallels the negative of the rate of substrate loss. Clearly, having a mechanism that yields the correct rate law does not guarantee that the mechanism is correct.

[2]While the quasi–steady-state assumption governing the formation of the reaction intermediate [Equation (10.4.9)] was made by Briggs and Haldane in 1925, the key insight of Michaelis and Menten that enshrined their names on the reaction mechanism was the proportionality between the rate of formation of product and that of the enzyme–substrate complex [Equation (10.4.8)] [10].

10.4.2 Application of the Quasi–Steady-State Assumption to Enzyme Kinetics

Let us now apply the concepts developed in Section 10.3 to enzyme kinetics. We want to determine the criteria for which the quasi–steady-state assumption is valid. To do this, we need to estimate the time constants for the fast and slow steps of the reaction. The following analysis was developed by Segel [11]. The fast step corresponds to the initial rapid phase of enzyme–substrate complex formation before the quasi–steady state is reached. During this period, we assume that the substrate concentration does not change appreciably (i.e., $C_S = C_{S_0}$). Further, the concentration of free enzyme is $C_E = C_{E_0} - C_{ES}$. Equation (10.4.6) for the enzyme–substrate complex now becomes

$$\frac{dC_{ES}}{dt} = k_1 C_{E_0} C_{S_0} - (k_{-1} + k_2 + k_1 C_{S_0}) C_{ES}. \tag{10.4.21}$$

For the initial condition that $C_{ES} = 0$ at $t = 0$, the solution of Equation (10.4.21) is

$$C_{ES} = \frac{C_{E_0} C_{S_0}}{K_M + C_{S_0}} \left\{ 1 - \exp\left[-k_1 (K_M + C_{S_0}) t \right] \right\}. \tag{10.4.22}$$

Equation (10.4.22) represents an approximate solution for the concentration of the enzyme–substrate complex during the rapid step before the quasi–steady-state is reached. The time constant for this fast phase of reaction is simply the reciprocal of the effective rate coefficient for the fast step, $k_1(K_M + C_{S_0})$. The magnitude of this rate coefficient increases as the substrate concentration increases.

The effective rate coefficient for the *slow* step can be approximated by examining the resulting rate expression when the quasi–steady-state assumption holds, namely,

$$\frac{dC_S}{dt} = -\frac{k_2 C_{E_0} C_S}{K_M + C_S}. \tag{10.4.23}$$

An upper limit for the time constant for the slow step is obtained by assuming again that the substrate concentration equals its initial value. From Equation (10.4.23), the rate coefficient for the slow step is then $k_2 C_{E_0}/(K_M + C_{S_0})$.

Thus, the quasi–steady-state approximation is valid as long as the effective rate coefficient for the fast step is much greater than the effective rate coefficient for the slow step—that is, as long as

$$\frac{k_2 C_{E_0}}{K_M + C_{S_0}} \ll k_1 (K_M + C_{S_0}) \tag{10.4.24a}$$

or

$$\frac{C_{E_0}}{K_M + C_{S_0}} \ll \left(1 + \frac{k_{-1}}{k_2}\right)\left(1 + \frac{C_{S_0}}{K_M}\right). \tag{10.4.24b}$$

For Equation (10.4.24b) to be valid, the change in substrate concentration must be small over the brief time when the enzyme–substrate complex is changing rapidly and approaching a steady state. As noted by Segel [11], the change in substrate concentration is small during the initial part of the reaction if the maximum rate of change of C_S (given by Equation (10.4.23)) divided by C_{S_0} is much smaller than the maximum relative rate of change of enzyme–substrate complex divided by C_{ES}:

$$\frac{1}{C_{S_0}}\left|\frac{dC_S}{dt}\right|_{\max} \ll \frac{1}{(C_{ES})_{\max}}\left|\frac{dC_{ES}}{dt}\right|_{\max}. \tag{10.4.25}$$

Initially, no enzyme–substrate complex is formed, and the enzyme and substrate concentrations equal their initial values. Thus, the maximum value for the time rate of change of the substrate is

$$\left|\frac{dC_S}{dt}\right|_{\max} \approx k_1 C_{E_0} C_{S_0}. \tag{10.4.26}$$

The derivative on the right-hand side of Equation (10.4.25) is obtained by differentiating Equation (10.4.22). Multiplying the result by the reciprocal of the maximum value of C_{ES} equals $k_1(K_M + C_{S_0})$. The criterion that C_S does not change appreciably can be stated as

$$k_1 C_{E_0} \ll k_1(K_M + C_{S_0}) \tag{10.4.27a}$$

or

$$\frac{C_{E_0}}{K_M + C_{S_0}} \ll 1. \tag{10.4.27b}$$

If Equations (10.4.24b) and (10.4.27b) are both satisfied, we can be assured that the quasi–steady-state assumption is valid. Clearly, Equation (10.4.27b) is the more restrictive requirement.

10.4.3 Determination of K_M and $R_{\max}$

The characterization of enzymatic reactions involves the determination of K_M and $R_{\max}$. Data for enzyme-catalyzed reactions are usually transformed to a form that is amenable to linear regression so that these parameters can be estimated. Three transformations are commonly used. For example, the reciprocal of Equation (10.4.2) yields

$$\frac{1}{R_S} = \frac{K_M}{R_{\max} C_S} + \frac{1}{R_{\max}}. \tag{10.4.28}$$

The concentration of S is set to its initial value. As a result, the analysis is strictly valid during the initial phase of the enzyme reaction, after the substrate is added to the enzyme solution. A plot of $1/R_S$ versus $1/C_S$ yields a straight line with slope $K_M/R_{\max}$ and intercept $1/R_{\max}$ (Figure 10.14). This type of transformation is known as the *Lineweaver–Burk equation.* Two other transformations of the Michaelis–Menten equation are

$$\frac{C_S}{R_S} = \frac{K_M}{R_{\max}} + \frac{C_S}{R_{\max}} \quad \text{(Scatchard equation)} \tag{10.4.29a}$$

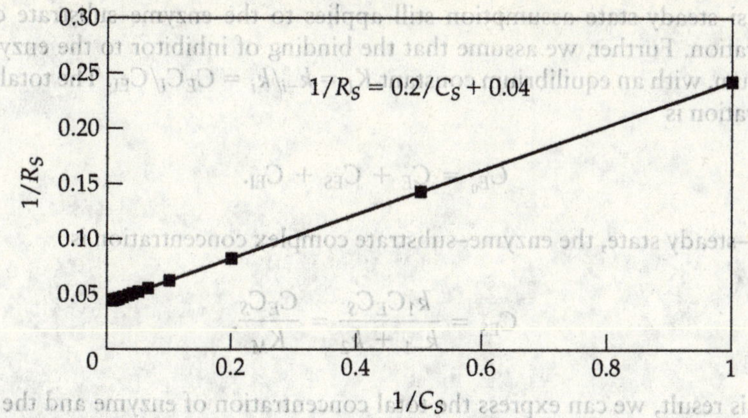

FIGURE 10.14 Lineweaver–Burk plot of the data in Figure 10.13.

and

$$R_S = R_{max} - K_M \frac{R_S}{C_S} \quad \text{(Eadie–Hofstee equation)}. \tag{10.4.29b}$$

Although these equations do present a linear relation and simplify the estimation of K_M and R_{max}, their practical application can lead to erroneous estimates of the constants.

10.5 Regulation of Enzyme Activity

Often, it is necessary to control the amount of product formed by a reaction. For example, the product may stimulate a specific biochemical pathway that is needed only at certain times (e.g., during growth) or the product of a biological reaction may be biologically active over a narrow range of concentrations. The rate of enzyme reactions can be regulated by the substrate, products, or other molecules that interact with the enzyme. Excess accumulation of products may require too much energy or interfere with other pathways. Substrate regulation can lead to either activation or inhibition of the reaction rate. Such inhibition can be classified, in the extreme, as either competitive or noncompetitive.

10.5.1 Competitive Inhibition

In competitive inhibition, the inhibitor molecule binds to the active site, preventing the substrate from binding. The following reactions occur:

$$E + S \underset{k_{-1}}{\overset{k_1}{\rightleftharpoons}} ES \xrightarrow{k_2} E + P, \tag{10.5.1}$$

$$E + I \underset{k_{-i}}{\overset{k_i}{\rightleftharpoons}} EI. \tag{10.5.2}$$

The quasi–steady-state assumption still applies to the enzyme–substrate complex concentration. Further, we assume that the binding of inhibitor to the enzyme is in equilibrium, with an equilibrium constant $K_I = k_{-i}/k_i = C_E C_I / C_{EI}$. The total enzyme concentration is

$$C_{E_0} = C_E + C_{ES} + C_{EI}. \tag{10.5.3}$$

At quasi–steady state, the enzyme–substrate complex concentration is

$$C_{ES} = \frac{k_1 C_E C_S}{k_{-1} + k_2} = \frac{C_E C_S}{K_M}. \tag{10.5.4}$$

Using this result, we can express the total concentration of enzyme and the concentration of the enzyme–substrate complex in terms of C_{E_0}:

$$C_E = \frac{C_{E_0}}{1 + (C_S/K_M) + (C_I/K_I)}, \tag{10.5.5a}$$

$$C_{ES} = \frac{C_{E_0} C_S}{K_M + C_S + (K_M C_I/K_I)}. \tag{10.5.5b}$$

The rate of formation of products is

$$\frac{dC_P}{dt} = k_2 C_{ES} = \frac{k_2 C_{E_0} C_S}{K_M(1+C_I/K_I) + C_S}. \tag{10.5.6}$$

An apparent Michaelis constant can be defined as

$$K'_M = K_M\left(1 + \frac{C_I}{K_I}\right). \tag{10.5.7}$$

Examination of Equations (10.5.6) and (10.5.7) indicates that a competitive inhibitor affects K'_M, but not the maximum rate. This is shown graphically for the rate as a function of concentration (Figure 10.15a) and in the Lineweaver–Burk plots (Figure 10.15b). Increasing the concentration of I leads to slower rates of formation of the product, although the rate eventually reaches the maximum value R_{max}.

10.5.2 Uncompetitive and Noncompetitive Inhibition

In addition to directly binding to the active sites, inhibitors can bind to other sites on the enzyme, reducing the reaction rate. In uncompetitive inhibition, the inhibitor does not directly block enzyme binding but interacts with the enzyme-bound substrate at another site, altering the rate of reaction [5]. Alternatively, the inhibitor may bind to the substrate bound to the enzyme. The effect of an uncompetitive inhibitor on the reaction rate can be determined by considering that the inhibitor only binds to the enzyme–substrate complex,

$$ES + I \underset{k_{-i}}{\overset{k_i}{\rightleftarrows}} ESI. \tag{10.5.8}$$

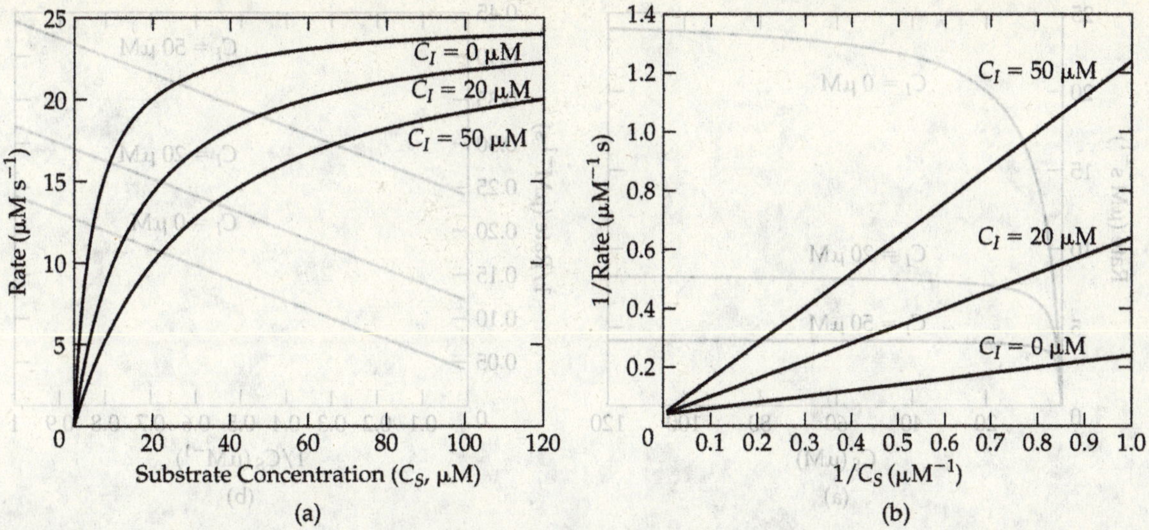

FIGURE 10.15 (a) Effect of competitive inhibition on the reaction rate for $R_{max} = 25\ \mu M\ s^{-1}$, $K_M = 5\ \mu M$, and $K_I = 10\ \mu M$. (b) Lineweaver–Burk plot for the data presented in panel (a).

Although the enzyme–substrate complex can still form, the complex ESI inhibits ES from forming products. The total concentration of enzyme then becomes

$$C_{E_0} = C_E + C_{ES} + C_{ESI}. \tag{10.5.9}$$

Assuming that the reaction represented by Equation (10.5.8) is in equilibrium (i.e., $C_{ESI} = C_{ES}C_I/K_I$), the free enzyme concentration and enzyme–substrate complex concentration are found in a manner analogous to those used for competitive inhibition. We get

$$C_E = \frac{C_{E_0}}{1+(C_S/K_M)(1+C_I/K_I)}, \tag{10.5.10a}$$

$$C_{ES} = \frac{C_{E_0}C_S}{K_M + C_S(1 + C_I/K_I)}. \tag{10.5.10b}$$

The rate of formation of product is

$$\frac{dC_P}{dt} = k_2 C_{ES} = \frac{K_2 C_{E_0} C_S}{K_M + C_S(1+C_I/K_I)}. \tag{10.5.11}$$

For uncompetitive inhibition, K_M is unaffected, but the maximum rate is reduced (Figure 10.16a). This can be detected in a Lineweaver–Burk plot as an increase in the intercept (Figure 10.16b).

In noncompetitive inhibition, the inhibitor binds to a site other than the active site of the enzyme as well as the enzyme–substrate complex [5]. Thus, both Equations (10.5.2) and (10.5.8) apply. In general, each inhibition reaction may be different and the equilibrium constants for binding the free enzyme and enzyme–substrate complex are denoted by K_I^{EI} and K_I^{ESI}, respectively. The total enzyme concentration is

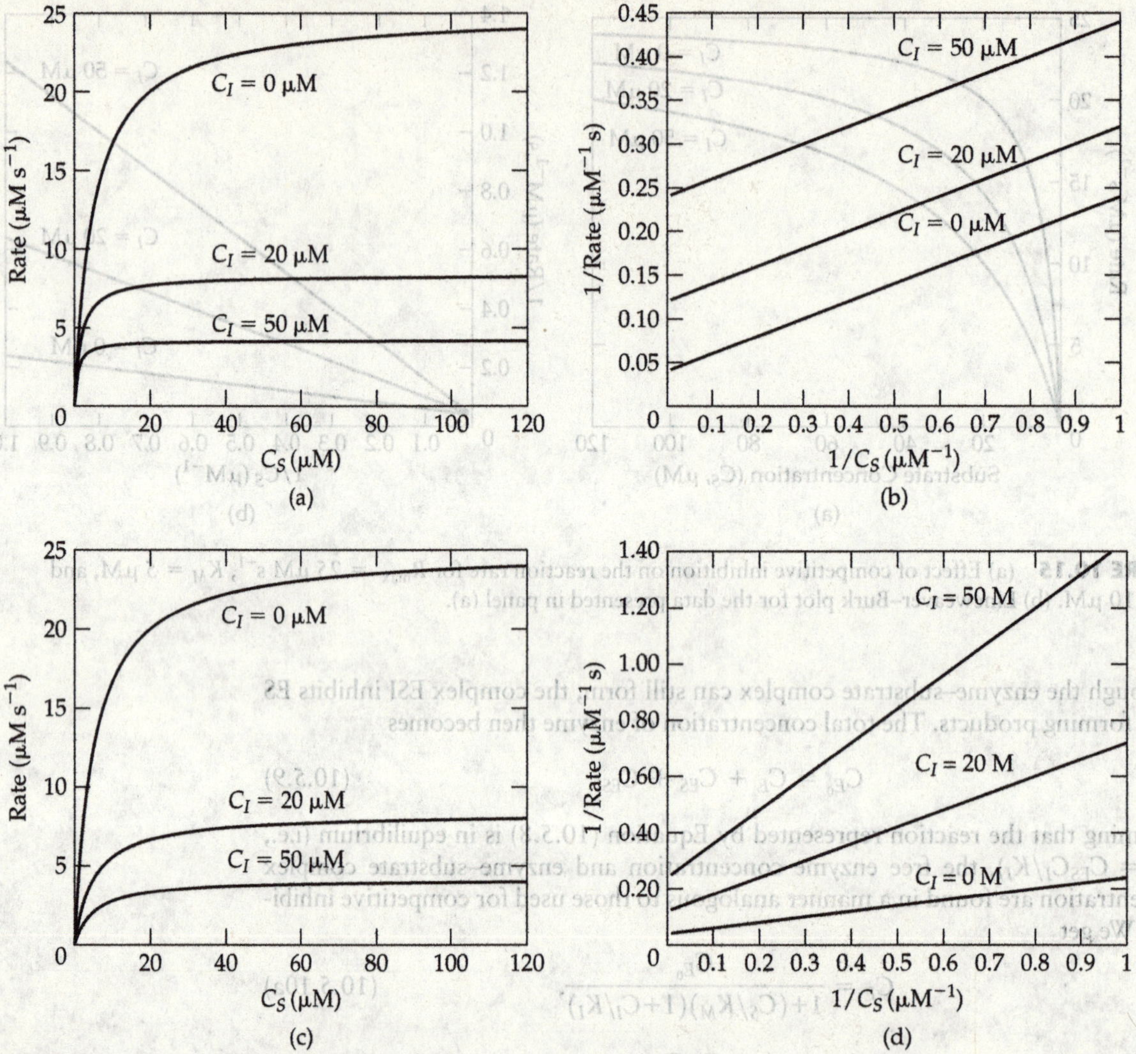

FIGURE 10.16 (a) Effect of an uncompetitive inhibitor on the initial rate of an enzymatic reaction for $R_{max} = 25$ μM s^{-1}, $K_M = 5$ μM, and $K_I = 10$ μM. (b) Lineweaver–Burk plot for the data presented in panel (a). (c) Effect of a noncompetitive inhibitor on the initial rate of an enzymatic reaction for $R_{max} = 25$ μM s^{-1}, $K_M = 5$ μM, and $K_I^{EI} = K_I^{ESI} = 10$ μM. (d) Lineweaver–Burk plot for the data presented in panel (c).

the sum of free enzyme, enzyme–substrate complex, inhibitor bound to the enzyme, and inhibitor bound to the enzyme–substrate complex. Applying the equilibrium relations for the inhibitor bound to the enzyme and inhibitor bound to the enzyme–substrate complex yields

$$C_{E_0} = C_E + C_{ES} + C_{EI} + C_{ESI} = C_E\left(1+\frac{C_I}{K_I^{EI}}\right)+C_{ES}\left(1+\frac{C_I}{K_I^{ESI}}\right). \quad (10.5.12)$$

Using this result and the quasi–steady-state assumption for the enzyme–substrate complex yields the following expression for the rate of product formation:

$$\frac{dC_P}{dt} = k_2 C_{ES} = \frac{k_2 C_{E_0} C_S}{K_M(1 + C_I/K_I^{EI}) + C_S(1 + C_I/K_I^{ESI})}. \tag{10.5.13}$$

For this case, inhibition affects both K'_M and the maximum rate (Figure 10.16c). The Lineweaver–Burk plot results in a change in slope and intercept (Figure 10.16d).

10.5.3 Substrate Inhibition

Substrates can either activate or inhibit the rate of reaction. Substrate inhibition can be a very important mechanism to limit the reaction rate. Inhibition can occur if the substrate binds to a second site on the enzyme. The binding of substrate to this second site does not lead to product formation and is likely to be of lower affinity because it has a larger dissociation constant than the substrate that binds to the active site on the enzyme. In many ways, substrate inhibition resembles uncompetitive inhibition. The reaction can be represented by Equation (10.5.1) and the following equation:

$$ES + SI \underset{k_{-11}}{\overset{k_{11}}{\rightleftharpoons}} ESS. \tag{10.5.14}$$

This reaction is assumed to be in equilibrium with $K_{D11} = C_{ES}C_S/C_{ESS} = k_{-11}/k_{10}$. The total enzyme concentration is

$$C_{E_0} = C_E + C_{ES} + C_{ESS}. \tag{10.5.15}$$

The enzyme–substrate complex C_{ES} is assumed to be in quasi–steady state, and ESS is assumed to be at equilibrium. Applying these assumptions yields the following relation for the rate of product formation:

$$\frac{dC_P}{dt} = k_2 C_{ES} = \frac{k_2 C_{E_0} C_S}{K_M + C_S(1 + C_S/K_{D11})}. \tag{10.5.16}$$

Figure 10.17 is a plot of the effect of substrate inhibition on the initial rates of product formation or substrate consumption. Substrate inhibition can severely reduce the maximum rate of reaction and slow down the rate at higher substrate concentrations. The maximum rate of reaction occurs at $C_S = (K_M K_{D11})^{1/2}$, which can be found by setting the derivative of the rate with respect to C_S equal to zero.

10.6 Effect of Diffusion and Convection on Chemical Reactions

Chemical reactions can occur either in a single phase or at the interface between two phases. The location of the reaction affects whether it appears in the conservation relation or the boundary conditions. If the reaction occurs in the phase through

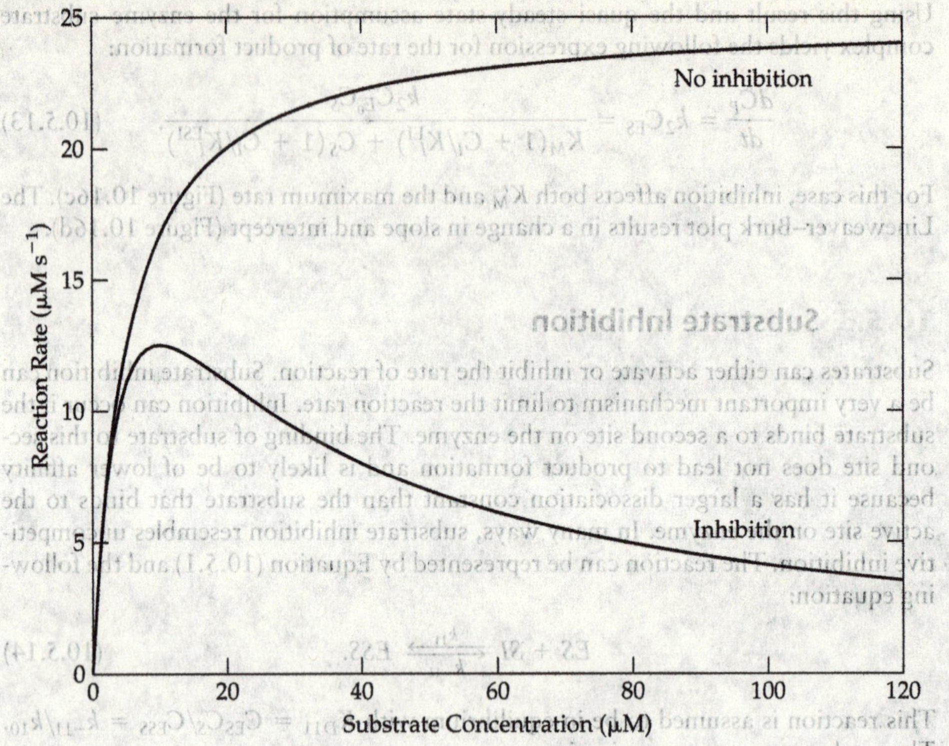

FIGURE 10.17 Effect of substrate inhibition on reaction rate for $K_M = 5\ \mu M$, $R_{max} = k_2$, $C_{E_0} = 25\ \mu M\ s^{-1}$, and $K_{D11} = 20\ \mu M$.

which the reacting molecules diffuse and are convected, *homogeneous reactions* affect the material balance. Reactions that occur on the surface of one phase in a system containing more than one phase are known as *heterogeneous reactions*. Such interfacial reactions occur as boundary conditions. A third possibility is a heterogeneous reaction that occurs throughout a porous material. As we discussed in Chapter 8, diffusion through porous media can be treated as diffusion through an effective single-phase medium. The binary diffusion coefficient is replaced with an effective diffusion coefficient. Using the same treatment, the rate of chemical reaction term appears in the conservation relation. Such reactions are known as *pseudohomogeneous reactions* if they are macroscopically homogeneous. Because the analyses of homogeneous reactions and pseudohomogeneous reactions are similar, we consider them together in Section 10.6.3.

Reactions in biological systems can occur in any of the three ways just described. The binding of small molecules to plasma proteins in blood or antigen–antibody binding in solution is an example of homogeneous reactions. Examples of heterogeneous reactions are clotting reactions on the surface of blood vessels, the hydrolysis of ATP by ectonucleases on the surface of endothelium, ion transport, and the release of nitric oxide. Heterogeneous reactions within tissues include aerobic metabolism, glucose consumption, and receptor-mediated endocytosis.

When studying the effect of diffusion and convection on the rate of reaction, one needs to distinguish between the *intrinsic reaction rate* and the *observed reaction rate*. The intrinsic reaction rate is the rate of reaction that would proceed if the reacting molecules were instantly brought into contact. This rate depends only upon the chemical properties of the reactants and is the fastest rate at which the reaction can occur at a given temperature. In contrast, the observed reaction rate is the rate that is actually measured. This rate is affected by diffusion and convection, as well as by the intrinsic kinetics between the reactants. In the current and subsequent sections, the factors that influence the observed rate are identified so that we can design conditions such that the observed rate equals the intrinsic rate.

The rates of homogeneous and heterogeneous reactions can be affected by diffusion and convection. In solution, the rate of a very fast reaction between two molecules depends upon how rapidly the molecules come together. Mixing by diffusion could be very slow, resulting in observed reaction rates much lower than the intrinsic rate. The observed or measured rates of reactions that occur on the lumenal surface of blood vessels are sensitive to the blood flow rate. Altering the flow rate alters the reaction rate, unless the intrinsic reaction rate is much less than the rates of diffusion and convection. Because the reactant must either diffuse or be convected to the surface prior to reaction, the mass-transfer process occurs in series with the reaction process. Reactions within tissues are often affected by the rate of diffusion to the cells. Because the diffusion process is occurring simultaneously with the reaction process, the effects of reaction and diffusion on the observed rate of reaction cannot be easily separated.

10.6.1 Reaction and Diffusion in Solution

In Section 6.9.1, we examined diffusion-limited binding in solution. In that analysis, we assumed that the rate of reaction was so fast that the reactant concentration at the reaction surface was zero. In this limit, the rate of reaction was dependent only upon the rate at which the reactant could diffuse to the reaction surface and was independent of the reaction kinetics. In general, however, the reaction rate for binding in solution can be affected by both the rate of diffusion and the intrinsic kinetics. To account for each of these factors, we need to reexamine the problem analyzed in Section 6.9.1.

The problem involves the diffusion of ligands to a central receptor of radius R_R. The ligands are assumed to be much smaller than the receptor. When the ligand reaches the surface, it reacts in accordance with a second-order rate coefficient k_1. Because the reaction occurs on the surface of the reactant molecule, the reaction is heterogeneous, and the conservation relation is the same as Equation (6.9.1):

$$\frac{D_L}{r^2}\frac{d}{dr}\left(r^2\frac{dC_L}{dr}\right) = 0. \qquad (6.9.1)$$

The boundary condition at the receptor surface is different from that used in Section 6.9.1. At the receptor surface, the steady-state rate of mass transfer by diffusion (flux times surface area) equals the rate of mass produced by the reaction. That is, at

$$r = R_R \quad -4\pi R_R^2 D_L\left(\frac{dC_L}{dr}\right)\bigg|_{r=R_R} = -k_1 C_L|_{r=R_R}, \qquad (10.6.1a)$$

where k_1 is the intrinsic rate coefficient of the reaction, with units of volume per molecule of receptor per unit time. Far from the surface of the receptor, the ligand concentration is uniform. That is, at

$$r \to \infty \quad C_L = C_{L_0}. \qquad (10.6.1b)$$

For this condition to be applicable, the receptor concentration must be dilute, so that the distance between receptors is large.

The solution of Equation (6.9.1), subject to the boundary conditions given in Equations (10.6.1a) and (10.6.1b), is

$$C_L = C_{L_0}\left(1 - \frac{k_1 R_R/r}{k_1 + D_L 4\pi R_R}\right). \tag{10.6.2}$$

In order to relate the reaction rate to the known concentration far from the receptor surface, the reaction rate is defined in terms of the *observable forward rate constant*:

$$\text{Rate} = -4\pi R_R^2 D_L\left(\frac{dC_L}{dr}\right)\bigg|_{r=R_R} = -k_f C_{L_0}. \tag{10.6.3}$$

Note that this quantity is referred to as an *observable rate coefficient* because it is determined on the basis of the observable bulk concentration.

Solving for k_f yields the relation

$$k_f = \frac{4\pi R_R D_L k_1}{4\pi R_R D_L + k_1}. \tag{10.6.4a}$$

For very fast surface reactions relative to diffusion ($k_1 \gg 4\pi R_R D_L$), Equation (10.6.4a) approaches the diffusion-limited result, Equation (6.9.5a). Alternatively, for very slow reactions relative to diffusion ($k_1 \ll 4\pi R_R D_L$), k_f approaches k_1 and the kinetics said to be *reaction limited*.

The reciprocal of Equation (10.6.4a) yields

$$\frac{1}{k_f} = \frac{1}{4\pi R_R D_L} + \frac{1}{k_1}. \tag{10.6.4b}$$

The quantity $4\pi R_R D_L$ represents the diffusion-limited rate coefficient. The rate coefficients are analogous to electrical conductances, and their reciprocals are equivalent to resistances. Equation (10.6.4b) indicates that diffusion and reaction act in series. Note that the units of k_1 and k_f are volume per molecule per unit time. To convert to more familiar units for second-order reactions, the rate coefficients are multiplied by Avogadro's number. Note that $4\pi R_R D_L$ is often represented as k_+ (see Equation (6.9.5a)).

Example 10.4 Effect of Diffusion on Biochemical Reaction Rates

Determine the contribution of diffusion to the overall reaction rate constant for the initial formation of a complex between transfer RNA (tRNA) and tRNA synthetase ([5], p. 341). The following data are available:

$$k_1 = 2.7 \times 10^8 \text{ M}^{-1}\text{ s}^{-1},$$
$$R_R = 1 \text{ nm},$$
$$D_L = 6 \times 10^{-7} \text{ cm}^2\text{ s}^{-1}.$$

Solution The rate coefficient, after converting to units of cm^3 $molecule^{-1}$ s^{-1} is 4.49×10^{-13}. The ratio $4\pi R_R D_L/(4\pi R_R D_L + k_1)$ in Equation (10.6.4) represents the fraction of the observed rate coefficient arising from the intrinsic kinetics. If this ratio is greater than 0.95, then diffusion can be assumed to have a negligible effect on the reaction rate. For the data provided, this ratio is 0.627. Diffusion reduces the reaction rate by 37.3%. Note that if the radius of the receptor were larger, then the ratio would be larger, and diffusion would contribute less to the reaction.

Most biological reactions in solution are not diffusion limited. Besides diffusion, electrostatic interactions between molecules can significantly influence the rate, making binding easier than if the molecules had to diffuse randomly to the correct conformation for binding to occur.

In order to examine the dissociation of the ligand from its receptor, the effective reverse rate constant k_r can be defined as $k_{-1}(1-\gamma)$, where k_{-1} is the *intrinsic dissociation rate coefficient* and γ is the *capture probability*, which equals $k_+/(k_+ + k_1)$ [12]. Thus,

$$k_r = \frac{k_{-1}k_1}{k_+ + k_1}. \tag{10.6.4c}$$

10.6.2 Interphase Mass Transfer and Reaction

When reactions occur on cell surfaces that are in contact with moving fluid phases, the analysis of the situation described in Section 10.6.1 is modified as follows. The physical situation is shown schematically in Figure 10.18. Fluid flows over a surface, and reactants in the fluid are transported to the surface by convection and diffusion. This analysis can be applied to any of the following processes occurring at the surface: a chemical reaction, the molecules binding to surface receptors, or molecules passing through a channel into cells.

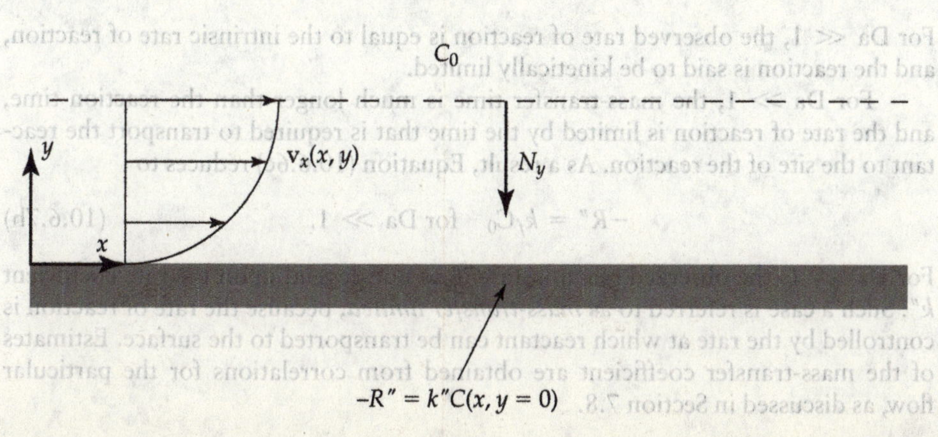

FIGURE 10.18 Interphase mass transfer and surface chemical reaction for the case of a solution in a moving fluid with velocity $v_x(x, y)$ and a surface reaction rate $-R''$. The solute flux to the surface is N_y.

Although the convection–diffusion equation can be solved for simple cases, an alternative approach is to use mass-transfer coefficients. The advantage of this approach is that we need only to solve an algebraic equation. First, we equate the flux of reactant molecules at the surface with the rate of production by reaction on the surface for a first-order reaction; that is,

$$N_y(y = 0) = k_f(C_0 - C(y = 0)) = k''C(y = 0), \tag{10.6.5}$$

where C_0 is the solute concentration in the bulk solution far from the surface, k_f is the fluid phase mass transfer coefficient (Equation 7.8.2), and k'' is the first-order rate coefficient for a heterogeneous reaction, with units of length per unit time. The choice of mass transfer coefficient (k_m, k_{loc} or k_{ln} as defined in Section 7.8, or (7.8.6), respectively) depends on the problem geometry and correlations available. Both the rate coefficient and the mass-transfer coefficient have the same units.

Solving for the surface concentration and calculating the surface reaction rate in terms of the bulk concentration yields

$$C(y = 0) = \frac{k_f C_0}{k'' + k_f'} \tag{10.6.6a}$$

$$-R'' = \frac{k'' k_f C_0}{k'' + k_f}. \tag{10.6.6b}$$

The Damkohler number $\text{Da} = k''/k_f$ represents the ratio of the mass-transfer time ($t_{\text{mt}} = L/k_f$, where L is a characteristic length) to the reaction time ($t_r = L/k''$). A typical characteristic length is the volume-to-surface ratio for the system. From the definition of the Damkohler number, Equation (10.6.6b) becomes

$$-R'' = \frac{k'' C_0}{1 + \text{Da}}. \tag{10.6.6c}$$

For $\text{Da} \ll 1$, the mass-transfer time is much less than the reaction time, and the rate of reaction is not limited by transport through the fluid phase. Consequently, the rate expression (Equation (10.6.6c)) becomes

$$-R'' = k''C_0 \quad \text{for Da} \ll 1. \tag{10.6.7a}$$

For $\text{Da} \ll 1$, the observed rate of reaction is equal to the intrinsic rate of reaction, and the reaction is said to be kinetically limited.

For $\text{Da} \gg 1$, the mass-transfer time is much longer than the reaction time, and the rate of reaction is limited by the time that is required to transport the reactant to the site of the reaction. As a result, Equation (10.6.6c) reduces to

$$-R'' = k_f C_0 \quad \text{for Da} \gg 1. \tag{10.6.7b}$$

For $\text{Da} \gg 1$, the observed reaction rate does not depend upon the rate coefficient k''. Such a case is referred to as *mass-transfer limited*, because the rate of reaction is controlled by the rate at which reactant can be transported to the surface. Estimates of the mass-transfer coefficient are obtained from correlations for the particular flow, as discussed in Section 7.8.

Example 10.5 Endothelial cells are present on the luminal surfaces of arteries and veins and are constantly exposed to blood flow. Laboratory studies of the response of endothelial cells to laminar flow have shown that the metabolism of ATP increases dramatically after the onset of shear stresses as low as 1 dyn/cm^2. ATP is metabolized by enzymes that are present on the cell surface. Some have speculated that the rate of ATP metabolism is influenced by mass transfer. To assess whether this hypothesis is correct, a mathematical model has been used to analyze mass transfer and heterogeneous reaction on an endothelial cell layer cultured on the bottom surface of a flow chamber [13]. For the concentrations of ATP in blood, the metabolism of ATP is a first-order surface reaction. Using the following data, determine whether the rate of ATP metabolism is mass-transfer limited:

a	0.02 cm	Gap height of the flow chamber
w	2.5 cm	Width of the flow chamber
L	8 cm	Length of surface on which endothelial cells are cultured
k''	0.005 cm s^{-1}	First-order heterogeneous rate constant
$\langle v \rangle$	5 cm s^{-1}	Average velocity
D_{ij}	2.6×10^{-6}cm^2 s^{-1}	Diffusion coefficient of ATP in water
k_f	$1.4674 D_{ij}\left(\dfrac{\langle v \rangle}{D_{ij}La}\right)^{1/3}$	Length-averaged mass-transfer coefficient

Solution The reaction rate is

$$-R'' = \frac{k'' C_0}{1+\text{Da}},$$

where Da $= k''/k_m$. For the data given, the mass-transfer coefficient is

$$k_m = 1.4674\,(2.6 \times 10^{-6}\ \text{cm}^2\text{s}^{-1})\left(\frac{5\ \text{cm s}^{-1}}{(2.6 \times 10^{-6}\ \text{cm}^2\ \text{s}^{-1})(8\ \text{cm})(0.02\ \text{cm})}\right)^{1/3}$$
$$= 0.000874\ \text{cm s}^{-1},$$

and Da $= 5.72$. Because Da is greater than unity, there is a significant reduction in the reaction rate due to external mass-transfer limitations. The rate is reduced by $1/(1+\text{Da})$, or 85%. Thus, the hypothesis is valid.

10.6.3 Intraphase Chemical Reactions

Heterogeneous chemical reactions that occur within a tissue (or a porous particle containing immobilized proteins) are said to be *intraphase*. Although the reaction takes place at the interface between the fluid and the cell surface or at extracellular matrix phases *within* the tissue, the reaction is treated as pseudohomogeneous if the reaction rate is macroscopically uniform. In this case, reaction and diffusion occur in parallel. As reactants enter the tissue, some react, reducing the concentration of reactants in the bulk phase. If the reaction proceeds rapidly relative to the rate of diffusion, then the reactant concentration may decrease to zero within the tissue. Such a reaction is inefficient because only part of the catalyst is used. Analyzing transport and reaction within tissues can assess the effect of diffusion on the reaction rate for biomolecules, drugs, or other therapeutic molecules. If diffusion affects the reaction rate significantly, strategies to minimize the effect of diffusion can be developed.

Applying Equation (8.4.5) for the case of steady diffusion without sources or sinks yields

$$D_{\text{eff}}\nabla^2 C = -R''a, \tag{10.6.8}$$

where D_{eff} is the effective diffusion coefficient within the porous media, R'' is the reaction rate per unit internal surface area, and a is the ratio of the internal surface area to the volume of the substrate. $R''a$ is an *effective homogeneous reaction rate* and is represented as R'''.

In general, we are interested neither in the spatial distribution of reactants within the porous substrate nor in the local rate of reaction as described by Equation (10.6.8). Instead, we want to know the macroscopic or observed rate of reaction. The macroscopic rate is the average of the local rate, R''', or

$$\overline{R}''' = \frac{1}{V}\int_V R''' dV. \tag{10.6.9a}$$

Equation (10.6.9a) can be rewritten in another way. First, by inserting Equation (10.6.8) into Equation (10.6.9a) and using the divergence theorem, we obtain the macroscopic, or global, reaction rate:

$$\overline{R}''' = \frac{1}{V}\int_V R''' dV = \frac{1}{V}\int_V -D_{\text{eff}}\nabla^2 C dV = \frac{1}{V}\int_S -D_{\text{eff}}\mathbf{n}\cdot\nabla C dS. \tag{10.6.9b}$$

The macroscopic or global reaction rate is simply the integral of the flux over the entire surface of the substrate in contact with the external fluid. (Note that S refers to the external surface area.)

The effect of diffusion on the reaction is assessed by comparing the observed reaction rate with the reaction rate that would occur if diffusion did not affect the reaction and the entire substrate were exposed to the bulk fluid with the solute concentration C_0. This ratio is known as the *effectiveness factor* and is given by

$$\eta = \frac{\overline{R}'''}{\overline{R}'''(C_0)}. \tag{10.6.9c}$$

The global reaction rate is obtained by solving Equation (10.6.8) for a particular geometry, rate expression, and boundary conditions. The flux at the surface is then computed, and the global reaction rate is obtained from Equation (10.6.9b).

An important application of these concepts is the case of diffusion and an irreversible first-order chemical reaction in a porous slab of thickness $2L$. Such a geometry might correspond to a rectangular piece of tissue immersed in a medium of uniform concentration. The surfaces ($x = \pm L$) are maintained at a concentration of C_0 (Figure 10.19). The reaction equation is

$$D_{eff}\frac{d^2C}{dx^2} = k'''C, \tag{10.6.10a}$$

with boundary conditions

$$x = \pm L \quad C = K_{av}C_0 \tag{10.6.10b}$$

and

$$x = 0 \quad \frac{dC}{dx} = 0, \tag{10.6.10c}$$

where $k''' = k''a$ and K_{av} is the available volume fraction.

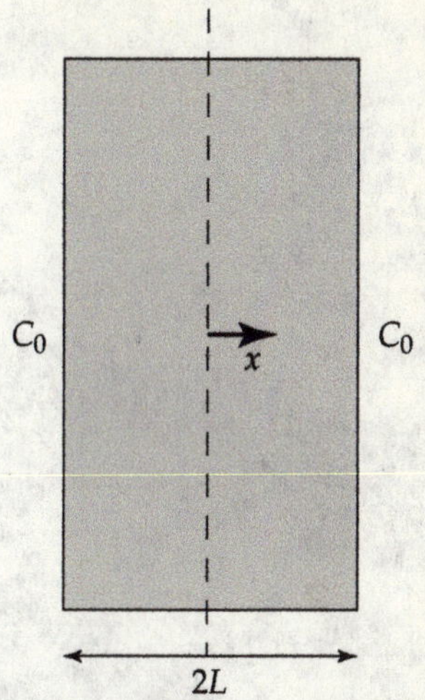

FIGURE 10.19 Diffusion and reaction in a rectangular region of tissue.

The solution of Equation (10.6.10a) subject to Equations (10.6.10b) and (10.6.10c) is facilitated by casting the equations in dimensionless form. To do so, let

$$\phi^2 = \frac{k'''L^2}{D_{eff}} \quad x^* = \frac{x}{L} \quad \theta = \frac{C}{K_{AV}C_0},$$

where ϕ^2 is the *Thiele modulus*, which represents the ratio of the diffusion time (L^2/D_{eff}) to the reaction time ($1/k'''$). Equations (10.6.10a) to (10.6.10c) then become

$$\frac{d^2\theta}{dx^{*2}} = \phi^2\theta, \tag{10.6.11a}$$

$$x^* = \pm 1 \quad \theta = 1, \tag{10.6.11b}$$

$$x^* = 0 \quad \frac{d\theta}{dx^*} = 0. \tag{10.6.11c}$$

Equation (10.6.11a) has the general solution

$$\theta = A\exp(\phi x^*) + B\exp(-\phi x^*). \tag{10.6.12a}$$

Applying the boundary condition at $x^* = 0$ yields $B = A$, and the expression for θ becomes

$$\theta = A\left[\exp(\phi x^*) + \exp(-\phi x^*)\right]. \tag{10.6.12b}$$

Next, we apply the boundary condition at $x^* = \pm 1$ and solve for A (due to the symmetry of the problem, both boundary conditions yield the same result):

$$A = \frac{1}{\exp(\phi) + \exp(-\phi)}. \tag{10.6.13}$$

Substituting Equation (10.6.13) into Equation (10.6.12b), the result is

$$\theta = \frac{\exp(\phi x^*) + \exp(-\phi x^*)}{\exp(\phi) + \exp(-\phi)}. \tag{10.6.14a}$$

Equation (10.6.14a) is often written in a somewhat different form using the definition of the hyperbolic cosine, namely,

$$\cosh(z) = \frac{\exp(z) + \exp(-z)}{2}. \quad (10.6.14b)$$

Thus,

$$\theta = \frac{\cosh(\phi x^*)}{\cosh(\phi)}. \quad (10.6.14c)$$

The global, or average, reaction rate for this problem is simply

$$-\overline{R}''' = \frac{2k'''}{2L}\int_0^L C dx = k''' K_{AV} C_0 \int_0^1 \theta dx^*. \quad (10.6.15a)$$

Substituting Equation (10.6.14c) into Equation (10.6.15a) and integrating yields

$$-\overline{R}''' = k''' C_0 K_{AV} \frac{\sinh(\phi)}{\phi \cosh(\phi)} = k''' C_0 K_{AV} \frac{\tanh(\phi)}{\phi}, \quad (10.6.15b)$$

where $\sinh(z) = (e^z - e^{-z})/2$.

When external mass-transfer limitations are not present, the surface concentration equals the bulk concentration, and the macroscopic rate is equal to $-k''' C_0 K_{AV}$. Thus, the effectiveness factor is

$$\eta = \frac{\tanh(\phi)}{\phi}. \quad (10.6.15c)$$

A graph of the effectiveness factor is presented in Figure 10.20. For small values of the Thiele modulus, reaction is slow relative to diffusion, and the effectiveness factor

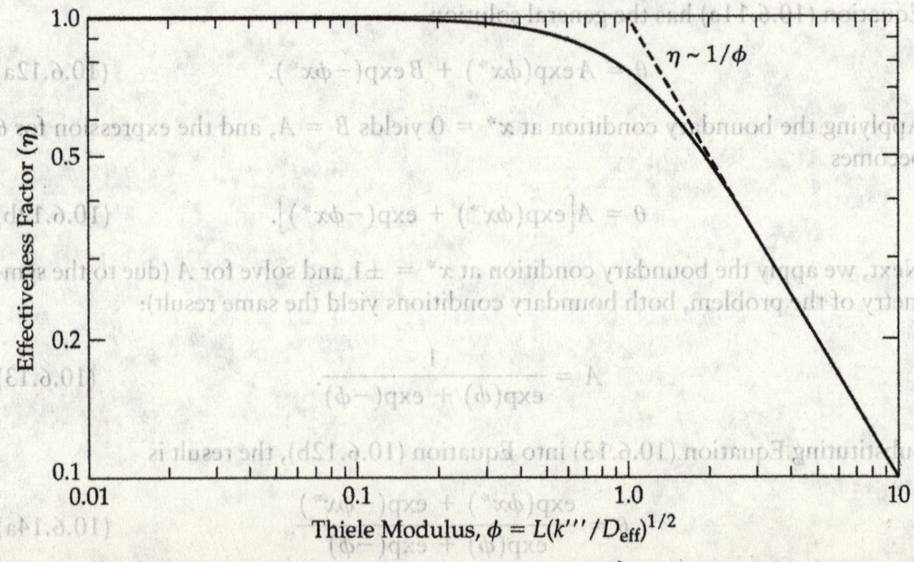

FIGURE 10.20 Effectiveness factor as a function of the Thiele modulus. The dashed line represents the asymptotic value of $\eta \sim 1/\phi$ for large values of ϕ.

approaches unity. As the Thiele modulus increases in size, diffusion limits the rate of reaction, and the effectiveness factor decreases in magnitude. For $\phi > 3$, $\tanh(\phi) \approx 1$ and $\eta \approx 1/\phi$.

For diffusion and first-order reaction in a sphere, the effectiveness factor has a slightly different form that can be determined by calculating the concentration profile. For steady-state radial diffusion and reaction in spherical coordinates, the mass balance (Equation (7.2.13c)) is

$$\frac{D_{\text{eff}}}{r^2}\frac{d}{dr}\left(r^2\frac{dC}{dr}\right) = k'''C. \tag{10.6.16}$$

The boundary conditions are

$$r = 0 \quad \frac{dC}{dr} = 0, \tag{10.6.17a}$$

$$r = R \quad C = K_{\text{av}}C_0. \tag{10.6.17b}$$

Using the definition of the Thiele modulus and letting $\xi = r/R$, we obtain, from Equation (10.6.16) and the associated boundary conditions,

$$\frac{1}{\xi^2}\frac{d}{d\xi}\left(\xi^2\frac{dC}{d\xi}\right) = \phi^2 C, \tag{10.6.18}$$

$$\xi = 0 \quad \frac{dC}{d\xi} = 0, \tag{10.6.19a}$$

and

$$\xi = 1 \quad C = K_{\text{AV}}C_0. \tag{10.6.19b}$$

To solve this problem, let $C = f/\xi$. Then the first derivative becomes

$$\frac{dC}{d\xi} = -\frac{f}{\xi^2} + \frac{df}{\xi d\xi}, \tag{10.6.20}$$

With this transformation, Equation (10.6.18) becomes

$$\frac{d^2f}{d\xi^2} = \phi^2 f. \tag{10.6.21}$$

Equation (10.6.21) is identical to Equation (10.6.11a) and has the general solution

$$f = C\xi = A\sinh(\phi\xi) + B\cosh(\phi\xi). \tag{10.6.22}$$

Evaluating the boundary conditions at $\xi = 0$ and $\xi = 1$ yields $B = 0$ and $A = K_{\text{AV}}C_0/\sinh(\phi)$. The concentration is

$$C = K_{\text{AV}}C_0\frac{\sinh(\phi\xi)}{\xi\sinh(\phi)}. \tag{10.6.23}$$

Defining the effectiveness factor as the average reaction rate in the sphere divided by the reaction rate if all of the sphere were exposed to $K_{AV}C_0$, we have

$$\eta = \frac{\int_0^R Cr^2 dr}{(R^3/3)K_{AV}C_0} = \frac{3\int_0^1 \sinh(\phi\xi)\xi d\xi}{\sinh(\phi)} = \frac{3}{\phi}\left(\frac{1}{\tanh(\phi)} - \frac{1}{\phi}\right). \quad (10.6.24)$$

Equations (10.6.24) for a sphere of radius R and (10.6.15c) for a rectangular slab of thickness $2L$ are shown in Figure 10.21a. Because the surface area decreases as the solute diffuses inward and reacts, the effectiveness factor is larger for a sphere than for a rectangular slab.

The graph for the slab geometry can be generalized to other geometries by using Equation (10.6.15c) with $L = R/2$ for a long cylinder of radius R and $L = R/3$ for a sphere of radius R. When this is done, the curves for a slab and sphere are almost coincident (Figure 10.21b).

10.6.4 Interphase and Intraphase Diffusion and Reaction

In general, the rates of biochemical reactions are affected both by external mass transfer from the blood or bulk fluid to the tissue surface and by diffusion within the tissue. The effectiveness factor can be generalized to include cases of external mass transfer. Consider Equation (10.6.10a), subject to the following boundary conditions:

$$x = \pm L \quad -D_{\text{eff}}\frac{dC}{dx} = \frac{k_f}{K_{AV}\left[K_{AV}C_0 - C(x = \pm L)\right]}, \quad (10.6.25a)$$

$$x = 0 \quad \frac{dC}{dx} = 0 \quad (10.6.25b)$$

In dimensionless form, with $\theta = C/(C_0K_{AV})$, Equations (10.6.10a) and (10.6.25a) become

$$\frac{d^2\theta}{dx^{*2}} = \phi^2\theta, \quad (10.6.26a)$$

$$x^* = \pm 1 \quad \frac{d\theta}{dx^*} = \text{Bi}[1 - \theta(x^* = \pm 1)], \quad (10.6.26b)$$

$$x^* = 0 \quad \frac{d\theta}{dx^*} = 0, \quad (10.6.26c)$$

where $\text{Bi} = k_f L/(D_{\text{eff}}K_{AV})$ is the *Biot number*, which represents the ratio of external mass transport to internal diffusion.

The solution of Equation (10.6.26a), subject to the boundary conditions given in Equations (10.6.26b) and (10.6.26c), parallels the solution of Equation (10.6.10a) until the boundary conditions at $x^* = \pm 1$ are evaluated. Thus, we can write

$$\theta = A[\exp(\phi x^*) + \exp(-\phi x^*)]. \quad (10.6.27a)$$

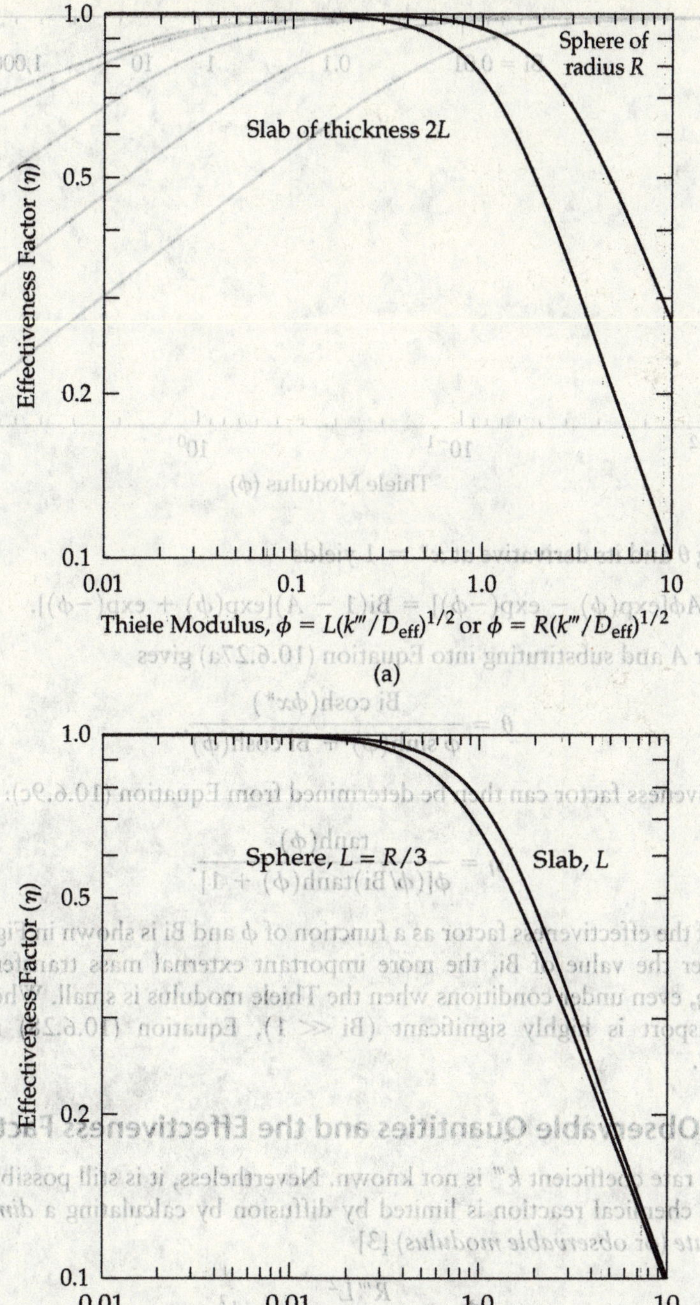

FIGURE 10.21 (a) Effectiveness factor for diffusion and first-order reaction in a slab and in a sphere. (b) Comparison of diffusion and reaction in a sphere with radius R using Equation (10.6.24), with $R = 3L$, and Equation (10.6.15c) for a rectangular slab with thickness $2L$.

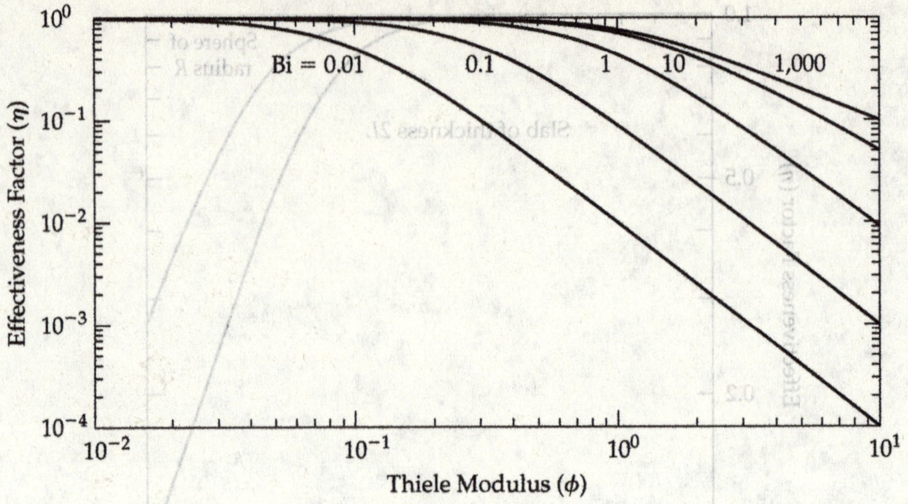

FIGURE 10.22 Effect of external mass transfer on the effectiveness factor from Equation (10.6.28).

Evaluating θ and its derivative at $x^* = 1$ yields

$$A\phi[\exp(\phi) - \exp(-\phi)] = \text{Bi}(1 - A)[\exp(\phi) + \exp(-\phi)]. \quad (10.6.27\text{b})$$

Solving for A and substituting into Equation (10.6.27a) gives

$$\theta = \frac{\text{Bi}\cosh(\phi x^*)}{\phi\sinh(\phi) + \text{Bi}\cosh(\phi)}. \quad (10.6.27\text{c})$$

The effectiveness factor can then be determined from Equation (10.6.9c):

$$\eta = \frac{\tanh(\phi)}{\phi[(\phi/\text{Bi})\tanh(\phi) + 1]}. \quad (10.6.28)$$

A graph of the effectiveness factor as a function of ϕ and Bi is shown in Figure 10.22. The smaller the value of Bi, the more important external mass transfer becomes, reducing η, even under conditions when the Thiele modulus is small. When external mass transport is highly significant (Bi $\ll$ 1), Equation (10.6.28) reduces to $\eta \approx \text{Bi}/\phi^2$.

10.6.5 Observable Quantities and the Effectiveness Factor

Often, the rate coefficient k''' is not known. Nevertheless, it is still possible to assess whether a chemical reaction is limited by diffusion by calculating a *dimensionless reaction rate* (or *observable modulus*) [3]

$$\Phi = \frac{\overline{R}'''L^2}{D_{\text{eff}}K_{AV}C_0} = \eta\phi^2, \quad (10.6.29)$$

where L is the half-thickness for a rectangular geometry, is $R/2$ for a cylinder, and is $R/3$ for a sphere. The dimensionless reaction rate is a measurable quantity and is directly related to the effectiveness factor. To see how Equation (10.6.29) is useful, consider the following cases of a first-order reaction with mass transfer occurring within and mass transfer occurring external to the tissue.

For reaction-limited conditions within the tissue, in which the Biot number is infinitely large (there are no external mass-transfer limitations), ϕ^2 is much less than unity and η is approximately equal to unity. Consequently, Equation (10.6.29) reduces to

$$\Phi = \frac{\overline{R}'''L^2}{D_{\text{eff}}K_{\text{AV}}C_0} \approx \phi^2 \ll 1 \text{ (diffusion not limiting)}, \tag{10.6.30a}$$

and the observed reaction rate is

$$\overline{R}''' = k'''K_{\text{AV}}C_0. \tag{10.6.30b}$$

For significant diffusion limitations, $\phi^2 \gg 1$ and $\eta \approx 1/\phi$. (*Note*: For $\phi > 3$, $\tanh(\phi) \approx 1$.) The observed modulus and observed reaction rate are

$$\Phi = \frac{\overline{R}'''L^2}{D_{\text{eff}}K_{\text{AV}}C_0} \approx \phi \gg 1 \text{ (significant diffusion limitations)}. \tag{10.6.31a}$$

Rearranging Equation (10.6.31a) and substituting for the definition of the Thiele modulus yields

$$\overline{R}''' = \frac{(k'''D_{\text{eff}})^{1/2}}{L}K_{\text{AV}}C_0. \tag{10.6.31b}$$

The effective rate coefficient $\sqrt{k'''D_{\text{eff}}}/L$ represents a weighting of reaction and diffusion.

For finite Bi, the observable quantity Φ is

$$\Phi = \frac{\phi\tanh(\phi)}{(\phi/\text{Bi})\tanh(\phi) + 1}. \tag{10.6.32a}$$

For rapid diffusion relative to reaction ($\phi \ll 1$), Equation (10.6.32a) reduces to

$$\Phi \approx \phi^2 \ll 1. \tag{10.6.32b}$$

For $\phi > 3$, $\tanh(\phi) \approx 1$ and the observable modulus reduces to

$$\Phi = \frac{\text{Bi}\,\phi}{\phi + \text{Bi}}, \tag{10.6.32c}$$

and the reaction rate is

$$\overline{R}''' = \frac{(k'''D_{\text{eff}})^{1/2}\text{Bi}}{L(\phi + \text{Bi})}K_{\text{AV}}C_0. \tag{10.6.32d}$$

Example 10.6 Cells metabolize oxygen in mitochondria, which are small organelles that contain the enzymes for respiration and that are uniformly dispersed throughout the cytoplasm of a cell. Oxygen diffuses in the cytoplasm until it reaches the mitochondria and is consumed. Using the following data for muscle cells and liver cells, determine whether the rate of oxygen reaction with mitochondrial enzymes is limited by diffusion in the cell cytoplasm (assume that $K_{AV} = 1$):

R_c	Cell radius	10 μm
R_m	Mitochondrion radius	0.45 μm
D_{eff}	Diffusion coefficient of O_2	1.92×10^{-5} cm^2 s^{-1}
	Muscle Cells under Exercise	
C_0	Extracellular concentration of O_2	1.1×10^{-8} mol cm^{-3}
$\overline{R}'''$	Observed rate of O_2 uptake by cells	2×10^{-6} mol cm^{-3} s^{-1}
	Liver Cells	
C_0	Extracellular concentration of O_2	10×10^{-8} mol cm^{-3}
$\overline{R}'''$	Observed rate of O_2 uptake by cells	7×10^{-8} mol cm^{-3} s^{-1}

Solution For each case, calculate the dimensionless reaction modulus given by Equation (10.6.29). In this problem, the cell can be treated as a sphere, and the cell radius is the appropriate length scale ($L = R_c/3$). For muscle cells under exercise,

$$\Phi = \frac{\overline{R}'''L^2}{D_{eff}K_{AV}C_0} = \frac{(2 \times 10^{-6}\text{mol cm}^{-3}\text{ s}^{-1})(0.001 \text{ cm}/3)^2}{(1.92 \times 10^{-5}\text{ cm}^2\text{ s}^{-1})(1.1 \times 10^{-8}\text{ mol cm}^{-3})} = 1.05.$$

A similar calculation for liver cells yields $\Phi = 0.004$.

For muscle cells under exercise, diffusion and reaction are of equal importance. So we would expect that the reaction rate would be less than the reaction limit. For liver cells, however, the small value of the modulus indicates that diffusion does not limit the rate of oxygen consumption *in vivo*.

Example 10.7 In order to determine the importance of diffusion within particles, a series of experiments was performed using various sizes of particles containing immobilized enzyme. The reaction may be assumed to be first order and irreversible ($r_{obs} = k'''C$). The surface concentration of reactant is 2×10^{-4} mol cm^{-3}.
The following data are given:

Diameter of the particle sphere (cm)	0.050	0.010	0.005	0.001	0.0005
r_{obs}(mol h^{-1} cm^{-3})	0.22	0.98	1.60	2.40	2.40

Determine the intrinsic rate coefficient k and the effective diffusivity D_{eff} from the data. Assume that $K_{AV} = 1$.

Solution Since neither the rate coefficient nor the effective diffusion coefficient is known, both the effectiveness factor and the reaction modulus given by Equations (10.6.15c) and (10.6.29) cannot be calculated immediately. Note that, as the size of the particle decreases, diffusion becomes less significant and $\eta \rightarrow 1$, and reaction rate is independent of the particle size. This is the case for the smallest two particles, from which we conclude that

$$2.40 \text{ mol h}^{-1} \text{ cm}^{-3} = k(2 \times 10^{-4} \text{ mol/cm}^3),$$

or $k = 12{,}000 \text{ h}^{-1} = 3.33 \text{ s}^{-1}$.

Conversely, for the largest particle size, the rate drops by a factor of 4.86 as the diameter increases from 0.01 cm to 0.05 cm, suggesting significant diffusion limitations. If we assume that for the largest particle size, $\eta \approx 1/\phi$, the rate expression is given by Equation (10.6.31b). Using the rate at a diameter of 0.05 cm, we find that the effective diffusion coefficient is 1.95×10^{-6} cm^2 s^{-1}.

To assess the validity of our assumptions, ϕ and η are determined at each particle diameter and the reaction rate is estimated. For a sphere, $L = R/3$, and we have the following results:

Diameter of sphere (cm)	0.050	0.010	0.005	0.001	0.0005
ϕ	10.89	2.178	1.089	0.218	0.109
η	0.092	0.448	0.731	0.985	0.996
Predicted rate (mol h^{-1} cm^{-3})	0.220	1.074	1.755	2.363	2.391
Percent error	0	−9.8	−9.7	1.54	0.38

The agreement is quite good, suggesting that the assumptions are reasonable. More accurate estimates of the rate coefficient and the diffusion coefficient can be obtained by nonlinear regression of Equation (10.6.29) to the data.

10.6.6 Transport Effects on Enzymatic Reactions

As the previous examples indicate, *in vivo* and *in vitro* enzymatic reactions may be subject to mass-transfer limitations. If blood flow cannot supply substrate to the enzymes that are present on the endothelial cell surface, then a concentration gradient occurs, and the substrate concentration reacting with the enzyme is less than the bulk concentration in the fluid. Likewise, concentration gradients occur in tissues if the diffusion time is larger than the characteristic reaction time.

Commercially, enzymes are used for the detoxification of blood, as biosensors, and in enzyme bioreactors for the generation of specialty chemicals [14]. In order to facilitate the separation of the enzyme from the product, the enzyme is immobilized on a solid support such as a porous bead. Alternatively, the reacting surface of the support is directly exposed to fluid, as is proposed for microfluidics devices [15], or the enzyme is immobilized in a porous material to provide a greater surface-to-volume ratio. A major limitation to the use of immobilized enzyme systems has been the random covalent modification of carboxyl and amine groups, which immobilizes the enzyme in a range of orientations, in some of which the active site is no longer accessible to the substrate. Further, the covalent modification may alter the active site. As a result, the activity of enzymes that are covalently immobilized is often significantly less than the activity of the enzyme in solution [16]. This limitation can be overcome by creating a fusion protein with a localized peptide sequence that is used in binding, such as biotin or a histidine tag [16,17].

Interphase transport and reaction. The results for first-order reactions can be extended to Michaelis–Menten kinetics. Two cases are considered: reactions limited by fluid-phase transport and reactions limited by diffusion within tissue. For fluid-phase mass transfer occurring in series with an enzymatic reaction, Equation (10.6.5) changes to

$$N_y(y=0) = k_f(C_0 - C(y=0)) = \frac{R''_{max}\, C(y=0)}{K_M + C(y=0)}, \tag{10.6.33}$$

where R''_{max} is the rate of surface reaction (e.g., moles cm^{-2} s^{-1}), is proportional to the enzyme concentration C_{E_0} at the surface. Before solving for $C(y = 0)$, we put the equations into dimensionless form by defining the following variables:

$$\theta = \frac{C(y=0)}{C_0} \qquad \beta = \frac{C_0}{K_M} \qquad \mathrm{Da} = \frac{R''_{max}}{K_M k_f} \tag{10.6.34}$$

The quantity β represents the dimensionless Michaelis constant. With these variables, Equation (10.6.33) becomes

$$(1-\theta) = \mathrm{Da}\frac{\theta}{1+\beta\theta}. \tag{10.6.35}$$

Equation (10.6.35) is a quadratic equation in θ, with only the positive root yielding physically reasonable values:

$$\theta = \frac{\beta - 1 - \mathrm{Da} + \sqrt{(1+\mathrm{Da}-\beta)^2 + 4\beta}}{2\beta}. \tag{10.6.36}$$

The effectiveness factor can be expressed in terms of θ [18] as

$$\eta = \frac{\text{Rate}}{\text{Rate}(C_0)} = \theta\left(\frac{1+\beta}{1+\beta\theta}\right). \tag{10.6.37}$$

When β is small, the value of θ approaches $1/(1 + \mathrm{Da})$, the value for first-order reactions (Figure 10.23). For larger values of β, the reaction rate is less sensitive to changes in concentration, and the effectiveness factor remains at unity even for larger values of Da. In the limit as β goes to infinity, the rate is insensitive to concentration, and the effectiveness factor is equal to unity. At large values of Da/β, the effectiveness factor declines as $\beta(1+\beta)/2\mathrm{Da}$. Thus, fluid phase mass transfer affects Michaelis–Menten kinetics only when the surface concentration declines below K_M.

Intraphase transport and reaction. For enzymatic reactions that behave according to Michaelis–Menten kinetics, a one-dimensional mass balance for diffusion and reaction in a rectangular region of tissue yields

$$D_{eff}\frac{d^2C}{dx^2} = \frac{R_{max}C}{K_M + C}, \tag{10.6.38}$$

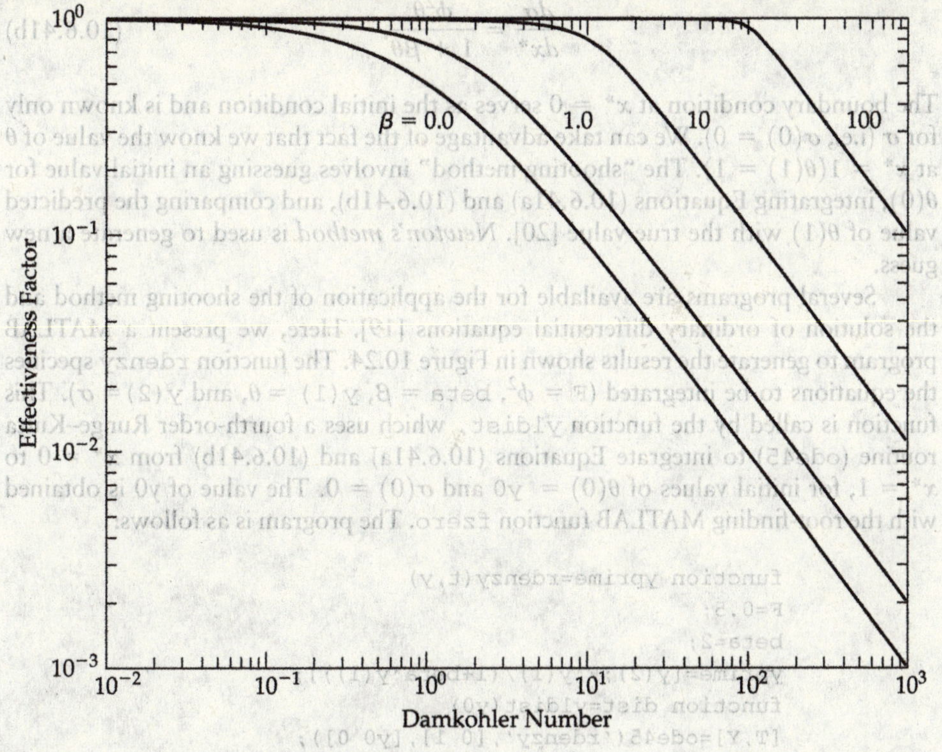

FIGURE 10.23 Effect of the dimensionless Michaelis constant $\beta = C_0/(K_M K_{AV})$ on the effectiveness factor.

where $R_{max} = R''_{max}a$ (see Equation (10.6.8)). The boundary conditions are the same as those used in the case of diffusion and a first-order reaction in rectangular coordinates (Equations (10.6.10b) and 10.6.10c)). The following dimensionless variables are defined:

$$\phi = L\sqrt{\frac{R_{max}}{K_M D_{eff}}} \qquad \beta = \frac{C_0 K_M}{K_{AV}} \qquad x^* = \frac{x}{L} \qquad \theta = \frac{C}{K_{AV} C_0}. \tag{10.6.39}$$

Note that R_{max}/K_M corresponds to the first-order rate coefficient. Placing Equation (10.6.38) into dimensionless form yields

$$\frac{d^2\theta}{dx^{*2}} = \frac{\phi^2\theta}{1 + \beta\theta}. \tag{10.6.40}$$

The dimensionless boundary conditions are given by Equations (10.6.11b) and (10.6.11c).

Equation (10.6.40) and the associated boundary conditions (Equations (10.6.11b) and (10.6.11c)) cannot be solved analytically. Several numerical techniques can be used, including the *finite-element*, or *finite-difference, method*. Alternatively, Equation (10.6.40) can be recast as a system of first-order ordinary differential equations and solved using a *Runge–Kutta method* [19]:

$$\frac{d\theta}{dx^*} = \sigma, \tag{10.6.41a}$$

$$\frac{d\sigma}{dx^*} = \frac{\phi^2\theta}{1 + \beta\theta}. \tag{10.6.41b}$$

The boundary condition at $x^* = 0$ serves as the initial condition and is known only for σ (i.e., $\sigma(0) = 0$). We can take advantage of the fact that we know the value of θ at $x^* = 1(\theta(1) = 1)$. The "shooting method" involves guessing an initial value for $\theta(0)$, integrating Equations (10.6.41a) and (10.6.41b), and comparing the predicted value of $\theta(1)$ with the true value [20]. *Newton's method* is used to generate a new guess.

Several programs are available for the application of the shooting method and the solution of ordinary differential equations [19]. Here, we present a MATLAB program to generate the results shown in Figure 10.24. The function `rdenzy` specifies the equations to be integrated (`F` = ϕ^2, `beta` = β, `y(1)` = θ, and `y(2)` = σ). This function is called by the function `y1dist,` which uses a fourth-order Runge–Kutta routine (`ode45`) to integrate Equations (10.6.41a) and (10.6.41b) from $x^* = 0$ to $x^* = 1$, for initial values of $\theta(0) =$ `y0` and $\sigma(0) = 0$. The value of y0 is obtained with the root-finding MATLAB function `fzero`. The program is as follows:

```
function yprime=rdenzy(t,y)
F=0.5;
beta=2;
yprime=[y(2);F*y(1)/(1+beta*y(1))];
function dist=y1dist(y0)
[T,Y]=ode45('rdenzy',[0 1],[y0 0]);
m=size(T);
z=m(1,1);
dist=1-Y(z,1);
```

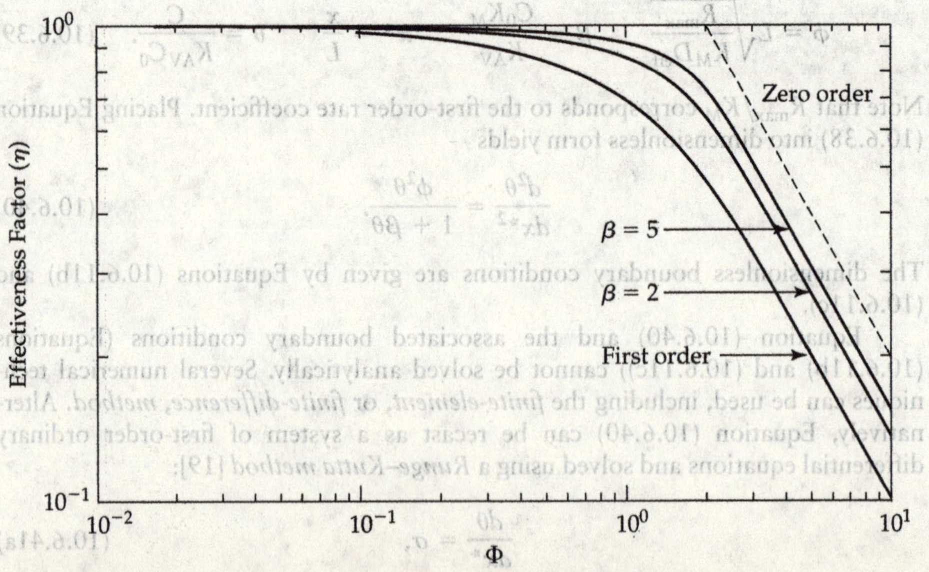

FIGURE 10.24 Effectiveness factor for Michaelis–Menten kinetics with $\beta = C_0/K_M$ and $\phi^2 = R_{max}L^2/K_M D_{eff}$ and $\Phi = \eta\phi^2/(1 + \beta)$. Also shown are results for first-order and zero-order reactions.

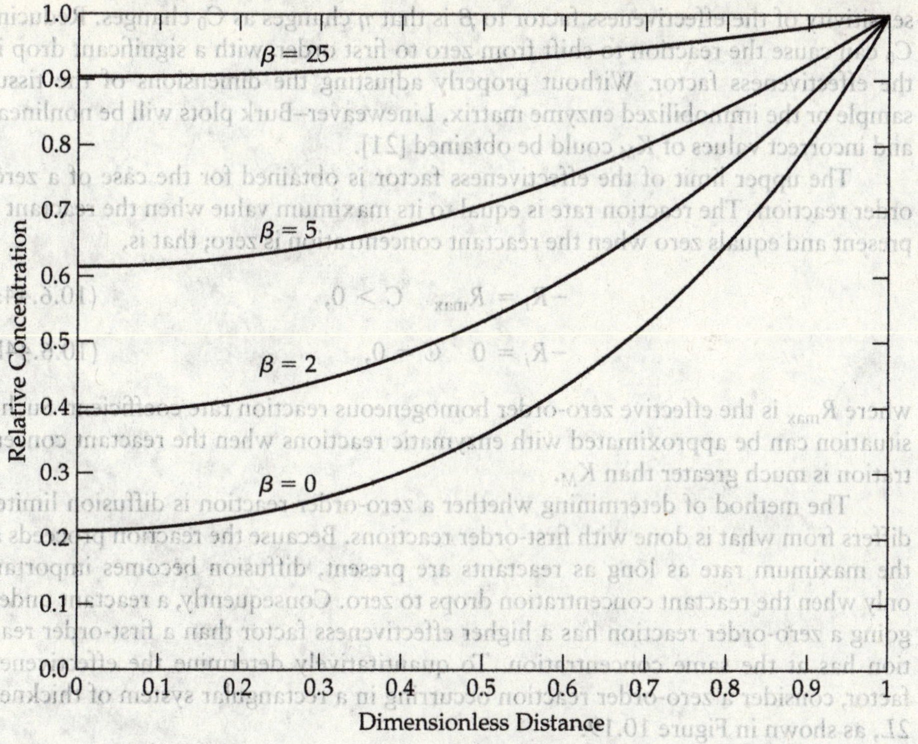

FIGURE 10.25 Dimensionless concentration profile for diffusion and reaction within a rectangular piece of tissue for $\phi^2 = 5$.

The user supplies an initial guess for `y0` and calls `fzero` to find the `y0` that satisfies the function `y1dist`. The subroutine `ode45` is called again to integrate the function `rdenzy` with the final value of `y0`. The effectiveness factor is found by evaluating the derivative[3] of θ at $x^* = 1$. We obtain

$$\eta = \frac{\frac{D_{\text{eff}}}{L}\frac{dC}{dx}\Big|_{x=L}}{R'''(C_0)} = \frac{\frac{D_{\text{eff}}}{L}\frac{dC}{dx}\Big|_{x=L}}{\frac{R_{\max} C_0 K_{AV}}{K_M + C_0 K_{AV}}} = \frac{(1+\beta)}{\phi^2}\frac{d\theta}{dx^*}\Big|_{x^*=1}. \quad (10.6.42)$$

The effectiveness factor is a function of the observable modulus Φ, and β is shown in Figure 10.24. For Michaelis–Menten kinetics, the observable modulus is

$$\Phi = \frac{\overline{R}''' L^2}{D_{\text{eff}} K_{AV} C_0} = \frac{\eta \phi^2}{1+\beta}. \quad (10.6.43)$$

Nonzero values of β lead to effectiveness factors greater than those obtained for first-order reactions (Figure 10.24). The reason is that the concentration decreases less for values of β greater than zero (Figure 10.25). An important implication of the

[3]Alternatively, one could calculate the average reaction rate in the tissue section. Both approaches yield the same value for the effectiveness factor. Since the program calculates the derivative at $x^* = 1$, no additional computation is needed.

sensitivity of the effectiveness factor to β is that η changes as C_0 changes. Reducing C_0 can cause the reaction to shift from zero to first order, with a significant drop in the effectiveness factor. Without properly adjusting the dimensions of the tissue sample or the immobilized enzyme matrix, Lineweaver–Burk plots will be nonlinear, and incorrect values of K_M could be obtained [21].

The upper limit of the effectiveness factor is obtained for the case of a zero-order reaction. The reaction rate is equal to its maximum value when the reactant is present and equals zero when the reactant concentration is zero; that is,

$$-R_i = R_{\max} \quad C > 0, \tag{10.6.44a}$$

$$-R_i = 0 \quad C = 0, \tag{10.6.44b}$$

where $R_{\max}$ is the effective zero-order homogeneous reaction rate coefficient. Such a situation can be approximated with enzymatic reactions when the reactant concentration is much greater than K_M.

The method of determining whether a zero-order reaction is diffusion limited differs from what is done with first-order reactions. Because the reaction proceeds at the maximum rate as long as reactants are present, diffusion becomes important only when the reactant concentration drops to zero. Consequently, a reactant undergoing a zero-order reaction has a higher effectiveness factor than a first-order reaction has at the same concentration. To quantitatively determine the effectiveness factor, consider a zero-order reaction occurring in a rectangular system of thickness $2L$, as shown in Figure 10.19.

For a zero-order reaction, Equation (10.6.38) is

$$D_{\text{eff}} \frac{d^2C}{dx^2} = R_{\max}. \tag{10.6.45}$$

The boundary conditions for this problem are the same as the ones used in the case of a first-order reaction (Equations (10.6.10b) and (10.6.10c)). Integrating Equation (10.6.45) twice yields

$$C = \frac{R_{\max}}{2D_{\text{eff}}} x^2 + Ax + B. \tag{10.6.46}$$

From the boundary condition at $x = 0$, it follows that $A = 0$. Applying the boundary condition at $x = L$ yields

$$B = K_{AV}C_0 - \frac{R_{\max}}{2D_{\text{eff}}} L^2. \tag{10.6.47}$$

The resulting concentration distribution is

$$C = K_{AV}C_0 - \frac{R_{\max}L^2}{2D_{\text{eff}}}\left(1 - \frac{x^2}{L^2}\right). \tag{10.6.48}$$

Equation (10.6.48) can be cast into dimensionless form by using the definitions of the Thiele modulus and β for Michaelis–Menten kinetics (Equation (10.6.39)). The dimensionless form of Equation (10.6.48) is

$$\theta = 1 - \frac{\phi^2}{2\beta}(1 - x^{*2}). \qquad (10.6.49)$$

For Thiele moduli greater than $\sqrt{2\beta}$, Equation (10.6.49) is no longer applicable, because the concentration is less than zero at some locations in the catalyst and the boundary condition at $x = 0$ is no longer valid [22]. Thus, this boundary condition is replaced by the derivative of the concentration with respect to x^* equal to zero at x_0^*. Using this relation in place of Equation (10.6.49) results in the following expression for the dimensionless concentration:

$$\theta = 1 + \frac{\phi^2}{\beta}\left(\frac{x^{*2}}{2} - \frac{1}{2} + x_0^* - x_0^* x^*\right). \qquad (10.6.50)$$

For $\phi > \sqrt{2\beta}$, the value of x_0^* is determined by setting the concentration at x_0^* equal to zero. Then

$$x_0^* = 1 - \sqrt{\frac{2\beta}{\phi^2}}. \qquad (10.6.51)$$

For a zero-order reaction, the effectiveness factor defined in Equation (10.6.9c) is

$$\eta = \frac{\overline{R}'''}{R'''(C_0)} = \frac{\int_V R'''(C_0)dV}{R'''(C_0)} = \frac{\int_{x_0*}^{1} R'''(C_0)dx^*}{R'''(C_0)} = 1 - x_0^*, \qquad (10.6.52a)$$

where the integral extends from x_0^* to unity because the rate is zero for $0 < x^* < x_0^*$. For $\phi < \sqrt{2\beta}$, $\eta = 1$. Substituting Equation (10.6.51) into Equation (10.6.52a) yields the following relation for the effectiveness factor for $\phi > \sqrt{2\beta}$

$$\eta = \sqrt{\frac{2\beta}{\phi^2}}. \qquad (10.6.52b)$$

For zero-order kinetics, $\beta \gg 1$, and the observable modulus is

$$\Phi = \frac{\overline{R}'''L^2}{D_{\text{eff}}K_{AV}C_0} = \phi\sqrt{\frac{2}{\beta}}. \qquad (10.6.53)$$

For values of Φ less than 2, the effectiveness factor is unity. For values of Φ greater than 2, the effectiveness factor is inversely related to ϕ. For zero-order reactions, all values of β fall on the same curve when η is plotted as a function of the observable modulus Φ (Figure 10.24).

Example 10.8 The following data were obtained for the rate of substrate reaction with an enzyme in a thin rectangular section of tissue 200 μm thick:

C_0 (μM)	R_{obs} (μM s^{-1})	Φ	η	Φ_{pred}
200	0.190	0.19	0.997	0.190
100	0.180	0.36	0.988	0.359
1	0.009	1.80	0.571	1.858
0.1	9.6×10^{-4}	1.92	0.480	1.921

The reactant diffusion coefficient in the tissue is 5×10^{-7} cm^2 s^{-1}. Find R_{max} and K_m.

Solution For each concentration, the observable modulus is calculated using Equation (10.6.29) and listed in the third column of the data. Examining Figure 10.24, we see that the reaction at $C_0 = 200$ μM has an effectiveness factor of approximately unity and is not diffusion limited. Initially, we also assume that 200 μM is much larger than K_M, so that the observed rate equals the maximum rate. At the two lowest concentrations, the observed rate is approximately proportional to the concentration. As a first guess, assume that K_M is greater than 1 μM and that the reaction is first order at $C_0 = 0.1$ μM. In this case, $\Phi = \phi\tanh(\phi)$. Solving iteratively, we find that ϕ is approximately 2.0. From this result and the modulus at a concentration of 200 μM, it follows that $K_M = 10$ μM. To verify these results, η is calculated numerically for each case and Φ is predicted. The agreement is very good, indicating that $R_{max} = 0.190$ μMs^{-1} and K_M is 10 μM.

10.7 QUESTIONS

10.1 Explain the difference between the reaction stoichiometry and the reaction mechanism.

10.2 Suppose that you are measuring the reaction of a sugar to form dimers. Discuss the experiments you need to perform to obtain the reaction order and the reaction rate coefficient at 37°C.

10.3 Component A reacts to form B and C, but the mechanisms are unknown. The following experiment was performed: Molecule A was added to solution at an initial concentration of 1 mM, and the concentrations of A, B, and C were measured with time. For the results shown in panels (a) and (b) of Figure 10.26, describe mechanisms that would be consistent with the results.

10.4 In parallel reactions, the same reactants produce two different products by two different reactions. For example, suppose A and B react as follows:

$$A + B \begin{array}{l} \xrightarrow{k_1} C \\ \xrightarrow[k_2]{} D \end{array}$$

Sketch concentrations of A, B, C, and D for the initial condition that $C_A = C_B = C_0$ and $C_C = C_D = 0$.

10.5 State appropriate conditions for which the quasi–steady-state assumption can be used.

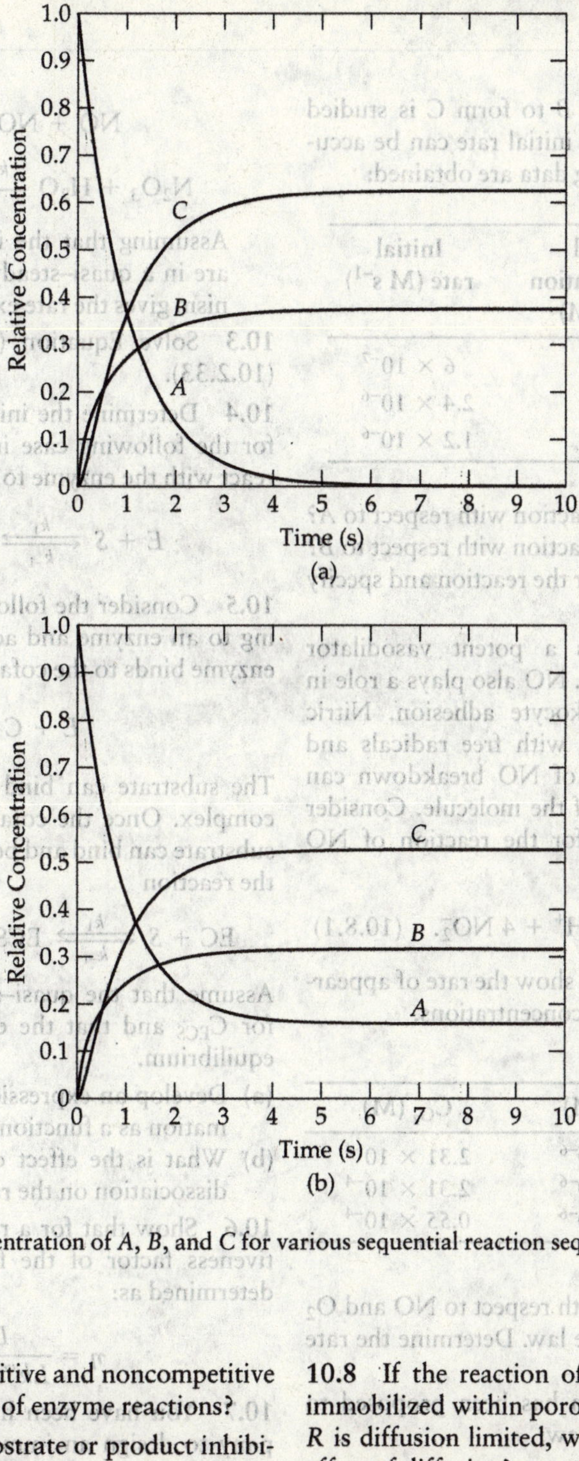

FIGURE 10.26 Change in concentration of A, B, and C for various sequential reaction sequences.

10.6 What effects do competitive and noncompetitive inhibition have on the kinetics of enzyme reactions?

10.7 Discuss the effect of substrate or product inhibition on the control of a reaction.

10.8 If the reaction of a substrate with an enzyme immobilized within porous spherical particles of radius R is diffusion limited, what can be done to reduce the effect of diffusion?

10.8 PROBLEMS

10.1 The reaction of A and B to form C is studied under conditions in which the initial rate can be accurately measured. The following data are obtained:

Initial concentration of A (M)	Initial concentration of B (M)	Initial rate (M s^{-1})
1.0	1.0	6×10^{-7}
2.0	1.0	2.4×10^{-6}
1.0	2.0	1.2×10^{-6}

(a) What is the order of the reaction with respect to A?
(b) What is the order of the reaction with respect to B?
(c) Find the rate coefficient for the reaction and specify the units.

10.2 Nitric oxide (NO) is a potent vasodilator released from endothelial cells. NO also plays a role in platelet aggregation and leukocyte adhesion. Nitric oxide is short lived, reacting with free radicals and other oxidants. The kinetics of NO breakdown can affect the biological activity of the molecule. Consider the following stoichiometry for the reaction of NO with oxygen:

$$4NO + O_2 + 2H_2O \longrightarrow 4H^+ + 4\,NO_2^-. \quad (10.8.1)$$

(a) The following kinetic data show the rate of appearance of NO_2^- for different concentrations.

Rate (M s^{-1})	C_{NO} (M)	C_{O_2} (M)
0.2218×10^{-6}	10×10^{-6}	2.31×10^{-4}
1.9958×10^{-6}	30×10^{-6}	2.31×10^{-4}
0.4752×10^{-6}	30×10^{-6}	0.55×10^{-4}

Find the reaction order with respect to NO and O_2 and state the resulting rate law. Determine the rate coefficient.

(b) The following mechanism has been proposed to explain the observed rate law:

$$2NO + O_2 \xrightarrow{k_1} 2NO_2 \quad (10.8.2a)$$

$$NO + NO_2 \underset{k_3}{\overset{k_2}{\rightleftharpoons}} N_2O_3 \quad (10.8.2b)$$

$$N_2O_3 + H_2O \xrightarrow{k_4} 2H^+ + 2NO_2^-. \quad (10.8.2c)$$

Assuming that the intermediates N_2O_3 and NO_2 are in a quasi-steady state, show that this mechanism gives the rate expression developed in part (a).

10.3 Solve Equation (10.2.32) to obtain Equation (10.2.33).

10.4 Determine the initial rate of product formation for the following case in which the product can also react with the enzyme to form ES:

$$E + S \underset{k_{-1}}{\overset{k_1}{\rightleftharpoons}} ES \underset{k_{-2}}{\overset{k_2}{\rightleftharpoons}} E + P. \quad (10.8.3)$$

10.5 Consider the following case for a cofactor binding to an enzyme and activating the enzyme: first, the enzyme binds to the cofactor via the reaction

$$E + C \underset{k_{-c}}{\overset{k_c}{\rightleftharpoons}} EC. \quad (10.8.4)$$

The substrate can bind only to the enzyme–cofactor complex. Once the cofactor binds to the enzyme, the substrate can bind and be converted to product through the reaction

$$EC + S \underset{k_{-1}}{\overset{k_1}{\rightleftharpoons}} ECS \overset{k_2}{\rightleftharpoons} EC + P. \quad (10.8.5)$$

Assume that the quasi–steady-state assumption holds for C_{ECS} and that the enzyme–cofactor binding is in equilibrium.

(a) Develop an expression for the rate of product formation as a function of cofactor concentration.
(b) What is the effect of cofactor concentration and dissociation on the rate of product formation?

10.6 Show that for a rectangular geometry the effectiveness factor of the first-order action can also be determined as:

$$\eta = \frac{D_{\text{eff}}}{Lk'''K_{AV}C_0} \left.\frac{dC}{dx}\right|_{x=L}.$$

10.7 You have been hired by a biotechnology company to design an immobilized enzyme bioreactor to remove a toxin from blood. A process has been developed to separate blood cells from the plasma and pass the blood plasma over a packed bed containing enzyme that is immobilized within porous beads. The following data have been obtained:

$$R_{max} = 1 \times 10^{-9} \text{ mol cm}^{-3} \text{ s}^{-1}$$
$$K_M = 1 \times 10^{-9} \text{ mol cm}^{-3}$$
$$D_{eff} = 5 \times 10^{-7} \text{ cm}^2 \text{ s}^{-1}$$

The company would like you to determine whether beads of radii 50 μm or 5 μm containing the immobilized enzyme will result in reaction-limited conditions. The concentration at which the toxin is lethal is 1×10^{-11} mol cm^{-3}.

10.8 There is considerable interest in delivering drugs to the brain for the treatment of cancer or degenerative nerve diseases. Delivering drugs to the brain via the circulation is extremely difficult because the microvascular endothelia are impermeable to most molecules. Implanting a polymeric capsule containing the drug of interest represents one possible approach to delivering drugs locally.

Consider a long cylindrical implant of radius R and length L. The implant dissolves at a constant rate R_0. The drug is released at the surface as a result of polymer dissolution. As the drug diffuses through the brain tissue, it is metabolized by a first-order reaction with a rate coefficient k_e. Assuming steady-state diffusion and assuming that the drug is diffusing into a region of length Λ, determine whether a reaction of the following drugs is diffusion limited:

$$R = 0.1 \text{ cm}$$
$$L = 1 \text{ cm}$$
$$\Lambda = 1 \text{ cm}$$

Anticancer drug

$$k_e = 2.8 \times 10^{-4} \text{ s}^{-1}$$
$$D_{eff} = 5 \times 10^{-6} \text{ cm}^2 \text{ s}^{-1}$$

Nerve growth factor

$$k_e = 1.1 \times 10^{-3} \text{ s}^{-1}$$
$$D_{eff} = 1 \times 10^{-6} \text{ cm}^2 \text{ s}^{-1}$$

10.9 The enzyme α-chymotrypsin was immobilized in porous spherical particles, and its activity was determined. The substrate was *N*-acetyl-L-tyrosine ethyl ester (ATEE). The following data were obtained:

$$R = 60 \ \mu\text{m}$$
$$D_{eff} = 1 \times 10^{-6} \text{ cm}^2 \text{ s}^{-1}$$
$$C_0 = 1 \times 10^{-6} \text{ mol cm}^{-3}$$
$$R_{obs} = 5 \times 10^{-6} \text{ moles cm}^{-3} \text{ s}^{-1}$$

K_M for this enzyme is 20×10^{-6} mol cm^{-3}. Determine the effectiveness factor and the value of R_{max}.

10.10 One approach to the development of new tissues is to grow the cells in a porous polymer matrix. The matrix provides support for the cells to adhere to, divide, and synthesize new tissue. One concern in the design of these tissue constructs is the optimum size that can be grown in culture. The size is determined, in part, by oxygen delivery to the cells.

Consider a polymer implant of thickness L filled with cells at density X (cells mL^{-1}). The cells consume O_2 at a constant rate of Q_{O_2} (moles O_2 (10^6 cells)$^{-1}$ s^{-1}). The oxygen consumption rate equals $Q_{O_2}X$. The oxygen concentration at the surface of the polymeric material is maintained uniformly at the value C_0. The diffusion coefficient of oxygen in cell-free polymers is the same as in water.

For these conditions, an oxygen consumption rate of 4×10^{-7}mol cm^{-3} s^{-1} was obtained for tissue-engineered matrices 200 μm thick. The following data are applicable:

$$D_{O_2} = 2 \times 10^{-5} \text{ cm}^2 \text{ s}^{-1}$$
$$Q_{O_2} = 1.1 \times 10^{-9} \text{ mol } (10^6 \text{ cells})^{-1} \text{ s}^{-1}$$
$$C_0 = 1 \times 10^{-7} \text{ mol mL}^{-1}$$
Cell diameter = 10 μm

Determine

(a) Whether oxygen consumption is diffusion- or reaction-limited
(b) The intrinsic reaction rate if all cells were exposed to C_0
(c) The cell density within the tissue

10.11 Adenosine triphosphate (ATP) can influence the relaxation of arteries. ATP is present in the blood and can be metabolized to adenosine diphosphate (ADP) by enzymes, known as ectonucleases, that are present on the endothelial cell surface. These enzymes follow Michaelis–Menten kinetics (Equation (10.4.2)), for which C_S is the concentration of ATP. Measured values are $R_{max} = 22 \times 10^{-9}$ mol min^{-1}(10^6 cells)$^{-1}$ and $K_M = 249 \times 10^{-6}$ M. The density of endothelial cells is 1.2×10^5 cells cm^{-2}. Generally, the ATP concentration is much less than K_M. Determine the effective first-order rate coefficient.

10.12 Diffusion and convection in blood occur in series with chemical reactions on the endothelial cell surface. For this problem, the reaction rate can be expressed in terms of the effective rate coefficient k'' and the Damkohler number $\text{Da} = k''/k_f$, where k_f is the mass-transfer coefficient. The rate is

$$-R'' = \frac{k'' C_S}{1 + \text{Da}}, \qquad (10.8.6)$$

where k_f is given by $k_f D/D_i = 1.615(\text{Pe}\, D/L)^{1/3}$, in which Pe is the Peclet number.

For the following values in the human aorta, determine the value of Da and assess the extent to which the reaction rate is reduced by blood flow:

$L = 60$ cm	(vessel length)
$D = 1.2$ cm	(vessel diameter)
$D_{ij} = 5.0 \times 10^{-6}$ cm^2 s^{-1}	(ATP diffusion coefficient)
Re = 750	(mean Reynolds number in the aorta)
$\nu = 0.03$ cm^2 s^{-1}	(kinematic viscosity of blood)

10.13 In Section 10.4.2, a qualitative analysis was used to assess the validity of the quasi–steady-state assumption. The criteria developed (Equations (10.4.24b) and (10.4.27b)) can be evaluated by numerical solution of the equations that describe enzyme kinetics. (In what follows, prime notation is used to denote a concentration that is based upon application of the quasi–steady-state assumption.) These scaled equations range in magnitude between zero and unity. From Equation (10.4.27b), the quasi–steady-state assumption is valid when

$$\frac{C_{E_0}}{K_M + C_{S_0}} \ll 1. \qquad (10.8.7)$$

To examine the assumption numerically, Equations (10.4.5) and (10.4.6) must be stated in a scaled form. The scaling normalizes the concentrations and provides relative references for the parameters. The substrate concentration is scaled by C_{S_0} so that $s = C_S/C_{S_0}$. Scaling the enzyme–substrate complex concentration by C_{S_0} would result in values that are always much less than unity, so instead we scale that concentration by $C_{E_0}C_{S_0}/(K_M + C_{S_0})$. Thus, the dimensionless complex concentration is

$$c = \frac{C_{E_0}C_S/(K_M + C_S)}{C_{E_0}C_{S_0}/(K_M + C_{S_0})}. \qquad (10.8.8)$$

Time is scaled by the slow time constant $k_2C_{E_0}/(K_M + C_{S_0})$ (i.e., $\tau = tk_2C_{E_0}/(K_M + C_{S_0})$). The other dimensionless groups are as follows:

$$\varepsilon = \frac{C_{E_0}}{K_M + C_{S_0}} \quad \sigma = \frac{C_{S_0}}{K_M} \quad \kappa = \frac{k_{-1}}{k_2}. \qquad (10.8.9\text{a,b,c})$$

Equations (10.4.5) and (10.4.6) are respectively cast into the following dimensionless form:

$$\frac{ds}{d\tau} = (\kappa + 1)(\sigma + 1)\left[-s + \frac{\sigma}{\sigma + 1}cs + \frac{\kappa}{(\kappa + 1)(\sigma + 1)}c\right], \qquad (10.8.10)$$

$$\frac{dc}{d\tau} = \frac{(\kappa + 1)(\sigma + 1)}{\varepsilon}\left[s - \frac{\sigma}{\sigma + 1}cs - \frac{1}{(\sigma + 1)}c\right]. \qquad (10.8.11)$$

Equations (10.4.13) and (10.4.14) are also placed into dimensionless form:

$$-\frac{ds'}{d\tau} = \frac{(1 + \sigma)s'}{1 + \sigma s'} \qquad (10.8.12)$$

$$c' = \frac{s'(\sigma + 1)}{s'\sigma + 1}, \qquad (10.8.13)$$

where the primes indicate that the quasi–steady-state assumption has been used.

(a) Using the definitions of the dimensionless groups s, es, τ, ε, κ, and σ, derive Equations (10.8.10) to (10.8.13) from Equations (10.4.5), (10.4.6), (10.4.13), and (10.4.14). Then show that

$$\frac{C_E}{C_{E_0}} = \varepsilon\frac{\sigma + 1}{\sigma}. \qquad (10.8.14)$$

(b) Using $\sigma = 0.1$, 1, and 10 and $\kappa = 10$, show that $\varepsilon \ll 1$ is suitable for the quasi–steady-state assumption. This can be demonstrated either by comparing the exact solution for s and c (Equations (10.8.10) and (10.8.11)) with the quasi–steady-state approximations (Equations (10.8.12) and (10.8.13)) or by varying individual parameters and making a phase plot of c as a function of s.

(c) Using simulations, show the effect of the parameter κ on the kinetics and the validity of the quasi–steady-state assumption.

For parts (b) and (c), the MATLAB M-file, *enzyme.m*, can be used to integrate Equations (10.8.10) to (10.8.12). In this program, `x(1)` represents s, `x(2)` denotes c, and `x(3)` designates s'. The array `dx` contains the three differential equations to be integrated (Equations (10.8.10) to (10.8.12)). The program is as follows:

```
function dx=enzyme(t,x)
global kappa sigma eps;
xs=sigma+1;
xk=kappa+1;
dx=[xk*xs*(-x(1)+sigma*x(2)*x(1)
/xs+kappa/(xs*xk)*x(2));...
xk*xs*(x(1)-sigma*x(2)*x(1)
/xs-x(2)/xs)/eps;...
-x(3)*xs/(sigma*x(3)+1)];
```

To integrate, the following commands are executed in the command window:

```
global kappa sigma eps;
kappa=10;
sigma=0.1;
eps=0.1;
[T,Y]=ode15s('enzyme',[0 2],[1;0;1]);
c=Y(:,3)*(sigma+1)/(1+sigma*Y(:,3));
```

The global command permits specification of the values external to the M-file. `Ode15s` is a numerical integration routine for "stiff" equations—equations that change their magnitude rapidly over a short timescale. The last expression solves Equation (10.8.13) for c', using the result for s' (x(3)).

10.14 Equation (10.6.53) represents a limiting case of the observable modulus for enzyme reactions when $\beta \gg 1$. When the reaction is diffusion limited, show that, using Equations (10.6.43) and (10.6.52b), the observable modulus becomes

$$\Phi = \frac{\eta\phi^2}{1+\beta} = \frac{\sqrt{2\beta}\phi}{1+\beta}.$$

Show that this result reduces to Equation (10.6.53) when $\beta \gg 1$.

10.15 The effect of a growth factor on hematopoietic stem cell function is being examined. The stem cells are grown in suspension in a small dish of radius 3.5 cm and fluid height H. Gas can only enter from the upper surface, which is in contact with air. You can assume rectangular coordinates.

Since researchers have observed that the effects of the growth factor are dependent upon the rate of binding to its cell-surface receptor, you have been asked to assess whether the growth factor binding is diffusion limited. The K_M is 10^{-7} M, C_0 is 5×10^{-7} M, $D_{\text{eff}} = 2 \times 10^{-6}$ cm^2 s^{-1}, and $K_{AV} = 1$. The observed rate of reaction is 6×10^{-12} M s^{-1} and $H = 1.0$ cm. Determine whether the reaction is diffusion limited and estimate the value of $R_{\max}$.

10.16 Diabetes involves damage to the insulin-producing islet cells in the pancreas. These cells do not grow and once damaged cannot be replaced by the body. One approach that is being considered to treat diabetes is to transplant cultured islet cells from a donor, encapsulate to prevent immune reactions, and implant to capsules. One challenge has been to create cultures of islet cells from the donor organ. The tissue must be well oxygenated. To do so, the tissue is cut into long cylinders and placed in a bath of culture medium and perfluorocarbon (PFC), which has a high solubility to oxygen (denoted by a solubility coefficient $K = C_{\text{PFC}}/C_{\text{aqueous solution}}$). In this way, high oxygen concentrations can be attained. The total oxygen concentration is the sum of the concentration of oxygen in solution and in PFC present at a volume fraction f.

(a) Consider a tissue cylinder of radius R_T and length L with $L \gg R_T$. The oxygen consumption rate is R_{O2}. The oxygen concentration at the surface is C_0. Assume zero-order reaction and solve for the steady state and radial concentration profile and obtain an expression for the maximum radius of the tissue.

(b) Use the following data to calculate the maximum radius:

$C_{aqueous} = 2.01 \times 10^{-7}$ mol cm^{-3}
$R_{O2} = 72 \times 10^{-9}$ mol cm^{-3} s^{-1}
$K = 20.4$
$D_{O2} = 2.0 \times 10^{-5}$ cm^2 s^{-1}
$f = 0.30$ (volume fraction of perfluorocarbon)

10.17 The BIACORE uses plasmon resonance on metal surfaces to measure the rate of reactions on surfaces. A receptor is immobilized on a gold surface, which forms the lower surface of a parallel plate flow chamber of height H, width W, and length L. A solution containing the ligand that binds to the receptor at concentration C_0 is injected between the parallel plates at a flow rate Q. The surface concentration of the receptor is large enough such that the reaction is effectively first order in the solution concentration, and few receptors on the surface are bound to the ligand. For a solution concentration of 2.5×10^{-8} M, the measured initial rate of reaction (R'') is 4.1×10^{-14} mol cm^{-2} s^{-1}. Use this result and the data below to determine the intrinsic rate coefficient for the reaction k'' and the extent to which transport through the liquid reduces the rate of reaction:

$D_{eff} = 1 \times 10^{-6}$ cm^2 s^{-1}
$Q = 0.00167$ cm^3 s^{-1}
$H = 0.005$ cm
$W = 0.05$ cm
$L = 0.24$ cm

10.18 Glucose isomerase is used in the commercial production of fructose from glucose. To enhance the purity of the final product and to recover the enzyme after reaction, the enzyme is immobilized in porous beads of radius R. The immobilization process alters the values for R_{max} and K_M, so these need to be measured. Use the following data to obtain the values for R_{max}, K_M, and D_{eff}. (Note: C_0 is the glucose concentration on the surface of the beads.) Briefly justify assumptions used.

	Observed isomerization rate (M s^{-1})		
Radius (μm)	$C_0 = 1 \times 10^{-4}$ M	$C_0 = 1 \times 10^{-7}$ M	$C_0 = 5 \times 10^{-8}$ M
5	9.26×10^{-5}	1.19×10^{-6}	5.98×10^{-7}
10	9.26×10^{-5}	1.17×10^{-6}	5.87×10^{-7}
50	4.63×10^{-4}	0.36×10^{-6}	1.82×10^{-9}

10.9 REFERENCES

1. Anderson, M.L.M., *Nucleic Acid Hybridization*. 1997, Oxford: Bios Scientific Publishers.
2. Britten, R.J., and Kohne, D.E., "Repeated sequences in DNA. Hundreds of thousands of copies of DNA sequences have been incorporated into the genomes of higher organisms." *Science*, 1968. 161: pp. 529–540.
3. Carberry, J.J., *Chemical and Catalytic Reaction Engineering*. 1976, New York: McGraw Hill.
4. Espenson, J.H., *Chemical Kinetics and Reaction Mechanisms*. 1981, New York: McGraw Hill.
5. Creighton, T.E., *Proteins, Structures and Molecular Principles*. 1984, New York: W.H. Freeman.
6. Price, N.C., and Stevens, L., *Fundamentals of Enzymology*. 3d ed. 1999, Avon, UK: Oxford University Press.
7. Tinico, I., Sauer, K., Wang, J.C., and Puglisi, J.D., *Physical Chemistry. Principles and Applications in Biological Sciences*. 4th ed. 2002, Upper Saddle River, NJ: Prentice Hall.
8. Barford, D., Das, A.K., and Egloff, M.K. "The structure and mechanism of protein phosphatases: insights into catalysis and regulation." *Ann. Rev. Biophys. Biomol. Struct.*, 1998. 27: pp. 133–164.
9. Bailey, J. and Ollis, D., *Biochemical Engineering Fundamentals*. 2d ed. 1986, New York: McGraw-Hill.
10. Segal, H., "The development of enzyme kinetics," in *The Enzymes*, P.D. Boyer, H. Lardy, and K. Myrback, editors. 1959, New York: Academic Press. pp. 1–48.

11. Segel, L. "On the validity of the steady state assumption of enzyme kinetics." *Bull. Math. Biol.*, 1988. **50**: 579–593.
12. Lauffenburger, D.A., and Linderman, J.J., *Receptors. Models for Binding, Trafficking, and Signaling*. 1993, New York: Oxford University Press.
13. Nollert, M.U., Diamond, S.L., and McIntire, L.V. "Hydrodynamic shear stress and mass transport modulation of endothelial cell metabolism." *Biotech. Bioeng.*, 1991. **38**: pp. 588–602.
14. Liang, J.F., Li, Y.T., and Yang, V.C. "Biomedical application of immobilized enzymes." *J. Pharm. Sci.*, 2000. **89**: pp. 979–990.
15. Mao, H., Yang, T., and Cremer, P.S. "Design and characterization of immobilized enzymes in microfluidic systems." *Anal. Chem.*, 2002. **74**: pp. 379–385.
16. Vishwanath, S.K., Watson, C.R., Huang, W., Bachas, L.G., and Bhattacharyya, D. "Kinetic studies of site-specifically and randomly immobilized alkaline phosphatase on functionalized membranes." *J. Chem. Tech. Biotechnol.*, 1997. **68**: pp. 294–302.
17. Eu, J., and Andrade, J. "Properties of firefly luciferase immobilized through a biotin carboxyl carrier protein domain." *Luminescence*, 2001. **16**: pp. 57–63.
18. Engasser, J.-M., and Horvath, C., "Diffusion and kinetics with immobilized enzymes," in *Applied Biochemistry and Bioengineering*, L.B. Wingard, E. Katchalski-Katzir, and L. Goldstein, editors. 1976, New York: Academic Press. pp. 127–220.
19. Pao, Y.C., *Engineering Analysis: Interactive Methods and Programs with FORTRAN, QuickBASIC, MATLAB, and Mathematica*. Chapter 6, "Ordinary Differential Equations—Initial and Boundary Value Problems." 1999, Boca Raton, FL: CRC Press.
20. Finlayson, B.A., *Nonlinear Analysis in Chemical Engineering*. 1980, New York: McGraw-Hill.
21. Hamilton, B.K., Gardner, C.R., and Colton, C.K. "Effect of diffusional limitations on Lineweaver–Burk plots for immobilized enzymes." *AIChE J.*, 1974. **20**: pp. 503–510.
22. Wheeler, A., *Reaction Rates and Selectivity in Porous Catalysts*, W.G. Frankenburg et al., editors. Vol. 3. 1951, New York: Academic Press. pp. 247–327.

CHAPTER 11

Cell-Surface Ligand–Receptor Kinetics and Molecular Transport Within Cells

11.1 Introduction

In this chapter, we examine both the application of kinetic models to processes involved in the transport of macromolecules into cells and the transmission of signals resulting from the binding of receptors and ligands. In Sections 11.2 and 11.3, we examine the kinetics of bimolecular receptor–ligand binding on the cell surface. The model presented is the simplest binding model possible that explains such interactions. We develop simplified expressions for the amount of ligand bound as a function of time and for the concentration of ligand in solution. In Section 11.4, we generalize these results to more complex receptor–ligand interactions on the cell surface. Receptor–ligand binding often leads to two complementary responses: *receptor-mediated endocytosis* and the generation of cellular signals. Receptor-mediated endocytosis (Section 11.5) involves the internalization and metabolism of ligands bound to cell-surface receptors. This process supplies cells with nutrients and regulates the cell-surface concentration of receptor-bound ligand. The regulation of receptor numbers and internalization dynamics (Section 11.6) provide mechanisms to control the number of molecules that are internalized. *Cellular signaling* involves the transmission of information from the extracellular environment to the intracellular environment as a result of binding and without the transport of the molecule into the cell. We use kinetic models (Section 11.7) to examine the effect of reaction cascades on signal molecule dynamics. In many cases, these signaling pathways affect the expression of genes; simplified models of such processes are presented in Section 11.8.

11.2 Receptor–Ligand Binding Kinetics

Although covalent bond formation is involved in synthetic and metabolic reactions and in the production of some regulatory molecules, cellular signaling and recognition often involve noncovalent interactions between molecules. *Receptors* are proteins that are present on the cell membrane, as well as on the membranes of organelles, including the nucleus. Their role is to transfer signals from the cell surface to the cell interior or between different parts of the cell. This transfer of information occurs as a result of the specific binding of a *ligand* to the receptor. A ligand can be a peptide, a protein, or a hormone. Based upon the types of signals that are generated, the receptors can be classified into four types [1]: G-protein–coupled receptors, ion-channel receptors, receptors that link to enzymes, and receptors with intrinsic enzymatic activity. Receptors that fall into each of these classes are listed in Table 11.1.

The binding of the receptor and ligand produces a conformational change in the cytoplasmic portion of the receptor (Figure 11.1). This conformational change initiates a cascade of signaling events, resulting in a functional change to the cell. Subsequently, several events may occur. Small-molecule second messengers may be activated. Alternatively, enzymes that modify proteins may be activated.

TABLE 11.1

Examples of Cell-Surface Receptors

Receptor	Classification	Effect
Epinephrine	G-protein-coupled receptor	Neurotransmitter and hormone that affects metabolic reactions
Glucagon	G-protein-coupled receptor	Glucose storage
Serotonin	G-protein-coupled receptor	Neurotransmitter that causes constriction of blood vessels in the brain
Acetylcholine	Ion-channel receptor	Neurotransmitter that stimulates or inhibits muscle activity
Cytokines	Tyrosine kinase[a] activator	Stimulates immune system cells to express receptors
Interferon	Tyrosine kinase activator	Cytokine that interferes with the replication of viruses
Insulin	Receptor exists as a monomer. Insulin binding stimulates dimerization, which results in tyrosine kinase activity	Increases glucose uptake by cells and functions as a growth factor
Growth factors	Receptors have intrinsic tyrosine kinase activity	Stimulates cell division

[a]A tyrosine kinase is an enzyme that adds a phosphate group to the amino acid tyrosine (i.e., it phosphorylates tyrosine).

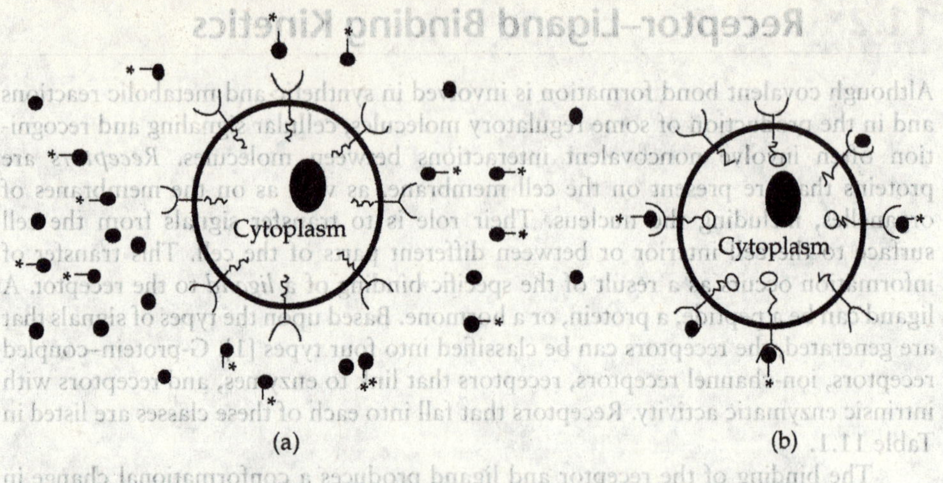

FIGURE 11.1 Schematic of ligand binding to receptors. (a) Binding produces a conformational change to the cytoplasmic portion of the receptor. (b) The ligand population consists of labeled (denoted with *) and unlabeled molecules in order to quantify the amount of ligand that is bound to receptors or to visualize binding.

In many cases, receptor–ligand binding can be treated as a reversible reaction (Figure 11.2). The accumulation of complex is due to the difference between receptor–ligand binding and the dissociation of the complex, according to the formula

$$\frac{dC_c}{dt} = k_1 C_R C_L - k_{-1} C_C, \tag{11.2.1}$$

where k_1 is the association rate coefficient (M^{-1} min^{-1} or M^{-1} s^{-1}) and k_{-1} is the dissociation rate coefficient (min^{-1} or s^{-1}). The ligand concentration is in molar units, whereas the concentrations of receptor and complex are in units of molecules per cell. When the receptor is in solution, the receptor concentration is represented as C_R. When binding occurs to receptors on the cell surface, the receptor concentration is represented as N_R, the number of receptors per cell. Assuming that the total number of receptors per cell on the cell surface is constant and equal to N_{R_T}, a mass balance on the receptor indicates that, at any time, the receptor is either free or bound to ligand. We have

$$N_{R_T} = N_R + N_C, \tag{11.2.2}$$

where N_C is the number of receptor–ligand complex molecules per cell. The cells can regulate the concentration of individual receptors, so the application of Equation (11.2.2) to a particular receptor–ligand interaction may be valid only under limited conditions. Receptor regulation is considered in Section 11.6.

$$L + R \underset{k_{-1}}{\overset{k_1}{\rightleftharpoons}} C$$

FIGURE 11.2 Reversible-binding reaction between receptor and ligand to form complex.

In experiments, the ligand is added in solution, usually in a labeled form, at an initial concentration C_{L_0}. If the ligand is not metabolized by the cells, then at some later time, the ligand is either in solution or bound to receptors. If the number of cells per volume is n, then the ligand concentration at any time is given by

$$C_{L_0} = C_L + \left(\frac{n}{N_A}\right) N_C, \tag{11.2.3}$$

where N_A is Avogadro's number.

Using Equations (11.2.2) and (11.2.3), we write the mass balance on the complex (Equation (11.2.1)) as a function of the complex concentration only:

$$\frac{dN_C}{dt} = k_1(N_{R_T} - N_C)\left(C_{L_0} - \frac{n}{N_A}N_C\right) - k_{-1}N_C. \tag{11.2.4}$$

Although Equation (11.2.4) can be solved exactly, experiments are often designed such that the ligand concentration does not change appreciably from its initial value. An examination of Equation (11.2.4) indicates that ligand depletion is not significant when

$$\left(\frac{n}{N_A}\right)N_C \ll C_{L_0}. \tag{11.2.5}$$

As a result, $C_L \approx C_{L_0}$. With this simplification, Equation (11.2.4) reduces to

$$\frac{dN_C}{dt} = k_1 N_{R_T} C_{L_0} - (k_{-1} + k_1 C_{L_0})N_C. \tag{11.2.6}$$

Equation (11.2.6) is a first-order ordinary differential equation with the following general solution when $N_C = N_{C_0}$ at $t = 0$:

$$N_C = N_{C_0}\exp[-(k_{-1} + k_1 C_{L_0})t] + \frac{k_1 C_{L_0} N_{R_T}}{k_1 C_{L_0} + k_{-1}}\{1 - \exp[-(k_{-1} + k_1 C_{L_0})t]\}. \tag{11.2.7}$$

Using the definition of the equilibrium dissociation constant $K_D = k_{-1}/k_1$, we write Equation (11.2.7) as

$$N_C = N_{C_0}\exp\left[-k_{-1}\left(1 + \frac{C_{L_0}}{K_D}\right)t\right] + \frac{C_{L_0}N_{R_T}}{C_{L_0} + K_D}\left\{1 - \exp\left[-k_{-1}\left(1 + \frac{C_{L_0}}{K_D}\right)t\right]\right\}. \tag{11.2.8}$$

The time required for the concentration to reach one-half of its maximum value for any given ratio of N_C/N_{R_T} is known as the *half-time*, $t_{1/2}$, and is found by setting the ratio of N_C to its maximum value for a given C_{L_0}, denoted $N_{C_{max}}$, to 0.5:

$$\frac{N_C}{N_{C_{max}}} = N_C\frac{C_{L_0} + K_D}{C_{L_0}N_{R_T}} = 0.5 = \left\{1 - \exp\left[-k_{-1}\left(1 + \frac{C_{L_0}}{K_D}\right)t_{1/2}\right]\right\}. \tag{11.2.9}$$

Solving Equation (11.2.9) for $t_{1/2}$ yields

$$t_{1/2} = \frac{\ln 2}{k_{-1}\left(1 + C_{L_0}/K_D\right)}. \tag{11.2.10}$$

For small ligand concentrations, the reaction is limited by dissociation and $k_{-1}t_{1/2}$ approaches ln(2). For large ligand concentrations, binding is rapid and the half-time approaches zero.

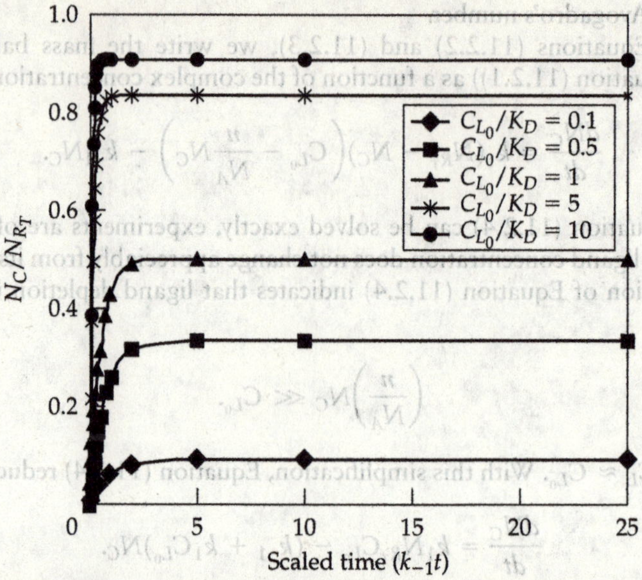

FIGURE 11.3 Changes in normalized concentration of bound ligand as a function of the scaled time, $k_{-1}t$.

Figure 11.3 shows results for the case when the initial concentration of complex is zero ($N_{C_0} = 0$). In general, the concentration rises rapidly and reaches a steady state that increases (and the time required to reach the steady state decreases) as the ratio C_{L_0}/K_D increases. Larger values of C_{L_0}/K_D produce larger values of $k_{-1}(1 + C_{L_0}/K_D)$, the rate coefficient for binding, and smaller values of $t_{1/2}$. As a result, binding approaches steady state more rapidly, and the amount bound at steady state increases.

Example 11.1 Determine the half-time for binding of endothelin to the endothelin 1 receptor [2] under the following conditions:

$$k_{-1} = 0.005 \text{ min}^{-1}$$

$$K_D = 16 \text{ pM} \quad (1 \text{ pM} = 10^{-12} \text{ M})$$

$$C_{L_0} = 1, 10, \text{ and } 100 \text{ pM}$$

Solution Inserting the data provided into Equation (11.2.10) yields the following values for $t_{1/2}$:

$$C_{L_0} = 1 \text{ pM} \qquad t_{1/2} = 130.5 \text{ min}$$

$$C_{L_0} = 10 \text{ pM} \qquad t_{1/2} = 85.3 \text{ min}$$

$$C_{L_0} = 100 \text{ pM} \qquad t_{1/2} = 19.1 \text{ min}$$

Clearly, the half-time to reach steady state decreases as the ligand concentration increases.

11.3 Determination of Rate Constants for Receptor–Ligand Binding

Three parameters are associated with the kinetic model for bimolecular receptor–ligand binding: k_1, k_{-1}, and N_{R_T}. These quantities are determined from kinetic and steady-state experiments at various values of the constant initial ligand concentration C_{L_0}, when no complex is present. The relevant formula is

$$N_C = \frac{C_{L_0} N_{R_T}}{C_{L_0} + K_D}\left\{1 - \exp\left[-k_{-1}\left(1 + \frac{C_{L_0}}{K_D}\right)t\right]\right\}. \tag{11.3.1}$$

Such measurements are performed with a labeled ligand (denoted $N^*_{L_0}$). The label may be either radioactive or fluorescent.

The simplest and most straightforward approach is to measure the amount of ligand bound as a function of time for several different ligand concentrations, as shown in Figure 11.3, and then to fit the data to the model (Equation (11.3.1)) via nonlinear regression. Alternatively, the data can be transformed at each ligand concentration by dividing N_C by the maximum value at each concentration ($N_{C_{max}} = C_{L_0}N_{R_T}/(C_{L_0} + K_D)$). Then, Equation (11.3.1) can be transformed into

$$\ln\left(1 - \frac{N_C}{N_{C_{max}}}\right) = -k_{-1}\left(1 + \frac{C_{L_0}}{K_D}\right)t. \tag{11.3.2}$$

A linear relation between $\ln(1 - N_C/N_{C_{max}})$ and time (Figure 11.4) is consistent with binding according to the bimolecular model. Next, a plot of the negative of the slope in Equation (11.3.2) (equal to $k_{-1}(1 + C_{L_0}/K_D)$) versus C_{L_0} should yield a straight line with slope equal to k_{-1}/K_D (= k_1) and an intercept equal to k_{-1} (Figure 11.5).

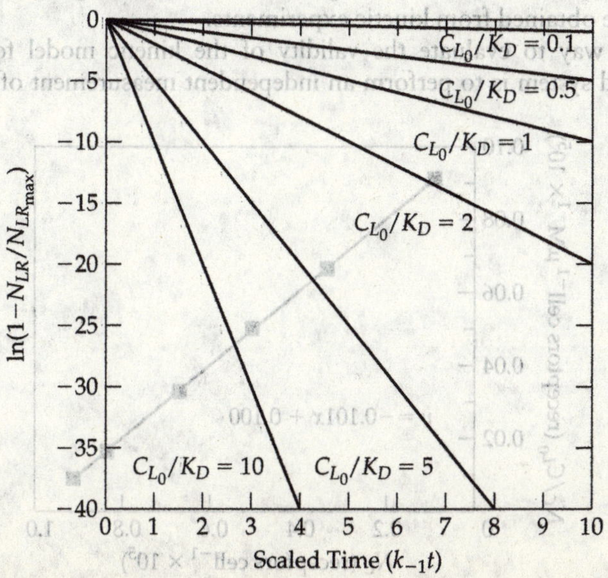

FIGURE 11.4 Plot of $\ln(1 - N_C/N_{C_{max}})$ versus time using data shown in Figure 11.3.

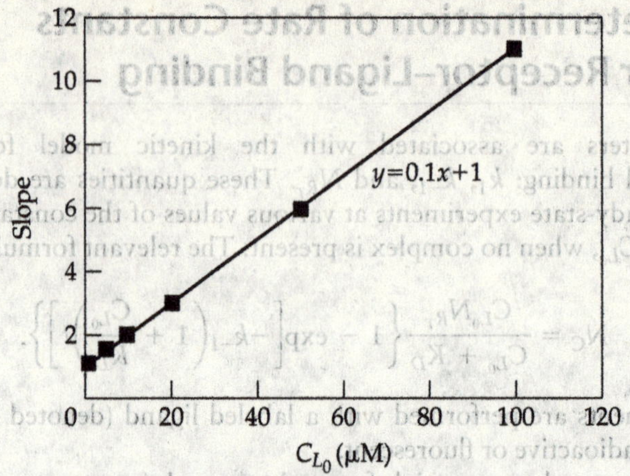

FIGURE 11.5 Plot of $k_{-1}(1 + C_{L_0}/K_D)$ (the slope in Equation (11.3.2)) versus C_{L_0} using data shown in Figures 11.3 and 11.4. The slope equals $k_1 = 0.1\ \mu\text{M min}^{-1}$ and the intercept equals $k_{-1} = 1\ \text{min}^{-1}$.

Again, a linear relation would support the idea that binding is a bimolecular interaction. Lastly, to obtain N_{R_T} and a separate estimate of K_D, the steady-state data can be analyzed. At steady state, Equation (11.3.1) becomes

$$N_C = \frac{C_{L_0} N_{R_T}}{C_{L_0} + K_D}. \tag{11.3.3}$$

This relation can be recast as the linear equation

$$\frac{N_C}{C_{L_0}} = \frac{N_{R_T}}{K_D} - \frac{N_C}{K_D}. \tag{11.3.4}$$

Data plotted according to Equation (11.3.4) is known as a *Scatchard plot*. A similar linearization of data is used to analyze enzyme kinetics in Section 10.5. The slope equals $-1/K_D$ and the intercept equals N_{R_T}/K_D. A Scatchard plot for the data shown in Figures 11.3 and 11.5 is shown in Figure 11.6. The resulting value of K_D is identical to the value obtained from kinetic experiments.

Another way to evaluate the validity of the kinetic model for a particular receptor–ligand system is to perform an independent measurement of some or all of

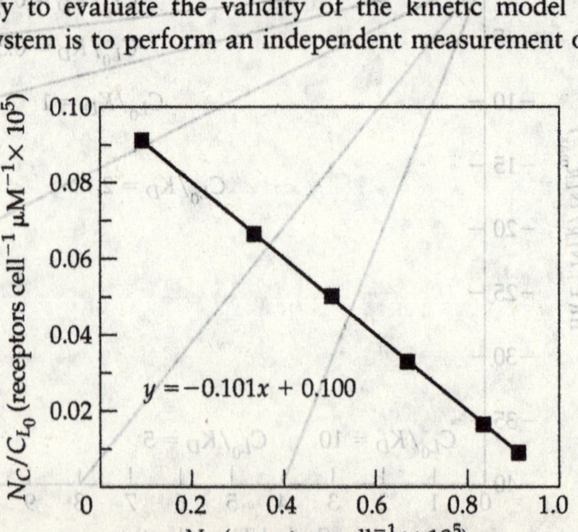

FIGURE 11.6 Scatchard plot of the steady-state data from Figure 11.3. $K_D = 10\ \mu\text{M}$ and $N_{R_T} = 1 \times 10^5$ receptor cell^{-1}.

the rate constants. One such experiment is a dissociation experiment. In this experiment, the labeled ligand is allowed to bind to receptors on cells for some period of time. The solution with the labeled ligand is removed, and the cells are rinsed to detach any loosely attached ligand. Then, the cells are incubated with a large concentration of unlabeled ligand. The labeled ligand dissociates from the receptor. Because there is a substantial amount of unlabeled ligand and the resulting solution concentration of labeled ligand is small, rebinding of labeled ligand does not occur. Since dissociation occurs, $C_{L_0} = 0$ during the dissociation period and Equation (11.2.4) reduces to

$$\frac{dN_C}{dt} = -k_{-1}N_C. \tag{11.3.5}$$

The initial condition is that $N_C = N_{C_0}$. The resulting solution of Equation (11.3.5) is simply

$$N_C = N_{C_0}\exp(-k_{-1}t). \tag{11.3.6}$$

For binding that conforms to the bimolecular model (Figure 11.2), a plot of $\ln(N_C/N_{C_0})$ versus time yields a straight line with slope equal to $-k_{-1}$.

In addition to binding specifically to receptors, ligand can bind nonspecifically to charged groups on the cell surface. A major difference between receptor binding and nonspecific binding is that nonspecific binding is *nonsaturable*. The amount that is bound nonspecifically is proportional to the ligand concentration. Thus, in a typical experiment, the measured amount of ligand that is bound to the cell surface is the sum of receptor-bound ligand and nonspecifically bound ligand (Figure 11.7). Since nonspecific binding is nonsaturable, it can be measured by incubating cells with labeled ligand and an excess of unlabeled ligand. The amount that is bound specifically can be determined by subtracting the amount bound nonspecifically from the total amount of bound ligand.

Example 11.2 Use the following data to determine the specific binding of a ligand to a receptor, the values of K_D, and the number of receptors per cell:

$C_{L_0}(M)$	Amount bound without unlabeled ligand	Amount bound with 100 excess unlabeled ligand
1×10^{-10}	15,000	5,000
5×10^{-10}	58,000	25,000
1×10^{-9}	100,000	50,000
5×10^{-9}	330,000	250,000
1×10^{-8}	590,000	500,000

Solution The amount specifically bound represents the difference between the amount bound without the excess unlabeled ligand and the amount bound with the excess unlabeled ligand. A Scatchard plot is linear, with K_D (= the negative of the slope) equal to 1×10^{-9} M and $N_{R_T}/K_D = 1 \times 10^{14}$ molecules cell^{-1} M^{-1}. From the values of N_{R_T}/K_D and K_D, N_{R_T} equals 1×10^5 molecules cell^{-1}.

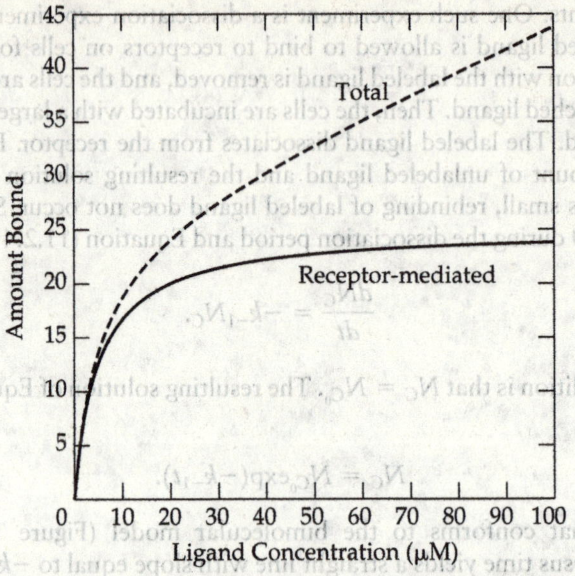

FIGURE 11.7 Receptor-specific binding and total binding.

11.4 Deviations from Simple Bimolecular Kinetics

Although it is appealing to assume that receptor–ligand binding is a simple bimolecular reversible reaction, this assumption must be examined for each receptor studied. Significant deviations from this idealized behavior can occur. Deviations in kinetic behavior consist of changes in the apparent association constants with ligand concentration and accelerated dissociation in the presence of unlabeled ligand. We briefly discuss the types of deviations that are observed and then present several general models that describe the mechanistic basis for empirical measurements.

Scatchard plots for equilibrium receptor–ligand binding are often used to assess whether binding departs from simple reversible behavior. In general, an apparent equilibrium dissociation constant can be defined from the reciprocal of the local slope of a Scatchard plot. When the curve is concave down, the apparent equilibrium dissociation constant K_D decreases as the amount of ligand that is bound increases (Figure 11.1). The decrease in K_D corresponds to an increase in affinity with an increasing amount bound and is known as *positive cooperativity.* When the curve is concave up, K_D increases as the amount of bound ligand increases (Figure 11.8). This property is known as *negative cooperativity.*

Cooperative behavior may be true or apparent. True cooperative behavior arises when binding of the receptor to the ligand alters the conformation of the receptor such that the affinity for ligand binding increases or decreases. For this to occur, the receptor usually consists of several subunits that have similar structures and that bind to a ligand molecule. Binding of the ligand to one subunit alters the conformation of the other subunits, either facilitating or hindering the binding of the ligand to those subunits.

The classic example of positive cooperativity is oxygen binding to hemoglobin. (This is not a receptor–ligand interaction, but the explanation is relevant to such

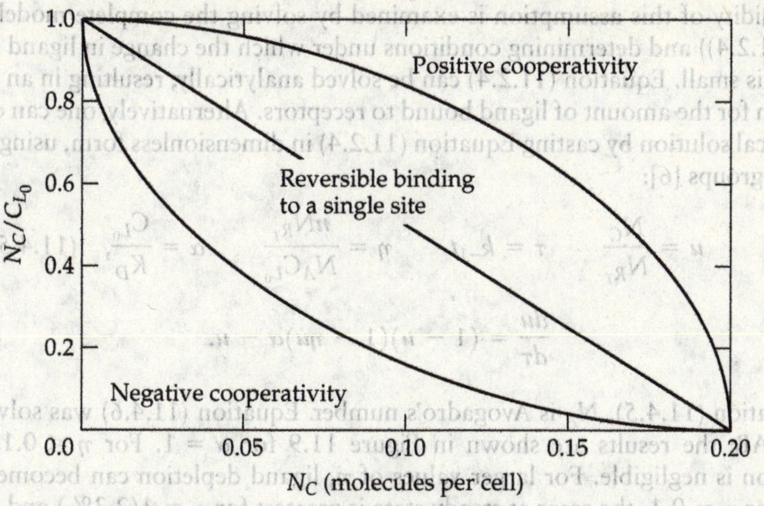

FIGURE 11.8 Scatchard plot showing deviations from equilibrium-binding behavior for complex formation.

behavior.) Hemoglobin has four subunits: two α chains and two β chains. Insulin binding to its receptor appears to represent a case of true negative cooperativity [3–5].

Apparent cooperativity can arise from a number of causes, some trivial and others quite significant. Among the causes of apparent cooperativity are the following:

- Ligand depletion
- Multiple-receptor subpopulations of differing affinity
- Multivalent ligand binding, in which more than one ligand binds to a receptor:

$$R + 2L \longleftrightarrow RL_2. \tag{11.4.1}$$

- Receptor aggregation, in which receptors may complex with each other before binding to ligand. For example, receptors may dimerize after colliding:

$$R + R \longleftrightarrow RR. \tag{11.4.2}$$

The dimer can bind ligand in a sequential fashion (e.g., erythropoietin binding to its receptor):

$$RR + L \longleftrightarrow LRR, \tag{11.4.3a}$$

$$LRR + L \longleftrightarrow LRRL. \tag{11.4.3b}$$

11.4.1 Ligand Depletion

As noted in Equation (11.2.5), if the concentration of complex is much less than the concentration of ligand, then the ligand concentration does not change appreciably during an experiment. This assumption implies that, throughout the binding process, the ligand concentration is approximately equal to the initial concentration of ligand:

$$C_L \approx C_{L_0} \quad \text{when} \left(\frac{n}{N_A}\right)\frac{N_{R_T}}{C_{L_0}} \ll 1. \tag{11.4.4}$$

The validity of this assumption is examined by solving the complete model (Equation (11.2.4)) and determining conditions under which the change in ligand concentration is small. Equation (11.2.4) can be solved analytically, resulting in an implicit solution for the amount of ligand bound to receptors. Alternatively, one can obtain a numerical solution by casting Equation (11.2.4) in dimensionless form, using the following groups [6]:

$$u = \frac{N_C}{N_{R_T}} \qquad \tau = k_{-1}t \qquad \eta = \frac{nN_{R_T}}{N_A C_{L_0}} \qquad \alpha = \frac{C_{L_0}}{K_D}, \tag{11.4.5a,b,c,d}$$

$$\frac{du}{d\tau} = (1-u)(1-\eta u)\alpha - u. \tag{11.4.6}$$

In Equation (11.4.5), N_A is Avogadro's number. Equation (11.4.6) was solved with MATLAB; the results are shown in Figure 11.9 for $\alpha = 1$. For $\eta < 0.1$, ligand depletion is negligible. For larger values of η, ligand depletion can become significant. For $\eta = 0.1$, the error at steady state is greatest for $\alpha = 1(2.3\%)$ and declines to 0.9% when $\alpha = 0.1$ or $\alpha = 10$. Maintaining η at less than 0.1 ensures that ligand depletion is not significant.

Ligand depletion produces nonlinear Scatchard plots [6]. To see this, note that the steady-state form of Equation (11.4.6) is

$$(1-u)(1-\eta u)\alpha - u = 0. \tag{11.4.7}$$

The solution of this quadratic equation can be plotted in the form of a Scatchard plot as u/α versus u (Figure 11.10). As a result, the curve is dependent only upon η. If η is much less than unity, then the solution of Equation (11.4.7) reduces to $u/\alpha = 1 - u$. As η increases in size, Scatchard plots exhibit negative cooperativity. Because there is less ligand present, binding is less likely to occur. This property appears as a reduction in the apparent K_D as the ligand concentration increases.

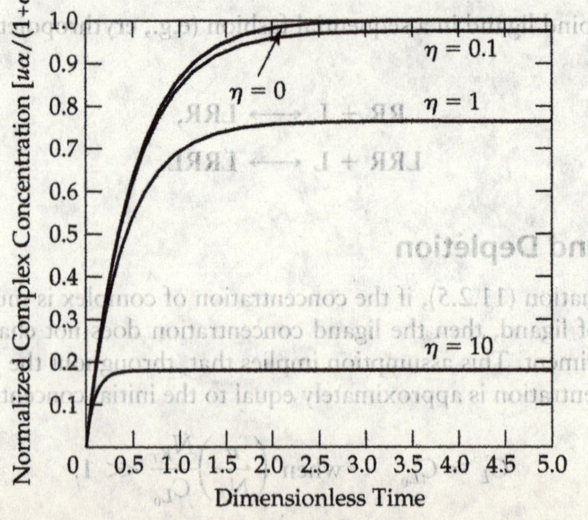

FIGURE 11.9 The effect of ligand depletion on the relative concentration as a function of time for $\alpha = C_{L_0}/K_D = 1$.

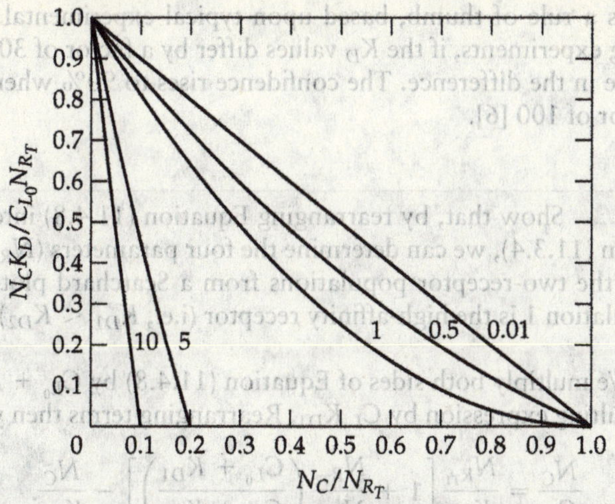

FIGURE 11.10 Normalized Scatchard plot showing the effect of ligand depletion for different values of $\eta = nN_{R_T}/N_A C_{L_0}$.

11.4.2 Two or More Receptor Populations

The binding of a ligand to two different populations of receptors produces nonlinear Scatchard plots. Each receptor acts independently, with equilibrium constants K_{D1} and K_{D2}. The total numbers of receptors in each population are $N_{R_{T1}}$ and $N_{R_{T2}}$. Since it is difficult to determine to which population a given ligand binds, the data have to be analyzed in terms of the total amount bound:

$$N_C = \frac{C_{L_0} N_{R_{T1}}}{C_{L_0} + K_{D1}} + \frac{C_{L_0} N_{R_{T2}}}{C_{L_0} + K_{D2}}. \tag{11.4.8}$$

The Scatchard plot is concave up (Figure 11.11), suggesting apparent negative cooperativity. The solid lines represent limiting curves for the high- and low-affinity populations of receptors. Note that unless the two values of K_D are more than an order of magnitude different, it may not be possible to distinguish two different receptor

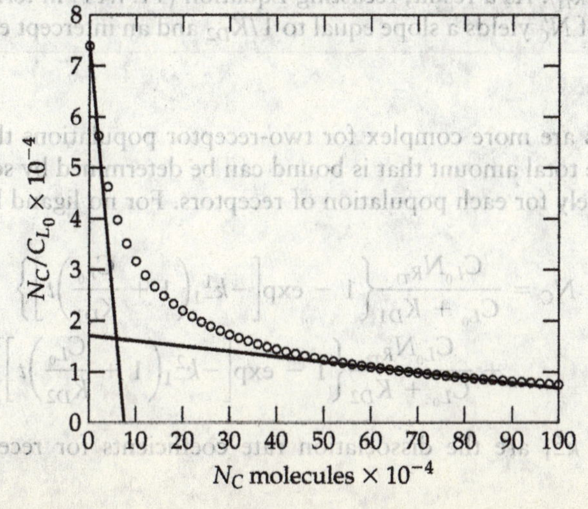

FIGURE 11.11 Scatchard plot for binding to two-receptor populations. The high-affinity population has $K_D = 1\ \mu$M. For the low-affinity population, $K_{D2} = 20\ \mu$M.

populations. As a rule of thumb, based upon typical experimental errors encountered in binding experiments, if the K_D values differ by a factor of 30, then there is a 50% confidence in the difference. The confidence rises to 90% when the K_D values differ by a factor of 100 [6].

Example 11.3 Show that, by rearranging Equation (11.4.8) into a form similar to Equation (11.3.4), we can determine the four parameters ($N_{R_{T1}}$, $N_{R_{T2}}$, K_{D1}, and K_{D2}) for the two-receptor populations from a Scatchard plot. Assume that receptor population 1 is the high-affinity receptor (i.e., $K_{D1} < K_{D2}$).

Solution We multiply both sides of Equation (11.4.8) by $C_{L_0} + K_{D1}$. Next, we divide the resulting expression by $C_{L_0}K_{D1}$. Rearranging terms then yields

$$\frac{N_C}{C_{L_0}} = \frac{N_{R_{T1}}}{K_{D1}}\left[1 + \frac{N_{R_{T2}}}{N_{R_{T1}}}\left(\frac{C_{L_0} + K_{D1}}{C_{L_0} + K_{D2}}\right)\right] - \frac{N_C}{K_{D1}}. \tag{11.4.9}$$

From this result, data can be presented in the form of a Scatchard plot. For low ligand concentrations ($C_{L_0} \ll K_{D1}$), the slope approaches $1/K_{D1}$. The intercept equals

$$\frac{N_C}{C_{L_0}} = \frac{N_{R_{T1}}}{K_{D1}}\left(1 + \frac{N_{R_{T2}}K_{D1}}{N_{R_{T1}}K_{D2}}\right) \quad (C_{L_0} \to 0). \tag{11.4.10}$$

If $K_{D2} \gg K_{D1}$, then the intercept equals $N_{R_{T1}}/K_{D1}$.

When $C_{L_0} \gg K_{D1}$, Equation (11.4.8) becomes

$$N_C = N_{R_{T1}} + \frac{C_{L_0}N_{R_{T2}}}{C_{L_0} + K_{D2}}. \tag{11.4.11}$$

Effectively, the amount bound behaves like receptor-mediated binding plus a constant, $N_{R_{T1}}$. Since $N_{R_{T1}}$ was determined by analyzing the low ligand concentration data, the amount bound can be modified by subtracting $N_{R_{T1}}$ (i.e., $N'_C = N_C - N_{R_{T1}}$). As a result, recasting Equation (11.4.11) in terms of N'_C/C_{L_0} as a function of N'_C yields a slope equal to $1/K_{D2}$ and an intercept equal to $N_{R_{T2}}$.

Binding kinetics are more complex for two-receptor populations than for a single population. The total amount that is bound can be determined by solving Equation (11.2.6) separately for each population of receptors. For no ligand bound to receptors at $t = 0$,

$$N_C = \frac{C_{L_0}N_{R_{T1}}}{C_{L_0} + K_{D1}}\left\{1 - \exp\left[-k_{-1}^1\left(1 + \frac{C_{L_0}}{K_{D1}}\right)t\right]\right\} + \frac{C_{L_0}N_{R_{T2}}}{C_{L_0} + K_{D2}}\left\{1 - \exp\left[-k_{-1}^2\left(1 + \frac{C_{L_0}}{K_{D2}}\right)t\right]\right\}, \tag{11.4.12}$$

where k_{-1}^1 and k_{-1}^2 are the dissociation rate coefficients for receptors 1 and 2, respectively.

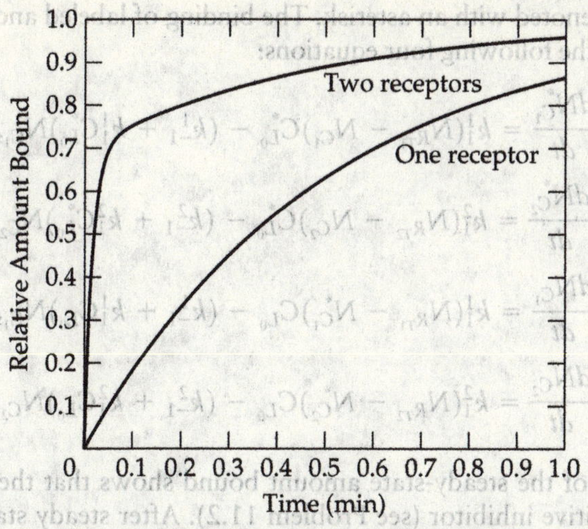

FIGURE 11.12 Binding kinetics for two-receptor populations. Parameter values are $N_{R_{T1}}/N_{C\max} = 0.3$, $K_{D1} = 1\ \mu\text{M}$ and $k_{-1}^{1} = 1\ \text{min}^{-1}$, $N_{R_{T2}}/N_{C_{\max}} = 0.7$, $K_{D2} = 100\ \mu\text{M}$, and $k_{-1}^{2} = 10\ \text{min}^{-1}$.

When the association rate coefficients are sufficiently different in magnitude, the two-receptor model exhibits *biphasic* kinetics, in which there is a rapid binding of ligand to the receptor population with the larger values of k_1 and K_D (Figure 11.12). The difference is more apparent when data are normalized by the maximum binding at a specific ligand concentration, $N_{C_{\max}}$, and plotted as $\ln(1 - N_C/N_{C_{\max}})$ versus time (Figure 11.13).

For binding to one or two populations of receptors, adding unlabeled ligand during the binding step reduces the amount that is bound but does not affect the kinetics of binding or the dissociation kinetics. To assess the effect of adding unlabeled ligand during binding, consider a dissociation experiment. The labeled population of ligand

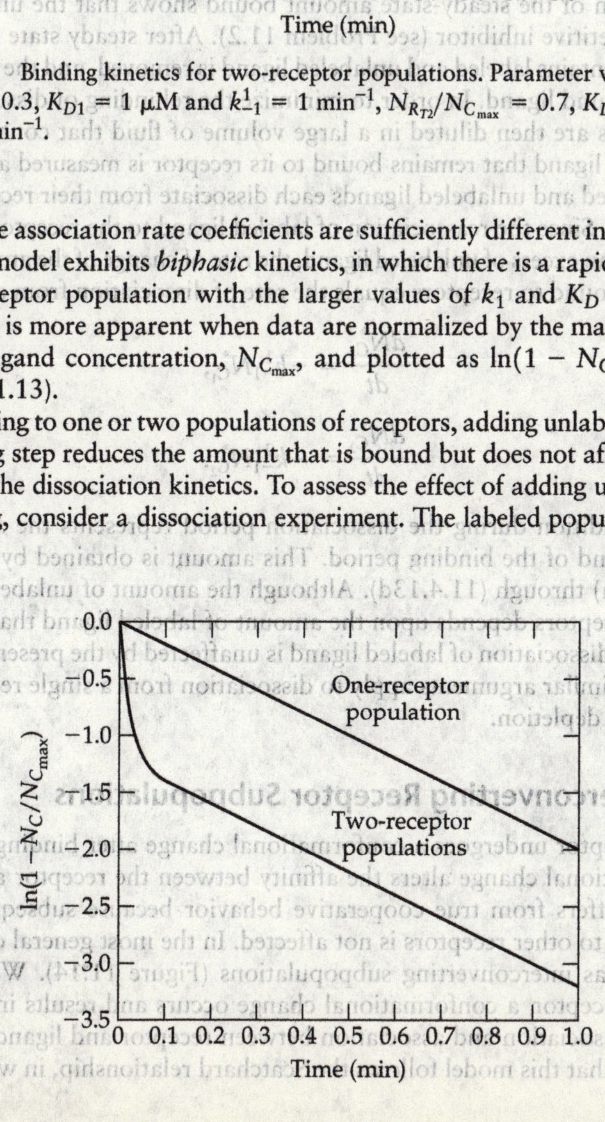

FIGURE 11.13 A plot of $\ln(1 - N_C/N_{C_{\max}})$ versus time for the data in Figure 11.12.

and complex is denoted with an asterisk. The binding of labeled and unlabeled ligand is represented as the following four equations:

$$\frac{dN^*_{C_1}}{dt} = k^1_1(N_{R_{T1}} - N_{C_1})C^*_{L_0} - (k^1_{-1} + k^1_1 C^*_{L_0})N^*_{C_1}, \tag{11.4.13a}$$

$$\frac{dN^*_{C_2}}{dt} = k^2_1(N_{R_{T2}} - N_{C_2})C^*_{L_0} - (k^2_{-1} + k^2_1 C^*_{L_0})N^*_{C_2}, \tag{11.4.13b}$$

$$\frac{dN_{C_1}}{dt} = k^1_1(N_{R_{T1}} - N^*_{C_1})C_{L_0} - (k^1_{-1} + k^1_1 C_{L_0})N_{C_1}, \tag{11.4.13c}$$

$$\frac{dN_{C_2}}{dt} = k^2_1(N_{R_{T2}} - N^*_{C_2})C_{L_0} - (k^2_{-1} + k^2_1 C_{L_0})N_{C_2}. \tag{11.4.13d}$$

An examination of the steady-state amount bound shows that the unlabeled ligand acts as a competitive inhibitor (see Problem 11.2). After steady state is reached, the solution that contains labeled and unlabeled ligand is removed, and the cells are rinsed to detach unbound ligand. In order to minimize the rebinding of dissociated labeled ligand, the cells are then diluted in a large volume of fluid that contains unlabeled ligand, and the ligand that remains bound to its receptor is measured as a function of time. The labeled and unlabeled ligands each dissociate from their receptors by first-order reactions. Since the reassociation of labeled ligand to the receptor is blocked by the presence of an excess of unlabeled ligand, the rate of change of the amount of labeled ligand that is bound to receptors equals the rate of dissociation from each receptor:

$$\frac{dN^*_{C_1}}{dt} = -k^1_{-1}N^*_{C_1}, \tag{11.4.14a}$$

$$\frac{dN^*_{C_2}}{dt} = -k^2_{-1}N^*_{C_2}. \tag{11.4.14b}$$

The initial condition during the dissociation period represents the amount that is bound at the end of the binding period. This amount is obtained by solving Equations (11.4.13a) through (11.4.13d). Although the amount of unlabeled ligand that is bound to receptors depends upon the amount of labeled ligand that is bound, the kinetics of the dissociation of labeled ligand is unaffected by the presence of the unlabeled ligand. Similar arguments apply to dissociation from a single receptor, with or without ligand depletion.

11.4.3 Interconverting Receptor Subpopulations

Often, the receptor undergoes a conformational change after binding to the ligand. The conformational change alters the affinity between the receptor and the ligand. The change differs from true cooperative behavior because subsequent receptor–ligand binding to other receptors is not affected. In the most general case, the receptor is present as interconverting subpopulations (Figure 11.14). When the ligand binds to the receptor, a conformational change occurs and results in differences in the rates of dissociation and association between receptor and ligand. Interestingly, one can show that this model follows the Scatchard relationship, in which [6]

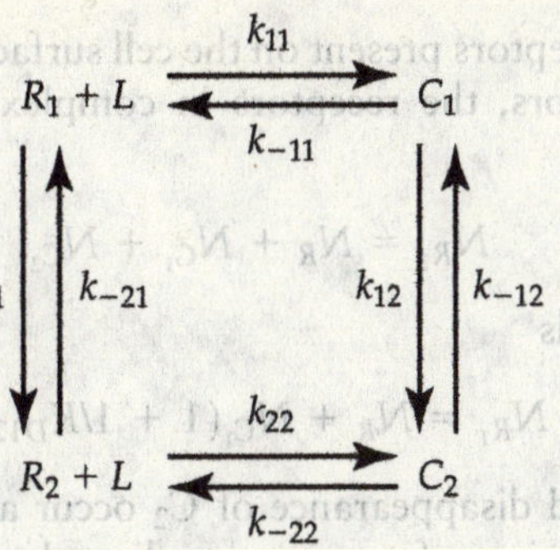

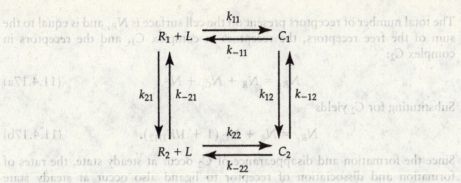

FIGURE 11.14 The most general model of interconverting receptor subpopulations [6].

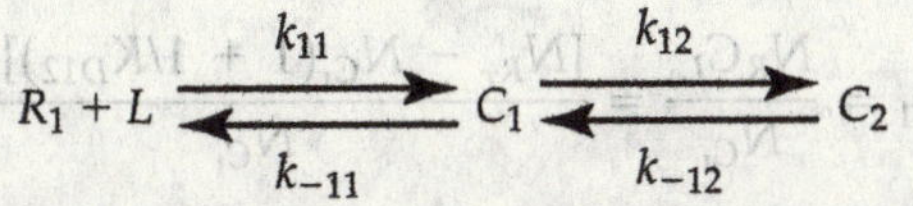

FIGURE 11.15 Interconverting receptor subpopulations in which C_2 is formed by conversion of C_1.

$$N_C = N_{C_1} + N_{C_2} = \frac{C_{L_0} N_{R_T}}{C_{L_0} + K_{Dapp}}, \tag{11.4.15}$$

where $N_{R_T} = N_{R_{T1}} + N_{R_{T2}} + N_{C_1} + N_{C_2}$ and

$$K_{Dapp} = K_{D11}\left(\frac{1 + 1/K_{D21}}{1 + 1/K_{D12}}\right) = K_{D22}\left(\frac{1 + K_{D21}}{1 + K_{D12}}\right), \tag{11.4.16}$$

in which $K_{D11} = k_{-11}/k_{11}$, $K_{D21} = k_{-21}/k_{21}$, $K_{D12} = k_{-12}/k_{12}$, and $K_{D22} = k_{-22}/k_{22}$.

An important special case of the model for two interconverting receptor subpopulations arises when $1/K_{D21} = 0$. In this case, the ligand cannot dissociate from the second population of receptors; the second state can be formed only after the receptor binds to ligand (Figure 11.15).

Both models exhibit linear Scatchard plots, but the dissociation of labeled ligand is influenced by the presence of unlabeled ligand. The reason is that, as unlabeled ligand forms C_1 or C_2, the conversion of labeled ligand is reduced, thereby resulting in greater dissociation of ligand from the receptors than would occur if no ligand were present.

Example 11.4 For the interconverting receptor model presented in Figure 11.15, determine the steady-state relationship between the total amount of ligand per cell and the ligand concentration in the extracellular fluid when ligand depletion is not important.

Solution At steady state, the number of complex C_2 molecules per cell is related to the number of complex C_1 molecules per cell, as $N_{C_2} = k_{12}/k_{-12}N_{C_1} = N_{C_1}/K_{D12}$.

The total number of receptors present on the cell surface is N_{R_T} and is equal to the sum of the free receptors, the receptors in complex C_1, and the receptors in complex C_2:

$$N_{R_T} = N_R + N_{C_1} + N_{C_2}. \qquad (11.4.17a)$$

Substituting for C_2 yields

$$N_{R_T} = N_R + N_{C_1}(1 + 1/K_{D12}). \qquad (11.4.17b)$$

Since the formation and disappearance of C_2 occur at steady state, the rates of formation and dissociation of receptor to ligand also occur at steady state ($k_{11}N_R C_{L_0} = k_{-11}N_{C_1}$). The steady-state relation for receptor–ligand binding and dissociation then becomes

$$K_{D11} = \frac{N_R C_{L_0}}{N_{C_1}} = \frac{[N_{R_T} - N_{C_1}(1 + 1/K_{D12})]C_{L_0}}{N_{C_1}}. \qquad (11.4.18)$$

Rearranging terms yields

$$N_{C_1} = \frac{N_R C_{L_0}}{K_{D11} + C_{L_0}(1 + 1/K_{D12})}. \qquad (11.4.19)$$

The total number of bound receptors is

$$N_C = N_{C_1} + N_{C_2} = \frac{N_{R_T} C_{L_0}(1 + 1/K_{D12})}{K_{D11} + C_{L_0}(1 + 1/K_{D12})} = \frac{N_{R_T} C_{L_0}}{[K_{D11}K_{D12}/(1 + K_{D12})] + C_{L0}}. \qquad (11.4.20)$$

Note that Equation (11.4.20) corresponds to Equations (11.4.15) and (11.4.16) for $K_{D21} = \infty$.

It is possible to discriminate between the various models for receptor–ligand binding by analyzing equilibrium binding (i.e., the kinetics of dissociation and the addition of unlabeled ligand during the binding step), the models for which are presented in Table 11.2 [6].

11.5 Receptor-Mediated Endocytosis

Only small nonpolar molecules permeate across the plasma membranes of cells. The entry of polar molecules, ions, and macromolecules is regulated by different mechanisms. Membrane proteins form transmembrane channels that regulate ion transport. Carrier molecules transport small polar molecules such as glucose. *Receptor-mediated endocytosis* is the major pathway by which macromolecules enter the cell. Receptor-mediated endocytosis acts to internalize proteins for metabolism, for the generation of signals, and for the regulation of receptor numbers. Molecules transported by this mechanism include nutrients (low-density lipoprotein [LDL] or transferrin), immunoglobulins, hormones (insulin), growth factors, and viruses.

TABLE 11.2

Behavior of Various Kinetic Models for Receptor–Ligand Binding

Type of binding	Scatchard plot	Dissociation kinetics	Effect of unlabeled ligand on dissociation[a]
Single receptor	Linear	Single exponential	No
Single receptor depletion–ligand	Nonlinear	Single exponential	No
Two populations of receptors	Nonlinear	Double exponential	No
Receptor–ligand interconversions	Linear	Two or more exponentials	Yes
Interconversion of ligand to a nondissociable form	Linear	Double exponential	Yes
True cooperativity	Nonlinear	Double exponential	Yes

Source: Adapted from Ref. [6], © 1993 by Oxford University Press, Inc. Used by permission of Oxford University Press, Inc.

[a]In this experiment, labeled and unlabeled ligand are added during the binding step. Then, after removing the solution and rinsing the cells to remove unbound receptor, the dissociation of labeled ligand is measured in the presence of excess unlabeled ligand.

Through the interaction of cytoplasmic proteins, cell-membrane components form vesicles and transport membrane, fluid, and solutes from one region of the cell to another [7]. *Endocytosis* is the process of vesicle formation from the plasma membrane. The vesicles fuse either with lysosomes, wherein protein degradation occurs, or with the Golgi apparatus. Some endocytotic vesicles are recycled back to the plasma membrane. *Exocytosis* is a process by which membrane vesicles form from the Golgi apparatus, fuse with the plasma membrane, and release newly synthesized proteins to the extracellular fluid. The processes of endocytosis and exocytosis produce a great amount of membrane traffic. Endocytosis can result in the turnover of the plasma-membrane lipids every 1–10 hours, a rate much faster than membrane synthesis. Most of the internalized membrane is recycled in an efficient manner, maintaining the polarity and composition of the plasma membrane. Vesicles do not diffuse in the cytoplasm but are transported along a network of microtubules.

The first step in receptor-mediated endocytosis is the binding of the protein to a specific receptor on the cell surface (Figure 11.16). Receptors migrate into specialized regions of the cell membrane known as *coated pits,* identifiable by their dark appearance in transmission electron micrographs. The pits contain *clathrin* and *adaptor proteins* (Figure 11.17). Clathrin is a cytoplasmic protein that initiates the formation of vesicles. Adaptor proteins bind to the cytoplasmic portion of the receptor and to clathrin. Several different cytoplasmic sequences regulate the binding of receptors to adaptor proteins. As a result, some receptors, such as the LDL receptor, bind to adaptor proteins regardless of whether a ligand is bound to the receptor. Other receptors, such as the epidermal growth factor (EGF) receptor, are activated after ligand binding and can then bind to adaptor proteins [8].

Coated pits typically cover a few percent of the cell membrane, although some cells have very high densities of coated pits (Figure 11.17). Coated pits continuously form coated vesicles every 2–3 minutes in a process referred to as *constitutive*. If a receptor-bound ligand is in the coated pit when a vesicle is formed, then it is internalized. In

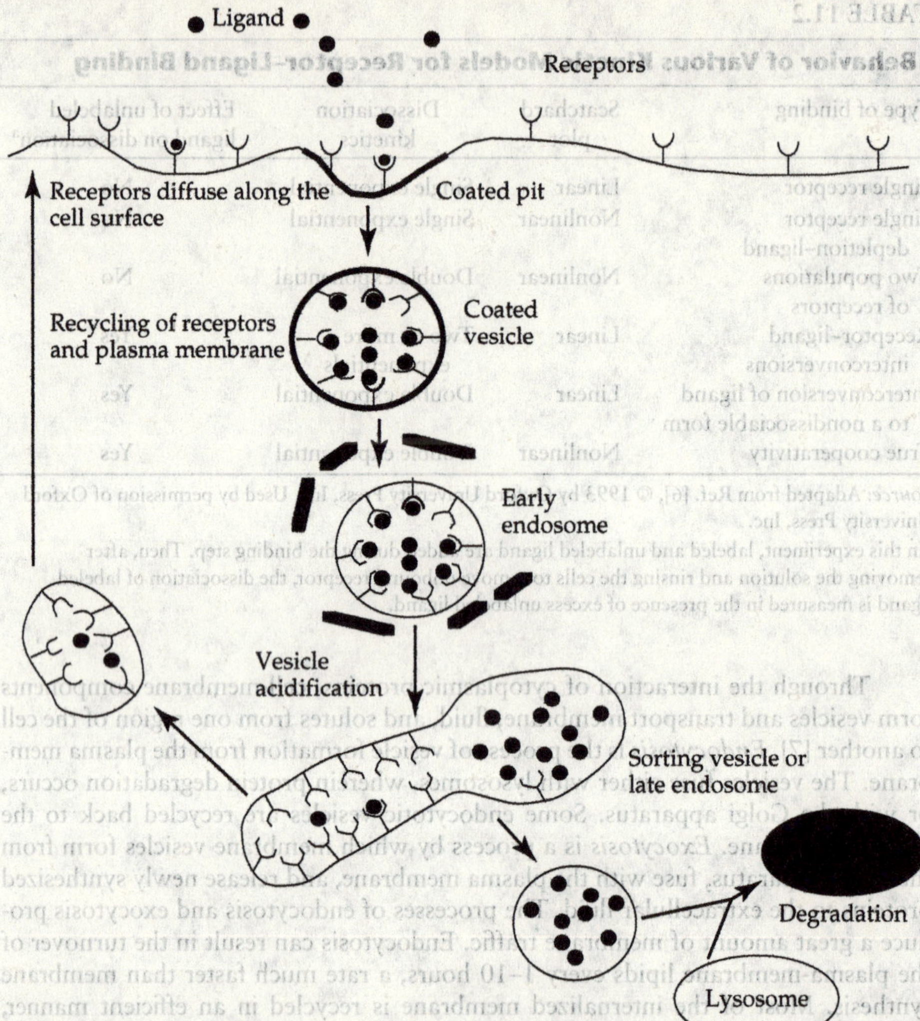

FIGURE 11.16 Schematic of receptor-mediated endocytosis and the fate of receptor and ligand.

addition, ligands can be internalized by fluid-phase endocytosis during vesicle formation, a process that is usually less efficient than receptor-mediated endocytosis.

Within 1–2 minutes after the internalization of vesicles, clathrin dissociates from the vesicle. The vesicle is now referred to as an *early endosome*. At this point, ion pumps in the vesicle membrane lower the pH within the endosome. In many receptor–ligand interactions, the receptor–ligand affinity decreases with decreasing pH. Dissociated receptor and ligand segregate by mechanisms that are poorly understood. Most of the ligand molecules remain in vesicles that are directed to *lysosomes*, where proteins are degraded to amino acids. It usually takes a molecule 15 minutes to reach the lysosome after it binds to the receptor. Most types of receptors are sorted and sent to tubular structures that return the receptors to the cell membrane. For example, the LDL receptor makes a round-trip in about 12 minutes. Each receptor makes about 150 round-trips before being degraded.

Receptors involved in receptor-mediated endocytosis are classified on the basis of whether they undergo endocytosis with or without binding to ligand. *Class 1*

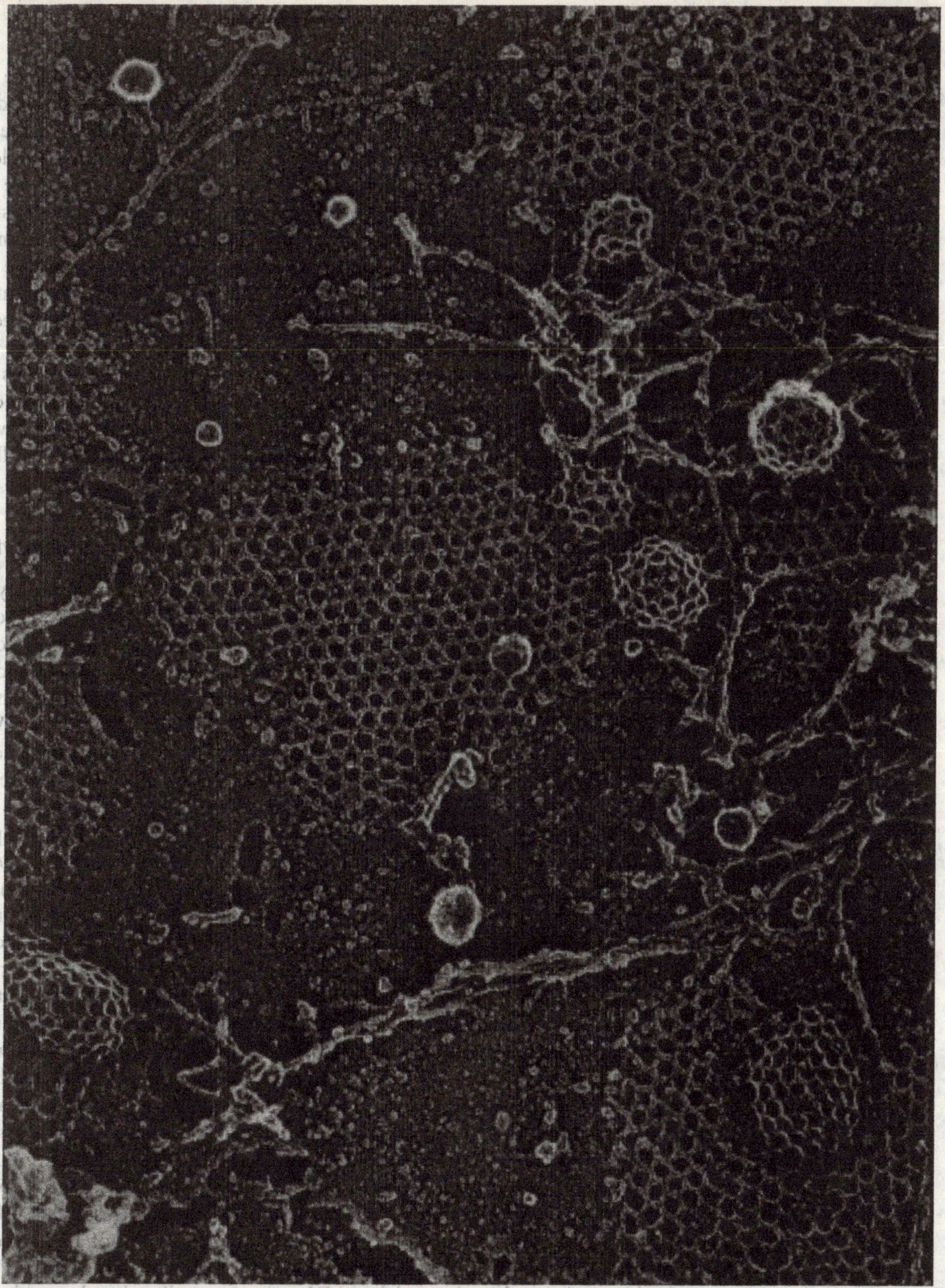

FIGURE 11.17 Cytoplasmic face of the plasma membrane, prepared using a rapid-freeze, deep-etch, rotary-shadowing method. Clathrin lattices consist mainly of hexagons, whereas clathrin-coated pits consist of pentagons and hexagons. (Reproduced from Ref. [9], with permission from Elsevier.)

receptors are receptors for hormones and nutritional molecules. In the absence of ligands, class 1 receptors are uniformly distributed over the cell surface. After binding to ligands, these receptors bind to coated-pit proteins and cluster rapidly in coated pits. Because the receptors are synthesized at a low rate (due to the low rate of endocytosis), the number of class 1 receptors on the cell surface declines rapidly after the receptors bind to ligands. *Class 2* receptors are receptors for nutritional molecules such as LDL and transferrin. Since cells have a continuous demand for molecules internalized by class 2 receptors, these receptors bind to coated-pit proteins in the absence of ligand. As a result, class 2 receptors tend to cluster in coated pits. Further, the number of receptor molecules per unit of surface area on the cell membrane is regulated only slowly by the addition of ligands, often by changes in receptor synthesis.

Another class of vesicles is known as *caveolae*—flask-like invaginations in the surface membrane [10]. These vesicles form more slowly than the clathrin-coated vesicles. Caveolin 1, the principal transmembrane protein found in the caveolae, binds to cholesterol and also interacts with signaling molecules. In addition, albumin binds specifically to caveolae.

11.5.1 A Kinetic Model for LDL Receptor-Mediated Endocytosis

In this section, we consider receptor-mediated endocytosis of LDL as a prototype for other molecules. LDL plays an important role in the delivery of cholesterol to cells. The LDL molecule itself consists of a core of cholesterol ester surrounded by cholesterol, phospholipids, and proteins. Both LDL and the less dense very-low-density lipoprotein (VLDL) have the same protein portion, known as *apoprotein B*. The liver uses cholesterol derived from food or synthesized by the liver to synthesize VLDL. The VLDL is transported through the blood, where it is converted to intermediate-density lipoprotein and LDL, which is internalized by a class 2 receptor. The receptor binds to a binding region on the protein portion of LDL.

An understanding of the dynamics of the LDL receptor is, in turn, helpful in understanding the basis for the genetic disease *familial hypercholesterolemia* (FH). This disease is characterized by extremely high levels of plasma cholesterol and the early onset of cardiovascular disease. Several mutations of the LDL receptor produce the disease. In some cases, the affinity between LDL and its receptor is reduced, so that K_D is much greater than normal, leading to impaired binding. In other cases, there is a defect in the internalization of the LDL receptor. Several rare mutations involve normal binding to the LDL receptor, but little localization in coated pits. The net effect is that the number of functional LDL receptors that undergo binding and internalization of LDL is reduced, with reduced LDL uptake by cells. In the heterozygous form of the disease, one copy of the LDL receptor gene is normal and the number of functional LDL receptors is 50% of normal. This form of FH is common, affecting 1 in 500 individuals. The homozygous form, in which there are no functional LDL receptors, is rare, affecting just 1 in 250,000 individuals. One of the most common treatments for elevated plasma cholesterol levels involves a class of drugs known as statins. These drugs cause an increase in the number of LDL receptors, particularly in the liver, thereby facilitating removal of LDL from the blood.

In this section, a kinetic model of LDL receptor-mediated endocytosis is presented to quantitatively describe LDL protein binding, internalization, and degradation [11], as well as to identify the effect of mutations on this process.

An important property of cells that facilitates the analysis of endocytosis is that, at 4°C, internalization does not occur. As a result, it is possible to separate the internalization process from the binding process and to deduce the kinetics of receptor-mediated endocytosis in a type of experiment known as the *pulse-chase experiment.* In this experiment, cells are incubated with radiolabeled LDL at 4°C. This period of incubation is known as the *pulse period.* Next, the solution is removed and the cells are rinsed at 4°C to remove LDL that is not bound to them. The solution is then replaced with a solution of unlabeled LDL and warmed to 37°C. Unlabeled LDL is used to prevent the rebinding of any LDL that dissociates from the receptor. During this period, known as the *chase period,* the disappearance of labeled LDL from the cell surface is measured. The appearance of labeled LDL in the medium is also measured, to determine the amount of LDL dissociation from its receptors. In addition, the amount of LDL that remains bound to the receptors, internalized, and degraded is measured.

The kinetics of labeled LDL disappearance from the surface of fibroblasts from a normal individual and from an individual who is homozygous for FH [12] are first-order reactions (see Section 11.4.3), but the rate constants differ widely (Figure 11.18). For normal cells, the rate constant is 0.257 min^{-1}, which represents the sum of the dissociation of LDL from its receptors and the internalization of LDL. From the measurements of the appearance of labeled LDL in the medium at 37°C and 4°C, the dissociation constant is 0.01 min^{-1}. For cells from a patient with FH, the dissociation constant is similar to the value obtained for the dissociation of LDL from its receptor, suggesting that, for this particular mutation, the LDL receptor is not internalized.

The amount of LDL bound to the surface, internalized, and degraded during the chase period (Figure 11.19) qualitatively resembles a sequential reaction (Section 10.3). These data are analyzed using the following general kinetic model for LDL metabolism during the chase period:

$$\text{LDL} + R \underset{k_{-1}}{\overset{k_2}{\rightleftharpoons}} \text{LDLR} \xrightarrow{k_2} \text{LDL}_i \xrightarrow{k_2} \text{LDL}_d \qquad (11.5.1)$$

Binding of LDL to receptor on cell surface — Internalization via coated pits — Degradation in lysosomes

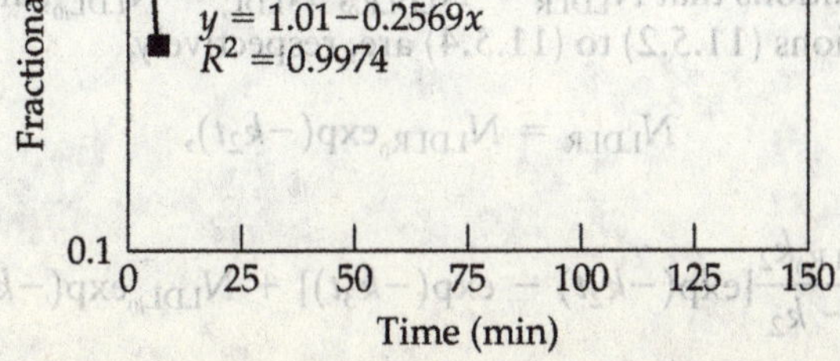

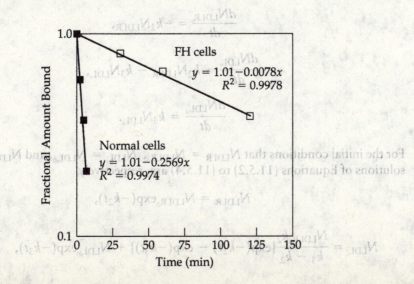

FIGURE 11.18 Disappearance of labeled LDL from the surface during the chase period of a pulse-chase experiment for normal fibroblasts and fibroblasts from a patient with FH. (Data from Ref. [12].)

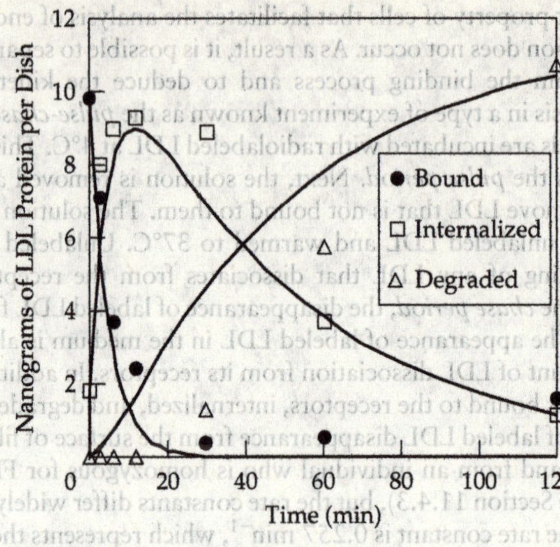

FIGURE 11.19 Pulse-chase experiment and kinetic model. (Data from Ref. [12].)

This model is a simplified representation of the many steps that take place in receptor-mediated endocytosis. In particular, receptors in coated pits and those that are not in coated pits are lumped together, and all of the intracellular ligands are lumped together. The conditions under which such simplifications can be made are discussed shortly.

The model given by Equation (11.5.1) can be used to analyze the pulse-chase experiment as follows: The association step does not occur during the chase period. On the basis of results obtained for LDL dissociation from its receptor and the dynamics of coated-pit internalization, the dissociation step is much slower than the internalization step ($k_{-1} \ll k_2$) and can be neglected.

Mass balances on surface-bound, internalized, and degraded LDL are as follows:

$$\frac{dN_{\text{LDLR}}}{dt} = -k_2 N_{\text{LDLR}}, \tag{11.5.2}$$

$$\frac{dN_{\text{LDL}_i}}{dt} = k_2 N_{\text{LDLR}} - k_3 N_{\text{LDL}_i}, \tag{11.5.3}$$

$$\frac{dN_{\text{LDL}_d}}{dt} = k_3 N_{\text{LDL}_i}. \tag{11.5.4}$$

For the initial conditions that $N_{\text{LDLR}} = N_{\text{LDLR}_0}$, $N_{\text{LDL}_i} = N_{\text{LDL}_{i0}}$, and $N_{\text{LDL}_d} = 0$, the solutions of Equations (11.5.2) to (11.5.4) are, respectively,

$$N_{\text{LDLR}} = N_{\text{LDLR}_0}\exp(-k_2 t), \tag{11.5.5}$$

$$N_{\text{LDL}_i} = \frac{N_{\text{LDLR}_0} k_2}{k_3 - k_2}[\exp(-k_2 t) - \exp(-k_3 t)] + N_{\text{LDL}_{i0}}\exp(-k_3 t), \tag{11.5.6}$$

and

$$N_{\mathrm{LDL}_d} = \frac{N_{\mathrm{LDLR}_0} k_2 k_3}{k_3 - k_2}\left[\left(\frac{1 - \exp(-k_2 t)}{k_2}\right) - \left(\frac{1 - \exp(-k_3 t)}{k_3}\right)\right] + N_{\mathrm{LDL}_{i0}}[1 - \exp(-k_3 t)]. \tag{11.5.7}$$

For $k_2 = 0.20\ \mathrm{min}^{-1}$ and $k_3 = 0.02\ \mathrm{min}^{-1}$, agreement between the model and the data is reasonable, although the model overestimates degradation for times between 5 and 60 minutes (Figure 11.19). Independent measurements show that $k_{-1} < 0.01\ \mathrm{min}^{-1}$.

In addition to performing pulse-chase experiments, researchers often measure binding, internalization, and degradation as a function of time. Based on the model represented by Equation (11.5.1), rate expressions for bound, internalized, and degraded ligands are

$$\frac{dN_{\mathrm{LDLR}}}{dt} = k_1 C_{\mathrm{LDL}} N_R - (k_2 + k_{-1}) N_{\mathrm{LDLR}}, \tag{11.5.8}$$

$$\frac{dN_{\mathrm{LDL}_i}}{dt} = k_2 N_{\mathrm{LDLR}} - k_3 N_{\mathrm{LDL}_i}, \tag{11.5.9}$$

and

$$\frac{dN_{\mathrm{LDL}_d}}{dt} = k_3 N_{\mathrm{LDL}_i}. \tag{11.5.10}$$

Initially, only ligand is present in the extracellular fluid, at a concentration C_{LDL_0}. Often, the ligand concentration can be assumed to be constant throughout the experiment. (The same criterion described for ligand depletion in receptor-binding experiments is applicable, except that now we must account for all cell-associated forms of ligand: bound, internalized, and degraded.) For the period when the concentrations are measured, receptor synthesis and degradation are in balance, so that the LDL receptor concentration on the cell surface is approximately constant. (This is true for class 2 receptors for times much shorter than the time required for their regulation.) Thus, receptors are either bound or free, and

$$N_{R_T} = N_R + N_{\mathrm{LDLR}}. \tag{11.5.11}$$

The solutions of Equations (11.5.8) to (11.5.10) are

$$N_{\mathrm{LDLR}} = \frac{C_{\mathrm{LDL}_0} N_{R_T}}{K + C_{\mathrm{LDL}_0}}(1 - e^{-kt}), \tag{11.5.12}$$

$$N_{\mathrm{LDL}_i} = \frac{k_2 C_{\mathrm{LDL}_0} N_{R_T}}{K + C_{\mathrm{LDL}_0}}\left(\frac{1 - e^{-k_3 t}}{k_3} - \frac{e^{-kt} - e^{-k_3 t}}{k_3 - k}\right), \tag{11.5.13}$$

and

$$N_{\mathrm{LDL}_d} = \frac{k_2 C_{\mathrm{LDL}_0} N_{R_T}}{K + C_{\mathrm{LDL}_0}}\left[t - \frac{1 - e^{-k_3 t}}{k_3} - \left(\frac{k_3}{k_3 - k}\right)\left(\frac{1 - e^{-kt}}{k} - \frac{1 - e^{-k_3 t}}{k_3}\right)\right], \tag{11.5.14}$$

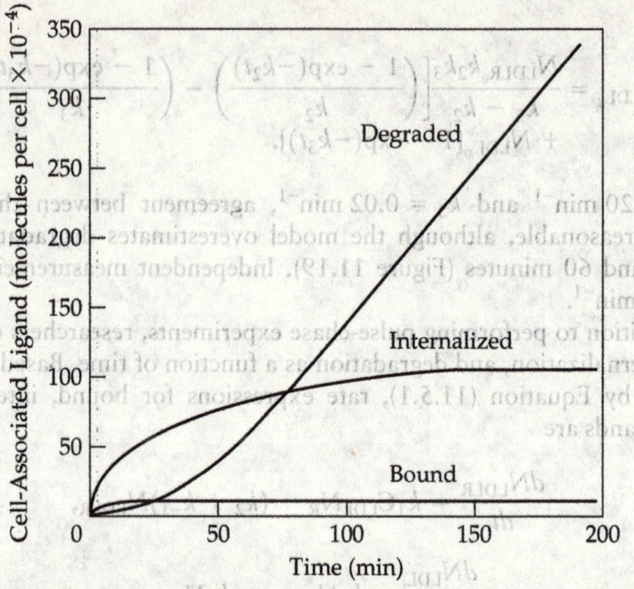

FIGURE 11.20 Bound, internalized, and degraded LDL after incubation of cells with 10 $\mu g\ mL^{-1}$ LDL. For these simulations, $k_2 = 0.20\ min^{-1}$, $K = 25\ \mu g\ mL^{-1}$, $k_3 = 0.02\ min^{-1}$, and $k_1 = 0.0084\ mL\ \mu g^{-1}\ min^{-1}$.

where $K = (k_{-1} + k_2)/k_1$ and $k = k_1(K + C_{LDL_0})$. The constant K is analogous to the Michaelis constant in enzyme kinetics (see Section 10.4).

The kinetics of bound, internalized, and degraded LDL are shown in Figure 11.20. Bound and internalized LDL kinetics resemble those for receptor binding to the cell surface. Note that the steady-state levels of bound and internalized LDL arise from a balance among binding, internalization, and degradation. The level of intracellular LDL is regulated by the balance between internalization and degradation; at steady state, it equals $N_{LDL_i} = (k_2/k_3)N_{LDLR}$. Because internalization is faster than degradation (i.e., $k_2 > k_3$), the amount that is internalized is significantly greater than the amount that is bound.

11.5.2 Receptor Interaction with Coated Pits

The previous model is the simplest that can be constructed to analyze receptor-mediated endocytosis. In this section, we examine the effect on the internalization process of receptor distribution over the cell-surface membrane. Only two-thirds of LDL receptors are present in coated pits, yet all bind to the receptor. This raises the following two questions: Can the distribution of LDL receptors be explained in terms of diffusion-limited binding to coated-pit proteins? What is the relationship between the observed internalization rate coefficient for the ligand k_2 and the rate coefficient for the formation of vesicles? To answer these questions, a model was developed (Figure 11.21) [13] to describe both receptor transport on the cell surface and internalization. In this model, receptors diffuse freely on the cell surface until

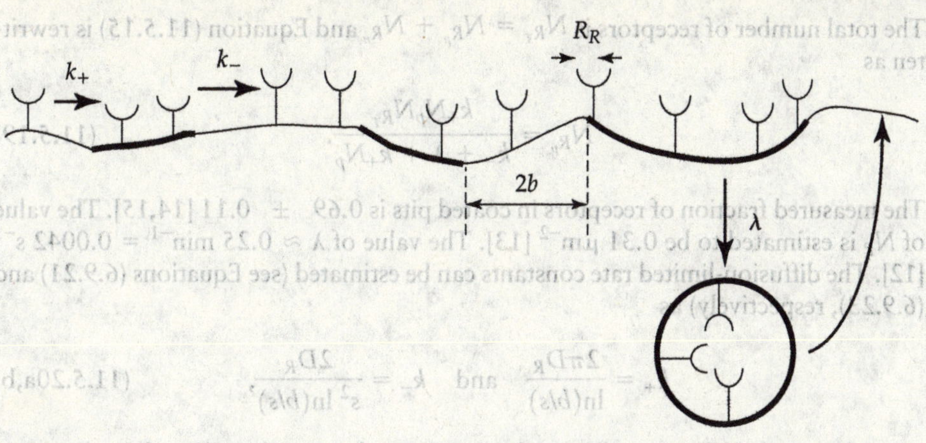

FIGURE 11.21 Receptor diffusion, binding to coated-pit proteins, and internalization.

they reach coated pits, where they bind rapidly and irreversibly to the adaptor proteins. In contrast, class 2 receptors bind to adaptor proteins independently of binding to ligands. The complex is then internalized when a vesicle forms. The sum of receptors on the cell surface and in coated pits is assumed to be constant. This latter simplification means that we do not need to include receptor synthesis and recycling. (We consider these processes in the next section.)

The diffusion-limited binding of receptors to adaptor proteins in coated pits can be viewed as a reversible reaction of receptors that are not in pits, denoted R_n, and those that are in coated pits, designated R_p:

$$R_n + P \underset{k_-}{\overset{k_+}{\rightleftharpoons}} R_p \xrightarrow{\lambda} R_i. \tag{11.5.15}$$

In this model of internalization, P is the density of coated pits (number per area), k_+ is the diffusion-limited association constant, k_- is the diffusion-limited dissociation constant, and λ is the rate constant for vesicle formation. The adaptor proteins are assumed to be uniformly distributed throughout the coated pit and are assumed not to be limiting for binding.

Mass balances on receptors outside of and within coated pits are as follows:

$$\frac{dN_{R_n}}{dt} = -k_+ N_{R_n} N_P + k_- N_{R_p}, \tag{11.5.16}$$

$$\frac{dN_{R_p}}{dt} = k_+ N_{R_n} N_P - (k_- + \lambda) N_{R_p}. \tag{11.5.17}$$

At steady state, the fraction of receptors in coated pits, $N_{R_p}/(N_{R_p} + N_{R_n})$, is

$$\frac{N_{R_p}}{N_{R_p} + N_{R_n}} = \frac{k_+ N_P}{k_- + \lambda + k_+ N_P}. \tag{11.5.18}$$

The total number of receptors is $N_{R_T} = N_{R_p} + N_{R_n}$ and Equation (11.5.15) is rewritten as

$$N_{R_p} = \frac{k_+ N_p N_{R_T}}{k_- + \lambda + k_+ N_p}. \tag{11.5.19}$$

The measured fraction of receptors in coated pits is 0.69 ± 0.11 [14,15]. The value of N_P is estimated to be 0.31 μm^{-2} [13]. The value of $\lambda \approx 0.25\ min^{-1} = 0.0042\ s^{-1}$ [12]. The diffusion-limited rate constants can be estimated (see Equations (6.9.21) and (6.9.23), respectively) as

$$k_+ = \frac{2\pi D_R}{\ln(b/s)} \quad \text{and} \quad k_- = \frac{2D_R}{s^2 \ln(b/s)}, \tag{11.5.20a,b}$$

where b is one-half the mean distance between coated pits, and s is the radius of a coated pit (Figure 6.33). For LDL receptors, $b = 1\ \mu m$, $s = 0.05\ \mu m$, and k_- is assumed to be zero because the reaction is irreversible. For LDL receptors (LDLR), $D_{LDLR} = 1.2 \times 10^{-10}\ cm^2\ s^{-1}$, $k_+ = 2.52 \times 10^{-10}\ cm^2\ s^{-1}$; as a result, $N_{R_p}/N_{R_T} = 0.65$, which is close to the experimental value. For EGF receptors (EGFR), all of the receptors are found in the coated pits after EGF binds to the receptors. The EGFR has a much higher diffusion coefficient than the LDL receptor does; the diffusion coefficient is $D_{EGFR} = 8 \times 10^{-10}\ cm^2\ s^{-1}$, yielding a value of $k_+ = 1.68 \times 10^{-9}\ cm^2\ s^{-1}$, almost 10 times larger than the value for the LDL receptor. The resulting fraction of receptors in coated pits after EGF binding is predicted to be 0.92. Again, this ratio is consistent with experimental data. More detailed analyses of receptor diffusion and binding to coated pits have considered the finite lifetime of coated pits and random appearance and disappearance on the cell membrane [13,16].

Example 11.5 For class 1 receptors, ligand binding to the receptor increases the binding between the receptor and adaptor proteins in coated pits. As a result, ligand binding to class 1 receptors perturbs the steady state, and the previous steady-state analysis is no longer applicable. For the scheme represented by Equation (11.5.15) and an initial distribution of ligand–receptor complexes on the cell surface produced by incubating cells with ligand at 4°C, determine whether a quasi–steady state can be assumed for EGFR internalization after ligand binding. Assuming diffusion-limited binding of the receptor to adaptor proteins in coated pits, find the value of the rate coefficient for receptor–ligand complex internalization.

Solution To assess whether receptor binding to coated pits or receptor internalization is rate limiting, time constants for the two steps need to be compared. For EGF binding to coated-pit proteins, the time constant is approximately $1/k_+N_P = 19.2$ s. The internalization time constant (238 seconds) is much larger than the time constant for receptor binding to coated pits, suggesting that internalization is the rate-limiting step. Thus, a quasi–steady state does not arise. EGFRs bound to ligand should rapidly diffuse into coated pits and are internalized more slowly.

The LDLR is continuously internalized at a rate λN_{R_p}. The timescale of internalization ($1/\lambda$ = 4 minutes) is comparable to the time constant for receptor diffusion into coated pits (2.1 minutes) but is significantly shorter than the timescale for receptor synthesis (5–10 hours). The distribution of receptors in coated pits can be assumed to be at steady state. Thus, the receptor population in coated pits is a fixed fraction of the receptors on the cell surface. As a result, the rate of internalization of receptors is

$$R_{\text{intern}} = \lambda N_{R_p} = \frac{\lambda k_+ N_p N_{R_T}}{k_- + \lambda + k_+ N_p} = k_2 N_{R_T}. \tag{11.5.21}$$

The rate coefficient for LDLR internalization is thus

$$k_2 = \frac{\lambda k_+ N_p}{k_- + \lambda + k_+ N_p}. \tag{11.5.22}$$

Equation (11.5.22) indicates that k_2 is dependent upon the density of coated pits as well as the rate of vesicle formation. These quantities are dependent upon cell type.

11.6 Receptor Regulation During Receptor-Mediated Endocytosis

The simple model considered in the previous section assumed that the number of receptors on the cell surface was constant. Even for class 2 receptors, however, this is true only for specific conditions. In general, cells regulate the number of receptors on the cell surface to control the rate of delivery of a specific ligand. Regulation can be rapid, due to the internalization and degradation of growth factor receptors, or slower, due to the regulation of LDL synthesis by the effect of intracellular cholesterol on LDL gene expression. To account for receptor regulation, the fate of the receptor must be analyzed. In a simple model, the major components are (a) reversible binding of receptors and ligands on the cell surfaces, (b) internalization by vesicle formation, (c) sorting, degradation, or recycling of receptors and ligands, and (d) receptor synthesis [6] (Figure 11.22).

Assuming that ligand depletion is negligible, the following mass balance can be written for the surface receptors:

$$\frac{dN_{R_S}}{dt} = k_1 C_{L_0} N_{R_S} + k_{-1} N_{C_S} - k_{eR} N_{R_S} + k_{\text{rec}}(1 - f_R) N_{Ri} + V_s. \tag{11.6.1}$$

The first two terms on the right-hand side of Equation (11.6.1) represent reversible binding and dissociation of cell-surface receptors and ligands. The third term represents the loss of surface receptors due to receptor-mediated endocytosis of the free receptor. The fourth term represents the addition of receptors due to receptor recycling, and the last term represents the addition of surface receptors due to receptor synthesis. For endocytosis of free receptors, k_{eR} is the internalization rate constant. As previously noted, the value of k_{eR} may differ from the value for the rate constant

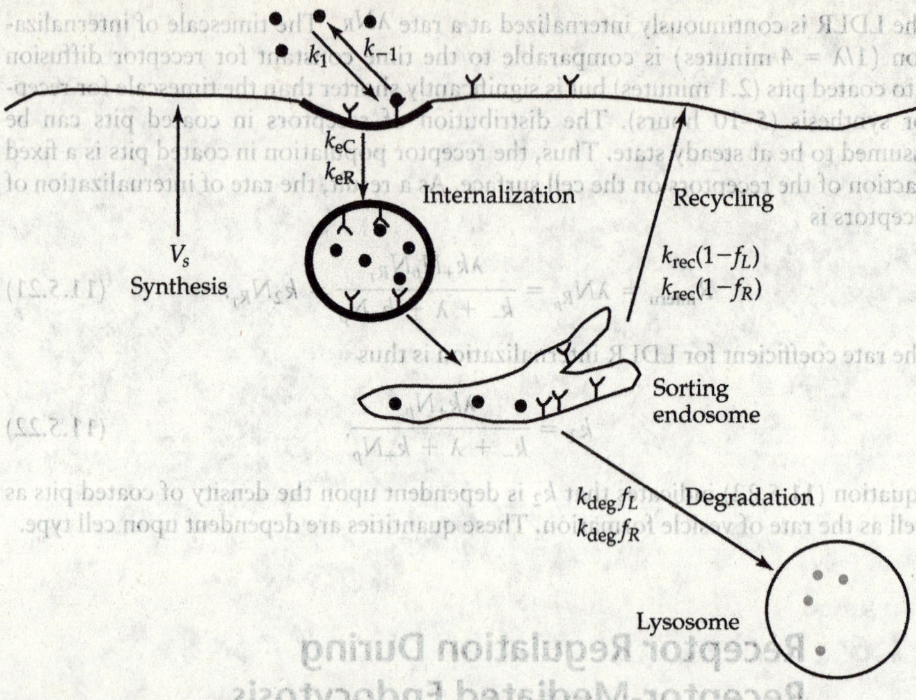

FIGURE 11.22 Schematic of the general model for receptor-mediated endocytosis and recycling of ligand and receptor. (Adapted from Ref. [6], with permission of Oxford University Press, Inc.)

for internalization of the complex, k_2, since some receptors do not localize in coated pits before they bind to ligands. Further, k_{eR} and the corresponding rate coefficient for the ligand-bound receptor k_{eC} are functions of the ligand concentration and time (see Problem 11.11). The rate constant k_{rec} is for recycling and represents the intrinsic rate at which endosomes segregate receptors and ligands and then return the receptors to the cell surface. The fraction of receptors that are sorted for degradation is f_R; $1 - f_R$ represents the fraction of receptors that are returned to the cell surface. V_s is the rate of receptor synthesis.

Complexes are formed as a result of receptor–ligand binding. Complexes disappear from the surface due to the dissociation of receptors and ligands, as well as to internalization. The accumulation of complex on the cell surface is given by the formula

$$\frac{dN_{C_s}}{dt} = k_1 C_{L_0} N_{R_s} - (k_{-1} + k_{eC}) N_{C_s}. \tag{11.6.2}$$

The rate coefficient k_{eC} is that for receptor internalization and is equal to the quantity k_2 that was presented in the model for LDL metabolism in Section 11.5.1.

Once within the cell, receptors and ligands are assumed to dissociate rapidly. Accordingly, only the free intracellular receptor (N_{R_i}) and ligand (N_{L_i}) concentrations are considered. Intracellular receptors arise from the internalization of free receptors and of complexes. These receptors are removed from the intracellular region by degradation and recycling. A mass balance on intracellular receptors yields

$$\frac{dN_{R_i}}{dt} = k_{eR}N_{R_s} + k_{eC}N_{C_s} - \left[k_{deg}f_R + k_{rec}(1 - f_R)\right]N_{R_i}. \quad (11.6.3)$$

The rate constant k_{deg} is the rate constant for receptor degradation.

Ligands are internalized by receptor-mediated endocytosis. (Nonspecific endocytosis is ignored in this analysis, a reasonable assumption for low ligand concentrations and short times.) Ligands are removed from the intracellular region by degradation and recycling. A mass balance for intracellular ligand yields

$$\frac{dN_{L_i}}{dt} = k_{eC}N_{C_s} - \left[k_{degL}f_L + k_{rec}(1 - f_L)\right]N_{L_i}. \quad (11.6.4)$$

Here, the rate constant k_{degL} is the rate constant for ligand degradation and is identical to k_3, as presented in the model for LDL. The quantity f_L is the fraction of ligand sorted for degradation.

This model can be used to examine receptor and ligand dynamics and to identify the various methods by which surface receptors are regulated. The analysis presented next follows that set forth in Ref. [6]. At steady state, in the presence of ligands, the number of receptors on the cell surface is obtained by setting the derivatives in Equations (11.6.1), (11.6.2), and (11.6.3) to zero and solving for the total number of receptors on the cell surface N_{R_T}. We obtain

$$N_{R_T} = N_{R_s} + N_C = \left(1 + \frac{K_D + k_{eC}/k_1}{C_{L_0}}\right)\left\{\frac{\left[1 + \left(k_{rec}(1 - f_{RL})/k_{deg}f_{RL}\right)\right]V_{SL}}{k_{eC} + k_{eR}\left((K_D + k_{eC}/k_1)/C_{L_0}\right)}\right\}. \quad (11.6.5)$$

Receptor sorting and synthesis are affected by ligands. For instance, some receptors may be degraded by remaining bound to ligands that are subsequently degraded. (Remember that this is a general model that accounts for multiple possibilities.) Hence, f_{RL} and V_{SL} are the values in the presence of ligands and f_{R0} and V_{S0} are the values in the absence of ligands. In the absence of ligands ($C_{L_0} = N_{C_s} = 0$), the steady-state solution of Equations (11.6.1) to (11.6.3) is

$$N_{R_{S0}} = \left[\frac{k_{deg}f_{R0} + k_{rec}(1 - f_{R0})}{k_{deg}f_{R0}k_{eR}}\right]V_{S0}. \quad (11.6.6)$$

Equations (11.6.5) and (11.6.6) can be used to describe several different ways in which receptors are regulated. First, if $k_{eC} = k_{eR}$, $f_{RL} = f_{R0}$, and $V_{SL} = V_{S0}$, then the ligand does not affect the number of receptors on the cell surface and $N_{R_S} = N_{R_{S0}}$.

Receptor regulation of class 1 receptors corresponds to the case for which $k_{eC} > k_{eR}$, $f_{RL} = f_{R0}$, and $V_{SL} = V_{S0}$. As a result, the ratio $N_{R_T}/N_{R_{S0}}$ becomes

$$\frac{N_{R_T}}{N_{R_{S0}}} = \frac{1 + \left[(K_D + k_{eC}/k_1)/C_{L_0}\right]}{k_{eC}/k_{eR} + \left[(K_D + k_{eC}/k_1)/C_{L_0}\right]}. \quad (11.6.7)$$

Since $k_{eC} > k_{eR}$, it follows that $N_{R_T}/N_{R_{S0}}$ is less than unity, and the number of receptors declines.

Regulation also occurs at the stage of sorting. The presence of ligands can cause a smaller fraction of receptors to be recycled, possibly because dissociation is not 100%. This case corresponds to $f_{RL} < f_{R0}$, $k_{eC} = k_{eR}$, and $V_{SL} = V_{S0}$. As a result, the ratio $N_{R_T}/N_{R_{S0}}$ is less than unity and equal to

$$\frac{N_{R_T}}{N_{R_{S0}}} = \frac{1 + \left[k_{rec}(1 - f_{RL})/k_{deg} f_{RL}\right]}{1 + \left[k_{rec}(1 - f_{R0})/k_{deg} f_{R0}\right]}. \quad (11.6.8)$$

If $f_{RL} = f_{R0}$, $k_{eC} = k_{eR}$, and $V_{SL} < V_{S0}$, then the ligand affects the rate of synthesis such that

$$\frac{N_{R_T}}{N_{R_{S0}}} = \frac{V_{SL}}{V_{S0}}. \quad (11.6.9)$$

The ratio $N_{R_T}/N_{R_{S0}}$ is less than unity, the same as is observed with the LDLR. Cholesterol esters, released by the degradation of the LDL protein in the lysosome, regulate the expression of the LDL receptor. A typical result is shown in Figure 11.23.

The decreased level of LDLR expression can be deduced empirically by fitting the data as shown in Figure 11.23 and rearranging terms to yield

$$\frac{N_{R_T}}{N_{R_{S0}}} = \frac{V_{SL}}{V_{S0}} = 0.2977 + \frac{2.238 \times 10^{-9}}{3.189 \times 10^{-9} + C_{L_0}}. \quad (11.6.10)$$

To determine the basal rate of synthesis of LDLRs V_{S0}, and the rate constant for receptor degradation, fibroblasts were incubated in the presence of cycloheximide, which blocks protein synthesis [17], and the number of receptors on the cell surface was measured at various times. Thus, $V_s = 0$. The disappearance of receptors from the cell surface was first order in the surface concentration of receptor–ligand complex, with a half-time of about 20 hours. Since this is much slower than the timescale for receptor internalization and recycling, we assume that receptors on the

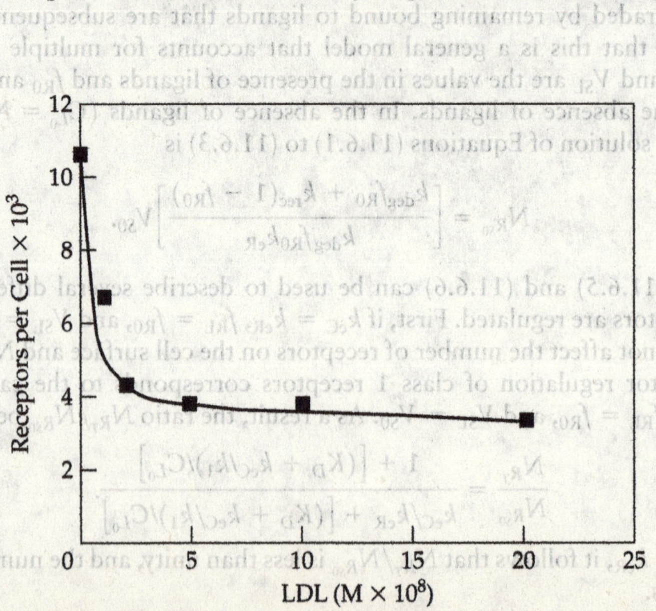

FIGURE 11.23 Effect of LDL concentration on the expression of LDL receptors on the surface of bovine aortic smooth muscle cells. The cells were incubated with varying concentrations of LDL for 24 hours and the number of receptors was determined by binding to radiolabeled LDL.

cell surface are in a quasi–steady state. For these conditions and $N_{C_S} = 0$, Equation (11.6.1) can be solved for N_{R_S} to yield

$$N_{R_S} = \frac{k_{\text{rec}}(1 - f_R)}{k_{\text{eR}}} N_{R_i}. \tag{11.6.11}$$

(Note that ligands are absent in these experiments.) Substituting Equation (11.6.11) into Equation (11.6.3) yields the following relation for the intracellular receptor concentration:

$$\frac{dN_{R_i}}{dt} = -k_{\text{deg}} f_R N_{R_i}. \tag{11.6.12}$$

The total cell-associated receptor concentration is the sum of the concentrations of the receptors on the cell surface and those within the cell, or

$$N_{R_T} = N_{R_S} + N_{R_i} = \left[1 + \frac{k_{\text{rec}}(1 - f_R)}{k_{\text{eR}}}\right] N_{R_i}. \tag{11.6.13}$$

Using the result from Equation (11.6.13), we can rewrite Equation (11.6.12) in terms of the total receptor concentration:

$$\frac{dN_{R_T}}{dt} = -k_{\text{deg}} f_R N_{R_T}. \tag{11.6.14}$$

From the half-time for LDLR disappearance, $k_{\text{deg}} f_R$ equals 5.8×10^{-4} min^{-1}. Since f_R is about 0.01, $k_{\text{deg}} = 0.058$ min^{-1}, which is approximately 3 times larger than the values obtained for LDL (see Table 11.3).

TABLE 11.3

Rate Constants for Receptor-Mediated Metabolism of LDL and EGF

Parameter	EGF[a]	LDL[b]
k_1(M^{-1} min^{-1})	7.2×10^7	0.32×10^7
k_{-1}(min^{-1})	0.34	0.01
k_{eR}(min^{-1})	0.03	0.25
k_{eC}(min^{-1})	0.3	0.25
k_{rec}(min^{-1})	0.058	0.02
k_{degL}(min^{-1})	0.0022	0.02
V_s (receptors min^{-1})	130	1.6–69
f_R	0.2–0.8	< 0.01
f_L	0.2–0.8	0.99

Source: [a]From Ref. [18]; [b]From analysis of data in Refs [11], [12], and [17], using the model presented in Section 11.5.1.

The value of the steady-state basal rate of receptor synthesis is now determined by resolving the steady-state form of Equation (11.6.1) in the absence of ligands:

$$N_{R_S} = \frac{k_{rec}(1 - f_R)}{k_{eR}} N_{R_i} + k_{eR}. \tag{11.6.15}$$

Inserting this result into Equation (11.6.3) yields

$$\frac{dN_{R_i}}{dt} = -k_{deg} f_R N_{R_i} + V_{S0}. \tag{11.6.16}$$

At the steady state, the intracellular receptor concentration of LDL is 20% of the total receptor concentration [17] and $N_{R_i} = 0.25 N_{R_S}$. Thus, from Equation (11.6.16), at the steady state, the rate of receptor synthesis can be determined to be

$$V_{S0} = k_{deg} f_R 0.25 N_{R_S} = 1.64 \text{ receptors min}^{-1}. \tag{11.6.17}$$

Other cells, that express higher levels of LDLRs, have higher levels of receptor synthesis. This analysis provides information about the rates of synthesis with and without ligands; it does not, however, consider the dynamics that govern the change in the number of receptors after exposure to LDL.

Rate constants and recycling parameters are available for LDL and EGF. The values are summarized in Table 11.3. There are considerable differences in the magnitudes of a number of the parameters, indicating that the dynamics of the two systems are different. Note in particular that k_{eR} for EGF (0.03 min^{-1}) is less than the value for the EGF complex (k_{eC} = 0.3 min^{-1}). Thus, the binding of EGF to its receptor leads to a rapid down-regulation of the number of receptors.

11.7 Signal Transduction

11.7.1 Qualitative Aspects of Signal Transduction

Cell-surface receptors have three principal functions: cell–cell communication, cell adhesion to other cells or to the extracellular matrix, and the delivery of nutrients. Cell–cell communication involves the transmission of information from one cell to another via hormones, peptide growth factors, or extracellular matrix molecules. Macromolecular nutrient delivery occurs via receptor-mediated endocytosis, as described in the previous section. In each of these cases, the receptors transmit signals to the cell.

The process of cellular signaling involves the transfer of information from the extracellular environment to the cell interior, resulting in the activation of a specific cellular response, such as cell division, cell migration (e.g., chemotaxis—movement against a concentration gradient), or protein synthesis [1]. Signaling molecules include extracellular molecules such as hormones, molecules linked to receptors that are stimulated after ligand binding (e.g., G-proteins), so-called second messengers, which are activated by signaling molecules produced by ligand–receptor binding

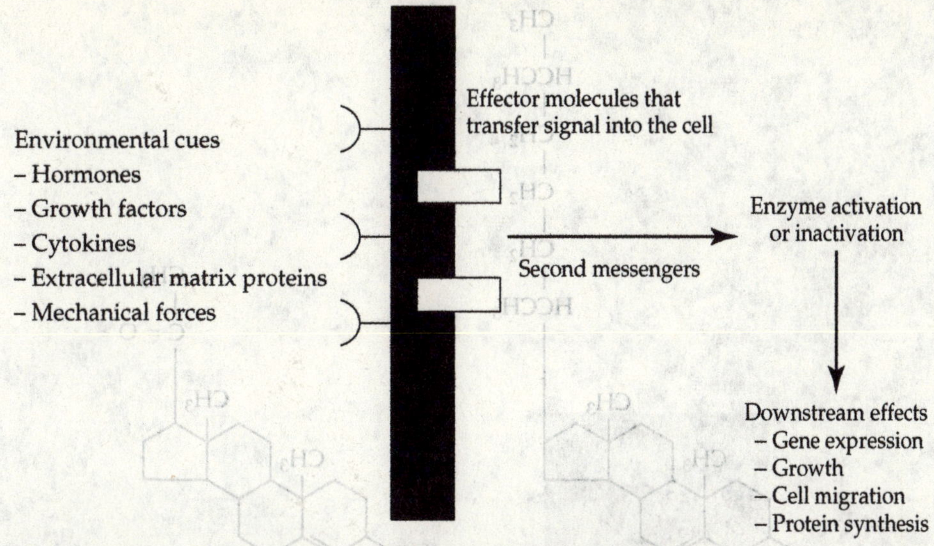

FIGURE 11.24 Overview of the steps involved in the transfer of information from the cell exterior to a functional change in the cell via signaling steps.

(e.g., calcium ions or cyclic adenosine monophosphate cAMP]), and the target molecules of second-messenger action (e.g., kinases) (Figure 11.24).

Signaling occurs in one of three ways. *Endocrine signaling* involves the release of hormones in one tissue, transport through the blood, and delivery to the target tissue. Such signaling requires minutes to deliver the hormone to its site of action. *Paracrine signaling* involves the release of growth factors or hormones from one cell and binding to receptors on cells nearby. The close proximity of the target and the secretory cells means that the hormone can diffuse rapidly between the two types of cells. In *autocrine signaling*, growth factors or hormones that are secreted by one cell bind to receptors on the same cell. Autocrine signaling is a common process in tumor cells.

Hormones can be classified in terms of their sites of action. Small lipophilic molecules (i.e., molecules that are soluble in lipids or organic molecules) diffuse across the plasma membrane and bind to receptors in the cytoplasm. The receptor–hormone complex is then transported into the cell nucleus, where it regulates the transcription of DNA. Examples of such lipophilic molecules include steroid hormones, thyroid hormones, and retinoids. *Steroid hormones* are derived from cholesterol (Figure 11.25). For example, estrogen stimulates the growth of the uterine wall. *Thyroid hormones* affect the expression of cytosolic enzymes that catabolize glucose, fats, and proteins. *Retinoids* regulate cell proliferation, differentiation, and death.

Receptors are classified in terms of the functional changes that arise on the cytoplasmic side of the cell after ligand binding to the receptor. The following are the major categories of receptors [1]:

- *G-protein-coupled receptors (GPCR).* G-proteins complex with guanidine triphosphate (GTP) or guanidine diphosphate (GDP) to regulate the generation of second messengers.

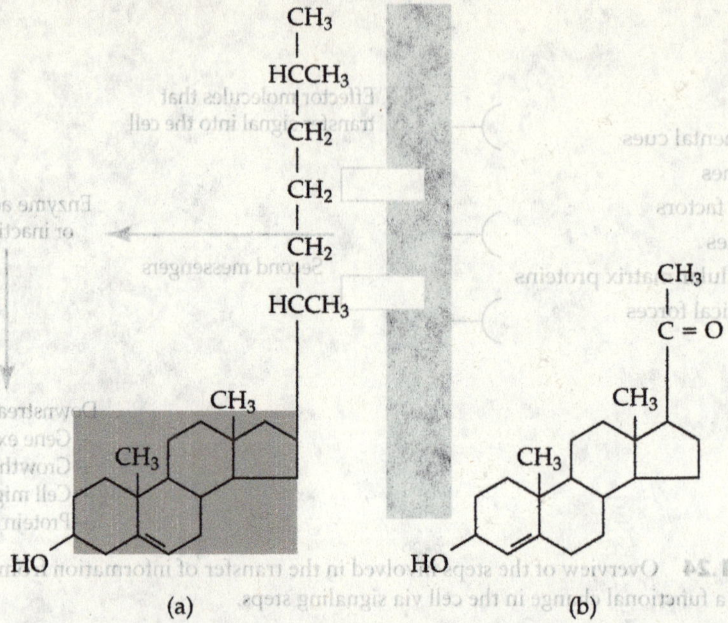

FIGURE 11.25 (a) The structure of cholesterol, precursor of steroid hormones. The steroid portion is shown in the shaded area. Most of the molecule is hydrophobic, but the hydroxyl group (−OH) is hydrophilic. (b) The structure of the steroid hormone progesterone.

- Receptors that couple with ion channels and open the channels transiently.
- *Tyrosine kinase–linked receptors*. Kinases are enzymes that add a phosphate group (i.e., they *phosphorylate*) to proteins at serine, threonine, or tyrosine residues.
- Receptors that function as enzymes.

These functional changes generate second messengers, small molecules that activate enzymes or transcription factors for genes. Examples of second messengers include Ca^{++}, cAMP, cyclic guanosine monophosphate (cGMP), diacylglycerol (DAG), and inositol trisphosphate (IP_3). In turn, second messengers alter enzyme activity by permitting cofactor binding or covalent modification of the enzyme.

A major benefit of signal transduction and second-messenger generation is the enormous amplification of the signal that can occur in a very short time. The process is further regulated by the transient release of hormones, receptor-mediated endocytosis of receptors and ligands after ligand binding, and rapid switching between active and inactive states of the receptors. Over the longer term, the ligand may regulate the expression of receptor molecules on the cell surface. As a result, a signaling event can be generated rapidly and the extent and duration of the response regulated.

G-protein receptor signaling. GPCRs are one of the most widely studied classes of signaling receptors. Over 450 drugs have been developed to alter signaling that is initiated by GPCRs. These drugs are either *agonists*, which bind to the receptor with a different affinity than the native ligand and stimulate G-protein activation, or *antagonists*, which bind to the receptor but do not stimulate G-protein activation.

A classic example of G-protein-mediated receptor activation is *epinephrine* (also known as *adrenaline*), a hormone released at nerve synapses and the adrenal gland. This hormone mediates the body's response to stress, such as fright or heavy exercise. Epinephrine acts in a coordinated fashion to increase heart rate and reduce blood flow and alters the metabolism of fats and carbohydrates [1]. In the heart, epinephrine binds to β-*adrenergic* receptors, which causes the cardiac muscle to contract, resulting in an increase in heart rate. In the tissues, epinephrine binding to α_2-adrenergic receptors causes vessels in the kidney and skin to contract, whereas binding to α_1-adrenergic receptors causes blood vessels in muscles to relax. The result is a redistribution of blood flow. These different effects after binding to epinephrine arise because the receptors activate different G-proteins.

The α_2- and β-*adrenergic* receptors are GPCRs that regulate the activity of the enzyme adenylyl cyclase, which produces cAMP in different ways. Binding to β-*adrenergic* receptors causes an increase in cAMP. In turn, cAMP activates *cAMP-dependent protein kinases* (i.e., enzymes that add a phosphate [phosphorylate] to specific proteins). In contrast, binding to α_2-adrenergic receptors causes a decrease in cAMP activity. The different modes of action of these receptors are due to differences in the types of G-proteins activated. Ligand-bound β-adrenergic receptors activate G_s proteins, which stimulate adenyl cyclase activity, whereas α_2-adrenergic receptors bind to G_i proteins, which inhibit adenyl cyclase activity.

The G_i proteins are linked to the plasma membrane and consist of a $G_{s\alpha}$ portion, which binds to GDP or GTP, and a $G_{\gamma\beta}$ portion, which serves to keep the $G_{s\alpha}$ in an inactive state. The G-proteins and other proteins diffuse in the plane of the cell membrane. In the absence of hormone binding to β-adrenergic receptors, $G_{s\alpha}$ is bound to GDP and $G_{\gamma\beta}$ and cannot bind to adenylyl cyclase when the two molecules collide (Figure 11.26a). When the hormone binds to the β-adrenergic receptors, a conformational change takes place in the cytoplasmic portion of the receptor that enables it to bind to the G-protein. As a result of this binding event, GDP dissociates from $G_{s\alpha}$, allowing GTP to bind (Figure 11.26b). Next, $G_{\gamma\beta}$ dissociates from the $G_{s\alpha}$–GTP complex. The $G_{s\alpha}$–GTP complex is now in an active form and binds to adenylyl cyclase. $G_{s\alpha}$–GTP acts as a cofactor for the enzyme, which enables the adenylyl cyclase to catalyze the conversion of ATP to cAMP (Figure 11.26c). The activity of adenylyl cyclase is decreased by the rapid hydrolysis of GTP to GDP, reforming $G_{s\alpha}$–GTP. The $G_{s\alpha}$–GTP complex dissociates from adenylyl cyclase, inactivating the enzyme (Figure 11.26d). $G_{s\alpha}$–GTP now binds to $G_{\gamma\beta}$, reestablishing the inactive form of the G-protein complex, which can then undergo another cycle of activation. Further regulation of the activity of the hormones occurs when the formation of $G_{s\alpha}$–GTP reduces the affinity of the hormone for its receptor (an increase in K_D).

The long-term activity of G-protein-linked receptors is regulated by the phosphorylation of the receptor, a process known as *desensitization*. The $G_{\gamma\beta}$–$G_{s\alpha}$–GDP complex can bind receptors bound to hormones (Figure 11.27a). Doing so, however, can lead to continuous stimulation, which may not be desirable. High levels of cAMP activate kinases that phosphorylate all G-proteins, reducing their affinity to the $G_{\gamma\beta}$–$G_{s\alpha}$–GDP complex. Since this kinase affects all G-proteins, the process is known as *heterologous desensitization*. In contrast, the long-term stimulation of β-adrenergic receptors with epinephrine leads to the phosphorylation of different sites of the β-adrenergic receptors by β-adrenergic receptor kinase (BARK) (Figure 11.27b). This receptor-specific phosphorylation is known as *homologous desensitization*. Subsequently, β-arrestin binds to the phosphorylated sites and blocks the binding of the G-protein with the receptor (Figure 11.27c). Further,

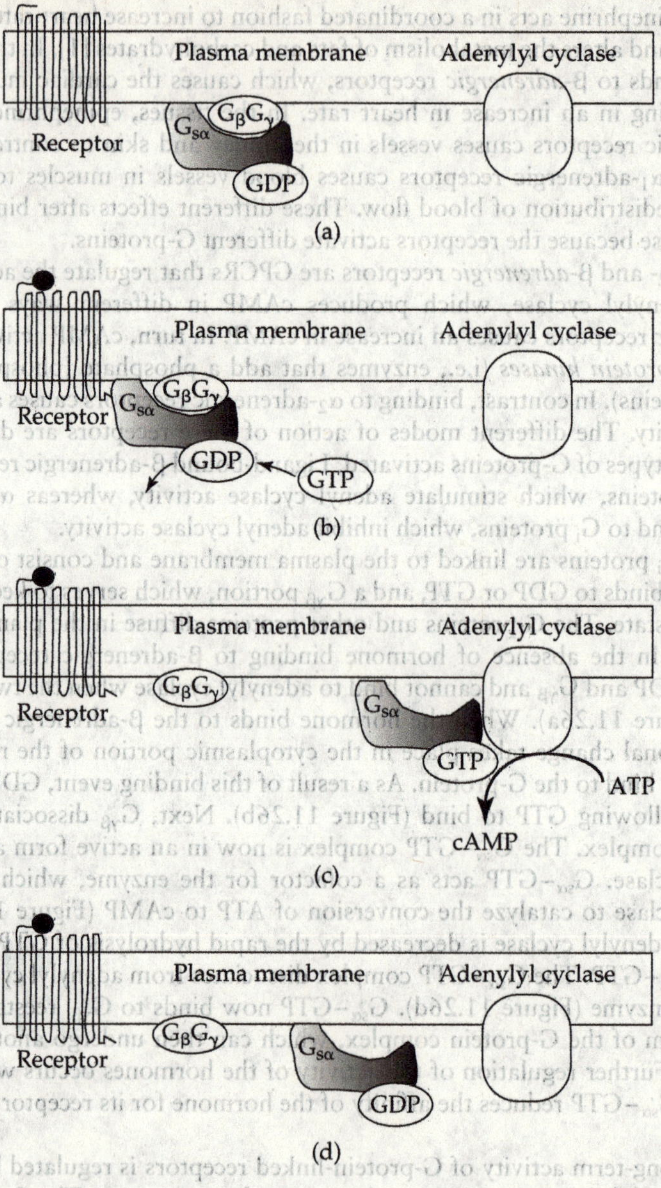

FIGURE 11.26 Cycle of G-protein activation and inactivation.

β-arrestin-bound receptors bind to coated-pit proteins and are internalized. Within the cell, β-arrestin dissociates from the receptor, and phosphodiesterases remove the phosphate groups. The receptor is recycled to the cell surface for reuse. If the presence of epinephrine persists, desensitization continues.

The G_α portion of G-proteins is present in several forms that regulate a variety of second-messenger pathways (Table 11.4). In addition, different isoforms of the $G_{\gamma\beta}$ subunit can increase the activity of phospholipase, which produces IP_3 or DAG or inhibits the activity of adenylyl cyclase.

During heart failure, epinephrine levels are elevated in order to strengthen the weakened heart. However, heterologous desensitization of the receptors leads to

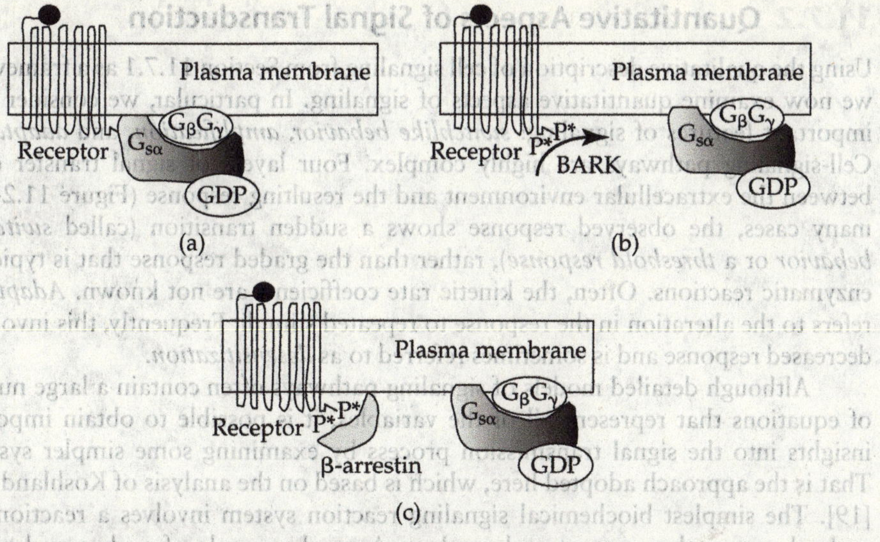

FIGURE 11.27 Phosphorylation of ligand-bound GPCR by β-adrenergic receptor kinase (BARK), leading to the binding of β-arrestin and inactivation of the receptor.

their reduced effectiveness. To compensate, the heart muscle increases in size (hypertrophy). Hypertrophy, however, results in complications that decrease the effectiveness of the heart. Clinical complications of heart failure are elevated lung pressure, edema, and reduced gas exchange.

The activity and function of second messengers are regulated in a number of ways, in addition to desensitization. Second-messenger degradation may be altered. For example, *cAMP phosphodiesterase* converts cAMP to adenosine monophosphate. The activity of cAMP phosphodiesterase is controlled by another second messenger, calcium ion. The downstream function of second messengers in various cells depends upon the specific cAMP-dependent kinases expressed. cAMP activates enzymes that stimulate glycogen breakdown and inhibit glycogen synthesis. This coordinate regulation effectively shifts carbohydrates into a form that can be rapidly metabolized.

TABLE 11.4

Classification of G-Protein Subunits

G_α Subclass	Effect
G_s	Increases the activity of adenylyl cyclase to produce cAMP. Other subforms open Ca^{++} channels or close Na^{+} channels.
G_i	Decreases the activity of adenylyl cyclase. Other subforms open K^{+} channels or close Ca^{++} channels.
G_q	Increases the activity of phospholipase, which produces IP_3 or DAG.
G_o	Increases the activity of phospholipase, which produces IP_3 or DAG. Other subclasses close Ca^{++} channels.
G_t	Increases the activity of cGMP phosphodiesterase, which produces cGMP from GTP.
$G_{\beta\gamma}$	Activates phospholipase and inhibits the activity of adenylyl cyclase.

11.7.2 Quantitative Aspects of Signal Transduction

Using the qualitative description of cell signaling from Section 11.7.1 as a framework, we now examine quantitative aspects of signaling. In particular, we consider three important features of signaling: *switchlike behavior, amplification*, and *adaptation*. Cell-signaling pathways are highly complex. Four layers of signal transfer occur between the extracellular environment and the resulting response (Figure 11.24). In many cases, the observed response shows a sudden transition (called *switchlike behavior* or a *threshold response*), rather than the graded response that is typical of enzymatic reactions. Often, the kinetic rate coefficients are not known. *Adaptation* refers to the alteration in the response to repeated stimuli. Frequently, this involves a decreased response and is sometimes referred to as *desensitization*.

Although detailed models of signaling pathways often contain a large number of equations that represent all of the variables, it is possible to obtain important insights into the signal transmission process by examining some simpler systems. That is the approach adopted here, which is based on the analysis of Koshland et al. [19]. The simplest biochemical signaling reaction system involves a reaction that cycles between the reactant and product. A good example of such a cycle is the process of phosphorylation and dephosphorylation of many intracellular proteins. Protein phosphorylation alters the activity of the protein. In the case of the G-protein receptors, phosphorylation inhibits binding. In other cases, phosphorylation can activate enzymes or promote binding reactions. The Koshland et al. analysis indicates that coupled enzyme reactions can exhibit large changes in product concentrations for small changes in maximum rates when the reactions are approximately of zeroth order. Before considering this analysis, the concept of *sensitivity* is defined.

The *logarithmic sensitivity* of a response such as a reaction rate R_s is defined in terms of the effect of the substrate concentration C_S on the reaction rate:

$$S = \frac{C_S \, dR_s}{R_s \, dC_S} = \frac{d \ln R_s}{d \ln C_S}. \tag{11.7.1}$$

The derivative dR_s/dC_s is referred to as the local sensitivity, s [20]. When the response is normalized in this fashion, differences in its absolute magnitude are not important. What is significant is the change in the response *relative* to a change in the input signal. The sign of the coefficient is not as critical as the magnitude of the *sensitivity coefficient*.

Example 11.6 Determine the sensitivity coefficients of the following expression for the reaction rate of a cooperative interaction:

$$R_S = \frac{R_{\max} C_S^n}{K_M^n + C_S^n}. \tag{11.7.2}$$

This empirical expression, known as the *Hill equation*, is frequently used to describe cooperative interactions. The coefficient n is known as the *Hill coefficient*. Ideally, for a cooperative interaction, n represents the number of interacting subunits. For apparent cooperativity, n represents the extent to which an enzymatic reaction exhibits cooperative behavior (Figure 11.28a). When $n = 1$, K_M is identical to the Michaelis constant. As n increases, the curve rises more rapidly over a shorter interval of concentration. When $n = 1$, the concentration changes by a factor of 81 as the enzyme reaction rate increases from $0.1R_{max}$ to $0.9R_{max}$. For other values of n, the concentration ratio for a change in rate from $0.1R_{max}$ to $0.9R_{max}$ is $(81)^{1/n}$. Thus, when $n = 3$, the concentration need change by a factor of only 4.326 for the rate to change from 10% to 90% of its maximum value. Compute the logarithmic sensitivity of Equation (11.7.2).

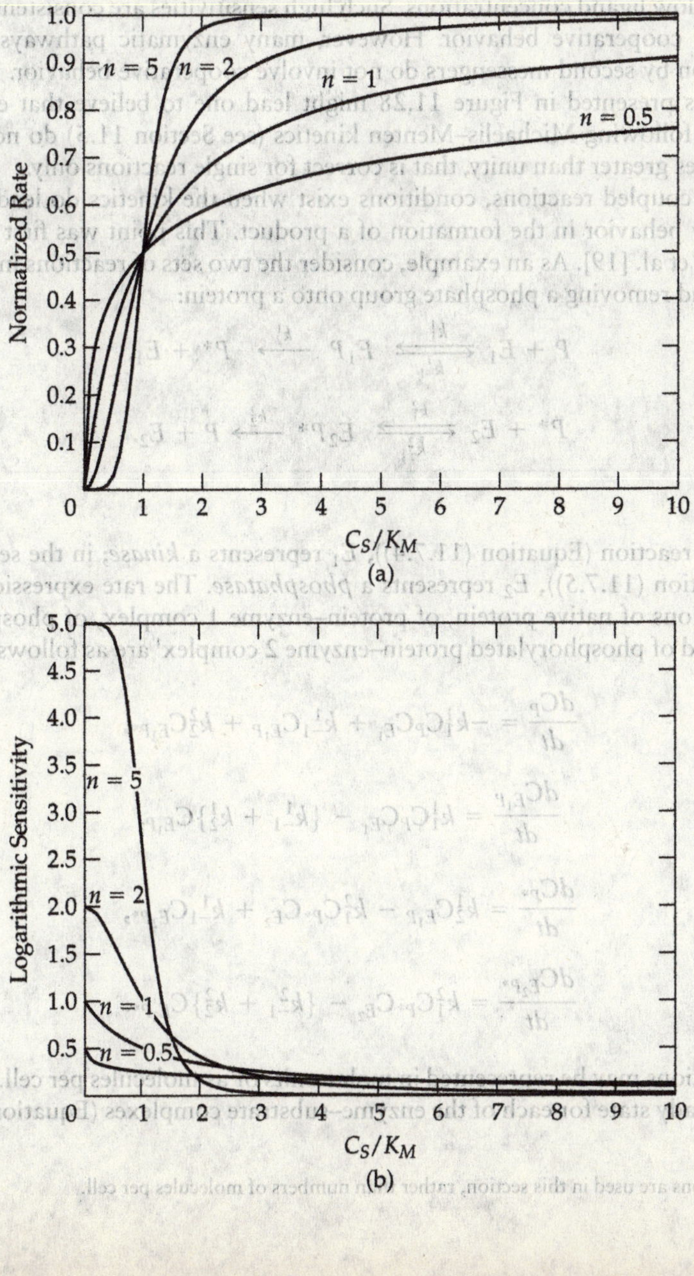

FIGURE 11.28 (a) Rate versus normalized concentration for Equation (11.7.3). (b) Sensitivity coefficients for the rate with respect to the substrate concentration.

Solution The resulting logarithmic sensitivity is obtained by first taking the derivative of R_s with respect to C_s and then multiplying by the ratio C_s/R_s to obtain

$$S = \frac{n}{1 + (C_S/K_M)^n}. \tag{11.7.3}$$

The logarithmic sensitivity is equal to the value of the Hill coefficient n at a concentration of zero and declines as the concentration increases (Figure 11.28b).

The preceding analysis shows that cooperative reactions lead to high sensitivities at low ligand concentrations. Such high sensitivities are consistent with the nature of cooperative behavior. However, many enzymatic pathways that are acted upon by second messengers do not involve cooperative behavior. Although the results presented in Figure 11.28 might lead one to believe that enzymatic reactions following Michaelis–Menten kinetics (see Section 11.5) do not exhibit sensitivities greater than unity, that is correct for single reactions only.

For coupled reactions, conditions exist when the kinetics do lead to high-sensitivity behavior in the formation of a product. This point was first noted by Koshland et al. [19]. As an example, consider the two sets of reactions involved in adding and removing a phosphate group onto a protein:

$$P + E_1 \underset{k_{-1}^1}{\overset{k_1^1}{\rightleftharpoons}} E_1P \xrightarrow{k_2^1} P^* + E_1, \tag{11.7.4}$$

$$P^* + E_2 \underset{k_{-1}^2}{\overset{k_1^2}{\rightleftharpoons}} E_2P^* \xrightarrow{k_2^2} P + E_2. \tag{11.7.5}$$

In the first reaction (Equation (11.7.4)), E_1 represents a *kinase*; in the second reaction (Equation (11.7.5)), E_2 represents a *phosphatase*. The rate expressions for the concentrations of native protein, of protein–enzyme 1 complex, of phosphorylated protein, and of phosphorylated protein–enzyme 2 complex[1] are as follows:

$$\frac{dC_P}{dt} = -k_1^1 C_P C_{E_1} + k_{-1}^1 C_{E_1P} + k_2^2 C_{E_2P^*}, \tag{11.7.6}$$

$$\frac{dC_{E_1P}}{dt} = k_1^1 C_P C_{E_1} - \{k_{-1}^1 + k_2^1\} C_{E_1P}, \tag{11.7.7}$$

$$\frac{dC_{P^*}}{dt} = k_2^1 C_{E_1P} - k_1^2 C_{P^*} C_{E_2} + k_{-1}^1 C_{E_2P^*}, \tag{11.7.8}$$

$$\frac{dC_{E_2P^*}}{dt} = k_1^2 C_{P^*} C_{E_2} - \{k_{-1}^2 + k_2^2\} C_{E_2P^*}. \tag{11.7.9}$$

Concentrations may be represented in molar units or as molecules per cell. Assuming a quasi-steady state for each of the enzyme–substrate complexes (Equations (11.7.7)

[1] Concentrations are used in this section, rather than numbers of molecules per cell.

and (11.7.9)), the following relation can be developed for the formation of the phosphorylated protein:

$$\frac{dC_{P^*}}{dt} = \frac{R_{\max_1} C_P}{K_{M_1} + C_P} - \frac{R_{\max_2} C_{P^*}}{K_{M_2} + C_{P^*}}. \tag{11.7.10}$$

Note that, since the concentration of the enzyme–substrate complexes is small (i.e., C_{E_1P} or $C_{E_2P^*} \ll C_{P_T}$), the sum of P and P^* is constant: $C_{P_T} = C_P + C_{P^*}$.

As discussed in Chapter 10, enzyme reactions have limiting behavior of zeroth order when K_M is much less than the substrate concentration and first order when K_M is much greater than the substrate concentration. Assuming that the K_M values for the phosphorylation and dephosphorylation reactions are equal ($K_{M_1} = K_{M_2}$), the kinetics of C_P^* formation exhibit different sensitivities to the ratio of reaction rates in the zeroth- and first-order regimes (Figure 11.29). In general, C_P^* is formed faster as the ratio $a = R_{\max_1}/R_{\max_2}$ increases. In the zeroth-order regime, there is an abrupt transition in product formation with $R_{\max_1}/R_{\max_2}$. When $R_{\max_1}/R_{\max_2} < 1$, there is little product formation. When $R_{\max_1}/R_{\max_2} > 2$, the reaction rapidly goes to completion (Figure 11.29a). For the first-order regime, there is a gradual transition from reactant to product with changes in the ratio $R_{\max_1}/R_{\max_2}$ (Figure 11.29b). The reason for this difference between the first- and zeroth-order regimes is that when the consumption of substrate is initially in the first-order regime, the reverse reaction is also in the first-order regime, and a decrease in the substrate concentration does not change the reaction order of the enzyme reaction. However, for a reaction in the zero-order regime, the reaction order changes once the substrate concentration has dropped about 20%. For larger values of $R_{\max_1}/R_{\max_2}$, the forward reaction undergoes a transition from zero- to first-order reaction rates, whereas the reverse reaction remains in the zero-order regime.

To gain additional insight into the sensitivity of the reaction to the ratio $R_{\max_1}/R_{\max_2}$, consider steady-state behavior. When the concentration of phosphorylated protein is at steady state, the following relation, derived from Equation (11.7.10), relates the relative rates to the concentration of the phosphorylated protein:

$$\frac{R_{\max_1}}{R_{\max_2}} = \frac{C_{P^*}(C_{P_T} - C_{P^*} + K_{M_1})}{(C_{P_T} - C_{P^*})(C_{P^*} + K_{M_2})}. \tag{11.7.11}$$

Because the maximum reaction rates 1 and 2 are proportional to the concentration of kinase and phosphatase, respectively, the ratio $R_{\max_1}/R_{\max_2}$ can be viewed as being proportional to the input signal C_{E_1}. Note that the total protein is the sum of the native form plus the phosphorylated form:

$$C_{P_T} = C_P + C_{P^*}. \tag{11.7.12}$$

Figure 11.30a shows the concentration of phosphorylated protein as a function of the maximum reaction rates and the Michaelis constants. When the maximum rate of the phosphorylation reaction is much less than the maximum rate of the dephosphorylation reaction, little phosphorylated protein is produced. As the ratio $R_{\max_1}/R_{\max_2}$ increases, more phosphorylated protein is produced. When the ratio is

FIGURE 11.29 Kinetics of formation of phosphorylated protein C_P^* for cases in which the enzymatic reaction is initially zero order (a) and first order (b). The quantity a equals $R_{\max_1}/R_{\max_2}$. At time 0, $C_{P_T} = C_P$ and $C_P^* = 0$.

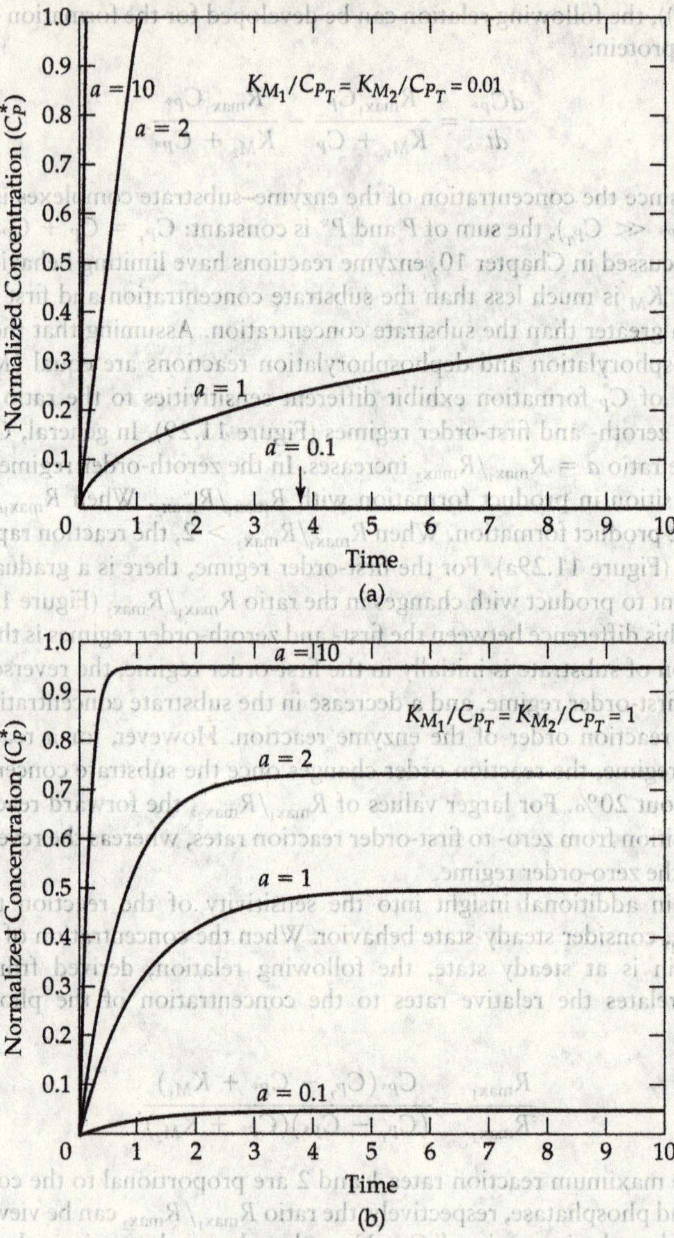

very large, almost all of the protein is in the phosphorylated form. The rate at which the phosphorylated protein changes relative to a change in $R_{\max_1}/R_{\max_2}$ depends upon the ratio of the Michaelis constant to the total protein concentration. The smaller the ratio K_{M_1}/C_T (or K_{M_2}/C_T), the more rapidly the concentration of phosphorylated protein changes with $R_{\max_1}/R_{\max_2}$. Values of K_{M_1}/C_T that are much less than unity correspond to zeroth-order kinetics (see Section 11.5). Likewise, the sensitivity coefficient changes dramatically when the reaction is zeroth order in nature (Figure 11.30b).

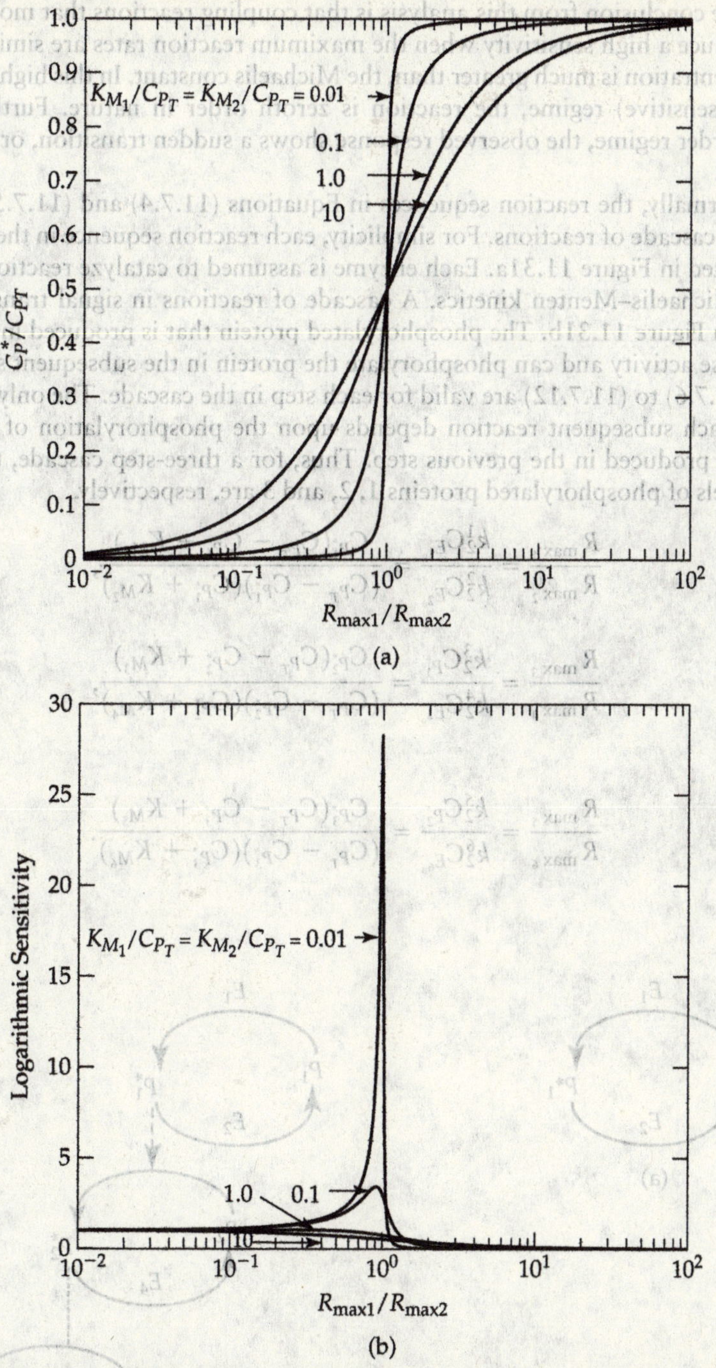

FIGURE 11.30 (a) The fractional amount of phosphorylated protein for the cyclic pair of reactions represented by Equations (11.7.4) and (11.7.5) as a function of the ratio of maximum reaction rates for various values of the Michaelis constants. (b) The corresponding sensitivity coefficients for the concentration (with respect to the reaction-rate ratio) as a function of the Michaelis constants.

The conclusion from this analysis is that coupling reactions that modify a protein produce a high sensitivity when the maximum reaction rates are similar to, and the concentration is much greater than, the Michaelis constant. In this high-sensitivity (or ultrasensitive) regime, the reaction is zeroth order in nature. Further, in the zeroth-order regime, the observed response shows a sudden transition, or switchlike behavior.

Normally, the reaction sequences in Equations (11.7.4) and (11.7.5) occur as part of a cascade of reactions. For simplicity, each reaction sequence in the cascade is represented in Figure 11.31a. Each enzyme is assumed to catalyze reactions according to Michaelis–Menten kinetics. A cascade of reactions in signal transduction is shown in Figure 11.31b. The phosphorylated protein that is produced in reaction 1 has kinase activity and can phosphorylate the protein in the subsequent step. Equations (11.7.6) to (11.7.12) are valid for each step in the cascade. The only difference is that each subsequent reaction depends upon the phosphorylation of the kinase that was produced in the previous step. Thus, for a three-step cascade, the steady-state levels of phosphorylated proteins 1, 2, and 3 are, respectively,

$$\frac{R_{\max 1}}{R_{\max 2}} = \frac{k_2^1 C_{E_{1_0}}}{k_2^2 C_{E_{2_0}}} = \frac{C_{P_1^*}(C_{P_T} - C_{P_1^*} + K_{M_1})}{(C_{P_T} - C_{P_1^*})(C_{P_1^*} + K_{M_2})}, \tag{11.7.13}$$

$$\frac{R_{\max 3}}{R_{\max 4}} = \frac{k_2^3 C_{P_1^*}}{k_2^4 C_{E_{4_0}}} = \frac{C_{P_2^*}(C_{P_T} - C_{P_2^*} + K_{M_3})}{(C_{P_T} - C_{P_2^*})(C_{P_2^*} + K_{M_4})}, \tag{11.7.14}$$

and

$$\frac{R_{\max 5}}{R_{\max 6}} = \frac{k_2^5 C_{P_2^*}}{k_2^6 C_{E_{6_0}}} = \frac{C_{P_3^*}(C_{P_T} - C_{P_3^*} + K_{M_5})}{(C_{P_T} - C_{P_3^*})(C_{P_3^*} + K_{M_6})}. \tag{11.7.15}$$

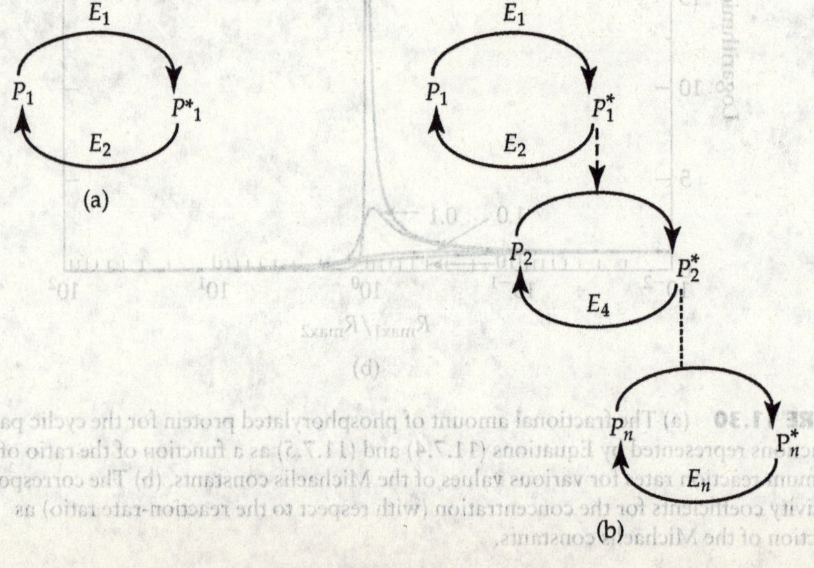

FIGURE 11.31 (a) A single sequence of enzymatic phosphorylation and dephosphorylation. (b) A cascade of signal-transduction reactions in which the modified protein (P_i^*) has catalytic activity.

The kinase concentrations in cascades 2 and 3 depend upon the values obtained from the previous step. The three equations can be solved sequentially for the kinase concentration that is produced in each step (Figure 11.32). For simplicity, the initial concentration of each of the proteins 1, 2, and 3 is assumed to be the same and equal to C_{P_T}. In order to determine the effect of the initial enzyme concentration $C_{E_{10}}$ and the phosphorylated protein concentrations $C_{P_2^*}$ and $C_{P_3^*}$, the maximum reaction rates are rewritten as follows:

$$\frac{R_{\max_1}}{R_{\max_2}} = \frac{k_2^1 C_{E_{1_0}}}{k_2^2 C_{E_{2_0}}} = \frac{k_2^1 C_{P_T} C_{E_{1_0}}}{k_2^2 C_{E_{2_0}} C_{P_T}}, \tag{11.7.16a}$$

$$\frac{R_{\max_3}}{R_{\max_4}} = \frac{k_2^3 C_{P_1^*}}{k_2^4 C_{E_{4_0}}} = \frac{k_2^3 C_{P_T} C_{P_1^*}}{k_2^4 C_{E_{4_0}} C_{P_T}}, \tag{11.7.16b}$$

$$\frac{R_{\max_5}}{R_{\max_6}} = \frac{k_2^5 C_{P_2^*}}{k_2^6 C_{E_{6_0}}} = \frac{k_2^5 C_{P_T} C_{P_2^*}}{k_2^6 C_{E_{6_0}} C_{P_T}}. \tag{11.7.16c}$$

When all of the Michaelis constants are the same and $k_2^3 C_{P_{1_T}}/k_2^4 C_{E_{4_0}} = k_2^5 C_{P_{2_T}}/k_2^6 C_{E_{6_0}} = 1$, then the maximum concentrations of $C_{P_2^*}$ and $C_{P_3^*}$ are less than the maximum value of $C_{P_1^*}$ (Figure 11.32a). Each subsequent reaction pair has a smaller ratio of reaction rates, and there is no increase in sensitivity with each subsequent reaction step. In order to produce an increase in sensitivity in each step, $k_2^3 C_{P_{1_T}}/k_2^4 C_{E_{4_0}}$ and $k_2^5 C_{P_{2_T}}/k_2^6 C_{E_{6_0}}$ must be less than unity. If the Michaelis constants are identical, then all of the curves pass through a normalized phosphorylated protein concentration at the same concentration of enzyme 1. In order to produce the situation in which steps 2 and 3 reach normalized concentrations of unity at lower concentrations of protein 1 (i.e., in order to obtain signal amplification), K_{M_3} and K_{M_5} must be smaller than K_{M_1}, K_{M_3} must be greater than K_{M_4}, and K_{M_5} must be greater than K_{M_6}. Thus, to produce ultrasensitive or switchlike behavior, the enzymatic reactions must be near the zeroth-order reaction limit.

For a set of sequential reactions without feedback, the overall local sensitivity is the product of the local sensitivities of the individual reactions [20]:

$$s_1^n = s_0^1 s_1^2 \ldots s_{n-1}^n. \tag{11.7.17}$$

The subscript refers to the previous step, the superscript to the current step. This result can be obtained by applying the chain rule (see Section A.1.E) to a cascade of reactions.

For example, consider a sequence of three cascades. If the individual sensitivities are each greater than unity, the overall sensitivity is greatly amplified. This result suggests why many signal-transduction reactions occur as cascades of coupled reactions. Each coupled reaction can have a sensitivity greater than unity, and the net product has a very high sensitivity.

Limited data support the basic approach of the preceding analysis [21]. The phosphorylation of mitogen-activated protein (MAP) kinase (MAPK) occurs at the end of a sequence of enzymatic reactions. For example, EGF binding to its receptor leads to receptor dimerization, which results in the activation of the membrane-bound protein *Ras*. In turn, Ras activates a MAP kinase kinase kinase (MAPKKK),

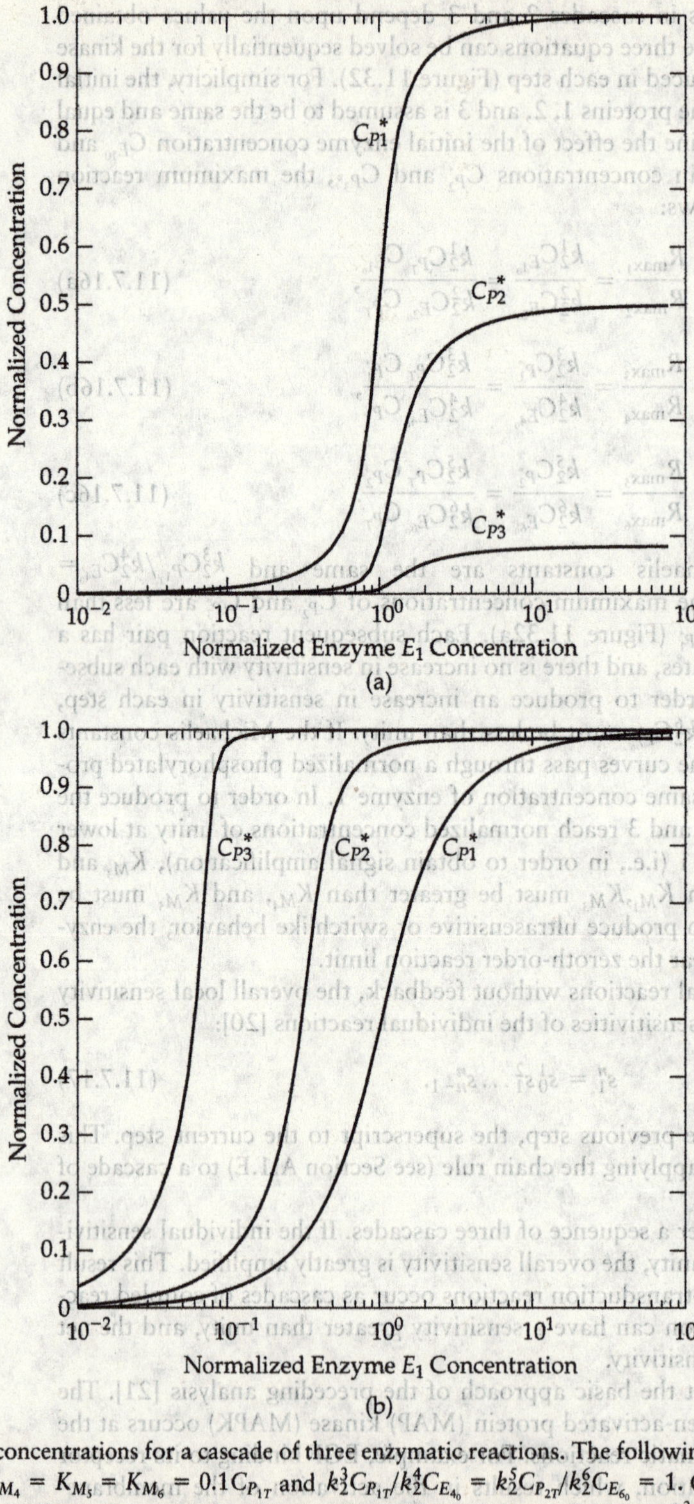

FIGURE 11.32 Relative concentrations for a cascade of three enzymatic reactions. The following parameters were used: (a) $K_{M_1} = K_{M_2} = K_{M_3} = K_{M_4} = K_{M_5} = K_{M_6} = 0.1C_{P_{1T}}$ and $k_2^3 C_{P_{1T}}/k_2^4 C_{E_{4_0}} = k_2^5 C_{P_{2T}}/k_2^6 C_{E_{6_0}} = 1$. Concentrations are normalized by C_{P_T}; (b) $K_{M_1} = K_{M_2} = K_{M_4} = K_{M_6} = 0.5\ C_{P_{1T}}$, $K_{m_3} = 0.1C_{P_{1T}}$, $K_{m_5} = 0.01C_{P_{1T}}$, $k_2^3 C_{P_{1T}}/k_2^4 C_{E_{4_0}} = 0.1$, and $k_2^5 C_{P_{1T}}/k_2^6 C_{E_{6_0}} = 0.01$.

which is a kinase that catalyzes two other phosphorylation reactions. In turn, MAPKKK enzymatically activates MAP kinase kinase (MAPKK), the enzyme that activates MAP kinase. MAPKKK is present at concentrations of 300 nM (C_{E_1}) [21]. The MAP kinase concentration ($C_{P_3^*}$) is 1.2 μM. Concentrations of the various phosphatases are not known. Michaelis constants are estimated to be about 60 nM [22]. Thus, it is likely that, for this system, $K_{M_3}/C_{P_{3T}} < 1$, which would produce a sensitivity increase in the latter steps of the cascade. *In vitro* studies indicate that some MAP kinases exhibit threshold behavior for different levels of MAPKKK (with a Hill coefficient of 5 when fit to Equation (11.7.2)), supporting the notion that ultrasensitive behavior can give rise to a coupled reaction that operates at near-zeroth-order kinetics [21].

The third major issue to consider is adaptation. A time-dependent phenomenon, adaptation cannot be demonstrated from a steady-state analysis. Feedback is essential for adaptation to occur [22]. Such feedback can take the form of noncompetitive inhibition of substrates by the product of a cascade or by the inactivation of the receptor involved in generating the second messenger.

Example 11.7 One way to attenuate the signal and amplify the response of a single site of an enzyme cascade of kinase and phosphatase reactions is to degrade the first kinase irreversibly and have $K_{M_1} = K_{M_2}$; that is,

$$\frac{dC_{P_1^*}}{dt} = \frac{R_{\max_1} C_{P_1}}{K_{M_1} + C_{P_1}} - \frac{R_{\max_2} C_{P_1^*}}{K_{M_2} + C_{P_1^*}} - k_3 C_{P_1^*}. \tag{11.7.18}$$

For $R_{\max_1} = R_{\max_2} = 1\ \mu\text{M s}^{-1}$, $C_{P_1}(t = 0) = 1\ \mu\text{M}$, $C_{P_1^*}(t = 0) = 0$, and $K_{M_1} = K_{M_2} = 1\ \mu\text{M}$, a maximum in the concentration of $C_{P_1^*}$ is reached at 3.22 seconds, with $C_{P_1} = 0.542\ \mu\text{M}$ and $C_{P_1^*} = 0.201\ \mu\text{M}$. Use these results to determine k_3.

Solution When the reaction reaches a maximum, the derivative is zero. All other terms besides k_3 are known. Solving Equation (11.7.18) yields a value of $k_3 = 0.916\ \text{s}^{-1}$.

11.8 Regulation of Gene Expression

Often, the final step in cellular signaling is the inhibition or activation of specific genes (Figure 11.24). Thus, a complete characterization of cell signaling needs to include the manner in which genes are regulated. Regulation can occur at the levels of mRNA synthesis (*transcription*) or degradation and protein synthesis (*translation*) or degradation. Synthesis and degradation of mRNA control the concentration of a specific mRNA within a cell. Although control can and does occur at all of these steps, transcriptional control is often a critical step, because it is the first step in the process. Regulation at the first step has the greatest overall effect, although regulation at a later step can control the dynamics of the process. This analysis focuses on transcriptional control.

Transcription is activated or repressed by protein binding to genetic regulatory sequences. Proteins and other molecules act as positive or negative regulators of gene transcription and are known as *transcription factors*. In bacteria, gene induction arises in response to the level of various nutrients. In eukaryotic cells, environmental regulation occurs, but it is less significant. Rather, gene regulation arises during embryological development and tissue differentiation. Although many of the processes of gene regulation identified in prokaryotes apply to eukaryotes, the processes in eukaryotes are more complex. The bacterial *lactose operon* system is used as a prototype for transcriptional regulation. The operon represents a group of functionally related genes under the same regulation.

Glucose metabolism in bacteria is controlled by three consecutive genes. The *lac Y* gene transcribes lactose permease, a membrane protein that pumps lactose into cells. The *lac Z* gene codes for β-galactosidase, which breaks lactose into glucose and galactose. The *lac A* gene codes for thiogalactoside transacetylase, an enzyme with a poorly understood function. The three genes are under the same genetic regulation (Figure 11.33).

Glucose metabolism by *E. coli* bacteria occurs as follows. *E. coli* preferentially metabolize glucose without the induction of any new enzymes. If glucose and lactose are both present, then the induction of enzymes for lactose metabolism occurs only after glucose is depleted. In the absence of glucose, *E. coli* synthesize cAMP before activating the lactose enzymes. In fact, β-galactosidase can be induced in the presence of glucose by adding dibutyryl cAMP, a stable analog of cAMP. Two levels of control are exerted to inhibit transcription in the presence of glucose: (a) RNA polymerase binding to the promoter site is dependent upon cAMP, which is synthesized when glucose levels are low; (b) in the absence of lactose, a repressor protein binds to the operator site, preventing the RNA polymerase from reaching the genes *Z*, *Y*, and *A*.

Repression and activation of the lactose genes occur as follows: RNA polymerase binds to a catabolite activator protein (CAP). Both molecules bind weakly to the lactose promoter site. As a result, low levels of lactose are metabolized, even in the absence of glucose. CAP can also bind to cAMP. When cAMP–CAP complex binds to RNA polymerase, the resulting complex binds to the promoter site with high affinity (Figure 11.34). As a result, high levels of β-galactosidase and the other enzymes are produced.

While control of RNA polymerase binding to the DNA is important, the primary means of transcriptional regulation of the *lac* operon is via repressor binding to

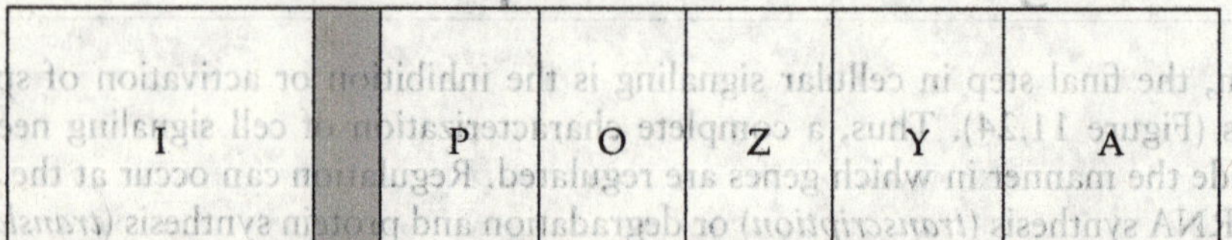

FIGURE 11.33 Schematic of the lactose operon. The lactose operon genes are Z, Y, and A. Regions to the left of Z are considered to be upstream. I encodes the repressor protein. P is the promoter binding site. RNA polymerase binds to this site. O is the operator or binding site of the repressor molecule. Binding of the repressor to the operator site O blocks the action of RNA polymerase.

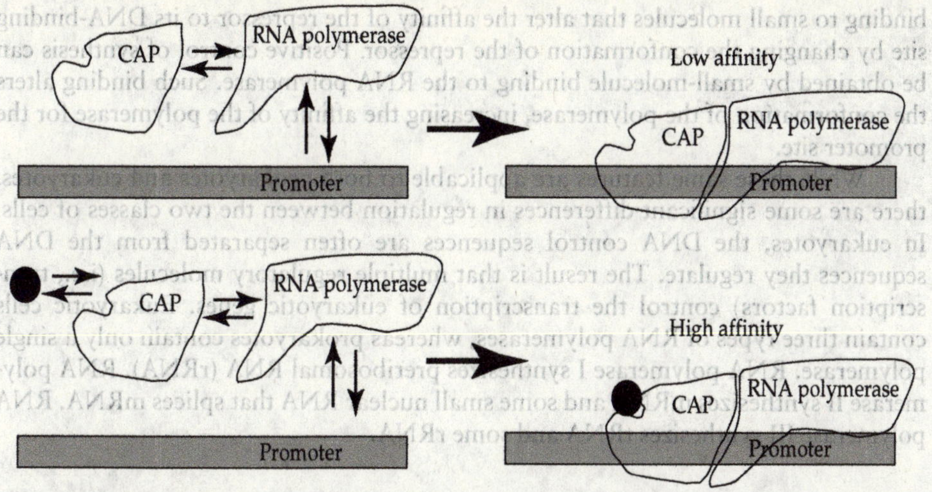

FIGURE 11.34 Schematic of the binding of RNA polymerase to the promoter site of the *lac* operon with and without cAMP. Binding can occur in the absence of cAMP, but the affinity is low.

the operator site (Figure 11.35). When glucose is high and cAMP is low, the repressor is synthesized and binds to the operator site with equilibrium constant K_A. No enzymes for lactose metabolism are synthesized. If lactose is now added, lactose binds to the repressor. The repressor may then dissociate from operator sites. Because of the weak interactions between RNA polymerase and DNA, little transcription occurs. In the absence of glucose, cAMP is synthesized, resulting in high-affinity binding of RNA polymerase–CAP–cAMP to the promoter site. The addition of lactose results in lactose binding to the repressor. The lactose-bound repressor binds to the operator site with reduced affinity. Consequently, RNA polymerase can initiate transcription.

Regulation of the *lac* operon highlights several features commonly observed in transcriptional regulation. Repressor proteins bind to specific sites upstream of a gene, inhibiting the synthesis of the gene. These repressor proteins are often regulated by

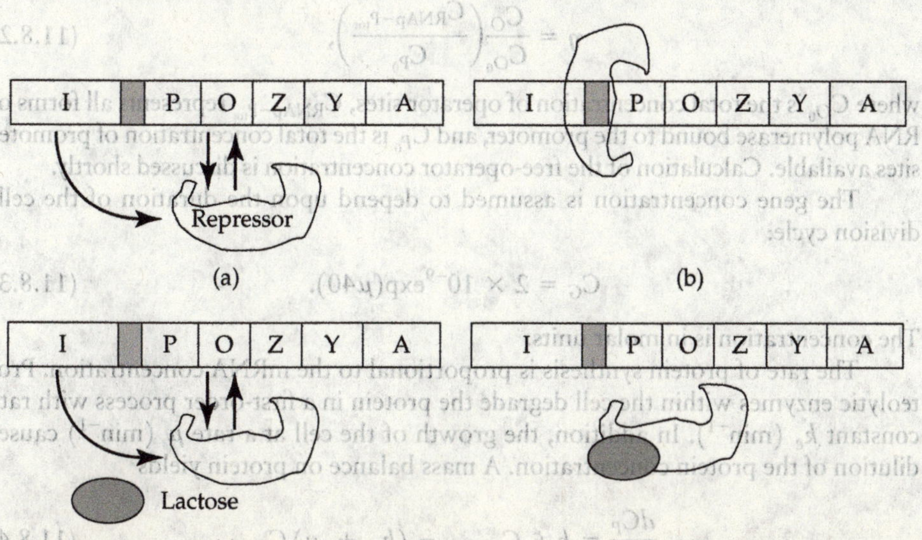

FIGURE 11.35 In the absence of lactose, repressor is synthesized (a) and binds tightly to the operator site (b). When lactose is present (c), it binds to the repressor (d), inhibiting repressor binding to the operator site.

binding to small molecules that alter the affinity of the repressor to its DNA-binding site by changing the conformation of the repressor. Positive control of synthesis can be obtained by small-molecule binding to the RNA polymerase. Such binding alters the conformation of the polymerase, increasing the affinity of the polymerase for the promoter site.

While these same features are applicable to both prokaryotes and eukaryotes, there are some significant differences in regulation between the two classes of cells. In eukaryotes, the DNA control sequences are often separated from the DNA sequences they regulate. The result is that multiple regulatory molecules (i.e., transcription factors) control the transcription of eukaryotic genes. Eukaryotic cells contain three types of RNA polymerases, whereas prokaryotes contain only a single polymerase. RNA polymerase I synthesizes preribosomal RNA (rRNA). RNA polymerase II synthesizes mRNA and some small nuclear RNA that splices mRNA. RNA polymerase III synthesizes tRNA and some rRNA.

11.8.1 Simplified Model for Gene Induction and Expression

Protein synthesis is modeled very simply [23,24]. First, mRNA is synthesized from the gene. The rate of synthesis is proportional to the concentration of the gene (C_G). mRNA is not stable and decays in a first-order process with a rate constant k_d (min^{-1}). In addition, growth of the cell at rate μ (min^{-1}) causes the mRNA concentration to become diluted. A mass balance on mRNA yields

$$\frac{dC_{\text{mRNA}}}{dt} = k_p \eta C_G - (k_d + \mu) C_{\text{mRNA}}, \tag{11.8.1}$$

where k_p is the rate constant for transcription, C_{mRNA} is the mRNA concentration, and η is the transcription efficiency, which depends upon repressor binding and RNA polymerase binding. The efficiency η is the product of the ratio of free operator sites to the total number of operator sites and the extent of RNA polymerase binding:

$$\eta = \frac{C_O}{C_{O_0}} \left(\frac{C_{\text{RNAp}-\text{P}_{\text{tot}}}}{C_{\text{P}_0}} \right), \tag{11.8.2}$$

where C_{O_0} is the total concentration of operator sites, $C_{\text{RNAp}-\text{P}_{\text{tot}}}$ represents all forms of RNA polymerase bound to the promoter, and C_{P_0} is the total concentration of promoter sites available. Calculation of the free-operator concentration is discussed shortly.

The gene concentration is assumed to depend upon the duration of the cell-division cycle:

$$C_G = 2 \times 10^{-9} \exp(\mu 40). \tag{11.8.3}$$

The concentration is in molar units.

The rate of protein synthesis is proportional to the mRNA concentration. Proteolytic enzymes within the cell degrade the protein in a first-order process with rate constant k_e (min^{-1}). In addition, the growth of the cell at a rate μ (min^{-1}) causes dilution of the protein concentration. A mass balance on protein yields

$$\frac{dC_P}{dt} = k_q \xi \, C_{\text{mRNA}} - (k_e + \mu) C_P, \tag{11.8.4}$$

where k_q is the rate constant for translation and ξ is the translation efficiency, which is influenced by specific codons that are preferentially translated. For simplicity, the translation efficiency is assumed to be equal to unity.

A simplified form of the *lac* operator region is shown schematically in Figure 11.36. All reactions are assumed to be at equilibrium. The repressor binds to the operator site, blocking RNA polymerase, in accordance with the following formula:

$$R + O \longleftrightarrow \text{RO}, \tag{11.8.5a}$$

$$K_A = \frac{C_{\text{RO}}}{C_R C_0}. \tag{11.8.5b}$$

Note that equilibrium constants are based upon the convention used by chemists. In addition, the repressor can bind to other sites on the DNA molecule, but at a much lower affinity:

$$R + \text{DNA} \longleftrightarrow \text{RDNA}, \tag{11.8.6a}$$

$$K_B = \frac{C_{\text{RDNA}}}{C_R C_{\text{DNA}}}. \tag{11.8.6b}$$

Transcription is increased when an inducer is added. For the *lac* operon, the inducer may be lactose or an analog such as isopropyl-β-D-thiogalactoside (IPTG) or 5-bromo-4-chloro-3-indolyl-β-D-thiogalactoside (X-gal). The inducer binds to free repressor or repressor bound to the operator sites. As a result of the high concentration of inducer, the RI complex dissociates from the operator site, reducing the amount of repressor bound to the operator sites. The relevant equations for inducer binding to repressor are as follows:

$$R + I \longleftrightarrow \text{RI}, \tag{11.8.7a}$$

$$K_C = \frac{C_{\text{RI}}}{C_R C_I}, \tag{11.8.7b}$$

$$\text{RI} + \text{O} \longleftrightarrow \text{RIO}, \tag{11.8.8a}$$

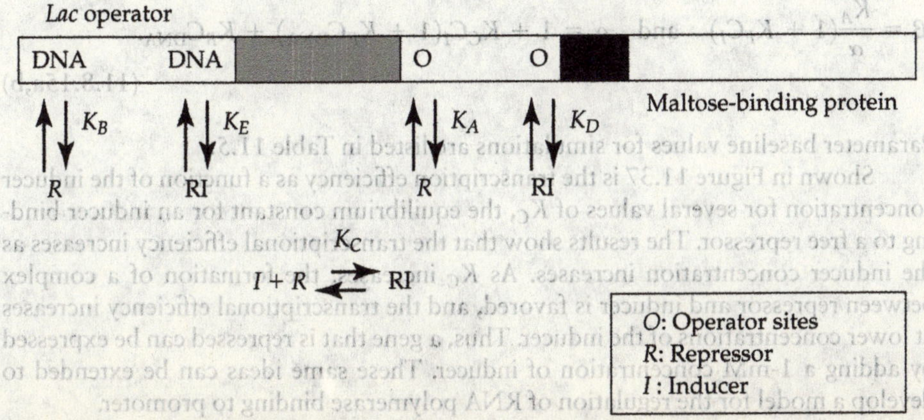

FIGURE 11.36 Simplified schematic of the *lac* operon transcription control system. (Adapted from Ref. [23].)

$$K_D = \frac{C_{RIO}}{C_{RI}C_O}. \tag{11.8.8b}$$

In addition, the inducer can bind operators that are bound to the repressor sites:

$$I + RO \longleftrightarrow RIO, \tag{11.8.9a}$$

$$K_F = \frac{C_{RIO}}{C_{RO}C_I}. \tag{11.8.9b}$$

Note that $K_D K_C = K_A K_F$.

Increasing the free-operator concentration increases the transcription efficiency (Equation (11.8.2)).

In addition to binding specifically, the repressor bound to inducer can bind nonspecifically to DNA:

$$RI + DNA = RIDNA, \tag{11.8.10}$$

$$K_E = \frac{C_{RIDNA}}{C_{RI}C_{DNA}}. \tag{11.8.11}$$

The total operator and repressor concentrations are, respectively,

$$C_{O_0} = C_O + C_{RO} + C_{RIO} \tag{11.8.12}$$

and

$$C_{R_T} = C_R + C_{RI} + C_{RO} + C_{RDNA} + C_{RIO} + C_{RIDNA}. \tag{11.8.13}$$

The subscript zero represents the initial concentration of a reactant or product. The concentration of nonspecific binding sites is large and can be assumed constant, so that $C_{DNA} = C_{DNA_0}$. After considerable algebraic manipulation of Equations (11.8.5) to (11.8.13), the following expression can be derived for the transcription efficiency:

$$\eta = 0.5\left[1 - \frac{1}{\beta C_{O_0}} - \frac{C_{R_T}}{C_{O_0}} + \sqrt{\left(1 - \frac{1}{\beta C_{O_0}} - \frac{C_{R_T}}{C_{O_0}}\right)^2 + \frac{4}{\beta C_{O_0}}}\right]. \tag{11.8.14}$$

In this equation,

$$\beta = \frac{K_A}{\alpha}(1 + K_F C_I) \quad \text{and} \quad \alpha = 1 + K_C C_I(1 + K_E C_{DNA}) + K_B C_{DNA}. \tag{11.8.15a,b}$$

Parameter baseline values for simulations are listed in Table 11.5.

Shown in Figure 11.37 is the transcription efficiency as a function of the inducer concentration for several values of K_C, the equilibrium constant for an inducer binding to a free repressor. The results show that the transcriptional efficiency increases as the inducer concentration increases. As K_C increases, the formation of a complex between repressor and inducer is favored, and the transcriptional efficiency increases at lower concentrations of the inducer. Thus, a gene that is repressed can be expressed by adding a 1-mM concentration of inducer. These same ideas can be extended to develop a model for the regulation of RNA polymerase binding to promoter.

TABLE 11.5

Baseline Parameter Levels for Transcriptional Efficiency

C_{DNA_0}	0.04 M
C_{O_0}	4×10^{-9} M
C_{R_0}	2×10^{-8} M
C_{I_0}	0–0.001 M
$\mu = \ln 2/t_d$*	0.0231 min^{-1}
K_A	$2 \times 10^{12} \text{M}^{-1}$
K_B	$1 \times 10^{3}\ \text{M}^{-1}$
K_C	$1 \times 10^{7} \text{M}^{-1}$
K_D	$2 \times 10^{9} \text{M}^{-1}$
K_E	$1.5 \times 10^{4} \text{M}^{-1}$
k_p	$\dfrac{2{,}400}{(233/\mu^2 + 78)}\ \text{min}^{-1}$
k_d	0.46 min^{-1}
k_e	0.01 min^{-1}
k_q	$\dfrac{3{,}600\delta}{(82.5/\mu + 145)}\ \text{min}^{-1}$
	$\delta = 1$ when $60\ \mu > \ln 2$
	$\delta = \dfrac{60\ \mu}{\ln 2}$ when $60\ \mu < \ln 2$

*t_d, doubling time.

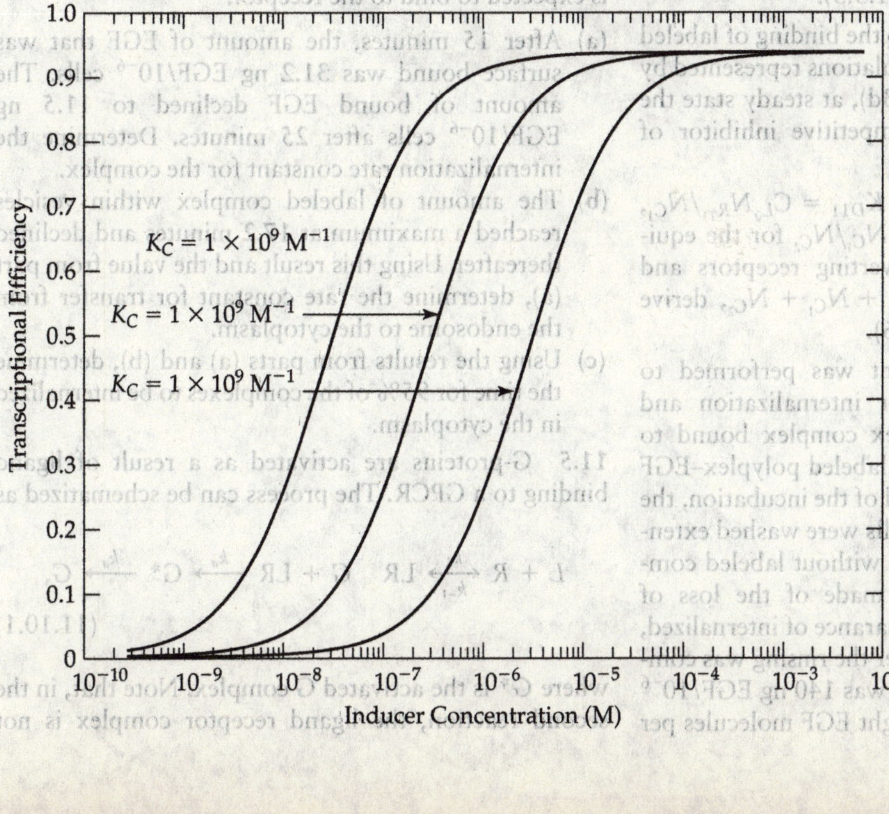

FIGURE 11.37 Effect of K_c for baseline parameters provided on in Table 11.5.

11.9 QUESTIONS

11.1 Explain how measurements of receptor–ligand binding at equilibrium can be used to determine whether a binding is a simple bimolecular reaction or involves the formation of a dimer.

11.2 How does cooperativity affect the binding between receptor and ligand?

11.3 Explain the difference between true cooperativity and apparent cooperativity.

11.4 Explain why, at steady state, the levels of internalized LDL are greater than the levels of receptor-bound LDL.

11.5 Discuss when the assumption of a constant ligand concentration is not appropriate to use.

11.6 What are the different ways in which receptor-mediated endocytosis can be regulated?

11.7 What is the advantage of using the logarithmic sensitivity coefficient versus the local derivative of the response with respect to the signal?

11.8 Explain how the signaling cascades in Figure 11.31 can exhibit switchlike behavior and amplification. What features need to be added for the cascade to exhibit adaptation?

11.9 What is the primary function of transcription factors?

11.10 PROBLEMS

11.1 Show that Equation (11.3.4) can be obtained by rearranging terms in Equation (11.3.3).

11.2 Show that, with respect to the binding of labeled and unlabeled ligand to two populations represented by Equations (11.4.13a) to (11.4.13d), at steady state the unlabeled ligand acts as a competitive inhibitor of labeled ligand binding.

11.3 Using the definitions $K_{D11} = C_{L_0}N_{R_{T1}}/N_{C_1}$, $K_{D21} = N_{R_{T1}}/N_{R_{T2}}$, and $K_{D12} = N_{C_1}/N_{C_2}$ for the equilibrium constants for interconverting receptors and the relation $N_{R_T} = N_{R_{T1}} + N_{R_{T2}} + N_{C_1} + N_{C_2}$, derive Equations (11.4.15) and (11.4.16).

11.4 The following experiment was performed to measure the rate constants for internalization and endosomal release of a polyplex complex bound to EGF: Cells were incubated with labeled polyplex–EGF for 45 minutes at 4°C. At the end of the incubation, the medium was removed and the cells were washed extensively. Fresh medium was added without labeled complex, and measurements were made of the loss of surface-bound EGF and the appearance of internalized, degraded, and recycled EGF. After the rinsing was complete, the amount of bound EGF was 140 ng EGF/10^{-6} cells. There were, on average, eight EGF molecules per complex, although only one EGF molecule per complex is expected to bind to the receptor.

(a) After 15 minutes, the amount of EGF that was surface bound was 31.2 ng EGF/10^{-6} cells. The amount of bound EGF declined to 11.5 ng EGF/10^{-6} cells after 25 minutes. Determine the internalization rate constant for the complex.

(b) The amount of labeled complex within vesicles reached a maximum at 17.2 minutes and declined thereafter. Using this result and the value from part (a), determine the rate constant for transfer from the endosome to the cytoplasm.

(c) Using the results from parts (a) and (b), determine the time for 95% of the complexes to be internalized in the cytoplasm.

11.5 G-proteins are activated as a result of ligand binding to a GPCR. The process can be schematized as

$$L + R \underset{k_{-1}}{\overset{k_1}{\rightleftharpoons}} \text{LR} \quad G + \text{LR} \xrightarrow{k_a} G^* \xrightarrow{k_a} G, \tag{11.10.1}$$

where G^* is the activated G complex. Note that, in the second reaction, the ligand receptor complex is not consumed, but merely activates the G-protein. Thus, a balance on the receptor is

$$C_{R_T} = C_R + C_{LR}, \qquad (11.10.2)$$

where C_{LR} is either free or bound to G-protein. The activated G-protein is deactivated with a rate constant k_i. In order to block desensitization, arrestin binding to LR is blocked by adding a drug that inactivates arrestin.

(a) A certain method was developed to measure the rate constants k_a and k_i. In the ensuing experiments, different amounts of ligand were bound to cells containing 5×10^4 receptors per cell. The number of activated G-proteins was measured after steady state was reached. The total G-protein concentration was 1×10^5 molecules per cell. Results from a typical experiment are shown in Table 11.6. Use these results to determine the ratio k_a/k_i.

(b) To determine k_i, G^* levels were allowed to reach steady state. At that point, the ligand rapidly dissociated from the receptor by incubating with a high-affinity antibody for the ligand. The activated G-protein was measured as a function of time (Table 11.7). Use this result to determine k_i.

(c) Using results from parts (a) and (b), determine the activation rate constant k_a.

11.6 Arrestin binds to a phosphorylated ligand–receptor complex, blocking the binding and activation of G-proteins, according to the following reaction:

$$LR + A \underset{k_{-1a}}{\overset{k_{1a}}{\rightleftharpoons}} LRA. \qquad (11.10.3)$$

Arrestin binds reversibly with LR with a dissociation constant of $K_A = C_{LR}C_A/C_{LRA}$, in units of molecules per cell. Only free LR can bind to G-proteins. A balance on LR yields $C_{LR_T} = C_{LR} + C_{LRA}$. To determine the effect of arrestin on G-protein activation, the experiment described in Problem 11.5 was repeated in the absence of the drug that inhibits arrestin binding.

(a) Assuming that $C_A \gg C_{LR_T}$, find an equilibrium expression for C_{LRA} in terms of K_A, C_{LR_T}, and C_{LR}. The total number of arrestin molecules is 5×10^5 molecules per cell. For $C_{LR} = 50{,}000$ molecules per cell, arrestin binding reduced the level of G^* to 40% of the value found in the absence of arrestin. Use this information to determine K_A.

(b) What is the effect of the amount of bound ligand on the relative inhibition in G-protein activation for an arrestin concentration of 5×10^5 molecules per cell?

TABLE 11.6

G-Protein Expression

C_{LR} (molecules per cell)	C_{G^*} (molecules per cell)
5,000	250
10,000	500
20,000	1,000
30,000	1,500
40,000	2,000
50,000	2,500

TABLE 11.7

Activated G-Protein Levels

Time (s)	C_{G^*} (molecules per cell)
0	2,500
1	2,047
2	1,676
5	920
10	338

11.7 *Receptor–Ligand Binding* Researchers have been studying the FY receptor, which appears to reverse the cell aging process. They performed equilibrium-binding experiments using 4×10^4 cells mL^{-1} and ligand concentrations of between 10^{-10} M and 10^{-6} M. After correcting for nonspecific binding, they appear to have identified two different classes of receptors. One class has 5×10^5 molecules per cell with $K_D = 1 \times 10^{-9}$ M. The other class has 5×10^8 molecules per cell with $K_D = 2 \times 10^{-7}$ M. To validate their results, the researchers performed a dissociation experiment in which 10^{-9} M of labeled ligand and 10^{-6} M of unlabeled ligand were bound to the receptors at 4°C for 2 hours, resulting in a steady state. The labeled medium was removed and the cells were rinsed. Dissociation from the receptor was measured after the cells were diluted in a medium without labeled ligand and a high concentration of unlabeled ligand. The experimenters found that dissociation was a first-order process with a rate constant of 3×10^{-3} s^{-1}. The presence of unlabeled ligand during the dissociation portion of the experiment did not affect the kinetics of receptor dissociation. When the binding portion of the experiment was performed without unlabeled ligand, it was found that dissociation involved two exponentials, one with a rate constant of $3 \times 10^{-3}\ s^{-1}$ and the other with a rate constant of $3 \times 10^{-5}\ s^{-1}$.

Determine whether the results are consistent. Also, determine whether the dissociation constants can be associated with one of the binding sites.

11.8 Find the logarithmic sensitivity coefficient for an enzymatic reaction in which the substrate is a noncompetitive inhibitor.

11.9 Derive Equation (11.7.10) from Equations (11.7.6) through (11.7.9).

11.10 *Receptor-Mediated Endocytosis of EGF* The EGFR is a class 1 receptor that undergoes increased internalization after binding to EGF. In this problem, a simplified kinetic model of EGFR-mediated endocytosis is used to explore the dynamics of EGF and its receptor with and without EGF. This model also accounts for ligand depletion. The model, which is presented in Section 11.6, incorporates ligand depletion and different degradation rate constants for the receptor (k_{deg}) and the ligand (k_{degL}).

A material balance on the ligand is

$$\frac{dC_L}{dt} = \frac{n_c}{N_A}[-k_1 C_L N_{R_S} + k_{-1} N_{C_S} + k_{rec}(1 - f_L) N_{L_i}], \qquad (11.10.4)$$

where n_c is the number of cells per mL and N_A is Avogadro's number.

Material balances on the surface receptor, the receptor–ligand complex, the intracellular receptor, and the intracellular ligand are, respectively,

$$\frac{dN_{R_S}}{dt} = -k_1 C_L N_{R_S} + k_{-1} N_{C_S} - k_{eR} N_{R_S} + k_{rec}(1 - f_R) N_{R_i} + V_s, \qquad (11.10.5)$$

$$\frac{dN_{C_S}}{dt} = k_1 C_L N_{R_S} - (k_{-1} + k_{eC}) N_{C_S}, \qquad (11.10.6)$$

$$\frac{dN_{R_i}}{dt} = k_{eR} N_{R_S} + k_{eC} N_{C_S} - [k_{deg} f_R + k_{rec}(1 - f_R)] N_{R_i}, \qquad (11.10.7)$$

$$\frac{dC_{L_i}}{dt} = k_{eC} N_{C_S} - [k_{deg\,L} f_L + k_{rec}(1 - f_L)] N_{L_i}. \qquad (11.10.8)$$

Baseline values of the parameters for the calculations are listed in Table 11.8.

(a) Solve the steady-state form of Equations (11.10.6) to (11.10.8) for $C_L = 0$, and determine the steady-state intracellular concentration N_{R_i} and the rate of new receptor synthesis V_s for $N_R = 0.5 \times 10^5$ receptors per cell.

(b) Use the differential-equation solver `ode45` in MATLAB to validate that, in the absence of ligand, the receptor levels on the cell surface and within the cell are constant for the values obtained in part (a).

(c) For the following values of EGF concentration, with the values listed in Table 11.8 for model parameters, use the differential-equation solver `ode45` in MATLAB to determine the concentration of bound and intracellular EGF as a function of time:

$$C_L = 1 \times 10^{-9}\text{M},\ 2 \times 10^{-9}\text{M},\ 5 \times 10^{-9}\text{M},\ 1 \times 10^{-8}\text{M},\ 5 \times 10^{-8}\text{M},\ 1 \times 10^{-7}\text{M}$$

Steady state should be achieved in about 240 minutes.

(d) Assess when ligand depletion becomes important.

(e) Explain the behavior of the binding curve, contrasting the results found here with those obtained for class 2 receptors.

TABLE 11.8

Baseline Values for Problem 11.10

Parameter	Value
$k_1 (M^{-1}\ min^{-1})$	7.2×10^7
$k_{-1} (min^{-1})$	0.3000
$k_{eR} (min^{-1})$	0.0300
$k_{eC} (min^{-1})$	0.3000
$k_{rec} (min^{-1})$	0.0800
$k_{deg} (min^{-1})$	0.0022
$k_{degL} (min^{-1})$	0.01
f_R	0.20
f_L	0.50
n_c (cells mL^{-1})	10^9

(f) Using simulations, explain how it would be possible to obtain k_{eC} and k_{eR}. If you wish to simulate a pulse-chase experiment, simulate the pulse portion first. Then use the endpoints from your simulation as the starting points for the chase period.

11.11 In order to obtain an expression for k_{eC}, assume that a class 1 receptor bound to ligand binds to adaptor proteins present in coated pits. The total number of adaptor proteins per cell is N_{A_T}. Assume that (a) the binding of receptors outside coated pits to adaptor proteins in coated pits is reversible and diffusion limited; (b) the fraction of receptors in coated pits (but not bound to adaptor proteins) equals the fractional area occupied by coated pits (2%), and (c) the internalization of ligand-bound receptors equals λN_{RA}, which is the amount of receptor bound to adaptor. Develop a relation for k_{eC} in terms of k_+, N_{A_T}, N_{R_T}, and λ. Evaluate how k_{eC} changes as a function of N_{A_T}/N_{R_T}.

11.12 For the reaction cascade represented by Equations (11.7.13) through (11.7.15) and the parameter values provided in the legend to Figure 11.32, determine the sensitivity of each reaction step. Use Equations (11.3.17a) and (11.3.17b) to show that the overall sensitivity can be greater than the individual sensitivities.

11.13 Two important features of signal transduction are amplification of the input signal and feedback inhibition of the signal. This problem examines the effect of feedback inhibition by the reaction product on a cascade of enzymatic reactions.

Consider a set of three phosphorylation–dephosphorylation reactions in which the phosphorylated protein produced in the first and second reactions serves as a kinase in the second and third reactions. Further, the product of the three reactions, P_3^*, noncompetitively inhibits the first reaction with an inhibition constant K_I. The cascade is shown schematically in Figure 11.38.

Each of the pairs of reactions in the sequence is assumed to follow Michaelis–Menten kinetics, so that, in the absence of inhibition, Equation (11.7.10) is valid for each step. For each protein i, the total concentration is divided between the native and phosphorylated forms:

$$C_{T_i} = C_{P_i} + C_{P_i}^* \qquad (11.8.9)$$

In steps 2 and 3, the phosphorylated protein that was produced in the previous step has kinase activity for the protein. Thus, $R_{\max_3} = k_2^3 C_{P_1}^*$ and $R_{\max_5} = k_2^5 C_{P_2}^*$. Note

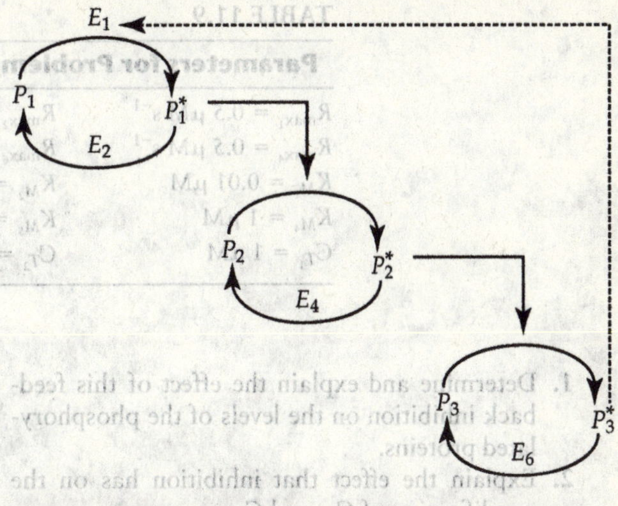

FIGURE 11.38 Schematic of a reaction cascade in which the end product P_3^* noncompetitively inhibits enzyme E_1.

that, in the absence of feedback, amplification arises when the concentration of P^* in the next step exceeds the value of P^* in the previous step.

(a) Use the parameter values listed in Table 11.9 as the baseline levels and the ordinary-differential-equation solver ode45 to determine the dynamics of P_1^*, P_2^*, and P_3^* in the absence of inhibition. The simulation time is 120 seconds. Address the following points:

1. Plot the concentrations of $C_{P_1^*}$, $C_{P_2^*}$, and $C_{P_3^*}$ versus time. Explain the dynamics.
2. From an analysis of the steady-state form of the equations describing the concentration of phosphorylated proteins, what is the expected relation between the various steady-state levels?
3. Amplification arises when the levels of $C_{P_2^*}$, and $C_{P_3^*}$ are higher than the levels of $C_{P_1^*}$. By adjusting the levels of parameters, identify which parameters have the greatest effect on amplification. Because of the large number of parameters that can be varied, carefully identify those parameters that should affect amplification, and adjust them rather than varying all of the parameters.

(b) Now assume that $C_{P_3^*}$ noncompetitively inhibits enzyme E_1. Use $K_I = 0.01\ \mu\text{M}$ for the inhibition constant.

TABLE 11.9

Parameters for Problem 11.13

$R_{max_1} = 0.5\ \mu M\ s^{-1}$	$R_{max_2} = 0.5\ \mu M\ s^{-1}$	$k_2^3 = 2.0\ s^{-1}$
$R_{max_4} = 0.5\ \mu M\ s^{-1}$	$R_{max_6} = 0.5\ \mu M\ s^{-1}$	$k_2^5 = 2.0\ s^{-1}$
$K_{M_1} = 0.01\ \mu M$	$K_{M_2} = 1\ \mu M$	$K_{M_3} = 1\ \mu M$
$K_{M_4} = 1\ \mu M$	$K_{M_5} = 1\ \mu M$	$K_{M_6} = 1\ \mu M$
$C_{T_1} = 1\ \mu M$	$C_{T_2} = 1\ \mu M$	$C_{T_3} = 1\ \mu M$

1. Determine and explain the effect of this feedback inhibition on the levels of the phosphorylated proteins.
2. Explain the effect that inhibition has on the amplification of C_{P_2} and C_{P_3}.

(c) In the cases you examined, the levels of C_{P_1}, C_{P_2}, and C_{P_3} do not return to zero. In many signaling cascades, however, the responses do return to zero and the overall response is attenuated. Propose and test a possible mechanism to cause all three responses to decay to zero.

11.11 REFERENCES

1. Lodish, H., Berk, A., Zipursky, A.L., Matsudaira, P., Baltimore, D., and Darnell, J., *Molecular Cell Biology*. 2000, New York: WH Freeman and Company.
2. Desmarets, J., Gresser, O., Guedin, D., and Frelin, C. "Interaction of endothelin-1 with cloned bovine ETA receptors: biochemical parameters and functional consequences." *Biochemistry*, 1996. 35: pp. 14858–14875.
3. DeMeyts, P. "The structural basis of insulin and insulin-like growth factor-I (IGF1) receptor binding and negative cooperativity, and its relevance to mitogenic versus metabolic signaling." *Diabetologica*, 1994. 37(S2): S135–S138.
4. Schaffer, L. "A model for insulin binding to the insulin receptor." *Eur. J. Biochem.*, 1994. 221: pp. 1127–1132.
5. DeMeyts, P., and Whittaker, J. "Structural biology of insulin and IGF1 receptors: implications for drug design." *Nat. Rev. Drug Discov.*, 2002. 1: pp. 769–783.
6. Lauffenburger, D.A., and Linderman, J.J., *Receptors: Models for Binding, Trafficking, and Signaling*. 1993, New York: Oxford University Press.
7. Brodsky, F.M., Chen, C.-Y., Knuehl, C., Towler, M.C., and Wakeham, D.E. "Biological basket weaving: formation and function of clathrin-coated vesicles." *Ann. Rev. Cell Dev. Biol.*, 2001. 17: pp. 516–568.
8. Sorkina, T., Huang, F., Beguinot, L., and Sorkin, A. "Effect of tyrosine kinase inhibitors on clathrin-coated pit recruitment and internalization of epidermal growth factor receptor." *J. Biol. Chem.*, 2002. 277: pp. 27433–27441.
9. Hirst, J., and Robinson, M.S. "Clathrins and adaptors." *Biochim. Biophys. Acta*, 1998. 1404: pp. 173–193.
10. Smart, E.J., Graf, G.A., McNiven, M.A., Sessa, W.C., Engelman, J.A., Scherer, P.E., Okamoto, T., and Lisanti, M.P. "Caveolins, liquid-ordered domains and signal transduction." *Mol. Cell Biol.*, 1999. 19: pp. 7289–7304.
11. Truskey, G.A., Colton, C.K., and Davies, P.F. "Kinetic analysis of receptor-mediated endocytosis and lysosomal degradation in cultured cells." *Ann. NY Acad. Sci.*, 1985. 435: pp. 349–351.
12. Brown, M.S., and Goldstein, J.L. "Analysis of a mutant strain of human fibroblasts with a defect in the internalization of receptor-bound low density lipoprotein." *Cell*, 1976. 9: pp. 663–674.
13. Goldstein, B., Wolfsky, C., and Bell, G. "Interactions of low density lipoprotein receptors with coated pits on human fibroblasts: Estimates of the forward rate constant and comparison with the diffusion limit." *Proc. Natl. Acad. Sci.*, 1981. 78: pp. 5695–5698.
14. Anderson, R.G.W., Goldstein, J.L., and Brown, M.S. "Localization of low density lipoprotein receptors on plasma membrane of normal human fibroblasts and their

absence in cells from a familial hypercholesterolemia homozygote." *Proc. Natl. Acad. Sci.*, 1976. **73**: pp. 2434–2438.

15. Anderson, R.G.W., Goldstein, J.L., and Brown, M.S. "Role of the coated endocytotic vesicle in the uptake of receptor bound low density lipoprotein in human fibroblasts." 1977. *Cell,* **10**: pp. 351–364.
16. Goldstein, B., Griego, R., and Wolfsky, C. "Diffusion-limited forward rate constants in two dimensions. Application to the trapping of cell surface receptors by coated pits." *Biophys. J.*, 1984. **46**: pp. 573–686.
17. Brown, M.S., and Goldstein, J.L. "Regulation of the activity of the low density lipoprotein receptor in human fibroblasts." *Cell,* 1975. **6**: pp. 307–316.
18. Starbuck, C., and Lauffenburger, D.A. "Mathematical model for the effects of epidermal growth factor receptor trafficking dynamics on fibroblast proliferation rates." *Biotechnol. Prog.*, 1992. **8**: pp. 132–143.
19. Koshland, D.E., Goldbeter, A., and Stock, J.B. "Amplification and adaptation in regulatory and sensory systems." *Science,* 1982. **217**: pp. 220–225.
20. Khodolenko, B.N., Hoek, J.B., Westerhoff, H.V., and Brown, G.C. "Quantification of information transfer via cellular signal transduction pathways." *FEBS Lett.*, 1997. **414**: pp. 430–434.
21. Huang, C.-Y.F., and Ferrell, J.E. "Ultrasensitivity in the mitogen-activated protein kinase cascade." *Proc. Natl. Acad. Sci.*, 1996. **93**: pp. 10078–10083.
22. Asthargari, A.R., and Lauffenburger, D.A. "A computational study of the feedback effects on signal dynamics in a mitogen-activated protein kinase (MAPK) pathway model." *Biotechnol. Prog.*, 2001. **17**: pp. 227–239.
23. Lee, S.B., and Bailey, J.E. "Genetically structured models for the *lac* promoter–operator function in the *Escherichia coli* chromosome and in multicopy plasmids: lac operator function." *Biotechnol. Bioeng.*, 1984. **26**: pp. 1372–1382.
24. Lee, S.B. and Bailey, J.E. "Analysis of growth rate effects on productivity of recombinant *Escherichia coli* populations using molecular mechanism model." *Biotechnol. Bioeng.*, 1984. **26**: pp. 66–73.

CHAPTER 12

Cell Adhesion

12.1 Introduction

Cell adhesion is the contact and firm interaction between cells and other cells, the extracellular matrix, or material surfaces. Most eukaryotic cells must adhere to other cells or to the extracellular matrix to function properly, to maintain the mechanical integrity of tissue, and to regulate cell migration. An understanding of the dynamics of cell adhesion is necessary in a variety of biomedical applications. Promoting or inhibiting cell adhesion to biomaterials is often crucial to the proper function of those materials [1]. When cancer cells metastasize, the adhesion of the cells to the endothelium is a primary step in invading and localizing in specific tissues. Cell adhesion and capsular tissue formation can impede the function of implanted sensors. Tissue-engineering applications are influenced by adhesion because adhesive events are involved in differentiation, migration, and ingrowth within natural or synthetic polymeric scaffolds.

In all cases of adhesion of cells to other cells or to the extracellular matrix, noncovalent receptor–ligand interactions are a prerequisite for firm adhesion. Adhesion that is induced by such interactions results in a number of functional changes to the receptors and in the initiation of signaling events. Binding to the extracellular matrix leads to the generation of signals that affect actin polymerization, cell migration, gene expression, and cell replication. When binding initiates these kinds of changes in cell function, the process is known as "outside in" signaling [2]. Conformational changes to a receptor can alter the affinity between the receptor and the ligand and can induce changes in affinity or localization in other receptors. Such conformational changes are induced either by ligand binding or by signals that are generated by *agonists*—molecules that bind to receptors on the cell surface in a process known as "inside out" signaling [2].

Because cell adhesion depends upon noncovalent receptor–ligand interactions, much of this chapter is focused upon these interactions. There are five classes of *adhesion molecules* [3, p. 970]: cadherins, the immunoglobulin (Ig) superfamily, integrins, selectins, and mucins. Adhesion between these molecules may be homotypic (i.e., between two similar types of adhesion molecules, such as cadherins adhering to cadherins) or heterotypic (i.e., between two different types of adhesion molecules or between an adhesion molecule and extracellular matrix proteins). Each class consists of a number of different molecules that share similar structures.

Cadherins are involved in cell-to-cell adhesion. These molecules are specific to given tissues and play a critical role in the development of the organism. Binding is the result of dimer formation of cadherins located on adjacent cells. For example, E-cadherins hold epithelial cells together and are found in the *adherens junctions* (see Section 1.5.1). The disruption of cadherin–cadherin bonds can produce an increase in molecular transport between these cells [4]. *Ig superfamily* molecules are involved in homotypic adhesion, such as that between nerve cell adhesion molecules, or in heterotypic adhesion with *integrins*—the principal adhesion molecules that bind to extracellular matrix proteins such as *fibronectin*, *laminin*, and *collagen*. *Selectins* are adhesion proteins that bind to *mucins*, which are adhesion molecules in which the adhesive region is a carbohydrate. Selectins play a critical role in the initial attachment of leukocytes to endothelial cells (see Section 12.4).

In this chapter, we examine the dynamics of adhesion. The effect of forces on adhesion bonds is discussed in Section 12.2. Cell-to-matrix adhesion and leukocyte and cancer adhesion to endothelium under flow conditions are examined in Sections 12.3 and 12.4, respectively. These two types of adhesion are considered separately because of differences in the steps leading to adhesion and in the functional consequences of adhesion. Receptor–ligand binding, as well as other biochemical reactions within cells, involves a small number of bonds, the effects of which on the distribution of binding and dissociation events are presented in Section 12.5.

12.2 Effect of Force on Bond Association and Dissociation

Tissues and the cells within them are constantly exposed to forces. Examples include the active stretching and contraction of skeletal and cardiac muscle, the dilation and relaxation of blood vessels in response to the pressure wave generated by the heart, and the compression of cartilage during body movements. In addition, cells generate stresses on the extracellular matrix, due to the contractile properties of the actin cytoskeleton.

Because cell adhesion requires noncovalent receptor–ligand interactions, the structure of tissues is maintained only when these interactions are able to resist the forces that are applied to the tissue. In this section, we examine how forces affect the kinetics and equilibria of such interactions and subsequent cell function.

12.2.1 The Influence of Energy Barriers on Molecular Interactions

Bond formation and disruption are dynamic processes. In the absence of an applied force on the molecules, the *bond distance* represents the energy minimum at which attractive and repulsive forces balance (Figure 12.1a). These forces include *van der Waals repulsion*; *London attraction* due to dipole–dipole interaction; and *Coulombic*

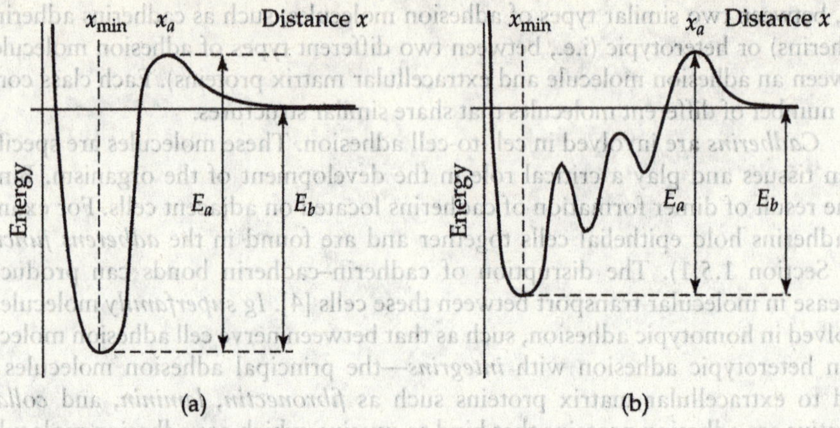

FIGURE 12.1 (a) The sum of attractive and repulsive interactions for a simple molecular interaction leading to an energy minimum at the bond distance (x_{min}). The activation energy barrier that must be overcome for bond dissociation is E_a at a distance x_a. The bond energy is E_b. (b) The bond-distance landscape for a ligand binding to a receptor. Such binding results in multiple interactions between molecules and ligands in the binding region.

interaction, which can be attractive or repulsive, depending on the charge. These forces also depend on bond stretching, bending, and torsion [5]. The deeper the energy minimum, the larger the barrier that must be overcome to break the interaction and the stronger the bond. Molecular collisions lead to a continuous exchange of energy between the reacting molecules, solvent molecules, and other solute molecules. As a result, interacting molecules are not static but move among various locations in the energy landscape (Figure 12.1b). If the energy minimum of a bond is shallow, then interacting molecules can easily overcome the barrier and break the bond.

For bonds formed by a ligand binding to a protein, the energy landscape involves multiple interactions between the ligand and amino-acid side chains in the binding pocket of the protein. Such a landscape is plotted in Figure 12.1b. A global energy minimum is present, as well as several other *metastable* states. These states result from interactions between the ligand and amino-acid side chains *other* than those that produce the most stable configuration. Because the energy barriers among the various local energy minima are much less than the activation barrier, dissociation is expected to be more likely in the scenario shown in Figure 12.2b than in the scenario shown in Figure 12.2a. The dynamics of escape from the energy minima for a protein–ligand interaction are more complex than those for a bond between two atoms.

Because there is a distribution of energies among the molecules in a cell and its surroundings, some molecules are likely to overcome the energy barriers. Macroscopically, this variation in energy levels is represented in terms of the dissociation rate coefficient k_{-1}, which has the general form [7]

$$k_{-1} = \nu \exp\left(\frac{E_a}{k_B T}\right), \tag{12.2.1}$$

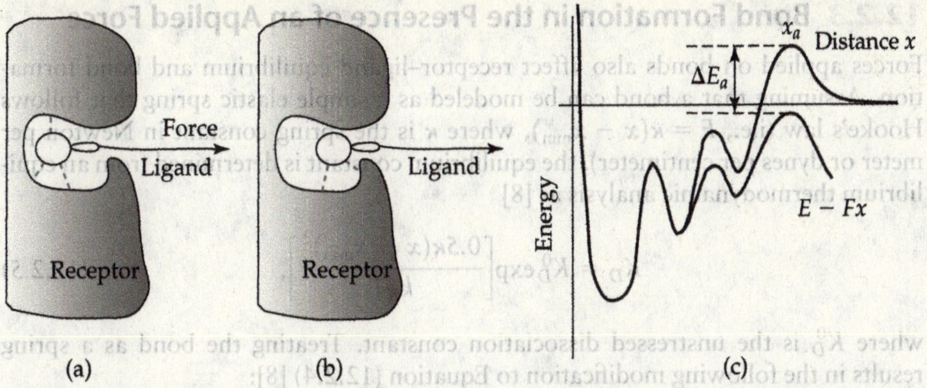

FIGURE 12.2 (a) Schematic of ligand bound to a receptor at energy minimum. Noncovalent interactions are indicated by the dashed line. (b) The displacement of ligand by applied force. A new set of noncovalent interactions have formed, which stabilizes the ligand in the binding site. (c) Distortion of the energy profile by an applied force [6].

where v is a prefactor, E_a is the activation energy that must be overcome for bond dissociation, k_B is Boltzmann's constant, and T is the absolute temperature. Raising the temperature increases the energy of the molecules, which enables a bound molecule to escape the energy barrier. The probability that a bond that is present at time 0 has dissociated at a later time t is

$$p = 1 - \exp(-k_{-1}t). \qquad (12.2.2)$$

12.2.2 Bond Disruption in the Presence of an Applied Force

When a force F is applied to one or both of the molecules, the energy landscape is distorted such that the energy barrier is reduced from E_a to $E_a - \Delta E_a$, where ΔE_a is the energy change that results from the applied force (see Figure 12.2c). The force reduces the energy barrier, which makes bond disruption easier. Bell applied this concept to the dissociation constant by postulating that the energy change equals x_aF [7]. As a result, Equation (12.2.1) becomes

$$k_{-1} = v\exp\left(\frac{E_a - x_aF}{k_BT}\right), \qquad (12.2.3)$$

where the prefactor is associated with the reciprocal of the natural frequency of oscillation in a solid. Defining the dissociation rate coefficient k_{-1} as the unstressed dissociation rate coefficient k_{-1}^0, we can write Equation (12.2.3) as

$$k_{-1} = k_{-1}^0\exp\left(\frac{x_aF}{k_BT}\right). \qquad (12.2.4)$$

This result clearly shows that an applied force can greatly increase the dissociation rate coefficient and shorten the time required for bond dissociation.

12.2.3 Bond Formation in the Presence of an Applied Force

Forces applied on bonds also affect receptor–ligand equilibrium and bond formation. Assuming that a bond can be modeled as a simple elastic spring that follows Hooke's law (i.e., $F = \kappa(x - x_{\min})$, where κ is the spring constant in Newton per meter or dynes per centimeter), the equilibrium constant is determined from an equilibrium thermodynamic analysis as [8]

$$K_D = K_D^0 \exp\left[\frac{0.5\kappa(x - x_{\min})^2}{k_B T}\right], \tag{12.2.5}$$

where K_D^0 is the unstressed dissociation constant. Treating the bond as a spring results in the following modification to Equation (12.2.4) [8]:

$$k_{-1} = k_{-1}^0 \exp\left[\frac{\kappa(x_a - x_{\min})(x - x_{\min})}{k_B T}\right], \tag{12.2.6}$$

where $x_{\min}$ is the location of the energy minimum.

According to Equation (12.2.6), the applied force changes with separation distance and lowers the energy barrier.

Given that the actual energy profile is complex and varies spatially, no simple analytical expression completely captures the behavior of noncovalent dissociation. In general, the dissociation rate coefficient can be written as

$$k_{-1} = k_{-1}^0 g(F), \tag{12.2.7}$$

where the function $g(F)$ may decrease, increase, or remain unchanged with the applied force [8]. The case considered so far, in which $g(F)$ increases with the applied force, is known as a *slip bond*. If $g(F)$ decreases with the applied force, the bond is referred to as a *catch bond*. A bond may change from one behavior to another as the force increases. The force stabilizes the bond by rearranging the structure of the binding pocket, exposing side chains that were buried. Slip-bond behavior has been observed with a number of receptor–ligand pairs [9–11]. Recently, catch-bond behavior was reported for P-selectin binding to its ligand at forces below 20 pN, with slip-bond behavior at higher forces [12].

12.2.4 The Effect of Loading Rates on Bond Forces

The previous section's analysis of the effect of force on bond dissociation indicates that the applied force affects the lifetime of the bond. Interestingly, the force required to break specific bonds depends upon the rate of loading, and there is no single force per bond that characterizes binding under all conditions [12]. The rate dependence of the force required to disrupt bonds arises from the interactions between the protein–ligand molecules and the surrounding solvent. The solvent surrounds the receptors and ligands and is present within the binding site. Collisions of water molecules with receptor–ligand pairs transfer energy to the system and constantly rearrange the molecular interactions. Such collisions occur on a *timescale* of 10^{-9} seconds [13]. When the timescale for the application of a force is longer than the timescale for collisions with the solvent, the ligands are exposed to a large number of energy levels and can locate lower energy pathways. If, however, the timescale for the application of a force is shorter than the timescale for collisions, then the ligands have few collisions with the solvent, and a larger force is needed to overcome the energy barrier. This analysis has two implications: (a) applying different loading rates of a force enables us to identify different energy barriers [6] and (b) the force in Equations (12.2.3) and (12.2.4) varies with the loading rate.

Several techniques are available for applying defined forces on receptor–ligand bonds: atomic force microscopy [14], micropipet aspiration [15], the surface force apparatus [16], the biointerface probe [17], laser tweezers, and flow chamber assays [18]. In general, ligands are immobilized onto beads or surfaces, and receptors can be expressed by cells or immobilized onto other surfaces. The techniques listed differ in the manner in which force is applied and in the loading rates applied to the bonds. The methods can measure separation distances of $10^{-10} - 10^{-9}$ m and forces as small as 1 pN (1 piconewton $= 1 \times 10^{-12}$N) (see Table 12.1). Further, by adjusting the density of ligands and receptors on the cell surface, it is possible to measure individual molecular binding events.

For a biointerface probe, the surfaces with ligand and receptor are brought into contact. An adhesive contact is noted when a sudden decrease in separation distance between the two surfaces arises from attractive forces (Figure 12.3). When the two surfaces are separated, a force must be added to overcome the bond force. When a single force and loading rate are applied, a distribution of forces is obtained, rather than a single force (see Figure 12.3a). The distribution arises from the complex energy profile and the thermal interactions between the solvent and the receptor molecules. The histogram of the forces also supports the view that adhesion is governed by a single bond, because multiple bonds would produce forces with maxima that are multiples of the lowest mean force [14]. (A more thorough examination of the presence of single bonds is carried out by Zhu and colleagues [19].) Increasing the loading rate leads to an increase in the force and a broadening of the force distribution.

When the mean force required to separate the bonds is plotted against the loading rate, the force increases with the loading rate [20]. A transition with protein–ligand interactions often occurs, due to the presence of multiple energy minima (Figure 12.3b). The transition occurs during the suppression of a lower energy barrier and the unmasking of a higher energy barrier. The applied force distorts the energy landscape so that the lower (or outer) energy barrier no longer limits dissociation. At lower

TABLE 12.1

Survey of Methods Used to Measure Molecular-Level Forces

Method	Forces (pN)	Resolvable distances (nm)	Loading rate (pN s^{-1})
Biointerface probe	0.01–1,000	10	10^{-2}–10^{5}
Laser tweezers	1–200	100	10^{-1}–10^{2}
Micropipet aspiration	10–1,000	100	10^{1}–10^{3}
Atomic force microscopy	10–10,000	0.1	10^{3}–10^{6}
Flow-chamber assays	1–1,000	100	10^{0}–10^{2}
Surface force apparatus	$10^{4}-10^{6}$	1	10^{4}–10^{6}

Source: Adapted from Refs [16,18].

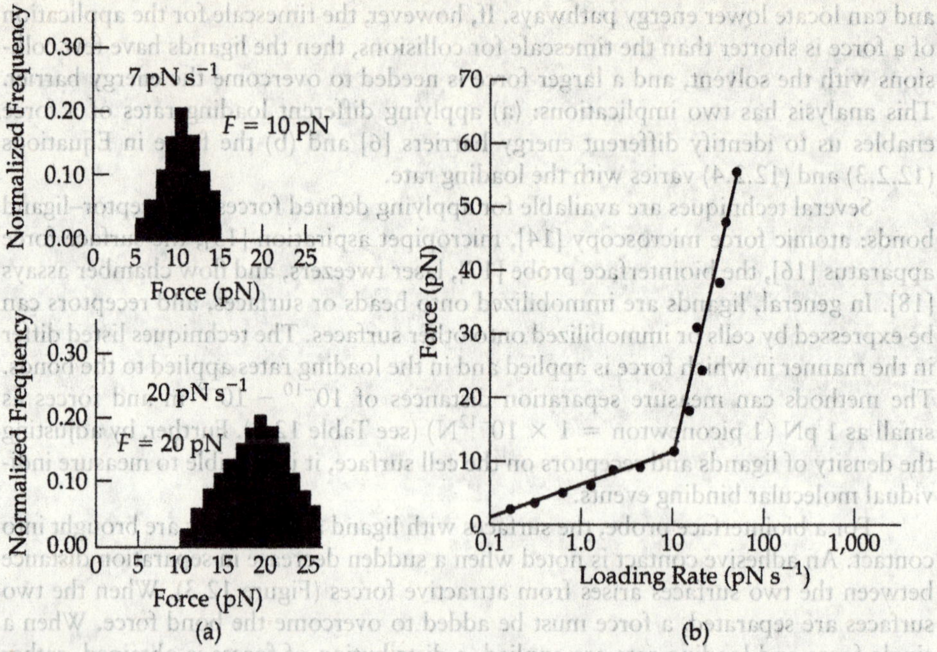

FIGURE 12.3 (a) Distribution of forces to break a receptor–ligand bond for two different loading rates. Increasing the loading rate increases the force needed to cause rupture. (b) The effect of loading rate on the mean force to break a receptor–ligand bond. The transition in the force at 10 pN s^{-1} indicates that a different energy well is involved in rupture. (Adapted from Ref. [6] by permission of The Royal Society of Chemistry.)

loading rates, the timescale for applying a force is much longer than the diffusion timescale. Collisions of the receptor with solvent in the binding region enable the ligand to move between the various energy states. Thus, only the lower energy state is a barrier to dissociation at the lower loading rates.

If a single barrier limits dissociation, then Equation (12.2.4) is valid at a given loading rate, and the mean force required to dissociate the bonds is related to the loading rate r_f as follows [6]:

$$F_b = \frac{k_B T}{x_a} \ln\left(\frac{r_f x_a}{k_{-1}^0 k_B T}\right). \tag{12.2.8}$$

12.3 Cell–Matrix Adhesion

Tissues consist of cells, extracellular matrix molecules, and interstitial fluid. The primary constituents of the extracellular matrix are collagen, proteoglycans, and elastin. In most tissues, the mechanical properties of the tissue depend extensively upon the matrix constituents; collagen and elastin provide elastic behavior, whereas proteoglycans have viscoelastic properties. The composition and density of matrix proteins affects the transport of molecules through the interstitial fluid. Other matrix molecules, such as fibronectin, vitronectin, and laminin, serve to connect the matrix proteins to cells.

Cells synthesize extracellular matrix molecules and maintain adhesive interactions with matrix proteins. Integrins are the primary cell-surface receptors involved in adhesion to extracellular matrix proteins. Integrins consist of two chains, labeled α and β. The 20 different arrangements between the various α and β chains result in the binding of integrins to a wide variety of ligands. Integrin binding is regulated by calcium binding.

A number of different types of cells, such as fibroblasts, liver cells, and muscle cells, exist within tissues and are surrounded by the extracellular matrix. In contrast, epithelial and endothelial cells interact with extracellular matrix proteins on one side only and are said to be *polarized*.

Cell adhesion to extracellular matrix proteins involves a defined set of molecular interactions at focal contacts that influence subsequent cell division and protein synthesis. *Focal contacts* are the sites of closest approach between the endothelium and a surface. Focal contacts are typically 0.5–2 μm wide and 5–10 μm long [21], with separation distances between 10 nm and 40 nm [22–24]. At focal contacts, adhesion proteins bind to integrins (Figure 12.4). The cytoplasmic side of the integrin β chain binds with the proteins talin and α-actinin. The association of focal adhesion kinase (FAK) with the developing focal contact results in the activation of FAK. In turn, FAK phosphorylates itself and the focal contact proteins vinculin and paxillin. The phosphorylation stimulates actin binding to the integrins via vinculin and α-actinin, resulting in actin polymerization (see Figure 12.5). Inhibition of FAK tyrosine kinase activity prevents the formation of focal contacts and stress fibers [25,26]. Further, tyrosine phosphorylation appears to activate mitogen-activated protein (MAP) kinase [27]. MAP kinase activation and nuclear translocation are two examples of stimuli that alter cell function upon integrin engagement with adhesion proteins.

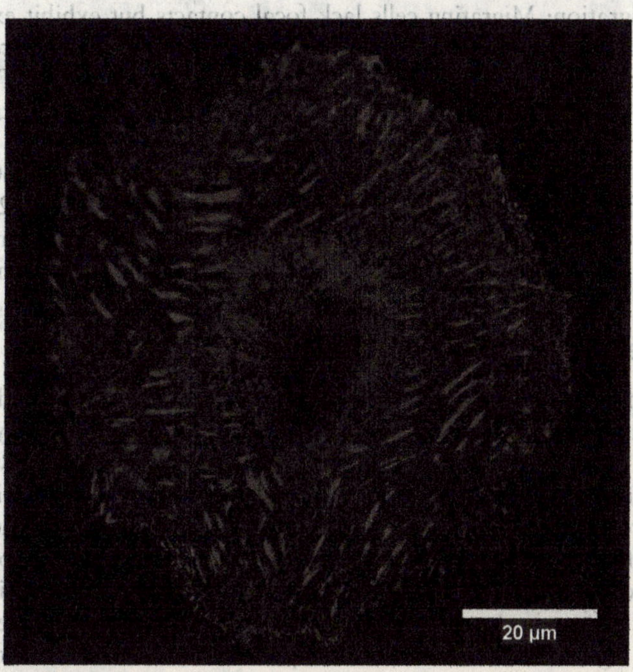

FIGURE 12.4 Focal contacts on adherent endothelial cells. Fibronectin adsorbed to glass slide, after which cells attached to the surfaces and were allowed to spread fully. Focal contacts were visualized using an antibody to vinculin. A second antibody containing a fluorescent molecule bound to the antivinculin antibody, making it possible to see the focal contacts. (Photograph courtesy of Dr. Charles Wallace.)

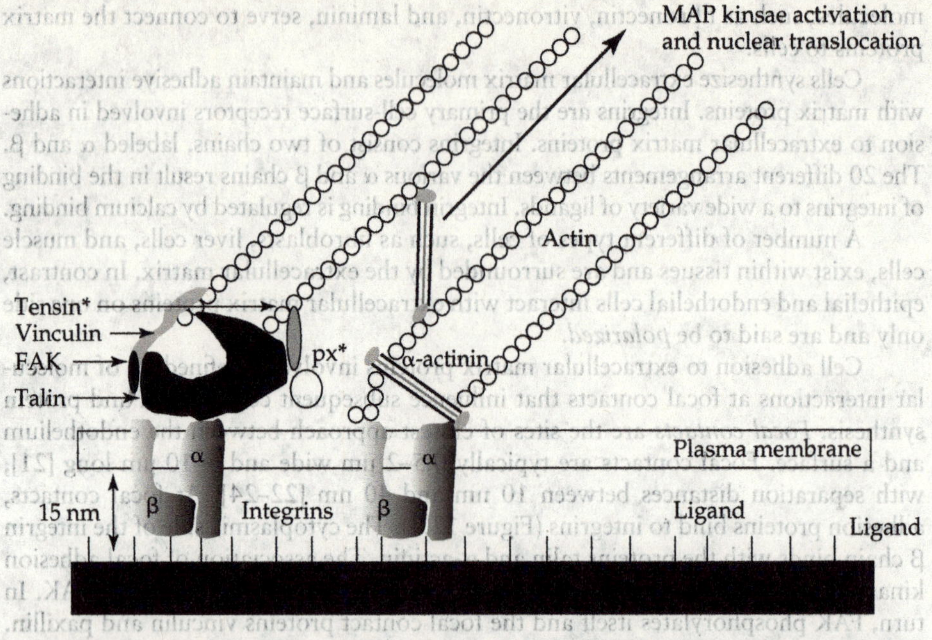

FIGURE 12.5 Schematic of proteins involved in focal contact formation and the signaling events initiated by focal contacts. Not shown are the molecules involved in signaling. The asterisk represents proteins that are phosphorylated, and px represents paxillin. (Adapted from Ref. [21], with permission of www.annualreviews.org, © 1996.)

Close contacts are membrane separations of 15 nm to 50 nm and contain many of the cytoplasmic proteins found in focal contacts [28]. Close contacts may be a factor in cell migration: Migrating cells lack focal contacts but exhibit extensive close contacts, as judged by interference reflection microscopy [28,29]. *Extracellular matrix contacts* are regions separated from the surface by 100 nm or more. These contacts occur after 24 or more hours of adhesion and contain fibrils of fibronectin. Bundles of actin filaments on the periphery of a cell that is spreading appear to maintain the spread shape of the cell but are not in direct contact with the surface [30]. Some cells do not form focal contacts; instead, they form *point contacts* 90 nm to 200 nm in size within 15 nm of the surface [31]. These point contacts contain integrin $\alpha_1\beta_1$, which binds to collagen and laminin, but they do not contain vinculin or f-actin [31].

12.3.1 Cell Attachment

The adhesion process can be divided into two phases: attachment and detachment. Under experimental conditions in vitro, the attachment process simply involves the formation of bonds between the adhesion receptors and their ligands when the cells are brought into contact. When cells attach under static conditions in vitro, the bond formation process does not depend upon force. When cells attach, the ligands and receptors diffuse in the plasma membrane and react in a small region of contact. When cells adhere to the extracellular matrix, the ligand proteins are immobile and the receptors diffuse in the plane of the cell membranes. Because adhesion receptor densities and ligand densities are often comparable when cells attach to one another or to the extracellular matrix, ligand depletion is often significant (see Section 11.4.1). Assuming that binding is a bimolecular interaction and that dissociation of the receptor–ligand complex is a first-order reaction (see Section 10.2.3), a mass balance on the number of complexes per unit area of cell membrane is

$$\frac{dN_C}{dt} = k'_1 N_L N_R - k_{-1} N_c. \tag{12.3.1}$$

For receptor and ligand, the total number of each type of molecule is assumed to be constant:

$$N_{L_0} = N_L + N_C, \tag{12.3.2a}$$

$$N_{R_0} = N_R + N_C. \tag{12.3.2b}$$

Substituting Equations (12.3.2a) and (12.3.2b) into Equation (12.3.1) yields

$$\frac{dN_C}{dt} = k_1(N_{L_0} - N_c)(N_{R_0} - N_c) - (k_{-1}N_c). \tag{12.3.3}$$

The solution of Equation (12.3.3) is presented in Equation (10.2.34). After a rearrangement of terms, that equation becomes

$$N_C = \frac{b + \sqrt{b^2 - 4a}}{2}\left\{\frac{1 - \exp\left[-k_1\left(\sqrt{b^2 - 4a}\right)t\right]}{1 - \left(\dfrac{b + \sqrt{b^2 - 4a}}{b - \sqrt{b^2 - 4a}}\right)\exp\left[-k_1\left(\sqrt{b^2 - 4a}\right)t\right]}\right\}, \tag{12.3.4}$$

where

$$a = N_{L_0} N_{R_0} \quad \text{and} \quad b = N_{L_0} + N_{R_0} + K_D, \tag{12.3.5a,b}$$

where $K_D = k_{-1}/k_1$. Note that as time becomes very large, Equation (12.3.4) approaches the steady-state value:

$$N_C = \frac{1}{2}\left[\left(N_{L_0} + N_{R_0} + K_D\right) - \sqrt{(N_{L_0} + N_{R_0} + K_D)^2 - 4N_{L_0}N_{R_0}}\right]. \tag{12.3.6}$$

A lower limit for the time required to reach steady state is obtained by assuming that k_1 is diffusion limited. This is a reasonable assumption for the membrane-bound receptor and ligand in the absence of convective transport of the cells [32]. The rate constant is estimated from Equation (6.9.21), which describes diffusion-limited binding in the cell membrane. Typical values for the number of integrin receptors per cell range from 500,000 for the $\alpha_5\beta_1$ receptor for fibronectin [33] to 3×10^6 for the $\alpha_v\beta_3$ receptor for vitronectin [34,35]. The resulting receptor densities range from 10^9 to 10^{11} molecules per cm^2. Ligand densities range from 5×10^8 [36] to 10^{10}

molecules per cm^2 [37]. The diffusion coefficient of integrin in the plasma membrane is 2×10^{-10} cm^2 s^{-1} [38]. These values indicate that binding is rapid and that the bond density reaches within 1% of the steady-state value in 0.2 seconds. The implication is that, as cells attach and spread, binding is in a quasi–steady state that depends upon the local receptor and ligand densities.

Subsequent to receptor–ligand binding, the receptor begins to interact with cytoskeletal proteins. Binding induces a conformational change to the cytoplasmic side of the integrin β chain, which causes the cytoplasmic side of the integrin to bind with the proteins talin and α-actinin (Figure 12.6). The binding reaction leads to an increase in the affinity between the ligand and the receptor, analogous to the formation of ternary complexes discussed in Section 11.6. Subsequent interactions with focal contacts produce signaling molecules (Figure 12.6).

The formation of *focal contacts*—localized sites of attachment—appears to be influenced by the diffusion-limited reversible binding of receptors and ligands and the formation of ternary complexes with the cytoskeleton proteins. Focal contacts are a periodic or quasi-periodic arrangement of clusters of receptor–ligand complexes. The binding kinetics that permit a nonuniform distribution of complexes at steady state are important in the formation of focal contacts. Although receptor–ligand complex diffusion coefficients are much smaller than the free-receptor diffusion coefficients in focal contacts [38], immobilization alone does not produce a nonuniform distribution of complexes. One way to produce a nonuniform distribution of complexes is to assume that the binding kinetics can be represented as reversible reactions, but that the rate coefficients are functions of the concentration of ligand–receptor complexes [39]. If the rate coefficients are independent of the density of complexes formed, then the complexes will be uniformly distributed. If, however, the rate coefficient for complex

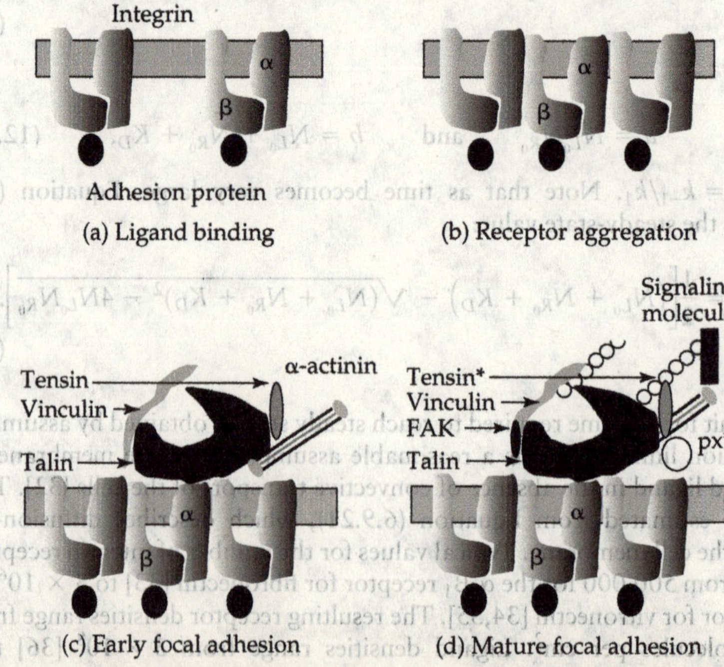

FIGURE 12.6 The sequence of events during focal contact formation and signal transduction. Integrins bind to adhesion proteins (a) and aggregate (b), resulting in binding of α-actinin, tensin, talin, and vinculin to cytoplasmic portions of integrins (c). FAK binding and tyrosine phosphorylation stimulate the interaction of cytoskeleton and signaling molecules with focal contact proteins (d). (Based on summary provided in Ref. [21] and data in Refs. [25–27].)

formation increases and the rate coefficient for complex dissociation decreases with concentration, then the complex distribution is nonuniform. Although ternary complex formation alone does not produce such positive cooperativity, the ternary complex permits receptors to bind to each other in a process known as *receptor cross-linking* that may permit a focal contact to form [40]. The ternary complex may be a simplification of the actual binding process, since a number of proteins (e.g., talin and vinculin) are required to stabilize the focal contact.

12.3.2 Cell Detachment

The molecular events that arise from receptor–ligand binding also lead to changes in the mechanics of the interaction between adherent cells. A force must be applied to separate the cells. Knowledge of the forces required to separate cells provides important information about the number and strength of receptor–ligand bonds formed. To measure adhesion quantitatively, a known force is applied to a cell in order to detach the cell. Several methods are available to measure the strength of cell adhesion, including flow chambers, micropipet aspiration, cell poking, and centrifugation. The choice of technique depends upon the objectives of the experiment and the specific measurements planned.

Net force on cells. A number of different kinds of flow chambers are available for measuring cell adhesion, including parallel-plate channels, variable-width and variable-height channels, cone-and-plate viscometers, and radial-disk systems. Parallel-plate channels and cone-and-plate viscometers expose all adherent cells to a single shear stress. With radial flow chambers, variable-width channels, and variable-height flow channels, a range of shear stresses can be generated in a single experiment.

Because cell-surface receptor density is heterogeneous, as is the contact size among the whole cell population, cells do not detach at a single fluid shear stress or centrifugal force. Rather, they detach over a range of stresses, which can be well described by a lognormal distribution [41]. The fraction of the population per unit area that detaches at a given shear stress τ and exposure time t equals the integral of $p(\mu_x, \sigma_x, \tau)$, the probability distribution of shear stresses at which cells detach:

$$1f(\tau_w, t) = 1 - \int_0^{\tau_w} p(\mu_x, \sigma_x, \tau)d\tau. \quad (12.3.7)$$

When $p(\mu_x, \sigma_x, \tau)$, is represented by a lognormal distribution, the mean (μ) and standard deviation (σ) are calculated from the mean (μ_x), median (M_x), and standard deviation (σ_x) of the untransformed data according to the following relations [42]:

$$\mu_x = \exp\left(\mu + \frac{\sigma^2}{2}\right), \quad (12.3.8)$$

$$M_x = \exp(\mu), \quad (12.3.9)$$

$$\sigma_x^2 = \mu_x^2\left[\exp\left(\sigma^2\right) - 1\right]. \quad (12.3.10)$$

The critical shear stress (τ_c), which is the shear stress required to detach 50% of the cells from the surface (i.e., $\tau_c = M_x$), is determined by applying nonlinear regression [43] of the lognormal distribution to cell detachment data. This simple statistical model can fit a wide range of detachment curves. Both μ_x and σ_x affect the critical shear stress. Small values of σ_x lead to detachment over a narrow range of shear stresses, whereas larger values lead to detachment over a wider range of shear stresses (Figure 12.7).

To use the results of the flow-channel measurements of the critical shear stress, it is necessary to calculate the force that is required to detach cells. The force is the integral of the stress over the surface of the cell that is exposed to flow. The cell shape is not uniform, varying as the cell spreads on the surface [44]. The stress distribution over the cell is obtained by numerical solution of either the Navier–Stokes equation or the Stokes equation, since Re is often less than 0.1. Accurate three-dimensional data on the cell surface can be obtained by using either atomic force microscopy [45] or confocal microscopy [46]. Once the stress distribution is known, the force and torque are computed numerically as

$$\mathbf{F} = \int_S \mathbf{n} \cdot \boldsymbol{\tau} dS, \tag{12.3.11a}$$

$$\mathbf{T} = \int_S \mathbf{F} \times \mathbf{x} dS, \tag{12.3.11b}$$

where **x** is a spatial vector between the center of mass of the cell and the cell surface.

As cells attach and spread, they change from a roughly spherical shape to a more flattened shape with a protrusion at the cell nucleus (see Figure 12.8). Several solutions are available for calculating flow over spheres and hemispheres. Comparing detailed numerical calculations for representative cell shapes with calculations for idealized shapes provides an indication about whether the idealized shapes are

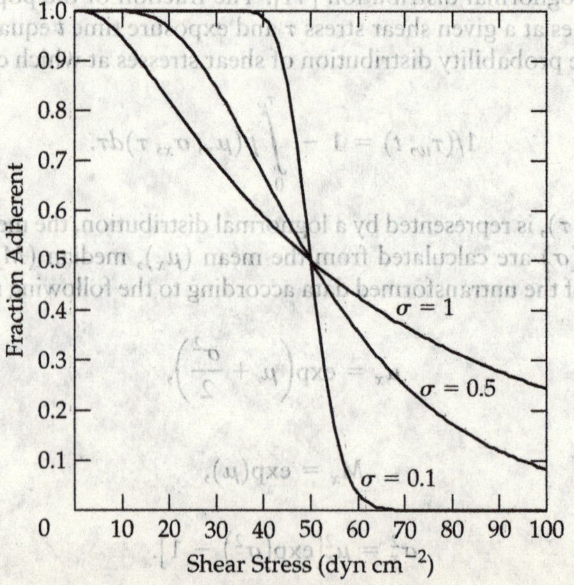

FIGURE 12.7 Effect of σ on the distribution of stresses that cause detachment when cells detach according to a lognormal distribution.

suitable for approximating flow over spreading endothelial cells. For a spherical cell just touching the surface, we can apply equations derived for a sphere touching the surface [47]. The drag force and moment are, respectively,

$$F_D = 6\pi\dot{\gamma}\mu a h F_x\left(\frac{a}{h}\right), \tag{12.3.12a}$$

and

$$T = 4\pi\dot{\gamma}\mu a^3 M_z\left(\frac{a}{h}\right), \tag{12.3.12b}$$

where h is the distance of the center of the sphere above the surface, and $F_x(a/h)$ and $M_z(a/h)$ account for hydrodynamic interactions between the sphere and the plane surface. $F_x(a/h)$ ranges from unity for a sphere that is infinitely far from the surface to 1.7005 for a sphere that touches the surface ($a/h = 1$) [47]. Likewise, $M_z(a/h)$ ranges from unity for a sphere infinitely far from the surface to 0.9439 for a sphere that touches the surface.

Numerical solution of the Stokes equation gives the following relations for the drag and moment on a hemisphere of radius a that touches the surface [48]:

$$F_D = 4.50\dot{\gamma}\mu\pi a^2, \tag{12.3.13a}$$

$$T = 2.58\dot{\gamma}\mu\pi a^3. \tag{12.3.13b}$$

For shear stresses of 30 dyn cm^{-2} or less, the Stokes flow solution is within 10% of the result obtained from the complete Navier–Stokes equation, which is consistent with the validity of the Stokes equation.

To assess the usefulness of these simplified geometries, Olivier performed a series of hydrodynamic calculations with spreading cells [48]. A straightforward way to compare the results for various geometries is to render the drag force and torque dimensionless by defining a drag coefficient C_D and a moment coefficient C_M. We have

$$C_D = \frac{|F_D|}{\tau_w A_p}, \tag{12.3.14a}$$

$$C_M = \frac{|T|}{\tau_w A_p h_c}, \tag{12.3.14b}$$

where A_p is the projected area of the cell obtained by conventional light microscopy, τ_w is the wall shear stress, and h_c is the cell height. Although the drag and moment

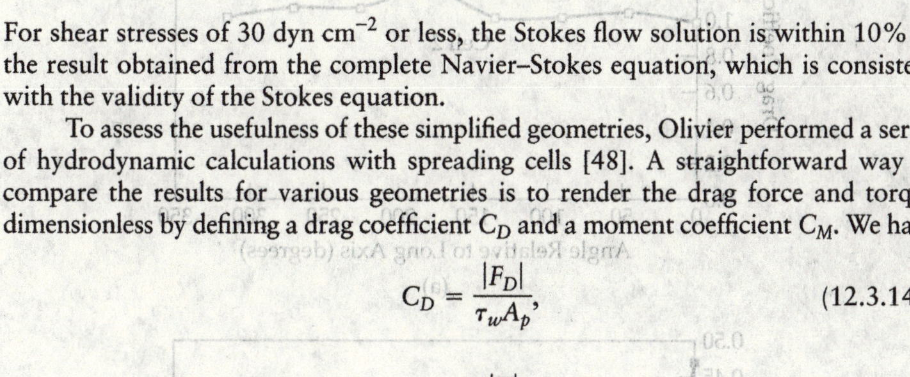

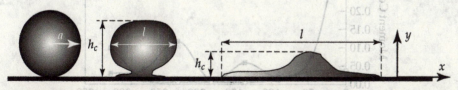

FIGURE 12.8 A side-view schematic of several representative shapes for spreading cells; a is the radius of the spherical cell, h_c is the height of the spreading cell above the surface, and l is the length of the cell.

coefficients are often based on the frontal area (i.e., the cross-sectional area presented to the flow [49]), the projected area is a useful reference that can be easily measured. If the cell were flat, then the force would be equal to the projected area times the wall shear stress; C_D would then be unity and C_M would be zero.

Results of the drag and moment coefficient calculations for two spreading cells indicate that the drag coefficient is not sensitive to the orientation of the cell in the flow field and that the moment coefficient is highly sensitive to the orientation of the cell relative to the flow field (see Figure 12.9). The drag coefficient can range from a value slightly greater than unity to values that are much larger (see Figure 12.9 and Table 12.2). The drag coefficient for the cells depends upon the cell height. The projected area A_p approximately accounts for cell shape. The drag coefficient for the cells lies below the value for the sphere and is generally less than the value for a hemisphere. The average moment coefficient for the cells is less than the value for a sphere or hemisphere.

The results shown in Figure 12.9 and Table 12.2 suggest that heterogeneity in cell detachment in flow chamber experiments is due, in part, to the variation in cell shape. Differences in cell shape lead to differences in the force acting on cells.

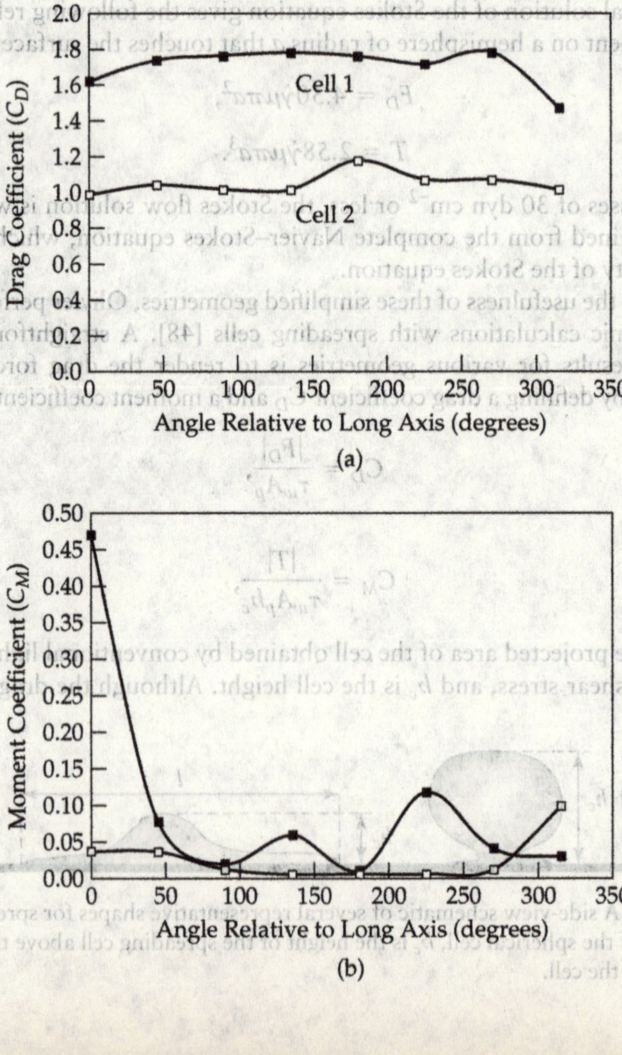

FIGURE 12.9 Drag (a) and moment (b) coefficients for two different cells as a function of the angle of the velocity vector relative to the long axis of the cell. An angle of 0 corresponds to flow approaching the cell in the same direction as the long axis of the cell [48].

TABLE 12.2

Drag and Moment Coefficients for Spreading Cells

Height (μm)	Length (μm)[a]	Width (μm)[a]	C_D	C_M
3.6	48	33	1.046	0.537
3.41	69	18	1.71	2.062
7.91	33	25	5.226	0.288
4.5	26.25	19.69	3.284	0.31
5	18.75	15.63	4.255	0.77
4	36.25	9.38	3.225	0.195
Hemisphere, finite element[b]			4.5	2.58
Sphere[c]			10.26	3.8

[a]The length corresponds to the major cell axis and the width corresponds to the minor cell axis.
Source: From [b]Ref. [48] and [c]Ref. [47].

Example 12.1 For a given cell, the drag coefficient was found to be 2.1. The projected cell area is 950 μm². Determine the drag force acting on the cell after it is exposed to a shear stress of 12 dyn cm^{-2}.

Solution From Equation (12.3.14a), $F_D = C_D \tau_w A_p$. Substituting, the drag force yields

$$F_D = 2.1(12 \text{ dyn cm}^{-2})(950 \times 10^{-8}\text{cm}^2) = 2.39 \times 10^{-4}\text{dyn}$$

$$= 2.39 \times 10^{-9}\text{N}.$$

Example 12.2 When cells detach from a surface, they assume a spherical shape, and the radius of the sphere can be measured, providing an upper limit on the cell height. When the cell is fully spread, the maximum cell height above the nucleus is 2.5 μm. Determine the drag and moment coefficients when the cells first attach and when they are fully spread.

Solution When the cells first attach, treat them as spheres. From Table 12.2, the drag and moment coefficients are $C_D = 10.26$ and $C_M = 3.8$, respectively. To determine the drag coefficient when the cells are fully spread, find the relation between the cell height and C_D. The data in the table are well fit by the cubic polynomial

$$C_D = 0.0273h_c^3 - 0.7697h_c^2 + 7.8171h_c - 12.046,$$

where h_c is measured in micrometers. Inserting the cell height, we find that $C_D = 1.112$. C_M does not exhibit a correlation with cell height. A likely upper limit can be obtained by averaging all of the data for spread cells in Table 12.2. The result is $C_M = 0.69$. These values indicate that when cells are exposed to the same shear stress, round cells are exposed to a much greater force than are spread cells. The greater force arises from the greater planar area that is exposed to flow and from torque that is applied to the cells.

The drag force and the torque can be used to determine the bond forces by applying (a) a force balance in the direction parallel to the flow (the x direction) and (b) a torque balance. In general, the bonds occur at different separation distances, because of the nonuniform topography beneath the cell. The local bond force varies spatially and equals the product of the bond density N_C, the force per bond f_{b_x}, and a differential area element. The force and torque balances are, respectively,

$$F_D = \int_{A_C} N_C f_{b_x} dA \tag{12.3.15a}$$

and

$$T + F_D h_c = \int_{A_C} N_c f_{b_y} r dA, \tag{12.3.15b}$$

where r is the distance from the center of mass of the cell to the bonds and A_C is the contact area. Note that the lift force in the y-direction has been neglected. This force is a few percent of the drag force [37]. The x-coordinate is parallel to the surface and the direction of flow; the y-coordinate is normal to the surface (Figure 12.8). To integrate Equations (12.3.15a) and (12.3.15b), we must know both the bond distribution and the separation distances between the cells and the surface. These distributions can be determined for known cell–substrate separation distances or for simple geometries (e.g., a sphere touching a surface). Otherwise, we need simplifying assumptions. When the equations are integrated, the total force on the bonds is

$$f_{b_T} = (f_{b_x}^2 + f_{b_y}^2)^{1/2}. \tag{12.3.15c}$$

Example 12.3 Develop an expression for the magnitude of the bond force when the bonds are uniformly distributed and stressed to the same extent. Assume that the cell contact area can be modeled as a circular disk with radius a. Use the resulting expression to compute the total force on the bonds ($f_{b_T} N_C A_C$) for a contact radius of 5 μm, a height of 5 μm, and a wall shear stress of 10 dyn cm^{-2}. Treat the cell as a hemispherical cap.

Solution For the conditions given, the terms f_{b_x} and f_{b_y} are constant, and Equations (12.3.15a) and (12.3.15b) can be integrated to give

$$F_D = N_C f_{b_x} A_C, \tag{12.3.16a}$$

$$T = \frac{2N_c f_{b_y} a A_C}{3}. \tag{12.3.16b}$$

Solving for f_{b_x} and f_{b_y} and inserting the result into Equation (12.3.15c) yields the desired expression:

$$f_{b_T} = \frac{1}{N_C A_C}\left[F_D^2 + \frac{9}{4}\left(\frac{T}{a}\right)^2\right]^{1/2}. \tag{12.3.16c}$$

Substituting Equations (12.3.14a) and (12.3.14b) for the force and torque into Equation (12.3.16c) and rearranging terms yields

$$f_{b_T} N_C A_C = \tau_w A_p \left[C_D^2 + \frac{9}{4}\left(\frac{h_C}{R}\right)^2 (C_M + C_D)^2 \right]^{1/2}. \quad (12.3.17)$$

Using the data provided, we find that the total force on all of the bonds is 4.66×10^{-5} dyn = 466 pN. Note that the force on the bonds is 5.9 times higher than the value obtained by treating the cell as a flat disk. Because the cell was modeled as a hemisphere, the torque contributed significantly to the total force on the bonds.

Factors influencing the dissociation of cells after exposure to a force. When an adherent cell is exposed to flow, Equation (12.3.3) is modified by including rate coefficients that are sensitive to the applied force. For now, we assume that the dissociation constant is sensitive to the applied force and can be described by Equation (12.2.4) or Equation (12.2.6). Using Equation (12.2.4), we can modify Equation (12.3.3) to read

$$\frac{dN_C}{dt} = k_1(N_{L_0} - N_C)(N_{R_0} - N_C) - k_{-1}^0 \exp\left(\frac{x_a F}{k_B T N_C}\right) N_C, \quad (12.3.18)$$

where F/N_C is the force per bond. The initial condition is the solution of Equation (12.3.4) after an attachment time t. Equation (12.3.18) cannot be integrated analytically.

Before we examine the kinetics of detachment, consider the steady-state solutions of Equation (12.3.18). As noted by Bell [7] and Cozens-Roberts et al. [50], steady-state solutions exist for specific ranges of bond density and applied force. This is most easily demonstrated by plotting each term on the right-hand side of Equation (12.3.18) for the case in which the ligand density is much greater than the receptor density ($N_{L_0} \gg N_{R_0}$ and $N_{L_0} - N_C \approx N_{L_0}$). The first term on the right-hand side decreases linearly as the number of complexes increases (see Figure 12.10). When the force is zero, the second term on the right-hand side of Equation (12.3.18)

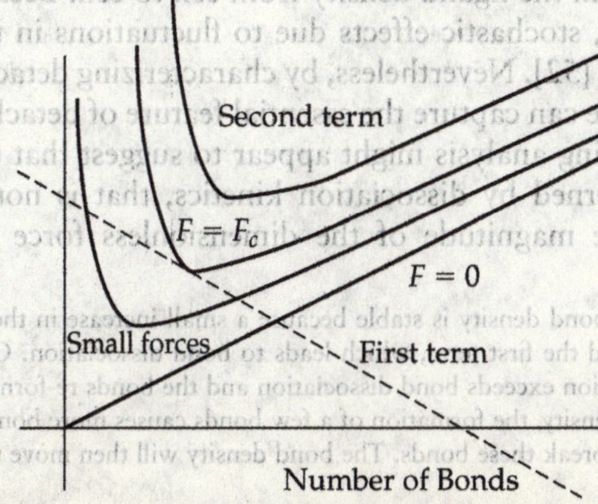

FIGURE 12.10 Comparison of the terms on the right-hand side of Equation (12.3.18). (Adapted with permission from Ref. [7], © 1978 American Association for the Advancement of Science.)

increases linearly with the bond density and intersects the first term at one value: the value that represents the steady-state bond density. When the force is nonzero, the second term on the right-hand side has various solutions, depending upon the force acting on the cell. At small forces, the curves intersect at two values. One value is a stable steady state at the higher bond density; the other is an unstable steady state at the lower bond density.[1] As the force increases, the two curves intersect at only one force, known as the *critical force*, F_c. Above F_c, the two curves do not intersect and a steady state does not exist. Thus, cells detach when the applied force exceeds the critical force.

The steady-state analysis indicates that, at the critical force, there is a rapid transition between adhesion and detachment. Analysis of the kinetics indicates that the application of a force equal to or greater than the critical force leads to a sudden loss of bonds and to cell detachment. Assuming that the ligand density is in excess, the solution of Equation (12.3.18) depends upon the following four dimensionless parameters [51]:

$$\tau = k_1 N_{L_0} t \qquad \kappa = \frac{K_D^0}{N_{L_0}} \qquad \alpha = \frac{x_a F}{k_B T N_{R_0}} \qquad \theta = \frac{N_C}{N_{R_0}} \tag{12.3.19a,b,c,d}$$

Equation (12.3.18) becomes:

$$\frac{d\theta}{d\tau} = 1 - \theta - \kappa\theta \exp\left(\frac{\alpha}{\theta}\right). \tag{12.3.20}$$

In detachment studies, the initial condition corresponds to $\theta = 1/(1 + \kappa)$, the equilibrium solution of Equation (12.3.20) prior to flow. When the force is below the critical force, a few bonds break and a new steady state is established (Figure 12.11). At or slightly above the critical force, bonds break slowly until the bond density reaches the value given by solution of the steady-state Equation (12.3.20). Then, the bond density drops dramatically to zero.

In cell-adhesion experiments, one often observes a rapid loss of cells after the onset of flow [41]. However, as previously noted, cells detach, not at a single shear stress, but over a range of shear stresses. This behavior is probably due to the differences in cell shape, which affect the force on the cell and the contact area with the surface, and to variations in the ligand density from cell to cell. Because large numbers of bonds are present, stochastic effects due to fluctuations in the numbers of bonds formed are minimal [52]. Nevertheless, by characterizing detachment in terms of a critical shear stress, we can capture the essential feature of detachment.

Although the foregoing analysis might appear to suggest that the detachment of spreading cells is governed by dissociation kinetics, that is not the case. The parameter controlling the magnitude of the dimensionless force α in Equation

[1]The steady state at the higher bond density is stable because a small increase in the number of bonds causes the second term to exceed the first term, which leads to bond dissociation. Conversely, if a few bonds are broken, bond association exceeds bond dissociation and the bonds re-form. However, for the steady state at the lower bond density, the formation of a few bonds causes more bonds to form because the force is not large enough to break these bonds. The bond density will then move to the stable steady state.

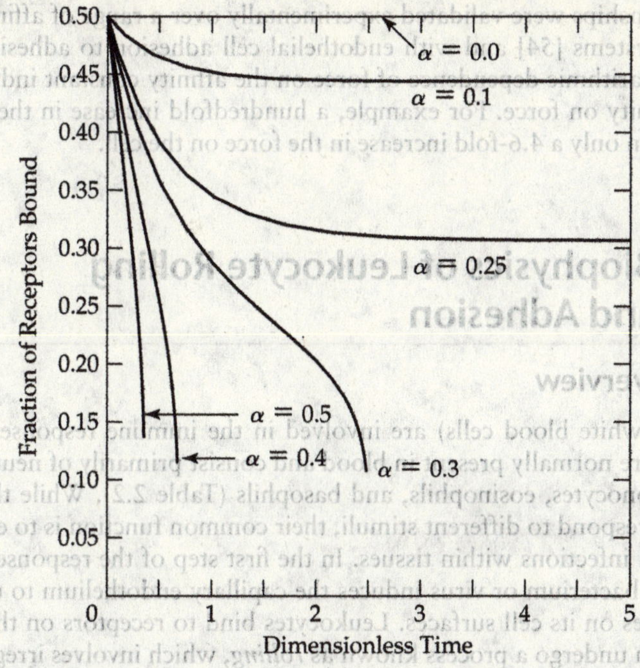

FIGURE 12.11 The dynamics of bond density when cells are exposed to a range of forces encompassing the critical force for $\kappa = 1$.

(12.3.20) is the dimensionless equilibrium constant κ. Using a more detailed model of forces acting or ligand-coated spherical beads bound to antibodies, Kuo et al. [53] found that detachment was governed by the equilibrium constant. The reason for this is that binding is rapid, and for spreading cells an equilibrium is established prior to the onset of fluid forces. Note that, as shown in Section 12.4, when cells are exposed to a force continuously, binding does not reach equilibrium and detachment is controlled by the dissociation constant.

An equilibrium analysis that was first developed by Dembo and colleagues can be used to determine the relation between force and the equilibrium constant [8]. Binding is assumed to be at equilibrium, and the generalized result when ligand and receptor depletion are both important is [37]

$$F_{Tb} = \frac{2k_B T A_c}{l_b}\left[-N_C + N_{R_0}\ln\left(\frac{N_{R_0}}{N_{R_0} - N_C}\right) + N_{L_0}\ln\left(\frac{N_{L_0}}{N_{L_0} - N_C}\right)\right], \tag{12.3.21}$$

where l_b is the *bond interaction distance*—the distance over which the noncovalent interaction between receptor and ligand occurs and A_c is the contact area. This distance is much smaller than the separation distance between the cell and substrate [8]. In their analysis, Kuo et al. [54] and Bell [7] used a value of 1×10^{-7}cm for l_b. If the ligand is in excess, Equation (12.3.21) reduces to [37]:

$$F_{Tb} = \frac{2k_B T A_c N_{R_0}}{l_b}\ln\left(1 + \frac{N_{L_0}}{K_D^0}\right). \tag{12.3.22}$$

These relationships were validated experimentally over a range of affinity constants in cell-free systems [54] and with endothelial cell adhesion to adhesion molecules [37]. The logarithmic dependence of force on the affinity constant indicates a weak effect of affinity on force. For example, a hundredfold increase in the affinity constant results in only a 4.6-fold increase in the force on the cell.

12.4 Biophysics of Leukocyte Rolling and Adhesion

12.4.1 Overview

Leukocytes (white blood cells) are involved in the immune response to infection. Leukocytes are normally present in blood and consist primarily of neutrophils, lymphocytes, monocytes, eosinophils, and basophils (Table 2.2). While these different types of cell respond to different stimuli, their common function is to eliminate bacterial or viral infections within tissues. In the first step of the response to infection, the infecting bacterium or virus induces the capillary endothelium to express adhesion molecules on its cell surfaces. Leukocytes bind to receptors on the endothelial cells and then undergo a process known as *rolling*, which involves irregular rotation and translation of the cells. The leukocytes subsequently adhere and then transmigrate between the endothelial cells to attack a site of infection (see Figure 12.12).

Both *selectins* and *integrins* are involved in adhesion. Selectins are involved in initial attachment and rolling (primary adhesion); they have a fast association constant and bind first. Selectin binding causes a redistribution of integrins, which facilitates firm integrin–ligand binding (secondary adhesion). Integrins have lower association-rate coefficients, so they bind more slowly.

In vitro studies using flow have proven useful in elucidating the paradigm of primary- and secondary-cell adhesion to activated endothelium. Such studies indicate that rolling occurs only with certain types of selectins: E- and P-selectin [55]. ICAM-1 [55] and IgE [56] bind to integrins and are involved only in firm adhesion. ICAM-1 requires the presence of selectin-mediated rolling. VCAM-1 appears unique

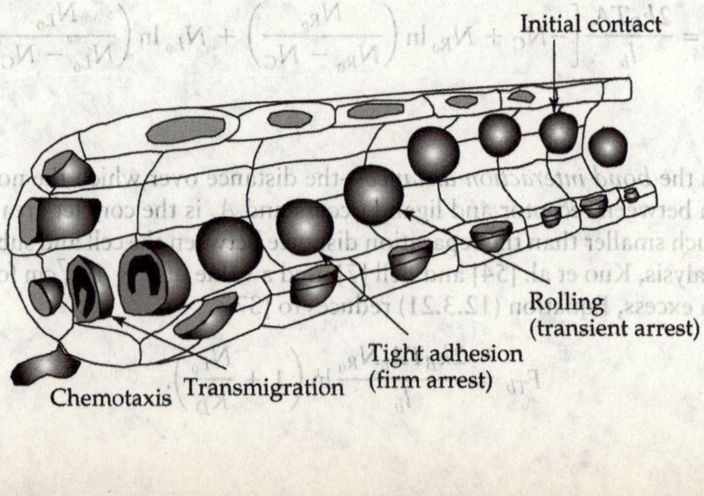

FIGURE 12.12 Schematic of the events leading to the attachment and transmigration of leukocytes to endothelium.

in that it can produce rolling and firm adhesion [57]. These different behaviors among ligands may be due to differences in the kinetics of bond formation and dissociation [8,58], to the mechanical properties of the bond [59], or to the location of ligands on the cell and the activation of the receptors [55].

At E- and P-selectin densities of 1–15 molecules cm^{-2}, and at low fluid velocities, neutrophils exhibit short periods of arrest followed by rolling at the hydrodynamic velocity [55,58]. The duration of the arrest period is a first-order reaction (see Section 11.4.3), suggesting that rolling is limited by bond dissociation and that a single bond is broken [8,58].[2] At ligand densities greater than 15 molecules cm^{-2}, cell dissociation is biphasic, suggesting that multiple bonds are broken [8] and reassociation may occur [58]. When the shear stress increases, the bond is stressed, which increases the likelihood of dissociation. According to the model proposed by Bell [7] for neutrophils binding to P-selectin (as well as to anti-P-selectin and anti-CD18 antibodies), the bond dissociation constant increases as the applied shear stress increases for P-selectin. In one study reported in the literature [8], the dissociation constant for P-selectin increased threefold when the force on the bond increased by a factor of 2. The characteristic distance over which the bond acted, x_a, was 0.049 ± 0.008 nm [8], which corresponds to hydrogen bond distances. The P-selectin bond thus exhibits a high tensile strength, which may explain how it can promote rolling. At higher ligand densities, rolling is continuous, although the velocity of rolling shows considerable variability [60]. Rolling velocities on E-selectin are less than those on P-selectin at the same ligand density. Further, rolling velocities on E-selectin have exhibited much less variability than rolling velocities on P-selectin [55]. These observations are consistent with the slower rate of E-selectin dissociation from its receptor ($k_r = 0.5\ s^{-1}$ [58]) relative to P-selectin-receptor dissociation ($k_r = 0.95\ s^{-1}$ [8]).

Extraction of receptors from the cell membrane is an alternative mechanism of rolling and detachment. The force required to extract receptors from the membrane is about 10–20 pN [61], whereas rolling cells may be acted upon by hydrodynamic forces well in excess of values in that range. Following activation, neutrophils and monocytes may shed a type of selectin called L-selectin [62], which could affect cell rolling. With inactivated cells, selectin-mediated rolling depends on an intact cytoskeleton, suggesting that selectins interact with the cytoskeleton. Such interaction reduces the likelihood that receptors are extracted. In support of this hypothesis, experiments involving the deletion of the cytoplasmic tail of α4 or treatment with cytochalasin B reduced the number of attached cells on VCAM-1 [9]. Cytochalasin B treatment also led to reduced rolling velocities on VCAM-1 [9], which may be due to higher mobility that could increase the likelihood of bond formation [63].

12.4.2 Modeling Leukocyte–Endothelial Cell Interactions

Mathematical models are able to predict rolling velocities [59] and variations in the instantaneous velocity [59]. The most sophisticated model [59] includes mechanical properties for bonds, stress-dependent association and dissociation rate constants, probabilistic binding and detachment, and a random distribution of receptors on the

[2] This case differs from the detachment of spreading cells discussed in Section 12.3 because the contact areas are small and new bonds do not form at these low ligand densities when the leukocytes detach.

cell surface. An important result from these models is that rolling is a stochastic process due to the formation of a small number of bonds, and bond dissociation is dependent upon the strain applied to the bond. Heterogeneity in the distribution of receptors among the cells may account for some of the variation in the velocity. In fact, the expression of adhesion molecules on activated endothelium is variable [64]. Although these models can help aid the analysis of experiments, they are limited by the large number of parameters, many of which have not been directly measured.

Kinetic models are based on the assumption that the adhesion and rolling of leukocytes on endothelium in the presence of flow are mediated by a small number of bonds. This assumption is supported by several experimental and theoretical studies. Leukocyte arrest durations at low ligand densities are described by a first-order process that suggests single-bond formation [8]. At higher densities, multiple-bond formation appears to occur. A small number of bonds appears to be involved in adhesion involving leukocytes rolling over purified selectins [65], cell–cell adhesion with micropipet adhesion assays [11,15], and neutrophil attachment to *tumor necrosis factor* (TNF)-α-activated endothelium under low shear stresses [58].

Bond formation is modeled by assuming that the cell is initially bound by a single bond and that multiple bonds subsequently form and dissociate while the cells are exposed to flow [58]. This model assumes that one receptor population is involved in binding to one population of ligands. The time interval is short enough such that, if i bonds are present at time t, then there will be i bonds, $i + 1$ bonds, or $i - 1$ bonds at time $t + \Delta t$. The probability $P_i(t)$ that a cell is bound by i bonds at time t is [66] governed by

$$P_i(t + \Delta t) - P_i(t) = k_f N_L[R - (i - 1)]P_{i-1}(t)\Delta t - k_f N_L[R - i]P_i(t)\Delta t$$
$$-k_r^i i P_i(t)\Delta t + k_r^{i+1}(i + 1)P_{i+1}(t)\Delta t, \quad (12.4.1)$$

where R is the receptor number and N_L is the ligand density. The first and second terms on the right-hand side of Equation (12.4.1) give the probability that $i - 1$ and i bonds, respectively, are present at time t and that one bond formed during Δt [66]. The third and fourth terms, respectively, give the probability that i and $i + 1$ bonds are present at time t and that one bond dissociated during Δt [66]. The forward rate constant k_f depends upon convective and diffusive mass transport between the receptor and the ligand. The effective association rate constant can be expressed as [32]

$$k_f = \frac{\pi D_{ij} \mathrm{Sh} \Lambda \mathrm{Da}}{1 + \Lambda \mathrm{Da}}, \quad (12.4.2)$$

where D_{ij} is the ligand diffusion coefficient, Sh is the Sherwood number (Section 7.7), Λ is the dimensionless encounter time ($\Lambda = \tau d_{ij}/a^2$, where τ is the average time to form a complex and a is the receptor radius), and Da is the Damkohler number ($\mathrm{Da} = k_{in}a^2/D_{ij}$, where k_{in} is the intrinsic rate constant) (Section 10.6.2). Both the Sherwood number and the dimensionless encounter time are functions of the Peclet number $\mathrm{Pe}(=av/D_{ij})$ [32].

Taking the limit as Δt goes to zero, in Equation (12.4.1), and assuming that few bonds are formed such that the ligand and receptor densities are approximately constant, we obtain the differential equations [58]

$$\frac{dP_0}{dt} = (k_r^1 - k_{ad})P_1, \tag{12.4.3a}$$

$$\frac{dP_n}{dt} = k_{ad}\,P_{n-1} - (k_{ad} + nk_r^n)P_n, \tag{12.4.3b}$$

and

$$\frac{dP_i}{dt} = \underbrace{k_{ad}P_{i-1}}_{i\text{th bond forms}} + \underbrace{(i+1)k_r^{i+1}P_{i+1}}_{(i+1)\text{th bond disassociates}} - \underbrace{(k_{ad} + ik_r^i)P_i}_{(i+1)\text{th bond forms and } i\text{th bond disassociates}} \quad \text{if } i > 1, \tag{12.4.4}$$

where the experimentally observed association rate constant (with units of per second) is

$$k_{ad} = k_f N_L R. \tag{12.4.5}$$

Assuming that all bonds are stressed equally, Equation (12.2.4) becomes

$$k_r^i = k_r^0 \exp\left(\frac{x_a F}{ik_B T}\right) \quad i = 1,2,\ldots,n, \tag{12.4.6}$$

where k_r^0 is the unstressed dissociation constant, x_a is the characteristic length of stretching, F is the net force on the cell, i is the number of bonds, k_B is Boltzmann's constant, and T is the absolute temperature. The total probability of forming bonds is $P = \sum_{i=0}^{\infty} P_i$.

To simplify the analysis of the detachment of spherical cells, N_L is assumed to be large relative to the number of bonds formed, and the probability of forming new bonds is assumed to be very low during the attachment time, so that Equation (12.4.4) simplifies to

$$\frac{dP_i}{dt} = (i+1)k_r^0 P_{i+1} - ik_r^0 P_i \quad i = 1,2,\ldots,n, \tag{12.4.7}$$

where n represents the number of bonds. Analytical solutions can be obtained for the case when the bonds are not exposed to a force (i.e., $k_r = k_r^0$). At time 0, $p_n = 1$ and $p_i = 0$ for i ranging from 1 to $n - 1$. Solving Equation (12.4.7) first for $i = n$ yields

$$P_n = \exp(-nk_r^0 t). \tag{12.4.8}$$

This result is then inserted into Equation (12.4.7) for $i = n - 1$. The resulting first-order equation can be solved by means of an integrating factor (Section A.1.D). The process is repeated for $i = n - 2$ to $i = 1$. Summing over all values of i yields the total probability distribution of bonds when a cell adheres with n initially formed bonds:

$$P(n,t) = 1 - P_0(t) = \sum_{i=1}^{n} P_i(n,t). \tag{12.4.9}$$

In a single bond ($n = 1$), Equation (12.4.9) reduces to the familiar equation for a first-order reaction. The formation of two or more bonds causes a significant change

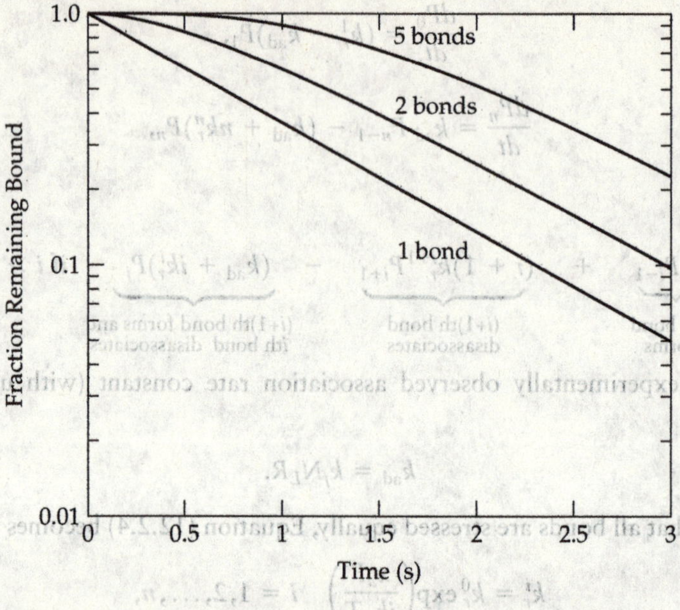

FIGURE 12.13 Effect of a number of bonds on the cell detachment curve for $k_r = 1\ \mathrm{s}^{-1}$ and no force on the bonds.

in the shape of the probability distribution (Figure 12.13). When more than one bond is present, the probability that a cell has more than one bond remains quite high initially. As more time elapses, each adherent cell has fewer bonds per cell. Eventually, all of the cells have a single bond, and detachment is a first-order process. Increasing the number of bonds per cell results in a longer time before the number of bonds per cell decreases and detachment becomes first order. Several systems for cell detachment can be explained in terms of this analysis, suggesting that single bonds are formed. However, single bonds have not been unambiguously identified, and small numbers of bonds can give the appearance of a single exponential [19]. The application of a force on the bonds causes the bonds to dissociate faster than they would in the absence of the force (Figure 12.14).

When a fluid shear stress is applied to an adherent, spherical leukocyte, all of the adhesion molecules are stretched as well as the microvilli to which the ligands are attached. When the force applied to the microvillus exceeds a critical value of 45 pN, a membrane tether is formed [67]. Tether formation and elongation significantly affect the force acting on cells. By calculating the x component of the force and performing a torque balance (Figure 12.15), one obtains [67]

$$F_b \cos\theta = 32.054 R^2 \tau_w, \tag{12.4.10}$$

$$F_b l \sin\theta = 43.916 R^3 \tau_w, \tag{12.4.11}$$

where $\tau_w = \mu\dot{\gamma}$ is the wall shear stress, l is the lever arm between the center of the cell and the microvillus, and

$$\theta = \tan^{-1}\left(\frac{R}{l}\right) + \cos^{-1}\left(\frac{l^2 + L^2}{2L\sqrt{l^2 + R^2}}\right), \tag{12.4.12}$$

where L is the length of the microvillus.

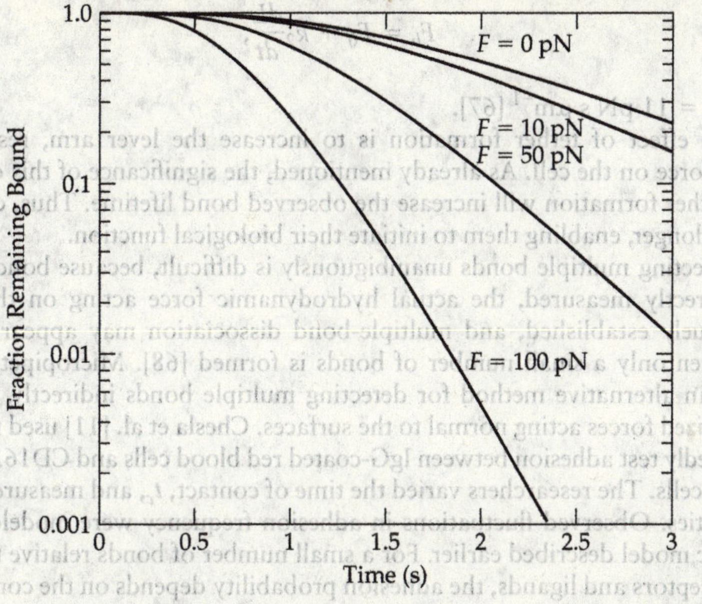

FIGURE 12.14 Solution of Equations (12.4.3a) and (12.4.3b) and Equation (12.4.4) to show the effect of force on cell detachment curve for the case of cells initially adherent by five bonds, $k_r = 1\ \text{s}^{-1}$, $x_a = 0.08$ nm, and $T = 310$ K.

After a tether forms, L and l increase, and the net force on the bonds declines. Because the mean lifetime of bonds declines at higher forces [8], tether formation results in longer bond lifetimes. Shao et al. [67] demonstrated that, for forces below 45 pN, the microvillus behaves as an elastic material, and the force on the bonds is given by

$$F_b = k_1(L - L_0), \tag{12.4.13}$$

where L_0 is 0.35 μm [67], the length of an unstressed microvillus. The spring constant for microvilli, k_1, was assumed to be 43 pN μm^{-1} [67]. When the critical force F_0 is exceeded, tethers form according to the relation

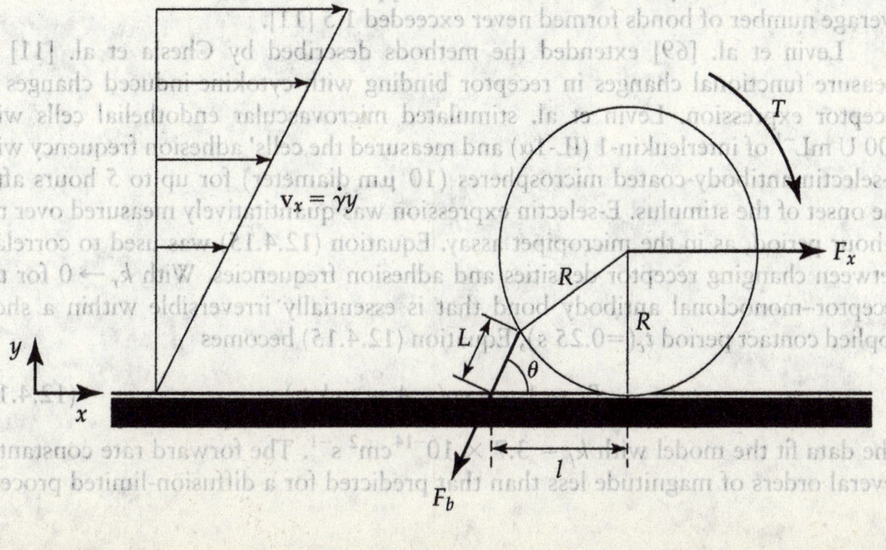

FIGURE 12.15 Schematic of the force and torque acting on an adherent cell.

$$F_b = F_0 + k_2\frac{dL}{dt}, \tag{12.4.14}$$

where $k_2 = 11$ pN s μm^{-1} [67].

The effect of tether formation is to increase the lever arm, resulting in a reduced force on the cell. As already mentioned, the significance of this observation is that tether formation will increase the observed bond lifetime. Thus, cells remain adherent longer, enabling them to initiate their biological function.

Detecting multiple bonds unambiguously is difficult, because bond formation is not directly measured, the actual hydrodynamic force acting on the bonds is not uniquely established, and multiple-bond dissociation may appear to be first order when only a small number of bonds is formed [68]. Micropipet techniques provide an alternative method for detecting multiple bonds indirectly, using well-characterized forces acting normal to the surfaces. Chesla et al. [11] used micropipets to repeatedly test adhesion between IgG-coated red blood cells and CD16A expressed on CHO cells. The researchers varied the time of contact, t_c, and measured adhesion probabilities. Observed fluctuations in adhesion frequency were modeled with the stochastic model described earlier. For a small number of bonds relative to the number of receptors and ligands, the adhesion probability depends on the contact time t_c and the densities of receptors and ligands, m_r and m_l, respectively, according to the relationship

$$P_a = 1 - \exp\left\{-A_c m_r m_l K_a^0\left[1 - \exp\left(-k_r^0 t_c\right)\right]\right\}, \tag{12.4.15}$$

where A_c is the contact area, K_a^0 is the equilibrium association constant ($=1/K_D^0$), and k_r^0 is the reverse rate constant. This solution assumes that the average number of bonds formed is in the form of a Poisson distribution, where

$$\langle n \rangle = A_c(m_r)^{v_r}(m_l)^{v_l}K_a^0\left[1 - \exp\left(-k_r^0 t_c\right)\right], \tag{12.4.16}$$

and v_r and v_l are the valence numbers for receptors and ligands, respectively. For the conditions studied by Chesla et al. and an apparent contact area of ~3 μm^2, the average number of bonds formed never exceeded 1.5 [11].

Levin et al. [69] extended the methods described by Chesla et al. [11] to measure functional changes in receptor binding with cytokine-induced changes in receptor expression. Levin et al. stimulated microvascular endothelial cells with 300 U mL^{-1} of interleukin-1 (IL-1α) and measured the cells' adhesion frequency with E-selectin antibody-coated microspheres (10 μm diameter) for up to 5 hours after the onset of the stimulus. E-selectin expression was quantitatively measured over the 5-hour period, as in the micropipet assay. Equation (12.4.15) was used to correlate between changing receptor densities and adhesion frequencies. With $k_r \rightarrow 0$ for the receptor–monoclonal antibody bond that is essentially irreversible within a short applied contact period $t_c(=0.25$ s), Equation (12.4.15) becomes

$$P_a = 1 - \exp(-A_c m_r m_l k_f t_c). \tag{12.4.17}$$

The data fit the model with $k_f = 3.7 \times 10^{-14}$cm^2 s^{-1}. The forward rate constant is several orders of magnitude less than that predicted for a diffusion-limited process, which suggests that the diffusion of receptors may be limited through steric hindrance or the immobilization of receptors within microdomains. Results suggest that cytokine-induced changes in microtopology or cytoskeletal linkages may affect adhesion.

12.4.3 Effect of Cell Deformation on Leukocyte Adhesion to Endothelium

The endothelium and the monocyte exhibit viscoelastic behavior. Fluid forces acting on the cells cause compression of the microvilli and deformation of the endothelium and monocyte. Such deformation has been observed in vivo during leukocyte adhesion to endothelium in capillaries. A side-view flow channel has recorded deformation of the leukocytes in vitro at shear stresses between 10 and 20 dyn cm^{-2} [70]. Leukocyte adhesion to planar surfaces or endothelium in flow chambers plateaus at higher shear stresses [55]. Mechanisms to explain this are the deformation of the leukocyte, leading to increased numbers of bonds, and the formation of membrane tethers that reduce the torque on the cell, leading to a reduction in the net force on the cell.

To address the effect of deformation on adhesion under flow conditions, Rinker et al. [71] separately varied the shear stress and shear rate by adjusting the flow rate and fluid viscosity. The contact time is inversely related to the shear rate. For flow between two long rectangular plates separated by a gap of thickness h and width w, the contact time can be computed as $wh^2/6Q$ from results presented in Section 2.7.2. The contact time was maintained constant by fixing the flow rate and the viscosity was adjusted so as to increase the shear stress (and hence the net force on the cell). At a fixed contact time, tethering frequencies increased, rolling velocities decreased (Figure 12.16), and median arrest durations increased with increasing shear stress. Increased adhesion frequency with force at a fixed shear rate is not consistent with adhesion dominated solely by transport effects (Equation (12.4.2)). Rolling and short arrests (<0.2 s) were well fit by a single exponential function consistent with adhesion via the formation of a single additional bond. The cell dissociation constant k_r increased when the shear stress was elevated at a constant shear rate.

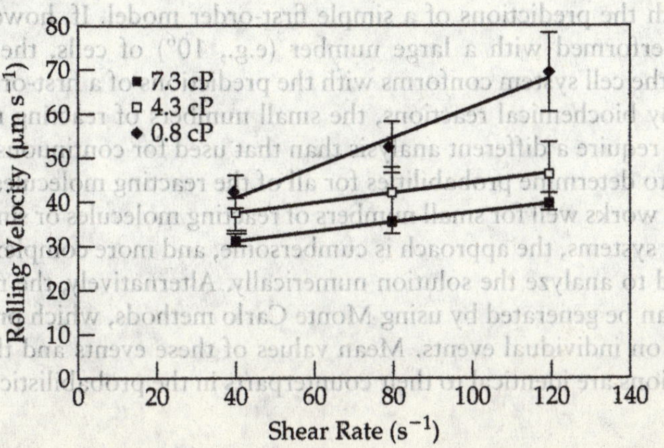

FIGURE 12.16 Effect of shear stress at constant shear rate on monocyte rolling velocities. (Adapted from Ref. [71], with permission.)

Firmly adherent cells arresting for at least 0.2 seconds were well fit by a model involving dissociation from multiple bonds (Equation (12.4.7)) and tether formation (Equations (12.4.10) to (12.4.14)). At a fixed contact time and increasing shear stress, bonds formed more frequently for rolling cells, resulting in more short arrests, and more bonds formed for firmly arresting cells, leading to longer arrest durations.

A spherical cell moving near a cell surface experiences forces normal and parallel to the surface. The drag force acts parallel to the surface. Due to the settling of the cells in the flow field, gravity cannot be neglected [72]. Because of gravity and cell rotation, a normal force acts on a microvillus as it contacts the endothelium. The normal force increases with shear stress when the shear rate is fixed.

Further support for the role of cell deformation and tether extension in modulating cell attachment is provided by a comparison of the adhesion of neutrophils or P-selectin glycoprotein ligand-1 (PSGL-1)-coated rigid beads onto rigid slides coated with P-selectin [73]. For comparable densities of cell-surface receptors, the neutrophils rolled more slowly than the beads. Cell deformation produces a larger contact area, facilitating bond formation and slowing rolling velocity.

12.5 Stochastic Effects on Chemical Interactions

Chemical reactions are abundant within cells. The number of reacting molecules may range from several hundred million per cell for oxygen to 1 per cell for a DNA molecule or a gene. When a small number of molecules react, the observed kinetics may actually differ dramatically from what is predicted for a model that assumes that the number of reactants is very large. The reason for this difference is simple: With a small number of reacting molecules, only a limited number of concentrations are possible. Take, for example, the case of a cell 10 μm in diameter that contains only four copies of a specific protein that reacts by a first-order process. Before any reaction proceeds, the concentration of this particular protein within the cell is 12.7×10^{-12} M. As each molecule reacts, the concentration drops to 9.52×10^{-12} M, 6.35×10^{-12} M, and, finally, 3.17×10^{-12} M. Measured as a function of time, the reaction exhibits discrete behavior that might not appear to conform with the predictions of a simple first-order model. If, however, measurements are performed with a large number (e.g., 10^6) of cells, then the average behavior of the cell system conforms with the predictions of a first-order model.

In many biochemical reactions, the small numbers of reacting molecules that are involved require a different analysis than that used for continuous systems. One approach is to determine probabilities for all of the reacting molecules as a function of time. This works well for small numbers of reacting molecules or simple reactions. For complex systems, the approach is cumbersome, and more complex mathematics must be used to analyze the solution numerically. Alternatively, the randomness of the system can be generated by using Monte Carlo methods, which provide detailed information on individual events. Mean values of these events and the statistics of the distributions are identical to their counterparts in the probabilistic analysis.

12.5.1 Kinetic Analysis of Stochastic Chemical Reactions

Consider the first-order irreversible reaction

$$A \xrightarrow{k_1} B.$$

Initially, there are n molecules of A and no molecules of B. The kinetic model is developed by computing the probability $P_x(t + \Delta t)$ that, among the n molecules, x have not reacted to form B by time $t + \Delta t$. The probability $P_x(t + \Delta t)$ equals the probability that there are x molecules at time t, plus the probability that $x + 1$ molecules have reacted to yield x unreacted molecules, minus the probability that x molecules have reacted to produce $x - 1$ unreacted molecules:

$$P_x(t + \Delta t) = P_x(t) + k_1(x + 1)P_{x+1}(t)\Delta t - k_1 x P_x(t)\Delta t. \quad (12.5.1)$$

Dividing by Δt and taking the limit as Δt goes to zero results in

$$\frac{dP_x}{dt} = k_1(x + 1)P_{x+1} - k_1 x P_x. \quad (12.5.2)$$

The value of x ranges from zero to n. The case of $x = 0$ represents no reacted molecules. For zero and n reacted molecules, Equation (12.5.2) reduces to

$$\frac{dP_0}{dt} = k_1 P_1 \quad x = 0, \quad (12.5.3)$$

$$\frac{dP_n}{dt} = -k_1 n P_n \quad x = n. \quad (12.5.4)$$

Initially, the probability that n molecules remain unreacted is unity (i.e., $P_n(t = 0) = 1$). Solving Equation (12.5.4) yields

$$P_n = \exp(-k_1 nt). \quad (12.5.5)$$

For $n - 1$ molecules, Equation (12.5.2) becomes

$$\frac{dP_{n-1}}{dt} = k_1 n P_n - k_1(n - 1)P_{n-1}. \quad (12.5.6)$$

Note that this equation is similar to Equation (12.4.7). The initial conditions are different, resulting in a different solution. Equation (12.5.6) is a first-order ordinary differential equation that can be solved by using an integrating factor (Section A.1.D):

$$P_{n-1} = \exp[-k_1(n - 1)t] \int_0^t k_1 n P_n \exp[k_1(n - 1)t]dt + C_{n-1}\exp[-k_1(n - 1)t]. \quad (12.5.7)$$

Substituting Equation (12.5.5) for P_n into Equation (12.5.7) and noting that, from the initial condition $P_{n-1}(t = 0) = 0$, it follows that $C_{n-1} = 0$. As a result, Equation (12.5.7) becomes

$$P_{n-1} = \exp[-k_1(n-1)t]\int_0^t k_1 n\exp(-k_1 nt)\exp[k_1(n-1)t]dt$$

$$= \exp[-k_1(n-1)t]\int_0^t k_1 n\exp(-k_1 t)dt. \qquad (12.5.8a)$$

Integrating yields

$$P_{n-1} = n\exp[-k_1(n-1)t][1-\exp(-k_1 t)]$$

$$= n\exp(-k_1 nt)[\exp(k_1 t) - 1]. \qquad (12.5.8b)$$

For $n - 2$ molecules remaining unreacted, the probability is

$$\frac{dP_{n-2}}{dt} = k_1(n-1)P_{n-1} - k_1(n-2)P_{n-2}. \qquad (12.5.9a)$$

Substituting Equation (12.5.8b) into Equation (12.5.9a) and integrating yields

$$P_{n-2} = \frac{n(n-1)}{2}\exp(-k_1 nt)[\exp(k_1 t) - 1]^2. \qquad (12.5.9b)$$

In general,

$$P_{n-j} = \frac{n!}{j!\,(n-j)!}\exp(-k_1 nt)[\exp(k_1 t) - 1]^j. \qquad (12.5.10)$$

Equation (12.5.10) is the binomial distribution. For the case of $n = 3$ bonds, the solutions embodied in that equation reduce to the following:

$$P_3 = \exp(-3k_1 t), \qquad (12.5.11a)$$

$$P_2 = 3\exp(-3k_1 t)[\exp(k_1 t) - 1], \qquad (12.5.11b)$$

$$P_1 = 3\exp(-3k_1 t)[\exp(k_1 t) - 1]^2, \qquad (12.5.11c)$$

$$P_0 = \exp(-3k_1 t)[\exp(k_1 t) - 1]^3. \qquad (12.5.11d)$$

Shown in Figure 12.17 are results for the time distribution of the probability that three, two, one, and no bonds are present. Initially, three bonds are present, but the situation rapidly changes to one in which there is a substantial probability of forming two, one, or no bonds. As time proceeds, the probability of forming no bonds increases until, at long times, it reaches unity.

Since the system behaves in a probabilistic manner, repeated experiments in which the number of molecules x at time t is measured produce a well-defined mean

$$\langle x\rangle = \sum_{x=0}^{n} xP_x. \qquad (12.5.12)$$

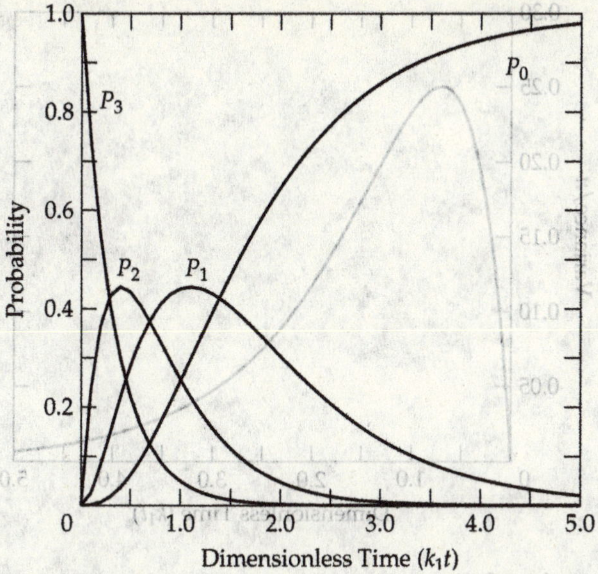

FIGURE 12.17 For a system consisting of three reacting molecules, the probability that 3, 2, 1, or 0 molecules have not reacted as a function of time.

For the model represented by Equations (12.5.3) through (12.5.6), the average behavior is simply a single exponential:

$$\langle x \rangle = n \exp(-k_1 t). \tag{12.5.13}$$

Thus, the probabilistic model yields the same mean behavior as the deterministic model for a first-order process.

The preceding results can be generalized to compute the average behavior and standard deviation for a reacting system. The mean value of the system is obtained by applying Equation (12.5.12) to the system of equations describing the change in probability with time. For a first-order system, applying Equation (12.5.12) to the system of equations represented by Equation (12.5.6) yields

$$\frac{d\langle x \rangle}{dt} = -k_1 \langle x \rangle. \tag{12.5.14}$$

At $t = 0$, there are n reacting molecules in the system.

The mean value of the probabilities equals the continuous distribution for the deterministic system. The variance of the distribution (the square of the standard deviation σ) is found from the following definition:

$$\sigma^2 = \sum_{x=0}^{n} x^2 P_x - \langle x \rangle^2 = \langle x^2 \rangle - \langle x \rangle^2. \tag{12.5.15}$$

Applying this definition to the system of equations describing a first-order reaction yields [74]

$$\frac{d\sigma^2}{dt} = -k_1 \langle x \rangle + 2k_1 \langle x \rangle^2. \tag{12.5.16}$$

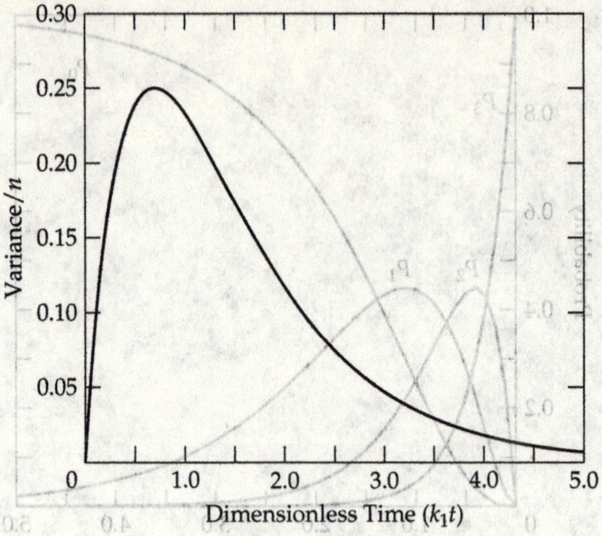

FIGURE 12.18 Variance in the probability of reaction for first-order reaction.

This equation has the solution

$$\sigma^2 = n\exp(-k_1t)\left[1 - \exp(-k_1t)\right]. \tag{12.5.17}$$

The variance is initially zero and increases as time progresses (see Figure 12.18). The function has a maximum at $k_1t = \ln(2)$. For large values of t, the variance decays to zero. This behavior is consistent with the progression of the reaction. At the beginning of the reaction, no molecules have reacted. As the reaction proceeds, some molecules have reacted and others have not. Figure 12.18 shows that, at this intermediate time, there is a wide distribution of states for the molecules. As the reaction progresses toward completion, the amount of unreacted molecules is small, resulting in a smaller variance.

For the reversible binding reaction represented by Equation (12.4.4) with excess ligand ($N_{L_0} \gg N_{R_0}$), the mean and variance for the probabilistic solution are [51], respectively,

$$\langle N_C \rangle = \frac{N_{R_0}N_{L_0}}{N_{L_0} + K_D}\left[1 - \exp(-kt)\right] \tag{12.5.18}$$

and

$$\frac{\sigma^2}{N_{R_0}} = \frac{K_D/N_{L_0}\left[1 - \exp(-kt)\right] + \exp(-kt) - \exp(-2kt)}{(1 + K_D/N_{L0})^2}, \tag{12.5.19}$$

where $k = k_1N_{L_0}(K_D/N_{L_0} + 1)$. The solution for the variance depends upon two dimensionless groups: $k_1N_{L_0}t$ and K_D/N_{L_0} (Figure 12.19). Unlike its behavior for irreversible reactions, the variance does not go to zero at long times. Rather, the solution approaches the steady-state value:

$$\sigma^2(kt \to \infty) = \frac{N_{R_0}K_D/N_{L_0}}{(1 + K_D/N_{L_0})^2}. \tag{12.5.20}$$

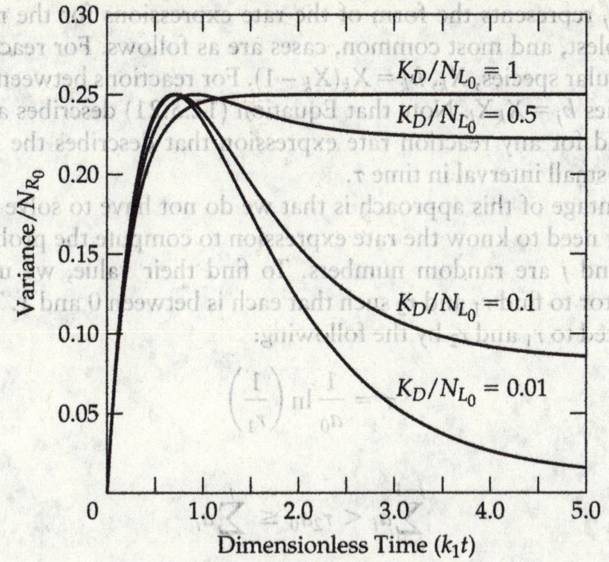

FIGURE 12.19 Variance in the probability for a reversible bimolecular reaction in the presence of excess ligand.

At low values of the ratio K_D/N_{L_0}, the reaction behaves as a first-order reaction, and the variance is similar to the value obtained for the first-order irreversible reaction. For larger values of the ratio K_D/N_{L_0}, the solution decays slowly with increasing time, eventually reaching the steady-state value. The reason for the larger variance for larger values of the ratio K_D/N_{L_0} is that the steady-state concentration increases with increasing values of the ratio. For the reversible reaction with large values of K_D/N_{L_0}, there are many possible states for the reactants, leading to greater variation in the observed response.

12.5.2 Monte Carlo Analysis of Stochastic Chemical Reactions

While stochastic analysis provides statistics on the mean and variance of a biochemical reaction, the simulations do not provide any details about the variation of individual reaction events or averages of small numbers of such events. Further, the kinetic analysis becomes unwieldy for complex reaction schemes or large numbers of different interactions. The Monte Carlo method was developed to compute the probability of individual reaction and transport events.

The Monte Carlo method is applied to a chemical reaction process as follows [74,75,76]. For the general case where there are N different reacting molecules and M kinetic expressions for the reactions, the probability, $P(\tau,j)$, that a reaction will occur in the time step t and involves reaction j is

$$P(\tau,j) = a_j\exp(-a_0\tau), \tag{12.5.21}$$

where $a_j = h_jk_j$ and

$$a_0 = \sum_{i=1}^{M} a_i. \tag{12.5.22}$$

The quantity h_j represents the form of the rate expressions for the reacting molecules. The simplest, and most common, cases are as follows. For reactions between the same molecular species, X_k, $h_k = X_k(X_k - 1)$. For reactions between two different molecular species $h_j = X_k X_l$. Note that Equation (12.5.21) describes a random process and is valid for any reaction rate expression that describes the probability of reaction over a small interval in time τ.

The advantage of this approach is that we do not have to solve a rate expression; we merely need to know the rate expression to compute the probability $P(\tau, j)$. In general, τ and j are random numbers. To find their value, we use a random-number generator to find r_1 and r_2 such that each is between 0 and 1. The quantities τ and j are related to r_1 and r_2 by the following:

$$\tau = \frac{1}{a_0} \ln\left(\frac{1}{r_1}\right) \tag{12.5.23}$$

$$\sum_{i=1}^{j-1} a_i < r_2 a_0 \leq \sum_{i=1}^{j} a_i \tag{12.5.24}$$

The randomly generated value of τ represents the time step over which a reaction occurs. The total elapsed time is the sum of all the values of τ for each generation of r_1. Equation (12.5.24) identifies the reaction step in which the reaction occurs. For the reaction $j - 1$, one molecule of each reactant would produce 1 molecule of product.

The following examples will show how this approach is applied. For a first-order reaction, there is only one reactant. Thus, $a_0 = a_1 = k_1 X_1$. Since $M = 1$, there is no need to compute r_2. For this case, the algorithm is simple. Start with an initial number of molecules N_1. Compute τ according to Equation (12.5.23). Then, reduce the number of molecules to $N_1 - 1$. Repeat this process until all of the molecules have reacted. A MATLAB program for such calculations based on the Gillespie algorithm [76] is

```
function y=unimolecular_gil
ii=1;
t(ii)=0;
X=20;
Z(1)=20;
y(1)=X;
k=1;
for j=1:1:19;
tau=(1/X)*log(1/rand(1,1));
ii=ii+1;
t(ii)=t(ii-1)+tau;
y(ii)=X-1;
X=X-1;
Z(ii)=20*exp(-t(ii));
end
plot(t,y,'o',t,Z)
```

Note that the rate coefficient was not stated in the expression for t since its value was set to 1. Alternatively, time can be treated as a dimensionless time equal to the product tk.

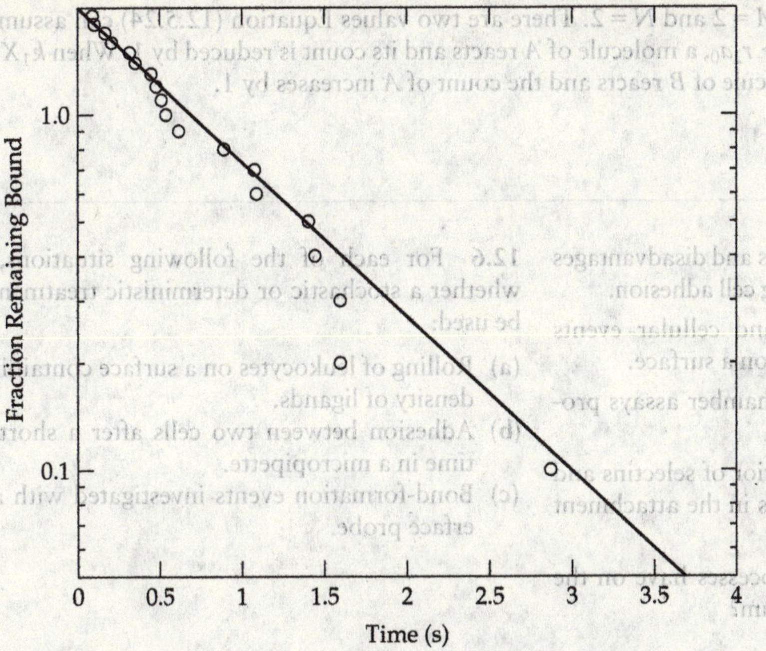

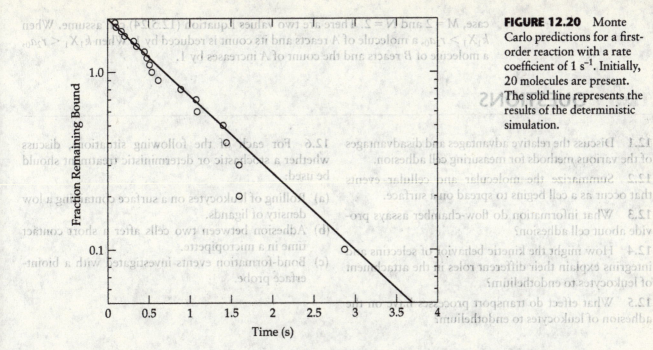

FIGURE 12.20 Monte Carlo predictions for a first-order reaction with a rate coefficient of 1 s^{-1}. Initially, 20 molecules are present. The solid line represents the results of the deterministic simulation.

Simulation results for one run with 20 molecules initially is shown in Figure 12.20. The solid line represents the deterministic results. While the Monte Carlo simulations follow the trend set by the deterministic relation, there are significant deviations. Since different random numbers are generated each time, repeated applications of the program will yield different results. Further, as the number of reacting molecules increases, the variance drops and the curves more closely following the deterministic results.

Deviation between the actual stochastic simulation and the deterministic model decreases with the number of molecules present. The variation can be as large as 25% for 20 particles. For 100 molecules of A initially, the variation can be as large as 12%; most of the variation is within 7%. Clearly, as the number of reacting molecules increases, the stochastic simulation approaches the deterministic model. By defining a root-mean-square error less than a specific tolerance, a criterion for convergence can be defined. Moreover, fluctuations still arise once steady state is reached. This is consistent with the concept of "dynamic equilibrium." That is, even at equilibrium, reaction still occurs, except that the forward and reverse reactions are, on average, balanced.

This approach can be extended to other reaction orders. For a bimolecular reaction, $h_1 = kX_1(X_1 - 1)$ and the quantity τ becomes `tau=(1/X(X-1))*log(1/rand(1,1));`. All other steps are the same. For a reversible first-order reaction, we have

$$A \underset{k_2}{\overset{k_1}{\rightleftharpoons}} B. \qquad (12.5.25)$$

Initially, only A is present with X_0 molecules. For this case, $h_1 = k_1X_1$ and $h_2 = k_2(X_0 - X_1)$. The quantity $X_0 - X_1$ is obtained from a mass balance. For this case, $M = 2$ and $N = 2$. There are two values Equation (12.5.24) can assume. When $k_1X_1 > r_2a_0$, a molecule of A reacts and its count is reduced by 1. When $k_1X_1 < r_2a_0$, a molecule of B reacts and the count of A increases by 1.

12.6 QUESTIONS

12.1 Discuss the relative advantages and disadvantages of the various methods for measuring cell adhesion.

12.2 Summarize the molecular and cellular events that occur as a cell begins to spread on a surface.

12.3 What information do flow-chamber assays provide about cell adhesion?

12.4 How might the kinetic behavior of selectins and integrins explain their different roles in the attachment of leukocytes to endothelium?

12.5 What effect do transport processes have on the adhesion of leukocytes to endothelium?

12.6 For each of the following situations, discuss whether a stochastic or deterministic treatment should be used:

(a) Rolling of leukocytes on a surface containing a low density of ligands.

(b) Adhesion between two cells after a short contact time in a micropipette.

(c) Bond-formation events investigated with a biointerface probe.

12.7 PROBLEMS

12.1 Estimate the force acting on a spreading cell of height 5 μm and length 20 μm exposed to a shear stress of 15 dyn cm^{-2}. Compare your estimate with results obtained by treating the cell as a flat disk of radius 10 μm.

12.2 Leukocyte adhesion in a parallel-plate flow channel is being studied in vitro under flow conditions. A leukocyte of radius 6 μm adheres to an endothelial cell via a single microvillus of length 0.35 μm. The location of the microvillus is not known. Calculate the net force acting on the cell when the cell is exposed to a wall shear stress of 2 dyn cm^{-2}. Assume that the microvillus is located in a region ranging from directly beneath the cell to a position that is 18.2° upstream relative to the position directly beneath the cell.

12.3 Examine the effect of ligand and receptor density on the steady-state behavior of Equation (12.3.18) for $N_{L_0} = N_{R_0}$.

12.4 Using different shear stress values, scientists detached adherent leukocytes from a surface containing a low density of antibody molecules. For each shear stress, the force was determined, and detachment was found to be a first-order process. Using the following data for the half-time of dissociation of the cell population, fit the values for k_r to the Bell model (Equation (12.4.6)) for $i = 1$ and determine k_r^0 and x_a:

Force (pN)	10	50	100	200
$t_{1/2}$ (s)	0.340	0.317	0.288	0.238

12.5 Show the effect of tether formation and elongation on the cell detachment curves for cells that adhere via a single bond attached to a microvillus of length 0.35 μm and exposed to a shear stress of 1 dyn cm^{-2}. The cell radius is 6 μm.

12.6 Using the approach outlined in Section 12.5.1, one can derive the following relation for the irreversible bimolecular decomposition of A_2 to $2A$ [77]:

$$\frac{dP_x}{dt} = 0.5k(x+2)(x+1)P_{x+2} - 0.5kx(x-1)P_x$$

In this equation, x represents the number of molecules of A_2 and k is the second-order rate coefficient. The indices of x are $x = 0, 2, 4, \ldots, x_0$, where x_0 is the initial number of molecules of x present in the system.

(a) Using MATLAB, present simulations of P_x as a function of dimensionless time, kx_0t for $x_0 = 6$.

(b) Compute $\langle x \rangle$ and compare it with the results obtained for a deterministic model.

12.7. Apply the Monte Carlo simulation approach to determine the time course of reactants in a sequential reaction of first-order reactions.

$$A \xrightarrow{k_1} B \xrightarrow{k_2} C$$

Perform simulations for 20, 50, and 100 molecules of A present initially. Assess the effect of initial number of molecules on the variation of simulation results.

12.8 REFERENCES

1. Hubbell, J.A., "Biomaterials in tissue engineering." *Biotechnology*, 1995. 13: pp. 565–576.
2. Miranti, C.K., and Brugge, J.S., "Sensing the environment: a historical perspective on integrin signal transduction." *Nature Cell Biol.*, 2002. 4: pp. 83–90.
3. Lodish, H., Berk, A., Zipursky, A.L., Matsudaira, P., Baltimore, D., and Darnell, J., *Molecular Cell Biology*. 2000, New York: WH Freeman and Company.
4. Corada, M., Mariotti, M., Thurston, G., Smith, K., Kunkel, R., Brockhaus, M., Lampugnani, M.G., Martin-Padura, I., Stoppacciaro, A., Ruco, L., Mcdonald, D., Ward, P.A., and Dejana, E., "Vascular endothelial-cadherin is an important determinant of microvascular integrity in vivo." *Proc. Natl. Acad. Sci.*, 1999. 96: pp. 9815–9820.
5. Tinico, I., Sauer, K., Wang, J.C., and Puglisi, J.D., *Physical Chemistry: Principles and Applications in Biological Sciences*. 4th ed. 2002, Upper Saddle River, NJ: Prentice Hall.
6. Evans, E., "Energy landscapes of biomolecular adhesion and receptor anchoring at interfaces explored with dynamic force spectroscopy." *Faraday Discuss.*, 1999. 111: pp. 1–16.
7. Bell, G., "Models for the specific adhesion of cells to cells." *Science*, 1978. 200: pp. 618–627.
8. Dembo, M., Torney, D.C., Saxman, K., and Hammer, D.A., "The reaction-limited kinetics of membrane-to-surface adhesion and detachment." *Proc. Royal Soc.*, 1988. 234: pp. 55–83.
9. Alon, R., Hammer, D.A., and Springer, T.A., "Lifetime of the P-selectin–carbohydrate bond and its response to tensile force in hydrodynamic flow." *Nature*, 1995. 374: pp. 539–542.
10. Alon, R., Kassner, P.D., Carr, M.W., Finger, E.B., Hemler, M.E., and Springer, T.A., "The integrin VLA-4 supports tethering and rolling in flow on VCAM-1." *J. Cell Biol.*, 1995. 128: pp. 1243–1253.
11. Chesla, S., Selvaraj, P., and Zhu, C., "Measuring two-dimensional receptor–ligand binding kinetics by micropipette." *Biophys. J.*, 1998. 75: pp. 1553–1572.
12. Marshall, B.T., Mian Long, M., Piper, J.W., Yago, T., McEver, R.P., and Zhu, C., "Direct observation of catch bonds involving cell-adhesion molecules." *Nature*, 2003. 423: pp. 190–193.
13. Evans, E., and Ritchie, K., "Dynamic strength of molecular adhesion bonds." *Biophys. J.*, 1997. 72: pp. 1541–1555.
14. Florin, E.L., Moy, V.T., and Gaub, H.E., "Adhesion forces between individual ligand–receptor pairs." *Science*, 1994. 264: pp. 415–417.
15. Shao, J.-Y., and Hochmuth, R., "Mechanical anchoring strength of L-selectin, β_2 integrins and CD45 to neutrophil cytoskeleton and membrane." *Biophys. J.*, 1999. 77: pp. 587–596.
16. Leckband, D., "Measuring the forces that control protein interactions." *Ann. Rev. Biophys. Biomol. Struct.*, 2000. 29: pp. 1–26.
17. Evans, E., Ritchie, K., and Merkel, R., "Sensitive force technique to probe molecular adhesion and structural linkages at biological interfaces." *Biophys. J.*, 1995. 68: pp. 2580–2587.
18. Pierres, A., Touchard, D., Benoliel, A.-M., and Bongrand, P., "Dissecting streptavidin–biotin interaction with a laminar flow chamber." *Biophys. J.*, 2002. 81: pp. 25–42.
19. Zhu, C., Long, M., Chesla, S.E., and Bongrand, P., "Measuring receptor/ligand interaction at the single-bond level: experimental and interpretative issues." *Ann. Biomed. Eng.*, 2002. 30: pp. 305–314.
20. Merkel, R., Nassoy, P., Leung, A., Ritchie, K., and Evans, E., "Energy landscapes of receptor–ligand bonds explored with dynamic force spectroscopy." *Nature*, 1999. 397: pp. 50–53.
21. Burridge, K., and Chrzanowska-Wodnicka, M., "Focal adhesions, contractility and signaling." *Ann. Rev. Cell Dev. Biol.*, 1996. 12: pp. 463–519.
22. Chen, W.T., and Singer, S.J., "Immunoelectron microscopic studies of the sites of cell–substratum and cell–cell contacts in cultured fibroblasts." *J. Cell Biol.*, 1982. 95: pp. 205–222.

23. Lanni, F., Waggoner, A.S., and Taylor, D.L., "Structural organization of interphase 3T3 fibroblasts studied by total internal reflection fluorescence microscopy." *J. Cell Biol.*, 1985. **100**: pp. 1091–1102.
24. Burmeister, J., Truskey, G., Yarbrough, J., and Reichert, W., "Imaging of cell/substrate contacts on polymers by total internal reflection fluorescence microscopy." *Biotechnol. Prog.*, 1994. **10**: pp. 26–31.
25. Burridge, K., Turner, C.E., and Romer, L.H., "Tyrosine phosphorylation of paxillin and pp125FAK* accompanies cell adhesion to extracellular matrix: a role in cytoskeleton assembly." *J. Cell Biol.*, 1992. **119**: pp. 893–903.
26. Chrzanowska-Wodnicka, M., and Burridge, K., "Tyrosine phosphorylation is involved in the reorganization of the actin cytoskeleton in response to serum or LPA stimulation." *J. Cell Sci.*, 1994. **107**: pp. 3643–3654.
27. Chen, Q., Kinch, M.S., Lin, T.H., Burridge, K., and Juliano, R.L., "Integrin-mediated cell adhesion activates mitogen-activated protein kinases." *J. Biol. Chem.*, 1994. **269**: pp. 26602–26605.
28. Lee, J., Leonard, M., Oliver, T., Ishihara, A., and Jacobson, K., "Traction forces generated by locomoting keratocytes." *J. Cell Biol.*, 1994. **127**: pp. 1957–1964.
29. Rinnerthaler, G., Geiger, B., and Small, J., "Contact formation during fibroblast locomotion: involvement of membrane ruffles and microtubules." *J. Cell Biol.*, 1988. **106**: pp. 747–760.
30. Zand, M., and Albrecht-Buehler, G., "What structures, besides adhesions, prevent spread cells from rounding up?" *Cell Motil. Cytoskel.*, 1989. **13**: pp. 195–211.
31. Tawil, N., Wilson, P., and Carbonetto, S., "Integrins in point contacts mediate cell spreading: factors that regulate integrin accumulation in point contacts vs. focal contacts." *J. Cell Biol*, 1993. **120**: pp. 261–271.
32. Chang, K.-C., and Hammer, D., "The forward rate of binding of surface-tethered reactants: effect of relative motion between two surfaces." *Biophys. J.*, 1999. **76**: pp. 1280–1292.
33. Akiyama, S.K., and Yamada, K.M., "The interaction of plasma fibronectin with fibroblastic cells in suspension." *J. Biol. Chem.*, 1985. **60**: pp. 4492–4500.
34. Conforti, G., Zanetti, A., Colella, S., Abbadini, M., Marchisio, P.C., Pytela, R., Giancotti, F., Tarone, G., Languino, L.R., and Dejana, E., "Interaction of fibronectin with cultured human endothelial cells: characterization of the specific receptor." *Blood*, 1989. **6**: pp. 1576–1585.
35. Preissner, K.T., Anders, E., Grulich-Henn, J., and Muller-Berghaus, G., "Attachment of human endothelial cells is promoted by specific association with S protein (vitronectin) as well as with the ternary S protein–thrombin–antithrombin III complex." *Blood*, 1988. **71**: pp. 1581–1589.
36. Massia, S.P. and Hubbell, J.A., "An RGD spacing of 440 nm is sufficient for integrin $\alpha_v\beta_3$-mediated fibroblast spreading and 140 nm for focal contact and stress fiber formation." *J. Cell Biol.*, 1991. **114**: pp. 1089–1100.
37. Xiao, Y., and Truskey, G.A., "The effect of receptor–ligand affinity on the strength of adhesion of endothelial cells to immobilized RGD peptides and adsorbed fibronectin." *Biophys. J.*, 1996. **70**: pp. 2869–2884.
38. Duband, J.L., Nuckolls, G.H., Ishihara, A., Hasegawa, T., Yamada, K.M., Thiery, J.P., and Jacobson, K., "Fibronectin receptor exhibits high lateral mobility in embryonic locomoting cells but is immobile in focal contacts and fibrillar streaks in stationary cells." *J. Cell Biol.*, 1988. **107**: pp. 1385–1396.
39. Segel, L.A., Volk, T., and Geiger, B., "On spatial periodicity in the formation of cell adhesions to a substrate." *Cell Biophys.*, 1983. **5**: pp. 95–104.
40. Ward, M.D., and Hammer, D.A., "Focal contact assembly through cytoskeletal polymerization: steady state analysis." *J. Math. Biol.*, 1994. **32**: pp. 677–704.
41. Truskey, G.A., and Pirone, J.S., "The effect of fluid shear stress upon cell adhesion to fibronectin-treated surfaces." *J. Biomed. Mat. Res.*, 1990. **24**: pp. 1333–1353.
42. Aitchison, J., and Brown, J., *The Lognormal Distribution*. 1957, Cambridge, UK: Cambridge University Press.
43. Bates, D.M., and Watts, D.G., *Nonlinear Regression Analysis and Its Applications*. 1988, New York: Wiley.
44. Olivier, L.A., and Truskey, G.A., "A numerical analysis of forces exerted by laminar flow on spreading cells in a parallel plate flow chamber assay." *Biotechnol. Bioeng.*, 1993. **42**: pp. 963–973.
45. Barbee, K.A., Mundel, T., Ratneshwar, L., and Davies, P.F., "Subcellular distribution of shear stress at the surface of flow-aligned and nonaligned endothelial monolayers." *Am. J. Physiol.*, 1995. **268**: pp. H1765–H1772.
46. Guilak, F., "Volume and surface area measurement of viable chondrocytes in situ using geometric modeling of serial confocal sections." *J. Microsc.*, 1994. **173**: pp. 245–256.
47. Goldman, A.J., Cox, R.G., and Brenner, H., "Slow viscous motion of a sphere parallel to a plane wall. I. Motion through a quiescent fluid." *Chem. Eng. Sci.*, 1967. **22**: pp. 637–651.
48. Olivier, L.A., "The response of endothelial cells to laminar flow in a parallel plate flow chamber." Ph.D. dissertation. 1996, Duke University.
49. White, F., *Fluid Mechanics*. 1986, New York: McGraw-Hill.
50. Cozens-Roberts, C., Quinn, J.A., and Lauffenburger, D.A., "Receptor-mediated adhesion phenomena: model studies

with the radial-flow detachment assay." *Biophys. J.*, 1990. **58**: pp. 107–125.

51. Cozens-Roberts, C., Lauffenburger, D.A., and Quinn, J.A., "Receptor-mediated cell attachment and detachment kinetics. I. Probabilistic model and analysis." *Biophys. J.*, 1990. **58**: pp. 841–856.
52. Saterbak, A., Kuo, S.C., and Lauffenburger, D.A., "Heterogeneity and probabilistic binding contributions to receptor-mediated cell detachment kinetics." *Biophys. J.*, 1993. **65**: pp. 243–252.
53. Kuo, S.C., Hammer, D.A., and Lauffenburger, D.A., "Simulation of detachment of specifically bound particles from surfaces by shear flow." *Biophys. J.*, 1997. **73**: pp. 517–531.
54. Kuo, S.C., and Lauffenburger, D.A., "Relationship between receptor/ligand binding affinity and adhesion strength." *Biophys. J.*, 1993. **65**: pp. 2191–2200.
55. Lawrence, M.B., and Springer, T.A., "Leukocytes roll on a selectin at physiologic flow rates: distinction from and prerequisite for adhesion through integrins." *Cell*, 1991. **65**: pp. 859–873.
56. Tempelman, L.A., and Hammer, D.A., "Receptor-mediated binding of IgE-sensitized rat basophilic leukemia cells to antigen-coated substrates under hydrodynamic flow." *Biophys. J.*, 1994. **66**: pp. 1231–1243.
57. Jones, D.A., McIntire, L.V., Smith, C.W., and Picker, L.J., "A two-step adhesion cascade for T cell/endothelial cell interactions under flow conditions." *J. Clin. Invest.*, 1994. **94**: pp. 2443–2450.
58. Kaplanski, G., Farnarier, C., Tissot, O., Pierres, A., Benoliel, A.M., Alessi, M.C., Kaplanski, S., and Bongrand, P., "Granulocyte–endothelium initial adhesion. Analysis of transient binding events mediated by E-selectin in a laminar shear flow." *Biophys. J.*, 1993. **64**: pp. 1922–1933.
59. Hammer, D.A., and Apte, S.M., "Simulation of cell rolling and adhesion on surfaces in shear flow: general results and analysis of selectin-mediated neutrophil adhesion." *Biophys. J.*, 1992. **63**: pp. 35–37.
60. Goetz, D.J., El-Sabban, M.E., Pauli, B.U., and Hammer, D.A., "Dynamics of neutrophil rolling over stimulated endothelium in vitro." *Biophys. J.*, 1994. **66**: pp. 2202–2209.
61. Evans, E., Berk, D., and Leung, A., "Detachment of agglutinin-bonded red blood cells. I. Forces to rupture molecular-point attachments." *Biophys. J.*, 1991. **59**: pp. 838–848.
62. Springer, T.A., "Adhesion receptors of the immune system." *Nature*, 1990. **346**: pp. 425–434.
63. Chan, P.Y., Lawrence, M.B., Dustin, M.L., Ferguson, L.M., Golan, D.E., and Springer, T.A., "Influence of receptor lateral mobility on adhesion strengthening between membranes containing LFA-3 and CD2." *J. Cell Biol.*, 1991. **115**: pp. 245–255.
64. Munn, L.L., Koenig, G.C., Jain, R.K., and Melder, R.J., "Kinetics of adhesion molecule expression and spatial organization using targeted sampling fluorometry." *BioTechniques*, 1995. **19**: pp. 622–631.
65. Chen, S., and Springer, T., "An automatic braking system that stabilizes leukocyte rolling by an increase in selectin bond number with shear." *J. Cell Biol.*, 1999. **144**: pp. 185–200.
66. Lauffenburger, D.A., and Linderman, J.J., *Receptors: Models for Binding, Trafficking, and Signaling*. 1993, New York: Oxford University Press.
67. Shao, J.-Y., Ting-Beall, H.P., and Hochmuth, R.M., "Static and dynamic lengths of neutrophil microvilli." *Proc. Natl. Acad. Sci.*, 1998. **95**: pp. 6797–6802.
68. Evans, E., "Probing the relation between force—lifetime—and chemistry in single molecular bonds." *Ann. Rev. Biophys. Biomol. Struct.*, 2001. **30**: pp. 105–128.
69. Levin, J.D., Ting-Beall, H.P., and Hochmuth, R.M., "Correlating the kinetics of cytokine-induced E-selectin adhesion and expression on endothelial cells." *Biophys. J.*, 2001. **80**: pp. 656–667.
70. Cao, J., Donell, B., Deaver, D.R., Lawrence, M.B., and Dong, C., "In vitro side-view imaging technique and analysis of human T-leukemic cell adhesion to ICAM-1 in shear flow." *Microvasc. Res.*, 1998. **55**: pp. 124–137.
71. Rinker, K.D., Prabhakar, V., and Truskey, G.A., "Effect of contact time and force on monocyte adhesion to vascular endothelium." *Biophys. J.*, 2001. **80**: pp. 1722–1732.
72. Zhao, Y., Chien, S., and Weinbaum, S., "Dynamic contact forces on leukocyte microvilli and their penetration of the endothelial glycocalyx." *Biophys. J.*, 2001. **80**: pp. 1124–1140.
73. Park, E.Y.H., Smith, M.J., Stropp, E.S., Snapp, K.R., DiVietro, J.A., Walker, W.F., Schmidtke, D.W., Diamond, S.L., and Lawrence, M.B., "Comparison of PSGL-1 microbead and neutrophil rolling: microvillus elongation stabilizes P-selectin bond clusters." *Biophys. J.*, 2002. **82**: pp. 1835–1847.
74. Bharucha-Reid, A.T., *Elements of the Theory of Markov Processes and Their Applications*. 1960, New York: McGraw-Hill.
75. Laurenzi, I.J., and Diamond, S.L., "Monte Carlo simulation for the heterotypic aggregation kinetics of platelets and neutrophils." *Biophys. J.*, 1999. **77**: pp. 1733–1746.
76. Gillespie, D.T. 1977. Exact stochastic simulation of coupled chemical reactions. *J. Phys. Chem.* **81**: 2340–2361.
77. McQuarrie, D.A., "Stochastic approach to chemical kinetics." *J. Appl. Prob.*, 1967. **4**: pp. 413–478.

Transport in Organs

PART IV

In this section, we apply the results of the previous three sections to specific mass transport problems in various tissues and organs. Among the examples we consider are oxygen and nitric oxide transport in Chapter 13; solute transport in the kidney in Chapter 14; transport in tumors in Chapter 15; and transport in organ systems and in the whole organism in Chapter 16. The relevant biology is introduced so that the importance of the transport problems can be appreciated. Although a few new concepts, such as facilitated and active transport, are presented, the focus is on developing models to provide a quantitative understanding of biological processes. As a result, the analysis often requires some level of simplification. This is accomplished by appropriate scaling of the equations describing solute transport and is necessarily specific to the solute and the organ under study. Although such scaling analysis does not always produce equations with analytical solutions, the scaling does isolate the most important factors affecting the transport process. Consequently, this approach often provides important insights into the biological and physical processes that affect the transport of molecules under physiological and pathological conditions.

Transport of Gases Between Blood and Tissues

CHAPTER 13

13.1 Introduction

In this chapter, we consider the transport of two important gases: molecular oxygen (O_2) and nitric oxide (NO). Oxygen is required in metabolic reactions that produce energy for cells. Many of these reactions occur within mitochondria. The efficient delivery of oxygen to the tissues ensures that the energetic needs of cells are satisfied. Oxygen transport within the body is a multistep process. First, during respiration, oxygen is transported by convection to the lungs. Next, oxygen passes across the lung capillaries by diffusion and binds to hemoglobin in red blood cells. Then, the oxygen in these cells moves throughout the blood by convection. Once oxygen reaches the tissues, it diffuses through them and within the cells to the mitochondria, where it is used in aerobic reactions that produce energy in the form of adenosine triphosphate (ATP). During heavy exercise, the demand for oxygen by the muscle rises tenfold from resting levels. Oxygen transport rises to meet this demand, through an increase in the breathing rate and the heart rate and via changes to the binding and dissociation of oxygen to hemoglobin in lungs and tissues.

Understanding the mechanisms for oxygen transport and chemical reaction within the body is a prerequisite for many clinical diagnoses and therapeutic interventions, including treating diseases in which the oxygen transport pathways are altered, designing efficient blood–gas exchangers and developing substitutes for blood. Understanding the dynamics of oxygen transport requires addressing several important questions: What are the kinetics and equilibria of oxygen binding to hemoglobin? What is the relationship between the structure and function of the lungs for efficient oxygen exchange from incoming air to the blood? What are the rate-limiting steps in oxygen transport in blood and in its binding to hemoglobin? How do oxygen diffusion in tissues and its consumption by mitochondria relate to the distance between capillaries in various tissues?

NO is a potent vasodilator that is released from endothelium in response to an increase in flow. NO diffuses through the blood vessel wall and then binds to the enzyme *guanylate cyclase* in smooth muscle cells to stimulate the production of cyclic guanine monophosphate (cGMP), which leads to smooth muscle cell relaxation.[1] NO release is stimulated by shear stress, acetylcholine, histamine, ATP, adenosine diphosphate, and hypoxia (low blood concentrations of oxygen).

NO is highly reactive. In addition to binding to guanylate cyclase, NO reacts with oxygen, hemoglobin, myoglobin, and free thiol groups on proteins. These reactions lead to the breakdown of NO or, in the case of binding to thiols, the formation of a reserve of bound NO that is released when the local NO concentration decreases. An important issue related to NO function is how the various reactions balance to yield sufficient NO to cause blood vessels to relax.

The material discussed in this chapter builds upon that presented in Chapters 6, 7, and 10, integrating the biological processes and transport analyses. In Section 13.2, we examine the equilibrium binding of oxygen to hemoglobin. This examination is followed by a discussion of the kinetics of oxygen–hemoglobin binding in Section 13.3. In Section 13.4, we examine the various steps involved in the oxygenation of red blood cells. Then, in Section 13.5, we present models for the delivery of oxygen to tissues. Finally, in Section 13.6, we consider the kinetics of NO formation and reaction and model NO transport in the microcirculation.

13.2 Oxygen–Hemoglobin Equilibria

The primary means by which oxygen is transported through blood is by binding to hemoglobin. The reason that oxygen must bind to hemoglobin is that oxygen is sparingly soluble in aqueous solutions. The solubility of oxygen is defined as

$$C_{O_2} = H \cdot P_{O_2}, \tag{13.2.1}$$

where C_{O_2} is the concentration of dissolved oxygen in solution, H is the Bunsen solubility coefficient, and P_{O_2} is the partial pressure of oxygen in the gas phase. Physiologists refer to this partial pressure as *oxygen tension*. The solubility of oxygen in blood plasma at standard temperature and pressure (0°C, 1.013×10^5 Pa, and 0% humidity) is 1.005×10^{-11} mol cm^{-3} Pa^{-1}. Thus, the partial pressure of oxygen present in the lung is approximately 13,328.9 Pa (or 100 mmHg),[2] and the concentration of oxygen in blood plasma is 1.34×10^{-4} M.[3] The solubility coefficient for oxygen in red blood cells at standard temperature and pressure is denoted H_{Hb} and is equal to 1.125×10^{-11} mol cm^{-3} Pa^{-1}, which is slightly higher than the value for plasma.

[1] The 1998 Nobel Prize in Medicine was awarded to Robert Furchgott, Louis Ignarro, and Ferid Murad for their discovery of nitric oxide as the molecule released by endothelium that causes blood vessels to dilate.

[2] This partial pressure of oxygen in the lungs is less than the partial pressure of oxygen in air (0.21 atm = 21,273 Pa), because the lungs are saturated with water vapor at 37°C.

[3] Physiologists often present pressure in units of torr (1 torr = 1 mmHg = 133.3 Pa), for which the solubility is 1.34×10^{-9} mol cm^{-3} torr^{-1} (1.005×10^{-11} mol cm^{-3} Pa^{-1}). In describing the oxygen capacity of blood and tissues, physiologists often compute the volume occupied by oxygen gas under ideal conditions and at standard temperature and pressure, in which case 1 mole of a gas occupies 22,400 cm^3 and the solubility can be written as 2.25×10^{-7} cm^3 O_2 cm^{-3} Pa^{-1}. Thus, the concentration of oxygen in plasma at 100 mmHg (13,328.9 Pa) is 0.0030 cm^3 O_2 per cm^3 plasma.

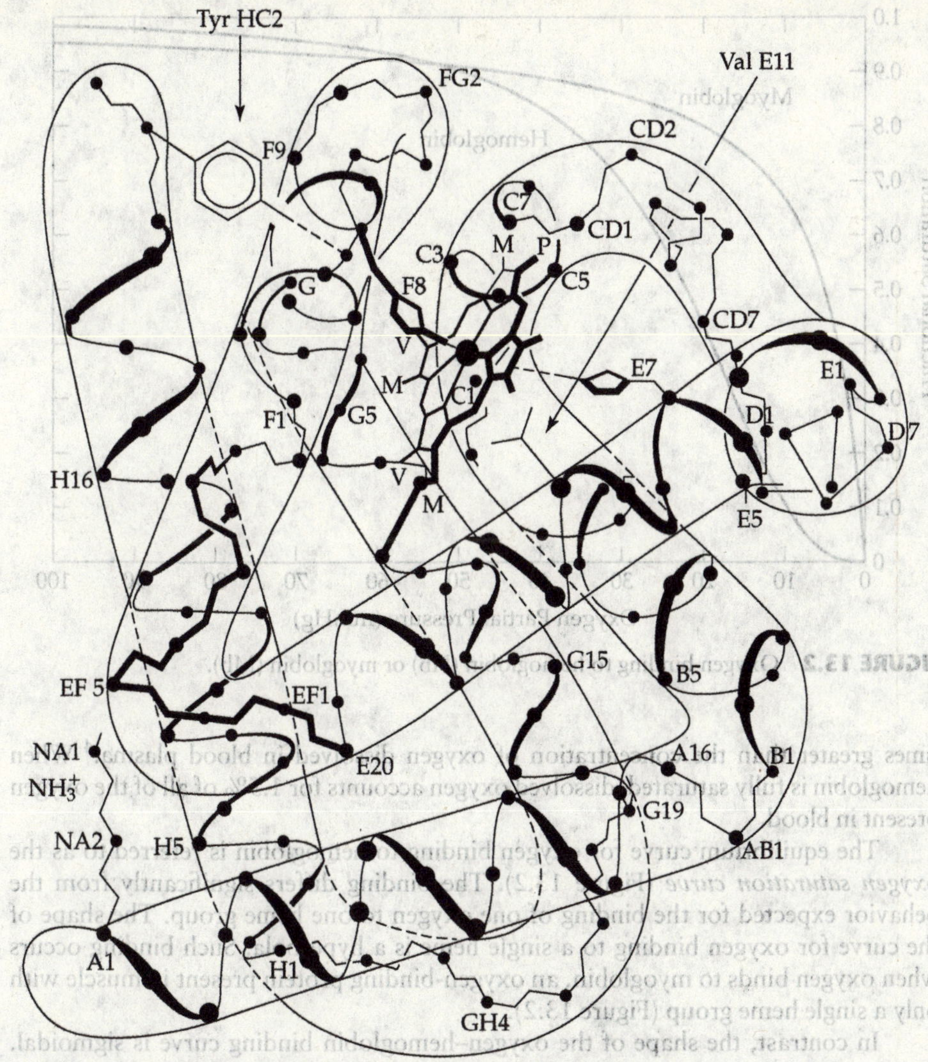

FIGURE 13.1 Secondary and tertiary structure of hemoglobin subunits. (Adapted from Ref. [1], with permission from *Annual Reviews*, © 1998.)

Hemoglobin is an oxygen-binding molecule that consists of four units (tetramers), each of which can bind one molecule of oxygen. The tetramers are all not of the same structure: the two α chains contain 141 amino acids and the two β chains have 146 amino acids. The molecular weight of hemoglobin is 64,500. Oxygen binds to the heme group that is present in each of the tetramers (Figure 13.1). The iron present in the heme group is also bound to histidines (shown by dashed lines in Figure 13.1). The heme group is planar, with the iron displaced above the heme, which lies between, and interacts with, several regions of the subunit.

The concentration of hemoglobin in blood at a hematocrit of 0.45 is 2.3×10^{-3} M. When oxygen is bound to all four heme groups, the concentration of oxygen bound to hemoglobin per total blood volume is 9.2×10^{-3} M, almost 70

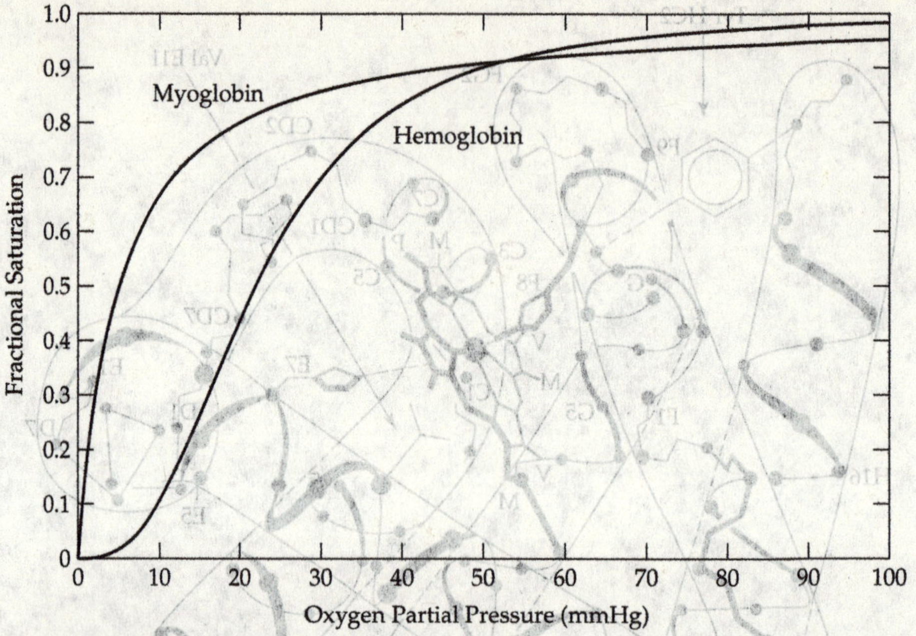

FIGURE 13.2 Oxygen binding to hemoglobin (Hb) or myoglobin (Mb).

times greater than the concentration of oxygen dissolved in blood plasma.[4] When hemoglobin is fully saturated, dissolved oxygen accounts for 1.5% of all of the oxygen present in blood.

The equilibrium curve for oxygen binding to hemoglobin is referred to as the *oxygen saturation curve* (Figure 13.2). The binding differs significantly from the behavior expected for the binding of one oxygen to one heme group. The shape of the curve for oxygen binding to a single heme is a hyperbola. Such binding occurs when oxygen binds to myoglobin, an oxygen-binding protein present in muscle with only a single heme group (Figure 13.2).

In contrast, the shape of the oxygen–hemoglobin binding curve is sigmoidal. At low concentrations, oxygen does not bind very efficiently, as evidenced by the low initial slope of the curve in Figure 13.2. As the partial pressure of oxygen increases, however, the slope increases, indicating that it becomes easier for oxygen to bind to hemoglobin. Then, as most of the heme sites become occupied with oxygen, the slope decreases.

Oxygen binding to hemoglobin is cooperative (see Section 11.7), in that the binding of one oxygen molecule to one heme group facilitates the binding of subsequent oxygen molecules to the other heme groups of the molecule. Cooperativity arises because oxygen binding to the ferrous ion moves the iron into the plane of the heme group. This motion pulls the histidine toward the heme group, resulting in a rearrangement of the subunit. As a consequence of this local movement, the α_2 and β_2 hemes move relative to the other two groups. This movement of the subunits makes

[4]This concentration of oxygen bound to hemoglobin corresponds to 0.201 cm^3 O_2 per cm^3 blood.

subsequent binding of oxygen to the heme easier. Crystallography shows that the deoxygenated and oxygenated forms of hemoglobin are in two different conformations, known as T (for tense) and R (for relaxed), respectively. However, the exact manner in which the local changes around the heme group lead to movement of the subunits is not completely understood [1].

Although models have been developed that explain the oxygen dissociation curves in terms of a low-affinity T state and a high-affinity R state [1], the most commonly used relation to describe the sigmoidal shape of the oxygen saturation curves is

$$S = \frac{(P/P_{50})^n}{1 + (P/P_{50})^n}, \tag{13.2.2}$$

where S is the fraction of hemoglobin saturated with oxygen, P is the partial pressure of oxygen, and P_{50} is the partial pressure at which the oxygen saturation is 50%. Equation (13.2.2), often referred to as the *Hill equation,* is derived by assuming that more than one oxygen binds to the hemoglobin molecule. However, $n = 4$ does not provide a good fit for oxygen binding to hemoglobin. For hemoglobin under normal physiological pH and temperature, best fits are found for $P_{50} = 26$ mmHg (3465.5 Pa) and $n = 2.7$. Equation (13.2.2) is valid for oxygen saturation between 20% and 80%. For myoglobin, $n = 1$, which is consistent with equimolar binding of oxygen and myoglobin, and $P_{50} = 5.3$ mmHg.

Example 13.1 For normal hemoglobin in red blood cells at a hematocrit of 0.45, determine the total concentration of oxygen in blood that is in equilibrium with oxygen at a partial pressure of 0.053 atm, corresponding to the partial pressure of oxygen in venous blood.

Solution Since 1 atm = 101,325 Pa, the oxygen partial pressure is 5,370 Pa. From Equation (13.2.1), the concentration of oxygen in plasma is 5.40×10^{-5} M. The total amount of oxygen in blood equals the sum of the oxygen in plasma plus the amount bound to hemoglobin. In terms of concentrations, the total concentration of oxygen in blood is

$$C_{B_T} = H \cdot P(1 - \text{Hct}) + \text{Hct}\left[H_{\text{Hb}}P + 4C_{\text{Hb}}\frac{(P/P_{50})^n}{1 + (P/P_{50})^n}\right]$$

$$= C_{\text{pl}} + \text{Hct} \cdot 4C_{\text{Hb}}\frac{(C_{\text{pl}}/C_{50})^n}{1 + (C_{\text{pl}}/C_{50})^n}, \tag{13.2.3}$$

where C_{Hb} is the concentration of hemoglobin in red blood cells, Hct is the hematocrit, and C_{pl} is the concentration of dissolved oxygen in plasma, equal to $H \cdot P - (H - H_{\text{Hb}})P \cdot \text{Hct}$. For $n = 2.7$, the fractional saturation S of hemoglobin is 0.765, corresponding to a concentration of oxygen bound to hemoglobin in red blood cells equal to 15.64×10^{-3} M. The total concentration of oxygen in blood is $(5.40 \times 10^{-5}\ \text{M} + 2.90 \times 10^{-6}\ \text{M}) + (15.64 \times 10^{-3}\ \text{M})0.45 = 7.09 \times 10^{-3}$ M.

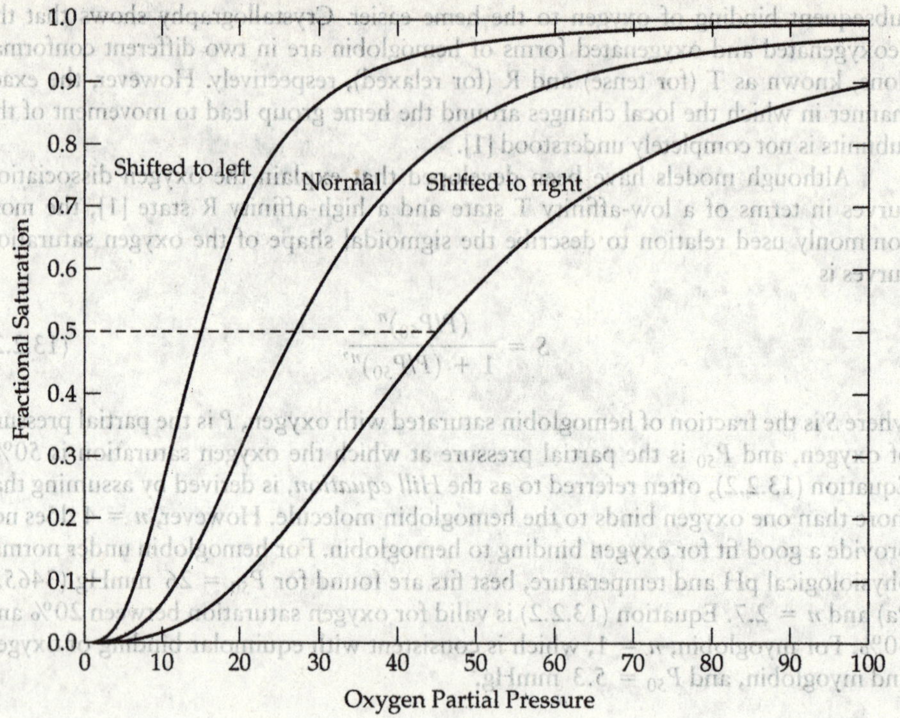

FIGURE 13.3 Changes in the oxygen–hemoglobin dissociation curve due to environmental factors.

The oxygen dissociation curve is sensitive to a number of environmental factors, such as pH, CO_2, 2,3-diphosphoglycerate (DPG), and temperature. An increase in P_{50} is referred to as a *shift to the right* (Figure 13.3) and indicates that a higher concentration of oxygen is needed to bind to the heme groups to achieve the same fractional saturation. A higher P_{50} makes it easier for oxygen to dissociate from hemoglobin in the capillaries of tissues. Such a shift to the right can be produced by an increase in CO_2 concentration, a decrease in pH, an increase in organic phosphates such as DPG, or an increase in temperature. Conversely, a lower P_{50} shifts the dissociation curve to the left. Such a reduced P_{50} is produced by lower CO_2 concentrations and an increase in pH.

Environmental effects on the oxygen–hemoglobin dissociation curve can be explained in relation to interactions with hemoglobin (Figure 1.16), which can bind hydrogen ions, CO_2, DPG, and NO at sites other than the heme group. The deoxygenated form of hemoglobin has a higher affinity for these ligands than do the oxygenated forms. When the ligands bind, they reduce the affinity of the heme group for oxygen. The effects of pH and CO_2 are interrelated, since CO_2 forms bicarbonate ion when dissolved in an aqueous solution:

$$CO_2 + H_2O \rightleftharpoons HCO_3^- + H^+. \tag{13.2.4}$$

This reaction is catalyzed by the enzyme carbonic anhydrase. Thus, an increase in dissolved CO_2 produces a reduction in pH.

The formation of HCO_3^- influences the dynamics of oxygen binding to hemoglobin. Although oxygen–hemoglobin binding may not be in equilibrium in tissues and the lungs, the following does provide a qualitative description of the effect of these factors on oxygen–hemoglobin binding: As blood moves through tissue capillaries, the CO_2 concentration increases from 40 to 46 mmHg, which reduces the pH from 7.4 to 7.2. As a result, the oxygen affinity for hemoglobin decreases along the capillary, resulting in a greater release of oxygen than would be released if the pH and CO_2 were constant. During exercise, the affinity of oxygen for hemoglobin is further reduced in muscle tissue, due to a rise in temperature and the release of DPG produced in metabolic reactions to supply the muscle with energy. In the lungs, the opposite effect occurs: the affinity of hemoglobin for oxygen increases as carbon dioxide diffuses from the blood into the lung alveoli and the CO_2 concentrations decline along the lung capillary.

13.3 Oxygen–Hemoglobin Binding Kinetics

Due to the conformational change that occurs in hemoglobin after oxygen binds to it, the kinetics describing the binding are complex. The initial binding of oxygen and hemoglobin can be modeled as a bimolecular reaction:

$$O_2 + Hb \underset{k'_{-1}}{\overset{k'_1}{\rightleftharpoons}} HbO_2. \tag{13.3.1}$$

The rate expression for the formation of the oxygenated hemoglobin is

$$\frac{dC_{HbO_2}}{dt} = k'_1 C_{O_2} C_{Hb} - k'_{-1}\, C_{HbO_2}. \tag{13.3.2}$$

The rate coefficients for oxygen binding to human hemoglobin at 37°C and pH 7.1 are $k'_1 = 3.5 \times 10^6\ M^{-1}\ s^{-1}$ and $k'_{-1} = 44\ s^{-1}$ (summarized in Refs [2] and [3]). For oxygen concentrations in the lung, the half-time for oxygenation is 2 ms (see Section 10.2). Although the model is simple, it does not yield an equilibrium relation of the form of Equation (13.2.2).

A more detailed model was developed by considering the sequential oxygenation of the four heme groups (reviewed in Ref. [2]):

$$O_2 + Hb_4(O_2)_{i-1} \underset{k'_{-i}}{\overset{k'_i}{\rightleftharpoons}} Hb_4(O_2)_i \qquad i = 1, 2, 3, 4, \tag{13.3.3}$$

$$\frac{dC_{Hb_4(O_2)_i}}{dt} = k_i C_{O_2} C_{Hb_4(O_2)_{i-1}} - k_{-i} C_{Hb_4(O_2)_i}. \tag{13.3.4}$$

Note that the case of $i - 1 = 0$ refers to deoxygenated hemoglobin.

The corresponding rate coefficients for human hemoglobin at 21.5°C and pH 7.0 are $k_1 = 17.7 \times 10^4\ M^{-1}\ s^{-1}$, $k_{-1} = 1900\ s^{-1}$, $k_2 = 33.2 \times 10^4\ M^{-1}\ s^{-1}$, $k_{-2} = 158\ s^{-1}$, $k_3 = 4.89 \times 10^4\ M^{-1}\ s^{-1}$, $k_{-3} = 539\ s^{-1}$, $k_4 = 33.0 \times 10^4 M^{-1}\ s^{-1}$, and $k_{-4} = 50\ s^{-1}$. A third approach, often found easier to use in transport models, is to use the model represented by Equation (13.3.1), but to represent k_1' as a function of saturation [3]; that is,

$$k_1' = \frac{k_{-1}'}{C_{50}}\left(\frac{S}{1-S}\right)^{1-1/n}, \tag{13.3.5}$$

where $C_{50} = H \cdot P_{50}$ and $n = 2.7$. Using Equation (13.3.5) for k_1' ensures that Equation (13.3.2) reduces to Equation (13.2.2) at steady state. The rate coefficients provided are valid for normal conditions. If the hemoglobin is modified by pH or DPG, then different values of the rate coefficients are needed. Meldon [4] provides a summary of measured values of k_1' and k_{-1}' for oxygen–hemoglobin binding with and without DPG.

13.4 Dynamics of Oxygenation of Blood in Lung Capillaries

The lungs are the site of oxygen exchange between air and blood. The alveolar oxygen partial pressure is approximately 100 mmHg (13.33 kPa). Blood enters the pulmonary circulation at 40 mmHg (or 5,333 Pa), which corresponds to 74% saturation of hemoglobin, and exits the lung in equilibrium with alveolar oxygen, which corresponds to 87% saturation. As noted in Chapter 1, the lungs are well suited for their function. An extremely large exchange area (126 m^2), a small capillary volume (213 cm^3), and short diffusion distances across the alveoli and lung capillaries enable efficient exchange. Under resting conditions, blood becomes fully oxygenated during transport along one-third of the length of the capillaries. The resulting excess capillary length provides a reservoir that allows complete oxygenation during exercise (Figure 13.4).

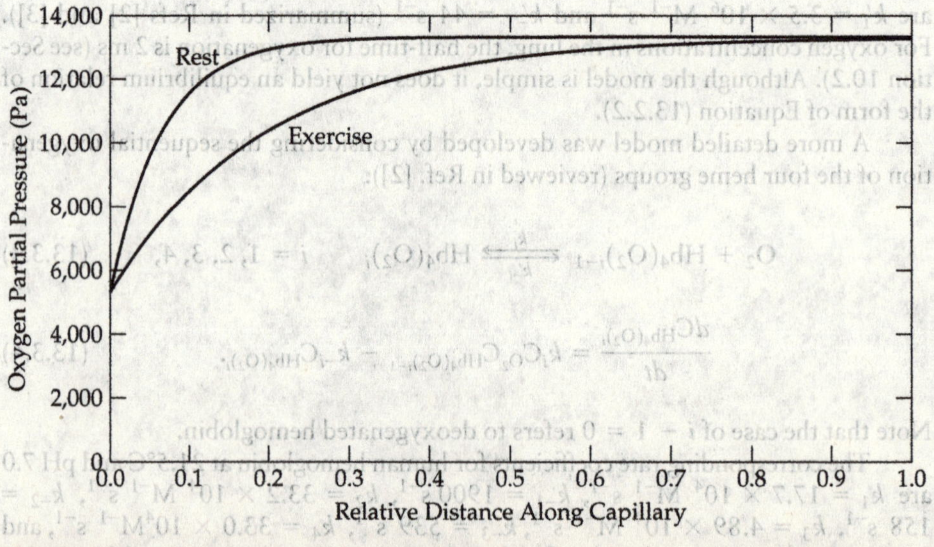

FIGURE 13.4 Oxygenation of blood along lung capillaries under rest and exercise.

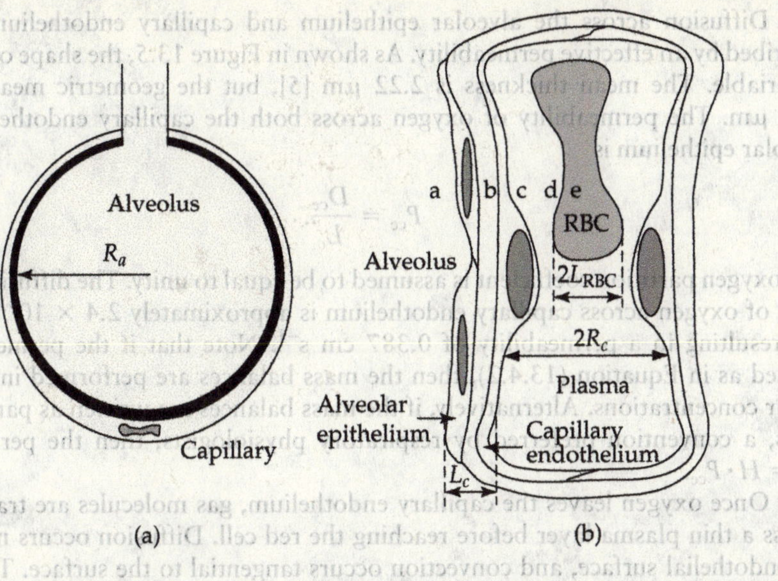

FIGURE 13.5 (a) Schematic of the alveolus and lung capillary. (b) Expanded view of capillary showing principal distances for oxygen transport.

Under resting conditions, the transit time of red blood cells through the lung capillaries is about 1 s, and oxygenation has occurred by 0.33 s. The following events take place as blood is oxygenated:

(a) Diffusion in the alveolar gas phase.
(b) Diffusion across the alveolar epithelium and capillary endothelium.
(c) Diffusion and convection in the blood plasma.
(d) Diffusion across the red-cell membrane.
(e) Diffusion in the red-cell cytoplasm and binding to hemoglobin.

The analysis that follows involves determining which events are significantly slower than others. Once this is done, the time to oxygenate blood is estimated and the validity of the simplifications assessed.

The anatomy of an alveolus and the surrounding capillary is shown schematically in Figure 13.5. The capillaries surround the alveolar surface, forming a continuous network that allows efficient gas exchange. Diffusion in the alveolar gas phase (Step a) involves determining whether concentration gradients exist in the alveolar gas. Since diffusion is the principal means of oxygen transport in the alveolus, and since the alveolus can be treated as a sphere, the characteristic time for diffusion is approximately

$$t_a = \frac{R_a^2}{D_a}. \tag{13.4.1}$$

The alveolar radius is typically 120 μm [5], and the diffusion coefficient of oxygen in the gas phase is 0.2 cm^2 s^{-1}. The characteristic diffusion time in the alveolus is 7.2×10^{-4} s. Because this time is much less than the time required for red cells to become oxygenated, the concentration of alveolar gas is uniform and the gas concentration in contact with the alveolar membrane does not vary with location.

Diffusion across the alveolar epithelium and capillary endothelium can be described by an effective permeability. As shown in Figure 13.5, the shape of the cells is variable. The mean thickness is 2.22 μm [5], but the geometric mean is only 0.62 μm. The permeability of oxygen across both the capillary endothelium and alveolar epithelium is

$$P_{ce} = \frac{D_{ce}}{L_c}. \tag{13.4.2}$$

The oxygen partition coefficient is assumed to be equal to unity. The diffusion coefficient of oxygen across capillary endothelium is approximately 2.4×10^{-5} cm^2 s^{-1} [6], resulting in a permeability of 0.387 cm s^{-1}. Note that if the permeability is defined as in Equation (13.4.2), then the mass balances are performed in terms of molar concentrations. Alternatively, if the mass balances are written as partial pressures, a convention preferred by respiratory physiologists, then the permeability $P^*_{ce} = H \cdot P_{ce}$.

Once oxygen leaves the capillary endothelium, gas molecules are transported across a thin plasma layer before reaching the red cell. Diffusion occurs normal to the endothelial surface, and convection occurs tangential to the surface. The thickness of the plasma layer ranges between 0.35 and 1.5 μm [5], and the diffusion coefficient in the plasma is 2.4×10^{-5} cm^2 s^{-1}. Convective transport is incorporated in terms of a mass-transfer coefficient (see Section 7.7). For transport in blood, the mass-transfer coefficient has been conventionally defined in terms of the difference in concentration between alveolar gas at the endothelial cell surface and the mean oxygen concentration in red blood cells [3,7]. The mass-transfer coefficient is presented as the dimensionless Sherwood number $\text{Sh} = k_{pl}L/D_{O_2,\text{plasma}}$. The Sherwood number for capillaries at a hematocrit of 0.25 is 1.3 [7]. The Sherwood number depends on hematocrit (see Section 2.8), in accordance with the following relation [3]:

$$\text{Sh}(\text{Hct}) = \text{Sh}_{\text{Hct}=0.25} + 0.84(\text{Hct} - 0.25). \tag{13.4.3}$$

At a normal hematocrit of 0.45, Sh = 1.468, which indicates that convection and diffusion contribute equally to the fluid-phase mass transfer of oxygen. For a fluid-phase thickness of 0.80 μm, the mass-transfer coefficient for the fluid phase is 0.44 cm s^{-1}.

Treating the transport barriers of the cells and the fluid phase as resistances in series yields the overall permeability for oxygen transport from the alveoli to the red-cell surface:

$$\frac{1}{P_{ce+pl}} = \frac{1}{P_{ce}} + \frac{1}{k_{pl}}. \tag{13.4.4}$$

On the basis of the preceding results, $P_{ce+pl} = 0.21$ cm s^{-1}.

Once oxygen reaches the red-cell surface, it diffuses across the red-cell membrane, which is 10 nm thick. The solubility of oxygen in the membrane is about twice the solubility in the blood plasma, and the diffusion coefficient is similar to that found in the blood. Consequently, the permeability of the membrane, P_m, is 48 cm s^{-1}. Thus, the red-cell membrane permeability is much larger than the combined permeability across the endothelium and plasma and the red-cell membrane contributes negligibly to the overall mass-transfer resistance.

Once within the red cell, oxygen diffuses through the hemoglobin solution in the cytoplasm and reacts with hemoglobin. As noted previously, the reaction is quite complex. The Thiele modulus provides an estimate of the relative importance of intracellular reaction and diffusion. For the initial binding of oxygen to hemoglobin, the average value of the association constant for the model represented by Equation (13.3.3) yields $k = 128\ s^{-1}$. A second estimate of the rate constant is obtained from the ratio of the rate of oxygen binding to red blood cells for oxygen saturations between 20% and 80%. This ratio is $1{,}555\ s^{-1}$ [8]. The average transport distance in the red cell is approximated by the red-cell thickness (L_{RBC}), which equals 1.35 μm. The diffusion coefficient of oxygen in a hemoglobin solution similar to that in red blood cells is $D_{O_2,RBC} = 6.0 \times 10^{-6}\ cm^2\ s^{-1}$. The Thiele modulus is thus

$$\phi = \sqrt{\frac{kL_{RBC}^2}{D_{O_2,\,RBC}}} = 0.62 \text{ to } 2.17. \tag{13.4.5}$$

Consequently, oxygen binding to hemoglobin is affected by diffusion within the red blood cell.

In order to determine the relative resistances offered by mass transfer across the alveolar epithelium, the capillary endothelium, the plasma, and the chemical reaction, a modified form of the effectiveness factor for combined internal and external mass transfer (Equation 10.6.28) is used:

$$\eta = \frac{\tanh(\phi)}{\phi\left[(\phi/Bi)\tanh(\phi) + 1\right]}, \tag{13.4.6}$$

where the Biot number represents the ratio of the diffusion time in the red blood cell to the transport time to the red blood cell:

$$\text{Bi} = \frac{P_{ce+pl}\,L_{RBC}}{D_{O_2,\,RBC}}. \tag{13.4.7}$$

Here, the diffusion coefficient in red blood cells is $6 \times 10^{-6}\ cm^2\ s^{-1}$. The resulting value of Bi is 5.1, and the effectiveness factor ranges from 0.33 to 0.83. Thus, external mass transfer and internal mass transfer each reduce the rate of reaction, with external mass transfer accounting for about 20% of the total resistance to oxygen delivery.

Having identified the limiting resistances as those due to transport from the alveolus to the red cell and reaction and diffusion in the red cell, it is now possible to examine overall balances to predict oxygenation along capillaries. To do this, we treat the capillary as a cylinder of radius R_c and length L. The effective length for mass transfer is equal to the circumference of an alveolus (754 μm). The average red-cell velocity is between 2,500 and 4,000 $\mu m\ s^{-1}$.

The simplest model that captures the details of oxygen–hemoglobin binding is to incorporate radial variations in concentration in the effectiveness factor and separately account for oxygen and hemoglobin transport. Separate mass balances are performed on oxygen and oxygen bound to hemoglobin. Balancing the rate of bound oxygen transport by convection in the axial direction with transport across the capillary endothelium, diffusion in the plasma and cytoplasm, and reaction in the red cell, we have

$$\pi R_c^2(v_z C_{O_2}|_z - v_z C_{O_2}|_{z+\Delta z}) + k(H\cdot P - C_{O_2})2\pi R_c \Delta z$$
$$- \eta(\text{rate of } O_2 \text{ reaction})\pi R_c^2 \Delta z = 0, \quad (13.4.8a)$$

where P is the alveolar pressure and is the free oxygen concentration in plasma. For oxygen bound to hemoglobin, the hemoglobin is confined in the red cell; hence, the permeation term is zero. The mass balance is

$$\pi R_c^2(v_z C_{HbO_2}|_z - v_z C_{HbO_2}|_{z+\Delta z}) + \eta(\text{rate of } O_2 \text{ reaction})\pi R_c^2 \Delta z = 0. \quad (13.4.8b)$$

The reaction term is represented by Equation (13.3.2), with k_1' modified to account for cooperative binding, as shown in Equation (13.3.5). Dividing each term of Equations (13.4.8a) and (13.4.8b) by $\pi R_c^2 \Delta z$ and taking the limit as Δz goes to zero, we have

$$v_z \frac{dC_{O_2}}{dz} = 2k/R_c(H\cdot P - C_{B_T}) - \eta(k_1' C_{O_2} 4C_{Hb} - k_{-1}' C_{HbO_2}) \quad (13.4.9)$$

$$v_z \frac{dC_{HbO_2}}{dz} = \eta(k_1' C_{O_2} 4C_{Hb} - k_{-1}' C_{HbO_2}). \quad (13.4.10)$$

The factor of 4 multiplying the hemoglobin concentration in Equations (13.4.9) and (13.4.10) accounts for the four oxygen-binding sites on hemoglobin. Equations (13.4.9) and (13.4.10) were solved numerically, and typical results are shown in Figure 13.6 for two different mean velocities and effectiveness factor values of 0.45

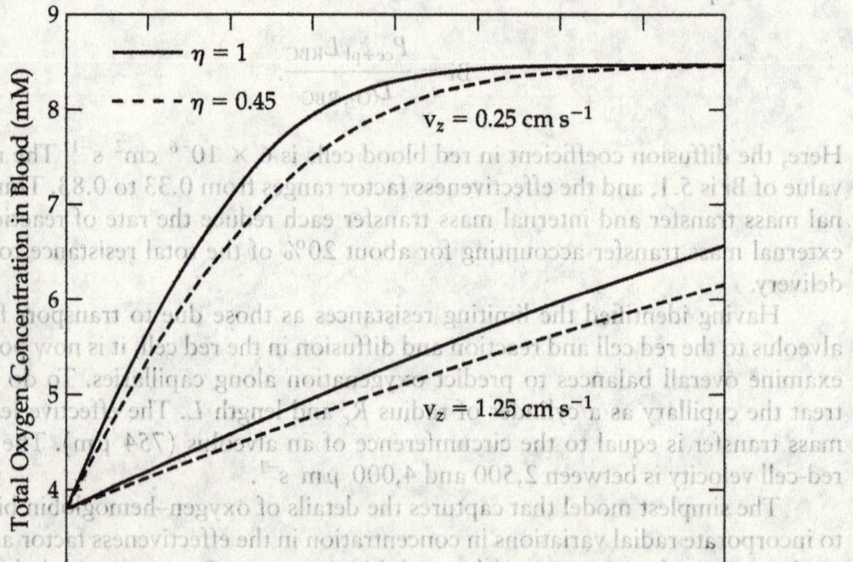

FIGURE 13.6 Model predictions for the oxygenation of blood in capillaries for average velocities representing resting conditions ($v_z = 0.25$ cm s^{-1}) and heavy exercise ($v_z = 1.25$ cm s^{-1}).

and 1.0. For the lower velocity, simulating rest, the blood is fully oxygenated before traversing 50% of the capillary. Decreasing the effectiveness factor to 0.45 does increase the distance required for full oxygenation, but the effect is not very large. During heavy exercise, in which the flow rate is increased fivefold, the model suggests that oxygenation is incomplete before blood exits the capillaries.

The preceding analysis indicates that transport from the alveolus to the red cell and reaction within the red cell are the limiting steps in oxygen delivery. In disease states, various resistances may change, affecting the time required to oxygenate red cells. For example, in chronic obstruction of the airways of the lung due to smoking or emphysema, the level of CO_2 in the alveoli increases. The rise in CO_2 causes the pH in the lung capillaries to decrease, raising P_{50}. Consequently, oxygen binds less tightly to hemoglobin, resulting in lower levels of oxygenation of pulmonary venous blood. In another example, in pulmonary edema, heart failure produces an increase in pulmonary blood pressure. This higher pressure increases fluid transport from the capillary to the tissue between the alveoli and capillaries. If the filtration flow rate exceeds the rate of lymphatic drainage, edema results. In mild cases, the edema may fill only the tissue between the alveolar epithelium and the capillary endothelium. In more extreme cases, the fluid accumulates in the alveoli. The net effect is a significant increase in mass-transfer resistance from the alveolus into the capillary and reduced levels of oxygenation of blood. Consequently, individuals suffering from pulmonary edema have shortness of breath and a limited capacity for physical exertion.

13.5 Oxygen Delivery to Tissues

When oxygen reaches the tissue, equilibrium favors the dissociation of the oxygen from the hemoglobin due to the lower oxygen levels and higher CO_2 levels in tissue. Oxygen transport from the oxygenated red cells to the tissue is not simply the opposite of the process occurring in the lungs because the kinetics of dissociation differs from that of binding, and the tissue thickness is much greater than the thickness of the alveolar epithelium and capillary endothelium.

Oxygen delivery to tissues is examined in two parts. First, oxygen transport into tissue is considered for the case of a uniform concentration of oxygen in the blood. This problem was first examined by the Danish physiologist August Krogh in the early part of the 20th century.[5] Then, we relax the assumption of a uniform oxygen concentration in blood and examine the variations in the concentration along the length of a capillary.

[5]Krogh was interested in respiration and developed accurate methods for measuring blood gases, including oxygen concentrations and respiration rates in tissues. He developed the model bearing his name to demonstrate that the net rate of oxygen transport to the tissue equaled the rate of consumption. He enlisted the help of a friend, Erlang, who was a mathematician, to solve the mathematical problem for him. In 1920, Krogh received the Nobel Prize in Medicine for his work on oxygen transport and the neural regulation of flow in small vessels.

13.5.1 The Krogh Cylinder Model of Oxygen Transport in Tissues

Oxygen delivery to tissues involves the dissociation of oxygen from hemoglobin, the diffusion and convection of oxygen through the plasma, diffusion across the endothelium, diffusion through the tissue and cells, and, finally, the reaction of oxygen in mitochondria as part of aerobic metabolism. In order to develop a model of oxygen transport to tissues, it is necessary to model a simplified organization of capillaries. In many tissues—particularly, skeletal muscle—capillaries are evenly spaced, with characteristic distances between them. This symmetry often leads to capillaries being located at the vertices surrounding a central capillary. The observed regularity of spacing led Krogh to develop the idealization that capillaries supply oxygen to a cylindrical region surrounding each capillary (Figure 13.7). Each capillary has a radius R_c. The tissue radius R_0 represents the half-distance between the center of two capillaries. Alternatively, R_0 can be computed as the reciprocal of the square root of the number of capillaries per unit area. Although this a reasonable first-order approximation, it is clear that the model fails to account for some area between two capillaries. Further, the model is not applicable in tissues such as brain tissue, where capillary organization does not fit this simple pattern.

The model consists of two parts: oxygen transport through blood, followed by oxygen transport and reaction in tissues. Let us consider transport and reaction in tissues first. Assuming a steady state, no convection in the tissue, and negligible axial diffusion, a mass balance on steady state, one-dimensional oxygen diffusion and reaction in the tissues yields the following relation:

$$\frac{D_{O_2}}{r}\frac{d}{dr}\left(r\frac{dC_{O_2}}{dr}\right) = R_{O_2}. \qquad (13.5.1)$$

Although oxygen consumption generally follows Michaelis–Menten kinetics, the reaction rate is assumed to be zero order. This assumption is based on the small value of the Michaelis constant ($K_M = 0.67\ \mu M$ to $2.7\ \mu M$ [9]) relative to average oxygen concentrations in tissue. That is, only when the oxygen concentrations are close to zero does the rate deviate from zero-order kinetics.

At the capillary surface, the tissue concentration is in equilibrium with the oxygen concentration in plasma, C_{R_c}:

$$r = R_c \qquad C_{O_2} = C_{R_c}. \qquad (13.5.2a)$$

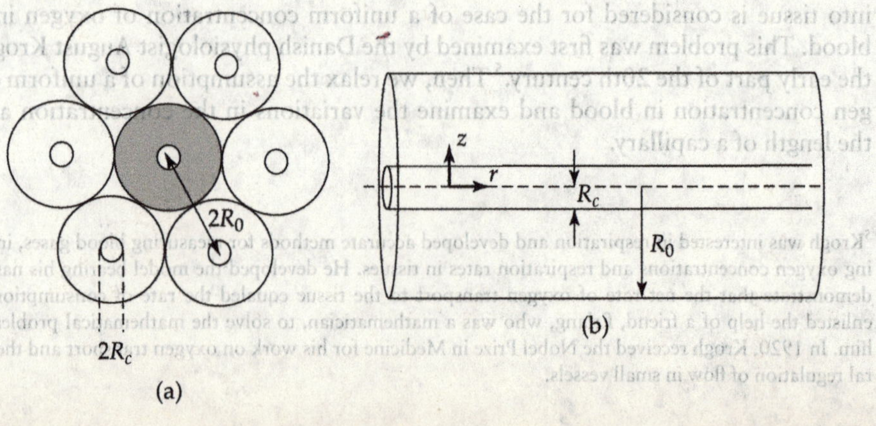

FIGURE 13.7 Krogh cylinder model. (a) Cross-sectional view; (b) side view showing the capillary and tissue cylinder radii.

As is noted subsequently, the oxygen concentration in the capillary is, in general, a function of axial distance along the capillary (Figure 13.7).

At R_0, however, there is a plane of symmetry between the two cylinders. Thus, the flux is zero:

$$\text{At } r = R_0, \quad -D_{O_2}\frac{dC_{O_2}}{dr} = 0. \tag{13.5.2b}$$

The solution of Equation (13.5.1) is

$$C_{O_2} = \frac{R_{O_2}r^2}{4D_{O_2}} + A\ln r + B. \tag{13.5.3}$$

After the boundary conditions (Equations (13.5.2a) and (13.5.2b)) are applied, the concentration profile in the tissue is

$$\frac{C_{O_2}}{C_{R_c}} = 1 + \frac{R_{O_2}R_0^2}{4C_{R_c}D_{O_2}}\left[\left(\frac{r^2}{R_0^2} - \frac{R_C^2}{R_0^2}\right) - 2\ln\left(\frac{r}{R_c}\right)\right]. \tag{13.5.4}$$

The group $R_{O_2}R_0^2/4C_{R_c}D_{O_2}$ represents a dimensionless reaction rate and is represented by the symbol Φ. With this definition and with $r^* = r/R_0$ and $R^* = R_c/R_0$, Equation (13.5.4) becomes

$$\frac{C_{O_2}}{C_{R_c}} = 1 + \Phi\left[(r^{*2} - R^{*2}) - 2\ln\left(\frac{r^*}{R^*}\right)\right]. \tag{13.5.5}$$

The effect of Φ on the concentration profile is shown in Figure 13.8. When Φ is 0.01 or less, the effect of metabolism leads to small drops in oxygen concentration

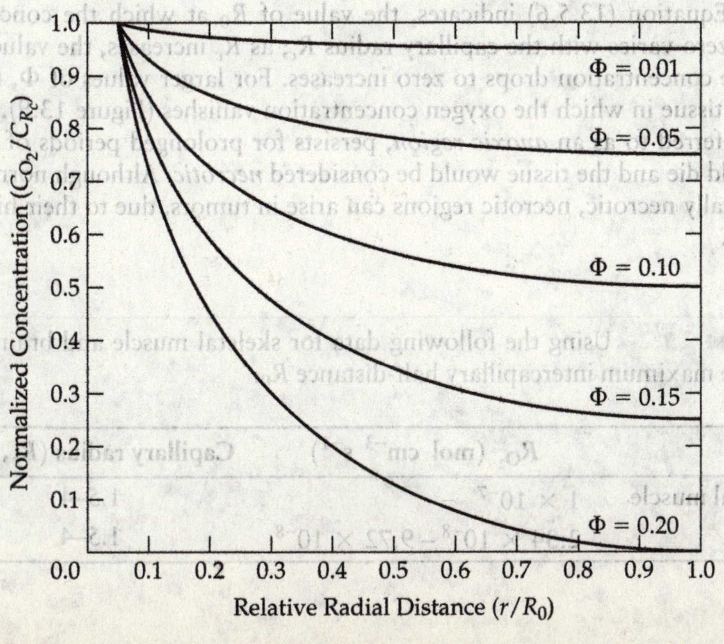

FIGURE 13.8 Oxygen concentration distribution for various values of the dimensionless metabolic parameter Φ for $R^* = R_c/R_0 = 0.05$.

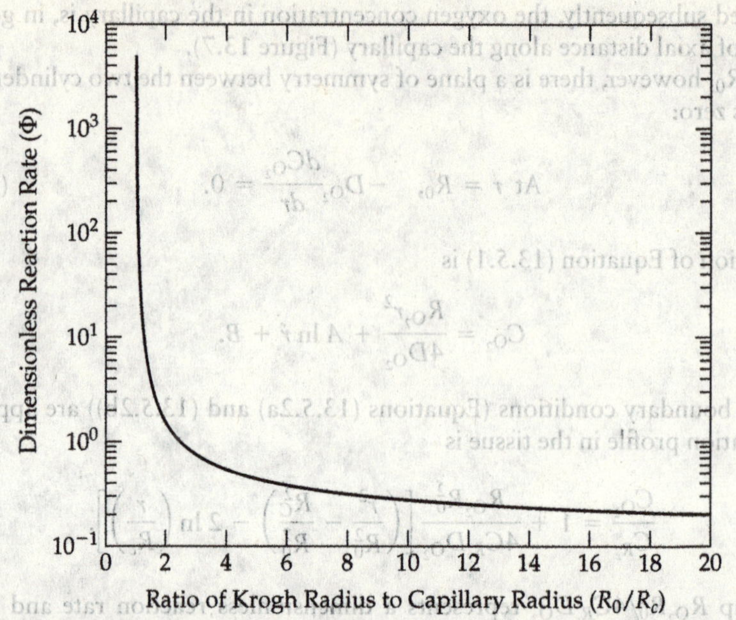

FIGURE 13.9 Ratio of maximum intercapillary distance to the capillary radius as a function of the oxygen consumption rate when $C_{O_2} = 0$ at $r = R_0$.

between the capillaries. As the value of Φ increases, the concentration drops. The concentration declines to zero at $r = R_0$ (or $r^* = 1$) when

$$\Phi = \frac{1}{R^{*2} - 1 - 2\ln(R^*)}. \tag{13.5.6}$$

For $R^* = R_c/R_0 = 0.05$, the concentration at $r = R_0$ (or $r^* = 1$) is zero for $\Phi = 0.20$. This represents the maximum distance attainable between two capillaries without causing oxygen starvation of cells.

As Equation (13.5.6) indicates, the value of R_0 at which the concentration drops to zero varies with the capillary radius R_c; as R_c increases, the value of Φ at which the concentration drops to zero increases. For larger values of Φ, there is a region of tissue in which the oxygen concentration vanishes (Figure 13.9). If such a region, referred to as an *anoxic region*, persists for prolonged periods of time, the cells would die and the tissue would be considered *necrotic*. Although normal tissue is not usually necrotic, necrotic regions can arise in tumors, due to their high metabolic rate.

Example 13.2 Using the following data for skeletal muscle and brain, determine the maximum intercapillary half-distance R_0:

Tissue	R_{O_2} (mol cm^{-3} s^{-1})	Capillary radius (R_c, μm)
Skeletal muscle	1×10^{-7}	1.5–4
Brain	$2.34 \times 10^{-8} - 9.72 \times 10^{-8}$	1.5–4

The diffusion coefficient of oxygen in tissue is 2.0×10^{-5} cm^2 s^{-1}. The oxygen concentration in plasma is 4.05×10^{-8} mol cm^{-3}.

Solution The maximum intercapillary half-distance represents the case when $C_{O_2} = 0$ at $r = R_c$. Since R_0 is not known, the values of Φ and R^* cannot be calculated. However, rearranging the dimensional form of Equation (13.5.6) gives the following relation:

$$R_C^2 - R_0^2 - 2R_0^2 \ln\left(\frac{R_C}{R_0}\right) = \frac{4C_{R_c}D_{O_2}}{R_{O_2}}. \quad (13.5.7)$$

For muscle, $R_c = 1.5 \times 10^{-4}$ cm. Substituting the values provided yields

$$2.25 \times 10^{-8}\ \text{cm}^2 - R_0^2 + 2R_0^2 \ln(6666.67R_0) = 3.24 \times 10^{-5}. \quad (13.5.8)$$

The value of R_0 is obtained by solving Equation (13.5.8) iteratively or by using a root-finding subroutine. Beginning with an initial guess of $R_0 = 50$ μm, we find that a value of $R_0 = 26.20$ μm ($R^* = 0.057$) results in the left-hand side of the equation coming within 0.10% of the value on the right-hand side. For a capillary radius of 4 μm, $R_0 = 31.95$ μm ($R^* = 0.125$).

For brain, with $R_{O_2} = 2.34 \times 10^{-8}$ mol cm^{-3} s^{-1} and $R_c = 1.5 \times 10^{-4}$ cm, Equation (13.5.7) becomes

$$2.25 \times 10^{-8}\ \text{cm}^2 - R_0^2 + 2R_0^2 \ln(6666.67R_0) = 1.385 \times 10^{-4}. \quad (13.5.9)$$

The left- and right-hand sides of Equation (13.5.9) are within 0.1% of each other for $R_0 = 48.25$ μm. For $R_c = 4$ μm, $R_0 = 56.71$ μm. For $R_{O_2} = 9.72 \times 10^{-8}$ mol cm^{-3} s^{-1} and $R_c = 1.5 \times 10^{-4}$ cm, Equation (13.5.7) becomes

$$2.25 \times 10^{-8}\ \text{cm}^2 - R_0^2 + 2R_0^2 \ln(6666.67R_0) = 3.33 \times 10^{-5}. \quad (13.5.10)$$

The left- and right-hand sides of Equation (13.5.10) are within 0.1% of each other for $R_0 = 26.50$ μm. For $R_c = 4$ μm, $R_0 = 32.33$ μm.

Although the original model developed by Krogh did not consider oxygen concentration gradients along the length of the capillary, axial variations in the oxygen concentration are included by applying a steady mass balance on oxygen in the blood. This balance is analogous to the balance presented for the lungs, except that the flux in the axial direction equals the rate of oxygen consumption by the tissue. Thus,

$$\pi R_c^2 v_z \frac{dC_{B_T}}{dz} = -\pi(R_0^2 - R_c^2)R_{O_2}. \quad (13.5.11)$$

Integrating Equation (13.5.11) from the inlet to the outlet yields

$$C_{B_T} = C_{B_a} - \frac{R_0^2 - R_c^2}{R_c^2 v_z} R_{O_2} z, \quad (13.5.12)$$

where C_{B_a} is the arterial concentration of C_{B_T} at $z = 0$. According to Equation (13.5.12), the concentration decreases linearly along the capillary and would decline to zero for a length

$$L = \frac{C_{B_a} R_c^2 v_z}{(R_0^2 - R_c^2) R_{O_2}}. \qquad (13.5.13)$$

Since the concentration decreases axially along the length of the capillary and radially in the tissue, for sufficiently high metabolic demand a region near R_0 at the venous end ($z \approx L$) could become anoxic. Such a condition is known as a *lethal corner* [2].

Example 13.3 For the data provided for muscle in Example 13.2, determine the capillary length at which the concentration at the outlet is zero. Oxygen enters the capillaries at a partial pressure of 13,329 Pa. At peak exercise, the average blood velocity in the capillaries is 0.20 cm s^{-1}.

Solution The inlet partial pressure of oxygen yields a value of C_{B_a} equal to 19.88×10^{-3} M. Using this value of C_{B_a}, the given blood velocity, $R_{O_2} = 1 \times 10^{-7}$ mol cm^{-3} s^{-1}, $R_c = 1.5 \times 10^{-4}$ cm, and $R_0 = 26.20 \times 10^{-4}$ cm, L is found from Equation (13.5.13):

$$L = \frac{(19.87 \times 10^{-6} \text{ mol cm}^{-3})(1.5 \times 10^{-4} \text{ cm})^2 (0.20 \text{ cm s}^{-1})}{\left[(26.20 \times 10^{-4} \text{ cm})^2 - (1.5 \times 10^{-4} \text{ cm})^2\right](1 \times 10^{-7} \text{ mol cm}^{-3} \text{ s}^{-1})}$$

$$= 0.168 \text{ cm}.$$

The predicted length is longer than the typical capillary length, 0.1150 cm. The model predicts that the outlet concentration of oxygen is 1.41×10^{-3} M, which corresponds to a saturation of 14%. Although this saturation is lower than the percentage typically observed, given the approximate nature of the model, the agreement is good.

13.5.2 Analysis of Assumptions Used in the Krogh Model

Because local oxygen concentrations are difficult to obtain in thick tissues, the theoretical analysis of oxygen transport provides an important complement to experimental studies. Consequently, there have been many theoretical investigations of oxygen transport. Two important issues are the accuracy of the Krogh model and the most realistic model of oxygen transport. The Krogh model was developed under the following set of assumptions [2,5,10]:

1. The capillaries are arranged in an ordered array that can be represented by a central capillary surrounded by a cylinder of tissue.
2. The oxygen–hemoglobin dissociation reaction is at equilibrium and occurs uniformly throughout the blood.

3. Oxygen release in plasma occurs uniformly, and the discrete nature of red blood cells is neglected.
4. The mass-transfer resistance offered by the cell-free fluid layer and endothelium is neglected.
5. Axial diffusion in blood is negligible.
6. Axial diffusion in tissue is negligible.
7. Oxygen uptake in tissue can be represented by zero-order kinetics, and the uptake rate is uniform throughout the tissue.
8. Other reactions of oxygen in tissue are negligible.

The significance of each assumption is examined next.

The use of a central capillary surrounded by a cylinder of tissue is an idealization that applies best to skeletal muscle. Nevertheless, the cylinder model does not include all of the tissue surrounding a central capillary. The closest packing of circular cylinders shown in Figure 13.7 excludes about 21% of the tissue, and the Krogh model underestimates oxygen consumption. Other geometries that have been considered include hexagonal, rectangular, and triangular cylinders. Geometries other than a circular cylinder require a numerical solution of the equations describing diffusion as a function of r and θ. Overall, the different geometries produce lower oxygen concentrations in tissue but have a minor effect on the distribution of oxygen [10]. In addition to the spatial arrangement of capillaries, the organization of arteriolar and venular ends of adjacent capillaries affects oxygen concentrations in tissues [2].

A comparison of oxygen delivery in models that consider either equilibrium between oxygen in solution and oxygen bound to hemoglobin or the kinetics of dissociation indicates that oxygen binding to hemoglobin in red cells is close to equilibrium. The equilibrium assumption applies because the time for oxygen diffusion in the red blood cell is much faster than the time for oxygen dissociation from hemoglobin. As a result, the oxygen concentration is uniform throughout most of the red cell and is in equilibrium with hemoglobin. Near the red-cell membrane, however, oxygen concentration gradients arise because deoxyhemoglobin cannot diffuse through the membrane. In this boundary layer region, oxygen is not in equilibrium with hemoglobin [11]. Assuming these conditions, Clark et al. [11] derived the following relation to describe the fractional saturation $S = C_{HbO_2}/(C_{Hb} + C_{HbO_2})$ as a function of contact time in the capillaries, $t = L/v_z$:

$$\frac{dS}{dt} = -\left(\frac{2nS^{(n+1)/n}}{t_u(n+1)(1-S)^{1/n}}\right)^{1/2}. \tag{13.5.14}$$

In Equation (13.5.14), the characteristic unloading time is [11]

$$t_u = \frac{V_{RBC}}{SA_{RBC}}\left(\frac{C_{Hb} + C_{HBO_2}}{D_{O_2}k'_{-1}C_{50}}\right)^{1/2}, \tag{13.5.15}$$

where V_{RBC} is the red-cell volume and SA_{RBC} is the red-cell surface area. The ratio V_{RBC}/SA_{RBC} equals 0.696 μm [11]. With this value and $k'_{-1} = 44\ s^{-1}$, $t_u = 0.053$ s. This time is much shorter than the red-cell transit time, in agreement with the view that unloading occurs rapidly and does not offer a significant transport barrier.

The Krogh model does not account for the discrete nature of red blood cells. As a result, oxygen release is not uniform, and the oxygen concentration exhibits radial and axial variations. Modeling diffusion and convection in the three-dimensional shape of the red blood cell in capillaries is a complex problem that must be solved numerically. As an approximation, the drop in concentration between red cells is neglected, and a radial variation in the fluid gap between the cells and the endothelium is assumed [10]. This approach involves adding an additional cell-free layer between the red cells and endothelium. There are, thus, three regions: a red-cell region in the capillaries ($0 < r < R_{RBC}$) in which Equation (13.5.12) applies, a cell-free region ($R_{RBC} < r < R_C$), and a tissue region ($R_C \leq r \leq R_0$) in which Equation (13.5.1) applies. In the cell-free region, there is oxygen diffusion in the radial direction but no reaction. Although a mass-transfer coefficient can be used (see Equation (13.4.4)), Groebe [10] took an alternate approach and examined the steady radial diffusion of oxygen in the plasma layer in the absence of reaction:

$$\frac{D_{O_2}}{r}\frac{d}{dr}\left(r\frac{dC'_{pl}}{dr}\right) = 0. \tag{13.5.16}$$

The prime denotes that the concentration is the oxygen concentration in the cell-free plasma region. The boundary conditions are

$$C'_{pl} = C_{pl} \qquad r = R_{RBC} \tag{13.5.17a}$$

and

$$-2\pi R_c l_{RBC} D_{O_2,pl}\left.\frac{dC'_{pl}}{dr}\right|_{r=R_C} = R_{O_2}(\pi R_0^2 - \pi R_c^2), \tag{13.5.17b}$$

where l_{RBC} is the fraction of the length of the capillary occupied by the red blood cells and is about 0.5 [10]. The resulting solution for the oxygen concentration in the cell-free layer is

$$C'_{pl} = C_{pl} - \frac{R_{O_2}(R_0^2 - R_c^2)}{2D_{pl}l_{RBC}}\ln\left(\frac{r}{R_{RBC}}\right). \tag{13.5.18}$$

Oxygen transport in capillaries and tissues is now described by the following three equations: Equation (13.5.12) for the red-cell–rich portion of blood with radius R_c replaced by R_{RBC}, Equation (13.5.18) for the cell-free layer between R_{RBC} and R_c, and Equation (13.5.4) for the tissue with $C_{R_C} = C'_{pl}$ at $R = R_c$. The cell-free layer provides an additional resistance to oxygen transport. For a capillary radius of 1.5 μm and a red-cell radius of 1.35 μm, the oxygen concentration at R_c is reduced by 3.2×10^{-9} mol cm^{-3}. This figure represents a 1% reduction in the surface concentration at the capillary entrance and a 10% reduction at the capillary exit. Near the exit, the drop can affect the onset of anoxic regions under exercise. For a capillary with a 3.5-μm radius, the resistance is much larger, leading to a 12.8% reduction in the surface concentration at the capillary entrance and a 26% reduction at the capillary exit.

The justification for neglecting the axial diffusion of oxygen in the blood is based on the Peclet number $Pe = \langle v\rangle 2R_c/D_{O_2}$. The value of Pe is 19.50, which indicates that convection is more important than diffusion. However, near the entrance to the capillary, diffusion may be important due to the large change in concentration [12].

Axial diffusion in the tissue is neglected because the tissue is generally very thin relative to the length of the capillary (i.e., $R_0 - R_c \ll L$). The scaling argument to neglect axial diffusion is thus similar to that used in boundary layer theory to neglect viscous transport in the direction parallel to the surface (Section 4.5).

In general, oxygen metabolism is well described by Michaelis–Menten kinetics. The value of K_M for cytochrome oxidase, the primary mitochondrial enzyme that reacts with oxygen, is around 9×10^{-8} M [9]. Mitochondrial metabolism is sensitive to oxygen when the oxygen concentration falls below 8×10^{-6} M [9], which corresponds to a value of K_M between 2×10^{-6} M and 3×10^{-6} M. Since oxygen levels in most tissues are well above the value of K_M, most investigators assume that the kinetics are zero order. McGuire and Secomb [13] used a more detailed model for oxygen consumption in tissues and compared oxygen concentrations and uptake for zero-order kinetics and Michaelis–Menten kinetics with $K_M = 3.4 \times 10^{-6}$ M. Zero-order kinetics results in a higher oxygen consumption rate than does Michaelis–Menten kinetics, since uptake changes only when the oxygen concentration reaches zero. The difference in oxygen uptake between zero-order kinetics and Michaelis–Menten kinetics for maximum oxygen consumption during exercise is only 3%.

The assumption of a uniform oxygen uptake rate in a tissue is not always appropriate. Nonuniformities in oxygen uptake may arise from the presence of more than one type of cell in a given tissue, such as the presence of glial and nerve cells in the brain or of hematopoietic cells in the bone marrow [14]. Due to the complex cell geometry, numerical models are needed to account for the different types of cells present. In some cases, the tissue can be treated as concentric rings of cells with different properties, and simple analytical solutions can be obtained [14].

In muscle tissue, the density of mitochondria is greater near the capillary than away from the capillary [5]. Thus, the oxygen consumption rate R_{O_2} is greatest near the capillary and declines with distance from the capillary. This variation can be modeled either by using concentric rings with different oxygen consumption rates or by multiplying the oxygen consumption rate by a function that decreases with distance.

In muscle tissue, oxygen reacts with myoglobin, increasing the total oxygen content in the muscle. As a result, the tissue can utilize more oxygen. Oxygen binding to myoglobin is assumed to be at equilibrium. Unlike binding to hemoglobin, oxygen binding to myoglobin is not cooperative and is well described by Equation (13.2.3) with $n = 1$.

Another function of myoglobin is to enhance oxygen transport [2]. Although myoglobin has a much smaller diffusion coefficient than oxygen, such enhanced transport is significant if the amount of oxygen bound to myoglobin in the muscle tissue is much greater than the amount of unbound oxygen in the tissue. Much controversy, however, has surrounded this topic. To address the issue [2], consider a mass balance on both free oxygen and oxygen bound to myoglobin. In this case, Equation (13.5.1) becomes

$$\frac{D_{O_2}}{r}\frac{d}{dr}\left(r\frac{dC_{O_2}}{dr}\right) + \frac{D_{MbO_2}}{r}\frac{d}{dr}\left(r\frac{dC_{MbO_2}}{dr}\right) = R_{O_2}, \quad (13.5.19)$$

where the concentration of oxygen bound to myoglobin is

$$C_{MbO_2} = \frac{C_{Mb_0} C_{O_2}}{K_{Mb} + C_{O_2}}, \tag{13.5.20}$$

in which K_{M_b} is the concentration at which oxygen binding to myoglobin is half maximal. Using the chain rule, we can write the derivative of C_{MbO_2} with respect to r as

$$\frac{dC_{MbO_2}}{dr} = \frac{dC_{MbO_2}}{dC_{O_2}} \frac{dC_{O_2}}{dr} = \frac{K_{Mb} C_{Mb_0}}{(K_{Mb} + C_{O_2})^2} \frac{dC_{O_2}}{dr}. \tag{13.5.21}$$

As a result, Equation (13.5.19) can be written in terms of C_{O_2} as

$$D_{O_2} \frac{1}{r} \frac{d}{dr} \left\{ r \left[1 + \frac{D_{MbO_2} K_{Mb} C_{Mb_0}}{D_{O_2} (K_{Mb} + C_{O_2})^2} \right] \frac{dC_{O_2}}{dr} \right\} = R_{O_2}. \tag{13.5.22}$$

The quantity in brackets on the left-hand side of Equation (13.5.22) represents the extent to which oxygen transport is increased due to the presence of myoglobin. This quantity is referred to as a *facilitation factor*, since multiplication by the oxygen diffusion coefficient yields an effective increase in the diffusion coefficient [2]. The facilitation factor is greatest at low oxygen concentrations. Typical values of these parameters in muscle are $K_{Mb} = 1.09 \times 10^{-8}$ mol cm^{-3}, $C_{Mb_0} = 3.8 \times 10^{-7}$ mol cm^{-3}, and $D_{MbO_2} = 2 \times 10^{-7}$ cm^2 s^{-1} [15]. For an oxygen concentration equal to the critical oxygen concentration, 1.6×10^{-8} mol cm^{-3}, the facilitation factor is 1.057, representing a rather modest effect. More detailed calculations of oxygen transport indicate that myoglobin-facilitated transport enhances transport 4% and 2% in maximally active skeletal muscle and cardiac muscle, respectively [15].

Oxygen is a substrate in a number of other reactions. One of the most significant is the formation of NO (see Section 13.6). The reaction occurs at the endothelial cell surfaces, and maximum NO production rates are as high as 20% of the maximum rate of oxygen consumption [16]. While this percentage may not normally represent significant consumption of oxygen, under certain conditions of high flow and muscular activity, NO formation may influence the overall distribution of oxygen in the tissue.

Finally, the Krogh model was developed to examine oxygen transport in capillaries. Recent experimental evidence, summarized in Ref. [3], indicates that a significant amount of oxygen consumption occurs in arterioles. In these vessels, 10–50 μm in radius, the blood behaves as a single-phase fluid, but the fluid mass-transfer resistance is quite significant. Oxygen uptake by the smooth muscle cells in the vessel wall is significant, limiting oxygen transport to the surrounding tissue [16].

13.6 Nitric Oxide Production and Transport in Tissues

As noted in Section 13.1, NO is an important vasodilator that also affects platelet and leukocyte adhesion to endothelium, hemoglobin function, neurotransmission, and smooth muscle cell growth. Disturbances in NO release during atherosclerosis and septic shock can affect the progression of those diseases. Because NO reacts rapidly in tissues, significant concentration gradients arise. Transport and reaction models have been developed to estimate the concentration of NO in various tissues and determine the distance over which NO is functionally active. After a brief overview of NO formation and breakdown reactions, we examine models for NO production, transport, and breakdown in tissues.

13.6.1 NO Formation and Reaction

Physiologically, NO is produced by the reaction of L-arginine with oxygen and nicotinamide adenine dinucleotide phosphate (NADPH) and is catalyzed by the enzyme nitric oxide synthase (NOS):

$$\text{L-arginine} + \text{NADPH} + O_2 \xrightarrow[\text{nitric oxide synthase}]{} \text{L-citrulline} + \text{NO} \quad (13.6.1)$$

Cofactors for this reaction include tetrahydrobiopterin and calcium bound to calmodulin. The value of K_M for the substrate L-arginine is between 2 μM and 5 μM.

Three different forms of the enzyme NOS are known: neuronal NOS (nNOS or NOS I), a form of NOS induced by cytokines (iNOS or NOS II), and endothelial NOS (eNOS or NOS III). NOS I functions in nerve cell communication and is also found in skeletal muscle, where it assists in the maturation of the muscle. NOS II aids in the immune response but has also been implicated in pathological conditions such as septic shock. NOS III is present in endothelium and is involved in smooth muscle cell relaxation as well as the inhibition of platelet and leukocyte adhesion. The three forms of NOS have very similar amino acid sequences, especially near the active site. The enzymes do differ, however, in the manner in which they are regulated. For instance, active eNOS is linked to caveolae that are present on the plasma membrane (see Chapter 1). In the absence of shear stress, NO release is low and NOS III levels are low. The onset of flow activates the enzyme by phosphorylation of a serine amino acid residue, resulting in increased NO production. After about 4 hours of continuous exposure to steady laminar flow, gene activation leads to increased NOS III production.

NO acts by reacting with heme complexes in proteins such as those found in hemoglobin, soluble guanylate cyclase (sGC), myoglobin, and cytochrome P450. In addition, NO reacts irreversibly with oxygen and reversibly with thiol groups (—SH).

In blood vessel relaxation, NO is synthesized by the endothelium and diffuses to the media, where it reacts with a heme group that is present in sGC (Figure 13.10).

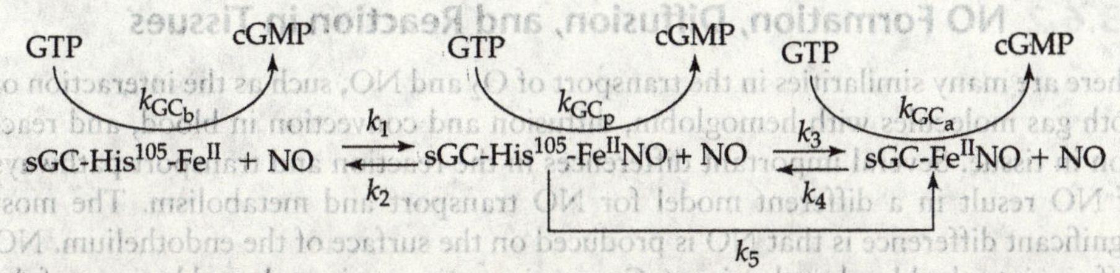

FIGURE 13.10 Mechanism by which NO reacts with soluble guanylate cyclase to produce active forms of the enzyme that catalyze cGMP [17].

The reaction involves a two-step process leading to the complete activation of sGC. Each of the intermediate forms can produce cGMP from guanine triphosphate (GTP), but the activity increases upon reaction with NO ($k_{GCa} > k_{GCp} > k_{GCb}$). In turn, cGMP activates cyclic guanine kinase, a cGMP-dependent kinase that decreases the activity of myosin light-chain kinase, resulting in smooth muscle cell relaxation.

The rate coefficients for the reaction of NO with sGC were measured at 4°C; the values obtained were $k_1 = 1.55 \times 10^8$ M^{-1} s^{-1}, $k_2 = 0.01$ s^{-1}, $k_3 = 1.3 \times 10^5$ M^{-1} s^{-1}, $k_4 = 0.001$ s^{-1}, and $k_5 = 5 \times 10^{-4}$ s^{-1} [17]. The effective Michaelis constant for the overall production of cGMP by NO is 23 nM [18]. This value is important in determining the distance over which NO is functional.

In addition to reacting with sGC in tissues, NO reacts with oxygen. The reaction stoichiometry is

$$4NO + O_2 + 2H_2O \longrightarrow 4H^+ + 4NO_2^- \tag{13.6.2}$$

The observed rate expression is

$$-\frac{dC_{NO}}{dt} = 4kC_{NO}^2 C_{O_2}. \tag{13.6.3}$$

The rate coefficient k equals 2×10^6 M^{-2} s^{-1} at 25°C [19].

NO reacts with hemoglobin in a complex manner. First, NO reacts with the heme group of deoxygenated hemoglobin to form a stable complex with iron. The reaction is first order in both NO and hemoglobin, with $k_1 = 1 \times 10^7$ M^{-1} s^{-1} at 20°C [20]. The reaction is essentially irreversible. In addition, NO reacts irreversibly with iron in oxygen bound to iron in a heme group to form nitrite:

$$NO + Hb(Fe^{2+})O_2 \xrightarrow{k_{Hb}} NO^{3-} + Hb(Fe^{2+}) \tag{13.6.4}$$

This reaction is first order in NO and $Hb(Fe^{2+})O_2$. The association rate coefficient for the reaction is $k_{Hb} = 8.9 \times 10^7$ M^{-1} s^{-1} at 25°C. In red blood cells, the reaction is diffusion limited with effectiveness factors of 0.002 or smaller. NO also reacts with free thiols in hemoglobin and other proteins to form nitrosothiol (SNOHb). Although the mechanism of the reaction has not yet been determined, the amount of SNOHb formed after oxygenated hemoglobin reacts with NO is one hundred times less than the amount of nitrite formed over a 6-minute period [21]. Other reactions of NO are presented in the review paper by Buerk [20].

13.6.2 NO Formation, Diffusion, and Reaction in Tissues

There are many similarities in the transport of O_2 and NO, such as the interaction of both gas molecules with hemoglobin, diffusion and convection in blood, and reaction in tissue. Several important differences in the reaction and transport pathways of NO result in a different model for NO transport and metabolism. The most significant difference is that NO is produced on the surface of the endothelium. NO diffuses into the blood or the tissue. Convective transport is neglected because of the uniform production of NO by the endothelium.

In this section, we develop a simple model for NO production by endothelium and for diffusion and reaction both in blood in the vessel lumen and in the vessel wall of small- and medium-size blood arterioles (R_c = 25 to 75 μm), to assess the distance over which NO acts in tissue. References to more detailed analyses are noted.

For steady-state radial diffusion and reaction, the conservation relation and the use of Fick's law of diffusion yield the following expression for NO transport in the vessel lumen and vessel wall:

$$\frac{D_{NO,i}}{r}\frac{d}{dr}\left(r\frac{dC_{NO,i}}{dr}\right) - R_{NO,i} = 0. \tag{13.6.5}$$

Here, the subscript i takes on the value B for the capillary lumen ($0 < r < R_C$) and the value T for $r > R_C$.

The reactions are lumped into reactions occurring in the blood and those taking place in tissue [22]. In the red cells, the reaction of NO with hemoglobin is the dominant reaction. Since the hemoglobin concentration (5.1 mM) is much higher than the NO concentration (≤0.002 mM) [23], the reaction can be treated as pseudo–first order in NO, and it follows that

$$R_{NO,B} = k_{Hb}C_{Hb}C_{NO} = k_B C_{NO}. \tag{13.6.6}$$

In the tissues, NO reacts with oxygen and sGC. An analysis of oxygen uptake at discrete locations from an endothelial cell source of NO indicates that the kinetics in vivo can be fit equally well by a model assuming first-order kinetics and one postulating second-order kinetics [23]. Uncertainty in the order of the reaction is due to the limited data available. For simplicity, we assume that the reaction is first order with a rate coefficient equal to $k_T = 0.01\ s^{-1}$ [23]. Since the reactions of NO in the blood and tissues are modeled as first order, Equation (13.6.5) can be rewritten as

$$\frac{D_{NO,i}}{r}\frac{d}{dr}\left(r\frac{dC_{NO,i}}{dr}\right) - k_i C_{NO,i} = 0. \tag{13.6.7}$$

The boundary conditions are as follows. At $r = 0$, the NO flux is zero, due to symmetry in the radial direction:

$$r = 0 \qquad \frac{dC_{NO,B}}{dr} = 0. \tag{13.6.8a}$$

At the endothelial cell surface ($r = R_C$), the rate of NO production is constant ($\dot{q}_{NO}$) and equal to the difference between the NO flux in the blood and tissue [22]:

$$r = R_C \qquad \dot{q}_{NO} = D_{NO,B}\frac{dC_{NO,B}}{dr} - D_{NO,T}\frac{dC_{NO,T}}{dr}. \tag{13.6.8b}$$

A more general analysis [22] has considered the release of NO from each endothelial cell surface, as well as diffusion and reaction of NO within the endothelium.

A second boundary condition at $r = R_C$ is that the concentration of NO is the same in blood and tissue (the partition coefficient is thus assumed to be unity):

$$C_{NO,B}(r = R_C) = C_{NO,T}(r = R_C). \tag{13.6.8c}$$

The boundary condition far from the endothelium is not well described. If the vessels are far apart, but uniformly spaced, then the flux far from the surface should be zero to ensure symmetry:

$$r \longrightarrow \infty \qquad \frac{dC_{NO,T}}{dr} = 0. \tag{13.6.8d}$$

Other boundary conditions that can be applied are that $C_{NO,T} = 0$ as r goes to infinity and that the flux is zero at a distance $r = R_0$, representing the half-distance between two blood vessels.

Since there is no characteristic concentration, Equation (13.6.7) and the associated boundary conditions (Equations (13.6.8a) to (13.6.8d)) are solved for the dimensional concentration as a function of dimensional variables. To solve these equations, we rewrite Equation (13.6.7) in the form

$$r\frac{d}{dr}\left(r\frac{dC_{NO,i}}{dr}\right) - r^2\frac{k_i}{D_{NO,i}}C_{NO,i} = 0. \tag{13.6.9}$$

Equation (13.6.9) has the form of Bessel's equation (see Equation A.2.31), with $\chi^2 = 0$ and $m_i^2 = -k_i/D_{NO,i}$. Since m_i is imaginary, the solution is presented in terms of the modified Bessel functions $I_\chi(m_i r)$ and $K_\chi(m_i r)$ [24]. Values of the modified Bessel functions for χ equal to zero and unity are shown in Figure 13.11.

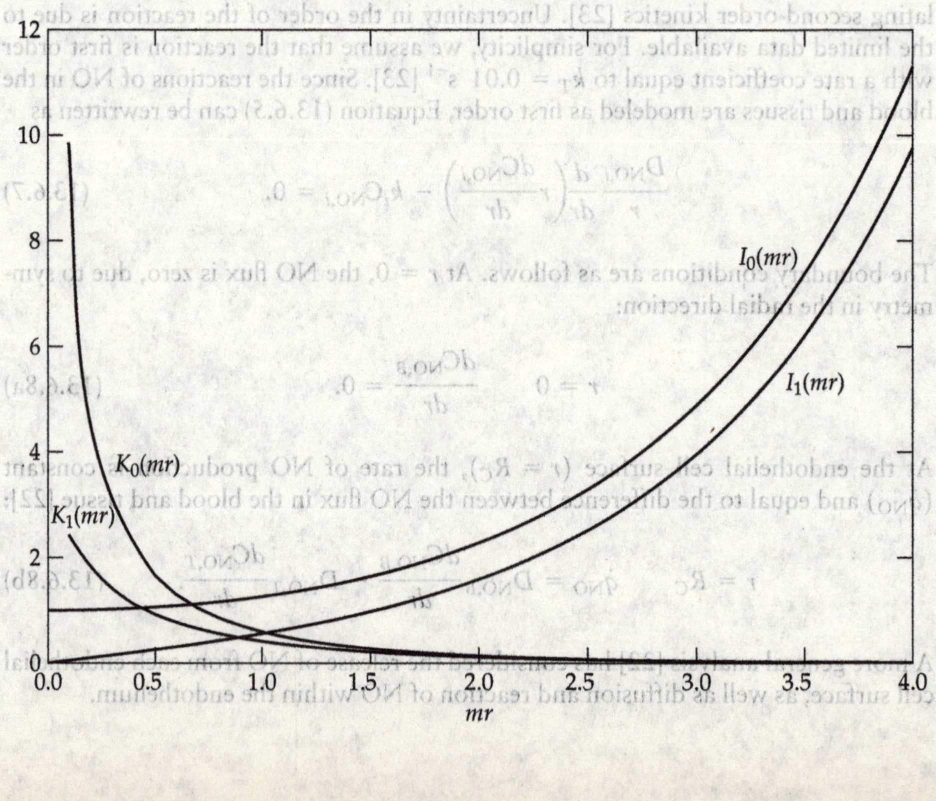

FIGURE 13.11 Modified Bessel functions $I_0(mr)$, $I_1(mr)$, $K_0(mr)$, and $K_1(mr)$ as a function of mr.

The modified Bessel function $I_0(mr)$ is 1 at $mr = 0$ and goes to infinity as mr becomes large. $I_1(mr)$ is zero at $mr = 0$ and also goes to infinity as mr becomes large. In contrast, both $K_0(mr)$ and $K_1(mr)$ are infinite at $mr = 0$ and go to zero as mr becomes large. Another important property of modified Bessel functions used in this problem is the following relation for the first derivative of $I_0(mr)$ and $K_0(mr)$ [24]:

$$\frac{d(I_0(mr))}{dr} = mI_1(mr) \qquad \frac{d(K_0(mr))}{dr} = -mK_1(mr). \tag{13.6.10a,b}$$

Although there is no characteristic concentration, the solution can be cast in terms of Thiele moduli, using the following definitions:

$$r^* = \frac{r}{R_C} \qquad \phi_B = R_C\sqrt{\frac{k_B}{D_{NO,B}}} \qquad \phi_T = R_C\sqrt{\frac{k_B}{D_{NO,T}}}. \tag{13.6.11a,b,c}$$

Thus, m_i can be rewritten as $\phi_i r^*$.

The solutions of Equation (13.6.9) have the following form:

$$C_{NO,i} = A_i I_0(\phi_i r^*) + B_i K_0(\phi_i r^*). \tag{13.6.12}$$

Applying the boundary conditions to evaluate the constants results in the following expressions for $C_{NO,B}$ and $C_{NO,T}$:

$$C_{NO,B} = \frac{\dot{q}_{NO} R_C I_0(\phi_B r^*) K_0(\phi_T)}{D_{NO,B}\phi_B I_1(\phi_B) K_0(\phi_T) + D_{NO,T}\phi_T I_0(\phi_B) K_1(\phi_T)}, \tag{13.6.13a}$$

$$C_{NO,T} = \frac{\dot{q}_{NO} R_C I_0(\phi_B) K_0(\phi_T r^*)}{D_{NO,B}\phi_B I_1(\phi_B) K_0(\phi_T) + D_{NO,T}\phi_T I_0(\phi_B) K_1(\phi_T)}. \tag{13.6.13b}$$

Parameter values are listed in Table 13.1. The value of k_T was obtained by the regression of measured NO concentrations at discrete locations in tissues [23]. The value of k_B includes the values with and without diffusion limitations within red blood cells and for a normal hematocrit [20–23]. Due to the Fahreus effect and the definition of $k_B(= k_{Hb}C_{Hb})$, the reduced hematocrit in capillaries may further decrease k_B by as much as 50%. These values were then used to simulate NO concentration profiles for different values of k_B and to determine the distance over which the concentration of NO is above 23 nM, the value of K_M required for binding to guanylate cyclase (Figure 13.12).

TABLE 13.1

Parameter Values for NO Transport and Reaction in Blood and Tissues

Parameter	Values	Parameter	Values
$\dot{q}_{NO}$	5.5×10^{-12} mol cm^{-2} s^{-1}	R_c	$25–75 \times 10^{-4}$ cm
k_B	$454–4.54 \times 10^5$ s^{-1}	k_T	0.01 s^{-1}
$D_{NO,B}$	4.5×10^{-5} cm^2 s^{-1}	$D_{NO,T}$	3.3×10^{-5} cm^2 s^{-1}
ϕ_B (R_c = 25 μm)	7.9–251	ϕ_T	0.0435–0.130

Source: Based on data summarized from Ref. [22].

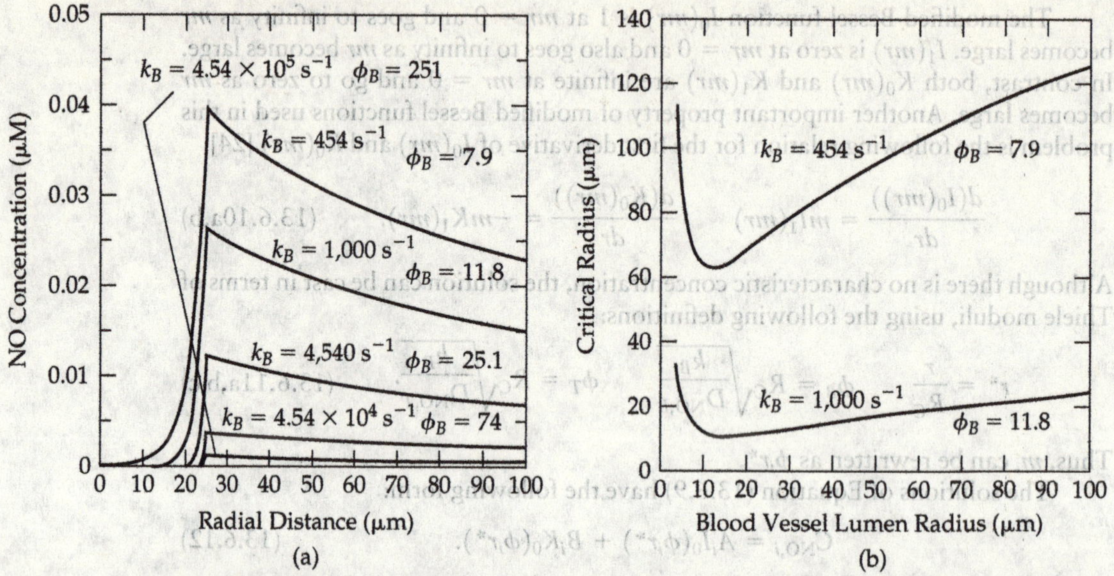

FIGURE 13.12 (a) Effect of vessel radius and value of k_B on the NO concentration profiles in blood and tissue. (b) Effect of k_B and blood vessel lumen radius on the critical radius for NO transport. For these simulations, $\phi_T = 0.0435$.

The value of k_B affects the NO concentration in both the blood and the tissue (Figure 13.12a). As more NO is consumed in the blood, the tissue concentration of NO decreases. Since ϕ_B is much larger than ϕ_T, the concentration at the endothelial cell surface is effectively controlled by the value of the rate coefficient k_B. The shape of the concentration profile in the tissue is affected by k_T. The results are qualitatively similar to those obtained from more complex models [22].

Defining the *critical concentration* as the concentration of NO above the concentration required for half-maximal binding to sGC [22], we note that the critical concentration (23 nM [18]) is exceeded only for values of k_B less than 1,000 s^{-1}. The critical concentration was determined as a function of the vessel lumen radius for k_B values of 454 s^{-1} and lower (Figure 13.12b). Two competing effects influence the distance at which the critical concentration is reached [22]. For small vessels between 2 and 10 μm in radius, the critical radius declines as the vessel radius increases. Increasing vessel radius leads to more NO consumption in the blood, which lowers the maximum NO concentration, causing the critical radius to decline with increasing vessel radius. However, the NO concentration in the blood drops to zero over a distance of 5–8 μm from the endothelial cell surface. For vessels larger than this size, there is no additional consumption of NO by blood and the maximum NO concentration does not change with increasing vessel radius. For vessels larger than 10 μm in radius, the critical radius increases due to a greater surface area for NO release. As a result, more NO is available for transport into the tissue, resulting in a longer diffusion distance before the critical radius is reached. The trends observed with the critical radius differ from those observed by Vaughn et al. [22].

The reason for the difference lies in the choice of the critical concentration. Vaughn et al. [22] used a value of 0.25 μM, based on the best estimate available at that time.

These simulations show that, on the basis of the current estimate of the concentration of NO needed to affect cGMP production, NO can diffuse sufficiently far in small arteries to stimulate relaxation throughout the vessel wall.

13.7 QUESTIONS

13.1 How does the structural change that takes place in hemoglobin after it binds to oxygen explain the hemoglobin–oxygen dissociation curve shown in Figure 13.3?

13.2 The hemoglobin of individuals who live at high altitude for long periods of time often exhibits a value of P_{50} below the normal value of 27 mmHg. Explain how this shift in P_{50} affects the oxygenation of blood at high altitudes.

13.3 Myoglobin is present in skeletal muscle and facilitates the delivery of oxygen in muscle. The molecule has a single heme group. Explain whether the reaction scheme represented by Equation (13.3.1) or Equation (13.3.3) would better describe the kinetics of oxygen binding to myoglobin.

13.4 Explain how carbon dioxide binding to hemoglobin would affect oxygen dissociation in capillaries.

13.5 Discuss how the oxidation of NO in blood vessels would affect the relaxation of smooth muscle cells.

13.6 Identify the factors that influence the NO concentration in the blood vessel wall.

13.8 PROBLEMS

13.1 Compare the reaction schemes represented by Equations (13.3.1) and (13.3.3) for the kinetics of oxygen binding to hemoglobin. For Equation (13.3.1), use both a constant value of k_1' and the variable form represented by Equation (13.3.5). Let $C_{Hb_0} = 2.3 \times 10^{-3}$ M and $C_{O_{2_0}} = 6 \times 10^{-5}$ M.

13.2 A person with anemia has a hemoglobin concentration of 0.1 g mL^{-1}. What is the total concentration of oxygen per milliliter of blood if the hemoglobin is saturated with oxygen?

13.3 Suppose an individual has edema that results in a 20-μm-thick fluid layer in the alveoli. By incorporating an additional mass-transfer resistance layer into the model of blood oxygenation presented in Section 13.4, determine the distance over which oxygenation occurs. Assume that the diffusion coefficient of oxygen in the fluid layer is the same as in plasma, and assume also that $k = 127\ s^{-1}$ and $v_z = 2{,}500\ \mu m\ s^{-1}$.

13.4 For muscle cells under exercise conditions, the observed rate of O_2 uptake by cells is 2×10^{-6} mol $cm^{-3}\ s^{-1}$. For the Krogh cylinder model, determine the maximum capillary spacing for which the oxygen concentration is zero at R_0. Use a capillary radius of $R_c = 3$ μm, $C_{O_2}(r = R_c) = 0.14$ mM, and an oxygen diffusion coefficient of $1.92 \times 10^{-5}\ cm^2\ s^{-1}$.

13.5 Repeat the analysis in Problem 13.4, but this time allow for axial variations in oxygen concentration. The arteriolar oxygen concentration is 20 mM and the venular concentration in RBC is 8 mM. Assume an average blood velocity in the arterioles and venules is either 0.2 cm s^{-1} or 2.0 cm s^{-1}.

13.6 Integrate Equation (13.5.14) numerically to find the time it takes for hemoglobin to change from a saturation of 0.95 to a saturation of 0.40.

13.7 In this problem, consider the effect of a nonuniform distribution of mitochondria on oxygen

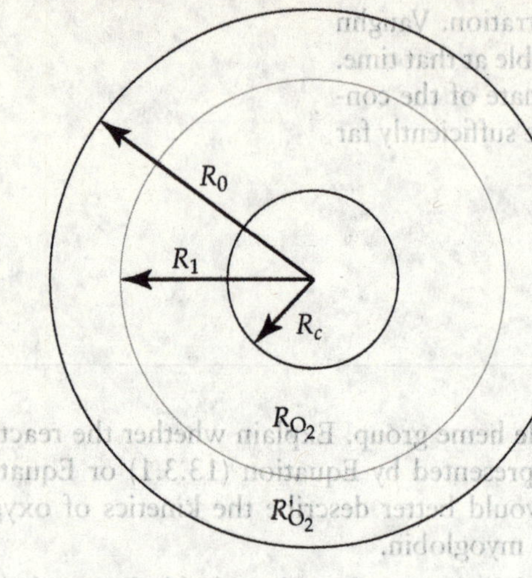

FIGURE 13.13 Schematic of a two-layer model to account for the nonuniform distribution of mitochondria.

consumption in muscle tissue. To simplify the analysis, consider the two-layer model shown in Figure 13.13. The muscle tissue consists of a region extending from $r = R_c$ to $r = R_1$ with an oxygen consumption rate R_{O_2}. In the region from $r = R_1$ to $r = R_0$, the oxygen consumption rate declines to R'_{O_2}. Assume that the oxygen diffusion coefficient and solubility are the same in both regions. Determine the oxygen concentration in the tissue for $C_{O_2} = C_{R_c}$ at $r = R_c$. At $r = R_1$, the oxygen concentrations and fluxes in each layer are the same.

13.8 For the following parameter values, use the result obtained in Problem 13.7 and plot the dimensionless oxygen concentration as a function of distance. Compare with the results for the Krogh model with a uniform oxygen consumption rate equal to $(R_{O_2} + R'_{O_2})/2$.

$R_c = 1.5\ \mu m \quad R_{O_2} = 5 \times 10^{-7}\ \text{mol cm}^{-3}\ \text{s}^{-1}$

$R_1 = 10\ \mu m \quad R_{O_2}' = 0.5 \times 10^{-7}\ \text{mol cm}^{-3}\ \text{s}^{-1}$

$R_0 = 25\ \mu m$

13.9 Determine the oxygen concentration profiles in tissue for oxygen transport in the presence of a reaction to form NO at the endothelial cell surface. The boundary condition at $r = R_C$ is

$$-D_{O_2}\frac{dC_{O_2}}{dr} = -k_{NO}C_{pl} + P_{ec}\left[C_{pl} - C(r = R_C)\right].$$

Let $k_{NO} = 10^{-3}$ cm s^{-1} and $P_{ec} = 0.10$ cm s^{-1}. Assess whether the concentration profile is significantly affected by the reaction.

13.10 Consider a spherical tumor of radius R_T that contains capillaries that provide oxygen at a concentration C_{pl}. The capillaries have a permeability P_{ec} and a surface-to-volume ratio S/V. Capillaries are present in the region from R_1 to R_T where $R_1 > 0$. In the region $0 < r < R_1$, only oxygen uptake occurs. Oxygen uptake is uniform in the tumor at a rate R_{O_2}.

(a) For these conditions, determine the oxygen concentration distribution as a function of position.
(b) Necrosis arises when the concentration of oxygen is zero. Determine a condition when necrosis occurs in the region without capillaries.

13.11 Determine the half-time for NO disappearance due to reaction, according to Equation (13.6.3), for an initial NO concentration of 0.1 μM and an oxygen concentration of 100 μM.

13.12 After 50 μM of oxygenated hemoglobin reacts with 5 μM of NO for 6 minutes, 0.04 μM of SNOHb is formed [21]. Assuming a bimolecular reaction, determine the rate constant for NO reacting with free sulfhydryls to form SNOHb.

13.13 Using a reaction that is second order in NO, examine the effect of the NO reaction order on NO consumption in tissues. Solve the equations numerically with MATLAB. Use a rate coefficient $k_T = 0.05\ \mu M^{-1}\ s^{-1}$ [23].

13.14 Consider a spherical tumor of radius R_T that is covered with capillaries that provide oxygen at a concentration C_B. There are no capillaries in the tumor itself. Oxygen uptake is uniform in the tumor at a rate R_{O_2}.

Necrosis arises when the concentration of oxygen is zero. For an oxygen consumption rate equal to 120 nmol cm^{-3} s^{-1}, determine the size of the tumor before necrosis occurs. Let $D_{eff} = 1.5 \times 10^{-5}$ cm^2 s^{-1} and $C_B = 13 \times 10^{-3}$ M.

13.15 Mitochondria are primary sites within the cells where oxygen is consumed. In the analysis of oxygen metabolism, the oxygen concentration within cells is assumed to be uniform. To test this, we will assume that mitochondria, which occupy a volume fraction f, can be treated as a single spherical mitochondrion of volume $V = fV_c$, where V_c is the cell volume (Figure 13.14). This

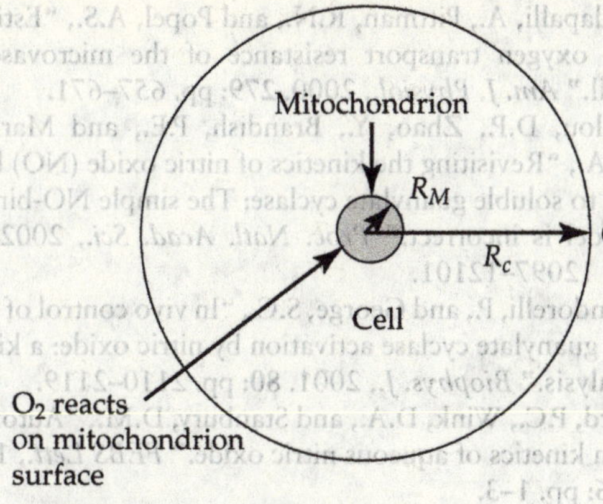

FIGURE 13.14 Schematic of a central mitochondrion in a spherical cell.

mitochondrion is located at the cell center and has a radius R_M. On this surface of the mitochondrion, the reaction rate per unit mitochondrial surface area (Γ, moles per square centimeter per second) equals the flux to the mitochondrial surface. Enzyme reactions on the mitochondrial surface are assumed to be zero order. On the surface of the cell, the oxygen concentration is C_0.

(a) Assuming that oxygen diffuses through the spherical cell and then reacts on the mitochondrion surface of radius R_M, determine the steady-state concentration profile.

(b) Use the data below to determine the concentration drop from the cell surface to the surface of the mitochondrion. Use this result to conclude whether there is a significant decline in oxygen concentration within the cell due to metabolism by the mitochondria.

(c) The K_M for the enzymatic metabolism of oxygen is 9×10^{-8} M. Is the assumption of zero-order reaction appropriate?

$R_c = 5 \times 10^{-4}$ cm	cell radius
$f = 0.085$	volume fraction of mitochondria
$\Gamma = 1.02 \times 10^{-10}$ mol cm^{-2} s^{-1}	oxygen consumption rate by mitochondria
$D_{eff} = 1.92 \times 10^{-5}$ cm^{-2} s^{-1}	oxygen diffusion coefficient within cells
$C_0 = 2.68 \times 10^{-8}$ mol cm^{-3}	oxygen concentration on the surface of cells

13.16 A hollow-fiber unit is used to grow liver cells to high density in the space between fibers. Separate fibers are used for gas and liquid nutrients. At the inlet, the oxygen partial pressure is 150 mmHg and the solubility is 1.34×10^{-6} M mmHg^{-1}. The spacing between hollow fibers that supply oxygen is 400 μm, and the reactor is 10 cm long. The exit partial pressure of oxygen is 10 mmHg. The oxygen consumption rate is 1×10^{-9} mole O_2 cm^{-3} s^{-1} per (1×10^6 cells). The hollow fibers are 50 μm in diameter and the gas velocity per tube is 0.3 cm s^{-1}. The hollow fibers are very permeable to gas, and the gas is in equilibrium with the extracellular solution with a partition coefficient of 1. Use the Krogh cylinder model with axial variations in concentration to determine the average number of cells per cm^3 assuming that the product of cell density times oxygen consumption rate per million cells is constant.

13.9 REFERENCES

1. Perutz, M.F., Wilkinson, A.J., Paoli, M., and Dodson, G.G., "The stereochemical mechanism of the cooperative effects in hemoglobin revisited." *Annu. Rev. Biophys. Biomol. Struct.*, 1998. 27: pp. 1–34.
2. Popel, A.S., "Theory of oxygen transport to tissue." *Crit. Rev. Biomed. Eng.*, 1989. 17: pp. 257–321.
3. Hellums, J.D., Nair, P.K., Huang, N.S., and Oshimara, N., "Simulation of intraluminal gas transport processes in the microcirculation." *Ann. Biomed. Eng.*, 1996. 24: pp. 1–24.
4. Meldon, J.H., "Blood–gas equilibria, kinetics and transport." *Chem. Eng. Sci.*, 1987. 42: pp. 199–211.
5. Weibel, E.R., *The Pathway for Oxygen*. 1984, Cambridge, MA: Harvard University Press.
6. Bentley, T.B., Meng, H., and Pittman, R.N., "Temperature dependence of oxygen diffusion and consumption in

mammalian arterial muscle." *Am. J. Physiol.*, 1993. **264**: pp. H1825–H1830.
7. Frank, A.O., Chuong, C.J., and Johnson, R.L., "A finite-element model of oxygen diffusion in the pulmonary capillaries." *J. Appl. Physiol.*, 1997. **82**: pp. 2036–2044.
8. Forster, R., "Diffusion of gases across the alveolar membrane," in *Handbook of Physiology: Section 3: The Respiratory System*, L.E. Farhi and S.M. Tenney, editors. 1987, Bethesda, MD: American Physiological Society. pp. 71–88.
9. Richmond, K.N., Shonat, R.D., and Pittman, R.M., "Critical P_{O_2} of skeletal muscle in vivo." *Am. J. Physiol.*, 1999. **277**: pp. H1831–H1840.
10. Groebe, K., "An easy-to-use model for O_2 supply to red muscle: validity of assumptions, sensitivity to errors in data." *Biophys. J.*, 1995. **68**: pp. 1246–1269.
11. Clark, A., Federspiel, W.J., Clark, P.A.A., and Cokelet, G.R., "Oxygen delivery from red cells." *Biophys. J.*, 1985. **47**: pp. 171–181.
12. Whiteley, J.P., Gavaghan, D.J., and Hahn, C.E.W., "Mathematical model of oxygen transport to tissue." *J. Math. Biol.*, 2002. **44**: pp. 503–522.
13. McGuire, B.J., and Secomb, T.W., "A theoretical model for oxygen transport in skeletal muscle under conditions of high oxygen demand." *J. Appl. Physiol.*, 2001. **91**: pp. 2255–2265.
14. Chow, D.C., Wenning, L.A., Miller, W.M., and Papoutsakis, E.T., "Modeling pO_2 distributions in the bone marrow hematopoietic compartment. II. Modified Kroghian models." *Biophys. J.*, 2001. **81**: pp. 685–696.
15. Papadopoulos, S., Endeward, W., Revesz-Walker, B., Jurgens, K.D., and Gros, G., "Radial and longitudinal diffusion of myoglobin in single living heart and skeletal muscle cells." *Proc. Natl. Acad. Sci.*, 2001. **98**: pp. 5904–5909.
16. Vadapalli, A., Pittman, R.N., and Popel, A.S., "Estimating oxygen transport resistance of the microvascular wall." *Am. J. Physiol.*, 2000. **279**: pp. 657–671.
17. Ballou, D.P., Zhao, Y., Brandish, P.E., and Marletta, M.A., "Revisiting the kinetics of nitric oxide (NO) binding to soluble guanylate cyclase: The simple NO-binding model is incorrect." *Proc. Natl. Acad. Sci.*, 2002. **99**: pp. 12097–12101.
18. Condorelli, P., and George, S.C., "In vivo control of soluble guanylate cyclase activation by nitric oxide: a kinetic analysis." *Biophys. J.*, 2001. **80**: pp. 2110–2119.
19. Ford, P.C., Wink, D.A., and Stanbury, D.M., "Autoxidation kinetics of aqueous nitric oxide." *FEBS Lett.*, 1993. **326**: pp. 1–3.
20. Buerk, D.G., "Can we model nitric oxide biotransport? A survey of mathematical models for a simple diatomic molecule with surprisingly complex biological activities." *Ann. Rev. Biomed. Eng.*, 2001. **3**: pp. 109–143.
21. Joshi, M.S., Ferguson, T.B., Han, T.H., Hyduke, D.R., Liao, J.C., Rassaf, T., Bryan, N., Feelisch, M., and Lancaster, J.R., "Nitric oxide is consumed, rather than conserved, by reaction with oxyhemoglobin under physiological conditions." *Proc. Natl. Acad. Sci.*, 2002. **99**: pp. 10341–10346.
22. Vaughn, M.W., Kuo, L., and Liao, J.C., "Effective diffusion distances of nitric oxide in the microcirculation." *Am. J. Physiol.*, 1998. **274**: pp. H1705–H1714.
23. Vaughn, M.W., Kuo, L., and Liao, J.C., "Estimation of nitric oxide production and reaction rates in tissue by use of a mathematical model." *Am. J. Physiol.*, 1998. **274**: pp. H2163–H2176.
24. Hildebrand, F.B., *Advanced Calculus for Applications*. 2d ed. 1976, Englewood Cliffs, NJ: Prentice-Hall.

Transport in the Kidneys

CHAPTER 14

14.1 Introduction

The kidney is a vital organ with three main functions: excretion, regulation, and secretion [1–3]. More specifically, the kidneys are involved in the regulation of the osmolality, volume, pH, and electrolyte balance in the body fluid. The kidneys regulate the amount of excretion of water, metabolic products, and foreign substances. In addition, the kidneys are endocrine organs that secrete hormones and enzymes important for bone formation, the regulation of blood pressure, and erythrocyte production in the bone marrow. These functions are closely linked to each other, making the kidneys critical in maintaining the stability of the chemical and physical environment in the body.

The functions of the kidneys can be affected by poisons, trauma, diseases in other organs (e.g., diabetes and hypertension), or genetic disorders in the kidneys themselves (e.g., the growth of cysts). Most kidney diseases damage the nephrons, which are the functional elements of the organ. Damage can occur quickly if it is caused by poison or trauma. In this case, however, kidney damage is reversible, provided that it is not severe. By contrast, other diseases may damage the nephrons slowly and silently over a number of years, causing the damaged nephrons to permanently lose their filtering capacity. The permanent damage is not life threatening if it occurs in only a small portion of the kidneys, and even if a large fraction of the kidneys is permanently damaged, patients can still stay alive through kidney dialysis or transplantation.

Kidney transplantation depends on the availability of donors, which can be either cadavers or healthy individuals. So far, the number of cadaver donors is extremely low. Thus, more kidneys from healthy donors are transplanted into patients, although the long-term health of the living donors is still a major concern. The donor availability problem may be mitigated in the future with advances in stem cell biology and tissue engineering.

Kidney dialysis has been used routinely to restore some of the kidney functions lost in patients with end-stage renal disease. Dialysis is an example of the major contributions of biomedical engineers to medicine. Kidney dialysis has two forms: hemodialysis and peritoneal dialysis. The former removes metabolic products from the body by circulating the blood through an external machine called an *artificial kidney*. (Details regarding the artificial kidney are discussed in Section 7.9.) Peritoneal dialysis removes metabolic products from the body by the direct infusion of a fluid, called *dialysate*, into the peritoneal cavity. The dialysate in the cavity is changed every few hours, often by the patients themselves.

Transport study in the kidneys is important in improving the design of kidney dialyzers. In addition, it is important in gaining an understanding of the pharmacokinetics of therapeutic agents in treating other diseases, since most drugs and their metabolites are removed from the body through the kidneys. Therefore, the quantitative analysis of molecular transport in the kidneys has a broad impact on medicine. In this chapter, we first review the mechanisms of transmembrane transport that are important in understanding renal reabsorption; we then discuss kidney physiology and introduce mathematical models of fluid and solute transport in glomerulus and renal tubules.

14.2 Mechanisms of Transmembrane Transport

As introduced in Chapter 1, transport across the plasma membrane of cells involves four fundamental mechanisms: direct diffusion, facilitated transport, active transport, and endocytosis. Here, we focus on the first three mechanisms. (Endocytosis is the transport mechanism mainly for large molecules and is not directly relevant to the key functions of the kidneys. The quantitative analysis of endocytosis is discussed in Chapter 11.)

14.2.1 Direct Diffusion

In the kidneys, the renal tubule wall consists of a thin layer of epithelial cells and the basal lamina. Pathways for transport across the epithelial layer can be either transcellular or paracellular. Because epithelial cells are connected by tight junctions, passive diffusion and convection of small solutes through the paracellular pathway are much slower than that across the walls of most blood capillaries. The direct diffusion of solutes through the plasma membrane of epithelial cells is possible, but the permeability of solutes across the membrane depends on the size and structure of the solutes. The permeability is high for small gas molecules (e.g., O_2 and CO_2) and low for uncharged polar molecules (e.g., H_2O, urea, and ethanol). When the molecular weight of uncharged polar molecules is increased, permeability decreases rapidly. For example, the permeability of glucose (180 daltons) is negligible compared with that of urea (64 daltons). The plasma membrane of cells is nearly impermeable to charged molecules because (a) the partition coefficient Φ of charged molecules between the membrane and solutions is extremely low and (b) the permeability coefficient P of any solute is proportional to the partition coefficient (see Problem 14.1); that is,

$$P = \frac{\Phi D_m}{\ell}, \tag{14.2.1}$$

where D_m is the diffusion coefficient of the solute in the membrane and l is the thickness of the membrane.

The partition coefficient of a solute in the membrane can be estimated on the basis of thermodynamic analysis. Transport between solution and membrane is an isothermal and isobaric process (i.e., temperature and pressure are maintained at constant levels). At equilibrium, the chemical potential of the solute in the solution, μ_s, should be the same as that in the membrane, μ_m:

$$\mu_s = \mu_m. \tag{14.2.2}$$

In general, the chemical potential of a solute at pressure p_r and temperature T_r is related to the activity a of the solute (see Section 6.4.2) [4,5],

$$\mu_j = \mu^0 + \mathrm{R}T_r \ln a_j, \tag{14.2.3}$$

where the subscript j indicates either s or m, R is the gas constant, μ^0 is the chemical potential of pure solute at a hypothetical reference state in which a_j is equal to unity, and the pressure and temperature are also maintained at p_r and T_r, respectively. For both neutral and charged solutes, the activity is approximately proportional to the mole fraction of the solutes, x_j, if x_j approaches zero or the mole fraction of solvent approaches unity (i.e., the Henry's law) [6]. The proportionality constant is denoted by f_j. In that case, solute–solute interactions are negligible, and Equation (14.2.3) becomes

$$\mu_j = \mu_j^c + RT_r \ln C_j, \tag{14.2.4}$$

with the chemical potential at the standard state (i.e., $C_j = 1$) defined by

$$\mu_j^c = \mu^0 + RT_r \ln(f_j) - RT_r \ln(C_j)_t \tag{14.2.5}$$

where $(C_j)_t$ is the total molar concentration in s or m. (Note that $C_j = x_j(C_j)_t$ and all molar concentrations are non-dimensionalized by 1M). Substituting Equation (14.2.4) into Equation (14.2.2), we have

$$\Phi \equiv \frac{C_m}{C_s} = \exp\left(-\frac{\mu_m^c - \mu_s^c}{RT_r}\right). \tag{14.2.6}$$

The partition coefficient can also be expressed in terms of the change in the partial molar Gibbs free energy at the standard state since the chemical potential μ_j of a solute in solution j is defined as the solution's partial molar Gibbs free energy, $\overline{G}_j$, or

$$\mu_j = \overline{G}_j = \left(\frac{\partial G_j}{\partial n}\right)_{T_r, p_r, n_i}, \tag{14.2.7}$$

where n is the number of moles of that particular solute in the solution, n_i represents the individual numbers of moles of all other solutes and solvent in the solution, and G_j is the total Gibbs free energy in the solution. Therefore, we have

$$\mu_j^c = \overline{G}_j^c \tag{14.2.8}$$

and

$$\mu_m^c - \mu_s^c = \overline{G}_m^c - \overline{G}_s^c = \Delta\overline{G}_{ms}^c. \tag{14.2.9}$$

$\Delta\overline{G}_{ms}^c$ can be divided into three parts: the electrostatic Gibbs energy change, $\Delta\overline{G}_E^c$; the hydrophobic Gibbs energy change, $\Delta\overline{G}_H^c$; and the Gibbs energy required to break hydrogen bonds between the polar molecules and water, $\Delta\overline{G}_{HB}^c$ [7] (i.e., $\Delta\overline{G}_{ms}^c = \Delta\overline{G}_E^c + \Delta\overline{G}_H^c + \Delta\overline{G}_{HB}^c$). For uncharged molecules (e.g., urea and glycerol), $\Delta\overline{G}_{ms}^c \approx \Delta\overline{G}_H^c + \Delta\overline{G}_{HB}^c$, where $\Delta\overline{G}_{HB}^c$ is approximately 63 kilojoules per mole (kJ mol^{-1}) for the breakage of hydrogen bonds between water molecules and a urea or glycerol molecule [7]. By contrast, $\Delta\overline{G}_E^c$ is the dominant term for small ions. In this case,

$$\Delta\overline{G}_{ms}^c \approx \Delta\overline{G}_E^c. \tag{14.2.10}$$

$\Delta\overline{G}_E^c$ can be estimated with the *Born model* [8]. The Born model assumes that the interactions between ion and solvent is solely electrostatic and that the solvent is a continuous medium, or continuum. In a continuum, the electric energy stored in a sphere is

$$e = \frac{q^2}{8\pi\varepsilon r}, \tag{14.2.11}$$

where ε is the dielectric permittivity constant of the continuum, and r and q are the radius and the total charge of the sphere, respectively. The derivation of Equation (14.2.11) can be found in most college physics textbooks. If both the solution and the membrane are considered as continuous media, the amount of work w required to move an ion from the solution into the membrane is equal to the change in the electric energy in the sphere, $\Delta e = e_m - e_s$; that is,

$$w = \Delta e = \frac{q^2}{8\pi r}\left(\frac{1}{\varepsilon_m} - \frac{1}{\varepsilon_s}\right) = \frac{q^2}{8\pi r\varepsilon_0}\left(\frac{1}{\kappa_m} - \frac{1}{\kappa_s}\right), \tag{14.2.12}$$

where ε_0 is the dielectric permittivity constant in the vacuum, and κ_m and κ_s are the dielectric constants in the membrane and solution, respectively. To move a mole of ions from the solution into the membrane, the amount of work required is

$$W = N_A w, \tag{14.2.13}$$

where N_A is Avogadro's number. If the movement of ions between the solution and the membrane is reversible (i.e., infinitesimally slow), isothermal, and isobaric, then the work performed on the ions is equal to the change in the molar Gibbs free energy of the ions. The latter is approximately equal to difference in the partial molar Gibbs free energies between the two phases at the standard state, i.e., $\Delta\overline{G}_E^c$, if changes in the total molar concentration (see Equation 14.2.5) in the membrane and solution, induced by the movement of ions, are negligible (since the mole fractions of the ions in both the membrane and solution approach zero). Taken together, Equations (14.2.10), (14.2.12), and (14.2.13) give

$$\Delta\overline{G}_{ms}^c \approx W = \frac{q^2 N_A}{8\pi r\varepsilon_0}\left(\frac{1}{\kappa_m} - \frac{1}{\kappa_s}\right) \tag{14.2.14}$$

Substituting Equations (14.2.9) and (14.2.14) into Equation (14.2.6), we obtain

$$\Phi = \exp\left[-\frac{1}{k_B T_r}\frac{q^2}{8\pi r \varepsilon_0}\left(\frac{1}{\kappa_m} - \frac{1}{\kappa_s}\right)\right], \quad (14.2.15)$$

where $k_B = R/N_A$ is the Boltzmann constant.

The Born model has been used to predict changes in Gibbs free energy and enthalpy at the standard state during ion–solvent interactions. However, the predicted values are always higher than the experimental data if r in the Born model is the crystallographic radius of ions. One reason for the over prediction is that the disturbances produced by ions on the dielectric properties of the membrane and solutions are neglected [8]. Despite the overestimation in Gibbs free-energy changes, the Born model is simple, and it is able to characterize some important features in ion partitions between different phases. For example, $\Delta\overline{G}^c_{ms}$ decreases with the size of ions and the dielectric permittivity constant of the membrane, but increases with the charge on ions (see Equation (14.2.14)).

Example 14.1 The dielectric constant of water is 78. Assume that the dielectric constant of a cell membrane is 2, the membrane thickness is 4 nm, and the diffusion coefficient of K^+ in the membrane is 2×10^{-5} cm^2 s^{-1}. The radius of K^+ is 0.149 nm.

(a) Determine the Gibbs free-energy difference and the partition coefficient of potassium ion between the plasma membrane and water at 30° C.

(b) Estimate the permeability coefficient of potassium ion across the cell membrane.

Solution The valence of K^+ is +1; thus, the charge is

$$q = 1 \times (1.602 \times 10^{-19}\text{ C}) = 1.602 \times 10^{-19}\text{ C}. \quad (14.2.16)$$

In addition, $k_B = 1.38 \times 10^{-23}$ J K^{-1}, $T = 273 + 30 = 303$°K, and $\varepsilon_0 = 8.9 \times 10^{-12}$ C^2 N^{-1}m^{-2}. Substituting these values into Equations (14.2.14) and (14.2.15), we have

$$\Delta\overline{G}^c_{ms} = \frac{6.023 \times 10^{23} \times (1.602 \times 10^{-19})^2 \times (1/2 - 1/78)}{8 \times 3.14 \times 0.149 \times 10^{-9} \times 8.9 \times 10^{-12} \times 1{,}000} = 226\text{ kJ mol}^{-1} \quad (14.2.17)$$

and

$$\Phi = \exp\left(-\frac{1.602^2 \times 10^{-38} \times (1/2 - 1/78)}{8 \times 3.14 \times 0.149 \times 10^{-9} \times 8.9 \times 10^{-12} \times 1.38 \times 10^{-23} \times 303}\right) = 1.04 \times 10^{-39} \quad (14.2.18)$$

indicating that the concentration of K^+ in the membrane is 39 orders of magnitude smaller than that in the water. Substituting the value of Φ into Equation (14.2.1), we obtain

$$P = 1.04 \times 10^{-39} \times 2 \times 10^{-5}/(4 \times 10^{-7}) = 5.19 \times 10^{-38}\text{ cm s}^{-1}. \quad (14.2.19)$$

The measured permeability of K^+ across a phosphatidylcholine (a type of phospholipid) membrane approximately 4 nm thick is 1.7×10^{-12} cm s^{-1} [7], which is much larger than the value predicted from Equation (14.2.19). The difference can be caused by several factors. First, the Born model overestimates $\Delta\overline{G}^{c}_{ms}$. Second, transport across the membrane may occur simultaneously through two different pathways. One is the hydrocarbon phase discussed previously. Another is the transient hydrophilic pores formed by thermal fluctuations of lipid molecules in the membrane [7]. The Gibbs free-energy change that is required to use the latter pathway can be several times smaller than that predicted by Equation (14.2.17). The mathematical modeling of ion transport through transient hydrophilic pores is beyond the scope of this book [7].

The Gibbs free-energy change required to transfer nonpolar molecules into the membrane is much smaller than that for ions. For example, the Gibbs free-energy change required to transfer a hydroxyl group is on the order of 20 kJ mol^{-1}. Substituting that quantity into Equation (14.2.6) and assuming that $T_r = 303$ K, we get $\Phi = 3.56 \times 10^{-4}$. Thus, the permeability coefficient of the hydroxyl group across the same membrane as that in Example 14.1 is 1.78×10^{-2} cm s^{-1}, which is 35 orders of magnitude larger than the permeability coefficient of K^+.

The discussion in this section indicates that direct diffusion through both transcellular and paracellular pathways is inefficient for the reabsorption of most solutes. Therefore, the reabsorption of these solutes in the kidneys must be mediated by *facilitated transport* and *active transport*. Both types of transport rely on *transporters*, which can be classified into two categories: *channels* (e.g., ion channels and water channels) and *carriers* (e.g., glucose carriers and the Na^+/K^+ pump) [9]. The carriers can be further divided into passive and active ones. Facilitated transport involves either channels or passive carriers, whereas active transport relies on active carriers. Most transporters are integral membrane proteins, although they may also be formed by some antibiotics. For example, gramicidin can form ion channels in the membrane that cause a loss of ions and thus the death of bacterial cells.

14.2.2 Facilitated Transport

Facilitated transport across the plasma membrane of tubular epithelial cells is similar to that in other cells. It occurs through either channels or passive carriers. To determine experimentally which transporters are involved in the transmembrane transport of a given solute, the rate and sensitivity of transport to membrane states are quantified in most studies. If the experimental data show that the rate of transport is about 10 million solute molecules per transporter per second or that the rate of transport is insensitive to membrane states, then it is likely that the transport is through channels. In contrast, carrier-mediated transport is suggested if the rate of transport is less than 1,000 molecules per transporter per second or the transport is sensitive to membrane states.

Facilitated transport is often called facilitated diffusion in the literature, but it is different from the classical concept of diffusion because some facilitated transport involves only chemical reactions of solutes with transporters and conformational changes of transporters. In these cases, solutes do not diffuse across the membrane during facilitated transport. The only thing in common between facilitated transport and diffusion is that they both depend on the concentration gradient of solutes and they always occur in the direction opposite that gradient.

Channels. *Channels* are transmembrane structures with a hydrophilic pore in the middle. More than a hundred channel structures have been identified in different cells. Transport through channels is driven by the concentration gradient, but it also depends on the electric potential difference across the membrane if the solutes contain charged groups. Compared with direct diffusion and ion migration through a uniform membrane, transport through channels has several unique features. First, channels are gated, or regulated, in most cases. Gating results from transitions of the channel structures between discrete conformational states. The transitions can be achieved through changing the transmembrane voltage, through the binding of ligands to either intracellular or extracellular regions, or through mechanical stretching of the membrane (see Figure 14.1). At the cellular level, gating can be modeled by kinetic equations; at the single-channel level, it has to be considered as a stochastic process [10].

Second, channels are selectively permeable to different molecules. For example, the Na channel is over 10 times more permeable to Na^+ than to K^+, whereas the K channel is over 100 times more permeable to K^+ than to Na^+. Mechanisms of selectivity are complicated and are related to the molecular properties of the solutes and the channels (e.g., size, charge, and configuration), to binding between solutes and the channel, to hydration and dehydration of solutes in the channel, and to solute-induced changes in channel structures [10]. The hydration–dehydration mechanism is unique for ion transport through channels, because unhydrated ions never exist in solutions. Instead, they are always surrounded by water molecules. Therefore, a hydrated ion will be able to enter a channel only if it is smaller than the channel or if the water molecules that are bound to the ion can be stripped off through channel–ion interactions. The dehydration mechanism partly explains why the K channel has a very low permeability to Na^+, although the unhydrated Na^+ (d = 0.27 nm) is smaller than the pore size in the K channel (d = 0.33 nm).

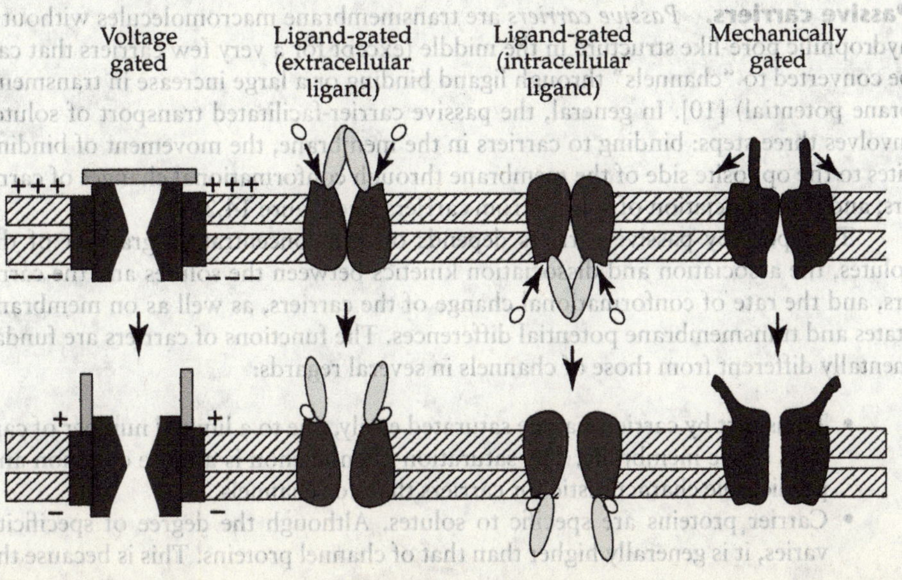

FIGURE 14.1 Regulation of channel opening and closing through different gating mechanisms. (Modified from Ref. [11], with permission.)

Data in the literature demonstrate that hydrated Na^+ and K^+ are both larger than the pore in the K channel. Thus, neither of them can pass the channel without dehydration. The selectivity filter pocket in the K channel is lined with oxygen atoms from carbonyl groups in surrounding amino acids [12]. The oxygen atoms can mimic water molecules that embrace K^+ ion in solutions, thus facilitating K^+ transfer from solution into the K channel. However, the oxygen atoms cannot embrace Na^+ ion in solutions, because Na^+ does not fit into the filter pocket. As a result, Na^+ dehydration is difficult, which in turn renders it difficult for Na^+ to enter the K channel. Some large channels in the plasma membrane of cells are less selective to ions. For example, acetylcholine receptors (AChR) are cation channels that are gated through binding with acetylcholine. They are nonselective to small metal ions, because their size is at least 0.65 nm. However, acetylcholine receptors are much less permeable to nonelectrolytes and nearly impermeable to anions.

Third, the electric conductance of channels can be saturated at a higher concentration gradient of ions or can be significantly reduced by other ions in some channels. The transport of ions through these channels is different from free diffusion in solutions, involving not only diffusion, but also transient binding and collision between ions and channels. When an ion binds to the inner wall of a narrow channel, the ion will either reduce or block the pathway of other ions behind it, due to steric exclusion (see Section 8.2). Electric repulsion will also play a role if the ions and the channel wall have the same sign of electrical charge. When the ion dissociates from the channel, the pathway becomes available for other ions. If the characteristic time of dissociation is much shorter than that of free diffusion through the channel, the effect of binding on ion transport should be small. The ratio of the characteristic time of free diffusion to the characteristic time of dissociation decreases with the increase in the concentration gradient of ions. At a very high concentration gradient, ion transport through channels is a binding-limited process. In this case, the flux of ions is independent of the concentration gradient and is reflected at the macroscopic level by the saturation of electric conductance of ion channels.

Passive carriers. *Passive carriers* are transmembrane macromolecules without a hydrophilic pore-like structure in the middle (except for a very few carriers that can be converted to "channels" through ligand binding or a large increase in transmembrane potential) [10]. In general, the passive carrier-facilitated transport of solutes involves three steps: binding to carriers in the membrane, the movement of binding sites to the opposite side of the membrane through conformational changes of carriers, and the dissociation of solutes from carriers (see Figure 14.2).

Transport by passive carriers depends on the concentration gradient of the solutes, the association and dissociation kinetics between the solutes and the carriers, and the rate of conformational change of the carriers, as well as on membrane states and transmembrane potential differences. The functions of carriers are fundamentally different from those of channels in several regards:

- Transport by carriers can be saturated easily, due to a limited number of carriers in the membrane. The saturation phenomenon is a more common and predictable characteristic for carriers than for channels.
- Carrier proteins are specific to solutes. Although the degree of specificity varies, it is generally higher than that of channel proteins. This is because the

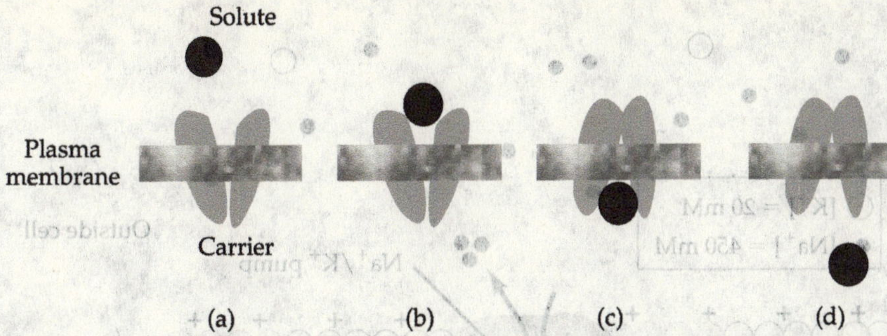

FIGURE 14.2 Passive-carrier-facilitated transport. (a) Initially, the solute is in the solution and a free carrier is available for transmembrane transport. (b) The first step in the transport is the binding of the solute to the carrier. (c) The second step is the conformational change of the carrier that moves the solute across the membrane. (d) The last step is the dissociation of the solute from the carrier.

carrier specificity is determined by both the geometrical and the physical complementarities between solutes and carriers, instead of by the molecular-sieving effect discussed above.

- Carriers can facilitate the transport of relatively large molecules (e.g., glucose and amino acids) across the membrane, whereas channels allow only small molecules to pass through.
- Transport via carriers is, in general, several orders of magnitude slower than transport via channels, since it depends on carrier protein rather than solute movement in the membrane.
- Carrier-mediated transport is more sensitive to changes in membrane structures than is transport through channels.
- Transmembrane transport via passive carriers is often asymmetric, due presumably to an asymmetric distribution of carrier proteins in the membrane and the difference in free energy of these proteins among distinct conformations. Thus, solute transport from one side to another of the membrane may involve less change in free energy than transport in the opposite direction.
- Carrier-facilitated transport can be inhibited by two different types of mechanisms [13]:
  - *Noncompetitive inhibition*, in which the inhibition is caused by the binding of inhibitors to either free carriers or carrier–solute complexes. In some rare cases, the inhibition depends only on the binding of inhibitors to solute–carrier complexes (called *uncompetitive inhibition* in the literature). In all of these cases, an increase in solute concentration has minimal effects on inhibitor functions.
  - *Competitive inhibition*, in which the inhibitors compete with solutes for the same binding sites in free carriers. This type of inhibition can be reversed by increasing the concentration of solutes.

Carrier-mediated transport can be classified into three different categories: *uniport*, *symport* (or *cotransport*), and *antiport* (or *exchange*). *Uniport* is defined as the

FIGURE 14.3 A sketch of the active transport of Na^+ and K^+ across the plasma membrane of cells through the Na^+/K^+ pump (or ATPase). (Modified from Figure 12.20 in Ref. [14], with permission.)

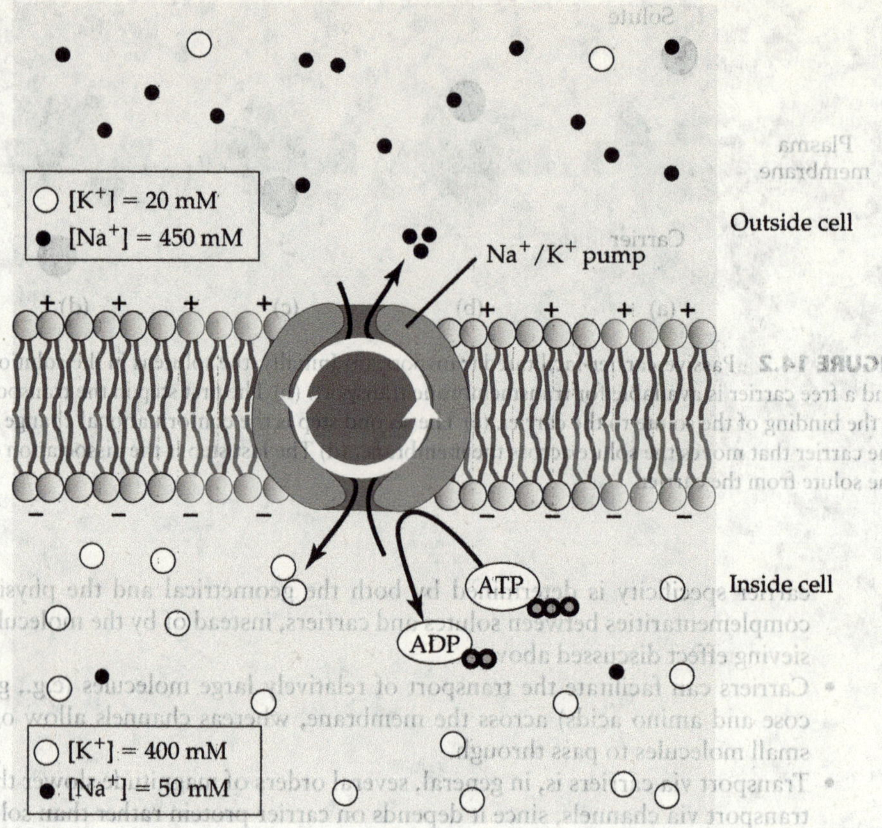

transport of one substrate (i.e., the molecules that can be transported) at a time through the membrane. *Symport* refers to the cotransport of two or more substrates through a single carrier in the same direction. If the transport direction of substrates differs, the transport phenomenon is called *antiport*. On the basis of this classification, the carriers that are responsible for uniport, symport, and antiport are defined as *uniporter*, *symporter* (or *cotransporter*), and *antiporter* (or *exchanger*), respectively. Comparing the three different transport phenomena mediated by passive carriers, we find that uniport is less common than symport or antiport. Uniport has been observed in the facilitated transport of glucose across the plasma membrane of some non-epithelial cells (e.g., red blood cells). Examples of symport and antiport will be discussed later in the chapter. Carriers are not limited to facilitating passive transport; active transport also involves carriers, as will be discussed in the next section.

14.2.3 Active Transport

Active transport refers to energy-dependent transport across the membrane. The energy is, in most cases, generated from chemical reactions, such as hydrolysis of the pyrophosphate bond (e.g., in adenosine triphosphate (ATP)), decarboxylation, methyl transfer, or oxidation–reduction reactions. There also exists light- or mechanical-driven transport [9]. Active transport can be further divided into two subcategories—*primary* and *secondary active transport*—depending on how energy is consumed during the transport. Primary active transport consumes energy directly. Secondary active transport is driven by the concentration gradients of other solutes that are established through the primary active transport; thus, it depends on energy indirectly. In comparison with passive diffusion and facilitated transport, one of the main features of active transport is that it can move solutes against the concentration gradient.

The most well-known mechanisms of primary active transport are related to ion pumps driven by ATP hydrolysis. Examples of these pumps are the Na^+/K^+ and Ca^{2+} pumps. In this chapter, we will focus on the Na^+/K^+ pump, which is found in the plasma membrane of all animal cells. Chemically, this pump also functions as an enzyme for hydrolyzing ATP, so it is also called Na^+/K^+ ATPase. In accordance with the classification of carriers discussed, an Na^+/K^+ pump is an active antiporter or exchanger (see Figure 14.3), simultaneously pumping out three Na^+ ions and transferring two K^+ ions into the cell. Meanwhile, an ATP molecule is hydrolyzed into an ADP and an orthophosphate to provide the chemical energy for the transport.

14.3 Renal Physiology

The basic functional unit of the kidney is the *nephron*, of which each human kidney contains about 10^6. A nephron consists of two portions: a *renal corpuscle* and a *renal tubule* (see Figure 14.4). The renal corpuscle is formed by a *glomerulus* (a group of blood capillary loops) and the surrounding double-layered epithelial cup called *Bowman's capsule*. The renal tubule has three portions: a *proximal convoluted tubule*, *Henle's loop*, and a *distal convoluted tubule*. The end of the distal convoluted tubule is connected to a serial structure that consists of a collecting duct, papillary ducts, the minor calyx, the major calyx, the renal pelvis, and the ureter. Transport phenomena in the kidneys are related to two major aspects of renal physiology: blood flow and urine formation.

14.3.1 Renal Blood Flow

The blood-flow rate into the kidney is about 1 L min^{-1}, accounting for about 20%–25% of the total cardiac output, although the kidney is less than 0.5% of the total body weight. Renal blood flow plays several important roles in physiological functions of the kidney. First, it delivers nutrients, hormones, and specific molecules to the kidneys for renal excretion and returns metabolic wastes and absorbed molecules to the systemic circulation. Second, it determines the glomerular filtration rate (GFR) by the microvascular pressure in glomerular capillaries. The microvascular pressure also affects the rate at which filtrate is reabsorbed through the peritubular capillaries. The filtration and reabsorption processes determine the amount of urine and the concentrations of major components in the urine.

Renal blood flow is autoregulated; that is, it is locally regulated. When the body's systemic blood pressure is increased from 100 to 200 mmHg, the total blood-flow rate in the kidney, Q, and GFR are nearly independent of the pressure (see Figure 14.5). The mechanisms of regulation may involve (a) the alternating contraction

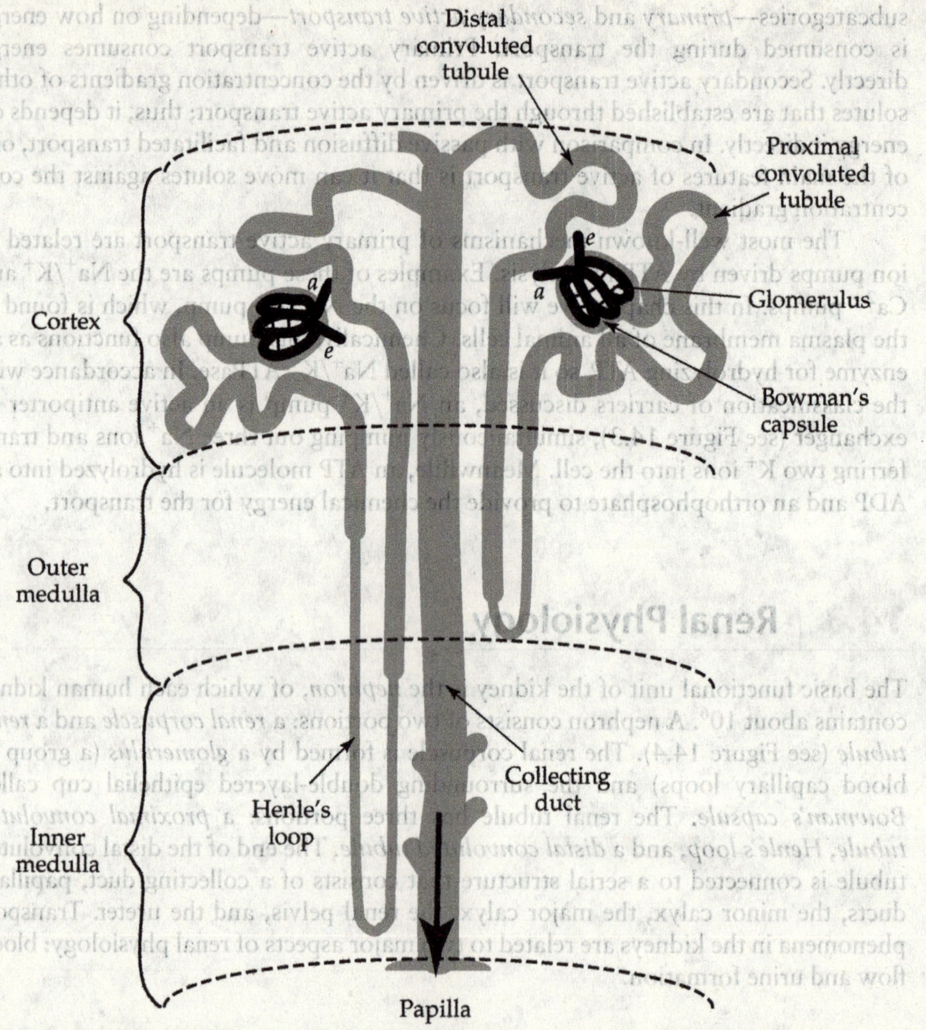

FIGURE 14.4 Schematic of a nephron. The letter *a* indicates afferent arterioles from which the blood enters the glomerulus, and the letter *e* indicates efferent arterioles that collect the blood from the glomerulus. Plasma is filtered in the glomeruli. The filtrate moves from the Bowman's capsules to the collecting ducts through the proximal convoluted tubules, the Henle's loops, and the distal convoluted tubules. (Modified from Figure 19.3 in Ref. [15], with permission.)

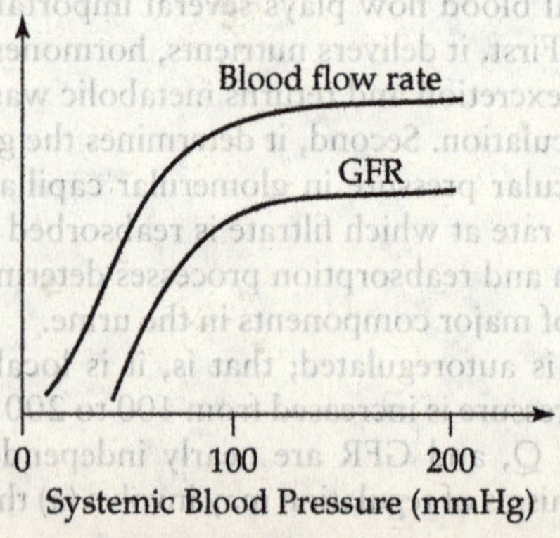

FIGURE 14.5 A schematic of the total blood flow rate in the kidneys and GFR as a function of the systemic blood pressure.

and relaxation of smooth muscle cells (SMCs) in the afferent arterioles (see Figure 14.4) and (b) tubuloglomerular feedback [2]. SMCs in the efferent arterioles (see Figure 14.4) may also contribute to the autoregulation of renal blood flow, but their contributions are likely to be small, for the following reasons.

The rate of blood flow, Q, depends on the ratio of the body's systemic pressure and the flow resistance R in the kidney. From Poiseuille's law (see Chapter 2), R in both afferent and efferent arterioles is proportional to the fourth power of the vessel diameter. Thus, R is highly sensitive to the contraction or dilation of the arterioles. As has been mentioned, the blood-flow rate does not change significantly when the systemic blood pressure increases from 100 to 200 mmHg (see Figure 14.5). This constancy indicates that the total R must be doubled in order to maintain the same flow rate. The change in the total R is the sum of the R's from the afferent and efferent arterioles. If the increase in total R were due mainly to the contraction of the efferent arteriole, then the microvascular pressure in the glomerular capillaries should also have been increased, because blood first passes these capillaries before entering the efferent arterioles. As a result, the GFR would have been increased as well. Recall, however, that the GFR also is not sensitive to the body's systemic pressure in the same pressure range as the blood-flow rate (see Figure 14.5). This contradiction demonstrates indirectly that efferent arterioles play a minimal role in regulating the renal blood flow when the systemic pressure is increased. Efferent arterioles may help to maintain the microvascular pressure in the glomerular capillaries when the systemic pressure is reduced from 100 mmHg to a lower level, but Figure 14.5 shows that both the blood flow rate in the kidneys and the GFR are decreased when the systemic pressure decreases. All these results together suggest that efferent arterioles contribute only minimally to the regulation of renal blood flow and GFR.

14.3.2 Urine Formation

Urine is formed primarily in the nephrons. Three processes are involved: *filtration*, *reabsorption*, and *secretion*.

Filtration. Filtration, in general, means the transport of solutions through a filter. The filter allows water and small solutes to pass through and blocks solutes larger than the size of the pores. In the kidney, the glomerulus is the filter. When blood flows through the glomerulus, water and small solutes in the plasma filtrate through the capillary walls to enter Bowman's capsule. Solutes can also enter Bowman's capsule by diffusion, but diffusion is a much slower process than filtration. Filtration is driven by the pressure difference across the microvessel wall. A solute with a molecular weight of less than 7,000 can move freely with water through the microvessel wall. However, large plasma proteins (e.g., albumin) are retained in the blood. The cutoff size of pores in the glomerulus is approximately 5.5 nm in diameter (see Figure 1.20), which is about the size of proteins with molecular weight of 70,000. Many kidney diseases increase the microvascular permeability of large proteins in the glomerulus, causing proteinuria.

The glomerular capillary wall consists of three layers: the *fenestrated endothelium*, the *basement membrane*, and the *epithelial cells* (*podocytes*), which have footlike structures called *pedicels* (see Figure 14.6). The podocytes form one of the two

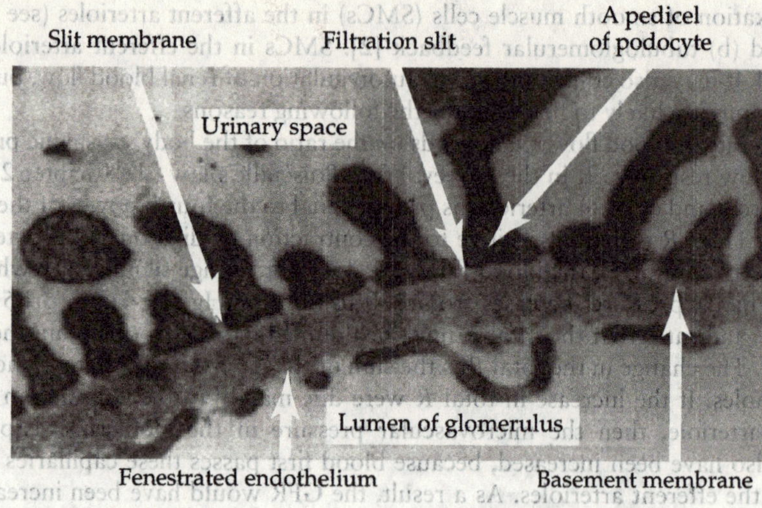

FIGURE 14.6 The structures of glomerular capillary wall. (Modified from Ref. [16], with permission of John Wiley & Sons, Inc.)

epithelial layers in the Bowman's capsule. The spaces between pedicels are called *filtration slits* or *slit pores*; they are covered by *slit membranes*.

Reabsorption. About 99% of the filtrate that passes through the glomeruli returns to the blood through *reabsorption* in the renal tubules (see Tables 1.11 and 14.1). Reabsorption determines the volume and composition of urine. The rate of filtration is nearly constant in normal adults (about 180 L, or 60 times the plasma volume, per day); it is the rate of reabsorption that accounts for variations in urine formation.

TABLE 14.1

Fraction of Major Substances that Are Reabsorbed in Renal Tubules

Molecules/ions	Fraction reabsorbed[a] (%)
Water	99
Na^+	99
HCO_3^-	99.9
K^+	86
Glucose	100
Ca^{2+}	98
Cl^-	99
Urea	50

Source: Modified from Ref. [1], with permission of Elsevier.

[a]The data are expressed as the percentage of filtered substances reabsorbed in the renal tubules. The reabsorption rates of the same substances are listed in Table 1.11.

More importantly, reabsorption controls the volume, osmolality, composition, and pH of the body fluid in both intracellular and extracellular spaces [1,2]. Most filtered water and solutes, including glucose, amino acids, urea, and ions (e.g., Na^+, K^+, Mg^{2+}, and Ca^{2+}), are reabsorbed in the proximal convoluted tubules of the nephrons. A small percentage of the total reabsorption occurs in Henle's loops and the distal convoluted tubules of the nephrons. Although the percentage is low, reabsorption in Henle's loops and the distal convoluted tubules is important for the delicate balances in the volume and composition of body fluid.

The mechanisms of reabsorption in the kidney are much more complicated than those of filtration and involve both passive and active transport of molecules across the epithelium of renal tubules and the endothelium of peritubular capillaries. The passive transport of solutes is spontaneous and driven by chemical and electrical potential gradients. As in filtration, the passive transport of solutes can occur by convection or by passive diffusion. For example, Mg^{2+} reabsorption is largely the result of passive diffusion through paracellular pathways. Unlike filtration, the passive transport of most solutes occurs as facilitated transport (see Section 14.2.2). In addition, both primary and secondary active transport occur in renal reabsorption (see Section 14.2.3). Although a complete review of these mechanisms is beyond the scope of the text, this section provides a brief discussion of the mechanisms of renal reabsorption of Na^+, water, glucose, and urea. The mechanisms of reabsorption of other molecules are discussed in textbooks related to renal physiology [1–3].

Na^+ reabsorption Na^+ reabsorption plays a key role in the regulation of the extracellular fluid volume in the body. Na^+ is usually transported together with Cl^- in the kidney. Mechanisms of Na^+ reabsorption involve passive diffusion through epithelial junctions and intracellular space, as well as facilitated transport and active transport across the plasma membrane of epithelial cells. The involvement of these mechanisms depends on the location in renal tubules. For Na^+ transport, a renal tubule can be divided into five distinct portions: (a) the early portion of the proximal tubule (see Figure 14.7a), (b) the late portion of the proximal tubule (see Figure 14.7b), (c) Henle's loop (see Figure 14.8), (d) the early portion of the distal tubule (see Figure 14.9a), and (e) the late portion of the distal tubule (see Figure 14.9b). Approximately 67% of filtrated Na^+ is reabsorbed in the proximal tubule, 25% in Henle's loop, and 7% in the distal tubule and the collecting duct. The remaining 1% is excreted with the urine. Na^+ transport from the peritubular interstitial space into the blood capillaries occurs spontaneously through passive diffusion driven by the concentration gradient of Na^+ and through convection driven by the hydrostatic pressure gradient.

In the early portion of the proximal tubule, Na^+ transport through epithelial cells is both facilitated and active. The facilitated transport is mediated by three types of carriers: (1) an Na^+/H^+ antiporter that internalizes Na^+ and removes the H^+ that is generated intracellularly from the chemical reaction of CO_2 and water in epithelial cells; (2) HCO_3^- transporters that remove the HCO_3^-, which is also formed in the reaction, from the epithelial cells; and (3) an Na^+/X symporter, where X can be glucose, inorganic phosphate, amino acids, or lactate. The primary active transport in epithelial cells uses the Na^+/K^+ pump or Na^+/K^+ ATPase to remove Na^+ and internalize K^+ ions.

In the late portion of the proximal tubule, Na^+ transport occurs through both paracellular and transcellular pathways. Paracellular transport takes place by

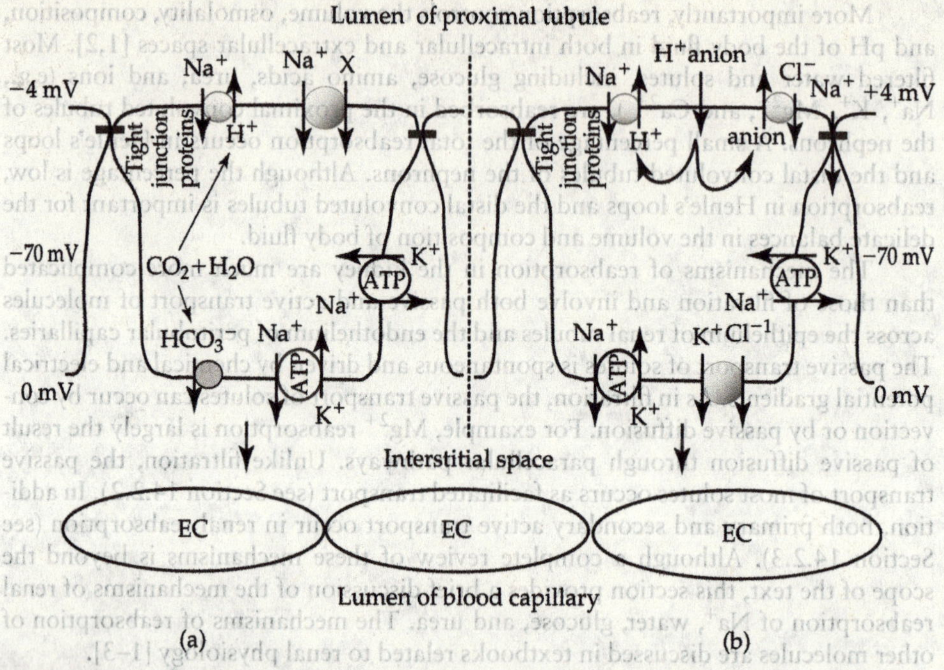

FIGURE 14.7 (a) Na^+ reabsorption in the early portion of the proximal tubule. (b) Na^+ reabsorption in the late portion of the proximal tubule. EC indicates endothelial cells.

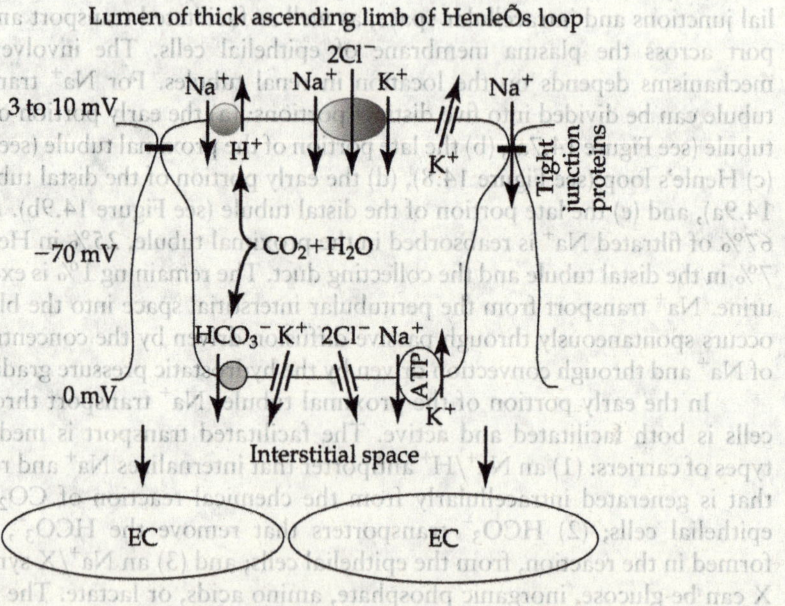

FIGURE 14.8 Na^+ reabsorption in the thick ascending limb of Henle's loop. EC indicates endothelial cells.

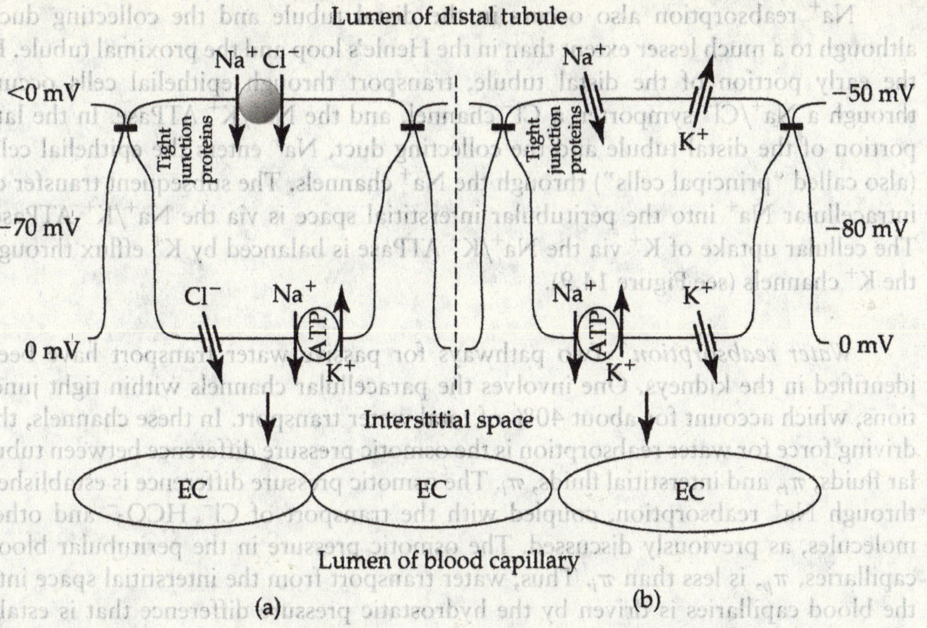

FIGURE 14.9 (a) Na^+ reabsorption in the early portion of the distal tubule. (b) Na^+ reabsorption in the late portion of the distal tubule and the collecting duct. EC indicates endothelial cells.

diffusion and convection; the transport pathway is regulated by tight-junctional proteins (e.g., members of the claudin family). The driving force for convective transport is coupled with the transcellular transport of Na^+, because the force is determined by the osmotic pressure difference, which, in turn, depends on the Na^+ concentration difference across the epithelial layer. The mechanisms of transcellular transport here are slightly different from those in the early portion of the proximal tubule, as shown in Figures 14.7a and 14.7b. The H^+ that is transported through the Na^+/H^+ antiporters is not generated within the cells; it originates outside and enters the epithelial cells with small organic anions (e.g., formate). The anions are then excreted through an anion/Cl^- antiporter. The Cl^- that enters the cells through this antiporter must be excreted across the basal-lateral membrane, via a K^+/Cl^- symporter.

In summary, Na^+ is reabsorbed either alone via direct diffusion and convection or with HCO_3^- and Cl^- as well as with other solutes (e.g., glucose, inorganic phosphate, amino acids, and lactate), in the proximal tubule via both facilitated and active transports.

Na^+ reabsorption in Henle's loop occurs mainly in the thick ascending limb of the loop (see Figure 14.4), through epithelial cell junctions and transcellular pathways. The transcellular transport involves Na^+/H^+ antiporters, $K^+$and Cl^- channels, $Na^+/2Cl^-/K^+$symporters, and Na^+/K^+ ATPase. An $Na^+/2Cl^-/K^+$ symporter simultaneously transfers one Na^+ ion, two Cl^- ions, and one K^+ ion across the plasma membrane (see Figure 14.8). The paracellular transport of Na^+ through epithelial cell junctions accounts for about half of the total Na^+ reabsorption in Henle's loop and is driven by the electric potential gradient (see Figure 14.8). Under normal physiological conditions, 25% of filtrated Na^+ is reabsorbed in the thick ascending limb. However, the percentage can be significantly increased if the proximal tubule absorbs less than two-thirds of the filtrated Na^+.

Na^+ reabsorption also occurs in the distal tubule and the collecting duct, although to a much lesser extent than in the Henle's loop and the proximal tubule. In the early portion of the distal tubule, transport through epithelial cells occurs through a Na^+/Cl^- symporter, a Cl^- channel, and the Na^+/K^+ ATPase. In the late portion of the distal tubule and the collecting duct, Na^+ enters the epithelial cells (also called "principal cells") through the Na^+ channels. The subsequent transfer of intracellular Na^+ into the peritubular interstitial space is via the Na^+/K^+ ATPase. The cellular uptake of K^+ via the Na^+/K^+ ATPase is balanced by K^+ efflux through the K^+ channels (see Figure 14.9).

Water reabsorption Two pathways for passive water transport have been identified in the kidneys. One involves the paracellular channels within tight junctions, which account for about 40% of total water transport. In these channels, the driving force for water reabsorption is the osmotic pressure difference between tubular fluids, π_t, and interstitial fluids, π_i. The osmotic pressure difference is established through Na^+ reabsorption, coupled with the transport of Cl^-, HCO_3^- and other molecules, as previously discussed. The osmotic pressure in the peritubular blood capillaries, π_p, is less than π_i. Thus, water transport from the interstitial space into the blood capillaries is driven by the hydrostatic pressure difference that is established when water is reabsorbed into the interstitial space. The second pathway for water reabsorption involves the transcellular channels that are formed by transmembrane proteins (i.e., members of the aquaporin family) on apical and basal–lateral membranes of epithelial cells. Water reabsorption induces the convective transport of solutes, one of the important means of all solute reabsorption through paracellular pathways, as has been mentioned for Na^+ transport from peritubular interstitial space into blood capillaries.

About two-thirds of water reabsorption occurs in the proximal tubule, due to the high hydraulic conductivity of the epithelial cell layer. Water reabsorption in Henle's loop occurs only in the descending thin limb; this reabsorption accounts for about 15% of the total water that is filtered. In the early portion of the distal tubule, the hydraulic conductivity of the tubular epithelial layer is extremely low. Thus, water reabsorption is negligible. In the late portion of the distal tubule and collecting duct, 8%–17% of the filtered water is reabsorbed. The exact amount of reabsorption in this region depends upon the plasma concentration of antidiuretic hormone (ADH). Without ADH, the epithelial layer in the late portion of distal tubule is impermeable to water.

Glucose and urea reabsorption Glucose is reabsorbed mainly in the proximal tubule. The mechanisms of glucose reabsorption involve secondary active transport across the apical membrane via glucose-Na^+ symporters (e.g., SGLT1 and SGLT2) and facilitated diffusion across the basal–lateral membrane via glucose uniporters (e.g., GLUT1 and GLUT2) (see Figure 14.10). Functionally, SGLT1 and SGLT2 differ by the number of Na^+ ions that are transported with one glucose molecule. SGLT1 symporters transport one Na^+ ion; SGLT2 symporters transport two Na^+ ions. In all cases, glucose reabsorption does not consume energy directly. The chemical energy (i.e., ATP) is required to maintain a very low intracellular concentration of Na^+ via the Na^+/K^+ ATPase (see Figure 14.3).

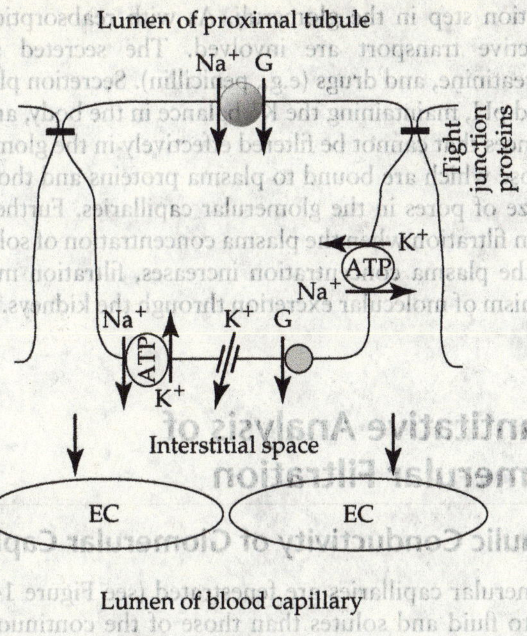

FIGURE 14.10 Glucose reabsorption in the proximal tubule. G is the glucose molecule. The Na^+/G symporter is either SGLT2 in the convoluted portion of the proximal tubule or SGLT1 in the straight portion of the proximal tubule. The glucose uniporter represents both GLUT1 and GLUT2. EC indicates endothelial cells.

The transport of glucose from the peritubular interstitial space to the lumen of blood capillaries occurs by passive diffusion, driven by the concentration gradient of glucose, and by convection, driven by the hydrostatic pressure gradient previously mentioned.

The reabsorption of urea in the kidney is much less than that of other molecules (see Table 14.1). As with glucose, urea is reabsorbed primarily in the proximal tubule. Urea, a relatively small, uncharged, polar molecule, can diffuse easily across the tubular epithelium, as well as through the peritubular interstitium and the capillary endothelium. Urea is one example of molecules that can be reabsorbed entirely by passive diffusion. In the case of urea, the diffusion is driven by the concentration gradient of the molecule, which depends upon the rate of water reabsorption and the flow rate of tubular fluid. The rate of water reabsorption has a positive effect, and the flow rate of the tubular fluid has a negative effect, on the urea concentration in the tubular fluid, which in turn controls the local concentration gradient of urea.

Urea can also be reabsorbed in the medullary collecting duct (see Figure 14.4), by facilitated transport through a urea uniporter in both apical and basal–lateral membranes of epithelial cells. The activity of the uniporter is ADH dependent. Urea reabsorption in this region raises its peritubular interstitial concentration, driving the diffusion of urea from the vicinity of the medullary collecting duct into Henle's loop and the interstitial space of papillae. Up to 50% of the papillary osmolality, which is important in water reabsorption through the papillary collecting duct, is maintained by urea.

Secretion. Renal reabsorption reduces the volume of the filtrate and the concentration of solutes in the filtrate. Meanwhile, the concentrations of other solutes, especially some wastes and excess substances, are increased by renal secretion in the tubules. Secretion is the direct transport of solutes from blood capillaries into renal tubules, without the filtration step in the glomeruli. As with reabsorption, both facilitated transport and active transport are involved. The secreted substances include K^+, H^+, NH^{4+}, creatinine, and drugs (e.g., penicillin). Secretion plays important roles in regulating blood pH, maintaining the K^+ balance in the body, and removing wastes and excess substances that cannot be filtered effectively in the glomeruli. Among these substances are those which are bound to plasma proteins and those which are larger than the cutoff size of pores in the glomerular capillaries. Furthermore, secretion is more effective than filtration when the plasma concentration of solutes is low (see Section 14.6.3). As the plasma concentration increases, filtration may become a more important mechanism of molecular excretion through the kidneys.

14.4 Quantitative Analysis of Glomerular Filtration

14.4.1 Hydraulic Conductivity of Glomerular Capillaries

The walls of glomerular capillaries are fenestrated (see Figure 14.6). The walls are more permeable to fluid and solutes than those of the continuous vessels in other tissues. The rate of fluid transport through the capillary walls can be modeled by Starling's law (Section 9.3),

$$J_v = L_P S(\Delta p - \sigma_s \Delta\pi), \tag{14.4.1}$$

where J_v is the flow rate, L_P is the hydraulic conductivity, S is the surface area of the endothelium, σ_s is the osmotic reflection coefficient, Δp is the difference between the microvascular pressure (p_{cap}) and the hydrostatic pressure in the Bowman capsule (p_{bc}), and $\Delta\pi$ is the osmotic pressure difference between the two compartments ($\pi_{\text{cap}} - \pi_{\text{bc}}$). Under normal physiological conditions, p_{cap} is maintained at approximately 45 mmHg [2]. The pressure drop from afferent arterioles to efferent arterioles in capillaries is negligible (see Figure 14.11). p_{bc} is about 10 mmHg. π_{cap} varies from 25 to 35 mmHg as the blood moves from afferent to efferent arteriole ends, due to a filtration-induced increase in the concentration of plasma proteins in the blood (see Figure 14.11). π_{bc} is close to zero and $\sigma_s \approx 1$.

The pressure distributions along microvessels in the kidney are different from those in non-renal tissues. For example, the hydrostatic pressure p_{cap} decreases significantly, while π_{cap} is maintained at a constant level, as the blood moves from arteriole to venule ends of capillaries in muscle tissues. These pressure distributions differ from those shown in Figure 14.11. The difference can be explained by the structures of capillary networks and the microvascular permeability of plasma proteins. In general, the net driving force for convective transport across the glomerular vessel wall is larger than that across nonrenal vessels, although the interstitial fluid pressure in muscles (approximately zero) is smaller than the hydrostatic pressure in Bowman's capsule(~10 mmHg).

The rate of filtration can be determined experimentally by infusing small inert molecules, such as inulin, which is a polysaccharide with a molecular weight of 5,000. In the glomerulus, inulin is transported primarily by convection; diffusion is

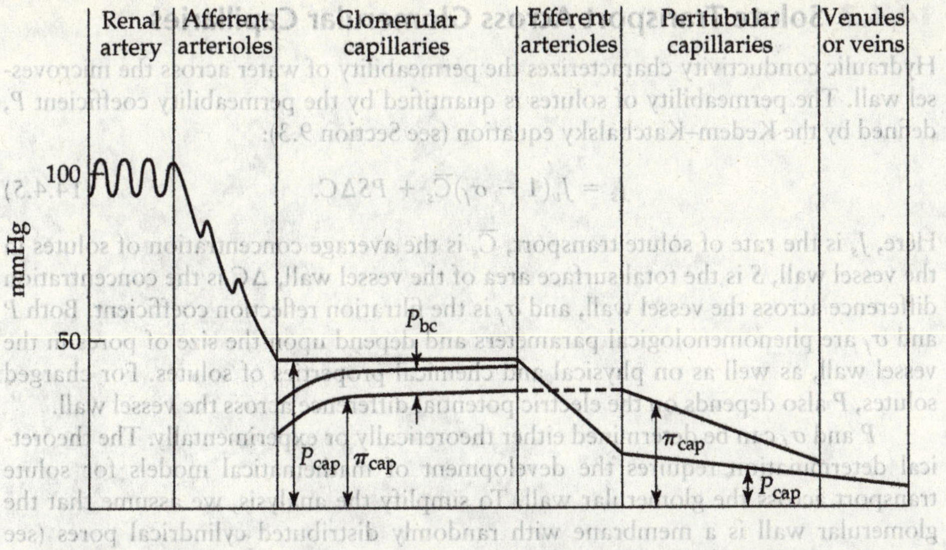

FIGURE 14.11 Schematic of pressure distributions in the kidneys. (Reprinted from Ref. [2], with permission of Springer Science + Business Media.)

negligible. Furthermore, because inulin is a small molecule, its convective transport through the microvessel wall is also negligibly retarded; the filtration reflection coefficient is close to zero (see the Kedem–Katchalsky equation in Section 9.3). Therefore, the filtration rate of inulin is

$$J_s \approx J_v C_p, \tag{14.4.2}$$

where C_p is the plasma concentration of inulin. Inulin is not metabolized, nor is it reabsorbed or secreted by the renal tubules after intravenous infusion. Therefore, inulin molecules found in the urine are originated entirely from filtration through the glomeruli, indicating that

$$J_v C_p = Q C_{\text{urine}}, \tag{14.4.3}$$

where Q is the rate of urine formation and C_{urine} is the urine concentration of inulin. Rearranging terms in Equation (14.4.3), one obtains

$$J_v = Q C_{\text{urine}} / C_p. \tag{14.4.4}$$

Equation (14.4.4) indicates that the flow rate can be calculated if C_p, Q, and C_{urine} are known. The concentrations of inulin can be determined by sampling blood and urine and then measuring the inulin concentration in the samples. Q can be measured by collecting the total urine produced per day by a person. Under normal conditions, Q may vary from time to time, due to changes in fluid reabsorption in renal tubules. However, data in the literature indicate that J_v in humans is nearly constant at about 125 mL min^{-1}, or 180 L day^{-1}, in a normal adult, out of which only 1–2 L day^{-1} is excreted as urine. Therefore, 99% of filtrate is reabsorbed into the blood through the peritubular capillaries as the filtrate passes through the renal tubules.

Substituting the flow rate and the net pressure difference across the microvessel wall into Equation (14.4.1), we obtain $L_p S \approx 7$ mL min^{-1} mmHg^{-1}, where S is the total surface area of microvessels in the glomeruli.

14.4.2 Solute Transport Across Glomerular Capillaries

Hydraulic conductivity characterizes the permeability of water across the microvessel wall. The permeability of solutes is quantified by the permeability coefficient P, defined by the Kedem–Katchalsky equation (see Section 9.3):

$$J_s = J_v(1 - \sigma_f)\overline{C}_s + PS\Delta C. \tag{14.4.5}$$

Here, J_s is the rate of solute transport, $\overline{C}_s$ is the average concentration of solutes in the vessel wall, S is the total surface area of the vessel wall, ΔC is the concentration difference across the vessel wall, and σ_f is the filtration reflection coefficient. Both P and σ_f are phenomenological parameters and depend upon the size of pores in the vessel wall, as well as on physical and chemical properties of solutes. For charged solutes, P also depends on the electric potential difference across the vessel wall.

P and σ_f can be determined either theoretically or experimentally. The theoretical determination requires the development of mathematical models for solute transport across the glomerular wall. To simplify the analysis, we assume that the glomerular wall is a membrane with randomly distributed cylindrical pores (see Figure 14.12) and neglect the details in the structure of the glomerular wall, which includes three layers: the endothelium, the glomerular basement membrane, and the epithelium (see Figure 14.6). In our analysis, the hindrances to solute transport presented by these layers are lumped together. More complicated models that consider structural details in the glomerular wall can be found in the literature [17].

Glomerular permeability: effect of solute size. The Kedem–Katchalsky equation is derived from the theory of irreversible thermodynamics. The transport parameters in the equation are phenomenological, being directly related neither to the structures of microvessels in tissues nor to the physical and chemical properties of solutes. Therefore, several other mathematical models have been developed to account for the effects of the structures of microvessels and the properties of solutes on transvascular transport. One such model is the *pore theory*, in which the microvessels wall is considered as a porous medium and the transport of solutes in

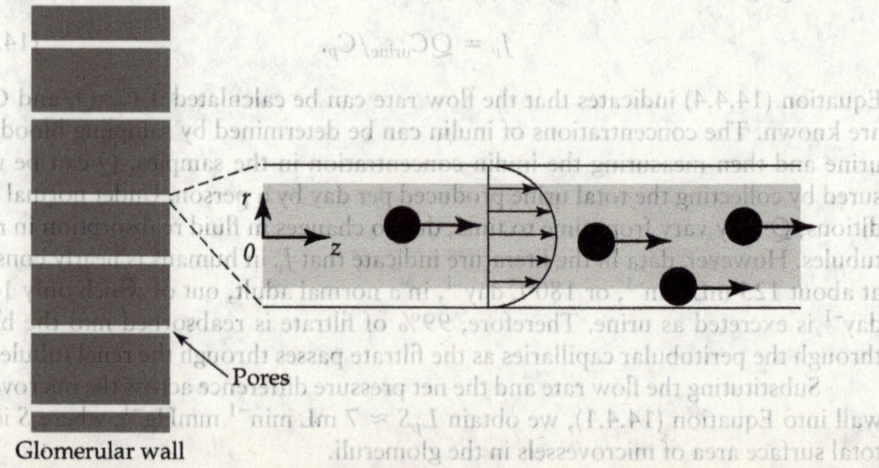

FIGURE 14.12 Geometry of mathematical models for solute transport across the glomerular wall. The transport includes both diffusion and convection. The black spheres represent solutes. The gray area inside the pore indicates the available space for solutes.

pores is retarded due to solute–pore interactions. Such interactions alter the fluid resistance to solute movement and the velocity profile of fluid in the pores. In addition, near the surfaces of pores is a region (see the white area in the pore shown in Figure 14.12) that is unavailable to solute molecules. A simple case in the pore theory is to assume that the pores are cylindrical channels and that solutes are spheres (see Figure 14.12). In this case, the average flux of solutes through the pore, $\overline{N}_s$, in the z direction was derived in Chapter 8 (see Equation (8.4.19)). We list it here as

$$\overline{N}_s = -HD_0\frac{dC}{dz} + WCv_m, \tag{14.4.6}$$

where v_m is the mean velocity of fluid in the pore, D_0 is the diffusion coefficient in the water, C is the solute concentration, z is the axial coordinate, and H and W are the *hydrodynamic resistance coefficients* for diffusion and convection, respectively. (W is also called the *retardation coefficient* in the pore.) Under the "centerline approximation" (see Chapter 8), H and W are functions only of λ. As $\lambda \rightarrow 0$, interactions between the pore and solutes become negligible. Thus, both H and W approach unity. As $\lambda \rightarrow 1$, both H and W approach zero. In this case, there is no transport of solutes across the microvessel wall. When $0 < \lambda < 1$, the dependence of H and W on λ can be determined numerically by solving the Stokes equation (3.5.3) for a rigid sphere at the centerline in a cylindrical channel. The numerical results are plotted in Figure 14.13. For $0 < \lambda < 0.4$, analytical expressions of H and W are also available (see Equations (8.4.25) and (8.4.26), respectively [18,19]).

Data shown in Figure 14.13 demonstrate that H decreases faster than W when λ is increased, suggesting that the retardation of solute transport is more significant for diffusion than for convection. Quantitatively, the ratio of H to W is always less than unity and approaches zero as λ approaches unity (see Figure 14.13). Thus, diffusion is negligible as λ approaches unity.

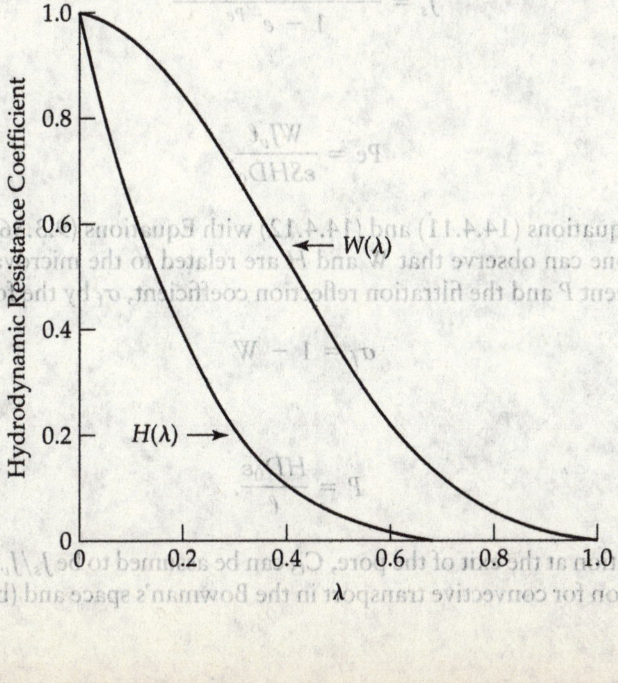

FIGURE 14.13 The dependence of H and W on λ. (Reprinted from Figure 2 in Ref. [20], with permission.)

At the steady state, $\overline{N}_s$ is a constant, independent of z. Thus, integrating Equation (14.4.6) gives a more convenient form of $\overline{N}_s$. The result is

$$\overline{N}_s = WC_0 v_m \frac{1 - (C_\ell/C_0)e^{-\mathrm{Pe}}}{1 - e^{-\mathrm{Pe}}}, \tag{14.4.7}$$

where C_ℓ and C_0 are the solute concentrations on $z = 0$ and $z = \ell$, respectively, ℓ is the length of the pore, and

$$\mathrm{Pe} = \frac{W v_m \ell}{HD_0} \tag{14.4.8}$$

is the Peclet number that characterizes the ratio of convection versus diffusion. In the pore, $\overline{N}_s$ and v_m can be used to predict the total rates of fluid transport, J_v, and solute transport, J_s, across the microvessel wall if all pores in the capillary wall are identical and there are no interactions between the pores. We have

$$J_s = \varepsilon S \overline{N}_s \tag{14.4.9}$$

and

$$J_v = \varepsilon S v_m, \tag{14.4.10}$$

where ε is the volume fraction of pores in the capillary wall and S is the total surface area of the capillary wall. (Note that the difference in the available volume fractions between solute and water has been considered in the derivation of H and W. Thus, it cannot be considered again. See Section 8.4.3) Substituting and rearranging terms in Equations (14.4.7) through (14.4.10) give

$$J_s = \frac{WJ_v(C_0 - C_\ell e^{-Pe})}{1 - e^{-Pe}} \tag{14.4.11}$$

and

$$\mathrm{Pe} = \frac{WJ_v\ell}{\varepsilon SHD_0}. \tag{14.4.12}$$

Comparing Equations (14.4.11) and (14.4.12) with Equations (9.3.36) and (9.3.30), respectively, one can observe that W and H are related to the microvascular permeability coefficient P and the filtration reflection coefficient, σ_f by the formulas

$$\sigma_f = 1 - \mathrm{W} \tag{14.4.13}$$

and

$$P = \frac{HD_0\varepsilon}{\ell}. \tag{14.4.14}$$

The concentration at the exit of the pore, C_ℓ, can be assumed to be J_s/J_v, since (a) there is no retardation for convective transport in the Bowman's space and (b) convection is the dominant mode of transport in glomerular filtration. Therefore, rearranging terms in Equation (14.4.11) gives

$$J_s = \frac{C_0 J_v W}{1 - e^{-Pe}(1 - W)}. \tag{14.4.15}$$

Glomerular permeability: effect of solute charge. Glomerular capillaries are selectively permeable to charged molecules because the vessel wall is covered with a layer of extracellular matrix (glycocalyx) with fixed negative charges. The glycocalyx provides an electrostatic barrier to the filtration of negatively charged solutes. Thus, these solutes are excreted more slowly than are positively charged solutes. In patients with proteinuric disorders, this barrier in the kidneys is damaged [17]. Consequently, the glomerular capillary wall becomes less selective to charged solutes. At present, there is no structure-based model for the transport analysis of charged molecules across the capillary wall [17]. In most studies, the rate of transport of these molecules in the x direction, J_s, is governed by the modified Nernst–Planck equation (see Chapter 7),

$$J_s = -\varepsilon SHD_0\left(\frac{dC}{dx} + \frac{zCF}{RT}\frac{d\Psi}{dx}\right) + WCJ_v, \tag{14.4.16}$$

where z is the net charge of solutes and Ψ is the electrical potential. However, Equation (14.4.16) does not take into account either the solute size and shape or the charge distributions in solutes and the structures of the capillary wall. Thus, the modified Nernst–Planck equation can be used only to compare changes in solute transport across the capillary wall after the surface charge of solutes is chemically modified.

Glomerular permeability: the molecular-sieving phenomenon. The transport of solutes can be selectively blocked when solutions pass through a filter or molecular sieve. Thus, the convective velocity of solutes is smaller than the fluid velocity. The retardation of convective transport causes an accumulation of solutes in a thin layer, called a *boundary layer*, near the retentive surface of the filter, leading to a higher concentration in this layer than in that away from the filter (see Figure 14.14).

The glomerular vessel wall can be considered a molecular sieve with a pore cutoff size of approximately 5.5 nm in diameter (see Figure 1.20). The concentration of solutes in the boundary layer near the luminal surface of the vessel wall is higher than the plasma concentration and the concentration in the Bowman's space.

The sieving effect is quantified by the *membrane-sieving coefficient*, defined as

$$\Theta = \frac{C_F}{C_M}, \tag{14.4.17}$$

where C_F is the concentration in the filtrate (e.g., in Bowman's space), and C_M is the concentration at the luminal surface of the membrane. The thickness of the boundary layer is on the order of a few microns. Thus, the quantification of C_M in such a small region is difficult. By contrast, the retentate concentration C_R (e.g, the concentration

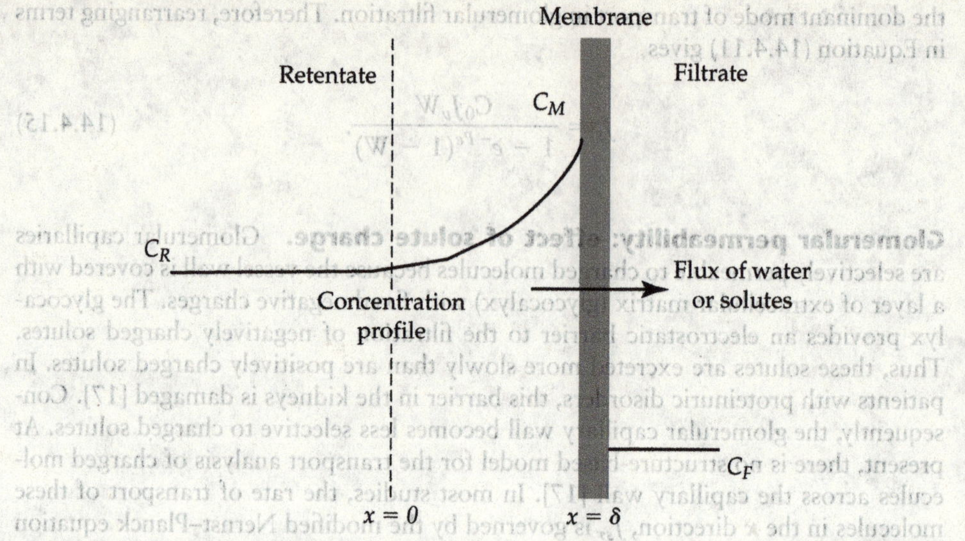

FIGURE 14.14 Molecular sieving phenomenon. (Modified from Figure 8 in Ref. [21]. Used with permission.)

in the plasma) can be determined in most cases. The overall sieving coefficient can be defined as

$$\Theta' = \frac{C_F}{C_R}. \qquad (14.4.18)$$

The relationship between Θ and Θ' can be established by analyzing solute transport in the microvessel wall. For neutral molecules, the flux of solutes on the retentive side of the vessel wall is

$$N_s = -D_0\frac{dC}{dx} + C\text{v}, \qquad (14.4.19)$$

where C is the concentration, and v is the fluid velocity. At the steady state, N_s is a constant. If the concentration outside the boundary layer is assumed to be uniform and equal to C_R, then integrating Equation (14.4.19) gives

$$N_s = C_R\text{v}\frac{1-(C_M/C_R)e^{-\text{Pe}}}{1-e^{-\text{Pe}}}, \qquad (14.4.20)$$

where

$$\text{Pe} = \frac{\text{v}\delta}{D_0}, \qquad (14.4.21)$$

in which δ is the thickness of the boundary layer (see Figure 14.14). On the Bowman's space side of the vessel wall,

$$N_s = C_F\text{v}. \qquad (14.4.22)$$

Substituting Equation (14.4.22) and the definitions of Θ and Θ' into Equation (14.4.20), one obtains

$$\Theta = \frac{\Theta'}{\Theta' + (1 - \Theta')\beta}, \tag{14.4.23}$$

where $\beta = \exp(\text{Pe})$ is called the *polarization factor* [21]. Equation (14.4.23) predicts that $\Theta' \geq \Theta$ and $C_M \geq C_R$, since $\text{Pe} \geq 0$. In general, Θ, Θ', and β depend on four types of variables: tissue structures, molecular properties of the solutes, the concentrations of the solutes, and the transmembrane differences of hydrostatic and osmotic pressures.

Equation (14.4.23) can be used to predict the membrane-sieving coefficient only if the overall sieving coefficient and the polarization factor are known. However, experimental data for Θ' and β for glomerular capillaries—or, indeed, for any microvessels—are limited in the literature. Thus, Θ has been determined only under some specific experimental conditions [21].

Example 14.2 Water and protein transport across the glomerular basement membrane (GBM) have been studied in vitro [21]. The investigators quantified the hydraulic conductivity of the GBM and the permeability of albumin and IgG across the GBM. From these data, the polarization factors β were estimated as 1.31 for albumin and 1.77 for IgG when the pressure difference across the GBM was 50 mmHg. In addition, the overall sieving coefficient Θ' was found to be 0.07 for albumin and 0.21 for IgG. Determine the membrane-sieving coefficients of albumin and IgG.

Solution Since β and Θ' are known, Θ can be easily calculated from Equation (14.4.23):

For albumin,

$$\Theta = \frac{0.07}{0.07 + (1 - 0.07)1.31} = 0.05.$$

For IgG,

$$\Theta = \frac{0.21}{0.21 + (1 - 0.21)1.77} = 0.13.$$

The membrane-sieving coefficient can also be predicted theoretically. One approach is to use the pore theory discussed above in this section. For transport across the microvessel wall, $C_0 = C_M$ and $J_s/J_v = C_F$. Substituting these relationships and the definition of Θ into Equation (14.4.15), one obtains,

$$\Theta = \frac{W}{1 - e^{-\text{Pe}}(1 - W)}. \tag{14.4.24}$$

Equation (14.4.24) predicts that the membrane-sieving coefficient depends on W and on Pe defined by Equation (14.4.12). When these parameters are known, the membrane-sieving coefficient can be calculated. Qualitatively, $\Theta \rightarrow 1$ when $W \rightarrow 1$ or $\text{Pe} \rightarrow 0$, suggesting that the sieving effect is negligible only if hydrodynamic retardation is low or if diffusion is the dominant mode of transport. Neither of these conditions can be satisfied, since convection is the dominant mode of transport during glomerular filtration, and the sieving effect in the glomerulus is caused primarily by the retardation of convective transport. In this case, the Peclet number Pe should be larger than unity, and $e^{-\text{Pe}}(1 - W)$ is much less than unity. Hence, Equation (14.4.24) becomes

$$\Theta \approx W. \tag{14.4.25}$$

As discussed earlier, the retardation coefficient W is approximately a function of the solute size relative to the pore diameter. Thus, Θ is independent of the filtration rate and the transmembrane pressure difference [21]. Furthermore, Equation (14.4.25) provides a method for estimating the retardation coefficient W of the microvessel wall, if $\text{Pe} \gg 1$. This is because Θ' and β can be measured experimentally, which determine Θ through Equation (14.4.23).

14.5 Quantitative Analysis of Tubular Reabsorption

The quantitative analysis of tubular reabsorption can be divided into two parts: transport within the lumen of tubules and transport across the tubular epithelium. The first part involves diffusion and convection in pure fluids, which were discussed fully in Chapter 6. The second part involves facilitated transport and active transport, which resemble chemical reactions near or inside the plasma membrane. These two kinds of transport can be modeled in the same manner as the kinetics of chemical reactions were modeled in Chapter 10. In general, the spatial distribution of solutes in the membrane of a cell is unknown and is neglected in reabsorption analysis. However, the spatial distributions of solutes in the tubular fluid and in peritubular tissues need to be considered.

14.5.1 Mass Balance Equations

Diffusion and convection in the lumen of a tubule are mainly in the axial direction, since the diameter of the tubule is much smaller than the length (see Figure 14.4). Transport across the epithelial layer is perpendicular to the tubule axis. (Transport between adjacent epithelial cells is negligible.) As a result, the transport processes can be assumed to be unidirectional (see Figure 14.15). In addition, the concentration distribution within a cell is assumed to be uniform, because the dimension of a cell is much smaller than the length of a tubule.

In this book, we consider reabsorption only in the steady state. In this case, the mass balance equations for the ith solute in the tubule are

$$\frac{dJ_{iL}}{dx} = -\frac{2}{\text{R}}J_{ic} + \frac{Q_{iL}}{\pi R^2}, \tag{14.5.1}$$

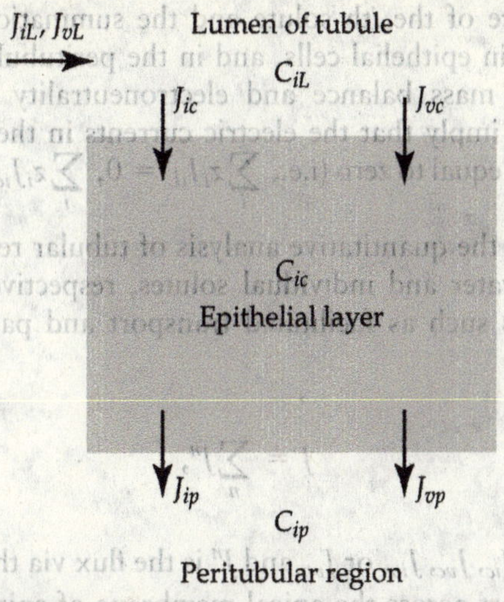

FIGURE 14.15 Water and solute transport in the lumen of tubules and across the tubular epithelial layer. J is the flux of either solute or water. C is the concentration of solutes. The subscript i represent the ith solute, v represents water. The subscripts L, c, and p represent the lumen of tubule, the intracellular region, and the peritubular region, respectively. To be consistent with the conventions in the literature, we use J instead of N in this section to represent fluxes. In other parts of this book, N is used to represent solute flux and v is used to represent the average fluid velocity which is equivalent to the fluid flux.

$$\frac{dJ_{vL}}{dx} = -\frac{2}{R}J_{vc}, \qquad (14.5.2)$$

$$J_{ic} = J_{ip} - \frac{Q_{ic}}{2\pi R}, \qquad (14.5.3)$$

and

$$J_{vc} = J_{vp}, \qquad (14.5.4)$$

where Q_{iL} and Q_{ic} are the rates of solute production per unit length of the tubule in the lumen and in the epithelial layer, respectively, R is the radius of the tubule, and x is the axial distance along the tubule. Equations (14.5.1) and (14.5.3) must be repeated for each solute of interest. For charged molecules, the concentrations in the lumen of the tubule, in the intracellular region, and in the peritubular region must satisfy the electroneutrality conditions

$$\sum_i z_i C_{iL} = 0, \qquad (14.5.5a)$$

$$\sum_i z_i C_{ic} = 0, \qquad (14.5.5b)$$

and

$$\sum_i z_i C_{ip} = 0, \qquad (14.5.5c)$$

where z_i is the valence of the ith solute and the summations are for all charged solutes in the lumen, in epithelial cells, and in the peritubular region, respectively. The requirements of mass balance and electroneutrality in Equations (14.5.1) through (14.5.5) also imply that the electric currents in the lumen and across the epithelial cells must be equal to zero (i.e., $\sum_i z_i J_{iL} = 0$, $\sum_i z_i J_{ic} = 0$, and $\sum_i z_i J_{ip} = 0$; see Problem 14.6).

The next step in the quantitative analysis of tubular reabsorption is to model the total flux J for water and individual solutes, respectively, based on different transport mechanisms such as facilitated transport and paracellular diffusion. In general,

$$J = \sum_n J^n, \tag{14.5.6}$$

where J can be either J_{ic}, J_{vc}, J_{ip}, or J_{vp} and J^n is the flux via the nth mechanisms. For example, Na^+ transport across the apical membrane of epithelial cells in the early portion of the proximal tubule is through Na^+/H^+ antiporters and Na^+/X symporters, where X can be glucose, inorganic phosphate, amino acids, or lactate (see Figure 14.7a). If the fluxes of Na^+ through the Na^+/H^+ antiporters and Na^+/X symporters can be determined, then the total flux of Na^+ across the apical membrane is equal to the summation of these fluxes. In the distal tubule, Na^+ transport through the apical surface of the epithelial layer is mediated by the Na^+/Cl^- symporters, the Na^+/H^+ antiporters, and the Na^+ channels. The rates of transport through these mechanisms are 304.6, 37.7, and 16.7 pmol min^{-1}, respectively. Thus, the total rate of transport is equal to 359.0 pmol min^{-1}. Across the basal–lateral surface of the epithelial cells, however, Na^+ transport is mediated only by Na^+/K^+ ATPase. On the basis of Equation (14.5.3), the rate of Na^+ transport via Na^+/K^+ ATPase should also equal 359.0 pmol min^{-1}, since $Q_c = 0$ (i.e., there is no Na^+ production in epithelial cells [22]).

Equation (14.5.6) also indicates that the total flux is location dependent, since the mechanisms of reabsorption depend on the location along the tubules, as previously discussed (see Figures 14.7 through 14.10). Furthermore, different solutes may be transported by the same symporters and antiporters, and different solutes in the same environments are subject to the electroneutrality conditions described previously. Thus, processes involving the reabsorption of different solutes must be considered simultaneously in the mathematical models. A complete description of the quantitative analysis of tubular reabsorption is complicated and beyond the scope of this book. Readers are directed to the literature [22–24].

In this chapter, we will describe some key models of transport, based on specific mechanisms in the tubules. When these models and Equations (14.5.1) through (14.5.6) are put together, one can develop a complete model of tubular reabsorption. In general, transport in the lumen of tubules is passive (see Section 14.5.2), whereas transport through epithelial cells may involve passive diffusion and convection (see Section 14.5.2), facilitated transport (see Sections 14.5.3 and 14.5.4), and active transport (see Section 14.5.5).

14.5.2 Fluxes of Passive Diffusion and Convection

Passive diffusion and convection are the mechanisms of transport in the lumen of renal tubules. Each mechanism can also affect both transcellular and paracellular transport across the epithelial layer. The transport of fluids and solutes in the lumen of tubules is, in general, three dimensional and time dependent. The fluid flow is governed by the Navier–Stokes equation (3.5.3), whereas the solute transport is governed by the general convective diffusion equation discussed in Chapters 6 and 7. Furthermore, the analyses of both fluid flow and solute transport are coupled with the analysis of transport across the epithelial layer.

The general analyses of passive diffusion and convection in the lumen of renal tubules are complicated and may not even be necessary. To simplify each analysis, the following assumptions can be made. First, the variation in the glomerular filtration rate is small in a normal person; thus, the time dependence of the transport is negligible. Second, the Reynolds number is small in the tubules, and the radius of a tubule is much smaller than its length. Therefore, one need consider only the average velocity at each cross-section of the tubules. The average velocity of fluid in a long, narrow tube can be determined approximately, based on the lubrication theory described in Chapter 4. In this approach, the fluid flux J_{vL}, which is equal to the average velocity in the axial (or x) direction, depends only on x. Similar to the way that Equation (4.7.19) for the fluid flow between two parallel porous plates was derived, the following equation can be derived for the fluid flow in a circular cylindrical pipe with a porous wall (see Problem 14.5):

$$J_{vL} = -\frac{R^2}{8\mu}\frac{dp}{dx}. \tag{14.5.7}$$

In this equation, R is the radius of the pipe or tubule, μ is the fluid viscosity, and dp/dx is the axial pressure gradient. Equation (14.5.7) is apparently the same as Poiseuille's law, discussed in Chapter 2, for the fluid flow in a circular cylindrical pipe with an impermeable wall. However, Equation (14.5.7) is not identical to Poiseuille's law, because J_{vL} is a constant in the latter but a function of x in the former. The derivative in Equation (14.5.2) is not equal to zero in renal tubules; rather, it depends on the fluid flux across the epithelium and the radius of the tubules.

Once the fluid flux is determined, the axial flux of the passive transport of solutes in the lumen of renal tubules can be modeled as

$$J_{iL} = C_{iL}J_{vL} - D_{iL}\frac{dC_{iL}}{dx}, \tag{14.5.8a}$$

where C_{iL} and D_{iL} are, respectively, the concentration and the dispersion coefficient of the ith solute in the lumen. In most studies, the first term in Equation (14.5.8a) is assumed to be much larger than the second term in the same equation. Thus,

$$J_{iL} \approx C_{iL}J_{vL}. \tag{14.5.8b}$$

The rates of passive transport across the epithelial layer through paracellular pathways are governed by the same equations as those for transvascular transport discussed in Section 14.4. Therefore, the analysis of paracellular transport will not be repeated here.

14.5.3 Goldman–Hodgkin–Katz Equation for Ion Channels

Ion transport across the cell membrane through ion channels is described in Chapter 7. If the electric field is assumed to be constant (i.e., if the electric potential is a linear function of the distance along the channel axis), then the ion flux J_i into the cell is determined by Equation (7.5.30b). More generally,

$$J_i = -P\frac{zFV_m}{RT}\frac{C_1 - C_2\exp(-zFV_m/RT)}{1-\exp(-zFV_m/RT)}, \tag{14.5.9}$$

where z is the valence of the ion, F is the Faraday constant, R is the gas constant, T is the absolute temperature, V_m is the transmembrane potential difference, and C is the concentration of the ion in solutions. The minus sign indicates that the flux is positive if it is in the direction from the extracellular spaces to the intracellular spaces. The subscripts 1 and 2 indicate intracellular and extracellular regions, respectively. P is the *Goldman permeability coefficient of ions*, which can be related to the ratio $\Phi_i D_{ij}/L$ in Equation (7.4.30b) on the basis of Equation (14.2.1). D_{ij} is the effective diffusion coefficient of ions in the channels and L is the length of the channels. In the literature of electrophysiology, Equation (14.5.9) is called the *Goldman–Hodgkin–Katz equation.*

Table 14.2 lists the values of the permeability coefficient in the Goldman–Hodgkin–Katz equation for Na^+, K^+, and Cl^- reabsorption in the distal tubule under normal conditions. The transmembrane potential difference in this region is indicated in Figure 14.9; the temperature can be assumed to be 310°K. With these values, the flux of ions across the apical and basal–lateral membranes of epithelial cells can be calculated from Equation (14.5.9).

Example 14.3 Determine the flux of Na^+ across the apical membrane of epithelium via Na^+ channels in the distal tubule if C_1 is 9.5 mM. P_{Na^+} is 1.19×10^{-6} cm s^{-1} in the early portion of the distal tube and 1.31×10^{-5} cm s^{-1}, in the late portion. C_2 is 28 mM and 20 mM in the early and late portions, respectively. $T = 310°K$, $R = 8.134 \times 10^{-3}$ J mmol^{-1} K^{-1}, and $F = 96.48$ C mmol^{-1}.

Solution The valence of Na^+ is + 1. In the early portion of the distal tubule, $V_m \approx -70$ mV (see Figure 14.9). Thus, $zFV_m/RT = -2.68$. Substituting these values into Equation (14.5.9), we obtain

$$J_i = 9.36 \times 10^{-5} \text{ cm mM s}^{-1} = 9.36 \times 10^{-8} \text{ mmol s}^{-1}\text{ cm}^{-2}. \tag{14.5.10a}$$

In the late portion of the distal tubule, $V_m = -30$ mV (see Figure 14.9). Thus, $zFV_m/RT = -1.15$. Substituting these values into Equation (14.5.9) yields

$$J_i = 4.77 \times 10^{-4} \text{ cm mM s}^{-1} = 4.77 \times 10^{-7} \text{ mmol s}^{-1}\text{ cm}^{-2}. \tag{14.5.10b}$$

TABLE 14.2

Goldman Permeability Coefficient of Na^+, K^+, and Cl^-

	Permeability in the early portion of the distal tubule (cm s^{-1})	Permeability in the late portion of the distal tubule (cm s^{-1})
$P_{Na^+}^{Lc}$ [a]	1.19×10^{-6}	1.31×10^{-5}
$P_{K^+}^{Lc}$	4.90×10^{-7}	3.35×10^{-5}
$P_{Cl^-}^{Lc}$	6.78×10^{-6}	1.85×10^{-5}
$P_{K^+}^{cp}$	4.74×10^{-4}	2.17×10^{-4}
$P_{Cl^-}^{cp}$	9.16×10^{-5}	2.33×10^{-5}

Source: Data from Ref. [25]

[a]The superscripts "Lc" and "cp" represent, respectively, apical and basal–lateral membranes of epithelial cells in the tubule. Ion channels are not involved in Na^+ transport across the basal–lateral membranes; thus, $P_{Na^+}^{cp}$ is not listed here.

These results demonstrate that Na^+ flux via the Na^+ channel in the early portion of the distal tubule is only about 20% of that in the late portion. This number is consistent with the experimental value of approximately ~30%. Furthermore, the Na^+ channels account for only about 5% of total Na^+ transport through epithelial cells in the early portion of the distal tubule. Most Na^+ ions enter the cells through an Na^+/Cl^- symporter (see Figure 14.9). From these results, one can conclude that Na^+ channels play a minor role in Na^+ reabsorption in the early portion of the distal tubule.

14.5.4 Mathematical Modeling of Carrier-Mediated Transport[1]

As mentioned in Section 14.2.2, carrier-mediated transport can be classified into three different types: uniport, symport, and antiport. Therefore, the mathematical models discussed in this section follow that sequence.

Uniport. As previously noted, uniport is carrier-mediated transport involving only one substrate. Uniport is based on *uniporters*, resulting from conformational change of transport proteins. Different conformational changes can occur, depending on the uniporters and the substrates. The same uniporter can mediate the transport of different substrates. For example, a glucose transporter can facilitate the transport of other sugar molecules, such as mannose and fructose [13]. A simple uniport model is shown in Figure 14.16.

We assume that the rate-limiting step in the uniport is the translocation of substrates between intracellular and extracellular spaces. Thus, we can assume that the binding between substrates and the uniporter is in the equilibrium state. Consequently, concentrations of S and its complexes with E and SE must satisfy the relationships

$$[SE_o] = [S_o][E_o]/K_o^S, \qquad (14.5.11)$$

[1]Note that, to simplify the notation, we will use [] instead of the letter "C" to represent the concentration in all analyses in Section 14.5.4.

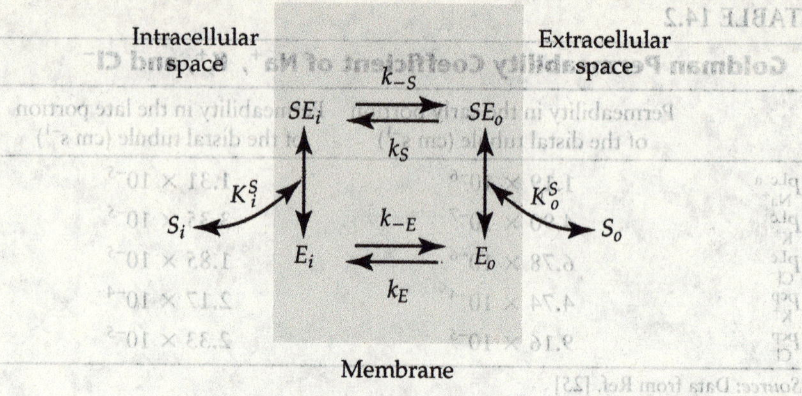

FIGURE 14.16 A simple kinetic scheme of uniport. S and E represent substrate and uniporter, respectively; other symbols are rate constants. The subscripts i and o represent intracellular and extracellular spaces, respectively.

and

$$[SE_i] = [S_i][E_i]/K_i^S, \tag{14.5.12}$$

where K is the ***equilibrium dissociation constant*** and the subscripts o and i indicate extracellular and intracellular, respectively. The total number of uniporters in the plasma membrane is assumed to be a constant, E_t. Hence,

$$[E_o] + [SE_o] + [E_i] + [SE_i] = E_t. \tag{14.5.13}$$

Furthermore, we assume that the total numbers of uniporters on both sides of the membrane are time independent. Thus, the total flux of E must be equal to zero:

$$k_{-S}[SE_i] - k_S[SE_o] + k_{-E}[E_i] - k_E[E_o] = 0. \tag{14.5.14}$$

Solving Equations (14.5.11) through (14.5.14) simultaneously, we have

$$[E_o] = \frac{E_t(k_{-E} + k_{-S}[S_i]/K_i^S)}{X} \tag{14.5.15}$$

and

$$[E_i] = \frac{E_t(k_E + k_S[S_o]/K_o^S)}{X}, \tag{14.5.16}$$

where

$$X = \left(1 + \frac{[S_o]}{K_o^S}\right)\left(k_{-E} + k_{-S}\frac{[S_i]}{K_i^S}\right) + \left(1 + \frac{[S_i]}{K_i^S}\right)\left(k_E + k_S\frac{[S_o]}{K_o^S}\right). \tag{14.5.17}$$

The net flux of S into the cell is

$$J_S = k_S[SE_o] - k_{-S}[SE_i] = \frac{E_t}{X}\left(k_S k_{-E}\frac{[S_o]}{K_o^S} - k_{-S}k_E\frac{[S_i]}{K_i^S}\right). \tag{14.5.18}$$

Equation (14.5.18) should also be valid when $[S_o] = [S_i]$. In this case, the net flux is equal to zero. Thus, the following relationship must be satisfied:

$$K_i^S k_S k_{-E} = K_o^S k_{-S} k_E. \tag{14.5.19}$$

Equation (14.5.19) indicates that these constants are not independent of each other.

Example 14.4 Consider a uniport experiment dealing with glucose transport into human red blood cells. The cells are mixed in a glucose solution with a concentration $[S_o]$. The intracellular concentration of glucose is negligible compared with $[S_o]$. Determine the net flux of glucose into the cell.

Solution We assume that $[S_i]$ is equal to zero. Thus, Φ and J_S become, respectively,

$$X = (k_{-E} + k_E) + (k_{-E} + k_S)\frac{[S_o]}{K_o^S} \tag{14.5.20}$$

and

$$J_S = \frac{E_t}{X} k_S k_{-E} \frac{[S_o]}{K_o^S}. \tag{14.5.21}$$

Now we define

$$K_m = K_o^S \frac{k_{-E} + k_E}{k_{-E} + k_S} \tag{14.5.22}$$

and

$$V_{\max} = E_t \frac{k_{-E} k_S}{k_{-E} + k_S}. \tag{14.5.23}$$

Therefore,

$$J_S = \frac{V_{\max}[S_o]}{K_m + [S_o]}. \tag{14.5.24}$$

Equation (14.5.24) is exactly the same as the Michaelis–Menten equation discussed in Chapter 10. K_m is equivalent to the Michaelis constant and $V_{\max}$ is equivalent to the maximum rate of reaction. K_m and $V_{\max}$ can be determined by fitting the experimental data of J_S versus $[S_o]$ to the curve represented by Equation (14.5.24). At 37°C, K_m and $V_{\max}$ are 4–10 mM and 600 mM min^{-1}, respectively [13].

In the derivation of Equation (14.5.24), we assume that the translocation of substrates is the rate-limiting step in the uniport. However, the equation is still valid if this assumption is violated. In that case, it can be shown that

$$K_m = \frac{(k_{-E} + k_E)(k_o k_{-S} K_o^S + k_o k_i K_o^S K_i^S + k_S k_i K_i^S)}{k_o[(k_{-S} + k_S + k_E)k_{-E} + k_i k_E K_i^S]} \tag{14.5.25}$$

and

$$V_{max} = \frac{k_i k_{-E} k_S E_t}{k_{-S}k_{-E} + k_S k_{-E} + (k_S + k_{-E})k_i K_i^S}, \tag{14.5.26}$$

where k_o and k_i are forward rate constants for S and E bindings on the extracellular and intracellular sides of the membrane, respectively (see Problem 14.11). Experimentally, one can always determine V_{max} and K_m, as long as J_S versus $[S_o]$ is quantified. Therefore, additional information on the uniport is required if one needs to determine whether the translocation is a rate-limiting step during the transport.

Symport. Mechanisms of symport—especially those involving Na^+ and another substrate—have been studied extensively in the literature. Many studies are based on the gradient hypothesis proposed by Crane [26], which is illustrated in Figure 14.17.

During symport, the binding of the two substrates to the carrier can be random, sequential (or ordered), or simultaneous (see Figure 14.17). The methods for kinetic analyses of the three processes are similar. Here, we will discuss only the random model with a stoichiometry of 1:1 [25]. The ordered model is analyzed in Problem 14.12.

As in the analysis of uniport, we again assume that the rate-limiting process in symport is the translocation of substrates between intracellular and extracellular spaces. Thus, we can assume that the binding between substrates and the symporter is in the equilibrium state. The effect of the binding of one substrate to the symporter on the binding of another substrate to the same symporter (i.e., the binding cooperativity) is accounted for by the factor α (see Figure 14.17a), which is equal to unity if there is no cooperativity. Experimental data in the literature suggest that α is less than unity [27]. Thus, there exists a negative cooperativity between the substrates. Under these assumptions, the concentrations of A, S, and their complexes with T (AT, ST, and AST) in the extracellular space must satisfy the following relationships:

$$[AT_o] = \frac{[A_o][T_o]}{K_o^A}, \tag{14.5.27}$$

$$[ST_o] = \frac{[S_o][T_o]}{K_o^S}, \tag{14.5.28}$$

$$[AST_o] = \frac{[AT_o][S_o]}{(\alpha_o K_o^S)}. \tag{14.5.29a}$$

If S binds to T first, then Equation (14.5.29a) needs to be replaced by

$$[AST_o] = \frac{[ST_o][A_o]}{(\alpha_o K_o^A)}. \tag{14.5.29b}$$

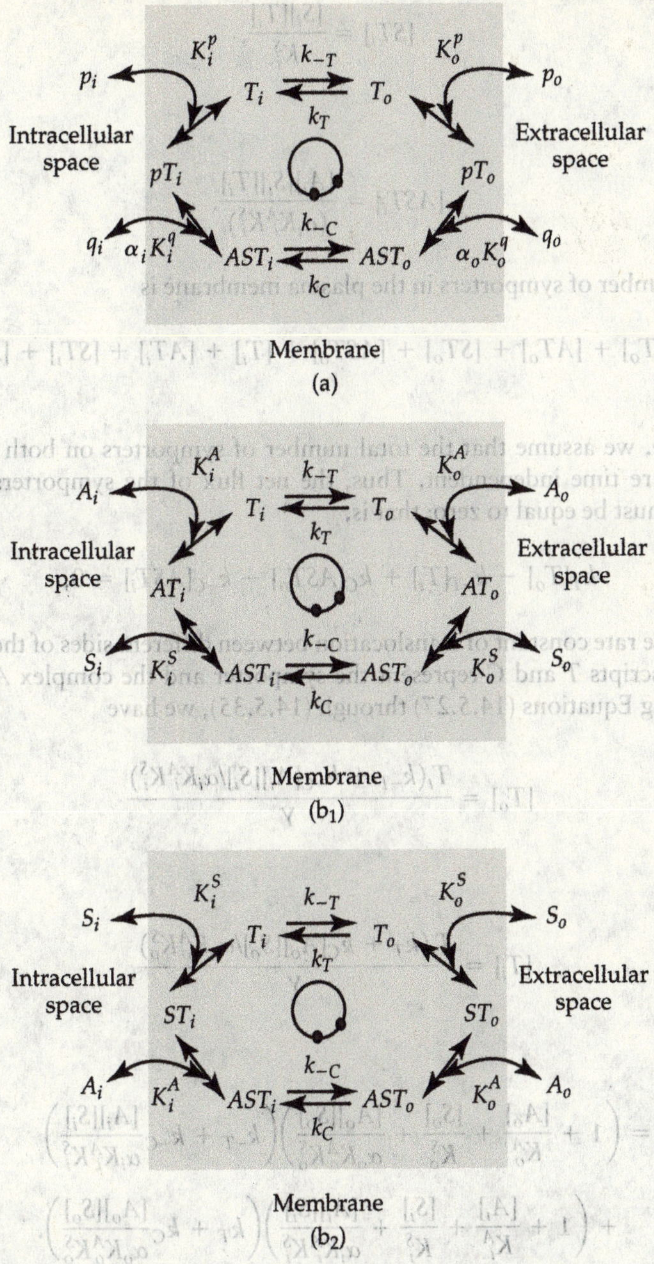

FIGURE 14.17 Kinetic schemes of symport. A and S represent substrates, T represents a symporter, and α accounts for the effect of binding of one substrate to the symporter on the binding of another substrate to the same symporter, i.e., the cooperativity in the substrate-to-symporter binding. K is the equilibrium dissociation constant, and k is the rate constant of translocation across the membrane. The subscripts i and o indicate intracellular and extracellular spaces, respectively. (a) The *random model*. The binding sequence of two substrates to the symporter is in a random manner. p and q represent different substrates and each of them can be either A or S. (b) The *ordered model*. The binding of two substrates to the symporter is in a sequential manner: (b_1) A binds to the symporter first on both sides of the membrane; (b_2) S binds to the symporter first. In the ordered model, α has been absorbed into the second equilibrium dissociation constant on both sides of the membrane. Thus, it is not considered explicitly. It is also possible that both substrates bind to the symporter simultaneously. In this case, AT and ST do not exist in the membrane.

In both sequences given by Equations (14.5.29a) and(14.5.29b), we have

$$[AST_o] = \frac{[A_o][S_o][T_o]}{(\alpha_o K_o^A K_o^S)}. \tag{14.5.30}$$

Similarly,

$$[AT_i] = \frac{[A_i][T_i]}{K_i^A}, \tag{14.5.31}$$

$$[ST_i] = \frac{[S_i][T_i]}{K_i^S}, \tag{14.5.32}$$

and

$$[AST_i] = \frac{[A_i][S_i][T_i]}{(\alpha_i K_i^A K_i^S)}. \tag{14.5.33}$$

The total number of symporters in the plasma membrane is

$$T_t = [T_o] + [AT_o] + [ST_o] + [AST_o] + [T_i] + [AT_i] + [ST_i] + [AST_i]. \tag{14.5.34}$$

Furthermore, we assume that the total number of symporters on both sides of the membrane are time independent. Thus, the net flux of the symporters across the membrane must be equal to zero; that is,

$$k_T[T_o] - k_{-T}[T_i] + k_C[AST_o] - k_{-C}[AST_i] = 0, \tag{14.5.35}$$

where k is the rate constant of translocation between different sides of the membrane and the subscripts T and C represent the symporter and the complex AST, respectively. Solving Equations (14.5.27) through (14.5.35), we have

$$[T_o] = \frac{T_t(k_{-T} + k_{-C}[A_i][S_i]/\alpha_i K_i^A K_i^S)}{Y} \tag{14.5.36}$$

and

$$[T_i] = \frac{T_t(k_T + k_C[A_o][S_o]/\alpha_o K_o^A K_o^S)}{Y}, \tag{14.5.37}$$

where

$$Y = \left(1 + \frac{[A_o]}{K_o^A} + \frac{[S_o]}{K_o^S} + \frac{[A_o][S_o]}{\alpha_o K_o^A K_o^S}\right)\left(k_{-T} + k_{-C}\frac{[A_i][S_i]}{\alpha_i K_i^A K_i^S}\right) + \left(1 + \frac{[A_i]}{K_i^A} + \frac{[S_i]}{K_i^S} + \frac{[A_i][S_i]}{\alpha_i K_i^A K_i^S}\right)\left(k_T + k_C\frac{[A_o][S_o]}{\alpha_o K_o^A K_o^S}\right). \tag{14.5.38}$$

The net fluxes of A and S into the cell are

$$J_A = J_S = k_C[AST_o] - k_{-C}[AST_i] = \frac{T_t}{Y}\left(k_C k_{-T}\frac{[A_o][S_o]}{\alpha_o K_o^A K_o^S} - k_{-C}k_T\frac{[A_i][S_i]}{\alpha_i K_i^A K_i^S}\right). \tag{14.5.39}$$

Equation (14.5.39) should also be valid when $[A_o] = [A_i]$ and $[S_o] = [S_i]$. In this case, the net flux is equal to zero. Thus, the following relationship must be satisfied:

$$\frac{k_C k_{-T}}{\alpha_o K_o^A K_o^S} = \frac{k_{-C} k_T}{\alpha_i K_i^A K_i^S}. \tag{14.5.40}$$

This means that the ten constants are not independent of each other; only nine of them can be specified independently.

Example 14.5 Equation (14.5.39) can be used to calculate Na^+ and Cl^- fluxes across the apical surface of the epithelial cells via the Na^+/Cl^- symporter in the distal tubule [22,25]. Assume that $\alpha_o = \alpha_i = 1$, $K_o^{Na} = K_i^{Na} = 51.1$ mM, $K_o^{Cl} = K_i^{Cl} = 19.2$ mM, $k_T = k_{-T} = k_C = k_{-C}$, and $J_{max} = k_T T_t = 1.04 \times 10^{-4}$ mmol cm^{-2} s^{-1}. In the early portion of the distal tubule, the concentrations of Na^+ and Cl^- in the lumen are the same, namely, $[Na_o]$. The corresponding concentrations in the cytoplasm of epithelial cells are 9.5 mM and 35.7 mM, respectively. In the late portion of the distal tubule, the intracellular concentrations of Na^+ and Cl^- are 9.7 mM and 17.7 mM, respectively. Determine the net flux of Na^+ as a function of $[Na_o]$, which varies from 20 to 60 mM.

Solution In the early portion of the distal tubule,

$$\begin{aligned}\frac{Y}{k_T} &= \left(1 + \frac{[Na_o]}{K_o^{Na}}\right)\left(1 + \frac{[Na_o]}{K_o^{Cl}}\right)\left(1 + \frac{[Na_i][Cl_i]}{K_o^{Na}K_o^{Cl}}\right) \\ &\quad + \left(1 + \frac{[Na_i]}{K_o^{Na}}\right)\left(1 + \frac{[Cl_i]}{K_o^{Cl}}\right)\left(1 + \frac{[Na_o]^2}{K_o^{Na}K_o^{Cl}}\right) \\ &= 4.75 + 0.097[Na_o] + 4.87 \times 10^{-3}[Na_o]^2\end{aligned} \tag{14.5.41}$$

and

$$\begin{aligned}J_{Na} &= \frac{T_t k_T^2 (1.02 \times 10^{-3}[Na_o]^2 - 0.346)}{Y} \\ &= \frac{1.06 \times 10^{-7}[Na_o]^2 - 0.360 \times 10^{-4}}{4.75 + 0.097[Na_o] + 4.87 \times 10^{-3}[Na_o]^2}.\end{aligned} \tag{14.5.42}$$

In the late portion of the distal tubule,

$$\begin{aligned}\frac{Y}{k_T} &= \left(1 + \frac{[Na_o]}{K_o^{Na}}\right)\left(1 + \frac{[Na_o]}{K_o^{Cl}}\right)\left(1 + \frac{[Na_i][Cl_i]}{K_o^{Na}K_o^{Cl}}\right) \\ &\quad + \left(1 + \frac{[Na_i]}{K_o^{Na}}\right)\left(1 + \frac{[Cl_i]}{K_o^{Cl}}\right)\left(1 + \frac{[Na_o]^2}{K_o^{Na}K_o^{Cl}}\right) \\ &= 3.46 + 0.084[Na_o] + 3.35 \times 10^{-3}[Na_o]^2\end{aligned} \tag{14.5.43}$$

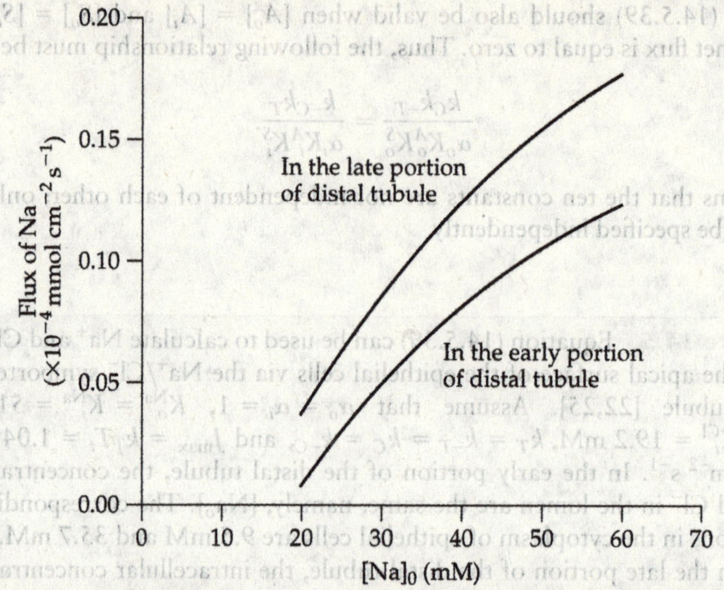

FIGURE 14.18 The flux of Na^+ across the apical membrane of the epithelial cells via the Na^+/Cl^- symporter in the distal tubule.

and

$$J_{Na} = \frac{T_t k_T^2(1.02 \times 10^{-3}[Na_o]^2 - 0.182)}{Y}$$

$$= \frac{1.06 \times 10^{-7}[Na_o]^2 - 0.182 \times 10^{-4}}{3.46 + 0.084[Na_o] + 3.35 \times 10^{-3}[Na_o]^2}. \qquad (14.5.44)$$

A plot of J_{Na} in both portions of the distal tubule is shown in Figure 14.18.

Under normal conditions, $[Na_o]$ depends on the location in the distal tubule. The average values of $[Na_o]$ are approximately 30 mM and 20 mM in the lumen of early and late portions of the distal tubule, respectively [25]. Thus, the corresponding fluxes are 4.93×10^{-6} and 3.73×10^{-6} mmol cm^{-2} s^{-1}, respectively.

Antiport. Antiport is often modeled by the so-called ping-pong mechanism [28]. In this model, the antiporter, *E*, has only one binding site for substrate *A* or substrate *S*, as shown in Figure 14.19. The ternary complex, consisting of *A*, *S*, and *E*, does not exist. *A* and *S* are translocated alternatively across the membrane via *E*. The ping-pong model is valid for both the Cl^-/HCO_3^- antiport and the Na^+/K^+ pump; each is involved in renal reabsorption. However, antiport cannot be modeled by the ping-pong mechanism in some cases, such as Cl^-/Cl^- self-exchange in a human myeloid leukemia cell line, HL60. The reason is that Cl^-/Cl^- antiport involves the ternary complex $Cl^-/E/Cl^-$ [28]. In this book, we will focus on the kinetic analysis of the ping-pong mechanism with a stoichiometry of 1:1. Such an analysis can be used to study Cl^-/HCO_3^- exchanges in the kidney [23].

As with the symport, we assume that in antiport the rate-limiting process is also the translocation of substrates between intracellular and extracellular spaces.

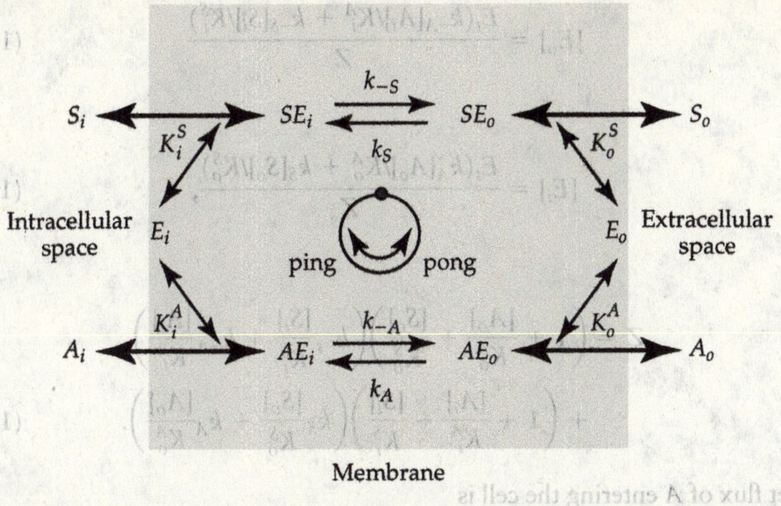

FIGURE 14.19 Kinetic scheme of antiport based on the ping-pong mechanism. A and S represent substrates, E represents the antiporter, K is the equilibrium dissociation constant, and k is the rate constant of translocation across the membrane. The subscripts i and o indicate intracellular and extracellular spaces, respectively.

Thus, we assume that the binding between substrates and the antiporter is again at the equilibrium state. The concentrations of A, S, and their complexes with E (AE and SE) must satisfy the relationships

$$[AE_o] = \frac{[A_o][E_o]}{K_o^A}, \tag{14.5.45}$$

$$[SE_o] = \frac{[S_o][E_o]}{K_o^S}, \tag{14.5.46}$$

$$[AE_i] = \frac{[A_i][E_i]}{K_i^A}, \tag{14.5.47}$$

and

$$[SE_i] = \frac{[S_i][E_i]}{K_i^S}, \tag{14.5.48}$$

where K is the equilibrium dissociation constant and the subscripts o and i indicate extracellular and intracellular spaces, respectively. The total number of antiporters in the plasma membrane is

$$E_t = [E_o] + [AE_o] + [SE_o] + [E_i] + [AE_i] + [SE_i]. \tag{14.5.49}$$

We assume that the total number of antiporters on both sides of the membrane are time independent. Thus, the net flux of the antiporter across the membrane must be equal to zero; that is,

$$k_A[AE_o] - k_{-A}[AE_i] + k_S[SE_o] - k_{-S}[SE_i] = 0, \tag{14.5.50}$$

where k is the rate constant of translocation between different sides of the membrane. Solving Equations (14.5.45) through (14.5.50) simultaneously, we have

$$[E_o] = \frac{E_t(k_{-A}[A_i]/K_i^A + k_{-S}[S_i]/K_i^S)}{Z} \tag{14.5.51}$$

and

$$[E_i] = \frac{E_t(k_A[A_o]/K_o^A + k_S[S_o]/K_o^S)}{Z}, \tag{14.5.52}$$

where

$$Z = \left(1 + \frac{[A_o]}{K_o^A} + \frac{[S_o]}{K_o^S}\right)\left(k_{-s}\frac{[S_i]}{K_i^S} + k_{-A}\frac{[A_i]}{K_i^A}\right) + \left(1 + \frac{[A_i]}{K_i^A} + \frac{[S_i]}{K_i^S}\right)\left(k_S\frac{[S_o]}{K_o^S} + k_A\frac{[A_o]}{K_o^A}\right). \tag{14.5.53}$$

The net flux of A entering the cell is

$$J_A = k_A[AE_o] - k_{-A}[AE_i] = \frac{E_t}{Z}\left(k_A k_{-S}\frac{[A_o][S_i]}{K_o^A K_i^S} - k_{-A}k_S\frac{[A_i][S_o]}{K_i^A K_o^S}\right). \tag{14.5.54}$$

The net flux of S exiting the cell is

$$J_S = -J_A = \frac{E_t}{Z}\left(-k_A k_{-S}\frac{[A_o][S_i]}{K_o^A K_i^S} + k_{-A}k_S\frac{[A_i][S_o]}{K_i^A K_o^S}\right). \tag{14.5.55}$$

When $[A_o] = [A_i]$ and $[S_o] = [S_i]$, there is no concentration gradient across the membrane. Thus, the net fluxes of A and S are equal to zero, and it follows that

$$\frac{k_A k_{-S}}{K_o^A K_i^S} = \frac{k_{-A}k_S}{K_i^A K_o^S} \quad \text{or} \quad \frac{k_A K_i^A}{K_o^A k_{-A}} = \frac{K_i^S k_S}{k_{-S}K_o^S}, \tag{14.5.56}$$

which means that the eight constants are not independent of each other; only seven can be specified independently. Furthermore, we can also show that

$$k_A[AE_o] = k_{-A}[AE_i], \tag{14.5.57}$$

when $[A_o] = [A_i]$ and $[S_o] = [S_i]$. Substituting Equations (14.5.45), (14.5.47), and (14.5.57) into Equation (14.5.56), we have

$$\frac{K_o^A k_{-A}}{k_A K_i^A} = \frac{k_{-S}K_o^S}{K_i^S k_S} = \frac{[E_o]}{[E_i]} = \lambda, \tag{14.5.58}$$

where λ is defined as an *asymmetry ratio* of unbound carriers between extracellular and intracellular spaces. Equation (14.5.58) indicates that λ is independent of substrates and is an intrinsic property of the antiporter. The value of λ is less than 0.1 for an anion exchanger, called AE1 or band 3, in the red blood cells [29], meaning that most free binding sites on the band 3 carrier are facing the intracellular, rather than extracellular, side of the plasma membrane.

Example 14.6 Consider Cl^- and HCO_3^- transport across the plasma membrane of epithelial cells via the Cl^-/HCO_3^- antiporter in the outer medullary collecting duct of the rat [23]. Here, $K_o^{Cl} = K_i^{Cl} = 50$ mM, $K_o^{HCO_3} = K_i^{HCO_3} = 198$ mM, $k_{Cl} = 562\ s^{-1}$, $k_{-Cl} = 61\ s^{-1}$, $k_{HCO_3} = 1247\ s^{-1}$, and $k_{-HCO_3} = 135\ s^{-1}$. The density of the antiporter is 9.6×10^{-8} mmol cm^{-2}, $[Cl]_o = 114$ mM, $[Cl]_i = 29$ mM, and $[HCO]_i = 26$ mM. Determine the net flux of HCO_3^- as a function of $[HCO_3]_o$.

Solution According to Equations (14.5.53) and (14.5.55),

$$\begin{aligned} Z &= (1 + 114/50 + [HCO_3]_o/198)(135/198 \times 26 + 61/50 \times 29) \\ &\quad + (1 + 29/50 + 26/198)(1{,}247/198 \times [HCO_3]_o + 562/50 \times 114) \\ &= 11.05 \times [HCO_3]_0 + 2{,}367 \end{aligned}$$

and

$$\begin{aligned} J_{HCO_3} &= 9.6 \times 10^{-8} \times (-562 \times 135 \times 114 \times 26/50/198 \\ &\quad + 61 \times 1{,}247 \times 29 \times [HCO_3]_0/50/198)/Z \\ &= \frac{[HCO_3]_0 - 101.94}{0.0052[HCO_3]_0 + 1.11} \times 0.01 (\text{nmol cm}^{-2}\ s^{-1}). \end{aligned}$$

A plot of the values of J_{HCO_3} is shown in Figure 14.20. Under normal conditions, $[HCO_3]_o \approx 10$ mM [23]. In this case, the efflux of HCO_3^- is 0.79 nmol cm^{-2} s^{-1} and the influx of Cl^- is also 0.79 nmol cm^{-2} s^{-1}.

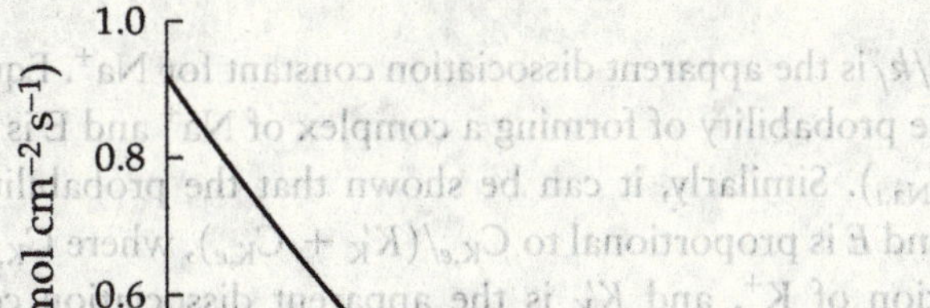

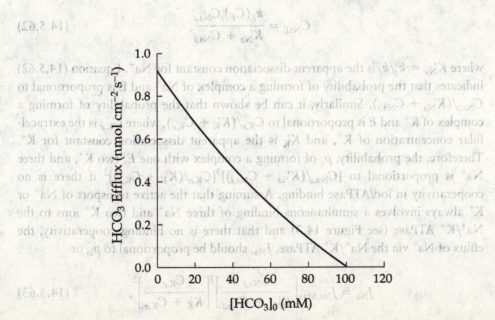

FIGURE 14.20 Efflux of HCO_3^- across the membrane of the epithelial cells via the Cl^-/HCO_3^- antiporter.

14.5.5 A Mathematical Model of Na^+/K^+ ATPase

Na^+/K^+ ATPase is an antiporter, and the antiport model previously described may be applied to it. Instead, however, we will introduce an alternative model of Na^+/K^+ ATPase [30]. The active transport of Na^+ via Na^+/K^+ ATPase involves the binding of Na^+ to the ATPase and the transmembrane translocation of Na^+, due to a conformational change of the ATPase. The binding process can be approximated by the reaction

$$Na,_i + E \underset{k_r}{\overset{k_f}{\longleftrightarrow}} NaE,$$

where $Na,_i$ is the intracellular Na^+, E is the ATPase, and NaE is the complex of Na^+ and E. The kinetic equation is

$$\frac{dC_{NaE}}{dt} = k_f C_{Na,i} C_E - k_r C_{NaE}. \tag{14.5.59}$$

In the steady state,

$$k_f C_{Na,i} C_E = k_r C_{NaE}. \tag{14.5.60}$$

The number of free ATPase molecules is equal to the total number of ATPase molecules minus those which have bound to Na^+; that is,

$$C_E = (C_E)_t - C_{NaE}, \tag{14.5.61}$$

where $(C_E)_t$ is the total concentration of ATPase in the membrane. Substituting Equation (14.5.61) into Equation (14.5.60), we have

$$C_{NaE} = \frac{k_f (C_E)_t C_{Na,i}}{K'_{Na} + C_{Na,i}}, \tag{14.5.62}$$

where $K'_{Na} = k_r/k_f$ is the apparent dissociation constant for Na^+. Equation (14.5.62) indicates that the probability of forming a complex of Na^+ and E is proportional to $C_{Na,i}/(K'_{Na} + C_{Na,i})$. Similarly, it can be shown that the probability of forming a complex of K^+ and E is proportional to $C_{K,e}/(K'_K + C_{K,e})$, where $C_{K,e}$ is the extracellular concentration of K^+, and K'_K is the apparent dissociation constant for K^+. Therefore, the probability, p_s of forming a complex with one E, two K^+, and three Na^+ is proportional to $[C_{Na,i}/(K'_{Na} + C_{Na,i})]^3[C_{K,e}/(K'_K + C_{K,e})]^2$ if there is no cooperativity in ion/ATPase binding. Assuming that the active transport of Na^+ or K^+ always involves a simultaneous binding of three Na^+ and two K^+ ions to the Na^+/K^+ ATPase (see Figure 14.3) and that there is no binding cooperativity, the efflux of Na^+ via the Na^+/K^+ ATPase, J_{Na}, should be proportional to p_s, or

$$J_{Na} = J_{Na,Max}\left[\frac{C_{Na,i}}{K'_{Na} + C_{Na,i}}\right]^3\left[\frac{C_{K,e}}{K'_K + C_{K,e}}\right]^2, \tag{14.5.63}$$

where $J_{Na,Max}$ is a constant that is equal to the maximum efflux of Na^+ at the steady state. Meanwhile, the influx of K^+ via the Na^+/K^+ ATPase must be

$$J_K = -\frac{2}{3} J_{Na}. \tag{14.5.64}$$

In general, $J_{Na,Max}$ increases with the intracellular concentration $C_{K,i}$ of K^+ and reaches a plateau when $C_{K,i}$ is greater than 30 mM [30]. Under normal conditions, $C_{K,i}$ is on the order of 150 mM. Thus, $J_{Na,Max}$ can be treated as a constant, with value ranging from 0.15×10^{-6} to 2.69×10^{-6} mmol cm^{-2} s^{-1} in the kidney [23,25]. However, K'_{Na} and K'_K in Equation (14.5.63) are linear functions of the concentrations of Na^+ and K^+ [25], respectively, or

$$K'_{Na} = a_{Na} + b_{Na} C_{K,i} \tag{14.5.65}$$

and

$$K'_K = a_K + b_K C_{Na,e}, \tag{14.5.66}$$

where $C_{Na,e}$ is the extracellular concentration of Na^+ and a_{Na}, a_K, b_{Na}, and b_K are constants approximately equal to 0.2 mM, 0.1 mM, 2.4×10^{-2}, and 5.4×10^{-3}, respectively [31]. The dependence of K'_{Na} on $C_{K,i}$ is due to a competitive inhibition of the binding of Na^+ to the Na^+/K^+ ATPase by K^+ in the cytosol of cells. Similarly, the dependence of K'_K on $C_{Na,e}$ is due to a competitive inhibition of the binding of K^+ to the Na^+/K^+ ATPase by Na^+ in the extracellular medium.

Example 14.7 Determine the efflux of Na^+ across an epithelial membrane via the Na^+/K^+ ATPase if $C_{Na,i} = 10$ mM, $C_{K,i} = 140$ mM, $C_{Na,e} = 145$ mM, and $C_{K,e} = 5$ mM. $J_{Na,Max}$ ranges from 0.15×10^{-6} to 2.69×10^{-6} mmol cm^{-2} s^{-1}.

Solution According to Equations (14.5.65) and (14.5.66), K'_{Na} and K'_K are 3.56 mM and 0.883 mM, respectively. Thus,

$$\left[\frac{C_{Na,i}}{K'_{Na} + C_{Na,i}}\right]^3 \left[\frac{C_{K,e}}{K'_K + C_{K,e}}\right]^2 = 0.29.$$

Substituting these values into Equation (14.5.63), we have the efflux of Na^+ ranging from 0.43×10^{-7} to 7.79×10^{-7} mmol cm^{-2} s^{-1}.

14.6 A Whole-Organ Approach to Renal Modeling

Renal reabsorption can be modeled at the organ level. In this approach, the kidney is treated as a black box. The input of the box is the plasma concentration of solutes, whereas the output is the rate of excretion of water or solutes through the urine [2]. The approach, in general, is called the *compartment model in pharmacokinetic analysis* and is discussed in detail in Chapter 16.

14.6.1 Filtration

Under normal conditions, the glomerular filtration rate (GFR) is a constant—approximately 125 mL min^{-1} in healthy adults. Thus, the rate of solute filtration can be estimated as

$$Q_f = (1 - \sigma_f)Q_v C_p, \tag{14.6.1}$$

where σ_f is the filtration reflection coefficient of solutes across the glomerular wall, Q_v is the GFR, and C_p is the plasma concentration of the solute. In this estimation, the diffusion of solutes across the glomerular wall is negligible compared with the convection. For small solutes (e.g., Na^+ and glucose), $\sigma_f \approx 0$.

14.6.2 Reabsorption

The average concentration of solutes in renal tubules is approximately proportional to their concentration in the Bowman's space, which is equal to Q_f/Q_v. According to Equation (14.6.1), Q_f/Q_v is proportional to C_p, since σ_f is a constant. Therefore, the average concentration is proportional to C_p. For solutes that are reabsorbed mainly through facilitated or active transport mechanisms, the rate of reabsorption will be saturated when C_p is increased beyond a threshold level. Such saturation is due to the saturation of transporters in epithelial cells. For glucose, the threshold level is approximately 4 mg mL^{-1} in healthy adults [2].

Although the mechanisms of facilitated and active transport vary for different solutes, the dependence of the reabsorption rate on the concentration of solutes in renal tubules can be modeled approximately by the Michaelis–Menten equation discussed in Chapter 10 for enzymatic reactions. On the basis of this equation, the rate of reabsorption in the kidneys is

$$Q_a = \frac{T_m C_p}{K_m + C_p}, \tag{14.6.2}$$

where K_m is a kinetic constant and T_m is the maximum capacity of reabsorption. In Equation (14.6.2), the solute concentration in renal tubules has been replaced by the plasma concentration of solutes, since, as has been mentioned, the two are proportional to each other. The proportionality constant has been absorbed into K_m, which, mathematically, is equal to a plasma concentration of solutes when Q_a is equal to one-half of T_m. Both K_m and T_m depend on the molecular properties of the solutes, on solute–transporter interactions, and on the total number of transporters in the cell membrane. For example, T_m is approximately 2 mmol min^{-1} for glucose and only 0.1 mmol min^{-1} for phosphate in healthy adults. The corresponding values of K_m are 8 mM and 0.5 mM, respectively. For comparison, the rates of filtration, reabsorption, and excretion for glucose and phosphate are plotted in Figure 14.21a and Figure 14.21b, respectively. The rate of excretion is calculated as the difference $Q_f - Q_a$.

In normal adults, C_p of glucose is between 60 and 100 mg/100mL [2], which is smaller than the threshold concentration for reabsorption saturation. In this case, filtration is balanced by reabsorption, and little glucose is excreted. The plasma concentration of glucose can be higher than the threshold level in diabetic patients,

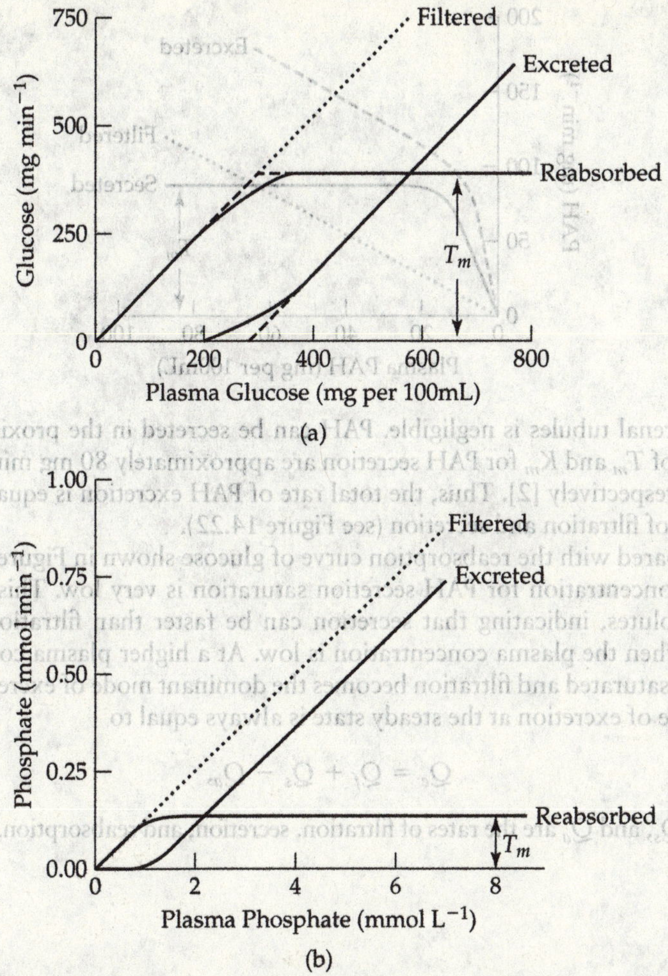

FIGURE 14.21 The plasma concentration-dependence of the rates of filtration, reabsorption, and excretion of (a) glucose and (b) phosphate. (Reprinted from Figures 5.3 and 5.5, respectively, in Ref. [2], with permission of Springer Science + Business Media.)

in which case filtered glucose cannot be completely reabsorbed. Thus, a large amount of glucose is found in the urine. The plasma concentration of phosphate is about 1 mM in normal adults, close to the threshold concentration. Under normal conditions, approximately 20% of the filtered phosphate is excreted.

14.6.3 Secretion

Like filtration and reabsorption, secretion can be analyzed quantitatively on the basis of two different approaches: (a) a distributed model that considers the details of transport mechanisms and (b) a whole-organ model that is phenomenological and neglects the details of transport mechanisms. The procedures for developing the distributed model are identical to those for reabsorption discussed in Section 14.5. At the organ level, Equation (14.6.2) can also be used to model the rate of solute secretion. For example, physiologists often use an organic acid, para-aminohippuric acid (PAH), to quantify renal blood-flow rate, because (i) the removal of PAH from the plasma is nearly complete in the kidney, as long as the plasma concentration of PAH in the systemic circulation, C_p, is very low (i.e., $C_p \ll 2K_m$), and (ii) the reabsorption

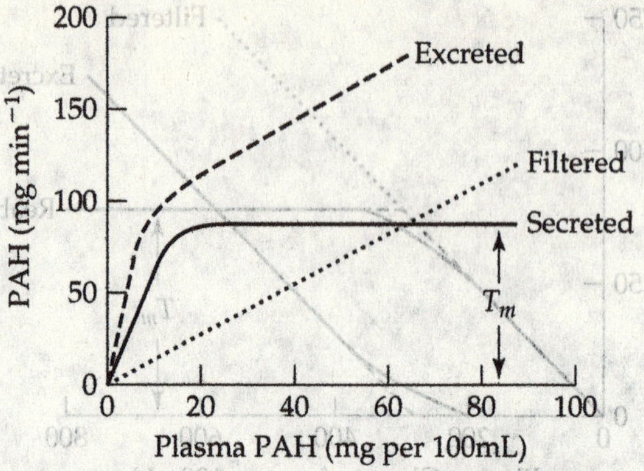

FIGURE 14.22 The plasma concentration-dependence of the rate of para-aminohippuric acid (PAH) secretion. The filtered and secreted curves can be approximated by Equations 14.6.1 and 14.6.2, respectively. (Reprinted from Figure 7.2 in Ref. [2], with permission of Springer Science + Business Media.)

of PAH in renal tubules is negligible. PAH can be secreted in the proximal tubules. The values of T_m and K_m for PAH secretion are approximately 80 mg min^{-1} and 0.07 mg min^{-1}, respectively [2]. Thus, the total rate of PAH excretion is equal to the sum of the rates of filtration and secretion (see Figure 14.22).

Compared with the reabsorption curve of glucose shown in Figure 14.21a, the threshold concentration for PAH secretion saturation is very low. This is also true for other solutes, indicating that secretion can be faster than filtration for solute excretion when the plasma concentration is low. At a higher plasma concentration, secretion is saturated and filtration becomes the dominant mode of excretion. In general, the rate of excretion at the steady state is always equal to

$$Q_e = Q_f + Q_s - Q_a, \tag{14.6.3}$$

where Q_f, Q_s, and Q_a are the rates of filtration, secretion, and reabsorption, respectively.

14.7 PROBLEMS

14.1 Derive Equation (14.2.1), $P = \Phi D_m/\ell$. P is defined as the flux of diffusion per unit concentration difference across the membrane. Assume that diffusion across the membrane is one dimensional and in the steady state.

14.2 Determine the permeability coefficient of a proton, H_3O^+, across a lipid membrane at 37°C. Assume that the dielectric constant of a cell membrane is 2, the membrane thickness is 4 nm, the radius of H_3O^+ is 0.11 nm, and the diffusion coefficient of H_3O^+ in the membrane is 1.4×10^{-4} cm^2 s^{-1}.

14.3 In an idealized glomerular capillary, one can assume that the blood pressure in the capillary is 45 mmHg, the hydrostatic pressure in the Bowman's space is 10 mmHg, and the osmotic pressure in the blood varies from 25 to 35 mmHg as a linear function of distance from the afferent arteriole end to the efferent arteriole end. The osmotic pressure in the Bowman's space is close to zero, and the osmotic reflection coefficient of solutes is close to unity. Determine the net pressure difference for filtration at both ends of the capillary and the average pressure difference across the capillary.

14.4 Derive Equations (14.4.7) and (14.4.15).

14.5 The average velocity, or the flux, of fluid in the axial direction of a long, narrow tube can be determined approximately on the basis of the lubrication theory described in Chapter 4. Similar to the way that Equation (4.7.19) for the fluid flow between two parallel porous plates was derived, derive Equation (14.5.7) for the fluid flow in a circular cylindrical pipe with a porous wall.

14.6 In general, total electric charge is conserved after chemical reactions, because highly charged solutes are unstable. Assuming this condition and the mass-balance and electroneutrality conditions in Equations (14.5.1) through (14.5.5), show that the electric currents in the lumen and across epithelial cells are equal to zero; that is, show that

$$\sum_i z_i J_{iL} = 0,$$

$$\sum_i z_i J_{ic} = 0,$$

and

$$\sum_i z_i J_{ip} = 0.$$

14.7 Inulin is excreted from the body by filtration only. Assume that the GFR is 125 mL min^{-1} and the plasma volume is 3 L. The half-life of solutes in the plasma is defined as the time when the plasma concentration is reduced to one-half of the initial value. If inulin is injected intravenously as a bolus (100 mg), determine the plasma concentration as a function of time and find the half-life of inulin.

14.8 PAH is excreted from the body by both filtration and secretion. Assume that the GFR is 125 mL min^{-1} and the plasma volume is 3 L. If 100 mg PAH is injected intravenously as a bolus, determine the plasma concentration as a function of time and find the half-life of PAH in the plasma.

14.9 Water reabsorption in renal tubules is passive, depending on the osmotic pressure difference across the epithelial layer, which is determined mainly by the concentration differences of several ions. To simplify the problem, consider only Cl^- and HCO_3^-. (This might be the case in the late portion of proximal tubules.) Experimental data have shown that the concentrations of Cl^- are approximately 120 mM and 100 mM in the tubular and interstitial fluids, respectively; the concentrations of HCO_3^- are 5 mM and 25 mM in the tubular and interstitial fluids, respectively. In addition, the epithelial layer is more permeable to Cl^- than HCO_3^-, and the hydrostatic pressure difference has a minimal contribution to water transport across the epithelial cell layer. Determine the direction of the driving force for water transport, and explain why the force is in that direction.

14.10 Consider the reabsorption of urea in renal tubules. The flux of reabsorption is proportional to the concentration difference across the epithelium. The permeability coefficient of urea across the epithelium is P. Assume that the rate of fluid flow in a tubule is Q, the length and the radius of the tubule are L and R, respectively, and the concentration of urea at the entrance of the tubule is C_0. The concentration of urea in the peritubular region is negligible. Assume a steady state. Find the concentration distribution of urea in the tubule and the total rate of urea reabsorption as a function of the flow rate.

14.11 In the derivation of Equation (14.5.24), we assumed that translocation of substrates is the rate-limiting process in uniport (i.e., the binding between substrate and uniporter occurs at equilibrium). Show that Equation (14.5.24) is also valid if this assumption is violated (i.e., derive Equations (14.5.25) and (14.5.26) for K_m and V_{max}, respectively, on the basis of the chemical reactions shown in Figure 14.23.

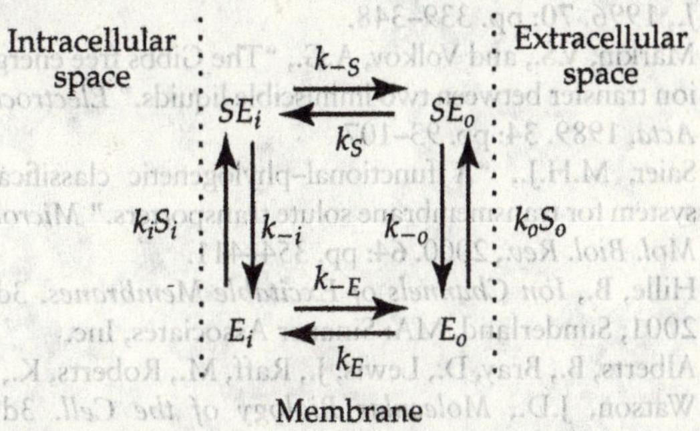

FIGURE 14.23 A kinetic scheme of uniport. S and E represent substrate and uniporter, respectively; other symbols are rate constants. The subscripts i and o represent intracellular and extracellular spaces, respectively.

14.12 During symport, the binding of the two substrates to the carrier can be random, sequential, or simultaneous (see Figure 14.17). We have derived the net fluxes of the substrates on the basis of the random model. Use a similar procedure to derive the net fluxes of the substrates on the basis of the sequential (or ordered) model. Assume that the substrate–symporter binding has a stoichiometry of 1:1 and that A binds to the symporter first on both sides of the membrane (see Figure 14.17b$_1$). Compare the net fluxes of the substrates with that derived from the random model (i.e., Equation (14.5.39)).

14.8 REFERENCES

1. Koeppen, B.M., and Stanton, B.A., *Renal Physiology*. 3d ed. 2001, St. Louis, MO: Mosby, Inc.
2. Lote, C.J., *Principles of Renal Physiology*. 4th ed. 2000, Boston, MA: Kluwer Academic Publishers.
3. Valtin, H., and Schafer, J.A., *Renal Function: Mechanisms Preserving Fluid and Solute Balance in Health*. 3d ed. 1995, Boston: Little, Brown.
4. Chang, R., *Physical Chemistry for the Chemical and Biological Sciences*. 2000, Sausalito, CA: University Science Books.
5. Tinoco, I.J., Sauer, K., and Wang, J.C., *Physical Chemistry: Principles and Applications in Biological Sciences*. 3d ed. 2002, NJ: Prentice Hall.
6. Newman, J.S., *Electrochemical Systems*. 1973, Englewood Cliffs, NJ: Prentice-Hall.
7. Paula, S., Volkov, A.G., Van Hoek, A.N., Haines, T.H., and Deamer, D.W., "Permeation of protons, potassium ions, and small polar molecules through phospholipid bilayers as a function of membrane thickness." *Biophys. J.*, 1996. 70: pp. 339–348.
8. Markin, V.S., and Volkov, A.G., "The Gibbs free energy of ion transfer between two immiscible liquids." *Electrochim. Acta*, 1989. 34: pp. 93–107.
9. Saier, M.H.J., "A functional–phylogenetic classification system for transmembrane solute transporters." *Microbiol. Mol. Biol. Rev.*, 2000. 64: pp. 354–411.
10. Hille, B., *Ion Channels of Excitable Membranes*. 3d ed. 2001, Sunderland, MA: Sinauer Associates, Inc.
11. Alberts, B., Bray, D., Lewis, J., Raff, M., Roberts, K., and Watson, J.D., *Molecular Biology of the Cell*. 3d ed. 1994, New York, Garland Publishing, Inc.
12. Zhou, Y., Morais-Cabral, J.H., Kaufman, A., and MacKinnon, R., "Chemistry of ion coordination and hydration revealed by a K+ channel-Fab complex at 2.0 A resolution." *Nature*, 2001. 414: pp. 43–48.
13. Stein, W.D., *Channels, Carriers, and Pumps: An Introduction to Membrane Transport*. 1990, San Diego: Academic Press.
14. Cooper, G.M., *The Cell: A Molecular Approach*. 1997, Washington, DC: ASM Press.
15. Ross, M.H., Reith, E.J., and Romrell, L.J., *Histology: A Text and Atlas*. 2d ed. 1989, Baltimore: Williams & Wilkins.
16. Tortora, G.J., and Grabowski, S.R., *Principles of Anatomy and Physiology*. 7th ed. 1993, New York, NY: HarperCollins College Publishers.
17. Deen, W.M., Lazzara, M.J., and Myers, B.D., "Structural determinants of glomerular permeability." *Am. J. Physiol. Renal Physiol.*, 2001. 281: pp. F579–F596.
18. Anderson, J.L., and Quinn, J.A., "Restricted transport in small pores: a model for steric exclusion and hindered particle motion." *Biophys. J.*, 1974. 14: pp. 130–150.
19. Deen, W.M., "Hindered transport of large molecules in liquid-filled pores." *AIChE J.*, 1987. 33: pp. 1409–1425.
20. Deen, W.M., Bohrer, M.P., and Brenner, B.M., "Macromolecule transport across glomerular capillaries: application of pore theory." *Kidney Intl.*, 1979. 16: pp. 353–365.
21. Daniels, B.S., Hauser, E.B., and Deen, W.M., "Glomerular basement membrane: in vitro studies of water and protein permeability." *Am. J. Physiol.*, 1992. 262: pp. F919–F926.
22. Chang, N., and Fujita, T., "A numerical model of acid–base transport in rat distal tubule." *Am. J. Physiol. Renal Physiol.*, 2001. 281: pp. F222–F243.
23. Weinstein, A.M., "A mathematical model of the outer medullary collecting duct of the rat." *Am. J. Physiol. Renal Physiol.*, 2000. 279: pp. F24–F45.
24. Layton, H.E., and Weinstein, A.M., eds. *Membrane Transport and Renal Physiology*. 2002, New York: Springer-Verlag.
25. Chang, H., and Fujita, T., "A numerical model of the renal distal tubule." *Am. J. Physiol.*, 1999. 276: pp. F931–F951.
26. Crane, R.K., "The gradient hypothesis and other models of carrier-mediated active transport." *Rev. Physiol. Biochem. Pharmacol.*, 1977. 78: pp. 99–159.
27. Monroy, A., Plata, C., Hebert, S.C., and Gamba, G., "Characterization of the thiazide-sensitive Na^+–Cl^- cotransporter: a new model for ions and diuretics interaction." *Am. J. Physiol. Renal Physiol.*, 2000. 279: pp. F161–F169.
28. Jennings, M.L., and Milanick, M.A., "Membrane transport in single cells." In *Section 14: Cell Physiology*, J. F. Hoffman and J.D. Jamieson, editors. 1997. New York: Oxford University Press, pp. 261–308.
29. Knauf, P.A., Raha, N.M., and Spinelli, L.J., "The noncompetitive inhibitor WW781 senses changes in erythrocyte anion exchanger (AE1) transport site conformation and substrate binding." *J. Gen. Physiol.*, 2000. 115: pp. 159–173.
30. Garay, R.P. and Garrahan, P.J., "The interaction of sodium and potassium with the sodium pump in red cells." *J. Physiol.*, 1973. 231: pp. 297–325.
31. Strieter, J., Stephenson, J.L., Giebisch, G., and Weinstein, A.M., "A mathematical model of the rabbit cortical collecting tubule." *Am. J. Physiol.*, 1992. 263: pp. F1063–F1075.

Drug Transport in Solid Tumors

CHAPTER 15

15.1 Introduction

15.1.1 Drug Delivery in Cancer Treatment

Tumors are abnormal masses of tissues in the body that arise when the mechanisms that regulate cell proliferation and differentiation are altered either by external mutagenic agents, such as chemicals, viruses, and radiation, or by internal errors in the duplication of genomic DNA during cell division. As a result, tumor cells grow in an unregulated manner. In general, tumors can be divided into two groups—malignant or benign—by the level of invasiveness and metastasis. Malignant tumors can invade and destroy surrounding normal tissues, and their cells can also disperse (i.e., *metastasize*) to distant normal organs through blood and lymph circulation systems. Benign tumors grow by expanding their volume without metastasizing and invading other tissues. Tumors can be further classified in terms of organs in which they grow, such as lung tumor, brain tumor, and liver tumor, or in terms of the types of tissue from which they originate, or in terms of gene expression profiles (or called gene expression signatures). Tumors that do not contain macroscopic regions of liquids are collectively called *solid* tumors.

Cancer is synonymous with *malignant tumor* and is a name for more than 100 different diseases. Epidemiologically, the incidence of cancer is due to environmental factors (e.g., smoking, pollution, viruses, and specific carcinogens to which individuals may be exposed) and an aging population. Some cancers (e.g., breast and colon cancers) can also be inherited from parents. Currently, cancer is treated primarily by surgery, targeted radiation, and the application of cytotoxic agents (i.e., chemotherapy). To treat certain tumors near the surface of the body or near the lining of internal organs, photodynamic therapy has also been clinically applied. Developments in molecular and cell biology, genetics, and genomics suggest that many new approaches to cancer treatment, such as antiangiogenesis therapy, immunotherapy, and gene therapy, will become available to the clinic in the next few decades.

Except for surgery, most of these approaches involve the delivery of therapeutic agents to tumor cells. The following is a brief summary of these agents and their roles in cancer treatment:

- *Chemotherapy* involves the delivery of cytotoxic agents into tumor cells. The concentration of these agents in every tumor cell must be higher than a therapeutic level before tumor cells can be killed. However, most anticancer drugs are toxic to both tumor and normal cells, so that the dose administered in the systemic delivery of drugs is restricted by normal tissue tolerance.
- *Radiation therapy* requires the presence of oxygen at targeted sites. Oxygen reacts with ionizing radiation to produce free radicals—molecules that have extra pairs of electrons. These electrons in turn can react with DNA and cause damage to tumor cells. Free radicals can enhance the radiation damage to tumor cells by up to about 300%.
- *Photodynamic therapy* involves the delivery of non-ionizing photons by a laser beam following the delivery of photosensitizers into tumor tissues. Photosensitizers are molecules that react with photons to produce free radicals, again damaging DNA in tumor cells.
- *Antiangiogenesis therapy* inhibits angiogenesis, the formation of new blood vessels in tumors. The inhibition of angiogenesis may reduce tumor growth, which depends upon nutrients supplied by the blood. Angiogenesis can be inhibited by the delivery of antiangiogenic agents into tumor tissues; these agents can either decrease the concentration ratio of angiogenic factors to inhibitors in tumor tissues or directly block the activities of the angiogenic agents.
- *Immunotherapy* currently involves two approaches. One consists of the delivery of toxins or radioisotopes to tumor cells by monoclonal antibodies that are able to recognize specific antigens on the plasma membranes of tumor cells. The conjugates of antibodies and toxins (called immunotoxins), or antibodies and radioisotopes, are able to act specifically on tumor cells. Another approach is to activate the immune system and cause it to attack tumor cells by delivering chemicals (e.g., lymphokines) to immune cells (e.g., lymphocytes and macrophages). The chemicals can be vaccines that help the immune system to recognize cancer cells, agents that stimulate the immune system to attack tumor cells, or molecules that are required to maintain the activities of the immune system. Once the immune cells are activated, they must reach tumors through the blood circulation. Within a tumor, immune cells must first adhere to the vascular endothelial cells and then migrate across the microvessel wall and through the interstitial space. The amount of activated immune cells delivered into the tumor tissues determines the clinical outcome of immunotherapy for cancer.
- *Gene therapy* involves the delivery of genes to the site of a tumor. Gene delivery, one of the major challenges in the development of this therapeutic approach, depends on *vehicles* (or *vectors*) to carry genes to the target sites in the body. Because vehicles are large molecules or nanoparticles, the transport of genes is severely hindered by various barriers in tumors.

In sum, various therapeutic agents must be delivered in the treatment of cancer. These agents can be as small as an oxygen molecule or as large as immune cells. To simplify the discussion in this book, all such agents are included under the term "drug."

15.1.2 Routes of Drug Administration

The routes by which drugs are administered can be divided into two categories: *local delivery* and *systemic delivery* (see Figure 15.1). Local delivery is often performed by the direct injection of drugs or by the implantation of controlled or sustained drug-release devices into tumors. In gene therapy, drugs can also be released locally from cells transfected with therapeutic genes. Local delivery bypasses the blood circulation and thus can significantly reduce the systemic toxicity of drugs. However, local delivery cannot be used to treat metastases—tumors formed by tumor cells disseminated from primary ones—when they are small and numerous in the body. Instead, the treatment of metastatic tumors requires systemic drug delivery.

In systemic delivery, drugs are either given orally or injected directly into a vein, the subcutaneous space, a muscle, or the peritoneal cavity. Except in intravenous injection, drugs enter the systemic blood circulation by being absorbed into the local blood microcirculation in the intestine or in the injected tissues. In the blood, drugs may bind to plasma proteins and be delivered to different organs. In the

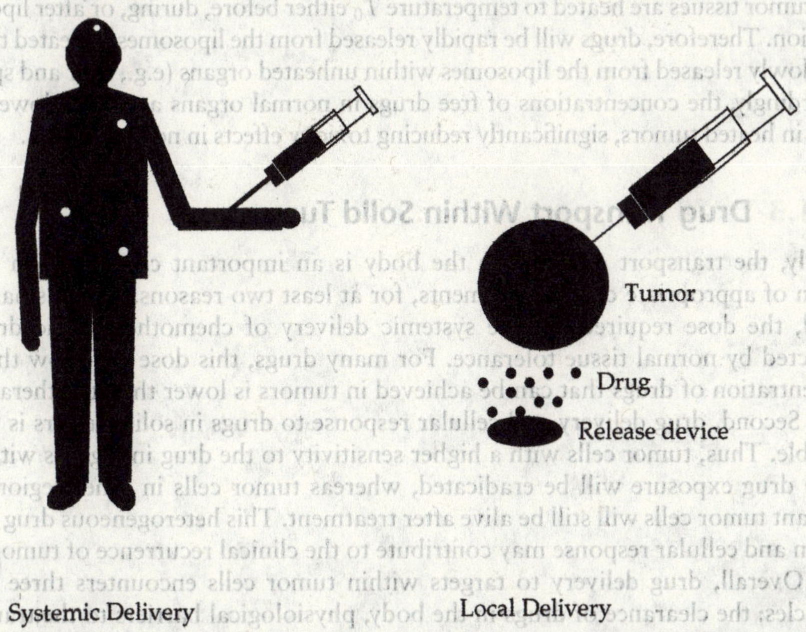

FIGURE 15.1 Routes for drug delivery to tumor cells. In systemic delivery, drugs are injected directly into either a vein, the subcutaneous space, a muscle, or the peritoneal cavity. Drugs can also be given orally. They will be absorbed in the intestine and then enter the systemic circulation. In local delivery, drugs are either injected directly into tumors or released from specific devices implanted in tumors or surrounding tissues. Drugs can also be released locally from cells transfected with therapeutic genes.

liver, drugs can be metabolized; in some cases, metabolites are the active forms of drugs. Drugs are removed from the body mainly through the liver and the kidney. The plasma half-life of drugs (i.e., the time required for the plasma concentration of a drug to be reduced to one-half of its initial value; see Chapter 16) varies from a few seconds to a few days. Any therapeutic agents that remain in the blood circulation retain a chance to be delivered to tumor tissues.

Sometimes systemic delivery and local release can be combined to improve the specificity and the efficiency of drug delivery. A case in point is the delivery of drugs via temperature-sensitive liposomes [1], which are fluid nanoparticles covered with one or more concentric phospholipid bilayers that are similar to the plasma membranes of mammalian cells. In addition to phospholipids, other molecules that alter the physical and chemical properties of liposomes can be incorporated into the bilayers. Liposomes used in most drug delivery studies have just one bilayer and are about 100 nm in diameter. The temperature-sensitive liposomes are those liposomes whose bilayer becomes leaky at a specific temperature T_0, due to either a phase transition or some other changes in the bilayer structure. Anticancer drugs encapsulated in these liposomes will be rapidly released only in tissues where the temperature is near T_0. The value of T_0, which can be predetermined via different liposome preparations, is usually between 39°C and 45°C. Liposomal drugs injected intravenously into the body will accumulate in both solid tumors and some normal organs (e.g., liver, spleen, and bone marrow); however, only tumor tissues are heated to temperature T_0 either before, during, or after liposome injection. Therefore, drugs will be rapidly released from the liposomes in heated tumors and slowly released from the liposomes within unheated organs (e.g., liver and spleen). Accordingly, the concentrations of free drugs in normal organs are much lower than those in heated tumors, significantly reducing toxicity effects in normal tissues.

15.1.3 Drug Transport Within Solid Tumors

Clearly, the transport of drugs in the body is an important consideration in the design of appropriate cancer treatments, for at least two reasons. First, as has been noted, the dose required for the systemic delivery of chemotherapeutic drugs is restricted by normal tissue tolerance. For many drugs, this dose is so low that the concentration of drugs that can be achieved in tumors is lower than the therapeutic level. Second, drug delivery and cellular response to drugs in solid tumors is highly variable. Thus, tumor cells with a higher sensitivity to the drug in regions with adequate drug exposure will be eradicated, whereas tumor cells in other regions and resistant tumor cells will still be alive after treatment. This heterogeneous drug distribution and cellular response may contribute to the clinical recurrence of tumors.

Overall, drug delivery to targets within tumor cells encounters three major obstacles: the clearance of drugs in the body, physiological barriers to the transport of therapeutic agents from the sites of administration to tumor cells, and the resistance of tumor cells to drugs. In order to improve the transport of therapeutic agents to tumor cells, an understanding of how drug molecules move in the body and within tumor tissues is required.

A tumor tissue can be divided into three compartments:[1] blood vessels, interstitium, and cells (see Figure 15.2). (In normal tissues, there are also lymph vessels, but

[1]See Chapter 16 for a complete discussion of compartment models.

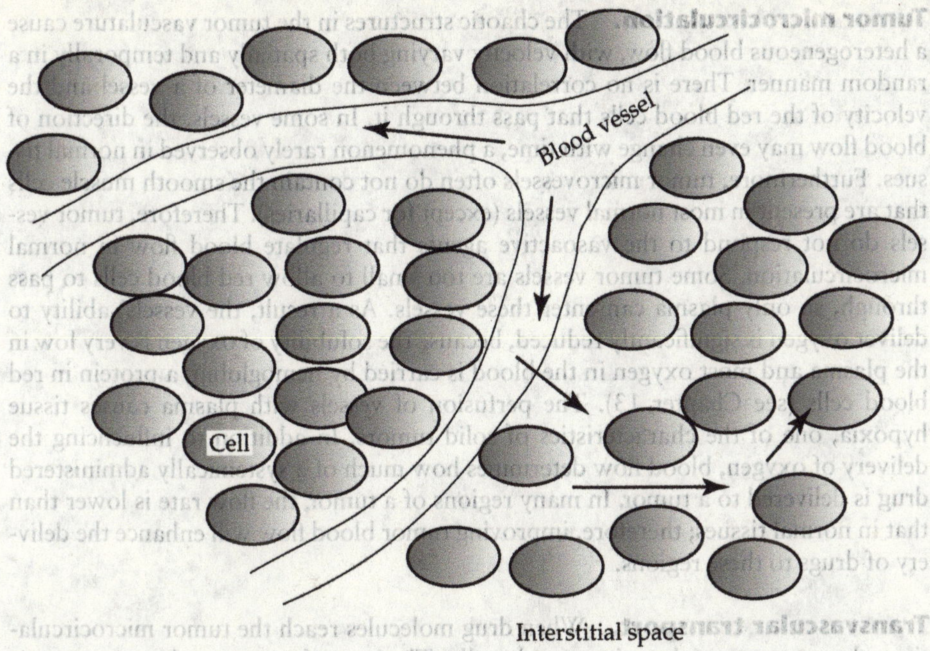

FIGURE 15.2 Schematic of the transport routes from blood microvessels to target sites in tumor cells. The arrows indicate directions of blood flow in microvessels, transvascular transport, interstitial transport, and transport into cells, respectively.

functional lymph vessels in solid tumors have not yet been observed.) The volume fraction of blood vessels varies from 0.01 in a rat fibrosarcoma (a tumor in fibrous tissues) to 0.28 in a dog lymphosarcoma (a tumor in lymphatic tissues) [2]. The range-of-volume fraction of interstitial space is between 0.13 in a human meningioma (tumor in the meninges, the membranes that cover and protect the brain and spinal cord) to 0.60 in a rat fibrosarcoma [3]. In normal tissues, the volume fraction of vessels varies from 0.019 in the skin to 0.300 in the lung, and the volume fraction of interstitial space varies from 0.06 in the brain to 0.30 in the skin [4].

Tumor vasculature. Vasculature in tumors is quite different from that in normal organs. Blood enters a normal organ through large arteries, which subsequently bifurcate into small arteries, arterioles, precapillary arterioles, and capillaries, the smallest blood vessels in the vascular network. Capillaries are responsible for nutrient and waste exchange between the blood and tissues. Blood from capillaries is collected by postcapillary venules, from which it then enters venules, small veins, and large veins, in turn. The large veins eventually send the blood from the organs back to the systemic circulation. In tumor vascular networks, there is no such order. Except for large arteries and veins that connect a tumor to the systemic circulation, structures in other vessels are heterogeneous and cannot be classified into any category. Therefore, they are collectively referred to as *tumor microvessels*. The tumor vascular network is chaotic. Unique features that rarely occur in normal vasculature include the formation of vessel loops and shunts, the bifurcation of a small vessel into two large vessels, and the trifurcation of vessels.

Tumor microcirculation. The chaotic structures in the tumor vasculature cause a heterogeneous blood flow, with velocity varying both spatially and temporally in a random manner. There is no correlation between the diameter of a vessel and the velocity of the red blood cells that pass through it. In some vessels, the direction of blood flow may even change with time, a phenomenon rarely observed in normal tissues. Furthermore, tumor microvessels often do not contain the smooth muscle cells that are present in most normal vessels (except for capillaries). Therefore, tumor vessels do not respond to the vasoactive agents that regulate blood flow in normal microcirculation. Some tumor vessels are too small to allow red blood cells to pass through, so only plasma can enter these vessels. As a result, the vessels' ability to deliver oxygen is significantly reduced, because the solubility of oxygen is very low in the plasma and most oxygen in the blood is carried by hemoglobin, a protein in red blood cells (see Chapter 13). The perfusion of vessels with plasma causes tissue hypoxia, one of the characteristics of solid tumors. In addition to influencing the delivery of oxygen, blood flow determines how much of a systemically administered drug is delivered to a tumor. In many regions of a tumor, the flow rate is lower than that in normal tissues; therefore, improving tumor blood flow will enhance the delivery of drugs to these regions.

Transvascular transport. When drug molecules reach the tumor microcirculation, they may cross the microvessel walls. The rate of transvascular transport is characterized by the *microvascular permeability coefficient* (see Section 9.3.3), which depends upon physical and chemical properties of the drug and structures of the vessel wall [5]. Tumor microvascular permeability is heterogeneous and is, on average, higher than that in normal tissues. In tumor vessels, the basement membranes are incomplete, and large pores exist in the vessel wall that allow the transport of macromolecules and nanoparticles. The cutoff pore size is heterogeneous and can be up to 2 microns in diameter. In most normal vessels, by contrast, the cutoff pore size is less than 20 nm. This difference between normal and tumor vessels suggests a strategy for delivering drugs specifically to solid tumors, since therapeutic agents larger than 20 nm and smaller than the cutoff size of pores in tumor vessels will accumulate preferentially in tumors. Such a strategy has been successfully applied to the delivery of doxorubicin, a potent anticancer drug, into solid tumors in patients. When doxorubicin is injected as a free drug, it may cause a severe cardiotoxicity. However, the encapsulation of doxorubicin into liposomes, mentioned earlier, can eliminate the cardiotoxicity because liposomal doxorubicin cannot cross the microvessel wall in the heart.

Tumor and normal microvascular permeabilities can be modulated either chemically or physically, although mechanisms of the modulation are not completely understood. Among all chemicals studied so far, the most potent factor for increasing microvascular permeability is *vascular endothelial growth factor* (VEGF), also called *vascular permeability factor* (VPF). VEGF can increase vascular permeability at a concentration of less than 1 nM, making it about 50,000 times more potent than histamine. Signal transduction in cells treated with VEGF [6] or histamine [7] involves the synthesis of nitric oxide (NO) in endothelial cells. Hence, inhibiting NO synthesis may abolish the effect of these agents on microvascular permeability [6–8]. VEGF, which can be released by various tumor cells and normal cells (e.g., fibroblasts, macrophages, and epidermal keratinocytes), is often overexpressed in the tumors. The expression of VEGF can be stimulated by hypoxia, growth factors, and tumor promoters. Eliminating VEGF in tumors can significantly reduce the microvascular permeability and the cutoff size of pores in the vessel wall. In addition, eliminating VEGF can cause vessel regression in tumors.

Vasoactive agents (e.g., leukotrienes and bradykinin) also increase microvascular permeability. These agents are especially useful in selectively enhancing the delivery of drugs to brain tumors. The reason is as follows: Normal brain capillaries, which are part of the blood–brain barrier (BBB), are protected by a biochemical barrier that blocks the effects of vasoactive agents. This biochemical barrier is absent in brain tumors. Therefore, vasoactive agents can increase tumor microvascular permeability while having minimal effects on the normal BBB.

Dexamethasone, a potent synthetic glucocorticoid, can be used to reduce microvascular permeability in tumors. Over the past three decades, dexamethasone has been utilized to treat cerebral edema caused by brain tumors or by brain surgery. In a recent study, the drug was used to decrease tumor interstitial fluid pressure by reducing tumor microvascular permeability [9]. The reduction is related to a down-regulation of VEGF expression in tumor cells and a decrease in the response of tumor endothelial cells to VEGF. A more direct approach to decreasing microvascular permeability is to treat tumors with antibodies that can block the binding of VEGF to its receptors on endothelial cells [5].

The regulation of microvascular permeability in normal tissues is also an important issue in cancer treatment [5]. For example, vasogenic cerebral edema remains a common problem for many brain cancer patients; there is a direct correlation between the edema, caused by the disruption of the BBB, and mortality. Another example is vascular leak syndrome (VLS), observed mainly in human patients. As one of the limiting factors in cancer treatment based on immunotoxin or interleukin-2 (IL-2), VLS is recognized by hypoalbuminemia (i.e., low albumin content in the plasma), peripheral edema, and fluid retention in the body. The mechanisms of VLS remain to be determined, but it has been hypothesized that IL-2-induced VLS is caused by the activation of complement and contact systems, as well as by leukocyte-mediated injury to endothelial cells. In contrast, immunotoxin-induced VLS is hypothesized to be the consequence of a disruption of the interactions between endothelial cells and the underlying extracellular matrix. Dexamethasone and inhibitors of complement and contact systems may, to a certain extent, reduce VLS.

Vascular permeability can be modulated, or at least changed, by physical means as well. Hyperosmolar solutions (e.g., mannitol solution at concentration of 1 M) have been used to open the interendothelial junctions and thereby improve drug delivery to tumors, especially those in the brain. However, the increased drug exposure in brain tumors is less than that in normal brain tissues. Consequently, the hyperosmolar solution treatment may significantly decrease the therapeutic index of many drugs. Some treatment modalities, such as ionizing radiation and hyperthermia, may also increase microvascular permeability in both normal and tumor tissues. Used in combination with chemotherapy, these modalities may improve the efficacy of drugs by increasing the amounts delivered to tumors.

Interstitial transport. When therapeutic agents cross the microvessel walls or are delivered locally from controlled-release devices, they move and react with other chemicals in the interstitial space of tumors. The interstitial space, a tissue compartment between blood vessels and cells (see Figure 15.2), consists of interstitial fluid and extracellular matrix. As noted previously, the matrix is formed by cross-linked fibers of collagens, elastins, glycosaminoglycans, proteoglycans, and structural glycoproteins [10]. The interstitial fluid is a solution of many compounds (e.g., nutrients, wastes, growth factors, and inhibitors). The main chemical contents of interstitial fluid in tumors are comparable to those in the plasma, except that tumor interstitial fluid is often acidic and contains less oxygen and glucose than the plasma does. The interstitial fluid can either flow freely or be immobilized by the cross-linked fibers to form hydrophilic gels.

Interstitial transport is one of the most difficult steps in both systemic and local drug delivery to tumor cells [11]. The difficulty is attributed to several unique characteristics at molecular, cellular, and tissue levels in solid tumors: (a) low convective transport due to elevated interstitial fluid pressure and the lack of functional lymphatic vessels; (b) outward gradients of the interstitial fluid pressure, which may cause convective transport of extravasated drugs from the interior to the periphery of tumors; (c) large diffusion distances in some regions of the interstitium; and (d) binding of drugs to various components in tissues, including soluble proteins released by the tumor and infiltrated host cells (e.g., macrophages and fibroblasts), the extracellular matrix, and receptors on cell membranes. Interstitial transport is especially difficult for the delivery of large therapeutic agents (e.g., monoclonal antibodies and genes) to cells in deeper layers of tumor tissues.

The main modes of drug transport in the interstitium are diffusion and convection. In normal tissues, diffusion over a distance between two adjacent blood vessels has been found to be dominant for small molecules and convection to be more important for large molecules. This is because the diffusion coefficient is molecular-size dependent (see Chapters 6 and 8). However, convection may not be important in the interstitial transport of any molecules in the interior of solid tumors, due to the absence of a driving force as previously discussed. Convection is important only at the periphery, where there is a sharp decrease in the interstitial fluid pressure. This convective transport oozes drug from the interior of the tumor to surrounding host tissues, thus reducing, rather than improving, drug delivery to tumor cells.

Intracellular transport. For drugs targeting molecules in the extracellular matrix or on the plasma membranes of cells, interstitial transport is the last step in their delivery. However, most anticancer drugs target intracellular molecules. Therefore, it is necessary for these drugs to move across cell membranes and within cells. Transport into the nuclei of cells may also be required if the targets are there. As drug molecules journey from cell membranes to targets, they interact with various intracellular molecules and organelles. These interactions hinder drug transport and cause drug inactivation or degradation. Furthermore, tumor cells often develop multidrug resistance (MDR), a normal defense system of cells, in response to drug treatment at sublethal doses. The development of MDR involves the expression of various transmembrane proteins that either redistribute intracellular drugs away from their targets or function as molecular pumps that move drugs from intracellular to extracellular spaces against the concentration gradient across cell membrane [12,13]. The removal of drugs via either of these mechanisms increases the difficulty involved in delivering drugs to intracellular targets and has contributed to the incomplete response of tumors to drug treatments in patients.

Overall, drug delivery in solid tumors involves various transport phenomena at molecular, cellular, and tissue levels. Understanding the mechanisms of drug transport can facilitate drug delivery and improve the efficacy of drugs in cancer treatment.

15.2 Quantitative Analysis of Transvascular Transport

The transport of molecules across tumor vessels can be analyzed by using the methods described in Chapter 9. Therefore, we will not repeat them here. On the other hand, transvascular transport in tumor tissues has three unique features. First, microvascular permeability is generally higher in tumors than in normal tissues, except for some primary brain tumors. (In these brain tumors, endothelial cells are connected by tight junctions similar to those found in the BBB. The BBB is impermeable even to small molecules such as glucose.) Second, the cutoff size of pores in the vessel wall is generally larger in tumors than in normal tissues. Particles up to the size of several hundred nanometers can extravasate in tumors, but not in normal tissues. Large pores have also been found in liver sinusoids (the blood capillaries in the liver), but the cutoff size of these pores is only about 100 nm. Third, convective transport across the microvessel wall can be neglected in the tumor except at the periphery due to elevated microvascular permeability and interstitial fluid pressure (see Section 15.4.1).

15.3 Quantitative Analysis of Interstitial Fluid Transport

15.3.1 Governing Equations

Like that in normal tissues, interstitial fluid transport in tumor tissues is governed by the mass conservation law and momentum balance equations in both fluid and solid phases. Therefore, the equations discussed in Chapter 8 can be used to investigate fluid flow in tumor tissues. The relevant equations, (8.5.6) through (8.5.8), are repeated here for convenience:

$$\nabla \cdot \left(\varepsilon \mathbf{v}_f + (1 - \varepsilon)\frac{\partial \mathbf{u}}{\partial t} \right) = \phi_B - \phi_L, \tag{15.3.1}$$

$$\varepsilon\left(\mathbf{v}_f - \frac{\partial \mathbf{u}}{\partial t} \right) = -K\nabla p_i, \tag{15.3.2}$$

$$\nabla \cdot \boldsymbol{\sigma} = 0. \tag{15.3.3}$$

In these equations, ϕ_B and ϕ_L are, respectively, the rate of fluid extravasation from blood vessels per unit tissue volume and the rate of the lymphatic drainage per unit tissue volume, $\mathbf{v}_f$ is the average fluid velocity in the fluid space, $\mathbf{u}$ is the average displacement in the solid phase, ε is the fractional volume of the interstitial fluid, ∇p_i

is the interstitial fluid pressure gradient, K is the hydraulic conductivity of tissues, and $\boldsymbol{\sigma}$ is the effective stress tensor.

In general, biological tissues are viscoelastic materials in which $\boldsymbol{\sigma}$ is a nonlinear function of the strain tensor in the tissue, E:

$$\boldsymbol{\sigma} = f(\mathrm{E}). \tag{15.3.4}$$

The function depends on the material properties of tissues, and Equation (15.3.4) is called the *constitutive equation* of tissues. The transvascular transport of fluid in solid tumors is governed by Starling's law (see Section 9.3.2), namely,

$$\phi_B = \frac{J_V}{V} = \frac{L_P S}{V}[p_B - p_i - \sigma_s(\pi_B - \pi_i)], \tag{15.3.5}$$

where p and π are the hydrostatic and osmotic pressures, respectively, the subscripts B and i represent microvascular and interstitial space, respectively, σ_s is the osmotic reflection coefficient, L_P is the hydraulic permeability of the microvessel wall in the tumor, and S/V is the vascular surface area per unit tissue volume.

There are no functional lymph vessels in solid tumors. Thus,

$$\phi_L = 0. \tag{15.3.6}$$

The hydraulic conductivity K depends upon the temperature and the structure of the tissue. The temperature dependence is determined mainly by the viscosity, μ, since the specific hydraulic permeability of tissues (see Section 8.3.1) is nearly independent of the temperature. Empirically, the values of K from different tissues are related to the tissue concentration of glycosaminoglycans (GAG), g/100 g tissue [14], or

$$K = 4.6 \times 10^{-13}[\mathrm{GAG}]^{-1.202}(\mu_{37}/\mu), \tag{15.3.7}$$

where μ_{37} is the viscosity of the interstitial fluid at 37°C and the unit of K is $\mathrm{cm^4\ dyn^{-1}\ s^{-1}}$. K is highly sensitive to changes in tissue structures [15]. A small tissue deformation can lead to changes in K of several orders of magnitude. Thus, the hydraulic conductivity in Equation (15.3.2) cannot be treated as a constant in quantitative analyses of the intratumoral infusion of drugs and genes.

15.3.2 Unidirectional Flow of Fluid at Steady State

At the steady state with $\phi_L = 0$, Equations (15.3.1) and (15.3.2) become, respectively,

$$\nabla \cdot (\varepsilon \mathbf{v}_f) = \phi_B \tag{15.3.8}$$

and

$$\mathbf{v}_f = -\frac{K}{\varepsilon}\nabla p_i. \tag{15.3.9}$$

Substituting Equation (15.3.9) into Equation (15.3.8) and assuming that K is a constant, we have

$$-K\nabla^2 p_i = \phi_B. \tag{15.3.10}$$

Equation (15.3.10) shows that the displacement of tissues has no direct effect on the interstitial fluid flow during the steady state. However, it may affect the flow indirectly, through changes in K that are mentioned in the previous section. If the proper boundary conditions are given and all constant values are known, Equation (15.3.10) can be used to predict the velocity and interstitial fluid pressure (IFP) profiles in tumor tissues.

To demonstrate the applications of Equation (15.3.10) to fluid flow analysis, we consider a unidirectional-flow problem [16]. In this application, we assume that the solid tumor is a sphere and that the transport of fluid is spherically symmetrical around the center of the tumor. These assumptions are based on a tissue-isolated tumor model, which is a mass derived from tumor tissue transplantation into the ovarian fat pads of mice or rats. The model has only one feeding artery and one vein that drains the blood from the tumor. The tissue-isolated tumor is approximately spherical, and IFP at the tumor surface is approximately zero. Substituting Equation (15.3.5) into Equation (15.3.10) we have in a spherical coordinate system

$$\frac{1}{r^2}\frac{\partial}{\partial r}\left(r^2\frac{\partial p_i}{\partial r}\right) = \frac{\alpha^2}{R^2}(p_i - p_e), \qquad (15.3.11)$$

where R is the radius of the tumor,

$$\alpha = \sqrt{\frac{L_pS}{KV}} \qquad (15.3.12)$$

is the ratio of the flow resistance in the interstitial space to that in the microvessel wall, and

$$p_e = p_B - \sigma_s(\pi_B - \pi_i) \qquad (15.3.13)$$

is the effective pressure.

When $p_i = p_e$, there is no fluid exchange between blood vessels and the interstitial space. The boundary conditions are

$$\frac{\partial p_i}{\partial r} = 0 \qquad \text{at } r = 0 \qquad (15.3.14a)$$

and

$$p_i = 0 \qquad \text{at } r = R. \qquad (15.3.14b)$$

Integrating Equation (15.3.11) directly and using the boundary conditions in Equations (15.3.14a) and (15.3.14b), we obtain the pressure profile as a function of r (see Figure 15.3a):

$$p_i = p_e\left(1 - \frac{R}{r}\frac{\sinh(r\alpha/R)}{\sinh(\alpha)}\right). \qquad (15.3.15)$$

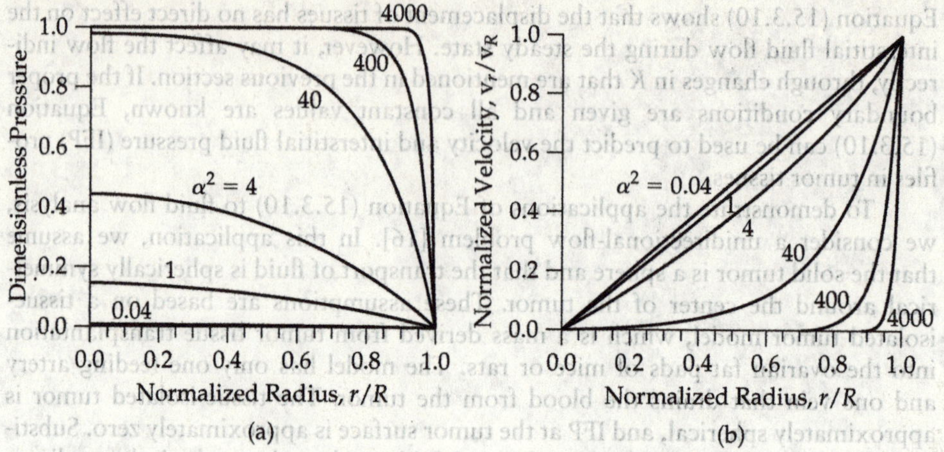

FIGURE 15.3 Pressure and velocity profiles as a function of r/R and α^2. The pressure is nondimensionalized by p_e; and the velocity is normalized by v_R, which is the velocity at $r = R$. (Modified from Ref. [16], with permission of Elsevier.)

Substituting Equation (15.3.15) into Equation (15.3.9), we get the profile of the r-component of v_f as a function of r (see Figure 15.3b):

$$v_r(r) = \frac{Kp_eR}{\varepsilon r^2 \sinh(\alpha)}\left[\frac{r\alpha}{R}\cosh\left(\frac{r\alpha}{R}\right) - \sinh\left(\frac{r\alpha}{R}\right)\right]. \qquad (15.3.16)$$

When the flow resistance of the interstitial space is much higher than that of the microvessel wall, $\alpha^2 \gg 1$. In this case, the IFP profile is uniform in the center of the tumor. The IFP gradient is negligible, which results in a low fluid velocity inside the tumor. When the flow resistance of the interstitial space is smaller than that of the microvessel wall, $\alpha^2 < 1$. As a result, the IFP is close to zero, and the fluid velocity decreases almost linearly from the surface to the center of the tumor (see Figure 15.3).

Experimentally, the IFP profiles in tumors are similar to those shown in Figure 15.3a, with large α values (see Figure 15.4). For a rat mammary adenocarcinoma (i.e., a breast cancer), $\alpha^2 = 1{,}200$, indicating that interstitial resistance is a rate-limiting factor, compared with the resistance across the microvessel wall, for fluid transport in tumors (see Equation (15.3.12)).

15.3.3 Unsteady-State Fluid Transport

The analysis of unsteady-state fluid flow is more complicated than that of steady-state flow, since tissue deformation is now coupled with the fluid flow. For unsteady-state flow problems, Equations (15.3.1), (15.3.2), (15.3.5), and (15.3.6) can still apply. These formulas govern the mass and momentum transport in the interstitial space, as well as the rates of fluid transport across the blood and lymph microvessel walls. There is no general constitutive equation for biological tissues. To simplify the analysis, we assume that tumor tissues are poroelastic materials and that tissue deformation is infinitesimal. Therefore, the stress–strain relationship in the solid phase obeys Hooke's law (see Section 8.5; i.e., the constitutive equation (15.3.4) becomes Equation (8.5.1)), and we have

$$\boldsymbol{\sigma} = 2\mu_G\mathbf{E} + \mu_\lambda e\mathbf{I} - p_i\mathbf{I}, \qquad (15.3.17)$$

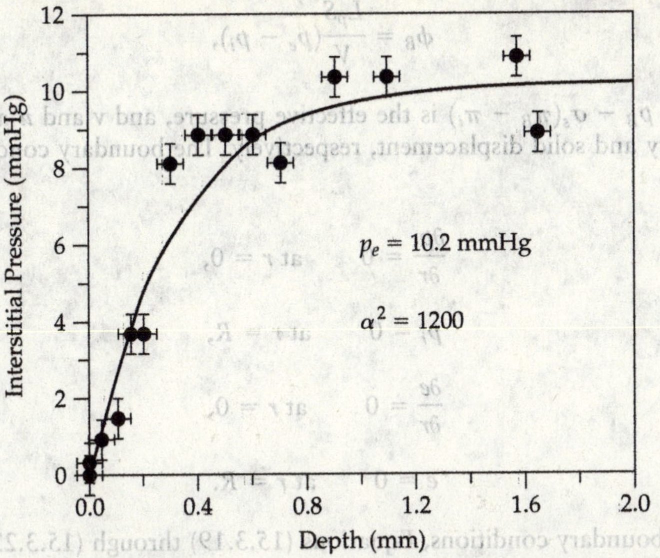

FIGURE 15.4 Experimental data of interstitial fluid pressure in a rat mammary adenocarcinoma. The solid line is generated from Equation (15.3.15), with p_e = 10.2 mmHg, R = 1.1 cm, and α^2 = 1,200. (Reprinted from Ref. [17], with permission.)

where $\mathbf{E} = [\nabla\mathbf{u} + (\nabla\mathbf{u})^T]/2$ and $e = \nabla\cdot\mathbf{u}$ is the *volume dilatation*. Physically, e is the change in volume per unit original volume. μ_G and μ_λ are Lamé constants. The momentum balance equation (15.3.3) can be replaced by Equation (8.5.13), which, from Section 8.5, is

$$\frac{\partial e}{\partial t} - K(2\mu_G + \mu_\lambda)\nabla^2 e = \phi_B. \tag{15.3.18}$$

Equation (15.3.18) is mathematically identical to the governing equation for the time-dependent diffusion of solutes, with the diffusion coefficient and the rate of chemical reaction being $K(2\mu_G + \mu_\lambda)$ and ϕ_B, respectively. The unsteady-state flow problems can be solved analytically or numerically on the basis of Equations (15.3.1), (15.3.2), (15.3.5), and (15.3.18).

If the fluid velocity is only in the radial direction and the transport constants are fixed, then Equations (15.3.1), (15.3.2), (15.3.5), and (15.3.18) can be simplified and expressed in spherical coordinates as

$$\varepsilon\frac{1}{r^2}\frac{\partial}{\partial r}(r^2 \mathrm{v}) + (1-\varepsilon)\frac{1}{r^2}\frac{\partial^2}{\partial r\partial t}(r^2 u) = \phi_B, \tag{15.3.19}$$

$$\varepsilon\left(\mathrm{v} - \frac{\partial u}{\partial t}\right) = -K\frac{\partial p_i}{\partial r}, \tag{15.3.20}$$

$$\frac{\partial e}{\partial t} - K(2\mu_G + \mu_\lambda)\frac{1}{r^2}\frac{\partial}{\partial r}\left(r^2\frac{\partial e}{\partial r}\right) = \phi_B, \tag{15.3.21}$$

and

$$\phi_B = \frac{L_p S}{V}(p_e - p_i), \tag{15.3.22}$$

where $p_e = p_B - \sigma_s(\pi_B - \pi_i)$ is the effective pressure, and v and u are the radial fluid velocity and solid displacement, respectively. The boundary conditions are as follows:

$$\frac{\partial p_i}{\partial r} = 0 \quad \text{at } r = 0, \tag{15.3.23a}$$

$$p_i = 0 \quad \text{at } r = R, \tag{15.3.23b}$$

$$\frac{\partial e}{\partial r} = 0 \quad \text{at } r = 0, \tag{15.3.24a}$$

$$e = 0 \quad \text{at } r = R. \tag{15.3.24b}$$

With these boundary conditions, Equations (15.3.19) through (15.3.22) have been solved analytically for two special cases [18], one involving a step increase in microvascular pressure, the other an abrupt stop in the transvascular exchange of fluid.

A step increase in microvascular pressure. If the microvascular pressure is increased as a step function of time, then the effective pressure p_e will also be increased as a step function of time (see Equation (15.3.13)). We assume that $p_e = p_{e0}$ when $t \leq 0$ and that $p_e = p_{e1}$ when $t > 0$. Using the separation-of-variables method, we can solve Equations (15.3.19) through (15.3.22) with the boundary conditions in Equations (15.3.23) and (15.3.24) (see Problem 15.3). The result is

$$p_i(r,t) = p_{e0}\left(1 - \frac{R}{r}\frac{\sinh(\alpha r/R)}{\sinh(\alpha)}\right) + \alpha^2(p_{e0} - p_{e1})\frac{R}{r}\sum_{n=1}^{\infty}\frac{2}{n\pi}(-1)^n c_n(1 - e^{-f_n t}), \tag{15.3.25}$$

where

$$c_n = \frac{\sin(n\pi r/R)}{(n\pi)^2 + \alpha^2},$$

$$f_n = [(n\pi)^2 + \alpha^2]\frac{K(2\mu_G + \mu_\lambda)}{R^2}, \tag{15.3.26}$$

$$\alpha^2 = R^2\frac{L_p S}{KV}.$$

The pressure profile predicted by Equation (15.23.25) has been verified experimentally [18] on the basis of the tissue-isolated tumor model introduced in Section 15.3.2. In this experiment, the tumor was derived from a type of human colon cancer cell transplanted into an immunodeficient mouse. During the experiment, the feeding artery was cannulated and infused with Krebs–Henseleit solution, a buffered saline developed in the 1930s for maintaining the viability of liver tissues. The infusion pressure was maintained at 60 mmHg. The IFP was monitored with a pressure transducer connected to a needle that was inserted into the middle of the tumor. At a specific point in time, the infusion pressure was suddenly increased from 60 mmHg to 94 mmHg and then reduced to 60 mmHg after 180 seconds. During the entire period, the IFP and the infusion pressure were recorded. The experimental results are shown in Figure 15.5. The IFP (i.e., p_i) was also predicted from Equation (15.3.25). The comparison between the experimental data and predictions from the model, also shown in the figure, indicates that they agree very well. Alternatively, the model's predictions and the experimental data can be compared in terms of the time constant τ_v, which determines how fast p_i will follow the changes in p_e:

$$\tau_v = \frac{1}{f_1} = \frac{R^2}{K(2\mu_G + \mu_\lambda)(\pi^2 + \alpha^2)}. \tag{15.3.27}$$

As previously discussed, the value of α^2 is, in general, much larger than unity in tumor tissues. Thus, Equation (15.3.27) can be written as

$$\tau_v = \frac{R^2}{K(2\mu_G + \mu_\lambda)\alpha^2} = \frac{V}{L_P S(2\mu_G + \mu_\lambda)}. \tag{15.3.28}$$

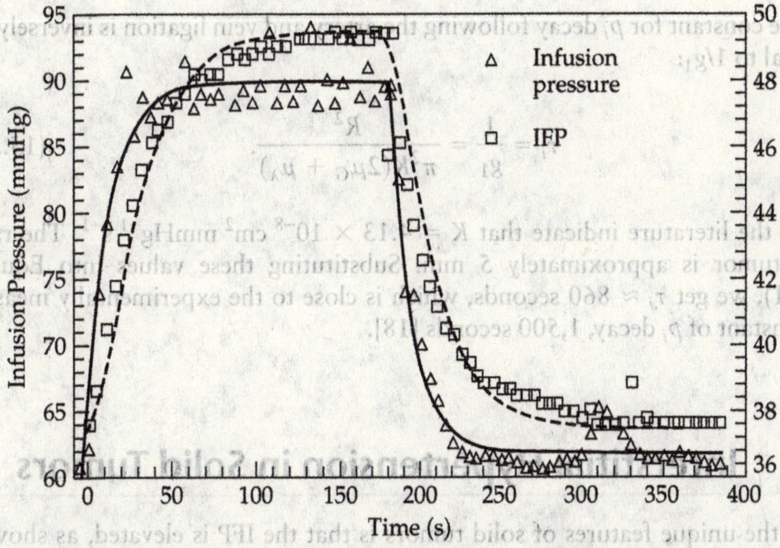

FIGURE 15.5 Comparison between experimental data (symbols) and model predictions (dashed line). The IFP and the infusion pressure were measured simultaneously when the infusion pressure was suddenly increased from 60 mmHg to 94 mmHg. After 180 seconds, the infusion pressure was reduce to 60 mmHg. The dashed line is the model prediction based on Equation (15.3.25) (Reprinted from Ref. [18], with permission.)

Data in the literature indicate that $L_P = 3.6 \times 10^{-7}$ cm mmHg^{-1} s^{-1}, $S/V = 200$ cm^{-1}, $\mu_G = 15.2$ mmHg, and $\mu_\lambda = 684$ mmHg [18]. Substituting these values into Equation (15.3.28), we get $\tau_v \approx 19$ seconds. Experimentally, it is difficult to produce a step increase in the arterial pressure. Thus, it is more appropriate to estimate τ_v by using the time delay between the change in IFP and the increase in arterial pressure. The data shown in Figure 15.5 indicate that the time delay is approximately 11 seconds, which is close to the prediction from the model.

An abrupt stop of the transvascular fluid exchange. If both the feeding artery and the draining vein are suddenly ligated, the transvascular fluid exchange stops immediately. In this case, the experimental condition can be modeled by setting ϕ_B to zero for $t > 0$. The analytical solution for p_i can then be obtained by using a method similar to that described in the previous section. The result is

$$p_i(r,t) = p_{e0}\left(1 - \frac{R}{r}\frac{\sinh(\alpha r/R)}{\sinh(\alpha)}\right) + \alpha^2 p_{e0}\frac{R}{r}\sum_{n=1}^{\infty}\frac{2}{n\pi}(-1)^n c_n(1 - e^{-g_n t}), \tag{15.3.29}$$

where c_n is defined in Equation (15.3.26) and

$$g_n = (n\pi)^2\frac{K(2\mu_G + \mu_\lambda)}{R^2}. \tag{15.3.30}$$

The time constant for p_i decay following the artery and vein ligation is inversely proportional to $1/g_1$:

$$\tau_i = \frac{1}{g_1} = \frac{R^2}{\pi^2 K(2\mu_G + \mu_\lambda)}. \tag{15.3.31}$$

Data in the literature indicate that $K = 4.13 \times 10^{-8}$ cm^2 mmHg^{-1} s^{-1}. The radius of the tumor is approximately 5 mm. Substituting these values into Equation (15.3.31), we get $\tau_i \approx 860$ seconds, which is close to the experimentally measured time constant of p_i decay, 1,500 seconds [18].

15.4 Interstitial Hypertension in Solid Tumors

One of the unique features of solid tumors is that the IFP is elevated, as shown in Figure 15.4. The first evidence of elevated tumor IFP was published by Young et al. in 1950 [17]. These authors found that, in a rabbit carcinoma, IFP = 19.3 ± 0.6 mmHg. In the 1990s, Jain and his collaborators demonstrated that the IFP was high in both animal and human tumors [11]. Table 15.1 lists typical values of IFP in human tumor and normal tissues.

TABLE 15.1

Interstitial Fluid Pressure (mmHg) in Human Tumor and Normal Tissues

Tumor or Normal Tissue	N[a]	Mean	Range
Brain tumors[b]	17	7	2 to 15
Brain tumors[b,c]	11	1	−0.5 to 8
Breast carcinomas	13	29	5 to 53
Breast carcinomas	8	15	4 to 33
Cervical carcinomas	26	23	6 to 94
Cervical tumors	102	19[d]	−3 to 48
Colorectal liver metastases	8	21	6 to 45
Head and neck carcinomas	27	19	1.5 to 79
Lung carcinomas	26	10	1 to 27
Lymphomas	7	4.5	1 to 12.5
Metastatic melanomas	14	21	0 to 60
Metastatic melanomas	12	14.5	2 to 41
Renal cell carcinomas	1	38	N/A
Normal breast	8	0	−0.5 to 3
Normal skin	5	0.4	−1 to 3

Source: Data collected from Ref. [11,19].
[a]*N* indicates the number of patients.
[b]The IFP was measured after patients were treated with antiedema agents.
[c]Some tumors are reported twice in this table, because they are obtained from different studies.
[d]Only the median of IFP was reported.

15.4.1 Effects of Interstitial Hypertension on Drug and Gene Delivery

Interstitial hypertension in tumors can affect the delivery of drugs and genes via four different mechanisms. First, the elevated IFP is nearly uniform (i.e., the IFP gradient is close to zero), except in a thin tissue layer (< 0.5 mm thick) at the periphery of the tumor (see Figure 15.4). As a result, convective transport is negligible in the interior of a tumor. Second, the IFP gradient is high and inward in the peripheral region, indicating that the direction of convection is outward. Drugs delivered into tumors can be pushed out, causing drugs to accumulate only at the periphery. Third, the elevated IFP reduces the pressure gradient across the microvessel wall. Boucher and his coworkers showed that the IFP is close to the microvascular pressure (MVP) [20] (see Figure 15.6). In addition, the osmotic pressure difference across the microvessel wall is minimal, due presumably to the elevated microvascular permeability [21]. Thus, according to Starling's law (see Section 9.3.2), convective transport across the microvessel wall is significantly reduced in the interior of a tumor. Finally, the interstitial hypertension, in combination with the hyperpermeability of tumor vessels (Section 15.2), may cause a significant reduction in blood flow in a tumor [22].

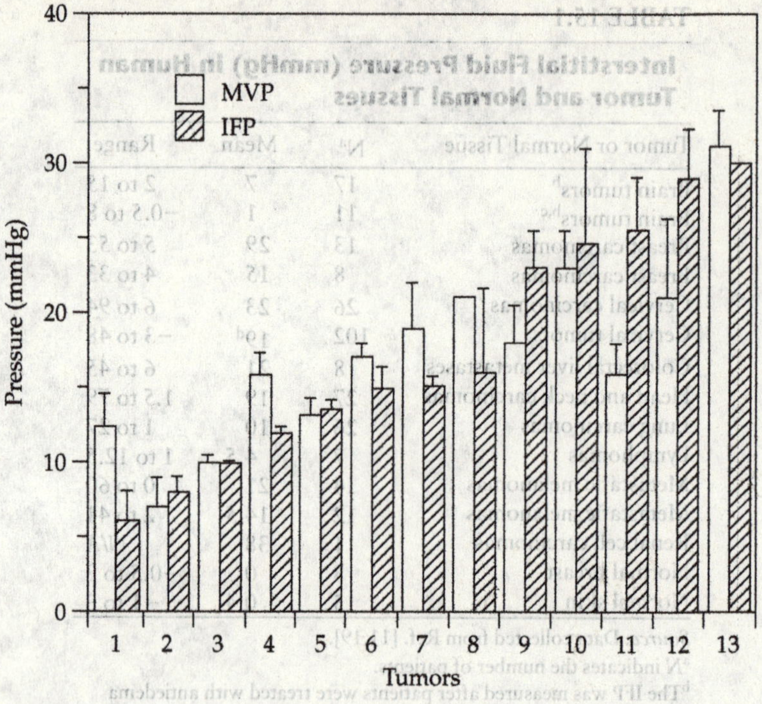

FIGURE 15.6 Comparison of microvascular pressure (MVP) and interstitial fluid pressure (IFP) measured in 13 tumors. The data show that IFP is correlated with MVP. (Reprinted from Ref. [20], with permission.)

15.4.2 Etiology of Interstitial Hypertension

The mechanisms of interstitial hypertension in tumors have been studied extensively. At the early stage of tumor growth, hypertension is caused by the vascular leakiness in solid tumors [23]. The leakiness increases fluid and plasma protein extravasation, which in turn results in an elevation of both hydrostatic and osmotic pressures in the interstitial space. At this stage, IFP increases with tumor size, although there is no correlation between IFP and tumor size when tumors are large.

Another important mechanism of the IFP elevation is the solid mechanical stress that is generated by tumor cell proliferation in a confined space. The solid stress has not been quantified directly; the only information in the literature is about the stress generated by a spheroidal tumor growing in agarose gels. This stress ranges from 45 to 120 mmHg [24]. The solid stress pushes surrounding tissues and may reduce the size, or cause the collapse, of blood vessels (see Figure 15.7). The compression of surrounding tissues can directly increase IFP, whereas the reduction in vessel size causes a higher resistance to blood flow. The increase in the flow resistance results in a higher MVP, which is a driving force for IFP. Thus, IFP is indirectly elevated by vessel compression.

In addition to the forces just mentioned, the absence of lymph vessels in tumors also contributes to the elevation of IFP. At present, no functional lymph vessels have been found in solid tumors, although histological staining with specific monoclonal antibodies has identified lymph endothelial cells in tumor tissues.

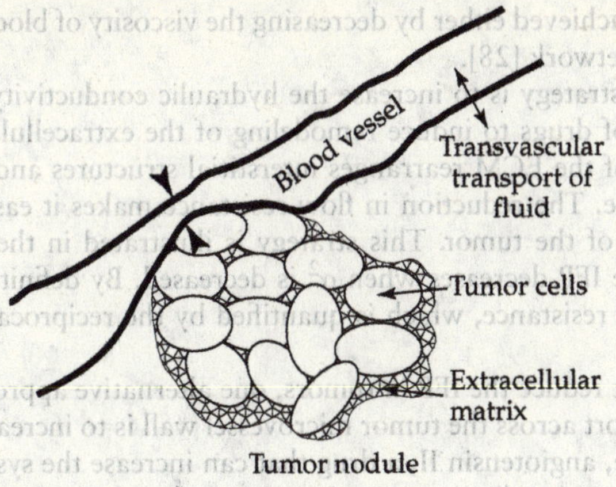

FIGURE 15.7 A mechanism of IFP elevation due to tumor cell proliferation in a confined tumor nodule. Cell proliferation exerts solid stress on the blood vessel to reduce its lumen and thus increase its resistance to blood flow. These changes in blood vessels will increase both MVP and IFP. (Modified from Ref. [25], with permission.)

15.4.3 Strategies for Reducing Interstitial Fluid Pressure

Elevated IFP is one of the major obstacles to the delivery of drugs and genes to solid tumors. As mentioned previously, elevated IFP reduces convective transport, both across the microvessel wall and in the interstitial space. In addition, it may cause drugs that are already inside the tumor to ooze out, because of the sharp IFP gradient at the periphery. Therefore, reducing IFP will undoubtedly improve the delivery of drugs and genes to solid tumors.

Four different strategies have been proposed for reducing IFP. One is to kill tumor cells directly with radiation or drugs [26,27]. Damaging the tumor cells can increase the interstitial fluid space and reduce solid stress, the changes that have a positive effect on reducing IFP. From another point of view, IFP can be used as a prognostic factor for determining the efficacy of radiation therapy or chemotherapy [26].

The second strategy is to reduce the microvascular permeability P and the hydraulic conductivity of microvessels, L_p. According to Starling's law (see Section 9.3.2),

$$\phi_B = \frac{J_v}{V} = \frac{L_P S}{V}[p_B - p_i - \sigma(\pi_B - \pi_i)]. \qquad (15.4.1)$$

Reducing L_P will directly reduce ϕ_B; reducing P will indirectly reduce ϕ_B, because a reduction in P decreases the interstitial osmotic pressure π_i, which in turn can either reduce fluid transport into the interstitial space or facilitate fluid absorption into blood vessels. Therefore, changes in both L_P and P will lower the IFP by decreasing fluid accumulation in tumors.

The third strategy is to reduce resistance to blood flow, which in turn will decrease the IFP by reducing the MVP (see Figure 15.6). The net resistance can be divided into two terms: the viscous resistance due to blood viscosity and the geometric resistance due to geometric parameters of the vascular network (e.g., the diameter and length of a vessel and the angle between two vessels). The reduction in flow resistance can be achieved either by decreasing the viscosity of blood or by normalizing the vascular network [28].

The fourth strategy is to increase the hydraulic conductivity of tumor tissues through the use of drugs to induce remodeling of the extracellular matrix (ECM). The remodeling of the ECM rearranges interstitial structures and thus may reduce the flow resistance. The reduction in flow resistance makes it easier for interstitial fluid to flow out of the tumor. This strategy is illustrated in the curves shown in Figure 15.3a. The IFP decreases when α^2 is decreased. By definition, α^2 is proportional to the flow resistance, which is quantified by the reciprocal of the hydraulic conductivity K.

If we cannot reduce the IFP in tumors, one alternative approach to improving convective transport across the tumor microvessel wall is to increase the blood pressure. For example, angiotensin II, a drug that can increase the systemic blood pressure and the microvascular pressure in tumors, has been used to improve the convective transport of drugs across the tumor microvessel wall [29]. However, the improvement is only transient, because the IFP will follow the increase in MVP, with a time delay of approximately 10 seconds, as shown in Figure 15.5. The amount of increase in drug delivery due to convection during this period can be estimated as [18]

$$V\frac{dC_t}{dt} = J_v C_B = L_p S \Delta p C_B, \tag{15.4.2}$$

where C is the drug concentration, Δp is the hydrostatic pressure difference across the microvessel wall, and J_v is the rate of fluid transport across the microvessel wall. The subscripts t and B represent tumor and plasma, respectively. Replacing the differential with a finite difference, we have

$$\Delta C = \frac{L_p S}{V}\Delta p C_B \tau_v, \tag{15.4.3}$$

where τ_v is the time delay (see Equation (15.3.28)). The relative change in the concentration of drugs in tumors is

$$\frac{\Delta C}{C_t} = \frac{C_B}{C_t}\frac{L_p S}{V}\Delta p \tau_v. \tag{15.4.4}$$

Equation (15.4.4) predicts that if $C_B/C_t = 3$, $L_pS/V = 7.2 \times 10^{-5}$ (mmHg s)$^{-1}$, $\Delta p = 10$ mmHg, and $\tau_v = 10$ seconds [18], then the relative change in the drug concentration is approximately 2%. To further increase C_t, the systemic blood pressure can be cycled by periodically injecting angiotensin II intravenously [11,30]. In each cycle, the blood pressure increases during the initial period, which will result in a 2% increase in drug delivery. In the rest period, the pressure difference across the vessel wall is either close to zero or reversed, due to the decrease in the blood pressure (see Figure 15.5). In this case, the rate of convection will be either zero or negative, causing the drug to flow back into the vessels. To prevent drug clearance, the drug molecules must be attached to molecules (e.g., monoclonal antibodies) that can bind to tumor tissues, before intravenous injection occurs. Therefore, the net effect of one pressure cycle is a 2% increase in drug delivery to the tumor. The percent increase rises to 20% if the blood pressure can be cycled 10 times. One drawback of this approach is that the cycling of the systemic blood pressure may cause damage to the cardiovascular system.

15.5 Quantitative Analysis of Interstitial Transport of Solutes

15.5.1 Governing Equations

A general approach to molecular transport in tumor tissues is based on the conservation laws for mass and momentum. In most situations, it is assumed that solute transport does not affect the fluid flow in tissues since the solution is dilute. Therefore, the governing equations for fluid flow and solute transport can be decoupled (i.e., one can first solve the problem of fluid flow as shown in Section 15.3 and then substitute the fluid velocity into the mass balance equation for solutes). The general equation for the mass balance of solutes in tissues is

$$\frac{\partial C}{\partial t} = -\nabla \cdot \mathbf{N} + (\Phi_B - \Phi_L) + R_{rxn}, \qquad (15.5.1)$$

where C is the solute concentration based on tissue volume, $\mathbf{N}$ is the solute flux in tissues, Φ_B is the rate of solute transport across a blood vessel wall per unit volume of tissue, Φ_L is the rate of solute transport across a lymph vessel wall per unit volume of tissue, and R_{rxn} is the rate of the chemical reaction. The transport of *neutral* solutes can be mediated by convection and diffusion. In dilute systems,

$$\mathbf{N} = \text{diffusion flux} + \text{convection flux} = -\mathbf{D}_{\text{eff}} \cdot \nabla C + \mathbf{v}_f f C, \qquad (15.5.2)$$

where $\mathbf{D}_{\text{eff}}$ and $\mathbf{v}_f$ are the effective diffusion tensor and the interstitial fluid velocity, respectively, averaged over the interstitial fluid space within the REV (see Section 8.3.1), and f is the *retardation coefficient*, which is equal to the ratio of the convective velocity of the solutes to the velocity of the interstitial fluid. Substituting Equation (15.5.2) into Equation (15.5.1) yields

$$\frac{\partial C}{\partial t} = \nabla \cdot (\mathbf{D}_{\text{eff}} \cdot \nabla C) - \nabla \cdot (\mathbf{v}_f f C) + (\Phi_B - \Phi_L) + R_{rxn}. \qquad (15.5.3)$$

If the diffusion in tissues is isotropic and uniform, then Equation (15.5.3) reduces to Equation (8.4.5), and we have

$$\frac{\partial C}{\partial t} = D_{\text{eff}} \nabla^2 C - \nabla \cdot (\mathbf{v}_f f C) + (\Phi_B - \Phi_L) + R_{rxn}. \qquad (15.5.4)$$

Since there are no functional lymphatics in solid tumors, the rate of solute transport across the lymph vessel wall is

$$\Phi_L = 0. \tag{15.5.5}$$

The rate of solute transport across the blood vessel wall can be modeled by different equations, as shown in Section 9.3.3. One of them is the Kedem–Katchalsky equation, modified from Equation (9.3.23) to the form

$$\Phi_B = \phi_B(1 - \sigma_f)\overline{C}_s + \frac{PS}{V}\left(C_p - \frac{C}{K_{AV}}\right), \tag{15.5.6}$$

where K_{AV} is the available volume fraction of solutes in tissues, ϕ_B is the fluid flow rate per unit tissue volume across the blood vessel wall, σ_f is the filtration reflection coefficient, P is the microvascular permeability coefficient, S/V is the vascular surface area per unit tissue volume, C_p is the solute concentration in the plasma, and $\overline{C}_s$ is a phenomenological parameter representing an average molar concentration of solutes in the vessel wall. Another model for Φ_B is the Patlak equation (9.3.38), modified to the form

$$\Phi_B = \phi_B(1 - \sigma_f)C_p + \frac{PS}{V}\left(C_p - \frac{C}{K_{AV}}\right)\frac{\text{Pe}}{e^{\text{Pe}} - 1}, \tag{15.5.7}$$

where

$$\text{Pe} = \frac{\phi_B(1 - \sigma_f)V}{PS}. \tag{15.5.8}$$

is the Peclet number, which is the ratio of convection versus diffusion across the microvessel wall. As discussed in Section 15.4.1, the pressure difference across the microvessel wall is close to zero in solid tumors. Therefore, convection can be neglected and $\text{Pe} \ll 1$. In this case, the Patlak equation is equivalent to the Kedem–Katchalsky equation, and Equations (15.5.6) and (15.5.7) both reduce to

$$\Phi_B = \frac{PS}{V}\left(C_p - \frac{C}{K_{AV}}\right). \tag{15.5.9}$$

15.5.2 Unidirectional Transport in a Solid Tumor

Equations (15.5.4), (15.5.5), and (15.5.9) are the most common forms used in solute transport simulations in solid tumors. These equations can be simplified further if the solid tumor is assumed to be a sphere and the transport is symmetric around the center of the tumor—that is, if the transport of both fluid and solutes is only in the radial direction and depends on r and t in a spherical coordinate system. In this case, Equation (15.5.4) becomes

$$\frac{\partial C}{\partial t} + \frac{1}{r^2}\frac{\partial(r^2 v_r f C)}{\partial r} = D_{\text{eff}}\frac{1}{r^2}\frac{\partial}{\partial r}\left(r^2\frac{\partial C}{\partial r}\right) + \frac{PS}{V}\left(C_p - \frac{C}{K_{AV}}\right) + R_{\text{rxn}}. \tag{15.5.10}$$

The boundary condition at $r = 0$ is

$$-D_{\text{eff}}\frac{\partial C}{\partial r} + v_r fC = 0. \tag{15.5.11}$$

At $r = R$, the boundary condition for C depends upon the solute transport in surrounding tissues. For the tissue-isolated tumor model described in Section 15.3.2, $C = 0$ at $r = R$. Equation (15.5.10) can be solved numerically by either the finite-difference or the finite-element method, provided that the velocity distribution v_r has been obtained by the methods discussed in Section 15.3.

15.5.3 Unidirectional Transport in the Krogh Cylinder

In the previous section, we discussed solute transport in the entire solid tumor, assuming symmetric transport about the center of the tumor. However, that analysis did not address solute distribution around a single microvessel, or *perivascular distribution*. The analysis of the perivascular distribution of solutes is an important tool in understanding the systemic delivery of drugs and genes into tumors. The most common way to model the perivascular distribution of solutes is based on the *Krogh cylinder* assumption. This model was originally developed to study oxygen delivery to cells around single capillaries (see Section 13.5.1). It was later adapted to analyze the transport of other solutes in biological tissues [31].

The Krogh cylinder model consists of two concentric cylinders. The inner cylinder represents a microvessel, and the outer cylinder represents the region in tissues that receives nutrients from the microvessel. If all microvessels were parallel to each other and were distributed uniformly throughout the tumor tissue, then the diameter of the outer cylinder would be equal to the distance between two adjacent vessels, and there would be no solute transport across the border of the outer cylinder. However, tumor vasculature is chaotic, and most vessels are *not* arranged in parallel. Consequently, the Krogh cylinder model is not valid, in general, for quantitative simulations of solute transport in tumors. Still, the model can be used to simulate drug delivery if the transport distance is much smaller than the intervessel distance. Under such conditions, the arrangement of vascular networks is not critical, because the adjacent vessels have minimal effects on the transport of the drugs around the vessel of interest.

One important application of the Krogh cylinder model in drug delivery is to simulate the transport of monoclonal antibodies in solid tumors [31], because the maximum depth of the interstitial penetration of IgG, which is only a few cell layers thick, is often smaller than the intervessel distance, which can be as large as 200 μm. To perform the simulation, we choose a cylindrical coordinate system and assume that the transport is axisymmetric. In addition, we assume that convection is negligible and that there is no metabolism of antibody in the tumor. Under these assumptions, the mass balance equation for free antibody in tumors is

$$\frac{\partial C_b}{\partial t} = D_{\text{eff}}\frac{1}{r}\frac{\partial}{\partial r}\left(r\frac{\partial C_b}{\partial r}\right) - R_b, \tag{15.5.12}$$

where C_b is the concentration of free antibody in tumors, D_{eff} is the effective diffusion coefficient of antibody in the interstitial space, and R_b is the rate of binding. In contrast to Equation (15.5.4), Equation (15.5.12) does not contain the source and sink terms, because the blood and lymph vessels are not part of the transport region in the Krogh cylinder. Thus, the fluid and solute exchange between the interstitial space and the blood or lymph vessels is accounted for in the boundary conditions.

The binding between antibody and antigen can be either monovalent or bivalent, depending upon the antibody and the distribution of antigens on the plasma membrane of the cells. In this example, we examine the Fab fragment of the antibody, which binds monovalently to the antigen, and the F(ab′)$_2$ fragment of the antibody and IgG, which bind bivalently to the antigen. The binding kinetics is modeled as

$$A_b + nA_g \underset{k_r}{\overset{k_f}{\longleftrightarrow}} A_b(A_g)_n,$$

where A_b is the free antibody, A_g is the free antigen, $A_b(A_g)_n$ is the antigen-bound antibody, and n is the number of binding sites on the antibody. For Fab, $n = 1$. For IgG or F(ab′)$_2$, $n = 2$. It is assumed that both binding sites of IgG or F(ab′)$_2$ are simultaneously occupied by the antigen (i.e., antigens are always paired and they can be treated as a single structure) and the binding follows a second-order reaction. Under this assumption, the net rate of binding is

$$R_b = k_f C_b C_g - k_r C_{bg}, \tag{15.5.13}$$

where C_g is the concentration of A_g, and C_{bg} is the concentration of $A_b(A_g)_n$ in tumors. The mass balance equations for A_g and $A_b(A_g)_n$ are, respectively,

$$\frac{\partial C_g}{\partial t} = nk_r C_{bg} - nk_f C_b C_g \tag{15.5.14}$$

and

$$\frac{\partial C_{bg}}{\partial t} = -k_r C_{bg} + k_f C_b C_g. \tag{15.5.15}$$

Experimental data in the literature suggest that the plasma concentration of macromolecules decreases biexponentially as a function of time and is given by

$$C_p = C_{p0}[\alpha \exp(-\lambda_1 t) + (1 - \alpha)\exp(-\lambda_2 t)], \tag{15.5.16}$$

where C_{p0} is the initial plasma concentration, which is equal to the dose of the bolus injection divided by the volume of the plasma; α is a constant; and λ_1 and λ_2 are called the *rate constants of plasma clearance*.

The boundary condition at $r = a$ is the continuity flux across the interface between the interstitial space and the microvessel wall; that is,

$$-D_{\text{eff}}\frac{\partial C_b}{\partial r} = P\left(C_p - \frac{C_b}{K_{\text{AV}}}\right) \approx PC_p \quad (\text{at } r = a), \qquad (15.5.17)$$

where P is the permeability coefficient. The approximation in Equation (15.5.17) is based upon the assumptions that (a) convection is negligible in tumors and (b) $C_p \gg C_b/K_{\text{AV}}$. The symmetry between different Krogh cylinders in the tumor requires a no-flux boundary condition at $r = R$, or

$$-D_{\text{eff}}\frac{\partial C_b}{\partial r} = 0 \quad (\text{at } r = R). \qquad (15.5.18)$$

The initial conditions are

$$C_b = 0, \; C_g = C_{g0}, \; \text{ and } C_{bg} = 0. \qquad (15.5.19)$$

The model constants from reference [31] are listed in Table 15.2. In addition, $C_{p0} = 100$ nM, $C_{g0} = 1$ μM, $a = 10$ μm, and $R = 75$ μm.

Equations (15.5.12) through (15.5.15), together with the boundary and initial conditions (i.e., Equations (15.5.16) through (15.5.19)), can be solved numerically by using either the finite-difference or the finite-element method. The simulation results are shown in Figures 15.8a through 15.8c for the model constants given in Table 15.2. One can see that Fab is more uniformly distributed than IgG, whereas the peak concentration and the total accumulation of IgG are higher than those of Fab, in tumor tissues. These differences are caused by the differences in molecule size, binding affinity, and plasma clearance rate between Fab and IgG, respectively.

To further understand how binding affinity affects antibody distributions in tumors, one can calculate the *specificity ratio*, defined as

$$\text{SR}_j = \frac{C_{sj}}{C_{nj}}, \qquad (15.5.20)$$

where C_{sj} is the total antibody concentration at location r_j in tumor tissues, which is equal to $C_b + C_{bg}$, and C_{nj} is the total concentration of nonspecific antibody at r_j.

TABLE 15.2

Model Parameters for Antibody Transport

	IgG	F(ab′)$_2$	Fab
α	0.27	0.61	0.68
λ_1 (s^{-1})	1.9×10^{-4}	1.5×10^{-4}	2.4×10^{-4}
λ_2 (s^{-1})	7.4×10^{-6}	6.3×10^{-6}	9.9×10^{-6}
D_{eff} (cm^2 s^{-1})	1.3×10^{-8}	2.0×10^{-8}	4.4×10^{-8}
P (cm s^{-1})	5.7×10^{-7}	7.9×10^{-7}	10.0×10^{-7}
k_f (M^{-1} s^{-1})	2.7×10^{4}	2.7×10^{4}	1.3×10^{4}
k_r (s^{-1})	5.3×10^{-5}	5.3×10^{-5}	15.3×10^{-5}

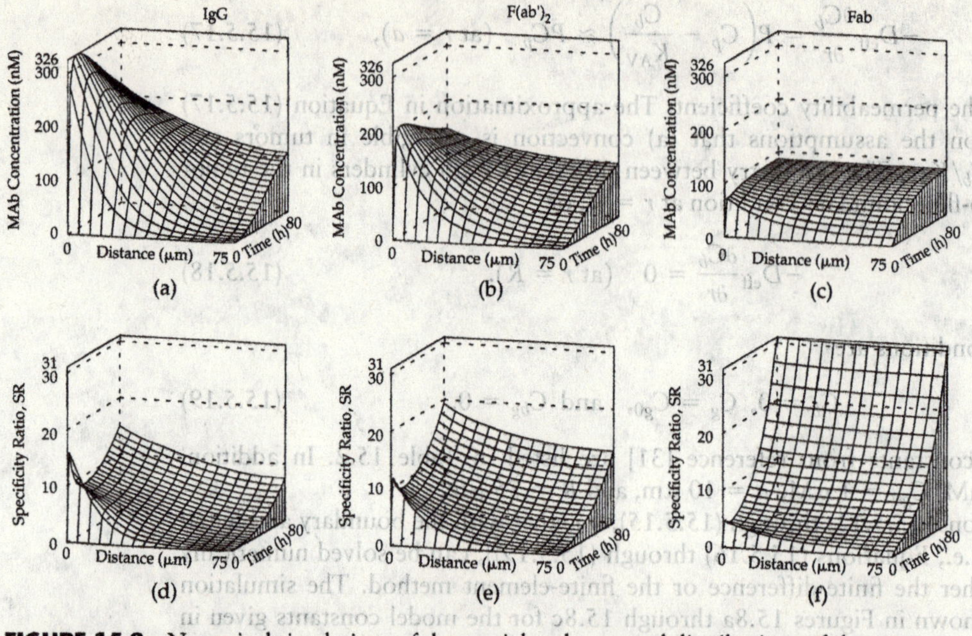

FIGURE 15.8 Numerical simulations of the spatial and temporal distributions of the monoclonal antibody concentrations in the perivascular region: (a) IgG, (b) F(ab$'$)$_2$, and (c) Fab. Numerical simulations of the spatial and temporal distributions of the antibody specificity ratios, SR$_j$, in the perivascular region: (d) IgG, (e) F(ab$'$)$_2$, and (f) Fab. (Modified from Ref. [31], with permission.)

C_{nj} can be calculated from Equation (15.5.12) with $R_b = 0$. The mean of SR$_j$ over the cylindrical space is defined as

$$\mathrm{SR} = \frac{\sum_{j=1}^{N} \mathrm{SR}_j r_j}{\sum_{j=1}^{N} r_j}, \qquad (15.5.21)$$

where N is the number of mesh points in the radial direction in the simulations. Similar to the concentration distributions, Figures 15.8d through 15.8f show that SR$_j$ is less uniform for IgG than Fab at any given time and that the absolute value of SR$_j$ for Fab is higher than that for IgG at later points in time, because of the fast clearance of free Fab from tumor tissues. In addition to being a function of the plasma clearance, the size of antibody molecules, and the binding affinity, SR$_j$ depends upon the dose of injection. For all of the antibodies, the higher the dose, the lower is SR$_j$ [31].

The spatial heterogeneity in antibody distribution can be characterized by the *index of spatial nonuniformity*, defined as

$$\mathrm{ISN} = \frac{\sqrt{\sum_{j=1}^{N} (C_{sj} - \overline{C})^2}}{\overline{C}\sqrt{N(N-1)}}, \qquad (15.5.22)$$

where $\overline{C}$ is the mean of C_{sj} in the cylinder. By definition, ISN is time dependent and ranges from zero to unity. ISN = 0 if C_{sj} is uniform and unity if C_{sj} is maximally nonuniform (i.e., a delta function of radial position r). ISN decreases with the dose of injection, but increases with the binding affinity for all antibodies [31].

15.6 PROBLEMS

15.1 If the diffusion coefficients of glucose and IgG are 2×10^{-6} and 2×10^{-8} cm^2 s^{-1}, respectively, and the convective velocity is 0.5 μm s^{-1}, determine which mode of transport is dominant in the following situations:

(a) glucose delivery over a distance of 100 μm.
(b) glucose delivery over a distance of 0.1 μm.
(c) IgG delivery over a distance of 100 μm.
(d) IgG delivery over a distance of 0.1 μm.

15.2 List at least four barriers to the systemic delivery of drugs that target a specific gene in tissues.

15.3 Derive Equations (15.3.25) and (15.3.29) for the interstitial fluid pressure p_i, and use those equations to calculate the volume dilatations e, tissue displacements u, and fluid velocities v in tumor tissues in those special cases described in Section 15.3.3.

15.4 For the unidirectional flow described in Section 15.3.3, the governing equations and the boundary conditions are Equations (15.3.19) through (15.3.24). If the interstitial fluid pressure p_i at r = o is p_0 initially, determine how p_i changes with time in tumors if the animal is sacrificed (i.e., p_e in Equation (15.3.22) is set to zero at $t = 0$). Compare the profile of p_i with others described in Section 15.3.3.

15.5 Consider the unidirectional transport of drugs in a tumor. The tumor is assumed to be a sphere, and the transport of fluid and drugs is spherically symmetrical around the center of the tumor. The velocity profile in tumors was solved in Section 15.3.2 (see Equation (15.3.16)). The concentration distribution is governed by Equations (15.5.10) and (15.5.11). To describe the solute transport problem completely, we need another boundary condition at $r = R$ and the initial concentration. In this exercise, we assume that there is no concentration gradient at $r = R$ and there is no drug in tumors at $t = 0$. The plasma concentration of drug is maintained at a constant level through a continuous intravenous infusion. Determine numerically the concentration distribution of solutes in tumors at 10, 20, 50, and 100 hours, based on the following model constants:

K	4.13×10^{-8} cm^2 mmHg^{-1} s^{-1}
D_{eff}	2×10^{-8} or 2×10^{-7} cm^2 s^{-1}
PS/V	1.26×10^{-4} s^{-1}
p_e	8 mmHg
C_p	1 μM
R	1 cm
f	1.0
ε	0.3
α	1, 10, or 100
R_{rxn}	0

15.6 Monoclonal antibodies, when conjugated with drugs, can improve the specificity of drug delivery because they are able to bind to specific antigens. Determine whether the following statements are true or false:

(a) Antibodies will enhance drug delivery in target tissues.
(b) Antibodies will improve the interstitial penetration of drugs.
(c) Antibodies will reduce drug accumulation in non-target tissues.
(d) Antibodies will increase drug retention in target tissues.

Explain your answers, based on the simulation results from the Krogh cylinder model.

15.7 Consider the unidirectional transport of drugs around a microvessel in tumors. The distribution of drugs can be simulated on the basis of the Krogh cylinder model, as shown in Section 15.5.3. Choose a cylindrical coordinate system. Assume that the transport is axisymmetric, that convection is negligible, and that there are no chemical reactions of drugs in the tumor. Initially, the concentration of drugs is zero in the tumor. The plasma concentration decays as a biexponential function of time; that is,

$$C_p = C_{p0}[\alpha \exp(-\lambda_1 t) + (1 - \alpha)\ \exp(-\lambda_2 t)]$$

(a) Determine the concentration distribution of drugs as a function of time (in a range from 0 to 100 hours) and radial distance, based on the following model constants:

D_{eff}	2×10^{-7} cm^2 s^{-1}
P	1.5×10^{-7} cm s^{-1}
K_{AV}	0.3
C_{p0}	1 μM
R	75 μm
a	5 μm
α	0.7
λ_1	4×10^{-4} min^{-1}
λ_2	7×10^{-6} min^{-1}

(b) Discuss how drug distribution changes with D_{eff} and P.

(c) Determine the index of spatial nonuniformity (ISN), defined in Equation (15.5.22), at 12 and 48 hours, respectively, as a function of D_{eff} ranging from 2×10^{-9} to 2×10^{-6} cm^2 s^{-1}.

15.8 Drugs infused directly into tumors can be cleared through absorption into microvessels. The rate of absorption can be estimated on the basis of the Krogh cylinder model discussed in Problem 15.7. Assume that the initial concentration of the drug is uniform and equal to 10 μM in the tumor. The plasma concentration is negligible compared with the drug concentration in the tumor. Other transport parameters are the same as those in Problem 15.7.

(a) Determine the drug concentration as a function of time and radial distance in tumors.

(b) Discuss how the rate of drug clearance depends on D_{eff} and P.

15.9 In Problem 15.8, we do not consider drug degradation in tumors. However, degradation cannot be neglected for some drugs. Assume that the rate of degradation follows a first-order chemical reaction with a rate constant $k = 0.5$ h^{-1}. Assume also that other conditions are the same as those in Problem 15.8.

(a) Determine the drug concentration as a function of time and radial distance in tumors.

(b) Discuss how the rate of drug clearance depends on k.

15.10 Polymer implants have been used for the controlled release of drugs in biological tissues. Consider a polymer membrane implanted in a tissue as described in Problem 8.12. Assume that the clearance of drug in the tissue is very slow so that the concentration of drug at the tissue–implant interface is equal to the solubility C_0 of the drug. The transport of drug in the tissue is one dimensional and in the direction normal to the membrane. Assume in addition that convection is negligible and the effective diffusion coefficient is D_{eff}. The drug molecules can be absorbed into the blood vessels at a rate governed by Equation (15.5.9).

(a) Determine the drug distribution in tumor tissues at the steady state.

(b) Define the penetration depth of the drug as the distance from the membrane at which the local concentration of the drug is equal to 1% of C_0. Find the penetration depth as a function of PS/V, and discuss how to improve drug penetration in tumors through changing the molecular properties of drugs or using drug carriers.

15.7 REFERENCES

1. Kong, G., and Dewhirst, M.W., "Hyperthermia and liposomes." *Int. J. Hyperthermia*, 1999. 15: pp. 345–370.
2. Jain, R.K., "Determinants of tumor blood flow: a review." *Cancer Res.*, 1988. 48: pp. 2641–2658.
3. Jain, R.K., "Transport of molecules in the tumor interstitium: a review." *Cancer Res.*, 1987. 47: pp. 3039–3051.
4. Khor, S.P., and Mayersohn, M., "Potential error in the measurement of tissue to blood distribution coefficients

in physiological pharmacokinetic modeling: residual tissue blood. I. Theoretical considerations." *Drug Metab. Disp.*, 1991. **19**: pp. 478–485.

5. Yuan, F., "Transvascular drug delivery in solid tumors." *Semin. Rad. Oncol.*, 1998. 8: pp. 164–175.
6. Wu, H.M., Huang, Q., Yuan, Y., and Granger, H.J., "VEGF induces NO-dependent hyperpermeability in coronary venules." *Am. J. Physiol.*, 1996. **271**: pp. H2735–H2739.
7. Yuan, Y., Granger, H.J., Zawieja, D.C., DeFily, D.V., and Chilian, W.M., "Histamine increases venular permeability via a phospholipase C–NO synthase–guanylate cyclase cascade." *Am. J. Physiol.*, 1993. **264**: pp. H1734–H1739.
8. Fukumura, D., Yuan, F., Endo, M., and Jain, R.K., "Role of nitric oxide in tumor microcirculation: blood flow, vascular permeability, and leukocyte–endothelial interactions." *Am. J. Pathol.*, 1997. **150**: pp. 713–725.
9. Kristjansen, P.E.G., Boucher, Y., and Jain, R.K., "Dexamethasone reduces the interstitial fluid pressure in a human colon adenocarcinoma xenograft." *Cancer Res.*, 1993. **53**: pp. 4764–4766.
10. Robert, L., "Cell–matrix interactions in cancer spreading—effect of aging: an introduction." *Semin. Cancer Biol.*, 2002. **12**: pp. 157–163.
11. Jain, R.K., "Delivery of molecules, particles, and cells to solid tumors." *Ann. Biomed. Eng.*, 1996. **24**: pp. 457–473.
12. Ambudkar, S.V., Dey, S., Hrycyna, C.A., Ramachandra, M., Pastan, I., and Gottesman, M.M., "Biochemical, cellular, and pharmacological aspects of the multidrug transporter." *Annu. Rev. Pharmacol. Toxicol.*, 1999. **39**: pp. 361–398.
13. Tan, B., Piwnica-Worms, D., and Ratner, L., "Multidrug resistance transporters and modulation." *Curr. Opin. Oncol.*, 2000. **12**: pp. 450–458.
14. Swabb, E.A., Wei, J., and Gullino, P.M., "Diffusion and convection in normal and neoplastic tissues." *Cancer Res.*, 1974. **34**: pp. 2814–2822.
15. Zhang, X.-Y., Luck, J., Dewhirst, M.W., and Yuan, F., "Interstitial hydraulic conductivity in a fibrosarcoma." *Am. J. Physiol.*, 2000. **279**: pp. H2726–H2734.
16. Baxter, L.T., and Jain, R.K., "Transport of fluid and macromolecules in tumors. I. Role of interstitial pressure and convection." *Microvasc. Res.*, 1989. **37**: pp. 77–104.
17. Boucher, Y., Baxter, L.T., and Jain, R.K., "Interstitial pressure gradients in tissue-isolated and subcutaneous tumors: implications for therapy." *Cancer Res.*, 1990. **50**: pp. 4478–4484.
18. Netti, P.A., Baxter, L.T., Boucher, Y., Skalak, R., and Jain, R.K., "Time-dependent behavior of interstitial fluid pressure in solid tumors: implications for drug delivery." *Cancer Res.*, 1995. **55**: pp. 5451–5458.
19. Milosevic, M., Fyles, A., Hedley, D., Pintilie, M., Levin, W., Manchul, L., and Hill, R., "Interstitial fluid pressure predicts survival in patients with cervix cancer independent of clinical prognostic factors and tumor oxygen measurements." *Cancer Res.*, 2001. **61**: pp. 6400–6405.
20. Boucher, Y., and Jain, R.K., "Microvascular pressure is the principal driving force for interstitial hypertension in solid tumors: implications for vascular collapse." *Cancer Res.*, 1992. **52**: pp. 5110–5114.
21. Stohrer, M., Boucher, Y., Stangassinger, M., and Jain, R.K., "Oncotic pressure in solid tumors is elevated." *Cancer Res.*, 2000. **60**: pp. 4251–4255.
22. Netti, P.A., Roberge, S., Boucher, Y., Baxter, L.T., and Jain, R.K., "Effect of transvascular fluid exchange on pressure–flow relationship in tumors: a proposed mechanism for tumor blood flow heterogeneity." *Microvasc. Res.*, 1996. **52**: pp. 27–46.
23. Boucher, Y., Leunig, M., and Jain, R.K., "Tumor angiogenesis and interstitial hypertension." *Cancer Res.*, 1996. **56**: pp. 4264–4266.
24. Helmlinger, G., Netti, P.A., Lichtenbeld, H.C., Melder, R.J., and Jain, R.K., "Solid stress inhibits the growth of multicellular tumor spheroids." *Nat. Biotechnol.*, 1997. **15**: pp. 778–783.
25. Yuan, F., "Stress is good and bad for tumors." *Nat. Biotechnol.*, 1997. **15**: pp. 722–723.
26. Roh, H.D., Boucher, Y., Kaliniki, S., Buchsbaum, R., Bloomer, W.D., and Jain, R.K., "Interstitial hypertension in cervical carcinomas in humans: possible correlation with tumor oxygenation and radiation response." *Cancer Res*, 1991. **51**: pp. 6695–6698.
27. Griffon-Etienne, G., Boucher, Y., Brekken, C., Suit, H.D., and Jain, R.K., "Taxane-induced apoptosis decompresses blood vessels and lowers interstitial fluid pressure in solid tumors: clinical implications." *Cancer Res.*, 1999. **59**: pp. 3776–3782.
28. Jain, R.K., "Normalizing tumor vasculature with anti-angiogenic therapy: a new paradigm for combination therapy." *Nat. Med.*, 2001. **7**: pp. 987–989.
29. Hori, K., Suzuki, M., Saito, S., Tanda, S., Zhang, Q.H., and Li, H.C., "Changes in vessel pressure and interstitial fluid pressure of normal subcutis and subcutaneous tumor in rats due to angiotensin II." *Microvasc. Res.*, 1994. **48**: pp. 246–256.
30. Netti, P.A., Hamberg, L.M., Babich, J.W., Kierstead, D., Graham, W., Hunter, G.J., Wolf, G.L., Fischman, A., Boucher, Y., and Jain, R.K., "Enhancement of fluid filtration across tumor vessels: implication for delivery of macromolecules." *Proc. Nat. Acad. Sci. USA*, 1999. **96**: pp. 3137–3142.
31. Fujimori, K., Covell, D.G., Fletcher, J.E., and Weinstein, J.N., "Modeling analysis of the global and microscopic distribution of immunoglobulin G, $F(ab')_2$, and Fab in tumors." *Cancer Res.*, 1989. **49**: pp. 5656–5663.

CHAPTER 16 Transport in Organs and Organisms

16.1 Introduction

The transport of molecules and particles in the body can be studied at four different levels: (a) transport to different organs through the systemic circulation (i.e., the blood and lymph circulation); (b) transport within individual organs through the local microcirculation and interstitial space; (c) transport around microvessels through the microvessel walls and surrounding tissues; and (d) transport within cells (see Figure 1.22). In previous chapters, we discussed blood flow, interstitial transport, transport across the microvessel wall, and transport within a cell. In this chapter, we focus on molecular transport to different organs.

Transport analysis at the organ level is straightforward, provided that we have information about blood and lymph flow distributions in the body and about how molecules are transported in each organ. In general, one can always perform a detailed analysis of blood flow, interstitial transport, transport across the microvessel wall, and transport within cells in each organ and then integrate the resulting information to obtain a complete view of the distribution of molecules as a function of time and location. However, this approach requires a great deal of effort and is often unnecessary, for a number of reasons. First, we generally do not know all of the transport properties for a given molecule in a specific organ, and these properties can be spatially heterogeneous and vary with time. Second, the distributions of blood and lymph vessels within individual organs are often unknown and are heterogeneous. In particular, the location of molecular exchange between the interstitial space and the blood and lymph microvessels in an organ cannot be determined precisely. Third, the governing equations of transport are, in general, three dimensional and time dependent, and they must be solved numerically at multiple length and timescales. The characteristic lengths can vary from a submicron to a meter—a range of more than 6 orders of magnitude—and the characteristic times can vary from a microsecond to several weeks—a range of more than 12 orders of magnitude. These variations render the numerical simulations expansive and time consuming—even impossible—with current technology. Therefore, most analyses of transport in organs and organisms are based on *compartment models*, in which the body is divided into several compartments, or, in some studies, the body itself is considered a compartment. Within each compartment, the molecules are assumed to be well mixed, so that the concentrations are uniform. This assumption simplifies the transport analysis significantly, because, rather than consider the movement of molecules between different locations within each compartment, we need only determine the temporal variations in the average concentrations of the molecules. The details of the compartmental analysis will be discussed later in this chapter.

Most applications of transport in organs and organisms are in the fields of pharmacokinetics and toxicology. The goal of the former is to determine distributions of drugs in the body; the goal of the latter is to investigate the uptake and accumulation of environmental toxins in the body. In this book, we will focus on pharmacokinetic analyses. In the literature, *pharmacokinetics* is defined as the study of mathematical relationships between the doses and the concentration distributions of drugs at various sites in the body (in other words, what the body does to the drug). Another term often seen in the pharmacology literature is *pharmacodynamics*—the study of mathematical relationships between the concentrations and the pharmacological effects of the drug in the body (in other words, what the drug does to the body).

16.2 General Considerations in Pharmacokinetic Analysis

In the compartmental approach to pharmacokinetic analysis, each compartment can represent a fraction of an organ (e.g., the cellular space), a single organ, or a group of organs. In some studies, the compartments have no direct correlation with any organs. In each compartment, molecules are well mixed, so there is no concentration gradient.

The concentration of drugs is often expressed in different units, depending on the study. For example, the concentration may be measured in milligrams (mg) or micrograms (μg) per unit weight of tissue or per unit volume of the plasma. *%I.D./g*, another unit of concentration, indicates the percentage of injected dose per gram of tissue. If the drug is labeled with a radionuclide, the unit of concentration becomes $cpm\ g^{-1}$ or $cpm\ mL^{-1}$, where *cpm* refers to radiation count per minute. The cumulative exposure of a compartment to a specific drug is quantified by the *area under the curve* (AUC), which is calculated as the area under the concentration-versus-time curve in the associated organ.

The dose of a drug is often expressed as the amount of drug per kilogram (kg) of body weight (B_w) or per square meter (m^{-2}) of body surface area (S). The amount of the drug is expressed in weight (e.g., μg and mg) or, if the drug is labeled with a radionuclide, in radioactivity (e.g., μC_i). In determining how to normalize a dose among different patients, body surface area is more accurate than body weight, because the rates of clearance of many drugs from the body are found to be proportional to a power function of the body weight, with an exponent close to 0.75 (see

Section 16.5). This relationship is similar to that between surface area and body weight:

$$S = 1.73\left(\frac{B_w}{70}\right)^{0.73}. \tag{16.2.1}$$

The similarity suggests that the drug clearance rate is approximately proportional to the surface area. The clearance rate and the dose of injection determine the plasma concentration of the drug. For drugs that are distributed uniformly in the body, different patients will have a similar drug exposure (approximately the same AUC in terms of the plasma concentration) if the amount of injection is normalized by the body surface area (see Equation (16.3.9b)).

The method of administering a drug varies with the treatment and can be oral uptake, local release from implanted devices, or injection. The route of injection can be intravenous (i.v.), intraperitoneal (i.p.), intramuscular (i.m.), or subcutaneous (s.c.). For each route, drugs can be injected as a *bolus* (a single dose) at different times, or they can be infused continuously.

After the drug is administered into the body, it will be cleared over a certain period of time by inactivation or excretion. Mechanisms of inactivation include metabolism, extracellular degradation, and binding to other molecules. Inactivation can happen in any organ. The liver is the main organ for drug metabolism, while excretion occurs mainly in the kidneys (see Chapter 14) but can also occur in the liver via the bile-secretion pathway. The secreted drugs will ultimately be excreted out of the body with feces. The rate of drug clearance depends upon the rate of excretion and inactivation. The *body clearance*, Cl_B, is a pharmacokinetic quantity defined as the rate of drug elimination in the body relative to its plasma concentration. The body clearance can be written as the summation

$$Cl_B = Cl_{kidney} + Cl_{liver} + Cl_{other}, \tag{16.2.2}$$

where the terms on the right-hand side represent the clearances through the kidneys, the liver, and other organs, respectively.

The goal of pharmacokinetic analysis is to establish mathematical relationships between the dose and the concentration distributions of drugs in all compartments. These relationships are generally derived mathematically, on the basis of a consideration of mass conservation and chemical reactions in each compartment. The constants in compartmental models depend upon the physical and chemical properties of the drugs, as well as the structures of tissues (i.e., the constants are drug and compartment dependent).

16.3 Simple Compartment Models in Pharmacokinetic Analysis

In pharmacokinetic analysis, the human body can be divided into several compartments, each is characterized by its volume, by the average concentration of drugs in the compartment, and by chemical reactions. Mass may be exchanged between different compartments, at rates characterized by the rate constants of mass transfer. The whole compartment system can also include input functions for drug administration and output functions for drug clearance.

16.3.1 One-Compartment Model

The simplest compartment model involves only one compartment (see Figure 16.1). This model is used in pharmacokinetic analysis if the drugs under study distribute uniformly in the body. The volume of the compartment is V and the concentration of drug in the compartment is C.

The input function is generally time dependent and, for a bolus i.v. injection, is a delta function $D\delta(t)$, where D is the amount of the injection. The output function is the rate of clearance, which is proportional to the concentration, with proportionality constant k_e. We assume that there is no drug in the body at $t < 0$ and that no chemical reactions occur within the compartment. In the one-compartment model, drug distribution within the compartment is uniform. Therefore, the compartment volume is equivalent to the distribution volume of the drug, and the concentration immediately after injection is equal to D/V. The mass balance equation for the drug is

$$\frac{dC}{dt} = -k_e C + \frac{D}{V}\delta(t), \tag{16.3.1}$$

and the initial boundary condition is

$$C = 0 \qquad \text{at } t = 0. \tag{16.3.2}$$

The mathematical model just described is equivalent to

$$\frac{dC}{dt} = -k_e C, \tag{16.3.3}$$

with the initial condition

$$C = C_0 \qquad \text{at } t = 0, \tag{16.3.4}$$

where $C_0 = D/V$. Solving Equation (16.3.3) gives

$$C = C_0 \exp(-k_e t). \tag{16.3.5}$$

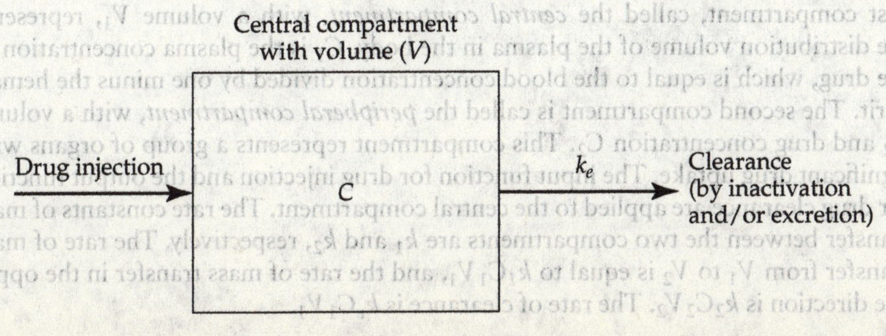

FIGURE 16.1 One-compartment model. C is the concentration of drugs in the compartment, and k_e is the rate constant of clearance.

Equation (16.3.5) indicates that drug concentration decreases exponentially with time. The half-life of the drug, $t_{1/2}$, is defined as the time when $C = C_0/2$. Substituting the definition into Equation (16.3.5), we have

$$t_{1/2} = \frac{\ln(2)}{k_e} = \frac{0.693}{k_e}. \qquad (16.3.6)$$

Equation (16.3.6) can be used to find k_e if $t_{1/2}$ is measured experimentally. The body clearance, Cl_B, is defined as the rate of drug elimination in the body relative to the drug's plasma concentration. In the one-compartment model, the plasma concentration is the same as the concentration in the central compartment shown in Figure 16.1. Thus,

$$Cl_B = \frac{Vk_eC}{C} = Vk_e, \qquad (16.3.7)$$

or

$$Cl_B = \frac{0.693V}{t_{1/2}}. \qquad (16.3.8)$$

The AUC can be obtained by integrating Equation (16.3.5) with respect to time:

$$AUC = \int_0^\infty C\,dt = \int_0^\infty C_0 \exp(-k_e t)\,dt. \qquad (16.3.9a)$$

The result is

$$AUC = \frac{C_0}{k_e} = 1.443C_0 t_{1/2}. \qquad (16.3.9b)$$

16.3.2 Two-Compartment Model

Few drugs can be analyzed with the one-compartment model, because the transport of most drugs in the plasma is significantly different from that in other tissues. As a result, at least two compartments are usually needed to characterize drug distribution in the body. One example of two-compartment models is given in Figure 16.2. The first compartment, called the *central compartment*, with a volume V_1, represents the distribution volume of the plasma in the body. C_1 is the plasma concentration of the drug, which is equal to the blood concentration divided by one minus the hematocrit. The second compartment is called the *peripheral compartment*, with a volume V_2 and drug concentration C_2. This compartment represents a group of organs with significant drug uptake. The input function for drug injection and the output function for drug clearance are applied to the central compartment. The rate constants of mass transfer between the two compartments are k_1 and k_2, respectively. The rate of mass transfer from V_1 to V_2 is equal to $k_1C_1V_1$, and the rate of mass transfer in the opposite direction is $k_2C_2V_2$. The rate of clearance is $k_eC_1V_1$.

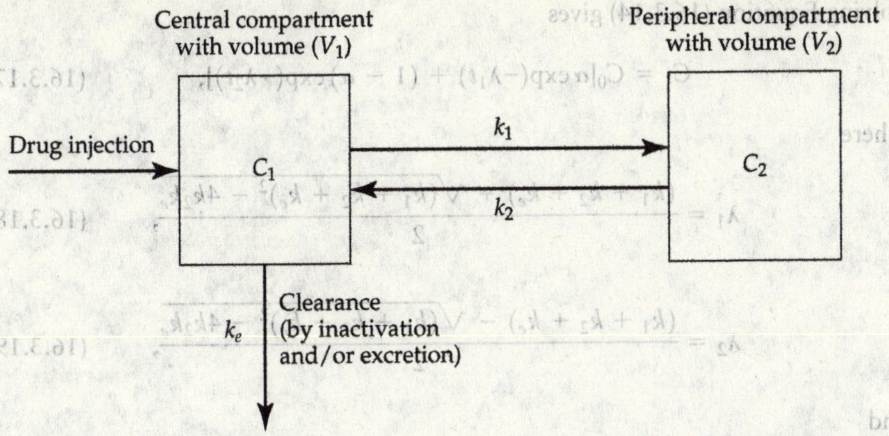

FIGURE 16.2 A two-compartment model. C represents concentration of drugs, k_e is the rate constant of clearance, k_1 and k_2 are the rate constants of mass transfer.

Again, we assume that the drug is administered via a bolus i.v. injection with an amount D. There is no drug in the peripheral compartment at $t = 0$, and chemical reactions are neglected in both compartments. Therefore, the mass balance equations are

$$\frac{dC_1}{dt} = -k_1C_1 + \zeta_{21}k_2C_2 - k_eC_1 \tag{16.3.10}$$

and

$$\zeta_{21}\frac{dC_2}{dt} = k_1C_1 - \zeta_{21}k_2C_2, \tag{16.3.11}$$

where $\zeta_{21} = V_2/V_1$. The boundary conditions are

$$C_1 = C_0 \quad \text{at } t = 0 \tag{16.3.12}$$

and

$$C_2 = 0 \quad \text{at } t = 0, \tag{16.3.13}$$

where $C_0 = D/V_1$, assuming that the drug in the plasma compartment reaches a uniform distribution instantaneously. Rearranging terms in Equations (16.3.10) and (16.3.11) to eliminate C_2, we obtain the governing equation for C_1:

$$\frac{d^2C_1}{dt^2} + (k_1 + k_2 + k_e)\frac{dC_1}{dt} + k_2k_eC_1 = 0. \tag{16.3.14}$$

The initial conditions for Equation (16.3.14) are

$$C_1 = C_0 \quad \text{at } t = 0 \tag{16.3.15}$$

and

$$\frac{dC_1}{dt} = -(k_1 + k_e)C_0 \quad \text{at } t = 0. \tag{16.3.16}$$

Solving Equation (16.3.14) gives

$$C_1 = C_0[\alpha \exp(-\lambda_1 t) + (1 - \alpha)\exp(-\lambda_2 t)], \tag{16.3.17}$$

where

$$\lambda_1 = \frac{(k_1 + k_2 + k_e) + \sqrt{(k_1 + k_2 + k_e)^2 - 4k_2 k_e}}{2}, \tag{16.3.18}$$

$$\lambda_2 = \frac{(k_1 + k_2 + k_e) - \sqrt{(k_1 + k_2 + k_e)^2 - 4k_2 k_e}}{2}, \tag{16.3.19}$$

and

$$\alpha = \frac{-\lambda_2 + k_1 + k_e}{\lambda_1 - \lambda_2}. \tag{16.3.20}$$

Alternatively, k_1, k_2, and k_e can be expressed as functions of λ_1, λ_2, and α:

$$k_1 = \frac{\alpha(1 - \alpha)(\lambda_2 - \lambda_1)^2}{(1 - \alpha)\lambda_1 + \alpha\lambda_2}, \tag{16.3.21}$$

$$k_2 = (1 - \alpha)\lambda_1 + \alpha\lambda_2, \tag{16.3.22}$$

$$k_e = \frac{\lambda_1\lambda_2}{(1 - \alpha)\lambda_1 + \alpha\lambda_2}. \tag{16.3.23}$$

Substituting Equation (16.3.17) into Equation (16.3.10) yields

$$C_2 = C_0 \frac{\alpha(1 - \alpha)(\lambda_2 - \lambda_1)}{\zeta_{21}[(1 - \alpha)\lambda_1 + \alpha\lambda_2]}[\exp(-\lambda_1 t) - \exp(-\lambda_2 t)]. \tag{16.3.24}$$

Equations (16.3.17) and (16.3.24) indicate that both C_1 and C_2 decrease biexponentially with time.

The difference between exponential versus biexponential decay is illustrated in Figure 16.3. The rate of exponential decay is uniform in the semilog plot, whereas the biexponential decay has two phases: a rapid decrease initially and a slow decrease later. The two phases are interpreted as the distribution phase and the elimination phase, respectively, of drugs in the body and have different half-lives:

$$t_{1/2\lambda_1} = \frac{0.693}{\lambda_1}, \tag{16.3.25}$$

$$t_{1/2\lambda_2} = \frac{0.693}{\lambda_2}. \tag{16.3.26}$$

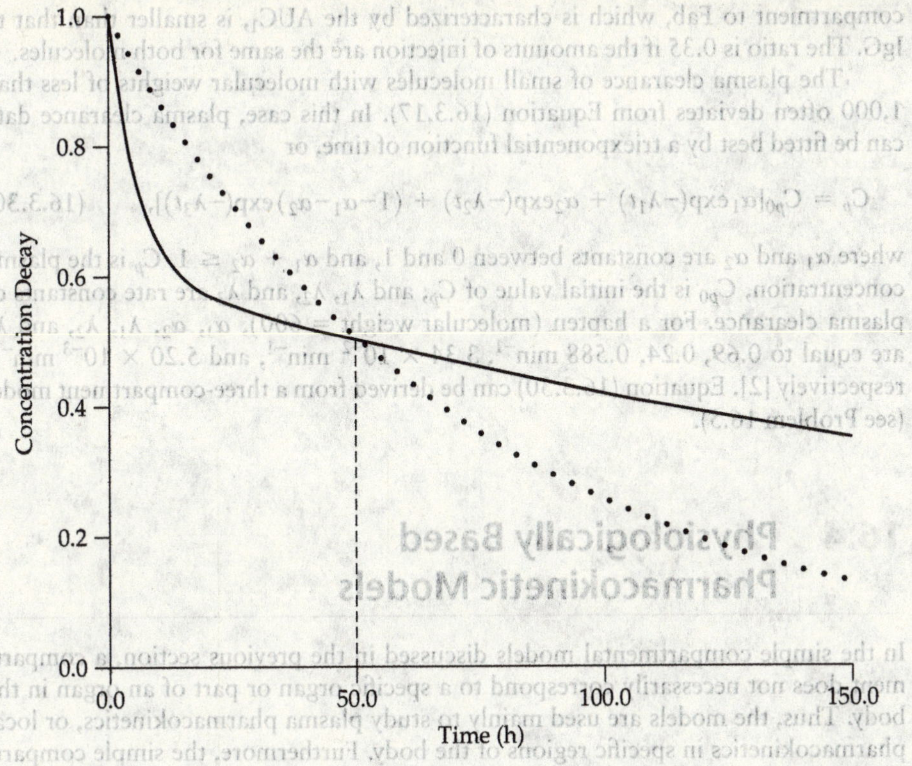

FIGURE 16.3 Exponential decays (dotted line) versus biexponential decays (solid line) of the concentration as a function of time. These curves are generated based upon Equations (16.3.5) and (16.3.17), respectively. The concentrations have been normalized by the initial concentration. α, λ_1, and λ_2 are equal to 0.4, $5 \times 10^{-5}\ s^{-1}$, and $1 \times 10^{-6}\ s^{-1}$, respectively. k_e is chosen to be $3.8 \times 10^{-6}\ s^{-1}$, so that both curves have the same half-life, which is approximately equal to 50 hours.

Based on the definition of drug clearance,

$$Cl_B = \frac{V_1 k_e C_1}{C_1} = V_1 k_e = \frac{V_1 \lambda_1 \lambda_2}{(1-\alpha)\lambda_1 + \alpha\lambda_2}. \tag{16.3.27}$$

The AUC in both compartments can be obtained by integrating Equations (16.3.17) and (16.3.24) with respect to time. The results are

$$AUC_1 = C_0\left[\frac{\alpha}{\lambda_1} + \frac{1-\alpha}{\lambda_2}\right] \tag{16.3.28}$$

and

$$AUC_2 = C_0 \frac{\alpha(1-\alpha)(\lambda_2 - \lambda_1)^2}{\zeta_{21}\lambda_1\lambda_2[(1-\alpha)\lambda_1 + \alpha\lambda_2]}. \tag{16.3.29}$$

For most macromolecules (e.g., monoclonal antibodies), the plasma concentration can be fitted by Equation (16.3.17). Curve fitting has been used to determine the values of α, λ_1, and λ_2 experimentally. For example, α, λ_1, and λ_2 for IgG (molecular weight = 150,000 are equal to 0.27, $1.9 \times 10^{-4}\ s^{-1}$, and $7.4 \times 10^{-6}\ s^{-1}$, respectively. For the Fab fragment of IgG (molecular weight = 50,000), the values are 0.68, $2.4 \times 10^{-4}\ s^{-1}$, and $9.9 \times 10^{-6}\ s^{-1}$, respectively [1]. These data indicate that Fab is eliminated faster than IgG in the plasma. Thus, the exposure of the central compartment to Fab, which is characterized by the AUC_1, is smaller than that to IgG. The ratio is 0.35 if the amounts of injection are the same for both molecules.

The plasma clearance of small molecules with molecular weights of less than 1,000 often deviates from Equation (16.3.17). In this case, plasma clearance data can be fitted best by a triexponential function of time, or

$$C_p = C_{p0}[\alpha_1 \exp(-\lambda_1 t) + \alpha_2 \exp(-\lambda_2 t) + (1-\alpha_1-\alpha_2)\exp(-\lambda_3 t)], \quad (16.3.30)$$

where α_1 and α_2 are constants between 0 and 1, and $\alpha_1 + \alpha_2 \leq 1$, C_p is the plasma concentration, C_{p0} is the initial value of C_p, and λ_1, λ_2, and λ_3 are rate constants of plasma clearance. For a hapten (molecular weight = 600), α_1, α_2, λ_1, λ_2, and λ_3 are equal to 0.69, 0.24, 0.588 min^{-1}, 3.34×10^{-2} min^{-1}, and 5.20×10^{-3} min^{-1}, respectively [2]. Equation (16.3.30) can be derived from a three-compartment model (see Problem 16.5).

16.4 Physiologically Based Pharmacokinetic Models

In the simple compartmental models discussed in the previous section, a compartment does not necessarily correspond to a specific organ or part of an organ in the body. Thus, the models are used mainly to study plasma pharmacokinetics, or local pharmacokinetics in specific regions of the body. Furthermore, the simple compartmental models cannot scale up a pharmacokinetic analysis from a small animal (e.g., a mouse) to a human being. (The scale-up issue is discussed in Section 16.5.)

To overcome these limitations, Bischoff and Dedrick proposed *physiologically based pharmacokinetic (PBPK) models* in the 1970s [3]. In such a model, each compartment represents an organ, and the number of compartments in the model corresponds to the number of organs in which drug accumulation is significant. In other words, the organs in which drug accumulation is insignificant can be neglected in a PBPK model. The general approach to PBPK analysis is as follows: (a) determine which organs need to be included in the analysis; (b) develop mathematical models of drug transport in each organ; (c) couple the models at the systemic level; (d) determine all model constants, based upon the data in the literature; and (e) solve the coupled mathematical equations simultaneously as functions of time. Applications of PBPK analysis include facilitating drug development, predicting the biodistribution of drugs and their metabolites in the body, scaling up the dose and schedule from mouse to human being, and performing risk assessments of environmental chemicals.

16.4.1 Transport in Individual Organs

Each organ in a PBPK model is divided into three subcompartments, representing vascular (or plasma), interstitial, and cellular spaces (see Figure 16.4). The plasma flows into the organ at the rate Q and out of the organ at the rate $Q - L$. The difference is the rate of lymph flow, L. The concentration of drugs in the feeding artery of the organ is C_p and is equal to that in the systemic circulation. C_v is the concentration in the vascular space and C_i is the interstitial concentration. The PBPK model

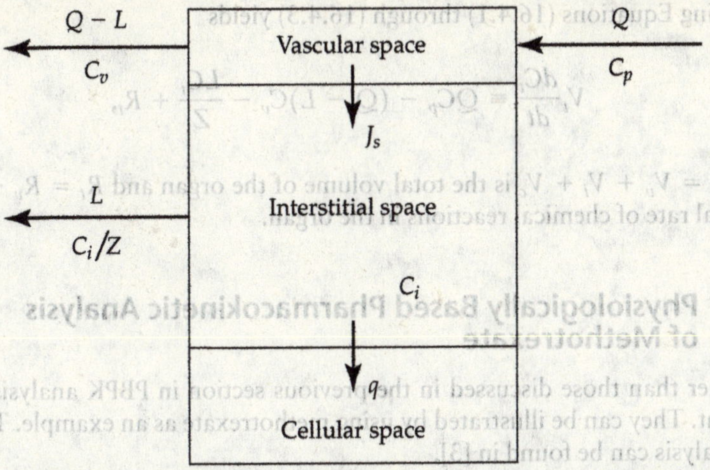

FIGURE 16.4 Three subcompartments in a typical organ in PBPK models: vascular (or plasma), interstitial, and cellular spaces.

assumes that the concentrations in the lymph vessel and the vein that drains the plasma from the organ are in equilibrium with those in the interstitial fluid and the vascular spaces, respectively. Thus, the concentrations in the lymph and the plasma in the vein are equal to C_i/Z and C_v, respectively. Here, Z is the equilibrium concentration ratio between the interstitium and the lymph or plasma. Z, which depends upon the steric exclusion of drugs and the binding or adsorption of drugs to different structures in tissues, can be either greater than or smaller than unity. When there is no binding and adsorption, Z is equivalent to K_{AV}, the available volume fraction introduced in Chapter 8. The rates of exchange are J_s between vascular and interstitial spaces and q between interstitial and cellular spaces.

The mass balance equations in the vascular, interstitial, and cellular spaces are, respectively,

$$V_v \frac{dC_v}{dt} = QC_p - (Q - L)C_v - J_s + R_v, \tag{16.4.1}$$

$$V_i \frac{dC_i}{dt} = J_s - \frac{LC_i}{Z} - q + R_i, \tag{16.4.2}$$

and

$$V_c \frac{dC_c}{dt} = q + R_c, \tag{16.4.3}$$

where V is the volume of the subcompartment, R is the rate of chemical reactions in the subcompartment. The subscripts v, i, and c represent vascular (or plasma), interstitial, and cellular spaces, respectively. The rate of exchange, J_s, can be modeled with the equations described in Chapter 9, whereas q can be modeled with the equations for molecular transport across the plasma membrane of cells (see Chapters 11 and 14). The rate q is equal to zero for molecules that cannot be internalized by cells. If we define the average concentration in an organ as

$$C_t = \frac{1}{V_t}(V_v C_v + V_i C_i + V_c C_c), \tag{16.4.4}$$

then adding Equations (16.4.1) through (16.4.3) yields

$$V_t \frac{dC_t}{dt} \equiv QC_p - (Q - L)C_v - \frac{LC_i}{Z} + R_t, \tag{16.4.5}$$

where $V_t = V_v + V_i + V_c$ is the total volume of the organ and $R_t = R_v + R_i + R_c$ is the total rate of chemical reactions in the organ.

16.4.2 Physiologically Based Pharmacokinetic Analysis of Methotrexate

Steps other than those discussed in the previous section in PBPK analysis are drug dependent. They can be illustrated by using methotrexate as an example. The details of the analysis can be found in [3].

Methotrexate is a drug used to treat cancer. The mechanism of the treatment is to inhibit DNA synthesis in tumor cells by blocking tetrahydrofolate synthesis. The molecular weight of methotrexate is 454. The delivery of small molecules like methotrexate can be assumed to be blood-flow limited (i.e., transport across the vessel wall is much faster than the convection of the drug in blood vessels). Methotrexate is secreted in the liver via the bile-secretion pathway and is subsequently transported into the gut lumen. Some methotrexate molecules in the gut lumen are absorbed into the gut tissue, and the rest is excreted with feces. In the systemic circulation, methotrexate can also be excreted through the kidney.

Given this information on methotrexate, the PBPK model must include the plasma, the liver, the gastrointestinal (GI) tract, the kidney, and the muscle (see Figure 16.5). The reason for including the muscle is that it is the largest organ in our bodies. Although the concentration in the muscle can be very low, the fraction of total injected dose of methotrexate in the muscle may not be negligible. Other organs may also contribute to the kinetics of methotrexate in the plasma and can be included in the PBPK model. However, their inclusion will render the model more complicated and may result in only a minor improvement in the accuracy of the model's predictions. For other drugs, organs besides those shown in Figure 16.5 have to be included in PBPK models (see Section 16.5.2).

The next step in the PBPK analysis is to establish the mass balance equation in each organ. In the muscle, it is assumed that there is no metabolism or cellular uptake of methotrexate and that the concentration has reached an equilibrium between the plasma and the interstitial fluid. Thus, $C_v = C_i/Z$ in Equation (16.4.5). The analysis also assumes that C_t is approximately equal to C_i. Under these assumptions, Equation (16.4.5) becomes

$$V_M \frac{dC_M}{dt} = Q_M\left(C_p - \frac{C_M}{Z_M}\right), \tag{16.4.6}$$

where the subscript M denotes muscle and V_M is the total volume of muscle. The model described in [3] also assumes a saturable binding between methotrexate and dihydrofolate reductase. This binding is neglected in the discussion in this section.

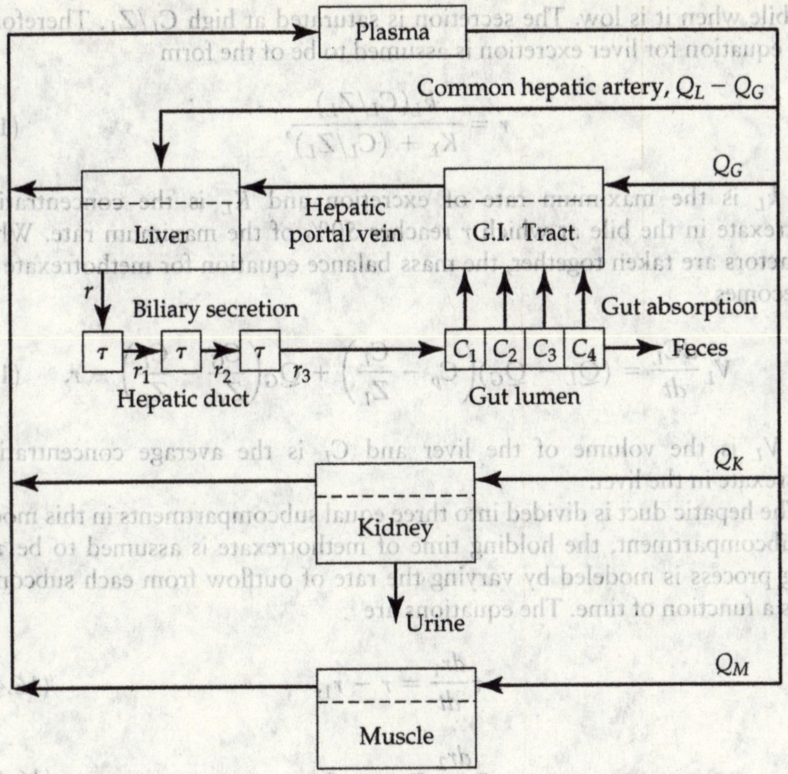

FIGURE 16.5 Diagram of the physiologically based pharmacokinetic model for methotrexate. (Adapted from Ref. [3], with permission.)

In the kidney, the mass balance equation is similar to that in the muscle, except that a rate of excretion is added and is assumed to be proportional to the concentration in the vascular space, i.e., equal to $k_K C_K/Z_K$, where the subscript K denotes the kidney and k_K is product of the kidney volume V_K and the rate constant of excretion (see Section 16.3.1). Therefore,

$$V_K \frac{dC_K}{dt} = Q_K\left(C_p - \frac{C_K}{Z_K}\right) - k_K \frac{C_K}{Z_K}. \tag{16.4.7}$$

The mass balance equation in the liver is a bit more complicated, since it involves two different blood supplies—one from the hepatic portal vein, the other from the common hepatic artery—as well as methotrexate excretion through the bile-secretion pathway. In accordance with Equation (16.4.6), the net rate of transport into an organ is equal to the plasma-flow rate times the concentration difference between the feeding artery and the draining vein. Thus, the net rates of methotrexate transport through the common hepatic artery and the hepatic portal vein are $(Q_L - Q_G)(C_p - (C_L/Z_L))$ and $Q_G((C_G/Z_G) - (C_L/Z_L))$, respectively, where Q is the plasma flow rate and C is the average concentration of methotrexate in tissues. The subscripts L and G denote the liver and the GI tract, respectively.

Methotrexate is excreted through the bile-secretion pathway. The rate of excretion through that pathway, r, is linearly dependent on the concentration C_L/Z_L

in the bile when it is low. The secretion is saturated at high C_L/Z_L. Therefore, the kinetic equation for liver excretion is assumed to be of the form

$$r = \frac{k_L(C_L/Z_L)}{K_L + (C_L/Z_L)}, \tag{16.4.8}$$

where k_L is the maximum rate of excretion and K_L is the concentration of methotrexate in the bile at which r reaches 50% of the maximum rate. When all these factors are taken together, the mass balance equation for methotrexate in the liver becomes

$$V_L\frac{dC_L}{dt} = (Q_L - Q_G)\left(C_p - \frac{C_L}{Z_L}\right) + Q_G\left(\frac{C_G}{Z_G} - \frac{C_L}{Z_L}\right) - r, \tag{16.4.9}$$

where V_L is the volume of the liver and C_L is the average concentration of methotrexate in the liver.

The hepatic duct is divided into three equal subcompartments in this model. In each subcompartment, the holding time of methotrexate is assumed to be τ. The holding process is modeled by varying the rate of outflow from each subcompartment as a function of time. The equations are

$$\tau\frac{dr_1}{dt} = r - r_1, \tag{16.4.10a}$$

$$\tau\frac{dr_2}{dt} = r_1 - r_2, \tag{16.4.10b}$$

and

$$\tau\frac{dr_3}{dt} = r_2 - r_3, \tag{16.4.10c}$$

where r_1, r_2, and r_3 are the outflow rates in the corresponding subcompartments. Alternatively, the holding process can be modeled as convection through the hepatic duct (see Problem 16.6).

The transport of methotrexate in the GI tract involves molecular exchange between the plasma and the gut tissue. As with the exchange in other organs, the net rate of exchange is $Q_G(C_p - (C_G/Z_G))$. In addition, a fraction of liver-excreted methotrexate is reabsorbed into the gut tissue from the gut lumen. To model the reabsorption, the gut lumen is divided into four equal subcompartments, in each of which the rate of absorption is given by

$$r_a = \frac{k_G C_j}{K_G + C_j} + bC_j, \tag{16.4.11}$$

where C_j is the concentration in the jth subcompartment and k_G, K_G, and b are constants. The first term on the right-hand side of Equation (16.4.11) represents saturable mechanisms of reabsorption and is similar to the rate of liver excretion (see Equation (16.4.8)), while the second term represents nonsaturable mechanisms of reabsorption. When these terms and the exchange rate are taken together, the mass balance equation in the gut tissue can be derived as

$$V_G \frac{dC_G}{dt} = Q_G\left(C_p - \frac{C_G}{Z_G}\right) + \sum_{j=1}^{4}\left(\frac{k_G C_j}{K_G + C_j} + bC_j\right), \qquad (16.4.12)$$

where V_G is the volume of gut tissue. In each subcompartment of the gut lumen, the mass balance equation can be derived as

$$\frac{V_{GL}}{4}\frac{dC_1}{dt} = r_3 - k_F\frac{V_{GL}}{4}C_1 - \left(\frac{k_G C_1}{K_G + C_1} + bC_1\right) \qquad (16.4.13)$$

and

$$\frac{V_{GL}}{4}\frac{dC_j}{dt} = k_F\frac{V_{GL}}{4}C_{j-1} - k_F\frac{V_{GL}}{4}C_j - \left(\frac{k_G C_j}{K_G + C_j} + bC_j\right) \qquad (j = 2,3,4), \qquad (16.4.14)$$

where V_{GL} is the total volume of gut lumen and k_F is the rate constant of methotrexate transport in the gut lumen. Physiologically, the first two terms on the right-hand side of Equations (16.4.13) and (16.4.14) represent rates of transport into and out of the subcompartment, respectively. The final form of the mass balance equation in the gut lumen can be written as

$$\frac{dC_{GL}}{dt} = \frac{1}{4}\sum_{j=1}^{4}\frac{dC_j}{dt} \qquad (16.4.15)$$

(i.e., the concentration in the gut lumen is equal to the average of methotrexate concentrations in four subcompartments). To close the loop in the development of the model, we must derive the mass balance equation in the plasma. The rate of accumulation in the plasma is equal to the rate of injection, plus the rate of transport from the veins in the liver, kidney, and muscle, minus the rate of transport into all organs through feeding arteries; that is,

$$V_p\frac{dC_p}{dt} = D\delta(t) + Q_L\frac{C_L}{Z_L} + Q_K\frac{C_K}{Z_K} + Q_M\frac{C_M}{Z_M} - (Q_L + Q_K + Q_M)C_p, \qquad (16.4.16)$$

where V_p is the plasma volume and $\delta(t)$ is the delta function, assuming that methotrexate is administered by a bolus injection with an amount D. As in our previous discussion if the initial plasma concentration is changed from zero to D/V_p, then Equation (16.4.16) becomes

$$V_p\frac{dC_p}{dt} = Q_L\frac{C_L}{Z_L} + Q_K\frac{C_K}{Z_K} + Q_M\frac{C_M}{Z_M} - (Q_L + Q_K + Q_M)C_P. \qquad (16.4.17)$$

The initial concentrations in other organs are equal to zero. Equations (16.4.6), (16.4.7), (16.4.9) through (16.4.15), and (16.4.17) can be solved simultaneously to

TABLE 16.1

Model Constants for Methotrexate Biodistribution in a 70-kg Human Body

V_p (mL)	3,000	Q_M (mL min^{-1})	420	Z_M	0.15
V_M (mL)	35,000	Q_K (mL min^{-1})	700	Z_K	3.0
V_K (mL)	280	Q_L (mL min^{-1})	800	Z_L	3.0
V_L (mL)	1,350	Q_G (mL min^{-1})	700	Z_G	1.0
V_G (mL)	2,100	k_L/K_L (mL min^{-1})	200	k_G (mg min^{-1})	1,900
V_{GL} (mL)	2,100	τ (min)	10	K_G (mg mL^{-1})	200
k_K (mL min^{-1})	190	k_F (min^{-1})	0.001	b (mL min^{-1})	0

Source: Data adapted from Ref. [3].

obtain the concentrations of methotrexate in each organ as functions of time, provided that the values of all constants in the model are known. The solution procedure involves numerical methods because some of these ordinary differential equations (ODEs) contain nonlinear terms. Such methods can easily be programmed or found in commercial software packages (e.g., MATLAB).

As an example, we will calculate the concentration distribution of methotrexate in a 70-kg human body. The model's constants are listed in Table 16.1. The liver excretion constants, k_L and K_L, are listed as k_L/K_L, assuming that $K_L \gg C_L/Z_L$. In this case, only the ratio of k_L to K_L in Equation (16.4.8) affects the biodistribution of methotrexate. The dose of injection is assumed to be 1 mg kg^{-1}. Thus, D is equal to 70 mg.

With the values in the table, Equations (16.4.6), (16.4.7), (16.4.9) through (16.4.15), and (16.4.17) are solved numerically with the ODE solver (`ode45`) in the MATLAB software. The results are presented in Figure 16.6. The predicted concentrations in different organs shown in the figure are consistent with experimental observations [3]. The concentration distributions of methotrexate in other animals (e.g., mouse, rat, dog, and monkey) have also been calculated in [3] (see Problem 16.7). Note that the predicted concentrations shown in Figure 16.6 are slightly smaller than those reported in [3], because we have neglected the binding of methotrexate to dihydrofolate reductase in all tissues.

16.5 Allometric Scaling Law and Its Application to Transport Properties

16.5.1 Scaling Laws

Many anatomical structures and physiological processes are related to the size or the weight of animals [4–9]. These relationships are called *scaling laws* and are governed by physical laws pertaining to structures and functions in animals, as well as by biological mechanisms involved in evolution, maturation, and seasonal adaptation [8,10]. The physical laws include all of those discussed in this book on the transport

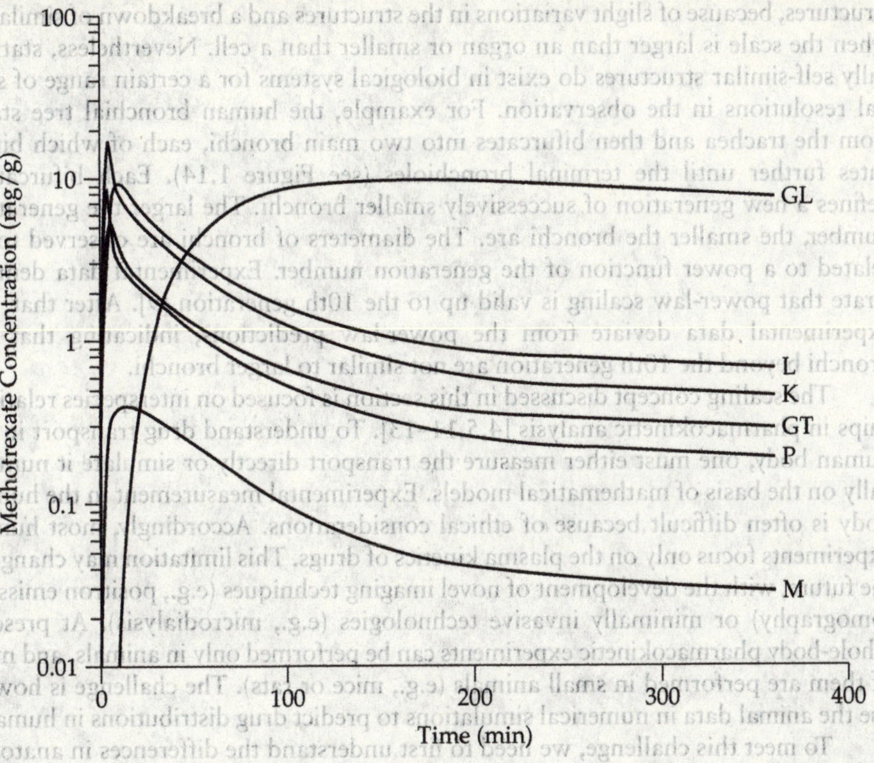

FIGURE 16.6 Biodistribution of methotrexate in gut lumen (GL), liver (L), kidney (K), gut tissue (GT), plasma (P), and muscle (M) of a 70-kg human. The dose of injection is 70 mg.

of mass and momentum, in addition to others on energy transport (see Chapter 17) and the mechanical strength of structures. The scaling laws have been used in many different applications. Here, we will introduce those dealing with predictions of drug distribution in one species (e.g., the human being) on the basis of distribution data of the same drug in another species (e.g., the mouse).

Scaling between large and small systems is not a new concept for engineers. As discussed in Section 3.5, one of the main applications of dimensional analysis and the Buckingham Pi theorem is the design of transport experiments based on model structures that are smaller than, but geometrically similar to, real structures. As a result, experimental data are not explicitly dependent on the size of structures, as long as all dimensionless groups are the same. In this case, the transport behaviors of large structures can be predicted by scaling up experimental data pertaining to small structures. For example, the fluid stresses exerted on a flying airplane or a running car can be predicted by wind-tunnel experiments using a model plane or a model car. In chemical engineering, a chemical reactor can be designed in the lab first and then scaled up in chemical plants.

Another scaling concept is related to the measured size of an object, which may depend on the scale of observation. For example, the measured length of the border between two countries depends on the spatial resolution in the measurement. The finer the resolution is, the longer the measured border will be. This relationship between the scale of the observation and the size of the object is found in any self-similar structures [9]. However, there is no absolute self-similarity in biological structures, because of slight variations in the structures and a breakdown of similarity when the scale is larger than an organ or smaller than a cell. Nevertheless, statistically self-similar structures do exist in biological systems for a certain range of spatial resolutions in the observation. For example, the human bronchial tree starts from the trachea and then bifurcates into two main bronchi, each of which bifurcates further until the terminal bronchioles (see Figure 1.14). Each bifurcation defines a new generation of successively smaller bronchi. The larger the generation number, the smaller the bronchi are. The diameters of bronchi are observed to be related to a power function of the generation number. Experimental data demonstrate that power-law scaling is valid up to the 10th generation [9]. After that, the experimental data deviate from the power-law predictions, indicating that the bronchi beyond the 10th generation are not similar to larger bronchi.

The scaling concept discussed in this section is focused on interspecies relationships in pharmacokinetic analysis [4,5,11–13]. To understand drug transport in the human body, one must either measure the transport directly or simulate it numerically on the basis of mathematical models. Experimental measurement in the human body is often difficult because of ethical considerations. Accordingly, most human experiments focus only on the plasma kinetics of drugs. This limitation may change in the future, with the development of novel imaging techniques (e.g., positron emission tomography) or minimally invasive technologies (e.g., microdialysis). At present, whole-body pharmacokinetic experiments can be performed only in animals, and most of them are performed in small animals (e.g., mice or rats). The challenge is how to use the animal data in numerical simulations to predict drug distributions in humans.

To meet this challenge, we need to first understand the differences in anatomical structures and physiological processes between small and large animals, which may help explain the differences in drug transport between small animals and humans. The most obvious difference is the size or the body weight (B_w). A wealth of experimental observations suggest that some structures and processes in the body depend strongly on B_w (e.g., blood volume and basal metabolic rate), whereas others are weakly dependent upon, or even independent of, B_w (e.g., cell size and capillary diameter and density). The dependence on B_w can be either linear (e.g., blood volume) or nonlinear (e.g., basal metabolic rate). The mathematical equation that describes the relationship between B_w and a physiological process or an anatomical structure is called the *allometric scaling law*. The law determines "the structural and functional consequences of a change in size or in scale among similarly organized animals" [14]. In general, the allometric scaling law can be described by the equation

$$Y = k(B_w)^\lambda, \tag{16.5.1}$$

where k and λ are constants and Y represents the rate of a physiological process or the size of an anatomical structure.

One of the most famous scaling laws states that the basal metabolic rate is dependent on B_w (see Chapter 17), with the exponent λ being $\frac{3}{4}$. The values of λ for other anatomical structures and physiological processes are listed in Table 16.2. The data indicate that the volume or mass of an organ is more sensitive to B_w than any physiological process is. Mechanisms that determine the value of λ vary with structures and processes. For example, drug clearance can occur in the kidney, the liver, and other organs. The rate of clearance in the kidney depends upon the rate of blood

TABLE 16.2

The Exponent in Scaling Laws for Different Anatomical Structures and Physiological Processes

Anatomical structure or physiological process	λ
Basal metabolic rate	0.75
O_2 consumption rate	0.75
Lung ventilation rate	0.75
Cardiac output	0.75
Respiration frequency	−0.26
Heart rate	−0.25
Lung volume	1.02
Tidal volume	1.01
Blood volume	0.99
Heart weight	0.99
Kidney mass	0.85
Liver mass	0.87
Creatinine clearance	0.69
Inulin clearance	0.77
Para-aminohippuric acid clearance	0.80
Gentamicin clearance	0.75
Methotrexate clearance	0.80

Source: Data are selected from references in Refs [4,5,14,15].

flow, which in turn is proportional to the cardiac output (see Chapter 14). The exponent for cardiac output scaling is approximately 0.75. Thus, for kidney clearance Cl_k, λ should also be close to 0.75 (see Table 16.2 for creatine, inulin, and para-aminohippuric acid). The scaling for drug clearance through the liver and other organs is complicated and depends upon the mechanisms of clearance. If the clearance is blood-flow limited, then λ is also approximately 0.75. However, λ may be meaningless if the clearance is capacity limited [15]. In this case, the allometric scaling law (Equation (16.5.1)) is invalid.

Another example of a physiological process that may not be scaled with B_w is the metabolic rate of drugs [4]. (This process is in contrast to the metabolism for energy production just discussed.) The inability to scale up between species or between animals with different sizes is likely caused by complex processes that involve biotransformation, genetic polymorphism, protein binding, capacity saturation, drug-induced alterations in physiology, enterohepatic recirculation, or renal tubular reabsorption that is sensitive to the pH of urine [15]. Even for anatomical structures or physiological processes that can be scaled allometrically, the values predicted from the scaling laws may differ from experimental data. Therefore, the predictions provide only general trends and order-of-magnitude estimations.

Different mechanisms of scaling have been proposed in the literature [7,10]. For example, the $\frac{3}{4}$-scaling law for the basal metabolic rate and its related anatomical structures and physiological processes has its origin in fractal-like networks in biological systems [7]. However, theories of scaling are still controversial [10], and allometric scaling in physiology is still empirical in most cases.

16.5.2 Applications of the Allometric Scaling Law in Pharmacokinetic Analysis

The allometric scaling law can be combined with the compartmental models discussed in this chapter to scale up pharmacokinetic analysis from small to large animals. Two main issues must be considered in this analysis. First, the PBPK model is a better choice in scale-up analysis (although other models have also been used), because the model's constants are directly related to either anatomical structures or physiological processes [2,4,5]. Thus, some of these constants can be scaled up directly, using Equation (16.5.1). Second, as we have said, not all physiological processes can be scaled up. Therefore, only a fraction of the constants in pharmacokinetic models can be predicted by the allometric scaling law; other constants must be determined independently in different species.

To illustrate how to perform a scale-up analysis based upon the PBPK model, we again study the biodistribution of methotrexate in the body. In this study, we compare the pharmacokinetics of this drug in a mouse (22 g) and a man (70 kg). The following model constants listed in Table 16.1 can be scaled allometrically: the organ volume (V), the plasma flow rate (Q), the kidney clearance rate (k_K), and the rate constant of methotrexate transport in the gut lumen (k_F). Other model constants cannot be scaled allometrically and must therefore be determined separately.

On the basis of the values of λ shown in Table 16.2, we can assume that

$$V \propto (B_w)^{1.0} \text{ (for plasma and muscle)}, \tag{16.5.2a}$$

$$V \propto (B_w)^{0.86} \text{ (for the liver, kidney, and GI tract)}, \tag{16.5.2b}$$

$$Q \propto (B_w)^{0.75}, \tag{16.5.3}$$

$$k_K \propto (B_w)^{0.75}, \tag{16.5.4}$$

and

$$k_F \propto (B_w)^{-0.25}, \tag{16.5.5}$$

where $\propto$ symbolizes "is proportional to." For these constants in mice, one can either predict their values, based on Equations (16.5.2) through (16.5.5) and the human data given in Table 16.1, or experimentally measure the constants. The results, shown in Table 16.3, indicate that the predicted values of V, Q, k_K, and k_F are close to the experimental data. The maximum difference is approximately twofold. The differences will affect the predictions of methotrexate biodistribution in the body slightly (see Problem 16.8).

Another example of allometric scale-up in pharmacokinetic analysis is the prediction of monoclonal antibody distribution in a human body, based on mouse data [13]. The prediction uses the PBPK model shown in Figure 16.7. The details of this model will not be discussed here; we recommend reading the original reference [13]. In brief, the model is a special case of the general approach discussed in Section 16.4.1. The model specifies the rates of transvascular transport and the chemical reactions in each subcompartment of an organ. The mass balance equation in the

TABLE 16.3

Differences in PBPK Model Constants between Predicted Values and Experimental Data in Mice

Model constant	Predicted values[a]	Experimental data[b]	Ratio
V_p (mL)	0.94	1.0	0.94
V_M (mL)	11.00	10.0	1.10
V_K (mL)	0.27	0.34	0.80
V_L (mL)	1.31	1.3	1.01
V_G (mL)	2.04	1.5	1.36
V_{GL} (mL)	2.04	1.5	1.36
Q_M (mL min^{-1})	0.99	0.5	1.98
Q_K (mL min^{-1})	1.65	0.8	2.07
Q_L (mL min^{-1})	1.89	1.1	1.72
Q_G (mL min^{-1})	1.65	0.9	1.84
k_K (mL min^{-1})	0.45	0.2	2.24
k_F (min^{-1})	0.008	0.01	0.75

Note: Four constants in the mouse PBPK model of methotrexate are predicted on the basis of corresponding human data.

[a]The predicted values are based on Equations (16.5.2) through (16.5.5) and the human data shown in Table 16.1.

[b]The experimental data in mice are obtained from Bischoff et al. [3].

vascular space is similar to Equation (16.4.1), except that $R_v = 0$ (i.e., there is no chemical reaction in vessels):

$$V_v \frac{dC_v}{dt} = QC_p - (Q - L)C_v - J_s. \tag{16.5.6}$$

Here, J_s is the rate of transvascular transport, modeled by a two-pore system that considers fluid recirculation within the microvessel wall. The details of the two-pore model are beyond the scope of this book and can be found in [16].

The mass balance equation relating to the interstitial space is similar to Equation (16.4.2), except that two antibody species must be considered in normal organs and three must be considered in the tumor compartment, because of the chemical reactions. Monoclonal antibodies can bind either nonspecifically in tissues or specifically to tumor-associated antigens. The nonspecific binding can be modeled as

$$a \underset{k_r^{NS}}{\overset{k_f^{NS}}{\longleftrightarrow}} b,$$

where a denotes free antibody, b represents nonspecifically bound antibody, and k_f^{NS} and k_r^{NS} are forward and reverse rate constants, respectively. The reaction of specific binding can be modeled as

$$a + G \underset{k_r^{S}}{\overset{k_f^{S}}{\longleftrightarrow}} aG,$$

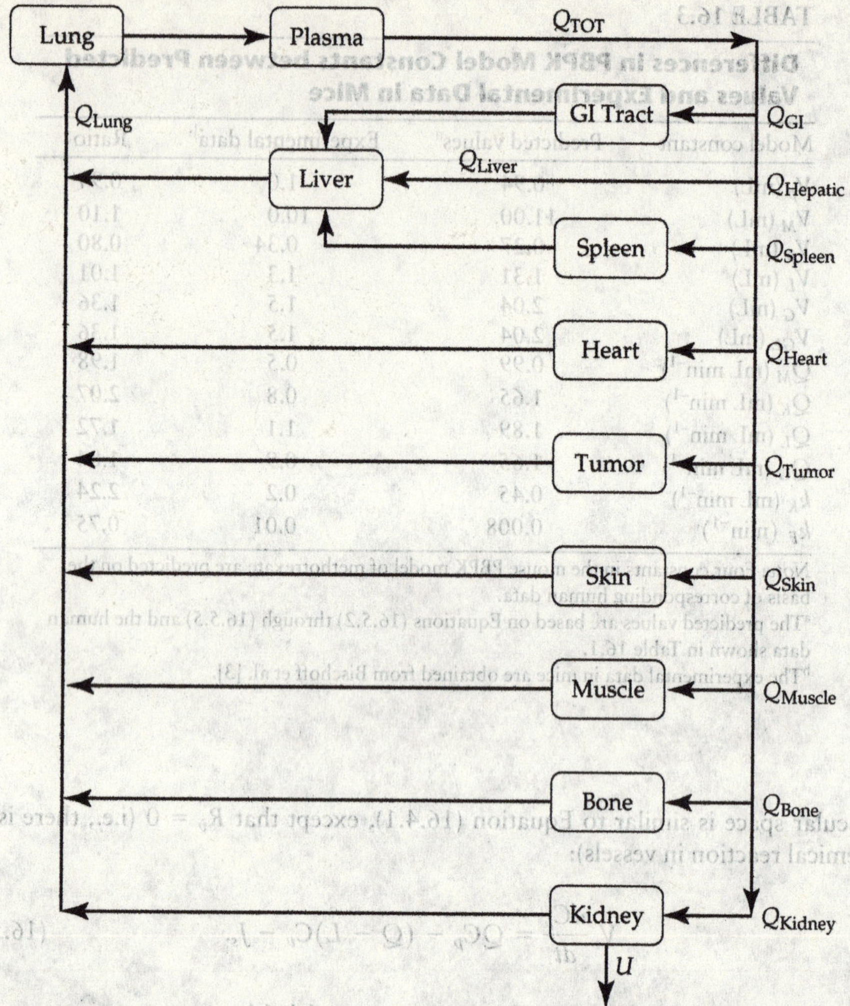

FIGURE 16.7 Diagram of the PBPK model for a monoclonal antibody that can bind to a tumor-associated antigen. Q represents the rate of plasma flow to individual organs. U is the rate constant of kidney clearance, assuming that the rate of clearance is proportional to the concentration in the kidney. (Adapted from Ref. [13], with permission.)

where G denotes free antigen, aG designates antibody–antigen complex, and k_f^S and k_r^S are forward and reverse rate constants, respectively, for specific binding in the tumor. On the basis of these reaction models, the following mass balance equations in the interstitial compartment can be derived for free antibody, nonspecifically bound antibody, and antibody–antigen complex, respectively:

$$V_i \frac{dC_i^a}{dt} = J_s - \frac{LC_i^a}{Z} - V_i(k_f^{NS}C_i^a - k_r^{NS}C_i^b) - V_i[k_f^S C_i^a(G_{max} - C_i^{aG}) - k_r^S C_i^{aG}], \tag{16.5.7a}$$

$$V_i \frac{dC_i^b}{dt} = V_i(k_f^{NS}C_i^a - k_r^{NS}C_i^b), \tag{16.5.7b}$$

$$V_i \frac{dC_i^{aG}}{dt} = V_i[k_f^S C_i^a(G_{max} - C_i^{aG}) - k_r^S C_i^{aG}]. \tag{16.5.7c}$$

In these equations, C_i represents the interstitial concentration, V_i is the interstitial volume, L is the total lymph flow rate from an organ, Z is the equilibrium ratio of antibody concentrations between tissue and lymph, and G_{max} is the total concentration of antigen in tumors. The superscripts a, b, and aG refer to free antibody, nonspecifically bound antibody, and antibody–antigen complex. Note that the antigen exists only in the tumor compartment. Therefore, Equation (16.5.7c) and the last term in Equation (16.5.7a) are omitted from normal organ compartments (i.e., k_f^S and k_r^S, as well as C_i^{aG}, are equal to zero in normal compartments). In addition,

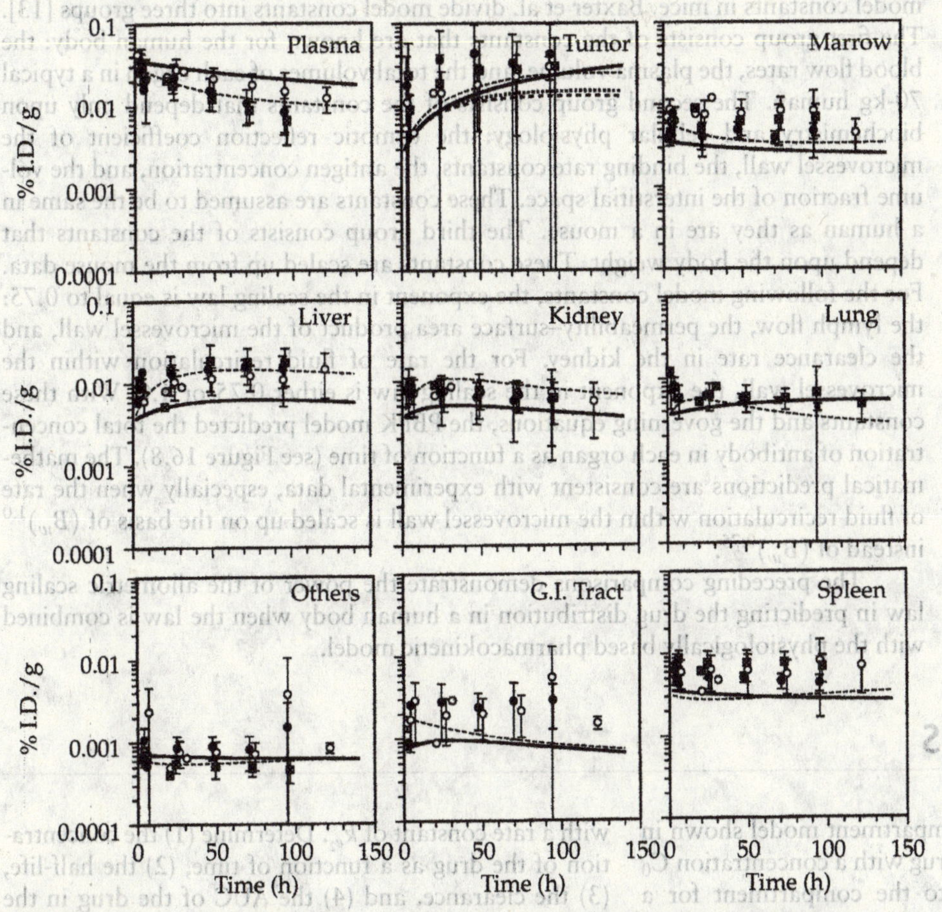

FIGURE 16.8 Comparisons of mathematical predictions (lines) and experimental data (symbols) of the biodistribution of a ^{111}In-labeled ZCE025 IgG against carcinoembryonic antigen in different organs. The "Others" compartment is the total of heart, skin, and muscle compartments. The antibody is injected intravenously with three different doses: 42 mg (○, ...), 21 mg (■, ...), or 2 mg (●, ______). The error bars indicate standard deviations of the experimental data. The model simulations assume that the human body weighs 70 kg and the tumor weight is 20 g. The mathematical predictions are nearly independent of the antibody dose; i.e., all three lines are indistinguishable except in the tumor. Furthermore, the mathematical predictions (.......) are closer to the experimental data in the plasma, kidney, and liver if the rate of fluid recirculation within the microvessel wall is scaled up based upon$(B_w)^{1.0}$. (Adapted from Ref. [13], with permission.)

$L = 0$ in the tumor compartment, because there is no functional lymphatic system in tumors (see Chapter 15). Cellular uptake is not considered in this model. Thus, there is no need to include Equation (16.4.3) for mass balance in the cellular space.

Equations (16.5.6) and (16.5.7a) through (16.5.7c) govern antibody transport in each organ. However, to determine the antibody distribution in the body, we also need to know the values of all constants in the model. Only a few such constants have been measured experimentally in the human body. Thus, the values of the rest of the model constants either must be assumed to be the same as the corresponding constants in mice or can be determined by using the allometric scaling law and the model constants in mice. Baxter et al. divide model constants into three groups [13]. The first group consists of the constants that are known for the human body: the blood flow rates, the plasma volume, and the total volumes of each organ in a typical 70-kg human. The second group consists of the constants that depend only upon biochemistry and cellular physiology: the osmotic reflection coefficient of the microvessel wall, the binding rate constants, the antigen concentration, and the volume fraction of the interstitial space. These constants are assumed to be the same in a human as they are in a mouse. The third group consists of the constants that depend upon the body weight. These constants are scaled up from the mouse data. For the following model constants, the exponent in the scaling law is equal to 0.75: the lymph flow, the permeability–surface area product of the microvessel wall, and the clearance rate in the kidney. For the rate of fluid recirculation within the microvessel wall, the exponent in the scaling law is either 0.75 or 1.0. With these constants and the governing equations, the PBPK model predicted the total concentration of antibody in each organ as a function of time (see Figure 16.8). The mathematical predictions are consistent with experimental data, especially when the rate of fluid recirculation within the microvessel wall is scaled up on the basis of $(B_w)^{1.0}$ instead of $(B_w)^{0.75}$.

The preceding comparisons demonstrate the power of the allometric scaling law in predicting the drug distribution in a human body when the law is combined with the physiologically based pharmacokinetic model.

16.6 PROBLEMS

16.1 Consider the one-compartment model shown in Figure 16.9. A solution of drug with a concentration C_0 is infused continuously into the compartment for a period of time T. The infusion starts at $t = 0$ and the flow rate is Q. Determine (1) the concentration as a function of time, (2) the clearance, (3) the area under the curve (AUC), and (4) the condition required for the concentration to reach the steady state.

16.2 Consider the two-compartment model shown in Figure 16.10. A drug is injected into the central compartment as a bolus and is excreted from the central compartment with a rate constant of k_{e_1}. Meanwhile, the drug is metabolized in the peripheral compartment with a rate constant of k_{e_2}. Determine (1) the concentration of the drug as a function of time, (2) the half-life, (3) the clearance, and (4) the AUC of the drug in the central compartment.

16.3 Consider the one-compartment model shown in Figure 16.11. A drug is injected into the compartment as a bolus and can nonspecifically bind to tissues via the reaction

$$D \underset{k_r}{\overset{k_f}{\longleftrightarrow}} B,$$

where D and B represent free and bound drugs, respectively, and k_f and k_r are forward and reverse rate

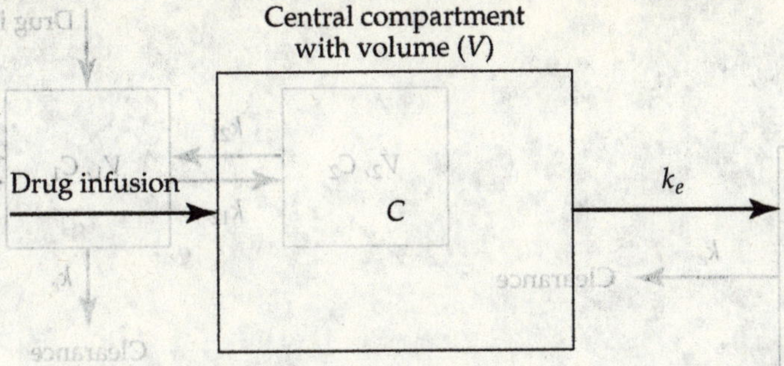

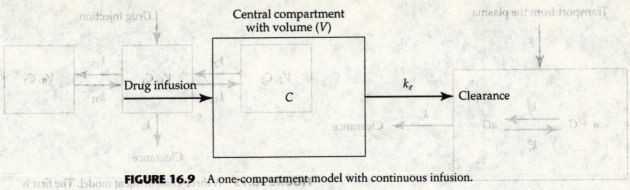

FIGURE 16.9 A one-compartment model with continuous infusion.

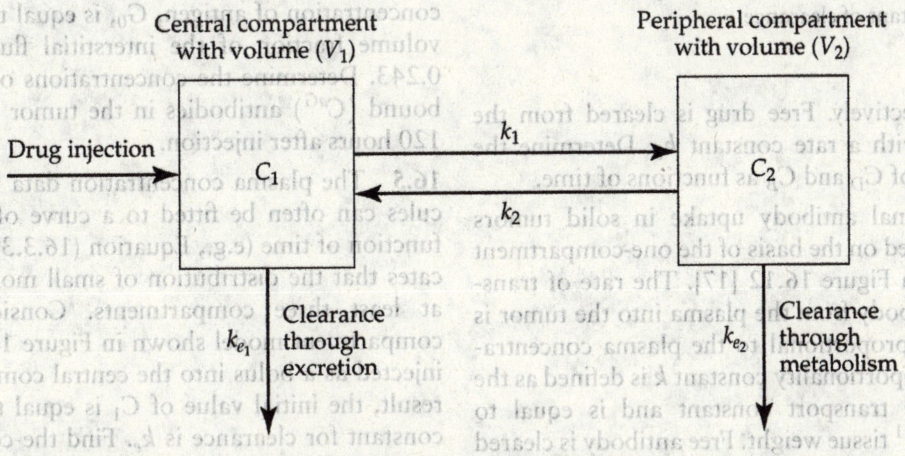

FIGURE 16.10 A two-compartment model with drug clearance in both compartments.

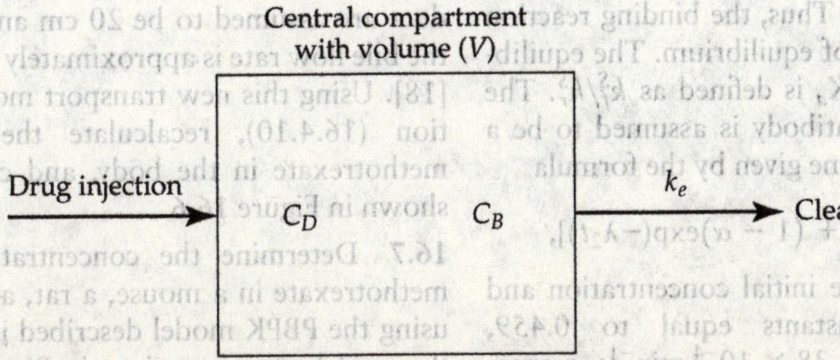

FIGURE 16.11 A one-compartment model involving the first-order chemical reaction.

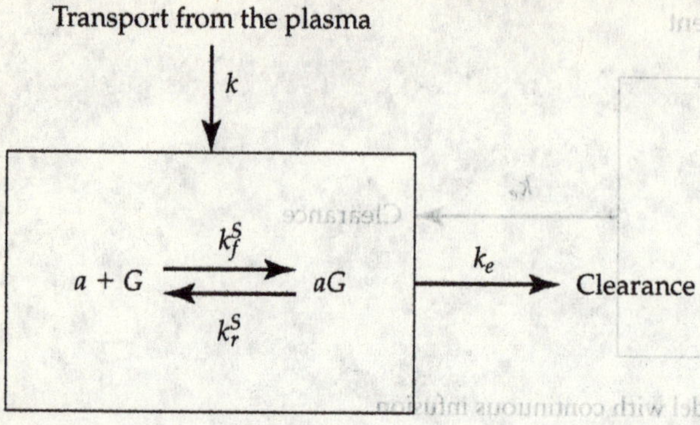

FIGURE 16.12 A one-compartment model of monoclonal antibody uptake in solid tumors. a denotes free antibody, G denotes free antigen, aG denotes antibody-antigen complex, and k_f^S and k_r^S are forward and reverse rateconstants, respectively. k is the plasma-to-tissue transport constant, and k_e is the rate constant of clearance.

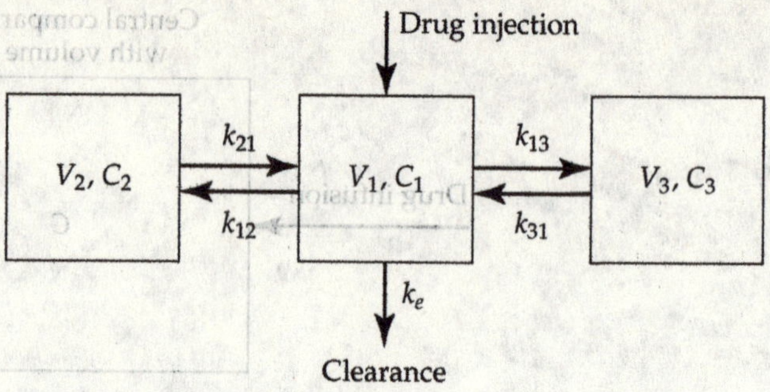

FIGURE 16.13 A three-compartment model. The first is the central compartment and the other two are peripheral compartments. Drug is injected into and cleared from the central compartment.

constants, respectively. Free drug is cleared from the compartment with a rate constant k_e. Determine the concentrations of C_D and C_B as functions of time.

16.4 Monoclonal antibody uptake in solid tumors has been analyzed on the basis of the one-compartment model shown in Figure 16.12 [17]. The rate of transport of the antibody from the plasma into the tumor is assumed to be proportional to the plasma concentration C_p. The proportionality constant k is defined as the plasma-to-tissue transport constant and is equal to 0.5 μL min^{-1} g^{-1} tissue weight. Free antibody is cleared from the tumor with a rate that is proportional to its concentration C_a. The rate constant of clearance is k_e and is equal to 0.8 μL min^{-1} g^{-1}. Within the tumor, antibody can bind to tumor-associated antigen in accordance with the reaction shown in Figure 16.12. The reaction is assumed to be much faster than the rate of change in concentration. Thus, the binding reaction is assumed to be in a state of equilibrium. The equilibrium association constant K_a is defined as k_f^S/k_r^S. The plasma concentration of antibody is assumed to be a biexponential function of time given by the formula

$$C_p = C_{p0}[\alpha \exp(-\lambda_1 t) + (1 - \alpha)\exp(-\lambda_2 t)],$$

where $C_{p0} = 100$ nM is the initial concentration and α, λ_1, and λ_2 are constants equal to 0.459, 1.11×10^{-2} min^{-1}, and 1.38×10^{-4} min^{-1}, respectively. Assume that K_a is equal to 10^9 M^{-1} and the total concentration of antigen, G_0, is equal to 100 nM. The volume fraction of the interstitial fluid space, ϕ, is 0.243. Determine the concentrations of free (C^a) and bound (C^{aG}) antibodies in the tumor during the first 120 hours after injection.

16.5 The plasma concentration data for small molecules can often be fitted to a curve of triexponential function of time (e.g., Equation (16.3.30)), which indicates that the distribution of small molecules involves at least three compartments. Consider the three-compartment model shown in Figure 16.13. A drug is injected as a bolus into the central compartment. As a result, the initial value of C_1 is equal to C_0. The rate constant for clearance is k_e. Find the concentration in the central compartment as a function of time.

16.6 An alternative approach to modeling the process of methotrexate transport in the hepatic duct is to consider the convection of methotrexate through the duct. Assume that diffusion is negligible and that transport is one dimensional. The length and the diameter of the duct are assumed to be 20 cm and 1 cm, respectively, the bile flow rate is approximately 20 mL h^{-1} in humans [18]. Using this new transport model to replace Equation (16.4.10), recalculate the biodistribution of methotrexate in the body, and compare it with that shown in Figure 16.6.

16.7 Determine the concentration distributions of methotrexate in a mouse, a rat, a dog, and a monkey, using the PBPK model described in Section 16.4.2 and the model constants given in Bischoff et al. [3]. Compare the concentration distributions in different species.

TABLE 16.4

Some Model Constants for Methotrexate Distribution in a 22-g Mouse

Z_M	0.15
Z_K	3.0
Z_L	10.0
Z_G	1.0
k_L/K_L (mL min^{-1})	0.4
τ (min)	2.0
k_G (mg min^{-1})	0.20
K_G (mg mL^{-1})	6.0
b (mL min^{-1})	0.001

16.8 The physiologically based pharmacokinetic model shown in Figure 16.5 can be used to scale up (or down) the biodistribution of methotrexate in the bodies of different species. However, the scaling law cannot predict the model constants precisely. Table 16.3 shows that the predicted values of V, Q, k_K, and k_F in mice based on human data, are different from the experimental data for mice. If other model constants are fixed as shown in Table 16.4, determine how the differences in predicted versus experimental values of V, Q, k_K, and k_F affect the simulated biodistribution of methotrexate in mice.

16.7 REFERENCES

1. Fujimori, K., Covell, D.G., Fletcher, J.E., and Weinstein, J.N., "Modeling analysis of the global and microscopic distribution of immunoglobulin G, F(ab')$_2$, and Fab in tumors." *Cancer Res.*, 1989. **49**: pp. 5656–5663.
2. Baxter, L.T., Yuan, F., and Jain, R.K., "Pharmacokinetic analysis of the perivascular distribution of bifunctional antibodies and haptens: comparison with experimental data." *Cancer Res.*, 1992. **52**: pp. 5838–5844.
3. Bischoff, K.B., Dedrick, R.L., Zaharko, D.S., and Longstreth, J.A., "Methotrexate pharmacokinetics." *J. Pharm. Sci.*, 1971. **60**: pp. 1128–1133.
4. Dedrick, R.L., "Animal scale-up." *J. Pharmacokinet. Biopharm.*, 1973. **1**: pp. 435–461.
5. Dedrick, R.L., and Bischoff, K.B., "Species similarities in pharmacokinetics." *Fed. Proc.*, 1980. **39**: pp. 54–59.
6. Nagy, K.A., Girard, I.A., and Brown, T.K., "Energetics of free-ranging mammals, reptiles, and birds." *Ann. Rev. Nutr.*, 1999. **19**: pp. 247–277.
7. West, G.B., Brown, J.H., and Enquist, B.J., "The fourth dimension of life: fractal geometry and allometric scaling of organisms." *Science*, 1999. **284**: pp. 1677–1679.
8. Schmidt-Nielsen, K., *Scaling: Why Is Animal Size So Important?* 1984, New York: Cambridge University Press.
9. Bassingthwaighte, J.B., Liebovitch, L.S., and West, B.J., *Fractal Physiology*. 1994, New York: Oxford University Press.
10. Feldman, H.A., "On the allometric mass exponent, when it exists." *J. Theor. Biol.*, 1995. **172**: pp. 187–197.
11. Zhou, S., Kestell, P., and Paxton, J.W., "Predicting pharmacokinetics and drug interactions in patients from in vitro and in vivo models: the experience with 5,6-dimethylxanthenone-4-acetic acid (DMXAA), an anti-cancer drug eliminated mainly by conjugation." *Drug Metab. Rev.*, 2002. **34**: pp. 751–790.
12. Lave, T., Coassolo, P., and Reigner, B., "Prediction of hepatic metabolic clearance based on interspecies allometric scaling techniques and in vitro-in vivo correlations." *Clin. Pharmacokinet.*, 1999. **36**: pp. 211–231.
13. Baxter, L.T., Zhu, H., Mackensen, D.G., Butler, W.F., and Jain, R.K., "Biodistribution of monoclonal antibodies: scale-up from mouse to human using a physiologically based pharmacokinetic model." *Cancer Res.*, 1995. **55**: pp. 4611–4622.
14. Schmidt-Nielsen, K., *Animal Physiology: Adaptation and Environment*. 1997, New York: Cambridge University Press.

15. Riviere, J.E., *Comparative Pharmacokinetics: Principles, Techniques, and Applications*. 1999, Ames, IA: Iowa State University Press.
16. Rippe, B., and Haraldsson, B., "Fluid and protein fluxes across small and large pores in the microvasculature: application of two-pore equations." *Acta Physiol. Scand.*, 1987. 131: pp. 411–428.
17. Sung, C., Shockley, T.R., Morrison, P.F., Dvorak, H.F., Yarmush, M.L., and Dedrick, R.L., "Predicted and observed effects of antibody affinity and antigen density on monoclonal antibody uptake in solid tumors." *Cancer Res.*, 1992. 52: pp. 377–384.
18. Chijiiwa, K., Mizuta, A., Ueda, J., Takamatsu, Y., Nakamura, K., Watanabe, M., Kuroki, S., and Tanaka, M., "Relation of biliary bile acid output to hepatic adenosine triphosphate level and biliary indocyanine green excretion in humans." *World J. Surg.*, 2002. 26: pp. 457–461.

Energy and Bioheat Transfer

PART V

Energy is necessary for all biological functions, and organisms have evolved to promote the efficient delivery of energy and removal of heat. The production of heat is a consequence of the second law of thermodynamics, and efficient heat transport is an important physiological function. In humans and other mammals, specialized processes have evolved to regulate heat production and removal and thereby maintain the body temperature in a very narrow range.

Like mass transport, energy transport is an important component in many biomedical technologies, surgical methods, and methods to preserve cells. Laser irradiation is now an indispensable component in many surgical procedures, used to open tissue and selectively remove and remodel tissue. Hyperthermia is an important component in the array of methods to treat cancer. Low temperatures are used to reduce the metabolic rate during surgery, and surgical instruments at subfreezing temperatures are used to selectively destroy diseased tissues. Cell preservation methods rely upon precise cooling to store cell and small tissue samples for later usage. Safety standards require an understanding of the ways that electromagnetic energy interacts with cells, tissues, and organisms.

In all of these physiological processes and biomedical applications, a quantitative description of energy transport is needed to understand the process, make predictions, and establish design guidelines. Although the underlying physics differs, mass and energy transport are similar in many respects. Both are scalar quantities and the conservation relations are similar. As a result, the governing equations that are solved are very similar. Energy transport by radiation and during phase changes does not have a counterpart in mass transport and must be handled separately. In Chapter 17, we focus on the transport of heat and its biomedical applications. Applications of the electromagnetic spectrum for imaging are important but are outside the scope of this book. The analogies between mass and energy transport are discussed extensively here, and we focus on a number of examples to formulate bioheat-transfer problems and to examine various physiological and biomedical applications.

Energy Transport in Biological Systems

CHAPTER 17

17.1 Introduction

Biological systems require energy to perform their normal functions and to perform work. As a consequence, heat is generated. Regulating the removal of heat is important for the survival of the organism. In plants, the energy supply is in the form of sunlight. Light is converted into chemical energy by the process of photosynthesis. Some bacteria living near the bottom of oceans can utilize Fe^{++} or Mn^{++} as an energy source. However, almost all bacteria, fungi, and animals derive their energy source from organic chemicals provided by plants, bacteria, or other animals. The energy is stored within cells in a number of chemicals with high-energy bonds. When these bonds are broken in metabolic processes, the energy released is used to perform the many functions of the organism. For example, cells use energy to maintain osmotic pressure gradients, to divide, to synthesize proteins, and to move. For animals, resting metabolic functions include autonomic motor activity for respiration and heart contraction, perfusion of blood and other fluids in the body, maintenance of body temperature, basic mental processes, food digestion, and simple body motion. In addition, animals must store enough energy for sudden increases in metabolic state arising from body responses to any number of environmental changes.

In order for biological systems to efficiently utilize chemical energy to do useful work, the reactions must be coupled. Otherwise, the energy produced from certain reactions could be transferred only as heat. Coupling usually involves utilization of energy from ATP or other high-energy small organic molecules to drive energy-dependent chemical reactions. Overall, the efficiency at which biological energy transfer occurs is the ratio of useful work, defined as the total work minus that done by the system to the energy input for volume expansion. For instance, aerobic glycolysis of 1 mole of glucose produces the equivalent of 36 moles of ATP. Under physiological conditions, the energy stored in each mole of ATP that is available to do useful work is 42 KJ. The complete breakdown of 1 mole of glucose to water and carbon dioxide releases 2823 kJ of energy, indicating that the maximum efficiency of energy transfer during glycolysis is 53%, which is much higher than that of many mechanical devices, even when they operate under a significant temperature gradient [1]. Energy from metabolism that is not converted to chemical energy or mechanical work is released as heat that is used to maintain the normal body temperature. The excess heat must be removed from the organism to avoid a significant increase in the body temperature. While the energetics of biochemical reactions is a branch of biochemistry, the transfer of heat is an important topic in biotransport phenomena.

In a number of biomedical and biotechnological processes, the addition or removal of heat is an important consideration. Large-scale culture of microorganisms generates considerable amounts of heat, which must be efficiently removed to prevent overheating of cultures, which could cause protein denaturation or cell death. A common treatment of cancer, known as hyperthermia, involves local heating of the tissue to temperatures between 40°C and 42°C, which damages cells locally and thereby facilitates radiation therapy and drug delivery [2]. More extreme heating, known as ablation, kills cells. Ablation is used to directly destroy tumors and cardiac tissue that exhibit irregular electrical conduction [3]. Energy is provided by radiofrequency radiation, lasers, microwaves, and focused ultrasound. For many biomedical devices, energy transfer is part of their normal operation. Therefore, strict guidelines exist to prevent overheating of these devices. Also, cooling of organs destined for transplantation reduces cell and tissue damage, and cooling of patients, or hypothermia, is used during heart or brain surgery. Preservation of cells involves the controlled cooling of cells suspended in a solution known as a cryopreservative, which is designed to minimize cell damage during cooling and thereby promote cell survival [4]. Cryosurgery is used to kill tumors or other tissues by exposing cells to subfreezing temperatures.

In this chapter, we examine the transfer of heat in organisms, produced by metabolism. We develop relations for the conservation of energy and apply these results to quantitatively understand energy transport and predict local temperature changes after heating or cooling. We begin in the next section by defining the first law of thermodynamics for an open system with energy and mass flow. As with mass transport, we will consider both differential and integral forms of these equations. We will examine special cases of these equations to consider energetics from biochemical reactions. Next, we will examine the most common forms of energy transfer, conduction and convection. This background enables us to quantitatively discuss metabolism and temperature distribution. Specific applications considered include cooling by evaporative heat loss (sweating) and cryopreservation.

17.2 First Law of Thermodynamics and Metabolism

17.2.1 Conservation Relations

The general statement of the law of conservation of energy[1] is that *energy can neither be created nor destroyed; it can only be changed from one form to another.*[2] Energy

[1]When considering energy transfer, it is important to define the region under study. This region is known as the *system*. Everything else is known as the *surroundings*. For an *open system*, there is exchange of mass and energy with the surroundings. *A closed system* does not exchange mass with the surroundings but does exchange energy with the surroundings. An *isolated system* exchanges neither mass nor energy with the surroundings.

[2]We are excluding the case of nuclear reactions that involve the interconversion of matter and energy.

exists in various forms: *kinetic energy,* which is one-half the product of the mass and the square of the velocity; *potential energy,* which is energy stored due to the position of an object in a gravitational field; *internal energy,* which represents the translational, rotational, and vibrational energy of molecules as well as the energy present in noncovalent bonds between molecules and covalent bonds between atoms in a molecule. Sometimes the bond energy within molecules is referred to as *chemical energy.* The energy of a system is a state variable because the value of the total energy depends only on the state of the system and not on the pathway taken to reach a specific state. Examples of other state variables include temperature and pressure.

Work equals the integral of the product of the force acting on a system, multiplied by the displacement of the system. The sign convention is that work is positive when work is done by the surroundings on the system. Thus,

$$W = \int \mathbf{F} \cdot \mathbf{n} dx, \tag{17.2.1}$$

where $\mathbf{n}$ is the unit outward normal and $\mathbf{F}$ is the force exerted on the surface of the system. If a compressive force is applied, $\mathbf{F} \cdot \mathbf{n}$ is negative and dx is also negative, resulting in a positive value for the work, which is in agreement with the sign convention.

Various types of work can be done on a system, including mechanical work, chemical work, and electrical work. Mechanical work may involve pushing a piston, raising the height of objects, stretching a spring, or turning a shaft. Humans and other animals can do work for lifting, walking, and running. Muscle fibers or cells perform work via contraction. Muscle contraction causes movement of bones, and cell contraction causes cell movement. At a molecular level, work is done by conformational change of macromolecules or motion of molecular motors (e.g., kinesin) along cytoskeletal filaments (e.g., microtubule). This work transports organelles or moves cytoskeletal elements such as actin or myosin.

Heat represents the transfer of energy between two bodies or between the system and its surroundings due to a temperature difference.[3] The temperature of the system represents the vibrational, rotational, and translational energies of the molecules as well as the intermolecular energy. Three mechanisms of heat transfer are considered in biological systems: conduction, convection, and thermal radiation. Conduction involves the direct transfer of energy via molecular contacts. In gases, the energy is transferred by molecular collisions and the thermal conductivity is proportional to the mean molecular speed, which is proportional to the square root of the absolute temperature [5]. In liquids, the molecules have limited translational motion but do collide with neighboring molecules, permitting energy exchange. The thermal conductivity depends on the square root of the isothermal compressibility [5]. In crystalline solids, heat conduction is mediated by exchange of vibrational and rotational energy among molecules. In metals, electrons transport energy, and the thermal conductivity is proportional to electrical conductivity. Biological tissues are a combination of liquids and fibrous solids and have values of thermal conductivity similar to that of water.

Convection mediates energy transfer through coupling of the bulk motion of fluids and heat conduction. For example, energy exchange between a solid object

[3]"heat" in *A Dictionary of Physics.* Ed. John Daintith. Oxford University Press, 2000. Oxford Reference Online. Oxford University Press. Duke University. 30 May 2008: <http://www.oxfordreference.com/views/ENTRY.html?subview=Main&entry=t83.e1361>

and a fluid flowing near its surface can occur through heat conduction. The rate of exchange depends on fluid velocity. At a microscopic level, energy transfer also occurs from hot to cold fluids mixed in a stream through conduction. The rate of heat transfer depends on the extent of mixing, which in turn is a function of fluid velocity and its distribution. In *forced convection*, fluid flow is independent of the temperature distribution, but does produce a temperature gradient. In the case of *natural convection*, a velocity field is generated by the buoyancy force induced by temperature gradient (see Section 17.4.2).

Thermal radiation is the transfer of energy in the form of electromagnetic waves, which causes temperature change in the object. The wavelength of thermal radiation is between 0.1 and 100 microns. When electromagnetic waves reach the system, part of the energy is reflected, part is transmitted through the system, and part is absorbed by the system. Energy absorption causes the temperature to rise. This is the major mechanism by which the sun's energy heats the earth. A system can also release electromagnetic waves from its surface. Radiation emission is one of the key mechanisms of heat removal from animals in cold environments. The extent of radiation emission and the resulting temperature reduction depend on the wavelength of radiation. The theory of thermal radiation is complicated and beyond the scope of this textbook. Therefore, effects of radiation absorption on internal energy are modeled in this chapter as external heat sources distributed in the system. The theory of radiation emission from an object surface is briefly introduced in Section 17.2.3 in the discussion of boundary conditions.

In general, organisms are *open systems*; that is, they exchange matter and energy with their surroundings. Further, organisms are multi-component systems, exchanging many different types of molecules with the surroundings. Each of the different molecular components entering or leaving a biological organism carries with it some energy in the form of internal molecular energy as well as kinetic energy. In this chapter, we define the open system such that molecular coupling between mass and heat transfer (e.g., the "Dufour effect") is negligible compared to other mechanisms of heat transfer. In addition, there are no molecular mixing, chemical/nuclear reaction, transport of charged particles (e.g., electrons and ions), or phase transition within the system. These phenomena occur only at boundaries of the system or in the surroundings. Although a more general equation of energy conservation can be derived without these assumptions, it is not very useful for determining the rate of heat exchange between the system and surroundings and the spatial and temporal distributions of temperature.

In general, the internal energy of the system depends on temperature, pressure, and material compositions. The dependence on composition is often unknown. For example, metabolism (i.e., chemical reactions) in a cell converts internal energy to heat and chemical work (e.g., synthesis of new molecules and pumping of ions across membrane). The rate of heat production can be measured experimentally, but is often difficult to quantify the work done. Therefore, we define the biological organisms as open systems in such a way that excludes all atoms and molecules involved in chemical/nuclear reactions. As a result, the internal energy depends only on temperature and pressure, which is adequate for heat-transfer studies. A more convenient way of thinking is to define the system as the solvent (e.g., water) plus nonreactive solid structures in cells or tissues, although it is not required in the following derivations.

In Section 4.4.2, we presented the equation for mechanical energy, which is one form of the conservation of energy. In this section, we state a more general form of energy conservation, which is a formal statement of the first law of thermodynamics. For an open system, this conservation relation can be stated as

$$\begin{bmatrix} \text{Rate of energy} \\ \text{accumulation} \\ \text{within the system} \end{bmatrix} = \begin{bmatrix} \text{Net rate of energy} \\ \text{transfer across} \\ \text{system surfaces} \end{bmatrix} + \begin{bmatrix} \text{Rate of work} \\ \text{done on the} \\ \text{system} \end{bmatrix} + \begin{bmatrix} \text{Rate of energy} \\ \text{production} \\ \text{within the system} \end{bmatrix} \tag{17.2.2}$$

The energy is the sum of kinetic and internal energy of the system. The net rate of energy transfer across surfaces represents the sum of heat conduction, transport of energy due to mass entering or leaving of the system, and electromagnetic radiation. In general, electromagnetic radiation can alter system energy through excitation of electrons, ionization of molecules, induction of chemical or nuclear reactions, and increase in temperature. Based on the definition of the open system used in this chapter, only thermal radiation needs to be considered here, since other changes do not occur in the system. The absorption of thermal radiation from surroundings is modeled as energy sources or sinks distributed throughout the system. Energy transfer, mediated by mass transport, can occur through convection, diffusion, and movement of changed or magnetic particles (e.g., electrons) driven by external fields. Also, based on the definition of the open system used in this chapter, only convective transport of energy needs to be considered. Therefore, the first term on the right side of Equation (17.2.2) includes only heat conduction and energy convection.

Although energy cannot be created or destroyed, the energy production term in Equation (17.2.2) is introduced to consider energy transfer into the system from external energy sources or sinks distributed within the same space as that occupied by the system. The energy production rate is positive at sources and negative at sinks. The energy sources or sinks are used to model various mechanisms of energy exchange between the system and the surroundings. The first mechanism is the absorption of thermal radiation. For example, absorption of photons by biological tissues during laser ablation produces a temperature increase. The thermal radiation absorbed at a location is converted to the internal energy of the system at the same location. The second mechanism represents interactions of the system with external molecules or particles (e.g., electrons). The energy transfer occurs when the internal and kinetic energies of these molecules or particles are reduced due to molecular mixing, chemical/nuclear reactions, or resistance to transport. For example, a biological tissue is heated around the electrodes when electric current is delivered into the tissue. The heating is due to electrical resistance in tissues. Chemical reactions can also heat tissues. These molecules/particles exist in external sources or sinks, since they do not belong to the system; and the nature of energy transfer is the same as that for heat conduction. The rate of heat production from chemical reactions is denoted by $\dot{Q}_m$. During reactions in biological systems (i.e., metabolism), a large fraction of the energy stored in reactants is dissipated as heat. Thus, $\dot{Q}_m$ represents the rate of heat production from metabolism (see Section 17.6). The third mechanism represents energy delivery into a tissue through convection in microvessels and will be discussed in the derivation of the bioheat-transfer equation in Section 17.7. In this model, the system is the extravascular space of the tissue, and the microvascular network is represented as external heat sources or sinks distributed in the system. Taken together, the energy production term in Equation (17.2.2) represents the sum of rates of energy transfer into the system, which are not considered explicitly in the first term on the right side. The total rate of production per unit volume is denoted by $\dot{Q}_p^*$.

The work done on the system is positive, since it adds energy to the system; the work done by the system is negative, since it transfers energy to the surroundings. This sign convention is consistent with Equation (17.2.1). Likewise, heat transfer is positive if it adds energy to the system and negative if it is released by the system.

17.2.2 Differential Forms of the Conservation of Energy

In this section, conservation of energy is expressed in differential form. The integral form will be derived in Section 17.6.2. With these mathematical equations, we will examine various problems in energy transport. Similar to the derivation of conservation equations for mass and momentum transport, we define a control volume that occupies the same space as the system does.

The system energy consists of kinetic energy ($\rho v^2/2$) and internal energy (U on an extensive or total basis or $\hat{U}$ on an intensive or mass basis),[4] where ρ and v are the mass density and the fluid velocity, respectively. Thus, on a mass basis, the system energy is

$$\hat{E} = \tfrac{1}{2}v^2 + \hat{U}. \tag{17.2.3}$$

For a multicomponent system, the density ρ is the total density of the mixture, v is the mass average velocity, and $\hat{U} = \sum \omega_i U_i$, where ω_i is the mass fraction (see Equation (6.2.2a)) and U_i is the partial specific internal energy of the ith component in the mixture.

Energy transfer between the system and the surroundings can be described in terms of a flux vector (e), which is analogous to that used in mass transport, as discussed in Sections 6.2 and 6.4.1. The flux includes convective transport of the energy ($\rho\hat{E}\mathbf{v}$) and the heat conduction (q), as discussed in the previous section. Heat transfer through thermal radiation is considered by heat sources or sinks in the system, as discussed above, and by specific boundary conditions discussed in the next section. Therefore, the energy flux e can be written as

$$\mathbf{e} = \rho\hat{E}\mathbf{v} + \mathbf{q}. \tag{17.2.4}$$

The types of work that can be done on the system or by the system include mechanical, electrical, and chemical work. According to the definition of the system adopted in this book, there is no chemical work, since there are no changes in the molecular compositions of the system. The rate of total work per unit volume is denoted by $\dot{W}_t$. Two types of mechanical work for fluid flow in the system are considered explicitly: work done by fluid stresses ($\boldsymbol{\sigma}$) and body forces (F) (e.g., gravitational

[4]Intensive variables are independent of the amount of mass present. Some variables, such as pressure, concentration, density, viscosity, and temperature, are naturally intensive, and they can vary locally. Other variables, such as energy, are extensive and depend on the amount of mass. By dividing by the mass (represented by the caret (^) over the variable), moles (represented by the overbar), or volume, the extensive variable becomes intensive.

force). The corresponding rates of work per unit volume are $\nabla \cdot (\boldsymbol{\sigma} \cdot \mathbf{v})$ and $\mathbf{F} \cdot \mathbf{v}$, respectively. The fluid stress tensor can be separated into two terms:

$$\boldsymbol{\sigma} = -p\mathbf{I} + \boldsymbol{\tau}, \qquad (17.2.5)$$

where p is the pressure, I is the unit tensor, and $\boldsymbol{\tau}$ is the shear stresses tensor. After subtracting the work for fluid flow, the rate of other work is denoted by $\dot{W}$ that is equal to $\dot{W}_t - \nabla \cdot (\boldsymbol{\sigma} \cdot \mathbf{v}) - \mathbf{F} \cdot \mathbf{v}$. The differential form of the conservation of energy is developed in the same manner as the conservation of momentum (Section 3.3) and mass in a multicomponent system (Section 7.3). Since the approach is analogous to that for the derivation of Equation (7.3.8), only the final result is presented here:

$$\rho\left(\frac{\partial \hat{E}}{\partial t} + \mathbf{v} \cdot \nabla \hat{E}\right) = -\nabla \cdot \mathbf{q} - \nabla \cdot (p\mathbf{v}) + \nabla \cdot (\boldsymbol{\tau} \cdot \mathbf{v}) + \mathbf{F} \cdot \mathbf{v} + \dot{W} + \dot{Q}_p^*, \qquad (17.2.6)$$

and the derivation is left as an exercise. In Equation (17.2.6), $\dot{Q}_p^*$ is the net rate of heat production per unit volume of the system. This is the equation for energy balance in the system. For heat-transfer analysis, it is more convenient to derive a conservation equation for internal energy ($\hat{U}$) balance because $\hat{U}$ is directly related to the temperature in the system. In the derivation, we first perform a dot product of $\mathbf{v}$ with the equation of linear momentum balance of fluid (e.g., Equation (3.3.16)). It results in the mechanical energy balance equation for fluid flow,

$$\frac{\rho}{2}\left(\frac{\partial \mathrm{v}^2}{\partial t} + \mathbf{v} \cdot \nabla \mathrm{v}^2\right) = -\mathbf{v} \cdot \nabla p + \mathbf{v} \cdot (\nabla \cdot \boldsymbol{\tau}) + \mathbf{F} \cdot \mathbf{v}, \qquad (17.2.7)$$

where $\mathbf{v}$ is the magnitude of the velocity vector. Subtracting Equation (17.2.7) from Equation (17.2.6) yields

$$\rho\left(\frac{\partial \hat{U}}{\partial t} + \mathbf{v} \cdot \nabla \hat{U}\right) = -\nabla \cdot \mathbf{q} - p(\nabla \cdot \mathbf{v}) + \dot{\Phi}_v + \dot{Q}_p^* + \dot{W}, \qquad (17.2.8)$$

We have used the conservation of mass (Equation (3.2.7)) in deriving Equation (17.2.8). $\dot{\Phi}_v$ is the rate of viscous dissipation [5], which is given by Equation 17.2.9,

$$\dot{\Phi}_v = \boldsymbol{\tau} : \nabla \mathbf{v}. \qquad (17.2.9)$$

This term represents the portion of the shear stress work that is dissipated as heat and thus not available for changing kinetic energy of fluid. Heat is either transferred to the surroundings or absorbed by the system to increase its temperature.

In biological systems, most energy transport phenomena occur in media (e.g., water) that are incompressible. In these systems,

$$\hat{C}_V = \hat{C}_p, \qquad \nabla \cdot \mathbf{v} = 0, \qquad \rho\frac{D\hat{U}}{Dt} = \rho\hat{C}_p\frac{DT}{Dt}, \qquad (17.2.10\text{a,b,c})$$

where $\hat{C}_V$ and $\hat{C}_p$ are the specific heat capacity at constant volume and pressure, respectively, and D/Dt is the material derivative, defined as

$$\frac{D}{Dt} = \frac{\partial}{\partial t} + \mathbf{v} \cdot \nabla. \qquad (17.2.10\text{d})$$

Equation (17.2.10c) is valid only if there are no changes in molecular compositions in the system, which is one of the key considerations above in the definition of the system. Substituting Equations (17.2.10b) and (17.2.10c) into Equation (17.2.8) yields

$$\rho \hat{C}_p \frac{DT}{Dt} = -\nabla \cdot \mathbf{q} + \dot{W} + \dot{Q}_p \qquad (17.2.11)$$

where $\dot{Q}_p$ is equal to $\dot{Q}_p^* + \dot{\Phi}_v$. This conservation equation is the starting point for the analysis of energy transport in biological systems. Both $\dot{Q}_p$ and $\dot{W}$ can vary with time and spatial location.

17.2.3 Constitutive Relations and Boundary Conditions

To use the conservation of energy in the analysis of biological energy transport problems, we need to specify a constitutive equation for the heat flux q. Jean Baptiste Joseph Fourier (1768–1830) first formulated the law of heat conduction, which he eventually published in 1822. Fourier also developed the Fourier series and the Fourier transform. Fourier's law of heat conduction is actually a constitutive relation and forms the basis for Adolf Fick's analysis of diffusion. In vector form, the relation for an isotropic medium is

$$\mathbf{q} = -k\nabla T, \qquad (17.2.12)$$

where k is the thermal conductivity with units of watts per meter per Kelvin or joules per meter per second per Kelvin. Substituting Equation (17.2.12) into Equation (17.2.11) yields

$$\rho \hat{C}_p \left(\frac{\partial T}{\partial t} + \mathbf{v} \cdot \nabla T \right) = k\nabla^2 T + \dot{Q}_p + \dot{W}. \qquad (17.2.13)$$

Table 17.1 lists Equation (17.2.13) in rectangular, cylindrical, and spherical coordinates.

TABLE 17.1

Energy Equation for Open System

Rectangular

$$\rho C_p \left(\frac{\partial T}{\partial t} + v_x \frac{\partial T}{\partial x} + v_y \frac{\partial T}{\partial y} + v_z \frac{\partial T}{\partial z} \right) = k \left(\frac{\partial^2 T}{\partial x^2} + \frac{\partial^2 T}{\partial y^2} + \frac{\partial^2 T}{\partial z^2} \right) + \dot{Q}_p + \dot{W} \qquad (17.2.14a)$$

Cylindrical

$$\rho C_p \left(\frac{\partial T}{\partial t} + v_r \frac{\partial T}{\partial r} + \frac{v_\theta}{r} \frac{\partial T}{\partial \theta} + v_z \frac{\partial T}{\partial z} \right) = k \left[\frac{1}{r} \frac{\partial}{\partial r} \left(r \frac{\partial T}{\partial r} \right) + \frac{1}{r^2} \frac{\partial^2 T}{\partial \theta^2} + \frac{\partial^2 T}{\partial z^2} \right] + \dot{Q}_p + \dot{W} \qquad (17.2.14b)$$

Spherical

$$\rho C_p \left(\frac{\partial T}{\partial t} + v_r \frac{\partial T}{\partial r} + \frac{v_\theta}{r} \frac{\partial T}{\partial \theta} + \frac{v_\phi}{r \sin\theta} \frac{\partial T}{\partial \phi} \right)$$

$$= k \left[\frac{1}{r^2} \frac{\partial}{\partial r} \left(r^2 \frac{\partial T}{\partial r} \right) + \frac{1}{r^2 \sin\theta} \frac{\partial}{\partial \theta} \left(\sin\theta \frac{\partial T}{\partial \theta} \right) + \frac{1}{r^2 \sin^2\theta} \frac{\partial^2 T}{\partial \phi^2} \right] + \dot{Q}_p + \dot{W} \qquad (17.2.14c)$$

The foregoing equations apply to both living and nonliving systems. An important limiting case of Equation (17.2.13) is pure conduction with no convection, energy production, or work. In this case,

$$\frac{\partial T}{\partial t} = \alpha \nabla^2 T, \tag{17.2.15}$$

where $\alpha = k/(\rho \hat{C}_p)$ is the thermal diffusivity in units of square meters per second or square centimeters per second. Equation (17.2.15) is analogous to the three-dimensional, unsteady diffusion equation considered in Sections 6.7 and 6.8. Some representative values of thermal properties are presented in Table 17.2. A more complete set of data can be found in the paper by Bowman et al. [6]. Many organs have thermal conductivities in the range of 0.4–0.7 W m^{-1} K^{-1}. The density of tissues is slightly greater than the density of water.

In general, these equations can be used to determine the energy or temperature distribution in the system, the heat released, and the work done, depending on the particular biological situation and available data. Boundary conditions need to be specified in order to solve heat-transfer problems. When two bodies are in direct contact without an energy source or sink at the interface, the heat fluxes are the same

TABLE 17.2

The Thermal Properties of Some Biological Tissues and Other Materials Relevant for Heat Transport in Biological Systems

Substance	T (K)	Thermal conductivity, (W m^{-1} K^{-1})	Specific heat capacity (kJ kg^{-1} K^{-1})	Thermal diffusivity (m^2 s^{-1})	Reference
Air (ρ = 1.177 kg m^{-3})	300	0.025	1.006	2.11×10^{-5}	[7]
Water (ρ = 996 kg m^{-3})	300	0.609	4.183	1.5×10^{-7}	[5,8]
Ice (ρ = 917.6 kg m^{-3})	273	2.22	2.050	1.06×10^{-6}	[9]
Ethanol (ρ = 783.5 kg m^{-3})	300	0.1676	2.454	4.01×10^{-5}	[5]
Copper (ρ = 19,320 kg m^{-3})	300	401	0.385	5.39×10^{-5}	[7]
Gold (ρ = 19,300 kg m^{-3})	298	318	0.129	0.128	[10]
Titanium (ρ = 4,540 kg m^{-3})	273	22.4	0.523	9.43×10^{-3}	[8]
Skin (ρ = 1,070 kg m^{-3})	310	0.442	3.471	1.19×10^{-7}	[6,11]
Fat (ρ = 937 kg m^{-13})	298	0.21	3.258	0.69×10^{-7}	[11]
Blood (ρ = 937 kg m^{-3})	298	0.642	3.889	1.76×10^{-7}	[11]
Bone (ρ = 1,920 kg m^{-3})	298	0.30–0.58	1.44	$1.085\text{–}2.097 \times 10^{-7}$	[10,12]
Tooth enamel (ρ =2,900 kg m^{-3})	310	0.92	0.75	4.2×10^{-7}	[10]
Hair	298	0.038–0.100			[13]
Woven fabric	298	0.040			[13]

and temperatures are in equilibrium. Thus, for objects 1 and 2 in contact, the boundary conditions are

$$\mathbf{q}\cdot\mathbf{n}|_1 + \mathbf{q}\cdot\mathbf{n}|_2 = 0 \qquad T|_1 = T|_2, \tag{17.2.16a,b}$$

where $\mathbf{n}$ is the unit outward normal from the interface. The temperature boundary condition at the interface is in contrast to mass transfer, in which the concentrations may differ due to solubility or partitioning. This assumes thermal equilibrium between the two materials. The temperatures may not be the same if the materials are rough or a barrier is placed between them. However, the temperatures at the interface may not be known *a priori* and may need to be determined as part of the solution. The fluxes may be unequal if there is an energy source or sink due to chemical reactions or surface friction [14].

Solid surfaces are often in contact with a flowing fluid, and convective heat transfer occurs. Much of the analysis is analogous to what is done for mass transfer, as discussed in Chapter 7. Correspondingly, we can define a local heat-transfer coefficient h_{loc} (Figure 17.1) by equating the flux due to conduction with the convective heat flux:

$$q_y(y=0) = -k\left(\frac{\partial T}{\partial y}\right)_{y=0} = h_{loc}(T_S - T_b), \tag{17.2.17}$$

where T_S is the temperature on the solid surface, T_b is the bulk temperature in the fluid, and its definition is application-dependent (see Equation 17.2.18d). The mean heat-transfer coefficient h_f is generally defined in terms of a spatially averaged heat flux [5,18]:

$$\langle q_y(y=0)\rangle = -\frac{1}{S}\int_S k\left(\frac{\partial T}{\partial y}\right)_{y=0} dS = h_f\Delta T, \tag{17.2.18a}$$

where S is the area of solid surface over which heat transfer occurs and ΔT is a measure of temperature difference between solid and fluid. In different situations, h_f and ΔT may take different forms. For flow over submerged objects, the suitable temperature difference is that between the temperature in the bulk fluid far away from the

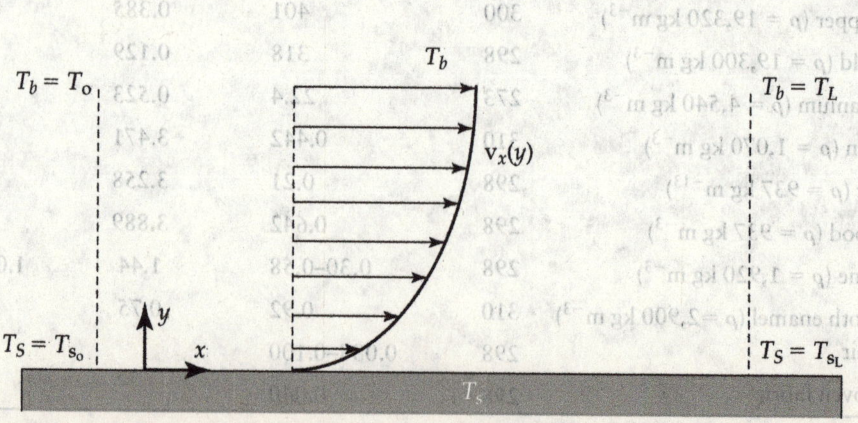

FIGURE 17.1 Heat-transfer coefficient from a solid surface into a moving fluid.

objects T_∞ and the average surface temperature $\langle T_S \rangle$. Thus, analogous to Equation (7.8.3), the mean heat-transfer coefficient for submerged objects h_m is given by

$$\langle q_y(y=0) \rangle = h_m(\langle T_S \rangle - T_\infty). \tag{17.2.18b}$$

Likewise, for confined flows in channels, the log-mean heat-transfer coefficient $h_{\ln}$ is defined as

$$\langle q_y(y=0) \rangle = h_{\ln}\left(\frac{(T_{S_0} - T_0) - (T_{S_L} - T_L)}{\ln\left((T_{S_0} - T_0)/(T_{S_L} - T_L)\right)}\right), \tag{17.2.18c}$$

where T_0 and T_L are the bulk temperature and T_{S_0} and T_{S_L} are the surface temperatures at two different locations in channels (see Figure 17.1). For confined flow in channels, the bulk temperature T_b used in the definition of local (h_{loc}) and log-mean ($h_{\ln}$) heat-transfer coefficients in Equations (17.2.17) and (17.2.18c), respectively, is given by

$$T_b = \frac{\int_A T\mathbf{v}\cdot\mathbf{n}dA}{\int_A \mathbf{v}\cdot\mathbf{n}dA} = \frac{\int_A T\mathrm{v}_z dA}{\int_A \mathrm{v}_z dA}, \tag{17.2.18d}$$

where A is the cross-sectional area. T_b in this case is also called the *flow average temperature*, and is location dependent; that is, T_0 can be different from T_L. Mass transfer correlations presented in Chapter 7 can be used to compute heat-transfer coefficients by replacing k_f and D_{ij} with the appropriate heat-transfer and heat-conduction coefficients, respectively.

Example 17.1 A runner has just completed a marathon on a warm, humid day. She steps into a cool room. Her body temperature initially is 39°C and the room air is at 28°C. The air flow is 1 m s^{-1}, for which the heat-transfer coefficient h_m is 16 W m^{-2} K^{-1} [15]. Determine the hear flux due to forced convection.

Solution From Equation (17.2.18b),

$$\langle q_y(y=0) \rangle = h_m(\langle T_S \rangle - T_\infty) = 16(11) = 176 \text{ W m}^{-2}$$

In addition to convection and conduction, energy can be transferred by thermal radiation which involves the emission of energy as electromagnetic waves. As such, the transfer does not have to involve matter, and energy can be transferred through a vacuum. This is the major mechanism by which solar energy reaches the earth. Although it is not a major mechanism of heat transfer within organisms, radiation can be a significant mechanism of heat transfer from organisms to the surroundings, or from the sun to organisms [15].

Thermal radiation emission often occurs at the surface of the system. The rate of radiation energy emitted per unit surface area is given by the Stefan–Boltzmann law [5]:

$$q|_{\text{radiation}} = \sigma e \mathrm{T}^4, \tag{17.2.19a}$$

where σ is the Stefan–Boltzmann constant ($\sigma = 5.67 \times 10^{-8}$ W m^{-2} K^{-4}) and e is the emissivity of the material in the object. The emissivity is the ratio of the radiation emitted by the object to the energy emitted by a perfectly emitting body, known as the black body. For a black body, emissivity is 1. Highly polished metal surfaces have small values of emissivity (e.g., 0.018 for highly polished copper at 28°C). Water has an emissivity of 0.95 and skin and fur have emissivities of 0.93. The net flux of radiation between two black bodies is

$$Q = q_y|_{\text{radiation}} A = \sigma A_1 F_{12}(T_1^4 - T_2^4) = \sigma A_2 F_{21}(T_1^4 - T_2^4), \quad (17.2.19b)$$

where F_{12} and F_{21} are the *view factors* or *configuration factors*, which are complex functions of the orientation of each object relative to the other, the shape of the surfaces, and the distance between the two objects, and A_1 and A_2 are the surface areas of the objects [5]. In general, the view factors range in magnitude from 0 to 1 and $A_1F_{12} = A_2F_{21}$ for black bodies. An object completely surrounded by a second object has a view factor of 1.

Most objects are not perfect absorbers or emitters and are known as gray bodies. For a small convex gray body in a large cavity, the emission of thermal radiation to the surroundings is governed by

$$Q = q_y|_{\text{radiation}} A = \sigma A_1(e_1 T_1^4 - a_1 T_2^4), \quad (17.2.19c)$$

where T_2 is the temperature in the cavity and a_1 is the absorptivity of the body at T_2. In general, e equals a at a given temperature, and a is not highly sensitive to the temperature. Thus, the absorptivity at T_2 is generally assumed to equal the absorptivity at T_1, which equals the emissivity at T_1 [5].

The energy transfer from body 1 to body 2 can occur in a vacuum, as in energy transfer from the sun to the earth, or through the atmosphere, which absorbs a negligible amount of radiation, except in the infrared region due to the presence of water and carbon dioxide. If the body emits radiation into the night sky, then T_2 can be assumed to be zero.

The thermal radiation reaching the earth's surface is about 1,000 W m^{-2} [16]. This energy must be reflected or removed to prevent overheating of plants and animals exposed to sunlight for long periods of time. Likewise, in the evening, organisms will radiate energy which can produce considerable cooling [16]. The maximum wavelength of energy emitted is given by the Wien displacement law

$$\lambda_{\max} = \frac{0.002884}{T}, \quad (17.2.19d)$$

where T is in Kelvin and $\lambda_{\max}$ is in meters. At normal terrestrial temperatures, Equation (17.2.19d) indicates that organisms radiate energy in the infrared range.

Example 17.2 After running the marathon on a warm, humid day, the winner steps into a large, cool room. The winner's body temperature has risen to 39°C. The surrounding air temperature is 28°C, and the walls are far from the individual. Determine the maximum energy flux from the runner by radiation and the total energy flux due to convection and radiation.

Solution Assuming that the absorptivity and emissivity are identical, the net energy flux by radiation is

$$q_y|_{radiation} = \sigma e_1(T_1^4 - T_2^4) = (5.67 \times 10^{-8})(0.93)(312^4 - 301^4) = 66.8 \text{ W m}^2.$$

The total flux is simply

$$q_y|_{total} = q_y|_{convection} + q_y|_{radiation} = 176 + 66.8 = 242.8 \text{ W m}^2.$$

Radiation increases the rate of heat transfer by 38%, above that due to convection alone. These results are upper limits, since the presence of clothing and other bodies nearby reduces both radiation and convection.

The radiant flux from the sun can be quite significant during the daytime. We all have had experience with high temperatures on a roadway, on sand, or on a roof. Interestingly, the radiant energy flux is greater at higher latitudes, especially during the summer. As a result, the surface temperature of fur, hair, or skin can rise significantly above body temperature [15].

17.2.4 The Dimensionless Form of Conservation Relations

Casting the energy equation in dimensionless form, we identify the key dimensionless groups and show the similarity to results for solute transport. Beginning with Equation (17.2.13) for the case of negligible work, the characteristic temperature drop is ΔT the characteristic length is L, and the characteristic time is (L^2/α). The characteristic velocity is the average velocity $\langle \mathrm{v} \rangle$. The corresponding dimensionless quantities are

$$T^* = \frac{T}{\Delta T} \qquad \mathbf{x}^* = \frac{\mathbf{x}}{L} \qquad \mathbf{v}^* = \frac{\mathbf{v}}{\langle \mathrm{v} \rangle} \qquad t^* = \frac{t\alpha}{L^2}, \tag{17.2.20a,b,c,d}$$

where $\mathbf{x}^*$ is the dimensionless position vector and t^* is known as the *Fourier number*, which scales time with the characteristic thermal diffusion time. As a result, Equation (17.2.3) becomes:

$$\frac{\partial T^*}{\partial t^*} + \frac{\langle v \rangle L}{\alpha}\mathbf{v}^* \cdot \nabla^* T^* = \nabla^{*2} T^* + \frac{\dot{Q}_p L^2}{\Delta T k} \tag{17.2.21}$$

The quantity $\langle \mathrm{v} \rangle L/\alpha$ represents the ratio of the time for thermal diffusion (L^2/α) to the time for convection $(L/\langle \mathrm{v} \rangle)$ and is defined as the thermal Peclet number:

$$Pe_T = \frac{\langle \mathrm{v} \rangle L}{\alpha} \tag{17.2.22}$$

The thermal Peclet number is the product of the Reynolds number times ν/α, which is known as the *Prandtl number*. The Prandtl number represents the ratio of fluid momentum transport to thermal diffusion. Its typical value is 0.6–0.8 for gases,

1–10 for liquids, and much less than 1 for metals. The Prandtl number is analogous to the Schmidt number in mass transfer, but it is much smaller than the Schmidt number, which means that for a given velocity, the thermal Peclet number will be less than the Peclet number for mass transfer. Consequently, the flow rates needed for significant convective heat transfer are higher than those for convective mass transport. The term $\dot{Q}_p L^2/(\Delta T k)$ represents the dimensionless heat production term. When this term is small, heat production does not affect the temperature distribution and can be neglected.

Applying dimensional analysis to Equation (17.2.18a) results in an expression for the dimensionless heat-transfer coefficient, or *Nusselt number* (Nu):

$$\mathrm{Nu} = \frac{h_f L}{k} = -\frac{1}{S}\int_S \left(\frac{\partial T^*}{\partial y^*}\right)\Bigg|_{y^*=0} dS. \tag{17.2.23}$$

The Nusselt number is analogous to the Sherwood number. In general, Nu = f (Re, Pr). As is discussed in more detail in Section 17.4, correlations developed for convective mass transfer generally apply to heat transfer, and vice versa.

17.3 Steady and Unsteady Heat Conduction

The conservation relations for energy and mass and the corresponding constitutive relations, Fourier's law and Fick's law, are of similar form. Both temperature and mass are scalars. As a result, many solutions developed for mass transport can be used in the solution of analogous heat-transfer problems. The next few problems explore this similarity, and the results are applied to bioheat transfer. The solution to many more heat-transfer problems with relevance to biomedical engineering can be found in Carslaw and Jaeger [17].

17.3.1 Insulation and Heat Conduction Through Layers of Different Thermal Conductivity

Different tissues have different values of thermal conductivity (see Table 17.2). Skin, blood, and fat have values similar to, but less than, water. The lower value for the thermal conductivity of fat is consistent with its role as an insulator. However, the difference is small, and the major effect on insulation is through the thickness of the fat layer. External layers of hair and fur provide additional insulation.

For steady-state one-dimensional conduction through one or more layers with no heat production, Equation (17.2.13) reduces to

$$\frac{d^2 T_i}{dx^2} = 0, \tag{17.3.1}$$

where the subscript i refers to the layer. As an example, consider a two-layer system of thicknesses L_1 and L_2 and conductivities k_1 and k_2 as shown in Figure 17.2. At $x = 0$, $T_1 = T_0$, and at $x = L_1 + L_2$, $T = T_L$. At $x = L_1$, $T_1 = T_2$

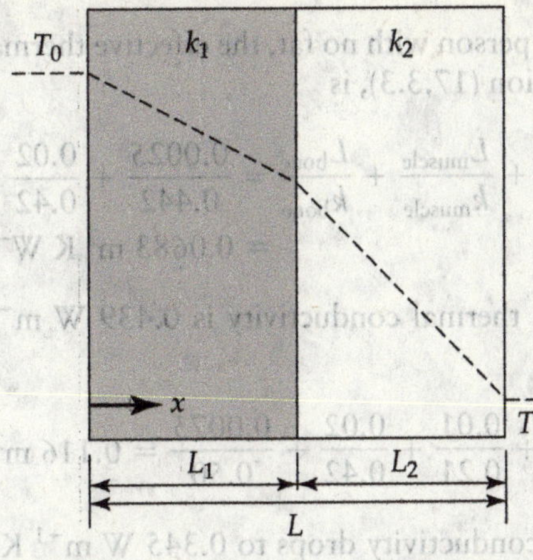

FIGURE 17.2 Heat conduction through a two-layer laminate.

and $-k_1(dT_1/dx)|_{x=L_1} = -k_2(dT_2/dx)|_{x=L_1}$. An important result is to determine the net flux across the layers and to determine an effective thermal conductivity. This situation is very similar to that discussed in Example 6.5 for diffusion through a laminate. The major difference is that temperatures in each layer must be the same at interfaces, whereas solute concentrations may differ. Consequently, we can use the results from Example 6.5 if we set $\Phi_1 = \Phi_2 = 1$ and equate the concentration with the temperature and the energy flux with the molar flux. Thus,

$$q_{1x} = q_{2x} = \frac{k_{\text{eff}}}{L}(T_0 - T_L), \tag{17.3.2}$$

and the effective thermal conductivity k_{eff} is

$$\frac{L}{k_{\text{eff}}} = \frac{L_1}{k_1} + \frac{L_2}{k_2}. \tag{17.3.3}$$

This approach can be extended to multiple layers, using an analogy to Equation (6.7.17).

Example 17.3 Using the following data for skin, fat, bone, and muscle layers, determine the effect of a layer of fat on the heat flux from the body. The thermal conductivity of muscle and bone are 0.42 W m^{-1} K^{-1} and 0.50 W m^{-1} K^{-1}, respectively, and values of the thermal conductivity for skin and fat are given in the following table:

Tissue	Thickness (cm)
Skin	0.25
Fat	1.0
Muscle	2.0
Bone	0.75

Solution For a lean person with no fat, the effective thermal conductivity, from the extension of Equation (17.3.3), is

$$\frac{L}{k_{\text{eff}}} = \frac{L_{\text{skin}}}{k_{\text{skin}}} + \frac{L_{\text{muscle}}}{k_{\text{muscle}}} + \frac{L_{\text{bone}}}{k_{\text{bone}}} = \frac{0.0025}{0.442} + \frac{0.02}{0.42} + \frac{0.0075}{0.50} = 0.0683 \text{ m}^2 \text{ K W}^{-1}. \qquad (17.3.4)$$

The resulting effective thermal conductivity is 0.439 W m^{-1} K^{-1}. With a layer of fat,

$$\frac{L}{k_{\text{eff}}} = \frac{0.0025}{0.442} + \frac{0.01}{0.21} + \frac{0.02}{0.42} + \frac{0.0075}{0.50} = 0.116 \text{ m}^2 \text{ K W}^{-1}. \qquad (17.3.5)$$

The effective thermal conductivity drops to 0.345 W m^{-1} K^{-1}, a 21% decrease. Assuming the same temperature drop, the flux is proportional to k_{eff}/L and is about 41% lower with the layer of fat.

Fur and hair can significantly reduce heat transfer from the body and provide insulation. The conductivity of hair is much less than that of tissue and is slightly greater than the value for air. The conductivity of hair (and fur) layer represents an average of the values of pure hair and air. Consider a layer of head hair that is 0.40 cm thick and has a thermal conductivity of 0.05 W m^{-1} K^{-1}. Adding this layer to the layers of bone, muscle, fat, and skin increases L/k_{eff} to 0.196 m^2 K W^{-1}, which reduces the flux by 35%.

To estimate the effective conductivity of hair layer, we use results for a composite of cylindrical fibers aligned parallel to the z-axis in a gas. The thermal conductivities of fiber and gas are k_0 and k_1, respectively. The effective thermal conductivities in the three directions are [5]

$$\frac{k_{\text{eff},zz}}{k_0} = 1 + \left(\frac{k_1 - k_0}{k_0}\right)\phi \qquad (17.3.6)$$

$$\frac{k_{\text{eff},xx}}{k_0} = \frac{k_{\text{eff},yy}}{k_0} = 1 + 2\phi \Big/ \left[\left(\frac{k_1 + k_0}{k_1 - k_0}\right) - \phi + \left(\frac{k_1 - k_0}{k_1 + k_0}\right) \times (0.30584\phi^4 + 0.013363\phi^8 + \cdots)\right], \qquad (17.3.7)$$

where ϕ is the volume fraction of fibers.

This analysis highlights the effect of fat and hair on the thermal conductivity, but heat transfer through tissues involves a combination of conduction through tissue and convection in blood. Thus, although the effect of fat on the conductive heat flux is modest, the lower capillary density in fat relative to muscle leads to a more significant reduction in overall heat transfer. Convective heat transfer by blood is discussed in Section 17.7.

The surface temperature is usually not equal to the air temperature due to convection in the surrounding air. A heat-transfer coefficient and correlation are used to relate the surface and air temperatures according to Equation (17.2.18b). Assuming that the surface temperature is constant, the local heat-transfer coefficient can be replaced by the average value, and the boundary condition at $x = L\,(= L_1 + L_2 + \cdots + L_n)$ becomes

$$q_x = -k\left(\frac{\partial T}{\partial x}\right)_{x=L} = h_m(T(x = L) - T_{air}). \tag{17.3.8}$$

For this case, Equation (17.3.3) becomes

$$\frac{L}{k_{\text{eff}}} = \frac{1}{h_m} + \sum_{i=1}^{n} \frac{L_i}{k_i}. \tag{17.3.9}$$

A lower-bound estimate for the value of h_m is 10 W (m^2 K)$^{-1}$. For the five-layer model of bone, muscle, fat, skin, and hair, the maximum effect of the thermal boundary layer is to raise L/k_{eff} to 0.296 m K W^{-1}, a 51% increase above the value calculated by assuming that the skin temperature equals the air temperature. This deviation drops as the wind speed increases.

17.3.2 Steady-State Conduction and Metabolic Energy Production

To maintain the body temperature at a constant value, heat production must balance heat removal. In the following problem, we consider the balance between these two effects to maintain a near-constant core body temperature.

Example 17.4 Consider a three-layer model (Figure 17.3). One layer is the body core with thickness $L_1 = 0.22$ m and thermal conductivity $k_1 = 0.42$ W (m K)$^{-1}$. $\dot{Q}_p = \dot{Q}_m$, the rate of metabolic heat production, that equals 880 W m^{-3}. The body core is surrounded by a fat and skin layer of thickness $L_2 = 0.08$ m, with an effective conductivity $k_2 = 0.235$ W (m K)$^{-1}$. Heat is removed from the body by convection in the air, which is characterized by the heat-transfer coefficient h_m. Determine the mean steady-state temperature in the core region and the maximum temperature drop in the core.

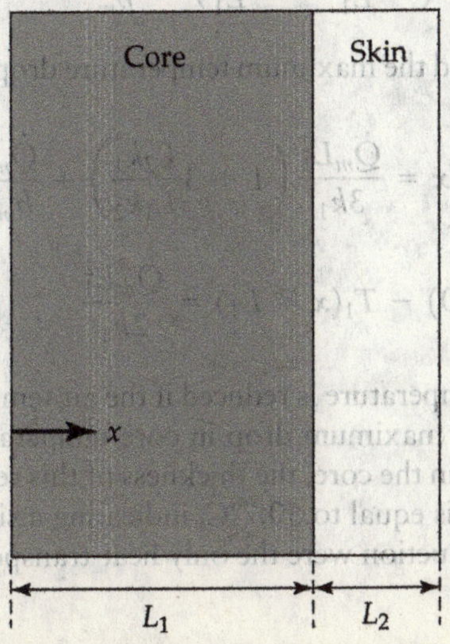

FIGURE 17.3 Conduction through a three-layer model consisting of body core, skin, and air.

Solution For steady-state, one-dimensional heat conduction with no work or heat production, Equation (17.2.14a) becomes, for region 1,

$$k_1 \frac{d^2T_1}{dx^2} = -\dot{Q}_m. \tag{17.3.10a}$$

For region 2, there is no heat production, so

$$\frac{d^2T_2}{dx^2} = 0. \tag{17.3.10b}$$

The boundary conditions are

$$x = 0, \quad \frac{dT_1}{dx} = 0 \tag{17.3.11a}$$

$$x = L_1, \quad k_1 \frac{dT_1}{dx} = k_2 \frac{dT_2}{dx} \tag{17.3.11b}$$

$$T_1 = T_2 \tag{17.3.11c}$$

$$x = L_1 + L_2 = L, \qquad -k_2 \frac{dT_2}{dx} = h_m(T_2(L) - T_{air}). \tag{17.3.11d}$$

The general solutions to Equations (17.3.10a) and (17.3.10b) are

$$T_1 = -\frac{\dot{Q}_m}{2k_1} x^2 + C_1 x + C_2 \tag{17.3.12a}$$

$$T_2 = C_3 x + C_4 \tag{17.3.12b}$$

After applying the boundary conditions, the resulting specific solutions are

$$T_1 = \frac{\dot{Q}_m L_1^2}{2k_1}\left(1 + 2\frac{L_2 k_1}{L_1 k_2} - \frac{x^2}{L_1^2}\right) + \frac{\dot{Q}_m}{h_m} L_1 + T_{air} \tag{17.3.13a}$$

$$T_2 = \frac{\dot{Q}_m L_1^2}{k_2}\left(\frac{L_1 + L_2}{L_1} - \frac{x}{L_1}\right) + \frac{\dot{Q}_m}{h_m} L_1 + T_{air}. \tag{17.3.13b}$$

The average temperature and the maximum temperature drop in the core region are

$$\langle T_1 \rangle = \frac{1}{L_1}\int_0^{L_1} T dx = \frac{\dot{Q}_m L_1^2}{3k_1}\left(1 + 3\frac{L_2 k_1}{L_1 k_2}\right) + \frac{\dot{Q}_m}{h_m} L_1 + T_{air} \tag{17.3.14}$$

$$\Delta T_1 = T_1(x = 0) - T_1(x = L_1) = \frac{\dot{Q}_m L_1^2}{2k_1}. \tag{17.3.15}$$

Note that the mean core temperature is reduced if the air temperature is lowered or h_m increases. However, the maximum drop in core temperature depends only on the rate of heat production in the core, the thickness of this region, and the thermal conductivity of the core. It is equal to 50.7°C, indicating a significant temperature drop across the core if conduction were the only heat-transport mechanism within the body. This is consistent with many experimental measurements. The temperature drop across the skin is

$$\Delta T_2 = T_2(x = L_1) - T_2(x = L_1 + L_2) = \frac{\dot{Q}_m L_1 L_2}{k_2}. \quad (17.3.16)$$

The temperature drop across the skin is 65.9°C. Thus, conduction is an inefficient method of heat removal from the body. As a result, the body temperatures would reach unacceptable levels. Fevers above 40°C are considered very serious, and sustained body temperatures above 42°C can cause death. These results clearly indicate that the body needs to have other mechanisms to establish a more uniform temperature at a value that can support life. These will be discussed in the next section.

17.3.3 Unsteady Heat Conduction

Unsteady heat conduction problems involve solution of Equation (17.2.15). The equation is analogous to the three-dimensional form of Fick's second law for unsteady diffusion, Equation (6.7.4b). As a result, we can make use of solutions obtained in Sections 6.8.2–6.8.4. The following problem provides an example of such an application.

Example 17.5 The polymerase chain reaction (PCR) is a technique to make multiple copies of DNA in vitro. A sample of DNA is mixed with nucleotides, fluors, and a DNA polymerase that is stable at high temperatures. The sample temperature is raised rapidly to 80°C–90°C, which is above the temperature needed to cause the DNA chains to separate (often called the "melting" temperature in the field.) The DNA polymerase then creates a copy of each chain. Each cycle lasts from 60 to 90 s. For a 10 μL volume, determine the time to reach steady-state temperature. Assume that the properties are the same as those of water.

Solution Since the liquid volume adopts a spherical shape, this problem involves unsteady diffusion in spherical coordinates. The corresponding diffusion problem is described by Equation (6.8.48), subject to the initial conditions of Equation (6.8.49a) and boundary conditions (6.8.49b) and (6.8.49c). This result can be used by replacing D_{ij} with α. Steady state occurs when $\alpha t/R^2 = 0.5$. The radius is found from the sample volume:

$$R = \left(\frac{3V}{4\pi}\right)^{1/3} = \left(\frac{3(0.01\ \text{cm}^3)}{4\pi}\right)^{1/3} = 0.134\ \text{cm} = 0.00134\ \text{m}$$

The time to reach steady state is

$$t = 0.5R^2/\alpha = 0.5(0.00134\ \text{m})^2/(1.5 \times 10^{-7}\ \text{m}^2\,\text{s}) = 5.95\ \text{s},$$

which is much less than the cycle time, indicating that the temperature is uniform in the DNA solution.

In Section 17.3.2, we noted that heat conduction alone produces large temperature gradients in the body. Further, the time to reach steady-state temperature can be very long. The solution for unsteady heat conduction in a cylinder of radius R is presented in Carslaw and Jaeger [17]. Steady state is reached when $\alpha t/R^2 \approx 1$. Using an effective thermal conductivity of $0.37\ \text{W}\ (\text{m K})^{-1}$, a density of $1100\ \text{kg m}^{-3}$, and a specific heat capacity of $3500\ \text{J}\ (\text{kg K})^{-1}$, $\alpha = k/\rho\hat{C}_p = 9.6\times10^{-8}\ \text{m}^2\ \text{s}^{-1}$. For $R = 0.30$ m, $t = 937{,}500$ s $= 206.4$ h. Such a large time constant makes conduction alone an inefficient process for producing thermal equilibrium within an organism.

17.3.4 Unsteady Heat Conduction with a Phase Change

An important class of unsteady heat-transfer problems relevant to the preservation of cells and tissues involves solidification and melting. Such problems have been treated extensively in the heat-transfer literature [17]. Most situations are complex and require numerical solutions, but an analytical solution can be obtained from melting or freezing in a semi-infinite medium. Such problems are known as Stefan problems, after Josef Stefan, who made numerous contributions to physics and the study of heat conduction. He published a solution for ice formation in 1891, although Franz Neumann actually published the first solutions to this problem in the 1860s [17].

Phase changes such as melting and boiling occur at a constant temperature for a pure material. For mixtures of several materials, the phase change may occur at several discrete temperatures or over a range of temperatures. The problem we consider here is for a pure material. (The application of this analysis to problems related to low-temperature preservation of cells and tissues is discussed in Section 17.8.) In some cases, the liquid state can exist below the freezing point, a situation known as supercooling. However, this state is unstable, and freezing can occur rapidly once ice crystals form locally.

A general feature of phase changes is that energy is used first to induce the change of phase and then to alter the temperature. In the problem we consider here, freezing and melting occur solely at the interface between the solid and liquid phases. This defines a front that moves in the direction of freezing or thawing, which is dictated by the temperatures of each phase.

Consider a solid layer initially occupying the region for $x \le 0$ and a liquid region occupying the region for $x > 0$. Since the temperature of the liquid is at or below the freezing temperature, freezing begins at $x = 0$ and proceeds in the positive x-direction (Figure 17.4). Freezing occurs as long as the solid is maintained at or below the freezing point. Heat transfer in each phase is governed by the equation for unsteady conduction. Thus, for each phase,

$$\frac{\partial T_i}{\partial t} = \alpha_i \frac{\partial^2 T_i}{\partial x^2}, \tag{17.3.17}$$

where $i = S$ for the solid phase and L for the liquid phase. The boundary and initial conditions are as follows:

$$x \to \infty \text{ and } \quad t \ge 0, \quad T_L = T_1 \tag{17.3.18a}$$

$$x = 0 \text{ and } \quad t \ge 0, \quad T_S = T_0. \tag{17.3.18b}$$

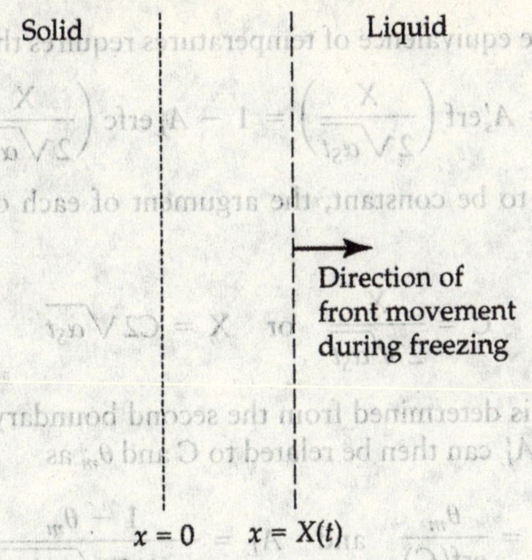

FIGURE 17.4 Schematic of advancing front during freezing. If thawing occurs, the front moves in the opposite direction.

The front is located at $x = X(t)$. The location is not known a priori but can be deduced along with its change in position with time. Two conditions must be satisfied at the front location. First, the liquid and solid temperatures must be equal to the melting temperature:

$$x = X, \qquad T_L = T_S = T_m. \tag{17.3.19a}$$

Second, the net flux of heat into the solid and liquid phases equals the heat generated or absorbed in the phase change. This is written as

$$-q_{Sx} + q_{Lx} = k_S \frac{\partial T_S}{\partial x} - k_L \frac{\partial T_L}{\partial x} = \Delta \hat{H} \rho \frac{dX}{dt}, \tag{17.3.19b}$$

where $\Delta \hat{H}$ is the latent heat per unit mass and ρ is the density of a solid if solidification occurs or ρ is the density of a liquid if melting occurs. Each of the solid and liquid phases is semi-infinite. Although the boundary conditions at $x = 0$ are different from those for diffusion in semi-infinite media, we can use the same approach that was used in Section 6.8.1. Thus, we can define a similarity variable η_i for each phase and a dimensionless temperature θ as

$$\eta_i = \frac{x}{\sqrt{4\alpha_i t}} \quad \text{and} \quad \theta_i = \frac{T_i - T_0}{T_1 - T_0}. \tag{17.3.20a,b}$$

As a result, Equations (17.3.15), (17.3.18a), and (17.3.18b) can be transformed into the following relations:

$$-2\eta_i \frac{d\theta_i}{d\eta_i} = \frac{d^2\theta_i}{d\eta_i^2}. \tag{17.3.21}$$

$$\eta_L \to \infty, \quad \theta_L = 1; \qquad \eta_S = 0, \quad \theta_S = 0 \tag{17.3.22a,b}$$

The general solution to equation (17.3.21) is

$$\theta_i = A_i' \mathrm{erf}(\eta_i) + B_i. \tag{17.3.23}$$

After applying the boundary conditions at $x = 0$ and $x \to \infty$,

$$\theta_S = A_s' \mathrm{erf}(\eta_S) \qquad \theta_L = 1 - A_L' \mathrm{erfc}(\eta_L) \tag{17.3.24a,b}$$

At the front $x = X$, the equivalence of temperatures requires that

$$\theta_m = A_s' \text{erf}\left(\frac{X}{2\sqrt{\alpha_S t}}\right) = 1 - A_L' \text{erfc}\left(\frac{X}{2\sqrt{\alpha_L t}}\right). \tag{17.3.25}$$

For the temperatures to be constant, the argument of each of the error functions must be constant. Let

$$C = \frac{X}{2\sqrt{\alpha_S t}} \quad \text{or} \quad X = C2\sqrt{\alpha_S t} \tag{17.3.26a,b}$$

where the constant C is determined from the second boundary condition at $x = X$. The constants A_s' and A_L' can then be related to C and θ_m as

$$A_s' = \frac{\theta_m}{\text{erf}(C)} \quad \text{and} \quad A_L' = \frac{1 - \theta_m}{\text{erfc}(C\sqrt{\alpha_S/\alpha_L})}. \tag{17.3.27a,b}$$

To evaluate the second boundary condition at $x = X$, the fluxes are rewritten in terms of the dimensionless variables:

$$q_{ix}(x = X) = -k_i \frac{\partial T_i}{\partial x} = -k_i(T_1 - T_0) \frac{d\theta_i}{d\eta_i} \frac{\partial \eta_i}{\partial x}\bigg|_{x=X}$$

$$= -\frac{k_i(T_1 - T_0)}{\sqrt{4\alpha_i t}} \frac{d\theta_i}{d\eta_i}\bigg|_{\eta_i = x}. \tag{17.3.28}$$

The derivatives for the solid and liquid phases at the front location are

$$\frac{d\theta_S}{d\eta_S}\bigg|_{x=X} = \frac{2\theta_m}{\sqrt{\pi}} \frac{\exp(-\eta_S^2)}{\text{erf}(C)}\bigg|_{x=X} = \frac{2\theta_m}{\sqrt{\pi}} \frac{\exp\left(-\frac{X^2}{4\alpha_S t}\right)}{\text{erf}(C)} = \frac{2\theta_m}{\sqrt{\pi}} \frac{\exp(-C^2)}{\text{erf}(C)} \tag{17.3.29a}$$

$$\frac{d\theta_L}{d\eta_L}\bigg|_{x=X} = \frac{2(1-\theta_m)}{\sqrt{\pi}} \frac{\exp\left(-\frac{X^2}{4\alpha_L t}\right)}{\text{erfc}\left(C\sqrt{\alpha_s/\alpha_L}\right)} = \frac{2(1-\theta_m)}{\sqrt{\pi}} \frac{\exp\left(-C^2 \frac{\alpha_s}{\alpha_L}\right)}{\text{erfc}\left(C\sqrt{\alpha_s/\alpha_L}\right)} \tag{17.3.29b}$$

After some algebraic manipulations, with the boundary condition at $x = X$, Equation (17.3.29b) is written as

$$\frac{k_S(T_m - T_0)\exp(-C^2)}{\text{erf}(C)} - \frac{k_L(T_1 - T_m)\sqrt{\alpha_S/\alpha_L}\exp(-C^2\alpha_S/\alpha_L)}{\text{erfc}(C\sqrt{\alpha_S/\alpha_L})}$$

$$= C\alpha_S\sqrt{\pi}\Delta\hat{H}\rho. \tag{17.3.30}$$

For ice formation, T_1 is equal to or less than T_m. When T_1 equals T_m, $\theta_m = 1$. Equation (17.3.30) reduces to

$$C = \frac{k_S(T_m - T_0)\exp(-C^2)}{\alpha_S\sqrt{\pi}\Delta\hat{H}\rho\,\text{erf}(C)} \tag{17.3.31}$$

and

$$A'_s = \frac{1}{\text{erf}(C)} \quad \text{and} \quad A'_L = 0. \tag{17.3.32a,b}$$

The liquid water temperature stays at the freezing point until ice forms; then the temperature drops. Values of C are listed in Table 17.3 for various temperature differences, $T_m - T_0$. If the ice is at the freezing point, $T_0 = T_m$, there are no changes in ice and liquid water temperatures.

Since the latent heat for ice formation is so large, C in Equation (17.3.31) can be approximated using [17]

$$C = \left[\frac{k_S(T_m - T_0)}{2\alpha_s\rho_s\Delta\hat{H}}\right]^{1/2} = \left[\frac{\hat{C}_{p,S}(T_m - T_0)}{2\Delta\hat{H}}\right]^{1/2}. \tag{17.3.33}$$

The term in parentheses represents the ratio of heat transfer into the solid phase to the latent heat. Generally, this ratio should be less than 1. For water, the ratio is quite small, making the approximation in Equation (17.3.33) reasonable. For $T_m - T_0 < 10$ K, the error is 1% or less.

Dimensionless temperatures in ice for $T_1 = T_m$ are presented in Figure 17.5 for temperature differences of 1, 5, and 10 K. The dimensionless temperature increases almost linearly with respect to η_S. The location of the front represents the distance at which the dimensionless temperature is 1.

TABLE 17.3

Values of the Root C for Different Values of $T_m - T_0$

$T_m - T_0$ (K)	C
0.1	0.0185
1.0	0.0584
2.0	0.0826
5.0	0.1301
10.0	0.1830

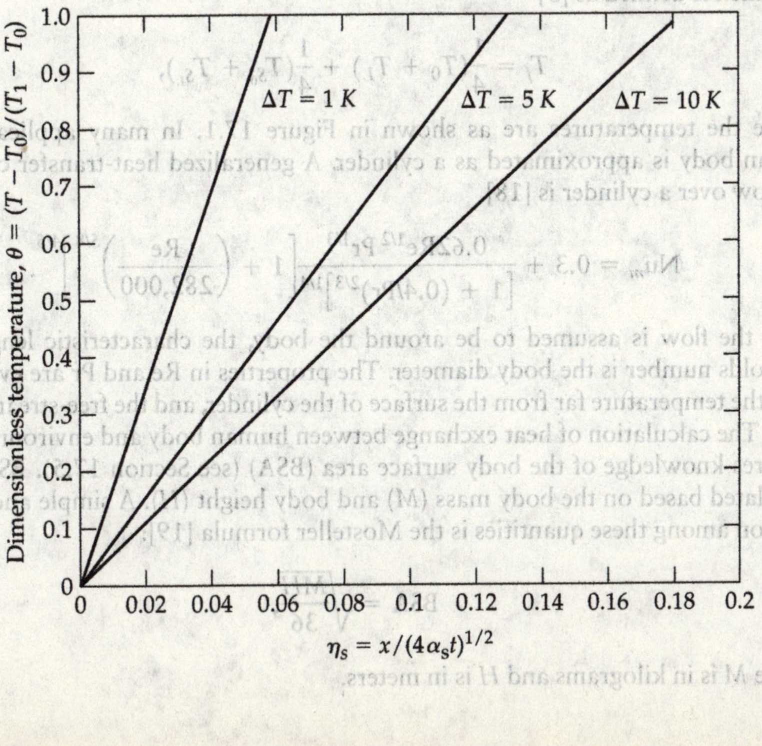

FIGURE 17.5 Dimensionless temperature profiles in ice for different values of $T_m - T_0$.

17.4 Convective Heat Transfer

As was the case with mass transfer, an imposed flow field can significantly increase the rate of heat transfer. Such a process is known as *forced convection.* The dimensionless heat-transfer coefficient, or Nusselt number (Nu), is a function of the Reynolds number and the Prandtl number. Since the fluid properties are dependent upon temperature, and the viscosity is rather sensitive to temperature, the Nusselt number may depend upon the variation in the viscosity due to the imposed temperature gradient. In addition, temperature differences can alter the density-induced fluid flow, a process known as *free convection.*

17.4.1 Correlations for Forced Convection

Due to the analogy between mass transfer and heat transfer, many of the mass transfer correlations reported in Chapter 7 (Table 7.5) were developed initially for heat transfer and either were found to apply to mass transfer or can be modified slightly to apply to heat-transfer problems. Thus, for laminar flow in a cylindrical tube of diameter D, in which the walls are heated at a constant flux, the dimensionless Nusselt number is [5]

$$\mathrm{Nu}_{\ln} = \frac{h_{\ln} D}{k} = 1.86\left(\mathrm{RePr}\frac{D}{L}\right)^{1/3}\left(\frac{\mu_b}{\mu_S}\right)^{0.14}, \tag{17.4.1}$$

where μ_b is the viscosity taken at the arithmetic mean of the bulk fluid temperature (T_b) and μ_S represents the viscosity at the arithmetic mean of the wall temperature (T_S) over the length L along the tube. The properties (viscosity, density, thermal diffusivity) in the Reynolds and Prandtl numbers are evaluated at the film temperature T_f, which is defined as [5]

$$T_f = \frac{1}{4}(T_0 + T_L) + \frac{1}{4}(T_{S_0} + T_{S_L}), \tag{17.4.2}$$

where the temperatures are as shown in Figure 17.1. In many applications, the human body is approximated as a cylinder. A generalized heat-transfer correlation for flow over a cylinder is [18]

$$\mathrm{Nu}_m = 0.3 + \frac{0.62\mathrm{Re}^{1/2}\,\mathrm{Pr}^{1/3}}{\left[1 + (0.4/\mathrm{Pr})^{2/3}\right]^{1/4}}\left[1 + \left(\frac{\mathrm{Re}}{282{,}000}\right)^{5/8}\right]^{4/5}. \tag{17.4.3}$$

Since the flow is assumed to be around the body, the characteristic length in the Reynolds number is the body diameter. The properties in Re and Pr are evaluated at T_∞, the temperature far from the surface of the cylinder, and the free-stream velocity.

The calculation of heat exchange between human body and environment often requires knowledge of the body surface area (BSA) (see Section 17.5). BSA is often calculated based on the body mass (M) and body height (H). A simple and accurate relation among these quantities is the Mosteller formula [19]:

$$\mathrm{BSA} = \sqrt{\frac{MH}{36}}, \tag{17.4.4}$$

where M is in kilograms and H is in meters.

In some applications, infants and small animals are treated as spheres for the purposes of calculating heat-transfer coefficients. The correlation for the Nusselt number around a sphere is [5]

$$\mathrm{Nu}_m = 2 + (0.4\mathrm{Re}^{1/2} + 0.06\mathrm{Re}^{2/3})\mathrm{Pr}^{0.4}\left(\frac{\mu_\infty}{\mu_S}\right)^{1/4}, \quad (17.4.5)$$

where the subscript ∞ refers to the fluid stream far from the object and the subscript S refers to the cylinder surface. The properties in Re and Pr are evaluated at ∞.

For the correlations for flow around the cylinder and the sphere, there are limiting values of the Nusselt number when the Reynolds number approaches zero (see Equations (17.4.3) and (17.4.5)). These values are obtained from the solution for steady-state conduction from the object when it is held at a constant surface temperature [5].

17.4.2 Natural Convection

While the flow of fluids over an object at a different temperature can enhance energy transfer by convection, the local heating or cooling of a fluid can generate a flow of fluid due to a density difference. Such a process is known as *natural* or *free convection*. This is an important mechanism by which air currents are generated near the surface, particularly along coastlines. The shimmering appearance of air near a hot surface, such as a sidewalk or roadway on a hot day or a radiator, is an example of natural convection. Likewise, after a period of strenuous exertion, heat transfer from a warm organism can produce natural convection, thereby facilitating cooling.

When natural convection arises, the local velocity field depends upon the temperature field. The temperature variation causes a variation in density, which induces a fluid flow. In general, the fluid density and viscosity are functions of temperature. We will introduce some simplifications to highlight the major features of free convection. Such simplifications are limited to small temperature differences and will be appropriate for natural convection around organisms. First, we neglect the temperature dependence of the viscosity. Second, the functional dependence of density upon temperature is approximated with a Taylor series expansion around the reference temperature T_0 [5]:

$$\rho = \rho_0 + \left(\frac{\partial \rho}{\partial T}\right)\bigg|_{T_0}(T - T_0) + \cdots = \rho_0 - \rho_0\beta(T_0)(T - T_0) + \cdots, \quad (17.4.6)$$

where ρ_0 is the density at the reference temperature and β is the coefficient of thermal expansion:

$$\beta(T) = \frac{1}{V}\left(\frac{\partial V}{\partial T}\right)_p = \frac{1}{\hat{V}}\left(\frac{\partial \hat{V}}{\partial T}\right)_p = \frac{1}{(1/\rho)}\left(\frac{\partial(1/\rho)}{\partial T}\right)_p = -\frac{1}{\rho}\left(\frac{\partial \rho}{\partial T}\right)_p. \quad (17.4.7)$$

In general, the coefficient of thermal expansion depends upon temperature. For many fluids, the volume increases and the density decreases with temperature so that $\beta > 0$ [14]. For an ideal gas, $\beta = 1/T$. For water, β is equal to $3.6 \times 10^{-4}\ \mathrm{K}^{-1}$ at

37°C, which is about 10 times smaller than the value for air at this temperature. Thus, natural convection is more important in gases than liquids.

Mathematically, the momentum and energy equations are coupled and should be solved together. To describe fluid motion, natural convection requires modification of the conservation of linear momentum (Equation (3.3.16)) or the Navier–Stokes equation (Equation (3.3.25)), which were derived based on the assumption that the temperature was the same everywhere. A commonly used approach is to examine only small temperature differences [5, 14], and the fluid is treated as incompressible. Thus, the Navier–Stokes equation (Equation (3.3.25)) is used in conjunction with Equation (17.4.6) and the energy transfer equation (Equation (17.2.13)). In many cases, a numerical solution is required. The following example shows the magnitude of free convection for a very simplified physical situation of biological relevance.

Example 17.6 A fluid is contained in a gap of width $2H$ between two long and wide, vertical parallel plates ($L \gg 2H$; $W \gg 2H$) maintained at temperatures T_0 and T_1, respectively (Figure 17.6). The temperature difference $T_0 - T_1$ is small, and $T_0 > T_1$. The heating induces a local change in density, which induces a velocity field in the y-direction. The velocity is assumed to be sufficiently small such that inertial effects are negligible. Both the velocity and temperature fields are at steady state. The velocity field is fully developed. The upper and lower ends are closed so that there is no net flow to or from the channel. The analysis is focused on the region away from the ends so that edge effects can be neglected. Further, we assume that the viscosity is constant. Based on these assumptions, determine the velocity profile $v_y(x)$.

Solution The fluid is warmer near $x = -H$ than at $x = H$. As a result, the fluid density is lower near the warmer surface and a flow is induced in the positive y-direction. Since there is no net flow, fluid will move in the negative y-direction near $x = H$.

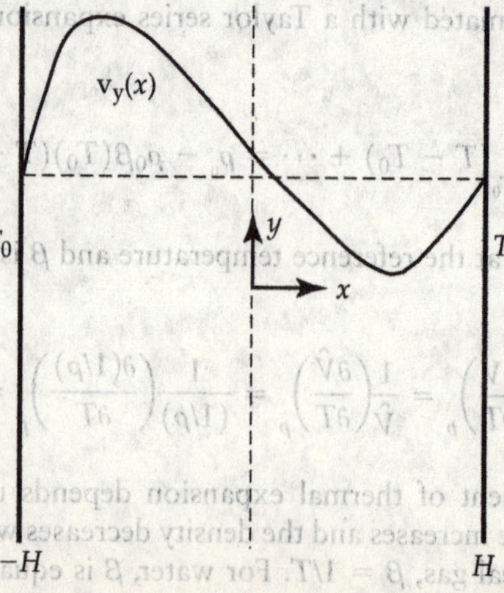

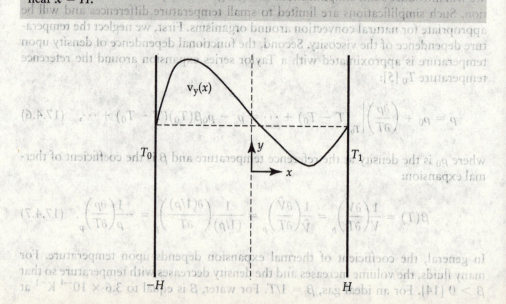

FIGURE 17.6 Sketch of the velocity field for free convection between two plates when $T_0 > T_1$.

For a steady, fully developed flow far from the ends of the channel, only $v_y(x)$ exists and all other velocity components are zero. Since the flow is induced by a density variation, gravity is important. As a result, the y-component of the Navier–Stokes equation is

$$\mu\frac{d^2 v_y}{dx^2} = \rho(T)g. \tag{17.4.8}$$

Next, differentiate Equation (17.4.8) with respect to x:

$$\mu\frac{d^3 v_y}{dx^3} = \frac{d\rho(T)}{dx}g. \tag{17.4.9}$$

Applying the chain rule to the derivative of density and using the definition of the coefficient of thermal expansion:

$$\frac{d\rho(T)}{dx} = \left(\frac{\partial\rho}{\partial T}\right)_p\frac{dT}{dx} = -\rho_0\beta\frac{dT}{dx}, \tag{17.4.10}$$

where T_0 is taken as the reference temperature for density (i.e., $\rho_0 = \rho(T_0)$). Replacing the derivative with respect to density in Equation (17.4.9) yields

$$\mu\frac{d^3 v_y}{dx^3} = -\rho_0 g\beta\frac{dT}{dx}. \tag{17.4.11}$$

The temperature field can be obtained by noting that since the plates are vertical and are each at a constant temperature, energy transfer is in the x-direction only, except near the ends. Thus, at steady state, with no heat production or work done, Equation (17.2.14a) reduces to

$$\frac{d^2 T}{dx^2} = 0, \tag{17.4.12}$$

subject to the boundary conditions that $T = T_0$ at $x = -H$ and $T = T_1$ at $x = H$. Integrating Equation (17.4.12) twice and applying the boundary conditions yields the following expression for $T(x)$:

$$T = T_0 - \frac{(x + H)}{2H}(T_0 - T_1). \tag{17.4.13}$$

Using Equation (17.4.13) to determine the derivative of T with respect to x, Equation (17.4.11) becomes

$$\mu\frac{d^3 v_y}{dx^3} = \rho_0 g\beta\frac{T_0 - T_1}{2H} = \rho_0 g\beta\frac{\Delta T}{2H}. \tag{17.4.14}$$

The right-hand side of Equation (17.4.14) is a constant. Integrating Equation (17.4.14) three times yields:

$$v_y = \frac{\rho_0 g\beta\Delta T}{12\mu H}x^3 + \frac{A}{2}x^2 + Bx + C. \qquad (17.4.15)$$

Three conditions are needed to determine the constants A, B, and C. Two conditions are the no-slip boundary conditions ($v_y = 0$ at $x = \pm H$). The third condition is that there is no net flow in the y-direction. Using the definition of the volumetric flow rate (Equation (2.2.6)), this condition can be expressed mathematically as

$$Q = W\int_{-H}^{H} v_y(x)dx = 0. \qquad (17.4.16a)$$

Applying this condition to Equation (17.4.15) yields

$$\int_{-H}^{H} v_y(x)dx = \frac{\rho_0 g\beta\Delta T}{48\mu H}x^4\Big|_{-H}^{H} + \frac{A}{6}x^3\Big|_{-H}^{H} + \frac{B}{2}x^2\Big|_{-H}^{H} + Cx\Big|_{-H}^{H} = 0 \qquad (17.4.16b)$$

$$\frac{A}{3}H^3 + C2H = 0 \quad \text{or} \quad C = -\frac{A}{6}H^2 \qquad (17.4.17a)$$

$$v_y = \frac{\rho_0 g\beta\Delta T}{12\mu H}x^3 + A\left(\frac{x^2}{2} - \frac{H^2}{6}\right) + Bx. \qquad (17.4.17b)$$

Applying the boundary conditions yields the following two relations:

$$0 = -\frac{\rho_0 g\beta\Delta T}{12\mu}H^2 + A\frac{H^2}{3} - BH \qquad (17.4.18a)$$

$$0 = \frac{\rho_0 g\beta\Delta T}{12\mu}H^2 + A\frac{H^2}{3} + BH. \qquad (17.4.18b)$$

Solving these two equations yields

$$A = 0 \qquad B = -\frac{\rho_0 g\beta\Delta TH}{12\mu} \qquad (17.4.19a,b)$$

$$v_y = \frac{\rho_0 g\beta\Delta T}{12\mu H}(x^3 - xH^2) = \frac{\rho_0 g\beta\Delta TH^2}{12\mu}\left(\frac{x^3}{H^3} - \frac{x}{H}\right). \qquad (17.4.20)$$

The velocity is a maximum at $x = -H/\sqrt{3}$ and a minimum at $x = H/\sqrt{3}$. The maximum velocity is

$$v_{max} = \frac{\sqrt{3}\rho_0 g\beta\Delta TH^2}{54\mu}. \qquad (17.4.21)$$

The previous analysis shows that the characteristic velocity in free convection is proportional to $(\rho g\beta\Delta TL^2)/\mu$, where L is the characteristic length, such as H in this example. Using this velocity, a new dimensionless group can be defined as the ratio of inertial forces to viscous forces. This group, known as the Grashof number, (Gr), is analogous to the Reynolds number, except that the characteristic velocity is the velocity induced by free convection. Thus,

$$\mathrm{Gr} = \frac{\rho v^2 L^3}{(\mu v/L)L^2} = \frac{\rho vL}{\mu} = \frac{\rho^2 g\beta\Delta TL^3}{\mu^2} = \frac{g\beta\Delta TL^3}{\nu^2}. \tag{17.4.22}$$

The maximum velocity can then be rewritten as

$$v_{max} = \frac{\sqrt{3}\nu}{54H}\mathrm{Gr}. \tag{17.4.23}$$

For a 15°C temperature difference between the body and air, and using the body temperature of 39°C as the reference temperature, the body height (2.0 m) as the characteristic length, and kinematic viscosity of air equal to 1.60×10^{-5} m^2 s^{-1} at 300 K [5], Gr $= 1.44 \times 10^7$. A correlation for the dimensionless heat-transfer coefficient near a vertical, heated plate is [5]

$$\mathrm{Nu}_m = C\mathrm{Pr}^{0.25}\mathrm{Gr}^{0.25}, \tag{17.4.24}$$

which can be used to study the free convection near the surface of the body if the body diameter is much larger than the thickness of the flow boundary layer in the air. The constant C equals 0.518 for a value of Pr $= 0.73$ corresponding to air [5]. (For water, $C \approx 0.66$.) The resulting value of the Nusselt number is 166. Based on Equation (17.4.3), the corresponding value of the Nusselt number for a light breeze of 5.0 miles h^{-1} is 130. When free and forced convection occur together, the net value of the Nusselt number is given by [5]

$$\mathrm{Nu}_m^{net} = \left[(\mathrm{Nu}_m^{free})^3 + (\mathrm{Nu}_m^{forced})^3\right]^{1/3}. \tag{17.4.25}$$

For the data provided, the net Nusselt number for the combined effect of free and forced convection is 189, 46% greater than the value due to forced convection alone. Additional correlations for free convection can be found in [18]. Free convection can be an important method of cooling with moderate temperature gradients in the absence of significant forced convection.

17.5 Energy Transfer Due to Evaporation

A common method by which mammals cool themselves involves the evaporation of water. This occurs during sweating by humans and panting by a number of animals, such as dogs. During sweating, an aqueous solution is secreted by the skin. Since the formation of water vapor from liquid is endothermic (i.e., energy must be added to the system), heat in the surrounding air is used to evaporate liquid, which reduces the temperature locally. This local absorption of heat to produce water vapor increases the flux of heat from the body.

The addition of heat can be used either to change the temperature of an object or to produce a phase change. At constant pressure without a phase transition, the absorption of heat produces a temperature change, which can be calculated by $Q = C_p(T - T_0)$, where Q is the heat added and T_0 is the initial temperature. For water, if the partial pressure of water in air is less than the vapor pressure, then the absorbed heat produces a phase change. The latent heat or enthalpy of vaporization $\Delta\hat{H}^{vap}$ is defined as the change in specific enthalpy between the gas phase and the liquid phase:

$$\Delta\hat{H}^{vap} = \hat{H}^{v} - \hat{H}^{l}, \tag{17.5.1}$$

where the superscripts v and l refer to the vapor and liquid phases, respectively.[5] The latent heat of vaporization of water at 100°C is 2,258 J g^{-1}. The value of $\Delta\hat{H}^{vap}$ for water varies with temperature and can be accurately estimated from the following correlation [20]:

$$\Delta\hat{H}^{vap}_{T_2} = \Delta\hat{H}^{vap}_{T_1}\left(\frac{1 - T_{R_2}}{1 - T_{R_1}}\right)^{0.38}, \tag{17.5.2a}$$

where T_R is the reduced temperature—that is, the ratio of the temperature to the temperature at the critical point, T_c. For water, $T_c = 647.3$ K [20]. Alternatively, the enthalpy of vaporization at any temperature can be determined by applying Kirchhoff's law for a closed system:

$$\Delta\hat{H}^{vap}_{T_2} = \Delta\hat{H}^{vap}_{T_1} + \int_{T_1}^{T_2} \Delta\hat{C}_p^{vap} dT, \tag{17.5.2b}$$

where $\Delta\hat{C}_p^{vap}$ represents the difference in specific heat capacity between the vapor and liquid phases. The following problem shows the effect of vaporization on the local air temperature.

Example 17.7 A total of 1.0 g of liquid water is vaporized at a temperature of 298.15 K (25°C) and a pressure of 1 atm. If the enthalpy of vaporization is completely converted to heat and transferred to 0.25 m^3 of air at the same pressure, determine the temperature change in the air. The molar heat capacity of air at constant pressure is 29.2 J K^{-1} mol^{-1}, and the mean molecular weight of air is 28.84 g mol^{-1}. The gas law constant is 8.314 J K^{-1} mol^{-1} and 1 atm = 101,325 Pa.

Solution The heat transfer at constant pressure is equal to the change in enthalpy. Thus, the total enthalpy change in water plus air should be equal to zero; that is,

$$m_{air}\Delta\hat{H}_{air} + m_{H_2O}\Delta\hat{H}^{vap}_{H_2O} = 0. \tag{17.5.3}$$

[5]This latent heat can also be expressed on a molar basis, $\Delta\overline{H}^{vap}$, or on a total energy basis, ΔH^{vap}.

The enthalpy change of the air at constant pressure is simply $\Delta\hat{H}_{air} = \hat{C}_{p,\,air}(T_2 - T_1)$. Using this relation and solving for T_2 yields

$$T_2 = T_1 - \frac{\Delta\hat{H}^{vap}_{H_2O} m_{H_2O}}{m_{air}\hat{C}_{p,\,air}}. \tag{17.5.4}$$

The enthalpy of vaporization at 25°C (298 K) can be calculated using Equation (17.5.2a) and its value at 100°C. The result is

$$\Delta\hat{H}^{vap}_{T_2} = \Delta\hat{H}^{vap}_{T_1}\left(\frac{1 - T_{R_2}}{1 - T_{R_1}}\right)^{1.38} = 2{,}257\left(\frac{1 - 298.15/647.3}{1 - 373.15/647.3}\right)^{1.38}$$

$$= 2{,}474.38 \text{ J g}^{-1}, \tag{17.5.5a}$$

where the temperature at the critical point is 647.3 K. If, instead, Equation (17.5.2b) is used in the calculation, then

$$\Delta\hat{H}^{vap}_{T_2} = \Delta\hat{H}^{vap}_{T_1} + \int_{T_1}^{T_2} \Delta\hat{C}^{vap}_p dT = 2{,}257 + 173.78 = 2{,}430.78 \text{ J g}^{-1}, \tag{17.5.5b}$$

where the specific heat capacity of water at constant pressure is assumed to be temperature independent and equal to 4.183 kJ kg^{-1} K^{-1} in the liquid phase and 1.866 kJ kg^{-1} K^{-1} in the vapor phase. In both estimations of the enthalpy of vaporization at 25°C, the errors are within 1.3% of the reported value of 2,444 J g^{-1} [21]. The mass of air can be obtained from the ideal gas law, $pV = nRT$, where n is the number of moles. At 298.15 K, n = 10.22 moles or 294.7 g of air. The specific heat capacity of air is 29.2 J K^{-1} mol^{-1}/28.84 g mol^{-1} = 1.012 J K^{-1} g^{-1}.

$$\Delta T = -\frac{\Delta\hat{H}^{vap}_{H_2O} m_{H_2O}}{m_{air}\hat{C}_{p,air}} = -\frac{(2{,}474 \text{ J g}^{-1})(1 \text{ g})}{(294.7 \text{ g})(1.012 \text{ J g}^{-1}\text{ K}^{-1})} = -8.30 \text{ K} \tag{17.5.6}$$

The previous calculation shows that the evaporation of water can produce a significant amount of cooling. As a result, the evaporation of water is used to cool the air in warm, dry climates—a process known as evaporative cooling. During sweating, the heat is removed from the air and the surface of the skin. Unlike the example provided above, the rates of sweat production and evaporation vary with time.

Evaporative cooling during sweating involves removing heat from a warm organism. As a result, body and air temperatures near the skin surface decline. The different temperature profiles from the surface of a warm organism are shown in Figure 17.7 for cases without and with sweating. Without sweating, the temperature declines monotonically to the ambient temperature. With sweating, there is removal

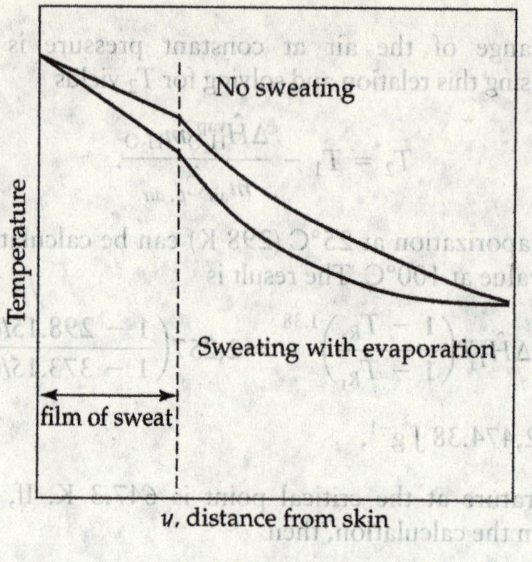

FIGURE 17.7 Schematic of the temperature profiles arising with and without sweating.

of heat from the air, reducing the temperature below the value of the surrounding air. As a result, the heat flux at the skin surface is greater for the case of evaporation, resulting in a faster rate of heat removal.

Sweating involves complex interactions among momentum, mass, and energy transfer. The coupling occurs because the rate of cooling or removal of heat via evaporation is equal to the product of heat removed per mole of sweat and the rate of evaporation. As a simple model for evaporation, consider a thin film of sweat attached to the body surface and in contact with air (Figure 17.8). The mass transfer process will be examined first, followed by an analysis of energy transfer to derive the temperature profile and energy flux.

Evaporation occurs if the local vapor pressure is higher than the partial pressure of water in the air. The gas-phase concentration is often expressed as a mole fraction or partial pressure, and ideal gas behavior is assumed. The mole fraction of

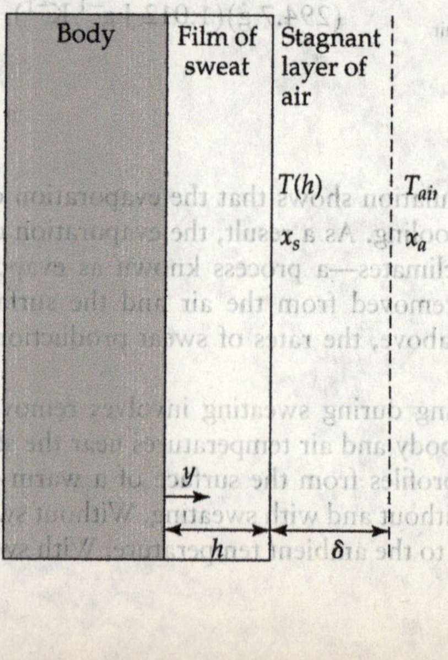

FIGURE 17.8 Schematic diagram of sweat film attached to the body and the surrounding air.

water (x_s) in the air is related to its partial pressure (p^{vap}) by $x_s = p^{vap}/p_{atm}$, where p_{atm} is the local atmospheric pressure. The evaporating liquid draws heat from the air or the body surface and locally reduces the temperature. As an example, consider that a person's body temperature has risen to 39°C and that the air far from the surface contains water at a mole fraction x_H and at a temperature of 25°C. We would like to predict the local temperature distribution and the rate of evaporation.

An important issue in the analysis of evaporation of sweat is the mechanism of mass transport in the gas. Because of the complexity of body's shape and uncertainties about air flow around the body, a common simplification is to assume that there is a stagnant air film of thickness δ. The thickness of the film is determined from convection, as described below.

In Section 6.7, we solved the problem for the concentration profile of an evaporating liquid. In that analysis, the vapor concentration at a certain distance away from the liquid surface is assumed to be zero. For the evaporation of water, the vapor concentration is x_a at $y = h + \delta$ and x_s at $y = h$. Based on these revised boundary conditions and the dimensions of this problem, the concentration profile of vapor in the stagnant layer of air is

$$\frac{1 - x}{1 - x_s} = \left(\frac{1 - x_a}{1 - x_s}\right)^{\frac{y-h}{\delta}}. \tag{17.5.7}$$

The rate of evaporation can be obtained from a quasi–steady-state analysis, since evaporation is slow relative to the time for diffusion. As a result, the concentration and temperature profiles adapt before the film thickness changes appreciably.

An important variable in this analysis is the film thickness. The thickness can be estimated from the mass transfer coefficient:

$$k_w = \frac{D_{w,\mathrm{air}}}{\delta} \tag{17.5.8}$$

In general, the mass and heat-transfer coefficients vary with the actual location around an organism due to local variations in the velocity field. Often, the average transfer coefficient is used, and this is obtained by treating a person as a cylinder, with the same surface area [15]! A generalized heat-transfer correlation for flow over a cylinder is given by Equation 17.4.3 [18]. The correlation can be used for mass transfer by replacing Nu with Sh and Pr with Sc.

We use these results to estimate the mass-transfer coefficient and δ as follows. For a female 5 ft 8 in. (1.7272 m) tall and 125 lb (56.75 kg), the BSA is 1.65 m^2 (see Equation 17.4.4). For an equivalent cylinder, the diameter is 0.304 m. For a light breeze of 0.25 miles h^{-1} (0.11 m s^{-1}) and a kinematic viscosity of air equal to 1.60×10^{-5} m^2 s^{-1} at 300 K, $D_{w,\mathrm{air}} = 2.6 \times 10^{-5}$ m^2 s^{-1}, Re = 2,120, and Sc = 0.62. As a result, Equation (17.4.3) predicts that the value of the Sherwood number is 22.3, which indicates that k_w is 1.91×10^{-3} m s^{-1}. Thus, $\delta = 0.0136$ m.

The rate of evaporation $\dot{R}_{evap}$ is simply the amount of sweat that forms water vapor per unit time. It can be represented as the rate of change in the number of moles of water molecules, which should equal the flux at the film surface, $y = h$ times the surface area of the film. Thus,

$$\text{Rate of evaporation (moles/time)} = \dot{R}_{evap} = N_{y=h}A, \tag{17.5.9}$$

where A is the surface area. The flux at $y = h$ can be determined with the aid of the following relationship for the derivative of $z = a^{u(x)}$:

$$\frac{dz}{dx} = z \ln a \frac{du}{dx}. \tag{17.5.10}$$

The flux at $y = h$ is found from Equation (17.5.7):

$$N_{y=h} = -\frac{cD_{w,\text{air}}}{1 - x}\frac{dx}{dy}\bigg|_{y=h} = \frac{cD_{w,\text{air}}}{\delta}\ln\left(\frac{1 - x_a}{1 - x_s}\right), \tag{17.5.11}$$

where c is the total concentration of gases in the air and is obtained from the ideal gas law ($c = n/V = p_{\text{tot}}/RT$). The rate of evaporation is then

$$\dot{R}_{\text{evap}} = \frac{cD_{w,\text{air}}}{\delta}\ln\left(\frac{1 - x_a}{1 - x_s}\right)A. \tag{17.5.12}$$

At normally encountered air temperatures (e.g., below 40°C), the water vapor partial pressure is much less than the atmospheric pressure. Thus, Equation (17.5.12) becomes

$$\dot{R}_{\text{evap}} = \frac{cD_{w,\text{air}}}{\delta}(x_s - x_a)A. \tag{17.5.13}$$

Therefore, the evaporation rate is proportional to the difference between the vapor pressure and the actual partial pressure of water in air (humidity). While this result does not depend explicitly upon the temperature, the water vapor pressure is a function of temperature [8]. As a result, the dependence is implicit.

In contrast, energy transfer depends explicitly upon mass transfer, since the rate of energy transfer from the sweat to the air equals the molar enthalpy of vaporization $\Delta\overline{H}^{\text{vap}}$ times the flux of water vapor into the air at the sweat/air interface, where $y = h$. This rate of energy change equals the sum of the fluxes in the air and liquid:

$$k_a \frac{dT_a}{dy}\bigg|_{y=h} - k_1 \frac{dT_1}{dy}\bigg|_{y=h} = \Delta\overline{H}^{\text{vap}} N_{y=h}, \tag{17.5.14}$$

where the subscript a refers to air and the subscript 1 refers to the liquid sweat film. Equation (17.5.14) is derived by equating the energy flux from the sweat to the flux of energy transferred to the air plus the energy needed to evaporate the fluid. The energy flux at $y = h$ represents one boundary condition for energy transfer. The other boundary conditions are that $T_l = T_b$ at $y = 0$, the body surface, $T_a = T_l$ at $y = h$, and $T_a = T_{\text{air}}$ at $y = h + \delta$. The steady-state energy flux in the air consists of both convection, due to vapor movement, and conduction. Based on Equation (17.2.11), the energy balance equation in the y-direction at the steady state is

$$\rho\hat{C}_p \mathbf{v} \cdot \nabla T_a = -\nabla \cdot \mathbf{q}_a \tag{17.5.15}$$

or

$$\rho\hat{C}_p v_y \frac{dT_a}{dy} = k_a \frac{d^2 T_a}{dy^2}, \tag{17.5.16}$$

where work and energy production by the air are assumed to be zero. The quantity v_y is the vapor velocity in the y direction and is related to the molar flux of vapor via the vapor concentration in the air by Equation (6.2.3b),

$$N_{y=h} = C_{vap}v_y \qquad (17.5.17)$$

where v_y is the vapor velocity and C_{vap} is the concentration of water vapor in the air. The term $\rho\hat{C}_p$ represents the product of mass density and the specific heat capacity of the humid air. Substituting Equation (17.5.17) into Equation (17.5.16) yields

$$\frac{\rho\hat{C}_p}{C_{vap}}N_{y=h}\frac{dT_a}{dy} = k_a\frac{d^2T_a}{dy^2}. \qquad (17.5.18)$$

To solve Equation (17.5.18), let $w = dT_a/dy$. Then Equation (17.5.18) is rewritten as

$$\frac{dw}{dy} = \frac{\rho\hat{C}_p}{C_{vap}k_a}N_{y=h}w. \qquad (17.5.19)$$

Integrating Equation (17.5.19) yields

$$w = \frac{dT_a}{dy} = a_1\exp\left(\frac{\rho\hat{C}_p}{C_{vap}k_a}N_{y=h}y\right). \qquad (17.5.20)$$

Integrating Equation (17.5.20) to find T_a yields the following expression in terms of two integration constants, a_1 and a_2:

$$T_a = \frac{a_1C_{vap}k_a}{N_{y=h}\rho\hat{C}_p}\exp\left(\frac{\rho\hat{C}_p}{C_{vap}k_a}N_{y=h}y\right) + a_2. \qquad (17.5.21)$$

In the sweat film, heat is transported by conduction and the movement of the sweat due to sweat production and evaporation. Assuming that the rate of sweat production balances the rate of evaporation, the velocity of sweat movement is

$$v_{yw} = N_{y=h}/C_w. \qquad (17.5.22)$$

The concentration of the water is 55.5 M or 5.55×10^4 mole m^{-3}. For air at 25°C and 60% relative humidity (i.e., the partial pressure of water vapor is 60% of the vapor pressure of water at that temperature), the vapor pressure is 23.756 mmHg [8]. The molar flux of water at $y = h$, from Equation (17.5.11), is 0.001 mole m^{-2} s^{-1}. Substituting the flux and water concentration into Equation (17.5.22), the fluid velocity is 1.8×10^{-8} m s^{-1}. Using a characteristic distance $h = 0.005$ m and the thermal diffusivity of water (Table 17.2), the thermal Peclet number in the liquid is 0.0006, which indicates that convection is not significant in the sweat. Thus, energy transport in the sweat is by conduction only:

$$\frac{d^2T_l}{dy^2} = 0, \qquad (17.5.23)$$

which has the general solution

$$T_l = a_3 y + a_4. \tag{17.5.24}$$

Applying the boundary conditions yields the following solution for T_a:

$$T = T_{\text{air}} + \left[\Delta T - \frac{\Delta \overline{H}^{vap} N_{y=h} h}{k_l}\right] \frac{\left[\exp\left(\frac{\rho \hat{C}_p N_{y=h}(y-h)}{C_{\text{vap}} k_a}\right) - \exp\left(\frac{\rho \hat{C}_p N_{y=h}\delta}{C_{\text{vap}} k_a}\right)\right]}{\left[1 - \left(\frac{k_a}{k_l}\right)\exp\left(\frac{\rho \hat{C}_p N_{y=h} h}{C_{\text{vap}} k_a}\right) - \exp\left(\frac{\rho \hat{C}_p N_{y=h}\delta}{C_{\text{vap}} k_a}\right)\right]}, \tag{17.5.25}$$

where $\Delta T = T_b - T_{\text{air}}$. Since $k_a \ll k_l$ (see Table 17.2), the terms $\left(\frac{k_a}{k_l}\right)\left(\frac{\rho \hat{C}_p N_{y=h} h}{C_{\text{vap}} k_a}\right) = 0.01$, and the temperature at $y = h$ is

$$T(y = h)_a = T_{\text{air}} + \left(\Delta T - \frac{\Delta \overline{H}^{\text{vap}} N_{y=h} h}{k_l}\right). \tag{17.5.26}$$

For the case of conduction only, the temperature profile is

$$T(y = h)_a = T_{\text{air}} + \Delta T\left[\frac{h + \delta - y}{\left(\frac{k_a}{k_l}\right)h + \delta}\right]. \tag{17.5.27}$$

The energy flux due to evaporation is $\Delta \overline{H}^{\text{vap}} N_{y=h}$ and equals 44.6 J m^{-2} s^{-1}. For reference, pure conduction in the air ($q = k\Delta T/\delta$) driven by a temperature difference from 39°C to 25°C is 25.7 J m^{-2} s^{-1}, about 58% of the value obtained by evaporation. From Equation (17.5.26), the air temperature at the surface of the sweat ($y = h = 0.005$ m) is 38.48°C, 0.31°C lower than the value calculated for conduction only (Equation (17.5.27) with $y = h$). So, although the temperature change is modest, evaporation removes a significant quantity of heat from the body, accelerating cooling.

17.6 Metabolism and Regulation of Body Temperature

17.6.1 Basal Metabolic Rate and Efficiency

Metabolism refers to all chemical reactions in organisms. It represents the expenditure of chemical energy to maintain life, serve useful biological functions, do work, and generate heat. In plants, the energy is most often supplied through sunlight, which is then converted to chemical energy. In animals, the energy is supplied through plants or other animals.

The minimal rate of metabolism to sustain the organism in a living but inactive state is referred to as the *basal metabolic rate*. This metabolic rate is measured under specific conditions in which the adult person or organism is awake but is in a resting and stress-free state. No digestion is occurring, and the external temperature does not produce heat or cause thermoregulation [22]. The basal metabolism generates energy to pump the blood through the body at rest, to produce breathing, maintain concentration gradients across cells, maintain normal electrical activity in the nervous system, and produce heat. Metabolic rate can be measured from oxygen consumption, based on assumptions about the energy liberated from oxygen metabolism.

Although physiologists often assume that the basal metabolic rate equals basal rate of heat production [23], that view is not correct, since some work must be done to maintain the minimal resting state. The energy for metabolism is supplied through the coupling of chemical reactions to produce a net decrease in free energy. The basal metabolic rate is 80 W (or 1,650 kcal day^{-1}) for adult males and 60.5 W (1,250 kcal day^{-1}) for adult females. The major organs involved in metabolism are the brain (20% of total heat production), skeletal muscle (20%), the liver (17%), the heart (11%), the gastrointestinal tract (10%), the kidneys (6%), and the lungs (4%) [22]. Resting skeletal muscle consumes much less energy, than that consumed during exercise. During sleep, the metabolic rate falls below the basal level because of the reduction in energy utilization, but under almost all other conditions, the rate of energy use is high, so that the metabolic rate exceeds the basal level. The additional energy required is used for normal functions, such as walking, sitting, eating, growth, and reproduction. During strenuous physical exertion, the metabolic rate can rise to 10–20 times the basal level, indicating a significant increase in energy use.

About 70% of basal oxygen consumption at the cellular level, or 70% of basal heat production, is associated with ATP production in the mitochondria [24]. The ATP molecules produced are then used for various biological functions, such as protein synthesis (~30%) and maintenance of transmembrane concentration gradients of Na^+ (~30%) and Ca^{++} (~7%). Other uses of ATP produced in basal metabolism include synthesis of glucose and urea and activation of actinomyosin ATPase. About 10% of basal oxygen consumption is nonmitochondrial; the remaining 20% of basal oxygen consumption is associated with metabolic reactions for maintaining proton gradients across mitochondria membrane [24].

Among animal species, the basal metabolic rate varies with body mass (M) as $M^{0.75}$, or the ratio of surface area to volume (S/V) as $(S/V)^{-3/2}$ (see also Table 16.2) [24, 25]. The basal metabolic rate per unit body mass scales as $M^{-0.25}$. That is, smaller animals have higher metabolic rates per unit mass than do larger animals. Variation within a species conforms to these relations, but other factors can produce wide variations.

Animals that regulate their body temperature, known as homeotherms or endotherms, have higher metabolic rates and less efficient utilization of energy for work, such as muscle contraction, than similarly sized cold-blooded animals, known as poikilotherms or ectotherms [26]. The result is that a considerable amount of the energy produced in homeotherms is converted to heat to maintain the body temperature. The heat production increases ATP usage and reduces the efficiency of biochemical reactions [22].

The thermodynamic efficiency η is defined as the ratio of useful work done by the system W to the maximal useful work that can be done by the system (W_{max}). Thus,

$$\eta = \frac{|W|}{|W_{max}|} \tag{17.6.1a}$$

For most biological systems in heat-transfer analysis, it can be assumed that temperature and pressure are maintained at constant levels. If mass transport into the system is negligible, then W_{max} is equal to the reduction in the total Gibbs energy (ΔG) of the system. Thus,

$$\eta = \frac{|W|}{\Delta G} \tag{17.6.1b}$$

The useful work done can be due to the direct conversion of enthalpy (i.e., $\Delta H < 0$) and entropy increase (i.e., $\Delta S > 0$). The work from entropy increase can arise when a molecule unwinds or tension is released. Such actions, observed in molecular motors within cells, cause motions of organelles or cells. Extracellular matrix molecules such as collagen can be compressed, producing a more ordered state, and do work during the recovery to a relaxed state. Thus, both the enthalpy and entropy changes contribute to the work done. Recall that at constant pressure and temperature, $\Delta G = \Delta H - T\Delta S$. As a result, $|W_{max}| = |\Delta G| > |\Delta H|$ if $\Delta S > 0$ and $\Delta H < 0$. On the other hand, it is possible to have $\Delta S < 0$ and $\Delta H < 0$. In this case, $|W_{max}| = |\Delta G| < |\Delta H|$. Based on the second law of thermodynamics, there must be a net heat transfer from system to surroundings, indicating that part of the ΔH consumed is transferred to the surroundings as heat in order to reduce the system entropy.

As an example of the efficiency of biological systems, consider the energetic requirements and efficiency of the heart. The heart comprises 0.4% of the body mass but contributes 11% of total oxygen consumption in the body at the resting level, more on a per kg basis than skeletal muscle or the brain. The energetic needs to pump blood account for only about 20% of the total energetic needs of the heart. The remaining 80% is lost as heat. The heat is generated during basal metabolism in the cardiac cells for pumping ions and activation of muscle for contraction, and also during cardiac muscle contraction [27]. The efficiency of cross-bridges ranges from 30% to 46% [28], consistent with the heat produced under resting conditions.

From thermodynamics, several general statements can be made about efficiency. The maximum efficiency (i.e., 100%) can be obtained only when the process is reversible. Reversible processes cannot happen in nature but can be approximated by moving the system from initial to final states through a large number of very small steps. For mechanical devices, such as engines, a thermal efficiency η_e of engines operating between a hot source with temperature T_1 and a cold sink with temperature T_2 can be defined as

$$\eta_e = \frac{|W|}{Q_h}, \tag{17.6.2a}$$

where Q_h is the heat transfer from the hot source into the engine and W is the work done by the engine. It can be shown that the maximum efficiency is [1]

$$\eta_{max} = 1 - \frac{T_2}{T_1}, \qquad (17.6.2b)$$

which occurs only when the engine operates through a reversible cycle. For example, an automobile engine operating at 220°C and releasing heat to the surrounding air at 25°C has an efficiency of

$$\eta_{max} = 1 - \frac{298.15}{493.15} = 0.396. \qquad (17.6.2c)$$

If the weather is very hot and the temperature rises to 40°C, the maximum efficiency drops to 0.365. Note that many engines operate at a much lower efficiency, typically around 0.20.

17.6.2 Regulation of Body Temperature

The existence of most organisms is limited to a narrow range of temperatures between freezing temperature, 0°C, and 42°C. Below freezing temperature, intracellular ice formation can destroy cells and extracellular ice can limit transport of nutrients. As discussed in Section 17.8, single-celled organisms can survive below 0°C, but special efforts are needed to limit ice formation and mechanical stresses. When the organism temperature is sustained above 42°C for prolonged periods of time, many important proteins undergo irreversible denaturation, resulting in cell death. Human beings maintain a mean body temperature of 37°C. Above 35°C, mechanisms are set in place to cool the body temperature. Conversely, only when the skin temperature falls below 23°C are mechanisms established to generate heat [26]. In this temperature range, normal mechanisms of heat generation and insulation are sufficient to maintain the core body temperature in a narrow range. The lower limit of body temperature before additional methods of heat generation are initiated is species specific. Small animals, such as rodents, begin to regulate their temperature below 30°C [26], in part because of their relatively large surface area to volume ratio.

Other than jumping into cool water, there are two mechanisms to cool the body when temperatures are elevated: vasodilation and evaporative cooling by sweating or panting. The reddish appearance of the skin after heavy exertion or exercise is an indication of vasodilation. Blood near the skin surface is cooled and goes back to the central veins and arteries, efficiently transporting heat. As a result, cooling occurs much faster than it would through conduction alone. As shown by the examples above, sweating can remove heat more efficiently than can convective heat-transfer in the air. As long as the partial pressure of water vapor in air is below the saturation level, sweating is an efficient mechanism to remove heat.

At temperatures below 32°C and above 23°C, the body reduces heat loss by reducing blood flow to the skin. This reduces convective transport within the body, resulting in reduced heat loss into the air by heat conduction and radiation. Below 23°C, vasoconstriction in the skin is insufficient, and additional heat is generated through shivering or physical activity. In small animals such as rodents, brown adipose tissue is metabolized to produce heat in cold environments [15].

Although there appears to be a set point about which the body temperature is regulated, the regulatory process is more complex, involving several different interacting control systems. Different transient receptor potential (TRP) ion channels are sensitive to hot and cold temperatures [29]. Activation of the TRP channels stimulates nerves that are controlled in the hypothalamus. Subsequent neural and hormonal signals then regulate vasodilation or vasoconstriction, blood flow, and changes in metabolic heat production.

17.6.3 Macroscopic Balance for Energy Transfer in Biological Systems

The macroscopic or integral forms of the energy equation are developed in order to consider the overall energy balance in organisms. The approach is similar to that used in Chapter 4 to derive the equation for mechanical energy balance. The definition of the system and the assumptions used in the following derivation are the same as those used in Section 17.2 to derive the differential form of energy conservation. By using the mass conservation equation and the incompressible condition for the system, Equation (17.2.8) becomes

$$\frac{D(\rho\hat{U})}{Dt} = -\nabla\cdot\mathbf{q} + \dot{W} + \dot{Q}_p. \tag{17.6.3}$$

Integrating Equation (17.6.3) over the volume of the system yields

$$\int_V \frac{\partial(\rho\hat{U})}{\partial t}dV + \int_V \nabla\cdot(\rho\mathbf{v}\hat{U})dV = -\int_V \nabla\cdot\mathbf{q}dV + \int_V (\dot{W} + \dot{Q}_p)dV. \tag{17.6.4}$$

Next, we apply the divergence theorem to the terms representing the flow across the system boundaries. This converts the volume integrals to surface integrals:

$$\frac{dU}{dt} + \int_S \rho\hat{U}\mathbf{v}\cdot\mathbf{n}dS = -\int_S \mathbf{q}\cdot\mathbf{n}dS + \frac{dQ_p^{\text{tot}}}{dt} + \frac{dW^{\text{tot}}}{dt}, \tag{17.6.5}$$

where dU/dt is the rate of change of total internal energy of the system; dW^{tot}/dt is the net rate of all works, except that for fluid flow; W^{tot} includes works done both *on* the system, which are positive, and *by* the system, which are negative; dQ_p^{tot}/dt is the net rate of heat production via different sources or sinks in the system. Among them, the dominant energy sources or sinks are formed by molecules involved in chemical reactions. In many biological applications, dQ_p^{tot}/dt is approximately equal to dQ_m^{tot}/dt, the net rate of metabolic heat production. Therefore, a reasonable simplification of Equation (17.6.5) is

$$\frac{dU}{dt} = (\dot{m}\hat{U})_{\text{in}} - (\dot{m}\hat{U})_{\text{out}} - \int_S \mathbf{q}\cdot\mathbf{n}dS + \frac{dQ_m^{\text{tot}}}{dt} + \frac{dW^{\text{tot}}}{dt}, \tag{17.6.6}$$

if the system has only one inlet and one outlet. In Equation (17.6.6), $\dot{m}$ is the rate of mass flow and dQ_m^{tot}/dt and dW^{tot}/dt can be constants or can vary with time.

Equation (17.6.6) indicates that the internal energy increase in the system can be mediated by four different mechanisms: (i) convective transport of internal energy, (ii) heat conduction, (iii) metabolic heat production, and (iv) works done on the system.

For a system with no net change in mass, $\dot{m}_{in} = \dot{m}_{out} = \dot{m}$. Equation (17.2.10c) indicates that $d\hat{U} = \hat{C}_p dT$ in incompressible media. Thus,

$$m\hat{C}_p\frac{dT}{dt} = \dot{m}\hat{C}_p(T_{in} - T_{out}) - \int_S \mathbf{q}\cdot\mathbf{n}dS + \frac{dQ_m^{tot}}{dt} + \frac{dW^{tot}}{dt}, \quad (17.6.7)$$

where T is the volume-averaged temperature in the system, T_{in} and T_{out} are the temperature at the inlet and outlet, respectively, and m is the total mass of the system. The left-hand side of Equation (17.6.7) represents the accumulation of internal energy in the organism, which leads to a change in body temperature. In certain applications, it can be assumed that the temperature distribution is uniform in the system. Additionally, the change in temperature from $T_{surroundings}$ to T is assumed to occur within a short distance at the inlet. In this case, T_{out} is equal to T in the system and T_{in} is equal to $T_{surroundings}$.

Example 17.8 Some athletes try to improve their performance by cooling their body prior to an endurance event, such as a marathon. Cooling can be achieved by wearing a vest with ice packs or very cool water. Consider an athlete wearing a vest around his chest and back containing 3 kg of ice. Assume that the whole-body heat capacity is 3.5 kJ kg^{-1} K^{-1}. Assume that the rate of basal metabolic energy production in a 70-kg male is 80 W and that 70% of it is used for heat production. Determine the body temperature if half the ice melts in 2 hours. The enthalpy of fusion of ice is 333.4 kJ kg^{-1} ice.

Solution Treat the body as the system under study and the ice as the surroundings. Assume that the body is in contact with ice at 273.15 K, and neglect heat exchange with the air. Since the individual is resting, no work is done. Thus, assume that the heat required to melt the ice causes cooling of the body temperature.

In Equation (17.6.7), the convection term is zero, since there is no mass exchange. Assuming efficient heat exchange, the heat removed from the organism via conduction (q) is the product of the latent heat or enthalpy of fusion ($\Delta\hat{H}_{fusion,H_2O}$) times the mass of ice melted (m_{H_2O}). Integrating the simplified form of Equation (17.6.7) over time yields

$$m\hat{C}_p(T - T_{initial}) = -m_{H_2O}\Delta\hat{H}_{fusion,H_2O} + \frac{dQ_m}{dt}\Delta t. \quad (17.6.8)$$

The heat transferred to melt the ice is negative, since it is transferred from the system to the surroundings. The final temperature is given by

$$T = T_{initial} - \left(\frac{m_{H_2O}\Delta\hat{H}_{fusion,H_2O} - \dfrac{dQ_m}{dt}\Delta t}{m\hat{C}_p}\right). \quad (17.6.9)$$

Inserting the values for the different quantities into Equation (17.6.9) yields

$$T = 310\text{ K} - \left(\frac{(3\text{ kg})(333.4\text{ kJ kg}^{-1}) - (0.080\text{ kJ s}^{-1})(0.70)(7{,}200\text{ s})}{(70\text{ kg})(3.5\text{ kJ (kg K)}^{-1})}\right)$$

$$T = 310.15\text{ K} - 2.44\text{ K} = 307.71\text{ K}$$

Example 17.9 Now consider a runner who has cooled down with the vest and is running the marathon without the vest. His time is 2 hours and 24 minutes. The surrounding air temperature is 30°C. During this time, the rate of metabolic heat production rises to 380 W. Assume that heat transfer between body and air is by convection only, with an effective heat-transfer coefficient of 25 W m^{-2} K^{-1}. Determine the rise in temperature and the time taken to reach a new steady state. Assess the effect of cooling. Neglect any enthalpy change due to breathing or drinking cold fluids during the race and the work done by the runner since they are small compared to metabolic heat production. Use a BSA of 1.8 m^2.

Solution Based on the assumptions stated above, Equation (17.6.7) becomes

$$m\hat{C}_p\frac{dT}{dt} = h(T_{\text{air}} - T)A + \frac{dQ_m}{dt}. \quad (17.6.10)$$

This is a first-order ordinary differential equation with the initial condition that $T = 307.71$ K. The solution to this equation, using the integrating factor, is

$$T = \left(T_{\text{air}} + \frac{1}{hA}\frac{dQ_m}{dt}\right)\left[1 - \exp\left(-\frac{hA}{m\hat{C}_p}t\right)\right] + T_0\exp\left(-\frac{hA}{m\hat{C}_p}t\right), \quad (17.6.11)$$

where T_0 is the initial body temperature. The time constant is $(m\hat{C}_p)/(hA) = 5{,}444$ s. The temperature at the end of the race is 310.80 K = 37.65°C. Without cooling, the body temperature would have risen to 311.30 = 38.15°C, 0.5°C higher. So cooling had a modest effect on body temperature.

Example 17.10 The culture of microorganisms for the production of recombinant proteins can generate significant heat, and large-scale cultures require extensive cooling. Consider an *Escherichia coli* culture in an aqueous broth at a density of 10 g L^{-1}. The bioreactor is insulated, so all of the heat released is absorbed by the water. The culture generates 5 kJ heat per gram of bacteria. Assuming that the broth has a density of 1,050 kg m^{-3} and a specific heat capacity of 4.3 kJ kg^{-1} K^{-1}, determine the steady-state temperature of the liquid if the initial temperature was 25°C.

Solution In this problem, the heat generated by the bacteria is transferred to the water, raising its temperature. Thus, an energy balance is

$$\frac{Q_{E.\,coli}}{V} = \rho \hat{C}_p \Delta T, \qquad (17.6.12)$$

where $Q_{E.\,coli}/V$ is the heat production per unit broth volume. Solving for the final temperature:

$$T = T_0 - \frac{Q_{E.\,coli}/V}{\rho \hat{C}_p} = 298.15 + \frac{(5\text{ kJ g}^{-1})(10\text{ g}/1{,}000\text{ cm}^3)(100\text{ cm m}^{-1})^3}{(1{,}050\text{ kg m}^{-3})(4.3\text{ kJ (kg K)}^{-1})}$$

$$= 298.15 + 11.08 = 309.22\text{ K}.$$

Example 17.11 Unlike the scenario discussed in Example 17.10, heat production in large-scale cell cultures is dynamic due to the growth and metabolism of microorganisms. Thus, if the total rate of heat production in the bioreactor is $\dot{Q}_p = \dot{Q}_m$, the amount of heat produced per unit mass of cells is Q_{cell}, the cell density is N_{cell} (mass of cells per unit volume), which can vary with time, the specific growth rate is μ (h^{-1}), and the bioreactor volume is V, then the rate of heat production is

$$\dot{Q}_m = Q_{cell} \mu N_{cell} V. \qquad (17.6.13)$$

The bioreactor volume is 6.3 m^3 and the specific growth rate is 1 h^{-1}. The cooling of the culture system is provided by flowing cold water at 20°C into the bioreactor. The water/bioreactor contact surface area is 12.5 m^2. When $N_{cell} = 10\text{ g L}^{-1}$, determine the heat-transfer coefficient needed to maintain the temperature of the broth no higher than 32°C if the mean water temperature in the heat exchanger is 25°C. In addition, calculate the flow rate if the outlet water temperature is 31.5°C.

Solution The rate of heat removal must balance the rate of heat generation to maintain the temperature at the desired level. Thus,

$$\dot{Q}_m = Q_{cell} \mu N_{cell} V = h_m (T_{bioreactor} - T_{water}) A. \qquad (17.6.14)$$

$$Q_{cell} \mu N_{cell} V = (5{,}000\text{ J g}^{-1})(1/3{,}600\text{ s})(10\text{ g}/1{,}000\text{ cm}^3)(100\text{ cm m}^{-1})^3(6.3\text{ m}^3)$$

$$= 87{,}500\text{ J s}^{-1}.$$

Solving for h_m,

$$h_m = \frac{Q_{cell} \mu N_{cell} V}{(T_{bioreactor} - T_{water})A} = \frac{87{,}500\text{ J s}^{-1}}{(7\text{ K})(12.5\text{ m}^2)} = 1{,}000\text{ W K}^{-1}\text{ m}^{-2}.$$

The flow rate can be determined from an energy balance on the heat removed:

$$Q_{cell} \mu N_{cell} V = \dot{m} \hat{C}_p (T_{exit} - T_{entrance}) \qquad (17.6.15)$$

$$\dot{m} = \frac{Q_{\text{cell}}\mu N_{\text{cell}} V}{\hat{C}_p(T_{\text{exit}} - T_{\text{entrance}})} = \frac{87{,}500 \text{ J s}^{-1}}{(4{,}183 \text{ J (kg K)}^{-1}(12 \text{ K})} = 1.74 \text{ kg s}^{-1}.$$

Based on the density of water, the volumetric flow rate is 0.00174 m^3 s^{-1}, which corresponds to 1.74 L s^{-1}. The results indicate that the heat exchange requirements are considerable.

17.7 The Bioheat-Transfer Equation

As noted in Section 17.3, conduction alone does not provide a mechanism for rapid and efficient heat transfer within the body. In this section, we examine how heat exchange between flowing blood and tissues affects the transfer of heat in tissues and the temperature profile. Local heat production and removal by blood can be considered by the soource/sink term in Equation (17.2.13) as discussed in section 17.2.1. A similar approach was used in Section 15.5.1. This model was first developed by Pennes [30,31]. To explicitly consider heat production by blood and from metabolism, the source or sink term in equation (17.2.13) is split into two terms as ahown in equation 17.7.1,

$$\rho \hat{C}_p \frac{\partial T}{\partial t} = k\nabla^2 \text{T} + \dot{Q}_m + \dot{Q}_b \tag{17.7.1}$$

The temperature and thermal properties refer to the extravascular portion of the tissue. $\dot{Q}_b$ represents the net rate of heat generation or removal due to blood flow. The model was developed from experimental data showing that the temperature in the forearm was a maximum at the center and declined with distance from the center [30]. One of the challenges in using Equation (17.7.1) has been to appropriate values of temperature to use for the term [32]. In general, arterial and venous blood may be at different temperatures than the tissue, and separate energy balance equations are needed to account for heat transfer to these regions.

The approach initially developed by Pennes was to assume that the venous blood was in thermal equilibrium with the tissue and that all arterial blood is at the same temperature. This is likely an oversimplification, because arterial blood will travel different distances before branching, and the arterial temperature varies with position. Nevertheless, these assumptions lead to a relatively simple expression for $\dot{Q}_b$:

$$\dot{Q}_b = \dot{V}_{\text{blood}} C_{p,\,\text{blood}}(T_a - T), \tag{17.7.2}$$

where $\dot{V}_{\text{blood}}$ is the perfusion rate of blood, $C_{p,\,\text{blood}}$ is the heat capacity of blood, and T_a is the arterial temperature. The steady-state solution of Equations (17.7.1) and (17.7.2) for a long cylindrical arm, subject to the boundary condition that the flux at the skin surface is the product of the heat-transfer coefficient h times the temperature difference between the skin surface and the ambient air (T_∞), has been derived in [32]. The temperature distribution in the radial (r) direction is

$$T(r) = T_a + \frac{\dot{Q}_m}{\dot{V}_{\text{blood}} C_{p,\text{blood}}} + \frac{T_\infty - (\dot{Q}_m/\dot{V}_{\text{blood}} C_{p,\text{blood}}) - T_a}{(k\lambda/h) I_1(\lambda R) + I_0(\lambda R)} I_0(\lambda r), \tag{17.7.3a}$$

where

$$\lambda = \sqrt{\frac{\dot{V}_{blood}C_{p,blood}}{k}}. \qquad (17.7.3b)$$

While Equation (17.7.3a) reproduces the shape of the data, the perfusion rate had to be estimated. As a result, agreement of the model was good, but artificial.

Several concerns have been raised about the use of Equation (17.7.2) [32]. First, using this equation results in a conservation relation that contains local temperatures. Second, the arterial temperature may not be independent of location. Third, there are three unknown temperatures, and conservation relations are needed for arterial and venous blood. The assumption that the venous blood temperature equals the tissue temperature may not be correct.

Efforts to address these shortcomings have involved adding separate energy balances for arterial and venous blood and examining the effect of the distribution of capillaries. These models are quite complex mathematically. Experimental results indicate that heat transfer from arteries is quite variable. When major arteries are present, the data of heat transfer conform to the original predictions of the Pennes model. Under other conditions, models involving separate energy balances for the blood and tissue are needed. Such analyses can be simplified to a one-dimensional equation with the following effective thermal conductivity [32]:

$$\frac{k_{eff}}{k_{tissue}} = 1 + \frac{n}{\sigma}\left(\frac{k_{blood}}{k_{tissue}}\right)^2\left(\frac{\pi a \mathrm{Pe}_T}{2}\right)^2, \qquad (17.7.4)$$

where a is the blood vessel radius, n is the number density of microvessels, Pe_T is the thermal Peclet number, and

$$\sigma = \frac{\pi}{\cosh^{-1}(l_s/2a)}, \qquad (17.7.5)$$

where $l_s/2a$ is the ratio of blood vessel spacing to vessel diameter. Equation (17.7.4) indicates that when the blood velocity in a highly perfused tissue increases, the effective thermal conductivity increases.

17.8 Cryopreservation

Many techniques that use cells for medical therapies involve the storage of cells at subfreezing temperatures, a process known as *cryopreservation*. Above the freezing point, the cellular metabolic rate slows as the temperature is reduced. If the cells are maintained at or below the freezing point, the metabolic rate is zero, but the cells are not dead. The low temperatures can preserve proteins, polynucleotides, and cell organelles so that upon thawing, cellular functions can be recovered. Cryopreservation is widely used to store blood cells, sperm, eggs, and many different types of cells that are used for research purposes. Cryopreservation of tissues is more challenging but is considered necessary for the successful application of engineered tissues.

The most critical problem affecting cell viability in cryopreservation is the formation of intracellular ice. The ice can damage cell membranes and denature proteins. High protein concentration regions within cells and membranes serve as nucleation sites for ice formation, whereas intracellular salts and other dissolved molecules inhibit ice formation by depressing the freezing point. The conditions during cooling must be such that intracellular ice formation is eliminated. Controlling the cooling rate in addition to adding cryopreservatives to inhibit ice formation or preventing freezing by vitrification can prevent intracellular ice formation.

When a suspension of cells is cooled slowly, ice first forms in the extracellular fluid at temperatures between –1°C and –4°C. If care is taken to eliminate spurious energy exchange (e.g., vibrations), homogeneous ice nucleation can be suppressed until the temperature reaches –40°C (233 K) [4] and liquid water can persist in a supercooled state, although that condition is thermodynamically unstable. Homogeneous nucleation occurs when several water molecules orient in a stable crystal pattern. Once a small stable nucleus of ice forms, the crystal grows, forming a solid state. However, heterogeneous nucleation often occurs around 18°C (255 K), so special treatment of the cell suspension is needed to prevent such ice nucleation. If ice formation can be suppressed to –138°C (135 K), water will form a glassy state with the release of some heat [4,33]. The process of cooling to form a glass material is known as vitrification, and the temperature at which the glassy state forms is known as the glass transition temperature. Vitrification is considered a very promising method for preserving tissues and organs, and research in this area is focused on ways to raise the glass transition temperature [4].

Ice formation can be suppressed by using high rates of cooling and adding cryoprotectants. Examples of cryoprotectants used with cells include glycerol and dimethyl sulfoxide. Often these agents are mixed together with high salt concentrations. Both approaches work by inhibiting ice nucleation. Recently, interest in the use of the disaccharide trehalose has increased because it can raise the glass transition temperature [34].

The cooling of a cell suspension to temperatures below the freezing point of water induces complex changes to the physicochemical environment that affects cell survival. A large number of experiments show that with or without cryoprotectants, there is an optimal cooling rate for cell survival, and the optimal cooling rate depends on the cell type [4,35]. Cooling rates for achieving 50% cell survival vary from a few degrees Celsius per minute for hepatocytes to over 500°C min^{-1} for red blood cells [4]. These optimal rates are due to a balance between mechanisms that protect cells and those that damage cells. When the cell suspension is cooled below the freezing point, the concentrations of dissolved molecules are initially greater outside the cell than within the cell, and extracellular ice forms first. Since the salts remain in the liquid phase when ice forms, salt concentrations increase as extracellular ice forms. An osmotic imbalance arises that favors secretion of water by the cell. Water is a small polar molecule that cannot permeate the lipid membrane. Rather, water transport occurs through special protein channels. The permeability is sensitive to temperature according to an Arrhenius-type relation [34]:

$$P_{H_2O} = P_0 \exp\left(-\frac{\Delta E_{act}}{R}\left(\frac{1}{T} - \frac{1}{T_0}\right)\right), \quad (17.8.1)$$

where P_0 is the permeability at a reference temperature T_0, R is the gas constant and ΔE_{act} is the activation energy. Thus, permeability increases with temperature for temperatures above the reference temperature and declines with decreasing temperature. If the water can leave the cell at a rate faster than the freezing rate, then the intracellular osmotic pressure increases and the freezing point is further suppressed. Consequently, very rapid cooling produces intracellular ice formation by heterogeneous nucleation, because insufficient water has left the cell and the salt concentration is too low to depress freezing. If the freezing rate is too slow, extracellular ice formation may become extensive, which will initiate intracellular ice formation through membrane damage or extension of ice into the cell [4]. As a result, cell survival represents a balance between dehydration of the cell and propagation of ice front.

In order to be effective, cryoprotectants must enter the cell and prevent nucleation. The levels of cryoprotectants can be high in the freezing medium, since the cryoprotectant can be removed after warming up the cell, causing little loss in cell viability. However, with tissues, the transport of cryoprotectants is much slower, potentially causing cell toxicity. Other factors affecting survival of tissues include thermal stresses generated due to the volume expansion of water near the freezing point and the inability to maintain uniform cooling rates for thick specimens.

17.9 QUESTIONS

17.1 Discuss the similarities and differences between heat transfer and mass transfer.

17.2 Explain why the thermodynamic efficiency is always less than 1.

17.3 Explain why the temperature of a metal surface on a sunny day can be warmer than the surrounding air.

17.4 Describe how the perfusion of blood through tissues enhances the transport of heat throughout the body.

17.5 Discuss how ice formation is deleterious to cell survival.

17.6 Identify problems affecting the cryopreservation of tissues and organs.

17.10 PROBLEMS

17.1 Complete the derivation of Equation (17.2.6).

17.2 Compute the work done to stretch a protein, assuming that the protein extension can be represented by the wormlike chain model relating force and extension:

$$F = \left(\frac{k_B T}{L_p}\right)\left(0.25\left(1 - \frac{x}{L}\right)^{-2} - 0.25 + \frac{x}{L}\right),$$

where k_B is Boltmann's constant and L_p is the persistence length.

17.3 For fully developed laminar flow in a cylindrical tube of radius R, show that Equation (17.2.9) is

$$\dot{\Phi}_v = \tau : \nabla \mathbf{v} = \mu\left(\frac{dv_r}{dr}\right)^2.$$

Use this result to estimate the viscous heating arising during steady blood flow in a vessel the size of the human aorta. Treat blood as a Newtonian fluid with a viscosity of 0.01 g cm^{-1}s^{-1} (=0.001 Pa s). Assume that the volumetric flow rate is 5 L min^{-1} and the radius is 1.5 cm. Assume that viscous heating produces a radial temperature profile, above the surrounding temperature T_0. Based on this result, estimate whether viscous heating is a significant source of heating in the body.

17.4 Consider a sphere of radius R maintained at constant temperature T_0 and placed in a large volume of a liquid. Natural convection is negligible and no forced convection occurs. Assume that the fluid properties are independent of temperature. The fluid temperature far from the sphere surface is T_∞. Determine the steady-state heat flux from the sphere surface and show that $\mathrm{Nu}_m = h_m D/k = 2$.

17.5 Show that for an ideal gas, $\beta = 1/T$.

17.6 Consider an arteriole and a venule in muscle separated by 250 μm, with mean temperatures T_a and T_v, respectively. Determine the time for temperature differences between the two vessels to reach steady state.

17.7 Determine the rate of growth of the ice front for the problem of unsteady ice formation discussed in Section 17.3.3. Explain how this result can be used to determine the freezing rate.

17.8 Derive Equation (17.5.7) for the diffusion of a vapor from a liquid surface to a gas with a relative humidity x_H.

17.9 Determine the effect of 20% and 80% humidity on the energy flux from sweat. Compare this with conduction through an air film when the temperature difference is either 15°C or 3°C.

17.10 Using the following data for the vapor pressure of water as a function of temperature [8], estimate the error in using Equation (17.5.13) to compute the flux for temperatures of 25°C and 40°C and a relative humidity of 75%.

Temperature (C)	Vapor pressure (Pa)
25	3,141
40	7,376

17.11 For the analysis of water vaporization, update the temperature and energy flux until the results change by less than 1%.

17.12 Determine the effect of wind speed upon the heat flux from the surface of the skin at 5°C. Assume that the skin is at 37°C.

17.13 Show that the Grashof number can also be written as

$$\mathrm{Gr} = \frac{\rho g \Delta\rho L^3}{\mu^2}$$

17.14 Compare the relative importance of radiation, conduction, forced convection, and free convection when an infant's head and an adult's head are exposed to a 10-mile-per-hour breeze on a cold winter day with an air temperature of 0°C. For the infant, use a head circumference of 39 cm for a 12-week-old girl (http://www.who.int/childgrowth/standards/en/); for the adult, use a head circumference of 56 cm.

17.15 Derive Equation (17.5.25) and Equation (17.5.27).

17.16 For values of $l_s/2a$ ranging from 1 to 100 and typical velocities in arterioles and capillaries (Table 2.4), determine the effect of convection on the effective thermal conductivity.

17.11 REFERENCES

1. Haynie, D.T., *Biological Thermodynamics*. 2000, Cambridge, England: Cambridge University Press.
2. Dewhirst, M.W., Viglianti, B.L., Lora-Michiels, M., Hanson, M., and Hoopes, P.J., "Basic principles of thermal dosimetry and thermal thresholds for tissue damage from hyperthermia." *Int. J. Hyperthermia*, 2003. 19: pp. 267–294.
3. Bischof, J.C., "Micro and nanoscale phenomena in bioheat transfer." *Heat Mass Transfer*, 2006. 42: pp. 955–966.
4. Fowler, A., and Toner, M., "Cryo-injury and biopreservation." *Ann. NY Acad. Sci.*, 2005. 1066: pp. 119–135.
5. Bird, R., Stewart, W., and Lightfoot, E., *Transport Phenomena*. 2d ed. 2002, New York: John Wiley and Sons. p. 895.
6. Bowman, H.F., Cravalho, E.G., and Woods, M., "Theory, measurement and application of thermal properties of biomaterials." *Ann. Rev. Biophys. Bioeng.*, 1975. 4: pp. 43–80.
7. Geankoplis, C.J., *Transport Processes and Separation Process Principles*. 4th ed. 2003, Upper Saddle River, NJ: Prentice Hall. p. 1026.
8. Weast, R., *CRC Handbook of Chemistry and Physics*. 67th ed. 1986, Boca Raton, FL: CRC Press.

9. Rubinsky, B., "Solidification processes in saline solutions." *J. Crystal Growth*, 1983. **62**: pp. 513–522.
10. Wong, K., Boyde, A., and Howell, P.G.T., "A model of temperature transients in dental implants." *Biomaterials*, 2001. **22**: pp. 2795–2797.
11. Dua, R., and Chakraborty, S., "A novel modeling and simulation technique of photo–thermal interactions between lasers and living biological tissues undergoing multiple changes in phase." *Comp. Biol. Med.*, 2005. **35**: pp. 447–462.
12. Davidson, S.R.H., and James, D.F., "Measurement of thermal conductivity of bovine cortical bone." *Med. Eng. Phys.*, 2000. **22**: pp. 741–747.
13. Cena, K., and Clark, J.A., "Thermal insulation of animal coats and human clothing." *Phys. Med. Biol.*, 1978. **23**: pp. 585–591.
14. Deen, W.M., *Analysis of Transport Phenomena*. 1998, New York: Oxford University Press. p. 597.
15. Blaxter, K., *Energy Metabolism in Animals and Man*. 1989, Cambdrige, England: Cambridge University Press. p. 336.
16. Vogel, S., "Living in a physical world IV. Moving heat around." *J. Biosci.*, 2005. **30**: pp. 449–460.
17. Carslaw, H.S., and Jaeger, J.C., *Conduction of Heat in Solids*. 2d ed. 1959, Oxford: Clarendon Press. p. 510.
18. Welty, J.R., Wicks, C.E., Wilson, R.E., and Rorrer, G.L., *Fundamentals of Momentum, Heat and Mass Transfer*. 2008, Hoboken, NJ: John Wiley and Sons.
19. Verbraecken, J., Van de Heyning, P., De Backera, W., and Van Gaal, L., "Body surface area in normal-weight, overweight, and obese adults. A comparison study." *Metabolism*, 2006. **55**: pp. 515–524.
20. Reid, R., Prausnitz, J., and Sherwood, T., *The Properties of Gases and Liquids*. 3rd ed. 1980, New York: McGraw-Hill.
21. Marsh, K.N., ed. *Recommended Reference Materials for the Realization of Physicochemical Properties*. 1987, Oxford: Blackwell.
22. Rolfe, D.F.S., and Brown, G.C., "Cellular energy utilization and molecular origin of standard metabolic rate in mammals." *Physiol. Rev.*, 1997. **77**: pp. 731–758.
23. Jessen, C., *Temperature Regulation in Humans and Other Mammals*. 2001, New York: Springer. p. 193.
24. Hulbert, A.J., and Else, P.L., Mechanisms underlying the cost of living in animals. *Ann. Rev. Physiol.*, 2000. **62**: pp. 207–235.
25. West, G.B., and Brown, J.H., "The origin of allometric scaling laws in biology from genomes to ecosystems: towards a quantitative unifying theory of biological structure and organization." *J. Exp. Biol.*, 2005. **208**: p. 1575–1592.
26. Silva, J.E., "Thermogenic mechanisms and their hormonal regulation." *Physiol. Rev.*, 2006. **86**: pp. 435–464.
27. Smith, N.P., Barclay, C.J., and Loisell, D.S., "The efficiency of muscle contraction." *Clin. Exp. Prog. Biophys. Molec. Biol.*, 2003. **88**: pp. 1–58.
28. Gibbs, C.L., and Barclay, C.J., "Cardiac efficiency." *Cardiovasc. Res.*, 1995. **30**: pp. 627–634.
29. Morrison, S.F., Nakamura, K., and Madden, C.J., "Central control of thermogenesis in mammals." *Exp. Physiol.*, 2008. **93**: pp. 773–797.
30. Pennes, H.H., "Analysis of tissue and arterial blood temperature in the resting human forearm." *J. Appl. Physiol.*, 1948. **1**: pp. 93–122.
31. Wissler, E.H., "Pennes' 1948 paper revisited." *J. Appl. Physiol.*, 1998. **85**: pp. 35–41.
32. Charny, C.K., *Mathematical models of bioheat transfer*, Y.I. Cho, editor. 1992, Boston: Academic Press. pp. 19–156.
33. Diller, K.R., "Modeling of bioheat transfer processes at high and low temperatures," in *Advances in Heat Transfer*, Y.I. Cho, editor. 1992, Boston: Academic Press. p. 157–358.
34. Chen, T., Fowler, A., and Toner, M., "Literature review: supplemented phase diagram of the trehalose-water binary mixture." *Cryobiology*, 2000. **40**: pp. 277–282.
35. Mazur, P., "The role of intracellular freezing in the death of cells cooled at supraoptimal rates." *Cryobiology*, 1977. **14**: pp. 251–272.

APPENDIX Mathematical Background

The analysis of transport and reaction processes relies heavily upon the use of mathematics. The objective of this appendix is to provide a brief overview of the mathematical procedures used throughout this text. We assume that readers are familiar with many of these procedures and that the overview will refresh their memory. For more detailed information, the reader should consult a text on calculus and mathematical analysis (e.g., Ref. [1]). This review covers topics in calculus, solution of ordinary and partial differential equations, and vectors and tensors.

A.1 Review of Calculus and Solution of Ordinary Differential Equations

The solution of some forms of the conservation equations requires one to solve ordinary differential equations. The following covers many of the techniques used in the solution of ordinary differential equations, integration, and evaluation of limits.

A.1.A Integration by Parts

Often a function u can be integrated by transposition of the function and the integrand dv using the following relation:

$$\int u dv = uv - \int v du \tag{A.1.1a}$$

Example A.1 Evaluate the following integral by parts

$$\int \sin^2(ax)dx.$$

Solution To solve, make the following substitutions:

$$u = \sin(ax) \quad dv = \sin(ax)dx$$

$$du = a\cos(ax)dx \quad v = -\frac{1}{a}\cos(ax)$$

Inserting these values into Equation (A.1.1), the integral becomes

$$\int \sin^2(ax)dx = \frac{1}{a}\sin(ax)\cos(ax) - \int \cos^2(ax)dx. \qquad \text{(A.1.1b)}$$

To simplify, note that $\cos^2(ax) = 1 - \sin^2(ax)$ and $\int \cos^2(ax)dx = \int [1 - \sin^2(ax)]dx$. After rearrangement, the solution is

$$\int \sin^2(ax)dx = -\frac{1}{2a}\sin(ax)\cos(ax) + \frac{x}{2}. \qquad \text{(A.1.1c)}$$

A.1.B Evaluation of Limits

L'Hôpital's rule. For two functions $f(z)$ and $g(z)$ that are zero at $z = a$ and have k derivatives equal to zero at $z = a$,

$$\underset{z\to a}{\text{limit}} \frac{f(z)}{g(z)} = \frac{f^{k+1}(a)}{g^{k+1}(a)}, \qquad \text{(A.1.2)}$$

where $f^{k+1}(a)$ is the $k + 1$ derivative of the function $f(z)$ evaluated at $z = a$ and $g^{k+1}(a) \neq 0$.

Example A.2 Determine the following limit:

$$\underset{z\to 0}{\text{limit}} \frac{1 - \cos(z)}{z^2}$$

Solution At $z = 0$, cos (z) is equal to 1 and both the numerator and denominator are zero. If we apply L'Hôpital's rule once, we obtain $\sin(z)/2z$, which at $z = 0$ is undefined. Applying L'Hôpital's rule a second time, we obtain $\cos(z)/2$, which is equal to one-half at $z = 0$. Thus,

$$\underset{z\to 0}{\text{limit}} \frac{1 - \cos(z)}{z^2} = \frac{\cos(0)}{2} = \frac{1}{2} \qquad \text{(A.1.3)}$$

A.1.C Taylor Series Approximations

The value of a function $f(z)$ about the point $z = a$ can be approximated by the following series expansion:

$$f(z) = f(a) + f'(a)(z-a) + \frac{f''(a)(z-a)^2}{2} + \frac{f'''(a)(z-a)^3}{6} + \cdots$$
$$= \sum_{n=0}^{\infty} \frac{f^n(a)(z-a)^n}{n!}, \tag{A.1.4}$$

where the prime denotes differentiation with respect to the independent variable z (e.g., $f(a) = (df(z)/dz)|_{z=a}$). Such an expansion is known as a *Taylor series approximation*.

A common application of the Taylor series approximation is to estimate the value of a function at a small perturbation away from a known value (Figure A.1). Consider a small perturbation $z + \varepsilon$, where $0 < \varepsilon \ll 1$. To simplify, truncate the Taylor series after the second term.

$$f(z+\varepsilon) \approx f(z) + \left.\frac{df(z)}{dz}\right|_{z=\varepsilon} \varepsilon. \tag{A.1.5}$$

For functions of two variables y and z, this approximation formula becomes

$$f(z+\varepsilon, y+\delta) \approx f(z,y) + \left.\frac{\partial f(z,y)}{\partial z}\right|_{z=\varepsilon} \varepsilon + \left.\frac{\partial f(z,y)}{\partial y}\right|_{y=\delta} \delta. \tag{A.1.6}$$

Example A.3 Derive the Taylor series approximation for $\sin(x)$.

Solution For this case, $a = 0$ and the Taylor series approximation is

$$\sin(x) = \sum_{n=0}^{\infty} \left.\frac{d^n \sin(x)}{dx^n}\right|_{x=0} \frac{x^n}{n!} \tag{A.1.7a}$$

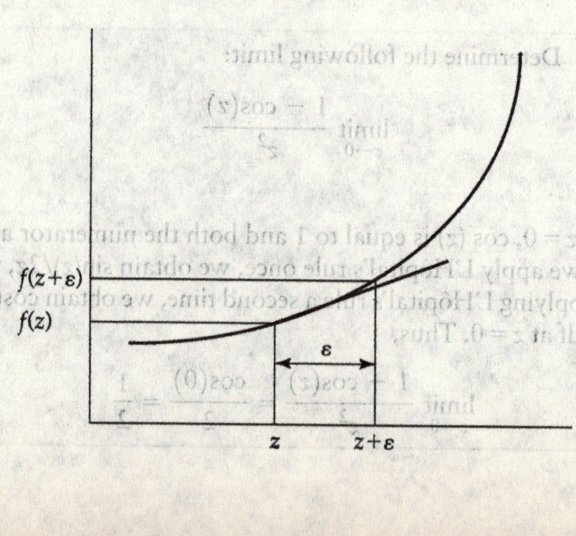

FIGURE A.1 Graphical representation of a Taylor series approximation.

For $n = 0, 2, 4, 6, \ldots$, the derivatives are $\sin(x), -\sin(x), \sin(x), -\sin(x)$, etc., which are equal to zero at $x = 0$. For $n = 1, 3, 5, 7, \ldots$, the derivatives are $\cos(x), -\cos(x), \cos(x), -\cos(x)$, etc., which are equal to 1 for $x = 0$. Thus,

$$\sin(x) = x - \frac{x^3}{3!} + \frac{x^5}{5!} - \frac{x^7}{7!} + \frac{x^9}{9!} - \cdots. \tag{A.1.7b}$$

Example A.4 If $f(x) = \ln(3 + x^2)$, estimate $f(x = 1 + \varepsilon)$ for $0 < \varepsilon \ll 1$.

Solution Because ε is a small parameter, we can use the truncated Taylor series approximation.

$$\ln\left[3 + (1 + \varepsilon)^2\right] \approx \ln(4) + \left.\frac{d\ln(3 + x^2)}{dx}\right|_{x=1} \varepsilon. \tag{A.1.7c}$$

Evaluating the derivative yields

$$\ln\left[3 + (1 + \varepsilon)^2\right] \approx \ln(4) + \frac{\varepsilon}{2}. \tag{A.1.8d}$$

A.1.D Solution of Linear Ordinary Differential Equations

First-order equations. First-order equations can be solved using an integrating factor. Consider

$$\frac{dy}{dx} + a(x)y = h(x). \tag{A.1.9}$$

Let $p = \exp\left(\int a(x)dx\right)$. Then $a(x)$ is equal to

$$\frac{1}{p}\frac{dp}{dx} = a(x), \tag{A.1.10}$$

and the first-order equation can be rewritten as

$$\frac{d(py)}{dx} = ph(x), \tag{A.1.11}$$

where p is known as the *integrating factor*. This equation has the following general solution for y:

$$y = \frac{1}{p}\int phdx + \frac{C}{p}, \tag{A.1.12}$$

where C is a constant that is determined by evaluating the condition $y = a$ at $x = b$.

Example A.5 Solve the following differential equation:

$$x\frac{dy}{dx} + (1 - x)y = xe^x.$$

Solution First, divide each term of the equation by x:

$$\frac{dy}{dx} + \left(\frac{1}{x} - 1\right)y = e^x. \tag{A.1.13a}$$

The integrating factor for this equation is

$$p = \exp\left[\int\left(\frac{1}{x} - 1\right)dx\right] = \exp(\ln(x) - x) = xe^{-x}. \tag{A.1.13b}$$

The quantity $h(x)$ in Equation (A.1.11) equals e^x. Thus, the solution is

$$y = \frac{e^x}{x}\int xe^{-x}e^x dx + \frac{Ce^x}{x} = \frac{xe^x}{2} + \frac{Ce^x}{2}. \tag{A.1.13c}$$

Linear differential equations with constant coefficients. Most of the linear differential equations you will solve are second-order equations and have the following form:

$$\frac{d^2y}{dx^2} + a_1\frac{dy}{dx} + a_2y = h(x), \tag{A.1.14}$$

where a_1 and a_2 are constants. The homogeneous solution (i.e., $h(x) = 0$), is of the form e^{rx}, where r is a constant, real or complex. This solution is chosen because all derivatives of e^{rx} are multiples of the function itself. The homogeneous solution is of the form

$$y = c_1e^{r_1x} + c_2e^{r_2x}. \tag{A.1.15}$$

The constants c_1 and c_2 are determined from the boundary conditions. The constants r_1 and r_2 can be determined by substituting the homogeneous solution back into the differential equation for $h(x) = 0$, which yields the following result:

$$c_1e^{r_1x}\left[r_1^2 + a_1r_1 + a_2\right] + c_2e^{r_2x}\left[r_2^2 + a_1r_2 + a_2\right] = 0. \tag{A.1.16}$$

Because the constants c_1 and c_2 are not necessarily zero, the above equation is zero only if the terms in brackets are zero. Because these terms are the same, we need only solve for one of them. The constants r_1 and r_2 are the roots of the quadratic equation

$$r_{1,2} = \frac{-a_1 \pm \sqrt{a_1^2 - 4a_2}}{2}. \tag{A.1.17}$$

The roots r_1 and r_2 may be real or imaginary. If the roots are real, then the solution involves exponentials. If the roots are imaginary and the coefficients a_1 and a_2 are real, then the roots must be conjugate pairs ($r_1 = a + ib$ and $r_2 = a - ib$, where $i = \sqrt{-1}$). Because the complex exponential can be written as $\exp(ibx) = \cos(bx) + i\sin(bx)$, the homogeneous solution is rewritten in terms of two constants:

$$y = e^{ax}\left[c_1^*\cos(bx) + c_2^*\sin(bx)\right]. \tag{A.1.18}$$

To determine the particular solution, assume it is of the form

$$y_p = h(x)c_3e^{r_1x}\left[c_3e^{r_1x} + c_4e^{r_2x} + c_5\right]. \tag{A.1.19}$$

The constants c_3, c_4, and c_5 can be determined by substituting the particular solution into the differential equation.

When both a_1 and a_2 in Equation (A.1.14) are equal to zero, the solution is straightforward. Consider

$$\frac{d^2y}{dx^2} = h, \tag{A.1.20}$$

subject to the boundary conditions, $x = 0$, $y = 0$ and $x = 1$, $y = A$. This equation is solved by integrating twice in succession. One integration yields

$$\frac{dy}{dx} = hx + A, \tag{A.1.21}$$

where A is a constant of integration. A second integration yields

$$y = \frac{h}{2}x^2 + Ax + B. \tag{A.1.22}$$

After evaluation of the boundary conditions, the complete solution is

$$y = \frac{hx}{2}(x - 1) + x. \tag{A.1.23}$$

Example A.6 Solve the following equation:

$$\frac{d^2y}{dx^2} = a^2y \tag{A.1.24a}$$

with the following boundary conditions: $x = 0$, $y = 1$ and $x = \infty$, $y = 0$.

Solution Since $h(x) = 0$, the particular solution equals zero and the homogeneous solution equals $y = c_1e^{r_1x} + c_2e^{r_2x}$. The coefficients are $a_1 = 0$ and $a_2 = -a^2$. The constants r_1 and r_2 are equal to $\pm a$. Thus,

$$y = c_1e^{ax} + c_2e^{-ax}. \tag{A.1.24b}$$

Because $y = 0$ as x approaches ∞, c_1 equals zero. At $x = 0$, $y = 1$, and $c_2 = 1$. The solution is

$$y = e^{-ax}. \tag{A.1.24c}$$

A.1.E Differentiation and the Chain Rule

In transport phenomena, dependent variables such as fluid velocity and concentration often depend upon more than one independent variable. For example, the fluid velocity is a function of the three spatial coordinates x, y, and z and, if the flow is unsteady, time. Thus, $\mathrm{v} = \mathrm{v}(x,y,z,t)$. The total differential is defined as

$$d\mathrm{v} = \frac{\partial \mathrm{v}}{\partial x}dx + \frac{\partial \mathrm{v}}{\partial y}dy + \frac{\partial \mathrm{v}}{\partial z}dz + \frac{\partial \mathrm{v}}{\partial t}dt. \tag{A.1.25}$$

If the spatial coordinates x, y, and z are also functions of time, then the dependent variable v is only a function of the variable t. Thus, the total differential or substantial derivative $D\mathrm{v}/Dt$ is equal to

$$\frac{D\mathrm{v}}{Dt} = \frac{\partial \mathrm{v}}{\partial x}\frac{dx}{dt} + \frac{\partial \mathrm{v}}{\partial y}\frac{dy}{dt} + \frac{\partial \mathrm{v}}{\partial z}\frac{dz}{dt} + \frac{\partial \mathrm{v}}{\partial t} = \mathrm{v}_x\frac{\partial \mathrm{v}}{\partial x} + \mathrm{v}_y\frac{\partial \mathrm{v}}{\partial y} + \mathrm{v}_z\frac{\partial \mathrm{v}}{\partial z} + \frac{\partial \mathrm{v}}{\partial t}. \tag{A.1.26}$$

Such differentiation is important in the calculation of fluid acceleration and mass transport because the local quantities change with time.

A.1.F Error Function

The integrand $e^{-x^2}dx$ appears frequently in transport problems but cannot be integrated analytically. This integral has been tabulated in terms of the error function, erf(n) (Figure A.2):

$$\mathrm{erf}(n) = \frac{2}{\sqrt{\pi}}\int_0^n e^{-x^2}dx \quad \text{where} \quad \frac{\sqrt{\pi}}{2} = \int_0^\infty e^{-x^2}dx \tag{A.1.27}$$

Limiting values of the error function are $\mathrm{erf}(0) = 0$ and $\mathrm{erf}(3) \approx 1$.

The complementary error function erfc(n) is written as $\mathrm{erfc}(n) = 1 - \mathrm{erf}(n)$. The error function and complimentary error function can be accessed in Matlab using the commands `erf(n)` and `erfc(n)`, respectively.

A.1.G Gamma Function

The gamma function is another integral that appears frequently in transport problems.

$$\Gamma(n) = \int_0^\infty x^{n-1}e^{-x}dx. \tag{A.1.28}$$

Some values of the gamma function are

$$\Gamma(n+1) = n\Gamma(n) \qquad \Gamma(1/2) = \pi^{1/2}$$
$$\Gamma(1/4) = 3.6256099082 \qquad \Gamma(2/3) = \mathrm{A}.3541179394$$
$$\Gamma(1/3) = 2.6789385347$$

For integer values of the argument of the gamma function, the general recursion relation is $\Gamma(n+1) = n!\Gamma(n)$. In Matlab, the command `gamma(b)` returns the value of the gamma function.

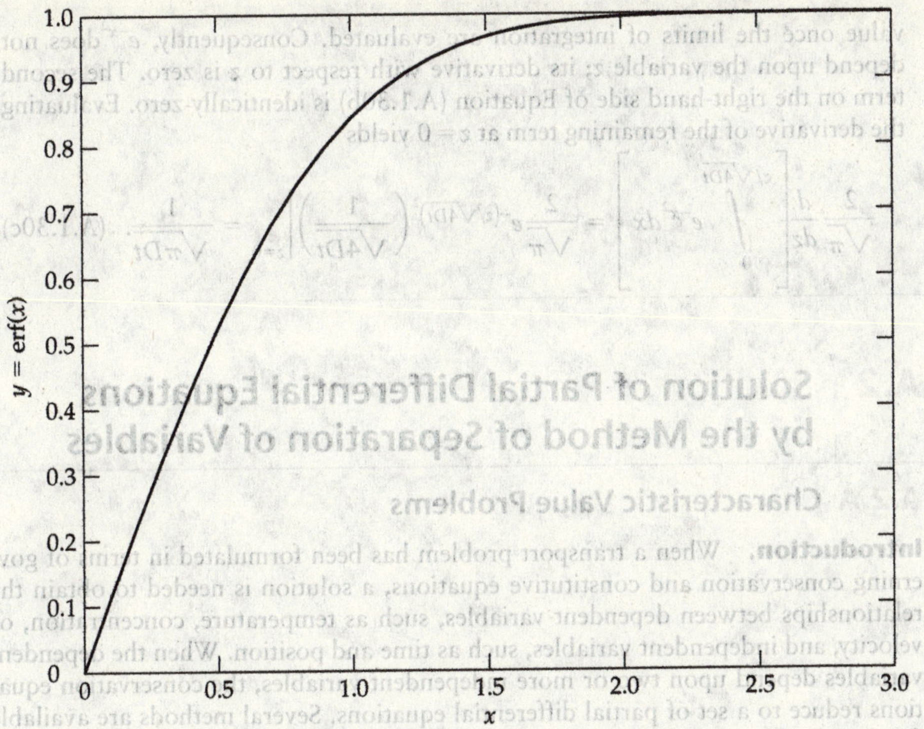

FIGURE A.2 Graph of the error function.

A.1.H Leibniz's Rule for Differentiating an Integral

$$\frac{d}{dt}\int_{a(t)}^{b(t)} f(x,t)dx = \int_{a(t)}^{b(t)} \frac{\partial f(x,t)}{\partial t}dx + f(b,t)\frac{db(t)}{dt} - f(a,t)\frac{da(t)}{dt} \tag{A.1.29}$$

Example A.7 Use Leibniz's rule to calculate the following derivative at $z = 0$:

$$\frac{d\,\mathrm{erf}(z/\sqrt{4Dt})}{dz} = \frac{2}{\sqrt{\pi}}\frac{d}{dz}\left[\int_0^{z/\sqrt{4Dt}} e^{-x^2}dx\right]. \tag{A.1.30a}$$

Solution Applying Leibniz's rule, we have

$$\frac{d}{dz}\left[\int_0^{z/\sqrt{4Dt}} e^{-x^2}dx\right] = \int_0^{z/\sqrt{4Dt}} \frac{\partial e^{-x^2}}{\partial z}dx + \frac{d0}{dz} + e^{-(z/\sqrt{4Dt})^2}\frac{d}{dz}\left(\frac{z}{\sqrt{4Dt}}\right). \tag{A.1.30b}$$

The first term on the right-hand side of Equation (A.1.30b) is zero because x is a *dummy* variable and is not a function of z. The dummy variable assumes a specific value once the limits of integration are evaluated. Consequently, e^{-x^2} does not depend upon the variable z; its derivative with respect to z is zero. The second term on the right-hand side of Equation (A.1.30b) is identically zero. Evaluating the derivative of the remaining term at $z = 0$ yields

$$\frac{2}{\sqrt{\pi}}\frac{d}{dz}\left[\int_0^{z/\sqrt{4Dt}} e^{-x^2}dx\right] = \frac{2}{\sqrt{\pi}}e^{-(z/\sqrt{4Dt})^2}\left(\frac{1}{\sqrt{4Dt}}\right)\bigg|_{z=0} = \frac{1}{\sqrt{\pi Dt}}. \quad \text{(A.1.30c)}$$

A.2 Solution of Partial Differential Equations by the Method of Separation of Variables

A.2.A Characteristic Value Problems

Introduction. When a transport problem has been formulated in terms of governing conservation and constitutive equations, a solution is needed to obtain the relationships between dependent variables, such as temperature, concentration, or velocity, and independent variables, such as time and position. When the dependent variables depend upon two or more independent variables, the conservation equations reduce to a set of partial differential equations. Several methods are available for the solution of such equations, including combination of variables, separation of variables, and transform techniques. Combination of variables is used for equations that lack an identifiable characteristic-length scale (e.g., unsteady diffusion into a semi-infinite medium); their solution is discussed in Chapter 6. Separation of variables can often be applied to finite-dimension steady and unsteady diffusion or fluid motion, such as the solution of the following equations [2,3]:

$$\text{Diffusion equation} \quad \frac{\partial C}{\partial t} = D\nabla^2 C, \quad \text{(A.2.1)}$$

$$\text{Laplace's Equation} \quad \nabla^2 C = 0. \quad \text{(A.2.2)}$$

Transform techniques are used to solve problems that cannot be solved by either the combination of variables method or the separation of variables method.

Separation of variables. Separation of variables applies to homogeneous boundary-value problems in two and three dimensions, as well as to homogeneous initial- and boundary-value problems. A boundary-value problem consists of a second-order differential equation in which conditions are specified at the boundaries of the problem. For example, the following equation is a boundary-value problem stated in terms of a second-order homogeneous ordinary differential equation:

$$\frac{d^2y}{dx^2} + a_1(x)\frac{dy}{dx} + a_2(x)y = 0, \quad \text{(A.2.3a)}$$

$$y(a) = 1 \qquad y(b) = 0. \quad \text{(A.2.3b)}$$

In contrast, an initial-value problem is one in which all conditions are specified at one boundary (i.e., the initial condition). Equations of this type are usually first-order equations in which the dependent variable is a function of time.

Homogeneous differential equations and boundary conditions. A *homogeneous linear differential equation* is one in which the equation is not altered by multiplying the dependent variable by an arbitrary constant. In other words, for a homogeneous equation, the variable and any linear combination of its derivatives vanish over the entire domain of the problem. Equation (A.2.3a) is homogeneous; if y is replaced by c_1y, the same equation results, because every term in the equation is multiplied by y.

The following is a *nonhomogeneous equation*:

$$\frac{d^2y}{dx^2} + a_1(x)\frac{dy}{dx} + a_2(x)y = g(x), \qquad \text{(A.2.4a)}$$

$$y(a) = 1 \qquad y(b) = 0. \qquad \text{(A.2.4b)}$$

Because the right-hand side of Equation (A.2.4a) is not linearly related to y, replacing y with c_1y in Equation (A.2.4a) results in a new equation in which $g(x)$ is divided by c_1

$$\frac{d^2y}{dx^2} + a_1(x)\frac{dy}{dx} + a_2(x)y = \frac{g(x)}{c_1}. \qquad \text{(A.2.4c)}$$

Similarly, a homogeneous boundary condition is one in which any linear combination of the function and its derivatives vanishes at the boundary. $y'(a) + y(a) = 0$ is a homogeneous boundary condition, whereas $y'(a) + y(a) = d$, where d is a constant, is not.

Characteristic-value problems. Characteristic-value problems are homogeneous boundary-value problems that yield solutions that are dependent upon periodic functions [3,4]. Not all boundary-value problems are characteristic-value problems, but all characteristic-value problems are boundary-value problems. For example, the following equation is a boundary-value problem:

$$\frac{d^2y}{dx^2} = \lambda^2 y, \qquad \text{(A.2.5a)}$$

$$y(0) = 1 \qquad y(\infty) = 0, \qquad \text{(A.2.5b)}$$

with the solution $y = \exp(-\lambda x)$. Equation (A.2.5a) is not, however, a characteristic-value problem, because the solution is not a periodic function. The following equation is both a boundary-value problem and a characteristic-value problem:

$$\frac{d^2y}{dx^2} = -\lambda^2 y, \qquad \text{(A.2.6a)}$$

$$y(0) = 0 \qquad y(L) = 0. \qquad \text{(A.2.6b)}$$

A solution to Equation (A.2.6a) subject to boundary conditions given by Equation (A.2.6b) is $y = \sin(\pi x/L)$ for $\lambda = \pi/L$. The solution of characteristic-value problems is best discussed in terms of the most general form of a second-order homogeneous differential equation

$$a_0(x)\frac{d^2y}{dx^2} + a_1(x)\frac{dy}{dx} + \left[a_2(x) + \lambda^2 a_3(x)\right]y = 0, \tag{A.2.7}$$

with boundary conditions specified at $y = a$ and $y = b$.

Equation (A.2.7) can be rewritten as

$$\frac{d}{dx}\left[p(x)\frac{dy}{dx}\right] + [q(x) + \lambda^2 w(x)]y = 0, \tag{A.2.8}$$

where $p = \exp\{\int[a_1(x)/a_0(x)]dx\}$, $q(x) = a_2(x)p(x)/a_0(x)$, and $w(x) = a_3(x)p(x)/a_0(x)$. A boundary-value problem of the form of Equation (A.2.8) with two homogeneous boundary conditions is called a *Sturm–Liouville* problem. The solution of a *characteristic-value problem* such as Equation (A.2.8) is expressed in terms of characteristic functions that are *periodic*. Examples of such functions are trigonometric functions (sin, cos, etc.) and Bessel functions. These functions are *orthogonal*. That is, two functions, $\phi_n(x)$ and $\phi_m(x)$, are orthogonal with respect to a weighting function $w(x)$ over the interval $[a,b]$ if

$$\int_a^b w(x)\phi_n(x)\phi_m(x)dx = 0 \quad n \neq m. \tag{A.2.9}$$

An arbitrary function $f(x)$ valid over the domain $[a,b]$ can be expanded as a series of orthogonal functions

$$f(x) = \sum a_n\phi_n(x), \tag{A.2.10}$$

where the summation goes from $n = 0$ to $n = \infty$. The coefficients a_n are obtained by multiplying both sides of Equation (A.2.10) by $w(x)\phi_m(x)$ and integrating from a to b:

$$\int_a^b w(x)f(x)\phi_m(x)dx = a_n\int_a^b w(x)\phi_n(x)\phi_m(x)dx. \tag{A.2.11}$$

From Equation (A.2.9), the integral on the right-hand side of Equation (A.2.11) equals zero when n is not equal to m. Consequently, the coefficients a_n are

$$a_n = \frac{\int_a^b w(x)f(x)\phi_n(x)dx}{\int_a^b w(x)[\phi_n(x)]^2dx}. \tag{A.2.12}$$

To use orthogonality, reconsider the problem represented by Equation (A.2.6a), subject to boundary conditions (A.2.6b). The general solution to this equation is

$$y = c_1\sin(\lambda x) + c_2\cos(\lambda x). \tag{A.2.13}$$

The boundary condition at $y = 0$ requires that $c_2 = 0$ and

$$y = c_1 \sin(\lambda x). \tag{A.2.14}$$

Application of the boundary condition at $y = L$ yields

$$0 = c_1 \sin(\lambda L). \tag{A.2.15}$$

Equation (A.2.15) has a trivial solution, $c_1 = 0$, which leads to $y = 0$. A nontrivial solution exists, however, for values of λ such that $\sin(\lambda L) = 0$—that is, when $\lambda = \pi/L$. Solutions also exist when $\lambda = 2\pi/L$, $3\pi/L$, etc. Therefore, $\lambda_n = n\pi/L$ and the characteristic functions of Equation (A.2.10) that satisfy the boundary conditions (A.2.6b) are

$$\phi_n(x) = a_n \sin(n\pi x/L). \tag{A.2.16a}$$

A nontrivial solution to Equation (A.2.15) results when λ_n assumes one of the characteristic values $n\pi/L$ for $n = 1, 2, 3$, etc. To determine the coefficients a_n, the solution is represented in terms of a series of orthogonal functions such as

$$f(x) = \sum a_n \sin\left(\frac{n\pi x}{L}\right). \tag{A.2.16b}$$

The function $f(x)$ satisfies Equation (A.2.6a), subject to the boundary conditions (A.2.6b).

The constants a_n are determined using the orthogonality of sine functions. Comparison of Equations (A.2.6a) and (A.2.7) indicates that the weighting function $w(x) = 1$ and the orthogonality relation reduces to

$$\int_0^L \sin\left(\frac{n\pi x}{L}\right)\sin\left(\frac{m\pi x}{L}\right)dx = 0 \quad n \neq m, \tag{A.2.17a}$$

$$\int_0^L \left[\sin\left(\frac{n\pi x}{L}\right)\right]^2 dx = \frac{L}{2} \quad n = m. \tag{A.2.17b}$$

The constants a_n are found by substitution of Equations (A.2.16b), (A.2.17a), and (A.2.17b) into Equation (A.2.12)

$$a_n = \frac{2}{L}\int_0^L f(x)\sin\left(\frac{n\pi x}{L}\right)dx. \tag{A.2.18}$$

Equation (A.2.18) is the Fourier sine series of $f(x)$ over the interval $[0,L]$.

To satisfy the boundary conditions (Equation (A.1.30a)), $f(x)$ must be an odd function of x over the domain $[0,L]$ because $f(-L) = -f(L)$. The function $f(x)$ can be determined by an additional condition that is usually obtained from the physical problem that the equations represent.

Separation of variables. The method of separation of variables is used to solve steady (or time-independent) two- and three-dimensional and unsteady (or time-dependent) problems by simplifying them to the solution of one or more *characteristic-value problems*. Consider steady-state two-dimensional diffusion in a rectangular region bordered by the lines $x = 0$, $y = 0$, $x = L$, and $y = L$. Initially the concentration is zero within the rectangular region. For $t > 0$, the concentration along the boundaries $x = 0$, $y = 0$, and $x = L$ is set to C_0, and the concentration along $y = L$ is set to $C = f(x)$.

The conservation relation for this problem is

$$\frac{\partial^2 C}{\partial x^2} + \frac{\partial^2 C}{\partial y^2} = 0. \tag{A.2.19a}$$

The boundary conditions are

$$y = 0 \quad C = C_0 \qquad y = L \quad C = f(x), \tag{A.2.19b}$$

$$x = 0 \quad C = C_0 \qquad x = L \quad C = C_0. \tag{A.2.19c}$$

Equation (A.2.19a) is homogeneous with nonhomogeneous boundary conditions. To apply the method of separation of variables, the boundary conditions must be made homogeneous. This is done by defining a new concentration $C^* = C - C_0$. Making this substitution, Equation (A.2.19a) becomes

$$\frac{\partial^2 C^*}{\partial x^2} + \frac{\partial^2 C^*}{\partial y^2} = 0, \tag{A.2.20a}$$

and the boundary conditions are

$$y = 0 \quad C^* = 0 \qquad y = L \quad C^*(L) = f(x) - C_0, \tag{A.2.20b}$$

$$x = 0 \quad C^* = 0 \qquad x = L \quad C^* = 0. \tag{A.2.20c}$$

As a result of this transformation, three of the four boundary conditions are homogeneous.

The method of separation of variables seeks a solution of the form

$$C^*(x,y) = X(x)(y). \tag{A.2.21}$$

Substituting Equation (A.2.21) into Equation (A.2.20a) and rearranging terms yields

$$-\frac{1}{X}\frac{d^2X}{dx^2} = \frac{1}{Y}\frac{d^2Y}{dy^2}. \tag{A.2.22}$$

Because the left-hand side of Equation (A.2.22) is a function of x only and the right-hand side of Equation (A.2.22) is a function of y only, Equation (A.2.22) must equal a constant, denoted as λ^2. Thus, the solution of the partial differential Equation (A.2.20a) reduces to the solution of the following two ordinary differential equations

$$\frac{d^2X}{dx^2} = -\lambda^2 X \qquad \frac{d^2Y}{dy^2} = \lambda^2 Y, \tag{A.2.23a,b}$$

with the boundary conditions $X = 0$ at $x = 0, L$ and $Y = 0$ at $y = 0$.

Equation (A.2.23a) and its associated boundary conditions are identical to the characteristic-value problem discussed above. Consequently, $\lambda = n\pi/L$ and

$$X = \sum A_n \sin\left(\frac{n\pi x}{L}\right). \tag{A.2.24}$$

The general solution to Equation (A.2.23b) is

$$Y = B_1 e(\lambda y) + B_2 e(-\lambda y). \tag{A.2.25}$$

Application of the boundary condition at $y = 0$ yields $B_2 = -B_1$ and

$$Y = 2B_1 \sinh(n\pi x/L), \tag{A.2.26}$$

where $\sinh(n\pi x/L) = 1/2[\exp(n\pi x/L) - \exp(-n\pi x/L)]$.

Substituting Equations (A.2.24) and (A.2.26) into Equation (A.2.21) and summing over all values of n yields

$$C^* = \sum_{n=1}^{\infty} a_n \sinh\left(\frac{n\pi y}{L}\right)\sin\left(\frac{n\pi x}{L}\right), \tag{A.2.27}$$

where $a_n = 2B_1A_n$. The variable a_n is evaluated using the boundary condition at $y = L$ and the orthogonality relations for the sine function (Equations (A.2.17a) and (A.2.17b)) over the domain $[0, L]$:

$$a_n \sinh(n\pi) = \frac{2}{L}\int_0^L C^*(L)\sin\left(\frac{m\pi x}{L}\right)dx. \tag{A.2.28}$$

If $f(x)$ in Equation (A.2.12) is a constant, then the coefficients are

$$a_n = \frac{2C^*(L)[1-(-1)^n]}{n\pi \sinh(n\pi)}. \tag{A.2.29}$$

The final solution to Equation (A.2.19a) is

$$C^* = \sum_{n=1}^{\infty} \frac{2C^*(L)[1-(-1)^n]\sinh(n\pi y/L)\sin(n\pi x/L)}{n\pi \sinh(n\pi)}. \tag{A.2.30}$$

In summary, for two-dimensional steady-state diffusion, the method of separation of variables can be used if all of the following are true:

1. The equations are linear and homogeneous.
2. The boundary conditions in one direction are homogeneous. This direction is the homogeneous direction, and the sign of λ^2 is chosen so that a characteristic-value problem results.
3. The other direction must possess a homogeneous and a nonhomgeneous boundary condition. This is the nonhomogeneous direction.

If the partial differential equation consists of an initial value and a boundary-value problem such as the diffusion equation (Equation (A.2.1)), then the third statement is replaced by the requirement that the initial condition be nonhomogeneous. Note that some nonhomogeneous boundary conditions can be transformed into homogeneous boundary conditions.

Bessel's equation. The following equation arises in problems dealing with transport in cylindrical coordinates, in which the independent variable varies with the radial direction and either the axial direction or time:

$$x\frac{d}{dx}\left(x\frac{dy}{dx}\right) + (m^2x^2 - \chi^2)y = 0. \tag{A.2.31}$$

Equation (A.2.31) is a Sturm–Liouville equation with $p(x) = x$, $q(x) = m^2x^2$, and $\lambda^2 w(x) = -x^2$. This equation has the following general solution:

$$y = C_1 J_\chi(mx) + C_2 Y_\chi(mx), \tag{A.2.32}$$

where $J_\chi(mx)$ and $Y_\chi(mx)$ are, respectively, Bessel functions of the first and second kind of order χ. Figure A.3 is a graph of $J_\chi(mx)$ and $Y_\chi(mx)$ for orders of 0, 1, and 2.

Properties of the Bessel functions and their derivatives are

$$\frac{dJ_0(mx)}{dx} = -mJ_1(mx) \qquad \frac{d(xJ_0(mx))}{dx} = -mxJ_1(mx), \tag{A.2.33a,b}$$

$$\frac{dY_0(mx)}{dx} = -mY_1(mx) \qquad \frac{d(xY_0(mx))}{dx} = -mxY_1(mx). \tag{A.2.34a,b}$$

Additional properties of Bessel functions are described in the *Handbook of Mathematical Functions* [1]. Bessel functions can be accessed in Matlab. For example, $J_\chi(mx)$ is `besselj`(χ, mx).

A.3 Basics of Vectors and Tensors [4,5]

A.3.A Notation

Vectors. *Vectors* associate a magnitude (scalar) with a direction (unit vector). Vectors are represented as lowercase symbols, either in boldface or with an underline:

$$\mathbf{v} = \mathbf{v}_1\mathbf{e}_1 + \mathbf{v}_2\mathbf{e}_2 + \mathbf{v}_3\mathbf{e}_3 = \sum_{i=1}^{3} \mathbf{v}_i\mathbf{e}_i, \tag{A.3.1}$$

where the subscripts 1, 2, and 3 refer to the x, y, and z directions and the unit vectors are represented as $\mathbf{e}_i$. Unit vectors define an orthogonal coordinate system.

Tensors. *Tensors* associate a vector with a direction. Tensors are represented by a lowercase Greek symbol, either in boldface or with a double underline. In some texts, an uppercase symbol is used:

$$\boldsymbol{\tau} = \mathbf{t}_1\mathbf{e}_1 + \mathbf{t}_2\mathbf{e}_2 + \mathbf{t}_3\mathbf{e}_3 = \sum_{i=1}^{3} \mathbf{t}_i\mathbf{e}_i. \tag{A.3.2}$$

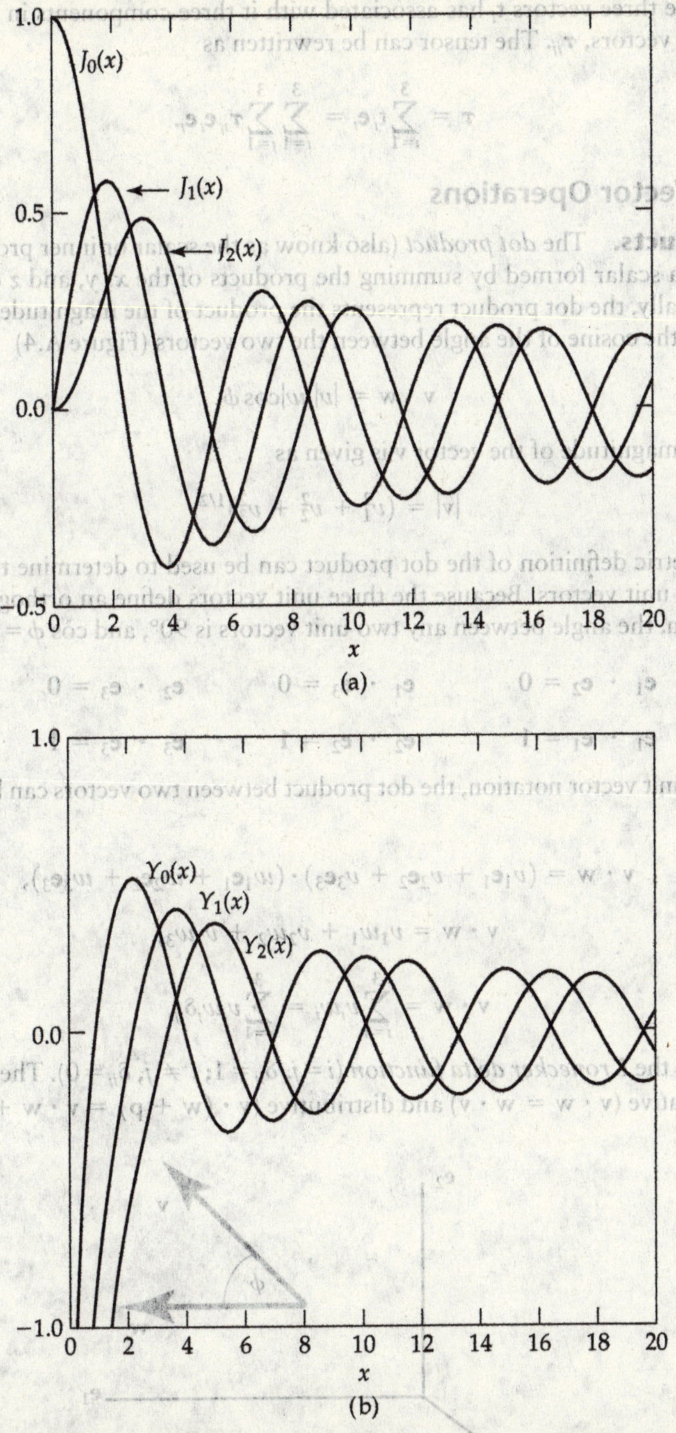

FIGURE A.3 Sketch of the Bessel functions $J_\chi(n)$ (a) and $Y_\chi(n)$ (b) for $x = 0$, 1, and 2. Results were generated using MATLAB functions for the Bessel functions.

Each of the three vectors $\mathbf{t}_i$ has associated with it three components in the direction of the unit vectors, τ_{ij}. The tensor can be rewritten as

$$\boldsymbol{\tau} = \sum_{i=1}^{3} \mathbf{t}_i \mathbf{e}_i = \sum_{i=1}^{3}\sum_{j=1}^{3} \tau_{ij}\mathbf{e}_i\mathbf{e}_j. \tag{A.3.3}$$

A.3.B Vector Operations

Dot products. The *dot product* (also know as the scalar or inner product) of two vectors is a scalar formed by summing the products of the x, y, and z components. Geometrically, the dot product represents the product of the magnitudes of two vectors times the cosine of the angle between the two vectors (Figure A.4)

$$\mathbf{v} \cdot \mathbf{w} = |v||w|\cos\phi, \tag{A.3.4}$$

where the magnitude of the vector $\mathbf{v}$ is given as

$$|\mathbf{v}| = (v_1^2 + v_2^2 + v_3^2)^{1/2}. \tag{A.3.5}$$

The geometric definition of the dot product can be used to determine the dot products of the unit vectors. Because the three unit vectors define an orthogonal coordinate system, the angle between any two unit vectors is 90°, and $\cos\phi = 0$. Thus,

$$\mathbf{e}_1 \cdot \mathbf{e}_2 = 0 \qquad \mathbf{e}_1 \cdot \mathbf{e}_3 = 0 \qquad \mathbf{e}_2 \cdot \mathbf{e}_3 = 0$$

$$\mathbf{e}_1 \cdot \mathbf{e}_1 = 1 \qquad \mathbf{e}_2 \cdot \mathbf{e}_2 = 1 \qquad \mathbf{e}_3 \cdot \mathbf{e}_3 = 1$$

Using the unit vector notation, the dot product between two vectors can be calculated as

$$\mathbf{v} \cdot \mathbf{w} = (v_1\mathbf{e}_1 + v_2\mathbf{e}_2 + v_3\mathbf{e}_3)\cdot(w_1\mathbf{e}_1 + w_2\mathbf{e}_2 + w_3\mathbf{e}_3),$$

$$\mathbf{v} \cdot \mathbf{w} = v_1w_1 + v_2w_2 + v_3w_3, \tag{A.3.6a}$$

$$\mathbf{v} \cdot \mathbf{w} = \sum_{i=1}^{3} v_iw_i = \sum_{i=1}^{3} v_iw_j\delta_{ij}, \tag{A.3.6b}$$

where δ_{ij} is the *Kronecker delta function* ($i = j$, $\delta_{ij} = 1$; $i \neq j$, $\delta_{ij} = 0$). The dot product is commutative ($\mathbf{v} \cdot \mathbf{w} = \mathbf{w} \cdot \mathbf{v}$) and distributive ($\mathbf{v} \cdot (\mathbf{w} + \mathbf{p}) = \mathbf{v} \cdot \mathbf{w} + \mathbf{v} \cdot \mathbf{p}$).

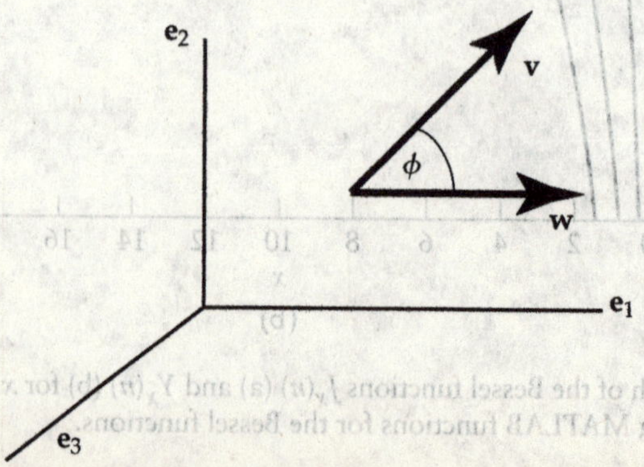

FIGURE A.4 Graphical representation of a dot product.

Cross products. The *cross product* or *vector product* ($\mathbf{v} \times \mathbf{w}$) of two vectors is a vector of magnitude $vw \sin \phi$, where ϕ is the angle formed between the two vectors. Its direction is perpendicular to the plane formed by $\mathbf{v}$ and $\mathbf{w}$. For the unit vectors that form an orthogonal coordinate system,

$$\mathbf{e}_1 \times \mathbf{e}_1 = 0 \qquad \mathbf{e}_2 \times \mathbf{e}_2 = 0 \qquad \mathbf{e}_3 \times \mathbf{e}_3 = 0 \quad \text{because } \sin 0° = 0$$

$$\mathbf{e}_1 \times \mathbf{e}_2 = \mathbf{e}_3 \qquad \mathbf{e}_2 \times \mathbf{e}_3 = \mathbf{e}_1 \qquad \mathbf{e}_3 \times \mathbf{e}_1 = \mathbf{e}_2$$

The cross product is *not commutative* ($\mathbf{v} \times \mathbf{w} = -\mathbf{w} \times \mathbf{v}$), but it is distributive ($\mathbf{v} \times (\mathbf{w}+\mathbf{n}) = \mathbf{v} \times \mathbf{w} + \mathbf{v} \times \mathbf{n}$). Thus, the cross product of two vectors is

$$\mathbf{v} \times \mathbf{w} = (v_1\mathbf{e}_1 + v_2\mathbf{e}_2 + v_3\mathbf{e}_3) \times (w_1\mathbf{e}_1 + w_2\mathbf{e}_2 + w_3\mathbf{e}_3),$$

$$\mathbf{v} \times \mathbf{w} = (v_1w_2 - v_2w_1)\mathbf{e}_3 + (v_3w_1 - v_1w_3)\mathbf{e}_2 + (v_2w_3 - v_3w_2)\mathbf{e}_1, \tag{A.3.7a}$$

$$\mathbf{v} \times \mathbf{w} = \sum_i^3 \sum_j^3 \sum_{k=1}^3 v_i w_j \mathbf{e}_k \varepsilon_{ijk}, \tag{A.3.7b}$$

where ε_{ijk} is known as the *permutation* and has the following values:

$$\varepsilon_{ijk} = 1 \text{ for } ijk = 123,\ 231,\ 312,$$

$$\varepsilon_{ijk} = -1 \text{ for } ijk = 321,\ 132,\ 213,$$

$$\varepsilon_{ijk} = 0 \text{ for } i = j \text{ or } i = k \text{ or } j = k.$$

A.3.C Vector Calculus

Vector differential operator or gradient. The gradient operator is a vector in which the spatial variation of a function is the same direction as the normal vector to a surface. The gradient is written as

$$\nabla = \mathbf{e}_1\frac{\partial}{\partial x_1} + \mathbf{e}_2\frac{\partial}{\partial x_2} + \mathbf{e}_3\frac{\partial}{\partial x_3} = \sum_{i=1}^{3}\mathbf{e}_i\frac{\partial}{\partial x_i}. \tag{A.3.8}$$

Gradient of a scalar. Consider a scalar quantity c that is a function of spatial location, $c = c(x_1, x_2, x_3)$. The gradient of the scalar (∇c) represents the local rate of change of c with respect to displacements in any direction. Suppose that $\mathbf{n}$ is a unit normal vector of magnitude one with a direction normal to a surface S and pointing outward and normal to the surface. Then the rate of change of c for a small displacement normal to the surface is

$$\mathbf{n} \cdot \nabla c = \sum_{i=1}^{3} n_i \frac{\partial}{\partial x_i}, \tag{A.3.9}$$

where n_i is the magnitude of the unit vector in the direction i.

Divergence of a vector field. The *divergence* is defined as the dot product of the gradient operator times the vector **v**:

$$\nabla \cdot \mathbf{v} = \sum_{i=1}^{3} \frac{\partial v_i}{\partial x_i}. \tag{A.3.10}$$

The Laplacian operator. The *Laplacian operator* is the divergence of the gradient vector:

$$\nabla \cdot \nabla = \sum_{i=1}^{3} \frac{\partial^2}{\partial x_i^2}. \tag{A.3.11}$$

You can apply the Laplacian operator to both scalar quantities and vector quantities.

Gradient of a vector field. A quantity that appears in the derivation of the conservation of linear momentum is the *gradient* of a vector:

$$\nabla \mathbf{v} = \sum_{i=1}^{3} \sum_{j=1}^{3} \frac{\partial v_j}{\partial x_i} \mathbf{e}_i \mathbf{e}_j. \tag{A.3.12}$$

The gradient of a vector is a tensor that associates one vector with another. The local rate of change of the vector **v** normal to a surface is another vector in the same direction as the vector **v**:

$$\mathbf{n} \cdot \nabla \mathbf{v} = \sum_{i=1}^{3} \sum_{j=1}^{3} \sum_{k=1}^{3} \frac{\partial v_k}{\partial x_j} \mathbf{e}_i \cdot \mathbf{e}_j \mathbf{e}_k = \sum_{i=1}^{3} \sum_{k=1}^{3} \mathbf{e}_k \frac{\partial v_k}{\partial x_i}. \tag{A.3.13}$$

Divergence theorem. If V is a closed region of space surrounded by a surface S, then the rate at which a quantity w accumulates within the enclosed region is equal to the net difference between the addition and the removal of w. If w is a vector, then the divergence theorem becomes

$$\int_V \nabla \cdot \mathbf{w} dV = \int_S \mathbf{n} \cdot \mathbf{w} dS. \tag{A.3.14}$$

The result of this operation is a scalar. If w is a scalar, then the divergence theorem is

$$\int_V \nabla w dV = \int_S \mathbf{n} w dS \tag{A.3.15}$$

and the result of the integration is a vector. The divergence theorem is very useful for converting surface integrals to volume integrals and vice versa.

A.3.D Tensors

Properties of tensors. A second-rank tensor, such as the shear-stress tensor, associates one vector with another vector (e.g., Equation (A.3.12)). Like matrices, tensors have the following properties:

- *Symmetric tensor.* A tensor is symmetric if, for every element τ_{ij} of a tensor, $\tau_{ij} = \tau_{ji}$.

- An *anti-symmetric tensor* would be $\tau_{ij} = -\tau_{ji}$.
- The *transpose* of a tensor is written as $\boldsymbol{\tau}^{\mathrm{T}}$. Each element ij is replaced with element ji. Thus,

$$\boldsymbol{\tau}^{\mathrm{T}} = \sum_{i=1}^{3}\sum_{j=1}^{3} \tau_{ji}\mathbf{e}_j\mathbf{e}_i. \qquad \text{(A.3.16a,b)}$$

For a symmetric tensor, $\boldsymbol{\tau} = \boldsymbol{\tau}^{\mathrm{T}}$.

Stress tensor. Consider a cubic control volume of fluid with volume $\Delta x \Delta y \Delta z$. The fluid is undergoing motion, but the control volume is fixed in space. A force is exerted on each face of the control volume. Consider face 1, which is a plane at constant x. The force acting on this face is the product of the stress times the area of the face, $\mathbf{t}_x \Delta y \Delta z$. The stress on a face of constant x is the result of compressive forces acting in the x direction (e.g., pressure), as well as tangential forces arising from the motion of the fluid (e.g., shear stresses). The components of the stress tensor on the x face are

$$\mathbf{t}_x = \sigma_{xx}\mathbf{e}_x + \sigma_{xy}\mathbf{e}_y + \sigma_{xz}\mathbf{e}_z. \qquad \text{(A.3.17a)}$$

Note that, in general, $\mathbf{t}_x$ is neither in the x direction nor normal to the surface. Only when the fluid is at rest are all of the forces normal to the surface. The forces acting on the faces of constants y and z are

$$\mathbf{t}_y = \sigma_{yx}\mathbf{e}_x + \sigma_{yy}\mathbf{e}_y + \sigma_{yz}\mathbf{e}_z, \qquad \text{(A.3.17b)}$$

$$\mathbf{t}_z = \sigma_{zx}\mathbf{e}_x + \sigma_{zy}\mathbf{e}_y + \sigma_{zz}\mathbf{e}_z. \qquad \text{(A.3.17c)}$$

The vectors $\mathbf{t}_i$ are sometimes referred to as the *surface traction vectors*.

The three stress vectors are the components of the stress tensor that is normal to each of the faces of constants x, y, and z. To see this, note that the stress vectors can be rewritten as

$$\mathbf{t}_i = \sum_{j=1}^{3} \sigma_{ij}\mathbf{e}_j. \qquad \text{(A.3.18)}$$

In general, the stress vector or surface traction vector is the dot product of the unit normal vector with the stress tensor

$$\boldsymbol{\tau}_i = \mathbf{n} \cdot \boldsymbol{\sigma} = \sum_{i=1}^{3}\sum_{j=1}^{3} \sigma_{ij}\mathbf{n} \cdot \mathbf{e}_i\mathbf{e}_j. \qquad \text{(A.3.19)}$$

Properties of the stress tensor. The stress tensor $\boldsymbol{\sigma}$ that is defined in Equations (A.3.17a) through (A.3.17c) actually consists of two terms: One represents the compressive or pressure forces that act normal to a surface, regardless of whether the fluid is static or in motion; the other represents shear stresses, or stresses arising from the motion of the fluid. The stress tensor can be rewritten to explicitly account for these two stresses:

$$\boldsymbol{\sigma} = -p\mathbf{I} + \boldsymbol{\tau}, \qquad \text{(A.3.20)}$$

where p is the fluid pressure, I is the identity matrix and $\boldsymbol{\tau}$ is the shear-stress tensor. The first term is referred to as the *isotropic stress* and represents the hydrostatic pressure. The second term is referred to as the *deviatoric stress*.

For most fluids, the shear-stress tensor is symmetric—that is, $\tau_{ij} = \tau_{ji}$. This assumption can be proved using the conservation of angular momentum. Simply stated, the conservation of angular momentum requires that the net torque on a fluid element goes to zero as the volume element shrinks to zero. This is true only if the shear stresses are symmetric. For some fluid, such as liquid crystals and highly polar materials, moments are generated by the molecular interactions between molecules (e.g., dipoles). For these fluids, the shear stresses are not symmetric. Nevertheless, for most fluids of interest, symmetry can be assumed. As a result, we need only specify six shear stresses, instead of nine surface forces acting on a control volume. The force exerted on a fluid element by the surrounding fluid is equal to the sum over the entire surface of the traction vectors that are normal to the surface $\mathbf{t}_n$. Using Equation (A.3.9), we see that the traction vector is equal to $\mathbf{n}\cdot\boldsymbol{\tau}$ and the force on the surface is

$$\mathbf{F} = \int_S \mathbf{t}_n dS = \int_S \mathbf{n}\cdot\mathbf{t}dS. \tag{A.3.21}$$

Example A.8 Show that for an incompressible Newtonian fluid,

$$\nabla\cdot\boldsymbol{\tau} = \mu\nabla^2\mathbf{v}. \tag{A.3.22}$$

Using the summation convention, we can write the left-hand side of Equation (A.3.22) as

$$\nabla\cdot\boldsymbol{\tau} = \sum_{i=1}^{3}\mathbf{e}_i\frac{\partial}{\partial x_i}\cdot\sum_{k=1}^{3}\sum_{j=1}^{3}\tau_{kj}\mathbf{e}_k\mathbf{e}_j = \sum_{i=1}^{3}\sum_{j=1}^{3}\frac{\partial\tau_{ij}}{\partial x_i}\mathbf{e}_j \tag{A.3.23}$$

because $\mathbf{e}_i\cdot\mathbf{e}_j = 1$ for $i = j$ and $\mathbf{e}_i\cdot\mathbf{e}_j = 0$ for $i \neq j$.

Solution For an incompressible Newtonian fluid, the constitutive equation is

$$\tau_{ij} = \mu\left(\frac{\partial v_i}{\partial x_j} + \frac{\partial v_j}{\partial x_i}\right). \tag{A.3.24}$$

Substituting Equation (A.3.24) into Equation (A.3.23) yields

$$\nabla\cdot\boldsymbol{\tau} = \sum_{i=1}^{3}\sum_{j=1}^{3}\frac{\partial\tau_{ij}}{\partial x_i}\mathbf{e}_j = \mu\sum_{i=1}^{3}\sum_{j=1}^{3}\left(\frac{\partial^2 v_i}{\partial x_j^2} + \frac{\partial^2 v_j}{\partial x_i\partial x_j}\right)\mathbf{e}_j. \tag{A.3.25}$$

Because the order of differentiation is unimportant, $\dfrac{\partial^2 v_i}{\partial x_i\partial x_j} = \dfrac{\partial^2 v_i}{\partial x_j\partial x_i}$ and

$$\nabla\cdot\boldsymbol{\tau} = \mu\sum_{i=1}^{3}\sum_{j=1}^{3}\left(\frac{\partial^2 v_i}{\partial x_j^2} + \frac{\partial}{\partial x_j}\left(\frac{\partial v_j}{\partial x_i}\right)\right)\mathbf{e}_j. \tag{A.3.26}$$

Using the summation notation for the divergence and the Laplacian operator (Equations (A.3.10) and (A.3.11), respectively), Equation (A.3.26) becomes

$$\nabla \cdot \tau = \mu(\nabla^2 \mathbf{v} + \nabla \cdot \mathbf{v}). \tag{A.3.27}$$

Because the fluid is incompressible, $\nabla \cdot \mathbf{v} = 0$, and Equation (A.3.27) reduces to Equation (A.3.22).

A.4 Physical Constants and Units

Force

$1\ \text{N} = 1\ \text{kg m s}^{-2}$

$1\ \text{dyn} = 1\ \text{g cm s}^{-2}$

$1\ \text{N} = 10^5\ \text{dyn}$

Energy

$1\ \text{J} = 1\ \text{N m} = 1\ \text{kg m}^2\ \text{s}^{-2}$

$1\ \text{erg} = 1\ \text{g cm}^2\ \text{s}^{-2} = 10^7\ \text{J}$

$1\ \text{cal} = 4.184\ \text{J}$

Pressure

$1\ \text{Pa} = 1\ \text{N m}^{-2} = 1\ \text{kg m}^{-1}\ \text{s}^{-2}$

$1\ \text{atm} = 101{,}325\ \text{Pa} = 101.325\ \text{kPa} = 760\ \text{mmHg}$

$1\ \text{Pa} = 10\ \text{dyn cm}^{-2}$

Density, ρ

$1\ \text{kg m}^{-3} = 0.001\ \text{g cm}^{-3}$

Viscosity, μ

$1\ \text{Pa s} = 1\ \text{kg m}^{-1}\ \text{s}^{-1}$

$1\ \text{Poise (P)} = 1\ \text{g cm}^{-1}\cdot\text{s}^{-1} = 1\ \text{dyn s} = 0.1\ \text{Pa s}$

Kinematic Viscosity, $\nu = \mu/\rho$

$1\ \text{m s}^{-2} = 10^4\ \text{Stokes (St)} = 10^4\ \text{cm s}^{-2}$

Reference

Water at 20 C, $\mu = 1$ cp

Universal Gas Law Constant, R

Values of R	Units
8.314472	$\text{J}\cdot\text{K}^{-1}\cdot\text{mol}^{-1}$
0.0820574587	$\text{L}\cdot\text{atm}\cdot\text{K}^{-1}\cdot\text{mol}^{-1}$
8.2057×10^{-5}	$\text{m}^3\cdot\text{atm}\cdot\text{K}^{-1}\cdot\text{mol}^{-1}$
62.36367	$\text{L}\cdot\text{mmHg}\cdot\text{K}^{-1}\cdot\text{mol}^{-1}$
1.987	$\text{cal}\cdot\text{K}^{-1}\cdot\text{mol}^{-1}$
8.314472×10^7	$\text{erg}\cdot\text{K}^{-1}\cdot\text{mol}^{-1}$

Avogadro's number, N_A

$N_A = 6.02214 \times 10^{23}$ molecules gmol^{-1}

Boltzmann constant, k_B

$1.380 \times 10^{-23}\ \text{J}\cdot\text{K}^{-1} = 1.3807 \times 10^{-16}\ \text{erg}\cdot\text{K}^{-1}$

$R = k_B N_A$

A.5 REFERENCES

1. Abramowitz, M., and Stegun, I.A., *Handbook of Mathematical Functions*. 1964, New York: Dover.
2. Crank, J., *The Mathematics of Diffusion*. 2d ed. 1975, Oxford: Clarendon Press. p. 414.
3. Hildebrand, F.B., *Advanced Calculus for Applications*. 2d ed. 1976, Englewood Cliffs, NJ: Prentice-Hall. pp. 186–268.
4. Deen, W., *Analysis of Transport Phenomena*. 1998, New York: Oxford University Press. p. 597.
5. Bird, R., Stewart, W., and Lightfoot, W., *Transport Phenomena*. 1960, New York: John Wiley and Sons. p. 780.

Index

A

D

H

I

J

K

Q

R

T

U